K

Klüver–Bucy syndrome: §12-4

Korsakoff's syndrome: Focus 14-4

L

Language disorders: §7-1, Focus 8-2, §8-3, §8-4, §10-3, §10-4, Focus 14-1, §15-1, §15-4

Learning disabilities: Focus 7-4, §8-4, Focus 14-1

Locked-in syndrome: Focus 11-1, §11-1

Lou Gehrig's disease (ALS): Focus 4-4, Focus 11-1

M

Major depression: §5-3, Focus 6-3, Focus 7-2, §12-4, Focus 16-2

Mania: §5-3, §6-2, §16-4

Memory deficits: §7-1, Focus 7-2, §14-2, §14-3, Focus 14-3

Ménière's disease: §11-4

Meningioma: Focus 3-2

Meningitis: §2-2

Mental retardation: see Developmental disability

Metabolic syndrome: Focus 13-1, §13-1

Metastatic tumor: Focus 3-2

Migraine: Focus 9-1

Mild cognitive impairment: Focus 14-3, §16-3

Minimally conscious state (MCS): §1-2

Monocular blindness: §9-5

Mood disorders: §12-4, §16-4

MPTP poisoning: Focus 5-4, §16-1

Multiple sclerosis (MS): Focus 3-4, §6-2, §16-3

Myasthenia gravis: Focus 4-2, §6-1

Myopia: §9-2

N

Narcolepsy: §13-6

Neglect: §15-2, §15-7

Neurodegenerative disease: §16-3

Neurotoxins: Focus 5-2, §5-3, §6-4, §16-1, §16-2, §16-3

Night terrors: §13-3

O

Obesity: §12-5

Obsessive compulsive disorder (OCD): §5-3, §12-4, §15-7, §16-2

Optic ataxia: §9-5

Osmotic thirst: §12-5

P

Pain: §6-2, §11-4, Focus 12-1, §16-2, §16-3

Panic disorder: Focus 12-3

Paralysis: Focus 2-4, §3-1

Paraplegia: §11-1, Focus 11-3

Parkinson's disease: §2-3, Focus 5-2, Focus 5-3, Focus 5-4, §5-3, §7-1, §7-5, §11-3, §16-3

Perseveration: §15-2

Persistent vegetative state: §1-2

Petit mal seizure: §16-3

Phantom-limb pain: Focus 11-5, §11-5

Phenylketonuria (PKU): §8-4, §16-1

Phobias: Focus 12-3, §16-2

Postictal depression: §16-3

Posttraumatic stress disorder (PTSD): §5-4, §6-5, §12-4, Focus

Presbyopia: §9-2

Psychosis: §6-2, Focus 6-4, Focus 7-3, §16-2, §16-3

Q

Quadrantanopia: §9-5

Quadripelegia: §11-1

R

Restless legs syndrome (RLS): Focus 13-4

Retrograde amnesia: Focus 14-4

'Roid rage: §6-5

S

Schizophrenia: §5-3, §6-1, §6-2, Focus 6-4, §7-1, §7-4, §7-5, Focus 7-3, Focus 8-5, §16-2, §16-4

Scotoma: Focus 9-1, §9-5

Seasonal affective disorder (SAD): Focus 13-2

Seizure: Focus 4-1, Focus 10-3, §16-3

Sleep apnea: Focus 13-5

Sleep paralysis: §13-6

Spina bifida: §8-4

Spinal-cord injury: §2-4, §6-2, Focus 11-3, §11-4, §12-3

Split-brain syndrome: Focus 15-1, §15-4

Stress: §2-5, §6-5, §7-5, §8-2, §8-4, §12-4, §16-1, §16-2, §16-3, §16-4

Stroke: §2-3, §7-1, §7-3, §7-5, §7-7, §16-3

Substance abuse: §2-3, §6-3, §12-4, §12-6

Sudden infant death syndrome (SIDS): §5-3, Focus 8-1, Focus 13-5

Suicide: Focus 6-3, §7-5, §16-4

Symptomatic seizure: §16-3

Synesthesia: §9-1, §15-5, Focus 15-6

T

Tardive dyskinesia: §16-2

Tay–Sachs disease: §3-3, §16-1

Tourette's syndrome: §2-3, Focus 11-4

Transsexuality: §12-5

Traumatic brain injury (TBI): Focus 1-1, §7-1, §7-5, §14-5, §16-3

Tumors: Focus 3-3, §7-3

V

Vertigo: §11-4

Visual-form agnosia: §9-5, §15-7

Visual illuminance: Focus 9-2

W

Water intoxication: §12-5

For a summary of general treatment categories, see the Chapter 16 Summary. For a summary of research techniques, see Table 7-1.

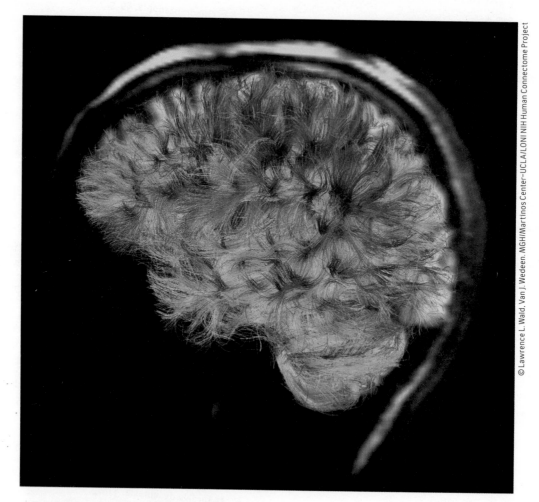

IMAGING THE BRAIN'S FIBER PATHWAYS

Commenting in *Nature* in 1993 on the "Backwardness of Human Neuroanatomy," Francis Crick and Edward Jones challenged the scientific community to map the human brain's connectivity. Rising to the challenge, scientists are exploiting MRI's sensitivity to motion to develop noninvasive methods for brain mapping in living humans, and the National Institutes of Health has launched the Human Connectome Project to catalyze new diagnostics based on connectivity (http://www.humanconnectomeproject.org/).

Diffusion MRI measures the microscopic, random movements of water molecules in living tissue. These movements act as a natural tracer of cellular architecture, including the orientations of axons, to create diffusion tractography: a virtual image representative of the brain's fiber pathways. The diffusion tractography shown on the cover was created under the aegis of NIH and employs the first in a new generation of MRI scanners designed and built to map human connectivity. The MGH/UCLA Siemens 3T Connectom affords gains in speed, sensitivity, and resolution up to tenfold over existing scanners.

On the cover, diffusion tractography images a typical human's left cerebral hemisphere, with the cerebellum at lower right and the nerve paths color-coded for clarity by their orientation in three-dimensional space. Each line represents hundreds of thousands of axons, primarily the short association pathways near the brain's surface. The image represents a milestone in answering the challenge of Crick and Jones and goes further—to the threshold of connectional imaging in mental health.

An Introduction to

BRAIN AND BEHAVIOR

FOURTH EDITION

BRYAN KOLB

University of Lethbridge

IAN Q. WHISHAW

University of Lethbridge

WORTH PUBLISHERS

*To the first neuron, our ancestors, our families, and
students who read this book*

Publisher: Kevin Feyen
Acquisitions Editor: Daniel DeBonis
Assistant Editor: Nadina Persaud
Marketing Manager: Jennifer Bilello
Developmental Editor: Barbara Brooks
Associate Media Editor: Betsy Block
Assistant Media Editor: Anthony Casciano
Supplements Project Editor: Edgar Bonilla
Senior Project Editor: Georgia Lee Hadler
Copy Editor: Penelope Hull
Production Manager: Sarah Segal
Art Director and Cover Design: Babs Reingold
Interior Design: Charles Yuen
Layout Designer and Chapter-Opening Photo Research: Lyndall Culbertson
Illustration Coordinator: Janice Donnola
Illustrations: Dragonfly Media Group and Northeastern Graphic, Inc.
Photo Editor: Cecilia Varas
Photo Researcher: Julie Tesser
Composition: Northeastern Graphic, Inc.
Printing and Binding: RR Donnelley

Library of Congress Control Number: 2012946361

ISBN-13: 9-781-4292-422-88
ISBN-10: 1-4292-4228-0

Credit is given to the following sources to use the chapter-opening images: Chapter 1: Jack Hollingsworth/Getty Images; Chapter 2: Dimitri Vervitsiotis/Getty Images; Chapter 3: (1) photosindia/Getty Images (2) Livet, Draft, Sanes, and Lichtman, Harvard University; Chapter 4: Take A Pix Media/Getty Images; Chapter 5: Diverse Images/Getty Images; Chapter 6: Image Source/Getty Images; Chapter 7: (1) Image Source/Getty Images (2) N. Kasthuri and J.W. Lichtman, Harvard University; Chapter 8: Priscilla Gragg/Getty Images; Chapter 9: (1) IMAGEMORE Co., Ltd./Getty Images (2) Novastock/StockConnection/PictureQuest; Chapter 10: Lane Oatey/Blue Jean/Getty Images; Chapter 11: Fotosearch/Getty Images; Chapter 12: Jupiterimages/Getty Images; Chapter 13: Brad Wilson/Getty Images; Chapter 14: Paula Hible/Getty Images; Chapter 15: (1) Nick Dolding/Getty Images (2) Image courtesy of Thomas Yeo and Fenna Krienen Yeo BTT, Krienen, F.M., Sepulcre, J., Sabuncu, M.R., Lashkari, L., Hollinshead, M., Roffman, J.L., Smoller, J.W., Zöllei, L., Polimeni, J.R., Fischl, B., Liu, H., Buckner, R.L.; Chapter 16: (1) Dimitri Vervitsiotis/Getty Images (2) Marcello Massimini/University of Wisconsin-Madison (3) Courtesy of Paul Thompson and Arthur W. Toga, University of California Laboratory of Neuro Imaging, Los Angeles, and Judith L. Rapoport, National Institute of Mental Health

Bryan Kolb received his Ph.D. from The Pennsylvania State University in 1973. He conducted postdoctoral work at the University of Western Ontario and the Montreal Neurological Institute. He moved to the University of Lethbridge in 1976, where he is currently Professor of Neuroscience and holds a Board of Governors Chair in Neuroscience. His current research examines how neurons of the cerebral cortex change in response to various factors, including hormones, experience, psychoactive drugs, neurotrophins, and injury, and how these changes are related to behavior in the normal and diseased brain. Kolb is a Fellow of the Royal Society of Canada, the Canadian Psychological Association (CPA), the American Psychological Association, and the Association of Psychological Science. He is a recipient of the Hebb Prize from CPA and from the Canadian Society for Brain, Behaviour, and Cognitive Science (CSBBCS) and has received honorary doctorates from the University of British Columbia and Thompson Rivers University. He is currently a member of the Experience-Based Brain and Behavioral Development program of the Canadian Institute for Advanced Research.

Ian Q. Whishaw received his Ph.D. from Western University and is a Professor of Neuroscience at the University of Lethbridge. He has held visiting appointments at the University of Texas, University of Michigan, Cambridge University, and the University of Strasbourg. He is a fellow of Clair Hall, Cambridge, the Canadian Psychological Association, the American Psychological Association, and the Royal Society of Canada. He is a recipient the Canadian Humane Society Bronze medal for bravery, the Ingrid Speaker Gold medal for research, and the distinguished teaching medal from the University of Lethbridge. He has received the Key to the City of Lethbridge and has honorary doctorates from Thompson Rivers University and the University of Lethbridge. His research addresses the neural basis of skilled movement and the neural basis of brain disease, and the Institute for Scientific Information includes him in its list of most cited neuroscientists. His hobby is training horses for western performance events.

Preface xvii
Media and Supplements xxiii

CHAPTER 1 What Are the Origins of Brain and Behavior? 1

CHAPTER 2 How Does the Nervous System Function? 33

CHAPTER 3 What Are the Functional Units of the Nervous System? 73

CHAPTER 4 How Do Neurons Use Electrical Signals to Transmit Information? 109

CHAPTER 5 How Do Neurons Use Electrochemical Signals to Communicate and Adapt? 139

CHAPTER 6 How Do Drugs and Hormones Influence the Brain and Behavior? 171

CHAPTER 7 How Do We Study the Brain's Structure and Functions? 211

CHAPTER 8 How Does the Nervous System Develop and Adapt? 245

CHAPTER 9 How Do We Sense, Perceive, and See the World? 281

CHAPTER 10 How Do We Hear, Speak, and Make Music? 319

CHAPTER 11 How Does the Nervous System Respond to Stimulation and Produce Movement? 353

CHAPTER 12 What Causes Emotional and Motivated Behavior? 397

CHAPTER 13 Why Do We Sleep and Dream? 443

CHAPTER 14 How Do We Learn and Remember? 481

CHAPTER 15 How Does the Brain Think? 523

CHAPTER 16 What Happens When the Brain Misbehaves? 565

Answers to Section Review Self-Tests A-1

Glossary G-1

References R-1

Name Index NI-1

Subject Index SI-1

PREFACE xvii

MEDIA AND SUPPLEMENTS xxiii

CHAPTER 1

What Are the Origins of Brain and Behavior? 1

CLINICAL FOCUS 1-1 Living with Traumatic Brain Injury 2

1-1 Neuroscience in the Twenty-First Century 2

Why Study Brain and Behavior? 3

What Is the Brain? 3

Gross Anatomy of the Nervous System 4

What Is Behavior? 5

1-2 Perspectives on Brain and Behavior 6

Aristotle and Mentalism 6

Descartes and Dualism 7

Darwin and Materialism 8

COMPARATIVE FOCUS 1-2 The Speaking Brain 9

Contemporary Perspectives on Brain and Behavior 12

1-3 Evolution of Brains and of Behavior 14

Origin of Brain Cells and Brains 15

Evolution of Animals Having Nervous Systems 15

THE BASICS Classification of Life 16

Chordate Nervous System 18

1-4 Evolution of the Human Brain and Behavior 19

Humans: Members of the Primate Order 19

Australopithecus: Our Distant Ancestor 20

The First Humans 21

Relating Brain Size and Behavior 22

Why the Hominid Brain Enlarged 23

RESEARCH FOCUS 1-3 Climate and the Evolving Hominid Brain 24

1-5 Modern Human Brain Size and Intelligence 27

Meaning of Human Brain-Size Comparisons 27

Culture 29

COMPARATIVE FOCUS 1-4 Evolution and Adaptive Behavior 30

SUMMARY 31

KEY TERMS 32

CHAPTER 2

How Does the Nervous System Function? 33

RESEARCH FOCUS 2-1 Evolution of Brain Size and Human Behavior 34

2-1 Overview of Brain Function and Structure 34

Plastic Patterns of Neural Organization 35

Functional Organization of the Nervous System 36

Surface Features of the Brain 37

THE BASICS Finding Your Way Around the Brain 38

CLINICAL FOCUS 2-2 Meningitis and Encephalitis 42

Internal Features of the Brain 43

CLINICAL FOCUS 2-3 Stroke 45

2-2 Evolutionary Development of the Nervous System 47

2-3 Central Nervous System: Mediating Behavior 50

Spinal Cord 50

Brainstem 51

Forebrain 54

Cortex 55

Basal Ganglia 57

Limbic System 58

Olfactory System 59

2-4 Somatic Nervous System: Transmitting Information 60

Cranial Nerves 60

Spinal Nerves 61

Connections of the Somatic Nervous System 62

CLINICAL FOCUS 2-4 Magendie, Bell, and Bell's Palsy 63

Integrating Spinal Functions 63

2-5 Autonomic Nervous System: Balancing Internal Functions 64

2-6 Ten Principles of Nervous-System Function 66

Principle 1: The Nervous System Produces Movement Within a Perceptual World the Brain Creates 66

Principle 2: The Hallmark of Nervous-System Functioning Is Neuroplasticity 66

Principle 3: Many of the Brain's Circuits Are Crossed 66

Principle 4: The Central Nervous System Functions on Multiple Levels 67

Principle 5: The Brain Is Both Symmetrical and Asymmetrical 67

Principle 6: Brain Systems Are Organized Both Hierarchically and in Parallel 68

Principle 7: Sensory and Motor Divisions Exist Throughout the Nervous System 68

Principle 8: Sensory Input to the Brain Is Divided for Object Recognition and Motor Control 69

Principle 9: Functions in the Brain Are Both Localized and Distributed 70

Principle 10: The Nervous System Works by Juxtaposing Excitation and Inhibition 70

SUMMARY 71
KEY TERMS 72

CHAPTER 3

What Are the Functional Units of the Nervous System? 73

RESEARCH FOCUS 3-1 A Genetic Diagnosis 74

3-1 Cells of the Nervous System 74

RESEARCH FOCUS 3-2 Brainbow: Rainbow Neurons 75

Neurons: The Basis of Information Processing 77

Five Types of Glial Cells 82

CLINICAL FOCUS 3-3 Brain Tumors 83

CLINICAL FOCUS 3-4 Multiple Sclerosis 86

3-2 Internal Structure of a Cell 87

THE BASICS Chemistry Review 88

The Cell As a Factory 90

Cell Membrane: Barrier and Gatekeeper 91

Nucleus: Site of Gene Transcription 92

Endoplasmic Reticulum: Site of RNA Translation 94

Proteins: The Cell's Product 94

Golgi Bodies and Microtubules: Protein Packaging and Shipment 95

Crossing the Cell Membrane: Channels, Gates, and Pumps 96

3-3 Genes, Cells, and Behavior 97

Mendelian Genetics and the Genetic Code 97

Applying Mendel's Principles 99

CLINICAL FOCUS 3-5 Huntington's Disease 101

Genetic Engineering 102

Phenotypic Plasticity and the Epigenetic Code 103

SUMMARY 107
KEY TERMS 108

CHAPTER 4

How Do Neurons Use Electrical Signals to Transmit Information? 109

CLINICAL FOCUS 4-1 Epilepsy 110

4-1 Searching for Electrical Activity in the Nervous System 111

Early Clues That Linked Electricity and Neuronal Activity 111

THE BASICS Electricity and Electrical Stimulation 112

Tools for Measuring a Neuron's Electrical Activity 114

How the Movement of Ions Creates Electrical Charges 116

4-2 Electrical Activity of a Membrane 118

Resting Potential 119

Maintaining the Resting Potential 119

Graded Potentials 121

Action Potential 122

Nerve Impulse 125

Refractory Periods and Nerve Action 126

Saltatory Conduction and the Myelin Sheath 127

4-3 How Neurons Integrate Information 128

Excitatory and Inhibitory Postsynaptic Potentials 129

CLINICAL FOCUS 4-2 Myasthenia Gravis 130

Summation of Inputs 131

Voltage-Sensitive Channels and the Action Potential 132

The Versatile Neuron 133

RESEARCH FOCUS 4-3 Optogenetics and Light-Sensitive Ion Channels 134

4-4 Into the Nervous System and Back Out 134

How Sensory Stimuli Produce Action Potentials 135

How Nerve Impulses Produce Movement 136

CLINICAL FOCUS 4-4 Lou Gehrig's Disease 137

SUMMARY 137

KEY TERMS 138

CHAPTER 5

How Do Neurons Use Electrochemical Signals to Communicate and Adapt? 139

RESEARCH FOCUS 5-1 The Basis of Neural Communication in a Heartbeat 140

5-1 A Chemical Message 140

CLINICAL FOCUS 5-2 Parkinson's Disease 142

Structure of Synapses 143

Neurotransmission in Four Steps 145

Varieties of Synapses 147

Excitatory and Inhibitory Messages 148

Evolution of Complex Neurotransmission Systems 149

5-2 Varieties of Neurotransmitters 150

Four Criteria for Identifying Neurotransmitters 150

Three Classes of Neurotransmitters 151

CLINICAL FOCUS 5-3 Awakening with L-Dopa 154

Two Classes of Receptors 155

5-3 Neurotransmitter Systems and Behavior 157

Neurotransmission in the Somatic Nervous System 158

Two Activating Systems of the Autonomic Nervous System 158

Four Activating Systems in the Central Nervous System 159

CLINICAL FOCUS 5-4 The Case of the Frozen Addict 162

5-4 Adaptive Role of Synapses in Learning and Memory 164

Habituation Response 164

Sensitization Response 166

Learning As a Change in Synapse Number 167

RESEARCH FOCUS 5-5 Dendritic Spines, Small but Mighty 169

SUMMARY 169

KEY TERMS 170

CHAPTER 6

How Do Drugs and Hormones Influence the Brain and Behavior? 171

CLINICAL FOCUS 6-1 Cognitive Enhancement 172

6-1 Principles of Psychopharmacology 173

Drug Routes into the Nervous System 173

Drug Action at Synapses: Agonists and Antagonists 175

An Acetylcholine Synapse: Examples of Drug Action 176

Tolerance 177

Sensitization 179

6-2 Grouping Psychoactive Drugs 181

Group I: Antianxiety Agents and Sedative Hypnotics 182

CLINICAL FOCUS 6-2 Fetal Alcohol Spectrum Disorder 184

Group II: Antipsychotic Agents 184

Group III: Antidepressants and Mood Stabilizers 186

Group IV: Opioid Analgesics 187

CLINICAL FOCUS 6-3 Major Depression 188

Group V: Psychotropics 190

6-3 Factors Influencing Individual Responses to Drugs 193

Behavior on Drugs 193

Addiction and Dependence 194

Sex Differences in Addiction 195

6-4 Explaining and Treating Drug Abuse 196

Wanting-and-Liking Theory 196

Why Doesn't Everyone Abuse Drugs? 198

Treating Drug Abuse 198

Can Drugs Cause Brain Damage? 199

CLINICAL FOCUS 6-4 Drug-Induced Psychosis 200

6-5 Hormones 202

Hierarchical Control of Hormones 202

Classes and Functions of Hormones 203

Homeostatic Hormones 203

Gonadal Hormones 204

Anabolic–Androgenic Steroids 204

Glucocorticoids and Stress 206

Ending a Stress Response 207

SUMMARY 209
KEY TERMS 210

CHAPTER 7

How Do We Study the Brain's Structure and Functions? 211

RESEARCH FOCUS 7-1 Tuning in to Language 211

7-1 Measuring Brain and Behavior 213

Linking Neuroanatomy and Behavior 213

Methods of Behavioral Neuroscience 215

Manipulating and Measuring Brain–Behavior Interactions 219

7-2 Measuring the Brain's Electrical Activity 223

EEG Recordings of Graded Potentials 224

Mapping Brain Function with Event-Related Potentials 225

Magnetoencephalography 226

CLINICAL FOCUS 7-2 Mild Head Injury and Depression 227

Recording Action Potentials from Single Cells 227

7-3 Static Imaging Techniques: CT and MRI 228

7-4 Dynamic Brain Imaging 231

Functional Magnetic Resonance Imaging 231

Positron Emission Tomography 231

Optical Tomography 233

7-5 Chemical and Genetic Measures of Brain and Behavior 235

Measuring the Brain's Chemistry 235

Measuring Genes in Brain and Behavior 236

Measuring Gene Expression 237

CLINICAL FOCUS 7-3 Cannabis Use, Psychosis, and Genetics 237

7-6 Comparing Neuroscience Research Methods 239

7-7 Using Animals in Brain–Behavior Research 240

Benefits of Creating Animal Models of Disease 240

Animal Welfare and Scientific Experimentation 241

RESEARCH FOCUS 7-4 Attention-Deficit/ Hyperactivity Disorder 241

SUMMARY 243
KEY TERMS 244

CHAPTER 8

How Does the Nervous System Develop and Adapt? 245

RESEARCH FOCUS 8-1 Linking Serotonin to SIDS 246

8-1 Three Perspectives on Brain Development 247

Predicting Behavior from Brain Structure 247

Correlating Brain Structure and Behavior 247

Influences on Brain and Behavior 248

8-2 Neurobiology of Development 248

Gross Development of the Human Nervous System 249

Origins of Neurons and Glia 251

Growth and Development of Neurons 253

CLINICAL FOCUS 8-2 Autism Spectrum Disorder 256

Unique Aspects of Frontal-Lobe Development 259

Glial Development 260

8-3 Correlating Behavior with Nervous-System Development 261

Motor Behaviors 261

Language Development 262

Development of Problem-Solving Ability 263

Caution about Linking Correlation to Causation 266

8-4 Brain Development and the Environment 267

Experience and Cortical Organization 267

RESEARCH FOCUS 8-3 Increased Cortical Activation for Second Languages 269

Experience and Neural Connectivity 269

Critical Periods for Experience and Brain Development 270

Abnormal Experience and Brain Development 271

Hormones and Brain Development 272

CLINICAL FOCUS 8-4 Romanian Orphans 273

Injury and Brain Development 275

Drugs and Brain Development 276

Other Kinds of Abnormal Brain Development 276

Developmental Disability 277

CLINICAL FOCUS 8-5 Schizophrenia 278

8-5 How Do Any of Us Develop a Normal Brain? 279

SUMMARY 280
KEY TERMS 280

CHAPTER 9

How Do We Sense, Perceive, and See the World? 281

CLINICAL FOCUS 9-1 Migraines and a Case of Blindsight 282

9-1 Nature of Sensation and Perception 283

Sensory Receptors 283

Neural Relays 285

Sensory Coding and Representation 285

Perception 286

9-2 Functional Anatomy of the Visual System 287

Structure of the Retina 287

THE BASICS Visible Light and the Structure of the Eye 288

Photoreceptors 291

CLINICAL FOCUS 9-2 Visual Illuminance 292

Retinal-Neuron Types 293

Visual Pathways 294

Dorsal and Ventral Visual Streams 296

9-3 Location in the Visual World 300

Coding Location in the Retina 300

Location in the Lateral Geniculate Nucleus and Region V1 301

The Visual Corpus Callosum 302

9-4 Neuronal Activity 303

Seeing Shape 303

Seeing Color 309

RESEARCH FOCUS 9-3 Color-Deficient Vision 310

Neuronal Activity in the Dorsal Stream 312

9-5 The Visual Brain in Action 313

Injury to the Visual Pathway Leading to the Cortex 313

Injury to the "What" Pathway 314

CLINICAL FOCUS 9-4 Carbon Monoxide Poisoning 314

Injury to the "How" Pathway 316

SUMMARY 317

KEY TERMS 318

CHAPTER 10

How Do We Hear, Speak, and Make Music? 319

RESEARCH FOCUS 10-1 Evolution of Language and Music 320

10-1 Sound Waves: Stimulus for Audition 321

Physical Properties of Sound Waves 321

Perception of Sound 325

Properties of Language and Music as Sounds 325

10-2 Functional Anatomy of the Auditory System 327

Structure of the Ear 327

Auditory Receptors 330

Pathways to the Auditory Cortex 331

Auditory Cortex 332

RESEARCH FOCUS 10-2 Seeing with Sound 333

10-3 Neural Activity and Hearing 334

Hearing Pitch 334

Detecting Loudness 336

Detecting Location 336

Detecting Patterns in Sound 336

10-4 Anatomy of Language and Music 337

Processing Language 338

CLINICAL FOCUS 10-3 Left-Hemisphere Dysfunction 342

CLINICAL FOCUS 10-4 Arteriovenous Malformations 344

Processing Music 345

CLINICAL FOCUS 10-5 Cerebral Aneurysms 345

RESEARCH FOCUS 10-6 The Brain's Music System 346

10-5 Auditory Communication in Nonhuman Species 348

Birdsong 348

Echolocation in Bats 350

SUMMARY 351

KEY TERMS 352

CHAPTER 11

How Does the Nervous System Respond to Stimulation and Produce Movement? 353

RESEARCH FOCUS 11-1 Neuroprosthetics 354

11-1 A Hierarchy of Movement Control 355

THE BASICS Relating the Somatosensory and Motor Systems 356

Forebrain and Initiation of Movement 358

Brainstem and Species-Typical Movement 360

CLINICAL FOCUS 11-2 Cerebral Palsy 362

Spinal Cord and Execution of Movement 363

CLINICAL FOCUS 11-3 Spinal-Cord Injury 364

11-2 Motor System Organization 365

Motor Cortex 365

Motor Cortex and Skilled Movement 367

Plasticity in the Motor Cortex 368

Corticospinal Tracts 369

Motor Neurons 370

Control of Muscles 371

11-3 Basal Ganglia, Cerebellum, and Movement 372

Basal Ganglia and Movement Force 372

CLINICAL FOCUS 11-4 Tourette's Syndrome 374

Cerebellum and Movement Skill 375

11-4 Organization of the Somatosensory System 378

Somatosensory Receptors and Perception 378

Dorsal-Root Ganglion Neurons 380

Somatosensory Pathways to the Brain 382

Spinal Reflexes 384

Feeling and Treating Pain 384

RESEARCH FOCUS 11-5 Phantom-Limb Pain 385

Vestibular System and Balance 388

11-5 Exploring the Somatosensory Cortex 390

Somatosensory Homunculus 391

RESEARCH FOCUS 11-6 Tickling 392

Effects of Damage to the Somatosensory Cortex 392

Somatosensory Cortex and Complex Movement 394

SUMMARY 395

KEY TERMS 396

CHAPTER 12

What Causes Emotional
and Motivated Behavior? 397

RESEARCH FOCUS 12-1 Pain of Rejection 398

12-1 Identifying the Causes of Behavior 399

Behavior for Brain Maintenance 400

Neural Circuits and Behavior 401

12-2 Chemical Senses 401

Olfaction 402

Gustation 404

12-3 Evolution, Environment, and Behavior 406

Evolutionary Influences on Behavior 406

Environmental Influences on Behavior 408

Inferring Purpose in Behavior: To Know a Fly 410

12-4 Neuroanatomy of Motivated and Emotional Behavior 411

Regulatory and Nonregulatory Behavior 411

Regulatory Function of the Hypothalamic Circuit 412

Organizing Function of the Limbic Circuit 417

Executive Function of the Frontal Lobes 419

CLINICAL FOCUS 12-2 Agenesis of the Frontal Lobes 421

Stimulating and Expressing Emotion 422

Amygdala and Emotional Behavior 423

Prefrontal Cortex and Emotional Behavior 424

Emotional Disorders 425

CLINICAL FOCUS 12-3 Anxiety Disorders 426

12-5 Control of Regulatory and Nonregulatory Behavior 427

Controlling Eating 428

CLINICAL FOCUS 12-4 Weight-Loss Strategies 429

Controlling Drinking 432

Controlling Sexual Behavior 433

CLINICAL FOCUS 12-5 Androgen-Insensitivity Syndrome and the Androgenital Syndrome 435

Sexual Orientation, Sexual Identity, and Brain Organization 436

Cognitive Influences on Sexual Behavior 437

12-6 Reward 438

SUMMARY 441

KEY TERMS 442

CHAPTER 13

Why Do We Sleep and
Dream? 443

CLINICAL FOCUS 13-1 Doing the Right Thing at the Right Time 444

13-1 A Clock for All Seasons 444

Origins of Biological Rhythms 445

Biological Clocks 446

Biological Rhythms 446

Free-Running Rhythms 447

Zeitgebers 449

CLINICAL FOCUS 13-2 Seasonal Affective Disorder 450

13-2 Neural Basis of the Biological Clock 452

Suprachiasmatic Rhythms 452

Keeping Time 453

RESEARCH FOCUS 13-3 Synchronizing Biorhythms at the Molecular Level 455

Pacemaking Circadian Rhythms 455

Pacemaking Circannual Rhythms 456

Cognitive and Emotional Rhythms 457

13-3 Sleep Stages and Dreaming 458

Measuring How Long We Sleep 459

Measuring Sleep in the Laboratory 459

Stages of Waking and Sleeping 459

A Typical Night's Sleep 460

Contrasting NREM Sleep and REM Sleep 461

CLINICAL FOCUS 13-4 Restless Legs Syndrome 462

Dreaming 463

What We Dream About 464

13-4 What Does Sleep Accomplish? 466

Sleep As a Biological Adaptation 467

Sleep As a Restorative Process 468

Sleep and Memory Storage 469

13-5 Neural Bases of Sleep 472

Reticular Activating System and Sleep 472

Neural Basis of the EEG Changes Associated with Waking 473

Neural Basis of REM Sleep 474

13-6 Sleep Disorders 475

Disorders of Non-REM Sleep 475

Disorders of REM Sleep 476

CLINICAL FOCUS 13-5 Sleep Apnea 477

13-7 What Does Sleep Tell Us About Consciousness? 479

SUMMARY 479

KEY TERMS 480

CHAPTER 14

How Do We Learn and Remember? 481

CLINICAL FOCUS 14-1 Remediating Dyslexia 482

14-1 Connecting Learning and Memory 483

Studying Learning and Memory in the Laboratory 483

Two Categories of Memory 485

What Makes Explicit and Implicit Memory Different? 487

What Is Special about Personal Memories? 488

14-2 Dissociating Memory Circuits 490

Disconnecting Explicit Memory 490

Disconnecting Implicit Memory 491

CLINICAL FOCUS 14-2 Patient Boswell's Amnesia 492

14-3 Neural Systems Underlying Explicit and Implicit Memories 493

Neural Circuit for Explicit Memories 493

CLINICAL FOCUS 14-3 Alzheimer's Disease 494

CLINICAL FOCUS 14-4 Korsakoff's Syndrome 498

Consolidation of Explicit Memories 499

Neural Circuit for Implicit Memories 500

Neural Circuit for Emotional Memories 500

14-4 Structural Basis of Brain Plasticity 502

Long-Term Potentiation 502

Measuring Synaptic Change 504

Enriched Experience and Plasticity 506

Sensory or Motor Training and Plasticity 507

RESEARCH FOCUS 14-5 Movement, Learning, and Neuroplasticity 510

Experience-Dependent Change in the Human Brain 510

Epigenetics of Memory 512

Plasticity, Hormones, Trophic Factors, and Drugs 512

Some Guiding Principles of Brain Plasticity 515

14-5 Recovery from Brain Injury 516

Donna's Experience with Traumatic Brain Injury 517

Three-Legged Cat Solution 517

New-Circuit Solution 518

Lost-Neuron-Replacement Solution 518

SUMMARY 521

KEY TERMS 522

CHAPTER 15

How Does the Brain Think? 523

RESEARCH FOCUS 15-1 Split Brain 524

15-1 Nature of Thought 525

Characteristics of Human Thought 525

Neural Unit of Thought 526

COMPARATIVE FOCUS 15-2 Animal Intelligence 527

15-2 Cognition and the Association Cortex 530

Knowledge about Objects 531

Multisensory Integration 531

Spatial Cognition 532

Attention 533

Planning 535

Imitation and Understanding 536

RESEARCH FOCUS 15-3 Consequences of Mirror-Neuron Dysfunction 537

15-3 Expanding Frontiers of Cognitive Neuroscience 538

Mapping the Brain 538

CLINICAL FOCUS 15-4 Neuropsychological Assessment 539

Social Neuroscience 541

Neuroeconomics 543

15-4 Cerebral Asymmetry in Thinking 544

Anatomical Asymmetry 544

Functional Asymmetry in Neurological Patients 544

Functional Asymmetry in the Normal Brain 546

Functional Asymmetry in the Split Brain 547

Explaining Cerebral Asymmetry 549

Left Hemisphere, Language, and Thought 550

15-5 Variations in Cognitive Organization 551

Sex Differences in Cognitive Organization 551

Handedness and Cognitive Organization 554

CLINICAL FOCUS 15-5 Sodium Amobarbital Test 555

Synesthesia 556

CLINICAL FOCUS 15-6 A Case of Synesthesia 556

15-6 Intelligence 557

Concept of General Intelligence 557

Multiple Intelligences 558

Divergent and Convergent Intelligence 559

Intelligence, Heredity, Epigenetics, and the Synapse 560

15-7 Consciousness 561

Why Are We Conscious? 561

What Is the Neural Basis of Consciousness? 562

SUMMARY 563

KEY TERMS 564

CHAPTER 16

What Happens When the Brain Misbehaves? 565

RESEARCH FOCUS 16-1 Posttraumatic Stress Disorder 566

16-1 Multidisciplinary Research on Brain and Behavioral Disorders 568

Causes of Abnormal Behavior 568

Investigating the Neurobiology of Behavioral Disorders 569

16-2 Classifying and Treating Brain and Behavioral Disorders 572

Identifying and Classifying Behavioral Disorders 572

Treatments for Disorders 574

RESEARCH FOCUS 16-2 Treating Behavioral Disorders with TMS 577

16-3 Understanding and Treating Neurological Disorders 581

Traumatic Brain Injury 581

CLINICAL FOCUS 16-3 Concussion 582

Stroke 584

Epilepsy 585

Multiple Sclerosis 587

Neurodegenerative Disorders 589

Are Parkinson's and Alzheimer's Aspects of One Disease? 594

Age-Related Cognitive Loss 595

16-4 Understanding and Treating Behavioral Disorders 596

Psychotic Disorders 597

Mood Disorders 599

RESEARCH FOCUS 16-4 Antidepressant Action and Brain Repair 601

Anxiety Disorders 602

16-5 Is Misbehavior Always Bad? 603

SUMMARY 603

KEY TERMS 604

ANSWERS TO SECTION REVIEW SELF-TESTS A-1

GLOSSARY G-1

REFERENCES R-1

NAME INDEX NI-1

SUBJECT INDEX SI-1

The Fourth Edition of *An Introduction to Brain and Behavior* continues to reflect the evolution of behavioral neuroscience. The major change in emphasis in this edition is the incorporation of epigenetics throughout. DNA, previously believed to be an unchanging template of heredity, is now known to respond to environmental events throughout life, leading to the concept of an *epigenome,* a record of the chemical changes to the DNA that regulate gene expression.

Epigenetics is especially important for understanding brain and behavior because environmentally induced modifications in gene expression alter the brain and ultimately behavioral development. Thus, experience—especially early experience—modifies how brain development unfolds. These modifications—of at least some behavioral traits—can be transferred across generations, as noted at the end of Section 3-3.

This edition fully addresses the advances in imaging technology, including the development and refinement of such MRI techniques as resting-state fMRI and diffusion tensor imaging, that are fueling the emerging field of *connectonomics.* As shown in the image on the book's cover, researchers are working to create a comprehensive map of neural connections—a "connectome" of the brain. These exciting advances are especially relevant in the second half of the book, where we review higher-level functions.

Imaging advances and epigenetics concepts and research are our prime focus in this revision but not our sole focus. The range of updates and new coverage in the Fourth Edition text and Focus features is listed, chapter-by-chapter, in the margins of these Preface pages. See for yourself the breadth and scope of the revision and then read on to learn more about the big-picture improvements in the Fourth Edition.

Our reviewers convinced us to move some material around. The chapter on Drugs and Hormones now follows Chapter 5, Neurotransmission. Long-term potentiation (LTP) has moved from Section 5-4 into an expanded discussion that begins Section 14-4 in the Learning and Memory chapter. And the Clinical Focus features on autism spectrum disorder and cerebral palsy have traded places: ASD now appears in Chapter 8, Development, and CP in Chapter 11, on Movement and Somatosensation.

In refining the book's learning apparatus, we have added sets of self-test questions at the end of the major sections in each chapter. These Section Reviews help students track their understanding as they progress. Answers appear at the back of the book.

We have expanded the popular margin notes, again thanks to feedback from readers, who confirm that the notes increase the reader's ease in finding information, especially when related concepts are introduced early in the text and then elaborated in later chapters. The addition of section numbers to each chapter's main headings makes it possible for readers to return quickly to an earlier discussion to refresh their knowledge or jump ahead to learn more. The margin notes also help instructors to move through the book to preview later discussions.

We have extended the appearance of illustrated Experiments that teach the scientific method visually from 8 to 13 chapters and the appearance of The Basics features that review scientific essentials from 4 to 6 chapters. The Experiments show readers how researchers design experiments—how they approach the study of brain–behavior relationships. The Basics let students brush up or get up to speed on their science foundation—knowledge that helps them comprehend behavioral neuroscience more fully.

We have made some big changes, yet much of the book remains familiar. Throughout, we continue to examine the nervous system with a focus on function, on how

FOURTH EDITION UPDATES CHAPTER-BY-CHAPTER

CHAPTER 1: ORIGINS

NEW Experiment 1-1 illustrates heritable factors and **genotype.**

UPDATED §1-2 defines **epigenetics**, adds case study on recovering consciousness post-TBI.

NEW The Basics: Classification of Life, in §1-3.

NEW Research Focus 1-3: climate change forced early hominids to develop more complex behaviors.

CHAPTER 2: NEUROANATOMY

NEW Figure 2-1 illustrates **phenotypic plasticity.**

NEW Experiment 2-1 demonstrates observational learning in the octopus.

NEW photo in §2-3: all sorts of activities can prove addictive.

CHAPTER 3: NEURONAL ANATOMY

NEW Research Focus 3-1: neuroscientists use genetics to investigate brain disorders.

NEW Experiment 3-1 tests how neural inhibition and excitation might produce behavior.

NEW Figure 3-6 contrasts early robots with contemporary social robots.

NEW coverage in §3-3: epigenetic mechanisms (Figure 3-25) and a case of "inheriting" experience.

CHAPTER 4: ELECTRICAL ACTIVITY IN THE NERVOUS SYSTEM

NEW Experiment 4-1 tests how motor neurons integrate information.

NEW Research Focus 4-3: the transgenic technique **optogenetics.**

- **CHAPTER 5: NEUROTRANSMISSION**

  NEW Figure 5-3: **gap junction**; expanded coverage of electrical synapses in §5-1.

  UPDATED Research Focus 5-5: evidence that dendritic spines form the structural basis of behavior.

- **CHAPTER 6: DRUGS AND HORMONES**

  Chapter now follows Neurotransmission.

  NEW Clinical Focus 6-1: Cognitive Enhancement.

  NEW Table 6-1: streamlined grouping for psychoactive drugs in §6-2.

  UPDATED Clinical Focus 6-3: evidence for ketamine as an acute treatment for patients with major depression.

  UPDATED §6-4: coverage of drug abuse includes epigenetics-based explanations.

  UPDATED §6-5: epigenetic effects of long-term stress on susceptibility to PTSD.

- **CHAPTER 7: RESEARCH METHODS**

  NEW Research Focus 7-1: **optical tomography.**

  NEW Experiment 7-1 demonstrates how researchers tested the idea that hippocampal neurons contribute to memory formation.

  NEW coverage of state-of-the-art methods: **optogenetics** in §7-1, **electrocorticography** in §7-2, **resting-state fMRI** in §7-4, and "epigenetic drift" in §7-5.

  NEW Figure 7-16: a series of **diffusion tensor images** links tractography to research on the *brain connectome*.

  NEW §7-6 and Table 7-1 summarize neuroscience research methods.

our behavior and our brains interact, by asking key questions that students and neuroscientists ask:

- Why do we have a brain?
- How is the nervous system organized?
- How do drugs affect our behavior?
- How does the brain learn?
- How does the brain think?

As it was when we wrote the First Edition, our goal in this new edition is to bring coherence to a vast subject by helping students understand the big picture. Asking fundamental questions about the brain has another benefit: it piques students' interest and challenges them to join us on the journey of discovery that is brain science.

Scientific understanding of the human brain and human behavior continues to grow at an exponential pace. We want to communicate the excitement of recent breakthroughs in brain science and to relate some of our own experiences from 45 years of studying brain and behavior both to make the field's developing core concepts and latest revelations understandable and meaningful and to transport uninitiated students to the frontiers of physiological psychology.

Every chapter centers on the relation between the brain and behavior. When we first describe how neurons communicate in Section 5-4, for example, we also describe how synaptic plasticity serves as the basis of learning. Later, in Section 14-4, we expand on plasticity as we explore learning and memory.

Emphasis on Evolution, Genetics and Epigenetics, Plasticity, and Psychopharmacology

To convey the excitement of neuroscience as researchers currently understand it, we interweave evolution, genetics and epigenetics, neural plasticity, and psychopharmacology throughout the book.

We address nervous-system evolution in depth in Chapters 1 and 2 and return to this perspective—neuroscience in an evolutionary context—in almost every chapter. Examples range from the evolution of the synapse in Section 5-1 to the evolution of visual pathways in Section 9-2, from ideas about how natural selection might promote overeating in Section 12-5 to evolutionary theories of sleeping and dreaming in Section 13-3, and from the evolution of sex differences in spatial cognition and language in Section 15-5 to links between our evolved reactions to stress and the development of anxiety disorders in Section 16-4.

We introduce the foundations of genetic and epigenetic research in Sections 1-3 and 2-1 and begin to elaborate on them in Section 3-3. Chapter 5 includes discussions of metabotropic receptors and DNA and of learning and genes. The interplay of genes and drug action is integral to Chapter 6, as is the role of genes and gene methylation in development to Chapter 8. Section 9-4 explains the genetics of color vision, and the genetics of sleep disorders anchors Section 13-6. Section 14-4 now includes the role of epigenetics in memory. Section 16-1 considers the role of genetics in understanding the causes of behavioral disorders.

Chapter 6 investigates drugs and behavior, a topic we revisit often through the book. You will find coverage of drugs and information transfer in Section 4-3, drugs and cellular communication in Section 5-3, drugs and motivation in Section 12-6, drugs and sleep disorders in Section 13-6, neuronal changes with drug use in Section 14-4, and drugs as treatments for a range of disorders in Section 16-4.

Neural plasticity continues as a hallmark of the book. We introduce the concept in Section 1-5, define it in Section 2-1, develop it in Section 2-6, and expand on it throughout, elaborating on the Basic Principles of Brain Plasticity at the conclusion of Section 14-4.

Scientific Background Provided

We describe the journey of discovery that is neuroscience in a way that students just beginning to study the brain and behavior can understand. The illustrated Experiments help students visualize the scientific method and how scientists think. We developed The Basics features to address the fact that this course can sometimes be daunting, largely because understanding brain function requires information from all the basic sciences. These encounters with basic science can prove both a surprise and a shock to introductory students, who often come to the course without the necessary background.

Our approach provides all the background students require to understand an introduction to brain science. The Basics features in Chapters 1 and 2 address the relevant evolutionary and anatomical background. In Chapter 3, The Basics provides a short introduction to chemistry before describing the chemical activities of the brain and, in Chapter 4, to electricity before exploring the brain's electrical activity. Readers already comfortable with the material can easily skip it; less experienced readers can learn it.

Similarly, we review such basic psychological facts as stages of behavioral development in Chapter 8 and the forms of learning and memory in Chapter 14. With this background students can tackle brain science with greater confidence. Students in social science disciplines often remark on the amount of biology and chemistry in the book, whereas an equal number of students in biological sciences remark on the amount of psychology. More than half the students enrolled in our Bachelor of Science in Neuroscience program have switched from biochemistry or psychology majors after taking this course. We must be doing something right!

Finally, Chapter 7 showcases the range of methods behavioral neuroscientists use to study brain and behavior—traditional methods and such cutting-edge techniques as optical tomography and resting-state fMRI. Expanded discussions of techniques appear where appropriate, especially in Research Focus features. Examples include Research Focus 3-2, "Brainbow: Rainbow Neurons"; Research Focus 4-3, "Optogenetics and Light-Sensitive Channels"; and Research Focus 16-1, "Posttraumatic Stress Disorder," which includes treatments based on virtual-reality exposure therapies.

Clinical Focus Maintained

We repeatedly emphasize that neuroscience is a human science. Everything in this book is relevant to our lives, and everything in our lives is relevant to neuroscience. Understanding neuroscience helps us understand how we learn, how we develop, and how we can help people who suffer brain and behavioral disorders. Knowledge of how we learn, how we develop, and the symptoms that people suffering brain and behavioral disorders display provides insights into neuroscience.

Clinical material also helps to make neurobiology particularly relevant to students who are going on to careers in psychology, social work, or other professions related to mental health, as well as to students pursuing careers in the biological sciences. We integrate clinical information throughout the text and Clinical Focus features, and we expand on it in Chapter 16, the book's capstone, as well.

In all four editions of *An Introduction to Brain and Behavior,* the placement of some topics is novel relative to traditional treatments. We include brief descriptions of brain diseases close to discussions of basic associated processes, as exemplified in the

FOURTH EDITION UPDATES CHAPTER-BY-CHAPTER

CHAPTER 8: DEVELOPMENT

UPDATED §8-2: how **gene (DNA) methylation** alters gene expression during brain development.

NEW Clinical Focus 8-2: Autism Spectrum Disorder.

NEW Figure 8-18 maps frontal-lobe development in a new discussion that points up frontal sensitivity to epigenetic influences.

UPDATED §8-4: benefits to infants of tactile stimulation beyond bonding.

NEW Figure 8-31 charts onset of behavioral disorders during adolescence.

CHAPTER 9: SENSATION, PERCEPTION, AND VISION

NEW in The Basics: the worldwide increase in myopia.

NEW in §9-2: the **retinohypothalamic tract**.

NEW Figure 9-19: temporal- and parietal-lobe areas specialized.

UPDATED discussion in §9-4: Seeing Color.

CHAPTER 10: AUDITION

NEW Research Focus 10-2: Seeing with Sound.

NEW research in §10-3 on the ventral and dorsal auditory streams.

NEW coverage: music as therapy concludes §10-4.

CHAPTER 11: MOVEMENT AND SOMATOSENSATION

UPDATED Research Focus 11-1: the promise of **neuroprosthetics**.

NEW The Basics: Relating the Somatosensory and Motor Systems, in §11-1.

CONDENSED §11-2: coverage of motor system organization and skilled movement, including plasticity in the motor system.

FOURTH EDITION UPDATES CHAPTER-BY-CHAPTER

● **CHAPTER 12: EMOTION, MOTIVATION, CHEMICAL SENSES**

NEW §12-2 is devoted to taste and smell; §12-4: neuroanatomy of motivated and emotional behaviors integrated to streamline coverage.

UPDATED Clinical Focus 12-4: Weight-Loss Strategies charts foods most likely to lead to weight gain or loss over time.

UPDATED §12-5: epigenetic influences on brain organization.

NEW Figure 12-19: prefrontal subregions important in motivated and emotional behavior.

NEW Figure 12-27: bilateral activity in the male hypothalamus during sexual arousal.

NEW Figure 12-30: enhanced connectivity in addiction-related limbic and prefrontal networks imaged by resting-state fMRI.

● **CHAPTER 13: SLEEP**

NEW Clinical Focus 13-1: disrupting the biological clock contributes to **metabolic syndrome**.

NEW Figure 13-6: path of the **retinohypothalamic tract**.

NEW coverage in §13-2: genetic and epigenetic influences on **chronotype**, biorhythmic influences on cognitive and emotional behavior.

UPDATED Research Focus 13-3 diagrams the cellular clock that paces SCN function.

NEW coverage in §13-4: sleep's possible roles in memory.

● **CHAPTER 14: LEARNING AND MEMORY**

NEW research in §14-3: memory **consolidation** and **reconsolidation**.

UPDATED §14-4 includes **LTP** and **LTD**, epigenetics of memory, **metaplasticity**.

NEW Figure 14-20 images hippocampal neurogenesis among London taxi drivers.

integrated coverage of Parkinson's disease through Chapter 5, "How Do Neurons Communicate and Adapt?" This strategy helps first-time students repeatedly forge close links between what they are learning and real-life issues.

In this new edition, the range of disorders we cover has expanded to nearly 150, all cross-referenced in the Index of Disorders inside the book's front cover. Chapter 16 expands on the nature of neuroscience research and the multidisciplinary treatment methods for neurological and psychiatric disorders described in preceding chapters, and it includes a discussion of causes and classifications of abnormal behavior.

Another area of emphasis is questions that relate to the biological bases of behavior. For us, the excitement of neuroscience lies in understanding how the brain explains what we do, whether it is talking, sleeping, seeing, or learning. Readers will therefore find nearly as many illustrations about behavior as illustrations about the brain. This emphasis on explaining the biological foundation of behavior is another reason that we include both Clinical Focus and Research Focus features throughout the text.

Abundant Chapter Pedagogy

In addition to the innovative teaching devices described so far, numerous in-text pedagogical aids adorn every chapter, beginning with an outline and an opening Focus feature that draws students into the chapter's topic. Clinical, Research, and Comparative Focus features dot each chapter to connect brain and behavior to relevant clinical or research experience. Within the chapters, boldface key terms are defined in the margins to reinforce their importance, margin notes link topics together, and end-of-section Review self-tests help students check their grasp of major points.

Each chapter ends with a Summary—several include summary tables or illustrations to help students visualize or review concepts—and a list of Key Terms, each referenced to the page number on which the term is defined. As in the Third Edition, additional resources can be found on the Companion Web Site at www.worthpublishers.com/kolbintro4e. Other material on the Web site will broaden students' understanding of chapter topics.

Superb Visual Reinforcement

You can see our most important learning aid simply by paging through the book: an expansive and, we believe, exceptional set of illustrations that, hand in hand with our words, describe and illuminate the nervous system. On the advice of instructors and readers, important anatomical diagrams have been enlarged to ease perusal, and we have retained and added new "applications" photos that range from Tweeting in Section 2-3 to a dance class for Parkinson's patients in Section 5-3 to a seniors' bridge game in Section 16-3.

Illustrations are consistent from chapter to chapter and reinforce one another. We consistently color-code diagrams that illustrate each aspect of the neuron, depict each structural region in the brain, and demark the divisions of the nervous system. We include many varieties of micrographic images to show what a particular neural structure actually looks like. These illustrations and images are included on our PowerPoint presentations and integrated as labeling exercises in our Study Guide and Testing materials.

Teaching Through Metaphors, Examples, and Principles

If a textbook is not enjoyable, it has little chance of teaching well. We heighten students' interest through abundant use of metaphors and examples. Students read about patients whose brain injuries are sources of insight into brain function, and we examine car engines, robots, and prehistoric flutes for the same purpose. Frequent comparative biology examples, illustrated Experiments, and representative Comparative Focus features help students understand how much we humans have in common with creatures as far distant from us as sea slugs.

We also facilitate learning by reemphasizing main points and by distilling sets of principles about brain function that can serve as a framework to guide students' thinking. Thus, Section 2-6 introduces ten key principles that explain how the various parts of the nervous system work together. Section 14-4 summarizes seven guiding principles of neuroplasticity. These principles form the basis of many discussions throughout the book, and marginal notes remind readers when they encounter the principles again—as well as where to review them in depth.

Big-Picture Emphasis

One challenge in writing an introductory book on any topic is deciding what to include and what to exclude. We organize discussions to focus on the bigger picture—a focus exemplified by the ten principles of nervous-system function introduced in Section 2-6 and echoed throughout the book. Any set of principles may be a bit arbitrary; nevertheless, it gives students a useful framework for understanding the brain's activities.

In Chapters 8 through 16 we tackle behavioral topics in a more general way than most contemporary books do. In Chapter 12, for instance, we revisit experiments and ideas from the 1960s to understand why animals behave as they do then we consider emotional and motivated behaviors as diverse as eating and anxiety attacks in humans. In Chapter 14, the larger picture of learning and memory is presented alongside a discussion of recovery from traumatic brain injury.

This broad focus helps students grasp the big picture that behavioral neuroscience is all about. While broadening our focus requires us to leave out some details, our experience with students and teachers through the three earlier editions confirms that discussion of the larger problems and issues in brain and behavior is of greater interest to students, especially those who are new to this field, and is more often remembered than are myriad details without context.

As in preceding editions, we are selective in our citation of the truly massive literature on the brain and behavior because we believe that numerous citations can disrupt the text's flow and distract students from the task of mastering concepts. We provide citations to classic works by including the names of the researchers and by mentioning where the research was performed. In areas where controversy or new breakthroughs predominate, we also include detailed citations to papers (especially reviews) from the years 2010–2013 when possible. An end-of-book References section lists all the literature used in developing the book, reflecting the addition of more than 200 new citations in this new edition and elimination of other, now superseded, research.

Acknowledgments

As in past editions of this text and *Fundamentals of Human Neuropsychology,* we have a special debt to Barbara Brooks, our development editor. She has learned how to extract the best from each of us by providing a firm guiding hand to our thinking. While we don't always initially agree with her (or our wives), we have learned to listen carefully and discover that she is usually right (as they are). Also, because we have now worked together on multiple editions, each of us has learned something of how the others think. Barbara's sense of humor is infectious and her commitment to excellence has again left a strong imprint on the entire book. Once again thank you, Barbara. You make this whole experience fun—believe it or not!

FOURTH EDITION UPDATES CHAPTER-BY-CHAPTER

CHAPTER 15: COGNITION

NEW Research Focus 15-1: experiment conducted with split-brain patients.

NEW Figure 15-2: prefrontal subregions important in cognition.

UPDATED Research Focus 15-3: evidence on mirror-neuron dysfunction in autism.

NEW in §15-3: mapping the **brain connectome**, the emerging fields of **social neuroscience** and **neuroeconomics**.

NEW Figures 15-17 and 15-18: sex differences in brain architecture.

UPDATED §15-6: efficiency of cortical networks and epigenetic factors in individual differences in intelligence.

CHAPTER 16: DISORDERS AND DYSFUNCTION

UPDATES chapterwide on epigenetic factors and neuroplasticity.

UPDATED Research Focus 16-1: increasing effectiveness of **virtual-reality exposure therapies**.

NEW Figure 16-10: healthy adult brain and brain shriveled by Alzheimer's disease.

NEW Clinical Focus 16-3: Concussion images **chronic traumatic encephalopathy**.

UPDATED §16-5 concludes with some pros and cons of **cognitive enhancement**.

We must sincerely thank the many people who contributed to the development of this edition. The staff at Worth Publishers is remarkable and makes revisions a joy to do. We thank our sponsoring editor Daniel DeBonis, assistant editor Nadina Persaud, our long-time project editor Georgia Lee Hadler, and our production manager Sarah Segal. Our manuscript editor Penelope Hull ensured the clarity and consistency of the text. Kate Scully and her team at Northeastern Graphic ensured that the page proofs were set properly and included the necessary revisions. We thank art director Babs Reingold for a striking cover and Charles Yuen and Lyndall Culbertson for a fresh, inviting, accessible new interior design. Thanks also to Cecilia Varas for coordinating photo research and to Julie Tesser, who found photographs and other illustrative materials that we would not have found on our own. We remain indebted to the illustrators at Dragonfly Media for their excellent work in creating new illustrations.

Our colleagues, too, have helped in the development of every edition. For their contributions to the Fourth Edition, we are especially indebted to the reviewers who provided extensive comments on selected chapters and illustrations: Mark Basham, *Regis University*; Pam Costa, *Tacoma Community College*; Russ Costa, *Westminster College*; Renee Countryman, *Austin College*; Kristen D'Anci, *Salem State University*; Trevor James Hamilton, *Grant MacGewn University*; Christian Hart, *Texas Woman's University*; Matthew Holahan, *Carleton University*; Chris Jones, *College of the Desert*; Joy Kannarkat, *Norfolk State University*; Jennifer Koontz, *Orange Coast College*; Kate Makerec, *William Paterson University of New Jersey*; Daniel Montoya, *Fayatteville State University*; Barbara Oswald, *Miami University of Ohio*; Gabriel Radvansky, *University of Notre Dame*; Jackie Rose, *Western Washington University*; Steven Schandler, *Chapman University*; Maharaj Singh, *Marquette University*; Manda Williamson, *University of Nebraska—Lincoln*.

Likewise, we continue to be indebted to the colleagues who provided extensive comments on selected chapters and illustrations during the development of the Third Edition: Chana Akins, *University of Kentucky*; Michael Anch, *Saint Louis University*; Maura Mitrushina, *California State University, Northridge*; Paul Wellman, *Texas A&M University*; and Ilsun White, *Morehead State University*. The methods chapter was new to the Third Edition and posed the additional challenge of taking what easily could read like a seed catalogue and making it engaging to readers. We therefore are indebted to Margaret G. Ruddy, *The College of New Jersey,* and Ann Voorhies, *University of Washington,* for providing extensive advice on the initial version of Chapter 7.

We'd also like to thank the reviewers who contributed their thoughts to the Second Edition: Barry Anton, *University of Puget Sound*; R. Bruce Bolster, *University of Winnipeg*; James Canfield, *University of Washington*; Edward Castañeda, *University of New Mexico*; Darragh P. Devine, *University of Florida*; Kenneth Green, *California State University, Long Beach*; Eric Jackson, *University of New Mexico*; Michael Nelson, *University of Missouri, Rolla*; Joshua S. Rodefer, *University of Iowa*; Charlene Wages, *Francis Marion University*; Doug Wallace, *Northern Illinois University*; Patricia Wallace, *Northern Illinois University*; and Edie Woods, *Madonna University.* Sheri Mizumori, *University of Washington,* deserves special thanks for reading the entire manuscript for accuracy and providing fresh ideas that proved invaluable.

Finally, we must thank our tolerant wives for putting up with sudden changes in plans as chapters returned from Barbara or Penny with hopes for quick turnarounds. We also thank our colleague Robbin Gibb, who uses the book and has provided much feedback, in addition to our graduate students, technicians, and postdoctoral fellows who kept our research programs moving forward when were engaged in revising the book.

Bryan Kolb and Ian Q. Whishaw

An Introduction to Brain and Behavior, Fourth Edition, features a wide array of supplemental materials designed exclusively for students and teachers of the text. For more information about any of the items, please visit Worth Publishers' online catalog at www.worthpublishers.com.

For Students

NEW! ONLINE NEUROSCIENCE TOOL KIT Available Spring 2013 at www.worth publishers.com/ntk The Neuroscience Tool Kit is a powerful Web-based tool for learning the core concepts of behavioral neuroscience—by witnessing them firsthand. These 30 interactive tutorials allow students to see the nervous system in action via dynamic illustrations, animations, and models that demystify the neural mechanisms behind behavior. Videos taken from actual laboratory research enhance student understanding of how we know what we know, and carefully crafted multiple-choice questions make it easy to assign and assess each activity. Based on Worth Publishers' groundbreaking *Foundations of Behavioral Neuroscience CD-ROM,* the Neuroscience Tool Kit is a valuable accompaniment to any biopsychology course.

AN INTRODUCTION TO BRAIN AND BEHAVIOR, FOURTH EDITION, COMPANION WEB SITE www.worthpublishers.com/kolbintro4e Created by Joe Morrissey of Binghamton University, the companion Web site is an online educational setting for students that provides a virtual study guide, 24 hours a day, 7 days a week. Best of all, the resources are free and do not require any special access codes or passwords. Tools on the site include chapter outlines; learning objectives; interactive flashcards; research exercises; selections from PsychSim 5.0 by Thomas Ludwig, Hope College; and online quizzes with immediate feedback and instructor notification.

For the instructor, the site offers access to a quiz gradebook for viewing student results, a syllabus posting service, lecture slides, illustration slides, electronic versions of illustrations in the book, and links to additional tools including course cartridges for Blackboard, WebCT, Angel, Desire2Learn, and others.

REVISED! STUDY GUIDE FOR *AN INTRODUCTION TO BRAIN AND BEHAVIOR* Written by Terrence J. Bazzett of the State University of New York at Geneseo, the revised Study Guide is carefully crafted to help students master each chapter of *An Introduction to Brain and Behavior,* Fourth Edition. To aid learning and retention, the Study Guide includes a review of key concepts and terms, practice tests, short-answer questions, illustrations for identification and labeling, Internet activities, and crossword puzzles. In one of the most difficult courses in psychology, the Study Guide is a great help to students at all levels.

COURSESMART E-BOOK A complete electronic version of *An Introduction to Brain and Behavior,* Fourth Edition, can be previewed and purchased at www.coursesmart.com. Students can choose to view the CourseSmart e-Book online or download it to a personal computer or a portable media player such as a smart phone or iPad. This flexible, easy-to-use format makes the text more portable than ever before!

PSYCHOLOGY AND THE REAL WORLD: ESSAYS ILLUSTRATING FUNDAMENTAL CONTRIBUTIONS TO SOCIETY A superb collection of essays by major researchers describing their landmark studies. Published in association with the not-for-profit FABBS Foundation, this engaging reader includes Bruce McEwen's work on the neurobiology of stress

and adaptation, Elizabeth Loftus's own reflections on her study of false memories, and Daniel Wegner on his study of thought suppression. A portion of the proceeds is donated to FABBS to support societies of cognitive, psychological, behavioral, and brain sciences.

IMPROVING THE MIND AND BRAIN: A SCIENTIFIC AMERICAN SPECIAL ISSUE On request, this reader is free of charge when packaged with the textbook. This single-topic issue from *Scientific American* magazine features the latest findings from the most distinguished researchers in the field.

Worth Publishers is pleased to offer cost-saving packages of *An Introduction to Brain and Behavior,* Fourth Edition, with our most popular supplements. Below is a list of some of the most popular combinations available for order through your local bookstore.

Hardcover Text & Neuroscience Tool Kit Access Card
ISBN-10: 1464143676

Hardcover Text & Study Guide
ISBN-10: 1464143641

Hardcover Text & *Improving the Mind and Brain: A* Scientific American *Special Issue*
ISBN-10: 1464143668

Hardcover Text & *Psychology and the Real World*
ISBN-10: 1464145180

For Instructors

REVISED! INSTRUCTOR'S RESOURCES Revised by Manda Williamson, University of Nebraska—Lincoln, this invaluable tool for new and experienced instructors alike, the resources include chapter-by-chapter learning objectives and chapter overviews, detailed lecture outlines, thorough chapter summaries, chapter key terms, in-class demonstrations and activities, springboard topics for discussion and debate, ideas for research and term-paper projects, homework assignments and exercises, and suggested readings from journals and periodicals. Course-planning suggestions and a guide to videos and Internet resources also are included. The Instructor's Resources can be downloaded from the Companion Web Site at www.worthpublishers.com/kolbintro4e.

NEW! FACULTY LOUNGE Faculty Lounge is an online forum provided by Worth Publishers where instructors can find and share favorite teaching ideas and materials, including videos, animations, images, lecture slides, news stories, articles, Web links, and in-class activities. Sign up to browse the site or upload your favorite materials for teaching psychology at http://psych.facultylounge.worthpublishers.com.

Assessment Tools

DIPLOMA COMPUTERIZED TEST BANK Prepared by Christopher Striemer of Grant MacEwan University, the revised Test Bank includes a battery of more than 1300 multiple-choice and short-answer test questions as well as diagram exercises—500 of which are new. Each item is keyed to the page in the textbook on which the answer can be found. All the questions have been thoroughly reviewed and edited for accuracy and clarity. The Test Bank is available on a dual-platform CD-ROM. Instructors are guided step-by-step through the process of creating a test and can quickly add,

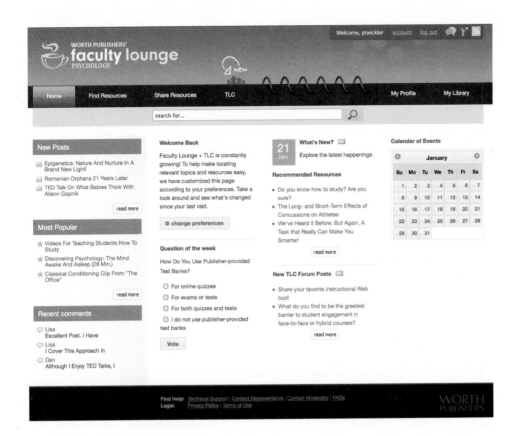

edit, scramble, or resequence items. The Test Bank will also allow you to export into a variety of formats that are compatible with many internet-based testing products. For more information on Diploma, please visit www.wimba.com/products/diploma.

ONLINE QUIZZES The Companion Web Site at www.worthpublishers.com/kolbintro4e features two multiple-choice quizzes to accompany each chapter of the text. Students can take each quiz multiple times and receive instant feedback. Instructors can then access student results in the online gradebook and view the results by quiz, student, or question, or they can obtain weekly results by email.

Presentation

ILLUSTRATION SLIDES AND LECTURE SLIDES Available at www.worthpublishers.com/kolbintro4e, these slides either can be used directly or customized to fit the needs of your course. There are two sets of slides for each chapter. One set features all the figures, photos, and tables, and the other features main points of the chapter with selected figures and illustrations.

Video

WORTH PUBLISHERS' NEUROSCIENCE VIDEO COLLECTION Edited by Ronald J. Comer of Princeton University and available on DVD, this video collection consists of dozens of short video segments to enhance the biopsychology classroom and taken from clinical documentaries, television news reports, archival footage, and other sources. Each segment has been created to bring a lecture to life, engaging students and enabling them to apply neuroscience theory to the real world. This collection offers powerful and memorable demonstrations of the links between the brain and behavior, neuroanatomical animations, cutting-edge neuroscience research, brain assessment in action, important historical events, interviews, and a wide sampling of brain phenomena and brain dysfunction. A special cluster of segments reveals the range of research methods used to study the brain. The accompanying Faculty Guide (available at www. worthpublishers.com/kolbintro4e) offers a description of each segment so that instructors can make informed decisions about how to best use the videos.

COURSE MANAGEMENT The various resources for this textbook are available in the appropriate format for users of Blackboard, WebCT, Angel, Desire2Learn, and other systems. Course outlines, prebuilt quizzes, links, and activities are included, eliminating hours of work for instructors. For more information, please visit our Web site at www.macmillanhighered.com/lms.

CHAPTER 1

What Are the Origins of Brain and Behavior?

CLINICAL FOCUS 1-1 LIVING WITH TRAUMATIC BRAIN INJURY

1-1 NEUROSCIENCE IN THE TWENTY-FIRST CENTURY

WHY STUDY BRAIN AND BEHAVIOR?

WHAT IS THE BRAIN?

GROSS ANATOMY OF THE NERVOUS SYSTEM

WHAT IS BEHAVIOR?

1-2 PERSPECTIVES ON BRAIN AND BEHAVIOR

ARISTOTLE AND MENTALISM

DESCARTES AND DUALISM

DARWIN AND MATERIALISM

COMPARATIVE FOCUS 1-2 THE SPEAKING BRAIN

CONTEMPORARY PERSPECTIVES ON BRAIN AND BEHAVIOR

1-3 EVOLUTION OF BRAINS AND OF BEHAVIOR

ORIGIN OF BRAIN CELLS AND BRAINS

EVOLUTION OF ANIMALS HAVING NERVOUS SYSTEMS

THE BASICS CLASSIFICATION OF LIFE

CHORDATE NERVOUS SYSTEM

1-4 EVOLUTION OF THE HUMAN BRAIN AND BEHAVIOR

HUMANS: MEMBERS OF THE PRIMATE ORDER

AUSTRALOPITHECUS: OUR DISTANT ANCESTOR

THE FIRST HUMANS

RELATING BRAIN SIZE AND BEHAVIOR

WHY THE HOMINID BRAIN ENLARGED

RESEARCH FOCUS 1-3 CLIMATE AND THE EVOLVING HOMINID BRAIN

1-5 MODERN HUMAN BRAIN SIZE AND INTELLIGENCE

MEANING OF HUMAN BRAIN-SIZE COMPARISONS

CULTURE

COMPARATIVE FOCUS 1-4 EVOLUTION AND ADAPTIVE BEHAVIOR

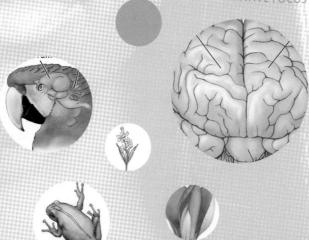

Living with Traumatic Brain Injury

Fred Linge, a clinical psychologist with a degree in brain research, wrote this description 12 years after his injury occurred:

> In the second it took for my car to crash head-on, my life was permanently changed, and I became another statistic in what has been called "the silent epidemic."
>
> During the next months, my family and I began to understand something of the reality of the experience of head injury. I had begun the painful task of recognizing and accepting my physical, mental, and emotional deficits. I couldn't taste or smell. I couldn't read even the simplest sentence without forgetting the beginning before I got to the end. I had a hair-trigger temper that could ignite instantly into rage over the most trivial incident. . . .
>
> Two years after my injury, I wrote a short article: "What Does It Feel Like to Be Brain Damaged?" At that time, I was still intensely focusing on myself and my own struggle. (Every head-injured survivor I have met seems to go through this stage of narcissistic preoccupation, which creates a necessary shield to protect them from the painful realities of the situation until they have a chance to heal.) I had very little sense of anything beyond the material world and could only write about things that could be described in factual terms. I wrote, for example, about my various impairments and how I learned to compensate for them by a variety of methods.
>
> At this point in my life, I began to involve myself with other brain-damaged people. This came about in part after the publication of my article. To my surprise, it was reprinted in many different publications, copied, and handed out to thousands of survivors and families. It brought me an enormous outpouring of letters, phone calls, and personal visits that continue to this day. Many were struggling as I had struggled, with no diagnosis, no planning, no rehabilitation, and most of all, no hope. . . . The catastrophic effect of my injury was such that I was shattered and then remolded by the experience, and I emerged from it a profoundly different person with a different set of convictions, values, and priorities. (Linge, 1990)

(*Left*) U.S. Representative Gabrielle Giffords (D-AZ) reenacts her swearing-in with House Speaker John Boehner in January 2011, days before a gunshot through the left side of her brain left her near death. (*Right*) One year later in Tucson, Rep. Giffords and her husband, former astronaut Mark Kelly, attend a candlelight vigil for all those who were shot, including the six who died. Giffords had regained limited speech, partly with the help of singing therapy, but mobility on her right side remained limited. Two weeks later she resigned her House seat to concentrate on recovering from the traumatic brain injury.

1-1 Neuroscience in the Twenty-First Century

In the years after his injury, Fred Linge made a journey. Before the car crash, he gave little thought to the relation between his brain and his behavior. After the crash, adapting to his injured brain and behavior dominated his life. On that journey he had to learn about his brain, he had to relearn many of his old skills, and he had learn to compensate for the impairments that his changed brain imposed on him.

The purpose of this book is to take *you* on a journey toward understanding the link between brain and behavior: how the brain is organized to create behavior. Evidence comes from studying three sources: (1) the evolution of brain and behavior in diverse animal species, (2) how the brain is related to behavior in normal people, and (3) how the brain changes in people who suffer brain damage or other brain abnormalities. The knowledge emerging from these lines of study is changing how we think about ourselves, how we structure education and our social interactions, and how we aid those with brain injury, disease, and disorder.

On our journey, we will learn about ourselves. We will learn how the brain stores and retrieves information, why we engage in the behaviors we engage in, and how we are able to read the lines on this page and generate ideas and thoughts. The coming decades will be exciting times for the study of brain and behavior. They will offer an opportunity for us to broaden our understanding of what makes us human.

We will marvel at the potential for future discoveries. We will begin to understand how genes control neural activity. The development of new imaging techniques will reveal how our own brains think. One day, we will be able to arrest the progress of brain disease. One day, we will be able to stimulate processes of repair in malfunctioning brains. One day, we will be able to make artificial brains that extend the functions of our own brains. One day, we will understand ourselves and other animals.

Why Study Brain and Behavior?

The *brain* is a physical object, a living tissue, a body organ. *Behavior* is action, momentarily observable, but fleeting. Brain and behavior differ greatly but are linked. They have evolved together: one is responsible for the other, which is responsible for the other, which is responsible for the other, and so on and on. There are three reasons for linking the study of brain to the study of behavior:

1. *How the brain produces behavior is a major unanswered scientific question.* Scientists and students study the brain for the purpose of understanding humanity. Understanding brain function will allow improvements in many aspects of our world, including educational systems, economic systems, and social systems. Many chapters in this book touch on the relation between psychological questions related to brain and behavior and philosophical questions related to humanity. For example, in Chapters 14 and 15, we address questions related to how we become conscious, how we speak, and how we remember.

2. *The brain is the most complex living organ on Earth and is found in many different groups of animals.* Students of the brain want to understand its place in the biological order of our planet. Chapter 1 describes the basic structure and evolution of brains, especially the human brain, Chapter 2 surveys its structures and functions, and Chapters 3 through 5 describe the functioning of brain cells—the building blocks of the brains of all animals.

3. *A growing list of behavioral disorders can be explained and cured by understanding the brain.* Indeed, more than 2000 disorders may in some way be related to brain abnormalities. As indexed inside the front cover of this book, we detail relations between brain disorders and behavioral disorders in every chapter, especially in the "Focus" features.

None of us can predict the ways in which knowledge about the brain and behavior may prove useful. A former psychology major wrote to tell us that she took our course because she was unable to register in a preferred course. She felt that, although our course was interesting, it was "biology, not psychology." After graduating and getting a job in a social service agency, she has found to her delight that an understanding of the links between brain and behavior is a source of insight into the disorders of many of her clients and the treatment options for them.

What Is the Brain?

For his postgraduate research, our friend Harvey chose to study the electrical activity given off by the brain. He said that he wanted to live on as a brain in a bottle after his body died. He expected that his research would allow his bottled brain to communicate with others who could "read" his brain's electrical signals. Harvey mastered the techniques of brain electrical activity but failed in his objective, not only because the goal was technically impossible but also because he lacked a full understanding of what "brain" means.

Brain is the Anglo-Saxon word for the tissue found within the skull, and it is this tissue that Harvey wanted to put into a bottle. As shown in **Figure 1-1,** the human brain comprises two major sets of structures, the cerebrum and the cerebellum. The **cerebrum** (*forebrain*), shown in Figure 1-1A, has two nearly symmetrical halves, called **hemispheres,**

cerebrum (forebrain) Major structure of the forebrain that consists of two virtually identical hemispheres (left and right) and is responsible for most conscious behavior.

hemisphere Literally, half a sphere, referring to one side of the cerebrum.

A series of illustrated Experiments appears through the book to reveal how neuroscientists conduct research. Section 7-7 frames the debates on the benefits and the ethics of conducting research using nonhuman animals.

Chapter 2 presents the brain's anatomical structures and functions in detail.

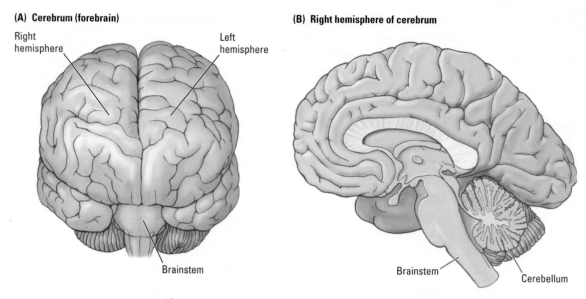

(A) Cerebrum (forebrain)

Right hemisphere

Left hemisphere

Brainstem

(B) Right hemisphere of cerebrum

Brainstem

Cerebellum

FIGURE 1-1 The Human Brain **(A)** Oriented within the human skull, the nearly symmetrical left and right hemispheres of the cerebrum shown head-on. **(B)** A cut through the middle of the brain from back to front reveals the right hemispheres of the cerebrum and cerebellum and the brainstem. The spinal cord emerges from the base of the brainstem.

FIGURE 1-2 Major Divisions of the Human Nervous System The brain and spinal cord together make up the central nervous system. All the nerve processes radiating out beyond the brain and spinal cord and all the neurons outside the CNS connect to sensory receptors, muscles, and internal body organs to form the peripheral nervous system.

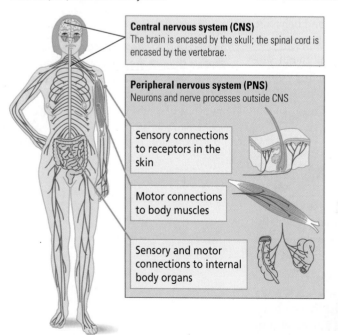

Central nervous system (CNS)
The brain is encased by the skull; the spinal cord is encased by the vertebrae.

Peripheral nervous system (PNS)
Neurons and nerve processes outside CNS

Sensory connections to receptors in the skin

Motor connections to body muscles

Sensory and motor connections to internal body organs

one on the left and one on the right. The cerebrum is responsible for most of our conscious behaviors. It enfolds the **brainstem** (Figure 1-1B), which is responsible for most of our unconscious behaviors. The second major brainstem structure, the **cerebellum,** is specialized for learning and coordinating skilled movements.

Harvey clearly wanted to preserve not just his brain but his *self*—his consciousness, his language, and his memory. This meaning of the term *brain* refers to something other than the organ found inside the skull. It refers to the brain as that which exerts control over behavior. It is what we intend when we talk of someone smart being "a brain" or speak of the computer that guides a spacecraft as being the vessel's "brain." The term *brain,* then, signifies both the organ itself and the fact that this organ produces behavior. Why could Harvey not manage to preserve his control-exerting self inside a bottle? Read on to learn one answer to this question.

Gross Anatomy of the Nervous System

The major divisions of the human nervous system, illustrated in **Figure 1.2,** are composed of cells, as is the rest of the body, and these nerve cells, or **neurons,** most directly control behavior. Neurons in the brain communicate with one another, with sensory receptors in the skin, with muscles, and with internal body organs. Most of the connections between the brain and the rest of the body are made through the **spinal cord,** which descends from the brainstem through a canal in the backbone.

Together, the brain and spinal cord make up the **central nervous system** (CNS). The CNS is encased in bone, the brain by the skull and the spinal cord by the vertebrae. The CNS is "central" both because it is physically located to be the core of the nervous system and because it is the core structure mediating behavior. All the processes radiating out beyond the brain and spinal cord as well as all the neurons outside the brain and spinal cord constitute the **peripheral nervous system** (PNS).

To return to Harvey's brain-in-a-bottle experiment, the effect of placing the brain or even the entire CNS in a bottle would be to separate it from the PNS and thus to separate it from the sensations and movements mediated by the PNS. Could the brain function without sensory information and without the ability to produce movement?

In the 1920s, Edmond Jacobson wondered what would happen if our muscles completely stopped moving, a question relevant to Harvey's experiment. Jacobson believed that, even when we think we are entirely motionless, we still make subliminal movements related to our thoughts. The muscles of the larynx subliminally move when we "think in words," for instance, and we make subliminal movements of our eyes when we imagine or visualize a scene. So, in Jacobson's experiment, people practiced "total" relaxation and were later asked what the experience was like. They reported a condition of "mental emptiness," as if the brain had gone blank (Jacobson, 1932).

In 1957, Woodburn Heron investigated another question related to Harvey's experiment. How would the brain cope without sensory input? He examined the effects of sensory deprivation, including feedback from movement, by having each subject lie on a bed in a bare, soundproof room and remain completely still. Padded tubes covered the subjects' arms so that they had no sense of touch, and translucent goggles cut off their vision. The subjects reported that the experience was extremely unpleasant, not just because of the social isolation but also because they lost their normal focus in this situation. Some subjects even hallucinated, as if their brains were somehow trying to create the sensory experiences that they suddenly lacked. Most asked to be released from the study before it ended.

One line of research and philosophical argument, called **embodied language,** proposes that the movements we make and the movements we perceive in others are central to communication with others (Prinz, 2008). That is, we understand one another not only by listening to words but also by observing gestures and other body language, and we think not only with silent language but also with overt gestures and body language.

Findings from these lines of research suggest that (1) the CNS needs ongoing sensory stimulation from the world and from its own body's movement and (2) the brain communicates by producing movement and observing the movements of others. Thus, when we use the term *brain* to mean an intelligent, functioning organ, we should refer to an active brain that is connected to the rest of the nervous system and engaged in doing its job of producing behavior. Unfortunately for Harvey, the normal functioning of a brain in a bottle, disconnected from the PNS, seems unlikely.

What Is Behavior?

Irenäus Eibl-Eibesfeldt began his textbook *Ethology: The Biology of Behavior,* published in 1970, with the following definition: "Behavior consists of patterns in time." These patterns can be made up of movements, vocalizations, or changes in appearance, such as the facial movements associated with smiling. The expression "patterns in time" includes thinking. Although we cannot directly observe someone's thoughts, techniques exist for monitoring changes in the brain's electrical and biochemical activity that may be associated with thought. So thinking, too, is a behavior that forms patterns in time.

The behavioral patterns of animals vary enormously. Animals produce behaviors that are inherited ways of responding, and they also produce behaviors that are learned. Most behaviors probably consist of a mix of inherited and learned actions. An example of the difference between a mainly inherited behavior and a mainly learned behavior is

brainstem Central structure of the brain responsible for most unconscious behavior.

cerebellum Major structure of the brainstem specialized for coordinating and learning skilled movements. In large-brained animals, the cerebellum may also have a role in coordinating other mental processes.

neuron Specialized nerve cell engaged in information processing.

spinal cord Part of the central nervous system encased within the vertebrae (spinal column) that provides most of the connections between the brain and the rest of the body.

central nervous system (CNS) The brain and spinal cord that together mediate behavior.

peripheral nervous system (PNS) All the neurons in the body located outside the brain and spinal cord; provides sensory and motor connections to and from the central nervous system.

embodied language Hypothesis that the movements we make and the movements we perceive in others are central to communication with others.

Specialized "mirror" neurons, described in Sections 11-1 and 15-2, facilitate this nonverbal communication.

Find out more about sensory deprivation in Section 12-1, where Figure 12-1 illustrates Heron's setting for the experiments.

A crossbill's beak is specifically designed to open pine cones. This behavior is innate.

A baby roof rat must learn from its mother how to eat pine cones. This behavior is learned.

FIGURE 1-3 Innate and Learned Behaviors Some animal behaviors are largely innate and fixed (*top*), whereas others are largely learned (*bottom*). This learning is a form of cultural transmission. (*Top*) Adapted from *The Beak of the Finch* (p. 183), by J. Weiner, 1995, New York: Vintage. (*Bottom*) Adapted from "Cultural Transmission in the Black Rat: Pinecone Feeding," by J. Terkel, 1995, *Advances in the Study of Behavior, 24*, p. 122.

François Gerard, *Psyche and Cupid* (1798)

the contrast in the eating behavior of two different animal species—crossbills and roof rats—illustrated in **Figure 1-3.**

A crossbill is a bird with a beak that seems to be awkwardly crossed at the tips; yet this beak is exquisitely evolved to eat pinecones. If the shape of a crossbill's beak is changed even slightly, the bird is unable to eat preferred pinecones until its beak grows back. Thus, eating, for crossbills, is a fixed behavioral pattern that is inherited and does not require much modification through learning. Roof rats, in contrast, are rodents with sharp incisor teeth that appear to have evolved to cut into anything. But roof rats can eat pinecones efficiently only if they are taught to do so by an experienced mother.

The mixture of inherited and learned behaviors varies considerably in different species. Generally, animals with smaller, simpler nervous systems have a narrow range of behaviors that depend on *heredity*. Animals with complex nervous systems have more behavioral options that depend on *learning*. We humans believe that we are the animal species with the most complex nervous system and the greatest capacity for learning new responses.

But even most human behaviors involve some mixture of inheritance and learning because we humans have not thrown away our simpler nervous systems. For this reason, although human behavior depends mostly on learning, we, like other species, still possess many inherited ways of responding. The sucking response of a newborn infant is an inherited eating pattern in humans, for example.

REVIEW 1-1
Neuroscience in the Twenty-First Century

Before you continue, check your understanding.

1. One major set of brain structures, the _____, or _____, whose nearly symmetrical left and right _____ enfold the _____, connects to the spinal cord.

2. The brain and spinal cord together make up the _____. All the nerve fibers radiating out beyond the brain and spinal cord as well as all the neurons outside the brain and spinal cord form the _____.

3. A simple definition of behavior is any kind of movement in a living organism. All behaviors have both a cause and a function, but they vary in complexity and in the degree to which they are _____, or automatic, and the degree to which they depend on _____.

4. Explain the concept of *embodied language* in a statement or brief paragraph.

Answers appear at the back of the book.

1-2 Perspectives on Brain and Behavior

Returning to the central topic in the study of brain and behavior—how the two are related—we now survey three classic theories about the cause of behavior: mentalism, dualism, and materialism. Then we explain why contemporary brain investigators subscribe to the materialist view. In reviewing these theories, you will recognize that some familiar "commonsense" ideas you might have about behavior are derived from one or another of these long-standing perspectives.

Aristotle and Mentalism

The hypothesis that the mind (or soul or psyche) is responsible for behavior can be traced back more than 2000 years to ancient Greece. In classical mythology, Psyche was a mortal who became the wife of the young god Cupid. Venus, Cupid's mother, opposed his marriage to a mortal, so she harassed Psyche with almost impossible tasks.

E. Lessing/Art Resource, New York

Psyche performed the tasks with such dedication, intelligence, and compassion that she was made immortal, thus removing Venus's objection to her. The ancient Greek philosopher Aristotle was alluding to this story when he suggested that all human intellectual functions are produced by a person's **psyche.** The psyche, Aristotle argued, is responsible for life, and its departure from the body results in death.

Aristotle's account of behavior had no role for the brain, which Aristotle thought existed to cool the blood. To him, the nonmaterial psyche was responsible for human consciousness, perceptions, and emotions and for such processes as imagination, opinion, desire, pleasure, pain, memory, and reason. The psyche was an entity independent of the body. Aristotle's view that a nonmaterial psyche governs our behavior was adopted by Christianity in its concept of the soul and has been widely disseminated throughout the world.

Mind is an Anglo-Saxon word for "memory," and when "psyche" was translated into English, it became "mind." The philosophical position that a person's mind (psyche) is responsible for behavior is called **mentalism.** Mentalism has influenced modern behavioral science because many terms—*consciousness, sensation, perception, attention, imagination, emotion, motivation, memory,* and *volition* among them—remain in use for patterns of behavior today. Indeed, these terms are frequently used as chapter titles in contemporary psychology and neuroscience textbooks.

Descartes and Dualism

In the first book on brain and behavior, René Descartes (1596–1650), a French philosopher, proposed a new explanation of behavior in which the brain played an important role. Descartes placed the seat of the mind in the brain and linked the mind to the body. In the first sentence of *Treatise on Man* (1664), he stated that mind and body "must be joined and united to constitute people."

To Descartes, most of the activities of the body and brain, such as motion, digestion, and breathing, could be explained by mechanical and physical principles. The nonmaterial mind, on the other hand, is responsible for rational behavior. Descartes's proposal that an entity called the mind directs a machine called the body was the first serious attempt to explain the role of the brain in controlling behavior.

The idea that behavior is controlled by two entities, a mind and a body, is called **dualism** (from Latin, meaning "two"). To Descartes, the mind receives information from the body through the brain. The mind also directs the body through the brain. The rational mind, then, depends on the brain both for information and to control behavior.

Descartes was also aware of the many new machines being built, including gears, clocks, and waterwheels. He saw mechanical gadgets on public display in parks. In the water gardens in Paris, one device caused a hidden statue to approach and spray water when an unsuspecting stroller walked past it. The statue's actions were triggered when the person stepped on a pedal hidden in the sidewalk. Influenced by these mechanical devices, Descartes developed mechanical principles to explain the functions of the body.

Descartes also developed a mechanical explanation of how the mind produces movement. He suggested that the mind works through a small structure in the center of the brain, the pineal body (now called the *pineal gland*), which is located beside fluid-filled cavities called *ventricles* (**Figure 1-4**). According to Descartes, the mind instructs the pineal body to direct fluid from the ventricles through nerves and into muscles. When the fluid expands the muscles, the body moves.

Descartes's dualistic theory faces many problems. It quickly became apparent to scientists that people who have damaged pineal bodies or even no pineal body at all still display normal intelligent behavior. Today, we understand that the pineal gland plays a role in behavior related to biological rhythms, but it does not govern human behavior. We now know that fluid is not pumped from the brain into muscles when they contract.

psyche Synonym for *mind,* an entity once proposed to be the source of human behavior.

mind Proposed nonmaterial entity responsible for intelligence, attention, awareness, and consciousness.

mentalism Explanation of behavior as a function of the nonmaterial mind.

dualism Philosophical position that holds that both a nonmaterial mind and a material body contribute to behavior.

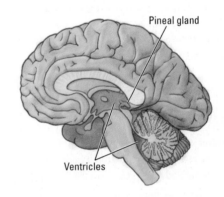

FIGURE 1-4 Dualist Hypothesis To explain how the mind controls the body, Descartes suggested that the mind resides in the pineal gland, where it directs the flow of fluid through the ventricles and into the body to investigate objects and to become informed about their properties.

Section 13-2 describes the pineal gland's function: it influences daily and seasonal biorhythms.

Placing an arm in a bucket of water and contracting the arm's muscles does not cause the water level in the bucket to rise, as it should if the volume of the muscle increased because fluid had been pumped into it. We now also know that there is no obvious way that a nonmaterial entity can influence the body, because doing so requires the spontaneous creation of energy, which violates the physical law of conservation of matter and energy.

The difficulty in Decartes's theory—how a nonmaterial mind and a physical brain might interact—has come to be called the **mind–body problem.** Nevertheless, Descartes proposed that his theory could be tested. To determine whether an organism possesses a mind, Descartes proposed the language test and the action test. To pass the language test, an organism must use language to describe and reason about things that are not physically present. The action test requires the organism to display behavior that is based on reasoning and is not just an automatic response to a particular situation. Descartes proposed that nonhuman animals and machines would be unable to pass the tests because they lack minds.

Descartes's theory of mind led to a number of unfortunate results. On the basis of it, some people argued that young children and the insane must lack minds, because they often fail to reason appropriately. We still use the expression "he's lost his mind" to describe someone who is "mentally ill." Some proponents of this view also reasoned that, if someone lacked a mind, that person was simply a machine not due normal respect or kindness. Cruel treatment of animals, children, and the mentally ill has been justified by Descartes's theory for centuries. It is unlikely that Descartes himself intended these interpretations. He was reportedly very kind to his own dog, Monsieur Grat.

Experimental research is also casting doubt on Descartes's view that only humans can pass the language and action tests. For example, studies of language in apes and other animals are partly intended to find out whether animals can describe and reason about things that are not present. Comparative Focus 1-2, "The Speaking Brain," summarizes a contemporary approach to studying language in animals. Computer specialists are also making progress on constructing robots with artificial intelligence—robots that think and remember.

Darwin and Materialism

By the mid-nineteenth century, another theory of brain and behavior emerged. This theory was **materialism**—the idea that rational behavior can be fully explained by the workings of the brain and the rest of the nervous system, without any need to refer to an immaterial mind. This perspective became prominent when supported by the evolutionary theory of Alfred Russel Wallace and Charles Darwin.

Evolution by Natural Selection

Wallace and Darwin independently arrived at the same conclusion—the idea that all living things are related. Both outlined this view in papers presented at the Linnaean Society of London in July 1858. Darwin further elaborated on the topic in his book *On the Origin of Species by Means of Natural Selection,* published in 1859. This book presented a wealth of supporting detail, which is why Darwin is regarded as the founder of modern evolutionary theory.

Both Darwin and Wallace had looked carefully at the structure of animals and at animal behavior. Despite the diversity of living animals, both men were struck by the myriad characteristics common to so many species. For example, the skeleton, muscles, and body parts of humans, monkeys, and other mammals are remarkably similar.

Such observations led first to the idea that living organisms must be related, an idea widely held even before Wallace and Darwin. But more important, these same observations led to Darwin's explanation of how the great diversity in the biological world

mind–body problem Quandary of explaining how a nonmaterial mind and a material body interact.

materialism Philosophical position that holds that behavior can be explained as a function of the brain and the rest of the nervous system without explanatory recourse to the mind.

natural selection Darwin's theory for explaining how new species evolve and how existing species change over time. Differential success in the reproduction of different characteristics (phenotypes) results from the interaction of organisms with their environment.

species Group of organisms that can interbreed.

phenotype Individual characteristics that can be seen or measured.

COMPARATIVE FOCUS ⬧ 1-2

The Speaking Brain

Language is such a striking characteristic of our species that it was once thought to be a trait unique to humans. Nevertheless, evolutionary theory predicts that language is unlikely to have appeared full-blown in modern humans. Language does have antecedents in other species. Many species lacking a cerebral cortex, including fish and frogs, are capable of elaborate vocalizations, and vocalization is still more elaborate in species having a cerebral cortex, such as birds, whales, and primates. But can nonhuman animals speak?

In 1969, Beatrice and Alan Gardner taught a version of American Sign Language to a chimpanzee named Washoe, showing that nonverbal forms of language might have preceded verbal language. Sue Savage-Rumbaugh and her coworkers (1999) then taught a pygmy chimpanzee named Malatta a symbolic language called Yerkish. (The pygmy chimpanzee, or *bonobo*, is a species thought to be an even closer relative of humans than the common chimp.)

Malatta and her son Kanzi were caught in the wild, and Kanzi accompanied his mother to class. It turned out that, even though he was not specifically trained, Kanzi learned more Yerkish than his mother did. Remarkably, Kanzi also displayed clear evidence of understanding complex human speech.

While recording vocalizations made by Kanzi interacting with people and eating food, Jared Taglialatela and coworkers found that Kanzi made many sounds associated with their meanings, or semantic context. For example, various peeps were associated with specific foods. The research group also found that chimps use a "raspberry" or "extended grunt" sound in a specific context to attract the attention of others, including people.

Imaging of blood flow in the brain associated with the use of "chimpanzeeish" indicates that the same brain regions that are activated when humans speak are also activated when chimpanzees speak (Taglialatela, 2011). These findings strongly support the idea that human language has antecedents in the vocalizations and gestures of nonhuman animals.

AP Images/Great Ape Trust/dapd

Kanzi

could have evolved from common ancestry. Darwin proposed that animals have traits in common because these traits are passed from parents to their offspring.

Natural selection is Darwin's theory for explaining how new species evolve and how existing species change over time. A **species** is a group of organisms that can breed among themselves but not with members of other species. Individual organisms within any species vary extensively in their **phenotype,** the characteristics we can see or measure. No two members of the species are exactly alike. Some are big, some are small, some are fat, some are fast, some are light-colored, and some have large teeth.

Individual organisms whose characteristics best help them to survive in their environment are likely to leave more offspring than are less fit members. This unequal ability of individual members to survive and reproduce leads to a gradual change in a species' population over time, with characteristics favorable for survival in a particular habitat becoming more prevalent in succeeding generations. Natural selection is nature's equivalent of the artificial selection practiced by plant and animal breeders to produce organisms with desirable traits.

Natural Selection and Heritable Factors

Neither Darwin nor Wallace understood the basis of the great variation in plant and animal species they observed. Another scientist, the monk Gregor Mendel, discovered one principle underlying phenotypic variation and how traits are passed from parents

Section 3-3 explains what constitutes a gene, how genes function, and how genes can change, or mutate.

to their offspring. Through experiments he conducted on pea plants in his monastery garden, beginning about 1857, Mendel deduced that heritable factors, which we now call *genes,* are related to various physical traits displayed by the species.

Members of a species that have a particular genetic makeup, or **genotype,** will *express* (turn on) that trait. If the gene or combination of genes for a trait, say, flower color, are passed on to offspring, the offspring will express the same trait, as illustrated in the Results section of **Experiment 1-1.** Two white-flowered pea plants produce white-flowered offspring in the first generation, or F1, and purple-flowered parents produce purple-flowered offspring. Observing this result, Mendel reasoned that two alternate heritable elements exist for the trait flower color.

Mendel then experimented with crossbreeding F1 purple and white pea plant flowers. The Results section of Experiment 1-1 shows that second-generation (F2) offspring all express the purple phenotype. Had the element that expresses white flowers disappeared? To find out, Mendel crossbred the F2 purple flowers. The third generation, F3, produced white flowers as well as purple in the ratio of roughly 1 white to 3 purple blooms.

EXPERIMENT 1-1

Question: How do parents transmit heritable factors to offspring?

Procedure
Mendel experimented by crossbreeding pea plants and then observing which traits parent plants passed on to their offspring in successive generations.

Results

White flower crosses produce white flowers in the first generation.

Purple flower crosses produce purple flowers in the first generation.

Parents   ×   Parents   ×

F1   F1

First-generation white flower crossed with first-generation purple flower produces all purple flowers in the second generation.

F1 Parents   ×

F2   ×

Second-generation purple flowers crossed with second-generation purple flowers produce, on average, three purple flowers and one white flower in the third generation.

F3

Conclusion: An individual inherits two factors, or genes, for each trait, but the effects of one gene may hide the other's effects in the individual's phenotype. The hidden gene can be reexpressed by crossbreeding.

This result suggested to Mendel that the trait for white flowers had not disappeared but rather was hidden by the trait for purple flowers. He concluded that individuals inherit two factors for each trait, but one factor may hide the other in the individual's phenotype.

The basic principles of inheritance that Mendel demonstrated through his experiments have led to countless discoveries about genetics. We now know, for example, that new traits appear because new gene combinations are inherited from parents, existing genes change or mutate, suppressed genes are reexpressed, expressed genes are suppressed, or genes or parts of genes are deleted or duplicated.

Thus, the unequal ability of individual organisms to survive and reproduce is related to the different genes they inherit from their parents and pass on to their offspring. By the same token, similar characteristics within or between species are usually due to similar genes. For instance, genes that produce the nervous system in different animal species tend to be very similar.

Interplay of Genes, Environment, and Experience

Genes alone cannot explain most traits. Mendel realized that environment plays a role in how genes express traits: planting tall peas in poor soil reduces their height, for example. Experience likewise plays a part. Children who are above average in intelligence but attend a substandard school, for example, have learning experiences that are far different from the experiences of similarly intelligent children who attend a model school.

Epigenetics is the study of differences in gene expression related to environment and experience. Epigenetic factors do not change your genes, but they do influence how your genes express the traits you've inherited from your parents. Epigenetic changes can persist throughout a lifetime, and the cumulative effects can make dramatic differences in how your genes work. Epigenetic studies described throughout this book promise to revolutionize our understanding of gene–brain interactions in normal brain development and in brain function. They will also help investigators develop new treatments for neurological disorders.

Summarizing Materialism

Darwin's theory of natural selection, Mendel's discovery of genetic inheritance, and the reality of epigenetics have three important implications for the study of the brain and behavior:

1. *Because all animal species are related, so too must be their brains.* A large body of research confirms that, first, the brain cells of all animals are so similar that these cells must be related and, second, all animal brains are so similar that they, too, must be related. Today, brain researchers study the nervous systems of animals as different as slugs, fruit flies, rats, and monkeys, knowing that they can extend their findings to human beings.

2. *Because all species of animals are related, so too must be their behavior.* Darwin was particularly interested in this subject. In his book *The Expression of the Emotions in Man and Animals,* he argued that emotional expressions are similar in humans and other animals because we inherited these expressions from a common ancestor. Evidence for such inheritance is illustrated in **Figure 1-5.** The fact that people in different parts of the world display the same behavior suggests that the trait is inherited rather than learned.

3. *Both the brain and behavior in complex animals such as humans evolved from the brain and behavior of simpler animals but also depend on learning.* In Section 1-3, we trace the steps in which the human nervous system and its repertoire of actions evolved from a simple netlike arrangement to a multipart nervous system with a brain that controls behavior.

genotype Particular genetic makeup of an individual.

epigenetics Differences in gene expression related to environment and experience.

Not all brain cells are neurons. Section 3-1 describes their varieties.

For more on emotions and their expression, see Sections 12-2, 12-4, and 14-3.

FIGURE 1-5 An Inherited Behavior
People from all parts of the world display the same emotional expressions that they recognize in others, as illustrated by these smiles. This evidence supports Darwin's suggestion that emotional expression is inherited.

minimally conscious state (MCS) Condition in which a person can display some rudimentary behaviors, such as smiling or uttering a few words, but is otherwise not conscious.

traumatic brain injury (TBI) Wound to the brain that results from a blow to the head.

persistent vegetative state (PVS) Condition in which a person is alive but unable to communicate or to function independently at even the most basic level.

clinical trial Consensual experiment directed toward developing a treatment.

deep brain stimulation (DBS) Neurosurgery in which electrodes implanted in the brain stimulate a targeted area with a low-voltage electrical current to facilitate behavior.

Contemporary Perspectives on Brain and Behavior

Where do modern students of the brain stand on the perspectives of mentalism, dualism, and materialism? In his influential book, *The Organization of Behavior,* published in 1949, psychologist Donald O. Hebb describes the scientific acceptance of materialism in a folksy manner:

> Modern psychology takes completely for granted that behavior and neural function are perfectly correlated, that one is completely caused by the other. There is no separate soul or life force to stick a finger into the brain now and then and make neural cells do what they would not otherwise. (Hebb, 1949, p. iii)

A contemporary philosophical school called *eliminative materialism* takes the position that if behavior can be described adequately without recourse to the mind, then the mental explanation should be eliminated. Daniel Dennett (1978) and other philosophers who have considered a number of mental entities, such as consciousness, pain, and attention, argue that once brain function is understood, the mental explanations of these "entities" can be replaced. Mentalism, by contrast, defines consciousness as an entity, an attribute, or a thing. Let us use the concept of consciousness to illustrate the argument for eliminative materialism.

Recovering Consciousness: A Case Study

One patient's case study offers insight into how the study of brain and behavior begins to describe consciousness. The patient, a 38-year-old man, had lingered in a **minimally conscious state** (MCS) for more than 6 years after an assault. He was occasionally able to communicate with single words, occasionally able to follow simple commands. He was able to make a few movements but could not feed himself despite 2 years of inpatient rehabilitation and 4 years in a nursing home.

This patient is one of approximately 1.4 million people each year in the United States who, as described by Fred Linge at the beginning of this chapter, contend with **traumatic brain injury** (TBI), a wound to the brain that results from a blow to the head. Among them, as many as 100,000 may become comatose, and only as few as 20 percent recover consciousness.

Among the remaining TBI patients, some are diagnosed as being in a **persistent vegetative state** (PVS), alive but unable to communicate or to function independently at even the most basic level because they have such extensive brain damage that no recovery can be expected. Others, such as the assault victim heretofore described, are diagnosed as being in an MCS because behavioral observation and brain-imaging studies suggest that they do have a great deal of functional brain tissue remaining.

Nicholas Schiff and his colleagues (Schiff & Fins, 2007) reasoned that, if they could stimulate the brain of their MCS patient, they could improve his behavioral abilities. As part of a **clinical trial** (a consensual experiment directed toward developing a treatment), they implanted thin wire electrodes into his brainstem via which they could administer a small electrical current.

Through these electrodes, which are visible in the X-ray image shown in **Figure 1-6,** the investigators applied the electrical stimulation for 12 hours each day. The procedure is called **deep brain stimulation** (DBS). The researchers found dramatic improvement in the patient's behavior and ability to follow commands, and he was, for the first time, able to feed himself and swallow food. He could even interact with his caregivers and watch television, and he showed further improvement in response to rehabilitation.

The experimenters' very practical measures of consciousness are formalized by the Glascow Coma Scale (GCS), an objective indicator of the degree of unconsciousness and of recovery from unconsciousness. The GCS rates eye movement, body movement, and speech on a 15-point scale. A low score indicates coma and a high score indicates consciousness. Thus, the ability to follow commands, to eat, to speak, and even to watch TV provide objective measures of consciousness that contrast sharply with the mentalistic description that sees consciousness as a single entity. Eliminative materialists would argue, therefore, that the measurably improved GCS score of behaviors in a brain-injured patient is more useful than a subjective, mentalistic explanation that consciousness has "improved."

Young people under the age of 19 can be especially vulnerable to head trauma and concussion when they participate in recreational activities and sporting competitions. The Centers for Disease Control and Prevention has cited bicycling and football injuries as leading causes of TBI in this population. TBI can also lead to accelerated brain aging, an area of growing concern for participants in athletic competitions that lead to repeated concussion or other head trauma.

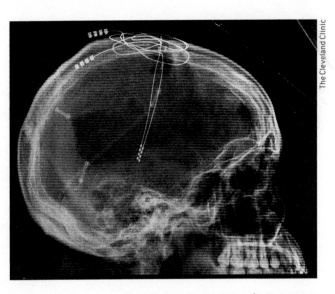

FIGURE 1-6 Deep Brain Stimulation X-ray of a human brain showing electrodes implanted in the thalamus for DBS. DBS also shows promise as a treatment for conditions such as Parkinson's disease, depression, and obsessive-compulsive disorder.

Sections 7-3, 14-5, 15-7 and Focus 16-3 detail research on and treatments for MCS, TBI, and concussion.

The Separate Realms of Science and Belief

Contemporary brain theory is materialistic. Although materialists, including the authors of this book, continue to use mentalistic words such as consciousness, pain, and attention to describe more complex sets of actions, at the same time materialists recognize that these words do not describe single mental entities. The materialistic view argues for *measurable* descriptions of behavior that can be referenced to the activity of the brain.

Some people may question materialism's tenet that only the brain is responsible for behavior because they think it denies religion. The theory, however, is neutral with respect to religion. Many of the world's major religions accept both evolution and the centrality of the brain in behavior as important scientific theories. Fred Linge has strong religious beliefs, as do the other members of his family. They used their religious strength to aid in his recovery. Yet, despite their religious beliefs, they realize that Linge's brain injury was the cause of his change in behavior and that the process of recovery that his brain underwent is the cause of his restored health.

Similarly, many behavioral scientists hold religious beliefs and see no contradiction between their beliefs and their engagement with science to examine the relations between the brain and behavior. Science is not a belief system but rather a set of procedures designed to allow investigators to confirm answers to a question independently.

Experiment 1-1 outlines the four-step procedure that allows scientists to *replicate*, or repeat, their original conclusions—or find that they cannot.

REVIEW 1-2
Perspectives on Brain and Behavior

Before you continue, check your understanding.

1. The view that behavior is the product of an intangible entity called the mind (psyche) is _____. The notion that the immaterial mind acts through the material brain to produce language and rational behavior is _____. _____, the view that brain function fully accounts for all behavior, guides contemporary research on the brain and behavior.

2. The implication that the brains and behaviors of complex animals such as humans evolved from the brains and behaviors of simpler animals draws on the theory of _____ _____ advanced by _____.

3. _____ is a wound to the brain that results from a blow to the head. The brain demonstrates a remarkable ability to recover, even after severe brain injury, but an injured person may linger in a _____, occasionally able to communicate with single words or to follow simple commands but otherwise not conscious. Those who suffer such extensive brain damage that no recovery can be expected remain in a _____, alive but unable to communicate or to function independently at even the most basic level.

4. Darwin and Mendel were nineteenth-century contemporaries. Briefly contrast the methods they used to reach their scientific conclusions.

Answers appear at the back of the book.

1-3 Evolution of Brains and of Behavior

The study of living organisms shows that nervous systems or brains are not common to all and that nervous systems and behavior built up and changed bit by bit as animals evolved. We trace the evolution of the human brain and behavior by describing (1) the animals that first developed a nervous system and muscles with which to move, (2) how the nervous system became more complex as the brain evolved to mediate complex behavior, and (3) how the human brain evolved its present complexity.

The popular interpretation of human evolution is that we are descended from apes. Actually, humans *are* apes. Other living apes are not our ancestors, although we are related to them through a **common ancestor,** a forebearer from which two or more lineages or family groups arise. To demonstrate the difference, consider the following story.

Two people named Joan Campbell are introduced at a party, and their names afford a good opening for a conversation. Although both belong to the Campbell lineage (family line), one Joan is not descended from the other. The two women live in different parts of North America, one in Texas and the other in Ontario, and both their families have been in those locations for many generations.

Nevertheless, after comparing family histories, the two Joans discover that they have ancestors in common. The Texas Campbells are descended from Jeeves Campbell, brother of Matthew Campbell, from whom the Ontario Campbells are descended. Jeeves and Matthew had both boarded the same fur-trading ship when it stopped for water in the Orkney Islands north of Scotland before sailing to North America in colonial times.

The Joan Campbells' common ancestors, then, were the mother and father of Jeeves and Matthew. Both the Texas and the Ontario Campbell family lines are descended

common ancestor Forebearer from which two or more lineages or family groups arise and so is ancestral to both groups.

nerve net Simple nervous system that has no brain or spinal cord but consists of neurons that receive sensory information and connect directly to other neurons that move muscles.

bilateral symmetry Body plan in which organs or parts present on both sides of the body are mirror images in appearance. For example, the hands are bilaterally symmetrical, whereas the heart is not.

segmentation Division into a number of parts that are similar; refers to the idea that many animals, including vertebrates, are composed of similarly organized body segments.

from this man and woman. If the two Joan Campbells were to compare their genes, they would find similarities that correspond to their common lineage.

In much the same way, humans and other apes are descended from common ancestors. But unlike the Joan Campbells, we do not know who those distant relatives were. By comparing the brain and behavioral characteristics of humans and related animals and by comparing their genes, however, scientists are tracing our lineage back farther and farther to piece together the origins of our brain and behavior.

Some living animal species display characteristics more similar to a common ancestor than do others. For example, with respect to humans, in some ways chimpanzees are more similar to the common ancestor of humans and chimpanzees than are modern humans. In the following sections, we trace some of the main evolutionary events that led to human brains and human behavior by looking at the nervous systems of living animal species and the fossils of extinct animal species.

Origin of Brain Cells and Brains

Earth formed about 4.5 billion years ago, and the first life-forms arose about a billion years later. About 700 million years ago, animals evolved the first brain cells, and by 250 million years ago, the first brain had evolved. A humanlike brain first developed only about 6 million years ago, and our modern human brain has been around for only the past 200,000 years.

Although life evolved very early in the history of our planet, brain cells and the brain evolved only recently, and large complex brains, such as ours, appeared only an eyeblink ago in evolutionary terms. If you are familiar with the general principles of taxonomic classification, which names and orders living organisms according to their relationships, read on. If you prefer a brief review before you continue, turn first to "The Basics: Classification of Life" on pages 16–17.

Evolution of Animals Having Nervous Systems

A nervous system is not essential for life. In fact, most organisms both in the past and at present have done without one. In animals that do have nervous systems, an examination of a wide variety of species presents the broad outlines of how the nervous system evolved. We summarize this evolution in the following general steps:

1. *Neurons and muscles.* Brain cells and muscles first evolved in animals, allowing the animals to move.

2. *Nerve net.* The nervous system representative of evolutionarily older phyla, such as jellyfishes and sea anemones, is extremely simple. It consists of a diffuse **nerve net,** which has no structure that resembles a brain or spinal cord but consists entirely of neurons that receive sensory information and connect directly to other neurons that move muscles. Look again at the human nervous system illustrated in Figure 1-2. Now imagine that the brain and spinal cord have been removed. The human PNS is reminiscent of the nerve net of phylogenetically simpler animals.

3. *Bilateral symmetry.* In more evolved animals such as flatworms, the nervous system is more organized and features **bilateral symmetry,** in which the nervous system on one side of the animal mirrors the nervous system on the other side. The human nervous system is bilaterally symmetrical (see Figure 1-2).

4. *Segmentation.* Animals such as earthworms feature **segmentation:** the body consists of a series of similar muscular segments. The nervous systems of these animals have similar repeating segments. The human spinal cord and brain display segmentation; in the vertebrae of the spinal column are the similar, repeating nervous system segments of the spinal cord.

For a visual recap, see "Evolution of the Nervous System" in The Basics on page 17.

Section 15-4 examines the asymmetric functions of each cerebral hemisphere in the context of the brain's bilateral symmetry.

Sections 2-3 and 2-4 outline the anatomy of the human spinal cord and describe how it functions.

⊕ THE BASICS

Classification of Life

Taxonomy is the branch of biology concerned with naming and classifying species by grouping organisms according to their common characteristics and their relationships to one another.

As shown in the left column of "Taxonomy of Modern Humans," which illustrates the human lineage, the broadest unit of classification is a kingdom, with more subordinate groups being phylum, class, order, family, genus, and species. This taxonomic hierarchy is useful in helping us trace the evolution of brain cells and the brain.

We humans belong to the animal kingdom, the chordate phylum, the mammalian class, the primate order, the great ape family, the *Homo* genus, and the *sapiens* species. Animals are usually identified by their genus and species names. So we humans are called *Homo sapiens,* meaning "wise humans."

The branches in "Cladogram," which shows the taxonomy of the animal kingdom, represent the evolutionary sequence (phylogeny) that connects all living organisms. Cladograms are read from left to right: the most recently evolved organism (animal) or trait (muscles and neurons) is located farthest to the right.

Of the five kingdoms of living organisms illustrated in this cladogram, only the one most recently evolved, Animalia, contains species with muscles and nervous systems. It is noteworthy that muscles and nervous systems evolved together to underlie the forms of movement (behavior) that distinguish members of the animal kingdom.

Taxonomy classifies groups of living organisms into increasingly specific, subordinate groups.

Living organisms

Classified in five main kingdoms: Monera (bacteria), Protista (single cells), Plantae (plants), Fungi (fungi), Animalia (animals)

Kingdom: Animals

Characteristics: Neurons and muscles used for locomotion

Phylum: Chordates

Characteristics: Brain and spinal cord

Class: Mammals

Characteristics: Large brains and social behavior

Order: Primates

Characteristics: Visual control of hands

Family: Great apes

Characteristics: Tool use

Genus: Human

Characteristics: Language

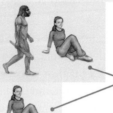

Modern humans are the only surviving species of the genus that includes numerous extinct species of humanlike animals.

Species: Modern human

Characteristics: Complex culture

Taxonomy of Modern Humans

---

5. *Ganglia.* In still more recently evolved phyla including clams, snails, and octopuses, are clusters of neurons called **ganglia** that resemble primitive brains and function somewhat like them in that they are "command centers." In some phyla, *encephalization,* meaning that the ganglia are found in the head, becomes distinctive. For example, insects have ganglia in the head that are sufficiently large to merit the term brain.

6. *Spinal cord.* In relatively highly evolved **chordates**—animals that have both a brain and a spinal cord—a single nervous system pathway connects the brain with sensory receptors and muscles. Chordates get their name from the *notochord,* a flexible rod that

ganglia Collection of nerve cells that function somewhat like a brain.

chordate Animal that has both a brain and a spinal cord.

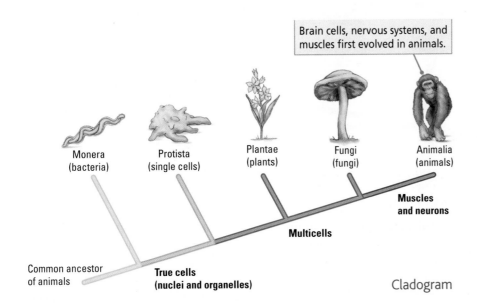

Brain cells, nervous systems, and muscles first evolved in animals.

Monera (bacteria)

Protista (single cells)

Plantae (plants)

Fungi (fungi)

Animalia (animals)

Muscles and neurons

Multicells

Common ancestor of animals

True cells (nuclei and organelles)

Cladogram

"Evolution of the Nervous System" shows the taxonomy of the 15 groups, or *phyla,* of animals, classified according to increasing complexity of nervous systems and movement. In proceeding to the right from the *nerve net,* we find that nervous systems in somewhat more recently evolved phyla, such as flatworms, are more complexly structured. These organisms have heads and tails, and their bodies show both *bilateral symmetry* (one half of the body is the mirror image of the other) and *segmentation* (the body is composed of similarly organized parts). This segmented nervous system resembles the structure of the human spinal cord.

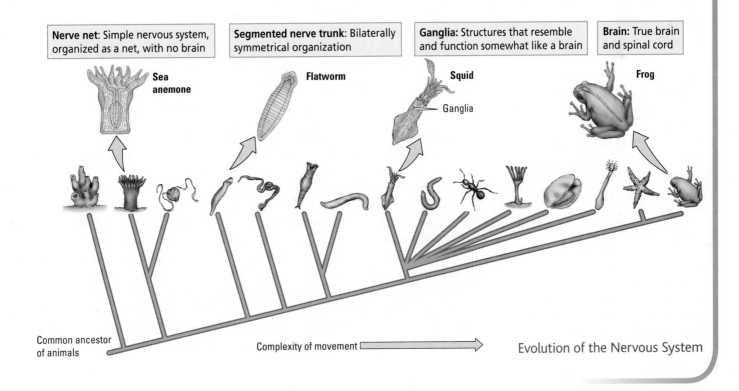

Nerve net: Simple nervous system, organized as a net, with no brain

Sea anemone

Segmented nerve trunk: Bilaterally symmetrical organization

Flatworm

Ganglia: Structures that resemble and function somewhat like a brain

Squid

Ganglia

Brain: True brain and spinal cord

Frog

Common ancestor of animals

Complexity of movement

Evolution of the Nervous System

runs the length of the back. In humans, the notochord is present only in the embryo; by birth, bony vertebrae encase the spinal cord.

7. *Brain.* The chordate phylum, of which frogs, reptiles, birds, and mammals are class members, display the greatest degree of encephalization: they have a true brain. Of all the chordates, humans have the largest brain relative to body size, but many other chordates have large brains as well. Although built to a common plan, the brain of each chordate species displays specializations related the control of the distinctive behaviors of the species.

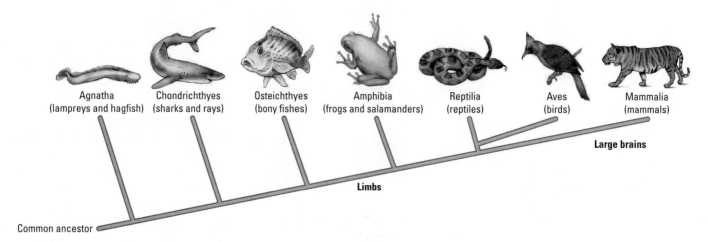

Agnatha (lampreys and hagfish)
Chondrichthyes (sharks and rays)
Osteichthyes (bony fishes)
Amphibia (frogs and salamanders)
Reptilia (reptiles)
Aves (birds)
Mammalia (mammals)

Large brains

Limbs

Common ancestor

FIGURE 1-7 Representative Classes of Chordates This cladogram illustrates the evolutionary relationship among animals that have a brain and a spinal cord. Brain size increased with the development of limbs in amphibia. Birds and mammals are the most recently evolved chordates, and large brains relative to body size are found in both classes.

Chordate Nervous System

A chart called a **cladogram** (from the Greek word *clados,* meaning "branch") displays groups of related organisms as branches on a tree. The cladogram in **Figure 1-7** represents seven of the nine classes to which the approximately 38,500 chordate species belong. Wide variation exists in the nervous systems of chordates, but the basic pattern of a structure that is bilaterally symmetrical, is segmented, and has a spinal cord and brain encased in cartilage or bone is common to all.

In addition, as chordates evolved limbs and new forms of locomotion, their brains became larger. For example, all chordates have a brainstem, but only the birds and mammals have a large forebrain.

The evolution of more complex behavior in chordates is closely related to the evolution of the *cerebrum* and of the *cerebellum.* Their increasing size in different classes of chordates is illustrated in **Figure 1-8.** The increases accommodate new behaviors, including new forms of locomotion on land, complex movements of the mouth and hands for eating, improved learning ability, and highly organized social behavior.

The cerebrum and the cerebellum are proportionately small and smooth in the earliest evolved classes (e.g., fish, amphibians, and reptiles). In later evolved chordates, especially the birds and mammals, these structures become much more prominent. In many large-brained mammals, both structures are extensively folded, which greatly increases their surface area while allowing them to fit into a small skull (just as folding a large piece of paper enables it to occupy a small container such as an envelope).

Increased size and folding become particularly pronounced in dolphins and primates, the animals with the largest brains relative to their body size. Because relatively large brains with a complex cerebrum and cerebellum have evolved in a number of animal lineages, humans are neither unique nor special in these respects. We humans are distinguished, however, in belonging to a lineage, the primates, that has large brains, and we are unique in having the largest brain of all animals relative to body size.

FIGURE 1-8 Brain Evolution The brains of representative chordates have many structures in common, illustrating a single basic brain plan across chordate species.

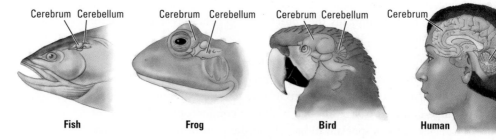

Cerebrum Cerebellum   Cerebrum Cerebellum   Cerebrum Cerebellum   Cerebrum   Cerebellum

Fish   **Frog**   **Bird**   **Human**

REVIEW 1-3
Evolution of Brains and of Behavior

Before you continue, check your understanding.

1. Because brain cells and muscles evolved only once in the animal kingdom, a similar basic pattern exists in the _____ of all animals.

2. Evolutionary relationships among the nervous systems of animal lineages are classified by increasing complexity and progress from the simplest _____ to a _____ and segmented nervous system to nervous systems controlled by _____ to, eventually, nervous systems featuring a brain and spinal cord in the _____ phylum.

3. Given that a relatively large brain with a complex cerebrum and cerebellum has evolved in a number of animal lineages, what if anything makes humans unique?

Answers appear at the back of the book.

cladogram Phylogenetic tree that branches repeatedly, suggesting a taxonomy of organisms based on the time sequence in which evolutionary branches arise.

1-4 Evolution of the Human Brain and Behavior

Anyone can see similarities among humans, apes, and monkeys, and if their brains are examined, similarities can also be seen. In this section, we consider only the brains and behaviors of some of the more prominent ancestors that link ancestral apes to our brain and our behavior. Then we consider the relation between brain size and behavior across different species. We conclude by surveying the leading hypotheses about how the human brain became so large as the nervous system became more complex and the brain evolved to mediate complex behavior. The evolutionary evidence from our ancestors shows that we humans are *specialized* in having an upright posture, making and using tools, and developing language but that we are not *special* because our ancestors also shared these traits, at least to some degree.

Humans: Members of the Primate Order

The human relationship to apes and monkeys places us in the primate order, a subcategory of mammals that includes not only apes and old world monkeys but new world monkeys, tarsiers, and lemurs as well (**Figure 1-9**). In fact, we humans are but 1 of about 275 species in the primate order. Primates have excellent color vision, with the eyes positioned at the front of the face to enhance depth perception, and they use this highly developed sense to deftly guide their hand movements.

FIGURE 1-9 Representatives of the Primate Order This cladogram illustrates hypothetical relationships among members of the primate order. Humans are members of the great ape family. In general, brain size increases across the groupings, with humans having the largest brain of all primates.

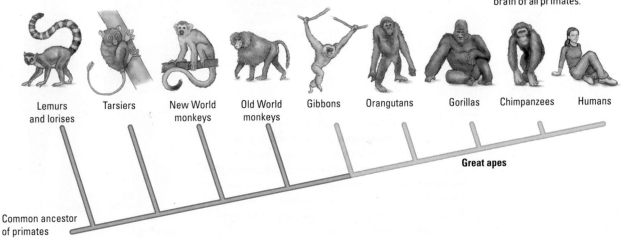

Lemurs and lorises   Tarsiers   New World monkeys   Old World monkeys   Gibbons   Orangutans   Gorillas   Chimpanzees   Humans

Great apes

Common ancestor of primates

hominid General term referring to primates that walk upright, including all forms of humans, living and extinct.

Female primates usually have only one infant per pregnancy, and they spend a great deal more time caring for their young than most other animals do. Associated with their skillful movements and their highly social nature, primates have brains that are on average larger than those of animals in other orders of mammals, such as rodents (mice, rats, beavers, squirrels) and carnivores (wolves, bears, cats, weasels).

Humans are members of the great ape family, which also includes orangutans, gorillas, and chimpanzees. Apes are arboreal animals with limber shoulder joints that allow them to brachiate in trees (swing from one handhold to another), a trait retained by humans, who generally do not live in trees these days. Nevertheless, freeing the arms at the shoulder joint is handy for all sorts of human activities, from traversing monkey bars on the playground to competing in the Olympic hammer toss to raising one's hand to ask a question in class. Apes are distinguished as well by their intelligence and large brains, traits that humans exemplify.

Among the apes, we are most closely related to the chimpanzee, having had a common ancestor between 5 million and 10 million years ago. In the past 5 million years, many **hominids**—primates that walk upright—evolved in our lineage. During most of this time, a number of hominid species coexisted. At present, however, we are the only surviving hominid species.

Australopithecus: **Our Distant Ancestor**

The name *Australopithecus* was coined by an Australian, Raymond Dart, for the skull of a child he found in a box of fossilized remains from a limestone quarry near Taung, South Africa, in 1924. Choosing a name to represent his native land probably is not accidental.

One of our hominid ancestors is probably *Australopithecus africanus.* (*Australopithecus* is from the Latin word *austral,* meaning "southern," and the Greek word *pithekos,* meaning "ape.") **Figure 1-10** shows reconstructions of the animal's face and body. We now know that many species of *Australopithecus* existed, some at the same time.

These early hominids were among the first primates to show distinctly human characteristics, including walking upright and using tools. Scientists have deduced their upright posture from the shape of their back, pelvic, knee, and foot bones and from a set of fossilized footprints that a family of australopiths left behind, walking through freshly fallen volcanic ash some 3.8 million years ago. The footprints feature

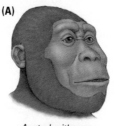

Australopithecus

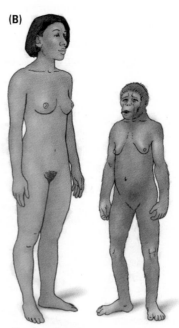

FIGURE 1-10 *Australopithecus* **(A)** The hominid *Australopithecus* walked upright with free hands, as do modern humans, but its brain was the size of that of a modern ape, about one-third the size of the modern human brain. **(B)** Human and *Australopithecus* figures compared (*right*) on the basis of the most complete *Australopithecus* skeleton yet found, a young female about 1 meter tall popularly known as Lucy, who lived 3 million years ago.

Homo sapiens   "Lucy"

the impressions of a well-developed arch and an unrotated big toe more like that of humans than of other apes. Evidence for tool use is implied in the structure of their hands (Pickering et al., 2011).

The First Humans

The oldest fossils designated as genus *Homo,* or human, are those found by Mary and Louis Leakey in the Olduvai Gorge in Tanzania in 1964, dated at about 2 million years. The primates that left these skeletal remains had a strong resemblance to *Australopithecus* but more closely resembled modern humans in one important respect: they made simple stone tools. The Leakeys named the species *Homo habilis* ("handy human") to signify that its members were toolmakers.

The first humans whose populations spread beyond Africa migrated into Europe and Asia. This species was *Homo erectus* ("upright human"), so named because of the mistaken notion that its predecessor, *H. habilis,* had a stooped posture. *Homo erectus* first shows up in the fossil record about 1.6 million years ago. As shown in **Figure 1-11,** its brain was bigger than that of any preceding hominid, overlapping in size the measurements of present-day human brains. The tools made by *H. erectus* were more sophisticated than those made by *H. habilis.*

Modern humans, *Homo sapiens,* appeared within about the past 200,000 years. Most anthropologists think that they also migrated from Africa. Until about 30,000 years ago in Europe and 18,000 years ago in Asia, they coexisted and interbred with other hominid species. An Asiatic species, *Homo floresienses,* found on the Indonesian island of Flores up to about 13,000 year ago, at about 3 feet tall, an especially small subspecies of *Homo erectus* (Gordon et al., 2008).

In Europe, *H. sapiens* coexisted with a subspecies of modern humans, Neanderthals, named after Neander, Germany, where the first Neanderthal skulls were found. As the first fossil ancestral humans to be discovered, the Neanderthals have maintained a preeminent place in the study of modern human ancestors. Neanderthals had brains as large as or larger than those of modern humans, used tools similar to those of early *H. sapiens* and possibly had a similar hunting culture, and wore jewelry and makeup.

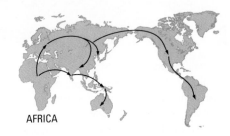

Proposed routes that early humans may have followed out of Africa, first to Asia and Europe, eventually to Australia, and finally to the Americas.

As Focus 10-1 reports, like us, Neanderthals apparently enjoyed music.

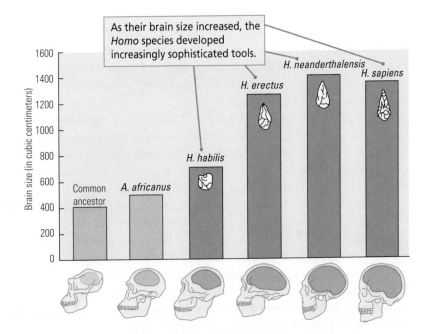

FIGURE 1-11 Increases in Hominid Brain Size The brain of *Australopithicus* was about the same size as that of living nonhuman apes, but succeeding members of the human lineage display a steady increase in brain size.

FIGURE 1-12 Does This Man Look Familiar? Modern Europeans acquired 4 percent of their genes from Neanderthals because their ancestors crossbred with Neanderthals. The sculpture was created by paleo-artist John Gurche.

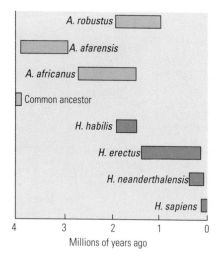

FIGURE 1-13 Origins of Humans The human lineage and a lineage of extinct *Australopithecus* probably arose from a common ancestor about 4 million years ago. Thus the ancestor of the human lineage *Homo* was probably an animal similar to *Australopithecus africanus*.

Section 14-4 details how changes in human brain size and complexity are related to the ability to learn.

We do not know how *H. sapiens* completely replaced other human species, but perhaps they had advantages in tool making, language use, or social organization. Contemporary genetic evidence shows that modern European humans who interbred with Neanderthals acquired genes that adapted them to the cold, to novel disease, and possibly to light skin that better absorbs vitamin D (Zhang et al., 2011). **Figure 1-12** is a recreation that shows what a Neanderthal man might have looked like.

One possible human lineage is shown in **Figure 1-13**. A common ancestor gave rise to the *Australopithecus* lineage, and one member of this group gave rise to the *Homo* lineage. Note that the bars in Figure 1-13 are not connected because many more hominid species have been discovered than are shown, and exact direct ancestors are uncertain. The bars overlap because many hominid species were alive at the same time until quite recently. The last of the australopith species disappeared from the fossil record about 1 million years ago.

Relating Brain Size and Behavior

Scientists who study brain evolution propose that a relative increase in the size and complexity of brains in different species enables the evolution of more complex behavior. Having a large brain clearly has been adaptive for humans, but many species of animals have large brains. Whales and elephants have brains much larger than ours. Of course, whales and elephants are much larger than humans. How is brain size measured and what does brain size signify?

In *The Evolution of the Brain and Intelligence,* published in 1973, Harry Jerison uses the *principle of proper mass* to sum up the idea that species exhibiting more complex behaviors must possess relatively larger brains than species whose behaviors are less complex. Jerison developed an index of brain size to compare the brains of different species relative to their differing body sizes. He calculated that, as body size increases, the size of the brain increases at about two-thirds the increase in body weight.

The diagonal trend line in **Figure 1-14** plots this expected brain–body size ratio. The graph also shows that some animals lie below the line: for these animals, brain size is smaller than would be expected for an animal of that size. Other animals lie above the line: for these animals, brain size is larger than would be expected for an animal of that size.

Using the ratio of actual brain size to expected size, Jerison developed a quantitative measure for brain size, the **encephalization quotient** (EQ). The lower an animal's brain falls below trend line in Figure 1-14, the smaller its EQs. The higher an animal's brain lies above the trend line, the larger its EQ. Notice that the rat's brain is a little smaller (lower EQ) and the elephant's brain a little larger (higher EQ) than the ratio predicts. A modern human is located farther above the line than any other animal, indicating that the human brain has the highest EQ of all the animals.

The top half of **Figure 1-15** lists the EQs for several familiar animal species, and the bottom half lists the EQs of representative species in the primate lineage. A comparison of the two lists reveals that, although the EQs for the primate lineage are large, there is overlap with other, quite different animals. The crow's EQ is similar to the monkey's, and the dolphin's EQ is comparable to that of *Homo erectus*.

A comparison of brain size and the complexity of behavior suggests that a larger brain is needed for increasingly complex behavior. People who study crows and dolphins would agree that they are intelligent and social animals. Thus, one property of the nervous system is that it can enlarge (or shrink), and it has done so in many species of animals to mediate more complex behavior or less complex behavior. Changes in brain size and complexity are related to the ability to learn, especially in humans. Next we pursue some ideas about how a large human brain evolved while other ape species showed no increase or even a decrease in brain size.

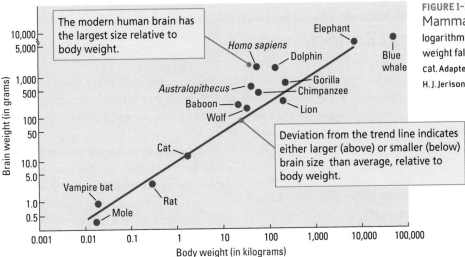

FIGURE 1-14 Brain-to-Body Size Ratios of Common Mammals A wide range of body and brain sizes is represented logarithmically on this graph. Average brain size relative to body weight falls along the diagonal trend line, where you find the cat. Adapted from *The Evolution of the Brain and Intelligence* (p. 175), by H. J. Jerison, 1973, New York: Academic Press.

Graph labels: The modern human brain has the largest size relative to body weight. Deviation from the trend line indicates either larger (above) or smaller (below) brain size than average, relative to body weight. Elephant, *Homo sapiens*, Dolphin, Blue whale, *Australopithecus*, Gorilla, Chimpanzee, Baboon, Wolf, Lion, Cat, Vampire bat, Mole, Rat. Brain weight (in grams); Body weight (in kilograms).

Why the Hominid Brain Enlarged

The evolution of modern humans—from the time when humanlike creatures first appeared until the time when humans like ourselves first existed—spans about 5 million years. As illustrated by the relative size differences of the hominid skulls pictured in **Figure 1-16**, much of this evolution was associated with increases in brain size, which were accompanied by changes in behavior. Brain-size changes were probably driven by many influences. Among the wide array of hypotheses that seek to explain why the modern human brain enlarged so much and so rapidly, we will examine four ideas.

The hypothesis detailed in Research Focus 1-3, "Climate and the Evolving Hominid Brain" on page 24 suggests that hominids were subjected to numerous, drastic climate changes that forced them to adapt and led to more complex behavior. Another hypothesis contends that the primate lifestyle favors an increasingly complex nervous system that humans capitalized on. A third links brain growth to brain cooling. And a fourth proposes that a slowed rate of maturation favors larger brains.

Primate Lifestyle

The primate lifestyle rests on living in large social groups. Robin Dunbar (1998), a British anthropologist, has assembled evidence that primates' group size is correlated with brain size. He concludes that the large average group size of about 150 favored by modern humans explains their large brain. His evidence for the number 150 is that

FIGURE 1-15 Comparing Encephalization Quotients The EQs of some familiar animals are ranked at the top of the chart, and members of the primate lineage are ranked at the bottom. Clearly, intelligence is widespread among animals.

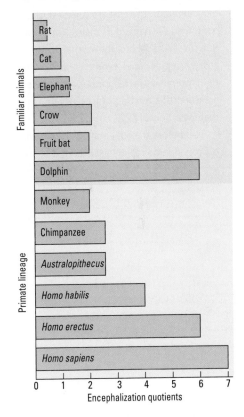

Chart labels: Familiar animals: Rat, Cat, Elephant, Crow, Fruit bat, Dolphin. Primate lineage: Monkey, Chimpanzee, *Australopithecus*, *Homo habilis*, *Homo erectus*, *Homo sapiens*. Encephalization quotients (0–7).

K. O'Farrell/Concepts

FIGURE 1-16 The Course of Human Evolution The relative size of the hominid brain has increased nearly threefold, illustrated here by a comparison of the skulls of *Australopithecus afarensis* (left), *Homo erectus* (center), and modern *Homo sapiens* (right). Missing parts of the *Australopithecus* skull, shown in blue, have been reconstructed. From *The Origin of Modern Humans* (p. 165), by R. Lewin, 1998, New York: Scientific American Library.

encephalization quotient (EQ) Jerison's quantitative measure of brain size obtained from the ratio of actual brain size to expected brain size, according to the principle of proper mass, for an animal of a particular body size.

Climate and the Evolving Hominid Brain

Anthropologist Rick Potts describes the human family tree as one littered with dead branches. Only a single living branch remains. Changes in climate may have driven many physical changes in hominids, including changes in the brain and the emergence of culture.

The nearly threefold increase in brain size from apes (EQ 2.5) to modern humans (EQ 7.0) appears to have taken place in a number of steps. Evidence suggests that each new hominid species appeared after climate changes devastated old environments and produced new environments.

About 8 million years ago, climate and a massive tectonic event (a deformation of Earth's crust) produced the Great Rift Valley, which runs from south to north across the eastern part of the African continent. The reshaped African landmass left a wet jungle climate to the west and a much drier savannah climate to the east. To the west, the apes continued unchanged in their former habitat. But the fossil record shows that in the drier eastern region, apes evolved rapidly into upright hominids in response to the selective pressures of a mixture of tree-covered and grassy regions that formed their new home.

Just before the appearance of *Homo habilis* 2 million years ago, the African climate rapidly grew even drier, with spreading grasslands and

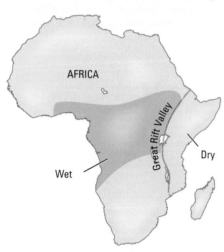

Africa's Great Rift Valley cut off ape species living in a wetter climate to the west from species that evolved into hominids to adapt to a drier climate to the east.

even fewer trees. Anthropologists speculate that the hominids that evolved into *H. habilis* adapted to this new habitat by becoming scavengers on the dead of the large herds of grazing animals that then roamed the open grasslands.

The appearance of *Homo erectus* 1 million years ago may have been associated with a further change in climate: a rapid cooling that lowered sea levels (by trapping more water as ice) and opened up land bridges into Europe and Asia. At the same time, the new hominid species upgraded their hunting skills and the quality of their tools for killing, skinning, and butchering animals.

Other climatic changes are associated with the disappearance of many other members of the human family. For example, the warming of Europe as recently as 30,000 years ago probably contributed to the migration of modern humans to the continent and to the disappearance of Neanderthals.

What makes modern humans special? Potts suggests that *Homo sapiens* has evolved to adapt to change itself and that this ability has allowed us to populate almost every climatic region on earth (Potts & Sloan, 2010). Potts also cautions that modern humans have been around only for a short time relative to the million years that *H. erectus* survived: our adaptability has yet to be severely tested.

it represents the estimated group size of hunter-gatherer groups and the average group size of many contemporary institutions—a company in the military, for instance. It also happens to be the number of people that each of us can gossip about.

Another way that the primate lifestyle favors a larger brain can be illustrated by examining how primates forage for food. Foraging is important for all animals, but some foraging activities are simple, whereas others are complex. Eating grass or vegetation is not difficult; an animal need only munch and move on. Vegetation eaters do not have especially large brains. Among the apes, gorillas, mainly vegetation eaters, have relatively small brains. In contrast, apes that eat fruit, such as chimpanzees and humans, have relatively large brains.

The relation between fruit foraging and larger brains is documented in a study by Katharine Milton (2003). She examined the feeding behavior and brain size of two South American (New World) monkeys that have the same body size—the spider monkey and the howler monkey. As illustrated in **Figure 1-17,** the spider monkey obtains nearly three-quarters of its nutrients from eating fruit and has a brain twice as large as that of the howler monkey, which obtains less than half of its nutrients from fruit.

What is it about eating fruit that favors a larger brain? The answer is not that fruit contains a brain-growth factor, although fruit is a source of sugar on which the brain depends for energy. The answer is that foraging for fruit is a far more complex behavior than grazing. Unlike plentiful vegetation within easy reach on the ground, fruit grows on trees, and only on certain trees in certain seasons. Among the many kinds of fruit, some are better for eating than others, and many different animals and insects compete for a fruit crop. Moreover, after a fruit crop has been eaten, it takes time for a new crop to grow. Each of these factors poses a challenge for an animal that eats mostly fruit.

Good sensory skills, such as color vision, are needed to recognize ripe fruit in a tree, and good motor skills are required to reach and manipulate it. Good spatial skills are needed to navigate to trees that contain fruit. Good memory skills are required to remember where fruit trees are, when the fruit will be ripe, and in which trees the fruit has already been eaten.

Fruit eaters have to be prepared to deal with competitors, including members of their own species, who also want the fruit. To keep track of ripening fruit, having friends who can help search also benefits a fruit eater. As a result, successful fruit-eating animals tend to have complex social relations and a means of communicating with others of their species. In addition, having a parent who can teach fruit-finding skills is helpful to a fruit eater; so being both a good learner and a good teacher is useful.

We humans are fruit eaters and we are descended from fruit eaters, so we are descended from animals with large brains. In our evolution, we also exploited and elaborated fruit-eating skills to obtain other temporary and perishable types of food as we scavenged, hunted, and gathered. These new food-getting efforts required navigating long distances, and they required recognition of a variety of food sources. At the same time, they required making tools for digging up food, killing animals, cutting skin, and breaking bones.

These tasks also require cooperation and learning. Humans distinguish themselves from other apes in displaying a high degree of male–male, female–female, and male–female cooperation in matters not related to sexual activity (Schuiling, 2005). The elaboration of all these lifestyle skills necessitated more brain cells over time. Added up, more brain cells produce an even larger brain.

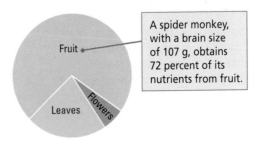

Spider monkey diet

A spider monkey, with a brain size of 107 g, obtains 72 percent of its nutrients from fruit.

Fruit

Leaves

Flowers

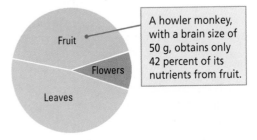

Howler monkey diet

A howler monkey, with a brain size of 50 g, obtains only 42 percent of its nutrients from fruit.

Fruit

Flowers

Leaves

FIGURE 1-17 Picky Eaters Katharine Milton examined the feeding behavior and brain size of two New World monkeys that have the same body size but different brain sizes and diets.

Changes in Hominid Physiology

One adaptation that may have given a special boost to greater brain size in our human ancestors was changes in the morphology, or form, of the skull. Dean Falk (2004), a neuropsychologist who studies brain evolution, developed the **radiator hypothesis** from her car mechanic's remark that, to increase the size of a car's engine, you have to also increase the size of the radiator that cools it.

Falk reasoned that, if the brain's radiator, the circulating blood, adapted into a more-effective cooling system, the brain could increase in size. Brain cooling is important because, although your brain makes up less than 2 percent of your body weight, it uses 25 percent of your body's oxygen and 70 percent of its glucose. As a result of all this metabolic activity, your brain generates a great deal of heat and is at risk of overheating under conditions of exercise or heat stress.

radiator hypothesis Idea that selection for improved brain cooling through increased blood circulation in the brains of early hominids enabled the brain to grow larger.

neoteny Process in which maturation is delayed and so an adult retains infant characteristics; idea derived from the observation that newly evolved species resemble the young of their common ancestors.

Falk argued that, unlike australopith skulls, *Homo* skulls contain holes through which cranial blood vessels pass. These holes suggest that *Homo* species had a much more widely dispersed blood flow from the brain than did earlier hominids, and this more widely dispersed blood flow would have greatly enhanced brain cooling.

A second adaptation, identified by Hansell Stedman and his colleagues (2004), stems from a genetic mutation associated with marked size reductions in individual facial-muscle fibers and entire masticatory muscles. The Stedman team speculates that smaller masticatory muscles in turn led to smaller and more delicate bones in the head. Smaller bones in turn allowed for changes in diet and an increase in brain size.

Stedman and his colleagues estimate that this mutation occurred 2.4 million years ago, coinciding with the appearance of the first humans. The methodology used by Stedman, in which human and ape genes are compared, will be a source of future insights into other differences between humans and apes, including those in brain size and function.

Neoteny

In the slowing of maturation, a process called **neoteny,** juvenile stages of predecessors become the adult features of descendants. Many features of human anatomy link us with the juvenile stages of other primates. These features include a small face, a vaulted cranium, an unrotated big toe, an upright posture, and a primary distribution of hair on the head, armpits, and pubic areas. Because the head of an infant is large relative to body size, neoteny has also led to adults with proportionally larger skulls to house larger brains.

Figure 1-18 illustrates that the shape of a baby chimpanzee's head is more similar to the shape of an adult human's head than it is to that of an adult chimpanzee's head. Humans also retain some behaviors of primate infants, including play, exploration, and an intense interest in novelty and learning. Neoteny is common in the animal world. Flightless birds are neotenic adult birds, domesticated dogs are neotenic wolves, and sheep are neotenic goats.

One aspect of neoteny related to human brain development is that slowing down human maturation would have allowed more time for brain cells to be produced (McKinney, 1998). Most brain cells in humans develop just before and after birth; so an extended prenatal and neonatal period would prolong the stage of life in which most brain cells are developing. This prolonged stage would, in turn, enable increased numbers of brain cells to develop.

There are a number of views about what promotes neoteny. One view is that, at times of abundant resources, less physiologically and behaviorally mature individual organisms can successfully reproduce, yielding offspring that have this trait in common. This "babies having babies" could lead to a population in which individual members have immature physical features and behavioral traits in common though at the same time being sexually mature. Another view is that, at times of food insufficiency, maturation and reproduction are slowed, allowing a longer time for development.

FIGURE 1-18 Neoteny The shape of an adult human's head more closely resembles that of a juvenile chimpanzee's head (*left*) than that of an adult chimp's head (*right*), leading to the hypothesis that we humans may be neotenic descendants of our more apelike common ancestors.

Brand X Pictures

R. Stacks/Index Stock

REVIEW 1-4
Evolution of the Human Brain and Behavior

Before you continue, check your understanding.

1. Modern humans share a _____ with the _____, our closest living relative.

2. Modern humans evolved from a _____ lineage that featured *Australopithicus*, *Homo habilis*, and *Homo erectus*, groups in which more than one species existed at the same time.

3. The large human brain evolved in response to a number of pressures and opportunities, including _____, _____, _____, and _____.

4. One hypothesis proposes that *Homo sapiens* has evolved to adapt to change itself. In a brief paragraph, explain, the reasoning behind this hypothesis.

Answers appear at the back of the book.

1-5 Modern Human Brain Size and Intelligence

The evolutionary approach that we have been using to explain how the large human brain evolved is based on comparisons *between* species. Special care attends the extension of evolutionary principles to physical comparisons *within* species, especially biological comparisons within or among groups of modern humans. We will illustrate the difficulty of within-species comparisons by considering the complexity of correlating human brain size with intelligence (Deary, 2000). Then we turn to another aspect of studying the brain and behavior in modern humans—the fact that, unlike the behavior of other animal species, so much of modern human behavior is culturally learned.

Meaning of Human Brain-Size Comparisons

We have documented parallel changes in brain size and behavioral complexity through the many species that form the human lineage. Some people have proposed that, because brain-size differences between species are related to behavioral complexity, brain-size differences might occur between individual members of a single species. Is there evidence to support this hypothesis?

There are large differences in the brains of individual people, but the reasons for the differences are numerous and complex. Consider some examples. Larger people are likely to have larger brains than smaller people. Men have somewhat larger brains than women, but they are proportionately physically larger. Nevertheless, girls mature more quickly than boys, so in adolescence the brain- and body-size differences may be absent. As people age, they generally lose brain cells, so their brains become smaller.

Neurological diseases associated with aging accelerate the age-related decrease in brain size. Brain injury before and or around birth often results in a dramatic reduction in brain size even in brain regions that are distant from the damage. Neurological disorders associated with a mother's abuse of alcohol or other drugs are associated with conditions such as fetal alcohol spectrum disorder, in which the brain can be greatly reduced in size. Autism spectrum disorder (ASD), a largely genetic condition affecting development, produces a wide variety of brain abnormalities, including either increases or decreases in brain size in different individuals. Neurologists are aware of potential effects of brain-size changes in disease states and may take a measurement of skull size with a tape measure from a subject as a preliminary investigation into a neurological disorder.

The size of the brain may also increase in individuals. For example, just as good nutrition in the early years of life can be associated with larger body size, good nutrition

For information on coverage of specific neurological disorders, consult the Index of Disorders inside the front cover of this book.

Sections 2-1 and 2-6 elaborate on plasticity, Section 8-4 on the effect of environment on brain development, Section 11-3 on skilled movement, and Section 14-1 on memory formation.

can also be associated with brain-size increase. The *plasticity* of the brain—its ability to change—in response to an enriched environment is associated with growth of existing brain cells and thus an increase in brain size. Furthermore, one way in which the brain stores new skills and memories is to form new connections among brain cells, and these connections in turn contribute to an increase in brain size.

A far different question related to brain size is Do differences in brain size in individuals correlate with their intelligence? Over a century ago some investigators promoted the simple conclusion that people with the largest brains display the most intelligent behavior. Stephen Jay Gould, in his 1981 book *The Mismeasure of Man,* reviews much of this early literature and is critical of this research on three counts.

First, measuring the size of a person's brain is difficult. If a tape measure is simply placed around a person's head, factoring out the thickness of the skull is impossible. There is also no agreement about whether volume or weight is a better measure of brain size. And no matter which indicator we use, we must consider body size. For instance, the human brain varies in weight from about 1000 grams to more than 2000 grams, but people also vary in body mass. To what extent should we factor in body mass in deciding if a particular brain is large or small? And how should we measure the mass of the body, given that a person's total weight can fluctuate widely over time?

Second, even if the problems of measurement could be solved, the question of what is causing what remains. For the many reasons listed earlier concerning why brain size varies, intelligence is likely to vary as well. Therefore, any potential correlation would not be especially meaningful.

Third, as if the preceding factors were not perplexing enough, we must also consider what is meant by "intelligence." When we compare the behavior of different species, we are comparing **species-typical behavior**—in other words, behavior displayed by all members of a species. For example, lamprey eels do not have limbs and cannot walk, whereas salamanders do have limbs and can walk; so the difference in brain size between the two species can be correlated with this trait. When we compare behavior *within* a species, however, we are usually comparing how well one individual member performs a certain task in relation to other members—how well one salamander walks relative to how well another salamander walks, for example.

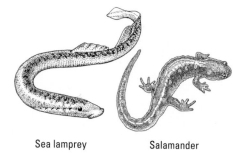

Sea lamprey   Salamander

We can make intraspecies comparisons for humans, but there are two problems. For one thing, individual performance on a task is influenced by many factors unrelated to inherent ability, among them opportunity, interest level, training, motivation, and health. For another, people vary enormously in their individual abilities, depending on the particular task. One person may have superior verbal skills but mediocre spatial abilities; another person may be adept at solving spatial puzzles but struggle with written work; still another may excel at mathematical reasoning and be average in everything else. Which of these people should we consider the most intelligent? Should certain skills carry greater weight as measures of intelligence? Clearly, it is difficult to say.

Early in the twentieth century, Charles Spearman carried out the first formal analysis of performance among various tests used to rate intelligence. He found a positive correlation among tests and suggested that a single common factor explained them. Spearman named it *g* for "general intelligence factor." Howard Gardner (2006), however, has proposed that a number of different intelligences (verbal, musical, mathematical, social, and so on) exist. Depending on whose definition is used, expectations about brain size can be quite different. Spearman's explanation might be consistent with a simple relationship, but Gardner's predicts that different regions of the brain may vary in size to mediate individual differences in traits and talents.

species-typical behavior Behavior that is characteristic of all members of a species.

culture Learned behaviors that are passed on from one generation to the next through teaching and experience.

Given the difficulty in measuring brain size and in defining intelligence, it is not surprising that very little research appears in the contemporary literature on the problem of gross brain size and intelligence. In case you are wondering whether having a larger brain might mean you could study a little less, consider this. The brains of people whom virtually everyone agrees are very intelligent have been found to vary in size from the low end to the high end of the range for our species. The brilliant physicist Albert Einstein had a brain of average size.

Section 8-3 correlates behavior with nervous-system development. Section 15-6 details the theories of Spearman and Gardner. Figure 15-19 is a photo of Einstein's brain.

Culture

The most remarkable thing that our brains have allowed us to develop is an extraordinarily rich **culture**—the complex learned behaviors passed on from generation to generation. Here is a list, in alphabetical order, of major categories of behavior that are part of human culture:

> Age-grading, athletic sports, bodily adornment, calendar [use], cleanliness training, community organization, cooking, cooperative labor, cosmology, courtship, dancing, decorative art, divination, division of labor, dream interpretation, education, eschatology, ethics, ethnobotany, etiquette, faith healing, family feasting, fire making, folklore, food taboos, funeral rites, games, gestures, gift giving, government, greetings, hair styles, hospitality, housing, hygiene, incest taboos, inheritance rules, joking, kin groups, kinship nomenclature, language, law, luck, superstitions, magic, marriage, mealtimes, medicine, obstetrics, penal sanctions, personal names, population policy, postnatal care, pregnancy usages, property rights, propitiation of supernatural beings, puberty customs, religious ritual, residence rules, sexual restrictions, soul concepts, status differentiation, surgery, tool making, trade, visiting, weaving, and weather control. (Murdock, 1965)

Not all the items in this list are unique to humans. Many other animal species display elements of some of these behaviors. For example, many other animals display age-grading (any age-related behavior or status), courtship behavior, rudimentary tool use, and elements of language. Furthermore, species that display well-developed traits have corresponding adaptions in brain structure, as is illustrated for tool use in birds in Comparative Focus 1-4, "Evolution and Adaptive Behavior" on page 30.

Despite such behavioral similarities across species, humans clearly have progressed much farther in the development of culture than other animals have. For humans, every category of activity on Murdock's list requires extensive learning from other members of the species, and exactly how each behavior is performed can differ widely from one group of people to another. Malcolm Gladwell (2000), author of *The Tipping Point* suggests that becoming an expert in many of these areas might require as many as 10,000 hours, or 10 years, of practice. You probably know this because you have spent more than 12 years learning in a formal educational setting.

Because of steady growth in cultural achievements, the behavior of *Homo sapiens* today is completely unlike that of *Homo sapiens* living 100,000 years ago. The earliest surviving art, such as carvings and paintings, dates back only some 30,000 years; agriculture appears still more recently, about 15,000 years ago; and reading and writing, the foundations of our modern literate and technical societies, were invented only about 7000 years ago.

Saint Ambrose, who lived in the fourth century, is reported to be the first person who could read silently. Most forms of mathematics, another basis of modern technology, were invented even more recently than reading and writing were. And many of our skills in using mechanical and digital devices are still more recent in origin.

These examples highlight a remarkable feature of the human brain: it now performs tasks that were not directly selected for in our early hominid evolution. The brains of early *Homo sapiens* certainly did not evolve to select smart-phone apps or travel to distant planets. And yet the same brains are capable of these tasks and more. Apparently, the things that the human brain did evolve to do contained the elements necessary for adapting to more sophisticated skills. Thus, humans and probably other large-brained species evolved a capacity for high flexibility in accommodating knowledge and culture.

Evolution and Adaptive Behavior

Tool use was once considered the exclusive domain of humans but is now recognized in all great apes, in other primates, and in 39 bird species, including parrots, corvids, herons, and raptors. Comparative research on nonhuman species allows for contrast among the behaviors and brains of many living species occupying widely different habitats.

The evolution of tool use in animals is correlated with significant increases in the relative size of the brain. Andrew Iwaniuk and his colleagues (2006) examined whether the size of the cerebellum and the extent of its foliation—its folding, shown on the left in the illustration—are related to tool use. Taking advantage of the many species of birds that do or do not use tools, they compared the volume and the degree of folding in the cerebellums of birds that use tools and those that do not. The investigators found that the extent of foliation was positively correlated with tool use.

Because all tool-using birds have been observed to use tools in their natural habitat, it is unclear whether the ability signifies the evolution of a narrow skill in manipulating objects or reflects a general intelligence marked by insight into the relationship between problems and tools.

To test the tool-use ability of non-tool–using birds in the laboratory, Christopher Bird and Nathan Emery

(2009) tested rooks (*Corvus frugilegus*), a bird that, although related to crows (shown on the right in the illustration), does not use tools in nature.

The rooks' problem was to obtain a worm located on a shelf in a tube. The birds quickly selected appropriately sized sticks and rocks to knock the worm from the shelf, used a different object to obtain the tool, removed branches from a stick in order to use it as a tool, and bent pieces of wire to hook the worm from the tube.

Bird and Emery propose that rooks have evolved brain structures and cognitive skills that give them insight into problem solving. The researchers further propose that such generalized intelligence may have evolved similarly in hominids to enable human tool use.

A. N. Iwaniuk et al., 2006 · *Gavin Hunt*

(*Left*) Cresyl-violet-stained section of the cerebellum of the Australian magpie, a tool user. (*Right*) Crows are among the animals that make and use tools, often, as here, to obtain food.

The acquisition of complex culture was a gradual, step-by-step process, with one achievement leading to another. Among our closest relatives, chimpanzees also have culture in the sense that some groups display tool-using skills that others have not acquired. In her book *The Chimpanzees of Gombe,* the primatologist Jane Goodall describes the process by which symbolic concepts, a precursor of language, might have developed in chimpanzees. She uses the concept of "fig" as an example, explaining how a chimp might progress from knowing a fig only as a tangible here-and-now entity to having a special vocal call that represents "fig" symbolically. Goodall writes:

> We can trace a pathway along which representations of . . . a fig become progressively more distant from the fig itself. The value of a fig to a chimpanzee lies in eating it. It is important that he quickly learn to recognize as fig the fruit above his head in a tree (which he has already learned to know through taste). He also needs to learn that a certain characteristic odor is representative of fig, even though the fig is out of sight. Food calls made by other chimpanzees in the place where he remembers the fig tree to be located may also conjure up a concept of fig. Given the chimpanzees' proven learning ability, there does not seem to be any great cognitive leap from these achievements to understanding that some quite new and different stimulus (a symbol) can also be representative of fig. Although chimpanzee calls are, for the most part, dictated by emotions, cognitive abilities are sometimes required to interpret them. And the interpretations themselves may be precursors of symbolic thought. (Goodall, 1986, pp. 588–589)

Presumably, in our own distant ancestors, the repeated acquisition of concepts, as well as the education of children in those concepts, gradually led to the acquisition of language and other aspects of a complex culture. The study of the human brain, then, is not just the study of the structure of a body organ. It is also the study of how that organ acquires cultural skills—that is, of how the human brain fosters behavior in to-day's world.

REVIEW 1-5
Modern Human Brain Size and Intelligence

Before you continue, check your understanding.

1. Behavior that is displayed by all members of a species is called _____.

2. We humans are distinguished in the animal kingdom by the development of _____, the vast proportion of our behavior that is passed on from generation to generation through learning and experience.

3. In a brief paragraph, explain the reasoning behind the following statement: What is true for evolutionary comparisons across different species may not be true for comparisons within a single species.

Answers appear at the back of the book.

SUMMARY

1-1 Neuroscience in the Twenty-First Century

Studying the brain and behavior leads us to understand our origins, to understand human nature, and to understand the causes of many behavioral disorders and their treatment.

The human nervous system is composed of the central nervous system (CNS), which includes the brain and the spinal cord, and the peripheral nervous system (PNS), through which the brain and spinal cord communicate with sensory receptors, with muscles and other tissues, and with the internal organs. The cerebrum and the cerebellum are the structures that have undergone the largest growth in large-brained species of animals.

Behavior can be defined as any kind of movement in a living organism, and, in many species, behavior is caused by the activity of the nervous system. The flexibility and complexity of behavior vary greatly among different species, as does the nervous system.

For some species, including humans, the brain is the organ that exerts control over behavior. The brain seems to need ongoing sensory and motor stimulation to maintain its intelligent activity.

1-2 Perspectives on Brain and Behavior

Mentalism is the view that behavior is a product of an intangible entity called the mind (psyche); the brain has little importance. Dualism is the notion that the immaterial mind acts through the material brain to produce language and rational behavior, whereas the brain alone is responsible for the "lower" kinds of actions that we have in common with other animal species.

Materialism, the view that brain function fully accounts for all behavior, language and reasoning included, guides contemporary research on the brain and behavior. Support for the materialistic view comes from the study of natural selection—the evolutionary theory that behaviors such as human language evolved from the simpler language abilities of human ancestors—and from discoveries about how genes function.

Traumatic brain injury (TBI) can be caused by a blow to the head. After severe brain injury, the brain demonstrates a remarkable ability to recover; but after either mild or severe injury, a person can be left with a permanent disability that prevents full recovery to former levels of function. The Glascow Coma Scale (GCS) objectively measures severe disabilities such as the minimal conscious state (MCS) and persistent vegetative state (PVS).

1-3 Evolution of Brains and of Behavior

Behavioral neuroscientists subscribe to the evolutionary principle that all living organisms are descended from a common ancestor. Brain cells and muscles are quite recent developments in the evolution of life on Earth. Because they evolved only once, a similar basic pattern exists in the nervous systems of all animals.

The nervous systems of some animal lineages have become more complex, with evolution featuring first a nerve net, followed by a bilaterally symmetrical and segmented nervous system, a nervous system controlled by ganglia, and eventually a nervous system featuring a brain and spinal cord.

A true brain and spinal cord evolved only in the chordate phylum. Mammals are a class of chordates characterized by large brains relative to body size. Modern humans belong to the primate order, an order distinguished by especially large brains, and to the family of great apes, whose members' limber shoulder joints allow them to brachiate.

1-4 Evolution of the Human Brain and Behavior

One of our early hominid ancestors was probably *Australopithecus*—or a primate very much like it—who lived in Africa several million years ago. From an australopith species, more humanlike species are likely to have evolved. Among them are *Homo habilis* and *Homo erectus*. Modern humans, *Homo sapiens,* appeared between 200,000 and 100,000 years ago.

Since *Australopithecus,* the hominid brain has increased in size almost threefold. Environmental challenges and opportunities that favored the natural selection of adaptability and more complex behavior patterns, changes in physiology, and neoteny stimulated brain evolution in human species.

1-5 Modern Human Brain Size and Intelligence

Principles learned in studying the evolution of the brain and behavior *across* species do not apply to the brain and behavior *within* a single species, such as *Homo sapiens*. As animals evolved, a larger brain was associated with more complex behavior; yet, within our species, the complexity of different brain regions are related to behavioral abilities. People vary widely in body size and in brain size as well as in having different kinds of intelligence, making a simple comparison of brain size and general intelligence impossible.

In the study of modern humans, recognizing the great extent to which our behavior is culturally learned rather than inherent in our nervous systems is paramount.

KEY TERMS

bilateral symmetry, p. 14

brainstem, p. 5

central nervous system (CNS), p. 5

cerebellum, p. 5

cerebrum (forebrain), p. 3

chordate, p. 16

cladogram, p. 19

clinical trial, p. 12

common ancestor, p. 14

culture, p. 28

deep-brain stimulation (DBS), p. 12

dualism, p. 7

embodied language, p. 5

encephalization quotient (EQ), p. 23

epigenetics, p. 11

ganglia, p. 16

genotype, p. 11

hemisphere, p. 3

hominid, p. 20

materialism, p. 8

mentalism, p. 7

mind, p. 7

mind–body problem, p. 8

minimally conscious state (MCS), p. 12

natural selection, p. 8

neoteny, p. 26

nerve net, p. 14

neuron, p. 5

peripheral nervous system (PNS), p. 5

persistent vegetative state (PVS), p. 12

phenotype, p. 8

psyche, p. 7

radiator hypothesis, p. 25

segmentation, p. 14

species, p. 8

species-typical behavior, p. 28

spinal cord, p. 5

traumatic brain injury (TBI), p. 12

How Does the Nervous System Function?

RESEARCH FOCUS 2-1 EVOLUTION OF BRAIN SIZE AND HUMAN BEHAVIOR

2-1 OVERVIEW OF BRAIN FUNCTION AND STRUCTURE

PLASTIC PATTERNS OF NEURAL ORGANIZATION

FUNCTIONAL ORGANIZATION OF THE NERVOUS SYSTEM

SURFACE FEATURES OF THE BRAIN

THE BASICS FINDING YOUR WAY AROUND THE BRAIN

CLINICAL FOCUS 2-2 MENINGITIS AND ENCEPHALITIS

CLINICAL FOCUS 2-3 STROKE

INTERNAL FEATURES OF THE BRAIN

2-2 EVOLUTIONARY DEVELOPMENT OF THE NERVOUS SYSTEM

2-3 THE CENTRAL NERVOUS SYSTEM: MEDIATING BEHAVIOR

SPINAL CORD

BRAINSTEM

FOREBRAIN

CORTEX

BASAL GANGLIA

LIMBIC SYSTEM

OLFACTORY SYSTEM

2-4 SOMATIC NERVOUS SYSTEM: TRANSMITTING INFORMATION

CRANIAL NERVES

SPINAL NERVES

CONNECTIONS OF THE SOMATIC NERVOUS SYSTEM

INTEGRATING SPINAL FUNCTION

CLINICAL FOCUS 2-4 MAGENDIE, BELL, AND BELL'S PALSY

2-5 AUTONOMIC NERVOUS SYSTEM: BALANCING INTERNAL FUNCTIONS

2-6 TEN PRINCIPLES OF NERVOUS-SYSTEM FUNCTION

PRINCIPLE 1: THE NERVOUS SYSTEM PRODUCES MOVEMENT WITHIN A PERCEPTUAL WORLD THE BRAIN CREATES

PRINCIPLE 2: THE HALLMARK OF NERVOUS-SYSTEM FUNCTIONING IS NEUROPLASTICITY

PRINCIPLE 3: MANY OF THE BRAIN'S CIRCUITS ARE CROSSED

PRINCIPLE 4: THE CENTRAL NERVOUS SYSTEM FUNCTIONS ON MULTIPLE LEVELS

PRINCIPLE 5: THE BRAIN IS BOTH SYMMETRICAL AND ASYMMETRICAL

PRINCIPLE 6: BRAIN SYSTEMS ARE ORGANIZED BOTH HIERARCHICALLY AND IN PARALLEL

PRINCIPLE 7: SENSORY AND MOTOR DIVISIONS EXIST THROUGHOUT THE NERVOUS SYSTEM

PRINCIPLE 8: SENSORY INPUT TO THE BRAIN IS DIVIDED FOR OBJECT RECOGNITION AND MOTOR CONTROL

PRINCIPLE 9: FUNCTIONS IN THE BRAIN ARE BOTH LOCALIZED AND DISTRIBUTED

PRINCIPLE 10: THE NERVOUS SYSTEM WORKS BY JUXTAPOSING EXCITATION AND INHIBITION

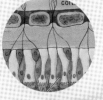

Evolution of Brain Size and Human Behavior

Compared with other mammals, primates have evolved larger brains than would be predicted from their body size. But within the primates, huge differences in brain size and structure exist, as displayed in the accompanying photographs. The human brain is much larger than the brain of the rhesus monkey or the chimpanzee, and the gap between human and chimpanzee is bigger than that between monkey and chimpanzee.

Brain size is related to body size, and humans are bigger than chimps. But when we factor in the relative sizes of brain and body as measured by the encephalization quotient (EQ), the human brain is nearly three times the size of the brain of a primate of our body size and nearly four times as big as that of a monkey of our body size.

This difference is huge. To put it in perspective, the gap in brain size between humans and chimpanzees is larger than the gap between chimpanzees and moles, which are tiny worm-eating mammals.

Increases in relative brain size are associated with increases in behavioral complexity and mental capacity. Moles live a simple life searching for worms and other terrestrial insects. Chimpanzees are not merely clever moles, however. Chimps make tools, lead a complex social life, and in captivity they can learn simple language. Given the large gap between humans and chimpanzees in brain size, we must consider additional capacities in the human brain that are not seen in the chimpanzee's brain.

One obvious difference is human language, a capacity that requires significant brainpower and is fundamental to understanding human brain organization. Richard Passingham (2008) has argued that communication by language leads to a novel form of understanding, not only of the world around us but also in the way that we think, reflect on our own thoughts, and imagine.

We have developed concepts of past and future as well as the capacity to see our autobiographical experiences as separate from those of

Monkey   Chimpanzee   Human

others. The implication is not that reasoning requires words but rather that the capacity to talk to others and, more important, to ourselves, has changed the nature of our mental life relative to that of other mammals.

Are our brains still "improving?" Douglas Fox (2011) concludes that we may be close to an evolutionary size limit that is set by physics. He concludes that the human brain has optimized its overall size, the size and number of neurons, the number and length of connections, and energy consumption. Changing any of these features would compromise the others and neutralize any performance improvements.

Engineers may be able to radically improve computers by redesigning chips or using new technologies, but evolution does not start over from scratch. It is constrained by the parts that have evolved from simpler brains over the past half-billion years. Our brains may well have reached an optimal neural blueprint that leaves little room for improvement.

On the surface, the brains of a monkey, chimpanzee, and human differ dramatically in size and in general appearance. The three brains are shown here to scale. With an EQ of 2.0, the monkey's brain is just over one-quarter the size of a human's (EQ 7.0), and the chimp's brain, EQ 2.5, is a bit more than one-third as large. **Wally Welker, University of Wisconsin Comparative Mammalian Brain Collection.**

Throughout this book, we examine the nervous system with a focus on function—on how our behavior and our brains interact. In this chapter, we consider the organization of the human nervous system and how its basic components function. We first focus on the biology of the brain and then elaborate on how the brain works in concert with the rest of the nervous system. This focus on function suggests ten basic principles of nervous-system organization that we note through the chapter and detail at its end, in Section 2-6. These "big ideas" apply equally to the micro and macro views of the nervous system presented in this chapter and to the broader picture of behavior that emerges in later chapters.

2-1 Overview of Brain Function and Structure

The brain's primary function is to produce behavior, or *movement*. To produce behavior as we search, explore, and manipulate our environments, the brain must get information about the world—about the objects around us: their sizes, shapes, and locations. Without such *stimulation,* the brain cannot orient and direct the body to produce an appropriate response.

The organs of the nervous system are designed to admit information from the world and to convert this information into biological activity that produces *perception,* or subjective experiences of reality. The brain thus produces what we believe is reality in order for us to move. These subjective experiences of reality are essential to carrying out any complex task.

When you answer the telephone, for example, your brain directs your body to reach for it as the nervous system responds to vibrating molecules of air by creating the subjective experience of a ringtone. We perceive this stimulation as sound and react to it as if it actually exists, when in fact the sound is merely a fabrication of the brain. That fabrication is produced by a chain reaction that takes place when vibrating air molecules hit the eardrum. Without the nervous system, especially the brain, there is no such thing as sound. Rather, there is only the movement of air molecules.

There is more to hearing a phone's ringtone than just the movement of air molecules, however. Our mental creation of reality is based not only on the sensory information received but also on the cognitive processes that each of us might use to interact with the incoming information. A telephone ringing when we are expecting a call has a different meaning from its ringing at three o'clock in the morning when we are not expecting a call.

The subjective reality created by the brain can be better understood by comparing the sensory realities of two different kinds of animals. You are probably aware that dogs perceive sounds that humans do not. This difference in perception does not mean that a dog's nervous system is better than ours or that our hearing is poorer. Rather, the perceptual world created by a dog brain simply differs from that created by a human brain. Neither subjective experience is "correct." The difference in experience is merely due to two differently evolved systems for processing physical stimuli.

When it comes to visual perception, dogs see very little color, whereas our world is rich with color because our brains create a different reality. Subjective differences in brains exist for good reason: they allow different animals to exploit different features of their environments. Dogs use their hearing to detect the movements of mice in the grass; early humans probably used color vision for identifying ripe fruit in trees. Evolution, then, fosters *adaptability:* it equips each species with a view of the world that helps it survive.

Plastic Patterns of Neural Organization

Although we tend to think of regions of the brain has having fixed functions, the brain is *plastic:* neural tissue has the capacity to adapt to the world by changing how its functions are organized. For example, a person blind from birth has enhanced auditory capacities because some of the usual visual regions have been co-opted for hearing. The brain is also plastic in the sense that connections among neurons in a given functional system are constantly changing in response to experience.

For us to learn anything new, neural circuits must change to represent and store this knowledge. As we learn to play a musical instrument or speak a new language, the cortical regions taking part can actually increase in size to accommodate the learning. An important aspect of human learning and brain plasticity is related to the development of language and to the expansion of the brain regions related to language discussed in Research Focus 2-1. We have learned to read, to calculate, to compose and play music, and to develop the sciences. Clearly, the human nervous system evolved long before we mastered these achievements.

In turn, culture now plays a dominant role in shaping our behavior. Because we drive cars and communicate electronically, we—and our nervous systems—must be different

Principle 1: The nervous system produces movement within a perceptual world the brain creates.

Section 9-1 elaborates on the nature of sensation and perception.

Principle 2: The hallmark of nervous-system functioning is neuroplasticity.

Epigenetics, defined in Section 1-2, studies differences in gene expression related to environment and experience.

FIGURE 2-1 Phenotypic Plasticity These two mice are genetically identical but express very different phenotypes because their mothers were fed different supplements when pregnant.

The chart in Figure 2-2A restates Figure 1-2, a diagram of the gross structure of the CNS and PNS in the human body.

FIGURE 2-2 Parsing the Nervous System The nervous system can be conceptualized **(A)** anatomically and **(B)** functionally. The functional approach employed in this book focuses on how the parts of the nervous system interact.

from those of our ancestors who did not engage in these activities. The basis for change in the nervous system is the fundamental property of **neuroplasticity,** the nervous system's potential for physical or chemical change that enhances its adaptability to environmental change and its ability to compensate for injury.

Although it is tempting to see neuroplasticity as a unique trait of animals' nervous systems, it is really part of a larger capacity called **phenotypic plasticity,** the individual's capacity to develop into more than one phenotype—characteristics that can be seen or measured. (See Gilbert & Epel, 2009, for a wonderful discussion of biological plasticity.) Stated simply, an individual's genotype (genetic makeup) interacts with the environment to elicit a specific phenotype from a large genetic repertoire of possibilities, a phenomenon that results from epigenetic influences.

Epigenetic factors do not change genes but rather influence how genes express the traits inherited from parents. The two mice pictured in **Figure 2-1** appear very different: one is fat, one thin; one has dark fur, the other is light-colored. Yet these mice essentially are clones and thus genetically identical. They appear so different because their mothers were fed different diets while pregnant. The diet supplements added chemical markers, or "epigenetic tags," on specific genes. The tags determine whether the gene is available to influence cells, including neurons, leading to differences in body structure and eating behavior.

Functional Organization of the Nervous System

From an anatomical standpoint, the brain and spinal cord together make up the central nervous system, and all the nerve fibers radiating out beyond the brain and spinal cord as well as all the neurons outside the brain and spinal cord form the peripheral nervous system. **Figure 2-2A** charts this *anatomical* organization. Nerves of the PNS carry sensory information into the CNS and motor instructions from the CNS to the body's muscles and tissues, including those that perform autonomic functions such as digestion and blood circulation.

In a *functional* organization, little changes; the focus is on how the parts of the system work together (Figure 2-2B). Neurons in the somatic division of the PNS connect through the cranial and spinal nerves to receptors on the body's surface and on its muscles to gather sensory information for the CNS and to convey information from the CNS to move muscles of the face, body, and limbs. Similarly, the autonomic division of the PNS enables the CNS to govern the workings of your body's internal organs—the beating of your heart, the contractions of your stomach, and the movement of your diaphragm to inflate and deflate your lungs.

From a functional standpoint, then, the major divisions of the PNS step up to constitute, along with the CNS, an interacting, three-part system:

- The CNS includes the brain and the spinal cord, the structures at the core of the nervous system that mediate behavior.

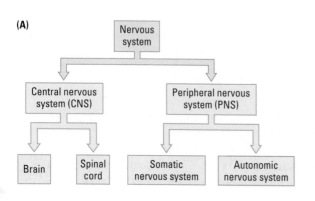

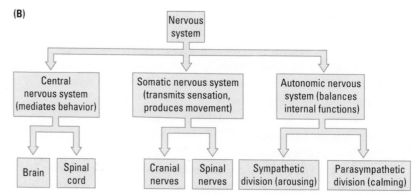

- The **somatic nervous system** (SNS), all the spinal and cranial nerves carrying sensory information to the CNS from the muscles, joints, and skin, also transmits outgoing motor instructions that produce movement.

- The **autonomic nervous system** (ANS) balances the body's internal organs to "rest and digest" through the *parasympathetic* (calming) *nerves* or to "fight or flee" or engage in vigorous activity through the *sympathetic* (arousing) *nerves*.

The direction of neural information flow is important. **Afferent** information is sensory information coming into the CNS or one of its parts (incoming information), whereas **efferent** information is information leaving the CNS or one of its parts (outgoing information). Thus, when you step on a tack, the sensory signals transmitted from the body into the brain are afferent. Efferent signals from the brain trigger a motor response: you lift your foot **(Figure 2-3)**.

Surface Features of the Brain

When buying a new car, people like to look under the hood and examine the engine, the part of the car responsible for most of its behavior—and misbehavior. All most of us can do is gaze at the maze of tubes, wires, boxes, and fluid reservoirs. What we see makes no sense, except in the most general way. We know that the engine burns fuel (whether it's gasoline, diesel, natural gas, or electricity) to make the car move and somehow generates power to run the sound system and lights. But this knowledge tells us nothing about what all the many parts of the engine do.

When it comes to behavior, the brain is the engine. In many ways, examining a brain for the first time is similar to looking under the hood of a car. We have a vague sense of what the brain does, but most of us have no sense of how the parts that we see accomplish these tasks. We may not even be able to identify the parts. If you are familiar with the anatomical terms and orientations used in brain drawings and images, read on. If you prefer to review this terminology before you continue, consult "The Basics: Finding Your Way Around the Brain" on pages 38–39.

Cerebral Security

The way to start our functional overview is to "open the hood" by observing the brain snug in its home within the skull. The first thing you encounter is not the brain but rather a tough, triple-layered, protective covering, the **meninges**, illustrated in **Figure 2-4.** The outer *dura mater* (from Latin, meaning "hard mother") is a tough double layer of fibrous tissue that encloses the brain and spinal cord in a kind of loose sack. In the middle is the *arachnoid* (from Greek, meaning "like a spider's web") *layer,* a very thin sheet of delicate connective tissue that follows the brain's contours. The inner layer, or *pia mater* (from Latin, meaning "soft mother"), is a moderately tough membrane of connective-tissue fibers that cling to the brain's surface.

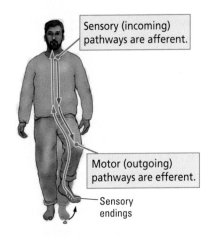

Sensory (incoming) pathways are afferent.

Motor (outgoing) pathways are efferent.

Sensory endings

FIGURE 2-3 Information Flow

neuroplasticity The nervous system's potential for physical or chemical change that enhances its adaptability to environmental change and its ability to compensate for injury.

phenotypic plasticity An individual's capacity to develop into more than one phenotype.

somatic nervous system (SNS) Part of the PNS that includes the cranial and spinal nerves to and from the muscles, joints, and skin that produce movement, transmit incoming sensory input, and inform the CNS about the position and movement of body parts.

autonomic nervous system (ANS) Part of the PNS that regulates the functioning of internal organs and glands.

afferent Conducting toward a central nervous system structure.

efferent Conducting away from a central nervous system structure.

meninges Three layers of protective tissue—dura mater, arachnoid, and pia mater—that encase the brain and spinal cord.

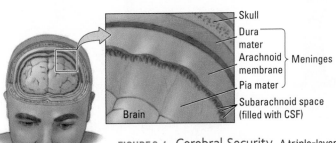

Skull
Dura mater
Arachnoid membrane   } Meninges
Pia mater
Subarachnoid space (filled with CSF)
Brain

FIGURE 2-4 Cerebral Security A triple-layered covering, the meninges, encases the brain and spinal cord, and the cerebrospinal fluid (CSF) cushions them.

⊕ THE BASICS

Finding Your Way Around the Brain

When the first anatomists began to examine the brain with the primitive tools of their time, the names they chose for brain regions often manifested their erroneous assumptions about how the brain works. They named one region of the brain the *gyrus fornicatus* because they thought that it had a role in sexual function, but most of this region actually has nothing to do with sexual activity.

A Wonderland of Nomenclature

As time went on, the assumptions and tools of brain research changed, but the naming continued to be haphazard and inconsistent. Many brain structures have several names, and terms are often used interchangeably. This peculiar nomenclature arose because research on brain and behavior spans several centuries and includes scientists of many nationalities and languages.

Early investigators named structures after themselves or objects or ideas. They used different languages, especially Latin, Greek, and English. More recently, investigators have often used numbers or letters, but even this system lacks coherence, because the numbers may be Arabic or Roman and are often used in combination with Greek or Latin letters.

Describing Locations in the Brain

Many names for nervous-system structures include information about their anatomical locations with respect to other body parts of the animal,

Brain–Body Orientation

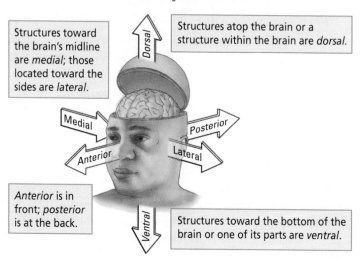

Structures toward the brain's midline are *medial*; those located toward the sides are *lateral*.

Structures atop the brain or a structure within the brain are *dorsal*.

Anterior is in front; *posterior* is at the back.

Structures toward the bottom of the brain or one of its parts are *ventral*.

with respect to their relative locations, and with respect to a viewer's perspective:

■ "Brain–Body Orientation" illustrates brain-structure location from the frame of reference of the face.

■ "Spatial Orientation" illustrates brain-structure location in relation to other body parts.

Spatial Orientation

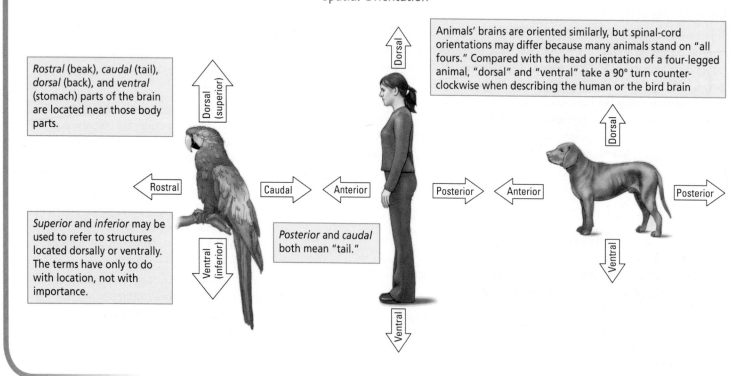

Rostral (beak), *caudal* (tail), *dorsal* (back), and *ventral* (stomach) parts of the brain are located near those body parts.

Superior and *inferior* may be used to refer to structures located dorsally or ventrally. The terms have only to do with location, not with importance.

Posterior and *caudal* both mean "tail."

Animals' brains are oriented similarly, but spinal-cord orientations may differ because many animals stand on "all fours." Compared with the head orientation of a four-legged animal, "dorsal" and "ventral" take a 90° turn counterclockwise when describing the human or the bird brain

Anatomical Orientation

"Anatomical Orientation" illustrates the direction of a cut, or section, through the brain (part A) from the perspective of a viewer (part B).

The orienting terms are derived from Latin. Consult the accompanying "Glossary of Anatomical Location and Orientation" for easy reference. It is common practice to combine orienting terms. A structure may be described as dorsolateral, for example, meaning that it is located "up and to the side."

Finally, the nervous system, like the body, is symmetrical, with a left side and a right side. Structures that lie on the same side are *ipsilateral*; if they lie on opposite sides, they are *contralateral* to each other. If a structure lies in each hemisphere, the structures are *bilateral*. Structures that are close to one another are *proximal*; those far from one another are *distal*.

(A) Plane of section

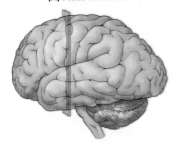

Coronal section

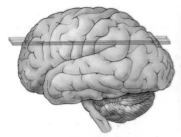

Horizontal section

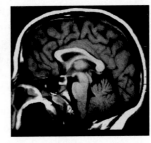

Sagittal section

(B) View of brain

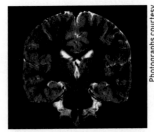

Frontal view

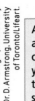
Photographs courtesy of Dr. D. Armstrong, University of Toronto/Lifeart.

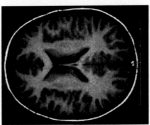

Dorsal view

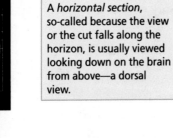

Medial view

A *coronal section* is cut in a vertical plane, from the crown of the head down, yielding a frontal view of the brain's internal structures.

A *horizontal section*, so-called because the view or the cut falls along the horizon, is usually viewed looking down on the brain from above—a dorsal view.

A *sagittal section* is cut lengthways from front to back and viewed from the side. (Imagine the brain split by an arrow—in Latin, *sagitta*.) Here, a cut in the *midsagittal plane* divides the brain into symmetrical halves, a medial view.

Glossary of Anatomical Location and Orientation

Term	Meaning with respect to the nervous system
anterior	Located near or toward the front of the animal or the front of the head (see also *frontal* and *rostral*)
caudal	Located near or toward the tail of the animal (see also *posterior*)
coronal	Cut vertically from the crown of the head down; used in reference to the plane of a brain section that reveals a frontal view
dorsal	On or toward the back of the animal or, in reference to human brain nuclei, located above; in reference to brain sections, a viewing orientation from above
frontal	"Of the front" (see also *anterior* and *rostral*); in reference to brain sections, a viewing orientation from the front
horizontal	Cut along the horizon; used in reference to the plane of a brain section that reveals a dorsal view
inferior	Located below (see also *ventral*)

Term	Meaning with respect to the nervous system
lateral	Toward the side of the body or brain
medial	Toward the middle, specifically the body's midline; in reference to brain sections, a side view of the central structures
posterior	Located near or toward the tail of the animal (see also *caudal*)
rostral	"Toward the beak" (front) of the animal (see also *anterior* and *frontal*)
sagittal	Cut lengthways from front to back of the skull; the plane that reveals a view into the brain from the side; a cut in the midsagittal plane divides the brain into symmetrical halves, a medial view.
superior	Located above (see also *dorsal*)
ventral	On or toward the belly or the side of the animal where the belly is located; in reference to brain nuclei, located below (see also *inferior*)

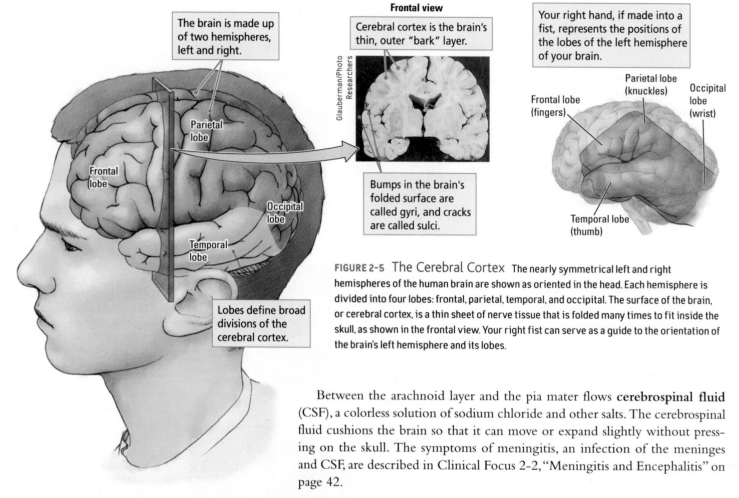

The brain is made up of two hemispheres, left and right.

Frontal view

Cerebral cortex is the brain's thin, outer "bark" layer.

Your right hand, if made into a fist, represents the positions of the lobes of the left hemisphere of your brain.

Glauberman/Photo Researchers

Parietal lobe

Frontal lobe

Occipital lobe

Parietal lobe (knuckles)

Occipital lobe (wrist)

Frontal lobe (fingers)

Temporal lobe

Bumps in the brain's folded surface are called gyri, and cracks are called sulci.

Temporal lobe (thumb)

Lobes define broad divisions of the cerebral cortex.

FIGURE 2-5 The Cerebral Cortex The nearly symmetrical left and right hemispheres of the human brain are shown as oriented in the head. Each hemisphere is divided into four lobes: frontal, parietal, temporal, and occipital. The surface of the brain, or cerebral cortex, is a thin sheet of nerve tissue that is folded many times to fit inside the skull, as shown in the frontal view. Your right fist can serve as a guide to the orientation of the brain's left hemisphere and its lobes.

Between the arachnoid layer and the pia mater flows **cerebrospinal fluid (CSF)**, a colorless solution of sodium chloride and other salts. The cerebrospinal fluid cushions the brain so that it can move or expand slightly without pressing on the skull. The symptoms of meningitis, an infection of the meninges and CSF, are described in Clinical Focus 2-2, "Meningitis and Encephalitis" on page 42.

Cerebral Geography

After removing the meninges, we can examine the brain's surface features, most prominently its two nearly symmetrical hemispheres, one on the left and one on the right. **Figure 2-5** shows the left hemisphere of a typical human forebrain oriented in the upright human skull. The entire outer layer of the forebrain consists of a thin, folded film of nerve tissue, the **cerebral cortex,** detailed in the frontal view in Figure 2-5. The word *cortex,* Latin for the bark of a tree, is apt, considering the cortex's heavily folded surface and its location, covering most of the rest of the brain. Unlike the bark on a tree, the brain's folds are not random but rather demarcate its functional zones.

Make a fist with your right hand and hold it up to represent the positions of the forebrain's broad divisions, or *lobes,* within the skull, as diagrammed on the right in Figure 2-5. Each lobe is named for the skull bone that it lies beneath.

• The forward-pointing **temporal lobe** is located at the side of the brain, approximately the same place as the thumb on your upraised fist. The temporal lobe functions in connection with hearing and with language and musical abilities

• Immediately above your thumbnail, your fingers correspond to the location of the **frontal lobe,** often characterized as performing the brain's "executive" functions, such as decision making.

• The **parietal lobe** is located at the top of the skull, behind the frontal lobe and above the temporal lobe. Parietal functions include directing our movements toward a goal or to perform a task, such as grasping an object.

• The area at the back of each hemisphere constitutes the **occipital lobe,** where visual processing begins.

Examining the Brain's Surface from All Angles

As we look at the dorsal view of the brain in **Figure 2-6**A, the wrinkled left and right hemispheres resemble a walnut meat taken whole from its shell. These hemispheres constitute the cerebrum, the major structure of the forebrain and most recently evolved feature of the central nervous system. From the opposite, ventral view in Figure 2-6B, the

cerebrospinal fluid (CSF) Clear solution of sodium chloride and other salts that fills the ventricles inside the brain and circulates around the brain and spinal cord beneath the arachnoid layer in the subarachnoid space.

cerebral cortex Thin, heavily folded film of nerve tissue composed of neurons that is the outer layer of the forebrain. Also called *neocortex.*

temporal lobe Part of the cerebral cortex that functions in connection with hearing, language, and musical abilities; lies below the lateral fissure, beneath the temporal bone at the side of the skull.

frontal lobe Part of the cerebral cortex often generally characterized as performing the brain's "executive" functions, such as decision making; lies anterior to the central sulcus and beneath the frontal bone of the skull.

parietal lobe Part of the cerebral cortex that functions to direct movements toward a goal or to perform a task, such as grasping an object; lies posterior to the central sulcus and beneath the parietal bone at the top of the skull.

occipital lobe Part of the cerebral cortex where visual processing begins; lies at the back of the brain and beneath the occipital bone.

(A) Dorsal view

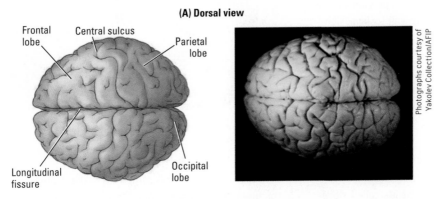

Frontal lobe
Central sulcus
Parietal lobe
Longitudinal fissure
Occipital lobe

(B) Ventral view

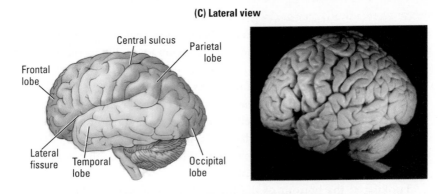

Temporal lobe
Cerebellum
Frontal lobe
Olfactory bulbs
Cranial nerves
Brainstem
Occipital lobe

Photographs courtesy of Yakolev Collection/AFIP

(C) Lateral view

Central sulcus
Parietal lobe
Frontal lobe
Lateral fissure
Temporal lobe
Occipital lobe

(D) Medial view

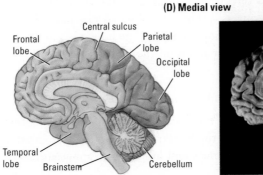
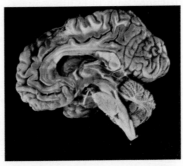

Central sulcus
Frontal lobe
Parietal lobe
Occipital lobe
Temporal lobe
Brainstem
Cerebellum

FIGURE 2-6 **Examining the Human Brain** Locations of the lobes of the cerebral hemispheres are shown in these top, bottom, side, and midline views, as are the cerebellum and the three major sulci.

Meningitis and Encephalitis

Harmful microorganisms can invade the layers of the meninges, particularly the pia mater and the arachnoid layer, as well as the CSF flowing between them, and cause a variety of infections that lead to a condition called *meningitis*. One symptom, inflammation, places pressure on the brain. Because the space between meninges and skull is slight, unrelieved pressure can lead to delirium and, if the infection progresses, to drowsiness, stupor, and even coma.

Usually, the earliest symptom of meningitis is severe headache and a stiff neck (cervical rigidity). Head retraction (tilting the head backward) is an extreme form of cervical rigidity. Convulsions, a common symptom in children, indicate that the brain also is affected by the inflammation.

Infection of the brain itself is called *encephalitis*. Some of the many forms of encephalitis have great historical significance. A century ago, in World War I, a form of encephalitis called sleeping sickness *(encephalitis lethargica)* reached epidemic proportions. Its first symptom is sleep disturbance. People sleep all day and become wakeful, even excited, at night. Subsequently, they show symptoms of Parkinson's disease including severe tremors, muscular rigidity, and difficulty in controlling body movements. Many are completely unable to make any voluntary movements, such as walking or even combing their hair. Survivors of sleeping sickness were immortalized by the neurologist Oliver Sacks in the book and movie *Awakenings*.

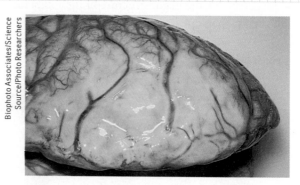

Pus is visible over the anterior surface of this brain infected with meningitis.

The cause of encephalitis symptoms is the death of an area deep in the brain, the *substantia nigra* ("black substance"), which you will learn about later in this chapter. Other forms of encephalitis may have different effects on the brain. For example, Rasmussen's encephalitis attacks one cerebral hemisphere in children. In most cases, the only effective treatment is radical: hemispherectomy, the surgical removal of the entire affected hemisphere.

Surprisingly, some young children who lose a hemisphere adapt rather well. They may even complete college, literally with half a brain. But retardation is a more common outcome of hemispherectomy as a result of encephalitis.

brainstem, including the wrinkly hemispheres of the smaller "little brain," or cerebellum, are visible. Both the cerebrum and the brainstem are visible in the lateral and medial views in Figure 2-6C and D.

Much of the crinkled-up cerebral cortex is invisible from the brain's surface. All we can see are bumps, or **gyri** (singular: gyrus), and cracks, or **sulci** (singular: sulcus). Some sulci are so deep that they are called *fissures*. The longitudinal fissure that runs between the cerebral hemispheres and the lateral fissure at the side of the brain are both shown in various views in Figure 2-6, along with the central sulcus that runs from the lateral fissures across the top of the cerebrum.

Looking at the bottom of the brain, the ventral view in Figure 2-6, we see in the midst of the wrinkled cerebrum and ventral to the cerebellum a smooth, whitish structure with little tubes attached. This central set of structures is the brainstem, the area responsible for most unconscious behavior. The little tubes in the illustration mark out the cranial nerves that run to and from the brain as part of the somatic nervous system.

Cerebral Circulation

One final gross feature is obvious: the brain's surface appears to be covered with blood vessels. Like the rest of the body, the brain receives blood through arteries and sends it back through veins to the kidneys and lungs for cleaning and oxygenation. The

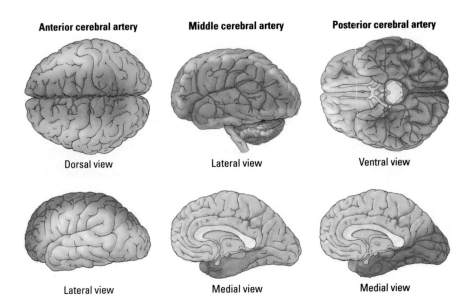

Anterior cerebral artery   **Middle cerebral artery**   **Posterior cerebral artery**

Dorsal view   Lateral view   Ventral view

Lateral view   Medial view   Medial view

FIGURE 2-7 Major Cerebral Arteries Each of the three major arteries that feed blood to the cerebral hemispheres branches extensively to service the regions shaded in pink.

cerebral arteries emerge from the neck to wrap around the outside of the brainstem, cerebrum, and cerebellum, finally piercing the brain's surface to nourish its inner regions.

Three major arteries send blood to the cerebrum—namely, the anterior, middle, and posterior cerebral arteries shown in **Figure 2-7**. Because the brain is very sensitive to loss of blood, a blockage or break in a cerebral artery is likely to lead to the death of the affected region, a condition known as **stroke,** the sudden appearance of neurological symptoms as a result of severely interrupted blood flow. Because the three cerebral arteries service different parts of the brain, strokes disrupt different brain functions, depending on the artery affected.

Because the brain's connections are crossed, stroke in the left hemisphere affects sensation and movement on the right side of the body. The opposite is true for those with strokes in the right hemisphere. Clinical Focus 2-3, "Stroke" on page 46, describes some disruptions that this condition causes, both to the person who experiences it and to those who care for stroke victims.

Section 16-3 elaborates on the effects of stroke and its treatment.

Principle 3: Many of the brain's circuits are crossed.

Internal Features of the Brain

The simplest way to examine the inside of something is to cut it in half. The orientation in which we cut makes a difference in what we see, however. Consider what happens when we slice through a pear. If we cut from side to side, we cut across the core, providing a dorsal view; if we cut from top to bottom, we cut parallel to the core, providing a medial view. Our impression of what the inside of a pear looks like is clearly influenced by how we slice it. The same is true of the brain.

Macro View

We can reveal the brain's inner features by slicing it downward through the middle, parallel to the front of the body, in a coronal section as shown in **Figure 2-8A**. The resulting frontal view, shown in Figure 2-8B, makes it immediately apparent that the interior is not homogeneous. Both dark and light regions of tissue are visible, and though these regions may not be as distinctive as the parts of a car's engine, they nevertheless represent different brain components.

The darker regions, called **gray matter,** are largely composed of cell bodies and capillary blood vessels. The neurons of the gray matter function either to collect and

gyrus (pl. gyri) A small protrusion or bump formed by the folding of the cerebral cortex.

sulcus (pl. sulci) A groove in brain matter, usually a groove found in the neocortex or cerebellum.

stroke Sudden appearance of neurological symptoms as a result of severely interrupted blood flow.

gray matter Areas of the nervous system composed predominantly of cell bodies and capillary blood vessels that function either to collect and modify information or to support this activity.

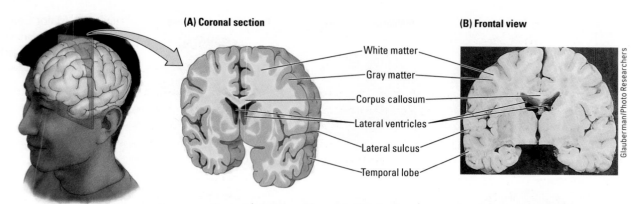

(A) Coronal section **(B) Frontal view**

White matter
Gray matter
Corpus callosum
Lateral ventricles
Lateral sulcus
Temporal lobe

Glauberman/Photo Researchers

FIGURE 2-8 Coronal Section Through the Brain **(A)** The brain is cut through the middle parallel to the front of the body and then viewed at a slight angle. **(B)** This frontal view displays white matter, gray matter, and the lateral ventricles. A large bundle of fibers, the corpus callosum, visible above the ventricles joins the hemispheres.

white matter Areas of the nervous system rich in fat-sheathed neural axons that form the connections between brain cells.

ventricle One of four cavities in the brain that contain cerebrospinal fluid that cushions the brain and may play a role in maintaining brain metabolism.

corpus callosum Band of white matter containing about 200 million nerve fibers that connects the two cerebral hemispheres to provide a route for direct communication between them.

modify information or to support this activity. The lighter regions, called **white matter,** are mostly nerve fibers with fatty coverings that produce the white appearance, much as fat droplets in milk make it appear white. The fibers of the white matter form the connections between the cells.

A second feature apparent at the middle of our frontal view in Figure 2-8B consists of two wing-shaped cavities—the **ventricles**—that contain cerebrospinal fluid. The brain contains four ventricles, shown in place in **Figure 2-9.** Cells that line the ventricles make the cerebrospinal fluid that fills them. The ventricles are connected; so the CSF flows from the two lateral ventricles to the third and fourth ventricles that lie on the brain's midline and into the cerebral aqueduct, a canal that runs the length of the spinal cord. Recall that CSF is also found in the space between the lower layers of the meninges wrapping around the brain and spinal cord (see Figure 2-4).

Although the functions of the ventricles are not well understood, researchers think that they play an important role in maintaining brain metabolism. The cerebrospinal fluid may allow certain compounds access to the brain, and it probably helps the brain excrete metabolic wastes. In the event of a traumatic brain injury or spinal trauma, CSF cushions the blow.

FIGURE 2-9 Cerebral Ventricles The four ventricles are interconnected. The lateral ventricles are symmetrical, one in each hemisphere. The third and fourth cerebral ventricles lie in the brain's midline and drain into the cerebral aqueduct that runs the length of the spinal cord.

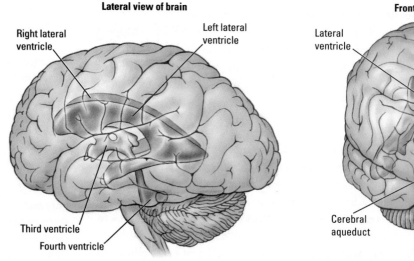

Lateral view of brain

Right lateral ventricle
Left lateral ventricle
Third ventricle
Fourth ventricle

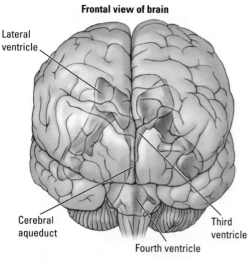

Frontal view of brain

Lateral ventricle
Cerebral aqueduct
Third ventricle
Fourth ventricle

CLINICAL FOCUS ✚ 2-3

Stroke

In the United States, someone suffers a stroke with obvious symptoms approximately every minute, producing more than a half million new stroke victims every year. Worldwide, stroke is the second leading cause of death. In addition to the visible strokes, at least twice as many "silent" strokes may occur. These small strokes of the white matter do not produce obvious symptoms.

Even with the best and fastest medical attention, most who endure stroke suffer some residual motor, sensory, or cognitive deficit. For every ten people who have a stroke, two die, six are disabled to varying degrees, and two recover to a degree but still endure a diminished quality of life. One in ten who survive risks further stroke.

The consequences of stroke are significant for victims, their families, and their lifestyles. Consider Mr. Anderson, a 45-year-old electrical engineer who took his three children to the movies one Saturday afternoon in 1998 and collapsed. Rushed to the hospital, he was diagnosed as having had a massive stroke of the middle cerebral artery of his left hemisphere. The stroke has impaired Mr. Anderson's language ever since and, because the brain's connections are crossed, his motor control on the right side as well.

Seven years after his stroke, Mr. Anderson remained unable to speak, but he could understand simple conversations. Severe difficulties in moving his right leg required him to use a walker. He could not move the fingers of his right hand and so had difficulty feeding himself, among other tasks. Mr. Anderson will probably never return to his engineering career or be able to drive or to get around on his own.

Like Mr. Anderson, most stroke survivors require help to perform everyday tasks. Their caregivers are often female relatives who give up their own careers and other pursuits. Half of these caregivers develop emotional illness, primarily depression or anxiety or both, in a year or so. Lost income and stroke-related medical bills have a significant effect on the family's standard of living.

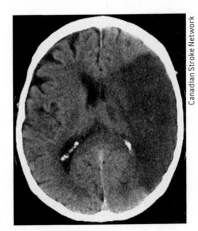

In this computer tomographic (CT) scan of a brain with a stroke, viewed dorsally, the dark area of the right hemisphere has been damaged by the loss of blood flow.

Canadian Stroke Network

Although we tend to speak of stroke as a single disorder, two major types of strokes have been identified. In the more common and often less severe *ischemic stroke,* a blood vessel is blocked (such as by a clot). The more severe *hemorrhagic stroke* results from a burst vessel bleeding into the brain.

The hopeful news is that ischemic stroke can be treated acutely with a drug called *tissue plasminogen activator* (t-PA) that breaks up clots and allows a return of normal blood flow to an affected region. Unfortunately, no treatment exists for hemorrhagic stroke, where the use of clot-preventing t-PA would be disastrous.

The results of clinical trials showed that, when patients are given t-PA within 3 hours of suffering an ischemic stroke, the number who make a nearly complete recovery increases by about 25 percent compared with those who are given a placebo (Hatcher and Starr, 2011). In addition, impairments are reduced in the remaining patients who survive the stroke. The risk of hemorrhage is about 6% in t-PA–treated patients relative to 0% in placebo-treated patients.

One difficulty is that many people are unable to get to a hospital soon enough for treatment with t-PA. Most stroke victims do not make it to an emergency room until about 24 hours after symptoms appear, too late for the treatment. Apparently, most people fail to realize that stroke is an emergency.

Other drugs producing an even better outcome than does t-PA are likely to become available. The hope is that these drugs will extend the 3-hour window for administering treatment after a stroke. There is also intense interest in developing treatments in the postacute period that will stimulate the brain to initiate reparative processes. Such treatment will facilitate the patient's functional improvement (see a review by Langhorne et al., 2011).

Another way to cut through the brain is perpendicularly from front to back, a sagittal section (**Figure 2-10A**). If we make our cut down the brain's midline, that is, in the midsagittal plane, we divide the cerebrum into its two hemispheres, revealing several distinctive brain components in a medial view (Figure 2-10B). One feature is a long band of white matter that runs much of the length of the cerebral hemispheres. This band, the **corpus callosum,** contains about 200 million nerve fibers that join the two hemispheres and allow communication between them.

FIGURE 2-10 Sagittal Section Through the Brain **(A)** This section in the midsagittal plane separates the hemispheres, allowing **(B)** a medial view of the midline structures of the brain, including the subcortical structures that lie below the corpus callosum.

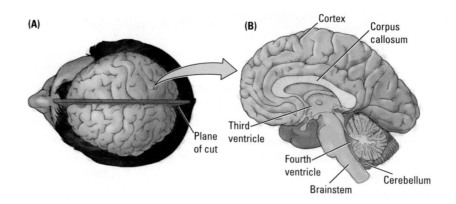

Figure 2-10B clearly shows that the cortex covers the cerebral hemispheres above the corpus callosum, whereas below it are various internal *subcortical regions.* The brainstem is a subcortical structure that generally controls basic physiological functions. But many subcortical regions are forebrain structures intimately related to the cortical areas that process motor, sensory, perceptual, and cognitive functions. This cortical–subcortical relation alerts us to the concept that redundancy of function exists at many different levels of the nervous system's organization.

If you were to compare medial views of the left and right hemispheres, you would be struck by their symmetry. The brain, in fact, has two of nearly every structure, one on each side. The few one-of-a-kind structures, such as the third and fourth ventricles, are found along the brain's midline (see Figure 2-9B). Another one-of-a-kind structure is the pineal gland that Descartes declared the seat of the mind in his dualistic theory about how the brain works.

Principle 4: The central nervous system functions on multiple levels.

Principle 5: The brain is both symmetrical and asymmetrical.

The human brain contains about 100 billion neurons and 100 billion glia. We examine their structures and functions in detail in Section 3-1.

Microscopic Inspection: Cells and Fibers

The fundamental units of the brain—its cells—are so small that they can be viewed only with the aid of a microscope. By using a microscope, we quickly discover that the brain has two main types of cells, illustrated in **Figure 2-11.** *Neurons* carry out the brain's major functions, whereas *glial cells* aid and modulate the neurons' activities—for example, by insulating neurons. Both neurons and glia come in many forms, each determined by the work that they do.

We can see the internal structures of the brain in much more detail by dyeing their cells with special stains (**Figure 2-12**). For example, if we use a dye that selectively stains cell bodies, we can see that the distribution of cells within the gray matter of the cerebral

FIGURE 2-11 Brain Cells A prototypical neuron (*left*) and glial cell (*right*) show that both have branches emanating from the cell body. This branching organization increases the surface area of the cell membrane. The neuron is called a pyramidal cell because the cell body is shaped somewhat like a pyramid; the glial cell is called an astrocyte because of its star-shaped appearance.

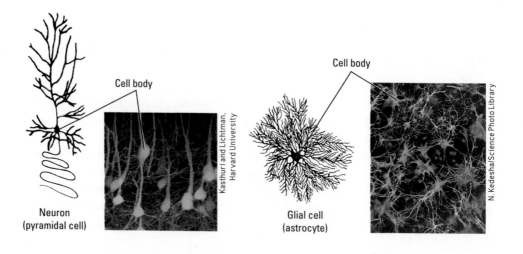

cortex is not homogeneous but rather forms layers, revealed by the bands of tissue in Figure 2-12A and C. Each layer contains cells that stain similarly. Stained subcortical regions are seen to be composed of clusters, or **nuclei,** of similar cells in Figure 2-12A and B.

Although layers and nuclei are very different in appearance, both form functional units within the brain. Whether a particular brain region has layers or nuclei is largely an accident of evolution. By using a stain that selectively dyes the fibers of neurons, as shown in Figure 2-12B and D, we can see the borders of the subcortical nuclei more clearly. In addition, we can see that the stained cell bodies lie in regions adjacent to the regions with most of the fibers.

A key feature of neurons is that they are connected to one another by fibers known as *axons.* When axons run along together, much like the wires that run from a car engine to the dashboard, they form a **nerve** or a **tract (Figure 2-13).** By convention, the term *tract* is usually used to refer to collections of nerve fibers found within the brain and spinal cord, whereas bundles of fibers located outside these CNS structures are typically referred to simply as *nerves.* Thus, the pathway from the eye to the brain is known as the optic nerve, whereas the pathway from the cerebral cortex to the spinal cord is known as the corticospinal tract.

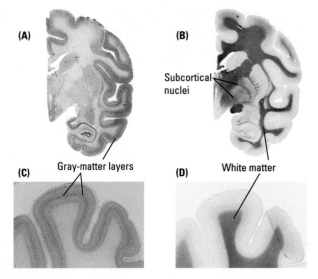

FIGURE 2-12 Cortical Layers and Glia
Brain sections from the left hemisphere of a monkey (midline is to the left in each image). Cells are stained with (**A** and **C**) a selective cell-body stain for neurons and (**B** and **D**) a selective fiber stain for insulating glial cells, or *myelin.* The images reveal very different pictures of the brain at a microscopic level (**C** and **D**).

REVIEW 2-1
Overview of Brain Function and Structure

Before you continue, check your understanding.

1. The function of the nervous system is to produce movement, or _____, within a perceptual world that is created by the _____.

2. The left and right cerebral hemispheres are each divided into four lobes: _____, _____, _____, and _____.

3. The human nervous system has evolved the potential to change, for example, to adapt to changes in the world or to compensate for injury. This attribute is called _____.

4. Neural tissue is of two main types: (1) _____ forms the connections among cells, and (2) _____ collects and processes incoming (afferent) sensory or outgoing (efferent) information.

5. The nerve fibers that lie within the brain form _____. Outside the brain they are called _____.

6. Chart the functional organization of the human nervous system.

Answers appear at the back of the book.

FIGURE 2-13 Neuronal Connections

Neuron 1

Axon

Neuron 2

Cell body

Terminal

Several axon fibers running together form a *nerve* when outside the CNS or a *tract* within the CNS.

2-2 Evolutionary Development of the Nervous System

The developing brain is less complex than the mature adult brain and provides a clearer picture of its basic structural plan. The biological similarity of embryos of vertebrate species as diverse as amphibians and mammals is striking in the earliest stages of development. In the evolution of complex nervous systems, simpler and evolutionarily more primitive forms have not been discarded and replaced but rather have been added to. As

nucleus (pl. nuclei) A group of cells forming a cluster that can be identified with special stains to form a functional grouping.

nerve Large collection of axons coursing together outside the central nervous system.

tract Large collection of axons coursing together within the central nervous system.

Section 1-3 outlines how the nervous system evolved. We explore biological and evolutionary similarities in development among humans and other species in Section 8-1.

Abnormalities associated with brain injury and brain disease that seem bizarre in isolation are but the normal manifestation of parts of a hierarchically organized brain. Our evolutionary history, our developmental history, and our own personal history are integrated at the various anatomical and functional levels of the nervous system.

FIGURE 2-14 Stages in Brain Evolution and Development The forebrain grows dramatically in the evolution of the mammalian brain.

a result, all anatomical and functional features of simpler nervous systems are present in the most complex nervous systems, including ours.

The bilaterally symmetrical nervous system of simple worms is common to complex nervous systems. Indeed, the spinal cord that constitutes most of the nervous system of the simplest fishes is recognizable in humans, as is the brainstem of more complex fishes, amphibians, and reptiles. The neocortex, although particularly complex in dolphins and humans, is nevertheless clearly the same organ found in other mammals.

The nervous system of a young vertebrate embryo begins as a sheet of cells that folds into a hollow tube and develops into three regions: forebrain, midbrain, and hindbrain (**Figure 2-14A**). These three regions are recognizable as a series of three enlargements at the end of the embryonic spinal cord. The adult brain of a fish, amphibian, or reptile is roughly equivalent to this three-part brain. The *prosencephalon* (front brain) is responsible for olfaction, the sense of smell; the *mesencephalon* (middle brain) is the seat of vision and hearing; and the *rhombencephalon* (hindbrain) controls movement and balance. The spinal cord is considered part of the hindbrain.

In mammals, the prosencephalon develops further to form the cerebral hemispheres, the cortex and subcortical structures known collectively as the *telencephalon* (endbrain), and the *diencephalon* (between brain) containing the thalamus, among other structures (Figure 2-14B). The hindbrain also develops further into the *metencephalon* (across brain), which includes the enlarged cerebellum, and the *myelencephalon* (spinal brain), including the medulla and the spinal cord.

The human brain is a more complex mammalian brain, possessing especially large cerebral hemispheres but retaining most of the features of other mammalian brains (Figure 2-14C). And, as illustrated in Research Focus 2-1, the human brain shows increases in selected cerebral areas when compared to the brains of other primates.

Most behaviors are the product not of a single locus in the brain but rather of many brain areas and levels. These several nervous-system layers do not simply replicate function; rather, each region adds a different dimension to the behavior. This hierarchical organization affects virtually every behavior in which humans engage.

Is the general organization of the vertebrate nervous system the only path to evolving intelligent behavior? Invertebrate animals, such as the octopus, have traveled on a separate evolutionary pathway from vertebrates for over 700 million years. Yet the octopus has

(A) Vertebrate

Prosencephalon (forebrain)
Mesencephalon (midbrain)
Rhombencephalon (hindbrain)
Spinal cord

(B) Mammalian embryo

Telencephalon
Diencephalon
Mesencephalon
Myelencephalon
Spinal cord
Metencephalon

(C) Fully developed human brain

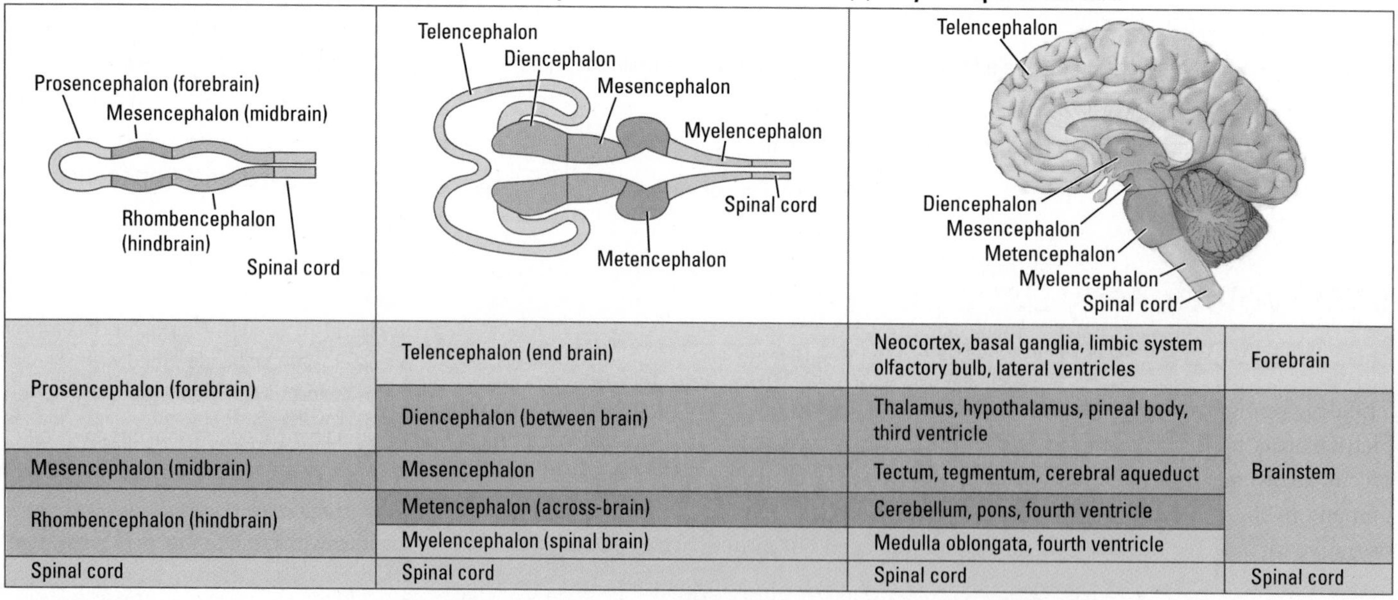

Telencephalon
Diencephalon
Mesencephalon
Metencephalon
Myelencephalon
Spinal cord

Prosencephalon (forebrain)	Telencephalon (end brain)	Neocortex, basal ganglia, limbic system olfactory bulb, lateral ventricles	Forebrain
	Diencephalon (between brain)	Thalamus, hypothalamus, pineal body, third ventricle	
Mesencephalon (midbrain)	Mesencephalon	Tectum, tegmentum, cerebral aqueduct	Brainstem
Rhombencephalon (hindbrain)	Metencephalon (across-brain)	Cerebellum, pons, fourth ventricle	
	Myelencephalon (spinal brain)	Medulla oblongata, fourth ventricle	
Spinal cord	Spinal cord	Spinal cord	Spinal cord

developed a complex nervous system that, while strikingly different from ours, may learn in strikingly similar ways, as **Experiment 2-1** demonstrates.

Italian biologists Graziano Fiorito and Pietro Scotto (1992) placed individuals of *Octopus vulgaris* (the common octopus) in separate tanks, each with an independent water supply, and allowed them to interact visually for 2 hours. As shown in the Procedures section of Experiment 2-1, during the observation phase that followed, the "observer" octopus watched the "demonstrator" octopus from an adjacent tank through a transparent wall. The demonstrator was being conditioned to learn that a red ball was associated with a reward, whereas a white ball was associated with a weak electric shock.

As noted in the Results section, the demonstrator animals quickly learned to distinguish between the colored balls. The observers then were placed in isolation. When tested later, they selected the same object as the demonstrators, responded faster than the demonstrators did during their conditioning, and performed the task correctly for 5 days without significant error or further conditioning.

EXPERIMENT 2-1

Question: Does intelligent behavior require a nervous system organized like those of vertebrate animals?

Procedure

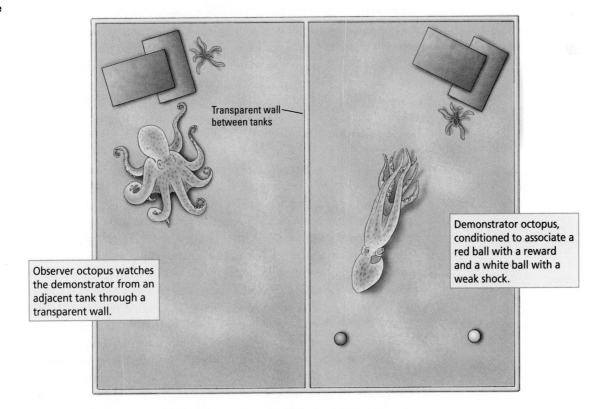

Transparent wall between tanks

Observer octopus watches the demonstrator from an adjacent tank through a transparent wall.

Demonstrator octopus, conditioned to associate a red ball with a reward and a white ball with a weak shock.

Results

1. The demonstrator animal quickly learns to distinguish between the colored balls.

2. When placed in isolation, then tested later, the observer animals selected the same object as the demonstrators, responded faster, and performed the task correctly for 5 days without significant error.

Conclusion: Invertebrate animals display intelligent behavior, such as learning by observation.

Adapted from "Observational learning in Octopus vulgaris," by G. Fiorito and P. Scotto (1992), Science, 256, 545–547.

REVIEW 2-2
Evolutionary Development of the Nervous System

Before you continue, check your understanding.

1. The brains of vertebrate animals have evolved into three regions: _____, _____, and _____.

2. The functional levels of the nervous system interact, each region contributing different aspects or dimensions, to produce _____.

3. In a brief paragraph, explain how the evolution of the forebrain region in mammals reinforces the principle that the CNS functions on multiple levels.

Answers appear at the back of the book.

2-3 Central Nervous System: Mediating Behavior

When we look under the hood, we can make some pretty good guesses about what each part of a car engine does. The battery must provide electrical power to run the radio and lights, for example, and, because batteries need to be charged, the engine must contain some mechanism for charging them. We can take the same approach to deduce the functions of the parts of the brain. The part connected to the optic nerve coming from each eye must have something to do with vision. Structures connected to the auditory nerve coming from each ear must have something to do with hearing.

From these simple observations, we can begin to understand how the brain is organized. The real test of inferences about the brain comes in analyzing actual brain function: how this seeming jumble of parts produces behaviors as complex as human thought. The place to start is the brain's anatomy, but learning the name of a particular CNS structure is pointless without also learning something about its function. In this section, therefore, we focus on the names and functions of the three major components of the CNS: the spinal cord, the brainstem, and the forebrain.

These three subdivisions reinforce the principle of levels of function, with newer levels partly replicating the work of older ones. A simple analogy to this evolutionary progress is learning to read. When you began to read, you learned simple words and sentences. As you progressed, you mastered new, more challenging words and longer, more complicated sentences, but you still retained the simpler skills that you had learned first. Much later, you encountered Shakespeare, with a complexity and subtlety of language unimagined in elementary school, taking you to a new level of reading comprehension.

Each new level of training adds new abilities that overlap and build on previously acquired skills. Yet all the functional levels deal with reading. Likewise, in the course of natural selection, the brain has evolved functional levels that overlap one another in purpose but allow for a growing complexity of behavior. For instance, the brain has functional levels that control movements. With the evolution of each new level, the complexity of movement becomes increasingly refined. We expand on the principle of evolutionary levels of function in Section 2-6.

Principle 6: Brain systems are organized both hierarchically and in parallel.

Spinal Cord

Although producing movement is the principal function of the brain, ultimately, the spinal cord produces most body movements, usually following instructions from the brain but at times acting independently. To understand how important the spinal cord is, think of the old saying "running around like a chicken with its head cut off." When a chicken's head is lopped off to provide dinner for the farmer's family, the chicken is still

brainstem Central structures of the brain, including the hindbrain, midbrain, thalamus, and hypothalamus, that are responsible for most unconscious behavior.

capable of running around the barnyard until it collapses from loss of blood. The chicken accomplishes this feat because the spinal cord is acting independently of the brain.

Grasping the complexity of the spinal cord is easier once you realize that it is not a single structure but rather a set of segmented "switching stations." As detailed in Section 2-4, each segment receives information from a discrete part of the body and sends out commands to that area. Spinal nerves, which are part of the somatic nervous system, carry sensory information to the cord from the skin, muscles, and related structures and, in turn, send motor instructions to control each muscle.

You can demonstrate movement controlled by the spinal cord in your own body by tapping your patellar tendon, just below your kneecap (the patella). The sensory input causes your lower leg to kick out and, try as you might, it is very hard to prevent the movement from occurring. Your brain, in other words, has trouble inhibiting this *spinal reflex,* which is automatic.

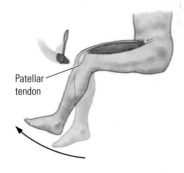

Patellar tendon

We explain reflexes in Section 11-4.

Brainstem

The **brainstem** begins where the spinal cord enters the skull and extends upward into the lower areas of the forebrain. The brainstem receives afferent nerves coming in from all of the body's senses, and it sends efferent nerves out to the spinal cord to control virtually all of the body's movements except the most complex movements of the fingers and toes. The brainstem, then, both directs movements and creates a sensory world.

In some animals, such as frogs, the entire brain is largely equivalent to the mammalian or avian brainstem. And frogs get along quite well, demonstrating that the brainstem is a fairly sophisticated piece of machinery. If we had only a brainstem, we would still be able to create a world, but it would be a far simpler, sensorimotor world, more like the world a frog experiences.

The brainstem, which is responsible for most unconscious behavior, can be divided into three regions: hindbrain, midbrain, and diencephalon, meaning "between brain" because it borders the brain's upper and lower parts. In fact, the "between brain" status of the diencephalon can be seen in a neuroanatomical inconsistency: some anatomists place it in the brainstem and others place it in the forebrain. **Figure 2-15A** illustrates the location of these three brainstem regions under the cerebral hemispheres, and Figure 2-15B compares the shape of the brainstem regions to the lower part of your arm held upright. The hindbrain is long and thick like your forearm, the midbrain is short and compact like your wrist, and the diencephalon at the end is bulbous like your hand forming a fist.

The hindbrain and midbrain are essentially extensions of the spinal cord; they developed first as vertebrate animals evolved a brain at the anterior end of the body. It makes

To keep the terms straight, remember that, alphabetically, afferent comes before efferent: sensory signals must come into the brain before an outgoing signal triggers a motor response.

Recall from Section 2-1 that an animal's perception of the external world depends on the complexity and organization of its nervous system.

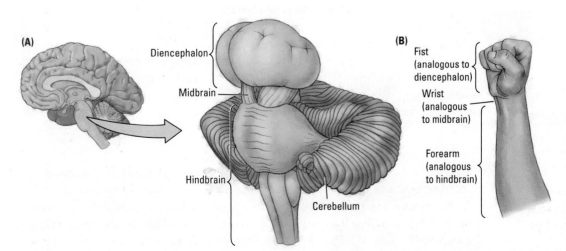

(A)

Diencephalon

Midbrain

Hindbrain

Cerebellum

(B)

Fist (analogous to diencephalon)

Wrist (analogous to midbrain)

Forearm (analogous to hindbrain)

FIGURE 2-15 Brainstem Structures **(A)** Medial view of the brain shows the relation of the brainstem to the cerebral hemisphere. **(B)** The shapes and relative sizes of the brainstem's parts can be imagined as analogous as to your fist, wrist, and forearm.

Principle 7: Sensory and motor divisions exist throughout the nervous system.

sense, therefore, that these lower brainstem regions should retain a division between structures having sensory functions and those having motor functions, with sensory structures located dorsally and motor ones ventrally.

Each brainstem region performs more than a single task. Each contains various subparts, made up of groupings of nuclei that serve different purposes. All three regions, in fact, have both sensory and motor functions. However, the hindbrain is especially important in motor functions, the midbrain in sensory functions, and the diencephalon in integrative tasks. Here we consider the central functions of these three regions; later chapters contain more detailed information about them.

Hindbrain

The **hindbrain** controls various motor functions ranging from breathing to balance to fine movements, such as those used in dancing. Its most distinctive structure, and one of the largest structures of the human brain, is the cerebellum. The relative size of the cerebellum increases with the physical speed and dexterity of a species, as shown in **Figure 2-16A**.

Animals that move relatively slowly (such as a sloth) have relatively small cerebellums for their body size, whereas animals that can perform rapid, acrobatic movements (such as a hawk or a cat) have very large cerebellums. The human cerebellum, which resembles a cauliflower in the medial view in Figure 2-16B, is important in controlling complex movements and apparently has a role in a variety of cognitive functions as well.

As we look below the cerebellum at the rest of the hindbrain, shown in **Figure 2-17**, we find three subparts: the reticular formation, the pons, and the medulla. Extending the length of the entire brainstem at its core, the **reticular formation** is a netlike mixture of neurons (gray matter) and nerve fibers (white matter) that gives this structure the mottled appearance from which its name derives (from the Latin *rete,* meaning "net"). The reticular formation's nuclei are localized along its length into small patches, each with a special function in stimulating the forebrain, such as in waking from sleep.

Not surprisingly, the reticular formation is sometimes called the *reticular activating system*.

The pons and medulla contain substructures that control many vital movements of the body. Nuclei within the pons receive inputs from the cerebellum and actually form a bridge from it to the rest of the brain (the Latin word *pons* means "bridge"). At the

FIGURE 2-16 The Cerebellum and Movement **(A)** Their relatively large cerebellums enable fine, coordinated movements such as flight and landing in birds and prey catching in cats. Slow-moving animals such as the sloth have smaller cerebellums relative to their brain size. **(B)** Like the cerebrum, the human cerebellum has an extensively folded cortex with gray and white matter and subcortical nuclei.

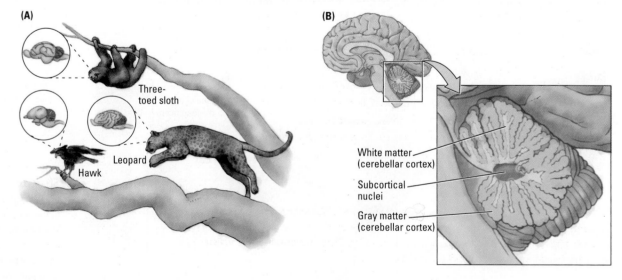

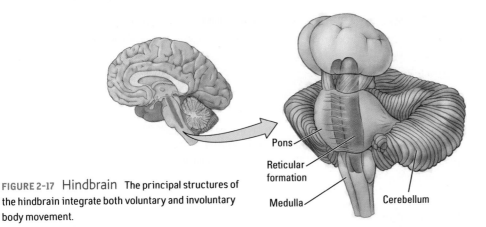

FIGURE 2-17 Hindbrain The principal structures of the hindbrain integrate both voluntary and involuntary body movement.

hindbrain Evolutionarily the oldest part of the brain; contains the pons, medulla, reticular formation, and cerebellum, structures that coordinate and control most voluntary and involuntary movements.

reticular formation Midbrain area in which nuclei and fiber pathways are mixed, producing a netlike appearance; associated with sleep–wake behavior and behavioral arousal.

midbrain Central part of the brain that contains neural circuits for hearing and seeing as well as orienting movements.

tectum Roof (area above the ventricle) of the midbrain; its functions are sensory processing, particularly visual and auditory, and the production of orienting movements.

tegmentum Floor (area below the ventricle) of the midbrain; a collection of nuclei with movement-related, species-specific, and pain-perception functions.

orienting movement Movement related to sensory inputs, such as turning the head to see the source of a sound.

rostral tip of the spinal cord, the medulla's nuclei control such vital functions as regulating breathing and the cardiovascular system. For this reason, a blow to the back of the head can kill you: your breathing stops if the control centers in the hindbrain are injured.

Midbrain

In the **midbrain,** shown in **Figure 2-18A**, the sensory component, the **tectum** (roof), is located dorsally, whereas a motor structure, the **tegmentum** (floor), is ventral. The tectum receives a massive amount of sensory information from the eyes and ears. The optic nerve sends a large bundle of nerve fibers to the *superior colliculus,* whereas the *inferior colliculus* receives much of its input from auditory pathways. The colliculi function not only to process sensory information but also to produce **orienting movements** related to sensory inputs, such as turning your head to see the source of a sound.

This orienting behavior is not as simple as it may seem. To produce it, the auditory and visual systems must share some sort of common "map" of the external world so that the ears can tell the eyes where to look. If the auditory and visual systems had different maps, it would be impossible to use the two together. In fact, the colliculi also have a tactile map. After all, if you want to look at the source of an itch on your leg, your visual and tactile systems need a common representation of where that place is so that you can scratch the itch with an arm and hand movement.

Lying ventral to the tectum, the tegmentum (shown in cross section in Figure 2-18B) is not a single structure but rather a composition of many nuclei, largely with movement-related functions. Several of its nuclei control eye movements. The so-called *red nucleus* controls limb movements, and the *substantia nigra* is connected to the forebrain, a connection especially important in initiating movements. (Recall from Clinical Focus 2-2 that the symptoms of Parkinson's disease are related to the destruction of the substantia nigra.) The *periacqueductal gray matter,* made up of cell bodies that surround the aqueduct joining the third and fourth ventricles, contains circuits controlling species-typical behaviors (e.g., female sexual behavior). These nuclei also play an important role in the modulation of pain by opioid drugs.

Principle 8: Sensory input to the brain is divided for sensory recognition and motor control.

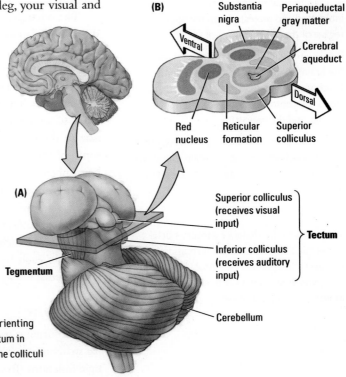

FIGURE 2-18 Midbrain (A) Structures in the midbrain are critical in producing orienting movements, species-specific behaviors, and the perception of pain. (B) The tegmentum in cross section, revealing various nuclei. Colliculus comes from *collis*, Latin for "hill." The colliculi resemble four little hills on the dorsal surface of the midbrain.

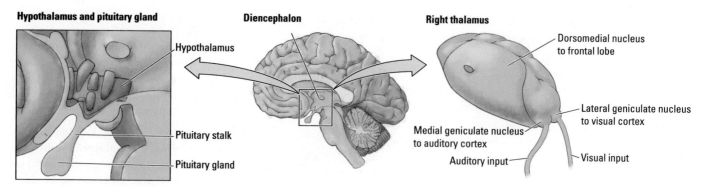

FIGURE 2-19 Diencephalon **The** diencephalon is composed of the thalamus, hypothalamus, and other structures. Thalamic regions connect to discrete regions of cortex. Below (hypo) the thalamus, at the base of the brain, the hypothalamus and pituitary lie above the roof of the mouth. The hypothalamus is composed of many nuclei, each with distinctly different functions.

diencephalon The "between brain" that integrates sensory and motor information on its way to the cerebral cortex.

hypothalamus Diencephalon structure that contains many nuclei associated with temperature regulation, eating, drinking, and sexual behavior.

thalamus Diencephalon structure through which information from all sensory systems is integrated and projected into the appropriate region of the neocortex.

forebrain Evolutionarily the newest part of the brain; coordinates advanced cognitive functions such as thinking, planning, and language; contains the limbic system, basal ganglia, and the neocortex.

neocortex (cerebral cortex) Newest, outer layer ("new bark") of the forebrain, composed of about six layers of gray matter; creates our reality.

We return to the thalamic sensory nuclei to examine how incoming information is processed in Sections 9-2, 10-2, 11-4, and 12-2, and in Section 14-3 to explore the brain's memory pathways.

Diencephalon

The **diencephalon,** shown in sagittal section in **Figure 2-19,** integrates sensory and motor information on its way to the cerebral cortex. The two principal structures of the diencephalon are the hypothalamus and the thalamus. The thalamus—one in each hemisphere—lies just to the left of the tip of the brainstem, and the hypothalamus lies to the left of the thalamus.

The **hypothalamus,** also found in each hemisphere lying bilaterally along the brain's midline, is composed of about 22 small nuclei, as well as nerve-fiber systems that pass through it. A critical function of the hypothalamus is to control the body's production of hormones, which is accomplished by its interactions with the pituitary gland, shown at the left in Figure 2-19. Although constituting only about 0.3 percent of the brain's weight, the hypothalamus takes part in nearly all aspects of behavior, including feeding, sexual behavior, sleeping, temperature regulation, emotional behavior, hormone function, and movement.

The hypothalamus is organized more or less similarly in different mammals largely because the control of feeding, temperature, and so on is carried out similarly. But there are sex differences in the structures of some parts of the hypothalamus, owing probably to differences between males and females in activities such as sexual behavior and parenting.

The other principal structure of the diencephalon, the **thalamus,** is much larger than the hypothalamus, as are its 20-odd nuclei. Perhaps most distinctive among the functions of the thalamus, shown at the right in Figure 2-19, is its role as a kind of gateway for channeling sensory information traveling to the cerebral cortex. All sensory systems send inputs to the thalamus for information integration and relay to the appropriate area in the cortex. The optic tract, for example, sends information through a large bundle of fibers to a region of the thalamus called the *lateral geniculate nucleus,* shown on the right side of the thalamus in Figure 2-19. In turn, the lateral geniculate nucleus processes some of this information and then sends it to the visual region of the cortex in each hemisphere.

The routes to the thalamus may be indirect. For example, the route for olfaction traverses several synapses before entering the *dorsomedial thalamic nucleus* on its way to the forebrain. Analogous sensory regions of the thalamus receive auditory and tactile information, which is subsequently relayed to the respective auditory and tactile cortical regions in each hemisphere. Some thalamic regions have motor functions or, like the dorsomedial nucleus that connects to most of the frontal lobe, perform integrative tasks.

Forebrain

The **forebrain** is the largest region of the mammalian brain; its major internal and external structures are shown in **Figure 2-20.** Each of its three principal structures has multiple functions. To summarize briefly, the *neocortex* (another name for the cerebral cortex) regulates a host of mental activities ranging from perception to planning; the *basal ganglia* control voluntary movement; and the *limbic system* regulates emotions and behaviors that create and require memory.

Extending our analogy between the brainstem and your forearm, imagine that the "fist" of the brainstem (the diencephalon) is thrust inside a watermelon. The watermelon represents the forebrain, with the cortex as the rind and the subcortical limbic system and basal ganglia as the fruit inside. By varying the size of the watermelon, we can vary the size of the brain, which in a sense is what evolution has done. The forebrain therefore varies considerably in size across species (recall the photographs in Research Focus 2-1).

Cortex

There are actually two types of cortex, the old and the new. The **neocortex** ("new bark") is the tissue that is visible when we view the brain from the outside, as in Figure 2-6 The neocortex is unique to mammals, and its primary function is to create a perceptual world and respond to that world. The older cortex, sometimes called *limbic cortex,* is more primitive than the neocortex. It is found in the brains of other chordates in addition to mammals, especially in birds and reptiles.

The limbic cortex is thought to play a role in controlling motivational states. Although anatomical and functional differences exist between the neocortex and the limbic cortex, the distinctions are not critical for most discussions in this book. Therefore, we will usually refer to both types of tissue simply as *cortex.*

Measured by volume, the cortex makes up most of the forebrain, comprising 80 percent of the human brain overall. It is the brain region that has expanded the most in the course of mammalian evolution. The human neocortex has a surface area as large as 2500 square centimeters but a thickness of only 1.5 to 3.0 millimeters. This area is equivalent to about four pages of this book. By contrast, a chimpanzee has a cortical area equivalent to about one page.

The pattern of sulci and gyri formed by the folding of the cortex varies across species. Some species, such as rats, have no sulci or gyri, whereas carnivores, such as cats, have gyri that form a longitudinal pattern. In primates, the sulci and gyri form a more diffuse pattern.

Cortical Lobes

The human cortex consists of two nearly symmetrical hemispheres, the left and the right, which are separated by the longitudinal fissure **(Figure 2-21, left).** Each hemisphere is subdivided into the four lobes introduced in Figure 2-5, corresponding to the skull

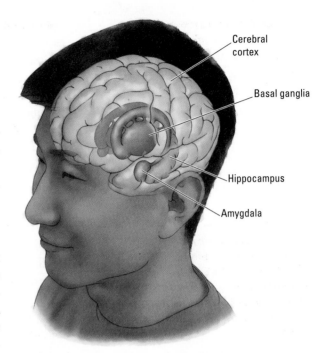

FIGURE 2-20 Forebrain Structures The major structures of the forebrain integrate sensation, emotion, and memory to enable advanced cognitive functions such as thinking, planning, and language.

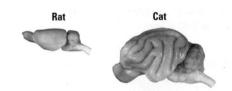

Courtesy of Wally Welker, University of Wisconsin Comparative Mammalian Brain Collection.

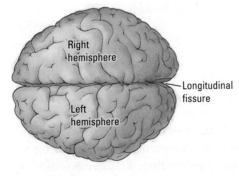

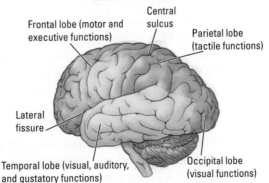

FIGURE 2-21 Cortical Boundaries

Principle 9: Functions in the brain are both localized and distributed.

bones overlying them: frontal, temporal, parietal, and occipital. Unfortunately, bone location and brain function are unrelated. As a result, the lobes of the cortex are rather arbitrarily defined regions that include many different functional zones.

Nevertheless, we can attach some gross functions to each lobe. The three posterior lobes have sensory functions: the occipital lobe is visual; the parietal lobe is tactile; and the temporal lobe is visual, auditory, and gustatory. In contrast, the frontal lobe is motor and is sometimes referred to as the brain's "executive" because it integrates sensory and motor functions and formulates plans of action. We can also predict some effects of injuries to each lobe:

- People with an injury to the occipital lobe have deficits in processing visual information. Although they may perceive light versus dark, they may be unable to identify either the shape or the color of objects.

- Injuries to the parietal lobe make it difficult to identify or locate stimulation on the skin. Deficits in making movements of the arms and hands to points in space may occur.

- Temporal lobe injuries result in difficulty recognizing sounds, although, unlike people with occipital injuries, they can still recognize that they are hearing some type of sound. Temporal lobe injuries can also produce difficulties in processing complex visual information, such as faces.

- Individuals with frontal-lobe injuries may have difficulties organizing their ongoing behavior as well as planning for the future.

Traditionally, the occipital lobes are defined on the basis of anatomical features that are presented in Section 9-2.

Fissures and sulci often establish the boundaries of cortical lobes (Figure 2-21, right). For instance, in humans, the central sulcus and lateral fissure form the boundaries of each frontal lobe. They also form the boundaries of each parietal lobe, which lies posterior to the central sulcus. The lateral fissure demarcates each temporal lobe, forming its dorsal boundary. The occipital lobes are not so clearly separated from the parietal and temporal lobes because no large fissure marks their boundaries.

Cortical Layers

The neocortex has six layers of gray matter atop a layer of white matter. (In contrast, the limbic cortex has three layers.) The layers of the neocortex have several distinct characteristics:

- Different layers have different types of cells.

- The density of cells in each layer varies, ranging from virtually no cells in layer I (the top layer) to very dense cell packing in layer IV of the neocortex (**Figure 2-22**).

- Other differences in appearance relate to the functions of cortical layers in different regions.

cytoarchitectonic map Map of the neocortex based on the organization, structure, and distribution of the cells.

basal ganglia Subcortical forebrain nuclei that coordinate voluntary movements of the limbs and body; connected to the thalamus and to the midbrain.

Parkinson's disease Disorder of the motor system correlated with a loss of dopamine in the brain and characterized by tremors, muscular rigidity, and a reduction in voluntary movement.

These visible differences led neuroanatomists of the early twentieth century to map the cortex. The map in **Figure 2-23** was developed by Korbinian Brodmann in about 1905. Because these maps are based on cell characteristics, the subject of cytology, they are called **cytoarchitectonic maps.** For example, viewed through a microscope, sensory cortex in the parietal lobe, shown in red in Figure 2-22, has a distinct layer IV. Motor cortex in the frontal lobe, shown in blue in Figure 2-22, has a distinctive layer V. Layer IV is afferent, whereas layer V is efferent. It makes sense that a sensory region would have a large input layer, whereas a motor region would have a large output layer.

Chemical differences in the cells in different cortical layers can be revealed by staining the tissue. Some regions are rich in one chemical, whereas others are rich in another. These differences presumably relate to functional specialization of different areas of the cortex.

The one significant difference between the organization of the cortex and the organization of other parts of the brain is its range of connections. Unlike most structures

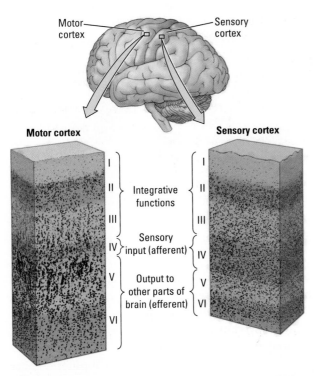

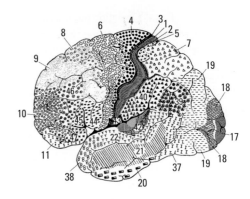

FIGURE 2-22 Layering in the Neocortex As this comparison of cortical layers in the sensory and motor areas shows, layer IV is relatively thick in the sensory cortex and relatively thin in the motor cortex. Afferents connect to layer IV (coming from the thalamus) as well as to layers II and III. Efferents in layers V and VI connect to other parts of the cortex and to the motor structures of the brain.

FIGURE 2-23 Early Brain Map In his cytoarchitectonic map of the cortex, Brodmann defined areas by the organization and characteristics of the cells that he examined. The regions shown in color are associated with the simplest sensory perceptions of touch (red), vision (purple), and hearing (orange). As we shall see, the areas of the cortex that process sensory information are far more extensive than Brodmann's basic areas.

that connect to only selective brain regions, the cortex is connected to virtually all other parts of the brain. The cortex, in other words, is the ultimate meddler. It takes part in everything. This fact not only makes it difficult to identify specific functions of the cortex but also complicates our study of the rest of the brain because the cortex's role in other brain regions must always be considered.

Consider your perception of clouds. You have no doubt gazed up at clouds on a summer day and imagined sailing ships, elephants, faces, and countless other objects. Although a cloud does not really look exactly like an elephant, you can concoct an image of one if you impose your frontal cortex—that is, your imagination—on the sensory inputs. This kind of cortical activity is known as "top-down processing" because the top level of the nervous system, the cortex, is influencing how information is processed in lower regions of the hierarchy—in this case, the midbrain and hindbrain.

The cortex influences many behaviors besides the perception of objects. It influences our cravings for foods, our lust for things (or people), and how we interpret the meaning of abstract concepts, words, and images. The cortex is the ultimate creator of our reality, and one reason it serves this function is that it is so well connected.

Basal Ganglia

A collection of nuclei that lie within the forebrain just below the white matter of the cortex, the **basal ganglia** consist of three principal structures: the *caudate nucleus,* the *putamen,* and the *globus pallidus,* all shown in **Figure 2-24.** Together with the thalamus and two closely associated structures, the substantia nigra and subthalamic nucleus, the basal ganglia form a system that functions primarily to control certain aspects of voluntary movement.

We can observe the functions of the basal ganglia by analyzing the behavior that results from the many diseases that interfere with the normal functioning of these nuclei. People afflicted with **Parkinson's disease,** a disorder of the motor system characterized by severe tremors, and one of the most common disorders of movement in the elderly, take short, shuffling steps, display bent posture, and often require a walker to get around. Many have an almost continual tremor of the hands and sometimes of the head as well.

Detailed coverage of Parkinson's disease appears in Focus features 5-2, 5-3, and 5-4, and in Section 16-3. Clinical Focus 11-4 details Tourette's syndrome.

FIGURE 2-24 Basal Ganglia This frontal section of the cerebral hemispheres shows the basal ganglia relative to surrounding structures. Two associated structures that are likewise instrumental in controlling and coordinating movement, the substantia nigra and subthalamic nucleus, also are shown.

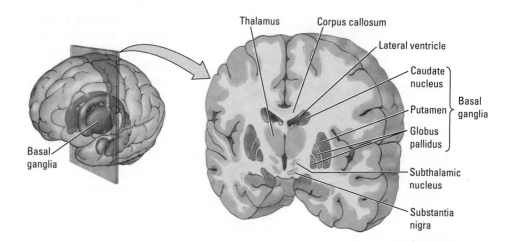

Another disorder of the basal ganglia is **Tourette's syndrome,** characterized by various motor tics, involuntary vocalizations (including curse words and animal sounds), and odd, involuntary movements of the body, especially of the face and head.

Neither Parkinsonism nor Tourette's syndrome is a disorder of *producing* movements, as in paralysis. Rather they are disorders of *controlling* movements. The basal ganglia, therefore, must play a role in the control and coordination of movement patterns rather than in activating the muscles to move.

Limbic System

In the 1930s, psychiatry was dominated by the theories of Sigmund Freud, who emphasized the roles of sexuality and emotion in understanding human behavior. At the time, regions in the brain controlling these behaviors had not been identified; at the same time, a group of brain structures collectively called the "limbic lobe" had no known function. It was a simple step to thinking that perhaps the limbic structures played a central role in sexuality and emotion.

Figure 16-1 charts a contemporary version of Freud's theory of mind.

The limbic system figures prominently in discussions of addiction in Section 6-3, motivation and emotion in Sections 12-3 and 12-4, memory in Section 14-3, and brain disorders in Research Focus 16-1 and Section 16-4.

One sign that this hypothesis might be correct came from James Papez (1937), who discovered that people with rabies have infections of limbic structures, and one of the symptoms of rabies is heightened emotionality. We now know that such a simple view is inaccurate. In fact, the **limbic system** is not a unitary system at all, and, although some limbic structures have roles in emotion and sexual behaviors, limbic structures serve other functions, too, including contributing to memory and motivation.

The principal limbic structures are shown in **Figure 2-25.** They include the *amygdala,* the *hippocampus,* and the *limbic,* or *cingulate, cortex,* which lies in the cingulate gyrus between the cerebral hemispheres. Recall that limbic cortex, structured in three or four layers of gray matter atop a layer of white matter, is evolutionarily older than the six-layered neocortex.

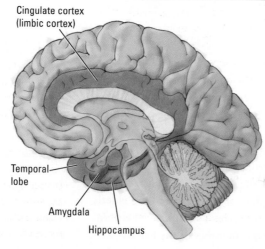

FIGURE 2-25 Limbic System This medial view of the right hemisphere illustrates the principal structures of the limbic system that play roles in emotional and sexual behaviors, motivation, and memory.

The hippocampus, the cingulate cortex, and associated structures have roles in certain memory functions, as well as in controlling navigation in space. Many limbic structures are also believed to be at least partly responsible for the rewarding properties of psychoactive drugs and other potentially addictive substances and behaviors. Repeated exposure to drugs such as amphetamine or nicotine produces both chemical and structural changes in the cingulate cortex and hippocampus, among other structures.

All sorts of behaviors can prove addictive—among them, eating, shopping, sex, video gaming, gambling, and, as pictured here, even Twitter! How else to explain these Canadian coeds tweeting in their winter coats when they could be defrosting on the Ft. Lauderdale beach in the Florida sun?

Removal of the amygdala produces truly startling changes in emotional behavior. A cat with the amygdala removed will wander through a colony of monkeys, completely undisturbed by their hooting and threats. No self-respecting cat would normally be caught anywhere near such bedlam.

Olfactory System

At the very front of the brain lie the olfactory bulbs, the organs responsible for our sense of smell. The olfactory system is unique among human senses, as **Figure 2-26** shows, because it is almost entirely a forebrain structure. Recall that the other sensory systems project most of their inputs from the sensory receptors to the midbrain and thalamus. Olfactory input takes a less direct route: the olfactory bulb sends most of its inputs to a specialized region, the *pyriform cortex* at the bottom of the brain, before progressing to the dorsal medial thalamus, which then provides a route to the frontal cortex.

A curious aspect of the olfactory system is that it is one of the first senses to have evolved in animals, yet it is found at the front of the human brain and considered part of the forebrain (see the ventral view in Figure 2-6). This is partly an "accident" of evolution. The olfactory bulbs lie near the olfactory receptors in the nasal cavity and, although they send their inputs to the pyriform cortex in mammals, the input to the brainstem is more direct in simpler brains.

Compared with the olfactory bulbs of animals such as rats, cats, and dogs, which depend more heavily on the sense of smell than we do, the human olfactory bulb is relatively small. Nonetheless, it is still sensitive and plays an important role in various aspects of our feeding and sexual behavior.

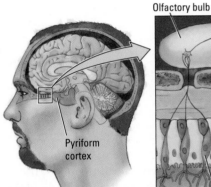

FIGURE 2-26 Sense of Smell Our small olfactory bulbs lie at the base of the forebrain, connected to receptor cells that lie in the nasal cavity.

We return to the olfactory system in Section 12-2 in considering the chemical senses of smell and taste in the context of emotional and motivated behavior.

Tourette's syndrome Disorder of the basal ganglia characterized by tics, involuntary vocalizations (including curse words and animal sounds), and odd, involuntary movements of the body, especially of the face and head.

limbic system Disparate forebrain structures lying between the neocortex and the brainstem that form a functional system controlling affective and motivated behaviors and certain forms of memory; includes cingulate cortex, amygdala, and hippocampus, among other structures.

REVIEW 2-3
Central Nervous System: Mediating Behavior

Before you continue, check your understanding.

1. The three functionally distinct sections of the CNS—spinal cord, brainstem, and forebrain—represent the evolution of multiple _____.

2. The _____ can perceive sensations from the skin and muscles and produce movements independent of the brain.

3. The brainstem includes three functional regions. The _____ is an extension of the spinal cord, the _____ is the first brain region to receive sensory inputs, and the _____ integrates sensory and motor information on its way to the cerebral cortex.

4. The forebrain's subcortical regions include the _____, which control voluntary movement, and the _____, which controls mood, motivation, and some forms of memory.

5. Briefly describe the functions performed by the forebrain.

Answers appear at the back of the book.

2-4 Somatic Nervous System: Transmitting Information

The somatic nervous system (SNS) is monitored and controlled by the CNS—the cranial nerves by the brain and the spinal nerves by the spinal cord segments.

Cranial Nerves

The linkages provided by the **cranial nerves** between the brain and various parts of the head and neck as well as various internal organs are illustrated and tabulated in **Figure 2-27.** Cranial nerves can have afferent functions, such as sensory inputs to the brain from the eyes, ears, mouth, and nose, or they can have efferent functions, such as motor control of the facial muscles, tongue, and eyes. Some cranial nerves have both sensory and motor functions, such as the modulation of both sensation and movement in the face.

FIGURE 2-27 Cranial Nerves Each of the 12 pairs of cranial nerves has a different function. A common mnemonic device for learning the order of the cranial nerves is "On old Olympus's towering top, a Finn and German vainly skip [and] hop." The first letter of each word (exept the second "and") is, in order, the first letter of the name of each nerve.

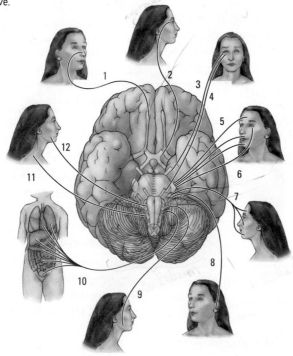

Cranial nerve	Name	Function
1	Olfactory	Smell
2	Optic	Vision
3	Oculomotor	Eye movement
4	Trochlear	Eye movement
5	Trigeminal	Masticatory movements and facial sensation
6	Abducens	Eye movement
7	Facial	Facial movement and sensation
8	Auditory vestibular	Hearing and balance
9	Glossopharyngeal	Tongue and pharynx movement and sensation
10	Vagus	Heart, blood vessels, viscera, movement of larynx and pharynx
11	Spinal accessory	Neck muscles
12	Hypoglossal	Tongue muscles

The 12 pairs of cranial nerves are known both by their numbers and by their names, as listed in Figure 2-27. One set of 12 controls the left side of the head, whereas the other set controls the right side. This arrangement makes sense for innervating duplicated parts of the head (such as the eyes), but why separate nerves should control the right and left sides of a singular structure (such as the tongue) is not so clear. Yet that is how the cranial nerves work. If you have ever received novocaine for dental work, you know that usually just one side of your tongue becomes anesthetized because the dentist injects the drug into only one side of your mouth. The rest of the skin and muscles on each side of the head are similarly controlled by cranial nerves located on the same side.

We consider many of the cranial nerves in some detail in later chapters in discussions on topics such as vision, hearing, olfaction, taste, and responses to stress. For now, you simply need to know that cranial nerves form part of the somatic nervous system, providing inputs to the brain from the head's sensory organs and muscles and controlling head and facial movements. The cranial nerves also contribute to maintaining autonomic functions by connecting the brain and internal organs and by influencing other autonomic responses, such as salivation.

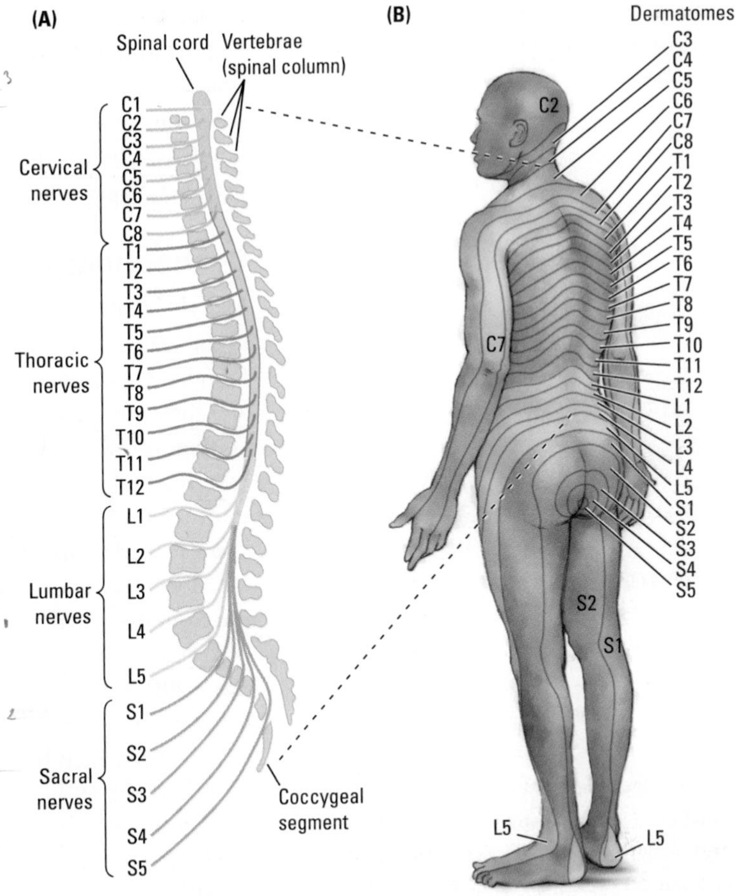

Spinal Nerves

The spinal cord lies inside the bony spinal column, which is made up of a series of small bones called **vertebrae,** categorized into five anatomical regions from top to bottom: cervical, thoracic, lumbar, sacral, and coccygeal, as diagrammed in **Figure** 2-28A. You can think of each vertebra within these five groups as a very short segment of the spinal column. The corresponding spinal-cord segment within each vertebral region functions as that segment's "minibrain."

This arrangement may seem a bit odd, but it has a long evolutionary history. Think of a simpler animal, such as a snake, that evolved long before humans did. A snake's body is a tube divided into segments. Within that tube is another tube, the spinal cord, which also is segmented. Each of the snake's nervous-system segments receives nerve fibers from sensory receptors in the part of the body adjacent to it, and that nervous-system segment sends fibers back to the muscles in that body part. Each segment, therefore, works independently.

A complication arises in animals such as humans, who have limbs that may originate at one spinal-segment level but extend past other segments of the spinal column. Your shoulders, for example, may begin at C5 (cervical segment 5), but your arms hang down well past the sacral segments. So, unlike the snake, which has spinal-cord segments that connect to body segments fairly directly adjacent to them, human body segments fall schematically into more of a patchwork pattern, as shown in Figure 2-28B. This arrangement makes sense if the arms are extended as they are when we walk on "all fours."

Regardless of their complex pattern, however, the segments of our bodies still correspond to segments of the spinal cord. Each of these body segments is called a **dermatome** (meaning "skin cut"). A dermatome has both a sensory nerve that sends information from the skin, joints, and muscles to the spinal cord and a motor nerve that controls the movements of the muscles in that particular segment of the body.

FIGURE 2-28 Spinal Segments and Dermatomes **(A)** Medial view of the spinal column, showing the five spinal-cord segments: cervical (C), thoracic (T), lumbar (L), sacral (S), and coccygeal. **(B)** Each spinal segment corresponds to a region of body surface (a dermatome) identified by the segment number (examples are C5 at the base of the neck and L2 in the lower back).

Spinal-cord segmentation and bilateral symmetry (one half of the body is the mirror image of the other) are two important structural features of human nervous-system evolution noted in Section 1-3.

cranial nerve One of a set of 12 nerve pairs that control sensory and motor functions of the head, neck, and internal organs.

vertebrae (sing. vertebra) The bones that form the spinal column.

dermatome Body segment corresponding to a segment of the spinal cord.

Sections 11-1 and 11-4 review the spinal cord's contributions to movement and to somatosensation.

These sensory and motor nerves, known as *spinal* (or *peripheral*) *nerves,* are functionally equivalent to the cranial nerves of the head. Whereas the cranial nerves receive information from sensory receptors in the eyes, ears, facial skin, and so forth, the spinal nerves receive information from sensory receptors in the rest of the body—that is, in the PNS. Similarly, whereas the cranial nerves move the muscles of the eyes, tongue, and face, the peripheral nerves move the muscles of the limbs and trunk.

Connections of the Somatic Nervous System

Like the central nervous system, the somatic nervous system is bilateral (two sided). Just as the cranial nerves control functions on the same side of the head on which they are found, the spinal nerves on the left side of the spinal cord control the left side of the body, and those on the right side of the spinal cord control the body's right side.

Figure 2-29A shows the spinal column in cross section. Look first at the nerve fibers entering the spinal cord's dorsal (back) side. These dorsal fibers are afferent: they carry in information from the body's sensory receptors. The fibers collect together as they enter a spinal-cord segment, and this collection of fibers is called a *dorsal root.*

Section 11-2 details the organization of the spinal tracts within the motor system.

Fibers leaving the spinal cord's ventral (front) side are efferent, carrying information out from the spinal cord to the muscles. They, too, bundle together as they exit the spinal cord and so form a *ventral root.* The outer part of the spinal cord, which is pictured in Figure 2-29B, consists of white matter, or CNS nerve tracts. These tracts are arranged so that, with some exceptions, dorsal tracts are sensory and ventral tracts are motor. The inner part of the cord, which has a butterfly shape, is gray matter composed largely of cell bodies.

Sections 11-1 and 11-4 explore causes of spinal-cord injuries and treatments for them. The link between spinal injury and loss of emotion is a topic in Section 12-3.

The observation that the dorsal spinal cord is sensory and the ventral side is motor is one of the nervous system's very few established laws, the **law of Bell and Magendie.** Combined with an understanding of the spinal cord's segmental organization, this law enables neurologists to make accurate inferences about the location of spinal-cord damage or disease on the basis of changes in sensation or movement that patients experience. For instance, if a person experiences numbness in the fingers of the left hand but can still move the hand fairly normally, one or more of the dorsal nerves in spinal-cord segments C7 and C8 must be damaged. In contrast, if sensation in the hand is normal but the person cannot move the fingers, the ventral roots of the same segments must be damaged. Clinical Focus 2-4, "Magendie, Bell, and Bell's Palsy," further explores the topic of diagnosing spinal-cord injury or disease.

FIGURE 2-29 Spinal-Nerve Connections **(A)** A cross section of the spinal cord, viewed from the front. The butterfly-shaped inner regions consist of neural cell bodies (gray matter), and the outer regions consist of nerve tracts (white matter) traveling to and from the brain. **(B)** A dorsal-view photograph shows the intact spinal cord exposed.

(A)

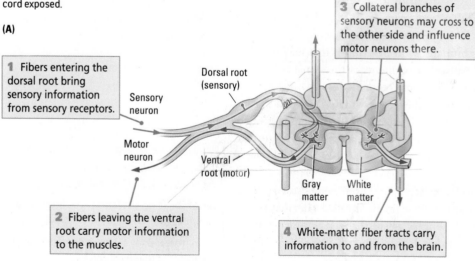

1 Fibers entering the dorsal root bring sensory information from sensory receptors.

Sensory neuron

Dorsal root (sensory)

3 Collateral branches of sensory neurons may cross to the other side and influence motor neurons there.

Motor neuron

Ventral root (motor)

Gray matter

White matter

2 Fibers leaving the ventral root carry motor information to the muscles.

4 White-matter fiber tracts carry information to and from the brain.

(B)

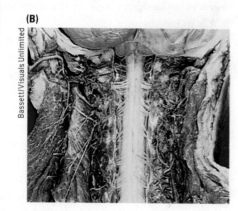

Bassett/Visuals Unlimited

Magendie, Bell, and Bell's Palsy

François Magendie, a volatile and committed French experimental physiologist, reported in a three-page paper in 1822 that he had succeeded in cutting the dorsal and ventral roots of puppies, animals in which the roots are sufficiently segregated to allow such surgery. Magendie found that cutting the dorsal roots caused loss of sensation, whereas cutting the ventral roots caused loss of movement.

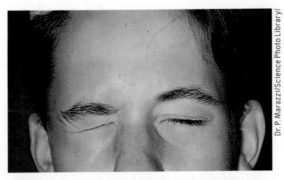

A young man suffering from Bell's palsy, a paralysis of the facial nerve that causes weakness over one side of the face. He was photographed during an involuntary tic (a nervous reaction) that affects the right side of the face, causing his right eye to close tightly.

Eleven years earlier, a Scot named Charles Bell had proposed functions for these nerve roots on the basis of anatomical information and the results of somewhat inconclusive experiments on rabbits. Although Bell's findings were not identical with those of Magendie, they were similar enough to ignite a controversy. Bell hotly disputed Magendie's claim to the discovery of dorsal- and ventral-root functions. As a result, the principle of sensory and motor segregation in the nervous system has been named after both researchers: the law of Bell and Magendie.

Magendie's conclusive experiment on puppies was considered extremely important because it enabled neurologists, for the first time, to localize nervous system damage from the symptoms that a patient displays. Bell went on to describe an example of such localized, cranial motor-nerve dysfunction that still bears his name—*Bell's palsy*, a facial paralysis that develops when the motor part of the facial nerve on one side of the head becomes inflamed (see the accompanying photograph).

The onset of Bell's palsy is typically sudden. Often the stricken person wakes up in the morning and is shocked to discover the face paralyzed on one side. He or she cannot open the mouth on that side of the head or completely close the eye on that side. Most people fully recover from Bell's palsy, although recovery may take several months. But in rare instances, such as that of Jean Chrétien, a former prime minister of Canada, partial paralysis of the mouth is permanent.

Photo credit (vertical): Dr. P. Marazzi/Science Photo Library/Photo Researchers

Integrating Spinal Functions

So far, we have emphasized the segmental organization of the spinal cord, but the spinal cord must also somehow coordinate inputs and outputs across different segments. For example, many body movements require the coordination of muscles that are controlled by different segments, just as many sensory experiences require the coordination of sensory inputs to different parts of the spinal cord. How is this coordination accomplished? The answer is that the spinal-cord segments are interconnected in such a way that adjacent segments can operate together to direct rather complex coordinated movements.

The integration of spinal-cord activities does not require the brain's participation, which is why the headless chicken can run around in a reasonably coordinated way. Still, a close working relation must exist between the brain and the spinal cord. Otherwise, how could we consciously plan and execute our voluntary actions?

Somehow information must be relayed back and forth, and examples of this information sharing are numerous. For instance, tactile information from sensory nerves in the skin travels not just to the spinal cord but also to the cerebral cortex through the thalamus. Similarly, the cerebral cortex and other brain structures can control movements through their connections to the ventral roots of the spinal cord. So even though the brain and spinal cord can function independently, the two are intimately connected in their CNS functions.

law of Bell and Magendie The general principle that sensory fibers are located dorsally and motor fibers are located ventrally.

REVIEW 2-4
Somatic Nervous System: Transmitting Information

Before you continue, check your understanding.

1. Two sets of SNS nerves, the _____ and the _____, function to receive sensory information or to send motor signals to muscles or both.

2. Both sets of SNS nerves are symmetrically organized, and each set controls functions on the _____ side of the body on which it is found.

3. The cranial nerves have both sensory and motor functions, receiving and sending information to the _____ and to the _____.

4. Define the law of Bell and Magendie and explain why it is important.

Answers appear at the back of the book.

2-5 Autonomic Nervous System: Balancing Internal Functions

The internal autonomic nervous system is a hidden partner in controlling behavior. Even without our conscious awareness, it stays on the job to keep the heart beating, the liver releasing glucose, the pupils of the eyes adjusting to light, and so forth. Without the ANS, which regulates the internal organs and glands via connections through the SNS to the CNS, life would quickly cease. Although learning to exert some conscious control over some of these vegetative activities is possible, such conscious interference is unnecessary. An important reason is that the ANS must keep working during sleep, when conscious awareness is off duty.

Although we might think that the autonomic system's organization must be pretty simple because it functions outside conscious awareness, the ANS, like the SNS, can be thought of as a collection of minibrains. The autonomic system has a surprisingly complex organization. For example, the gut responds to a range of hormones and other chemicals with exquisite neural responses. Some scientists have even proposed that the central nervous system evolved from the gut of very simple organisms.

The two divisions of the ANS, sympathetic and parasympathetic, work in opposition. The **sympathetic division** arouses the body for action, for example, by stimulating the heart to beat faster and inhibiting digestion when we exert ourselves during exercise or times of stress—the familiar "fight or flight" response. The **parasympathetic division** calms the body down, for example, by slowing the heartbeat and stimulating digestion to allow us to "rest and digest" after exertion and during quiet times.

Like the SNS, the ANS interacts with the rest of the nervous system. Activation of the sympathetic division starts in the thoracic and lumbar spinal-cord regions. But the spinal nerves do not directly control the target organs. Rather, the spinal cord is connected to autonomic control centers, which are collections of neural cells called *ganglia*. The ganglia control the internal organs, and each acts as a minibrain for specific organs.

The sympathetic ganglia are located near the spinal cord, forming a chain that runs parallel to the cord, as illustrated at the left in **Figure 2-30.** The parasympathetic division also is connected to the spinal cord—specifically, to the sacral region—but the greater part of it derives from three cranial nerves: the vagus nerve, which calms most of the internal organs, and the facial and oculomotor nerves, which control salivation and pupil dilation, respectively (review Figure 2-27). In contrast with the sympathetic division, the parasympathetic division connects with ganglia that are near the target organs, as shown at the right in Figure 2-30.

Section 5-3 explains how the CNS and ANS communicate, Section 6-5 diagrams the autonomic stress response, and Section 16-4 discusses how mood affects an individual's reactivity to stress.

Principle 10: The nervous system works by juxtaposing excitation (increased neural activity) and inhibition (decreased neural activity).

sympathetic division Part of the autonomic nervous system; arouses the body for action, such as mediating the involuntary fight-or-flight response to alarm by increasing heart rate and blood pressure.

parasympathetic division Part of the autonomic nervous system; acts in opposition to the sympathetic division—for example, preparing the body to rest and digest by reversing the alarm response or stimulating digestion.

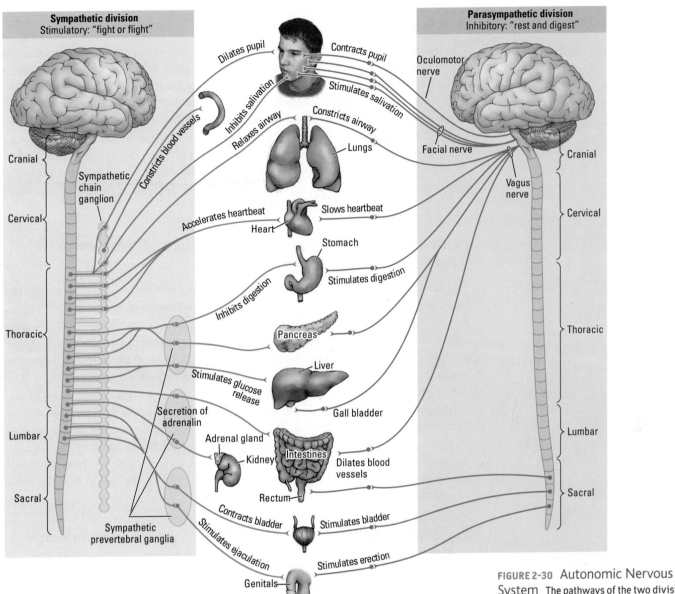

Sympathetic division
Stimulatory: "fight or flight"

Dilates pupil

Inhibits salivation

Constricts blood vessels

Relaxes airway

Cranial

Sympathetic chain ganglion

Cervical

Accelerates heartbeat

Heart

Thoracic

Inhibits digestion

Stimulates glucose release

Secretion of adrenalin

Lumbar

Adrenal gland

Kidney

Sacral

Sympathetic prevertebral ganglia

Contracts bladder

Stimulates ejaculation

Genitals

Parasympathetic division
Inhibitory: "rest and digest"

Contracts pupil

Oculomotor nerve

Stimulates salivation

Constricts airway

Lungs

Facial nerve

Cranial

Vagus nerve

Cervical

Slows heartbeat

Stomach

Stimulates digestion

Thoracic

Pancreas

Liver

Gall bladder

Lumbar

Intestines

Dilates blood vessels

Rectum

Sacral

Stimulates bladder

Stimulates erection

FIGURE 2-30 Autonomic Nervous System The pathways of the two divisions of the ANS exert opposing effects on the organs that they innervate. All autonomic fibers connect at "stops" en route from the CNS to their target organs. (*Left*) Arousing sympathetic fibers connect to a chain of ganglia near the spinal cord. (*Right*) Calming parasympathetic fibers connect to individual parasympathetic ganglia near the target organs.

REVIEW 2-5
Autonomic Nervous System: Balancing Internal Functions

Before you continue, check your understanding.

1. The ANS interacts with the CNS and SNS via sets of autonomic control centers called _____, which act as minibrains to control the internal organs.

2. The _____ division of the ANS arouses the body for action, and the _____ division calms the organs. The two divisions work _____ to allow for quick defensive responses (fight or flight) or to induce a calming (rest and digest) state.

3. Why is the ANS essential to life?

Answers appear at the back of the book.

2-6 Ten Principles of Nervous-System Function

The balance created within the whole nervous system, within the functioning brain, and within individual cells works in concert to produce behavior. Knowing the parts of the nervous system and some general notions about what they do is only the beginning. Learning how the parts work together allows us to proceed to a closer look, in the chapters that follow, at how the brain produces behavior.

In this chapter, we have identified ten principles related to the nervous system's functioning. Here, we elaborate on each one. As you progress through the book, review these ideas regularly with an eye toward understanding the concept rather than simply memorizing the principle. Soon you will find yourself applying them as you encounter new information about the brain and behavior.

Principle 1: The Nervous System Produces Movement Within a Perceptual World the Brain Creates

The fundamental function of the nervous system is to produce behavior, or movement. But movements are not made in a vacuum. They are related to objects, places, memories, and myriad other forces and factors. Your representation of the world depends on the nature of the information sent to your brain. The representation of the world of people who are color-blind is very different from the representation of the world of those who perceive color. The perceptual world of people who have perfect pitch is different from that of those who do not.

Although we tend to think that the world that we perceive is what is actually there, individual realities, both between and within species, are clearly but rough approximations of what is actually present. A special function of the brain of each animal species is to produce a reality that is adaptive for that species to survive. In other words, the behavior that the brain produces is directly related to the world that the brain has created.

Principle 2: The Hallmark of Nervous-System Functioning Is Neuroplasticity

Experience alters the brain's organization, and neuroplasticity is required for learning and memory functions as well as for survival. In fact, information is stored in the nervous system only if neural connections change. Forgetting is presumably due to a loss of the connections that represented the memory.

As Experiment 2-1 on page 49 demonstrates, neuroplasticity is a characteristic not just of the mammalian brain; it is found in the nervous systems of all animals, even the simplest worms. Nonetheless, larger brains have more capacity for change and are thus likely to show more plastic neural organization.

Neuroplasticity is not always beneficial. Analyses of the brains of animals given addicting doses of drugs such as cocaine or morphine reveal large changes in neural connectivity suspected of underlying some maladaptive behaviors related to addiction. Among the many other examples of pathological neuroplasticity are pathological pain, epilepsy, and dementia.

Principle 3: Many of the Brain's Circuits Are Crossed

A most peculiar organizational feature of the brain is that most of its inputs and outputs are "crossed." Each hemisphere receives sensory stimulation from the opposite (contralateral) side of the body and controls muscles on the contralateral side as well. Crossed organization explains why people who experience strokes or other damage to the left

Section 14-4 explores the neural bases of plasticity and drug addiction. Section 11-4 explains how we feel pain and can treat it. Clinical Focus 4-1 describes the symptoms of epilepsy; Section 16-3 details its diagnosis and treatment. Section 16-3 also describes the spectrum of dementias that neuroscientists have identified.

cerebral hemisphere may have difficulty in sensing stimulation to the right side of the body or in moving body parts on the right side. The opposite is true of people with strokes in the right cerebral hemisphere.

A crossed nervous system must somehow join the two sides of the perceptual world together. To do so, innumerable neural connections link the left and right sides of the brain. The most prominent connecting cable is the corpus callosum, which joins the left and right cerebral hemispheres with about 200 million nerve fibers.

Two important exceptions to the crossed-circuit principle are olfactory sensation and the somatic nervous system. Olfactory information does not cross but rather projects directly into the same (ipsilateral) side of the brain. The cranial and spinal nerves of the SNS do not cross but are connected ipsilaterally.

As illustrated in Figure 9-10, the human visual system has evolved a fascinating solution to the challenge of representing the world seen through two eyes as a single perception: both eyes connect with both hemispheres.

Principle 4: The Central Nervous System Functions on Multiple Levels

Sensory and motor functions are carried out at many places in the brain, in the spinal cord and the brainstem as well as in the forebrain. This multiplicity of functions results from the nature of brain evolution.

Simple animals such as worms have a spinal cord, more complex animals such as fish have a brainstem as well, and yet more complex animals have evolved a forebrain. Each new addition to the CNS has added a new level of behavioral complexity without discarding previous levels of control. As animals evolved legs, for example, brain structures had to be added to move the legs. Later, the development of independent digit movements also required more brainpower.

The addition of new brain areas can be viewed as the addition of new levels of nervous-system control. The new levels are not autonomous but rather must be integrated into the existing neural systems. Each new level can be conceived of as a way of refining and elaborating the control provided by the earlier levels.

The idea of levels of function can be seen not only in the addition of forebrain areas to refine the control of the brainstem but also within the forebrain itself. As mammals evolved, they developed an increased capacity to represent the world in the cortex, an ability that is related to the addition of more "maps." The new maps must be related to the older ones, however, and again are simply an elaboration of the sensory world that was there before.

Principle 5: The Brain Is Both Symmetrical and Asymmetrical

Although the left and the right hemispheres look like mirror images, they have some dissimilar features. Cortical asymmetry is essential for integrative tasks, language and body control among them.

Consider speaking. If a language zone existed in both hemispheres, each connected to one side of the mouth, we would actually be able to talk out of both sides of our mouths at once. That would make talking awkward, to say the least. One solution is to locate language control of the mouth on one side of the brain only. Organizing the brain in this way allows us to speak with a single voice.

A similar problem arises in controlling the body's movement in space. We would not want the left and the right hemispheres each trying to take us to a different place. Again, the problem can be solved if a single brain area controls this sort of spatial processing.

In fact, functions such as language and spatial navigation are localized on only one side of the brain. Language is usually on the left side, and spatial functions are usually on the right. The brains of many species have both symmetrical and asymmetrical features. The control of singing is located in one hemisphere in the bird brain. Like human language, birdsong is usually located on the left side. It is likely that the control of song by two sides of the brain would suffer the same problems as the control of language, and it is likely that birds and humans evolved the same solution independently—namely, to assign the control to only one side of the brain.

Principle 6: Brain Systems Are Organized Both Hierarchically and in Parallel

When we consider that the CNS comprises multiple levels of function, these levels clearly must be extensively interconnected to integrate their processing and create unified perceptions or movements. The nature of this connectivity leads to the next principle of brain function: the brain has both serial (or hierarchical) and parallel circuitry.

A hierarchical circuit hooks up a linear series of all regions concerned with a particular function. Consider vision. In a serial system, the information from the eyes goes to regions that detect the simplest properties, such as color or brightness. This information would then be passed to another region that determines shape and then to another region that measures movement and so on until, at the most complex level, the information is understood to be, say, your grandmother. Information therefore flows in a hierarchical manner sequentially from simpler to more complex regions as illustrated in **Figure 2-31A**.

One difficulty with hierarchical models, however, is that functionally related structures in the brain are not always linked in a linear series. Although the brain has many serial connections, many expected connections are missing. For example, within the visual system, not all cortical regions are connected to one another. The simplest explanation is that the unconnected regions must have very different functions.

Parallel circuits operate on a different principle, which also is illustrated by the visual system. Imagine looking at a car. As we look at a car door, one set of visual pathways processes information about its nature, such as its color and shape, whereas another set of pathways processes information about door-related movements, such as movements required to open the door.

These two visual systems are independent of each other, yet they must interact in some manner. Your perception when you pull the door open is not one of two different representations—the door's size, shape, and color on the one hand and the opening movements on the other. When you open the door, you have the impression of unity in your conscious experience.

Figure 2-31B illustrates the flow of information in such a distributed hierarchy. If you trace the information flow from the primary area to levels 2, 3, and 4, you can see the parallel pathways. These multiple parallel pathways are also connected to one another. However, the connections are more selective than those that exist in a purely serial circuit.

Interestingly, the subsystems of the brain are organized into multiple parallel pathways. Yet our conscious experiences are always unified. As we explore this conundrum throughout the book, keep in mind that your commonsense impressions of how the brain works may not always be correct.

Principle 7: Sensory and Motor Divisions Exist Throughout the Nervous System

The segregation of sensory and motor functions described by the Bell and Magendie law exists throughout the nervous system. However, distinctions between motor and sensory functions become subtler in the forebrain.

Sensory and Motor Divisions in the SNS

The spinal nerves are either sensory or motor in function. Some cranial nerves are exclusively sensory; some are exclusively motor; and some have two parts, one sensory and one motor, much like spinal nerves serving the skin and muscles.

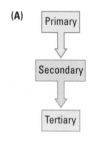

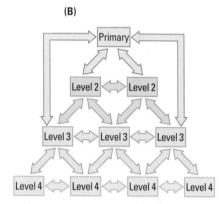

FIGURE 2-31 Models of Neural Information Processing (A) Simple serial hierarchical model of cortical processing similar to that first proposed by Alexandre Luria in the 1960s. (B) In the distributed hierarchical model of Daniel Felleman and David van Essen (1991), multiple levels exist in each of several processing streams. Areas at each level interconnect.

You can review cranial-nerve and spinal-nerve connections in Figures 2-27 and 2-28.

Sensory and Motor Divisions in the CNS

The lower brainstem regions—hindbrain and midbrain—are essentially extensions of the spinal cord and retain the spinal-cord division, with sensory structures located dorsally and motor structures ventrally. An important function of the midbrain is to orient the body to stimuli. Orienting movements require both sensory input and motor output. The midbrain's colliculi, which are located dorsally in the tectum, are the sensory component, whereas the tegmentum, which is ventral, is a motor structure that plays a role in controlling various movements, including orienting ones.

Distinct sensory nuclei are present in the thalamus, too, although they are no longer located dorsally. Because all sensory information reaches the forebrain through the thalamus, to find separate nuclei associated with vision, hearing, and touch is not surprising. Separate thalamic nuclei also control movements. Other nuclei have neither sensory nor motor functions but rather connect to cortical areas, such as the frontal lobe, that perform more integrative tasks.

Finally, sensory and motor functions are divided in the cortex in two ways:

1. Separate sensory and motor cortical regions process particular sensory inputs, such as vision, hearing, or touch. Others control detailed movements of discrete body parts, such as the fingers.

2. The entire cortex is organized around the sensory and motor distinction. Layer IV of the cortex always receives sensory inputs, layers V and VI always send motor outputs, and layers I, II, and III integrate sensory and motor operations.

Principle 8: Sensory Input to the Brain Is Divided for Object Recognition and Motor Control

Sensory systems evolved first for controlling motion, not for recognizing things. Simple organisms can detect stimulation such as light and move to or from it. It is not necessary to "perceive" an object to direct movements toward or away from it. As animals and their behaviors became more complex, they began to evolve ways of representing their environment. Animals with complex brains, such as ourselves, evolved separate systems for producing movement toward objects and for recognizing them. The visual system exemplifies this separation.

Visual information travels from the eyes to the thalamus to visual regions of the occipital lobe. From the occipital cortex it follows one of two routes: one route, known as the *ventral stream,* leads to the temporal lobe for object identification, whereas the other route, known as the *dorsal stream,* goes to the parietal lobe to guide movements relative to objects **(Figure 2-32).** People with ventral-stream injuries are "blind" for the recognition of objects, yet they nevertheless shape their hands appropriately when asked to reach for the objects that they cannot identify.

Consider reaching for a cup. When a normal participant reaches for a cup, his or her hand forms a shape that is different from the shape it forms when reaching for a spoon. People with ventral-stream injuries can make appropriate hand shapes, yet they do not consciously recognize the object. In contrast, people with dorsal-stream injuries can recognize objects but make clumsy reaching movements because they do not form appropriate hand postures until they contact objects. Only then do they shape the hand on the basis of tactile information.

The recognition that perception for movement and perception for object recognition are independent processes has three important implications for understanding brain organization:

1. The dorsal and ventral systems provide an excellent example of parallel information processing in the brain.

Brainstem structures are illustrated in Figures 2-15 through 2-19.

To review the layered structure of the cortex, see Figure 2-22.

In Sections 9-2 and 9-3, we review the evidence that led to understanding the functions of the visual streams and how each stream processes visual information.

FIGURE 2-32 Neural Streams The dorsal and ventral streams mediate vision for action and recognition, respectively.

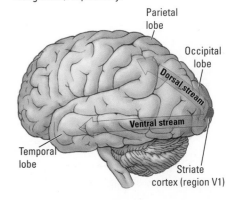

Alzheimer's disease Degenerative brain disorder related to aging that first appears as progressive memory loss and later develops into generalized dementia.

excitation Increase in the activity of a neuron or brain area.

inhibition Decrease in the activity of a neuron or brain area.

Focus 1-1 reports that singing therapy helped Gabrielle Giffords regain limited speech after a gunshot wound disrupted major language areas in her brain. Section 10-4 explains how the brain processes the sounds we perceive as language and music.

Figure 14-5 illustrates the extensive distribution of memory through the brain.

You will find information about the neurochemistry of Alzheimer's disease in Section 5-3, its incidence and possible causes in Section 14-3, and treatments in Section 16-4.

2. Although we may think we are aware of our entire sensory world, the sensory analysis required for some movements is clearly not conscious.

3. The presence of unconscious and conscious brain processing underlies an important difference in our cognitive functions. The unconscious-movement system is always acting in the present and in response to ongoing sensory input. In contrast, the recognition system allows us to escape the present and bring to bear information from the past. Thus, the object-recognition system forms the neural basis of enduring memory.

Principle 9: Functions in the Brain Are Both Localized and Distributed

One of the great debates in the history of brain research has concerned what aspects of different functions are actually localized in specific brain regions. Perhaps the fundamental problem is defining a function. Language, for example, includes the comprehension of spoken words, written words, signed words (as in American Sign Language), and even touched words (as in Braille). Language also includes processes of producing words orally, in writing, and by signing, as well as constructing whole linguistic compositions, such as stories, poems, songs, and essays.

Because the function that we call language has many aspects, it is not surprising that these aspects reside in widely separated areas of the brain. We see evidence of this widespread distribution in language-related brain injuries. People with injuries in different locations may selectively lose the abilities to produce words, understand words, read words, write words, and so forth. Specific language-related abilities, therefore, are found in specific locations, but language itself is distributed throughout a wide region of the brain.

Memory provides another example of this same distributed pattern. Memories can be extremely rich in detail and can include sensual feelings, words, images, and much more. Like language, then, aspects of memory are located in many brain regions distributed throughout a vast area of the brain.

Because many functions are both localized and distributed in the brain, damage to a small brain region produces only focal symptoms. Massive brain damage is required to completely remove some functions. Thus, a small injury could impair some aspect of language functioning, but it would take a widespread injury to completely remove all language abilities. In fact, one characteristic of dementing diseases is that people can endure widespread deterioration of the cortex yet maintain remarkably normal language functions until late stages of the disease. **Alzheimer's disease** is a degenerative brain disorder related to aging that first appears as progressive memory loss and only much later develops into generalized dementia.

Principle 10: The Nervous System Works by Juxtaposing Excitation and Inhibition

Although we have emphasized the brain's role in *making* movements, we must also recognize that the brain acts to *prevent* movements. To make a directed movement, such as picking up a glass of water, we must refrain from other movements, such as waving the hand back and forth. In producing movement, then, the brain produces some action through **excitation** (increased neural activity) and, through **inhibition** (decreased neural activity), prevents other action.

Brain injury or disease can produce either a *loss* or a *release* of behavior through changes in the balance between excitation and inhibition. A brain injury to a person in a region that normally initiates speech may render the person unable to talk. This result is a loss of behavior. A person with an abnormality in a region that inhibits inappropriate language (such as swearing) may be unable to inhibit this form of talking. This result is a release of behavior that can be seen in Tourette's syndrome.

Patients with Parkinson's disease may endure uncontrollable shaking of the hands because the neural system that inhibits such movements has failed. Paradoxically, they often have difficulty initiating movements and appear frozen because they are unable to generate the excitation needed to produce movements.

This juxtaposition of excitation and inhibition is central to how the brain produces behavior and can be seen at the level of individual neurons. All neurons have a spontaneous rate of activity that can be either increased (excitation) or decreased (inhibition). Some neurons excite others, whereas other neurons inhibit. Both effects are produced by specific neurochemicals by which neurons communicate.

> Tourette's syndrome and Parkinson's disease are disorders representative of dysfunction in the basal ganglia, which coordinates voluntary movement.

> Chapter 3 details the structure of nervous-system cells. Chapter 4 reveals how neurons transmit and integrate information. Chapter 5 explains neural communication and adaptation through learning.

REVIEW 2-6
Ten Principles of Nervous-System Function

Before you continue, check your understanding.

1. Many of the brain's input and output circuits are crossed. Within the nervous system, two exceptions to this principle are the _____ and the _____.

2. The vertebrate brain has evolved three regions—hindbrain, midbrain, and forebrain—leading to _____ and flexibility in controlling behavior.

3. One aspect of neural activity that resembles the "on–off" language of digital devices is the juxtaposition of _____ and _____.

4. Explain this statement: Perception is not reality.

Answers appear at the back of the book.

SUMMARY

2-1 Overview of Brain Function and Structure

The primary function of the brain is to produce behavior, or movement, in a perceptual world that is created by the brain. This perceptual world is ever-changing, and thus, to adapt the brain must also change, a property referred to as neuroplasticity.

To study how the nervous system functions, we abandon the anatomical divisions between the central nervous system and the peripheral nervous system to focus instead on function—on how the CNS interacts with the divisions of the PNS: the somatic and autonomic nervous systems.

2-2 Evolutionary Development of the Nervous System

The vertebrate nervous system has evolved from a relatively simple structure mediating reflexlike behaviors to the complex human brain mediating advanced cognitive processes. Primitive forms have not been replaced but rather have been adapted and modified as new structures have evolved to allow for more complex behavior in an increasingly sophisticated perceptual world.

The principles of nervous-system organization and function generalize across the three regions of the vertebrate brain—hindbrain, midbrain, and forebrain—leading to multiple levels of functioning. The evolution of levels of control thus adds flexibility to the control of behavior.

2-3 Central Nervous System: Mediating Behavior

The CNS includes the brain and the spinal cord. The spinal cord can perceive sensations from the skin and muscles and produce movements independent of the brain. The brain can be divided into the brainstem and forebrain, each made up of hundreds of parts. The brainstem both directs movements and creates a sensory world through its connections with the sensory systems, spinal cord, and forebrain. The forebrain modifies and elaborates basic sensory and motor functions, regulates cognitive activity, including thought and memory, and holds ultimate control over movement. The most elaborate part of the forebrain is the cerebral cortex, which grows disproportionately large in the human brain.

2-4 Somatic Nervous System: Transmitting Information

The SNS consists of the spinal nerves that enter and leave the spinal column, going to and from muscles, skin, and joints in the body, and the cranial nerves that link the muscles of the face and some internal organs to the brain. Both sets of SNS nerves have a symmetrical organization, with one set controlling each side of the body. Some cranial nerves are sensory, some are motor, and some combine both functions. The spinal cord functions as a kind of minibrain for the peripheral (spinal) nerves that enter and leave its five segments. Each spinal segment works independently, although CNS fibers interconnect them and coordinate their activities.

2-5 Autonomic Nervous System: Balancing Internal Functions

The ANS controls the body's glands and internal organs and operates largely outside conscious awareness. Its sympathetic (arousing) and parasympathetic (calming) divisions work in opposition. The parasympathetic division directs the organs to "rest and digest," whereas the sympathetic division prepares for "fight or flight."

2-6 Ten Principles of Nervous-System Function

The ten principles that form the basis for many discussions throughout this book are listed in the right column. Understanding these principles fully will place you at an advantage in your study of brain and behavior.

Ten Principles of Nervous-System Function

Principle 1	The nervous system produces movement within a perceptual world the brain creates.
Principle 2	The hallmark of nervous-system functioning is neuroplasticity.
Principle 3	Many of the brain's circuits are crossed.
Principle 4	The central nervous system functions on multiple levels.
Principle 5	The brain is both symmetrical and asymmetrical.
Principle 6	Brain systems are organized both hierarchically and in parallel.
Principle 7	Sensory and motor divisions exist throughout the nervous system.
Principle 8	Sensory input to the brain is divided for object recognition and motor control.
Principle 9	Functions in the brain are both localized and distributed.
Principle 10	The nervous system works by juxtaposing excitation and inhibition.

KEY TERMS

afferent, p. 37

Alzheimer's disease, p. 70

autonomic nervous system (ANS), p. 37

basal ganglia, p. 56

brainstem, p. 50

cerebral cortex, p. 41

cerebrospinal fluid (CSF), p. 41

corpus callosum, p. 44

cranial nerve, p. 61

cytoarchitectonic map, p. 56

dermatome, p. 61

diencephalon, p. 54

efferent, p. 37

excitation, p. 70

forebrain, p. 54

frontal lobe, p. 41

gray matter, p. 43

gyrus (pl. gyri), p. 43

hindbrain, p. 53

hypothalamus, p. 54

inhibition, p. 70

law of Bell and Magendie, p. 63

limbic system, p. 59

meninges, p. 37

midbrain, p. 53

neocortex (cerebral cortex), p. 54

nerve, p. 47

neuroplasticity, p. 37

nucleus (pl. nuclei), p. 47

occipital lobe, p. 41

orienting movement, p. 53

parasympathetic division, p. 64

parietal lobe, p. 41

Parkinson's disease, p. 56

phenotypic plasticity, p. 37

reticular formation, p. 53

somatic nervous system (SNS), p. 37

stroke, p. 43

sulcus (pl. sulci), p. 43

sympathetic division, p. 64

tectum, p. 53

tegmentum, p. 53

temporal lobe, p. 41

thalamus, p. 54

Tourette's syndrome, p. 59

tract, p. 47

ventricle, p. 44

vertebrae (sing. vertebra), p. 61

white matter, p. 44

Please refer to the Companion Web Site at www.worthpublishers.com/kolbintro4e *for Interactive Exercises and Quizzes.*

CHAPTER

3

What Are the Functional Units of the Nervous System?

RESEARCH FOCUS 3-1 A GENETIC DIAGNOSIS

3-1 CELLS OF THE NERVOUS SYSTEM

RESEARCH FOCUS 3-2 BRAINBOW: RAINBOW NEURONS

NEURONS: THE BASIS OF INFORMATION PROCESSING

FIVE TYPES OF GLIAL CELLS

CLINICAL FOCUS 3-3 BRAIN TUMORS

CLINICAL FOCUS 3-4 MULTIPLE SCLEROSIS

3-2 INTERNAL STRUCTURE OF A CELL

THE BASICS CHEMISTRY REVIEW

THE CELL AS A FACTORY

CELL MEMBRANE: BARRIER AND GATEKEEPER

NUCLEUS: SITE OF GENE TRANSCRIPTION

ENDOPLASMIC RETICULUM: SITE OF RNA TRANSLATION

PROTEINS: THE CELL'S PRODUCT

GOLGI BODIES AND MICROTUBULES: PROTEIN PACKAGING AND SHIPMENT

CROSSING THE CELL MEMBRANE: CHANNELS, GATES, AND PUMPS

3-3 GENES, CELLS, AND BEHAVIOR

MENDELIAN GENETICS AND THE GENETIC CODE

APPLYING MENDEL'S PRINCIPLES

CLINICAL FOCUS 3-5 HUNTINGTON'S DISEASE

GENETIC ENGINEERING

PHENOTYPIC PLASTICITY AND THE EPIGENETIC CODE

A Genetic Diagnosis

Fraternal twins Alexis and Noah Beery seemingly acquired cerebral palsy perinatally (at or near birth). They had poor muscle tone and could barely walk or sit. Noah drooled and vomited, and Alexis suffered from tremors.

Typically, children with cerebral palsy, a condition featuring perinatal brainstem damage, do not get worse with age, but the twins' condition deteriorated. Their mother, Retta Beery, observed as well that Alexis's symptoms fluctuated: for example, they improved after she slept or napped.

In searching the literature for similar cases, Retta found a photocopy of a 1991 news report that described a child first diagnosed with

Courtesy of Retta Beery

Noah, Retta, Joe, and Alexis Beery at Baylor College with the Solid Sequencer that decoded the genomes of twins Noah and Alexis.

cerebral palsy, then found to have a rare condition, dopa-responsive dystonia ("dystonia" for abnormal muscle tone). The condition stems from a deficiency of a neurochemical, dopamine, produced by a relatively small cluster of cells in the midbrain.

When Alexis and Noah received a daily dose of L-dopa, a chemical that brain cells convert into dopamine, they displayed remarkable improvement. "We knew that we were witnessing a miracle," Retta recalled.

A few years later, in 2005, Alexis began to experience new symptoms marked by difficulty in breathing. At this time the twins' father, Joe, worked for Life Technologies, a biotech company that makes equipment used for sequencing DNA, the molecule found in every cell that codes our genetic makeup. Joe arranged for samples of the twins' blood to be sent to the Baylor College of Medicine's sequencing center.

The twin's genome was sequenced and compared with that of their parents and close relatives. The analysis showed that the twins had an abnormality in a gene for an enzyme that enhances the production not only of dopamine but another chemical made by brainstem cells, serotonin (Bainbridge et al., 2011).

When the twins' doctors added the chemical that is converted to serotonin to the L-dopa, both twins showed remarkable improvement. Alexis was able to compete in junior high school track, and Noah was able to compete in volleyball in the Junior Olympics. This is the first diagnosis established through genome sequencing that has led to a treatment success, a scientific miracle indeed.

The Beery twins' remarkable story highlights how neuroscientists are taking advantage of advances in genetics to investigate brain disorders. Understanding genes, proteins, and cellular function allows us to understand normal brain function as well.

We begin this chapter by describing the structure of nervous system cells and relating their structure to their functions. Brain cells not only give the nervous system and its many parts their structure but also mediate its moment-to-moment changes. These changes underlie our behavior. We conclude the chapter by elaborating on Mendelian genetics and how Mendel's theory contrasts with epigenetics.

3-1 Cells of the Nervous System

Cells have been likened to factories in that they make a product, proteins. But cells are more than factories. They use the proteins they produce to play a dynamic role in orchestrating our behavior.

Nervous-system cells are small, are packed tightly together, and have the consistency of jelly. To see a brain cell, it must first be distinguished from surrounding cells and then magnified using a microscope. Scientists, including anatomists who study tissue, have

Brainbow: Rainbow Neurons

Were it not for the discovery of different stains that can highlight the features of brain cells, their complexity and connections would remain unknown. Jean Livet (2007) and his colleagues at Harvard University developed a transgenic technique that labels many different neurons by highlighting them with distinct colors, a technique called "brainbow," a play on the word "rainbow." (Transgenic techniques are a form of genetic engineering, discussed in Section 3-3.)

In the same way a television monitor produces the full range of colors that the human eye can see by mixing only red, green, and blue, the brainbow scientists introduced genes that produce cyan (blue), green, yellow, and red fluorescent proteins into the cells of mice. The red gene is obtained from coral, and the blue and green genes are obtained from jellyfish. (The 2008 Nobel Prize in chemistry was awarded to Roger Tsien, Osamu Shimomura, and Martin Chalfie for their discovery of fluorescent proteins in coral and jellyfish.)

The mice also received a bacterial gene called *Cre*. This gene activates the color genes inside each cell, but owing to chance factors, the extent to which each gene is activated varies. As the mice develop, the variable expression of the color-coding genes results in cells that fluoresce in one hundred or more different hues. When viewed through a fluorescent microscope that is sensitive to these wavelengths, individual brain cells and their connections can be visualized because they have slightly different hues, as illustrated in the accompanying micrographs.

Because many individual cells can be visualized, brainbow offers a way to describe where each neuron sends its processes and how it interconnects with other neurons. You have probably seen an electrical power cord in which the different wires have different colors (black, white, red) that signify what they do and how they should be connected. By visualizing living brain tissue in a dish, brainbow provides a method of examining changes in neural circuits with the passage of time.

In the future, brainbow will prove useful for examining specific populations of cells—for example, cells that are implicated in specific brain diseases. In principle, brainbow could be turned on at specific times, as a child develops or ages or even as the child solves a particular problem, for example. Brainbow could reveal the pattern of changes associated with behavioral changes in each case.

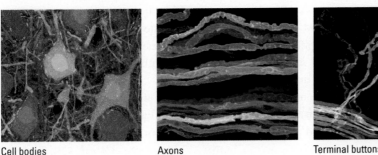
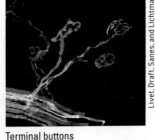

Livet, Draft, Sanes, and Lichtman, Harvard University

Cell bodies   Axons   Terminal buttons

developed many techniques for viewing and describing both dead and living brain cells. The pictures of brain cells that emerge from these techniques provide insights into what cells do.

The first anatomists to study brain cells developed methods of highlighting individual cells in nervous system tissue. To make it firm, they soaked it in formaldehyde, which removes the water from the tissue. The firmed-up tissue was then sliced in thin sheets that could be placed under a microscope for viewing. Scientists continue to develop ways of visualizing cells so that an individual cell or its parts stand out from the dense background of surrounding cells. Visualization is aided by using dyes that either color an individual cell completely, color some of the cells' components, or, as described in Research Focus 3-2, "Brainbow: Rainbow Neurons," color the cell only when it is engaged in a particular activity.

Scientists have also developed techniques of viewing living cells in the nervous system or viewing cells that are "cultured" in a dish with fluids that nurture cells and so keep them alive. Some of these staining techniques present not only an image of the cell but also allow its activity to be viewed and controlled. These experiments are aided using miniature microscopes fitted with cameras that are mounted over living areas of the nervous system (Ghosh et al., 2011).

FIGURE 3-1 Two Views of a Cell **(A)** Tissue preparation revealing human pyramidal cells stained by using the Golgi technique. **(B)** Cajal's drawing of a single Purkinje neuron made from Golgi stained tissue. **(B)** From *Histologie du système nerveux de l'homme et des vertebres*, by S. Ramón y Cajal, 1909–1911, Paris: Maloine.

(A)

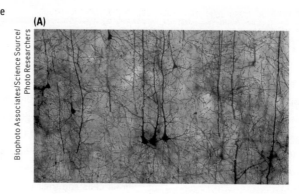

(B)

There remains, however, the problem of making sense of what you see. Different brain samples can yield different images, and different people can interpret the images in different ways. So began a controversy between two great scientists—the Italian Camillo Golgi and the Spaniard Santiago Ramón y Cajal—that resulted in our understanding of what neurons really are. Both men were awarded the Nobel Prize for medicine in 1906 for their descriptions of brain cells.

Imagine that you are Camillo Golgi hard at work in your laboratory staining and examining cells of the nervous system. You immerse a thin slice of brain tissue in a solution containing silver nitrate and other chemicals, a technique used at the time to produce black-and-white photographic prints. A contemporary method, shown in **Figure 3-1**A, produces a color-enhanced microscopic image that resembles the images seen by Golgi.

The image is beautiful and intriguing, but what do you make of it? To Golgi, this structure suggested that the nervous system is composed of a network of interconnected fibers. He thought that information, like water running through pipes, somehow flowed around this "nerve net" and produced behavior. His theory was reasonable, given what he saw.

Golgi never revealed just how he came to develop his technique.

But Santiago Ramón y Cajal came to a different conclusion. He used Golgi's stain to study the brain tissue of chick embryos. He assumed that their nervous systems would be simpler and easier to understand than an adult person's nervous system. Figure 3-1B shows one of the images that he rendered from the neural cells of a chick embryo using the Golgi stain. Cajal concluded that the nervous system is made up of discrete cells that begin life with a rather simple structure that becomes more complex with age. When mature, each cell consists of a main body with extensions projecting from it.

The structure looks something like a plant, with branches coming out of the top and roots coming out of the bottom. In addition, Cajal showed that cells with these plantlike features came in many different shapes and sizes. Cajal's *neuron theory*—that neurons are the functional units of the nervous system—is now universally accepted. The neuron theory includes the idea that it is the interactions between these discrete cells that enables behavior.

Figure 3-2 shows the three basic subdivisions of a neuron. The core region is called the **cell body,** or **soma** (Greek, meaning "body"; the root of words such as "somatic"). A neuron's branching extensions, or **dendrites** (from Greek for "tree"), collect information from other cells, and its main "root" is the single **axon** (Greek for "axle") that carries messages to other neurons. So a neuron has only one axon, but most have many dendrites. Some small neurons have so many dendrites that they look like a garden hedge.

The human nervous system contains more than 100 billion neurons. How can we explain how 100 billion cells cooperate, make connections, and produce behavior? Fortunately, brain cells have a common plan: examining how one cell works can be a source

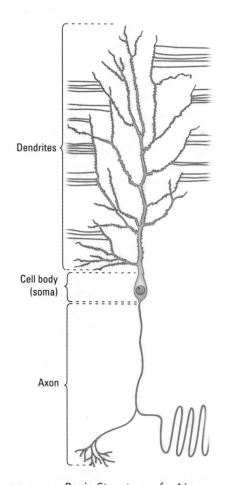

FIGURE 3-2 Basic Structure of a Neuron Dendrites gather information from other neurons, the cell body (soma) integrates the information, and the axon sends the information to other cells. Note that, though it may have many dendrites, there is only one axon.

Dendrites

Cell body (soma)

Axon

of insight that we can generalize to other cell types. As you learn to recognize some of their different types, you will also see how their specialized structures contribute to their functions in your body.

Neurons: The Basis of Information Processing

As the information-processing units of the nervous system, neurons acquire information from sensory receptors, pass the information on to other neurons, and make muscles move to produce behaviors. They encode memories and produce our thoughts and emotions. At the same time, they regulate body processes such as breathing, heartbeat, and body temperature, to which we seldom give a thought.

Some scientists think that a specific function can be assigned to an individual neuron. For example, Einat Adar and colleagues (2008) studied how birds learn songs and proposed a relation between the number of neurons produced for singing and the number of notes in the song that is sung. For most behaviors in most species, however, scientists think that neurons work together in groups of many hundreds to many thousands to produce some aspect of the behavior.

In support of the idea of group function, the loss of a neuron or two is no more noticeable than the loss of one or two voices from a cheering crowd. It is the crowd that produces the overall sound, not each person. In much the same way, although neuroscientists say that neurons are the information-processing units of the brain, they really mean that large teams of neurons serve this function.

Scientists also speak informally about the structure of a particular neuron as if the structure never changes. But neurons are the essence of plasticity. If fresh brain tissue is viewed through a microscope, the neurons reveal themselves to be surprisingly active, producing new branches, losing old ones, and making and losing connections with each other. It is this dynamic activity that underlies both the constancies and the changes in our behavior.

Another important property of most neurons is their longevity. At a few locations in the human nervous system, the ongoing production of new neurons does take place throughout life, and some behavior does depend on the production of new neurons. But most of your CNS neurons are with you for life and are never replaced. If the brain or spinal cord is damaged, the neurons that are lost are not replaced, and functional recovery is poor.

In the peripheral nervous system and in some other species of animals, neurons can be lost and replaced. The variability in longevity and replacement of neurons are properties that are important to scientists who study diseases of aging associated with age-related changes in the nervous system and diseases or injuries that damage the nervous system.

Structure and Function of the Neuron

Figure 3-3 details the external and internal features common to neurons. The surface area of the cell is increased immensely by its extensions into dendrites and an axon (Figure 3-3A and B). The dendritic area is further increased by many small protrusions called **dendritic spines** (Figure 3-3C). A neuron may have from 1 to 20 dendrites, each may have from one to many branches, and the spines on the branches may number in the thousands. The dendrites collect information from other cells, and the spines are the points of contact with other neurons. The extent of a cell's branches corresponds to its information processing capacity.

Each neuron has only a single axon that carries messages to other neurons. The axon begins at one end of the cell body at an expansion known as the **axon hillock** (little hill) shown in Figure 3-3D. The axon may branch out into one or many **axon collaterals** that usually emerge from it at right angles, as shown at the bottom of Figure 3-3B.

Note also that the lower tip of an axon may divide into a number of smaller branches (*teleodendria,* or end branches). At the end of each teleodendrion is a knob called an *end foot* or **terminal button.** The terminal button sits very close to a dendritic spine or

No one has counted all 100 billion neurons. Scientists have estimated the total number by counting the cells in a small sample of brain tissue and then multiplying by the brain's volume.

cell body (soma) Core region of the cell containing the nucleus and other organelles for making proteins.

dendrite Branching extension of a neuron's cell membrane that greatly increases the surface area of the cell and collects information from other cells.

axon "Root," or single fiber, of a neuron that carries messages to other neurons.

dendritic spine Protrusion from a dendrite that greatly increases the dendrite's surface area and is the usual point of dendritic contact with the axons of other cells.

axon hillock Juncture of soma and axon where the action potential begins.

axon collateral Branch of an axon.

terminal button (end foot) Knob at the tip of an axon that conveys information to other neurons.

FIGURE 3-3 Major Parts of a Neuron (A) Typical neuron stained by using the Golgi technique to reveal its dendrites and cell body. (B) The neuron's basic structures identified. (C) An electron micrograph captures the synapse between an axon from another neuron and a dendritic spine. (D) A high-power light-microscopic view inside the cell body. Note the axon hillock at the junction of the soma and axon.

(A)

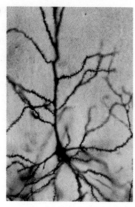

(B)

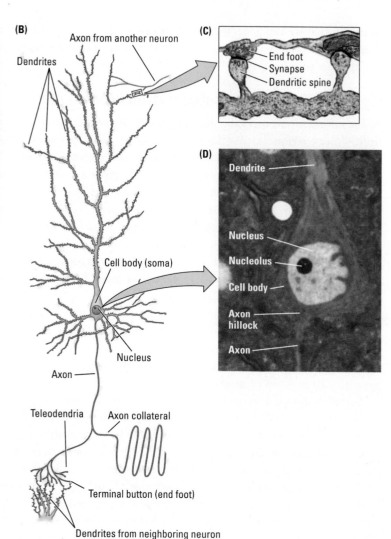

Axon from another neuron

Dendrites

Cell body (soma)

Nucleus

Axon

Teleodendria

Axon collateral

Terminal button (end foot)

Dendrites from neighboring neuron

(C)

End foot
Synapse
Dendritic spine

(D)

Dendrite

Nucleus

Nucleolus

Cell body

Axon hillock

Axon

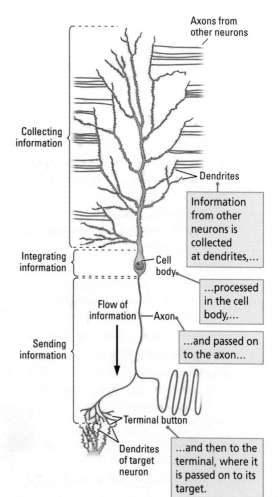

Axons from other neurons

Collecting information

Dendrites

Information from other neurons is collected at dendrites,...

Integrating information

Cell body

...processed in the cell body,...

Flow of information — Axon

...and passed on to the axon...

Sending information

Terminal button

Dendrites of target neuron

...and then to the terminal, where it is passed on to its target.

FIGURE 3-4 Information Flow Through a Neuron

some other part of another cell, although it usually does not touch it (see Figure 3-3C). This "almost connection," which includes the surfaces of the end foot and the neighboring dendritic spine as well as the space between them, is called a **synapse.** The synapse is the information-transfer site between neurons.

Chapter 4 describes how neurons transmit information; here, we simply generalize about neuronal function by examining shape. Imagine looking at a river system from an airplane. You see many small streams merging to make creeks, which join to form tributaries, which join to form the main river channel. As the river reaches its delta, it breaks up into a number of smaller channels again before discharging its contents into the sea.

The general shape of a neuron suggests that it works in a broadly similar way. As illustrated in **Figure 3-4,** the neuron collects information from many different sources on its dendrites. It channels the information onto its axon, which can send out only a single message. The eventual branching of the axon then allows the message to be sent to many different target surfaces.

Three Types of Neurons

The nervous system contains an array of neurons in varying shapes and sizes that are structured differently to perform specialized tasks. Some appear quite simple and others very complex. **Sensory neurons** (**Figure 3-5A**) are designed to bring

(A) Sensory neurons

Bring information to the central nervous system

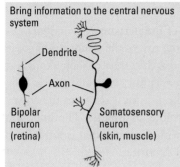

Dendrite

Axon

Bipolar neuron (retina)

Somatosensory neuron (skin, muscle)

(B) Interneurons

Associate sensory and motor activity in the central nervous system

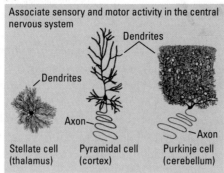

Dendrites

Dendrites

Axon

Axon

Stellate cell (thalamus)

Pyramidal cell (cortex)

Purkinje cell (cerebellum)

(C) Motor neurons

Send signals from the brain and spinal cord to muscles

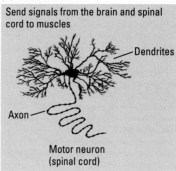

Dendrites

Axon

Motor neuron (spinal cord)

FIGURE 3-5 Neuron Shape and Function (A) Sensory neurons collect information from a source and pass it on to (B) an interneuron. The many branches of interneurons collect information from many sources and link to (C) motor neurons, which are distinctively large and pass information on to command muscles to move. Note that these cells are not drawn to scale.

information into the brain from sensory receptors, **interneurons** (Figure 3-5B) to associate sensory and motor activity in the central nervous system, and **motor neurons** (Figure 3-5C) to carry information out of the brain and spinal cord to the body's muscles.

SENSORY NEURONS Sensory neurons are the simplest neurons structurally. A **bipolar neuron** found in the retina of the eye, for example, has a single short dendrite on one side of its cell body and a single short axon on the other side. Bipolar neurons transmit afferent (incoming) sensory information from the retina's light receptors to the neurons that carry information into the visual centers of the brain.

A sensory neuron that is a bit more complicated is the **somatosensory neuron,** which brings sensory information from the body into the spinal cord. Structurally, the somatosensory dendrite connects directly to its axon, so the cell body sits to one side of this long pathway.

INTERNEURONS Also called *association cells* because they link up sensory and motor neurons, interneurons branch extensively, the better to collect information from many sources. A major difference between animals with small brains and animals with large brains is that large-brained animals have more interneurons. A specific interneuron, the *stellate* (star-shaped) *cell,* is characteristically small, with many dendrites extending around the cell body. Its axon is difficult to see in the maze of dendrites.

A **pyramidal cell** has a long axon, a pyramid-shaped cell body, and two sets of dendrites, one set projecting from the apex of the cell body and the other from its sides. Pyramidal interneurons carry information from the cortex to the rest of the brain and spinal cord. A **Purkinje cell** (named for its discoverer) is a distinctive output cell with extremely branched dendrites that form a fan shape. It carries information from the cerebellum to the rest of the brain and spinal cord.

MOTOR NEURONS To collect information from many sources, motor neurons have extensive networks of dendrites, large cell bodies, and long axons that connect to muscles. Motor neurons are located in the lower brainstem and spinal cord. All efferent (outgoing) neural information must pass through them to reach the muscles.

Neural Networks

Sensory neurons collect afferent information from the body and connect to interneurons that process the information and then pass it on to motor neurons whose efferent connections move muscles and so produce behavior. So neurons are "networkers," and the physical appearance of each neuron reveals its role in the network. Figure 3-5 illustrates the relation between form and function in neurons but does not illustrate their relative sizes.

Neurons that project for long distances, such as somatosensory neurons, pyramidal neurons, and motor neurons, are large relative to other neurons. In general, neurons with

synapse Junction between one neuron and another that forms the information-transfer site between neurons.

sensory neuron Neuron that carries incoming information from sensory receptors into the spinal cord and brain.

interneuron Association neuron interposed between a sensory neuron and a motor neuron; thus, in mammals, interneurons constitute most of the neurons of the brain.

motor neuron Neuron that carries information from the brain and spinal cord to make muscles contract.

bipolar neuron Sensory neuron with one axon and one dendrite.

somatosensory neuron Brain cell that brings sensory information from the body into the spinal cord.

pyramidal cell Distinctive interneuron found in the cerebral cortex.

Purkinje cell Distinctive interneuron found in the cerebellum.

large cell bodies have extensions that are long, whereas neurons with small cell bodies, such as stellate interneurons, have short extensions.

Long extensions carry information to distant parts of the nervous system; short extensions are engaged in local processing. For example, the tips of the dendrites of some somatosensory neurons are located in your big toe, whereas the target of their axons is at the base of your brain. These sensory neurons send information over a distance as long as 2 meters, or more. The axons of some pyramidal neurons must reach from the cortex as far as the lower spinal cord, a distance that can be as long as a meter. The imposing size of this pyramidal cell body is therefore in accord with the work that it must do in providing nutrients and other supplies for its axons and dendrites.

The Language of Neurons: Excitation and Inhibition

Neurons are networkers with elaborate interconnections, but how do they communicate? Simply put, neurons either excite other neurons (turn them on) or inhibit other neurons (turn them off). Like computers, neurons send "yes" or "no" signals to one another; the "yes" signals are excitatory, and the "no" signals are inhibitory. Each neuron receives thousands of excitatory and inhibitory signals every second.

The neuron's response to all those inputs is democratic: it sums them. A neuron is spurred into action only if its excitatory inputs exceed its inhibitory inputs. If the reverse is true and inhibitory inputs exceed excitatory inputs, the neuron does not activate.

By exciting or inhibiting one another, a network of neurons can detect sensory information and "decide" what kind of motor response to make to that information. To confirm whether they understand how a neural network produces behavior, scientists might make a model, such as a robot, intended to function in the same way. Robots, after all, engage in goal-oriented actions, just as animals do. A robot's computer must guide and coordinate those actions, doing much the same work that an animal's nervous system does.

The construction of humanlike machines is one objective of the science of artificial intelligence (AI) (Riva et al., 2008). Barbara Webb's cricket robot, constructed from Lego blocks, wires, and a motor **(Figure 3-6, left),** is designed to mimic a female cricket that listens for the source of a male's chirping song and travels to it (Reeve et al., 2007). This is the beginning step in constructing intelligent robots Figure 3-6, right).

In approaching a male, a female cricket must avoid open, well-lit places where a predator could detect her. The female must often choose between competing males, preferring, for example, the male that makes the longest chirps. All these behaviors must be "wired into" a successful cricket robot, making sure that one behavior does not interfere with another. In simulating cricket behavior in a robot, Webb is duplicating the rules of a cricket's nervous system, which are "programmed" by its genes.

The Procedures sections of **Experiment 3-1** illustrate some simple ways that neural inhibition and excitation might produce the cricket robot's behavior. The Results section confirms that this hypothetical arrangement mimics the function of sensory and motor neurons and the principle of summating excitatory and inhibitory signals—but with only six neurons and each neuron connected to only one other neuron. Imagine how infinitely more complex a human nervous system is with its hundred billion neurons, most of which are interneurons, each with thousands of connections.

Recall Principle 10 from Section 2-6: The nervous system works through a combination of excitatory and inhibitory signals. Section 4-3 explains how neurons sum up excitatory and inhibitory signals.

FIGURE 3-6 Nervous System Mimics (*Left*) Rules obtained from the study of crickets' behavior can be programmed into robots to be tested. (*Right*) Advances made by social roboticists like Heather Knight, pictured here with her companion, promise human-like robots that can hold a conversation, convey emotional responses, and make short work of Rubik's cubes. (*Left*) From "A Cricket Robot," by B. Webb, 1996, *Scientific American, 214*(12), p. 99.

Robert P. Carr/Bruce Coleman (animal); Barbara Webb (model)

Louise Stein/Heather Knight

EXPERIMENT 3-1

Question: Can the principles of neural excitation and inhibition control the activity of a simple robot?

Procedure A

If we insert sensory neurons between the microphone for sound detection on each side of this hypothetical robot and the motor on the opposite side, we need only two rules to instruct the female robot to seek out a chirping male cricket.

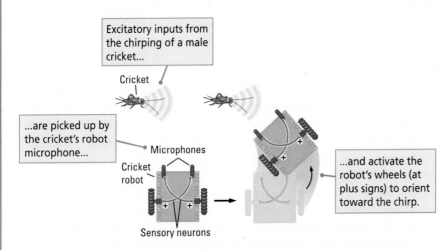

Excitatory inputs from the chirping of a male cricket...

Cricket

...are picked up by the cricket's robot microphone...

Microphones

Cricket robot

...and activate the robot's wheels (at plus signs) to orient toward the chirp.

Sensory neurons

Rule 1 When a microphone detects a male cricket's song, an excitatory message is sent to the opposite wheel's motor, activating it so the robot turns toward the cricket.

Rule 2 If the chirp is coming from the robot's left side, it will be detected as being louder by the microphone on the left, which will make the right wheel turn a little faster, swinging the robot to the left.

Procedure B

We add two more sensory neurons, coming from photoreceptors on the robot. When activated, these light-detecting sensory neurons inhibit the motor neurons leading to the wheels and prevent the robot from moving toward a male cricket. Now the female cricket robot will move only when it is dark and "safe."

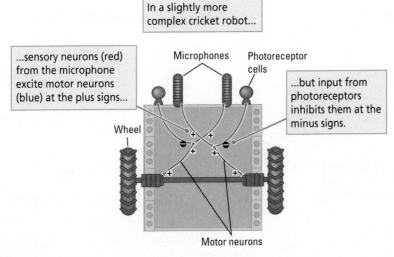

In a slightly more complex cricket robot...

...sensory neurons (red) from the microphone excite motor neurons (blue) at the plus signs...

Microphones   Photoreceptor cells

...but input from photoreceptors inhibits them at the minus signs.

Wheel

Motor neurons

Result

This hypothetical arrangement illustrates the biological function of sensory and motor neurons and the neural principle of summating excitatory and inhibitory signals.

Conclusion: A simple robot will operate following the principles of neural excitation and inhibition. To make the robot more "intelligent" requires more neurons.

Five Types of Glial Cells

Neurons, the information-processing cells of the nervous system, are aided in their function by another group of cells. **Glial cells** (from the Greek word for "glue") are often described as the support cells of the nervous system. Although they do not transmit information themselves, they help neurons carry out this task, binding them together (some *do* act as glue) and providing support, nutrients, and protection, among other functions.

Table 3-1 lists the five major types of glia. Each has a characteristic structure and function. Glial cells are different from neurons in that they can replace themselves, and errors in the way they do so can result in abnormal growth. Clinical Focus 3-3, "Brain Tumors," describes the results of uncontrolled glial cell growth.

Glia form the fatty coverings around neurons that show up as white matter in brain images such as Figure 2-12B and D.

Ependymal Cells

On the walls of the ventricles, the cavities inside your brain, are **ependymal cells** that produce and secrete the cerebrospinal fluid that fills the ventricles. CSF is constantly being formed and flows through the ventricles toward the base of the brain, where it is absorbed into the blood vessels. Cerebrospinal fluid serves several purposes. It acts as a shock absorber when the brain is jarred, it provides a medium through which waste products are eliminated, it assists the brain in maintaining a constant temperature, and it is a source of nutrients for parts of the brain located adjacent to the ventricles.

As CSF flows through the ventricles, it passes through some narrow passages, especially from the cerebral aqueduct into the fourth ventricle, which runs through the brainstem. If the fourth ventricle is fully or partly blocked, the fluid flow is restricted. Because CSF is continuously being produced, this blockage causes a buildup of pressure that begins to expand the ventricles, which in turn push on the surrounding brain.

You can review the location of the cerebral aqueduct and the ventricles in Figure 2-9.

If such a blockage develops in a newborn infant, before the skull bones are fused, the pressure on the brain is conveyed to the skull and the baby's head consequently swells.

glial cell Nervous-system cell that provides insulation, nutrients, and support and that aids in repairing neurons and eliminating waste products.

ependymal cell Glial cell that makes and secretes cerebrospinal fluid; found on the walls of the ventricles in the brain.

tumor Mass of new tissue that grows uncontrolled and independent of surrounding structures.

hydrocephalus Buildup of pressure in the brain and, in infants, swelling of the head caused if the flow of cerebrospinal fluid is blocked; can result in retardation.

astrocyte Star-shaped glial cell that provides structural support to neurons in the central nervous system and transports substances between neurons and blood vessels.

blood–brain barrier Tight junctions between the cells that compose blood vessels in the brain, providing a barrier to the entry of an array of substances, including toxins, into the brain.

TABLE 3-1 Types of Glial Cells

Type	Appearance	Features and function
Ependymal cell		Small, ovoid; secretes cerebrospinal fluid (CSF)
Astrocyte		Star shaped, symmetrical; nutritive and support function
Microglial cell		Small, mesodermally derived; defensive function
Oligodendroglial cell		Asymmetrical; forms myelin around CNS axons in brain and spinal cord
Schwann cell		Asymmetrical; wraps around peripheral nerves to form myelin

CLINICAL FOCUS ✛ 3-3

Brain Tumors

One day while she was watching a movie in a neuropsychology class, R. J., a 19-year-old college sophomore, collapsed on the floor and began twitching, displaying symptoms of a brain seizure. The instructor helped her to the university clinic, where she recovered, except for a severe headache. She reported that she had suffered from severe headaches on a number of occasions.

A few days later, computer tomography (CT) was used to scan her brain; the scan showed a tumor over her left frontal lobe. She underwent surgery to have the tumor removed and returned to classes after an uneventful recovery. She successfully completed her studies, finished law school, and has been practicing law without any further symptoms.

A **tumor** is a mass of new tissue that undergoes uncontrolled growth and is independent of surrounding structures. No region of the body is immune, but the brain is a common site. The incidence of brain tumors in the United States is about 20 per 100,000 according to the Central Brain Tumor Registry of the United States (2011). Brain tumors do not grow from neurons but rather from glia or other supporting cells. The rate of growth depends on the type of cell affected.

Some tumors are benign, as R. J.'s was, and not likely to recur after removal; others are malignant, likely to progress, and apt to recur after removal. Both kinds of tumors can pose a risk to life if they develop in sites from which they are difficult to remove.

The earliest symptoms usually result from increased pressure on surrounding brain structures and can include headaches, vomiting, mental dullness, changes in sensory and motor abilities, and seizures such as R. J. experienced. Many symptoms depend on the precise location of the tumor. The three major types of brain tumors are classified according to how they originate:

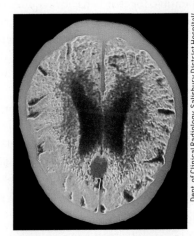

Dept. of Clinical Radiology, Salisbury District Hospital/ Science Photo Library/Photo Researchers

The red area in this colored CT scan is a meningioma, a noncancerous tumor arising from the arachnoid membrane covering the brain. A meningioma may grow large enough to compress the brain but usually does not invade brain tissue.

1. *Gliomas* arise from glial cells, are slow growing, not often malignant, and relatively easy to treat if they arise from astrocytes. In contrast, gliomas that arise from the precursor *blast* or *germinal* cells that grow into glial cells are much more often malignant, grow more quickly, and often recur after treatment. Senator Edward Kennedy was diagnosed with a malignant glioma in his left parietal cortex in 2008. He died a year later. Like R. J., his first symptom was an epileptic seizure.

2. *Meningiomas,* the type of tumor that R. J. had, attach to the meninges and so grow entirely outside the brain, as shown in the accompanying CT scan. These tumors are usually well encapsulated, and if they are located in places that are accessible, recovery after surgery is good.

3. The *metastatic tumor* becomes established by a transfer of tumor cells from one region of the body to another (which is what the term *metastasis* means). Typically, metastatic tumors are present in multiple locations, making treatment difficult. Symptoms of the underlying condition often first appear when the tumor cells reach the brain.

Treatment for a brain tumor is usually surgery, which also is one of the main means of diagnosing the type of tumor. Radiotherapy (treatment with X-rays) is more useful for destroying brain tumor cells. Chemotherapy, although common for treating tumors in other parts of the body, is less successful in the treatment of brain tumors because getting the chemicals across the blood–brain barrier is difficult.

This condition, called **hydrocephalus** (literally, "water brain"), can cause severe mental retardation and even death. To treat it, doctors insert one end of a tube, called a *shunt*, into the blocked ventricle and the other end into a vein. The shunt allows the CSF to drain into the bloodstream.

Astroglia

Astrocytes (star-shaped glia shown in Table 3-1), also called *astroglia*, provide structural support within the central nervous system. Their extensions attach to blood vessels and to the brain's lining, creating scaffolding that holds neurons in place. These same extensions provide pathways for the movement of certain nutrients between blood vessels and neurons. Astrocytes also secrete chemicals that keep neurons healthy and help them heal if injured.

At the same time, astrocytes play a role in contributing to a protective partition between blood vessels and the brain, the **blood–brain barrier.** As shown in **Figure 3-7,**

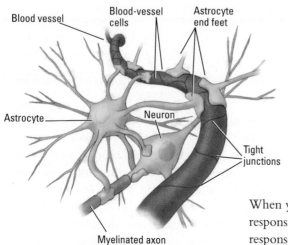

FIGURE 3-7 Blood–Brain Barrier Astrocyte processes attach to neurons and to blood vessels to provide support between different structures in the brain, stimulate the cells on blood vessels to form tight junctions and so form the blood–brain barrier, and transport chemicals excreted by neurons to blood vessels.

Growth factors, introduced in Section 8-2, are one class of neurotrophic compounds, chemicals that stimulate and support the growth, survival, and, as described in Section 14-4, perhaps even the plasticity of brain cells.

the end feet of astrocytes attach to the blood-vessel cells, causing them to bind tightly together. These tight junctions prevent an array of substances, including many toxins, from entering the brain through the blood-vessel walls.

The molecules (smallest units) of these substances are too large to pass between the blood-vessel cells unless the blood–brain barrier is somehow compromised. But the downside to the blood–brain barrier is that many useful drugs, including antibiotics such as penicillin, cannot pass through it to the brain either. As a result, brain infections are very difficult to treat. Scientists can bypass the blood-brain barrier and introduce drugs into the brain by inserting small tubes that allow the delivery of a drug directly to a targeted brain region.

Yet another important function of astrocytes is to enhance brain activity. When you engage in any behavior, whether it's reading or running, the neural network responsible for that behavior requires more fuel in the form of oxygen and glucose. In response to the activity in the network, the blood vessels that supply the network expand, allowing greater oxygen- and glucose-carrying blood flow. But what triggers the blood vessels to dilate? This is where the astrocytes come in. They receive signals from the neurons that they pass on to the blood vessels, stimulating them to expand and so provide more fuel.

Astrocytes also contribute to the process of healing damaged brain tissue. If the brain is injured by a blow to the head or penetrated by some object, astrocytes form a scar to seal off the damaged area. Although the scar tissue is beneficial in healing the injury, it can unfortunately act as a barrier to the regrowth of damaged neurons. One experimental approach to repairing brain tissue seeks to get the axons and dendrites of CNS neurons to grow around or through a glial scar.

Microglia

Unlike other glial cells, which originate in the brain, **microglia** originate in the blood as an offshoot of the immune system and migrate throughout the nervous system, where they make up about 20% of all glial cells. The brain is largely "immune privileged" because the blood–brain barrier prevents most immune system cells from entering the brain. Microglia monitor the health of brain tissue and play the role of its immune system. They identify and attack foreign tissue. When brain cells are damaged, microglia invade the area to provide *growth factors* that aid in repair.

There are several different kinds of microglia, and microglia take different shapes depending upon the role they are performing. Microglia engulf any foreign tissue and dead brain cells, an immune process called *phagocytosis*. When full they take on a distinctive appearance. The stuffed and no-longer-functioning microglia can be detected as dark bodies in and near regions of the brain that have been damaged, as illustrated in **Figure 3-8**.

Because microglia are frontline players in protecting the nervous system and removing nervous system waste, considerable research is directed toward the extent to

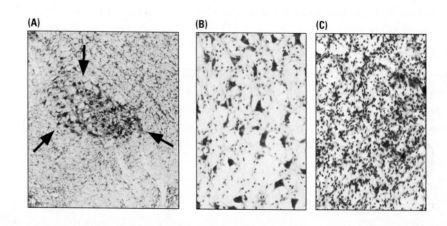

FIGURE 3-8 Detecting Brain Damage
(A) Arrows indicate a brain area called the red nucleus in a rat. **(B)** Closeup of cresyl-violet-stained neurons in the healthy red nucleus. **(C)** After exposure to a neurotoxin, only microglia are present.

which microglia are involved in protecting the nervous system from disease. A characteristic of Alzheimer's disease, the degenerative brain disorder commonly associated with aging, is the deposit of distinctive bodies called *plaques* in regions of damage. Microglia consume plaques and may play a role in slowing or halting the disease (Naert and Rivest, 2011).

It is unclear why Alzheimer's disease can continue to progress, but possibly the disease process disarms microglia and prevents them from doing their job. Microglia may actually play a harmful role, by consuming inflamed tissue rather than protecting it. Although small, as their name suggests, microglia play a mighty role in maintaining the brain's health.

Oligodendroglia and Schwann Cells

Two kinds of glial cells insulate the axons of neurons. Like the rubber insulation on electrical wires, **myelin** prevents adjacent neurons from short-circuiting. **Oligodendroglia** myelinate axons in the brain and spinal cord by sending out large, flat branches that enclose and separate adjacent axons (the prefix *oligo* means "few" and here refers to the fact that these glia have few branches in comparison with astrocytes; see Table 3-1).

Schwann cells myelinate axons in the peripheral nervous system. Each Schwann cell wraps itself repeatedly around a part of an axon, forming a structure somewhat like a bead on a string. In addition to the myelination, Schwann cells and oligodendroglia contribute to a neuron's nutrition and function by absorbing chemicals that the neuron releases and releasing chemicals that the neuron absorbs.

In Section 4-2, you will learn how myelin speeds up the flow of information along a neuron. Neurons that are heavily myelinated send information much faster than neurons having little or no myelin. Neurons that send messages over long distances quickly, including sensory and motor neurons, are heavily myelinated.

If myelin is damaged, a neuron may be unable to send any messages over its axons. In **multiple sclerosis** (MS), the myelin formed by oligodendroglia is damaged, and the functions of the neurons whose axons it encases are disrupted. Clinical Focus 3-4, "Multiple Sclerosis" on page 86, describes the course of the disease.

Glial Cells and Neuron Repair

A deep cut on your body—on your arm or leg for instance—may cut the axons connecting your spinal cord to muscles and to sensory receptors. Severed motor-neuron axons will render you unable to move the affected part of your body, whereas severed sensory fibers will result in loss of sensation from that body part. Cessation of both movement and sensation is **paralysis.** In a period of weeks to months after motor and sensory axons are severed, movement and sensation will return. The human body can repair this kind of nerve damage, so the paralysis is not permanent.

Both microglia and Schwann cells play a part in repairing damage to the peripheral nervous system. When a PNS axon is cut, it dies back to the cell body, as shown at the top of **Figure 3-9.** Microglia remove all the debris left by the dying axon. Meanwhile, the Schwann cells that provided the axon's myelin shrink and then divide, forming numerous smaller glial cells along the path the axon formerly took. The neuron then sends out axon sprouts that search for the path made by the Schwann cells and follow it.

Eventually, one sprout reaches the intended target, and this sprout becomes the new axon; all other sprouts retract. The Schwann cells envelop the new

microglia Glial cells that originate in the blood, aid in cell repair, and scavenge debris in the nervous system.

myelin Glial coating that surrounds axons in the central and peripheral nervous systems; prevents adjacent neurons from short-circuiting.

oligodendroglia Glial cells in the central nervous system that myelinate axons.

Schwann cell Glial cell in the peripheral nervous system that myelinates sensory and motor axons.

multiple sclerosis (MS) Nervous-system disorder that results from the loss of myelin (glial-cell covering) around neurons.

paralysis Loss of sensation and movement due to nervous-system injury.

For statistics on the incidence of MS as well as possible causes and treatments, see Section 16-3.

FIGURE 3-9 **Neuron Repair** Schwann cells aid the regrowth of axons in the somatic division of the peripheral nervous system.

1 When a peripheral axon is cut, the axon dies.

2 Schwann cells first shrink and then divide, forming glial cells along the axon's former path.

3 The neuron sends out axon sprouts, one of which finds the Schwann-cell path and becomes a new axon.

4 Schwann cells envelop the new axon, forming new myelin.

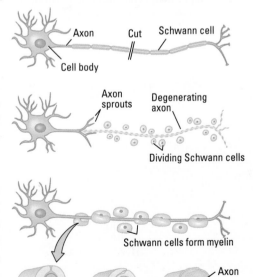

Multiple Sclerosis

One day J. O., who had just finished university requirements to begin work as an accountant, noticed a slight cloudiness in her right eye that did not go away when she wiped her eye. The area of cloudiness grew over the next few days. Her optometrist suggested that she see a neurologist, who diagnosed optic neuritis, a symptom that can be a flag for multiple sclerosis (MS).

MS is caused by a loss of myelin (see illustration), both on pathways bringing sensory information to the brain and on pathways taking commands to muscles. This loss of myelin occurs in patches, and scarring is frequently left in the affected areas.

Eventually, a hard scar, or *plaque,* forms at the site of myelin loss. (MS is called a "sclerosis" from the Greek word meaning "hardness.") Associated with the loss of myelin is impairment of neuron function, causing characteristic MS symptoms of sensory loss and difficulty in moving.

Fatigue, pain, and depression are common related symptoms. Bladder dysfunction, constipation, and sexual dysfunction all complicate the condition. Multiple sclerosis greatly affects a person's emotional, social, and vocational functioning. As yet, it has no cure.

J. O.'s eye cleared over the next few months, and she had no further symptoms until after the birth of her first child 3 years later, when she felt a tingling in her right hand that spread up her arm, until gradually she lost movement in the arm. Movement was restored 5 months later. Then 5 years later, after her second child was born, she felt a tingling in her left big toe that spread along the sole of her foot and then up her leg, eventually leading again to loss of movement. J. O. received corticosteroid treatment, which helped, but the condition rebounded when she stopped treatment. Then it subsided and eventually disappeared.

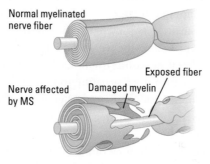

Normal myelinated nerve fiber

Nerve affected by MS

Damaged myelin

Exposed fiber

Adapted from Mayo Foundation for Medical Education and Research.

Since then, J. O. has had no major outbreaks of motor impairment, but she reports enormous fatigue, takes long naps daily, and is ready for bed early in the evening. Her sister and a female cousin have experienced similar symptoms, and recently a third sister began to display similar symptoms in middle age. One of J. O.'s grandmothers had been confined to a wheelchair, although the source of her problem was never diagnosed.

The symptoms of MS are difficult to diagnose; they usually appear in adulthood, and their onset is quite sudden and swift. Initial symptoms may be loss of sensation in the face, limbs, or body or loss of control over movements or loss of both sensation and control. Motor symptoms usually appear first in the hands or feet.

Often early symptoms go into remission and do not appear again for years. In some forms, however, the disease progresses rapidly over a period of just a few years until the person is bedridden.

MS is common in the most northern and most southern latitudes, suggesting that it may be related to a lack of vitamin D, which is usually obtained by the action of summer sunlight on the skin. The disease may also be related to genetic succeptibility, as is likely in J. O.'s case. Vitamin D_3 and vitamin B_{12} are frequently taken by MS patients. Patients are also treated with anti-inflammatory agents.

It has been suggested that blood flow from the brain is reduced in MS, allowing a buildup of toxic iron in the brain. Widening veins that drain the brain is suggested as a treatment. Clinical trials evaluating both the venous cause and venous widening as a treatment are still inconclusive (Zamboni et al., 2011).

axon, forming new myelin and restoring normal function. In the PNS, then, Schwann cells serve as signposts to guide axons to their appropriate end points. Axons can get lost, however, as sometimes happens after surgeons reattach a severed limb. If axons destined to innervate one finger end up innervating another finger instead, the wrong finger will move when a message is sent along that neuron.

When the CNS is damaged, as happens, for example, when the spinal cord is cut, regrowth and repair do not occur, even though the distance that damaged fibers must bridge is short. That recovery should take place in the peripheral nervous system but not in the central nervous system is both a puzzle and a challenge to treating people with brain and spinal-cord injuries. In nonmammalian vertebrates (fish, amphibians, and reptiles) and in birds CNS neurons can regrow. Regrowth in mammals may be lacking in part because, as mammalian neuronal circuits mature, they become exquisitely tuned to mediate individualized behavior and so are protected from the proliferation of new cells or the regrowth of existing cells. In the CNS, the oligodendrocytes themselves play a role in inhibiting neuron regrowth (McDonald et al., 2011).

The absence of recovery after spinal-cord injury is especially frustrating because the spinal cord contains many axon pathways, just like those found in the PNS. Researchers investigating how to encourage the regrowth of CNS neurons have focused on the spinal cord. They have placed tubes across an injured area, trying to get axons to regrow through the tubes. They have also inserted immature glial cells into injured areas to facilitate axon regrowth, and they have used chemicals to stimulate the regrowth of axons. Some success has been obtained with each of these techniques, but none is as yet sufficiently advanced to restore function in people with spinal-cord injuries. Research in which inhibitory factors secreted by oligodendroglia are themselves turned off have been only modestly successful, perhaps in part because not all the inhibitory factors have been identified.

Sections 11-1 and 11-4 detail causes of and treatments for spinal-cord injury.

REVIEW 3-1
Cells of the Nervous System

Before you continue, check your understanding.

1. The two classes of nervous-system cells are _____ and _____.

2. Neurons, the information-conducting units of the nervous system, either _____ or _____ one another through their connecting synapses.

3. The three types of neurons are _____, _____, and _____.

4. The five types of glial cells are _____, _____, _____, _____, and _____. Their functions include _____, _____, _____, _____, and _____ neurons.

5. What is the main obstacle to producing a robot with all the behavioral abilities displayed by a mammal?

Answers appear at the back of the book.

3-2 Internal Structure of a Cell

What is it about the structure of neurons that gives them their remarkable ability to receive, process, store, and send a seemingly limitless amount of information? To answer this question, we must look inside a neuron to see what its components are and understand what they do. Although neurons are very small, they can be viewed with an electron microscope, in which electrons take the place of the photons of the light microscope. Packed inside are hundreds of interrelated parts that do the cell's work.

Figure 5-1 illustrates the difference in magnification between a light microscope and an electron microscope.

To a large extent, the characteristics and functions of a cell are determined by its proteins. Each cell can manufacture thousands of proteins, which variously take part in building the cell and in communicating with other cells. When a neuron malfunctions or contains errors, proteins are implicated and so are involved in many kinds of brain disease. In this section, we explain how the different parts of a cell contribute to protein manufacture, describe what a protein is, and detail some major functions of proteins.

Water, salts, and ions play prominent parts in the cell's functions, as you will learn in this and the next few chapters. If you already understand the structure of water and you know what a salt is and what ions are, read on. If you prefer a brief chemistry review first, turn to "The Basics: Chemistry Review," on pages 88–89.

⊕ THE BASICS

Chemistry Review

The smallest unit of a protein or any other chemical substance is the molecule. Molecules and the even smaller elements and atoms that make them up are the raw materials for the cellular factory.

Elements, Atoms, and Ions

Chemists represent each element, a substance that cannot be broken down into another substance, by a symbol—for example, O for oxygen, C for carbon, and H for hydrogen. Of Earth's 92 naturally occurring elements, the 10 listed in "Chemical Composition of the Brain" constitute virtually the entire makeup of an average living cell. Many other elements are vital to the cell but present in minute quantities.

An atom is the smallest quantity of an element that retains the properties of that element. The basic structures of a cell's most common atoms are shown in "Chemical Composition of the Brain." Ordinarily, atoms are electrically neutral, as illustrated in part A of "Ion Formation."

Atoms of chemically reactive elements such as sodium and chlorine can easily lose or gain electrons. When an atom gives up electrons, it becomes positively charged; when it takes on extra electrons, it becomes negatively charged, as illustrated in part B of "Ion Formation." In either case, the charged atom is now an *ion*. The positive and negative charges of ions allow them to interact, a property central to cell function.

Chemical Composition of the Brain

Percentage of weight	Element name and symbol	Nucleus and electrons (not to scale)
65.0	Oxygen, O	
18.5	Carbon, C	
9.5	Hydrogen, H	
3.5	Nitrogen, N	
1.5	Calcium, Ca	
1.0	Phosphorus, P	

Together, oxygen, carbon, and hydrogen account for more than 90 percent of a cell's makeup.

Percentage of weight	Element name and symbol	Nucleus and electrons (not to scale)
0.4	Potassium, K	
0.2	Sulfur, S	
0.2	Sodium, Na	
0.2	Chlorine, Cl	

Some symbols derive from an element's Latin name—K for *kalium* (Latin for "potassium") and Na for *natrium* (Latin for "sodium"), for example.

Ions Critical to Cell Function

Na+	Sodium
K+	Potassium
Ca²+	Calcium
Cl−	Chloride

Ions formed by losing electrons are represented by an element's symbol followed by one or more plus signs.

Ions formed by gaining electrons are represented by an element's symbol followed by a minus sign.

Ion Formation

(A) Atoms

Total positive (+) and negative (–) charges in atoms are equal. The nucleus contains *neutrons* (no charge) and *protons* (positive charge). Orbiting the nucleus are *electrons* (negative charge).

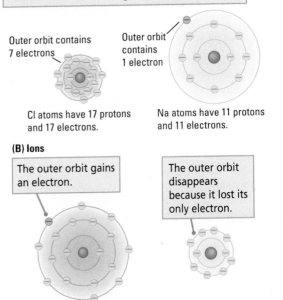

Outer orbit contains 7 electrons

Outer orbit contains 1 electron

Cl atoms have 17 protons and 17 electrons.

Na atoms have 11 protons and 11 electrons.

(B) Ions

The outer orbit gains an electron.

The outer orbit disappears because it lost its only electron.

Charged chloride ion (Cl⁻)

Charged sodium ion (Na⁺)

Chemistry of Water

(A) Water molecule

Two hydrogen atoms share electrons unequally with one oxygen atom, creating a polar water molecule positively charged on the hydrogen end and negatively charged on the oxygen end.

H   H

H   H

O

O

H_2O

(B) Hydrogen bonds

Hydrogen bonds join water molecules to a maximum of four partners.

H^+ and O^{2-} ions in each water molecule are attracted to nearby water molecules.

Because water molecules are polar, they are attracted to other electrically charged substances and to one another. Part B of "Chemistry of Water" illustrates this attracting force, called a *hydrogen bond*. Hydrogen bonding enables water to dissolve electrically neutral salt crystals into their component ions. Salts thus cannot retain their shape in water. As illustrated in "Salty Water," the polar water molecules muscle their way into the Na^+ and Cl^- lattice, surrounding and separating the ions.

Essentially, it is salty water that bathes our brain cells, provides the medium for their activities, supports their communications, and constitutes the brain's cerebrospinal fluid. Sodium chloride and many other dissolved salts, including KCl (potassium chloride) and $CaCl_2$ (calcium chloride) are constituents of the brain's salty water.

Molecules: Salts and Water

Salt crystals form bonds through the electrical attraction between ions. The formula for table salt, NaCl (sodium chloride), means that this molecule consists of one sodium ion and one chloride ion. KCl, the formula for the salt potassium chloride, is composed of one potassium ion and one chloride ion.

Atoms bind together to form molecules, the smallest units of a substance that contain all of that substance's properties. A water molecule (H_2O) is the smallest unit of water that still retains the properties of water. Breaking down water any further would release its component elements, the gases hydrogen and oxygen. The formula H_2O indicates that a water molecule is the union of two hydrogen atoms and one oxygen atom.

Charged ionic bonds hold salt molecules together, but water molecules share electrons. As you can see in part A of "Chemistry of Water," the electron sharing is not equal: H electrons spend more time orbiting the O atom than orbiting each H atom. This structure gives the oxygen region of the water molecule a slight negative charge and leaves the hydrogen regions with a slight positive charge. Like atoms, most molecules are electrically neutral, but water is a polar molecule: it has opposite charges on opposite ends, just as Earth does at the North and South Poles.

Salty Water

Negative Cl^- ions attract the positive poles of water molecules. Positive Na^+ ions attract the negative poles of water molecules.

Salt (NaCl)

Water (H_2O)

Polar water molecules surround Na^+ and Cl^- ions in a salt crystal, dissolving it.

The Cell as a Factory

We have compared a cell to a miniature factory, with work centers that cooperate to make and ship the cell's products—proteins. To investigate the internal parts of a cell—the *organelles*—and how they function, we begin with a quick overview of the internal structure of a cell. **Figure 3-10** displays many cellular components.

A factory's outer wall separates it from the rest of the world and affords some security. Likewise, a cell's double-layered outer wall, or *cell membrane,* separates the cell from its surroundings and allows it to regulate what enters and leaves its domain. The cell membrane surrounds the neuron's cell body, its dendrites and their spines, and its axon and its terminals and so forms a boundary around a continuous intracellular compartment.

Very few substances can enter or leave a cell spontaneously because the cell membrane is almost *impermeable* (impenetrable). Proteins made by the cell are embedded in the cell membrane to facilitate the transport of substances into and out of the cell. Some proteins thus serve as the cellular factory's gates.

FIGURE 3-10 Typical Nerve Cell This view inside a neuron reveals its organelles and other internal components.

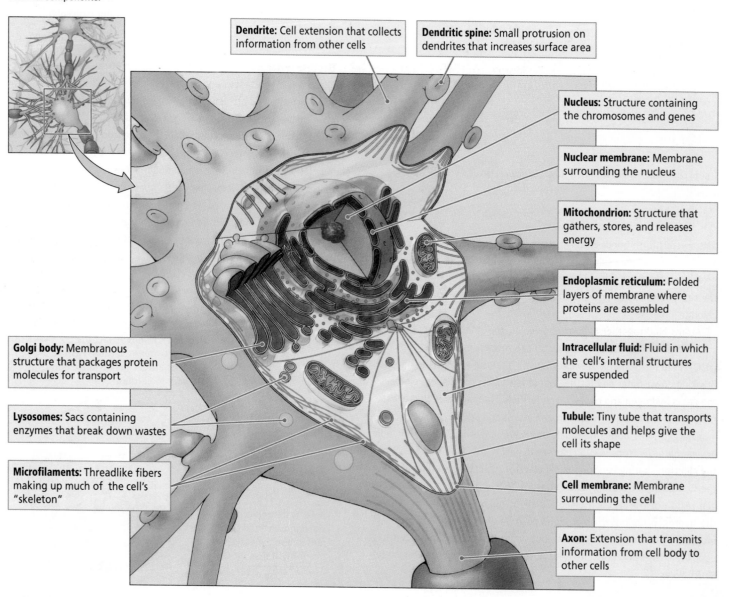

Dendrite: Cell extension that collects information from other cells

Dendritic spine: Small protrusion on dendrites that increases surface area

Nucleus: Structure containing the chromosomes and genes

Nuclear membrane: Membrane surrounding the nucleus

Mitochondrion: Structure that gathers, stores, and releases energy

Endoplasmic reticulum: Folded layers of membrane where proteins are assembled

Intracellular fluid: Fluid in which the cell's internal structures are suspended

Tubule: Tiny tube that transports molecules and helps give the cell its shape

Cell membrane: Membrane surrounding the cell

Axon: Extension that transmits information from cell body to other cells

Golgi body: Membranous structure that packages protein molecules for transport

Lysosomes: Sacs containing enzymes that break down wastes

Microfilaments: Threadlike fibers making up much of the cell's "skeleton"

Although the neurons and glia appear to be tightly packed together, they, like all cells, are separated by *extracellular fluid* composed mainly of water with dissolved salts and many other chemicals. A similar *intracellular fluid* is found inside a cell as well. The important point is that the relative impermeability of a cell's membrane ensures that concentrations of substances inside and outside the cell are different.

Within the cell shown in Figure 3-10 are membranes that surround its organelles, similar to the work areas demarcated by the inner walls of a factory. The membrane of each organelle is also relatively impermeable and so concentrates needed chemicals while keeping out unneeded ones.

The prominent *nuclear membrane* surrounds the cell's *nucleus,* where the genetic blueprints for the cell's proteins are stored, copied, and sent to the "factory floor." The *endoplasmic reticulum* (ER) is an extension of the nuclear membrane; the cell's protein products are assembled in the ER in accord with instructions from the nucleus.

When those proteins are assembled, they are packaged and sent throughout the cell. Parts of the cell called the *Golgi bodies* provide the "packaging rooms" where proteins are wrapped, addressed, and shipped.

Other cell components are called *tubules;* there are several kinds of tubules. Some (*microfilaments*) reinforce the cell's structure, others aid in the cell's movements, and still others (*microtubules*) form the transportation network that carries the proteins to their destinations, much as roads allow a factory's trucks and forklifts to deliver goods to their destinations.

Two other important parts of the cellular factory shown in Figure 3-10 are the *mitochondria,* the cell's power plants that supply its energy needs, and *lysosomes,* vesicles that transport incoming supplies and move and store wastes. Interestingly, more lysosomes are found in old cells than in young ones. Cells apparently have trouble disposing of their garbage, just as we do.

Cell Membrane: Barrier and Gatekeeper

The cell membrane separates the intracellular from the extracellular fluid and so allows the cell to function as an independent unit. Its double-layered structure, shown in **Figure 3-11**A, regulates the movement of substances into and out of the cell. One of these

In the CNS, the extracellular fluid is cerebrospinal fluid.

FIGURE 3-11 Structure of the Cell Membrane (A) Double-layered cell membrane close up. **(B)** Detail of a phospholipid molecule's polar head and electrically neutral tails. **(C)** Space-filling model shows why the phosphate head's polar regions (positive and negative poles) are hydrophilic, whereas its fatty acid tail, having no polar regions, is hydrophobic.

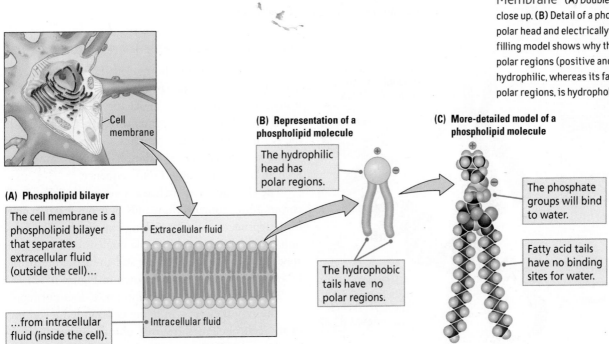

(A) Phospholipid bilayer

The cell membrane is a phospholipid bilayer that separates extracellular fluid (outside the cell)...

...from intracellular fluid (inside the cell).

Cell membrane

Extracellular fluid

Intracellular fluid

(B) Representation of a phospholipid molecule

The hydrophilic head has polar regions.

The hydrophobic tails have no polar regions.

(C) More-detailed model of a phospholipid molecule

The phosphate groups will bind to water.

Fatty acid tails have no binding sites for water.

gene DNA segment that encodes the synthesis of a particular protein.

substances is water. If too much water enters a cell, it will burst, and, if too much water leaves a cell, it will shrivel. The cell membrane's structure helps ensure that neither will happen.

The cell membrane also regulates the differing concentrations of salts and other chemicals on its inner and outer sides. This regulation is important because, if the concentrations of chemicals within a cell become unbalanced, the cell will not function normally. What properties of a cell membrane allow it to regulate water and salt concentrations within the cell? One property is its special molecular construction. These molecules, called *phospholipids,* are named for their structure, shown close up in Figure 3-11B.

Figure 3-11C shows a chemical model of the phosopholipid molecule. The molecule has a "head" containing the element phosphorus (P) bound to some other atoms, and it has two "tails," which are lipids, or fat molecules. The head has a polar electrical charge, with a positive charge in one location and a negative charge in another, like water molecules. The tails consist of hydrogen and carbon atoms that are tightly bound to one another by their shared electrons; hence there are no polar regions in the fatty tail.

The polar head and the nonpolar tails are the underlying reasons that a phospholipid molecule can form cell membranes. The heads are hydrophilic (Greek *hydro,* meaning "water," and *philia,* meaning "love") and so are attracted to one another and to polar water molecules. The nonpolar tails have no such attraction for water. They are *hydrophobic,* or water hating (from the Greek word *phobia,* meaning "fear").

Quite literally, then, the head of a phospholipid loves water and the tails hate it. To avoid water, the tails of phospholipid molecules point toward each other, and the hydrophilic heads align with one another and point outward to the intracellular and extracellular fluid. In this way, the cell membrane consists of a bilayer (two layers) of phospholipid molecules (see Figure 3-11A).

The bilayer cell membrane is flexible while still forming a remarkable barrier to a wide variety of substances. It is impenetrable to intracellular and extracellular water, because polar water molecules cannot pass through the hydrophobic tails on the interior of the membrane. Ions in the extracellular and intracellular fluid also cannot penetrate this membrane, because they carry charges and thus cannot pass the polar phospholipid heads. In fact, only a few small molecules, such as oxygen (O_2), carbon dioxide (CO_2), and glucose, can pass through a phospholipid bilayer.

Nucleus: Site of Gene Transcription

In our factory analogy, the nucleus is the cell's executive office where the blueprints for making proteins are stored, copied, and sent to the factory floor. These blueprints are called **genes,** segments of DNA that encode the synthesis of particular proteins. Genes are contained within the *chromosomes,* the double-helix structures that hold an organism's entire DNA sequence.

The chromosomes are like a book of blueprints. Each chromosome contains thousands of genes. Each gene is the blueprint, or code, for making one protein. The location of the chromosomes in the nucleus of the cell, the appearance of a chromosome, and the structure of the DNA in a chromosome are shown in **Figure 3-12.**

This static picture of chromosomes does not represent the way they look in living cells. Video recordings of the cell nucleus show that chromosomes are constantly changing shape and moving in relation to one

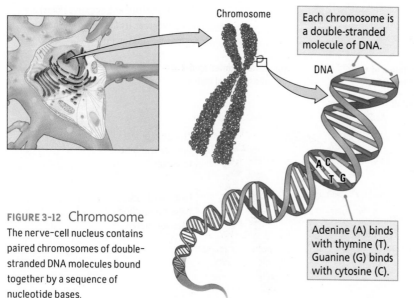

Chromosome

Each chromosome is a double-stranded molecule of DNA.

DNA

A C T G

Adenine (A) binds with thymine (T). Guanine (G) binds with cytosine (C).

FIGURE 3-12 Chromosome The nerve-cell nucleus contains paired chromosomes of double-stranded DNA molecules bound together by a sequence of nucleotide bases.

another so as to occupy the best locations within the nucleus for collecting the molecular building blocks of proteins and making proteins. By changing shape, chromosomes expose different genes to the surrounding fluid, thus allowing the processes of protein formation to take place.

A human somatic (body) cell has 23 pairs of chromosomes, or 46 chromosomes in all (in contrast, the 23 chromosomes within a reproductive cell are not paired). Each chromosome is a double-stranded molecule of *deoxyribonucleic acid* (DNA), which is capable of replicating and determining the inherited structure of a cell's proteins. The two strands of a DNA molecule coil around each other, as shown in Figure 3-12. Each strand possesses a variable sequence of four *nucleotide bases,* the constituent molecules of the genetic code: *adenine* (A), *thymine* (T), *guanine* (G), and *cytosine* (C).

Adenine on one strand always pairs with thymine on the other, whereas guanine on one strand always pairs with cytosine on the other. The two strands of the DNA helix are bonded together by the attraction that the bases in each pair have for each other, as illustrated in Figure 3-12. Sequences of hundreds of nucleotide bases within the chromosomes spell out the genetic code. Scientists represent this code by the letters of the nucleotide bases, for example ATGCCG, and so forth.

A gene is a segment of a DNA strand, and it encodes the synthesis of a particular protein. The code is contained in the sequence of the nucleotide bases, much as a sequence of letters spells out a word. The sequence of bases "spells out" the particular order in which *amino acids,* the constituent molecules of proteins, should be assembled to construct a certain protein.

To initiate the process, the appropriate gene segment of the DNA strands first unwinds. The exposed sequence of nucleotide bases on one of the DNA strands then serves as a template to attract free-floating molecules called *nucleotides.* The nucleotides thus attached form a complementary strand of *ribonucleic acid* (RNA), the single-stranded nucleic acid molecule required for protein synthesis. This process, called *transcription,* is shown in steps 1 and 2 of **Figure 3-13.** (To transcribe means "to copy," as in copying part of a message you receive in a text.)

The word *chromosome* means "colored body," referring to the fact that chromosomes can be readily stained with certain dyes.

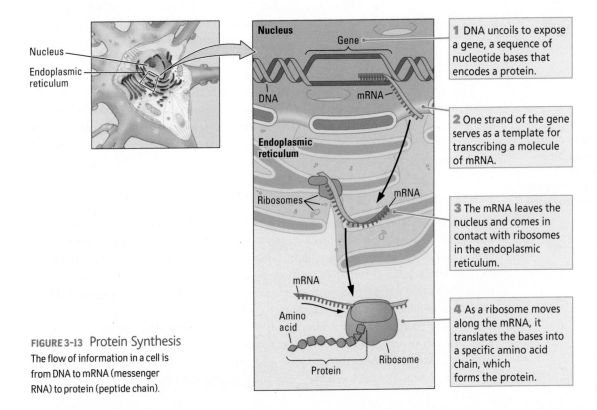

FIGURE 3-13 Protein Synthesis
The flow of information in a cell is from DNA to mRNA (messenger RNA) to protein (peptide chain).

1 DNA uncoils to expose a gene, a sequence of nucleotide bases that encodes a protein.

2 One strand of the gene serves as a template for transcribing a molecule of mRNA.

3 The mRNA leaves the nucleus and comes in contact with ribosomes in the endoplasmic reticulum.

4 As a ribosome moves along the mRNA, it translates the bases into a specific amino acid chain, which forms the protein.

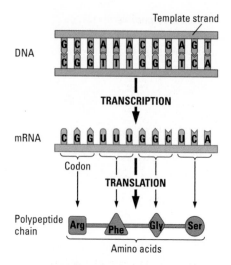

FIGURE 3-14 Transcription and Translation In protein synthesis (see Figure 3-12), a strand of DNA is transcribed into mRNA. Each sequence of three bases in the mRNA strand (a codon) encodes one amino acid. Directed by the codons, the amino acids link together to form a polypeptide chain. The amino acids illustrated are tryptophan (Trp), phenylalanine (Phe), glycine (Gly), and serine (Ser).

(A) Amino acid structure

The chemical composition of the R group distinguishes one amino acid from another.

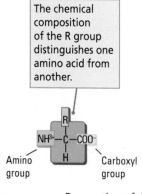

(B) Polypeptide chain

Peptide bond

A chain of amino acids forms a protein.

FIGURE 3-15 Properties of Amino Acids (A) Each amino acid consists of a central carbon atom (C) attached to an amine group (NH³⁺), a carboxyl group (COO⁻), and a distinguishing side chain (R). (B) The amino acids are linked by peptide bonds to form a polypeptide chain.

protein Folded-up polypeptide chain.

Endoplasmic Reticulum: Site of RNA Translation

The RNA produced through transcription is much like a single strand of DNA except that the base *uracil* (U), which also is attracted to adenine, takes the place of thymine. The transcribed strand of RNA is called *messenger RNA* (mRNA) because it carries the genetic code out of the nucleus to the endoplasmic reticulum, where proteins are manufactured.

Steps 3 and 4 in Figure 3-13 show that the ER consists of membranous sheets folded to form numerous channels. A distinguishing feature of the ER is that it may be studded with *ribosomes,* protein structures that act as catalysts in the building of proteins. When an mRNA molecule reaches the ER, it passes through a ribosome, where its genetic code is "read."

In this process, called *translation,* a particular sequence of nucleotide bases in the mRNA is transformed into a particular sequence of amino acids. (To translate means to convert one language into another, in contrast to transcription, in which the language remains the same.) *Transfer RNA* (tRNA) assists in translation.

As shown in **Figure 3-14,** each group of three consecutive nucleotide bases along an mRNA molecule encodes one particular amino acid. These sequences of three bases are called *codons.* For example, the codon uracil, guanine, guanine (UGG) encodes the amino acid tryptophan (Trp), whereas the codon uracil, uracil, uracil (UUU) encodes the amino acid phenylalanine (Phe). Codons also direct the placement of particular amino acids into a polypeptide (meaning "many peptides") chain.

Humans use 20 different amino acids to synthesize polypeptide chains. All 20 are structurally similar, as illustrated in **Figure 3-15A.** Each consists of a central carbon atom (C) bound to a hydrogen atom (H), an *amino group* (NH³⁺), a *carboxyl group* (COO⁻), and a *side chain* (represented by the letter R). The side chain varies in chemical composition from one amino acid to another, which helps give different polypeptide molecules their distinctive biochemical properties.

Amino acids in a polypeptide chain are linked together chemically by a special *peptide bond* (Figure 3-15B). Just as a remarkable number of words can be made from the 26 letters of the English alphabet, a remarkable number of peptide chains can be made from the 20 different amino acids. These amino acids can form 400 (20 × 20) different dipeptides (two-peptide combinations), 8000 (20 × 20 × 20) different tripeptides (three-peptide combinations), and almost countless polypeptides.

In summary, the flow of information contained in the genetic code is conceptually quite simple: a gene, or portion of a DNA strand, is transcribed into an mRNA strand, and the mRNA strand is translated by ribosomes into a molecular chain of amino acids, a polypeptide chain. Thus the sequence of events in building a protein is DNA → mRNA → protein.

Proteins: The Cell's Product

A polypeptide chain and a protein are related, but they are not the same. The relation is analogous to the relation between a length of ribbon and a bow of a particular size and shape that can be made from the ribbon. Long polypeptide chains have a strong tendency to twist into a helix (a spiral) or to form pleated sheets, which, in turn, have a strong tendency to fold together to form more complex shapes, as shown in **Figure 3-16.** A **protein** is a folded-up polypeptide chain.

Any one neuron may use as many as 10,000 protein molecules. The number of proteins that can be produced by the cell's factory is far larger than the number of genes. Although each gene codes for one protein, the protein can be cut into pieces to form still other proteins. Proteins can also combine, forming still other proteins.

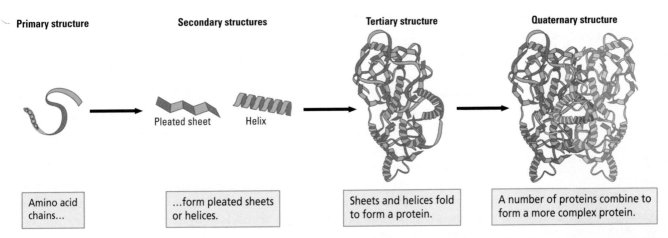

| Primary structure | Secondary structures | Tertiary structure | Quaternary structure |

Pleated sheet   Helix

| Amino acid chains... | ...form pleated sheets or helices. | Sheets and helices fold to form a protein. | A number of proteins combine to form a more complex protein. |

FIGURE 3-16 Four Levels of Protein Structure Whether a polypeptide chain forms a pleated sheet or a helix and its ultimate three-dimensional shape are determined by the sequence of amino acids in the primary structure.

A protein's shape and ability to change shape and to combine with other proteins are central to the protein's function. Through their shapes and changes in shape, they can combine with other proteins in chemical reactions. They can modify the length and shape of other proteins and so act as *enzymes.* They can be embedded into a cell's membrane to form channels and gates that regulate the flow of various substances through the membrane. They can be exported to travel to other cells and so act as messenger molecules.

Golgi Bodies and Microtubules: Protein Packaging and Shipment

Getting proteins to the right destinations is the task of the cell components that package, label, and ship them. These components operate much like a postal service.

To reach their appropriate destinations, the protein molecules that have been synthesized in the cell are wrapped in membranes and marked with their destination addresses to indicate where they are to go. This wrapping and labeling takes place in the organelles called Golgi bodies. The packaged proteins are then loaded onto motor molecules that "walk" along the many microtubules radiating through the cell, thus carrying each protein to its destination. The work of exporting proteins is illustrated in **Figure 3-17.**

If a protein is destined to remain within the cell, it is unloaded into the intracellular fluid. If it is to be incorporated into the cell membrane, it is carried to the membrane, where it inserts itself. Some proteins are exported by the cell. In this process, called *exocytosis,* the membrane, or *vesicle,* in which the protein is wrapped first fuses with the membrane of the cell. Now the protein inside the vesicle can be expelled into the

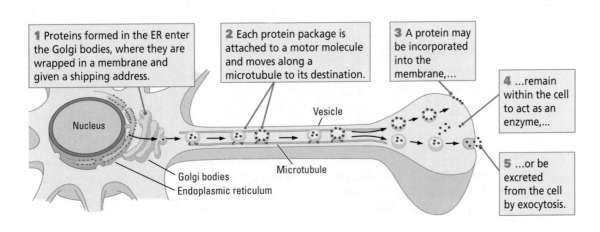

1 Proteins formed in the ER enter the Golgi bodies, where they are wrapped in a membrane and given a shipping address.

2 Each protein package is attached to a motor molecule and moves along a microtubule to its destination.

3 A protein may be incorporated into the membrane,...

4 ...remain within the cell to act as an enzyme,...

5 ...or be excreted from the cell by exocytosis.

Nucleus

Vesicle

Microtubule

Golgi bodies
Endoplasmic reticulum

FIGURE 3-17 Protein Export Exporting a protein entails packaging, transport, and its function at the destination.

Chapter 4 explains how neurons process information. Chapter 5 explores how neural communication can determine behavior.

extracellular fluid. The role that proteins play when they are embedded into the cell's membrane or exported from the cell are central to understanding how neurons process information and determine behavior.

Crossing the Cell Membrane: Channels, Gates, and Pumps

Proteins embedded in the cell membrane serve many functions. One is to help transport substances across the membrane. We will consider how three such membrane proteins work: channels, gates, and pumps. In each case, notice how the function of the particular protein is an emergent property of its shape.

Both the shape of a protein and its ability to change shape are emergent properties of the precise sequence of amino acids that compose the protein molecule. Some proteins change shape when other chemicals bind to them, others change shape as a function of temperature, and still others change shape in response to changes in electrical charge. The ability of a protein molecule to change shape is analogous to a lock in a door. When a key of the appropriate size and shape is inserted into the lock and turned, the locking device activates and changes shape, allowing the door to be closed or opened.

Such a shape-changing protein is illustrated in **Figure 3-18.** The surface of this protein molecule has a groove, called a *receptor,* which is analogous to a keyhole. Small molecules, such as glucose, or other proteins can bind to a protein's receptors and cause the protein to change shape. Changes in shape then allow the proteins to serve some new function.

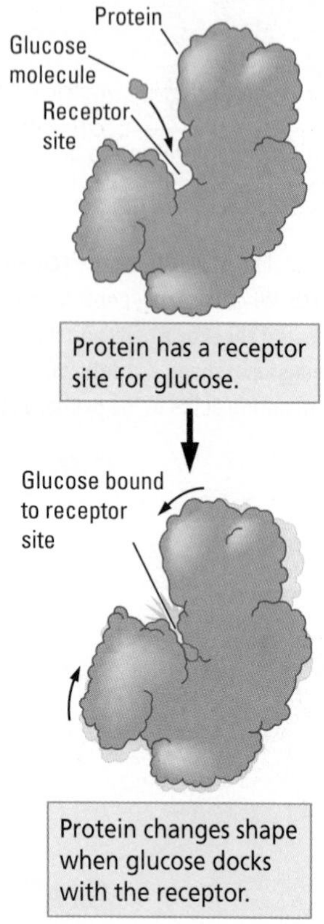

Some membrane proteins create **channels** through which substances can pass. Different-sized channels in different proteins allow the passage of different substances. **Figure 3-19**A illustrates a protein with a particular shape forming a small channel in the cell membrane that is large enough for potassium (K^+) ions, but not other ions, to pass through. Other protein channels allow sodium ions or chloride ions to pass into or out of the cell. Still others allow the passage of many various substances.

Figure 3-19B shows a protein molecule that acts as a **gate** to regulate the passage of substances across the cell membrane. Like the protein in Figure 3-18, it changes its shape in response to some trigger. The protein allows the passage of substances when its shape forms a channel and prevents the passage of substances when its shape leaves the channel closed. Thus, a part of this protein acts as a gate.

Changes in the shape of a protein can also allow it to act as a **pump.** Figure 3-19C shows a protein that, when Na^+ and K^+ ions bind to it, changes its shape to carry

Protein

Glucose molecule

Receptor site

Protein has a receptor site for glucose.

Glucose bound to receptor site

Protein changes shape when glucose docks with the receptor.

FIGURE 3-18 Receptor Binding When substances bind to a protein's receptors, the protein changes shape, which may change its function.

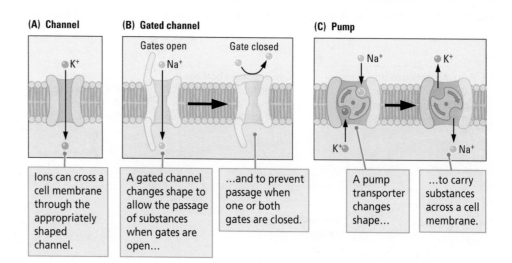

(A) Channel

K⁺

Ions can cross a cell membrane through the appropriately shaped channel.

(B) Gated channel

Gates open

Na⁺

Gate closed

A gated channel changes shape to allow the passage of substances when gates are open…

…and to prevent passage when one or both gates are closed.

(C) Pump

Na⁺

K⁺

K⁺

Na⁺

A pump transporter changes shape…

…to carry substances across a cell membrane.

FIGURE 3-19 Transmembrane Proteins Channels, gates, and pumps are proteins embedded in the cell membrane.

("pump") the substances across the membrane, exchanging the Na$^+$ on one side of the membrane for the K$^+$ on the other side of the membrane.

Channels, gates, and pumps play an important role in allowing substances to enter and leave a cell. This passage of substances is critical in explaining how neurons send messages. Chapter 4 explores how neurons use electrical activity to communicate.

REVIEW 3-2
Internal Structure of a Cell

Before you continue, check your understanding.

1. The constituent parts of the cell include the _____, _____, _____, _____, _____, and _____.

2. The product of the cell is _____, which serve many functions that include acting at the cell membrane as _____, _____, and _____ to allow substances to cross the membrane.

3. The basic sequence of events in building a protein is _____ makes _____ makes _____.

4. Once proteins are formed in the _____, they are wrapped in membranes by _____ and transported to their designated sites in the neuron, to its membrane, or for export from the cell by _____.

5. Why is a cell more than a factory for making proteins?

Answers appear at the back of the book.

3-3 Genes, Cells, and Behavior

Your *genotype* (genetic makeup) influences your physical and behavioral traits, which combine to form your *phenotype*. Genetic analysis conducted by the Human Genome Project has cataloged the human genome—all 20,000 or so genes in our species—and today individuals' genomes are routinely documented (recall the Beery twins in Research Focus 3-1). James Watson, the co-discoverer of DNA, was the first person to have his genome catalogued. Researchers have succeeded in sequencing the long-extinct Neanderthal human genome as well. The genomes of James Watson and the Neanderthal are surprisingly similar, as might be expected for close hominid relatives.

Studying the effects that genes have in influencing our traits is the objective of Mendelian genetics, named for Gregor Mendel, whose research led to the concept of the gene. Studying how the environment influences gene expression is the objective of epigenetics. In this section we describe how these two factors influence our phenotypes.

Mendelian Genetics and the Genetic Code

Recall that the nucleus of each human somatic cell contains 23 pairs of chromosomes, or 46 in all. One member of each pair of chromosomes comes from the mother, and the other member comes from the father. The chromosome pairs are numbered from 1 to 23, roughly according to size, with chromosome 1 being the largest **(Figure 3-20)**.

channel Opening in a protein embedded in the cell membrane that allows the passage of ions.

gate Protein embedded in a cell membrane that allows substances to pass through the membrane on some occasions but not on others.

pump Protein in the cell membrane that actively transports a substance across the membrane.

Figure 1-12 shows a paleo-artist's recreation of an adult male Neanderthal.

FIGURE 3-20 Human Chromosomes The nucleus of a human cell contains 23 chromosomes derived from the father and 23 from the mother. Sexual characteristics are determined by the 23rd pair, the X and Y sex chromosomes.

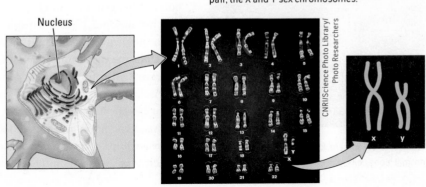

Nucleus

allele Alternate form of a gene; a gene pair contains two alleles.

homozygous Having two identical alleles for a trait.

heterozygous Having two different alleles for the same trait.

wild type Refers to a normal (most common in a population).

mutation Alteration of an allele that yields a different version of the allele.

Tay-Sachs disease Inherited birth defect caused by the loss of genes that encode the enzyme necessary for breaking down certain fatty substances; appears 4 to 6 months after birth and results in retardation, physical changes, and death by about age 5.

Chromosome pairs 1 through 22 are called *autosomes,* and they contain the genes that contribute most to our physical appearance and behavioral functions. The 23rd pair comprises the *sex chromosomes,* which contribute to our physical and behavioral sexual characteristics. The two types of mammalian sex chromosomes are referred to as X and Y because of their appearance, shown at the right in Figure 3-20. Female mammals have two X chromosomes, whereas males have an X and a Y.

Because all but your sex chromosomes are "matched" pairs, a cell contains two copies of every gene, one inherited from your mother, the other from your father. These two matching copies of a gene are called **alleles.** The term "matched" here does not necessarily mean identical. The nucleotide sequences in a pair of alleles may be either identical or different. If they are identical, the two alleles are **homozygous** (*homo* means "the same"). If they are different, the two alleles are **heterozygous** (*hetero* means "different").

The nucleotide sequence that is most common in a population is called the **wild-type** allele, whereas a less frequently occurring sequence is called a **mutation.** While mutant genes can be beneficial, more often they determine genetic disorders.

Dominant and Recessive Alleles

If both alleles in a pair of genes are homozygous, the two encode the same protein, but if the two alleles in a pair are heterozygous, they encode two different proteins. Three possible outcomes attend the heterozygous condition when these proteins express a physical or behavioral trait: (1) only the allele from the mother may be expressed, (2) only the allele from the father may be expressed, or (3) both alleles may be expressed simultaneously.

A member of a gene pair that is routinely expressed as a trait is called a *dominant* allele; a routinely unexpressed allele is *recessive.* Alleles can vary considerably in their dominance. In complete dominance, only the allele's own trait is expressed in the phenotype. In incomplete dominance, the expression of the allele's own trait is only partial. In *codominance,* both the allele's own trait and that of the other allele in the gene pair are expressed completely.

Each gene makes an independent contribution to the offspring's inheritance, even though the contribution may not always be visible in the offspring's phenotype. When paired with a dominant allele, a recessive allele is often not expressed. Still, it can be passed on to future generations and influence their phenotypes when not masked by the influence of some dominant trait.

Genetic Mutations

The mechanism for reproducing genes and passing them on to offspring is fallible. Errors can arise in the nucleotide sequence when reproductive cells make gene copies. The altered alleles are mutations.

A mutation may be as small as a change in a single nucleotide base. Because the average gene has more than 1200 nucleotide bases, an enormous number of mutations can potentially occur on a single gene. For example, the *BRCA1* (breast cancer) gene, found on chromosome 17, is a caretaker gene that contributes to preventing breast cancer, but more than 1000 different mutations have already been found on this gene. Thus, in principle, there are more than 1000 different ways in which to inherit a predisposition to breast cancer from just this gene.

A change in a nucleotide or the addition of a nucleotide in a gene sequence can be either beneficial or disruptive. An example of a mutation that is both causes sickle-cell anemia, a condition in which blood cells have an abnormal sickle shape. The sickle shape offers some protection against malaria but sickle cells also have poor oxygen-carrying capacity, thus weakening the person who possesses them.

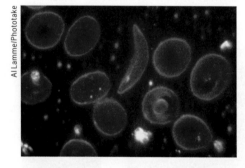

Al Lamme/Phototake

In this micrograph a sickle cell is surrounded by normal blood cells.

Other genetic mutations are more purely beneficial in their results, and still others are seemingly neutral to the functioning of the organism that carries them. Most mutations, however, have a negative effect. If not lethal, they produce in their carriers debilitating physical and behavioral abnormalities. Each of us carries a surprisingly large number of genetic mutations that, because of cell division, may differ in different parts of our bodies and brains (Charney, 2012). These localized mutations may contribute to individual variations in brain function.

Even though neuroscientists cannot yet explain human behavior in relation to genes and neurons, we know the severe behavioral consequences of about 2000 genetic abnormalities that affect the nervous system. For example, an error in a gene could produce a protein that should be an ion channel but will not allow the appropriate substance to pass, it may produce a pump that will not pump, or it may produce a protein that the transportation system of the cell refuses to transport.

Applying Mendel's Principles

Gregor Mendel introduced the concept of dominant and recessive alleles in the nineteenth century, when he studied pea plants. Today, scientists studying genetic variation are gaining insight into how genes, neurons, and behaviors are linked. This knowledge may help to reduce the negative effects of genetic abnormalities, perhaps someday even eliminating them completely, and will contribute to our understanding of normal brain function.

Experiment 1-1 reviews one of Mendel's experiments on "heritable factors" that we now call genes. Epigenetics studies differences in gene expression related to environment and experience.

Allele Disorders That Affect the Brain

Some disorders caused by mutant genes illustrate Mendel's principles of dominant and recessive alleles. One is **Tay-Sachs disease,** caused by a dysfunctional protein that acts as an enzyme known as HexA (hexosaminidase A) that fails to break down a class of lipids (fats) in the brain.

The disorder is named for Warren Tay and Bernard Sachs, the scientists who first described it.

Symptoms usually appear a few months after birth and rarely at later ages. The baby begins to suffer seizures, blindness, and degenerating motor and mental abilities. Inevitably, the child dies within a few years. Tay-Sachs mutations appear with high frequency among certain ethnic groups, including Jews of European origin and French Canadians, but the mutation in different populations is different.

The dysfunctional Tay-Sachs HexA enzyme is caused by a recessive allele of the HexA gene on chromosome 15. Distinctive inheritance patterns result from recessive alleles, because two copies (one from the mother and one from the father) are needed for the disorder to develop. A baby can inherit Tay-Sachs disease only when both parents carry the recessive allele.

Because both parents have survived to adulthood, both must also possess a corresponding dominant normal HexA allele for that particular gene pair. The egg and sperm cells produced by this man and woman will therefore contain a copy of one or the other of these two alleles. Which allele is passed on is determined completely by chance.

This situation gives rise in any child produced by two Tay-Sachs carriers to three different potential gene combinations, as diagrammed in **Figure 3-21**A. The child may have two normal alleles, in which case he or she will be spared the disorder and cannot pass on the disease. The child may have one normal and one Tay-Sachs allele, in which case he or she, like the parents, will be a carrier of the disorder. Or the child may have two Tay-Sachs alleles, in which case he or she will develop the disease.

In such a recessive condition, the chance of a child of two carriers being normal is 25 percent, the chance of being a carrier is 50 percent, and the chance of having Tay-Sachs disease is 25 percent. If one parent is a Tay-Sachs carrier and the other is normal,

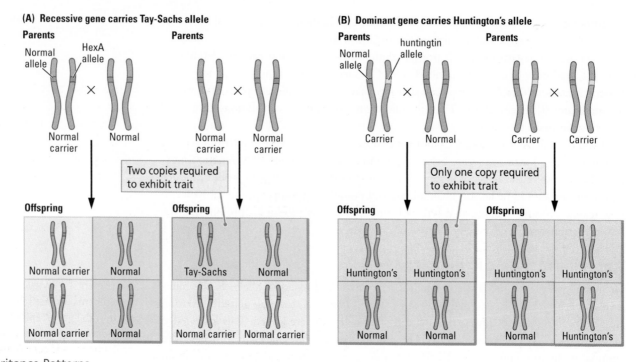

FIGURE 3-21 Inheritance Patterns
(A) Recessive condition: If a parent has one mutant allele, the parent will not show symptoms of the disease but will be a carrier. If both parents carry a mutant allele, each of their offspring stands a 1 in 4 chance of developing the disease. **(B)** Dominant condition: A person with a single allele will develop the disease. If this person mates with a normal partner, offspring have a 50-50 chance of developing the disease. If both parents are carriers, both will develop the disease, and offspring have a 75 percent chance of developing it.

then any of their children has a 50-50 chance of being either normal or a carrier. Such a couple has no chance of conceiving a baby with Tay-Sachs disease.

The Tay-Sachs allele operates independently of the dominant allele. As a result, it still produces the defective HexA enzyme, so the person who carries it has a higher-than-normal lipid accumulation in the brain. Because this person also has a normal allele that produces a functional enzyme, the abnormal lipid accumulation is not enough to cause Tay-Sachs disease.

Fortunately, a blood test can detect whether a person carries the recessive Tay-Sachs allele. People found to be carriers can make informed decisions about conceiving children. If they avoid having children with another Tay-Sachs carrier, none of their children will have the disorder, although some will probably be carriers. Where genetic counseling has been effective, the disease has been eliminated.

The one normal allele that a carrier of Tay-Sachs possesses produces enough functional enzyme to enable the brain to operate in a satisfactory way. That would not be the case if the normal allele were dominant, however, as happens with the genetic disorder **Huntington's disease.** Here, the buildup of an abnormal version of a protein known as *huntingtin* kills brain cells, especially cells in the basal ganglia and the cortex.

Symptoms can begin at any time from infancy to old age, but they most often start in midlife and include abnormal involuntary movements, which is why the disorder was once called a *chorea* (from the Greek, meaning "dance"). Other symptoms are memory loss and eventually a complete deterioration of behavior, followed by death. The abnormal huntingtin allele is dominant, the recessive allele normal, so only one defective allele is needed to cause the disorder, as discussed further in Clinical Focus 3-5, Huntington's Disease.

Figure 3-21B illustrates the inheritance patterns associated with a dominant allele on chromosome 4 that produces Huntington's disease. If one parent carries the defective allele, offspring have a 50 percent chance of inheriting the disorder. If both parents have the defective allele, the chance of inheriting it increases to 75 percent. Because the

Huntington's Disease

Woody Guthrie, whose protest songs made him a spokesman for farm workers during the Great Depression of the 1930s, is revered as one of the founders of American folk music. His best-known song is "This Land Is Your Land." Bob Dylan was instrumental in reviving Woody's popularity in the 1960s.

Guthrie died in 1967 after struggling with the symptoms of what was eventually diagnosed as Huntington's disease. His mother had died of a similar condition, although her illness was never diagnosed. Two of Guthrie's five children, from two marriages, developed the disease, and his second wife, Marjorie, became active in promoting its study.

Huntington's disease is devastating, characterized by memory impairment, abnormal uncontrollable movements, and marked changes in personality, eventually leading to virtually total loss of normal behavioral, emotional, and intellectual functioning. Fortunately, it is rare, with an incidence of only 5 to 10 victims in 100,000 people. It is most common in people of European ancestry.

The symptoms of Huntington's disease result from the degeneration of neurons in the basal ganglia and cortex. Those symptoms can appear at any age but typically start in midlife. In 1983, the *huntingtin* gene responsible for the disease was located on chromosome 4, and subsequently the abnormality in the base pairs of the gene has been described.

Woody Guthrie, whose unpublished lyrics and artwork are archived at woodyguthrie.org.

With permission of the Woody Guthrie Archives, photograph by Robin Carson

The study of the *huntingtin* gene has been a source of fascinating insights into the transmission of genetic disorders. A part of the gene contains a number of CAG repeats, and this combination of nucleotide bases encodes the amino acid glutamine. If the number of CAG repeats exceeds about 40, the carrier will likely display Huntington symptoms. As the number of CAG repeats increases, the onset of symptoms occurs earlier in life, and the progression of the disease becomes more rapid. Typically, non-Europeans have fewer repeats than do Europeans, so the disease is more common in Europeans. The number of repeats can also increase with transmission from the father but not from the mother.

Why brain cells containing the abnormal huntingtin protein die and why symptom onset takes so long are unknown, but the answers to these questions may be sources of insight into other brain diseases with onsets later in life, including degenerative diseases such as Alzheimer's. Insights into the neuroscience of Huntington's disease are now coming from animal models of mice, rats, and monkeys that have received the huntingtin gene and display the symptoms of the disease (McFarland and Cha, 2011). Although Huntington's disease is quite rare (5 to 10 victims in 100,000 people), Alzheimer's disease, which is most prevalent in people older than 65, can affect about 1 in 70 people.

abnormal *huntingtin* allele usually is not expressed until midlife, after the people who possess it have already had children, it is passed from generation to generation even though it is lethal.

As with the allele causing Tay-Sachs disease, there is now a genetic test for determining if a person possesses the allele that causes Huntington's disease. If a person is found to have the allele, he or she can elect not to procreate. A decision not to have children in this case will reduce the incidence of the abnormal *huntingtin* allele in the human gene pool.

Chromosome Abnormalities

Genetic disorders are not caused only by single defective alleles. Some nervous-system disorders are caused by aberrations in a part of a chromosome or even an entire chromosome.

In humans, one condition due to a change in chromosome number is **Down syndrome,** which affects approximately 1 in 700 children. Down syndrome is usually the result of an extra copy of chromosome 21. One parent (usually the mother) passes on two of these chromosomes to the child, rather than the normal single chromosome. Combining

Huntington's disease Hereditary disease characterized by chorea (ceaseless, involuntary, jerky movements) and progressive dementia, ending in death.

Down syndrome Chromosomal abnormality resulting in mental retardation and other abnormalities, usually caused by an extra chromosome 21.

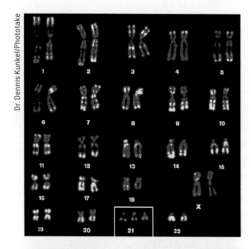

FIGURE 3-22 Chromosome Aberration (*Top*) Down syndrome, also known as trisomy 21, is caused by an extra copy of chromosome 21. (*Bottom*) Chris Burke, who lives with Down syndrome, attends a Broadway opening in 2008. Burke played a leading role on the television series *Life Goes On* in the 1990s.

Sections 7-1 and 7-5 review genetic methods used in neuroscience research.

these two chromosomes with one from the other parent yields three chromosomes 21, an abnormal number called a *trisomy* (**Figure 3-22**).

Although chromosome 21 is the smallest human chromosome, its trisomy severely alters a person's phenotype. People with Down syndrome have characteristic facial features and short stature. They also endure heart defects, susceptibility to respiratory infections, and mental retardation. They are prone to developing leukemia and Alzheimer's disease. Although people with Down syndrome usually have a much shorter-than-normal life span, some live to middle age or beyond. Improved education for children with Down syndrome shows that they can learn to compensate greatly for the brain changes that cause their mental handicaps.

Genetic Engineering

Despite advances in understanding the structure and function of genes, there remains a gap in understanding how genes produce behavior. To investigate gene structure and behavior relations, geneticists have invented a number of methods to influence the traits that genes express. This approach collectively defines the science of *genetic engineering*. In its simplest forms, genetic engineering entails manipulating a genome, removing a gene from a genome, modifying a gene, or adding a gene to the genome. Genetic engineering techniques include selective breeding, cloning, and transgenics.

Selective Breeding

The oldest means of influencing genetic traits is the selective breeding of animals and plants. Beginning with the domestication of wolves into dogs more than 15,000 years ago, many species of animals have been domesticated by selectively breeding males and females that display particular traits. The selective breeding of dogs, for example, has produced breeds that can run fast, haul heavy loads, retrieve prey, dig for burrowing animals, climb rocky cliffs in search of sea birds, herd sheep and cattle, or sit on an owner's lap and cuddle.

Selective breeding is an effective way to alter gene expression. As is described by Erik Karlsson (Karlsson & Lindblad-Toh, 2008) in regard to the dog genome, insights into the relations among genes, behavior, and disease can be usefully examined because, as the result of selective breeding, dogs display the most diverse traits of any animal species.

Maintaining spontaneous mutations is one objective of selective breeding. By using this method, researchers create whole populations of animals possessing some unusual trait that originally arose as an unexpected mutation in only one individual animal or in a few of them. In laboratory colonies of mice, for example, large numbers of spontaneous mutations have been discovered and maintained in various mouse strains.

There are strains of mice that have abnormal movements, such as reeling, staggering, and jumping. Some have diseases of the immune system; others have sensory deficits and are blind or cannot hear. Some mice are smart, some mice are not, some have big brains, some small, and many display distinctive behavioral traits. Many of these genetic variations can also be found in humans. As a result, the neural and genetic bases of the altered behavior in the mice can be studied systematically to develop treatments for human disorders.

Cloning

More direct approaches to manipulating the expression of genetic traits include altering early embryonic development. One such method is cloning—producing an offspring that is genetically identical to another animal.

To clone an animal, scientists begin with a cell nucleus containing DNA, usually from a living animal, place it into an egg cell from which the nucleus has been removed, and after stimulating the egg to start dividing, implant the new embryo into the uterus of a female. Because each individual animal that develops from these cells is genetically identical with the donor of the nucleus, clones can be used to preserve valuable traits, to study the relative influences of heredity and environment, or to produce new tissue or organs for transplant to the donor. Dolly, a female sheep, was the first mammal to be cloned.

Cloning has matured from an experimental manipulation to a commercial enterprise. The first horse to be cloned was Charmayne James's horse Scamper, the mount she rode to 11 world championships in barrel racing. The first cat to be cloned was called Copycat, shown in **Figure 3-23.** The first rare species cloned was an Asian gaur, an animal related to the cow. One group of investigators anticipates cloning the mastodon, an extinct elephant species, using cells from carcasses found frozen in the Arctic tundra.

Transgenic Techniques

Transgenic technology enables scientists to introduce genes into an embryo or remove genes from it. For example, introducing a new gene can enable goats to produce medicines in their milk. The medicines can be extracted from the milk to treat human diseases (Kues & Niemann, 2011).

Chimeric animals are composites formed when an embryo of one species receives cells from a different species. The resulting animal has cells with genes from both parent species and behaviors that are a product of those gene combinations. The chimeric animal may display an interesting mix of the behaviors of the parent species. For example, chickens that have received Japanese quail cells in early embryogenesis display some aspects of quail crowing behavior rather than chicken crowing behavior, providing evidence for the genetic basis of some bird vocalization (Balaban, 2005). The chimeric preparation provides an investigative tool for studying the neural basis of crowing because quail neurons can be distinguished from chicken neurons when examined under a microscope.

In *knock-in technology,* a number of genes or a single gene from one species is added to the genome of another species and is passed along and expressed in subsequent generations of **transgenic animals.** Brainbow technology, described in Research Focus 3-2, is an application of transgenic technology. Another application is in the study and treatment of human genetic disorders. For instance, researchers have introduced into a line of mice the human gene that causes Huntington's disease (Gill and Rego, 2009). The mice express the abnormal *huntingtin* allele and display symptoms similar to the disorder in humans. This mouse line is being used to study potential therapies for the Huntington's disorder in humans.

Knockout technology is used to inactivate a gene so that a line of mice fails to express it (Eisener-Dorman et al., 2008). The line of mice can then be examined to determine whether the targeted gene is responsible for a human disorder and to examine possible therapies for the disorder. It is potentially possible to knock out genes that are related to certain kinds of memory, such as emotional memory, social memory, or spatial memory. Such technology would provide a useful way of investigating the neural basis of memory.

Phenotypic Plasticity and the Epigenetic Code

Our genotype is not sufficient to explain our phenotype. We all know that if we expose ourselves to the sun, our skin becomes darker; if we exercise, our muscles become larger. Our phenotype also changes depending on our diet and as we age. In short, the extent of our phenotypic variation, given the same genotype, can be dramatic.

Dolly was cloned in 1996 by a team of researchers in Scotland. As an adult, she mated and bore a lamb.

Photographs used with permission of Texas A&M College of Veterinary Medicine and Biochemical Sciences

FIGURE 3-23 A Clone and Her Mom
Copycat (*left*) and Rainbow (*right*), the cat that donated the cell nucleus for cloning. Although the cats' genomes are identical, their phenotypes, including fur color, differ. One copy of the X chromosome is randomly inactivated in each cell, which explains the color differences. Even clones are subject to phenotypic plasticity: they retain the capacity to develop into more than one phenotype.

Neural basis of memory is the topic of Section 14-3.

transgenic animal Product of technology in which number of genes or a single gene from one species is introduced into the genome of another species and passed along and expressed in subsequent generations.

The mice shown in Figure 2-1 are genetic clones but express different phenotypes because their pregnant mothers' diets contained different supplements.

The corpus callosum, shown in Figure 2-10, is the large band of nerve fibers that connects the two hemispheres and allows them to communicate.

FIGURE 3-24 Gene Expression Identical coronal sections through the brains of mice with identical genotypes reveal frontal views of distinctly different phenotypes. Mouse (A) has a corpus callosum, whereas mouse (B) does not. Adapted from "Defects of the fetal forebrain in acallosal mice," by D. Wahlsten and H. W. Ozaki, 1994, in *Callosal Agenesis* (p. 126), edited by M. Lassonde and M. A. Jeeves, New York: Plenum Press.

Section 8-2 describes the origins, growth, and development of neurons and glia.

Several countries have combined their research resources to form the International Human Epigenome Consortium (IHEC). Its mandate: to describe the epigenetic code, as the Human Genome Project has described the genetic code.

Every individual has a capacity to develop into more than one phenotype. This *phenotypic plasticity* is due in part to the capacity of the genome to express a large number of phenotypes and in part to epigenetics, the influence of environment in selecting one or another phenotype.

Seemingly puzzling features in the expression of genomes in relation to phenotypes are illustrated in strains of genetically identical mice, some of which develop a brain with no corpus callosum **(Figure 3-24)**. The absence of a corpus callosum results from an epigenetic influence on whether the trait is expressed in a particular mouse and occurs in the embryo at about the time at which the corpus callosum should form. This lack of concordance (incidence of similar behavioral traits) is also observed in patterns of disease incidence in human identical twins, who share the same genome.

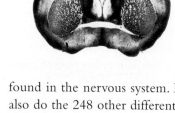

(A) Corpus callosum

Anterior commissure

(B)

The concordance rate between identical twins for a vast array of diseases—including schizophrenia (in which a person experiences hallucinations and delusions), Alzheimer's disease; multiple sclerosis (see Clinical Focus 3-3). Crohn's disease (a form of inflammatory bowel disease), asthma, diabetes, and prostate cancer—are between 30 and 60 percent. For cleft palate and breast cancer, the concordance rate is about 10 percent. These less-than-perfect concordance rates point to contributing factors other than Mendelian genetic principles.

Phenotypic plasticity is in evidence not only in adult organisms but also in cells. In Section 3-1, we described the many kinds of neurons and glia found in the nervous system. Each of these cells usually has the same genotype. So also do the 248 other different cell types of the body. How then do they become so different?

The genes that are expressed in a cell are influenced by factors within the cell and in the cell's environment. Once a fertilized egg begins to divide, each new cell finds itself in a somewhat different environment from that of its parent cell. The cell's environment will determine which genes are expressed and so what kind of tissue it becomes, including what kind of nervous system cell it becomes.

Applying the Epigenetic Code

Epigenetic mechanisms create phenotypic variation without altering the base-pair nucleotide sequence of the genes. Through these mechanisms the environment can allow a gene to be expressed or prevent its expression. Epigenetics is viewed as a second code; the first code is the genome. Epigenetics describes how a single genetic code produces each different somatic cell type, explains how a single genome can code for many different phenotypes, and describes how cells go astray in their function to produce diseases ranging from cancer to brain dysfunction.

Epigenetic mechanisms can influence protein production either by blocking a gene so that it cannot be transcribed or by unlocking a gene so that it can be transcribed. This is where environmental influences come into play. To review, each of your chromosomes consists of a long, double-stranded chain of nucleotide bases that forms your DNA. Each gene on a chromosome is a segment of DNA that encodes the synthesis of a particular protein (see Figure 3-13).

Chromosomes are wrapped around supporting molecules of a protein called histone. This histone wrapping allows the many yards of a chromosome to be packaged in a small space, as yards of thread are wrapped around a spool. For any gene to be transcribed into messenger RNA, its DNA must be unspooled from the histones. Once unspooled, each gene must be instructed to transcribe mRNA. Then the mRNA must be translated into an amino acid chain that forms the protein. **Figure 3-25** illustrates some of the ways that each of these three steps can be either enabled or blocked:

1. *Histone modification* DNA may unwrap or be stopped from unwrapping from the histone. In Figure 3-25, a methyl group (CH_3) or other molecule that binds to the tails of histones blocks DNA from unspooling, and genes cannot be exposed for transcription.

2. *DNA modification* Transcription of DNA into mRNA may be enabled or blocked. In Figure 3-25, one or more methyl groups bind to CG base pairs to block transcription.

3. *mRNA modification* mRNA translation may be enabled or blocked. In Figure 3-25, noncoding RNA (ncRNA) binds to mRNA, blocking translation.

An environmental influence can either induce or remove one or more blocks, thus allowing the environment to regulate gene expression (Charney, 2012). It is through these epigenetic mechanisms that cells are instructed to differentiate into different body

This process of *methylation* dramatically alters gene expression during brain development (see Sections 8-2 and 12-5) and can affect memory and brain plasticity (see Section 14-4).

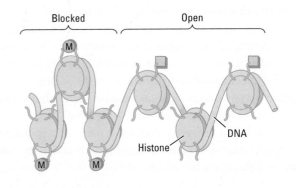

1 Histone modification
A methyl group (CH_3) or other molecules bind to the tails of histones, either blocking them from opening (orange circles) or allowing them to open for transcription (green squares).

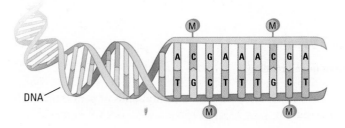

2 DNA modification
Methyl groups (M) bind to CG base pairs to block transcription.

3 mRNA modification
ncRNA binds to mRNA, preventing translation.

FIGURE 3-25 Epigenetic Mechanisms

tissues and that our unique environments and experiences induce changes in our brains that make us unique individuals. Some experientially induced experiences can also be passed from one generation to the next, as the following study illustrates.

A Case of Inheriting Experience

The idea that traits are passed from parent to child through genes is a cornerstone of Mendelian genetics. Mendel's theory also predicts that individual life experience cannot be inherited. Lars Olov Bygren and colleagues (Kaati et al., 2007), however, found that an individual's nutritional experiences *can* affect the health of their offspring.

The investigators focused on Norrbotten, a sparsely populated, northern Swedish region. In the nineteenth century, Norrbotten was virtually isolated from the outside world. If the harvest there was bad, people starved. According to historical records, the years 1800, 1812, 1821, 1836 and 1856 saw total crop failure. The years 1801, 1822, 1828, 1844 and 1863 brought good harvests and abundance.

Bygren and colleagues identified at random individuals who had been subjected to famine or to plenty in the years just before they entered puberty. Then the researchers examined the health records and longevity of these people's children and grandchildren.

The findings seem to defy logic. The descendants of the plenty group experienced higher rates of cardiovascular disease and diabetes and had a reduced life expectancy of more than seven years compared to the famine group! Notably, these effects were found only in male offspring of males and female offspring of females.

Bygren and colleagues propose that diet during a *critical period* can modify the genetic expression of sex chromosomes—the Y chromosome in males and the X chromosome in females. Further, this change can be passed on to subsequent generations. The dietary experience in the prepubertal period, just before the onset of sexual maturity, is important: this is the time at which gene expression on the sex chromosomes begins.

The seminal findings of Bygren and coworkers are supported by many other studies that, together, make a strong argument for epigenetics and for the idea that some epigenetic influences can be passed on for at least a few generations. Evidence that epigenetic influences play a demonstrable role in determining gene expression is disclosing how our experiences shape our brains to influence who we become.

Section 8-4 examines critical periods, time spans during which events have long-lasting influences on development.

REVIEW 3-3
Genes, Cells, and Behavior

Before you continue, check your understanding.

1. Each of our _____ chromosome pairs contains thousands of genes, and each gene contains the code for one _____.

2. The genes that we receive from our mothers and fathers may include slightly different _____ of particular genes, which will be expressed in slightly different _____.

3. Abnormalities in a gene, caused by a(n) _____, can result in an abnormally formed protein that, in turn, results in the abnormal cell function. Chromosome abnormality can result in the abnormal function of many genes. _____, for example, is caused by an extra copy of chromosome 21.

4. The neurological disorder Tay-Sachs disease results from a(n) _____ allele being expressed; Huntington's disease results from the expression of a(n) _____ allele.

5. _____ is the oldest form of genetic manipulation. Genetic engineering techniques manipulate or alter the genome of an animal. _____ produces an animal that is genetically identical with to a parent or sibling; _____ contain new, altered, or inactivated genes.

6. What distinguishes Mendelian genetics from epigenetics?

Answers appear at the back of the book.

SUMMARY

3-1 Cells of the Nervous System

The nervous system is composed of two kinds of cells: neurons that transmit information and glia that support brain function. Sensory neurons send information from the body's sensory receptors to the brain, motor neurons send commands enabling muscles to move, and interneurons link sensory and motor activities.

Like neurons, glial cells can be grouped by structure and function. Ependymal cells produce cerebrospinal fluid. Astrocytes structurally support neurons, help to form the blood–brain barrier, and seal off damaged brain tissue. Microglia aid in the repair of brain cells. Oligodendroglia and Schwann cells myelinate axons in the CNS and the somatic division of the PNS, respectively.

A neuron is composed of three basic parts: a cell body, or soma; branching extensions called dendrites, designed to receive information; and a single axon that passes information along to other neurons. A dendrite's surface area is greatly increased by numerous dendritic spines. An axon may have branches called axon collaterals, which are further divided into teleodendria, each ending at a terminal button, or end foot. A synapse is the "almost connection" between a terminal button and the membrane of another cell.

3-2 Internal Structure of a Cell

A surrounding cell membrane protects the cell and regulates what enters and leaves it. Within the cell are a number of compartments, also enclosed in membranes. These compartments include the nucleus (which contains the cell's chromosomes and genes), the endoplasmic reticulum (where proteins are manufactured), the mitochondria (where energy is gathered and stored), the Golgi bodies (where protein molecules are packaged for transport), and lysosomes (which break down wastes). A cell also contains a system of tubules that aid its movements, provide structural support, and act as highways for transporting substances.

To a large extent, the work of cells is carried out by proteins. The nucleus contains chromosomes—long chains of genes, each gene encoding a specific protein needed by the cell. Proteins perform diverse tasks by virtue of their diverse shapes. Some act as enzymes to facilitate chemical reactions; others serve as membrane channels, gates, and pumps; and still others are exported for use in other parts of the body.

A gene is a segment of a DNA molecule and is made up of a sequence of nucleotide bases. Through a process called transcription, a copy of a gene is produced in a strand of messenger RNA. The mRNA then travels to the endoplasmic reticulum, where a ribosome moves along the mRNA molecule, translating it into a sequence of amino acids. The resulting chain of amino acids is a polypeptide. Polypeptides fold and combine to form protein molecules with distinctive shapes that are used for specific purposes in the body.

3-3 Genes, Cells, and Behavior

From each parent, we inherit one of each of the chromosomes in our 23 chromosome pairs. Because all but the sex chromosomes are "matched" pairs, a cell contains two alleles of every gene. Sometimes the two alleles of a pair are homozygous (the same), and sometimes they are heterozygous (different).

An allele may be dominant and expressed as a trait, recessive and not expressed, or co-dominant, in which case both it and the other allele in the pair are expressed in the individual organism's phenotype. One allele of each gene is designated the wild type—that is, the most common one in a population—whereas the other alleles of the gene are called mutations. A person might inherit any of these alleles from a parent, depending on the parent's genotype.

Genes can potentially undergo many mutations, in which their codes are altered by one or more changes in the nucleotide sequence. Most mutations are harmful and may produce abnormalities in nervous-system structure and behavioral function. Genetic research seeks to prevent the expression of genetic and chromosomal abnormalities and to find cures for those that are expressed.

Selective breeding is the oldest form of genetic manipulation. Genetic engineering is a new science in which the genome of an animal is artificially altered. The genetic composition of a cloned animal is identical with that of a parent or sibling; transgenic animals contain new or altered genes; and knockouts have genomes from which a gene has been deleted.

The genome encodes a range of phenotypes. The phenotype eventually produced is determined by epigenetics and influenced by the environment. Epigenetic mechanisms can influence whether genes are transcribed or transcription is blocked, without changing the genetic code itself.

KEY TERMS

allele, p. 98

astrocyte, p. 82

axon, p. 77

axon collateral, p. 77

axon hillock, p. 77

bipolar neuron, p. 79

blood–brain barrier, p. 82

cell body (soma), p. 77

channel, p. 97

dendrite, p. 77

dendritic spine, p. 77

Down syndrome, p. 101

ependymal cell, p. 82

gate, p. 97

gene, p. 92

glial cell, p. 82

heterozygous, p. 98

homozygous, p. 98

Huntington's disease, p. 101

hydrocephalus, p. 82

interneuron, p. 79

microglia, p. 85

motor neuron, p. 79

multiple sclerosis (MS), p. 85

mutation, p. 98

myelin, p. 85

oligodendroglia, p. 85

paralysis, p. 85

protein, p. 94

pump, p. 97

Purkinje cell, p. 79

pyramidal cell, p. 79

Schwann cell, p. 85

sensory neuron, p. 79

somatosensory neuron, p. 79

synapse, p. 79

Tay-Sachs disease, p. 98

terminal button (end foot),
    p. 77

transgenic animal, p. 103

tumor, p. 82

wild type, p. 98

How Do Neurons Use Electrical Signals to Transmit Information?

CLINICAL FOCUS 4-1 EPILEPSY

4-1 SEARCHING FOR ELECTRICAL ACTIVITY IN THE NERVOUS SYSTEM

EARLY CLUES THAT LINKED ELECTRICITY AND NEURONAL ACTIVITY

THE BASICS ELECTRICITY AND ELECTRICAL STIMULATION

TOOLS FOR MEASURING A NEURON'S ELECTRICAL ACTIVITY

HOW THE MOVEMENT OF IONS CREATES ELECTRICAL CHARGES

4-2 ELECTRICAL ACTIVITY OF A MEMBRANE

RESTING POTENTIAL

MAINTAINING THE RESTING POTENTIAL

GRADED POTENTIALS

ACTION POTENTIAL

NERVE IMPULSE

REFRACTORY PERIODS AND NERVE ACTION

SALTATORY CONDUCTION AND THE MYELIN SHEATH

4-3 HOW NEURONS INTEGRATE INFORMATION

EXCITATORY AND INHIBITORY POSTSYNAPTIC POTENTIALS

CLINICAL FOCUS 4-2 MYASTHENIA GRAVIS

SUMMATION OF INPUTS

VOLTAGE-SENSITIVE CHANNELS AND THE ACTION POTENTIAL

THE VERSATILE NEURON

RESEARCH FOCUS 4-3 OPTOGENETICS AND LIGHT-SENSITIVE ION CHANNELS

4-4 INTO THE NERVOUS SYSTEM AND BACK OUT

HOW SENSORY STIMULI PRODUCE ACTION POTENTIALS

HOW NERVE IMPULSES PRODUCE MOVEMENT

CLINICAL FOCUS 4-4 LOU GEHRIG'S DISEASE

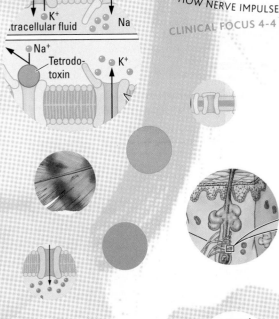

Epilepsy

J. D. worked as a disc jockey for a radio station and at parties in his off-hours. One evening, he set up on the back of a truck at a rugby field to emcee a jovial and raucous rugby party. Between musical sets, he made introductions, told jokes, and exchanged toasts.

About one o'clock in the morning, J. D. suddenly collapsed, making unusual jerky motions, and then passed out. He was rushed to a hospital emergency room, where he gradually recovered. The attending physician noted that he was not intoxicated, released him to his friends, and recommended a series of neurological tests for the next day. Neuroimaging with state-of-the-art brain scans can usually reveal brain abnormalities (Bano et al., 2011), but it did not do so in J. D.'s case.

When the electrical activity in J. D.'s brain was recorded while a strobe light was flashed before his eyes, an electroencephalogram, or EEG, displayed a series of abnormal electrical patterns characteristic of epilepsy. The doctor prescribed Dilantin (diphenylhydantoin), an

The EEG detects electrical signals given off by the brain in various states of consciousness, as explained in Sections 7-2 and 13-3. Section 16-3 details the diagnosis and treatment of epilepsy.

anesthetic agent given in low doses, and advised J. D. to refrain from drinking. He was required to give up his driver's license to prevent the possibility that an attack while driving could cause an accident. And he lost his job at the radio station.

After 3 months of uneventful drug treatment, he was taken off medication and his driver's license was restored. J. D. convinced the radio station that he could resume work, and subsequently he has been seizure free.

Epilepsy is the most common neurological disease worldwide: 1 person in 20 experiences an epileptic seizure in his or her lifetime. Synchronous stimuli can trigger a seizure; thus, a strobe light is often used in diagnosis. Some epileptic seizures can be linked to a specific symptom, such as infection, trauma, tumor, or other damage to a part of the brain. Others appear to arise spontaneously. Their cause is poorly understood.

Three symptoms are common to many kinds of epilepsy:

1. An *aura,* or warning, of an impending seizure, which may take the form of a sensation, such as an odor or sound, or may simply be a "feeling"

2. Abnormal movements such as repeated chewing or shaking; twitches that start in a limb and spread across the body; and in some cases, a total loss of muscle tone and postural support causes the person to collapse

3. Loss of consciousness and later unawareness that the seizure happened

If seizures occur repeatedly and cannot be controlled by drug treatment, surgery may be performed. The goal of surgery is to remove damaged or scarred tissue that serves as the focal point of a seizure. Removing this small area prevents seizures from starting and spreading to other brain regions. The condition of epilepsy reveals that the brain is normally electrically active and that if this activity becomes abnormal, the consequences are severe.

The most reproduced drawing in behavioral neuroscience is nearly 350 years old, predating our understanding of the electrical basis of epilepsy by centuries. Taken from René Descartes's book *Treatise on Man* and reproduced in **Figure 4-1,** it illustrates the first serious attempt to explain how information travels through the nervous system. Descartes proposed that the carrier of information is cerebrospinal fluid flowing through nerve tubes.

Descartes reasoned that, when the fire burns the man's toe, it stretches the skin, which tugs on a nerve tube leading to the brain. In response to the tug, a valve in a ventricle of the brain opens and CSF flows down the tube, filling the leg muscles and causing them to contract and pull the toe back from the fire. The flow of fluid through other tubes to other muscles of the body (not shown in Figure 4-1) causes the head to turn toward the painful stimulus and the hands to rub the injured toe.

Descartes proposed the idea behind dualism—that the nonmaterial mind controls body mechanics—described in Section 1-2.

Descartes's theory was inaccurate, yet it is remarkable because he isolated the three basic questions that underlie a behavioral response to stimulation:

1. How do our nerves detect a sensory stimulus and inform the brain about it?

2. How does the brain decide what response should be made?

3. How does the brain command muscles to move to produce a behavioral response?

Descartes was trying to explain the very same things that scientists have sought to explain in the intervening centuries. If not by stretched skin tugging on a nerve tube initiating the message, the message must still be initiated somehow. If not by opening valves to initiate the flow of CSF to convey information, the information must still be sent. If not by filling the muscles with fluid that produces movements, the muscles must be caused to contract by some other mechanism.

What all these mechanisms in fact are is the subject of this chapter. We examine how neurons convey information from the environment throughout the nervous system and ultimately activate muscles to produce movement. We begin by describing the clues and tools that explained the electrical activity of the nervous system.

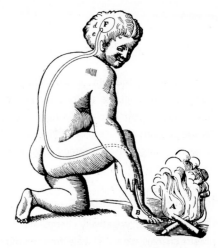

FIGURE 4-1 Descartes's Theory of Information Flow   From Descartes, 1664.

4-1 Searching for Electrical Activity in the Nervous System

The first hints about how the nervous system conveys its messages came in the eighteenth century, following the discovery of electricity. Early discoveries about the nature of electricity quickly led to proposals that it plays a role in conducting information in the nervous system. We describe a few milestones that lead from this idea to an understanding of how the nervous system really conveys information. If you have a basic understanding of how electricity works and how it is used to stimulate neural tissue, read on. If you prefer to brush up on electricity and electrical stimulation first, turn to "The Basics: Electricity and Electrical Stimulation" on page 112.

Early Clues That Linked Electricity and Neuronal Activity

In a dramatic demonstration in 1731, Stephen Gray, an eighteenth-century amateur English scientist, rubbed a rod with a piece of cloth to accumulate electrons on the rod. Then he touched the charged rod to the feet of a boy suspended on a rope and brought a metal foil to the boy's nose. The foil was attracted to the boy's nose and bent on approaching it, and as foil and nose touched, electricity passed from the rod through the boy to the foil.

Yet the boy was completely unaware that the electricity had passed through his body. Therefore, Gray speculated that electricity might be the messenger that spreads information through the nervous system. Two other lines of evidence, drawn from electrical stimulation and electrical recording studies, implicated electrical activity in the nervous system's flow of information.

Electrical Stimulation Studies

When the eighteenth-century Italian scientist Luigi Galvani, a contemporary of Gray, observed that frogs' legs hanging on a wire in a market twitched during a lightning storm, he surmised that sparks of electricity from the storm were activating the muscles. Investigating this possibility, he found that, if an electrical current is applied to a dissected nerve, the muscle connected to that nerve contracts. It was unclear how the process worked, but Galvani had discovered the technique of **electrical stimulation:** passing an electrical current from the uninsulated tip of an electrode onto a nerve produces behavior—a muscular contraction.

Among the many researchers who used Galvani's technique to produce muscle contraction, two mid-nineteenth-century Prussian scientists, Gustave Theodor Fritsch and

Gray's experiment is similar to accumulating electrons by combing your hair. If you hold a piece of paper near the comb, the paper will bend in its direction. Negative charges on the comb have pushed the paper's negative charges to its backside, leaving the front side positively charged. Because opposite charges attract, the paper bends toward the comb.

electrical stimulation Passage of an electrical current from the uninsulated tip of an electrode through tissue, resulting in changes in the electrical activity of the tissue.

⊕ THE BASICS
Electricity and Electrical Stimulation

Electricity powers the lights in your home and the batteries that run so many gadgets, including smart phones. *Electricity* is the flow of electrons from a body that contains a higher charge (more electrons) to a body that contains a lower charge (fewer electrons). This electron flow can perform work, such as lighting an unlit bulb. If biological tissue contains an electrical charge, the charge can be recorded; if tissue is sensitive to an electrical charge, the tissue can be stimulated.

How Electricity Works

In "Power Source," negatively charged electrons are attracted to the positive pole because opposite charges attract. The electrons on the negative pole have the potential to flow to the positive pole. This *electrical potential,* or electrical charge, is the ability to do work through the use of stored electrical energy.

Electrical charge is measured in *volts,* the difference in charge between the positive and the negative poles. The positive and negative poles in a battery, like the poles in each wall socket in your home, when not connected, have a voltage between the poles.

Electrical Activity in Cells

If the bare tip of an insulated wire, or *electrode,* from each pole of a battery is brought into contact with biological tissue, current will flow from the electrode connected to the negative pole into the tissue and then from the tissue into the electrode connected to the positive pole. The most intense stimulation comes from the tip of the electrode. Microelectrodes can record from or stimulate tissue as small as a single living cell.

Electrical stimulation, illustrated in part A of "Studying Electrical Activity in Animal Tissue," is most effective when administered in brief pulses. A timer in the stimulator turns the current on and off to produce the pulses. In electrical recording, voltage can be displayed by the dial on a voltmeter, a recording device that measures the voltage of a battery or of biological tissue (part B).

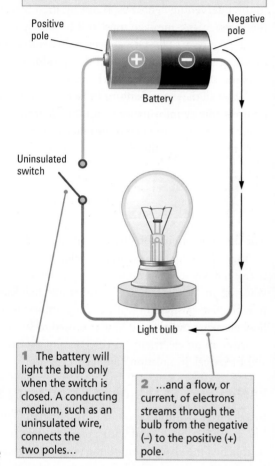

Because electrons are negatively charged, the negative pole has a higher electrical charge (more electrons) than the positive pole.

Positive pole

Negative pole

Battery

Uninsulated switch

1 The battery will light the bulb only when the switch is closed. A conducting medium, such as an uninsulated wire, connects the two poles…

2 …and a flow, or current, of electrons streams through the bulb from the negative (–) to the positive (+) pole.

Light bulb

Power Source

(A) Electrical stimulation

Current leaves the stimulator through a wire lead (red) that attaches to an electrode. From the uninsulated tip of the electrode, the current enters the tissue and stimulates it. The current flows back to the stimulator through a second lead (green) connected to a reference electrode.

1 A stimulating electrode delivers current (electrons) ranging from 2 to 10 millivolts, intensities sufficient to produce a response without damaging cells.

2 The reference electrode contacts a large surface area that spreads out the current and thus does not excite the tissue here.

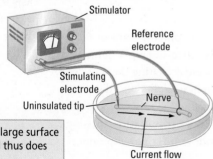

Stimulator

Reference electrode

Stimulating electrode

Uninsulated tip

Nerve

Current flow

(B) Electrical recording

The difference in voltage between the tip of a recording electrode and a reference electrode deflects a needle that indicates the current's voltage.

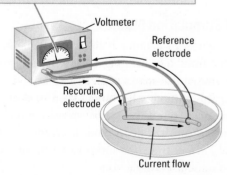

Voltmeter

Reference electrode

Recording electrode

Current flow

Studying Electrical Activity in Animal Tissue

Eduard Hitzig, demonstrated that electrical stimulation of the neocortex causes movement. They studied several animal species, including rabbits and dogs, and may even have stimulated the neocortex of a person whom they were treating for head injuries sustained on a Prussian battlefield. They observed movements of the arms and legs of their subjects in response to the stimulation of specific parts of the neocortex.

In 1874, Roberts Bartholow, a Cincinnati physician, wrote the first report describing the effects of human brain stimulation. His patient, Mary Rafferty, had a skull defect that exposed part of her neocortex. Bartholow stimulated her exposed brain tissue to examine the effects. In one of his observations he wrote:

> Passed an insulated needle into the left posterior lobe so that the non-insulated portion rested entirely in the substance of the brain. The reference was placed in contact with the dura mater. When the circuit was closed, muscular contraction in the right upper and lower extremities ensued. Faint but visible contraction of the left eyelid, and dilation of the pupils, also ensued. Mary complained of a very strong and unpleasant feeling of tingling in both right extremities, especially in the right arm, which she seized with the opposite hand and rubbed vigorously. Notwithstanding the very evident pain from which she suffered, she smiled as if much amused. (Bartholow, 1874)

As you might imagine, Bartholow's report was not well received. An uproar after its publication forced him to leave Cincinnati. Nevertheless, he had demonstrated that the brain of a conscious person could be stimulated electrically to produce movement of the body.

Electrical Recording Studies

Another, less invasive line of evidence that the flow of information in the brain is partly electrical in nature came from the results of electrical recording experiments. Richard Caton, a Scottish physician who lived in the early twentieth century, was the first to measure the electrical currents of the brain with a sensitive **voltmeter**, a device that measures the flow and the strength of electrical voltage by recording the difference in electrical potential between two bodies. Caton reported that, when he placed electrodes on the skull of a human subject, he could detect fluctuations in his voltmeter recordings. Today, this type of brain recording, the **electroencephalogram** (EEG), is a standard tool used to monitor sleep stages and record waking activity as well as to diagnose disruptions such as those that occur in epilepsy.

The results of all these studies provided evidence that neurons send electrical messages, but concluding that nerves and tracts carry conventional electrical currents proved problematic. Hermann von Helmholtz, a nineteenth-century German scientist, stimulated a nerve leading to a muscle and measured the time the muscle took to contract. The nerve conducted information at the rate of only 30 to 40 meters per second, whereas electricity flows along a wire at the much faster speed of light (3×10^8 meters per second).

The flow of information in the nervous system, then, is much too slow to be a flow of electricity. To explain the electrical signals of a neuron, Julius Bernstein suggested in 1886 that the chemistry of neurons produces an electrical charge. He also proposed that the charge can change and so act as a signal. Bernstein's idea was that successive waves of electrical change constitute the message conveyed by the neuron.

Notice that it is not the electrical *charge* but the *wave* that travels along the axon. To understand the difference, consider other kinds of waves. If you drop a stone into a pool of still water, the contact produces a wave that travels away from the site of impact, as shown in **Figure 4-2**. The water itself does not travel. Only the change in pressure moves, changing the height of the surface of the water and creating the wave effect.

Similarly, when you speak, you induce pressure waves in air, and these waves carry the "sound" of your voice to a listener. If you flick a towel, a wave

voltmeter Device that measures the flow and the strength of electrical voltage by recording the difference in electrical potential between two bodies.

electroencephalogram (EEG) Graph that records electrical activity through the skull or from the brain and represents graded potentials of many neurons.

In the twentieth century, the scientific community established ethical standards for research on human and nonhuman subjects (see Section 7-7), and brain stimulation became a standard part of many neurosurgical procedures, such as those described in Section 16-2.

Detail on these applications of the EEG appears in Sections 7-2, 13-3, and 16-3.

FIGURE 4-2 Wave Effect Waves created by dropping a stone into still water do not entail the forward movement of the water but rather differences in pressure that change the height of the surface of the water.

Young-Wolff/PhotoEdit

Section 9-1 explains how we perceive sounds.

travels to the other end of the towel. Just as waves through the air send a spoken message, Bernstein's idea was that waves of chemical change travel along an axon to deliver a neuron's message.

Tools for Measuring a Neuron's Electrical Activity

The waves that carry nervous-system messages are very small and are restricted to the surfaces of neurons. Still, we can measure these waves using electrical stimulation and electrical recording techniques and determine how they are produced. If a single axon is stimulated, it produces a wave of excitation, and if an electrode connected to a voltmeter is placed on a single axon, the electrode can detect a change in electrical charge on that axon's membrane as the wave passes, as illustrated in **Figure 4-3**.

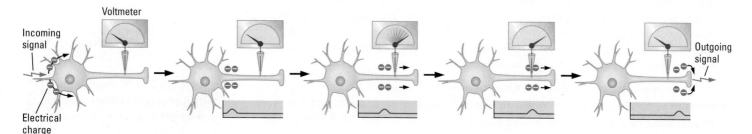

FIGURE 4-3 Wave of Information
Neurons can convey information as a wave induced by stimulation on the cell body traveling down the axon to its terminal. A voltmeter detects the passage of the wave.

As simple as this process may seem, recording a wave and explaining how it is produced requires a neuron large enough to record, a recording device sensitive enough to detect a small electrical impulse, and an electrode small enough to place on the surface of a single neuron. The fortuitous discovery of the giant axon of the squid, the invention of the oscilloscope, and the development of microelectrodes met all these requirements.

1 micrometer = one-millionth of a meter or one-thousandth of a millimeter.

Giant Axon of the Squid

The neurons of most animals, including humans, are tiny, on the order of 1 to 20 micrometers in diameter, too small to be seen by the eye and too small on which to perform experiments easily. The British zoologist J. Z. Young, when dissecting the North Atlantic squid, *Loligo vulgaris,* noticed that it has giant axons, as much as a millimeter (1000 micrometers) in diameter. **Figure 4-4** illustrates *Loligo* and the giant axons leading to its body wall, or mantle, which contracts to propel the squid through the water.

Loligo is not a giant squid. It is only about a foot long. But its axons are giants as axons go. Each axon is formed by the fusion of many smaller axons. Because larger axons send messages faster than smaller axons do, these giant axons allow the squid to jet propel away from predators.

In 1936, Young suggested to Alan Hodgkin and Andrew Huxley, two neuroscientists at Cambridge University in England, that *Loligo*'s axons were large enough to use for electrical recording studies. A giant axon could be dissected out of the squid and kept

FIGURE 4-4 Laboratory Specimen **(A)** The North Atlantic squid propels itself both with fins and by contracting its mantle to force water out for propulsion. **(B)** The stellate ganglion projects giant axons to contract the squid's mantle.

(A)

(B)

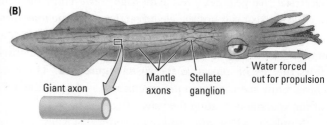

Giant axon   Mantle axons   Stellate ganglion   Water forced out for propulsion

functional in a bath of salty liquid that approximates body fluids. In this way, Hodgkin and Huxley described the neuron's electrical activity. In 1963 they received the Nobel Prize for their accomplishment.

Oscilloscope

Hodgkin and Huxley's experiments were made possible by the invention of the **oscilloscope,** a device that serves as a voltmeter sensitive enough to record the very small electrical signals from a nerve. You probably have seen one form of oscilloscope, a boxy analog television set that uses glass vacuum tubes. Today, the digital oscillocope (**Figure 4-5**A) has replaced cathode-ray oscilloscopes.

oscilloscope Device that serves as a sensitive voltmeter by registering the flow of electrons to measure voltage.

microelectrode A microscopic insulated wire or a salt-water-filled glass tube of which the uninsulated tip is used to stimulate or record from neurons.

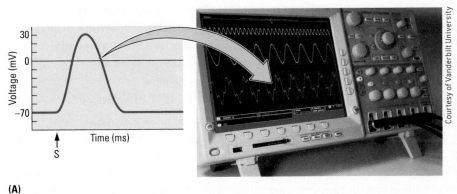

(A)

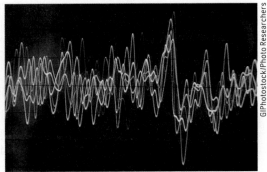

Courtesy of Vanderbilt University

GJPhotostock/Photo Researchers

(B)

As graphed at the left in Figure 4-5A, the scales used when recording the electrical charge from a nerve are millivolts (1 millivolt is 1/1000 of a volt) and milliseconds (1 millisecond is 1/1000 of a second). In an oscilloscope, an electron beam leaves a trace on a screen (Figure 4-5B). Deflections of the beam can be used to record voltage changes on an axon.

Microelectrodes

The final ingredient needed to measure a neuron's electrical activity is an electrode small enough to place on or into an axon—a **microelectrode.** Microelectrodes can deliver an electrical current to a single neuron or record from it. One way to make a microelectrode is to etch the tip of a piece of thin wire to a fine point of about 1 micrometer in size and insulate the rest of the wire. The tip is placed on or into the neuron, as shown in the left image in **Figure 4-6**A.

Microelectrodes can also be made from a thin glass tube tapered to a very fine tip (Figure 4-6A, right image). The tip of a glass microelectrode can be as small as 1 micrometer, even though it still remains hollow. When the glass tube is then filled with salty water, which provides the conducting medium through which an electrical current can travel, it acts as an electrode. A wire placed in the salt solution connects the electrode to a stimulation or recording device.

Microelectrodes are used to record from an axon in a number of different ways. Placing the tip

FIGURE 4-5 Oscilloscope Recording
(A) Basic wave shapes are displayed on a digital oscilloscope, a versatile electronic instrument used to visualize and measure electrical signals changing in time. To the left of the photo, in the graph of a trace produced by an oscilloscope, S stands for stimulation. The horizontal axis measures time, and the vertical axis measures voltage. The voltage of the axon is represented as −70 millivolts (mV). **(B)** Neuron traces displayed on a digital oscilloscope screen.

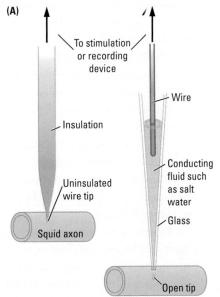

(A)

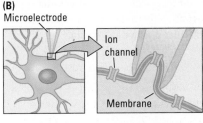

FIGURE 4-6 Uses of Microelectrodes
(A) A squid axon is larger than the tip of either a wire (*left*) or a glass (*right*) microelectrode. Both can be placed on an axon or into it. (Drawings are not to scale.) **(B)** A microelectrode can record from only a small area of an axon by suctioning the membrane up into the glass electrode.

diffusion Movement of ions from an area of higher concentration to an area of lower concentration through random motion.

concentration gradient Differences in concentration of a substance among regions of a container that allow the substance to diffuse from an area of higher concentration to an area of lower concentration.

voltage gradient Difference in charge between two regions that allows a flow of current if the two regions are connected.

"The Basics: Chemistry Review" on pages 88–89 covers ions. The "Salty Water" illustration there shows how water molecules dissolve salt crystals.

of a microelectrode on an axon provides an extracellular measure of the electrical current from a very small part of the axon. If a second microelectrode is used as the reference, one tip can be placed on the surface of the axon and the other inserted into the axon. This technique provides a measure of voltage across the cell membrane.

A still more refined use of a glass microelectrode is to place its tip on the neuron's membrane and apply a little back suction until the tip becomes sealed to a patch of the membrane, as shown in Figure 4-6B. This technique is analogous to placing the end of a soda straw against a piece of plastic wrapping and sucking to grasp the plastic. This method allows a recording to be made from only the small patch of membrane that is sealed within the perimeter of the microelectrode tip.

Using the giant axon of the squid, an oscilloscope, and microelectrodes, Hodgkin and Huxley recorded the electrical voltage on an axon's membrane and explained the *nerve impulse* as changes in ion concentration across the cell membrane. The basis of this electrical activity is the movement of intracellular and extracellular ions, which carry positive and negative charges. To understand Hodgkin and Huxley's results, you first need to understand the principles underlying the movement of ions.

How the Movement of Ions Creates Electrical Charges

The intracellular and extracellular fluids of a neuron are filled with various ions, including positively charged Na^+ (sodium) and K^+ (potassium) ions and negatively charged Cl^- (chloride) ions. These fluids also contain numerous negatively charged protein molecules (A^- for short). Positively charged ions are called *cations,* and negatively charged ions, including protein molecules, are called *anions.* Three factors influence the movement of anions and cations into and out of cells: diffusion, concentration gradient, and charge.

Because molecules move constantly, they spontaneously tend to spread out from where they are more concentrated to where they are less concentrated. This spreading out is **diffusion.** Requiring no work, diffusion results from the random motion of molecules as they move and bounce off one another to gradually disperse in a solution. When diffusion is complete, a dynamic equilibrium, with an equal number of molecules everywhere, is created.

Smoke from a fire gradually diffuses into the air of a room until every bit of air contains the same number of smoke molecules. Dye poured into water diffuses in the same way—from its point of contact to every part of the water in the container. Salts placed in water dissolve into ions surrounded by water molecules. Carried by the random motion of the water molecules, these ions diffuse throughout the solution to equilibrium, when every part of the container has exactly the same salt concentration.

Concentration gradient describes the relative concentration of a substance in space or in a solution. As illustrated in **Figure 4-7A**, when you drop a little ink into a beaker of water, the dye starts out concentrated at the site of contact and then diffuses. The ink spreads out from a point of higher concentration to points of lower concentration until it is equally distributed, and all the water in the beaker is the same color.

A similar process takes place when a salt solution is put into water. The salt concentration is initially high in the location where it enters the water, but it then diffuses from that location until its ions are in equilibrium. You are familiar with other kinds of gradients. A car parked on a hill will roll down the grade if the car is taken out of gear, a skier will slide down a mountain, and a dropped ball falls to the ground.

Because ions carry an electrical charge and like charges repel one another, ion movement can be described either by a concentration gradient, the difference in the number of ions between two regions, or by a **voltage gradient,** the difference in charge between two regions. Ions will move down a voltage gradient from an area of higher charge to

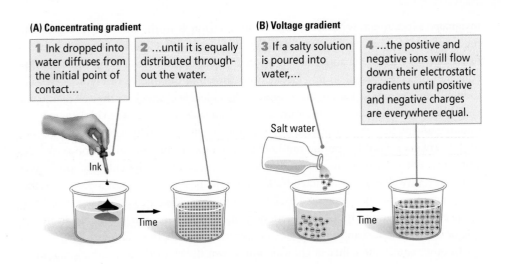

FIGURE 4-7 Moving to Equilibrium

(A) Concentrating gradient

1 Ink dropped into water diffuses from the initial point of contact...

2 ...until it is equally distributed throughout the water.

Ink

Time

(B) Voltage gradient

3 If a salty solution is poured into water,...

4 ...the positive and negative ions will flow down their electrostatic gradients until positive and negative charges are everywhere equal.

Salt water

Time

an area of lower charge, just as they move down a concentration gradient from an area of higher concentration to an area of lower concentration.

Figure 4-7B illustrates this process: when salt is dissolved in water, its diffusion can be described either as movement down a concentration gradient (for sodium and chloride ions) or movement down a voltage gradient (for the positive and negative charges). In a container, such as a beaker, that allows unimpeded movement of ions, the positive and negative charges eventually balance.

An imaginary experiment illustrates how a cell membrane influences the movement of ions. **Figure 4-8A** shows a container of water divided in half by a solid, impermeable membrane. If we place a few grains of salt (NaCl) in the left half of the container, the salt dissolves. The ions diffuse down their concentration and voltage gradients until the water in the left compartment is in equilibrium.

In the left side of the container, there is no longer a gradient for either sodium or chloride ions because the water everywhere is equally salty. There are no gradients for these ions on the other side of the container either because the solid membrane prevents the ions from entering that side. But there are concentration and voltage gradients for both sodium and chloride ions *across* the membrane—that is, from the salty side to the freshwater side.

Recall that protein molecules embedded in a cell membrane form channels that act as pores to allow certain kinds of ions to pass through the membrane. Returning to our

The cell membrane is impermeable to salty solutions because the salt ions, surrounded by water molecules, will not pass through the membrane's hydrophobic tails (review Figure 3-11).

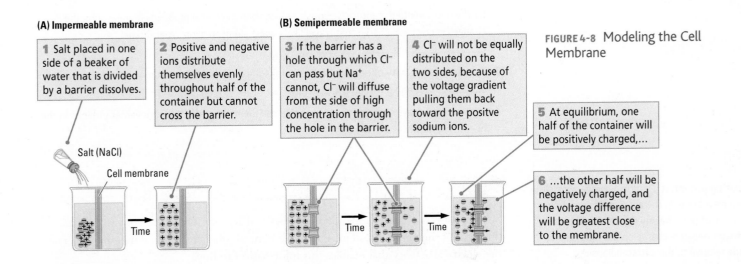

(A) Impermeable membrane

1 Salt placed in one side of a beaker of water that is divided by a barrier dissolves.

2 Positive and negative ions distribute themselves evenly throughout half of the container but cannot cross the barrier.

Salt (NaCl)

Cell membrane

Time

(B) Semipermeable membrane

3 If the barrier has a hole through which Cl⁻ can pass but Na⁺ cannot, Cl⁻ will diffuse from the side of high concentration through the hole in the barrier.

4 Cl⁻ will not be equally distributed on the two sides, because of the voltage gradient pulling them back toward the positve sodium ions.

Time   Time

5 At equilibrium, one half of the container will be positively charged,...

6 ...the other half will be negatively charged, and the voltage difference will be greatest close to the membrane.

FIGURE 4-8 Modeling the Cell Membrane

Dissolved sodium ions are smaller than chloride ions but have a greater tendency to stick to water molecules and so are bulkier and will not pass through a small chloride channel.

imaginary experiment, we place a few chloride channels in the membrane that divides the container of water, as illustrated at the left in Figure 4-8B. Chloride ions will now diffuse across the membrane and move down their concentration gradient on the side of the container that previously had no chloride ions, shown in the middle of Figure 4-8B. The sodium ions, in contrast, cannot cross through the chloride channels or the cell membrane.

If the only factor affecting the movement of chloride ions were the chloride concentration gradient, the efflux (outward flow) of chloride from the salty to the freshwater side of the container would continue until chloride ions were in equilibrium on both sides. But this is not what happens. Because opposite charges attract, the chloride ions, which carry a negative charge, are attracted back toward the positively charged sodium ions they left behind. Because they are pulled back toward the Na$^+$ ions, the Cl$^-$ ions cannot diffuse completely. Consequently, the concentration of chloride ions remains higher in the left side of the container than in the right, as illustrated at the right in Figure 4-8B.

In other words, the efflux of chloride ions down the chloride concentration gradient is counteracted by the influx (inward flow) of chloride ions down the chloride voltage gradient. At some point, equilibrium is reached: the concentration gradient of chloride ions on the right side of the beaker is balanced by the voltage gradient of chloride ions on the left. In brief:

$$\text{concentration gradient} = \text{voltage gradient}$$

At this equilibrium, there is a differential concentration of the chloride ions on the two sides of the membrane, the difference in ion concentration produces a difference in charge, so a voltage exists across the membrane. The left side of the container is positively charged because some chloride ions have migrated, leaving a preponderance of positive (Na$^+$) charges. The right side of the container is negatively charged because some chloride ions (Cl$^-$) have entered that chamber where no ions were before. The charge is highest on the surface of the beaker membrane, the area at which positive and negative ions accumulate. This is much the same as what happens in a real cell.

REVIEW 4-1

Searching for Electrical Activity in the Nervous System

Before you continue, check your understanding.

1. Experimental results obtained over hundreds of years from electrical _____ and, more recently, from electrical _____ implicated electrical activity in the nervous system's flow of information.

2. By the mid-twentieth century, scientists solved three technical problems in measuring the changes in electrical charge that travel like a wave along an axon's membrane: _____, _____, and _____.

3. The electrical activity of neuronal axons entails the diffusion of ions. Ions may move down a(n) _____ and down a(n) _____.

4. In what three ways does the semipermeable cell membrane affect the movement of ions in the nervous system?

Answers appear at the back of the book.

resting potential Electrical charge across the cell membrane in the absence of stimulation; a store of potential energy produced by a greater negative charge on the intracellular side relative to the extracellular side.

4-2 Electrical Activity of a Membrane

Specific aspects of the cell membrane's electrical activity interact to convey information throughout the nervous system. The movement of ions across neural membranes creates the electrical activity that enables this information to flow.

Resting Potential

Figure 4-9 graphs the voltage difference recorded when one microelectrode is placed on the outer surface of an axon's membrane and another is placed on its inner surface. In the absence of stimulation, the difference is about 70 millivolts. Although the charge on the outside of the membrane is actually positive, by convention it is given a charge of zero. Therefore, the inside of the membrane at rest is −70 millivolts *relative to* the extracellular side.

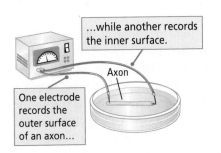

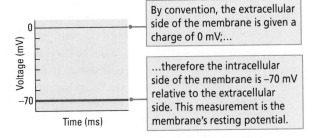

FIGURE 4-9 Resting Potential
The electrical charge across a resting cell membrane creates a store of potential energy.

If we were to continue to record for a long period of time, the charge across the unstimulated membrane would remain much the same. The charge can change, given certain changes in the membrane, but at rest, the difference in charge on the inside and outside of the membrane creates an electrical *potential*—the ability to use its stored power. The charge is thus a store of potential energy called the membrane's **resting potential.**

We might use the term "potential" in the same way to talk about the financial potential of someone who has money in the bank—the person can spend the money at some future time. The resting potential, then, is a store of energy that can be used at a later time. Most of your body's cells have a resting potential, but it is not identical on every axon. A resting potential can vary from −40 to −90 millivolts on axons of different animal species. The exact potential on an axon does not influence the neuron's ability to participate in generating brain activity.

Four charged particles take part in producing the resting potential: ions of sodium (Na^+) and potassium (K^+), chloride ions (Cl^-), and large protein molecules (A^-). These are the cations and anions defined in Section 4-1. As **Figure 4-10** shows, these charged particles are distributed unequally across the axon's membrane, with more protein anions and K^+ ions in the intracellular fluid and more Cl^- and Na^+ ions in the extracellular fluid. How do the unequal concentrations arise and how does each contribute to the resting potential?

Maintaining the Resting Potential

The cell membrane's channels, gates, and pumps maintain the resting potential. **Figure 4-11,** which shows the resting membrane close up, details how these three features of the cell membrane contribute to its resting charge:

1. Because the membrane is relatively impermeable, large negatively charged protein molecules remain inside the cell.

FIGURE 4-10 Ion Distribution Across the Resting Membrane The number of ions distributed across the resting cell membrane is unequal. Protein ions are represented by A^-.

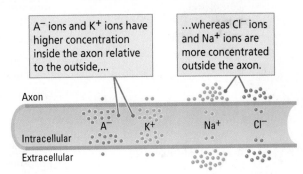

A^- ions and K^+ ions have higher concentration inside the axon relative to the outside,...

...whereas Cl^- ions and Na^+ ions are more concentrated outside the axon.

Axon

Intracellular

Extracellular

A^- K^+ Na^+ Cl^-

FIGURE 4-11 Maintaining the Resting Potential Channels, gates, and pumps in the cell membrane contribute to the transmembrane charge.

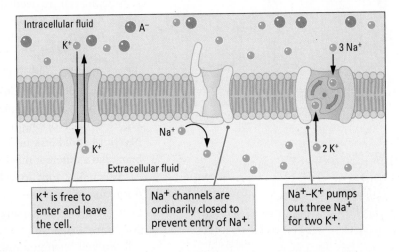

Intracellular fluid A^-

K^+

$3 Na^+$

Na^+

K^+

$2 K^+$

Extracellular fluid

K^+ is free to enter and leave the cell.

Na^+ channels are ordinarily closed to prevent entry of Na^+.

Na^+–K^+ pumps out three Na^+ for two K^+.

graded potential Small voltage fluctuation in the cell membrane restricted to the vicinity on the axon where ion concentrations change to cause a brief increase (hyperpolarization) or decrease (depolarization) in electrical charge across the cell membrane.

2. Ungated potassium (and chloride) channels allow K⁺ (and Cl⁻) ions to pass more freely, but gates on sodium channels keep out positively charged Na⁺ ions.

3. Na⁺–K⁺ pumps extrude Na⁺ from the intracellular fluid and inject K⁺.

Inside the Cell

Large protein anions are manufactured inside cells. No membrane channels are large enough to allow these proteins to leave the cell, and their negative charge alone is sufficient to produce a transmembrane voltage or resting potential. Because most cells in the body manufacture these large, negatively charged protein molecules, most cells have a charge across the cell membrane.

To balance the negative charge created by large protein anions in the intracellular fluid, cells accumulate positively charged K⁺ ions to the extent that about 20 times as many K⁺ ions cluster inside the cell as outside it. Potassium ions cross the cell membrane through open K⁺ channels, as shown in Figure 4-11. With this high concentration of K⁺ ions inside the cell, however, the potassium concentration gradient across the membrane limits the number of K⁺ ions entering the cell. In other words, not all the K⁺ ions that could enter do enter. Because the internal concentration of K⁺ ions is much higher than the external K⁺ concentration, potassium ions are drawn out of the cell by the potassium concentration gradient.

A few residual K⁺ ions on the outside of the membrane are enough to contribute to the charge across the membrane. They add to the negative charge on the intracellular side of the membrane relative to the extracellular side. You may be wondering whether you read the last sentence correctly. If there are 20 times as many positively charged K⁺ ions inside the cell as there are outside, why should the inside of the membrane have a negative charge? Should not all those K⁺ ions in the intracellular fluid give the inside of the cell a positive charge instead? No, because not quite enough K⁺ ions are able to enter the cell to balance the negative charge of the protein anions.

Think of it this way: if the number of K⁺ ions that could accumulate on the intracellular side of the membrane were unrestricted, the positively charged K⁺ ions inside would exactly match the negative charges on the intracellular protein anions. There would be no charge across the membrane at all. But there is a limit to the number of K⁺ ions that accumulate inside the cell because, when the intracellular K⁺ ion concentration becomes higher than the extracellular concentration, further K⁺ ion influx is opposed by the potassium concentration gradient.

Outside the Cell

The equilibrium of the potassium voltage gradient and the potassium concentration gradient results in some K⁺ ions remaining outside the cell. Only a few K⁺ ions staying outside the cell are needed to maintain a negative charge on the inner side of the membrane. As a result, K⁺ ions contribute to the charge across the membrane.

Sodium (Na⁺) and chloride (Cl⁻) ions also take part in producing the resting potential. If positively charged Na⁺ ions were free to move across the membrane, they would diffuse into the cell and eliminate the transmembrane charge produced by the unequal distribution of K⁺ ions inside and outside the cell. This diffusion does not happen because a gate on the sodium ion channels in the cell membrane is ordinarily closed (see Figure 4-11), blocking the entry of most Na⁺ ions. Still, given enough time, sufficient Na⁺ ions could leak into the cell to neutralize its membrane potential. The cell membrane has a different mechanism to prevent this neutralization from happening.

When Na⁺ ions do leak into the neuron, they are immediately escorted out again by the action of a *sodium–potassium pump*, a protein molecule embedded in the cell membrane. A membrane's many thousands of pumps continually exchange three intracellular Na⁺ ions for two K⁺ ions, as shown in Figure 4-11. The K⁺ ions are free to leave the cell through open potassium channels, but closed sodium channels slow the reentry of the Na$^+$ ions. In this way, Na$^+$ ions are kept out to the extent that about 10 times as many Na$^+$ ions reside on the extracellular side of the axon membrane as on the membrane's intracellular side. The difference in Na$^+$ concentrations also contributes to the membrane's resting potential.

Now consider the chloride ions. Unlike Na$^+$ ions, Cl$^-$ ions move in and out of the cell through open channels in the membrane. The equilibrium at which the chloride concentration gradient equals the chloride voltage gradient is approximately the same as the membrane's resting potential, and so Cl$^-$ ions ordinarily contribute little to the resting potential. At this equilibrium point, there are about 12 times as many Cl$^-$ ions outside the cell as inside it.

The cell membrane's semipermeability and the actions of its channels, gates, and pumps thus create a voltage across the cell membrane: its resting potential (**Figure 4-12**).

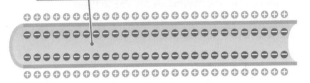

FIGURE 4-12 Resting Transmembrane Charge

Graded Potentials

The resting potential provides an energy store that can be used somewhat like the water in a dam, where small amounts can be released by opening gates for irrigation or to generate electricity. If the concentration of any of the ions across the unstimulated cell membrane changes, the membrane voltage changes. Conditions under which ion concentrations across the cell membrane change produce **graded potentials,** small voltage fluctuations that are restricted to the vicinity on the axon where ion concentrations change.

Just as a small wave produced by dropping a stone into the middle of a large, smooth pond decays before traveling very far, graded potentials produced on a cell membrane decay before traveling very far. But an isolated axon will not undergo a spontaneous change in charge. For a graded potential to arise, an axon must somehow be stimulated.

Stimulating an axon electrically through a microelectrode mimics the way in which membrane voltage changes to produce a graded potential in the living cell. If the voltage applied to the inside of the membrane is negative, the membrane potential increases in negative charge by a few millivolts. As illustrated in **Figure 4-13A**, it may change from a resting potential of -70 millivolts to a new, slightly greater potential of -73 millivolts.

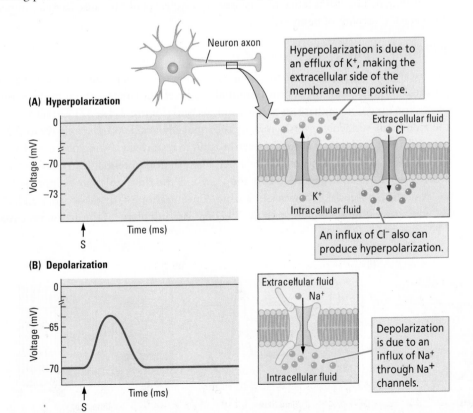

FIGURE 4-13 Graded Potentials
(A) Stimulation (S) that increases relative membrane voltage produces a hyperpolarizing graded potential. (B) Stimulation that decreases relative membrane voltage produces a depolarizing graded potential.

hyperpolarization Increase in electrical charge across a membrane, usually due to the inward flow of chloride or sodium ions or the outward flow of potassium ions.

depolarization Decrease in electrical charge across a membrane, usually due to the inward flow of sodium ions.

action potential Large, brief reversal in the polarity of an axon.

threshold potential Voltage on a neural membrane at which an action potential is triggered by the opening of Na⁺ and K⁺ voltage-sensitive channels; about −50 millivolts relative to extracellular surround.

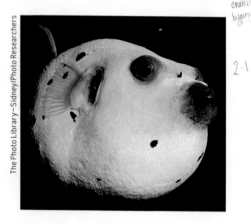

Pufferfish

The Photo Library–Sidney/Photo Researchers

This change is a **hyperpolarization** because the charge (polarity) of the membrane increases. Conversely, if positive voltage is applied inside the membrane, its potential decreases by a few millivolts. As illustrated in Figure 4-13B, it may change from a resting potential of −70 millivolts to a new, slightly lower potential of −65 millivolts. This change is a **depolarization** because the membrane charge decreases. Graded potentials are usually brief, lasting only milliseconds.

Hyperpolarization and depolarization typically take place on the soma (cell-body) membrane and on the dendrites of neurons. These areas contain channels that can open and close, causing the membrane potential to change as illustrated in Figure 4-13. Three channels—for potassium, chloride, and sodium ions—underlie graded potentials:

1. *Potassium channels* For the membrane to become hyperpolarized, its extracellular side must become more positive, which can be accomplished with an efflux of K⁺ ions. But if potassium channels are ordinarily open, how can a greater-than-normal efflux of K⁺ ions take place? Apparently, even though potassium channels are open, some resistance remains to the outward flow of K⁺ ions. Reducing this resistance enables hyperpolarization.

2. *Chloride channels* The membrane can also become hyperpolarized if there is an influx of Cl⁻ ions. Even though chloride ions can pass through the membrane, more ions remain on the outside than on the inside, so a decreased resistance to Cl⁻ flow can result in brief increases of Cl⁻ inside the cell.

3. *Sodium channels* Depolarization can be produced by an influx of sodium ions and is produced by the opening of normally closed gated sodium channels.

Evidence that potassium channels have a role in hyperpolarization comes from the fact that the chemical tetraethylammonium (TEA), which blocks potassium channels, also blocks hyperpolarization. The involvement of sodium channels in depolarization is indicated by the fact that the chemical tetrodotoxin, which blocks sodium channels, also blocks depolarization. The pufferfish, which is considered a delicacy in some countries, especially Japan, secretes this potentially deadly poison; so skill is required to prepare this fish for dinner. The fish is lethal to the guests of careless cooks because its toxin impedes the electrical activity of neurons.

Action Potential

Electrical stimulation of the cell membrane at resting potential produces localized graded potentials on the axon. An **action potential** is a brief but larger reversal in the polarity of an axon's membrane that lasts about 1 millisecond (**Figure 4-14A**). The voltage across the membrane suddenly reverses, making the intracellular side positive relative to the extracellular side, and then it abruptly reverses again to restore the resting potential. Because the duration of the action potential is brief, many action potentials can occur within a second, as illustrated in Figure 4-14B and C, where the time scales are compressed.

An action potential occurs when a large concentration of, first, Na⁺ ions and, then, K⁺ ions crosses the membrane rapidly. The depolarizing phase of the action potential is

FIGURE 4-14 Measuring Action Potentials **(A)** Phases of a single action potential. The time scales on the horizontal axes are compressed to chart **(B)** each action potential as a discrete event and **(C)** the ability of a membrane to produce many action potentials in a short time.

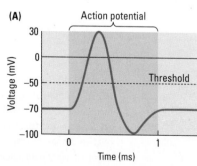
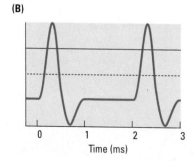
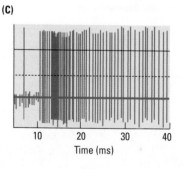

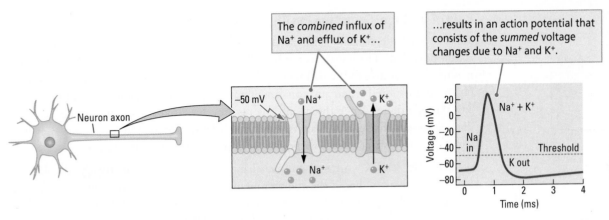

FIGURE 4-15 Triggering an Action Potential

due to Na¹ influx, and the hyperpolarizing phase to K^+ efflux. In short, Na^+ ions rush in and then K^+ ions rush out. As shown in **Figure 4-15**, the *combined* flow of Na^+ and K^+ ions underlies the action potential.

An action potential is triggered when the cell membrane is depolarized to about -50 millivolts. At this **threshold potential,** the membrane charge undergoes a remarkable further change with no additional stimulation. The relative voltage of the membrane drops to zero and then continues to depolarize until the charge on the inside of the membrane is as great as $+30$ millivolts—a total voltage change of 100 millivolts. Then the membrane potential reverses again, becoming slightly hyperpolarized—a reversal of a little more than 100 millivolts. After this second reversal, the membrane slowly returns to its resting potential at -70 millivolts.

The action potential normally consists of the summed current changes caused first by the inflow of Na^+ ions and then by the outflow of K^+ ions on an axon. Experimental results reveal that, if an axon membrane is stimulated to produce an action potential while the solution surrounding the axon contains the chemical TEA (to block potassium channels), a smaller-than-normal action potential due entirely to a Na^+ influx is recorded. Similarly, if an axon's membrane is stimulated to produce an action potential while the solution surrounding the axon contains tetrodotoxin (to block sodium channels), a slightly different action potential due entirely to the efflux of potassium is recorded. **Figure 4-16** illustrates these experimental results. Note that the graphs in Figure 4-16 represent ion flow, not voltage change.

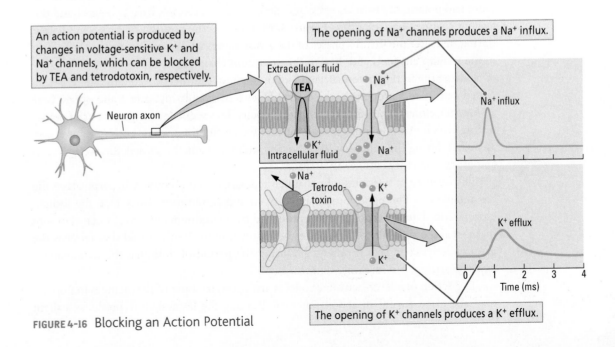

FIGURE 4-16 Blocking an Action Potential

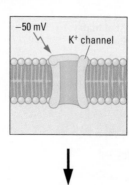

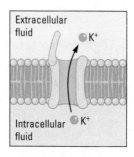

FIGURE 4-17 Voltage-Sensitive Potassium Channel

Exceptions do exist: some CNS neurons can discharge again during the repolarizing phase.

Role of Voltage-Sensitive Ion Channels

What cellular mechanisms underlie the movement of Na^+ and K^+ ions to produce an action potential? The answer is the behavior of a class of gated Na^+ and K^+ ion channels that are sensitive to the membrane's voltage (**Figure 4-17**). These **voltage-sensitive channels** are closed when an axon's membrane is at its resting potential, so ions cannot pass through them. When the membrane reaches threshold voltage, the configuration of the voltage-sensitive channels alters: they open briefly, enabling ions to pass through, then close again to restrict their flow. Following is the sequence of their actions:

1. Both Na^+ and K^+ voltage-sensitive channels are attuned to the threshold voltage of −50 millivolts. Thus, if the cell membrane changes to reach this voltage, both types of channels open to allow ion flow across the membrane.

2. The voltage-sensitive sodium channels are more sensitive than the potassium channels and so open first. As a result, the voltage change due to Na^+ ion influx takes place slightly before the voltage change due to K^+ ion efflux can begin.

3. The Na^+ channels have two gates. Once the membrane depolarizes to about +30 millivolts, its voltage triggers the closing of one of the Na^+ gates. Thus, Na^+ influx begins quickly and as quickly ends.

4. The K^+ channels open more slowly than the Na^+ channels and they remain open longer. Thus, the efflux of K^+ reverses the depolarization produced by Na^+ influx and even hyperpolarizes the membrane.

Action Potentials and Refractory Periods

There is an upper limit to how frequently action potentials occur. Sodium and potassium channels are responsible for this property of the action potential. If the axon membrane is stimulated during the depolarizing phase of the action potential, another action potential will not occur. Nor is the axon able to produce another action potential when it is repolarizing. During these times, the membrane is described as being **absolutely refractory.**

If on the other hand the axon membrane is stimulated during hyperpolarization, another action potential can be induced, but the stimulation must be more intense than that which initiated the first action potential. During this phase, the membrane is **relatively refractory.**

Refractory periods result from the way in which gates of the voltage-sensitive sodium and potassium channels open and close. Sodium channels have two gates, and potassium channels have one gate. **Figure 4-18** illustrates the position of these gates before, during, and after the various phases of the action potential. We will describe changes first in the sodium channels and then in the potassium channels.

During the resting potential, gate 1 of the sodium channel depicted in Figure 4-18 is closed; only gate 2 is open. At the threshold level of stimulation, gate 1 also opens. Gate 2, however, closes very quickly after gate 1 opens. This sequence produces a brief period during which both sodium gates are open. During the time when both gates are open as well as during the time that gate 2 of the sodium channel is closed, the membrane is absolutely refractory.

The opening of the potassium channels repolarizes and eventually hyperpolarizes the cell membrane. The potassium channels open and close more slowly than the sodium channels do. The hyperpolarization produced by a continuing efflux of potassium ions makes it more difficult to depolarize the membrane to the threshold that reopens the gates underlying an action potential. During the period of time that the membrane is hyperpolarizing, it is relatively refractory.

The action of a lever-activated toilet is analogous to some of the changes in polarity that take place during an action potential. Pushing the lever slightly produces a slight

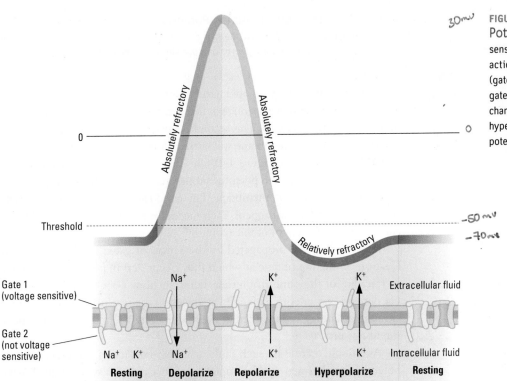

FIGURE 4-18 **Phases of an Action Potential** Initiated by changes in voltage-sensitive sodium and potassium channels, an action potential begins with a depolarization (gate 1 of the sodium channel opens and then gate 2 closes). The slower-opening potassium-channel gate contributes to repolarization and hyperpolarization until the resting membrane potential is restored.

flow of water, which stops when the lever is released. This activity is analogous to a graded potential. A harder lever press brings the toilet to threshold and initiates flushing, a response that is out of all proportion to the lever press. This activity is analogous to the action potential.

During the flush, the toilet is absolutely refractory: another flush cannot be induced at this time. During the refilling of the bowl, in contrast, the toilet is relatively refractory, meaning that reflushing is possible but harder to bring about. Only after the cycle is over and the toilet is once again "resting," can the usual flush be produced again.

Nerve Impulse

Suppose you place two recording electrodes at a distance from one another on an axon membrane and then electrically stimulate an area adjacent to one of these electrodes. That electrode would immediately record an action potential. A similar recording would register on the second electrode in a flash. Thus, an action potential has arisen near this second electrode also, even though it is some distance from the original point of stimulation.

Is this second action potential simply an echo of the first that passes down the axon? No, it cannot be, because the size and shape of the action potential are exactly the same at the two electrodes. The second is not just a faint, degraded version of the first but is equal in magnitude. Somehow the full action potential has moved along the axon. This propagation of an action potential along an axon is called a **nerve impulse**.

Why does an action potential move? Remember that the total voltage change during an action potential is 100 millivolts, far beyond the 20-millivolt change needed to bring the membrane from its resting state of −70 millivolts to the action potential threshold level of −50 millivolts. Consequently, the voltage change on the part of the membrane at which an action potential first occurs is large enough to bring adjacent parts of the membrane to a threshold of −50 millivolts.

voltage-sensitive channel Gated protein channel that opens or closes only at specific membrane voltages.

absolutely refractory Refers to the state of an axon in the repolarizing period during which a new action potential cannot be elicited (with some exceptions), because gate 2 of sodium channels, which is not voltage sensitive, is closed.

relatively refractory Refers to the state of an axon in the later phase of an action potential during which increased electrical current is required to produce another action potential; a phase during which potassium channels are still open.

nerve impulse Propagation of an action potential on the membrane of an axon.

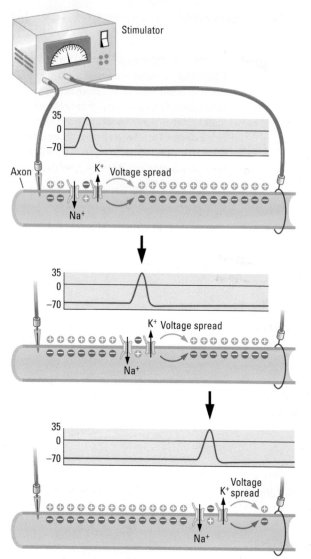

Stimulator

Axon

35
0
−70

K⁺ Voltage spread

Na⁺

35
0
−70

K⁺ Voltage spread

Na⁺

35
0
−70

Voltage
K⁺ spread

Na⁺

The domino effect

FIGURE 4-19 Propagating an Action Potential Voltage sufficient to open Na⁺ and K⁺ channels spreads to adjacent sites of the axon membrane, inducing voltage-sensitive gates to open. Here, voltage changes are shown on only one side of the membrane.

When the membrane of an adjacent part of the axon reaches −50 milli-volts, the voltage-sensitive channels at that location pop open to produce an action potential there as well. This second occurrence, in turn, induces a change in the voltage of the membrane still farther along the axon, and so on and on, down the axon's length. **Figure 4-19** illustrates this process. The nerve impulse occurs because each action potential propagates another action potential on an adjacent part of the axon membrane. The word *propagate* means "to give birth," and that is exactly what happens. Each successive action potential gives birth to another down the length of the axon.

Because they are propagated by the action of gated ion channels at the location on the membrane at which they occur, action potentials on a nerve or tract are of the same magnitude wherever they occur. An action potential depends on energy expended at the site where it occurs, and the same amount of energy is expended at every site along the membrane as a nerve impulse is propagated.

As a result, there is no such thing as a dissipated action potential: an action potential is either generated completely or it is not generated at all. A nerve impulse always maintains a constant size, and the action potential—the nerve's "message"—arrives unchanged to every terminal of the nerve that receives it.

Think of the voltage-sensitive channels along the axon as a series of dominoes. When one domino falls, it knocks over its neighbor, and so on down the line. There is no decrement in the size of the fall. The last domino travels exactly the same distance and falls just as hard as the first one did.

Essentially, this "domino effect" happens when voltage-sensitive channels open. The opening of one channel produces a voltage change that triggers its neighbor to open, just as one domino knocks over the next. The channel-opening response does not grow any weaker as it moves along the axon, and the last channel opens exactly like the first, just as the domino action stays constant to the end of the line.

Refractory Periods and Nerve Action

Refractory periods are determined by the position of the gates that mediate ion flow in the voltage sensitive channels. The refractory phase of the action potential has two practical uses for nerves conducting information. First, because of refractory periods, there is about a 5-millisecond limit on how frequently action potentials can occur. In other words, refractory periods limit the maximum rate of action potentials to about 200 per second. Variations in the sensitivity of voltage sensitive channels in different kinds of neurons likewise determine how frequently the neurons can fire.

Second, although an action potential can travel in either direction on an axon, refractory periods prevent it from reversing direction and returning to the point from which it came. Thus, refractory periods create a single, discrete impulse that travels away from the point of initial stimulation. When an action potential begins at the cell body, it usually travels down the axon to the terminals.

To return to our domino analogy, once a domino falls, it takes time to set it back up. This is its refractory period. Because each domino falls as it knocks down its neighbor, the sequence cannot reverse until the domino is set upright again. Thus, the dominos can fall in only one direction. The same principle determines the direction of the action potential.

Saltatory Conduction and the Myelin Sheath

Because the giant axons of squid are so large, they can transmit nerve impulses very quickly, much as a large-diameter pipe can deliver a lot of water at a rapid rate. But large axons take up substantial space, so a squid cannot accommodate many of them or its body would become too bulky. For us mammals, with our many axons producing repertoires of complex behaviors, giant axons are out of the question. Our axons must be extremely slender because our complex behaviors require a great many of them.

Our largest axons are only about 30 micrometers wide, so the speed with which they convey information should not be especially fast. And yet, like most vertebrate species, we humans are hardly sluggish creatures. We process information and generate responses with impressive speed. How do we manage to do so if our axons are so thin? The vertebrate nervous system has evolved a solution that has nothing to do with axon size.

Glial cells play a role in speeding nerve impulses in the vertebrate nervous system. Schwann cells in the human peripheral nervous system and oligodendroglia in the central nervous system wrap around each axon, forming the myelin that insulates it (**Figure 4-20**). Action potentials cannot occur where myelin is wrapped around an axon. For one thing, the myelin creates an insulating barrier to the flow of ionic current. For another, regions of an axon that lie under myelin have few channels through which ions can flow, and channels are essential to generating an action potential.

But axons are not totally encased in myelin. Unmyelinated gaps on the axon between successive glial cells are richly endowed with voltage-sensitive channels. These tiny gaps in the myelin sheath, the **nodes of Ranvier,** are sufficiently close to one another that an action potential occurring at one node can trigger the opening of voltage-sensitive gates at an adjacent node. In this way, a relatively slow action potential jumps at the speed of light from node to node, as shown in **Figure 4-21**. This flow of energy is called **saltatory conduction** (from the Latin verb *saltare,* meaning "to dance").

Jumping from node to node speeds the rate at which an action potential can travel along an axon. On larger, myelinated mammalian axons, nerve impulses can travel at a

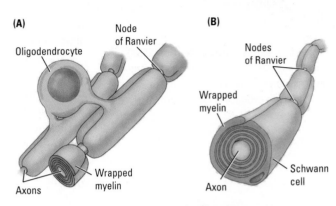

FIGURE 4-20 Myelination An axon is insulated by **(A)** oligodendroglia in the CNS and **(B)** Schwann cells in the PNS. Each glial cell is separated by a gap, or node of Ranvier.

You can review the appearance and function of all five types of glial cells in Table 3-1.

node of Ranvier The part of an axon that is not covered by myelin.

saltatory conduction Propagation of an action potential at successive nodes of Ranvier; saltatory means "jumping" or "dancing."

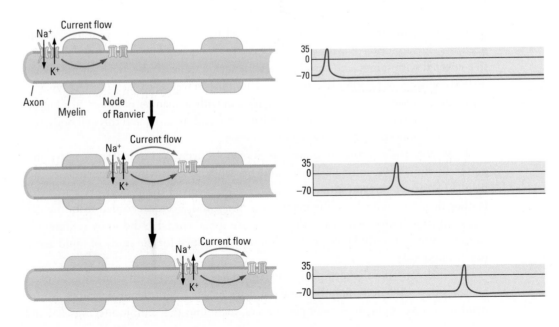

FIGURE 4-21 Saltatory Conduction Myelinated stretches of axon are interrupted by nodes of Ranvier, rich in voltage-sensitive channels. In saltatory conduction, the action potential jumps from node to node rapidly.

excitatory postsynaptic potential (EPSP)
Brief depolarization of a neuron membrane in response to stimulation, making the neuron more likely to produce an action potential.

rate as high as 120 meters per second. On smaller, uninsulated axons they travel only about 30 meters per second.

Spectators at sporting events once performed a "wave" that traveled around a stadium. As one person rose, the adjacent person rose, producing the wave effect. This human wave is like conduction along an unmyelinated axon. Now think of how much faster the wave would complete its circuit around the field if only spectators in the corners rose to produce it, which is analogous to a nerve impulse that travels by jumping from one node of Ranvier to the next. The quick reactions that humans and other mammals are capable of are due in part to this saltatory conduction in their nervous systems.

REVIEW 4-2
Electrical Activity of a Membrane

Before you continue, check your understanding.

1. The _____ results from the unequal distribution of _____ inside and outside the cell membrane.

2. Because it is _____, the cell membrane prevents the efflux of large protein anions and pumps sodium ions out of the cell to maintain a slightly _____ charge in the intracellular fluid relative to the extracellular fluid.

3. For a graded potential to arise, an axon must be stimulated to the point that the transmembrane charge increases slightly to cause a(n) _____ or decreases slightly to cause a(n) _____.

4. The voltage change associated with a(n) _____ is sufficiently large to stimulate adjacent parts of the axon membrane to the threshold for propagating it along the length of an axon as a(n) _____.

5. Briefly explain why nerve impulses travel faster on myelinated than on unmyelinated axons.

Answers appear at the back of the book.

4-3 How Neurons Integrate Information

A neuron is more than just an axon connected to microelectrodes by some curious scientist who stimulates it with electrical current. A neuron has an extensive dendritic tree covered with spines, and through these dendritic spines, it can establish more than 50,000 connections to other neurons. Nerve impulses traveling to each of these synapses from other neurons bombard the receiving neuron with all manner of inputs. In addition, a neuron has a cell body between its dendritic tree and its axon, and this cell body, too, can receive connections from many other neurons.

At the cellular level, the neurons that receive more than one kind of input sum up the information that they get.

How does the neuron integrate this enormous array of inputs into a nerve impulse? In the 1960s, John C. Eccles and his students performed experiments that helped to answer this question, for which Eccles received the Nobel Prize in 1963. Rather than recording from the giant axon of a squid, Eccles recorded from the cell bodies of large motor neurons in the vertebrate spinal cord. He did so by refining the electrical stimulating and recording techniques developed for the study of squid axons (see Section 4-1).

A spinal-cord motor neuron has an extensive dendritic tree with as many as 20 main branches that subdivide numerous times and are covered with dendritic spines. Motor neurons receive input from multiple sources, including the skin, joints, muscles, and brain, which is why they are ideal for studying how a neuron responds to diverse inputs. Each motor neuron sends its axon directly to a muscle. The motor neuron, then, is the path by which the nervous system produces behavior. Clinical Focus 4-2, "Myasthenia Gravis" on page 130, explains what happens when a motor neuron can no longer deliver messages to a muscle.

Excitatory and Inhibitory Postsynaptic Potentials

To study the activity of motor neurons, Eccles inserted a microelectrode into the spinal cord of a vertebrate animal until the tip was located in or right beside a motor neuron's cell body. He then placed stimulating electrodes on the axons of sensory nerve fibers entering the spinal cord. By teasing apart the fibers of the incoming sensory nerves, he was able to stimulate one fiber at a time.

Experiment 4-1 diagrams the experimental setup Eccles used. As shown at the left in the Procedures section, stimulating some incoming sensory fibers produced a depolarizing graded potential (reduced the charge) on the membrane of the motor neuron to which these fibers were connected. Eccles called these graded potentials **excitatory postsynaptic potentials** (EPSPs). As graphed at the left in the Results section, EPSPs

Figure 2-29 diagrams the spinal cord in cross section. Chapter 11 details the motor system and how we move.

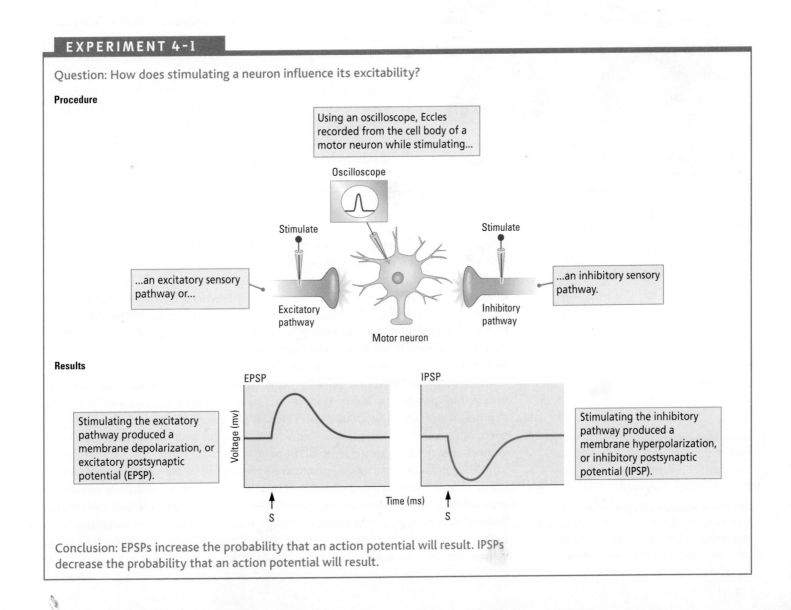

EXPERIMENT 4-1

Question: How does stimulating a neuron influence its excitability?

Procedure

Using an oscilloscope, Eccles recorded from the cell body of a motor neuron while stimulating...

Oscilloscope

Stimulate

Stimulate

...an excitatory sensory pathway or...

Excitatory pathway

...an inhibitory sensory pathway.

Inhibitory pathway

Motor neuron

Results

EPSP

IPSP

Stimulating the excitatory pathway produced a membrane depolarization, or excitatory postsynaptic potential (EPSP).

Voltage (mv)

Stimulating the inhibitory pathway produced a membrane hyperpolarization, or inhibitory postsynaptic potential (IPSP).

Time (ms)

S

S

Conclusion: EPSPs increase the probability that an action potential will result. IPSPs decrease the probability that an action potential will result.

Myasthenia Gravis

R. J. was 22 years old when she noticed that her eyelid drooped. In the course of the next few years, she experienced some difficulty in swallowing, general weakness in her limbs, and terrific fatigue.

Many of the symptoms would disappear for days and then suddenly reappear. About 3 years after the symptoms first appeared, she was diagnosed with myasthenia gravis, a condition that affects the communication between motor neurons and muscles.

In myasthenia gravis, the receptors of muscles are insensitive to the chemical messages passed from axon terminals. Na+ and K+ ions do not move through the end-plate pore, and the muscle does not receive the signal to contract. As a result, the muscles do not respond to commands from motor neurons.

Myasthenia gravis is rare, with a prevalence of 14/100,000, and the disorder is more common in women than in men. The age of onset is usually in the thirties or forties for women and after age 50 for men. In about 10 percent of cases, the condition is limited to the eye muscles, but, for the majority of patients, the condition gets worse.

At the time when R. J. contracted the disease, about a third of myasthenia gravis patients died from the disease or from complications such as respiratory infections. A specialist suggested that R. J. undergo a treatment in which the thymus gland is removed. Within the next 5 years, all her symptoms gradually disappeared, and she remained symptom free thereafter.

The thymus is an immune system gland that takes part in producing antibodies to foreign material and viruses that enter the body. In myasthenia gravis, the thymus may start to make antibodies to the end-plate receptors on muscles.

Myasthenia gravis is one of nearly 80 **autoimmune diseases,** disorders in which the immune system makes antibodies to a person's own body (Rezania et al., 2011). Others include neuromyalgia, multiple sclerosis, and diabetes. Contemporary treatments besides the removal of the thymus gland include drug treatments, such as those that increase the release of the chemical transmitter acetylcholine at muscle receptors. As a result, most myasthenia gravis conditions are controlled.

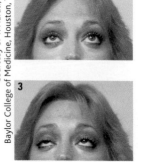

Courtesy of Y. Harati, M.D., Baylor College of Medicine, Houston, Texas

(1) This myasthenia gravis patient was asked to look up.
(2, 3) Her eyelids quickly become fatigued and droop.
(4) Her eyelids open normally after a few minutes rest.

reduce the charge on the membrane toward the threshold level and increase the probability that an action potential will result.

In contrast, as shown at the right in the Procedures section, when Eccles stimulated other incoming sensory fibers, he produced a hyperpolarizing graded potential (increased the charge) on the receiving motor-neuron membrane. Eccles called these graded potentials **inhibitory postsynaptic potentials** (IPSPs). As graphed at the right in the Results section, IPSPs increase the charge on the membrane away from the threshold level and decrease the probability that an action potential will result.

Both EPSPs and IPSPs last only a few milliseconds before they decay and the neuron's resting potential is restored. EPSPs are associated with the opening of sodium channels, which allows an influx of Na$^+$ ions. IPSPs are associated with the opening of potassium channels, which allows an efflux of K$^+$ ions (or with the opening of chloride channels, which allows an influx of Cl$^-$ ions).

Although the size of a graded potential is proportional to the intensity of the stimulation, an action potential is not produced on the motor neuron's cell-body membrane even when an EPSP is strongly excitatory. The reason is simple: the cell-body membrane of most neurons does not contain voltage-sensitive channels. The stimulation must reach the axon hillock, the area of the cell where the axon begins. The hillock is rich in voltage-sensitive channels.

autoimmune disease Illness resulting from the loss of the immune system's ability to discriminate between foreign pathogens in the body and the body itself.

inhibitory postsynaptic potential (IPSP) Brief hyperpolarization of a neuron membrane in response to stimulation, making the neuron less likely to produce an action potential.

Summation of Inputs

Remember that a motor neuron has myriad dendritic spines, allowing for myriad inputs to its membrane, both EPSPs and IPSPs. How do these incoming graded potentials interact? For example, what happens if there are two EPSPs in succession? Does it matter if the time between them is increased or decreased? And what happens when an EPSP and an IPSP arrive together?

Temporal Summation

If one excitatory pulse of stimulation is delivered and is followed some time later by a second excitatory pulse, one EPSP is recorded and after a delay, a second identical EPSP is recorded, as shown at the top left in **Figure 4-22.** These two widely spaced EPSPs are independent and do not interact. If the delay between them is shortened so that the two occur in rapid succession, however, a single, large EPSP is produced, as shown in the left-center panel of Figure 4-22.

Here, the two excitatory pulses are summed (added together to produce a larger depolarization of the membrane than either would induce alone). This relation between two EPSPs occurring close together or even at the same time (bottom left panel) is called **temporal summation.** The right side of Figure 4-22 illustrates that equivalent results are obtained with IPSPs. Therefore, temporal summation is a property of both EPSPs and IPSPs.

Spatial Summation

What happens when inputs to the cell body's membrane are located close together or are spaced far apart? By using two recording electrodes (R_1 and R_2) we can see the effects of spatial relations on the summation of inputs.

If two EPSPs are recorded at the same time but on widely separated parts of the membrane (**Figure 4-23**A), they do not influence one another. If two EPSPs occurring close together in time are also located close together, however, they sum to form a larger EPSP (Figure 4-23B). This **spatial summation** indicates that two separate inputs occurring very close to one another on the cell membrane in time sum. Similarly, two IPSPs produced at the same time sum if they occur at approximately the same place on the cell-body membrane but not if they are widely separated.

temporal summation Graded potentials that occur at approximately the same time on a membrane are summed.

spatial summation Graded potentials that occur at approximately the same location and time on a membrane are summed.

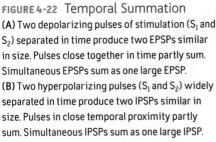

FIGURE 4-22 Temporal Summation **(A)** Two depolarizing pulses of stimulation (S_1 and S_2) separated in time produce two EPSPs similar in size. Pulses close together in time partly sum. Simultaneous EPSPs sum as one large EPSP. **(B)** Two hyperpolarizing pulses (S_1 and S_2) widely separated in time produce two IPSPs similar in size. Pulses in close temporal proximity partly sum. Simultaneous IPSPs sum as one large IPSP.

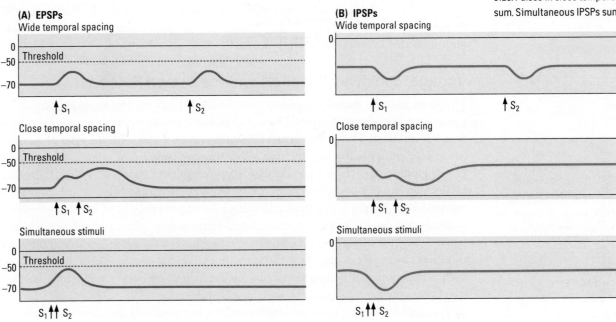

FIGURE 4-23 Spatial Summation
Illustrated here is the process for EPSPs; the process for IPSPs is equivalent.

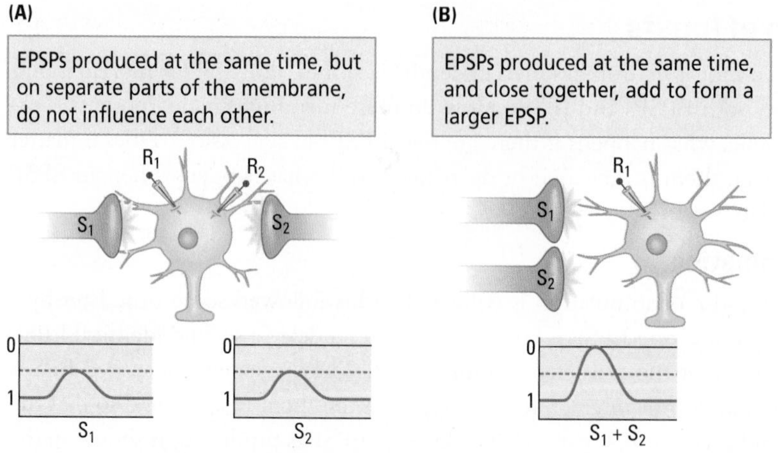

(A)

EPSPs produced at the same time, but on separate parts of the membrane, do not influence each other.

(B)

EPSPs produced at the same time, and close together, add to form a larger EPSP.

FIGURE 4-24 Triggering an Action Potential If the summated graded potentials—the EPSPs and IPSPs—on the dendritic tree and cell body of a neuron charge the membrane to threshold level at the axon hillock, an action potential travels down the axon membrane.

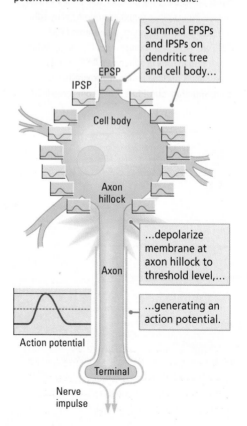

Role of Ions in Summation

Summation is a property of both EPSPs and IPSPs in any combination. The interactions between EPSPs and IPSPs make sense when you consider that the influx and the efflux of ions are being summed. The influx of sodium ions accompanying one EPSP is added to the influx of sodium ions accompanying a second EPSP if the two occur close together in time and space. If the two influxes of sodium ions are remote in time or in space or in both, no summation is possible.

The same is true regarding effluxes of potassium ions. When they occur close together in time and space, they sum; when they are far apart in either or both of these ways, there is no summation. The patterns are identical for an EPSP and an IPSP. The influx of sodium ions associated with the EPSP is added to the efflux of potassium ions associated with the IPSP, and the difference between them is recorded as long as they are spatially and temporally close together. If, on the other hand, they are widely separated in time or in space or in both, they do not interact and there is no summation.

A neuron with thousands of inputs responds no differently from one with only a few inputs. It democratically sums all inputs that are close together in time and space. The cell-body membrane, therefore, always indicates the summed influences of multiple inputs. Because of this temporal and spatial summation, a neuron can be said to analyze its inputs before deciding what to do. The ultimate decision is made at the axon hillock, the region that initiates the action potential.

Voltage-Sensitive Channels and the Action Potential

The axon hillock, shown emanating from the cell body in **Figure 4-24,** is rich in voltage-sensitive channels. These channels, like those on the squid axon, open at a particular membrane voltage. The actual threshold voltage varies with the type of neuron, but to keep things simple, we will stay with a threshold level of −50 millivolts.

To produce an action potential, the summed graded potentials—the IPSPs and EPSPs—on the cell-body membrane must depolarize the membrane at the axon hillock to −50 millivolts. If that threshold voltage is obtained only briefly, voltage-sensitive channels open, and just one or a few action potentials may occur. If the threshold level is maintained for a longer period, however, action potentials will follow one another in rapid succession, just as quickly as the gates on the voltage-sensitive channels can reset. Each action potential is then repeatedly propagated to produce a nerve impulse that travels from the axon hillock down the length of the axon.

Neurons often have extensive dendritic trees, but dendrites and dendritic branches do not have many voltage-sensitive channels and ordinarily do not produce action potentials. And distant branches of dendrites may have less influence in producing action potentials initiated at the axon hillock than do the more proximal branches of the dendrites. Consequently, inputs close to the axon hillock are usually much more dynamic in their influence than those occurring some distance away. Those distant inputs usually have a modulating effect. As in all democracies, some inputs have more say than others (Debanne, 2011).

The Versatile Neuron

Dendrites collect information in the form of graded potentials (EPSPs and IPSPs), and the axon hillock initiates discrete action potentials that are delivered to other target cells via the axon. But exceptions to this picture of how a neuron works do exist. For example, some cells in the hippocampus can produce additional action potentials, called *depolarizing potentials,* when the cell would ordinarily be refractory. The function of depolarizing potentials is not fully understood.

The hippocampus, a structure in the subcortical limbic system, controls some aspects of memory (see Figure 2-25).

A typical neuron does not initiate action potentials on its dendrites because the cell-body membrane does not contain voltage-sensitive channels. In some neurons, however, voltage-sensitive channels on dendrites do enable action potentials. The reverse movement of an action potential from the axon hillock into the dendritic field of a neuron is called **back propagation.** Back propagation signals the dendritic field that the neuron is sending an action potential over its axon and may play a role in plastic changes in the neuron that underlie learning. For example, back propagation may make the dendritic field refractory to incoming inputs, set the dendritic field to an electrically neutral baseline, or reinforce signals coming in to certain dendrites (Legenstein and Maass, 2011).

We explore the neuronal basis of learning in Sections 5-4 and 14-4.

Additionally, the neurons of some nonmammalian species do not have dendritic branches. And some species' ion channels respond to light rather than to voltage changes. The many differences among neurons suggest that the nervous system capitalizes on modifications of structure and function to produce adaptive behavior in each species. These variations do not exhaust the adaptability of neuronal mechanisms, because neuroscientists have engineered some of their own adaptions, as described in Research Focus 4-3, "Optogenetics and Light-Sensitive Channels," on page 134.

Section 7-1 describes the promise of optogenetics for neuroscience research and for clinical applications.

REVIEW 4-3

How Neurons Integrate Information

Before you continue, check your understanding.

1. Graded potentials that decrease the charge on the cell membrane, moving it toward the threshold level, are called _____ because they increase the likelihood that an action potential will occur. Graded potentials that increase the charge on the cell membrane, moving it away from the threshold level, are called _____ because they decrease the likelihood that an action potential will result.

2. EPSPs and IPSPs that occur close together in _____ and/or in _____ are summed. This is how a neuron _____ the information it receives from other neurons.

3. The membrane of the _____ does not contain voltage-sensitive ion channels, but if summed inputs excite the _____ to a threshold level, action potentials are triggered and then propagated as they travel along the cell's _____ as a nerve impulse.

4. Explain what happens during back propagation.

Answers appear at the back of the book.

back propagation Reverse movement of an action potential into the dendritic field of a neuron; postulated to play a role in plastic changes that underlie learning.

Optogenetics and Light-Sensitive Ion Channels

Membrane channels that are responsive to light have been discovered in nonmammalian animal species. Using the transgenic technique of **optogenetics,** researchers have successfully introduced light-sensitive channels into worms, fruit flies, and mice.

Optogenetics combines genetics and light to control targeted cells in living tissue. Here we examine how introducing different light-sensitive channels into a species excites the organism's movements with one light wavelength and inhibits them with another light wavelength.

One class of light-activated ion channels in the green algae, *Chlamydomonas reinhardtii,* is channelrhodopsin-2. The ChR2 channel absorbs blue light and in doing so, opens briefly to allow the passage of cations, including sodium and potassium ions. These light-sensitive channels allow the passage of Na$^+$ and K$^+$ ions when a cell is illuminated with blue light. The resulting depolarization excites the cell to generate action potentials.

Halorhodopsin (NpHR) is a light-driven ion pump, specific for chloride ions and found in phylogenetically ancient bacteria (archaea) known as halobacteria. When illuminated with green-yellow light, the halorhodopsin pumps Cl$^-$ anions into the cell, hyperpolarizing it and so inhibiting its activity.

The movements of worms, fruit flies, and mice with genetically introduced light-sensitive channels have been controlled when their nervous-system cells have been illuminated with appropriate wavelengths of light. One such species, *Caenorhabditis elegans,* is a roundworm about 1 millimeter long that lives in soil.

C. elegans is popular in neuroscience experiments because it is transparent and has a simple nervous system. It is also the first species to have all its neurons and synapses and its genome described. The accompanying illustration shows **(A)** the normal movements of *C. elegans* and diagrams the light-sensitive membrane channel responses that **(B)** excite or **(C)** inhibit movement.

Using optogenetic techniques, light-sensitive channels can be incorporated into specific neural circuits so that only a subset of neurons is controlled by light stimulation. The promise is that investigating specific neuron populations can provide insight into brain disease, including conditions such as addiction (Cao et al., 2011).

Also encouraging are the results of a study suggesting that the impaired vision suffered by people due to the loss of the light-sensitive retina of the eye could be restored with light-sensitive channels (Fenno et al., 2011).

(A) *C. elegans*

Carolina Biological/
Visuals Unlimited

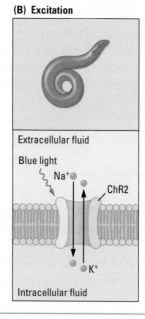

(B) Excitation

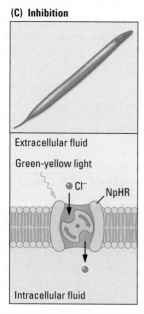

(C) Inhibition

Light-Sensitive Channels

(A) Normal movements of *C. elegans.* **(B)** When light-sensitive channelrhodopsin-2 ion channels are introduced into its neurons, *C. elegans* becomes active and coils when exposed to blue light. **Adapted from Zhang et al., 2007. (C)** With light-sensitive halorhodopsin ion pumps introduced into its muscles, *C. elegans* elongates and becomes immobile when exposed to green-yellow light. **Adapted from Liewald et al., 2008.**

4-4 Into the Nervous System and Back Out

The nervous system allows us to respond to afferent sensory stimuli by detecting them and sending messages about them to the brain. The brain interprets the information, triggering efferent responses that contract muscles and produce behavior. Until now, we have been dealing only with the middle of this process—how neurons convey information to one another, integrate the information, and generate action potentials. Now we explore the beginning and end of the journey.

To fill in the missing pieces, we explain how a sensory stimulus initiates a nerve impulse and how a nerve impulse produces a muscular contraction. You will see that ion channels are again important and that these channels are modifications of those described so far.

How Sensory Stimuli Produce Action Potentials

We receive information about the world through bodily sensations (touch and balance), auditory sensations (hearing), visual sensations (sight), and chemical sensations (taste and olfaction). Each sensory modality has one or more separate functions. For example, the body senses include, in addition to touch, pressure, joint sense, pain, and temperature. Receptors for audition and balance are modified touch receptors. The visual system has receptors for light and for different colors. And taste and olfactory senses respond to myriad chemical compounds.

To process all these different kinds of sensory inputs requires a remarkable array of different sensory receptors. But one thing that neurons related to these diverse receptors have in common is the presence of ion channels on their cell membranes. These ion channels initiate the chain of events that produces a nerve impulse.

Touch provides an example. Each hair on the human body allows us to detect even a very slight displacement. You can demonstrate this sensitivity to yourself by selecting a single hair on your arm and bending it. If you are patient and precise in your experimentation, you will discover that some hairs are sensitive to displacement in one direction only, whereas others respond to displacement in any direction. What enables this very finely tuned sensitivity?

The dendrite of a touch neuron is wrapped around the base of each hair. As shown in **Figure 4-25,** when a hair is mechanically displaced, as when you bend it, the encircling dendrite is stretched. The displacement opens **stretch-sensitive channels** in the dendrite's membrane. When these channels open, they allow an influx of Na$^+$ ions sufficient to depolarize the dendrite to its threshold level. At threshold, the voltage-sensitive sodium and potassium channels are activated to open and initiate a nerve impulse that conveys touch information to your brain.

Other kinds of sensory receptors have similar mechanisms for *transducing* (transforming) the energy of a sensory stimulus into nervous-system activity. When displaced, the *hair receptors* that provide information about hearing and balance likewise activate stretch-sensitive channels. In the visual system, light particles strike chemicals in receptors within the eye, and the resulting chemical change activates ion channels in the membranes of *relay neurons.* An odorous molecule in the air that lands on an olfactory receptor and fits itself into a specially shaped compartment opens chemical-sensitive ion channels. When tissue is damaged, injured cells release chemicals that activate channels on a pain nerve. The point here is that, in all our sensory systems, ion channels begin the process of information conduction.

optogenetics Transgenic technique that combines genetics and light to control targeted cells in living tissue.

stretch-sensitive channel Ion channel on a tactile sensory neuron that activates in response to stretching of the membrane, initiating a nerve impulse.

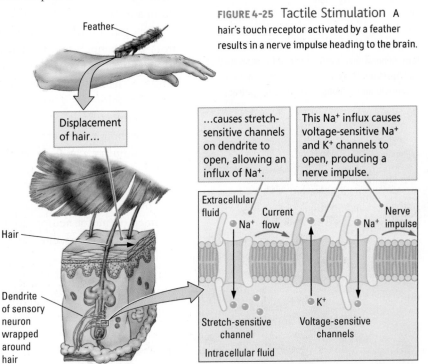

FIGURE 4-25 Tactile Stimulation A hair's touch receptor activated by a feather results in a nerve impulse heading to the brain.

Later chapters detail how sensory receptors transduce energy from the external world into action potentials:

Sensation, perception, and vision in Sections 9-1 and 9-2

Hearing in Section 10-1

Touch, pain, and balance in Section 11-4

Smell and taste in Section 12-2

end plate On a muscle, the receptor–ion complex that is activated by the release of the neurotransmitter acetylcholine from the terminal of a motor neuron.

transmitter-sensitive channel Receptor complex that has both a receptor site for a chemical and a pore through which ions can flow.

Sections 5-2 and 5-3 describe the varieties of chemical transmitters, including acetylcholine, and how they function.

FIGURE 4-26 Muscle Contraction **(A)** A motor-neuron's axon collaterals contact muscle end plates in this microscopic view. Dark patches are end plates, axon terminal buttons are not visible. **(B)** When an axon terminal contacts an end plate, **(C)** acetylcholine attaches to receptor sites on transmitter-sensitive channels, opening them. These large membrane channels allow simultaneous influx of Na+ ions and efflux of K+ ions, generating current sufficient to activate voltage-sensitive channels, triggering action potentials, and causing the muscle to contract.

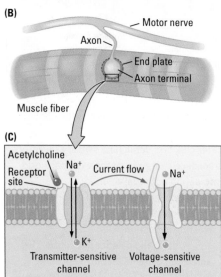

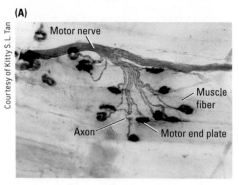

How Nerve Impulses Produce Movement

What happens at the end of the neural journey? How, after sensory information has traveled to the brain and been interpreted, is a behavioral response that includes the contraction of muscles generated? Behavior, after all, is movement, and for movement to take place, muscles must contract. Motor neurons in the spinal cord are responsible for activating muscles. Without them movement becomes impossible and muscles atrophy, as described in Clinical Focus 4-4, "Lou Gehrig's Disease."

Motor neurons send nerve impulses to synapses on muscle cells. The synapses are instrumental in making the muscle contract. The axon of each motor neuron makes one or a few synapses with its target muscle, similar to those that neurons make with one another **(Figure 4-26)**. The part of the muscle membrane that is contacted by the axon terminal is a specialized area called an **end plate,** shown in Figure 4-26A and B. The axon terminal releases onto the end plate a chemical transmitter called *acetylcholine*.

Acetylcholine does not enter the muscle but rather attaches to **transmitter-sensitive channels** on the end plate (Figure 4-26C). When these channels open in response, they allow a flow of Na+ and K+ ions across the muscle membrane sufficient to depolarize the muscle to the threshold for its action potential. Yes, muscles generate action potentials in order to contract. At this threshold, adjacent voltage-sensitive channels open. They, in turn, produce an action potential on the muscle fiber, as they do in a neuron.

The transmitter-sensitive channels on muscle end plates are somewhat different from the channels on axons and dendrites. A single end-plate channel is larger than two sodium and two potassium channels on a neuron combined. So when transmitter-sensitive channels open, they allow both Na+ influx and K+ efflux through the same pore. Therefore, to generate a sufficient depolarization on the end plate to activate neighboring voltage-sensitive channels requires the release of an appropriate amount of acetylcholine.

In summary, a wide range of neural events can be explained by the actions of membrane channels. Some channels generate the transmembrane charge. Others mediate graded potentials. Still others trigger the action potential. Sensory stimuli activate channels on neurons to initiate a nerve impulse, and the nerve impulse eventually activates channels on motor neurons to produce muscle contractions.

These various channels and their different functions probably evolved over a long period of time in the same way that new species of animals and their behaviors evolve. We have not described all the different ion channels that neural membranes possess, but you will learn about some additional ones in subsequent chapters.

REVIEW 4-4
Into the Nervous System and Back Out

Before you continue, check your understanding.

1. Different sensory stimuli initiate nerve impulses for each _____ in a similar way.

2. The membrane of a(n) _____ contains a mechanism for transducing sensory energy into changes in ion channels that, in turn, allow ion flow to alter the voltage of the membrane to the point that _____ channels open, initiating a nerve impulse.

3. Sensory stimuli activate ion channels to initiate a nerve impulse that activates channels on _____ neurons that in turn contract _____.

4. Why have so many different kinds of ion channels evolved on cell membranes?

Answers appear at the back of the book.

CLINICAL FOCUS ⊕ 4-4

Lou Gehrig's Disease

Baseball legend Lou Gehrig played for the New York Yankees from 1923 until 1939. He was a member of numerous World Series championship teams; set a host of individual records, some of which still stand today; and was immensely popular with fans, who knew him as the "Iron Man." His record of 2130 consecutive games was untouched until 1990, when Cal Ripkin, Jr., played his 2131st consecutive game.

In 1938, Gehrig started to lose his strength. In 1939, he played only eight games and then retired from baseball. Shortly afterward he was diagnosed with amyotrophic lateral sclerosis (ALS), a disorder first described by French physician Jean-Martin Charcot in 1869. Once Lou Gehrig developed ALS, however it became known as Lou Gehrig's disease. Gehrig died in 1941 at the age of 38.

There are about 5000 new cases in the United States each year. ALS strikes most commonly between the ages of 50 and 75, although its onset can be as early as the teenage years. About 10 percent of victims have a family history of the disorder. The disease begins with general weakness, at first in the throat or upper chest and in the arms and legs. Gradually, walking becomes difficult and falling common.

The patient may lose use of the hands and legs, have trouble swallowing, and have difficulty speaking. The disease does not usually affect any sensory systems, cognitive functions, bowel or bladder control, or even sexual function. Death often occurs within 5 years of diagnosis.

ALS is due primarily to the death of spinal motor neurons but can affect brain neurons as well. The technical term, amyotrophic lateral sclerosis, describes its consequences, both on muscles (*amyotrophic* means "muscle weakness") and on the spinal cord (*lateral sclerosis* means "hardening of the lateral spinal cord").

Several theories have been advanced to explain why motor neurons suddenly start to die in ALS victims. Recent evidence suggests that ALS can result from head trauma that activates the cell's DNA to produce

Lou Gehrig in his prime, jumping over Yankee teammate Joe DiMaggio's bat.

signals that initiate the neuron's death (*apoptosis*, or programmed cell death). Genetic factors are also suspected (Allen et al., 2011). At the present time, there is no cure for ALS, although some newly developed drugs appear to slow its progression and offer some hope for future treatments.

SUMMARY

4-1 Searching for Electrical Activity in the Nervous System

Electrical-stimulation studies, dating far back as the eighteenth century, show that stimulating a nerve with electrical current induces a muscle contraction. More recently, electrical-recording studies, in which the brain's electrical current is measured with a voltmeter, show that electrical activity is continually taking place within the nervous system.

To measure the electrical activity of a single neuron, researchers used giant axons of the squid. They recorded small, rapid electrical changes with an oscilloscope through microelectrodes that they could place on or into the cell.

The electrical activity of neurons is generated by the flow of electrically charged ions across the cell membrane. These ions flow both down a concentration gradient (from an area of relatively high concentration to an area of lower concentration) and down a voltage gradient (from an area of relatively high voltage to an area of lower voltage). The distribution of ions is also affected by the opening and closing of ion channels in neural cell membranes.

4-2 Electrical Activity of a Membrane

The neuron's resting potential results from an unequal distribution of ions on a membrane's two sides, with the intracellular side registering about −70 millivolts relative to the extracellular side. Negatively charged protein anions are too large to leave the neuron, and the cell membrane actively pumps out positively charged sodium ions. In addition, unequal distributions of potassium cations and chloride anions contribute to the resting potential.

Graded potentials result when the neuron is stimulated because ion channels in the membrane are affected, which in turn changes the distribution of ions across the membrane, suddenly increasing or decreasing the transmembrane voltage by a small amount. A slight increase in the voltage is called hyperpolarization, whereas a slight decrease is called depolarization.

An action potential is a brief but large change in the polarity of an axon membrane that is triggered when the transmembrane voltage drops to a threshold level of about −50 millivolts. For an action potential, the transmembrane voltage suddenly reverses (with the intracellular side becoming positive relative to the extracellular side) and then abruptly reverses again, after which the resting potential is gradually restored. The membrane changes are due to the behavior of voltage-sensitive channels—sodium and potassium channels that are sensitive to the membrane's voltage.

When an action potential is triggered at the axon hillock, it can propagate along the axon as a nerve impulse. Nerve impulses travel more rapidly on myelinated axons because of saltatory conduction: the action potentials jump between the nodes separating the glial cells that form the axon's myelin sheath.

4-3 How Neurons Integrate Information

The inputs to neurons from other cells can produce both excitatory postsynaptic potentials and inhibitory postsynaptic potentials. EPSPs and IPSPs are summed both temporally and spatially, which integrates the incoming information. If the resulting sum moves the voltage of the membrane at the axon hillock to the threshold level, an action potential is produced on the axon.

The neuron is a versatile cell. Some species' ion channels respond to light rather than to voltage changes, an attribute that genetic engineers are exploiting. Most of our neurons do not initiate action potentials on dendrites because the cell-body membrane does not contain voltage-sensitive channels. But some voltage-sensitive channels on dendrites do enable action potentials. Back propagation, the reverse movement of an action potential from the axon hillock into the dendritic field of a neuron, may play a role in plastic changes that underlie learning.

4-4 Into the Nervous System and Back Out

Sensory-receptor cells in the body contain mechanisms for transducing sensory energy into energy changes in ion channels. These changes, in turn, alter the transmembrane voltage to the point at which voltage-sensitive channels open, triggering an action potential and propagating a nerve impulse that transmits sensory information to relevant parts of the nervous system.

Ion channels again come into play to activate muscles because the chemical transmitter acetylcholine, released at the axon terminal of a motor neuron, activates channels on the end plate of a muscle-cell membrane. The subsequent flow of ions depolarizes the muscle-cell membrane to the threshold for its action potential. This depolarization, in turn, activates voltage-sensitive channels, producing an action potential on the muscle fiber. The action potential induces the contraction of muscle fibers that enable movement.

KEY TERMS

absolutely refractory, p. 125

action potential, p. 122

autoimmune disease, p. 130

back propagation, p. 133

concentration gradient, p. 116

depolarization, p. 122

diffusion, p. 116

electrical stimulation, p. 111

electroencephalogram (EEG), p. 113

end plate, p. 136

excitatory postsynaptic potential (EPSP), p. 128

graded potential, p. 120

hyperpolarization, p. 122

inhibitory postsynaptic potential (IPSP), p. 130

microelectrode, p. 115

nerve impulse, p. 125

node of Ranvier, p. 127

optogenetics, p. 135

oscilloscope, p. 115

relatively refractory, p. 125

resting potential, p. 118

saltatory conduction, p. 127

spatial summation, p. 131

stretch-sensitive channel, p. 135

temporal summation, p. 131

threshold potential, p. 122

transmitter-sensitive channel, p. 136

voltage gradient, p. 116

voltage-sensitive channel, p. 125

voltmeter, p. 113

CHAPTER

5

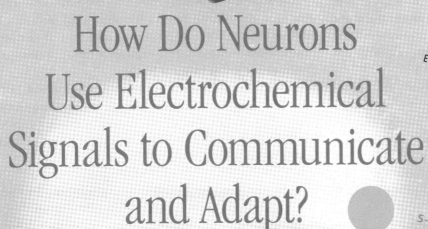

How Do Neurons Use Electrochemical Signals to Communicate and Adapt?

RESEARCH FOCUS 5-1 THE BASIS OF NEURAL COMMUNICATION IN A HEARTBEAT

5-1 A CHEMICAL MESSAGE

CLINICAL FOCUS 5-2 PARKINSON'S DISEASE

STRUCTURE OF SYNAPSES

NEUROTRANSMISSION IN FOUR STEPS

VARIETIES OF SYNAPSES

EXCITATORY AND INHIBITORY MESSAGES

EVOLUTION OF COMPLEX NEUROTRANSMISSION SYSTEMS

5-2 VARIETIES OF NEUROTRANSMITTERS

FOUR CRITERIA FOR IDENTIFYING NEUROTRANSMITTERS

THREE CLASSES OF NEUROTRANSMITTERS

CLINICAL FOCUS 5-3 AWAKENING WITH L-DOPA

TWO CLASSES OF RECEPTORS

5-3 NEUROTRANSMITTER SYSTEMS AND BEHAVIOR

NEUROTRANSMISSION IN THE SOMATIC NERVOUS SYSTEM

TWO ACTIVATING SYSTEMS OF THE AUTONOMIC NERVOUS SYSTEM

FOUR ACTIVATING SYSTEMS IN THE CENTRAL NERVOUS SYSTEM

CLINICAL FOCUS 5-4 THE CASE OF THE FROZEN ADDICT

5-4 ADAPTIVE ROLE OF SYNAPSES IN LEARNING AND MEMORY

HABITUATION RESPONSE

SENSITIZATION RESPONSE

LEARNING AS A CHANGE IN SYNAPSE NUMBER

RESEARCH FOCUS 5-5 DENDRITIC SPINES, SMALL BUT MIGHTY

139

The Basis of Neural Communication in a Heartbeat

Discoveries about how neurons communicate stem from experiments designed to study what controls an animal's heart rate. Like that of any animal, your heartbeat quickens if you are excited or exercising; if you are resting, it slows. Heart rate changes to match energy expenditure—that is, to meet the body's nutrient and oxygen needs.

Heartbeat undergoes a most dramatic change when you dive beneath water: it almost completely stops. This drastic slowing, called diving *bradycardia,* conserves the body's oxygen when you are not breathing. Bradycardia (*brady,* meaning "slow," and *cardia,* meaning "heart") is a useful survival strategy. This energy-conserving response under water is common to many animals. But what controls it?

Otto Loewi, a great storyteller, recounted that his classic experiment, which earned him a Nobel Prize in 1936, came to him in a dream. As shown in the Procedure section of Experiment 5-1, Loewi first maintained a frog's heart in a salt bath and then electrically stimulated the vagus nerve—the cranial nerve that leads from the brain to the heart. At the same time, he channeled some of the fluid bath from the vessel containing the stimulated heart through a tube to another vessel in which a second heart was immersed but not electrically stimulated.

Loewi recorded the beating rates of both hearts. His findings are represented in the Results section of Experiment 5-1. The electrical stimulation decreased the beating rate of the first heart, but more important, the second heartbeat also slowed. This finding suggested that the fluid transferred from the first to the second container carried the message "slow down."

But where did the message come from originally? Loewi reasoned that a chemical released from the stimulated vagus nerve must have diffused into the fluid to influence the second heart. The experiment therefore demonstrated that the vagus nerve contains a chemical that tells the heart to slow its rate of beating.

Loewi subsequently identified the messenger chemical. Later, he identified a chemical that tells the heart to speed up. Apparently, the heart adjusts its rate of beating in response to at least two different messages: an excitatory message that says "speed up" and an inhibitory message that says "slow down."

Puffins fish by diving underwater, propelling themselves by flapping their short stubby wings, as if flying. During these dives, their hearts display the diving-bradycardia response, just as our hearts do. Here, a puffin emerges from a dive.

Acetylcholine (ACh)

Epinephrine (EP)   Norepinephrine (NE)

Acetylcholine inhibits heartbeat; epinephrine and norepinephrine excite the heart in frogs and humans, respectively.

In this chapter, we explain how excitatory and inhibitory signals enable neurons to communicate, the chemicals that carry the neuron's signal, and the receptors on which those chemicals act to produce behavior. Then we explore the neural bases of simple *learning*—that is, how neural synapses adapt physically as a result of an organism's experience.

5-1 A Chemical Message

Loewi's successful heartbeat experiment, diagrammed in **Experiment 5-1,** marked the beginning of research into how chemicals carry information from one neuron to another. Loewi was the first to isolate a chemical messenger. We now know the chemical as **acetylcholine** (ACh), the same transmitter that activates skeletal muscles, as described in Section 4-4. Yet here, ACh acts to inhibit heartbeat, to slow it down. It turns out that ACh activates skeletal muscles in the somatic nervous system and may either excite or inhibit internal organs in the autonomic system. And, yes, ACh is the chemical messenger that slows the heart in diving bradycardia.

In further experiments, Loewi stimulated another nerve to the heart, the accelerator nerve, and obtained a speeded-up heart rate. As before, the fluid that bathed the accelerated heart increased the rate of beating of a second heart that was not electrically

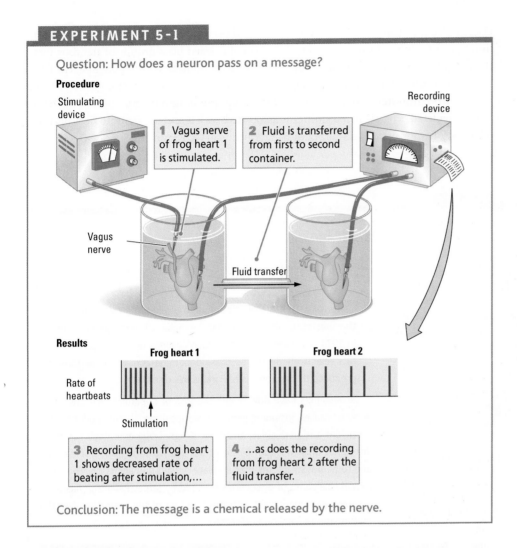

EXPERIMENT 5-1

Question: How does a neuron pass on a message?

Procedure

Stimulating device

Recording device

1 Vagus nerve of frog heart 1 is stimulated.

2 Fluid is transferred from first to second container.

Vagus nerve

Fluid transfer

Results

Frog heart 1

Frog heart 2

Rate of heartbeats

Stimulation

3 Recording from frog heart 1 shows decreased rate of beating after stimulation,...

4 ...as does the recording from frog heart 2 after the fluid transfer.

Conclusion: The message is a chemical released by the nerve.

acetylcholine (ACh) First neurotransmitter discovered in the peripheral and central nervous systems; activates skeletal muscles in the somatic nervous system and may either excite or inhibit internal organs in the autonomic system.

epinephrine (EP, *or* adrenaline) Chemical messenger that acts as a hormone to mobilize the body for fight or flight during times of stress and as a neurotransmitter in the central nervous system.

norepinephrine (NE, *or* noradrenaline) Neurotransmitter found in the brain and in the sympathetic division of the autonomic nervous system; accelerates heart rate in mammals.

neurotransmitter Chemical released by a neuron onto a target with an excitatory or inhibitory effect.

stimulated. Loewi identified the chemical that carries the message to speed up heart rate in frogs as **epinephrine** (EP), also known as *adrenaline*. Adrenaline (Latin) and epinephrine (Greek) are the same substance, produced by the adrenal glands located atop the kidneys. Adrenaline is the name more people know, in part because a drug company used it as a trade name, but EP is common parlance in the neuroscience community.

Further experimentation eventually demonstrated that the chemical that accelerates heart rate in mammals is **norepinephrine** (NE, also *noradrenaline*), a chemical closely related to EP. The results of Loewi's complementary experiments showed that ACh from the vagus nerve inhibits heartbeat, and EP from the accelerator nerve excites it.

Messenger chemicals released by a neuron onto a target to cause an excitatory or inhibitory effect are referred to as **neurotransmitters.** Outside the central nervous system, many of the same chemicals, EP among them, circulate in the bloodstream as *hormones.* Under control of the hypothalamus, the pituitary gland directs hormones to excite or inhibit targets such as the organs and glands in the autonomic nervous system. In part because hormones travel through the bloodstream to distant targets, their action is slower than that of CNS neurotransmitters prodded by the lightning-quick nerve impulse.

Loewi's discoveries led to the search for more neurotransmitters and their functions. How many transmitters there are is an open question, with the number of 100 given for the maximum number and the number 50 given for the confirmed number. Whether a chemical is accepted as neurotransmitter depends on the extent to which it meets certain

Among the cranial nerves illustrated in Figure 2-29, the vagus nerve influences the heart and many other internal body processes.

Section 6-5 explains how hormones influence the brain and behavior.

criteria. As this chapter unfolds, you will learn those criteria along with the names and functions of many neurotransmitters. You will also learn how groups of neurons form neurotransmitter systems throughout the brain to modulate, or temper, aspects of behavior. The three Clinical Focus boxes in this chapter tell the fascinating story of one such neurotransmitter system that has yielded deep insight into the function of the brain. When depleted, this neurotransmitter is associated with a specific neurological disorder. The story begins with Clinical Focus 5-2, "Parkinson's Disease."

CLINICAL FOCUS ✛ 5-2

Parkinson's Disease

. . . is seventy-two years of age. . . . About eleven or twelve, or perhaps more, years ago, he first perceived weakness in the left hand and arm, and soon after found the trembling to commence. In about three years afterwards the right arm became affected in a similar manner: and soon afterwards the convulsive motions affected the whole body and began to interrupt speech. In about three years from that time the legs became affected. (James Parkinson, 1817/1989)

In the 1817 essay from which this case study is taken, James Parkinson, a British physician, reported similar symptoms in six patients, some of whom he observed only in the streets near his clinic. Shaking was usually the first symptom, and it typically began in a hand. Over a number of years, the shaking spread to include the arm and then other parts of the body.

As the disease progressed, patients had a propensity to lean forward and walk on the balls of their feet. They also tended to run forward to prevent themselves from falling. In the later stages of the disease, patients had difficulty eating and swallowing. They drooled and their bowel movements slowed. Eventually, the patients lost all muscular control and were unable to sleep because of the disruptive tremors.

More than 50 years after James Parkinson's description, French neurologist Jean-Martin Charcot named the condition **Parkinson's disease.** Three findings have helped researchers understand its neural basis:

1. In 1919, Constantin Tréatikoff studied the brains of nine Parkinson patients on autopsy and found that the substantia nigra, a small nucleus in the midbrain, had degenerated. In the brain of one patient who had experienced symptoms of Parkinson's disease on one side of the body only, the substantia nigra had degenerated on the side opposite that of the symptoms.

2. Chemical examination of the brains of Parkinson patients showed that symptoms of the disease appear when the level of **dopamine,** then a proposed

neurotransmitter, was reduced to less than 10 percent of normal in the basal ganglia (Ehringer & Hornykiewicz, 1960).

3. Confirming the role of dopamine in a neural pathway connecting the substantia nigra to the basal ganglia, Urban Ungerstedt found in 1971 that injecting a neurotoxin called 6-hydroxydopamine into rats selectively destroyed these dopamine-containing neurons and produced the symptoms of Parkinson's disease.

Researchers have now linked the loss of dopamine neurons to an array of causes, including genetic predisposition, the flu, pollution, insecticides and herbicides, and toxic drugs. Dopamine itself has been linked not only to motor behavior but also to some forms of learning and to neural structures that mediate reward and addiction. Thus, this remarkable series of discoveries initiated by James Parkinson has been a source of more insight into the function of the brain than has the investigation of any other disease.

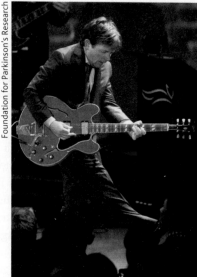

Universal Pictures/Photofest

Mike Coppola/Getty Images for the Michael J. Fox Foundation for Parkinson's Research

Actor Michael J. Fox gained wide fame in the 1980s for his starring role in the *Back to the Future* movie series, which included his rendition of Chuck Berry's pop classic, "Johnny B. Goode" (*left*). In 1991 at age 30, Fox was diagnosed with young-onset Parkinson's disease. When he performed the song 20 years later to benefit the Michael J. Fox Foundation for Parkinson's Research, he labored but still had the moves (*right*).

Structure of Synapses

Loewi's discovery about the regulation of heart rate by chemical messengers was the first of two important findings that form the foundation for current understanding of how neurons communicate. The second had to wait nearly 30 years, for the invention of the electron microscope, which enabled scientists to see the structure of a synapse.

The electron microscope, shown at the right in **Figure 5-1**, uses some of the principles of both an oscilloscope and a light microscope, shown at the left. The electron microscope works by projecting a beam of electrons through a very thin slice of tissue. The varying structure of the tissue scatters the beam onto a reflective surface where it leaves an image, or shadow, of the tissue.

The resolution of an electron microscope is much higher than that of a light microscope because electron waves are smaller than light waves, so there is much less scatter when the beam strikes the tissue. If the tissue is stained with substances that reflect electrons, very fine structural details can be observed. Compare the images at the bottom of Figure 5-1.

Chemical Synapses

The first good electron micrographs, made in the 1950s, revealed the structure of a synapse for the first time. In the center of the micrograph in **Figure 5-2A**, the upper part of the synapse is the axon terminal, or "end foot"; the lower part is the receiving dendrite. Note the round granular substances in the terminal. They are the **synaptic vesicles** containing the neurotransmitter.

The dark patches on the dendrite consist mainly of protein receptor molecules that receive chemical messages. Dark patches on the axon terminal membrane are protein molecules that serve largely as ion channels and pumps to release the transmitter or to recapture it after its release. The terminal and the dendrite are separated by a small space, the **synaptic cleft.** The synaptic cleft is central to synapse function because neurotransmitter chemicals must bridge this gap to carry a message from one neuron to the next.

Figure 4-5 shows a digital oscilloscope and describes how it measures voltage in biological tissue.

Parkinson's disease Disorder of the motor system correlated with a loss of dopamine in the brain and characterized by tremors, muscular rigidity, and reduction in voluntary movement.

dopamine (DA) Amine neurotransmitter that plays a role in coordinating movement, in attention and learning, and in behaviors that are reinforcing.

synaptic vesicle Organelle consisting of a membrane structure that encloses a quantum of neurotransmitter.

synaptic cleft Gap that separates the presynaptic membrane from the postsynaptic membrane.

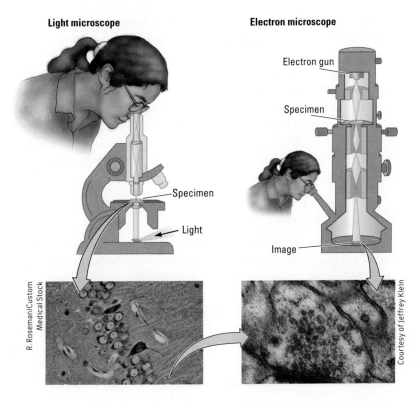

Light microscope

Specimen

Light

Electron microscope

Electron gun

Specimen

Image

R. Roseman/Custom Medical Stock

Courtesy of Jeffrey Klein

FIGURE 5-1 Microscopic Advance Whereas a light microscope (*left*) can be used to see the general features of a cell, an electron microscope (*right*) can be used to examine the details of a cell's organelles.

(A)

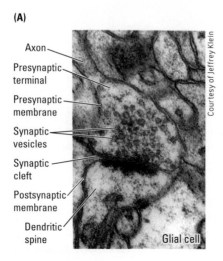

Courtesy of Jeffrey Klein

Axon

Presynaptic terminal

Presynaptic membrane

Synaptic vesicles

Synaptic cleft

Postsynaptic membrane

Dendritic spine

Glial cell

FIGURE 5-2 **Chemical Synapse**

(A) Surrounding the centrally located synapse in this electron micrograph are glial cells, axons, dendrites, and other synapses. **(B)** Within a chemical synapse, storage granules hold vesicles containing neurotransmitter that, when released, travels to the presynaptic membrane. It is expelled into the synaptic cleft through the process of exocytosis, crosses the cleft, and binds to receptor proteins on the postsynaptic membrane.

(B)

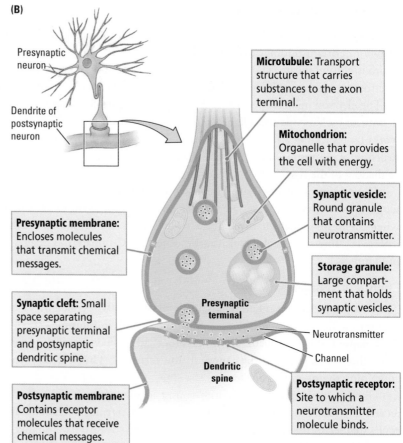

Presynaptic neuron

Dendrite of postsynaptic neuron

Microtubule: Transport structure that carries substances to the axon terminal.

Mitochondrion: Organelle that provides the cell with energy.

Synaptic vesicle: Round granule that contains neurotransmitter.

Storage granule: Large compartment that holds synaptic vesicles.

Presynaptic membrane: Encloses molecules that transmit chemical messages.

Synaptic cleft: Small space separating presynaptic terminal and postsynaptic dendritic spine.

Postsynaptic membrane: Contains receptor molecules that receive chemical messages.

Presynaptic terminal

Dendritic spine

Neurotransmitter

Channel

Postsynaptic receptor: Site to which a neurotransmitter molecule binds.

FIGURE 5-3 **Gap Junction**

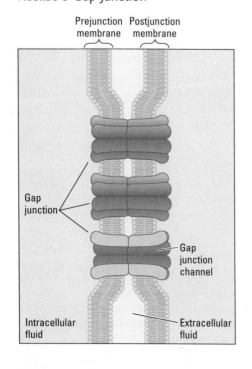

Prejunction membrane

Postjunction membrane

Gap junction

Gap junction channel

Intracellular fluid

Extracellular fluid

You can also see in the micrograph that the synapse is sandwiched by many surrounding structures. These structures include glial cells, other axons and dendritic processes, and other synapses. The surrounding glia contribute to chemical neurotransmission in a number of ways—by supplying the building blocks for the synthesis of neurotransmitters or by mopping up excess neurotransmitter molecules, for example.

The drawing in Figure 5-2B details the process of neurotransmission at a **chemical synapse,** the junction where messenger molecules are released from one neuron to excite the next neuron. Here the **presynaptic membrane** forms the axon terminal, the **postsynaptic membrane** forms the dendritic spine, and the space between the two is the synaptic cleft. Within the axon terminal are specialized structures, including mitochondria, the organelles that supply the cell's energy needs; **storage granules,** large compartments that hold several synaptic vesicles; and microtubules that transport substances, including the neurotransmitter, to the terminal.

Electrical Synapses

Chemical synapses are the rule in mammalian nervous systems, but they are not the only kind of synapse. Some neurons influence each other electrically through a **gap junction,** or *electrical synapse,* where the "prejunction" and "postjunction" cell membranes are fused **(Figure 5-3).** Ion channels in one cell membrane connect to ion channels in the other membrane, forming a pore that allows ions to pass directly from one neuron to the next.

This fusion eliminates the brief delay in information flow—about 5 milliseconds per synapse—of chemical transmission (compare Figure 5-3 with Figure 5-2B). For example, the crayfish's gap junctions activate its tail flick, a response that allows it to quickly escape from a predator. Gap junctions are found in the mammalian brain, where in some regions they allow groups of interneurons to synchronize their firing rhythmically. Gap junctions also allow glial cells and neurons to exchange substances (Dere & Zlomuzica, 2011).

Why, if chemical synapses transmit messages more slowly, do mammals rely on them more than on gap junctions? The answer is that chemical synapses are flexible in controlling whether a message is passed from one neuron to the next, they can amplify or diminish a signal sent from one neuron to the next, and they can change with experience to alter their signals and so mediate learning.

Neurotransmission in Four Steps

The four-step process of transmitting information across a chemical synapse is illustrated in **Figure 5-4** and explained in this section. In brief, the neurotransmitter must be

1. synthesized and stored in the axon terminal.

2. transported to the presynaptic membrane and released in response to an action potential.

3. able to activate the receptors on the target-cell membrane located on the postsynaptic membrane.

4. inactivated, or it will continue to work indefinitely.

Step 1: Neurotransmitter Synthesis and Storage

Neurotransmitters are derived in two general ways, and these origins define two broad classes of neurotransmitters. Some are synthesized in the cell body according to instructions contained in the neuron's DNA, packaged in membranes on the Golgi bodies and transported on microtubules to the axon terminal. Cell-derived neurotransmitters may also be manufactured within the presynaptic terminal from mRNA that is transported to the terminal.

Other neurotransmitters are synthesized in the axon terminal from building blocks derived from food. **Transporters,** protein molecules that pump substances across the cell membrane, absorb the required precursor chemicals from the blood supply. (Sometimes transporter proteins absorb the neurotransmitter ready-made.) Mitochondria in the axon terminal provide the energy needed both to synthesize precursor chemicals into the neurotransmitter and to wrap them in membranous vesicles.

Regardless of their origin, neurotransmitters in the axon terminal can usually be found in three locations, depending on the type of neurotransmitter. Some vesicles are warehoused in granules, some are attached to microfilaments in the terminal, and still others are attached to the presynaptic membrane. These sites represent the steps in which a transmitter is transported from a granule to the membrane, ready to be released into the synaptic cleft.

Step 2: Neurotransmitter Release

When an action potential is propagated on the presynaptic membrane, voltage changes on the membrane set the release process in motion. Calcium cations (Ca^{2+}) play an important role. The presynaptic membrane is rich in voltage-sensitive calcium channels, and the surrounding extracellular fluid is rich in Ca^{2+}. As illustrated in **Figure 5-5**, the action potential's arrival opens these calcium channels, allowing an influx of calcium ions into the axon terminal.

The incoming Ca^{2+} binds to the protein *calmodulin,* and the resulting complex takes part in two chemical reactions: one releases vesicles bound to the presynaptic membrane, and the other releases vesicles bound to microfilaments in the axon terminal. The vesicles released from the presynaptic membrane empty their contents

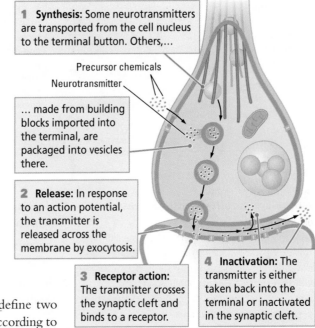

1 Synthesis: Some neurotransmitters are transported from the cell nucleus to the terminal button. Others,...

Precursor chemicals

Neurotransmitter

... made from building blocks imported into the terminal, are packaged into vesicles there.

2 Release: In response to an action potential, the transmitter is released across the membrane by exocytosis.

3 Receptor action: The transmitter crosses the synaptic cleft and binds to a receptor.

4 Inactivation: The transmitter is either taken back into the terminal or inactivated in the synaptic cleft.

FIGURE 5-4 Synaptic Transmission

For a refresher on protein export, review Figure 3-17.

chemical synapse Junction at which messenger molecules are released when stimulated by an action potential.

presynaptic membrane Membrane on the transmitter-output side of a synapse (axon terminal).

postsynaptic membrane Membrane on the transmitter-input side of a synapse (dendritic spine).

storage granule Membranous compartment that holds several vesicles containing a neurotransmitter.

gap junction (electrical synapse) Fused prejunction and postjunction cell membrane in which connected ion channels form a pore that allows ions to pass directly from one neuron to the next.

transporter Protein molecule that pumps substances across a membrane.

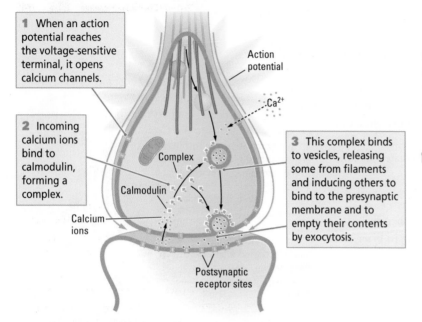

1 When an action potential reaches the voltage-sensitive terminal, it opens calcium channels.

Action potential

Ca²⁺

2 Incoming calcium ions bind to calmodulin, forming a complex.

Complex

Calmodulin

Calcium ions

3 This complex binds to vesicles, releasing some from filaments and inducing others to bind to the presynaptic membrane and to empty their contents by exocytosis.

Postsynaptic receptor sites

FIGURE 5-5 Neurotransmitter Release

into the synaptic cleft through the process of exocytosis. The vesicles from storage granules and on filaments then move up to replace the vesicles that just emptied their contents.

Step 3: Receptor-Site Activation

After the neurotransmitter has been released from vesicles on the presynaptic membrane, it diffuses across the synaptic cleft and binds to specialized protein molecules embedded in the postsynaptic membrane. These **transmitter-activated receptors** have binding sites for the transmitter substance. Through the receptors, the postsynaptic cell may be affected in one of three ways, depending on the type of neurotransmitter and the kind of receptors on the postsynaptic membrane. The transmitter may

1. depolarize the postsynaptic membrane and so have an excitatory action on the postsynaptic neuron.

2. hyperpolarize the postsynaptic membrane and so have an inhibitory action on the postsynaptic neuron.

3. initiate other chemical reactions that modulate either effect, inhibitory or excitatory, or that influence other functions of the receiving neuron.

In addition to interacting with the postsynaptic membrane's receptors, a neurotransmitter may interact with receptors on the presynaptic membrane. That is, it may influence the cell that just released it. Presynaptic receptors that may be activated by a neurotransmitter are called **autoreceptors** (self-receptors) to indicate that they receive messages from their own axon terminals.

How much neurotransmitter is needed to send a message? Bernard Katz was awarded a Nobel Prize in 1970 for providing an answer. Recording electrical activity from the postsynaptic membranes of muscles, he detected small, spontaneous depolarizations now called *miniature postsynaptic potentials.* The potentials varied in size, but each size appeared to be a multiple of the smallest potential.

Katz concluded that the smallest postsynaptic potential is produced by the release of the contents of just one synaptic vesicle. This amount of neurotransmitter is called a **quantum.** To produce a postsynaptic potential that is large enough to initiate a postsynaptic action potential requires the simultaneous release of many quanta from the presynaptic cell.

The results of subsequent experiments show that the number of quanta released from the presynaptic membrane in response to a single action potential depends on two factors: (1) the amount of Ca²⁺ that enters the axon terminal in response to the action potential and (2) the number of vesicles docked at the membrane, waiting to be released. Both factors are relevant to synaptic activity during learning, which we consider at the end of the chapter.

Step 4: Neurotransmitter Deactivation

Chemical transmission would not be a very effective messenger system if a neurotransmitter lingered within the synaptic cleft, continuing to occupy and stimulate receptors. If this happened, the postsynaptic cell could not respond to other messages sent by the presynaptic neuron. Therefore, after a neurotransmitter has done its work, it is quickly removed from receptor sites and from the synaptic cleft. Deactivation is accomplished in at least four ways:

1. **Diffusion:** Some of the neurotransmitter simply diffuses away from the synaptic cleft and is no longer available to bind to receptors.

2. **Degradation** by enzymes in the synaptic cleft

3. **Reuptake:** Membrane transporter proteins specific to that transmitter may bring the transmitter back into the presynaptic axon terminal for subsequent reuse. The by-products of degradation by enzymes also may be taken back into the terminal to be used again in the cell.

4. **Glial uptake:** Some neurotransmitters are taken up by neighboring glial cells. Potentially, the glial cells can also store transmitters for re-export to the axon terminal.

As part of the flexibility of synaptic function, an axon terminal has chemical mechanisms that enable it to respond to the frequency of its own use. If the terminal is very active, the amount of neurotransmitter made and stored there increases. If the terminal is not often used, however, enzymes located within the terminal buttons may break down excess transmitter. The by-products of this breakdown are then reused or excreted from the neuron. Axon terminals may even send messages to the neuron's cell body requesting increased supplies of the neurotransmitter or the molecules with which to make it.

Varieties of Synapses

So far, we have considered a generic chemical synapse, with features possessed by most synapses. In the nervous system, synapses vary widely, and each type is specialized in location, structure, function, and target. **Figure 5-6** illustrates this diversity on a single hypothetical neuron.

transmitter-activated receptor Protein that has a binding site for a specific neurotransmitter and is embedded in the membrane of a cell.

autoreceptor "Self-receptor" in a neural membrane that responds to the transmitter released by the neuron.

quantum (pl. quanta) Amount of neurotransmitter, equivalent to the contents of a single synaptic vesicle, that produces a just observable change in postsynaptic electric potential.

reuptake Deactivation of a neurotransmitter when membrane transporter proteins bring the transmitter back into the presynaptic axon terminal for subsequent reuse.

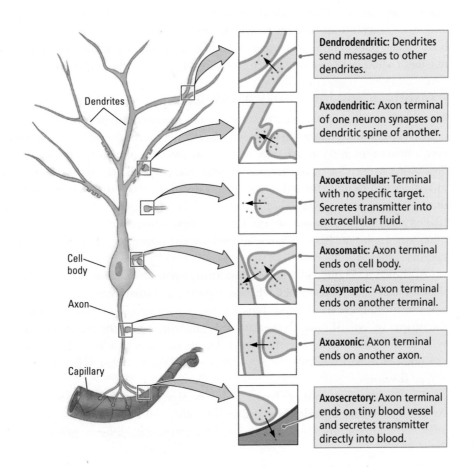

Dendrodendritic: Dendrites send messages to other dendrites.

Axodendritic: Axon terminal of one neuron synapses on dendritic spine of another.

Axoextracellular: Terminal with no specific target. Secretes transmitter into extracellular fluid.

Axosomatic: Axon terminal ends on cell body.

Axosynaptic: Axon terminal ends on another terminal.

Axoaxonic: Axon terminal ends on another axon.

Axosecretory: Axon terminal ends on tiny blood vessel and secretes transmitter directly into blood.

FIGURE 5-6 The Versatile Synapse

Figure 4-26 shows both microscopic and schematic views of an axomuscular synapse.

You have already encountered two kinds of synapses. One is the *axomuscular synapse,* in which an axon synapses with a muscle end plate, releasing acetylcholine. The other synapse familiar to you is the *axodendritic synapse* detailed in Figure 5-2B, in which the axon terminal of a neuron ends on a dendrite or dendritic spine of another neuron.

Figure 5-6 diagrams the axodendritic synapse as well as the *axosomatic synapse,* an axon terminal ending on a cell body; the *axoaxonic synapse,* an axon terminal ending on another axon; and the *axosynaptic synapse,* an axon terminal ending on another presynaptic terminal—that is, at the synapse between some other axon and its target. *Axoextracellular synapses* have no specific targets but instead secrete their transmitter chemicals into the extracellular fluid. In the *axosecretory synapse,* a terminal synapses with a tiny blood vessel called a capillary and secretes its transmitter directly into the blood. Finally, synapses are not limited to axon terminals. Dendrites also may send messages to other dendrites through *dendrodendritic synapses.*

This wide variety of connections makes the synapse a versatile chemical delivery system. Synapses can deliver transmitters to highly specific sites or diffuse locales. Through connections to the dendrites, cell body, or axon of a neuron, transmitters can control the actions of the neuron in different ways.

Through axosynaptic connections, they can also provide exquisite control over another neuron's input to a cell. By excreting transmitters into extracellular fluid or into the blood, axoextracellular and axosecretory synapses can modulate the function of large areas of tissue or even the entire body. Recall that many transmitters secreted by neurons act as hormones circulating in your blood, with widespread influences on your body.

Gap junctions, shown in Figure 5-3, further increase the diversity of signaling between one part of a neuron and another part of the same neuron. Intraneuronal communication may occur via dendrodendritc and axoaxonic gap junctions. Gap junctions also allow neighboring neurons to synchronize their signals through *somasomatic* (cell body to cell body) connections, and they allow glial cells, especially astrocytes, to pass nutrient chemicals to neurons and to receive waste products from them.

Excitatory and Inhibitory Messages

Once again we see that the nervous system works through a combination of excitatory and inhibitory signals. Each neuron receives thousands of excitatory and inhibitory signals every second.

A neurotransmitter can influence the function of a neuron through a remarkable number of mechanisms. In its direct actions in influencing a neuron's electrical excitability, however, a neurotransmitter acts in only one of two ways. It influences transmembrane ion flow either to increase or to decrease the probability that the cell with which it comes in contact will produce an action potential. Thus, despite the wide variety of synapses, they all convey messages of only these two types, excitatory or inhibitory, and they are labeled as such. *Type I synapses* are excitatory in their actions, whereas *Type II synapses* are inhibitory. Each type has a different appearance and is located on different parts of the neurons under its influence.

As shown in **Figure 5-7,** Type I (excitatory) synapses are typically located on the shafts or the spines of dendrites, whereas Type II (inhibitory) synapses are typically located on a cell body. In addition, Type I synapses have round synaptic vesicles, whereas the vesicles of Type II synapses are flattened. The material on the presynaptic and postsynaptic membranes is denser in a Type I synapse than it is in a Type II, and the Type I synaptic cleft is wider. Finally, the active zone on a Type I synapse is larger than that on a Type II synapse.

The different locations of Type I and Type II synapses divide a neuron into two zones: an excitatory dendritic tree and an inhibitory cell body. You can think of excitatory and inhibitory messages as interacting from these two different perspectives.

Viewed from an inhibitory perspective, you can picture excitation coming in over the dendrites and spreading to the axon hillock to trigger an action potential. If the message is to be stopped, it is best stopped by applying inhibition on the cell body, close to the axon hillock where the action potential originates. In this model of excitatory–inhibitory interaction, inhibition blocks excitation by a "cut 'em off at the pass" strategy.

Another way to conceptualize excitatory–inhibitory interaction is to picture excitation overcoming inhibition. If the cell body is normally in an inhibited state, the only way to generate an action potential at the axon hillock is to reduce the cell body's inhibition. In this "open the gates" strategy, the excitatory message is like a racehorse ready to run down the track, but first the inhibitory starting gate must be removed.

Evolution of Complex Neurotransmission Systems

Considering all the biochemical steps required for getting a message across a synapse and the variety of synapses, you may well wonder why—and how—such a complex communication system ever evolved. How did chemical transmitters originate?

If you think about the feeding behaviors of simple single-celled creatures, the origin of chemical secretions for communication is easier to imagine. The earliest unicellular creatures secreted juices onto bacteria to immobilize and prepare them for ingestion. These digestive juices were probably expelled from the cell body by exocytosis, in which a vacuole or vesicle attaches itself to the cell membrane and then opens into the extracellular fluid to discharge its contents. The prey thus immobilized is captured through the reverse process of endocytosis.

The mechanism of exocytosis for digestion parallels its use to release a neurotransmitter for communication. Quite possibly the digestive processes of single-celled animals were long ago adapted by evolution into processes of neural communication in more complex organisms.

Behaviors are lost when a disorder prevents excitatory instructions; behaviors are released when a disorder prevents inhibitory instructions.

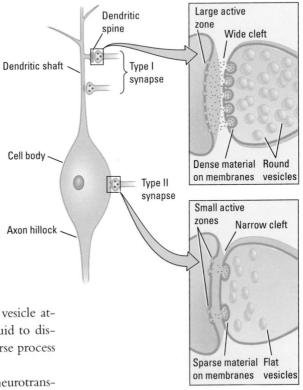

FIGURE 5-7 Excitatory and Inhibitory Zones Type I synapses occupy the spines and dendritic shafts of the neuron, creating an excitatory zone. Type II synapses found on the cell body create an inhibitory zone.

REVIEW 5-1

A Chemical Message

Before you continue, check your understanding.

1. In mammals, the principal form of communication between neurons occurs via _____, even though this structure is slower and more complex than the fused _____.

2. To generate an action potential that travels from the _____ across the synaptic cleft onto a _____ requires the simultaneous release of many _____ of chemical transmitter.

3. The nervous system has evolved a variety of synapses:
   _____ between axon terminals and dendrites,
   _____ between axon terminals and cell bodies,
   _____ between axon terminals and muscles,
   _____ between axon terminals and other axons,
   _____ between axon terminals and other synapses.
   A(n) _____ synapse releases chemical transmitters into extracellular fluid, a(n) _____ synapse releases transmitter into the bloodstream as hormones, and still another, the _____ synapse, connects dendrites to other dendrites.

small-molecule transmitter Quick-acting neurotransmitter synthesized in the axon terminal from products derived from the diet.

4. Excitatory synapses, known as Type I, are usually located on a(n) _____, whereas inhibitory synapses, known as Type II, are usually located on a(n)_____.

5. Describe the four steps in chemical neurotransmission.

Answers appear at the back of the book.

5-2 Varieties of Neurotransmitters

Subsequent to Otto Loewi's discovery, in 1921, that excitatory and inhibitory chemicals control heart rate, many researchers thought that the brain must work under much the same type of dual control. They reasoned that there must be excitatory and inhibitory brain cells and that norepinephrine and acetylcholine were the transmitters through which these neurons worked. They did not imagine what we know today: the human brain employs a wide variety of neurotransmitters. These chemicals operate in even more versatile ways: some may be excitatory at one location and inhibitory at another location, for example, and two or more may team up in a single synapse so that one makes the other more potent.

In this section, you will learn how neurotransmitters are identified and how they fit within three broad categories on the basis of their chemical structure. The functional aspects of neurotransmitters interrelate and are intricate, with no simple one-to-one relation between a single neurotransmitter and a single behavior.

Four Criteria for Identifying Neurotransmitters

Among the many thousands of chemicals in the nervous system, which are neurotransmitters? **Figure** 5-8 presents four identifying criteria:

1. The chemical must be synthesized in the neuron or otherwise be present in it.

2. When the neuron is active, the chemical must be released and produce a response in some target.

3. The same response must be obtained when the chemical is experimentally placed on the target.

4. A mechanism must exist for removing the chemical from its site of action after its work is done.

The criteria for identifying a neurotransmitter are fairly easy to apply when examining the somatic nervous system, especially at an accessible nerve–muscle junction with only one main neurotransmitter, acetylcholine. But identifying chemical transmitters in the central nervous system is not so easy. In the brain and spinal cord, thousands of synapses are packed around every neuron, preventing easy access to a single synapse and its activities. Consequently, a number of techniques, including staining, stimulating, and collecting, are used to identify substances thought to be CNS neurotransmitters. A suspect chemical that has not yet been shown to meet all the criteria is called a *putative* (supposed) *transmitter.*

Researchers trying to identify new CNS neurotransmitters can use microelectrodes to stimulate and record from single neurons. A glass microelectrode is small enough to be placed on specific targets on a neuron. It can be filled with a chemical of interest and, when a current is passed through the electrode, the chemical can be ejected into or onto the neuron to mimic the release of a neurotransmitter onto the cell.

FIGURE 5-8 Criteria for Identifying Neurotransmitters

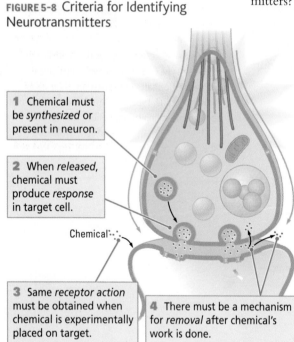

1 Chemical must be *synthesized* or present in neuron.

2 When *released*, chemical must produce *response* in target cell.

Chemical

3 Same *receptor action* must be obtained when chemical is experimentally placed on target.

4 There must be a mechanism for *removal* after chemical's work is done.

Figure 4-6 illustrates the use of a glass microelectrode.

Many staining techniques can identify specific chemicals inside the cell. Methods have also been developed for preserving nervous-system tissue in a saline bath while experiments are performed to determine how the neurons in the tissue communicate. The use of "slices of tissue" simplifies the investigation by allowing the researcher to view a single neuron through a microscope while stimulating it or recording from it.

Acetylcholine was not only the first substance identified as a neurotransmitter but also the first substance identified as a CNS neurotransmitter. A logical argument that predicted its presence even before experimental proof was gathered greatly facilitated the process. As you know, all motor-neuron axons leaving the spinal cord use ACh as a transmitter. Each of these axons has an axon collateral within the spinal cord that synapses on a nearby CNS interneuron. The interneuron, in turn, synapses back on the motor neuron's cell body. This circular set of connections, called a *Renshaw loop* after the researcher who first described it, is shown in **Figure 5-9**.

Because the main axon to the muscle releases acetylcholine, investigators suspected that its axon collateral also might release ACh. For two terminals of the same axon to use different transmitters seemed unlikely. Knowing what chemical to look for made it easier to find and obtain the required proof that ACh is in fact a neurotransmitter in both locations.

The loop made by the axon collateral and the interneuron in the spinal cord forms a feedback circuit that enables the motor neuron to inhibit itself from becoming overexcited if it receives a great many excitatory inputs from other parts of the CNS. Follow the positive and negative signs in Figure 5-9 to see how the Renshaw loop works. If the Renshaw loop is blocked, as can be done with the toxin strychnine, motor neurons become overactive, resulting in convulsions that can choke off respiration and so cause death.

The term "neurotransmitter" is used more broadly now than it was when researchers began to identify these chemicals. Today, the term applies to chemicals that

- carry a message from one neuron to another by influencing the voltage on the postsynaptic membrane.

- have little effect on membrane voltage but rather have a common message-carrying function, such as changing the structure of a synapse.

- communicate not only by *delivering* a message from the presynaptic to the postsynaptic membrane but by *sending* messages in the opposite direction as well. These reverse-direction messages influence the release or reuptake of transmitters.

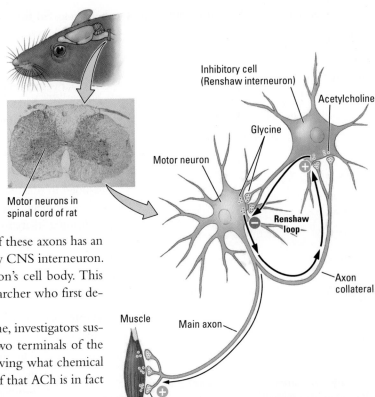

FIGURE 5-9 Renshaw Loop At top left, some spinal-cord motor neurons project to the muscles of the rat's forelimb. At right, in a Renshaw loop the main motor axon (green) projects to a muscle, and its axon collateral remains in the spinal cord to synapse with a Renshaw interneuron (red) that contains the inhibitory transmitter glycine (Gly). Both the main motor axon and its collateral terminals contain acetylcholine. When the motor neuron is highly excited, it can modulate its activity level through the Renshaw loop (plus and minus signs).

Three Classes of Neurotransmitters

Some order can be imposed on the diversity of neurotransmitters by classifying them into three groups on the basis of their chemical composition: (1) small-molecule transmitters, (2) peptide transmitters, and (3) transmitter gases.

Small-Molecule Transmitters

The first neurotransmitters that were identified are the quick-acting, **small-molecule transmitters** such as acetylcholine. Typically, they are synthesized from dietary nutrients and packaged ready for use in axon terminals. When a small-molecule transmitter has

TABLE 5-1 Small-Molecule Neurotransmitters

Acetylcholine (ACh)
Histamine (H)
Amines
Dopamine (DA)
Norepinephrine (NE, or noradrenaline, NA)
Epinephrine (EP, or adrenaline)
Serotonin (5-HT)
Amino acids
Glutamate (Glu)
Gamma-aminobutyric acid (GABA)
Glycine (Gly)

As explained in Section 6-1, taking drugs orally is the safest and easiest method, but not all drugs can pass through the walls of the digestive tract.

been released from a terminal button, it can quickly be replaced at the presynaptic membrane.

Because small-molecule transmitters or their main components are derived from the food that we eat, their level and activity in the body can be influenced by diet. This fact is important in the design of drugs that act on the nervous system. Many neuroactive drugs are designed to reach the brain by the same route that small-molecule transmitters or their precursor chemicals follow: the digestive tract.

Table 5-1 lists some of the best-known and most extensively studied small-molecule transmitters. In addition to acetylcholine, four amines (related by a chemical structure that contains an amine, or NH, group) and three amino acids are included in this list. A few other substances, including histamine, also are classified as small-molecule transmitters.

Among its many functions, which include the control of arousal and of waking, the transmitter **histamine** (H) can cause the constriction of smooth muscles. When activated in allergic reactions, histamine contributes to asthma, a constriction of the airways. You are probably familiar with antihistamine drugs used to treat allergies.

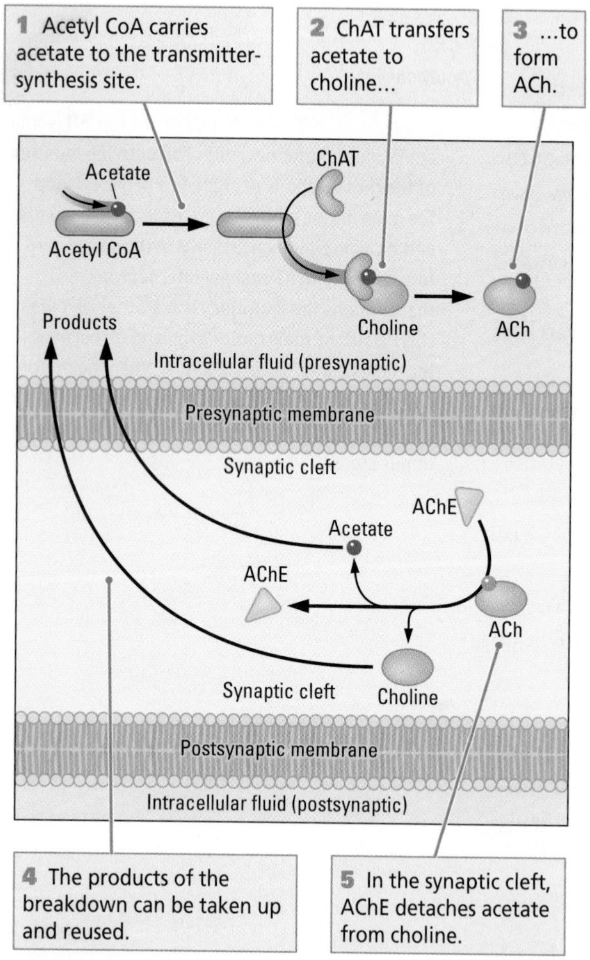

ACETYLCHOLINE SYNTHESIS **Figure 5-10** illustrates how acetylcholine molecules are synthesized and broken down. As you know, ACh is present at the junction of neurons and muscles, including the heart, as well as in the CNS. The molecule is made up of two substances, choline and acetate.

Choline is among the breakdown products of fats in foods such as egg yolk, avocado, salmon, and olive oil; acetate is a compound found in acidic foods, such as vinegar and lemon juice. As depicted in Figure 5-10, inside the cell, acetyl coenzyme A (acetyl CoA) carries acetate to the synthesis site, and the transmitter is synthesized as a second enzyme, choline acetyltransferase (ChAT), transfers the acetate to choline to form ACh. After ACh has been released into the synaptic cleft and diffuses to receptor sites on the postsynaptic membrane, a third enzyme, acetylcholinesterase (AChE), reverses the process by detaching acetate from choline. These breakdown products can then be taken back into the presynaptic terminal for reuse.

AMINE SYNTHESIS Some of the transmitters grouped together in Table 5-1 have common biochemical pathways to synthesis and so are related. You are familiar with the amines dopamine (DA), norepinephrine (NE), and epinephrine (EP). To review, DA loss has a role in Parkinson's disease, EP is the excitatory transmitter at the amphibian heart, and NE is the excitatory transmitter at the mammalian heart.

Figure 5-11 charts the biochemical sequence that synthesizes these amines in succession. The precursor chemical is tyrosine, an amino acid abundant in food. (Hard cheese and bananas are good sources.) The enzyme tyrosine hydroxylase (enzyme 1 in Figure 5-11) changes tyrosine into L-dopa, which is sequentially converted by other enzymes into dopamine, norepinephrine, and, finally, epinephrine.

An interesting fact about this biochemical sequence is that the supply of the enzyme tyrosine hydroxylase is limited. Consequently, so is the rate at which dopamine, norepinephrine, and epinephrine can be produced, regardless of how much tyrosine is present or ingested. This **rate-limiting factor** can be bypassed by the oral administration of L-dopa, which is why L-dopa is a medication used in the treatment of Parkinson's disease, as described in Clinical Focus 5-3, "Awakening with L-Dopa" on page 154.

1 Acetyl CoA carries acetate to the transmitter-synthesis site.

2 ChAT transfers acetate to choline...

3 ...to form ACh.

4 The products of the breakdown can be taken up and reused.

5 In the synaptic cleft, AChE detaches acetate from choline.

FIGURE 5-10 **Chemistry of Acetylcholine** Two enzymes combine the dietary precursors of ACh within the cell, and a third breaks them down in the synapse for reuptake.

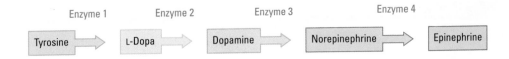

FIGURE 5-11 Sequential Synthesis of Three Amines A different enzyme is responsible for each successive molecular modification in this biochemical sequence.

SEROTONIN SYNTHESIS The amine transmitter **serotonin** (5-HT, for 5-hydroxytryptamine) is synthesized differently. Serotonin plays a role in regulating mood and aggression, appetite and arousal, respiration, and the perception of pain. Serotonin is derived from the amino acid tryptophan, which is abundant in turkey, milk, and bananas, among other foods.

AMINO ACID SYNTHESIS Two amino acid transmitters, **glutamate** (Glu) and **gamma-aminobutyric acid** (GABA), also are closely related. GABA is formed by a simple modification of the glutamate molecule, as shown in **Figure 5-12**. These two transmitters are the workhorses of the brain because so many synapses use them.

In the forebrain and cerebellum, glutamate is the main excitatory transmitter and GABA is the main inhibitory transmitter. Type I excitatory synapses thus have glutamate as a neurotransmitter, and Type II inhibitory synapses have GABA as a neurotransmitter. So the appearance of a synapse provides information about the neurotransmitter and its function (review Figure 5-7). Interestingly, glutamate is widely distributed in CNS neurons, but it becomes a neurotransmitter only if it is appropriately packaged in vesicles in the axon terminal. The amino acid transmitter glycine (Gly) is a much more common inhibitory transmitter in the brainstem and spinal cord, where it acts within the Renshaw loop, for example (review Figure 5-9).

Focus 3-1 describes how genome sequencing led to a diagnosis for the Beery twins, who lack an enzyme that enhances the production of dopamine and serotonin.

histamine (H) Neurotransmitter that controls arousal and waking; can cause the constriction of smooth muscles and so, when activated in allergic reactions, contributes to asthma, a constriction of the airways.

rate-limiting factor Any enzyme that is in limited supply, thus restricting the pace at which a chemical can be synthesized.

serotonin (5-HT) Amine neurotransmitter that plays a role in regulating mood and aggression, appetite and arousal, the perception of pain, and respiration.

glutamate (Glu) Amino acid neurotransmitter that excites neurons.

gamma-aminobutyric acid (GABA) Amino acid neurotransmitter that inhibits neurons.

neuropeptide Multifunctional chain of amino acids that acts as a neurotransmitter; synthesized from mRNA on instructions from the cell's DNA. Peptide neurotransmitters can act as hormones and may contribute to learning.

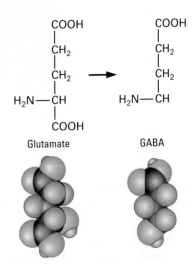

FIGURE 5-12 Amino Acid Transmitters (*Top*) Removal of a carboxyl (COOH) group from the bottom of the glutamate molecule produces GABA. (*Bottom*) Their different shapes thus allow these amino acid transmitters to bind to different receptors.

Peptides are proteins (molecular chains of amino acids) connected by peptide bonds, which accounts for the name (see Figure 3-15).

Peptide Transmitters

More than 50 amino acid chains of various lengths form the families of the peptide transmitters listed in **Table 5-2**. **Neuropeptides,** synthesized through the translation of mRNA from instructions contained in the neuron's DNA, are multifunctional chains of amino acids that act as neurotransmitters.

In some neurons, peptide transmitters are made in the axon terminal, but most are assembled on the neuron's ribosomes, packaged in a membrane by Golgi bodies, and

TABLE 5-2 Peptide Neurotransmitters

Family	Example
Opioids	Enkephaline, dynorphin
Neurohypophyseals	Vasopressin, oxytocin
Secretins	Gastric inhibitory peptide, growth-hormone-releasing peptide
Insulins	Insulin, insulin growth factors
Gastrins	Gastrin, cholecystokinin
Somatostatins	Pancreatic polypeptides
Corticosteroids	Glucocorticoids, mineralocorticoids

Awakening with L-Dopa

He was started on L-dopa in March 1969. The dose was slowly raised to 4.0 mg a day over a period of three weeks without apparently producing any effect. I first discovered that Mr. E. was responding to L-dopa by accident, chancing to go past his room at an unaccustomed time and hearing regular footsteps inside the room. I went in and found Mr. E., who had been chair bound since 1966, walking up and down his room, swinging his arms with considerable vigor, and showing erectness of posture and a brightness of expression completely new to him. When I asked him about the effect, he said with some embarrassment: "Yes! I felt the L-dopa beginning to work three days ago—it was like a wave of energy and strength sweeping through me. I found I could stand and walk by myself, and that I could do everything I needed for myself—but I was afraid that you would see how well I was and discharge me from the hospital." (Sacks, 1976)

In this case history, neurologist Oliver Sacks describes administering L-dopa to a patient who acquired Parkinsonism as an aftereffect of severe influenza in the 1920s. The relation between the influenza and the symptoms of Parkinson's disease suggests that the flu virus entered the brain and selectively attacked dopamine neurons in the substantia nigra. L-Dopa, by increasing the amount of DA in remaining synapses, relieved the patient's symptoms.

Two separate groups of investigators had quite independently given L-dopa to Parkinson patients beginning in 1961 (Birkmayer & Hornykiewicz, 1961; Barbeau et al., 1961). Both research teams knew that the chemical is catalyzed into dopamine at DA synapses (see Figure 5-11). The L-dopa turned out to reduce the muscular rigidity that the patients suffered.

The movie *Awakenings* recounts the L-dopa trials conducted by Oliver Sacks and described in his book of the same title.

This work was the first demonstration that a neurological condition can be relieved by a drug that aids in replacing a neurotransmitter. L-Dopa has since become a standard treatment for Parkinson's disease. Its effects have been improved by the administration of drugs that prevent L-dopa from being broken down before it gets to dopamine neurons in the brain.

L-Dopa is not a cure. Parkinson's disease still progresses during treatment, and as more and more dopamine synapses are lost, the treatment becomes less and less effective. Eventually, L-dopa begins to produce dyskinesias—involuntary, unwanted movements, such as tremors. When these side effects eventually become severe, the treatment must be discontinued.

---

transported by the microtubules to the axon terminals. The entire process of neuropeptide synthesis and transport is relatively slow compared with the nearly ready-made formation of small-molecule neurotransmitters. Consequently, peptide transmitters act slowly and are not replaced quickly.

Neuropeptides, however, perform an enormous range of functions in the nervous system, as might be expected from the large number that exist there. They act as hormones that respond to stress, enable a mother to bond with her infant, regulate eating and drinking and pleasure and pain, and probably contribute to learning. Opium and related synthetic chemicals such as morphine, long known to both produce euphoria and reduce pain, appear to mimic the actions of three natural brain neurotransmitter peptides: met-enkephalin, leu-enkephalin, and beta-endorphin. (The term *enkephalin* derives from the phrase "in the cephalon," meaning "in the brain or head," whereas the term *endorphin* is a shortened form of "endogenous morphine.")

A part of the amino acid chain in each of these three peptide transmitters is structurally similar to the others, as illustrated for two of these peptides in **Figure 5-13**. Presumably, opium mimics this part of the chain. The discovery of naturally occurring opium-like peptides suggested that one or more of them might take part in the management of pain. Opioid peptides, however, appear in a number of locations and perform a variety of functions in the brain, including the inducement of nausea. Therefore opium-like drugs are still preferred for pain management.

Met-enkephalin

(Tyr)(Gly)(Gly)(Phe)(Met)

Leu-enkephalin

(Tyr)(Gly)(Gly)(Phe)(Leu)

FIGURE 5-13 Opioid Peptides Parts of the amino acid chains of some neuropeptides that act on brain centers for pleasure and pain are similar in structure and are also similar to drugs such as opium and morphine, which mimic their functions.

Some CNS peptides take part in specific periodic behaviors, each month or each year perhaps. For instance, neuropeptide transmitters act as hormones to prepare a female deer for the fall mating season (luteinizing hormone). Come winter, a different set of biochemicals facilitates the development of the deer fetus. The mother gives birth in the spring, and yet another set of highly specific neuropeptide hormones—such as oxytocin, which enables her to bond to her fawn, and prolactin, which enables her to nurse—takes control.

The same neuropeptides serve similar, specific hormonal functions in humans. Others, such as neuropeptide growth hormones, have much more general functions in regulating growth. And neuropeptide corticosteroids mediate general responses to stress.

Unlike small-molecule transmitters, neuropeptides do not bind to ion channels, so they have no direct effects on the voltage of the postsynaptic membrane. Instead, peptide transmitters activate synaptic receptors that indirectly influence cell structure and function. Because peptides are amino acid chains that are degraded by digestive processes, they generally cannot be taken orally as drugs, as the small-molecule transmitters can.

Transmitter Gases

The gases **nitric oxide** (NO) and **carbon monoxide** (CO) further expand the biochemical strategies that transmitter substances display. As water-soluble gases, they are neither stored in synaptic vesicles nor released from them; instead, they are synthesized in the cell as needed. After synthesis, each gas diffuses away, easily crossing the cell membrane and immediately becoming active. Both NO and CO activate metabolic (energy-expending) processes in cells, including processes modulating the production of other neurotransmitters.

Nitric oxide serves as a chemical messenger in many parts of the body. It controls the muscles in intestinal walls, and it dilates blood vessels in brain regions that are in active use, allowing these regions to receive more blood. Because it also dilates blood vessels in the sexual organs, NO is active in producing penile erections. Viagra, a drug used to treat erectile dysfunction in men, acts by enhancing the chemical pathways influenced by NO. Note that NO does not of itself produce sexual arousal.

Two Classes of Receptors

When a neurotransmitter is released from any of the wide varieties of synapses onto a wide variety of targets, as illustrated in Figure 5-6, it crosses the synaptic cleft and binds to a receptor. What happens next depends on the receptor type. Each of the two general classes of receptor proteins has a different effect. One directly changes the electrical potential of the postsynaptic membrane; the other induces cellular change indirectly.

Ionotropic receptors allow the movement of ions such as Na^+, K^+, and Ca^{2+}, across a membrane (the suffix *tropic* means "to move toward"). As **Figure 5-14** illustrates, an ionotropic receptor has two parts: (1) a binding site for a neurotransmitter and (2) a pore, or channel. When the neurotransmitter attaches to the binding site, the receptor changes shape, either opening the pore and allowing ions to flow through it or closing the pore and blocking the flow of ions. Because the binding of the transmitter to the receptor is quickly followed by the opening or closing of the receptor pore that affects the flow of ions, ionotropic receptors bring about very rapid changes in membrane voltage. Ionotropic receptors are usually excitatory: they trigger an action potential.

In contrast, a **metabotropic receptor** has a binding site for a neurotransmitter but lacks its own pore through which ions can flow. Through a series of

Section 6-5 introduces the major classes and functions of hormones, Section 8-4 describes their role in brain development, Sections 12-4 and 12-5 explain hormonal influence over emotional and motivated behaviors, and Section 16-4 details their role in understanding and treating mood disorders.

nitric oxide (NO) Gas that acts as a chemical neurotransmitter—for example, to dilate blood vessels, aid digestion, and activate cellular metabolism.

carbon monoxide (CO) Gas that acts as a neurotransmitter in the activation of cellular metabolism.

ionotropic receptor Embedded membrane protein that acts as (1) a binding site for a neurotransmitter and (2) a pore that regulates ion flow to directly and rapidly change membrane voltage.

metabotropic receptor Embedded membrane protein, with a binding site for a neurotransmitter but no pore, linked to a G protein that can affect other receptors or act with second messengers to affect other cellular processes.

Structurally, ionotropic receptors are similar to the voltage-sensitive channels, discussed in Section 4-3, that propagate the action potential.

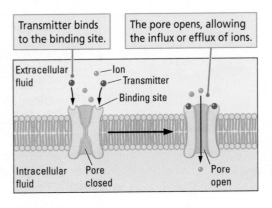

| Transmitter binds to the binding site. | The pore opens, allowing the influx or efflux of ions. |

Extracellular fluid — Ion — Transmitter — Binding site

Intracellular fluid — Pore closed — Pore open

FIGURE 5-14 Ionotropic Receptor When activated, these embedded proteins bring about direct, rapid changes in membrane voltage.

steps, activated metabotropic receptors indirectly produce changes in nearby ion channels or in the cell's metabolic activity.

Figure 5-15A shows the first of these two indirect effects. The metabotropic receptor consists of a single protein that spans the cell membrane, its binding site facing the synaptic cleft. Receptor proteins are each coupled to one of a family of guanyl-nucleotide-binding proteins, **G proteins** for short, shown on the inner side of the cell membrane in Figure 5-15A.

A G protein consists of three subunits: alpha, beta, and gamma. The alpha subunit detaches when a neurotransmitter binds to the G protein's associated metabotropic receptor. The detached alpha subunit can then bind to other proteins within the cell membrane or within the cytoplasm of the cell.

If the alpha subunit binds to a nearby ion channel in the membrane as shown at the bottom of Figure 5-15A, the structure of the channel changes, modifying the flow of ions through it. If the channel is open, it may be closed by the alpha subunit or, if closed, it may open. Changes in the channel and the flow of ions across the membrane influence the membrane's electrical potential.

The binding of a neurotransmitter to a metabotropic receptor can also trigger more complicated cellular reactions, summarized in Figure 5-15B. All these reactions begin

FIGURE 5-15 Metabotropic Receptors When activated, these embedded membrane proteins trigger associated G proteins, thereby exerting indirect effects (**A**) on nearby ion channels or (**B**) in the cell's metabolic activity.

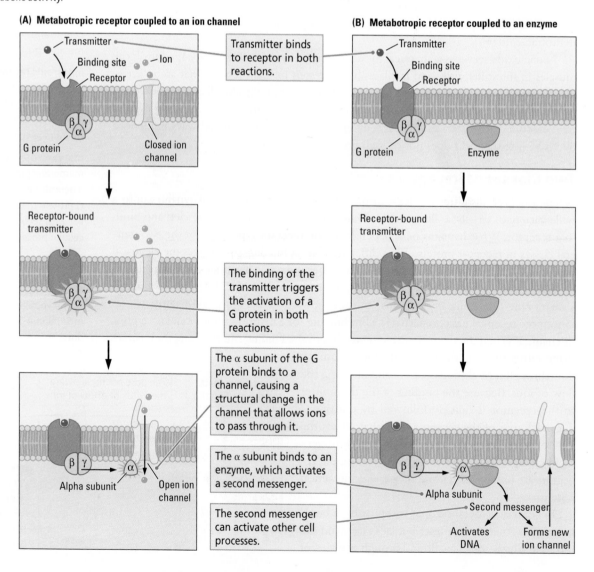

when the detached alpha subunit binds to an enzyme. The enzyme in turn activates a **second messenger** (the neurotransmitter is the first messenger) that carries instructions to other structures inside the cell. As illustrated at the bottom of Figure 5-15B, the second messenger can

- bind to a membrane channel, causing the channel to change its structure and thus alter ion flow through the membrane.

- initiate a reaction that causes protein molecules within the cell to become incorporated into the cell membrane, for example, resulting in the formation of new ion channels.

- instruct the cell's DNA to initiate or cease the production of a protein.

In addition, metabotropic receptors allow for the possibility that a single neurotransmitter's binding to a receptor can activate an escalating sequence of events called an *amplification cascade.* The cascade effect causes many downstream proteins (second messengers or channels or both) to be activated or deactivated. Ionotropic receptors do not have such a widespread "amplifying" effect.

No one neurotransmitter is associated with a single receptor type. A neurotransmitter may bind either to an ionotropic receptor and have an excitatory effect on the target cell or to a metabotropic receptor and have an inhibitory effect.

Recall that acetylcholine has an excitatory effect on skeletal muscles. Here it activates an ionotropic receptor. You know that ACh has an inhibitory effect on the heart rate. Here it activates a metabotropic receptor. In addition, each transmitter may bind with several different kinds of ionotropic or metabotropic receptors. Elsewhere in the nervous system, for example, acetylcholine may activate a wide variety of either receptor type.

G protein Guanyl-nucleotide-binding protein coupled to a metabotropic receptor that, when activated, binds to other proteins.

second messenger Chemical that carries a message to initiate a biochemical process when activated by a neurotransmitter (the first messenger).

REVIEW 5-2
Varieties of Neurotransmitters

Before you continue, check your understanding.

1. Neurotransmitters are identified using four experimental criteria: _____, _____, _____, and _____.

2. The three broad classes of chemically related neurotransmitters are _____, _____, and _____. All three classes, encompassing the approximately 100 likely neurotransmitters active in the nervous system, are associated with both _____ and _____ receptors.

3. Contrast the major characteristics of ionotropic and metabotropic receptors.

Answers appear at the back of the book.

5-3 Neurotransmitter Systems and Behavior

When researchers began to study neurotransmission, you'll recall, they reasoned that any given neuron would contain only one transmitter at all its axon terminals. New methods of analysis, however, reveal that this hypothesis is a simplification.

A single neuron may use one transmitter at one synapse and a different transmitter at another synapse. Moreover, different transmitters may coexist in the same terminal or synapse. Neuropeptides have been found to coexist in terminals with small-molecule transmitters, and more than one small-molecule transmitter may be found in a single synapse. In some cases, more than one transmitter may even be packaged within a single vesicle.

All such findings allow for a number of combinations of neurotransmitters and receptors for them. They caution as well against the assumption of a simple cause-and-effect relation between a neurotransmitter and a behavior. What are the functions of so many combinations? The answer will likely vary, depending on the behavior that is controlled. Generally, neurotransmission is simplified by concentrating on the dominant transmitter located within any given axon terminal. The neuron and its dominant transmitter can then be related to a function or behavior.

We now consider some of the links between neurotransmitters and behavior. We begin by exploring the two parts of the peripheral nervous system: SNS and ANS. Then we investigate neurotransmission in the central nervous system.

Neurotransmission in the Somatic Nervous System

Motor neurons in the brain and spinal cord send their axons to the body's skeletal muscles, including the muscles of the eyes and face, trunk, limbs, fingers, and toes. Without these SNS neurons, movement would not be possible. Motor neurons are also called **cholinergic neurons** because acetylcholine is their main neurotransmitter. At a skeletal muscle, cholinergic neurons are excitatory and produce muscular contractions.

Just as a single main neurotransmitter serves the SNS, so does a single main receptor, an ionotropic, transmitter-activated channel called a *nicotinic ACh receptor* (nAChr). When ACh binds to this receptor, its pore opens to permit ion flow, thus depolarizing the muscle fiber. The pore of a nicotinic receptor is large and permits the simultaneous efflux of K^+ and influx of Na^+. The molecular structure of nicotine, a chemical found in tobacco, activates the nAChr in the same way that ACh does, which is how this receptor got its name. The molecular structure of nicotine is sufficiently similar to ACh that nicotine acts as a mimic, fitting into acetylcholine-receptor binding sites.

Although acetylcholine is the primary neurotransmitter at skeletal muscles, other neurotransmitters also are found in these cholinergic axon terminals and are released onto the muscle along with ACh. One of these neurotransmitters is a neuropeptide called calcitonin-gene-related peptide (CGRP) that acts through CGRP metabotrophic receptors to increase the force with which a muscle contracts.

Two Activating Systems of the Autonomic Nervous System

The complementary divisions of the ANS, sympathetic and parasympathetic, regulate the body's internal environment. The sympathetic division rouses the body for action, producing the fight-or-flight response. Heart rate ramps up, digestive functions ramp down. The parasympathetic division calms the body down, producing an essentially opposite rest-and-digest response. Digestive functions are turned up, heart rate is turned down, and the body is made ready to relax.

Figure 5-16 shows the neurochemical organization of the ANS. Both ANS divisions are controlled by acetylcholine neurons that emanate from the CNS at two levels of the spinal cord. The CNS neurons synapse with parasympathetic neurons that also contain acetylcholine and with sympathetic neurons that contain norepinephrine. In other words, ACh neurons in the CNS synapse with sympathetic NE neurons to prepare the body's organs for fight or flight. Cholinergic (ACh) neurons in the CNS synapse with autonomic ACh neurons in the parasympathetic division to prepare the body's organs to rest and digest.

Whether acetylcholine synapses or norepinephrine synapses are excitatory or inhibitory on a particular body organ depends on that organ's receptors. During sympathetic arousal, norepinephrine turns up heart rate and turns down digestive functions because NE receptors on the heart are excitatory, whereas NE receptors on the gut

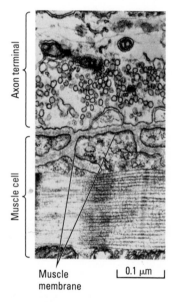

Axon terminal

Muscle cell

Muscle membrane

⊢ 0.1 μm ⊣

Nicotinic Acetylcholine Receptor From *The Nervous System*, Vol. 1, *Handbook of Physiology*, by J. E. Heuser and T. Reese, 1977, edited by E. R. Kandel, Oxford University Press, p. 266.

To review the divisions of the autonomic nervous system in detail, see Figure 2-30.

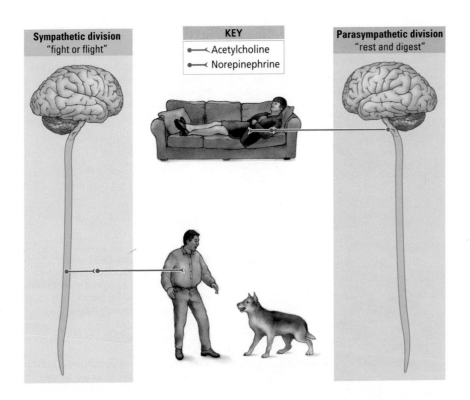

FIGURE 5-16 Controlling Biological Functions in the Autonomic Nervous System   The neurotransmitter of all the neurons leaving the spinal cord is acetylcholine. (*Left*) In the sympathetic division, these ACh neurons activate autonomic norepinephrine neurons, which stimulate organs required for fight or flight and suppress the activity of organs used to rest and digest. (*Right*) In the parasympathetic division, ACh neurons from the spinal cord activate ACh neurons in the ANS, which suppress activity in organs used for fight or flight and stimulate organs used to rest and digest.

are inhibitory. Similarly, acetylcholine turns down heart rate and turns up digestive functions because its receptors on these organs are different. Acetylcholine receptors on the heart are inhibitory, whereas those on the gut are excitatory. The activity of neurotransmitters, excitatory in one location and inhibitory in another, mediate the sympathetic and parasympathetic divisions to form a complementary autonomic regulating system that maintains the body's internal environment under differing circumstances.

Four Activating Systems in the Central Nervous System

Just as there is an organization to the neurochemical systems of the PNS, there is an organization of neurochemical systems in the CNS. These systems are remarkably similar across a wide range of animal species, which allowed for their identification, first in the rat brain and then in the human brain (Hamilton et al., 2010).

For each of the four **activating systems** that we describe here, a relatively small number of neurons grouped together in one or a few brainstem nuclei send axons to widespread CNS regions, suggesting that these nuclei and their terminals play a role in synchronizing activity throughout the brain and spinal cord. You can envision an activating system as analogous to the power supply to a house. The fuse box is the source of the house's power and from it, lines go to each room.

Just as in the ANS, the precise action of the CNS transmitter depends on the region of the brain that is innervated and on the types of receptors on which the transmitter acts at that region. To continue our analogy, precisely what the activating effect of the power is in each room depends on the electrical devices in the room.

Each of four small-molecule transmitters participates in its own neural activating system—the cholinergic, dopaminergic, noradrenergic, and serotonergic systems. **Figure 5-17** maps the location of each system's nuclei, with arrow shafts mapping the pathways of axons and arrowheads indicating axon-terminal locales.

cholinergic neuron Neuron that uses acetylcholine as its main neurotransmitter. The term *cholinergic* applies to any neuron that uses ACh as its main transmitter.

activating system Neural pathways that coordinate brain activity through a single neurotransmitter; cell bodies are located in a nucleus in the brainstem and axons are distributed through a wide region of the brain.

FIGURE 5-17 Major Activating Systems
Each system's cell bodies are gathered into nuclei (shown as ovals) in the brainstem. The axons project diffusely through the brain and synapse on target structures. Each activating system is associated with one or more behaviors or diseases.

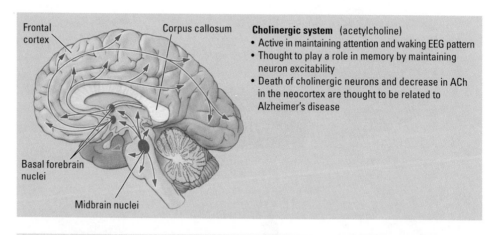

Cholinergic system (acetylcholine)
- Active in maintaining attention and waking EEG pattern
- Thought to play a role in memory by maintaining neuron excitability
- Death of cholinergic neurons and decrease in ACh in the neocortex are thought to be related to Alzheimer's disease

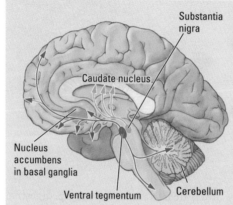

Dopaminergic system (dopamine)
Nigrostriatal pathways (orange projections)
- Active in maintaining normal motor behavior
- Loss of DA is related to muscle rigidity and dyskinesia in Parkinson's disease

Mesolimbic pathways (purple projections)
- Dopamine release causes feelings of reward and pleasure
- Thought to be the neurotransmitter system most affected by addictive drugs and behavioral addictions
- Increases in DA activity may be related to schizophrenia
- Decreases in DA activity may be related to deficits of attention

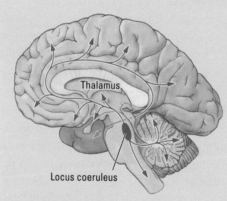

Noradrenergic system (norepinephrine)
- Active in maintaining emotional tone
- Decreases in NE activity are thought to be related to depression
- Increases in NE are thought to be related to mania (overexcited behavior)
- Decreased NE activity is associated with hyperactivity and attention-deficit/hyperactivity disorder

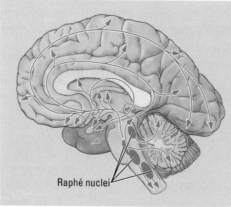

Serotonergic system (serotonin)
- Active in maintaining waking EEG pattern
- Changes in serotonin activity are related to obsessive–compulsive disorder, tics, and schizophrenia
- Decreases in serotonin activity are related to depression
- Abnormalities in brainstem 5-HT neurons are linked to disorders such as sleep apnea and SIDS

Alzheimer's disease Degenerative brain disorder related to aging that first appears as progressive memory loss and later develops into generalized dementia.

As summarized on the right in Figure 5-17, each activating system is associated with a number of behaviors. With the exception of dopamine's clear link to Parkinson's disease, however, associations among activating systems, behavior, and brain disorders are far less certain. All these relations are subjects of ongoing research.

The difficulty in making definitive correlations between activating systems and behavior or activating systems and a disorder is that the axons of these systems connect to almost every part of the brain. They likely have both specific functions and modulatory roles. We will detail some of the documented relations between the systems and behavior and disorders here and in many subsequent chapters.

Cholinergic System

Figure 5-18 shows a cross section of a rat brain stained for the enzyme acetylcholinesterase (AChE) that breaks ACh down in synapses, as diagrammed earlier in Figure 5-10. The darkly stained areas have high AChE concentrations, indicating the presence of cholinergic terminals. Note that AChE is located throughout the cortex and is especially dense in the basal ganglia. Many of these ACh synapses are connections from ACh nuclei in the brainstem as illustrated in the top panel of Figure 5-17.

The cholinergic system plays a role in normal waking behavior and is thought to function in attention and in memory. For example, cholinergic neurons take part in producing one form of waking EEG activity. People who suffer from the degenerative **Alzheimer's disease** that begins with minor forgetfulness, progresses to major memory dysfunction, and later develops into generalized dementia, show a loss of cholinergic neurons at autopsy. One treatment strategy currently being pursued for Alzheimer's is to develop drugs that stimulate the cholinergic system to enhance alertness. But the beneficial effects of these drugs are not dramatic (Herrmann et al., 2011). Recall that ACh is synthesized from nutrients in food; thus, the role of diet in maintaining ACh levels also is being investigated.

The brain abnormalities associated with Alzheimer's disease are not limited to the cholinergic neurons, however. Autopsies reveal extensive damage to the neocortex and other brain regions. As a result, what role the cholinergic neurons play in the progress of the disorder is not yet clear. Perhaps their destruction causes degeneration in the cortex or perhaps the cause-and-effect relation is the other way around, with cortical degeneration causing cholinergic cell death. Then, too, the loss of cholinergic neurons may be just one of many neural symptoms of Alzheimer's disease.

Dopaminergic System

Note in Figure 5-17 that the dopaminergic activating system operates in two distinct pathways. The *nigrostriatal dopaminergic system* plays a role in coordinating movement. As described throughout this chapter in relation to Parkinsonism, when DA neurons in the substantia nigra are lost, the result is a condition of extreme muscular rigidity. Opposing muscles contract at the same time, making it difficult for an affected person to move.

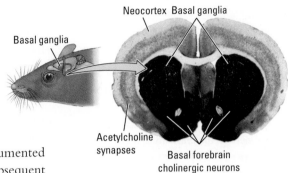

Basal ganglia · Neocortex · Basal ganglia · Acetylcholine synapses · Basal forebrain cholinergic neurons

FIGURE 5-18 **Cholinergic Activation** The drawing shows the cortical location of the micrograph stained to reveal the enzyme AChE. Cholinergic neurons of the basal forebrain project to the neocortex, and the darkly stained bands in the cortex show areas rich in cholinergic synapses. The darker central parts of the section, also rich in cholinergic neurons, are the basal ganglia.

The EEG, or electroencephalogram, detects electrical signals the brain emits during various states of consciousness (see Sections 7-2 and 13-3). Section 4-1 relates how the EEG helped explain how the nervous system conveys information.

The current state of research on Alzheimer's disease and treatment for dementias are detailed in Focus 14-3 and Section 16-3.

For Parkinson's patients, rhythmic movement apparently helps to restore the balance between neural excitation and inhibition—between the loss and the release of behavior. Patients who attend dance classes, as pictured here, report that moving to music helps them regain muscle control. Exercise and music are helpful additions to treatments directed toward replacing depleted dopamine.

Parkinson patients also exhibit rhythmic tremors, especially of the limbs, which signals a release of formerly inhibited movement. Although Parkinson's disease usually arises for no known cause, it can actually be triggered by the ingestion of certain "toxic" drugs, as described in Clinical Focus 5-4, "The Case of the Frozen Addict." Those drugs may act as selective neurotoxins that kill dopamine neurons.

Dopamine in the *mesolimbic dopaminergic system* may be the neurotransmitter most affected in addiction—to food, drugs, and to other behaviors that involve a loss of impulse control. A common feature of addictive behaviors is that stimulating the mesolimbic DA system enhances responses to environmental stimuli, thus making the stimuli attractive and rewarding.

Excessive mesolimbic DA activity is proposed as well to play a role in **schizophrenia,** a behavioral disorder characterized by delusions, hallucinations, disorganized speech, blunted

Drug effects on the mesolimbic dopaminergic system are described in Sections 6-3, 6-4, and 12-3. We examine schizophrenia's possible causes in Sections 6-2 and 7-4 and its neurobiology in Section 16-4.

CLINICAL FOCUS ⊕ 5-4

The Case of the Frozen Addict

Patient 1: During the first 4 days of July 1982, a 42-year-old man used 4½ grams of a "new synthetic heroin." The substance was injected intravenously three or four times daily and caused a burning sensation at the site of injection. The immediate effects were different from heroin, producing an unusual "spacey" high as well as transient visual distortions and hallucinations. Two days after the final injection, he awoke to find that he was "frozen" and could move only in "slow motion." He had to "think through each movement" to carry it out. He was described as stiff, slow, nearly mute, and catatonic during repeated emergency room visits from July 9 to July 11. (Ballard et al., 1985, p. 949)

Patient 1 was one of seven young adults hospitalized at about the same time in California. All showed symptoms of severe Parkinson's disease that appeared very suddenly after drug injection. These symptoms are extremely unusual in this age group. All those affected reportedly injected a synthetic heroin that was being sold on the streets in the summer of 1982.

J. William Langston (2008) and his colleagues found that the heroin contained a contaminant called MPTP (1-methyl-4-phenyl-1,2,3,6-tetrahydropyridine) resulting from poor preparation during its synthesis. The results of experimental studies in rodents showed that MPTP was not itself responsible for the patients' symptoms but was metabolized into MPP+ (1-methyl-4-phenylpyridinium), a neurotoxin.

In one autopsy of a suspected case of MPTP poisoning, the victim suffered a selective loss of dopamine neurons in the substantia nigra. The rest of the brain was normal. Injection of MPTP into monkeys, rats, and mice produced similar symptoms and a similar selective loss of DA neurons in the substantia nigra. Thus, the combined clinical and experimental evidence indicates that Parkinson's disease can be induced by a toxin that selectively kills dopamine neurons.

In 1988, Patient 1 received an experimental treatment at University Hospital in Lund, Sweden. Dopamine neurons taken from human fetal brains at autopsy were implanted into the caudate nucleus and putamen. Extensive work with rodents and nonhuman primates in a number

Dr. Hakan Widner, M.D., PhD., Lund University, Sweden

Positron emission tomographic (PET) images of Patient 1's brain before the implantation of fetal dopamine neurons (*left*) and 12 months after the operation (*right*). The increased areas of red and gold show that the transplanted neurons are producing DA. From **"Bilateral Fetal Mesencephalic Grafting in Two Patients with Parkinsonism Induced by 1-Methyl-4-phenyl-1,2,3,6-tetrahydropyradine (MPTP),"** by H. Widner, J. Tetrud, S. Rehngrona, B. Snow, P. Brundin, et al., 1992, *New England Journal of Medicine, 327,* p. 151.

of laboratories had demonstrated that fetal neurons, which have not yet developed dendrites and axons, can survive transplantation and grow into mature neurons that can secrete neurotransmitters.

Patient 1 had no serious postoperative complications and was much improved 24 months after the surgery. He could dress and feed himself, visit the bathroom with help, and make trips outside his home. He also responded much better to medication.

The transplantation of fetal neurons to treat Parkinson's disease does not work in all patients and can produce unwanted side effects, including frequent involuntary movements. Because Parkinson's disease can affect as many as 20 people per 100,000, scientists continue to experiment with new approaches to brain transplantation and with genetic approaches for modifying remaining dopamine neurons (Lane et al., 2010).

emotion, agitation or immobility, and a host of associated symptoms. Schizophrenia is one of the most common and debilitating psychiatric disorders, affecting 1 in 100 people.

Noradrenergic System

The term *noradrenergic neuron* describes a neuron using noradrenaline as its transmitter. The term *noradrenaline* is derived from *adrenaline,* the Latin name for epinephrine. Norepinephrine (noradrenalin) may play a role in learning by stimulating neurons to change their structure. It may also facilitate normal development of the brain and play a role in organizing movements.

In the main, behaviors and disorders related to the noradrenergic system concern the emotions. Some symptoms of **major depression**—a mood disorder characterized by prolonged feelings of worthlessness and guilt, the disruption of normal eating habits, sleep disturbances, a general slowing of behavior, and frequent thoughts of suicide—may be related to decreases in the activity of noradrenergic neurons. Conversely, some symptoms of **mania** (excessive excitability) may be related to increases in the activity of these same neurons. Decreased NE activity has also been associated both with hyperactivity and attention-deficit/hyperactivity disorder (ADHD).

Serotonergic System

The serotonergic activating system maintains a waking EEG in the forebrain when we move and thus plays a role in wakefulness, as does the cholinergic system. Like norepinephrine, serotonin also plays a role in learning, as described in Section 5-4. Some symptoms of depression may be related to decreases in the activity of serotonin neurons, and drugs commonly used to treat depression act on serotonin neurons. Consequently, two forms of depression may exist, one related to norepinephrine and another related to serotonin.

Likewise, the results of some research suggest that various symptoms of schizophrenia also may be related to increases in serotonin activity, which implies that there may be different forms of schizophrenia. Increased serotonergic activity is also related to symptoms observed in **obsessive-compulsive disorder** (OCD), a condition in which a person compulsively repeats acts (such as hand washing) and has repetitive and often unpleasant thoughts (obsessions). Evidence also points to a link between abnormalities in serotonergic nuclei and conditions such as sleep apnea and sudden infant death syndrome (SIDS).

schizophrenia Behavioral disorder characterized by delusions, hallucinations, disorganized speech, blunted emotion, agitation or immobility, and a host of associated symptoms.

noradrenergic neuron From adrenaline, Latin for "epinephrine"; a neuron containing norepinephrine.

major depression Mood disorder characterized by prolonged feelings of worthlessness and guilt, the disruption of normal eating habits, sleep disturbances, a general slowing of behavior, and frequent thoughts of suicide.

mania Disordered mental state of extreme excitement.

obsessive-compulsive disorder (OCD) Behavior disorder characterized by compulsively repeated acts (such as hand washing) and repetitive, often unpleasant, thoughts (obsessions).

Details on the causes of SIDS appear in Research Focus 8-1. Clinical Focus 13-5 details sleep apnea. Information on causes and treatments for OCD appears in Sections 6-2, 12-4, and 16-3.

REVIEW 5-3

Neurotransmitter Systems and Behavior

Before you continue, check your understanding.

1. Although neurons can synthesize more than one _____, they are usually identified by the principal _____ in their axon terminals.

2. In the peripheral nervous system, the neurotransmitter at somatic muscles is _____; in the autonomic nervous system, _____ neurons from the spinal cord connect to _____ neurons for parasympathic activity and with _____ neurons for sympathetic activity.

3. The four main activating systems of the brain are _____, _____, _____, and _____.

4. How would you respond to the comment that a behavior is caused solely by a "chemical imbalance in the brain"?

Answers appear at the back of the book.

Adaptive Role of Synapses in Learning and Memory

Experiment 2-1, which relates how investigators first demonstrated observational learning in the octopus, points to the ubiquity of neuroplasticity, the nervous system's potential for change that enhances its ability to adapt.

Aplysia californica

In Section 14-4 we investigate in detail the neural bases of brain plasticity in conscious learning and in memory.

learning Relatively permanent change in behavior that results from experience.

habituation Learning behavior in which a response to a stimulus weakens with repeated stimulus presentations.

Among our most cherished abilities are learning and remembering. Neuroplasticity is both a requirement for learning and memory and a characteristic not only of the mammalian brain but also of the nervous systems of all animals, even the simplest worms. Larger brains with more connections are more plastic, however, and thus likely to show more adaptability in neural organization.

Greater adaptability happens because experience alters the synapse. Not only are synapses versatile in structure and function, they are also plastic: they can change. The synapse, therefore, provides a site for the neural basis of **learning,** a relatively permanent change in behavior that results from experience.

Donald O. Hebb (1949) was not the first to suggest that learning is mediated by structural changes in synapses. But the change that he envisioned in his book *The Organization of Behavior* was novel 65 years ago. Hebb theorized, "When an axon of cell A is near enough to excite a cell B and repeatedly or persistently takes part in firing it, some growth process or metabolic change takes place in one or both cells such that A's efficiency, as one of the cells firing B, is increased" (Hebb, 1949, p. 62). A synapse that physically adapts in this way is called a *Hebb synapse* today.

Eric Kandel was awarded a Nobel Prize in 2000 for his descriptions of the synaptic basis of learning in a way that Hebb envisaged: learning in which the conjoint activity of nerve cells serves to link them. Kandel's subject, the marine snail *Aplysia californica,* is an ideal subject for learning experiments. Slightly larger than a softball and lacking a shell, *Aplysia* has roughly 20,000 neurons. Some are quite accessible to researchers, who can isolate and study circuits having very few synapses.

When threatened, *Aplysia* defensively withdraws its more vulnerable body parts—the gill (through which it extracts oxygen from the water to breathe) and the siphon (a spout above the gill that excretes seawater and waste). By stroking or shocking the snail's appendages, Kandel and his coworkers produced enduring changes in its defensive behaviors. They used these behavioral responses to study underlying changes in the snail's nervous system.

We now illustrate the role of synapses in two kinds of learning that Kandel has studied: habituation and sensitization. For humans, both are called "unconscious" because they do not depend on a person's knowing precisely when and how they occur. In learning how unconscious learning takes place you will recognize each of these behaviors as part of your experience. You will also recognize the neural basis of unconscious learning because it entails changes in synaptic function and structure with which you are familiar.

Habituation Response

In **habituation,** the response to a stimulus weakens with repeated presentations of the stimulus. If you are accustomed to living in the country and then move to a city, you might at first find the sounds of traffic and people extremely loud and annoying. With time, however, you stop noticing most of the noise most of the time. You have habituated to it.

Habituation develops with all our senses. When you first put on a shoe, you "feel" it on your foot, but very soon it is as if the shoe were not there. You have not become insensitive to sensations, however. When people talk to you, you still hear them; when someone steps on your foot, you still feel the pressure. Your brain simply has habituated to the customary "background" sensation of a shoe on your foot.

Aplysia habituates to waves in the shallow tidal zone where it lives. These snails are constantly buffeted by the flow of waves against their bodies, and they learn that waves are just the "background noise" of daily life. They do not flinch and withdraw every time a wave passes over them. They habituate to this stimulus.

A sea snail that is habituated to waves remains sensitive to other touch sensations. Prodded with a novel object, it responds by withdrawing its siphon and gill. The animal's reaction to repeated presentations of the same novel stimulus forms the basis for **Experiment 5-2,** studying its habituation response.

Neural Basis of Habituation

The Procedure section of Experiment 5-2 shows the setup for studying what happens to the withdrawal response of *Aplysia*'s gill after repeated stimulation. A gentle jet of water is sprayed on the siphon while movement of the gill is recorded. If the jet of water is presented to *Aplysia*'s siphon as many as 10 times, the gill-withdrawal response is weaker some minutes later when the animal is again tested with the water jet. The decrement in the strength of the withdrawal is habituation, which can last as long as 30 minutes.

The Results section of Experiment 5-2 starts by showing a simple representation of the pathway that mediates *Aplysia*'s gill-withdrawal response. For purposes of illustration, only one sensory neuron, one motor neuron, and one synapse are shown; in actuality, about 300 neurons may take part in this response. The jet of water stimulates the sensory neuron, which in turn stimulates the motor neuron responsible for the gill withdrawal. But exactly where do the changes associated with habituation take place? In the sensory neuron? In the motor neuron? In the synapse between the two?

Habituation does not result from an inability of either the sensory or the motor neuron to produce action potentials. In response to direct electrical stimulation, both the sensory neuron and the motor neuron retain the ability to generate action potentials even after habituation. Electrical recordings from the motor neuron show that, as habituation develops, the excitatory postsynaptic potentials in the motor neuron become smaller.

The most likely way in which these EPSPs decrease in size is that the motor neuron is receiving less neurotransmitter from the sensory neuron across the synapse. And if less neurotransmitter is being received, then the changes accompanying habituation must be taking place in the presynaptic axon terminal of the sensory neuron.

Calcium Channels Habituate

Kandel and his coworkers measured neurotransmitter output from a sensory neuron and verified that less neurotransmitter is in fact released from a habituated neuron than from a nonhabituated one. Recall from Figure 5-5 that the release of a neurotransmitter in response to an action potential requires an influx of calcium ions across the presynaptic membrane. As habituation takes place, that Ca^{2+} influx decreases in response to

---

EXPERIMENT 5-2

Question: What happens to gill response after repeated stimulation?

Procedure

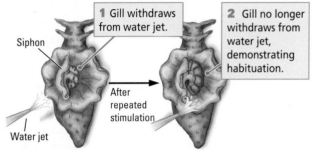

1 Gill withdraws from water jet.

2 Gill no longer withdraws from water jet, demonstrating habituation.

Siphon

After repeated stimulation

Water jet

Results

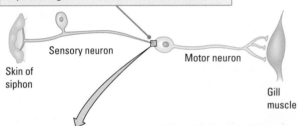

The sensory neuron stimulates the motor neuron to produce gill withdrawal before habituation.

Sensory neuron

Motor neuron

Skin of siphon

Gill muscle

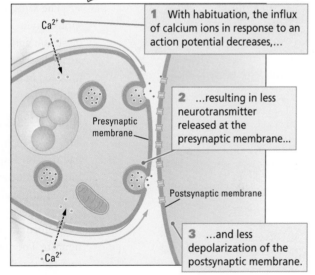

Ca^{2+}

1 With habituation, the influx of calcium ions in response to an action potential decreases,…

Presynaptic membrane

2 …resulting in less neurotransmitter released at the presynaptic membrane…

Postsynaptic membrane

Ca^{2+}

3 …and less depolarization of the postsynaptic membrane.

Conclusion: Withdrawal response weakens with repeated presentation of water jet (habituation) owing to decreased Ca^{2+} influx and subsequently less neurotransmitter release from the presynaptic axon terminal.

the voltage changes associated with an action potential. Presumably, with repeated use, voltage-sensitive calcium channels become less responsive to voltage changes and more resistant to the passage of calcium ions.

The neural basis of habituation lies in the change in presynaptic calcium channels. Its mechanism, which is summarized close up in the Results section of Experiment 5-2, is a reduced sensitivity of calcium channels and a consequent decrease in the release of a neurotransmitter. Thus, habituation can be linked to a specific molecular change, as summarized in the experiment's Conclusion.

Sensitization Response

A sprinter crouched in her starting blocks is often hyperresponsive to the starter's gun: its firing triggers in her a rapid reaction. The stressful, competitive context in which the race takes place helps to sensitize her to this sound. **Sensitization,** an enhanced response to some stimulus, is the opposite of habituation. The organism becomes hyperresponsive to a stimulus rather than accustomed to it.

Sensitization occurs within a context. Sudden, novel stimulation heightens our general awareness and often results in larger-than-normal responses to all kinds of stimulation. If you are suddenly startled by a loud noise, you become much more responsive to other stimuli in your surroundings, including some to which you had been previously habituated. In **posttraumatic stress disorder** (PTSD), physiological arousal related to recurring memories and dreams surrounding a traumatic event persist for months or years after the event. One characteristic of PTSD is a heightened response to stimuli, suggesting that the disorder is in part related to sensitization.

The same thing happens to *Aplysia*. Sudden, novel stimuli can heighten a snail's responsiveness to familiar stimulation. When attacked by a predator, for example, the snail displays heightened responses to many other stimuli in its environment. In the laboratory, a small electric shock to *Aplysia*'s tail mimics a predatory attack and effects sensitization, as illustrated in the Procedure section of **Experiment 5-3.** A single electric shock to the snail's tail enhances its gill-withdrawal response for a period that lasts from minutes to hours.

Neural Basis of Sensitization

The neural circuits participating in sensitization differ from those that take part in a habituation response. The Results section of Experiment 5-3 again shows one of each kind of neuron: the heretofore-described sensory and motor neurons that produce the gill-withdrawal response and adds an interneuron that is responsible for sensitization.

An interneuron that receives input from a sensory neuron in the tail (and so carries information about the shock) makes an axoaxonic synapse with a siphon sensory neuron. The interneuron's axon terminal contains serotonin. Consequently, in response to a tail shock, the tail sensory neuron activates the interneuron, which in turn releases serotonin onto the axon of the siphon sensory neuron. Information from the siphon still comes through the siphon sensory neuron to activate the motor neuron leading to the gill muscle, but the gill-withdrawal response is amplified by the interneuron's action in releasing serotonin onto the presynaptic membrane of the sensory neuron.

At the molecular level, shown close up in Experiment 5-3, the serotonin released from the interneuron binds to a metabotropic serotonin receptor on the axon of the siphon sensory neuron. This binding activates second messengers in the sensory neuron. Specifically, the serotonin receptor is coupled through its G protein to the enzyme adenyl cyclase. This enzyme increases the concentration of the second messenger cyclic adenosine monophosphate (cAMP) in the presynaptic membrane of the siphon sensory neuron.

Through a number of chemical reactions, cAMP attaches a phosphate molecule (PO_4) to potassium channels, and the phosphate renders the potassium channels less

The role of stress in fostering and prolonging the effects of PTSD is a topic in Section 6-5 and 12-4. Section 16-4 covers treatment strategies.

sensitization Learning behavior in which the response to a stimulus strengthens with repeated presentations of that stimulus because the stimulus is novel or because the stimulus is stronger than normal—for example, after habituation has occurred.

posttraumatic stress disorder (PTSD) Syndrome characterized by physiological arousal symptoms related to recurring memories and dreams related to a traumatic event for months or years after the event.

responsive. The closeup in Experiment 5-3 sums it up. In response to an action potential traveling down the axon of the siphon sensory neuron (such as one generated by a touch to the siphon), the potassium channels on that neuron are slower to open. Consequently, K^+ ions cannot repolarize the membrane as quickly as is normal, so the action potential lasts longer than it usually would.

Potassium Channels Sensitize

The longer-lasting action potential that occurs because potassium channels are slower to open prolongs the inflow of Ca^{2+} into the membrane. You know that Ca^{2+} influx is necessary for neurotransmitter release. Thus, more Ca^{2+} influx results in more neurotransmitter being released from the sensory synapse onto the motor neuron.

This increased release of neurotransmitter produces greater activation of the motor neuron and thus a larger-than-normal gill-withdrawal response. The gill withdrawal may also be enhanced by the fact that the second messenger cAMP may mobilize more synaptic vesicles, making more neurotransmitter ready for release into the sensory–motor synapse.

Sensitization, then, is the opposite of habituation at the molecular level as well as at the behavioral level. In sensitization, more Ca^{2+} influx results in more transmitter being released, whereas in habituation, less Ca^{2+} influx results in less neurotransmitter being released. The structural basis of cellular memory in these two forms of learning is different, however. In sensitization, the change takes place in potassium channels, whereas in habituation, the change takes place in calcium channels.

Learning As a Change in Synapse Number

The neural changes associated with learning must last long enough to account for a relatively permanent change in an organism's behavior. The changes at synapses described in the preceding sections develop quite quickly, but they do not last indefinitely, as memories often do. How, then, can synapses be responsible for the long-term changes associated with learning and memory?

Repeated stimulation produces habituation and sensitization that can persist for months. Brief training produces short-term learning, whereas longer training periods produce more enduring learning. If you cram for an exam the night before you take it, you might forget the material quickly, but if you study a little each day for a week, your learning may tend to endure. What underlies this more persistent form of learning?

Craig Bailey and Mary Chen (see Miniaci et al., 2008) found that the number and size of sensory synapses change in well-trained, habituated, and sensitized *Aplysia*. Relative to a control neuron, the number and size of synapses decrease in habituated animals and increase in sensitized animals, as represented in **Figure 5-19**. Apparently, synaptic events associated with habituation and sensitization can also trigger processes in the sensory cell that result in the loss or formation of new synapses.

A mechanism through which these processes can take place begins with calcium ions that mobilize second messengers to send instructions to nuclear DNA. The transcription and translation of nuclear DNA, in turn, initiate structural changes at synapses, including the formation of new synapses and new

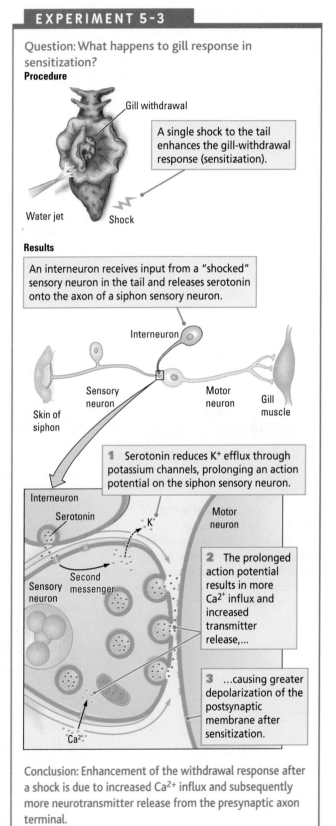

EXPERIMENT 5-3

Question: What happens to gill response in sensitization?

Procedure

Gill withdrawal

A single shock to the tail enhances the gill-withdrawal response (sensitization).

Water jet   Shock

Results

An interneuron receives input from a "shocked" sensory neuron in the tail and releases serotonin onto the axon of a siphon sensory neuron.

Interneuron

Sensory neuron   Motor neuron   Gill muscle

Skin of siphon

1 Serotonin reduces K^+ efflux through potassium channels, prolonging an action potential on the siphon sensory neuron.

Interneuron

Serotonin K^+

Motor neuron

Second messenger

Sensory neuron

2 The prolonged action potential results in more Ca^{2+} influx and increased transmitter release,...

3 ...causing greater depolarization of the postsynaptic membrane after sensitization.

Ca^{2+}

Conclusion: Enhancement of the withdrawal response after a shock is due to increased Ca^{2+} influx and subsequently more neurotransmitter release from the presynaptic axon terminal.

FIGURE 5-19 Physical Basis of Memory
Relative to a control neuronal connection (*left*), the number of synapses between *Aplysia*'s sensory neuron and a motor neuron decline as a result of habituation (*center*) and increase as a result of sensitization (*right*). Such structural changes may underlie enduring memories.

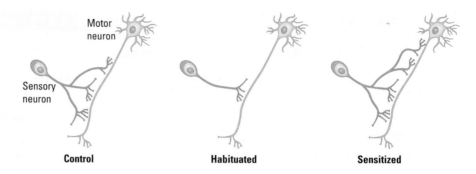

Motor neuron

Sensory neuron

Control   **Habituated**   **Sensitized**

Section 8-2 explains where neuronal cells originate and how they grow and develop.

Drosophila

	cAMP	
No learning	High levels	*dunce*
Learning	Normal levels	No mutation
No learning	Low levels	*rutabaga*

FIGURE 5-20 Genetic Disruption of Learning   Two mutations in the fruit fly, *Drosophila*, inactivate the second messenger cAMP by moving its level above or below the concentration range at which it can be regulated, thus disrupting learning.

dendritic spines. Research Focus 5-5, "Dendritic Spines, Small but Mighty," summarizes experimental evidence about structural changes in dendritic spines.

The second messenger cAMP plays an important role in carrying instructions regarding these structural changes to nuclear DNA. The evidence for cAMP's involvement comes from studies of fruit flies. In the fruit fly, *Drosophila*, two genetic mutations can produce the same learning deficiency. Both render the second messenger cAMP inoperative, but in opposite ways. One mutation, called *dunce*, lacks the enzymes needed to degrade cAMP, so the fruit fly has abnormally high cAMP levels. The other mutation, called *rutabaga*, reduces levels of cAMP below the normal range for *Drosophila* neurons.

Significantly, fruit flies with either of these mutations are impaired in acquiring habituated and sensitized responses because their levels of cAMP cannot be regulated. New synapses seem to be required for learning to take place, and the second messenger cAMP seems to be needed to carry instructions to form them. **Figure 5-20** summarizes these research findings.

More lasting habituation and sensitization are mediated by relatively permanent changes in neuronal structure—by fewer or more synaptic connections—and the effects can be difficult to alter. As a result of sensitization, for example, symptoms of posttraumatic stress disorder can persist indefinitely.

REVIEW 5-4
Adaptive Role of Synapses in Learning and Memory

Before you continue, check your understanding.

1. Experience alters the _____, the site of the neural basis of _____, a relatively permanent change in behavior that results from experience.

2. *Aplysia*'s synaptic function mediates two basic forms of learning: _____ and
   _____.

3. Changes that accompany habituation take place within the _____ of the _____ neuron, mediated by _____ channels that grow _____ sensitive with use.

4. The sensitization response is amplified by _____ that release serotonin onto the presynaptic membrane of the sensory neuron, changing the sensitivity of presynaptic _____ channels and increasing the influx of _____.

5. One characteristic of _____, defined as physiological arousal related to recurring memories and dreams surrounding a traumatic event that persist for months or years after the event, is a heightened response to stimuli. This suggests that the disorder is in part related to
   _____.

6. Describe the benefits and/or drawbacks of permanent habituation and sensitization.

Answers appear at the back of the book.

Dendritic Spines, Small but Mighty

Dendritic spines measure from about 1 to 3 micrometers long, are less than 1 micrometer in diameter, and protrude from the dendrite shaft. Each neuron may have many thousands of spines. The number of dendritic spines in the human cerebral cortex may be 10^{14}.

Dendritic spines have their origins in filopodia (from the Latin *file,* for "thread," and the Greek *podium,* for "foot") that bud out of neurons, especially their dendrites. Microscopic observation of dendrites shows that filopodia are constantly emerging and retracting, over times on the order of seconds.

The budding of filopodia is much more pronounced in developing neurons and in the developing brain. Because filopodia can become dendritic spines, the budding suggests that they are searching for contacts from axon terminals, so as to form synapses. When contact is made, some synapses so formed may have only a short life; others will endure.

A permanent dendritic spine tends to have a large head, giving it a large area of contact with a terminal button, and a long stem, giving it an identity apart from that of its dendrite. The heads and the terminals of presynaptic connections serve as biochemical compartments that can generate huge electrical potentials and so influence the neuron's electrical messages.

Dendritic spines provide the biostructural basis of our behavior, our individual skills, and our memories (Bosch & Hayashi, 2011). Impairments in forming spines characterize some kinds of mental disability, and the loss of spines is associated with the dementia of Alzheimer's disease.

Dendritic spines mediate learning that lasts, unconscious habituation and sensitization included. To mediate learning, each spine must be able to act independently, undergoing changes that its neighbors do not undergo.

Examination of dendritic spines in the nervous system shows that some are simple and others complex. The cellular mechanisms that allow synapses to appear on spines and to change shape include a number of different microfilaments linked to the membrane receptors, transport of proteins from the cell body, and the incorporation of nutrients from the extracellular space.

This variety suggests that all this activity changes the appearance of both presynaptic and postsynaptic structures. The accompanying illustration summarizes synaptic structures that can be measured and related to learning and behavior and to the structural changes that may subserve learning.

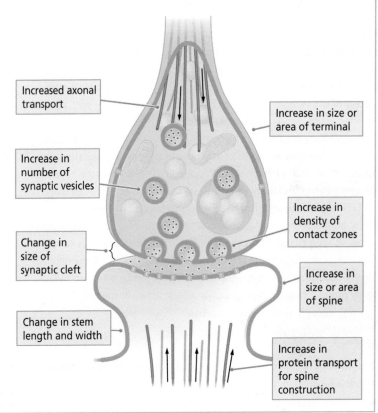

Increased axonal transport

Increase in size or area of terminal

Increase in number of synaptic vesicles

Increase in density of contact zones

Change in size of synaptic cleft

Increase in size or area of spine

Change in stem length and width

Increase in protein transport for spine construction

SUMMARY

5-1 A Chemical Message

In the 1920s, Otto Loewi suspected that nerves to the heart secrete a chemical that regulates its rate of beating. His subsequent experiments with frogs showed that acetylcholine slows heart rate, whereas epinephrine increases it. This observation provided the key to understanding the basis of chemical neurotransmission.

A synapse consists of the sending neuron's axon terminal (surrounded by a presynaptic membrane), a synaptic cleft (a tiny gap between the two neurons), and a postsynaptic membrane on the receiving neuron. Systems for the chemical synthesis of an excitatory or inhibitory neurotransmitter are located in the presynaptic neuron's axon terminal or its soma, whereas systems for storing the neurotransmitter are in its axon terminal. Receptor systems on which that neurotransmitter acts are located on the postsynaptic membrane. Neurons also make direct connections with each other through gap junctions, channel-forming proteins that allow direct sharing of ions or nutrient substances.

The four major stages in neurotransmission are (1) synthesis and storage, (2) release from the axon terminal, (3) action on postsynaptic receptors, and (4) inactivation. After synthesis, the neurotransmitter is wrapped in a membrane to form synaptic vesicles in the axon terminal. When an action potential is propagated on the presynaptic membrane, voltage changes set in motion the attachment of vesicles to the presynaptic membrane and the release of the neurotransmitter by exocytosis.

One synaptic vesicle releases a quantum of neurotransmitter into the synaptic cleft, producing a miniature potential on the postsynaptic membrane. To generate an action potential on the postsynaptic cell requires the simultaneous release of many quanta of transmitter. After a transmitter has done its work, it is inactivated by such processes as diffusion out of the synaptic cleft, breakdown by enzymes, and reuptake of the transmitter or its components into the axon terminal (or sometimes uptake into glial cells).

5-2 Varieties of Neurotransmitters

Small-molecule transmitters, neuropeptides, and transmitter gases are broad classes for ordering the roughly 100 neurotransmitters that investigators propose might exist. Neurons containing these transmitters make a variety of connections with various parts of other neurons as well as with muscles, blood vessels, and extracellular fluid.

Functionally, neurons can be both excitatory and inhibitory, and they can participate in local circuits or in general brain systems. Excitatory synapses, known as Type I, are usually located on a dendritic tree, whereas inhibitory synapses, known as Type II, are usually located on a cell body.

Each neurotransmitter may be associated with both ionotropic and metabotropic receptors. An ionotropic receptor quickly and directly produces voltage changes on the postsynaptic cell membrane as its pore opens or closes to regulate the flow of ions through the cell membrane. Slower-acting metabotropic receptors activate second messengers to indirectly produce changes in the function and structure of the cell.

5-3 Neurotransmitter Systems and Behavior

Because neurotransmitters are multifunctional, scientists find it impossible to isolate single-neurotransmitter–single-behavior relations. Rather, activating systems of neurons that employ the same principal neurotransmitter influence various general aspects of behavior. For instance, acetylcholine, the main neurotransmitter in the somatic nervous system, controls movement of the skeletal muscles, whereas acetylcholine and norepinephrine, the main transmitters in the autonomic system, control the body's internal organs.

The central nervous system contains not only widely dispersed glutamate and GABA neurons but also neural activating systems that employ acetylcholine, norepinephrine, dopamine, or serotonin as their main neurotransmitter. All these systems ensure that wide areas of the brain act in concert, and each is associated with various classes of behaviors and disorders.

5-4 Adaptive Role of Synapses in Learning and Memory

Changes in synapses underlie learning and memory. In habituation, a form of learning in which a response weakens as a result of repeated stimulation, calcium channels become less responsive to an action potential. Consequently, less neurotransmitter is released when an action potential is propagated.

In sensitization, a form of learning in which a response strengthens as a result of stimulation, changes in potassium channels prolong the duration of the action potential, resulting in an increased influx of calcium ions and, consequently, release of more neurotransmitter. With repeated training, new synapses can develop, and both forms of learning can become relatively permanent.

In *Aplysia*, the number of synapses connecting sensory neurons and motor neurons decreases in response to repeated sessions of habituation. Conversely, in response to repeated sessions of sensitization, the number of synapses connecting sensory and motor neurons increases. These changes in synapse number are related to long-term learning.

⊞ KEY TERMS

acetylcholine (ACh), p. 141

activating system, p. 159

Alzheimer's disease, p. 160

autoreceptor, p. 147

carbon monoxide (CO), p. 155

chemical synapse, p. 145

cholinergic neuron, p. 159

dopamine (DA), p. 143

epinephrine (EP), p. 141

G protein, p. 157

gamma-aminobutyric acid (GABA), p. 153

gap junction (electrical synapse), p. 145

glutamate (Glu), p. 153

habituation, p. 164

histamine (H), p. 153

ionotropic receptor, p. 155

learning, p. 164

major depression, p. 163

mania, p. 163

metabotropic receptor, p. 155

neuropeptide, p. 153

neurotransmitter, p. 141

nitric oxide (NO), p. 155

noradrenergic neuron, p. 163

norepinephrine (NE), p. 141

obsessive-compulsive disorder (OCD), p. 163

Parkinson's disease, p. 143

postsynaptic membrane, p. 145

posttraumatic stress disorder (PTSD), p. 166

presynaptic membrane, p. 145

quantum (pl. quanta), p. 147

rate-limiting factor, p. 153

reuptake, p. 147

schizophrenia, p. 163

second messenger, p. 157

sensitization, p. 166

serotonin (5-HT), p. 153

small-molecule transmitter, p. 150

storage granule, p. 145

synaptic cleft, p. 143

synaptic vesicle, p. 143

transmitter-activated receptor, p. 147

transporter, p. 145

How Do Drugs and Hormones Influence the Brain and Behavior?

CLINICAL FOCUS 6-1 COGNITIVE ENHANCEMENT

6-1 PRINCIPLES OF PSYCHOPHARMACOLOGY

DRUG ROUTES INTO THE NERVOUS SYSTEM

DRUG ACTION AT SYNAPSES: AGONISTS AND ANTAGONISTS

AN ACETYLCHOLINE SYNAPSE: EXAMPLES OF DRUG ACTION

TOLERANCE

SENSITIZATION

6-2 GROUPING PSYCHOACTIVE DRUGS

GROUP I: ANTIANXIETY AGENTS AND SEDATIVE HYPNOTICS

CLINICAL FOCUS 6-2 FETAL ALCOHOL SPECTRUM DISORDER

GROUP II: ANTIPSYCHOTIC AGENTS

GROUP III: ANTIDEPRESSANTS AND MOOD STABILIZERS

GROUP IV: OPIOID ANALGESICS

CLINICAL FOCUS 6-3 MAJOR DEPRESSION

GROUP V: PSYCHOTROPICS

6-3 FACTORS INFLUENCING INDIVIDUAL RESPONSES TO DRUGS

BEHAVIOR ON DRUGS

ADDICTION AND DEPENDENCE

SEX DIFFERENCES IN ADDICTION

6-4 EXPLAINING AND TREATING DRUG ABUSE

WANTING-AND-LIKING THEORY

WHY DOESN'T EVERYONE ABUSE DRUGS?

TREATING DRUG ABUSE

CAN DRUGS CAUSE BRAIN DAMAGE?

CLINICAL FOCUS 6-4 DRUG-INDUCED PSYCHOSIS

6-5 HORMONES

HIERARCHICAL CONTROL OF HORMONES

CLASSES AND FUNCTIONS OF HORMONES

HOMEOSTATIC HORMONES

GONADAL HORMONES

ANABOLIC–ANDROGENIC STEROIDS

GLUCOCORTICOIDS AND STRESS

ENDING A STRESS RESPONSE

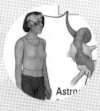

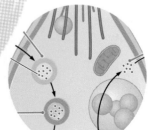

Cognitive Enhancement

A new name for an old game? An article in the preeminent science publication *Nature* floated the idea that certain "cognitive-enhancing" drugs improve school and work performance in otherwise healthy individuals by improving brain function (Greely et al., 2008). The article was instigated in part by reports that up to 20 percent—and in some schools up to 80 percent—of high school and university students were using the combination of Adderall (mainly dextroamphetamine) and Ritalin (methylphenidate) as a study aid to help meet deadlines and to cram for examinations.

Both drugs are prescribed as a treatment for **attention-deficit/hyperactivity disorder** (ADHD), a developmental disorder characterized by core behavioral symptoms of impulsivity, hyperactivity, and/or inattention. Methylphenidate and dextroamphetamine are Schedule II drugs, signifying that they carry the potential for abuse and require a prescription when used medically. Their main illicit source is through falsified prescriptions or purchase from someone who has a prescription. Both drugs share the pharmacological properties of cocaine of stimulating dopamine release and also blocking its reuptake (see Section 6-2).

The use of cognitive enhancers is not new. In his classic paper on cocaine, Viennese psychoanalyst Sigmund Freud stated in 1884, "The main use of coca [cocaine] will undoubtedly remain that which the Indians [of Peru] have made of it for centuries . . . to increase the physical capacity of the body." Freud later withdrew his endorsement when he realized that cocaine is addictive.

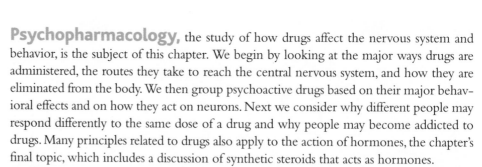

Andrew Jacobs/Redux

In 1937, an article in the *Journal of the American Medical Association* reported that a form of amphetamine, Benzedrine, improved performance on mental-efficiency tests. This information was quickly disseminated among students, who began to use the drug as an aid to study for examinations. In the 1950s, dextroamphetamine, marketed as Dexedrine, was similarly prescribed for narcolepsy and used illicitly by students as a study aid.

The complex neural effects of amphetamine stimulants center on learning at the synapse by means of habituation and sensitization. With repeated use for nonmedicinal purposes, the drugs can also begin to produce side effects including sleep disruption, loss of appetite, and headaches. Some people develop cardiovascular abnormalities and/or become addicted to amphetamine.

Treating ADHD with prescription drugs is itself controversial, despite their widespread use for this purpose. Aagaard and Hansen (2011) note that assessing the adverse effects of cognitive enhancement medication is hampered because many participants drop out of studies and the duration of the studies is short.

Despite their contention that stimulant drugs can improve school and work performance by improving brain function in otherwise healthy individuals, Greely and coworkers (2008) call for more research into the legal implications related to using cognitive enhancers, their beneficial effects, and the long-term neural consequences of their use.

Psychopharmacology, the study of how drugs affect the nervous system and behavior, is the subject of this chapter. We begin by looking at the major ways drugs are administered, the routes they take to reach the central nervous system, and how they are eliminated from the body. We then group psychoactive drugs based on their major behavioral effects and on how they act on neurons. Next we consider why different people may respond differently to the same dose of a drug and why people may become addicted to drugs. Many principles related to drugs also apply to the action of hormones, the chapter's final topic, which includes a discussion of synthetic steroids that acts as hormones.

Before we examine how drugs produce their effects on the brain for good or for ill, we must raise a caution: the sheer number of neurotransmitters, receptors, and possible sites of drug action is astounding. Most drugs act at many sites in the body and brain and affect more than one activating system, and most receptors on which drugs act have many variations. Individual differences among people—their sex, genetic make-up, and age, height, and weight—all influence how drugs affect them. Considering all the variables, psychopharmacological research has made important advances in understanding drug action. And yet it remains safe to say that neuroscientists do not know everything there is to know about any drug.

6-1 Principles of Psychopharmacology

Drugs are chemical compounds administered to bring about some desired change in the body. Drugs are usually used to diagnose, treat, or prevent illness, to relieve pain and suffering, or to improve some adverse physiological condition. In this chapter, we focus on **psychoactive drugs**—substances that act to alter mood, thought, or behavior, are used to manage neuropsychological illness, and may be abused. We also consider psychoactive drugs that can act as toxins, producing sickness, brain damage, or death.

Drug Routes into the Nervous System

To be effective, a psychoactive drug has to reach its nervous-system target. The way in which a drug enters and passes through the body to reach its target is called its *route of administration*. Drugs can be administered orally, inhaled into the lungs, administered through rectal suppositories, absorbed from patches applied to the skin or mucous membranes, or injected into the bloodstream, into a muscle, or even into the brain. **Figure 6-1** illustrates some of these routes of drug administration and summarizes the characteristics of drugs that allow them to pass through various barriers to reach their targets.

Oral administration is easy and convenient but also the most complex route. To reach the bloodstream, an ingested drug must first be absorbed through the lining of the stomach or small intestine. If the drug is liquid, it is absorbed more readily. Drugs taken in solid form are not absorbed unless they can be dissolved by the stomach's gastric juices. Some drugs may be destroyed or altered by enzymes in the gastrointestinal tract. Whether a drug is an acid or a base influences its absorption.

Once absorbed by the stomach or intestine, the drug must next enter the bloodstream. This part of the journey requires that the drug have additional properties. Because blood has a high water concentration, the drug must be soluble in water. In the blood, it is then diluted by the approximately 6 liters of blood that circulate through an adult's body. When the drug leaves the bloodstream, the body's roughly 35 liters of extracellular fluid further dilutes it.

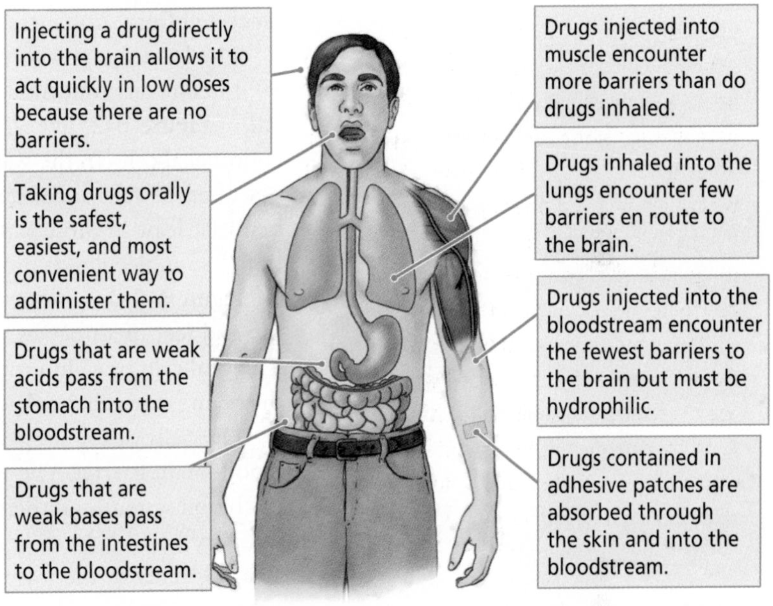

Drugs that are administered as gases or aerosols penetrate the cell linings of the respiratory tract easily and are absorbed across these membranes into the bloodstream nearly as quickly as they are inhaled. Thus, they reach the bloodstream by circumventing the barriers posed by the digestive system. When administered as a gas or in smoke, drugs of abuse, including nicotine, cocaine, and marijuana, are absorbed in a similar way.

Our largest organ, the skin, has three layers of cells designed to be a protective body coat. Some small-molecule drugs (e.g., nicotine when used in a drug patch) penetrate the skin's barrier almost as easily as they penetrate the cell lining of the respiratory tract. There are still fewer obstacles to a drug destined for the brain if that drug is injected directly into the bloodstream. The fewest obstacles are encountered if a psychoactive drug is injected directly into the brain.

With each obstacle eliminated en route to the brain, the dosage of a drug can be reduced by a factor of 10. For example, 1000 micrograms of *amphetamine,* a psychomotor stimulant and major component in the drugs described in Clinical Focus 6-1, produces a noticeable behavioral change when ingested orally. If inhaled into the lungs or injected into the blood, thereby circumventing the stomach, a dose of just 100 micrograms produces the same results. If amphetamine is injected into the cerebrospinal fluid, thus

attention-deficit/hyperactivity disorder (ADHD) Developmental disorder characterized by core behavioral symptoms of impulsivity, hyperactivity, and/or inattention.

psychopharmacology Study of how drugs affect the nervous system and behavior.

psychoactive drug Substance that acts to alter mood, thought, or behavior; is used to manage neuropsychological illness; or is abused.

Injecting a drug directly into the brain allows it to act quickly in low doses because there are no barriers.

Taking drugs orally is the safest, easiest, and most convenient way to administer them.

Drugs that are weak acids pass from the stomach into the bloodstream.

Drugs that are weak bases pass from the intestines to the bloodstream.

Drugs injected into muscle encounter more barriers than do drugs inhaled.

Drugs inhaled into the lungs encounter few barriers en route to the brain.

Drugs injected into the bloodstream encounter the fewest barriers to the brain but must be hydrophilic.

Drugs contained in adhesive patches are absorbed through the skin and into the bloodstream.

FIGURE 6-1 Routes of Drug Administration

1000 micrograms = 1 milligram

bypassing both the stomach and the blood, 10 micrograms is enough to produce an identical outcome, as is merely 1 microgram if dilution in the cerebrospinal fluid also is skirted and the drug is injected directly onto target neurons.

This math is well known to sellers and users of illicit drugs. Drugs that can be prepared to be inhaled or injected intravenously are much cheaper per dose because the amount required is so much smaller than that needed for an effective oral dose.

Revisiting the Blood–Brain Barrier

The body presents a number of barriers to the internal movement of drugs: cell membranes, capillary walls, and the placenta. The passage of drugs across capillaries in the brain is made difficult by the *blood–brain barrier,* the tight junctions between the cells of blood vessels in the brain that prevent the passage of most substances. The blood–brain barrier protects the brain's ionic balance and denies neurochemicals from the rest of the body passage into the brain where they can disrupt communication between neurons. It protects the brain from the effects of many circulating hormones and from various toxic and infectious substances. Injury or disease can sometimes rupture the blood–brain barrier, letting pathogens through. For the most part, however, the brain is very well protected from substances potentially harmful to its functioning.

The brain has a rich capillary network. None of its neurons is farther than about 50 micrometers (one-millionth of a meter) away from a capillary. As you can see at the left of **Figure 6-2,** like all capillaries, brain capillaries are composed of a single layer of endothelial cells. In the walls of capillaries in most parts of the body, endothelial cells are not fused, so substances can pass through the clefts between the cells. In contrast, in the brain (at least in most parts of it), endothelial cell walls are fused to form "tight junctions," so molecules of most substances cannot squeeze between them.

Figure 6-2 also shows that the endothelial cells of brain capillaries are surrounded by the end feet of astrocytes attached to the capillary wall, covering about 80 percent of it. The glial cells provide a route for the exchange of food and waste between capillaries and the brain's extracellular fluid and from there to other cells, shown at the right in Figure 6-2.

The cells of capillary walls in the three brain regions shown in **Figure 6-3** lack a blood–brain barrier. The pituitary is a source of many hormones that are secreted into the blood system, and their release is triggered in part by other hormones carried to the pituitary by the blood (see Section 6-5). The absence of a blood–brain barrier in the area postrema of the lower brainstem allows toxic substances in the blood to trigger a vomiting response. The pineal gland also lacks a blood–brain barrier, enabling hormones to reach it and modulate the day–night cycles controlled by this structure.

Figures 4-7 and 4-8 illustrate ion diffusion and concentration and voltage gradients.

Section 13-2 details the pacemaking function of the pineal gland.

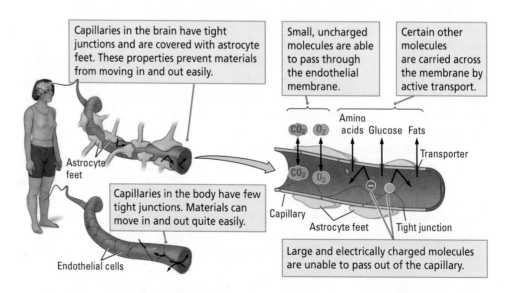

Capillaries in the brain have tight junctions and are covered with astrocyte feet. These properties prevent materials from moving in and out easily.

Small, uncharged molecules are able to pass through the endothelial membrane.

Certain other molecules are carried across the membrane by active transport.

Capillaries in the body have few tight junctions. Materials can move in and out quite easily.

Large and electrically charged molecules are unable to pass out of the capillary.

Astrocyte feet

Endothelial cells

CO_2 O_2 Amino acids Glucose Fats

CO_2 O_2

Transporter

Capillary

Astrocyte feet

Tight junction

FIGURE 6-2 Blood–Brain Barrier
Capillaries in most of the body allow for the passage of substances between capillary cell membranes, but those in the brain, stimulated by the actions of astrocytes, form the tight junctions of the blood–brain barrier.

To carry out its work, the brain needs, among other substances, oxygen and glucose for fuel and amino acids to build proteins. The fuel molecules reach brain cells from the blood, just as carbon dioxide and other waste products are excreted from brain cells into the blood. Molecules of these vital substances cross the blood–brain barrier in two ways:

1. Small molecules such as oxygen and carbon dioxide can pass through the endothelial membrane.

2. Molecules of glucose, amino acids, and other food components are carried across the membrane by active-transport systems or ion pumps—transporter proteins specialized to convey a particular substance.

The blood–brain barrier has relevance for understanding drug actions on the nervous system. Because few psychoactive drug molecules are small or have the correct chemical structure, few can gain access to the CNS. An estimated 98% of all drugs that may affect brain function and so may have some therapeutic use, cannot cross the blood-brain barrier.

How the Body Eliminates Drugs

After a drug has been administered, the body soon begins to break it down (*catabolize*) and remove it. Drugs are diluted throughout the body and are sequestered in many regions, including fat cells. They are also catabolized throughout the body, including the kidneys, liver and in the intestine by bile. They are excreted in urine, feces, sweat, breast milk, and exhaled air. Drugs that are developed for therapeutic purposes are usually designed not only to increase their chances of reaching their targets but also to enhance their survival time in the body.

The liver is especially active in catabolizing drugs. Owing to a family of enzymes called the *cytochrome P450 enzyme family* (some are also present in the gastrointestinal tract) that are involved in drug catabolism, the liver is capable of catabolizing many different drugs into forms that are more easily excreted from the body. Substances that cannot be catabolized or excreted can build up in the body and become poisonous. For instance, the metal mercury is not easily eliminated and can produce severe neurological conditions.

Drugs that are eliminated from the body can remain problematic. They may be reingested in food and water by many animal species, including humans. Some may then affect fertility, the development of embryos, and even the physiology and behavior of adult organisms. This problem can be limited by the redesign of waste-management systems to remove byproducts eliminated by humans as well as by other animals (Radjenović et al., 2009).

Drug Action at Synapses: Agonists and Antagonists

Most drugs that have psychoactive effects do so by influencing the chemical reactions at synapses. So to understand how drugs work, we must explore the ways in which they modify synaptic actions. **Figure 6-4** summarizes the seven major steps in neurotransmission at a synapse—each a site of drug action:

1. *Synthesis* of the neurotransmitter can take place in the cell body, the axon, or the terminal.

2. *Storage* of the neurotransmitter in granules or in vesicles or in both

3. *Release* of the transmitter from the terminal's presynaptic membrane into the synapse

4. *Receptor interaction* in the postsynaptic membrane, as the transmitter acts on an embedded receptor

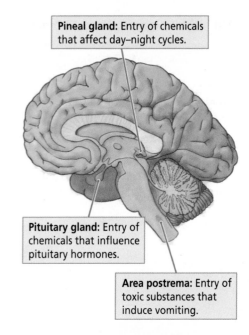

FIGURE 6-3 Barrier-Free Brain Sites The pituitary gland is a target for many blood-borne hormones, the pineal gland is a target for hormones that affect circadian rhythms, and the area postrema initiates vomiting in response to noxious substances.

Catabolic processes break down; metabolic processes build up.

FIGURE 6-4 Points of Influence In principle, a drug can modify seven major chemical processes, any of which results in enhanced or reduced synaptic transmission, depending on the drug's action as an agonist or an antagonist.

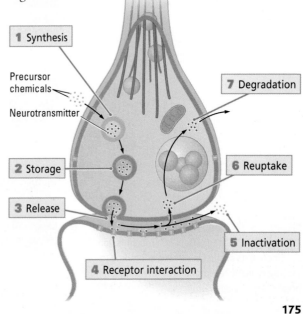

agonist Substance that enhances the function of a synapse.

antagonist Substance that blocks the function of a synapse.

tolerance Decrease in response to a drug with the passage of time.

5. *Inactivation* of excess neurotransmitter at the synapse

6. *Reuptake* into the presynaptic terminal for reuse

7. *Degradation* of excess neurotransmitter by synaptic mechanisms and removal of un-needed by-products from the synapse

A drug that affects any of these synaptic functions ultimately has one of two effects: it either increases or diminishes neurotransmission. Drugs that increase neurotransmission are called **agonists,** whereas drugs that decrease neurotransmission are called **antagonists.** To illustrate, consider the acetylcholine synapse between motor neurons and muscles.

An Acetylcholine Synapse: Examples of Drug Action

Figure 6-5 shows how selected drugs and toxins act as agonists or antagonists at the acetylcholine synapse on skeletal muscles. Acetylcholine agonists excite muscles, increasing muscle tone, whereas acetylcholine antagonists inhibit muscles, decreasing muscle tone. Some of these substances may be new to you, but you have probably heard of others. Knowing their effects at the ACh synapse allows you to understand the behavioral effects that they produce.

Figure 4-26 details the structure of and ACh action at a neuromuscular synapse.

Figure 6-5 shows two toxins that influence the release of ACh from the axon terminal. Black widow spider venom acts as an agonist by promoting the release of acetylcholine to excess. A black widow spider bite does not inject enough drug to paralyze a person, though a victim may feel some muscle weakness.

Botulin toxin, the poisonous agent in tainted foods, such as canned goods that have been improperly processed, is an antagonist. It blocks the release of ACh, an effect that can last from weeks to months. A severe case of poisoning can result in the paralysis of both movement and breathing and so cause death.

Focus 11-2 describes the causes and range of outcomes for cerebral palsy.

Botulin toxin has medical uses. Injected into a muscle, it can selectively paralyze the muscle. This action makes it useful in blocking excessive and enduring muscular twitches or contractions, including the spasms that make movement difficult, for example, in people with cerebral palsy. Under the trade name Botox, botulin toxin is also used cosmetically to paralyze facial muscles that cause facial wrinkling.

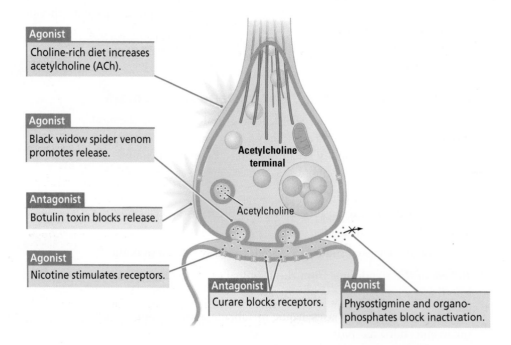

FIGURE 6-5 Acetylcholine Agonists and Antagonists Drugs affect ACh transmission by affecting its synthesis, release, or binding to the postsynaptic receptor and by affecting its breakdown or inactivation.

Figure 6-5 also shows two drugs that act on receptors for acetylcholine. Nicotine's molecular structure is similar enough to that of ACh to allow nicotine to fit into the receptors' binding sites where it acts as an agonist. Curare acts as an antagonist by occupying cholinergic receptors and so preventing acetylcholine from binding to them. After having been introduced into the body, curare acts quickly, and it is cleared from the body in a few minutes. Large doses, however, arrest movement and breathing for a period sufficient to result in death.

Early European explorers of South America discovered that the Indians along the Amazon River killed small animals by using arrowheads coated with curare prepared from the seeds of a plant. The hunters themselves did not become poisoned when eating the animals, because ingested curare cannot pass from the gut into the body. Many curarelike drugs have been synthesized. Some are used to briefly paralyze large animals so that they can be examined or tagged for identification. You have probably seen this use of these drugs in wildlife programs on television. Skeletal muscles are more sensitive to curarelike drugs than are respiratory muscles; an appropriate dose paralyzes an animal's movement temporarily but still allows it to breathe.

The final drug action shown in Figure 6-5 is that of physostigmine, an agonist that inhibits acetylcholinesterase, the enzyme that breaks down acetylcholine, thus increasing the amount of ACh available in the synapse. Physostigmine is obtained from an African bean and is used as a poison for hunting.

Large doses of physostigmine can be toxic because they produce excessive excitation of the neuromuscular synapse and so disrupt movement and breathing. In small doses, however, physostigmine is used to treat myasthenia gravis, a condition of muscular weakness in which muscle receptors are less than normally responsive to acetylcholine. Physostigmine's action is short lived, lasting only a few minutes or, at most, a half hour.

Organophosphates, another class of compounds, bind irreversibly to acetylcholinesterase and consequently are extremely toxic by allowing a buildup of ACh in the synaptic space. Many insecticides and chemical weapons are organophosphates. Insects use glutamate as a neurotransmitter at the nerve–muscle junction, but elsewhere in their nervous systems, they have nicotine receptors. Thus, organophosphates poison insects by acting centrally, but they poison chordates by acting peripherally as well.

Does a drug or toxin that affects neuromuscular synapses also affect acetylcholine synapses in the brain? That depends on whether the substance can cross the blood–brain barrier. For example, physostigmine and nicotine can readily pass the blood–brain barrier; curare cannot. Nicotine is the active ingredient in cigarette smoke and its actions on the brain account for its addictive properties (see Section 6-4). Physostiginelike drugs are reported to have some beneficial effects in treating memory disorders.

Tolerance

College roommates B. C. and A. S. went to a party, then to a bar, and by 3 A.M., were in a restaurant ordering pizza. A. S. decided that he wanted to watch the chef make his pizza, and off he went to the kitchen. Shortly, both were asked to leave the restaurant.

The roommates were headed away in A. S.'s car when a police officer, called by the manager, drove up. The officer, uncertain who was driving, gave both a breathalyzer test that estimates blood-alcohol content. A. S. failed the test, and B. C. passed, even though both had consumed the same amount of alcohol. Why this difference in their responses to the drinking bout?

The reason could be that B. C. had developed greater tolerance for alcohol than A. S. had. **Tolerance** is a decrease in response to a drug with the passage of time. Harris Isbell and coworkers (1955) show how such tolerance comes about. These researchers gave

As illustrated in Section 5-3, a single main receptor serves the somatic nervous system: the nicotinic ACh receptor (nAChr).

Figure 5-10 illustrates ACh synthesis and how acetylcholinesterase breaks down ACh.

Focus 4-3, "Myasthenia Gravis," explains what happens when muscle receptors lose their sensitivity to motor-neuron messages.

The Basics in Section 1-3 charts the evolution of animals' nervous systems, from nerve net to segmentation to ganglia (insects) to a brain and spinal cord (chordates, including humans).

In tolerance, as in habituation, a learned behavior results when a response to a stimulus weakens with repeated presentations (see Experiment 5-2).

EXPERIMENT 6-1

Question: Will the consumption of alcohol produce tolerance?

Procedure

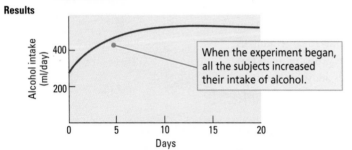

Subjects were given alcohol every day for 13 weeks—enough to keep them intoxicated.

Results

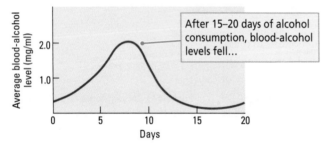

When the experiment began, all the subjects increased their intake of alcohol.

After 15–20 days of alcohol consumption, blood-alcohol levels fell...

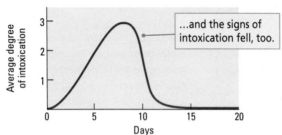

...and the signs of intoxication fell, too.

Conclusion: Because of tolerance, much more alcohol was required by the end of the study to obtain the same level of intoxication that was produced at the beginning.

Adapted from "An Experimental Study of the Etiology of 'Rum Fits' and Delirium Tremens," by H. Isbell, H. F. Fraser, A. Winkler, R. E. Belleville, and A. J. Eisenman, 1955, *Quarterly Journal of Studies on Alcohol, 16*, pp. 1–21.]

volunteers in a prison enough alcohol daily in a 13-week period to keep them in a constant state of intoxication. Yet they found that the subjects did not stay drunk for 3 months straight.

When the experiment began, all the participants showed rapidly rising levels of blood alcohol and behavioral signs of intoxication, as shown in the Results section of **Experiment 6-1.** Between the twelfth and twentieth days of alcohol consumption, however, blood alcohol and the signs of intoxication fell to very low levels, even though the subjects maintained a constant alcohol intake. Although blood-alcohol levels and signs of intoxication fluctuated in subsequent days of the study, one did not always correspond with the other. A relatively high blood-alcohol level was sometimes associated with a low outward appearance of intoxication. Why?

The results were the products of three different kinds of tolerance:

1. In the development of *metabolic tolerance,* the number of enzymes needed to break down alcohol in the liver, blood, and brain increases. As a result, any alcohol that is consumed is metabolized more quickly, so blood-alcohol levels are reduced.

2. In the development of *cellular tolerance,* the activities of brain cells adjust to minimize the effects of alcohol present in the blood. This kind of tolerance can help explain why the behavioral signs of intoxication may be very low despite a relatively high blood-alcohol level.

3. *Learned tolerance,* too, can help explain a drop in the outward signs of intoxication. As people learn to cope with the daily demands of living while under the influence of alcohol, they may no longer appear to be intoxicated.

It may surprise you that learning plays a role in tolerance to alcohol, but this role has been confirmed in many studies. In an early description of the effect, John Wenger and his coworkers (1981) trained rats to walk on a narrow conveyor belt to prevent electric shock to their feet from a grid over which the belt slid. One group of rats received alcohol after training in walking the belt, whereas another group received alcohol before training. A third group received training only, and a fourth group received alcohol only.

After several days of exposure to their respective conditions, all groups were given alcohol before a walking test. The rats that had received alcohol before training performed well, whereas those that had received training and alcohol separately performed just as poorly as those that had never had alcohol before or those that had not been trained. Apparently, animals can acquire the motor skills needed to balance on a narrow belt despite alcohol intoxication. With motor experience, in other words, they can learn to compensate for being intoxicated.

All forms of tolerance are much more likely to develop with repeated drug use than with periodic drug use. B. C. came from a small town where he was the acclaimed local pool shark. He was accustomed to "sipping a beer" both while waiting to play and during play, which he did often. B. C.'s body, then, was prepared to metabolize alcohol, and his experience in drinking while engaging in a skilled sport had prepared him to display controlled behavior under the influence of alcohol. By contrast, A. S. was unaccustomed to the effects of alcohol.

Sensitization

Repeated exposure to the same drug does not always result in tolerance, which resembles habituation in that the response to the drug weakens with repeated presentations. The occasional drug taker may experience the opposite reaction, sensitization—increased responsiveness to successive equal doses. Whereas tolerance generally develops with repeated use of a drug, sensitization is much more likely to develop with occasional use.

To demonstrate drug sensitization, Terry Robinson and Jill Becker (1986) isolated rats in observation boxes and recorded their reactions to an injection of amphetamine, which stimulates dopamine receptors. Every 3 or 4 days, the investigators injected the rats and found that their motor activities—sniffing, rearing, and walking—were more vigorous with each administration of the same dose of the drug, as graphed in Results 1 of **Experiment 6-2.**

The increased motor activity on successive tests was not due to the animals becoming comfortable with the test situation. Control animals that received no drug did not display a similar escalation. Administering the drug to rats in their home cages did not affect activity in subsequent tests either. Moreover, the sensitization to amphetamine was enduring. Even when two injections of amphetamine were separated by months, the animals still showed an escalation of motor behavior. Even a single exposure to the drug produced sensitization.

Sensitization is not always characterized by an increase in an emitted behavior but may also manifest as a progressive decrease in behavior. One of us and his coworkers (Whishaw et al., 1989) administered Flupentixol, a drug that blocks dopamine receptors, to rats that had been well trained in a swimming task. As illustrated in Results 2 of Experiment 6-2, the rats displayed such a decrease in swimming speed when escaping from the water that they completely stopped swimming over successive trials. The trial-dependent decrease in swimming was similar when the trials were massed on the same day or spaced over days or weeks. Administering the drug to rats left in their home environment did not influence performance in subsequent swim tests.

The neural basis of sensitization lies in part in changes at the synapse. Studies on the dopamine synapse after sensitization to amphetamine show that more dopamine is released from the presynaptic terminal in sensitized animals. Sensitization can also be associated with changes in the number of receptors on the postsynaptic membrane, changes in the rate of transmitter metabolism in the synaptic space, changes in transmitter reuptake by the presynaptic membrane, and changes in the number and size of synapses.

Part of the basis of sensitization is that animals are showing a change in learned responses to the cues in their environment as sensitization progresses. Consequently, sensitization is difficult to achieve in an animal that is tested in its home cage. Sabina Fraioli and her coworkers (1999) gave amphetamine to two groups of rats and recorded the rats' behavioral responses to successive injections. One group of rats lived in the test apparatus, so for that group home was the test box. The other group was taken out of their normal home cage and placed in the test box for each day's experimentation. The "home" group showed no sensitization to amphetamine, whereas the "out" group displayed robust sensitization.

At least part of the explanation of the "home–out" effect is that the animals are accustomed to engaging in a certain repertoire of behaviors in their home environment, so it is difficult to get them to change the behavior to "home cues" even in response to

Experiment 5-3 describes sensitization at the level of neurons and synapses. Section 14-4 relates drug-induced behavioral sensitization to neuroplasticity and learned addictions.

EXPERIMENT 6-2

Question: Does the injection of a drug always produce the same behavior?

Procedure 1

In the Robinson and Becker study, animals were given periodic injections of the same dose of amphetamine. Then the researchers measured the number of times each rat reared in its cage.

Procedure 2

In the Whishaw study, animals were given different numbers of swims after being injected with Flupentixol. Then the researchers measured their speed to escape to a platform in a swimming pool.

Agonist	Antagonist
Amphetamine	Flupentixol

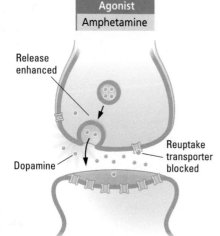

Release enhanced

Dopamine

Reuptake transporter blocked

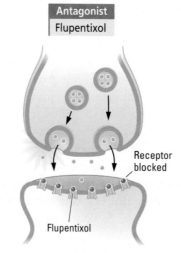

Receptor blocked

Flupentixol

Results 1

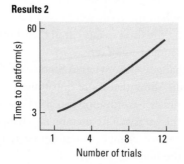

Number of incidents of rearing

24

12

1 3 5 9

Number of injections

Results 2

Time to platform(s)

60

3

1 4 8 12

Number of trials

Conclusion 1: Sensitization, indicated by increased rearing, develops with periodic repeated injections.

Conclusion 2: Sensitization depends on the occurrence of a behavior: the number of swims increases the time that it takes the rat to reach the platform.

(*Left*) Adapted from "Enduring Changes in Brain and Behavior Produced by Chronic Amphetamine Administration: A Review and Evaluation of Animal Models of Amphetamine Psychosis," by T. E. Robinson and J. B. Becker, 1986, *Brain Research Reviews, 397*, pp. 157–198. (*Right*) Adapted from "Training-Dependent Decay in Performance Produced by the Narcoleptic cist(Z)-Flupentixol on Spatial Navigation by Rats in a Swimming Pool," by I. Q. Whishaw, G. Mitt Elman, and J. L. Evend, 1989, *Pharmacology, Biochemistry, and Behavior, 32*, pp. 211–220.]

a drug. When away from home the "out cues" are novel and so favor conditioning of new responses.

The phenomenon of sensitization is relevant to understanding at least two psychopharmacological effects of drugs. First, for many drug therapies, including therapies for the psychiatric disorder schizophrenia, a drug must be taken for a number of weeks before it has beneficial effects. Possibly, sensitization underlies the development of the drug's beneficial effects. Second, before a person becomes dependent on or addicted to a drug, the person must have a number of experiences with the drug away from the home environment. Possibly, sensitization is related to the development of drug dependence.

REVIEW 6-1
Principles of Psychopharmacology

Before you continue, check your understanding.

1. _____, substances that produce changes in behavior by acting on the nervous system, are one subject of _____, the study of how drugs affect the nervous system and behavior.

2. Perhaps the most important obstacle on a psychoactive drug's journey between its entry into the body and its action at a target is the _____, which generally allows only substances needed for nourishment to pass from the capillaries into the _____.

3. Most drugs that have psychoactive effects influence chemical reactions at neuronal _____. Drugs that influence communication between neurons do so by acting either as _____ (increasing the effectiveness of neurotransmission) or as _____ (decreasing the effectiveness of neurotransmission).

4. Behavior may change in a number of ways with the repeated use of a psychoactive drug. These changes include _____ and _____, in which the effect of the drug decreases or increases, respectively, with repeated use.

5. The body eliminates drugs through _____, _____, _____, _____, and _____.

6. Describe briefly how tolerance and sensitization might affect someone who uses cognitive enhancers occasionally (a) at home or (b) at work.

Answers appear at the back of the book.

6-2 Grouping Psychoactive Drugs

You may be surprised to learn that most psychoactive drugs and their effects were discovered by accident. Subsequently, scientists and pharmaceutical companies have experimented to explain drug action, to synthesize alternate forms for therapeutic treatments, and to modify drugs to reduce side effects. A full appreciation of any drug's action requires a multifaceted description, such as can be found in compendiums of drug action.

Creating unambiguous groupings of psychoactive drugs is virtually impossible because most drugs influence many behaviors. Descriptions of behavior undergo constant review, as illustrated by the continuing revisions of the *Diagnostic and Statistical Manual of Mental Disorders* (DSM), a classification system for diagnosing neurological and behavioral disorders, published by the American Psychiatric Association. Further, drugs with similar chemical structures can have different effects, whereas drugs with different structures can have similar effects. Finally, a single drug usually acts on many different neurochemical systems and thus has many different effects.

Table16-3 summarizes the classification system used in the DSM.

TABLE 6-1 Grouping Psychoactive Drugs

Group I: Antianxiety agents and sedative hypnotics	Group IV: Opioid analgesics
Benzodiazepines: diazepam (Valium, Xanax, Clonapin)	Morphine, codeine, heroin
Barbiturates (anesthetic agents); alcohol	Endomorphins, enkephlins and dynorphins
Other anesthetics: gamma-hydroxybuterate (GHB), ketamine (Special K), phencyclidine (PCP, angel dust)	**Group V: Psychotropics**
Group II: Antipsychotic agents	Behavioral stimulants: amphetamine, cocaine
First generation: phenothiazines: chlorpromazine (Thorazine); butyrophenones: haloperidol (Haldol)	Psychedelic and hallucinogenic stimulants (listed by neurotransmitter affected)
Second generation: clozapine (Clozaril), aripiprazole (Abilify, Aripiprex)	Acetylcholine psychedelics: atropine, nicotine
Group III: Antidepressants and mood stabilizers	Anandamide psychedelics: tetrahydrocannabinol (THC)
Antidepressants	Glutamate psychedelics: phencyclidine (PCP, angel dust), ketamine (Special K)
MAO inhibitors	Norepinephrine psychedelics: mescaline
Tricyclic antidepressants: imipramine (Tofranil)	Serotonin psychedelics: Lysergic acid diethylamide (LSD), psilocybin, MDMA (ecstasy)
SSRIs (atypical antidepressants): fluoxetine (Prozac); sertraline (Zoloft); paroxetine (Paxil, Seroxat)	General stimulants: caffeine
Mood stabilizers	
Lithium, sodium valproate, carbamazepine (Tegretol)	

The grouping of psychoactive drugs in **Table 6-1** is based on their most-pronounced behavioral or psychoactive effects (Julien et al., 2011). Each of the five groups may contain from a few to many thousands of different chemicals in its subcategories. In the following sections we highlight drug actions on neurochemical systems in the brain and on synaptic function.

Most psychoactive drugs have three names: a chemical name, a generic name, and a brand name. The chemical name describes a drug's chemical structure, the generic name is nonproprietary, and the capitalized brand name is proprietary and given by the pharmaceutical company that sells it. Some drugs also sport "street" names or are known as "club drugs."

Group I: Antianxiety Agents and Sedative Hypnotics

At low doses, antianxiety drugs and sedative hypnotics reduce anxiety; at medium doses, they sedate; at high doses, they anesthetize or induce coma. At very high doses, they can kill **(Figure 6-6)**. However, antianxiety drugs are safer at high doses than sedative hypnotics are, and the use of prescribed sedative hypnotics for all purposes is decreasing. The best-known **antianxiety agents,** also known as minor tranquilizers, are the benzodiazepines such as diazepam, which is marketed as the widely prescribed brand-name drugs Valium, Xanax, and Clonapin. Benzodiazepines are often used by people who are having trouble coping with a major life stress, such as a traumatic accident or a death in the family. They are given to aid sleep and also used as presurgical relaxation agents.

The sedative hypnotics include alcohol and barbiturates. Alcohol is well known to most people because it is so widely consumed. **Barbiturates** are sometimes still prescribed as a sleeping medication but are used mainly to induce anesthesia before surgery. Both alcohol and barbiturates induce sleep, anesthesia, and coma at doses only slightly higher than those that sedate.

A characteristic feature of sedative hypnotics is that the user who takes repeated doses develops a tolerance for them. A larger dose is then required to maintain the drug's initial effect. **Cross-tolerance** results when the tolerance developed for one drug is carried over to a different member of the drug group.

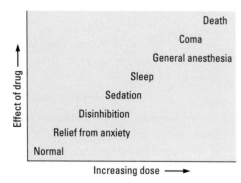

FIGURE 6-6 Behavioral Continuum of Sedation Increasing doses of sedative-hypnotic and antianxiety drugs affect behavior: low doses reduce anxiety and very high doses result in death.

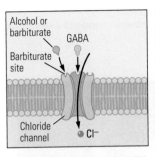
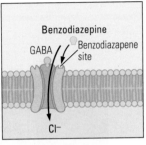
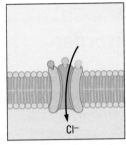
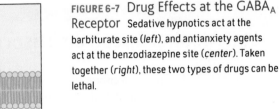

Sedative-hypnotic drugs (alcohol or barbiturates) increase the binding of GABA by maximizing the time the pore is open.

Antianxiety drugs (benzodiazepines) influence the frequency of pore opening.

Because of their different actions, these drugs should not be taken together.

FIGURE 6-7 Drug Effects at the GABA$_A$ Receptor Sedative hypnotics act at the barbiturate site (*left*), and antianxiety agents act at the benzodiazepine site (*center*). Taken together (*right*), these two types of drugs can be lethal.

Cross-tolerance suggests that antianxiety and sedative-hypnotic drugs act on the nervous system in similar ways. One target common to both drug types is a receptor for the inhibitory neurotransmitter gamma-aminobutyric acid, or GABA. The GABA$_A$ receptor, illustrated in **Figure 6-7,** contains a chloride ion channel.

Excitation of the receptor by GABA$_A$ produces an influx of Cl$^-$ ions through its pore. An influx of Cl$^-$ ions increases the concentration of negative charges inside the cell membrane, hyperpolarizing it, and making it less likely to propagate an action potential. The inhibitory effect of GABA, therefore, is to decrease a neuron's firing rate. Widespread reduction of neuronal firing in part underlies the behavioral effects of drugs that affect the GABA$_A$ synapse.

The GABA$_A$ receptor illustrated in Figure 6-7 also has binding sites for chemicals other than GABA, including, shown in the panel at left, a barbiturate site and, shown in the center panel, a benzodiazepine site. Activation of each site also promotes an influx of Cl$^-$ ions, but in different ways. Activating the barbiturate site increases the binding of GABA, and of benzodiazepines if present, and maximizes the time the pore is open. Activating the benzodiazepine site enhances the natural action of GABA by influencing the frequency that the ion pore opens in response to GABA. Because the effects of actions at these three sites summate, sedative-hypnotics, including alcohol, and antianxiety drugs should not be taken together. Combined doses of drugs reportedly contribute to as many deaths as occur annually from automobile accidents in the United States. Such was the case in 2012, when singer Whitney Houston succumbed to drowning.

The GABA$_A$ receptor also has binding sites that, when active, block the ion pore. Picrotoxin is a compound that blocks the pore, producing overexcitation and epileptic discharges in postsynaptic neurons. Administration of GABA$_A$ agonists can block the action of picrotoxin. Sedative-hypnotic and antianxiety drugs are thus useful in treating epileptic discharges and may act in part through the GABA$_A$ receptor.

Drugs that act on GABA receptors may affect brain development, because GABA is one substance that regulates brain development. Alcohol's potentially devastating effects on developing fetuses are explored in Clinical Focus 6-2, "Fetal Alcohol Spectrum Disorder" on page 184.

Many other psychoactive drugs have sedative-hypnotic and antianxiety actions. They include phencyclidine (PCP, angel dust) and two drugs—gamma-hydroxybutyric acid (GHB) and ketamine (Special K)—that have gained notoriety as "date rape" drugs, are soluble in alcohol, act quickly, and, like other sedative hypnotics, impair memory for recent events. Because they can be dissolved in a drink, party-goers and clubbers should never accept drinks from anyone, drink from punch bowls, or leave drinks unattended.

GABA, an amino acid, is the main inhibitory neurotransmitter in the central nervous system. Figure 5-12 shows its chemical structure.

antianxiety agent Drug that reduces anxiety; examples are minor tranquillizers such as benzodiazepines and sedative-hypnotic agents.

barbiturate Drug that produces sedation and sleep.

cross-tolerance Reduction of response to a novel drug because of tolerance developed in response to a chemically related drug.

Fetal Alcohol Spectrum Disorder

The term **fetal alcohol spectrum disorder** (FASD) was coined in 1973 to describe a pattern of physical malformation and mental retardation observed in some children born to alcoholic mothers. Children with FASD may have abnormal facial features, such as unusually wide spacing between the eyes. Their brains display a range of abnormalities, from small size with abnormal gyri to abnormal clusters of cells and misaligned cells in the cortex.

Related to these brain abnormalities are certain behavioral symptoms that FASD children tend to have in common. They display varying degrees of learning disabilities and low intelligence test scores as well as hyperactivity and other social problems. Individuals with FASD are nineteen times more likely to be incarcerated than those without the disorder (Popova et al., 2011).

The identification of FASD stimulated widespread interest in the effects of alcohol consumption by pregnant women. The offspring of approximately 6 percent of alcoholic mothers suffer from pronounced FASD. In major cities, the incidence of FASD is about 1 in 700 births (Popova et al., 2011). Its incidence is especially high among Native Americans on reservations in Canada, some other minority groups, and single mothers.

A major problem is that women who are most at risk for bearing FASD babies are poor and not well educated, their alcohol-consumption problems predate pregnancy, and they have little access to prenatal care. It is often difficult to inform these women about the dangers that alcohol poses to a fetus and to encourage them to abstain from drinking before and while they are pregnant.

Alcohol-induced abnormalities can vary from hardly noticeable physical and psychological effects to full-blown FASD. The severity of effects is related to when, how much, and how frequently alcohol is consumed over the course of pregnancy. The effects are worse if alcohol is consumed in the first trimester, a time when many women do not yet realize that they are pregnant.

Severe FASD is also more likely to coincide with binge drinking, which produces high blood-alcohol levels. Other factors related to a more severe outcome are poor nutritional health of the mother and the mother's use of other drugs, including the nicotine in cigarettes.

A major question raised by FASD is how much alcohol is too much to drink during pregnancy. To be completely safe, it is best not to drink

George Steinmetz

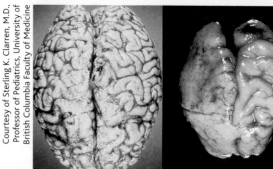

Courtesy of Sterling K. Clarren, M.D., Professor of Pediatrics, University of British Columbia Faculty of Medicine

(*Top*) Effects of fetal alcohol syndrome are not merely physical; many FAS children are severely retarded. (*Bottom*) The convolutions characteristic of the brain of a normal child at age 6 weeks, shown at the left, are grossly underdeveloped in the brain of a child who suffered from fetal alcohol syndrome, shown at the right.

at all in the months preceding as well as during pregnancy. This conclusion is supported by findings that as little as a single drink of alcohol per day during pregnancy can lead to a decrease in intelligence test scores of children.

Group II: Antipsychotic Agents

Focus 8-5 relates the possible origin of schizophrenia and the progress of the disease.

The term *psychosis* is applied to behavioral disorders such as schizophrenia, which is characterized by hallucinations (false sensory perceptions) and delusions (false beliefs), among a host of symptoms. The use of antipsychotic drugs has improved the functioning of schizophrenia patients and contributed to sharply reducing the number housed in institutions, as **Figure 6-8** graphs. The success of antipsychotic agents is an important

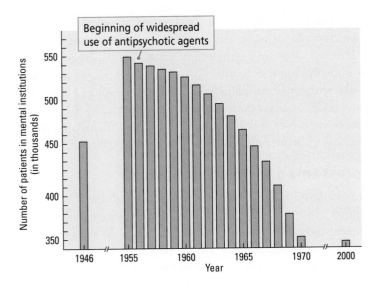

Beginning of widespread use of antipsychotic agents

FIGURE 6-8 **Trends in Resident Care** The dramatic decrease in the number of resident patients in state and municipal mental hospitals in the United States began after 1955, when psychoactive drugs were introduced into widespread therapeutic use. Adapted from *A Primer of Drug Action* (8th ed., p. 499), by R. M. Julien, 1998. New York: W. H. Freeman and Company

therapeutic achievement because the incidence of schizophrenia is high, about 1 in every 100 people.

Antipsychotic agents have been widely used since the mid-1950s, beginning with the development of what are now called first-generation antipsychotics (FGAs) that include a class of drugs called the phenothiazines (e.g., chlorpromazine, Thorazine) and a class called the butyrophenones (e.g., haloperidol, Haldol). FGAs act mainly by blocking the dopamine D_2 receptor. Beginning in the 1980s, newer drugs, such as clozapine (Clozaril) and a number of other compounds, were developed and became the second generation antipsychotics (SGAs). SGAs weakly block D_2 receptors but also block serotonin $5\text{-}HT_2$ receptors. Antipsychotic drugs now in development will likely form a third generation.

One negative side effect of FGAs can be to produce symptoms reminiscent of Parkinson's disease, in which control over movement is impaired. SGA agents can affect motivation and reduce agitation, and one of their side effects, weight gain, may be related to this action. Although for some patients, FGAs and SGAs seem to be equally effective, subgroups of patients that do not respond to FGAs may respond to SGAs.

The therapeutic actions of antipsychotic agents are not understood fully. The **dopamine hypothesis of schizophrenia** holds that some forms of the disease may be related to excessive dopamine activity—especially in the frontal lobes. Other support for the dopamine hypothesis comes from the schizophrenialike symptoms of chronic users of amphetamine, a stimulant. As **Figure 6-9** shows, amphetamine is a dopamine agonist. It fosters the release of dopamine from the presynaptic membrane of D_2 synapses and blocks the reuptake of dopamine from the synaptic cleft. If amphetamine causes schizophrenia-like symptoms by increasing dopamine activity, perhaps naturally occurring schizophrenia is related to excessive dopamine action too. Both FGAs and SGAs block the D_2 receptor for dopamine, which has an immediate effect in reducing motor activity and alleviates the excessive agitation of some schizophrenia patients.

fetal alcohol spectrum disorder (FASD) Range of physical and intellectual impairments observed in some children born to alcoholic mothers.

dopamine hypothesis of schizophrenia Idea that excess activity of the neurotransmitter dopamine causes symptoms of schizophrenia.

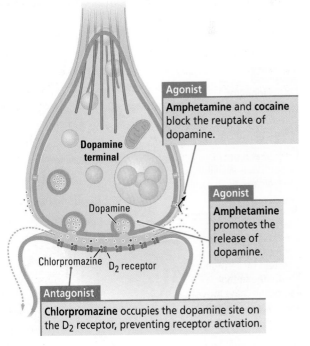

Agonist
Amphetamine and **cocaine** block the reuptake of dopamine.

Dopamine terminal

Agonist
Amphetamine promotes the release of dopamine.

Dopamine

Chlorpromazine D_2 receptor

Antagonist
Chlorpromazine occupies the dopamine site on the D_2 receptor, preventing receptor activation.

FIGURE 6-9 **Drug Effects at D_2 Receptors** That the antipsychotic agent chlorpromazine can lessen schizophrenia symptoms, whereas the abuse of amphetamine or cocaine can produce them, suggests that excessive activity at the D_2 receptor is related to schizophrenia.

Other drug models of schizophrenia include the psychotropic stimulant LSD (lysergic acid diethylamide), which produces hallucinations and is a serotonin agonist that acts at the 5-HT$_2$ receptor. Hallucinations are a symptom of schizophrenia, suggesting that excess serotonin action is involved in the disorder. Two other psychotropic drugs that produce schizophrenia-like symptoms, including hallucinations and out of body experiences, are phencyclidine (PCP or angel dust) and ketamine (Special K). Both drugs, formerly used as anesthetics, exert part of their action by blocking glutamate receptors, suggesting the involvement of excitatory glutamate synapses in schizophrenia as well.

Group III: Antidepressants and Mood Stabilizers

Major depression—a mood disorder characterized by prolonged feelings of worthlessness and guilt, the disruption of normal eating habits, sleep disturbances, a general slowing of behavior, and frequent thoughts of suicide—is very common. At any given time, about 6 percent of the U.S. adult population experiences major depression, and, in the course of a lifetime, 30 percent may experience at least one episode that lasts for months or longer. Depression is diagnosed in twice as many women as men.

Inadequate nutrition, stress from difficult life conditions, acute changes in neuronal function, and damage to brain neurons are among the factors implicated in depression. These factors may be related: nutritional deficiencies may increase vulnerability to stress, stress may change neuronal function, and if unrelieved, altered neuronal function may lead to neuron damage. Section 6-5 has more information on stress.

Among the nutrients that may be related to symptoms of depression (Smith et al., 2010) are folic acid and other B vitamins and omega-3 fatty acids, a rich source of vitamin D obtained from fish. Our skin synthesizes vitamin D on exposure to sunlight, but our bodies cannot store it. Vitamin D deficiency is reportedly widespread in people living in northern climates due to inadequate consumption of fish and lack of exposure to sunlight in winter months. Although Hoang and coworkers (2011) note an association between vitamin D deficiency and depressive symptoms, little is known about the relationship between long-term deficiencies in nutrients, depression and associated brain changes, and the effectiveness of dietary supplements.

Not surprisingly, alongside improved nutrition, a number of pharmacological approaches to depression are available, including normalizing stress hormones, modifying neuronal responses, and stimulating processes of neuronal repair.

Antidepressant Medications

Three different types of drugs have antidepressant effects: the **monoamine oxidase (MAO) inhibitors;** the **tricyclic antidepressants,** so called because of their three-ringed chemical structure; and the **second-generation antidepressants,** sometimes called *atypical antidepressants* (see Table 6-1). Second-generation antidepressants do not have a three-ringed structure, but they share some similarities to the tricyclics in their actions.

Antidepressants are thought to act by improving chemical neurotransmission at serotonin, noradrenaline, histamine, and acetylcholine synapses, and perhaps at dopamine synapses as well. **Figure 6-10** shows the actions of MAO inhibitors and second-generation antidepressants at a serotonin synapse, the synapse on which most research is focused. MAO inhibitors and the tricyclic and second-generation antidepressants all act as agonists but have different mechanisms for increasing the availability of serotonin.

Glutamate is the main excitatory neurotransmitter in the forebrain and cerebellum. Figure 5-12 shows its chemical structure.

Section 12-4 explores the neuroanatomy of emotional disorders such as depression. Section 16-4 reviews the neural consequences of and treatments for depression.

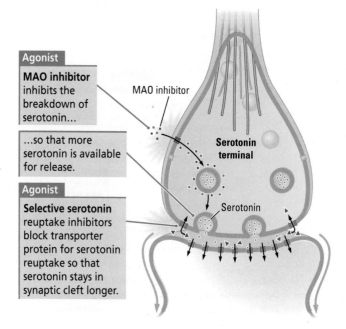

Agonist

MAO inhibitor inhibits the breakdown of serotonin…

…so that more serotonin is available for release.

Agonist

Selective serotonin reuptake inhibitors block transporter protein for serotonin reuptake so that serotonin stays in synaptic cleft longer.

MAO inhibitor

Serotonin terminal

Serotonin

FIGURE 6-10 Drug Effects at Serotonin Receptors Different antidepressant drugs act on the serotonin synapse in different ways to increase the availability of serotonin.

An MAO inhibitor provides for more serotonin release with each action potential by inhibiting monoamine oxidase, an enzyme that breaks down serotonin within the axon terminal. In contrast, the tricyclics and second-generation antidepressants block the reuptake transporter that takes serotonin back into the axon terminal. The second-generation antidepressants are thought to be especially selective in blocking serotonin reuptake; consequently, some are also called **selective serotonin reuptake inhibitors** (SSRIs). Because the transporter is blocked, serotonin remains in the synaptic cleft, prolonging its action on postsynaptic receptors.

Although these drugs begin to affect synapses very quickly, their antidepressant actions take weeks to develop. One explanation is that antidepressants, especially SSRIs, stimulate second messengers in neurons to activate the repair of those damaged by stress. Of interest in this respect one SSRI, fluoxetine (Prozac), increases the production of new neurons in the hippocampus, a limbic structure in the temporal lobes. As detailed in Section 6-5, the hippocampus is vulnerable to stress-induced damage, and its restoration by fluoxetine is proposed to underlie one of the drug's antidepressant effects (DeCarolis & Eisch, 2010).

Most people recover from depression within a year of its onset. If the illness is left untreated, however, the incidence of suicide is high, as described in Clinical Focus 6-3, "Major Depression" on page 188. Of all psychological disorders, major depression is one of the most treatable, and cognitive and intrapersonal therapies are as effective as drug therapies (Comer, 2011).

Even so, about 20 percent of patients with depression fail to respond to antidepressant drugs. Accordingly, it is likely that depression can have many other causes, including dysfunction in other transmitter systems and even brain damage, including frontal-lobe damage. Some people have difficulty tolerating the side effects of antidepressants—increased anxiety, sexual dysfunction, sedation, dry mouth, blurred vision, and memory impairment among them.

Mood Stabilizers

Bipolar disorder, once referred to as manic–depressive illness, is characterized by periods of depression alternating with normal periods and periods of intense excitation, or *mania*. According to the National Institute of Mental Health, bipolar disorder can affect as much as 2.6% of the adult population of the United States.

The difficulty in treating bipolar disorder with drugs is related to the difficulty in understanding how a disease produces symptoms that appear to be opposites: mania and depression. Consequently, bipolar disorder often is treated with a number of drugs, each directed toward a different symptom. **Mood stabilizers,** which include the salt lithium, mute the intensity of one pole of the disorder, thus making the other less likely to occur. Lithium does not directly affect mood and so may act by stimulating mechanisms of neuronal repair, such as the production of neuron growth factors.

A variety of drugs effective in treating epilepsy (carbamazepine, valproate) have positive effects, perhaps by muting the excitability of neurons during the mania pole. And antipsychotic drugs that block D_2 receptors effectively control the hallucinations and delusions associated with mania. It is important to note that all these treatments have side effects: enhancing beneficial effects while minimizing side effects is a major focus of new drug development (Severus et al., 2011).

Group IV: Opioid Analgesics

An *opioid* is any compound that binds to a group of brain receptors that are also sensitive to morphine. The term *narcotic analgesics* was first used to describe these drugs because **opioid analgesics** have sleep-inducing (narcotic) and pain-relieving (analgesic) properties. There are two natural sources of opioids.

Reuptake is part of transmitter deactivation, the last in the four-step process of neurotransmission, illustrated in Figure 5-4.

We consider the neurobiology of depression in Section 16-4.

major depression Mood disorder characterized by prolonged feelings of worthlessness and guilt, the disruption of normal eating habits, sleep disturbances, a general slowing of behavior, and frequent thoughts of suicide.

monoamine oxidase (MAO) inhibitor Antidepressant drug that blocks the enzyme monoamine oxidase from degrading neurotransmitters such as dopamine, noradrenaline, and serotonin.

tricyclic antidepressant First-generation antidepressant drug with a chemical structure characterized by three rings that blocks serotonin reuptake transporter proteins.

second-generation antidepressant Drug whose action is similar to that of tricyclics (first-generation antidepressants) but more selective in its action on the serotonin reuptake transporter proteins; also called *atypical antidepressant.*

selective serotonin reuptake inhibitor (SSRI) Tricyclic antidepressant drug that blocks the reuptake of serotonin into the presynaptic terminal.

bipolar disorder Mood disorder characterized by periods of depression alternating with normal periods and periods of intense excitation, or *mania.*

mood stabilizer Drug for treatment of bipolar disorder that mutes the intensity of one pole of the disorder, thus making the other pole less likely to recur.

opioid analgesic Drug like morphine, with sleep-inducing (narcotic) and pain-relieving (analgesic) properties; originally *narcotic analgesic.*

CLINICAL FOCUS ⬩ 6-3

Major Depression

P. H. was a 53-year-old high school teacher who, although popular with his students, was deriving less and less satisfaction from his work. His marriage was foundering because he was growing apathetic and no longer wanted to socialize or go on holidays. He was having difficulty getting up in the morning and arriving at school on time.

P. H. eventually consulted a physician, complaining of severe chest pains, which he thought signified that he was about to have a heart attack. He informed his doctor that a heart attack would be a welcome

relief because it would end his problems. The physician concluded that P. H. was suffering from depression and referred him to a psychiatrist.

Since the 1950s, depression has been treated with antidepressant drugs, a variety of cognitive-behavior therapies (CBTs), and electroconvulsive therapy (ECT), a treatment in which electrical current is passed briefly through one hemisphere of the brain. Of the drug treatments available, tricyclic antidepressants and the selective serotonin reuptake inhibitors (SSRIs) are favored.

The risk of suicide and self-injurious behaviors is high in depression, especially among depressive adolescents who are resistant to treatment with SSRIs (Asarnow et al., 2011). Even for patients who do respond positively to SSRI treatment, the benefits may not occur for weeks.

Matthew and coworkers (2012) report that the glutamate antagonist ketamine, when given in subanesthetic doses, can produce rapid beneficial effects that last for weeks, even in patients who are resistant to SSRI medication. Ketamine is thus proposed to be useful as an acute treatment for patients with major depression who are at risk for suicide and even for patients with bipolar depression who are at risk for suicide.

Prompted by complaints from family members that antidepressant drug treatments have caused suicide, especially in children, the U.S. Food and Drug Administration has advised physicians to monitor the side effects of SSRIs including fluoxetine (Prozac), sertraline (Zoloft), and paroxetine (Paxil, Seroxat). Findings from a number of studies show no difference in the rate of suicide between groups receiving these SSRIs and a placebo, and the incidence of suicide after prescriptions were reduced subsequent to the FDA warning actually increased (Kutcher & Gardner, 2008).

David Braun/Masterfile

Depressed? Virtually everyone who exercises will tell you that it can work wonders to brighten your mood.

One source is *opium,* an extract of the seeds of the opium poppy, *Papaver somniferum,* shown in **Figure 6-11.** Opium has been used for thousands of years to produce euphoria, analgesia, sleep, and relief from diarrhea and coughing. In 1805, German chemist Friedrich Sertürner synthesized two chemicals from opium: codeine and morphine. Codeine is often an ingredient in prescription cough medicine and pain relievers: it is converted into morphine by the liver. Morphine, shown in Figure 6-11 and named for Morpheus, the Greek god of dreams, is a very powerful pain reliever. Despite decades of research, no other drug has been found that exceeds morphine's effectiveness as an analgesic.

The second natural source of opioids is the brain. In the 1970s, several groups of scientists injected radioactive opiates into the brain and identified special receptors to which the opiates bind. At roughly the same time, other groups of investigators identified a number of brain peptides as the neurotransmitters that naturally affect these receptors. The peptides in the body that have opioid-like effects are collectively called **endorphins** (endogenous morphines).

Research has identified three classes of endorphins: *endomorphins, enkephalins* (meaning in the head), and *dynorphins.* There are also three receptors—*mu, kappa,* and *delta*—on which each endorphin is relatively specific. All endorphins and their receptors are

Peptides, including the endorphins illustrated in Figure 5-13, are molecular chains of amino acids connected by peptide bonds. Table 5-2 lists the families of peptide neurotransmitters.

FIGURE 6-11 Potent Poppy Opium is obtained from the seeds of the opium poppy (*left*). Morphine (*center*) is extracted from opium, and heroin (*right*) is in turn synthesized from morphine.

found in many regions of the brain and spinal cord as well as in other parts of the body, including the digestive system. Morphine most closely mimics the endomorphins and binds most selectively to the *mu* receptors.

In addition to the natural opioids, many synthetic opioids, such as heroin, affect *mu* receptors. Heroin is synthesized from morphine. It is more fat-soluble than morphine and penetrates the blood–brain barrier more quickly, allowing it to produce very rapid but shorter-acting relief from pain. Although heroin is a legal drug in some countries, it is illegal in others, including the United States.

Among the synthetic opioids that are prescribed for clinical use in pain management are hydromorphone, levorphanol, oxymorphone, methadone, meperidine, oxycodone, and fentanyl. All opioids are potently addictive, and abuse of prescription opioids is growing more common. Opioids are also illegally modified, manufactured, and distributed. People who suffer from chronic pain and who use opioids for pain relief also can become addicted; some obtain multiple prescriptions and sell their prescriptions illicitly.

A number of drugs act as antagonists at opioid receptors. They include *nalorphine* (Lethidrone, Nalline) and *naloxone* (Narcan, Nalone). These drugs are **competitive inhibitors:** they compete with opioids for neuronal receptors. Because they can enter the brain quickly, they can quickly block the actions of morphine and so are essential aids in treating morphine overdoses. Many people addicted to opioids carry a competitive inhibitor as a treatment for overdosing. Because they can also be long-acting, competitive inhibitors can be used to treat opioid addiction after the addicted person has recovered from withdrawal symptoms.

Researchers have extensively studied whether endorphins that exist in the brain can be used as drugs to relieve pain without producing the addictive effects of morphine. The answer is so far mixed, and the objectives of pain research in producing an analgesic that does not produce addiction may be difficult to realize.

Opioid drugs, such as heroin, are addictive and are abused worldwide. The hypodermic needle was developed in 1853 and used in the American Civil War for the intravenous injection of morphine for pain treatment. This practice reportedly produced 400,000 sufferers of the "soldiers disease" of morphine addiction. Morphine can be administered by many routes, but intravenous injection is preferred because it produces euphoria described as a "rush." Morphine does not readily cross the blood–brain barrier, whereas heroin does, so the latter is even more likely to produce a rush.

If opioids are used repeatedly, they produce tolerance such that, within a few weeks, the effective dose may increase tenfold. Thereafter, many of the desired effects with respect to both pain and addiction are no longer realized. An addicted person cannot simply stop using the drug, however. A severe sickness called "withdrawal" results if drug use is stopped.

endorphin Peptide hormone that acts as a neurotransmitter and may be associated with feelings of pain or pleasure; mimicked by opioid drugs such as morphine, heroin, opium, and codeine.

competitive inhibitor Drug such as nalorphine and naloxone that acts quickly to block the actions of opioids by competing with them for binding sites; used to treat opioid addiction.

Feeling and treating pain are topics in Section 11-4. Focus 12-1 reports that emotional pain activates the same neural areas that physical pain activates.

Because morphine results in both tolerance and sensitization, the morphine user is always flirting with the possibility of an overdose. The unreliability of appropriate information on the purity of "street" forms of morphine contributes to the risk of overdosing. A lack of sterile needles for injections also leaves the morphine user at risk for many other diseases, including AIDS (acquired immunodeficiency syndrome) and hepatitis.

The ingestion of opioids produces a wide range of physiological changes in addition to pain relief, including relaxation and sleep, euphoria, and constipation. Other effects include respiratory depression, decreased blood pressure, pupil constriction, hypothermia, drying of secretions (e.g., dry mouth), reduced sex drive, and flushed warm skin. Withdrawal is characterized by sicknesslike symptoms that are physiologically and behaviorally opposite those produced by the drug. Thus, a major part of the addiction syndrome is the drive to prevent withdrawal symptoms.

Group V: Psychotropics

Psychotropic drugs are stimulants that mainly affect mental activity; motor activity; and arousal, perception, and mood. Behavioral stimulants affect motor activity and mood. Psychedelic and hallucinogenic stimulants affect perception and produce hallucinations. General stimulants mainly affect mood.

Behavioral Stimulants

Behavioral stimulants increase motor behavior as well as elevating a person's mood and alertness. Rapid administration of behavioral stimulants is most likely to be associated with addiction. As explained in Section 6-1 and Figure 6-1, the quicker a drug reaches its target—in this case, the brain—the quicker it takes effect. Further, with each obstacle eliminated en route to the brain, drug dosage can be reduced by a factor of 10, making it cheaper per dose. Two behavioral stimulants are amphetamine and cocaine.

Amphetamine is a synthetic compound that was discovered in attempts to synthesize the CNS neurotransmitter epinephrine, which also acts as a hormone to mobilize the body for fight or flight in times of stress (see Figure 6-22). Both amphetamine and cocaine are dopamine agonists that act first by blocking the dopamine reuptake transporter. Interfering with the reuptake mechanism leaves more dopamine available in the synaptic cleft. Amphetamine also stimulates the release of dopamine from presynaptic membranes. Both mechanisms increase the amount of dopamine available in synapses to stimulate dopamine receptors. As noted in Focus 6-1, amphetamine-based drugs are widely prescribed to treat attention-deficit/hyperactivity disorder (ADHD).

A form of amphetamine was first used as a treatment for asthma: Benzedrine was sold in inhalers as a nonprescription drug through the 1940s. Soon people discovered that they could open the container and ingest its contents to obtain an energizing effect. Amphetamine was widely used in World War II—and is still used today to help troops and pilots stay alert, increase confidence and aggression, and boost morale—and was used then to improve the productivity of wartime workers. Today, amphetamine is also used as a weight-loss aid. Many over-the-counter compounds marketed as stimulants or weight-loss aids have amphetamine-like pharmacological actions.

An illegal amphetamine derivative, methamphetamine (also known as meth, speed, crank, smoke, or crystal ice) continues in widespread use. Lifetime prevalence of methamphetamine use in the U.S. population, estimated to be as high as 8% (Durell et al., 2008), is related to its ease of manufacture in illicit laboratories and to its potency, thus making it a relatively inexpensive, yet potentially devastating, drug.

Cocaine is a powder extracted from the Peruvian coca shrub, shown in **Figure 6-12.** The indigenous people of Peru have chewed coca leaves through the generations to increase their stamina in the harsh environment and high elevations where they live.

Section 5-1 describes the experiments Otto Loewi performed to identify epinephrine (EP, or adrenaline). Section 7-7 describes some symptoms of and outcomes for ADHD and the search for an animal model of the disease.

amphetamine Drug that releases the neurotransmitter dopamine into its synapse and, like cocaine, blocks dopamine reuptake.

psychedelic drug Drug that can alter sensation and perception; examples are lysergic acid dielthylmide, mescaline, and psilocybin.

FIGURE 6-12 Behavioral Stimulant Cocaine (*left*) is obtained from the leaves of the coca plant (*center*). Crack cocaine (*right*) is chemically altered to form "rocks" that vaporize when heated at low temperatures.

Refined cocaine powder can either be sniffed ("snorted") or injected. Cocaine users who do not like to inject cocaine intravenously or cannot afford it in powdered form, sniff or smoke "rocks" or "crack," a potent, highly concentrated form. Crack is chemically altered so that it vaporizes at low temperatures, and the vapors are inhaled.

Sigmund Freud originally popularized cocaine in the late 1800s as an antidepressant. It was once widely used in the manufacture of soft drinks and wine mixtures that were promoted as invigorating tonics. It is the origin of the trade name Coca-Cola, because this soft drink once contained cocaine **(Figure 6-13)**. The addictive properties of cocaine soon became apparent, however.

Freud also recommended that cocaine could be used as a local anesthetic. Cocaine did prove valuable for this purpose, and many derivatives, such as Novocaine, are used today. These local anesthetic agents reduce a cell's permeability to Na$^+$ ions and so reduce nerve conduction.

Psychedelic and Hallucinogenic Stimulants

Psychedelic drugs alter sensory perception and cognitive processes and can produce hallucinations. We categorize the major groups of psychedelics by their actions on specific neurotransmitters, here and in Table 6-1 on page 182.

ACETYLCHOLINE PSYCHEDELICS These drugs either block (atropine) or facilitate (nicotine) transmission at acetylcholine synapses.

ANANDAMIDE PSYCHEDELICS Results from numerous lines of research suggest that this endogenous neurotransmitter plays a role in enhancing forgetting. Anandamide prevents the brain's memory systems from being overwhelmed by all the information to which we are exposed each day. Tetrahydrocannabinol (THC), the active ingredient in marijuana obtained from the hemp plant *Cannabis sativa* and shown in **Figure 6-14,** acts on endogenous THC receptors for anandamide, the CB1 and CB2 receptors. Thus, THC use may have a detrimental effect on memory or a positive effect on mental overload.

Evidence points to the usefulness of THC as a therapeutic agent for a number of clinical conditions. It relieves nausea and emesis (vomiting) in patients undergoing cancer chemotherapy who are not helped by other treatments, for example, and stimulates the appetite in AIDS patients suffering from anorexia–cachexia (wasting) syndrome. THC has been found helpful for treating chronic pain through mechanisms that appear to be different from those of the opioids. It has also proved useful for treating glaucoma (increased pressure in the eye), for spastic disorders such as multiple sclerosis

FIGURE 6-13 Warning Label Cocaine was formerly an ingredient in a number of invigorating beverages, including Coca-Cola, as this advertisement suggests.

Anandamide (from the Sanskrit word for "joy" or "bliss") acts on a THC receptor that naturally inhibits adenyl cyclase, part of a second-messenger system active in sensitization and noted in Section 5-4.

FIGURE 6-14 *Cannabis sativa* The hemp plant is an annual herb that reaches a height between 3 and 15 feet. Hemp grows in a wide range of altitudes, climates, and soils and has myriad practical uses, including in the manufacture of rope, cloth, and paper.

and disorders associated with spinal-cord injury, and it may have some neuroprotective properties (see Section 6-4).

Synthetic and derived forms of THC have been developed in part to circumvent legal restrictions on THC use. Nevertheless, legal restrictions against THC use hamper investigations into its useful medicinal effects.

GLUTAMATE PSYCHEDELICS Phencyclidine (PCP, angel dust) and ketamine (Special K) can produce hallucinations and out of body experiences. Both drugs, formally used as anesthetics (see Table 6-1, Group I), exert part of their action by blocking glutamate NMDA receptors, receptors that are involved in learning. Other NMDA receptor antagonists include dextromethorphan and nitrous oxide. Although the primary psychoactive effects of PCP last for a few hours, its total elimination rate from the body can extend its action for eight days or longer.

NOREPINEPHRINE PSYCHEDELICS Mescaline, obtained from the peyote cactus, is legal in the United States for use by Native Americans for religious practices. Mescaline produces pronounced psychic alterations including a sense of spatial boundlessness and visual hallucinations. The effects of a single dose last up to ten hours.

SEROTONIN PSYCHEDELICS The synthetic drug lysergic acid diethylamide (LSD) and naturally occurring psilocybin (obtained from a certain mushroom) stimulate some serotonin receptors and block the activity of other serotonergic neurons through serotonin autoreceptors.

Serotonin psychedelics may stimulate other transmitter systems, including norepinephrine receptors. MDMA (ecstasy) is one of several synthetic amphetamine derivatives. It induces a sense of well-being and disembodiment as well as visual distortions. Repeated use of MDMA is associated with sleep, mood, and anxiety disorders and may also be associated with memory and attention deficits.

General Stimulants

General stimulants are drugs that cause an overall increase in the metabolic activity of cells. Caffeine, a widely used stimulant, inhibits an enzyme that ordinarily breaks down the second messenger cyclic adenosine monophosphate (cAMP). The resulting increase in cAMP leads to an increase in glucose production within cells, thus making more energy available and allowing higher rates of cellular activity.

A cup of coffee contains about 100 milligrams of caffeine, and many common soft drinks contain almost as much—some energy drinks pack as much as 500 milligrams. You may be using more caffeine than you realize. Excess levels can lead to the jitters. Regular caffeine users who quit may experience headaches, irritability, and other withdrawal symptoms.

Section 14-4 describes the roles of glutamate and NMDA receptors in long-term learning.

This enzyme plays a role in sensitization by increasing the concentration of cAMP in the presynaptic membrane. The action potential lasts longer than it usually would. Caffeine lowers cAMP concentrations, action potentials are briefer, and we get the coffee jitters.

REVIEW 6-2
Grouping Psychoactive Drugs

Before you continue, check your understanding.

1. Because of their diverse actions, it is useful to group drugs in terms of their most pronounced _____ or _____ effects.

2. Antianxiety and sedative-hypnotic drugs affect the _____ receptor, which through _____ influx hyperpolarizes neurons.

3. Among the antidepressant drug types, _____ increase the amount of serotonin available in the presynaptic terminal while _____ block serotonin reuptake at the synapse.

4. Opioids mimic the action of _____ by binding to the same receptors.

5. Amphetamine stimulates _____ and cocaine blocks _____ at the _____ synapse.

6. On which neurotransmitters do drugs that produce psychotropic effects act?

Answers appear at the back of the book.

6-3 Factors Influencing Individual Responses to Drugs

Many behaviors trigger predictable results. You strike the same piano key repeatedly and hear the same note each time. You flick a light switch each day, and the bulb glows exactly as it did yesterday. This cause-and-effect consistency does not extend to the effects of psychoactive drugs. Individuals respond to drugs in remarkably different ways at different times.

Behavior on Drugs

Ellen is a healthy, attractive, intelligent 19-year-old university freshman who knows the risks of unprotected sexual intercourse. She learned about the transmission of HIV and other sexually transmitted diseases (STDs) in her high-school health class. More recently, a seminar about the dangers of unprotected sexual intercourse was part of her college orientation, in which senior students provided the freshmen in her residence with free condoms and "safe sex" literature. Ellen and her former boyfriend were always careful to use latex condoms during intercourse.

At a homecoming party in her residence hall, Ellen has a great time, drinking and dancing with her friends and meeting new people. She is particularly taken with Brad, a sophomore at her college, and the two of them decide to go back to her room to order a pizza. One thing leads to another, and Ellen and Brad have sexual intercourse without using a condom. The next morning, Ellen wakes up, dismayed and surprised at her behavior and very concerned that she may be pregnant or may have contracted an STD. She is terrified that she may have contracted AIDS (MacDonald et al., 2000).

What happened to Ellen? What is it about drugs, especially alcohol, that makes people do things that they would not ordinarily do? Alcohol is associated with many harmful behaviors that are costly both to individual persons and to society. These harmful behaviors include not only unprotected sexual activity but also driving while intoxicated, date rape, spousal or child abuse, and other forms of aggression and crime.

An early and still widely held explanation of the effects of alcohol is the **disinhibition theory.** It holds that alcohol has a selective depressant effect on the cortex, the region of the brain that controls judgment, while sparing subcortical structures, those areas of the brain responsible for more-primitive instincts, such as desire. Stated differently, alcohol presumably depresses learned inhibitions based on reasoning and judgment while releasing the "beast" within.

This theory often excuses alcohol-related behavior with such statements as "She was too drunk to know better" or "The boys had a few too many and got carried away." Does disinhibition explain Ellen's behavior? Not entirely. Ellen had used alcohol in the past and managed to practice safe sex despite the effects of the drug. The disinhibition theory cannot explain why her behavior was different on this occasion. If alcohol is a disinhibitor, why is it not always so?

Craig MacAndrew and Robert Edgerton (1969) questioned disinhibition theory along just these lines in their book *Drunken Comportment.* They cite many instances in

disinhibition theory Explanation holding that alcohol has a selective depressant effect on the cortex, the region of the brain that controls judgment, while sparing subcortical structures responsible for more primitive instincts, such as desire.

which behavior under the influence of alcohol changes from one context to another. People who engage in polite social activity at home when consuming alcohol may become unruly and aggressive when drinking in a bar.

Even behavior at the bar may be inconsistent. Take Joe, for example. While drinking one night at a bar, he becomes obnoxious and gets into a fight. On another occasion, he is charming and witty, even preventing a fight between two friends, whereas, on a third occasion, he becomes depressed and only worries about his problems. MacAndrew and Edgerton also cite examples of cultures in which people are disinhibited when sober only to become inhibited after consuming alcohol and cultures in which people are inhibited when sober and become more inhibited when drinking. How can all these differences in alcohol's effects be explained?

MacAndrew and Edgerton suggested that behavior under the effects of alcohol represents learned behavior. Learned behavior is specific to culture, group, and setting and can in part explain Ellen's decision to sleep with Brad. Where alcohol is used to facilitate social interactions, behavior while intoxicated represents a "time out" from more conservative rules regarding dating.

But Ellen's lapse in judgment regarding safe sex is more difficult to explain by learning theory. Ellen had never practiced unsafe sex before and had never made it a part of her time-out social activities. So why did she engage in it with Brad?

Alcohol myopia theory (Griffin et al., 2010) suggests an explanation for alcohol-related lapses in judgment like Ellen's. **Alcohol myopia** (nearsightedness) is the tendency for people under the influence of alcohol to respond to a restricted set of immediate and prominent cues while ignoring more remote cues and potential consequences. Immediate and prominent cues are very strong and obvious and are close at hand.

People like this air surfer, who enjoy high-risk adventure, may have a genetic predisposition toward experimenting with drugs, but people with no interest in risk taking are just as likely to use drugs. Section 6-4 discusses genetic influences on drug taking.

In an altercation, the person with alcohol myopia will be quicker than normal to throw a punch because the cue of the fight is so strong and immediate. Similarly, at a raucous party, the myopic drinker will be more eager than usual to join in because the immediate cue of boisterous fun dominates the person's view. Once Ellen and Brad arrived at Ellen's room, the sexual cues of the moment were far more immediate than concerns about long-term safety. As a result, Ellen responded to those immediate cues and behaved as she normally would not.

Alcohol myopia can explain many lapses in judgment that lead to risky behavior, including aggression, date rape, and reckless driving while intoxicated. After drinking, individuals may also have poor insight into their level of intoxication: they may assume that they are less impaired than they actually are (Quinn & Fromme, 2011).

Addiction and Dependence

B. G. started smoking when she was 13 years old. Now a university lecturer, she has one child and is aware that smoking is not good for her own health or for the health of her family. She has quit smoking many times without success. Recently, she used a nicotine patch that absorbs the nicotine through the skin, without the smoke.

After successfully abstaining from cigarettes for more than 6 months on this treatment, B. G. began to smoke again. Because the university where she works has a no-smoking policy, she has to leave the campus and stand across the street from the building in which she works to smoke. Her voice has developed a rasping sound, and she has an almost chronic "cold." She says that she used to enjoy smoking but does not any more. Concern about quitting dominates her thoughts.

Oliver Furrer/Brand X/Corbis

B. G. has a drug problem. She is one of more than 20 percent of North Americans who smoke. Most begin between the ages of 15 and 35, consuming an average of about 18 cigarettes daily, nearly a pack-a-day habit. Like B. G., most smokers realize that smoking is a health hazard, have experienced unpleasant side effects from it, and have attempted to quit but cannot. B. G. is exceptional only in her white-collar occupation. Today, most smokers are found in blue-collar occupations rather than among professional workers.

Substance abuse is a pattern of drug use in which people rely on a drug chronically and excessively, allowing it to occupy a central place in their lives. A more advanced state of abuse is *substance dependence,* popularly known as **addiction.** Addicted people are physically dependent on a drug in addition to abusing it. They have developed tolerance for the drug, so an addict requires increased doses to obtain the desired effect.

Drug addicts may also experience unpleasant, sometimes dangerous, physical **withdrawal symptoms** if they suddenly stop taking the abused drug. Symptoms can include muscle aches and cramps, anxiety attacks, sweating, nausea, and even, for some drugs, convulsions and death. Withdrawal symptoms from alcohol or morphine can begin within hours of the last dose and tend to intensify over several days before they subside.

Although B. G. abuses the drug nicotine, she is not physically dependent on it. She smokes approximately the same number of cigarettes each day (she has not developed tolerance to nicotine), and she does not get sick if she is deprived of cigarettes (she does not suffer severe withdrawal symptoms from lack of nicotine but does display some physical symptoms—irritability, anxiety, and increases in appetite and insomnia). B. G. illustrates that the power of psychological dependence can be as influential as the power of physical dependence.

Many different kinds of abused or addictive drugs—including sedative hypnotics, antianxiety agents, opioids, and stimulants—have a common property: they produce **psychomotor activation** in some part of their dosage range. That is, at certain levels of consumption, these drugs make the user feel energetic and in control. This common effect has led to the hypothesis that all abused drugs may act on the same target in the brain: dopamine in the mesolimbic pathways of the dopaminergic activating system. Drugs that are abused increase mesolimbic dopamine activity, either directly or indirectly (**Figure 6-15**), and drugs that blunt abuse and addiction decrease mesolimbic dopamine activity.

alcohol myopia "Nearsighted" behavior displayed under the influence of alcohol: local and immediate cues become prominent, and remote cues and consequences are ignored.

substance abuse Use of a drug for the psychological and behavioral changes it produces aside from its therapeutic effects.

addiction Desire for a drug manifested by frequent use of the drug, leading to the development of physical dependence in addition to abuse; often associated with tolerance and unpleasant, sometimes dangerous, withdrawal symptoms on cessation of drug use. Also called *substance dependence.*

withdrawal symptom Physical and psychological behavior displayed by an addict when drug use ends.

psychomotor activation Increased behavioral and cognitive activity; at certain levels of consumption, the drug user feels energetic and in control.

(A)

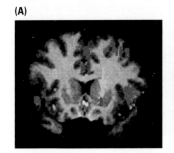

(B)

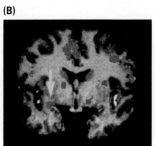

(C)

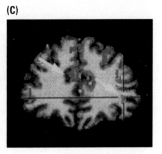

FIGURE 6-15 Targeted by Nicotine Functional MRI imaging of brain regions (yellow arrows) in the basal ganglia **(A)** and limbic system **(B)** and **(C)** reveal the neural areas that nicotine activates. All are mesolimbic dopamine projections (shown in red), consistent with the idea that stimulation of the dopaminergic activating system is related to addiction. From "Nicotine-induced limbic cortical activation in the human brain: A functional MRI study," by F. A. Stein, J. Pankiewicz, H. H. Harsch, J. K. Cho, S. A. Fuller, et al., 1998, *American Journal of Psychiatry, 155,* pp. 1009-1015.

Sex Differences in Addiction

Vast differences in individual responses to drugs are due to differences in age, body size, metabolism, and sensitivity to a particular substance. Larger people, for instance, are generally less sensitive to a drug than smaller people are: their greater volume of body fluids dilutes drugs more. Old people may be twice as sensitive to drugs as young people are. The elderly often have less effective barriers to drug absorption as well as less effective processes for metabolizing and eliminating drugs from their bodies. Individuals also respond to drugs in different ways at different times.

wanting-and-liking theory When a drug is associated with certain cues, the cues themselves elicit desire for the drug; also called *incentive-sensitization theory.*

Females are about twice as sensitive to drugs as are males, on average, owing in part to their relatively smaller body size but also to hormonal differences. The long-held general assumption that human males are more likely to abuse drugs than are human females led investigators to neglect researching drug use and abuse in human females. But the results of more recent research support a view quite the opposite of this common wisdom: females are less likely to become addicted to some drugs than are males, but females are catching up and for some drugs are surpassing males in the incidence of addiction.

Although the general pattern of drug use is similar in males and females, the sex differences are striking (Becker and Hu, 2008). Females are more likely than men are to abuse nicotine, alcohol, cocaine, amphetamine, opioids, cannabinoids, caffeine, and phencyclidine (PCP, or "angel dust"). Females begin to regularly self-administer licit and illicit drugs of abuse at lower doses than do males, use escalates more rapidly to addiction, and females are at greater risk for relapse after abstinence. This susceptibility may be explained by the role that the dopamine system plays in nurturing behavior, a behavior that is related mother–infant bonding.

REVIEW 6-3

Factors Influencing Individual Responses to Drugs

Before you continue, check your understanding.

1. Of the three explanations for the effects of alcohol on behavior, _____ and _____ are less explicative than _____.

2. _____ is a condition in which people rely on drugs chronically and to excess, whereas _____ is a condition in which people are physically dependent on a drug as well.

3. The evidence that many abused or addictive drugs produce _____, which makes the user feel energetic and in control, suggests that activation in the _____ plays a role in drug abuse and addiction.

4. Common wisdom is incorrect in suggesting that _____ are less likely to abuse drugs than _____ are.

5. Why can alcohol-related behavior vary widely in a single individual from time to time?

Answers appear at the back of the book.

6-4 Explaining and Treating Drug Abuse

Why do people become addicted to drugs? Early explanations centered on pleasure and dependence: habitual drug users initially experience pleasure but then endure psychological and physiological withdrawal symptoms as the drug wears off. They feel anxious, insecure, or just plain sick in the absence of the drug, so they take it again to alleviate those symptoms. In this way, they get "hooked" on the drug.

Although this "dependency hypothesis" may account for part of drug-taking behavior, it has shortcomings as a general explanation. For example, an addict may abstain from a drug for months, long after any withdrawal symptoms have abated, and yet still be drawn back to using it. In addition, some psychoactive drugs, such as the tricyclic antidepressants, produce withdrawal symptoms when discontinued, but these drugs are not abused.

Wanting-and-Liking Theory

To account for all the facts about drug abuse and addiction, Terry Robinson and Kent Berridge (2008) proposed the *incentive-sensitization theory.* This perspective is also called the **wanting-and-liking theory** because wanting and liking are produced by two different brain systems. "Wanting" is equivalent to cravings for a drug, whereas "liking" is the pleasure that drug taking produces. With repeated use, tolerance for liking develops, and the expression of liking (pleasure) decreases as a consequence **(Figure 6-16)**. In contrast, the system that mediates wanting sensitizes, and wanting the drug (craving) increases.

The first step on the proposed road to drug dependence is the initial experience, when the drug affects a neural system associated with "pleasure." At this stage, the user may experience liking the substance—including liking to take it within a social context. With repeated use, liking the drug may decline from its initial level. At this stage, the user may also begin to show tolerance to the drug's effects and so may begin to increase the dosage to increase liking.

With each use, the drug taker increasingly associates the cues related to drug use—be it a hypodermic needle, the room in which the drug is taken, or the people with whom the drug is taken—with the drug-taking experience. The user at this stage makes this association because the drug enhances classically conditioned cues associated with drug taking. Later encounters with these wanting cues, rather than the expected liking—the pleasure derived from the drug's effects—initiates wanting, or craving.

The neural basis of addiction is proposed to involve a number of brain systems. First, the decision to take a drug is made in the frontal cortex, an area proposed to be involved in most daily decisions. Next, when a drug is taken, it activates opioid systems in the brainstem that are generally related to pleasurable experiences. Third, wanting drugs may spring from activity in the mesolimbic pathways of the dopaminergic activating system.

In the mesolimbic pathways diagrammed in **Figure 6-17,** the axons of dopamine neurons in the midbrain project to the nucleus accumbens, the frontal cortex, and the limbic system. When drug takers encounter cues associated with drug taking, the mesolimbic system becomes active, releasing dopamine. Dopamine release is the neural correlate of subjectively experiencing wanting.

A fourth brain system may be responsible for conditioning drug-related cues to drug taking. Everitt and colleagues (2008) propose that the repeated pairing of drug-related cues to drug taking forms neural associations, or learning, in the dorsal *striatum,* a region in the basal ganglia consisting of the caudate nucleus and putamen. As the user engages in repeated drug-taking experiences, voluntary control of drug taking gives way to unconscious processes—a "habit." The result: drug users lose control of decisions related to drug taking, and the wanting—the voluntary control over drug taking—gives way to the craving of addiction.

A number of findings align with the wanting-and-liking explanation of drug addiction. Ample evidence confirms that abused drugs and the context in which they are taken initially has a pleasurable effect and that habitual users continue to use their drug of choice, even when taking it no longer produces any pleasure. Heroin addicts sometimes report that they are miserable, their lives are in ruins, and the drug is not even pleasurable anymore. But they still want it. What's more, desire for the drug often is greatest just when the addicted person is maximally high, not when he or she is undergoing withdrawal. Finally, cues associated with drug taking—the social situation, the sight of the drug, and drug paraphernalia—strongly influence decisions to take, or continue taking, a drug.

We can extend wanting-and-liking theory to many life situations. Cues related to sexual activity, food, and even sports can induce a state of wanting, sometimes in the absence of liking. We frequently eat when prompted by the cue of other people eating, even though we may not be hungry and derive little pleasure from eating at that time. The similarities between exaggerating normal behaviors and drug addiction suggest that both depend on the same learning and brain mechanisms. For this reason, any addiction is extremely difficult to treat.

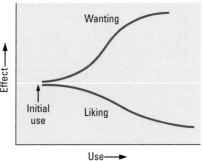

FIGURE 6-16 Wanting-and-Liking Theory Wanting a drug and liking a drug go in opposite directions with repeated drug use. Wanting (craving) is associated with drug cues.

In classical, or Pavlovian, conditioning, learning to associate some formerly neutral stimulus (the sound of a bell) with a stimulus (food) elicits an involuntary response (salivation).

Analogously, when a rat is placed in an environment where it anticipates a favored food or sex, investigators record dopamine increases in the striatum (see Section 7-5).

FIGURE 6-17 Mesolimbic Dopamine Pathways Axons of neurons in the midbrain ventral tegmentum project to the nucleus accumbens, frontal cortex, and hippocampus.

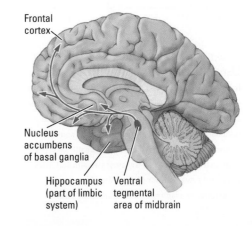

Why Doesn't Everyone Abuse Drugs?

Observing that some people are more prone than other people to drug abuse and dependence, scientists have investigated and found three lines of evidence suggesting a genetic contribution to differences in drug use. First, if one twin abuses alcohol, the other twin is more likely to abuse it if the twins are identical (have the same genetic makeup) than if they are fraternal (have only some of their genes in common). Second, people adopted shortly after birth are more likely to abuse alcohol if their biological parents were alcoholic, even though they have had almost no contact with those parents. Third, although most animals do not care for alcohol, the selective breeding of mice, rats, and monkeys can produce strains that consume large quantities of it.

Each line of evidence presents problems, however. Perhaps identical twins show greater concordance rates (incidence of similar behavioral traits) for alcohol abuse because they are exposed to more similar environments than fraternal twins are. And perhaps the link between alcoholism in adoptees and their biological parents has to do with nervous-system changes due to prenatal exposure to the drug. Finally, the fact that animals can be selectively bred for alcohol consumption does not mean that all human alcoholics have a similar genetic makeup. The evidence for a genetic basis of alcohol abuse will become compelling only when a gene or set of genes related to alcoholism is found.

Epigenetics offers an alternative to the "inherited" explanations of susceptibility to addiction (Robison & Nestler, 2011). The epigenetic explanation posits that addictive drugs can influence gene regulation relatively directly. By determining which genes are expressed, addictive drugs can selectively turn off genes related to voluntary control and turn on genes related to behaviors susceptible to addiction. Epigenetic changes in an individual's gene expression are relatively permanent and can be passed along, perhaps through the next few generations. For these reasons, epigenetics can also account both for the enduring behaviors that support addiction and for the tendency of drug addiction to be inherited.

Treating Drug Abuse

Figure 6-18 charts the percentage of people in the United States who reported using at least one psychoactive drug during the year preceding a 2010 survey by the Department of Health and Human Services. The two most-used drugs, alcohol and tobacco, are legal. Drugs against which laws are most harsh, cocaine and heroin, are used by far fewer people. But criminalizing drugs clearly is not a solution to drug use or abuse, as illustrated by the widespread use of marijuana, the third most-used drug on the chart.

Treating drug abuse is difficult in part because legal proscriptions in relation to drug use are irrational. In the United States, the Harrison Narcotics Act of 1914 made heroin and a variety of other drugs illegal and made the treatment of addicted people by physicians in their private offices illegal. The Drug Addiction Treatment Act of 2000 partly reversed this prohibition, allowing the treatment of patients but with a number of restrictions. In addition, legal consequences attending drug use vary greatly with the drug that is abused and the jurisdiction in which it is used.

From a health standpoint, using tobacco has much greater proved health risks than does using marijuana. The moderate use of alcohol is likely benign. The moderate use of opioids is likely impossible. Social coercion is useful in reducing tobacco use: witness the marked decline in smoking as a result of prohibitions against smoking in public places. Medical intervention is necessary for providing methadone and other drug treatment of opioid abusers.

The numerous approaches to treating drug abuse vary, depending on the drug to which a person is addicted. Many Web sites support self-help groups and professional groups that address the treatment of various drug addictions. Importantly, because addiction is associated with unconscious conditioning to drug-related cues, relapse remains an enduring risk for people who have "kicked their habit."

Section 3-3 observes that the less-than-perfect concordance rates between identical twins for a vast array of diseases, from schizophrenia to asthma, point to the effects of epigenetic factors on all sorts of behaviors and their persistence into the next generation.

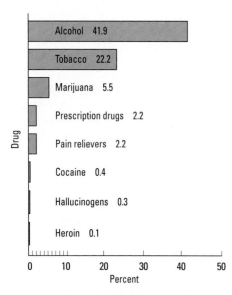

FIGURE 6-18 Relative Incidence of Drug Use in the United States, 2010 Results from a national survey of Americans who reported using the drug during the past year. Results from the *2010 National Survey on Drug Use and Health: Summary of National Findings*, U.S. Department of Health and Human Services, Substance Abuse and Mental Health Services Administration, Center for Behavioral Health Statistics and Quality. Available for downloading from the U.S. Department of Health and Human Services Web site.

Neuroscience research will continue to lead to a better understanding of the neural basis of drug use and to better treatment. It is likely that the best approach to any drug treatment recognizes that addiction will be a lifelong problem for most people. Thus, drug addiction must be treated in the same way as chronic behavioral addictions and medical problems—analogous to recognizing that controlling weight with appropriate diet and exercise is a lifelong problem for many people.

Can Drugs Cause Brain Damage?

Many natural substances can act as neurotoxins; **Table 6-2** lists some of them. Extensive investigations of the neurotoxicity of these substances and other drugs are ongoing in animal models. These studies show that many substances can cause brain damage. Whether drugs of abuse cause brain damage in humans, and especially whether they can do so at the doses that humans take, are more difficult to determine. It is difficult to sort out other life experiences from drug-taking experiences. It is also difficult to obtain the brains of drug users for examination at autopsy.

In the late 1960s, many reports circulated linking monosodium glutamate, MSG, a salty-tasting, flavor-enhancing food additive, to headaches in some people. In the process of investigating this effect, scientists placed large doses of MSG on cultured neurons and noticed that the neurons died. Subsequently, they injected MSG into the brains of experimental animals, where it also produced neuron death.

These findings raised the question of whether large doses of the neurotransmitter glutamate, which MSG resembles structurally, might also be toxic to neurons. It turned out that they are. Glutamate-receptor activation results in an influx of Ca^{2+} into the cell, and the influx of excessive Ca^{2+} may, through second messengers, activate a "suicide gene" in a cell's DNA leading to *apoptosis* (cell death). These discoveries led to the

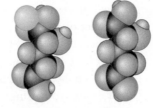

Monosodium glutamate   Glutamate
(MSG)

understanding that a drug might not be toxic only because of its general effect on cell function but also as an epigenetic agent that activates cell process related to apoptosis.

Many glutamatelike chemicals, including domoic acid and kainic acid, toxins in seaweed; and ibotenic acid, which is found in some poisonous mushrooms, are now known to similarly kill neurons **(Figure 6-19)**. Many of these glutamate analogs are now used to make experimental lesions in the brains of research animals. Some drugs, such as phencyclidine and ketamine, also act as glutamate agonists, leaving open the possibility that at high doses they, too, can cause neuronal death.

What about the many recreational drugs that affect the nervous system? Are any neurotoxic? Sorting out the effects of the drug itself from the effects of other factors related to taking the drug is a major problem. Chronic alcohol use, for instance, can be associated with damage to the thalamus and limbic system, producing severe memory disorders. Alcohol itself does not seem to cause this damage; rather, it stems from complications related to alcohol abuse, including vitamin deficiencies due to poor diet.

Alcoholics typically obtain reduced amounts of thiamine (vitamin B_1) in their diets, and alcohol interferes with the absorption of thiamine by the intestine. Thiamine plays a vital role in maintaining cell-membrane structure. Nevertheless, drugs of abuse do produce many direct effects on the body (Milroy and Parai, 2011).

Similarly, among the many reports of people who suffer some severe psychiatric disorder subsequent to their abuse of certain recreational drugs, in most cases, determining

TABLE 6-2 Some Neurotoxins, Their Sources, and Their Actions

Substance	Origin	Action
Tetrodotoxin	Pufferfish	Blocks membrane permeability to Na^+ ions
Magnesium	Natural element	Blocks Ca^{2+} channels
Reserpine	Tree	Destroys storage granules
Colchicine	Crocus plant	Blocks microtubules
Caffeine	Coffee bean	Blocks adenosine receptors, blocks Ca^{2+} channels
Spider venom	Black widow spider	Stimulates ACh release
Botulin toxin	Food poisoning	Blocks ACh release
Curare	Plant berry	Blocks ACh receptors
Rabies virus	Infected animal	Blocks ACh receptors
Ibotenic acid	Mushroom	Similar to domoic acid, mimics glutamate
Strychnine	Plant	Blocks glycine
Apamin	Bees and wasps	Blocks Ca^{2+} channels

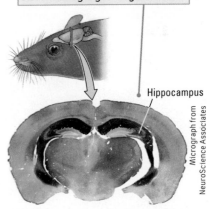

Domoic acid produces hippocampal damage, as shown by a dark silver stain that highlights degeneration.

Hippocampus

Micrograph from NeuroScience Associates

FIGURE 6-19 Neurotoxicity Domoic acid damage in this rat's hippocampus and to a lesser extent in many other brain regions, indicated by darker coloring. Domoic acid is a glutamate antagonist and the causative agent in amnesic shellfish poisoning, which can result in permanent short-term memory loss, brain damage, and in severe cases, death.

Focus 5-4 reports the chilling case of heroin addicts who developed Parkinson's disease after using synthetic heroin, owing to a contaminant (MPTP) in the drug.

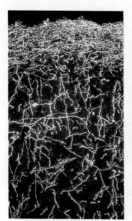

FIGURE 6-20 Drug Damage The administration of MDMA changes the density of serotonin axons in the neocortex of a squirrel monkey: (*left*) normal monkey; (*right*) monkey 18 months after administration of MDMA. From "Long-Lasting Effects of Recreational Drugs of Abuse on the Central Nervous System," by U. D. McCann, K. A. Lowe, and G. A. Ricaurte, 1997, *The Neurologist, 3*, p. 401.

Focus 7-3 explores the hypothesis that a genetic vulnerability predisposes some adolescents to develop a psychotic condition when exposed to cannabis.

whether the drug initiated the condition or just aggravated an existing problem is difficult. Exactly determining whether the drug itself or some contaminant in the drug is related to a harmful outcome also is difficult. With the increasing sensitivity of brain-imaging studies, however, there is increasing evidence that many drugs used "recreationally" can cause brain damage.

The strongest evidence that some recreational drugs can cause brain damage and cognitive impairments comes from the study of the synthetic amphetaminelike drug, MDMA, also called *ecstasy* (Büttner, 2011). Although MDMA is structurally related to amphetamine, it produces hallucinogenic effects and is referred to as a "hallucinogenic amphetamine." Findings from animal studies show that doses of MDMA approximating those taken by human users result in the degeneration of very fine serotonergic nerve terminals. In monkeys, the terminal loss may be permanent, as shown in **Figure 6-20**.

Memory impairments and damage revealed by brain imaging have been reported in users of MDMA, which may be a result of similar neuronal damage (Cowan et al., 2008). MDMA may also contain a contaminant, paramethoxymethamphetamine (PMMA). This notoriously toxic amphetamine is often called "Dr. Death" because the difference between a dose that causes behavioral effects and a dose that causes death is minuscule (Vevelstad et al., 2012). Contamination by unknown compounds can occur in any drug purchased on the street.

The psychoactive properties of cocaine are similar to those of amphetamine, and the possible deleterious effects of cocaine use have been subjected to intense investigation. The results of many studies show that cocaine use is related to the blockage of cerebral blood flow and other changes in blood circulation. Brain-imaging studies also suggest that cocaine use can be toxic to neurons because a number of brain regions are found to be reduced in size in cocaine users (Barrós-Loscertales et al., 2011).

A number of cases of chronic marijuana use have been associated with psychotic attacks. Clinical Focus 6-4, "Drug-Induced Psychosis," describes one. The marijuana plant contains at least 400 chemicals, 60 or more of which are structurally related to its active ingredient, tetrahydrocannabinol. Determining whether a psychotic attack is related to THC or to some other ingredient contained in marijuana or to aggravation of an existing condition is almost impossible.

Regardless of whether THC can cause psychosis, there is no evidence that the disease is a result of brain damage (DeLisi, 2008). Indeed, beyond the therapeutic applications of TCH cited in Section 6-2, recent studies suggest that THC may have neuroprotective properties. It can aid brain healing after traumatic brain injury and slow the progression of diseases associated with brain degeneration, including Alzheimer's disease and Huntington's disease (Sarne et al., 2011).

REVIEW 6-4

Explaining and Treating Drug Abuse

Before you continue, check your understanding.

1. The wanting-and-liking theory of addiction suggests that, with repeated use, _____ the drug decreases as a result of _____ while _____ increases as a result of _____.

2. At the neural level, the decision to take a drug is made in the brain's _____. Once taken, the drug activates opioid systems related to pleasurable experiences in the _____. Drug cravings may originate in the _____, and the repeated pairing of drug-related cues and drug taking forms neural associations in the _____ that loosen voluntary control over drug taking.

CLINICAL FOCUS ⊹ 6-4

Drug-Induced Psychosis

At age 29, R. B. S. smoked marijuana chronically. For years, he had been selectively breeding a particularly potent strain of marijuana in anticipation of the day when it would be legalized. R. B. S. made his living as a pilot, flying small-freight aircraft into coastal communities in the Pacific Northwest.

One evening, R. B. S. had a sudden revelation: he was no longer in control of his life. Convinced that he was being manipulated by a small computer implanted in his brain when he was 7 years old, he confided in a close friend, who urged him to consult a doctor. R. B. S. insisted that he had undergone the surgery when he participated in an experiment at a local university. He also claimed that all the other children who participated in the experiment had been murdered.

The doctor told R. B. S. that the computer implantation was unlikely but called the psychology department at the university and got confirmation that children had in fact taken part in an experiment conducted years before. The records of the study had long since been destroyed. R. B. S. believed that this information completely vindicated his story. His delusional behavior persisted and eventually cost him his pilot's license.

Employees fill prescriptions at a medical marijuana clinic in San Francisco. California is one of several states that have decriminalized the use of medical marijuana in recent years.

Jim Wilson/The New York Times/Redux

R. B. S. seemed to compartmentalize the delusion. When asked why he could no longer fly, he intently recounted the story of the implant and the murders, asserting that its truth had cost him the medical certification needed for a license. Then he happily discussed other topics in a normal way.

R. B. S. was suffering from a mild focal psychosis: he was losing contact with reality. In some cases, this break is so severe and the capacity to respond to the environment so impaired and distorted that the person can no longer function. People in a state of psychosis may experience hallucinations or delusions or they may withdraw into a private world isolated from people and events around them.

A variety of drugs can produce psychosis, including LSD, amphetamine, cocaine, and, as shown by this case, marijuana. At low doses, the active ingredient in marijuana, tetrahydrocannabinol, has mild sedative-hypnotic effects similar to those of alcohol. At the high doses that R. B. S. used, THC can produce euphoria and hallucinations.

Marijuana comes from the leaves of the hemp plant, *Cannabis sativa*. Humans have used hemp for thousands of years to make rope, paper, cloth, and a host of products. And marijuana has a number of beneficial medical effects. In the Pacific Northwest, marijuana is the largest agricultural crop and makes a larger contribution to the economy than does forestry.

R. B. S.'s heavy marijuana use certainly raises the suspicion that the drug had some influence on his delusional condition. *Cannabis* use has been reported to moderately increase the risk of psychotic symptoms in young people and has a much stronger effect in those with a predisposition for psychosis, who have been abused in childhood, or who have other childhood disabilities (Henquet et al., 2008). Although there is evidence that heavy marijuana use may be associated alterations in brain development, it is unclear whether brain abnormalities are a result of marijuana use or a causal factor in its use.

R. B. S.'s delusions might have eventually arisen anyway, even if he had not used marijuana. Furthermore, any of the 400 or so compounds besides THC present in marijuana could trigger psychotic symptoms. Approximately 10 years after his initial attack, R. B. S.'s symptoms subsided, and he returned to flying.

---

3. As an alternative to explanations of susceptibility to addiction based on genetic _____, _____ can account both for the enduring behaviors that support addiction and for the tendency of drug addiction to be inherited.

4. It is hard to determine whether recreational drugs cause brain damage in humans because it is difficult to distinguish the effects of _____ from the effects of _____.

5. Briefly describe the basis for a reasonable approach to treating drug addiction.

Answers appear at the back of the book.

6-5 Hormones

In 1849, European scientist A. A. Berthold removed the testes of a rooster and found that the rooster no longer crowed; nor did it engage in sexual or aggressive behavior. Berthold then reimplanted one testis in the rooster's body cavity. The rooster began crowing and displaying normal sexual and aggressive behavior again. The reimplanted testis did not establish any nerve connections, so Berthold concluded that it must release a chemical into the rooster's circulatory system that influenced the animal's behavior.

That chemical, we now know, is **testosterone,** the sex hormone secreted by the testes and responsible for the distinguishing characteristics of the male. The effect that Berthold produced by reimplanting the testis can be mimicked by administering testosterone to a castrated rooster, or capon. The hormone is sufficient to make the capon behave like a rooster with testes.

Testosterone's influence on the rooster illustrates some of the ways that this hormone produces male behaviors. Testosterone also initiates changes in the size and appearance of the mature male body. In a rooster, for example, testosterone produces the animal's distinctive plumage and crest, and it activates other sex-related organs.

Hierarchical Control of Hormones

Figure 6-21 shows that hormones operate within a hierarchy that begins when the brain responds to sensory experiences and cognitive activity. The hypothalamus produces neurohormones that stimulate the pituitary gland to secrete "releasing hormones" into the circulatory system. The pituitary hormones, in turn, influence the remaining endocrine glands to release appropriate hormones into the bloodstream. These hormones then act on various targets in the body and send feedback to the brain about the need for more or less hormone release.

Berthold's experiment demonstrated the existence and function of *hormones*—chemicals released by *endocrine glands* into the bloodstream—that circulate to a body target and affect it. The endocrine glands operate under the influence of the CNS and the ANS (see Figure 2-30).

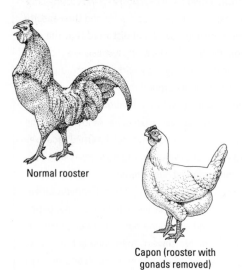

Normal rooster

Capon (rooster with gonads removed)

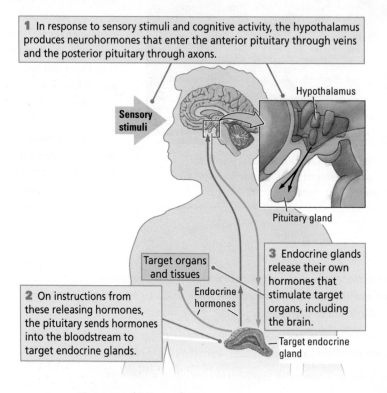

1 In response to sensory stimuli and cognitive activity, the hypothalamus produces neurohormones that enter the anterior pituitary through veins and the posterior pituitary through axons.

Hypothalamus

Sensory stimuli

Pituitary gland

3 Endocrine glands release their own hormones that stimulate target organs, including the brain.

Target organs and tissues

Endocrine hormones

2 On instructions from these releasing hormones, the pituitary sends hormones into the bloodstream to target endocrine glands.

Target endocrine gland

FIGURE 6-21 Hormonal Hierarchy

Hormones not only affect body organs but also target the brain and neurotransmitter-activating systems there. Almost every neuron in the brain contains receptors on which various hormones can act. In addition to influencing sex organs and physical appearance in a rooster, testosterone may have neurotransmitterlike effects on the brain cells that it targets, especially neurons that control crowing, male sexual behavior, and aggression.

In these neurons, testosterone is transported into the cell nucleus, where it activates genes. The genes, in turn, trigger the synthesis of proteins needed for cellular processes that produce the rooster's male behaviors. Thus, the rooster receives not only a male body but a male brain as well.

Although many questions remain about how hormones produce complex behavior, the diversity of testosterone's functions clarifies why the body uses hormones as messengers: their targets are so widespread that the best possible way of reaching all of them is to travel in the bloodstream, which goes everywhere in the body.

Classes and Functions of Hormones

Hormones can be used as drugs to treat or prevent disease. People take synthetic hormones as replacement therapy if the glands that produce the hormones are removed or malfunction. People also take hormones, especially sex hormones, to counteract the effects of aging and to increase physical strength and endurance and to gain an advantage in sports. In the human body, as many as 100 hormones are classified chemically as either steroids or peptides.

Steroid hormones, such as testosterone and cortisol, are synthesized from cholesterol and are lipid (fat) soluble. Steroids diffuse away from their sites of synthesis in glands, including the gonads, adrenal cortex, and thyroid, easily crossing the cell membrane. They enter target cells in the same way and act on the cells' DNA to increase or decrease the production of proteins.

Peptide hormones, such as insulin, growth hormone, and the endorphins, are made by cellular DNA in the same way other proteins are made. They influence their target cell's activity by binding to metabotropic receptors on the cell membrane, generating a second messenger that affects the cell's physiology.

Steroid and peptide hormones fall into one of three main functional groups with respect to behavior, and they may function in more than one group:

1. **Homeostatic hormones** maintain a state of internal metabolic balance and regulate physiological systems in an organism. Mineralocorticoids (e.g., aldosterone) control both the concentration of water in blood and cells and the levels of sodium, potassium, and calcium in the body, and they promote digestive functions.

2. **Gonadal (sex) hormones** control reproductive functions. They instruct the body to develop as male (e.g., the hormone testosterone) or female (e.g., the hormone estrogen); influence sexual behavior and the conception of children; and, in women, control the menstrual cycle (e.g., estrogen and progesterone), the birthing of babies, and the release of breast milk (e.g., prolactin, oxytocin).

3. **Glucocorticoids** (cortisol and corticosterone are examples), a group of steroid hormones secreted in times of stress, are important in protein and carbohydrate metabolism and in controlling sugar levels in the blood and the absorption of sugar by cells. Hormones activated in psychologically challenging events or emergency situations prepare the body to cope by fighting or fleeing.

Homeostatic Hormones

The homeostatic hormones are essential to life. The body's internal environment must remain within relatively constant parameters for us to function. An appropriate balance of sugars, proteins, carbohydrates, salts, and water is required in the bloodstream,

We take up the story of hormones again as we examine motivation and emotion, including eating in Section 12-5 and the relation between learning and memory in Section 14-4.

To refresh your understanding of metabotropic receptors, review Figure 5-15.

testosterone Sex hormone secreted by the testes and responsible for the distinguishing characteristics of the male.

steroid hormone Fat-soluble chemical messenger synthesized from cholesterol.

peptide hormone Chemical messenger synthesized by cellular DNA that acts to affect the target cell's physiology.

homeostatic hormone One of a group of hormones that maintain internal metabolic balance and regulate physiological systems in an organism.

gonadal (sex) hormone One of a group of hormones, such as testosterone, that control reproductive functions and bestow sexual appearance and identity as male or female.

glucocorticoid One of a group of steroid hormones, such as cortisol, secreted in times of stress; important in protein and carbohydrate metabolism.

The term *homeostasis* comes from the Greek words *stasis* (standing) and *homeo* (in the same place).

The normal concentration of glucose in the bloodstream varies between 80 and 130 milligrams per 100 milliliters of blood.

in the extracellular compartments of muscles, in the brain and other body structures, and within all body cells. Homeostasis of the internal environment must be maintained regardless of a person's age, activities, or conscious state. As children or adults, at rest or in strenuous work, when we have overeaten or when we are hungry, to survive we need a relatively constant internal environment.

A typical homeostatic function is the control of blood-sugar levels. One group of cells in the pancreas releases insulin, a homeostatic hormone that causes blood sugar to fall by instructing the liver to start storing glucose rather than releasing it and by instructing cells to increase glucose uptake. The resulting decrease in glucose then decreases the stimulation of pancreatic cells so that they stop producing insulin.

Diabetes mellitus is caused by a failure of the pancreatic cells to secrete enough insulin or any at all. As a result, blood-sugar levels can fall (hypoglycemia) or rise (hyperglycemia). In hyperglycemia, blood-glucose levels rise because insulin does not instruct cells of the body to take up glucose. Consequently, cell function, including neural function, can fail through glucose starvation, even in the presence of high levels of glucose in the blood. In addition, chronic high blood-glucose levels cause damage to the eyes, kidneys, nerves, heart, and blood vessels.

In hypoglycemia, inappropriate diet can lead to low blood sugar, which can be severe enough to cause fainting. Eric Steen and his coworkers (2005) propose that insulin resistance in brain cells may be related to Alzheimer's disease. They raise the possibility that Alzheimer's disease may be a third type of diabetes.

Gonadal Hormones

We are prepared for our adult reproductive roles by the gonadal hormones that give us our sexual appearance, mold our identity as male or female, and allow us to engage in sex-related behaviors. Sex hormones begin to act on us even before we are born and continue their actions throughout our lives.

Section 8-2 reviews the neurobiology of development and Section 8-4 the role of hormones in brain development.

The male Y chromosome contains a gene called the sex-determining region, or *SRY* gene. If cells in the undifferentiated gonads of the early embryo contain an *SRY* gene, they develop into a testis, and if they do not, they develop into an ovary. In the male, the testes produce the hormone testosterone, which in turn masculinizes the body, producing the male body and genital organs and the male brain.

The **organizational hypothesis** proposes that actions of hormones in the course of development alter tissue differentiation. Thus, testosterone masculinizes the brain early in life by being taken up in brain cells, where it is converted into estrogen by the enzyme aromatase. Estrogen then acts on estrogen receptors to initiate a chain of events that include the activation of certain genes in the cell nucleus. These genes then contribute to the masculinization of brain cells and their interactions with other brain cells.

That estrogen, a hormone usually associated with the female, masculinizes the male brain may seem surprising. Estrogen does not have the same effect on the female brain, because females have a blood enzyme that binds to estrogen and prevents its entry into the brain. Hormones play a somewhat lesser role in producing the female body as well as the female brain, but they control the mental and physical aspects of menstrual cycles, regulate many facets of pregnancy and birth, and stimulate milk production for breast-feeding babies.

Section 12-5 describes the effects of gonadal hormones on sexual behavior. Section 15-5 recounts sex differences in thinking patterns.

Hormones contribute to surprising differences in the brain and in cognitive behavior, and, as noted in Section 6-3, they play a role in male–female differences in drug dependence and addiction. The male brain is slightly larger than the female brain after corrections are made for body size, and the right hemisphere is somewhat larger than the left in males. The female brain has a higher rate both of cerebral blood flow and of glucose utilization. There are also a number of differences in brain size in different regions of the brain, including nuclei in the hypothalamus, that are related to sexual function, parts of the corpus callosum that are larger in females, and a somewhat larger language region in the female brain.

Three lines of evidence, summarized by Elizabeth Hampson and Doreen Kimura (2005), support the conclusion that sex-related cognitive differences result from these brain differences. These cognitive differences also depend in part on the continuing circulation of the sex hormones. The evidence:

1. The results of spatial and verbal tests given to females and males in many different settings and cultures show that males tend to excel in the spatial tasks tested and females in the verbal tasks.

2. The results of similar tests given to female subjects in the course of the menstrual cycle show fluctuations in test scores with various phases of the cycle. During the phase in which the female sex hormones estradiol (metabolized from estrogen) and progesterone are at their lowest levels, women do comparatively better on spatial tasks, whereas, during the phase in which levels of these hormones are high, women do comparatively better on verbal tasks.

3. Tests comparing premenopausal and postmenopausal women, women in various stages of pregnancy, and females and males with varying levels of circulating hormones all provide some evidence that hormones affect cognitive function.

Sex-hormone-related differences in cognitive function are not huge. A great deal of overlap in performance scores exists between males and females. Yet statistically, the differences are reliable. Similar influences of sex hormones on behavior are found in other species. Berthold's rooster experiment described earlier shows the effects of testosterone on the rooster's behavior. Findings from a number of studies demonstrate that motor skills in female humans and other animals improve at estrus, a time when progesterone levels are high.

Anabolic–Androgenic Steroids

A class of synthetic hormones related to the male sex hormone testosterone has both muscle-building (anabolic) and masculinizing (androgenic) effects. These *anabolic–androgenic steroids,* commonly known simply as **anabolic steroids** (in common parlance, "roids"), were synthesized originally to build body mass and enhance endurance. Russian weight lifters were the first to use them, in 1952, to enhance performance and win international competitions.

Synthetic steroid use rapidly spread to other countries and sports, eventually leading to a ban from track and field and then from many other sports as well. These bans are enforced by drug testing. Testing policy has led to a cat-and-mouse game in which new anabolic steroids and new ways of taking them and masking them are devised to evade detection.

Today, the use of anabolic steroids is about equal among athletes and nonathletes. More than 1 million people in the United States have used anabolic steroids not only to enhance athletic performance but also to enhance physique and appearance. Anabolic steroid use in high schools may be as high as 7 percent for males and 3 percent for females.

The use of anabolic steroids carries health risks. Their administration results in the body reducing its manufacture of the male hormone testosterone, which in turn reduces male fertility and spermatogenesis. Muscle bulk is increased and so is male aggression. Cardiovascular effects include increased risk of heart attacks and stroke. Liver and kidney function may be compromised, and the risk of tumors may increase. Male-pattern baldness may be enhanced, and females may experience clitoral enlargement, acne, an increase in body hair, and a deepened voice.

organizational hypothesis Proposal that actions of hormones in development alter tissue differentiation; for example, testosterone masculinizes the brain.

anabolic steroid Belongs to a class of synthetic hormones related to testosterone that have both muscle-building (anabolic) and masculinizing (androgenic) effects; also called *anabolic–androgenic steroid.*

Aggressive episodes attributed to anabolic-steroid use are called "'roid rage."

Anabolic steroids also have approved clinical uses. Testosterone replacement is a treatment for hypogonadal males. It is also useful for treating muscle loss subsequent to trauma and for the recovery of muscle mass in malnourished people. In females, anabolic steroids are used to treat endometriosis and fibrocystic disease of the breast.

Glucocorticoids and Stress

"Stress" is a term borrowed from engineering to describe a process in which an agent exerts a force on an object. Applied to humans and other animals, a *stressor* is a stimulus that challenges the body's homeostasis and triggers arousal. Stress responses are not only physiological but also behavioral and include both arousal and attempts to reduce stress. A stress response can outlast a stress-inducing incident and may even occur in the absence of an obvious stressor. Living with constant stress can be debilitating.

Surprisingly, the body's response is the same whether the stressor is exciting, sad, or frightening. Robert Sapolsky (1992) uses the vivid image of a hungry lion chasing down a zebra to illustrate the stress response. The chase elicits very different reactions in the two animals, but their physiological stress responses are exactly the same. The stress response begins when the body is subjected to a stressor and especially when the brain perceives a stressor and responds with arousal. The response consists of two separate sequences, one fast and the other slow.

The fast response is shown at the left in **Figure 6-22.** The sympathetic division of the autonomic nervous system is activated to prepare the body and its organs for "fight or flight," and the parasympathetic division for "rest and digest" is turned off. In addition, the sympathetic division stimulates the medulla on the interior of the adrenal gland to release epinephrine. The epinephrine surge (often called the "adrenaline surge" after epinephrine's original name) prepares the body for a sudden burst of activity. Among

FIGURE 6-22 Activating a Stress Response Two pathways to the adrenal gland control the body's response to stress. The fast-acting pathway primes the body immediately for fight or flight. The slow-acting pathway both mobilizes the body's resources to confront a stressor and repairs stress-related damage. Abbreviations: CRH, corticotropin-releasing hormone; ACTH, adrenocorticotropic hormone.

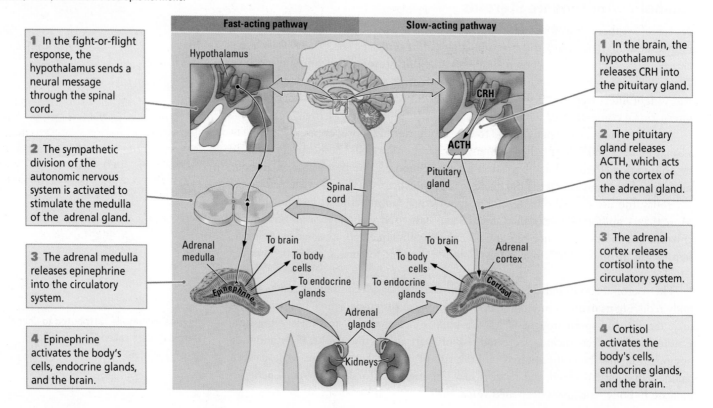

its many functions, epinephrine stimulates cell metabolism so that the body's cells are ready for action.

The hormone controlling the slow response is the steroid cortisol, a glucocorticoid released from the outer layer (cortex) of the adrenal gland, as shown at the right in Figure 6-22. The cortisol pathway is activated more slowly, taking from minutes to hours. Cortisol has a wide range of functions, which include turning off all bodily systems not immediately required to deal with a stressor. For example, cortisol turns off insulin so that the liver starts releasing glucose, thus temporarily increasing the body's energy supply. It also shuts down reproductive functions and inhibits the production of growth hormone. In this way, the body's energy supplies can be concentrated on dealing with the stress.

Ending a Stress Response

Normally, stress responses are brief. The body mobilizes its resources, deals with the challenge physiologically and behaviorally, and then shuts down the stress response. Just as the brain is responsible for turning on the stress reaction, it is also responsible for turning it off. Consider what can happen if the stress response is not shut down:

- The body continues to mobilize energy at the cost of energy storage.

- Proteins are used up, resulting in muscle wasting and fatigue.

- Growth hormone is inhibited, so the body cannot grow.

- The gastrointestinal system remains shut down, reducing the intake and processing of nutrients to replace used resources.

- Reproductive functions are inhibited.

- The immune system is suppressed, contributing to the possibility of infection or disease.

Sapolsky (2005) argued that the hippocampus plays an important role in turning off the stress response. The hippocampus contains a high density of cortisol receptors, and it has axons that project to the hypothalamus. Consequently, the hippocampus is well suited to detecting cortisol in the blood and instructing the hypothalamus to reduce blood-cortisol levels.

There may, however, be a more-insidious relation between the hippocampus and blood-cortisol levels. Sapolsky and his coworkers observed wild-born vervet monkeys that had become agricultural pests in Kenya and had therefore been trapped and caged. They found that some of the monkeys became sick and died of a syndrome that appeared to be related to stress. Those that died seemed to have been subordinate animals housed with particularly aggressive, dominant monkeys. Autopsies showed high rates of gastric ulcers, enlarged adrenal glands, and pronounced hippocampal degeneration. The hippocampal damage may have been due to prolonged high cortisol levels produced by the unremitting stress of being caged with the aggressive monkeys.

Cortisol levels are usually regulated by the hippocampus, but if these levels remain elevated because a stress-inducing situation continues, cortisol eventually damages the hippocampus. The damaged hippocampus is then unable to do its work of reducing the level of cortisol. Thus, a vicious circle is set up in which the hippocampus undergoes progressive degeneration and cortisol levels are not controlled (**Figure 6-23**).

Because stress-response circuits in monkeys are very similar to those in humans, the possibility exists that excessive stress in humans also can lead to damaged hippocampal neurons. Because the hippocampus is thought to play a role in memory, stress-induced damage to the hippocampus is postulated to result in impaired memory as

FIGURE 6-23 **Vicious Circle** Unrelieved stress promotes an excessive release of cortisol that damages neurons in the hippocampus. The damaged neurons are unable to detect cortisol and therefore cannot signal the adrenal gland to stop producing it. The result is a feedback loop in which the enhanced secretion of cortisol further damages hippocampal neurons

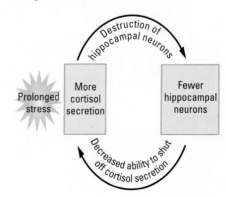

PTSD, introduced in Section 5-4 in relation to sensitization, is among the anxiety disorders detailed in Section 12-4. Focus 16-1 and Section 16-4 consider treatments.

well as in posttraumatic stress disorder. PTSD is characterized by physiological arousal symptoms related to recurring memories and dreams related to a traumatic event—for months or years after the event. People with PTSD feel as if they are re-experiencing the trauma, and the accompanying physiological arousal enhances their belief that danger is imminent.

Research has not yet led to a clear-cut answer on whether the cumulative effects of stress damage the human hippocampus. For example, research on women who were sexually abused in childhood and were diagnosed as suffering from PTSD yields some reports of changes in memory or in hippocampal volume, as measured with brain-imaging techniques. Other studies report no differences in abused and nonabused subjects (Landré et al., 2010). The fact that such apparently similar studies can obtain different results can be explained in a number of ways.

First, the amount of damage to the hippocampus that must occur to produce a stress syndrome is not certain. Second, brain-imaging techniques may not be sensitive to subtle changes in hippocampal-cell function or to moderate cell loss. Third, wide individual and environmental differences influence how people respond to stress. Finally, preexisting injury to the hippocampus or other brain regions could influence the probability of developing PTSD (Gilbertson et al., 2002).

Humans are long-lived and gather many life experiences that complicate simple extrapolations from a single stressful event. Nevertheless, Patrick McGowan and his coworkers (2009) report that the glucocorticoid-receptor density in the hippocampi of suicide victims who had been sexually abused in childhood was decreased compared with that of suicide victims who had not been abused and with that of control subjects.

The decrease in receptors and in glucocorticoid mRNA suggests that childhood abuse induces epigenetic changes in the expression of glucocorticoid genes. The decrease in glucocorticoid receptors presumably renders the hippocampus less able to depress stress responses. The importance of the McGowan study is its suggestion of a mechanism through which stress can influence hippocampal function without necessarily being associated with a decrease in hippocampal volume.

REVIEW 6-5

Hormones

Before you continue, check your understanding.

1. The hypothalamus produces _____ that stimulate the _____ to secrete _____ into the circulatory system. Hormone levels circulating in the bloodstream send feedback to the _____.

2. Hormones are classified chemically as _____ or _____.

3. Broadly speaking, _____ hormones regulate metabolic balance, _____ hormones regulate reproduction, and _____ regulate stress.

4. One class of synthetic hormones is _____, which increase _____ and have _____ effects.

5. The stress response has a fast-acting pathway mediated by _____ and a slow-acting pathway mediated by _____.

6. Describe the proposed relationship among stress, cortisol, and the hippocampus.

Answers appear at the back of the book.

SUMMARY

6-1 Principles of Psychopharmacology

Psychoactive drugs—substances that alter mood, thought, or behavior—produce their effects by acting on neuronal receptors or on chemical processes in the nervous system, especially on processes of neurotransmission at synapses. Drugs act either as agonists to stimulate neuronal activity or as antagonists to depress it. Psychopharmacology is the study of drug effects on the brain and behavior.

Drugs are administered by mouth, by inhalation, by absorption through the skin, and by injection. To reach a target in the nervous system, a psychoactive drug must pass through numerous barriers posed by digestion and dilution, the blood–brain barrier, and cell membranes. Drugs are diluted by body fluids as they pass through successive barriers; metabolized in the body; and excreted through sweat glands and in feces, urine, breath, and breast milk.

A common misperception about psychoactive drugs is that they act specifically and consistently, but learning also affects individual reactions to drugs. The body and brain may rapidly become tolerant (habituated) of many drugs, so the dose must be increased to produce a constant effect. Alternatively, people may become sensitized to a drug: the same dose produces increasingly greater effects. These forms of unconscious learning also play an important role in a person's behavior under the influence of a drug.

6-2 Grouping Psychoactive Drugs

Psychoactive drugs can be organized according to their major behavioral effects into five groups: antianxiety agents and sedative hypnotics, antipsychotic agents, antidepressants and mood stabilizers, opioid analgesics, and psychotropics. Each group, summarized in Table 6-1 on page 182 contains natural or synthetic drugs or both, and they may produce their actions in different ways.

6-3 Factors Influencing Individual Responses to Drugs

A drug does not have a uniform action on every person. Physical differences—in body weight, sex, age, or genetic background—influence the effects of a given drug on a given person, as do behaviors, such as learning, and the cultural and environmental contexts.

The influence of drugs on behavior varies widely with the situation and as a person learns drug-related behaviors. Alcohol myopia, for example, can influence a person to focus primarily on prominent cues in the environment. These cues may encourage the person to act in ways in which he or she would not normally behave.

Females are more sensitive to drugs than males are and may become addicted more quickly than males to lower doses of drugs. The incidence of female abuse of many kinds of drugs currently equals or exceeds male abuse of those drugs.

6-4 Explaining and Treating Drug Abuse

The neural mechanisms implicated in addiction are the same neural systems responsible for wanting and liking more generally. So anyone is likely to be a potential drug abuser. Addiction develops in a number of stages as a result of repeated drug taking.

Initially, drug taking produces pleasure (liking), but with repeated use, the behavior becomes conditioned to associated objects, events, and places. Eventually, the conditioned cues motivate the drug user to seek them out (wanting), which leads to more drug taking. These subjective experiences associated with prominent cues and drug seeking promote craving for the drug. As addiction proceeds, the subjective experience of liking decreases while wanting increases.

Drug treatment varies by the drug that is abused. Whatever the treatment approach, it is likely that success depends on permanent lifestyle changes. Considering how many people use tobacco, drink alcohol, use recreational drugs, or abuse prescription drugs, to find someone who has not used a drug when it was available is probably rare. But some people do seem vulnerable to drug use and addiction due either to genetic or epigenetic influences.

Excessive alcohol use can be associated with damage to the thalamus and hypothalamus, but the cause of the damage is poor nutrition rather than the direct actions of alcohol. Cocaine can harm brain circulation, producing brain damage by reducing blood flow or by bleeding into neural tissue. MDMA (ecstasy) use can result in the loss of fine axon collaterals of serotonergic neurons and the associated impairments in cognitive function.

Psychedelic drugs, such as marijuana and LSD, can be associated with psychotic behavior. Whether this behavior is due to the direct effects of the drugs or to the aggravation of preexisting conditions is not clear.

6-5 Hormones

Steroid and peptide hormones produced by endocrine glands circulate in the bloodstream to affect a wide variety of targets. Interacting to regulate hormone levels is a hierarchy of sensory stimuli and cognitive activity in the brain that stimulate the pituitary gland through the hypothalamus. The pituitary stimulates or inhibits the endocrine glands, which, through other hormones, send feedback to the brain.

Homeostatic hormones regulate the balance of sugars, proteins, carbohydrates, salts, and other substances in the body. Gonadal hormones regulate the physical features and behaviors associated with sex characteristics and behaviors, reproduction, and caring for offspring. Glucocorticoids are steroid hormones that regulate the body's ability to cope with stress—with arousing and challenging situations.

The hippocampus plays an important role in ending the stress response. Failure to turn stress responses off after a stressor has passed can contribute to susceptibility to PTSD and other psychological and physical diseases. Stress may activate epigenetic changes that modify the expression of genes that regulate hormonal responses to stress and produce brain changes that persist long after the stress-provoking incident has passed.

Synthetic anabolic steroids, used by both athletes and nonathletes, mimic the effects of testosterone and so increase muscle bulk, stamina, and aggression but can have deleterious side effects.

KEY TERMS

addiction, p. 195

agonist, p. 176

alcohol myopia, p. 195

amphetamine, p. 190

anabolic steroid, p. 205

antagonist, p. 176

antianxiety agent, p. 183

attention deficit/hyperactivity disorder (ADHD), p. 173

barbiturate, p. 183

bipolar disorder, p. 187

competitive inhibitor, p. 189

cross-tolerance, p. 183

disinhibition theory, p. 193

dopamine hypothesis of schizophrenia, p. 185

endorphin, p. 189

fetal alcohol spectrum disorder (FASD), p. 185

glucocorticoid, p. 203

gonadal (sex) hormone, p. 203

homeostatic hormone, p. 203

major depression, p. 187

monoamine oxidase (MAO) inhibitor, p. 187

mood stabilizer, p. 187

opioid analgesic, p. 187

organizational hypothesis, p. 205

peptide hormone, p. 203

psychedelic drug, p. 190

psychoactive drug, p. 173

psychomotor activation, p. 195

psychopharmacology, p. 173

second-generation antidepressant, p. 187

selective serotonin reuptake inhibitor (SSRI), p. 187

steroid hormone, p. 203

substance abuse, p. 195

testosterone, p. 203

tolerance, p. 176

tricyclic antidepressant, p. 187

wanting-and-liking theory, p. 196

withdrawal symptom, p. 195

Please refer to the Companion Web Site at www.worthpublishers.com/kolbintro4e *for Interactive Exercises and Quizzes.*

CHAPTER

7

How Do We Study the Brain's Structure and Functions?

RESEARCH FOCUS 7-1 TUNING IN TO LANGUAGE

7-1 MEASURING BRAIN AND BEHAVIOR

LINKING NEUROANATOMY AND BEHAVIOR

METHODS OF BEHAVIORAL NEUROSCIENCE

MANIPULATING AND MEASURING BRAIN–BEHAVIOR INTERACTIONS

7-2 MEASURING THE BRAIN'S ELECTRICAL ACTIVITY

EEG RECORDINGS OF GRADED POTENTIALS

MAPPING BRAIN FUNCTION WITH EVENT-RELATED POTENTIALS

CLINICAL FOCUS 7-2 MILD HEAD INJURY AND DEPRESSION

MAGNETOENCEPHALOGRAPHY

RECORDING ACTION POTENTIALS FROM SINGLE CELLS

7-3 STATIC IMAGING TECHNIQUES: CT AND MRI

7-4 DYNAMIC BRAIN IMAGING

FUNCTIONAL MAGNETIC RESONANCE IMAGING

POSITRON EMISSION TOMOGRAPHY

OPTICAL TOMOGRAPHY

7-5 CHEMICAL AND GENETIC MEASURES OF BRAIN AND BEHAVIOR

MEASURING THE BRAIN'S CHEMISTRY

MEASURING GENES IN BRAIN AND BEHAVIOR

MEASURING GENE EXPRESSION

CLINICAL FOCUS 7-3 CANNABIS USE, PSYCHOSIS, AND GENETICS

7-6 COMPARING NEUROSCIENCE RESEARCH METHODS

7-7 USING ANIMALS IN BRAIN–BEHAVIOR RESEARCH

BENEFITS OF CREATING ANIMAL MODELS OF DISEASE

ANIMAL WELFARE AND SCIENTIFIC EXPERIMENTATION

RESEARCH FOCUS 7-4 ATTENTION-DEFICIT/HYPERACTIVITY DISORDER

Tuning In to Language

The search to understand the organization and operation of the human brain is driven partly by emerging technologies. Over the past decade, neuroscience researchers have developed dramatic new, noninvasive ways to image the brain's activity in subjects who are awake. One technique, **functional near-infrared spectroscopy** (fNIRS), gathers light transmitted through cortical tissue to image blood–oxygen consumption, or oxygenated hemoglobin, in the brain. NIRS, a form of *optical tomography,* is detailed in Section 7-4.

fNIRS allows investigators to measure oxygen consumption in relatively select regions of the cerebral cortex, even in newborn infants. In one study (May et al., 2011), newborns (0–3 days old) wore a mesh cap containing the NIRS apparatus, made up of optical fibers, as they listened to the sounds of a familiar or unfamiliar language.

When newborns listened to a familiar language, their brains showed a generalized increase in oxygenated hemoglobin; when they heard an unfamiliar language, oxygenated hemoglobin decreased overall. But when the babies heard the same sentences played backwards, there was no difference in brain response.

The opposing responses to familiar and unfamiliar languages mean that prenatal language experience shapes how the newborn brain responds to familiar and unfamiliar tongues. This finding leads to many questions, among them: How does prenatal exposure to language influence later language learning? Do children who are exposed to multiple languages prenatally show better language acquisition than those exposed just to one? How much prenatal language exposure is necessary, and do premature infants show the same results as full-term babies?

Whatever the answers, this study shows that the prenatal brain is tuned in to the language environment into which it will be born.

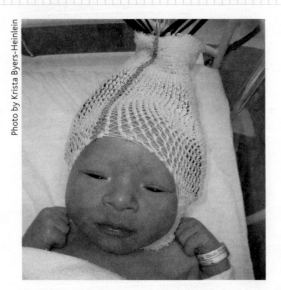

Newborn with probes placed on the head.

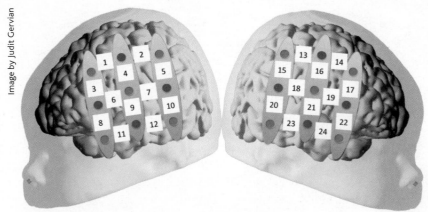

Configuration of probes overlaid on schematics of an infant's left and right hemispheres. The red dots indicate light emitting fibers; the blue dots indicate light detectors. The light detectors in the outer strips in both hemispheres sit over the regions specialized for language in adults.

The simple and noninvasive nature of fNIRS is likely to yield new insights not only into brain development but also into adult brain function. Over the coming decades, our understanding of the brain–behavior relationship will continue to be driven in part by applying novel research methods and in part by exploiting existing ones.

To understand how far the methods of neuroscience research have progressed, imagine that it is the year 1800. You are a neurologist interested in studying how the brain works. The challenge is how to begin. The two most obvious choices are to dissect the brains of dead people and other animals or to study people who have sustained brain injuries. Indeed, this was the how the relationship between brain and behavior was studied well into the twentieth century.

functional near-infrared spectroscopy (fNIRS) Noninvasive technique that gathers light transmitted through cortical tissue to image blood-oxygen consumption; form of optical tomography.

neuropsychology Study of the relations between brain function and behavior.

Techniques for studying the brain's physiological processes began to develop in the years between World War I and World War II, when emerging methods of research began to record electrical activity emitted from the brain. One breakthrough was the *electroencephalograph* (EEG), developed by Hans Berger in the 1930s. Advances in understanding genetics and the analysis of behavior in the early 1950s set the stage for neuroscience as we know it today.

An explosion of knowledge in neuroscience that began around 1970, driven by the analysis of brain and behavior using new research methods, has occurred across multiple and disparate disciplines. Today, brain–behavior analyses combine the efforts of anatomists and geneticists, psychologists and physiologists, chemists and physicists, endocrinologists and neurologists, pharmacologists and psychiatrists. For the aspiring brain researcher in the twenty-first century, the range of available research methods is breathtaking.

We begin this chapter by reviewing how investigators measure behavior in both human and nonhuman subjects and how neuroscientists can manipulate behavior by perturbing the brain. We then consider electrical techniques, including the EEG, for recording brain activity; noninvasive procedures, such as fNIRS, that produce both static and dynamic images of the brain; and chemical and genetic methods for measuring brain and behavior. At the chapter's end, we review issues that surround using nonhuman animals in research.

Section 4-1 reviews the history of the EEG and how it enabled investigators to explain electrical activity in the nervous system.

7-1 Measuring Brain and Behavior

During a lecture at a meeting of the Anthropological Society of Paris in 1861, Ernest Auburtin, a French physician, argued that language functions are located in the brain's frontal lobes. Five days later a fellow French physician, Paul Broca, observed a brain-injured patient who had lost his speech and was able to say only "tan" and utter a swear word. The patient soon died. Broca and Auburtin examined the man's brain and found that the left frontal lobe was the focus of his injury.

By 1863 Broca had collected eight more similar cases and concluded that speech is located in the third frontal convolution of the left frontal lobe—a region now called *Broca's area*. Broca's findings attracted others to study brain–behavior relationships in patients. The field that developed we now call **neuropsychology,** the study of the relations between brain function and behavior. Today, measuring brain and behavior increasingly includes noninvasive imaging, complex neuroanatomical measurement, and sophisticated behavioral analyses.

Section 10-4, which explores the anatomy of language and music, describes Broca's contributions.

Linking Neuroanatomy and Behavior

At the beginning of the twentieth century, neuroanatomy's primary tools were *histological:* brains were sectioned postmortem and the tissue (histo) stained with different dyes. As shown in **Figure 7-1,** staining sections of brain tissue can identify the cell bodies in the brain when viewed with a light microscope (A), and selectively staining individual neurons reveals their complete structure (B). An electron microscope (C) makes it possible to view synapses in detail. These techniques allowed researchers such as Korbinian Brodmann to divide the cerebral cortex into many distinct zones. These zones, investigators presumed, had specific functions.

By the dawn of the twenty-first century, dozens of techniques for labeling neurons and their connections, as well as glial cells, had developed. These techniques allow researchers to identify molecular, neurochemical, and morphological (structural) differences among neuronal types and ultimately to relate these characteristics to behavior. One fast-developing technique, shown in Figure 7-1D, uses a multiphoton microscope that makes it possible to image living brain tissue in a three-dimensional view.

Compare Brodmann's map of the cortex, based on staining and shown in Figure 2-23, to the multiphoton microscopic image in Figure 7-1D.

FIGURE 7-1 Staining Cerebral Neurons Viewed through a light microscope, **(A)** Nissl-stained section of parietal cortex shows all cell bodies but no cell processes (axons and dendrites). **(B)** At higher magnification, an individual Golgi-stained pyramidal cell from the parietal cortex is visible. The cell body (dark triangular shape at center) and spiny dendrites A and B are visible in detail. These dendrites and spines are shown at right at an even higher magnification. **(C)** The view through an electron microscope shows neuronal synapses in detail. **(D)** Multiple images from a multiphoton microscope, merged to generate a 3-D image of living tissue.

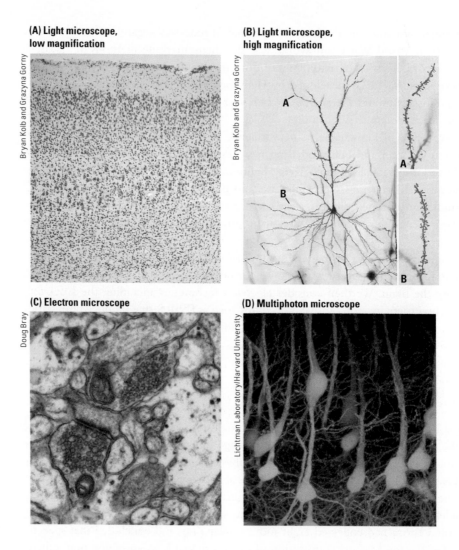

(A) Light microscope, low magnification

Bryan Kolb and Grazyna Gorny

(B) Light microscope, high magnification

Bryan Kolb and Grazyna Gorny

A

B

(C) Electron microscope

Doug Bray

(D) Multiphoton microscope

Lichtman Laboratory/Harvard University

Parkinson's disease offers one of the clearest examples of neuronal/behavioral relations. Early analysis of postmortem brain tissue from humans afflicted with the disease showed that cells in the brainstem's midbrain region, the substantia nigra, had died **(Figure 7-2)**. Later studies using laboratory animals showed that if the substantia nigra was killed experimentally, the animals showed symptoms remarkably similar to those in human Parkinson's patients, including tremors, muscular rigidity, and a reduction in voluntary movement.

FIGURE 7-2 Pathology in Parkinson's Disease A variety of motor disturbances appear when enough dopamine-producing cells in the substantia nigra die. Adapted from healthguide.howstuffworks.com/substantia-nigra-and-parkinsons-disease-picture.htm

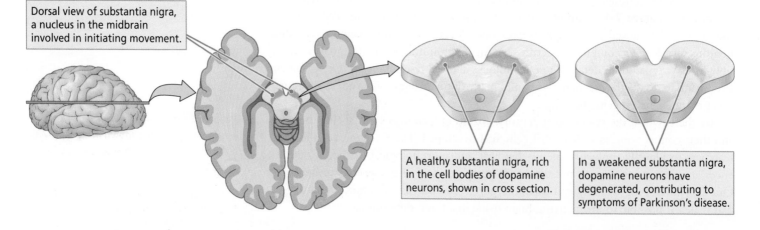

Dorsal view of substantia nigra, a nucleus in the midbrain involved in initiating movement.

A healthy substantia nigra, rich in the cell bodies of dopamine neurons, shown in cross section.

In a weakened substantia nigra, dopamine neurons have degenerated, contributing to symptoms of Parkinson's disease.

The connection between anatomy and behavior can also be seen in studies of animals trained on various types of learning tasks, such as spatial mazes. Such learning can be correlated with a variety of neuronanatomical changes, such as modifications in the synaptic organization of cells in specific cortical regions—the visual cortex in animals trained in visually guided mazes is one example—or in the number of newly generated cells that survive in the hippocampus. The hippocampus is necessary for remembering the context in which we encounter information.

Although the changes in synaptic organization or cell survival have not yet been proved to be the basis of the new learning, experimental evidence reveals that preventing the growth of new hippocampal neurons leads to memory deficits. To test the idea that hippocampal neurons contribute to memory formation, researchers tested normal rats and "ADX" rats—rats with adrenal glands removed, which results in an absence of the hormone corticosterone. Without corticosterone, neurons in the hippocampus die.

Procedure 1 in **Experiment 7-1** contrasts the appearance of, at left, a normal rat hippocampus and, at right, the neuronal degeneration in an ADX rat after surgery. The behavior of normal and ADX rats was studied in the object/context mismatch task diagrammed in Procedure 2. During the training phase, the rats were placed in two distinct contexts, A and B, for 10 minutes on each of 2 days. Each context contained a different type of object. On the test day, the rats were placed in either context A or context B but with two different objects, one from that context and a second from the other context.

As noted in the Results section of the experiment, when normal rats encounter objects in the correct context, they spend little time investigating because the objects are familiar. If, however, they encounter an object in the wrong context, they are curious and spend about three-quarters of their time investigating, essentially treating the mismatched object as new. But the ADX rats with cell loss in the hippocampus treated both the object in context and the mismatched object the same, spending about half of their investigation time with each object.

Another group of ADX rats given treatment known to increase neuron generation in hippocampus—enriched housing and exercise in running wheels—was not impaired at the context/mismatch task. Experiment 7-1 concludes that the cellular changes in the hippocampus and behavioral changes are closely linked: hippocampal neurons are necessary for contextual learning to take place.

Methods of Behavioral Neuroscience

The ultimate function of any brain region is to produce behavior (movement). It follows that brain dysfunction will alter behavior in some way. The study of brain–behavior relationships began to be called neuropsychology in the 1940s, although the term is now often confined to the study of humans. The broader field, including both human and laboratory animals, is now referred to as **behavioral neuroscience,** the study of the biological bases of behavior.

A major challenge to behavioral neuroscientists is to develop methods for studying both normal and abnormal behavior. Measuring behavior in humans and laboratory animals is different, in large part because humans speak: investigators can ask them about their symptoms. It is also possible to use both paper-and-pencil and computer-based tests to identify specific symptoms in people.

Measuring behavior in laboratory animals is more complex. Researchers must learn to speak "ratese" to "talk" to rat subjects or "monkeyese" to "talk" to monkeys. In short, researchers must develop ways to enable the animals to reveal their symptoms. The development of the fields of animal learning and *ethology,* the study of animal behavior, provided the basis for modern behavioral neuroscience (see Whishaw & Kolb, 2005). To illustrate the logic of behavioral neuroscience, we describe some measurement tools used with humans and some used with rats.

The hippocampus is part of the limbic system (see Figure 2-25). The adrenal glands sit atop the kidneys (see Figure 2-30). Corticosterone, a steroid hormone secreted in times of stress, is important in protein and carbohydrate metabolism (see Section 6-5).

behavioral neuroscience Study of the biological bases of behavior.

EXPERIMENT 7-1

Question: Do hippocampal neurons contribute to memory formation?

Procedure 1
Rat hippocampus before (*left*) and after (*right*) surgical removal of the adrenal glands. Arrows point to neuronal degeneration resulting from a lack of corticosterone.

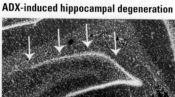

Normal rat hippocampus

ADX-induced hippocampal degeneration

Hippocampus

Procedure 2
The behavior of normal and ADX rats was studied in an object context mismatch task in which two distinct contexts each contained a different type of object.

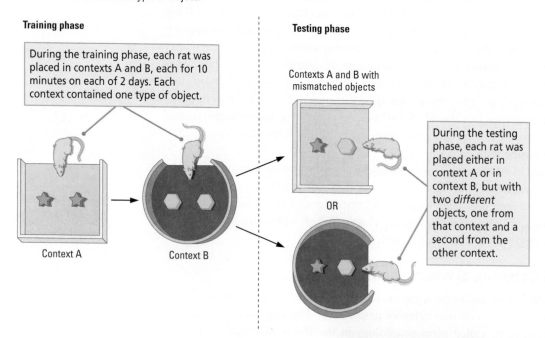

Training phase

During the training phase, each rat was placed in contexts A and B, each for 10 minutes on each of 2 days. Each context contained one type of object.

Context A

Context B

Testing phase

Contexts A and B with mismatched objects

OR

During the testing phase, each rat was placed either in context A or in context B, but with two *different* objects, one from that context and a second from the other context.

Results
Normal rats investigate the mismatch object more than the object that is in context, but the ADX rats performed at chance. The rats in another ADX group were given treatments known to increase neuron generation in the hippocampus (enriched housing and exercise in running wheels). The rats with hippocampal regeneration were not impaired at the mismatch task.

The confocal photo at right shows a rat hippocampus. A specific stain was used to identify new neurons, which appear yellow.

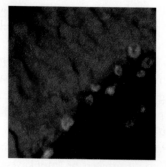

Conclusion: Hippocampal neurons are necessary for the contextual learning.

Adapted from "Object/Context-Specific Memory Deficits Associated with Loss of Hippocampal Granule Cells after Adrenalectomy in Rats," by S. Spanswick and R. J. Sutherland, 2010, *Learning and Memory, 17,* 241–245, and from "A Novel Animal Model of Hippocampal Cognitive Deficits, Slow Neurodegeneration, and Neuroregeneration," by S. Spanswick, H. Lethman, and R. J. Sutherland, 2011, *Journal of Biomedicine and Biotechnology,* Article ID 527201.

Neuropsychological Testing of Humans

The brain has exquisite control of an amazing array of functions ranging from control of movement and sensory perception to memory, emotion, and language. As a consequence, any analysis of behavior must be tailored to the particular function(s) under investigation. Consider the analysis of memory.

People with damage to the temporal lobes often complain of memory disturbance. But memory is not a single function. We have memory for events, colors, names, places, and motor skills, among other categories, and each must be measured separately. It would be rare indeed for someone to be impaired in *all* forms of memory. Neuropsychological tests of three distinct forms of memory are illustrated in **Figure 7-3**.

The Corsi block-tapping test shown in Figure 7-3A requires participants to observe an experimenter tap a sequence of blocks—blocks 4-6-1-8-3, for instance. The task is to repeat the sequence correctly. Note that the subject does not see numbers on the blocks but rather must remember the location of the blocks tapped.

The Corsi test provides a measure of the short-term recall of spatial position, an ability we can call *block span*. The test can be made more difficult by determining the maximum block span of an individual (say, 6 blocks) and then adding one (*span + 1*). By definition, the participant will fail on the first presentation, but given the span +1 repeatedly, the participant will eventually learn it.

Span + 1 identifies a different form of memory from block span. Different types of neurological dysfunction interfere differentially with tasks that superficially appear to be quite similar. Block span measures the short-term recall of information, whereas the span + 1 task reflects the learning and longer-term memory storage of information.

The *mirror-drawing task* (Figure 7-3B) requires a person to trace a pathway, such as a star, by looking in a mirror. This motor task initially proves quite difficult because our movements are backward in the mirror. With practice, participants learn how to accomplish the task accurately, and they show considerable recall of the skill when retested days later. Curiously, subjects with certain types of memory problems have no recollection of learning the task on the previous day but nevertheless perform it flawlessly.

In the *recency memory task* (Figure 7-3C), participants are shown a long series of cards, each bearing two stimulus items that are words or pictures. On some trials a question mark appears between the items. The subjects' task is to indicate whether they have seen the items before and if so, which item they saw most recently. People may be able to recall that they have seen items before but be unable to recall which was most recent.

Sections 14-1 through 14-3 analyze the many varied categories of memory.

In this book, we refer to humans in research studies who are "normal" as *participants* and to people who have a brain or behavioral impairment as *subjects*.

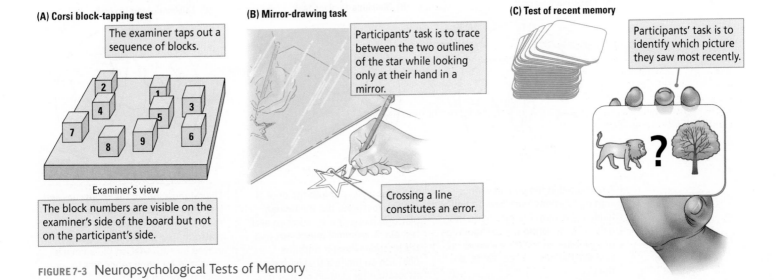

(A) Corsi block-tapping test

The examiner taps out a sequence of blocks.

Examiner's view

The block numbers are visible on the examiner's side of the board but not on the participant's side.

(B) Mirror-drawing task

Participants' task is to trace between the two outlines of the star while looking only at their hand in a mirror.

Crossing a line constitutes an error.

(C) Test of recent memory

Participants' task is to identify which picture they saw most recently.

FIGURE 7-3 Neuropsychological Tests of Memory

To illustrate the power of neuropsychological assessment, Focus 15-3 compares the effects that injuries to different brain regions have on performing particular tasks.

Conversely, they may not be able to identify the items as being familiar, but when forced to choose the most recent one, they can correctly identify it.

The latter result is counterintuitive and reflects the need for behavioral researchers to develop ingenious ways of identifying memory abilities. It is not enough simply to ask people to recall information verbally, although this, too, measures a form of memory.

Behavioral Analysis of Rats

Over the past 100 years, psychologists interested in the neural basis of memory have devised a vast array of mazes to investigate different forms of memory in laboratory animals. **Figure 7-4** illustrates three different tests based on a task originally devised by Richard Morris in 1980. Researchers place rats in a large swimming pool where an escape platform lies just below the surface of the water, invisible to the rats.

FIGURE 7-4 Swimming Pool Tasks

General arrangement of the swimming pool used in three different visuospatial learning tasks for rats. The red lines in parts A, B, and C mark the rat's swimming path on each trial (T). (A) Adapted from "Spatial Localization Does Not Require the Presence of Local Cues," by R. G. M. Morris et al., 1981, *Learning and Motivation, 12,* 239–260. (B) Adapted from "Dissociating Performance and Learning Deficits in Spatial Navigation Tasks in Rats Subjected to Cholinergic Muscarinic Blockade," by I. Q. Whishaw, 1989, *Brain Research Bulletin, 23,* 347–358. (C) Adapted from "Behavioural and Anatomical Studies of the Posterior Parietal Cortex of the Rat," by B. Kolb and J. Walkey, 1987, *Behavioural Brain Research, 23,* 127–145.

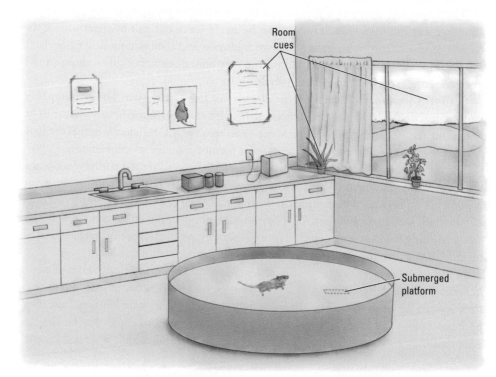

(A) Place-learning task

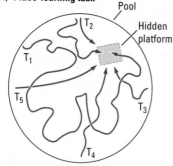

A rat placed in the pool at various starting locations must learn to find a hidden platform. The rat can do this only by considering the configuration of visual cues in the room—windows, wall decorations, potted plants, and the like.

(B) Matching-to-place task

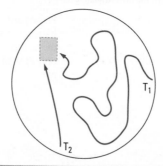

The rat is again put into the pool at random locations, but the hidden platform is in a new location on each test day. The animal must learn that the location where it finds the platform on the first trial each day is its location for all trials on that day.

(C) Landmark-learning task

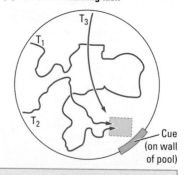

The rat must ignore the room cues and learn that only the cue on the wall of the pool signals the location of the platform. The platform and cue are moved on each trial, so the animal is penalized for using room cues to try to solve the problem.

In one version of the task, *place learning,* the rat must find the platform from any starting location in the pool (Figure 7-4A). The only cues available are outside the pool, so the rat must learn the relation between several cues in the room and the platform's location. In a second version of the task, *matching-to-place learning,* the rat has already learned that a platform always lies somewhere in the pool but is moved to a different location every day. The rat is released and searches for the platform (Figure 7-4B). Once the rat finds the platform, the animal is removed from the pool and after a brief delay (such as 10 seconds) is released again. The rat's task is to swim directly to the platform.

The challenge for the rat in the matching-to-place test is to develop a strategy for finding the platform consistently: it is always in the same location on each trial, each day, but each new day brings a new location. In the *landmark* version of the task, the platform's location is identified by a cue on the pool wall (Figure 7-4D). The platform moves on every trial, but the relation to the cue is constant. In this task the brain is learning that the distant cues outside the pool are irrelevant; only the local cue is relevant. Rats with different neurological perturbations are selectively impaired in the three different versions of the swimming pool task.

Another type of behavioral analysis in rats is related to movement. A major problem facing people with stroke is a deficit in the control of hand and limb movements, prompting considerable interest in devising ways to analyze such motor behaviors for the purpose of testing new therapies for facilitating recovery. Ian Whishaw (Whishaw & Kolb, 2005) has devised both novel tasks and novel scoring methods to measure the fine details of skilled reaching movements in rats.

In one test, rats are trained to reach through a slot to obtain a piece of sweet food. The movements are remarkably similar to the movements people make in a similar task and can be broken down into segments. Investigators can score the segments, which are differentially affected by different types of neurological perturbation, separately.

The photo series in **Figure 7-5** details how a rat orients its body to the slot (A), aims its paw through the slot (B), rotates the paw horizontally to grasp the food (C), then rotates the paw vertically and withdraws the paw to obtain the food (D). Contrary to reports common in neurology textbooks, primates are not the only animals to make fine digit movements, but because the rodent paw is small and moves so quickly, digit dexterity can be seen in rodents only by using high-speed photography.

Experiment 14-1 demonstrates fear conditioning in rats, Experiment 14-2 plasticity in the monkey's motor cortex, and Experiment 14-3 neuronal effects of amphetamine sensitization in rats. Figure 14-2 demonstrates instrumental conditioning in cats, Figure 14-10 visuospatial learning in monkeys, and Figure 14-12 monkeys' short-term memory.

FIGURE 7-5 Skilled Reaching in Rats
Movement series displayed by rats trained to reach through a narrow vertical slot to obtain sweet food: **(A)** aim the paw, **(B)** reach over the food, **(C)** grasp the food, **(D)** withdraw and move food to the mouth.

(A) 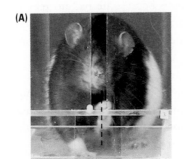 **(B)** 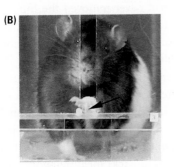 **(C)** 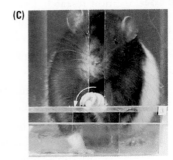 **(D)**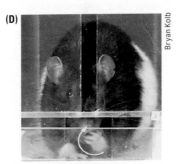

Bryan Kolb

Manipulating and Measuring Brain–Behavior Interactions

One strategy for studying brain–behavior relationships is to modify the brain and see how behavior is altered. There are two reasons to do so. The first is to develop hypotheses about how the brain affects behavior; the second is to test the hypotheses.

Neuroscientists can hypothesize about the functions of brain regions by studying how their absence affects behavior. Recall that Broca studied patients with naturally occurring injuries and made inferences about the organization of language. Similarly, Parkinson's patients have motor disturbances and associated cell death in the substantia

Focus features 5-2, 5-3, and 5-4 detail many aspects of Parkinson's disease.

stereotaxic apparatus Surgical instrument that permits the researcher to target a specific part of the brain.

akinesia Slowness or absence of movement.

Sections 2-6 and 5-3 introduce factors that contribute to dementias such as Alzheimer's disease. Focus 14-3 and Section 16-3 describe, respectively, research on and treatments for Alzheimer's.

Scoville's patient, H. M., who is profiled in Section 14-2, became the most-studied case in neuroscience.

To refresh your memory for anatomical locations and orientations in the brain, review The Basics in Section 2-1.

Figure 8-15 shows a brain-imaging atlas that tracks cortical thickness over time.

nigra (see Figure 7-2). It is a small step experimentally to produce specific injuries to different regions in the brains of laboratory animals and study their behavior. Such studies tell investigators not only about the function of the injured region but also what the remaining brain can do in the absence of the injured region.

A second reason to manipulate the brain is to develop animal models of neurological and psychiatric disorders. The general presumption in neurology and psychiatry is that it ought to be possible to restore at least some normal functioning by pharmacological, behavioral, or other interventions. A major problem for developing such treatments is that, like most new treatments in medicine, they must be developed in nonhuman subjects first. (In Section 7-7, we take up scientific and ethical issues surrounding the use of animals in research.)

For brain disorders, researchers must develop models of different diseases that then can be treated. Consider dementia, a condition of progressive memory impairment that appears to be related to neuronal death in specific brain regions. The goal for treatment is to reverse or prevent cell death, but developing effective treatments requires an animal model that mimics dementia.

Brains can be manipulated in a variety of ways, the precise manner depending on the specific research question being asked. The principal techniques are to lesion the brain or to electrically or chemically (with drugs) stimulate or inactivate the brain.

Brain Lesions

The first—and the simplest—technique used was to ablate (remove or destroy) tissue. Beginning in the 1920s Karl Lashley, a pioneer of neuroscience research, used *ablation,* and for the next 30 years he tried to find the location of memory in the brain. He trained monkeys and rats on various mazes and motor tasks and then removed bits of cerebral cortex with the goal of producing amnesia for specific memories.

To his chagrin, Lashley failed to produce amnesia. He discovered instead that memory loss was related to the amount of tissue he removed. The only conclusion Lashley could reach was that memory is distributed throughout the brain and not located in any single place.

Ironically, just as Lashley was retiring in the 1950s, William Scoville and Brenda Milner (1957) described a patient from whose brain Scoville had removed the hippocampus as a treatment for epilepsy. The surgery rendered this patient amnesic. During his ablation research, Lashley had never removed the hippocampus because he had no reason to believe that this structure had any role in memory. And because the hippocampus is not accessible on the surface of the brain, other techniques had to be developed before subcortical lesions could be used.

The solution to accessing subcortical regions is to use a **stereotaxic apparatus,** a device that permits a researcher or a neurosurgeon to target a specific part of the brain for ablation, as shown in **Figure 7-6.** The head is held in a fixed position, and because brain structures hold a fixed relationship with the location of the junction of the skull bones, it is possible to imagine a three-dimensional map of the brain.

Rostral–caudal (front to back) measurements, corresponding to the x axis in Figure 7-6, are made relative to the junction of the frontal and parietal bones (the *bregma*). Dorsal–ventral (top to bottom) measurements, the y axis, are made relative to the surface of the brain. Medial–lateral measurements, the z axis, are made relative to the midline junction of the cranial bones. Atlases of the brains of humans and laboratory animals have been created from postmortem tissue so that the precise location of any structure can be specified in three-dimensional space.

Consider the substania nigra. To ablate this region to produce a rat that displays symptoms of Parkinson's disease, the structure and its three-dimensional location in the brain atlas is located. A small hole is then drilled in the skull, as shown in Figure 7-6, and

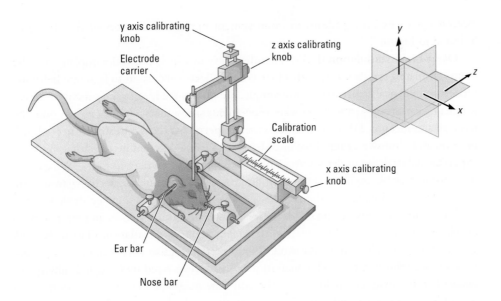

FIGURE 7-6 **Stereotaxic Apparatus** This instrument allows the precise positioning of electrodes for lesioning or for stimulating brain regions.

an electrode lowered to the substantia nigra. If a current is passed through the electrode, the tissue in the region of the electrode tip is killed, producing an *electrolytic lesion*.

A problem with electrolytic lesions is that not only are the neurons of the tissue (in this case substantia nigra) killed but any nerve fibers passing through the region die as well. One solution is to lower a narrow, metal tube (a cannula) instead of an electrode, infuse a neuron-killing chemical, and thus produce a *neurotoxic lesion* (Figure 7-22 diagrams this procedure). A selective toxin can be injected that kills only neurons, sometimes only certain types of neurons, and spares the fibers.

To create a Parkinsonian rat, a toxin can be injected that is selectively taken up by dopaminergic neurons, thus leading to a condition that mimics human Parkinson pathology. Animals with such neurotoxic lesions have a variety of motor symptoms including **akinesia** (slowness or absence of movement), short footsteps, and tremor. Drugs such as ʟ-dopa, an agonist that enhances dopamine production, and atropine, an antagonist that blocks acetylcholine production, relieve these symptoms in human patients. Ian Whishaw and his colleagues (Schallert et al., 1978) thus were able to selectively lesion the substantia nigra in rats to produce a behavioral model of Parkinson's disease.

Shuffling gait of a Parkinsonian rat, captured in prints left by its ink-stained hind feet.

Brain Stimulation

The brain operates on both electrical and chemical energy, so it is possible to selectively turn brain regions "on" or "off" by using electrical or chemical stimulation, usually delivered via a stereotaxic apparatus. Electrical stimulation was first used by Wilder Penfield to stimulate the cerebral cortex of humans directly during neurosurgery. Later researchers used the stereotaxic instrument to place an electrode or a cannula in specific brain locations with the objective of enhancing or blocking the activity of neurons and observing the behavioral effects.

You can read more about Penfield's dramatic discoveries in Sections 10-4 and 11-2.

Perhaps the most dramatic research example comes from stimulating specific regions of the hypothalamus. Rats with electrodes placed in the lateral hypothalamus will eat whenever the stimulation is turned on. If the animals have the opportunity to press a bar that briefly turns on the current, they quickly learn to press the bar to obtain the current, a behavior known as *electrical self-stimulation*. It appears that the stimulation is affecting a neural circuit that involves both eating and pleasure.

Figure 2-19 diagrams the location of the hypothalamus within the brainstem. Section 12-3 details its role in motivated and emotional behavior.

Brain stimulation can also be used as a therapy. Stimulation can be applied directly to the cortex. When the intact cortex adjacent to cortex injured by a stroke is stimulated

Section 1-2 describes how DBS (Figure 1-6) dramatically improved responsiveness in a man who had lingered in a minimally conscious state for six years following a traumatic brain injury.

electrically, for example, it leads to improvement in motor behaviors such as those illustrated in Figure 7-5.

Deep-brain stimulation (DBS) is neurosurgery in which electrodes implanted in the brain stimulate a targeted area with a low-voltage electrical current to facilitate behavior. DBS to subcortical structures—for example, the globus pallidus in the subcortical basal ganglia in Parkinson's patients—makes movements smoother. Medications can often be reduced significantly. DBS has also been used experimentally as a treatment in both depression and obsessive-compulsive disorder using several different neural targets. Further experimental trials are underway to identify the optimal brain regions for DBS as a treatment for intractable psychiatric disorders and possibly for stimulating recovery from TBI.

Electrical stimulation of the brain is invasive because holes must be drilled in the skull. Researchers took advantage of the relation between magnetism and electricity to develop a noninvasive technique, **transcranial magnetic stimulation** (TMS). A small wire coil is placed adjacent to the skull, as illustrated in **Figure 7-7**. A high-voltage current is pulsed through the coil, which in turn produces a rapid increase and subsequent decrease in the magnetic field around the coil. The magnetic field easily passes through the skull and causes a population of neurons in the cerebral cortex to depolarize and fire.

Research Focus 16-2 describes how TMS is used to treat depression and other behavioral disorders.

If the motor cortex is stimulated, movement is evoked, or if a movement is in progress, a disruption of movement occurs. Similarly, if the visual cortex is stimulated, the participant sees dots of light (*phasphenes*). The effects of brief pulses of TMS do not outlive the stimulation, but *repetitive TMS* (rTMS), which involves continuous stimulation for up to several minutes, produces more long-lasting effects. TMS and rTMS can be used to study brain–behavior relationships in normal participants, and rTMS has been used as a potential treatment for a variety of behavioral disorders. A growing body of research supports its antidepressant actions.

Brain activity can also be stimulated by administering drugs that pass into the bloodstream and eventually enter the brain. The drugs influence the activity of specific neurons in specific brain regions. For example, the drug haloperidol, which is used to treat schizophrenia, reduces dopaminergic neuron function and makes normal rats dopey and inactive (*hypokinetic*).

Section 6-1 describes routes of administration and how drugs act at synapses either as agonists or antagonists.

In contrast, drugs that increase dopaminergic activity, such as amphetamine, produce *hyperkinetic rats*—rats that are hyperactive. The advantage of administering drugs through the bloodstream is that their effects wear off in time as the drugs are metabolized. It thus is possible to study the effects of drugs on learned behaviors, such as skilled reaching (see Figure 7-5), and then to reexamine the behavior after the drug effect has worn off.

Claudia Gonzalez and her colleagues (2006) administered nicotine to rats as they learned a skilled reaching task and then studied their later acquisition of a new skilled reaching task. The researchers found that the later motor learning was impaired by the earlier, nicotine-enhanced motor learning. This finding surprised the investigators, but

FIGURE 7-7 Transcranial Magnetic Stimulation

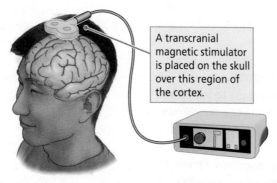

A transcranial magnetic stimulator is placed on the skull over this region of the cortex.

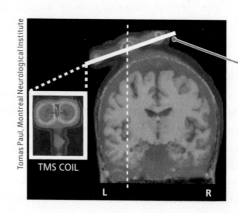

Tomas Paul, Montreal Neurological Institute

TMS COIL

L   R

The TMS coil, shown here in a composite MRI and PET-scan photograph, interferes with brain function in the adjacent area.

it now appears that repeated exposure to psychomotor stimulants such as amphetamine, cocaine, and nicotine can produce long-term effects on the brain's later plasticity, its ability to change in response to experience, including the learning of specific tasks.

Optogenetics

Optogenetics, a transgenic technique that combines genetics and light to control targeted cells in living tissue, is based on the discovery that light can activate proteins. The proteins can occur naturally or can be inserted into cells. For example, opsins, proteins derived from microorganisms, combine a light-sensitive domain with an ion channel. The first opsin to be used for the optogenetic technique was channelrhodopsin-2 (ChR2).

When ChR2 is expressed in a neuron and exposed to blue light, the ion channel opens and immediately depolarizes the neuron, causing excitation. In contrast, stimulation of halorhodopsin (NpHR) with a green-yellow light activates a chloride pump, hyperpolarizing the neuron and causing inhibition. A fiber-optic light can be delivered to selective regions of the brain such that all neurons exposed to the light respond immediately.

Optogenetics is being applied to behavioral studies. For example, within the limbic system, the amygdala is a key structure in generating fear in animals. If the amygdala is targeted with opsins and then exposed to an inhibitory light, rats immediately show no fear and wander about in a novel open space. As soon as the light is turned off, they scamper back to the safety of a hiding place.

Optogenetics has tremendous potential not only as a research tool but also in clinical practice, with uses much like DBS. It is now possible to insert light-sensitive proteins into specific neuron types, such as pyramidal cells, and use light to selectively activate a single cell type.

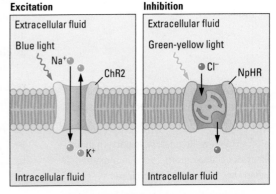

Lighting Up Neurons Specific wavelengths activate light-sensitive proteins expressed in neurons. Research Focus 4-3 explains how scientists engineered the optogenetic technique.

REVIEW 7-1
Measuring Brain and Behavior

Before you continue, check your understanding.

1. Neuropsychology, the study of relations between _____ and _____, employs memory tests such as _____ in humans and _____ in nonhuman animals such as rats.

2. Anatomical studies rely on techniques such as _____ tissue postmortem or visualizing living tissue with a _____.

3. Methods developed to manipulate the brain include _____, _____, and _____.

4. Outline the various methods of brain stimulation that either activate or inhibit neural activity.

Answers appear at the back of the book.

7-2 Measuring the Brain's Electrical Activity

The brain is always electrically active, even when we sleep. Electrical measures of brain activity are important for studying brain function, for medical diagnosis, and for monitoring the effectiveness of therapies used to treat brain disorders. The four major techniques for tracking the brain's electrical activity are electroencephalography (EEG), event-related potentials (ERP), magnetoencephalography (MEG), and single-cell recording.

In part, these techniques are used to record electrical activity in different parts of neurons, such as action potentials and graded potentials. The electrical behavior of cell

deep-brain stimulation (DBS) Neurosurgery in which electrodes implanted in the brain stimulate a targeted area with a low-voltage electrical current to facilitate behavior.

transcranial magnetic stimulation (TMS) Procedure in which a magnetic coil is placed over the skull to stimulate the underlying brain; used either to induce behavior or to disrupt ongoing behavior.

optogenetics Transgenic technique that combines genetics and light to control targeted cells in living tissue.

Figure 4-11 diagrams the electrical activity of a cell membrane at rest, Figure 4-13 during graded potentials, and Figure 4-15 generating the action potential.

electrocorticography (ECoG) Graded potentials recorded with electrodes placed directly on the brain's surface.

alpha rhythm Regular wave pattern in an electroencephalogram; found in most people when they are relaxed with closed eyes.

event-related potentials (ERPs) Complex electroencephalographic waveforms related in time to a specific sensory event.

Amplitude is the height of a recorded brain wave. *Frequency* is the number of brain waves recorded per second.

Section 13-3 describes how sleep researchers use the EEG to measure sleep stages and dreaming. Focus features 4-1 and 10-3 and Section 16-3 detail the diagnosis of and treatments for various types of epilepsy.

bodies and dendrites tends to be much more varied than that of axons, which conduct action potentials. Graded potentials are recorded from cell bodies and dendrites, as these regions normally do not produce action potentials.

EEG Recordings of Graded Potentials

In the early 1930s, Hans Berger discovered that electrical activity in the brain could be recorded simply by placing electrodes on the scalp. Recording this electrical activity, popularly known as "brain waves," produces an "electrical record from the head"—an electroencephalogram. The EEG measures the summed graded potentials from many thousands of neurons. EEG waves are usually recorded with a special kind of oscilloscope called a polygraph (meaning "many graphs"), illustrated in **Figure 7-8**. Electrodes can also be placed directly on the cerebral cortex during neurosurgery, a method referred to as **electrocorticography**, or ECoG.

EEGs reveal some remarkable features of the brain's electrical activity. The EEG recordings in **Figure 7-9** illustrate three:

1. The EEG changes as behavior changes.

2. An EEG recorded from the cortex displays an array of patterns, some of which are rhythmical.

3. The living brain's electrical activity is never silent, even when a person is asleep or comatose.

When a person is aroused, excited, or even just alert, the EEG pattern has a low amplitude and a fast frequency, as shown in Figure 7-9A. This pattern is typical of an EEG taken from anywhere on the skull of an alert subject, not only humans but other animals too. In contrast, when a person is calm and quietly relaxed, especially with eyes closed, the rhythmical brain waves shown in Figure 7-9B often emerge. These **alpha rhythms** are extremely regular, with a frequency of approximately 11 cycles per second and amplitudes that wax and wane as the pattern is recorded. In humans, alpha rhythms are generated in the region of the visual cortex at the back of the brain. If a relaxed person is disturbed or opens his or her eyes, the alpha rhythms abruptly stop.

The EEG is a sensitive indicator of behaviors beyond simple arousal and relaxation. Parts C, D, and E of Figure 7-9 illustrate EEG changes as a person moves from drowsiness to sleep and finally into deep sleep. EEG rhythms become progressively slower and larger in amplitude. Still slower waves appear during anesthesia, after brain trauma, or when a person is in a coma (illustrated in Figure 7-9F). In brain death, the EEG becomes a flat line.

These distinctive brain-wave patterns make the EEG a reliable tool for monitoring sleep stages, estimating the depth of anesthesia, evaluating the severity of head injury, and searching for other brain abnormalities. The brief periods of unconsciousness and

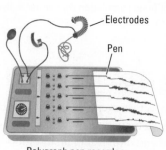

Michael Rosenfeld/Stone Images

SIU/Photo Researchers

FIGURE 7-8 Polygraph Recording EEG
A simple, noninvasive method for recording the brain's electrical activity. EEG waves are recorded via computer today (see Figure 4-5) and can match wave activity to specific regions of the brain..

Electrodes

Pen

Polygraph pen recorder

1 Electrodes are attached to the scalp, corresponding to specific areas of the brain.

2 Polygraph electrodes are connected to magnets, which are connected to pens...

3 ...that produce a paper record of electrical activity in the brain. This record indicates a relaxed person.

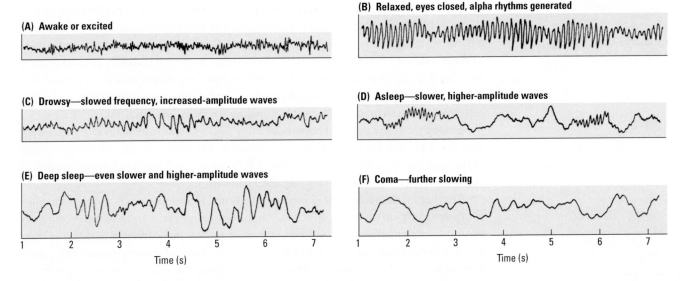

(A) Awake or excited

(B) Relaxed, eyes closed, alpha rhythms generated

(C) Drowsy—slowed frequency, increased-amplitude waves

(D) Asleep—slower, higher-amplitude waves

(E) Deep sleep—even slower and higher-amplitude waves

(F) Coma—further slowing

Time (s)

Time (s)

FIGURE 7-9 Characteristic EEG Recordings Brain-wave patterns reflect different states of consciousness in humans. Adapted from *Epilepsy and the Functional Anatomy of the Human Brain* (p. 12), by W. Penfield and H. H. Jasper, 1954, Boston: Little, Brown.

involuntary movements that characterize epileptic seizures are associated with highly abnormal spike-and-wave patterns in the EEG.

The important point here is that EEG recording provides a useful tool both for research and for diagnosing brain abnormalities. EEG can also be used in combination with the brain-imaging techniques described in Sections 7-3 and 7-4 to provide a more accurate identification of the source of abnormal EEG waves in epilepsy.

Mapping Brain Function with Event-Related Potentials

Brief changes in an EEG signal in response to a discrete sensory stimulus are called **event-related potentials** (ERPs). ERPs are largely the graded potentials on dendrites that a sensory stimulus triggers. You might think that they should be easy to detect, but they are not.

The problem is that ERPs are mixed in with so many other electrical signals in the brain that they are difficult to spot just by visually inspecting an EEG record. One way to detect ERPs is to produce the stimulus repeatedly and average the recorded responses. Averaging tends to cancel out any irregular and unrelated electrical activity, leaving in the EEG record only the potentials that the stimulus generated.

To clarify this procedure, imagine throwing a small stone into a lake of choppy water. Although the stone produces a splash, the splash is hard to see among all the ripples and waves. This splash made by a stone is analogous to an event-related potential caused by a sensory stimulus. Like the splash surrounded by choppy water, the ERP is hard to detect because of all the other electrical activity around it.

A solution is to throw a number of stones exactly the same size, always hitting the same spot in the water and producing the same splash over and over. If a computer is then used to calculate an average of the water's activity, random wave movements will tend to average one another out, and you will see the splashes produced by the stones as clearly as if a single stone had been thrown into a pool of calm water.

Figure 7-10 shows an ERP record (top) that results when a person hears a tone. Notice that the EEG record is very irregular when the tone is first presented. But after averaging over 100 stimulus presentations, a distinctive wave pattern appears, as shown in the bottom panel of Figure 7-10. This ERP pattern consists of a number of negative (N) and positive (P) waves that occur over a period of a few hundred milliseconds after the stimulus.

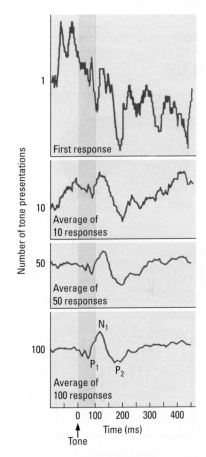

Number of tone presentations

First response

Average of 10 responses

Average of 50 responses

N_1

P_1 P_2

Average of 100 responses

Time (ms)

Tone

FIGURE 7-10 Detecting ERPs In the averaging process for an auditory ERP, a tone is presented at time 0, and EEG activity in response is recorded. After many successive presentations of the tone, the averaged EEG wave sequence develops a distinctive shape that becomes extremely clear after averaging 100 responses, as shown in the bottom panel. Positive and negative waves that appear at different times after the stimulus presentation are used for analysis.

The waves may also be labeled as N$_{100}$ and P$_{200}$.

The waves are numbered in relation to the time at which they occur. For instance, in Figure 7-10, N$_1$ is a negative wave occurring about 100 milliseconds after the stimulus, whereas P$_2$ is a positive wave occurring about 200 milliseconds after the stimulus. Not all these waves are unique to this particular stimulus. Some are common to any auditory stimulus that might be presented. Other waves, however, correspond to important differences in this specific tone. ERPs to spoken words even contain distinctive peaks and patterns that differentiate such similar-sounding words as "cat" and "rat."

Among the many practical reasons for using ERPs to study the brain is the advantage that, like EEG, the ERP technique is noninvasive. Electrodes are placed on the surface of the skull, not into the brain. Therefore, ERPs can be used to study humans, including college students—the most frequently used participants.

Another advantage is cost. Compared to other techniques, such as brain imaging, ERPs are inexpensive and can be recorded from many brain areas simultaneously by pasting an array of electrodes (sometimes more than 200) onto different parts of the scalp. Because certain brain areas respond only to certain kinds of sensory stimuli (e.g., auditory areas respond to sounds and visual areas to sights), the relative responses at different locations can be used to map brain function.

Figure 7-11 shows a multiple-recording method that uses 128 electrodes simultaneously to detect ERPs at many cortical sites. Computerized averaging techniques reduce the masses of information obtained to simpler comparisons between electrode sites. For example, if the focus of interest is P$_3$, a positive wave occurring about 300 milliseconds after the stimulus, the computer can display a graph of the skull showing only the amplitude of P$_3$. A computer can also convert the averages at different sites into a color code, creating a graphic representation that shows the brain regions most responsive to the signal.

ERPs can be used not only to detect which areas of the brain are processing particular stimuli but also to study the order in which different regions play a role. This second use of ERPs is important because we want to know the route information takes as it travels through the brain. In Figure 7-11, the participant is viewing a picture of a rat that appears repeatedly in the same place on a computer screen. The P$_3$ recorded on the posterior right side of the head is larger than any other P$_3$ occurring elsewhere, meaning that this region is a "hot spot" for processing the visual stimulus. Presumably, for this particular participant, the right posterior part of the brain is central in decoding the picture of the rat 300 milliseconds after it is presented.

Many other interesting research areas can benefit from investigation using ERPs, as illustrated in Clinical Focus 7-2, "Mild Head Injury and Depression." They also can be used to study how children learn and process information differently as they mature. ERPs can be used to examine how a person with a brain injury compensates for the impairment by using other, undamaged regions of the brain. ERPs can even help reveal which brain areas are most sensitive to the aging process and therefore contribute most to declines in behavioral functions among the elderly. All these areas can be addressed with this simple, inexpensive research tool.

FIGURE 7-11 Using ERPs to Image Brain Activity

Electrodes attached to the scalp of a research subject are connected to...

Electrodes in geodesic sensor net

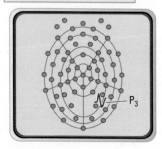

...a computer display of electrical activity, showing a large positive (P$_3$) wave at the posterior right side of the head.

This electrical activity can be converted into a color representation showing the hot spot for the visual stimulus.

Resting   300 ms after viewing

Magnetoencephalography

A magnetic field passing across a wire induces an electrical current in the wire. When a current flows along a wire, it induces a magnetic field around the wire. Neural activity, by generating an electrical field, also produces a magnetic field. Although the magnetic field produced by a single neuron is vanishingly small, the field produced by many neurons is sufficiently strong to be recorded on the scalp. The record of this phenomenon, a **magnetoencephalogram** (MEG), is the magnetic counterpart of the EEG or ERP.

Calculations based on MEG measurements not only provide a description of the electrical activity of neurons but also permit a three-dimensional localization of the cell groups generating the measured field. Magnetic waves conducted through living tissue

Mild Head Injury and Depression

B. D. was an industrial tool salesman who suffered an accident when a pallet of boxed tools tipped and part of the load struck his head. He did not lose consciousness but did have a serious cut to his scalp as well as damage to two vertebrae in his spine. The attending physician at the hospital emergency room suspected a mild concussion, but no further neurological workup was done at the time.

B. D.'s spinal symptoms gradually cleared, but he had persisting symptoms of irritability, anxiety, and depression that did not resolve even two years later. B. D. was unable to work, and his behavioral change put a major strain on his family. His emotional problems led him to withdraw from the world, which only worsened his predicament.

A neuropsychological exam was administered to B. D. about two years after the injury. His general cognitive ability was found to be well above average, with an IQ score of 115. B. D. had significant attentional and short-term memory deficits, however. A subsequent MRI of his brain failed to find any injury that could explain his symptoms. B. D.'s serious emotional symptoms are common following mild head injury, even when no other neurological or radiological signs of brain injury present themselves.

One way to investigate brain functioning in such cases is to use ERP. Reza and colleagues (2007) compared healthy control participants to groups of subjects with mild head injuries, with or without depression. The investigators found that all subjects with the head injury had a delayed P_3 wave, but only those with depression as well also had a delayed N_2 wave. These findings demonstrate that ERP can identify abnormalities in cerebral processing in people with depression after mild head injury, even when MRI scans are negative. Such evidence can be important for people like B. D. who are seeking long-term disability support following what appears to be a "mild" head injury.

---

undergo less distortion than electrical signals do, so an MEG can have a higher resolution than an ERP. Thus, a major advantage of the MEG over the EEG and ERP is its ability to more precisely identify the source of the activity being recorded. For example, the MEG has proved useful in locating the source of epileptic discharges. The disadvantage of the MEG is its cost. The equipment for producing it is expensive in comparison with the apparatus used to produce EEGs and ERPs.

Recording Action Potentials from Single Cells

An EEG is recorded from the scalp and reflects the summed activity of thousands of graded potentials. But what do individual cells do? By the early 1950s it was becoming possible to record the activity of individual cells by measuring the action potentials of single neurons with fine electrodes inserted into the brain. These microelectrodes can be placed next to cells (*extracellular recording*) or inside cells (*intracellular recording*). Modern extracellular recording techniques make it possible to distinguish the activity of as many as 40 neurons at once. Intracellular recording allows direct study and recording of a single neuron's electrical activity. The disadvantage is that inserting the electrode into the cell can kill the cell.

We now know that cells in the sensory regions of the brain are highly specific in what excites them. Some cells in the visual system fire vigorously to specific wavelengths of light (a color) or to specific orientations of bars of light (vertical, for example). More interesting, other cells respond to more complex patterns such as faces or hands. Similarly, cells in the auditory system respond to specific sound frequencies (a low or high pitch) or to more complex combinations of sounds, such as speech (the syllable "ba," for example).

But cells may also have much more complex interests that can tell us much about brain–behavior relationships. John O'Keefe and his colleagues (1971) showed that certain cells in the hippocampus of the rat and mouse brain vigorously fire when an animal is in a specific place in the environment, as illustrated in **Figure 7-12**. These cells are thought to code spatial location by helping to create some type of spatial map in the brain. O'Keefe called them **place cells.**

magnetoencephalogram (MEG) Magnetic potentials recorded from detectors placed outside the skull.

place cells Neurons maximally responsive to specific locations in the world.

Figure 4-6 illustrates the structure and use of two types of microelectrodes.

FIGURE 7-12 Place-Cell Specificity
The increasing "heat" of the colors—from dark (slowest) to light blue to green, yellow, and red—shows increasing cell firing rate. Place-cell firing is **(A, B)** highly specific in young control and transgenic animals, **(C)** slightly less specific in the old control animals, and **(D)** nonspecific in the aged transgenic mouse. From "Place Cell Firing Correlates with Memory Deficits and Amyloid Plaque Burden in Tg2576 Alzheimer Mouse Model," by F. Cacucci, M. Yi, T. J. Wills, P. Chapmans, and J. O'Keefe, 2008, *Proceedings of the National Academy of Sciences (USA), 105,* 7863–7868.

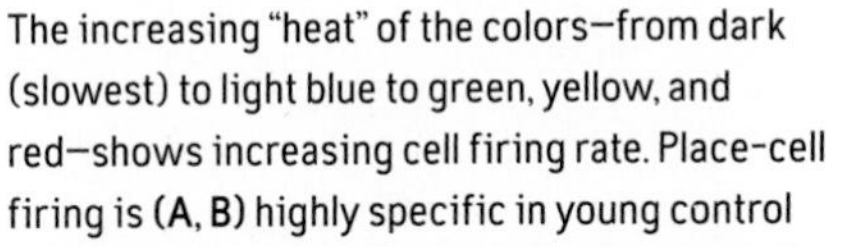

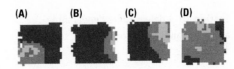

(A)   **(B)**   **(C)**   **(D)**

Section 13-4 discusses the role of place cells in storing memories.

More recently, O'Keefe's group (Cacucci et al., 2008) demonstrated that, in mice with a genetically engineered mutation that produces deficits in spatial memory, place cells lack specificity: the cells fire to a very broad region of their world. As a result, these mice have difficulty finding their way around their world, much as human patients with Alzheimer's dementia tend to get lost. One of the reasons may be that a change similar to the engineered mutation takes place in human brain cells.

REVIEW 7-2

Measuring the Brain's Electrical Activity

Before you continue, check your understanding.

1. The four major techniques for tracking the brain's electrical activity are _____, _____, _____, and _____.

2. EEG measures _____ on the cell membrane.

3. Single-cell recording measures _____ from a single neuron.

4. Magnetoencephalography measures the _____ and also provides a _____.

5. What is the advantage of ERP over EEG?

Answers appear at the back of the book.

7-3 Static Imaging Techniques: CT and MRI

Until the early 1970s, the only way to actually image the living brain was procedures such as X-ray. The modern era of brain imaging began in the early 1970s, when Allan Cormack and Godfrey Hounsfield independently developed an X-ray approach now called **computerized tomography:** the *CT scan*. Cormack and Hounsfield both recognized that a narrow X-ray beam could be passed through the same object at many different angles, creating many different images, and then the images could be combined with the use of computing and mathematical techniques to create a three-dimensional image of the brain.

Tomo comes from the Greek word for "section," so tomography yields a picture through a single brain section.

The CT method resembles the way in which our two eyes (and our brains) work in concert to perceive depth and distance to locate an object in space. The CT scan, however, coordinates many more than two images, analogous perhaps to our walking to several new vantage points to obtain multiple views. X-ray absorption varies with tissue density. High-density tissue, such as bone, absorbs a lot of radiation. Low-density material, such as ventricular fluid or blood, absorbs little. Neural tissue absorption lies between these two extremes. CT scanning software translates these differences in absorption into an image of the brain in which dark colors indicate low-density regions and light colors indicate high-density regions.

Figure 7-13A shows a typical CT scan. The dense skull forms a white border. The density of the brain's gray matter does not differ sufficiently from that of white matter for a CT scan to distinguish between the two clearly, so the cortex and its underlying white matter show up as a more or less homogeneous gray. Ventricles can be visualized, however, because the fluid in them is far less dense: they, as well as some of the major fissures in the cortex, are rendered darker in the CT scan. Each point on the image in Figure 7-13A represents about a 1-millimeter-diameter circle of tissue, a resolution sufficient to distinguish two objects about 5 millimeters apart and appropriate for localizing brain tumors and lesions.

Section 10-4 delves more deeply into aphasias that result from damage to speech areas in the brain.

The lesion revealed in Figure 7-13A is a damaged region where the presence of fewer neurons and more fluid produces a contrast that appears as a dark area in the CT scan. This subject presented with symptoms of *Broca's aphasia,* the inability to speak fluently despite

(A)
Lesion

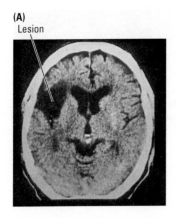

(B)
Anterior

Lesion

Posterior

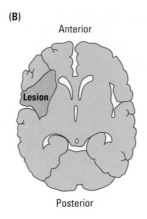

(C)
Lesion

Plane of section
in parts
A and B

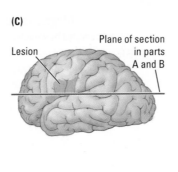

FIGURE 7-13 CT Scan and Brain Reconstruction **(A)** Dorsal view of a horizontal CT scan of a subject who presented with Broca's aphasia. The dark region at the left anterior is the location of the lesion. **(B)** A schematic representation of the horizontal section, with the area of the lesion shown in blue. **(C)** A reconstruction of the brain, showing a lateral view of the left hemisphere with the lesion shown in blue. Adapted from *Lesion Analysis in Neuropsychology* (p. 56), by H. Damasio and A. R. Damasio, 1989, New York: Oxford University Press.

the presence of normal comprehension and intact vocal mechanisms. The location of the lesion in the left frontal cortex (adjacent to the butterfly-shaped lateral ventricles) confirms this diagnosis. Figure 7-13B, a drawing of the same horizontal section, uses color to portray the lesion. Figure 7-13C is a lateral-view drawing of the left hemisphere reconstructed from a series of horizontal CT scans and showing the extent of the lesion.

A more recent alternative to the CT scan, **magnetic resonance imaging** (MRI), is based on the principle that hydrogen atoms behave like spinning bar magnets in the presence of a magnetic field. The MRI procedure is illustrated in **Figure 7-14.** The dorsal-view brain image portrays density differences among the hydrogen atoms in different neural regions as colors on the horizontal slice through the head.

Normally, hydrogen atoms point randomly in different directions, but when placed in a magnetic field, they line up in parallel as they orient themselves with respect to the field's lines of force. In MRI, radio pulses are applied to a brain whose atoms have been aligned in this manner, and the radio pulses form a second magnetic field. The second field causes the spinning atoms to wobble irregularly, thus producing a tiny electrical current that the MRI measures.

computerized tomography (CT) X-ray technique that produces a static, three-dimensional image of the brain in cross section—a *CT scan.*

magnetic resonance imaging (MRI) Technique that produces a static, three-dimensional brain image by passing a strong magnetic field through the brain, followed by a radio wave, then measuring the radiation emitted from hydrogen atoms.

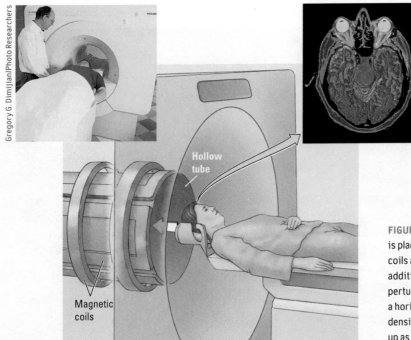

Hollow tube

Magnetic coils

FIGURE 7-14 Magnetic Resonance Imaging The subject is placed in a long metal cylinder that has two sets of magnetic coils arranged at right angles, as shown in the drawing. An additional radiofrequency coil (not shown) surrounds the head, perturbing the static magnetic fields to produce an MRI image of a horizontal section through the head, shown in dorsal view. The density differences in hydrogen atoms in different regions show up as colors in the brain image.

FIGURE 7-15 Magnetic Resonance Image Electrical currents emitted by wobbling atoms are recorded by MRI to represent different types of tissue—cerebrospinal fluid, brain matter, and bone, for example—as lighter or darker depending on the density of the hydrogen atoms in the tissue.

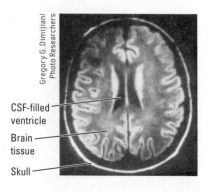

Gregory G. Dimijian/ Photo Researchers

CSF-filled ventricle

Brain tissue

Skull

When the currents are recorded, MRI images can be made based on the density of the hydrogen atoms in different brain regions. For example, areas of the brain with high water (H_2O) content (neuron-rich areas) will stand out from areas with lower water content (axon-rich areas). **Figure 7-15** shows such a magnetic resonance image.

Diffusion tensor imaging (DTI) is an MRI method that detects the directional movements of water molecules to image nerve fiber pathways in the brain. DTIs can be used to delineate abnormalities in neural pathways. Diffusion tensor images are also used to identify changes in the myelination of fibers, such as the damage that leads to loss of myelin in multiple sclerosis.

Clinical Focus 3-3 describes how myelin loss in MS disrupts neuronal function.

Each scan in the series of DTIs shown in **Figure 7-16** represents a dorsal view at increasing depths through the brain. Although the images appear to show real fibers, the DTIs are virtual and are based on computer reconstructions of bits of actual fibers. Nonetheless, abnormalities, such as occur in MS, stroke, or concussion, are easily detected in the imaged fiber pathways and in their myelin sheaths.

Focus 16-3 explores the relationship between concussion and degenerative brain disease.

Magnetic resonance spectroscopy (MRS) is an MRI method that uses the hydrogen proton signal to determine the concentration of brain metabolites such as N-acetyl aspartate (NAA) in brain tissue. This measurement is especially useful in detecting persisting abnormalities in brain metabolism in disorders such as concussion.

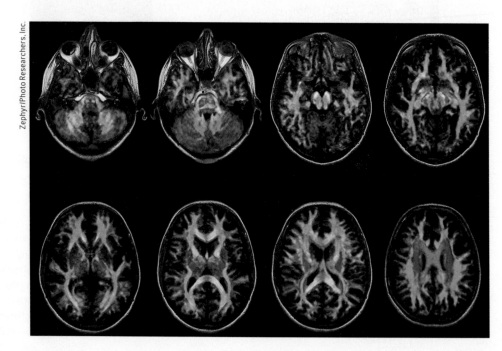

Zephyr/Photo Researchers, Inc.

FIGURE 7-16 Diffusion Tensor Imaging MRI can be used to measure the diffusion of water molecules in white matter, allowing the visualization of nerve fiber tracts. The front of the brain is at the top in these scans of sections through a healthy brain. The axons are colored according to orientation: fibers running left–right are red, front–back are blue, and up–down are green. Section 15-3 outlines how DTI is helping researchers develop a *brain connectome* to map functional connections in the living brain.

REVIEW 7-3
Static Imaging Techniques: CT and MRI

Before you continue, check your understanding.

1. The principal static methods of imaging the brain are _____ and _____.

2. Diffusion tensor imaging identifies _____ whereas magnetic resonance spectroscopy determines _____.

3. In addition to imaging the density of different brain regions, CT and MRI can be used to assess _____.

4. Explain briefly how the development of the CT scan ushered in the brain-imaging techniques used today in neuroscience research.

Answers appear at the back of the book.

diffusion tensor imaging (DTI) Magnetic resonance imaging method that, by detecting the directional movements of water molecules, can image fiber pathways in the brain.

magnetic resonance spectroscopy (MRS) Magnetic resonance imaging method that uses the hydrogen proton signal to determine the concentration of brain metabolites

functional magnetic resonance imaging (fMRI) Magnetic resonance imaging in which changes in elements such as iron or oxygen are measured during the performance of a specific behavior; used to measure cerebral blood flow during behavior or resting.

7-4 Dynamic Brain Imaging

Advances in MRI and computing technologies led from static to dynamic brain-imaging techniques that allow investigators to measure the amount of blood, oxygen, and glucose the brain uses as subjects solve cognitive problems. When a region of the brain is active, the amount of blood, oxygen, and glucose flowing to the region increases. It therefore is possible to infer changes in brain activity by measuring either blood flow or levels of the constituents of blood such as oxygen, glucose, and iron. Three techniques developed around this logic are functional MRI, positron emission tomography, and optical tomography.

Functional Magnetic Resonance Imaging

As neurons become active, they use more oxygen, resulting in a temporary dip in the amount of oxygen in the blood. At the same time, active neurons signal the blood vessels to dilate to increase blood flow and bring more oxygen to the area. Peter Fox and colleagues (1986) discovered that when human brain activity increases, the increase in oxygen produced by increased blood flow actually exceeds the tissue's need for oxygen. As a result, the amount of oxygen in an activated brain area increases.

Changes in the oxygen content of the blood alter the magnetic properties of the water in the blood. In 1990, Segi Ogawa and his colleagues showed that MRI could accurately match these changes in magnetic properties to specific locations in the brain (Ogawa et al., 1990). This process, **functional magnetic resonance imaging** (fMRI), signals which areas are displaying change in activity.

Figure 7-17 shows changes in the fMRI signal in the visual cortex of a person who is being stimulated with light. When the light is turned on, the visual cortex (bottom of the brain images) becomes more active than it was during baseline (no light). In other words, from increases and decreases in the MRI signal produced by changes in oxygen levels, functional changes in the brain are inferred.

When superimposed on MRI-produced brain images, fMRI changes in activity can be attributed to particular structures. The dense blood-vessel supply to the cerebral cortex allows for a spatial resolution of fMRI on the order of 1 millimeter, affording good spatial resolution of the source of brain activity. On the other hand, because changes in blood flow take as long as a third of a second, the temporal (time) resolution of fMRI is not as precise as that obtained with EEG recordings and ERPs.

fMRI also has the disadvantage that subjects must lie motionless in a long, noisy tube, an experience that can prove claustrophobic. The confined space and lack of mobility

Figure 2-7 diagrams the extent of the major cerebral arteries.

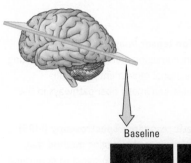

FIGURE 7-17 Imaging Changes in Brain Activity Functional MRI sequence of a horizontal section at mid-occipital lobe (bottom of each image) in a normal human brain during visual stimulation. A baseline acquired in darkness (*far left*) was subtracted from the subsequent images. The participant wore tightly fitting goggles containing light-emitting diodes that were turned on and off as a rapid sequence of scans was obtained over a period of 270 seconds. Note the prominent activity in the visual cortex when the light is on and the rapid cessation of activity when the light is off, all measured in the graph of signal intensity below the images. Adapted from "Dynamic Magnetic Resonance Imaging of Human Brain Activity During Primary Sensory Stimulation," by K. K. Kwong et al., 1992, *Proceedings of the National Academy of Sciences (USA), 89*, 5678.

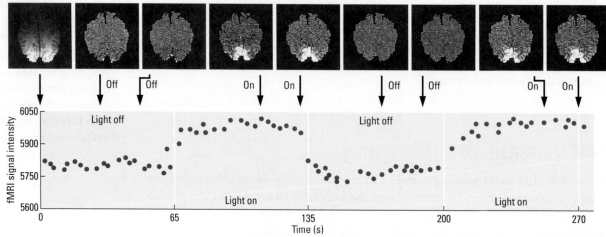

resting-state fMRI (rs-fMRI) Magnetic resonance imaging method that measures changes in elements such as iron or oxygen when the individual is resting (not engaged in a specific task).

positron emission tomography (PET) Imaging technique that detects changes in blood flow by measuring changes in the uptake of compounds such as oxygen or glucose; used to analyze the metabolic activity of neurons.

Most of the oxygen in air we breathe is the stable ¹⁶O molecule.

also restricts the types of behavioral experiments that can be performed. Nonetheless, fMRI has become a major tool in cognitive neuroscience.

The living brain is always active, and researchers have succeeded in inferring brain function and connectivity by studying fMRI signals when participants are "resting," that is, not engaged in any specific task. This signal, **resting-state fMRI** (rs-fMRI), is collected when participants are asked to look at a fixation cross and to keep their eyes open.

The scanner collects brain activity, typically for at least 4-minute-long blocks. Researchers are attempting to shorten the period. Statistical analysis of the data involves correlating activity in different brain regions over time. Although rs-fMRI is still in development, investigators already have identified many consistent networks of brain activity and abnormalities in disease states such as dementia and schizophrenia (Van den Heuvel & Hulshoff Pol, 2010).

Positron Emission Tomography

Researchers use **positron emission tomography** (PET) to study the metabolic activity of brain cells engaged in processing brain functions such as language. PET imaging detects changes in the brain's blood flow by measuring changes in the uptake of compounds such as oxygen and glucose (Posner & Raichle, 1997). A PET camera, like the one shown in **Figure 7-18,** is a doughnut-shaped array of radiation detectors that encircles a person's head. A small amount of water, labeled with radioactive molecules, is injected into the bloodstream. The person injected with these molecules is in no danger because the molecules, such as the radioactive isotope oxygen-15 (¹⁵O), are very unstable. They break down in just a few minutes and are quickly eliminated from the body.

The radioactive ¹⁵O molecules release tiny, positively charged, subatomic particles known as positrons (electrons with a positive charge). Positrons are emitted from an atom that is unstable because it is deficient in neutrons. The positrons are attracted to the negative charge of electrons in the brain, and the subsequent collision of the two particles leads to both of them being annihilated, thus creating energy.

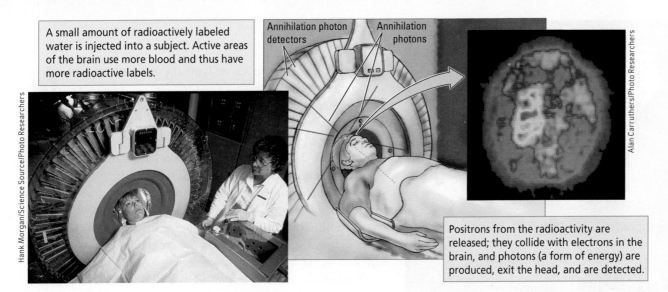

A small amount of radioactively labeled water is injected into a subject. Active areas of the brain use more blood and thus have more radioactive labels.

Annihilation photon detectors

Annihilation photons

Positrons from the radioactivity are released; they collide with electrons in the brain, and photons (a form of energy) are produced, exit the head, and are detected.

This energy, in the form of two photons (a photon is a unit of light energy), leaves the head at the speed of light and is detected by the PET camera. The photons leave the head in exactly opposite directions from the site of positron–electron annihilation, so their source can be identified by annihilation photon detectors, as illustrated in Figure 7-18. A computer identifies the coincident photons and locates the annihilation source to create the PET image.

The PET system enables the measurement of blood flow in the brain because the unstable radioactive molecules accumulate in the brain in direct proportion to the rate of local blood flow. Local blood flow, in turn, is related to neural activity because potassium ions released from stimulated neurons dilate adjacent blood vessels. The greater the blood flow, the higher the radiation counts recorded by the PET camera.

With the use of sophisticated computer imaging, blood flow in the brain when a person is at rest with closed eyes can be mapped **(Figure 7-19).** The map shows where the blood flow is highest in a series of frames. Even though the distribution of blood is not uniform, it is still difficult to conclude very much from such a map because the entire brain is receiving oxygen and glucose.

FIGURE 7-18 PET Scanner and Image
Subject lying in a PET scanner, the design of which is illustrated in the drawing. In the scan, the bright red and yellow areas are regions of high blood flow.

Section 10-4 reports on methods used, predating PET, to observe and measure blood flow in the brain.

The development of rs-fMRI suggests that a parallel resting-state PET analysis may one day emerge.

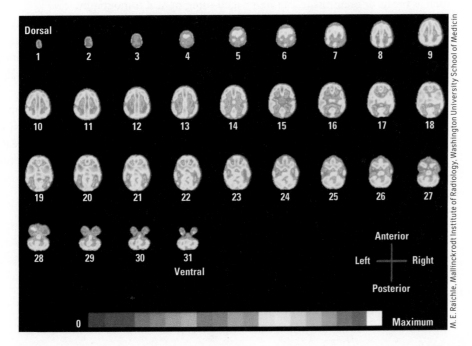

FIGURE 7-19 Resting State PET images of blood flow obtained while a single subject rested quietly with eyes closed. Each scan represents a horizontal section of the brain, from the dorsal surface (1) to the ventral surface (31).

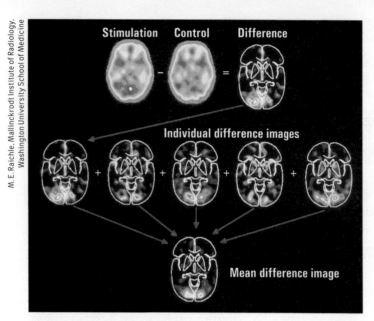

Stimulation   **Control**   **Difference**

Individual difference images

Mean difference image

FIGURE 7-20 The Procedure of Subtraction In the upper row of scans, the control condition of resting while looking at a static fixation point (control) is subtracted from the experimental condition of looking at a flickering checkerboard (stimulation). The subtraction produces a different scan for each of five experimental subjects, shown in the middle row, but all show increased blood flow in the occipital region. The difference scans are averaged to produce the representative image at the bottom.

PET researchers who are studying the link between blood flow and mental activity resort to a statistical trick. They subtract the blood-flow pattern when the brain is in a carefully selected control state from the pattern of blood flow imaged when the subject is engaged in the experimental task under study, as illustrated in the top row of **Figure 7-20.** This subtraction process images the change in blood flow in the two states. The change can be averaged across subjects (middle row) to yield a representative, average image difference that reveals which areas of the brain are selectively active during the task (bottom). Note that PET does not measure local neural activity directly; rather, it infers activity on the assumption that blood flow increases where neuron activity increases.

A significant disadvantage of PET is that radioactive materials must be prepared in a cyclotron located close to the scanner. Generating these materials is very expensive. But in spite of the expense, PET has important advantages over other imaging methods:

- PET can detect the decay of literally hundreds of radiochemicals, which allows the mapping of a wide range of brain changes and conditions, including changes in pH, glucose, oxygen, amino acids, neurotransmitters, and proteins.

- PET can detect relative amounts of a given neurotransmitter, the density of neurotransmitter receptors, and metabolic activities associated with learning, brain poisoning, and degenerative processes that might be related to aging.

- PET is widely used to study cognitive function, with great success. For example, PET confirms that various regions of the brain perform different functions.

Optical Tomography

Research Focus 7-1 describes a brain-imaging study that used functional near-infrared spectroscopy to investigate newborn infants' responses to language. fNRIS is a form of *optical tomography,* a dynamic imaging technique that operates on the principle that an object can be reconstructed by gathering light that was transmitted through the object. One requirement is that the object must at least partially transmit light. Soft body tissue, such as breast or brain tissue, can be imaged using optical tomography.

In fNRIS, reflected infrared light infers blood flow because hemoglobin, a protein in the blood that carries oxygen, absorbs light at a particular wavelength. By measuring the blood's light absorption it is possible to measure the brain's average oxygen consumption. To do so, an array of optical transmitter and receiver pairs are fitted across the scalp, as illustrated in **Figure 7-21**A.

The obvious advantage of fNRIS is that it is relatively easy to hook subjects up and record from them throughout life, from infancy to senescence. The disadvantage is that the light does not penetrate the brain very far, so researchers are restricted to measuring cortical activity (Figure 7-21B). The spatial resolution is also not as good as other noninvasive methods, although NIRS equipment now uses over 100 light detectors on the scalp, which allows acceptable spatial resolution in the image.

REVIEW 7-4
Dynamic Brain Imaging

Before you continue, check your understanding.

1. The principal methods of dynamic brain imaging are _____, _____, and _____.

(A)

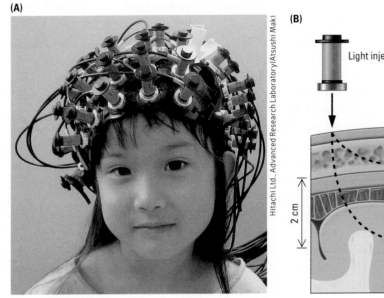

(B)

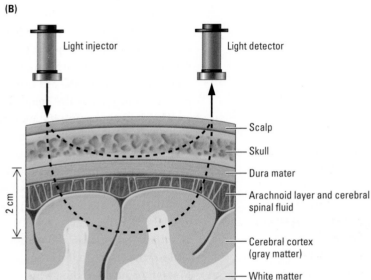

Light injector

Light detector

Scalp

Skull

Dura mater

Arachnoid layer and cerebral spinal fluid

Cerebral cortex (gray matter)

White matter

2 cm

Hitachi Ltd., Advanced Research Laboratory/Atsushi Maki

FIGURE 7-21 How NIRS Works **(A)** Light injectors (red) and detectors (blue) are distributed in an array across the head. **(B)** Light injected through the scalp and skull penetrates the brain to a depth of about 2 centimeters. A small fraction of the light is reflected and captured by a detector on the scalp surface. Light is reflected from as deep as 2 centimeters but also from the tissue above it, as illustrated by the banana-shaped curves. **(B)** Adapted from "Optical Topography and the Color of Blood," by L. Spinney, *The Scientist, 19,* 25–27.

2. PET uses _____ to measure brain processes and to identify _____ changes in the brain.

3. fMRI and optical imaging measure changes in _____.

4. Why are resting-state measurements useful to researchers?

Answers appear at the back of the book.

7-5 Chemical and Genetic Measures of Brain and Behavior

Our focus so far has been on how neuroscientists study the activity of neurons, individually and collectively, and how neuronal activity relates to behavior. Neurons are regulated by *genes,* segments of DNA that encode the synthesis of particular proteins within cells. Genes control the cell's production of chemicals, so it is possible to relate behavior to genes and to chemicals inside and outside the cell. Chemical and genetic approaches require sophisticated technologies that have seen major advances in the past decade.

Section 3-2 investigates how neurons function. Section 3-3 describes how genes determine cell function, genetic engineering, and epigenetic mechanisms.

Measuring the Brain's Chemistry

The brain contains a wide mixture of chemicals ranging from neurotransmitters and hormones to glucose and carbon monoxide, among many others. Abnormalities in these chemicals can cause serious disruptions in behavior. Prime examples are Parkinson's disease, characterized by low dopamine levels, and depression, correlated with low serotonin and/or noradrenaline production. The simplest way to measure brain chemistry in these types of diseases is to extract tissue postmortem from humans or laboratory animals and undertake traditional biochemical techniques, such as high-performance liquid chromatography (HPLC), to measure specific chemical levels.

Fluctuations in brain chemistry are associated not only with abnormalities in behavior but also with ongoing normal behavior. For example, the research of at least the last 30 years shows that dopamine levels fluctuate in the nucleus accumbens (a structure in the subcortical basal ganglia) in association with stimuli related to rewarding behaviors such as food and sex. Changes in brain chemistry can be measured in

Section 12-6 explores the neural effects of rewarding events on behavior.

Section 4-1 explains diffusion and concentration gradients in detail.

This result mirrors wanting-and-liking theory described in Section 6-4.

Section 6-4 investigates why glutamate and chemically similar substances can act as neurotoxins at very high doses.

microdialysis Technique used to determine the chemical constituents of extracellular fluid.

striatum Caudate nucleus and putamen of the basal ganglia.

cerebral voltammetry Technique used to identify the concentration of specific chemicals in the brain as animals behave freely.

FIGURE 7-22 **Microdialysis** Adapted from "Cerebral Microdialysis: Research Technique or Clinical Tool," by M. M. Tisdall and M. Smith, 2006, *British Journal of Anaesthesia, 97,* 18–25.

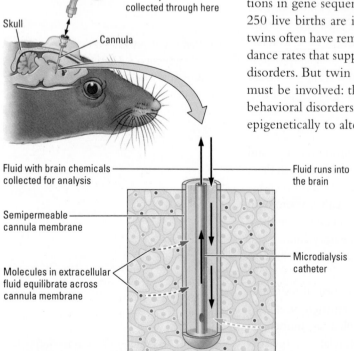

freely moving animals using two different methods, cerebral microdialysis and cerebral voltammetry.

Microdialysis has been widely used in the laboratory and in the past decade has begun to find clinical application. A catheter with a semipermeable membrane at its tip is placed in the brain as illustrated in **Figure 7-22.** A fluid flows through the cannula where it passes along the cell membrane. Simple diffusion drives the passage of extracellular molecules across the membrane along their concentration gradient.

The fluid containing the molecules from the brain exits through tubing to be collected for analysis. The fluid is removed at a constant rate so that changes in brain chemistry can be correlated with behavior. For example, if a rat is placed in an environment in which it anticipates sex or a favored food, microdialysis will record an increase in dopamine in a region of the basal ganglia known as the **striatum.**

Microdialysis is used in some medical centers to monitor brain chemistry in the injured brain. The effects of TBI or stroke can be worsened by secondary events such as a drastic increase in the neurotransmitter glutamate. Such biochemical changes can lead to irreversible cell damage or death. Physicians are beginning to use microdialysis to monitor such changes, which then can be treated.

Cerebral voltammetry works on a different principal. A small carbon fiber electrode and a metal electrode are implanted in the brain, and a small current is passed through the metal electrode. The current causes electrons to be added to or removed from the surrounding chemicals, and these changes can be translated into a measure of extracellular levels of specific neurotransmitters that are measured as they occur.

Because different currents lead to changes in different compounds, it is possible to identify levels of different transmitters, such as serotonin or dopamine, and related chemicals. Voltammetry has the advantage of not requiring the chemical analysis of fluid removed from the brain, as microdialysis does, but it has the disadvantage of being destructive. That is, the measurement of chemicals requires the degradation of one chemical into another. Thus this technique is not well suited to clinical uses.

Measuring Genes in Brain and Behavior

Most human behaviors cannot be explained by genetic inheritance alone, but variations in gene sequences do contribute significantly to brain organization. About 1 in 250 live births are identical twins, people who share an identical genome. Identical twins often have remarkably similar behavioral traits. Twin studies show strong concordance rates that support genetic contributions to drug addiction and other psychiatric disorders. But twin studies also show that environmental factors and life experience must be involved: there is generally far less than 100 percent concordance for most behavioral disorders, such as schizophrenia and depression. Life experiences are acting epigenetically to alter gene expression.

Genetic factors can also be studied by comparing people who were adopted early in life and normally would not have a close genetic relationship to their adoptive parents. Here, a high concordance rate for behavioral traits would imply a strong environmental influence on behavior. Ideally, an investigator would be able to study both the adoptive and biological parents to tease out the relative heritability of behavioral traits.

With the development of relatively inexpensive methods of identifying specific genes in people, it is now possible to relate the *alleles* (different forms) of specific genes to behaviors. A gene related to the production of a compound called brain-derived neurotrophic factor (BDNF) is representative. BDNF plays an important role in stimulating neural plasticity, and low levels of

BDNF have been revealed in mood disorders such as depression. The two alleles of this gene are BDNF Val 66Met and BDNF Val 66Val.

Joshua Bueller and his colleagues (2006) showed that the Met allele is associated with an 11 percent reduction in hippocampal volume in healthy subjects. Other studies have associated the Met allele with poorer memory for specific events (*episodic memory*) and a higher incidence of dementia later in life. However, the Val allele is by no means the "better" variant: although Val carriers have better episodic memory, they also have a higher incidence of neuroticism and anxiety disorders, as illustrated in Clinical Focus 7-3, "Cannabis Use, Psychosis, and Genetics." The two alleles simply produce different phenotypes because they influence brain structure and functions differently. Note too that other genes that were not measured also differed among Bueller's subjects and may have contributed to the observed difference.

Measuring Gene Expression

An individual's genotype exists in an environmental context that is fundamental to *gene expression,* the way genes become active or not. While epigenetic factors do not change the DNA sequence, the genes that are expressed can change dramatically in response to environment and experience. Epigenetic changes can persist throughout a lifetime and even across multiple generations.

Changes in gene expression can result from a wide range of experience, including chronic stress, traumatic events, drugs, culture, and disease. A study by Mario Fraga and

Section 8-2 explains how neurotrophic factors, nourishing chemical compounds, support growth and differentiation in developing neurons and may act to keep certain neurons alive in adulthood.

Focus 6-4 relates a case where chronic cannabis use may have led to psychosis.

See "A Case of Inheriting Experience" in Section 3-3.

CLINICAL FOCUS ✛ 7-3

Cannabis Use, Psychosis, and Genetics

Cannabis is the most widely used illicit drug in the world. Although it is usually considered "safe," it is not completely free of side effects. There is a modest risk for the emergence of psychosis, especially when adolescents use cannabis. Given that most adolescents who use it do not develop psychosis, however, it is likely that some sort of genetic vulnerability predisposes certain individuals to develop a psychotic condition when exposed to cannabis.

A working hypothesis contends that the COMT gene may be the culprit because this gene has been associated with schizophrenia. The COMT gene product is an enzyme involved in metabolizing dopamine in the synapse, and abnormalities in dopaminergic activity are associated with psychosis. The hypothesis predicts that adolescents who develop psychosis after cannabis use have an abnormality in the COMT gene.

Avshalom Caspi and colleagues (2005) analyzed the COMT gene in nearly 1000 26-year-old subjects who had participated in a long-term health study in New Zealand. As shown in the adjoining figure, although no genotype was more likely to use cannabis in adolescence, carriers of the Val allele were far more likely to develop psychotic symptoms if they used cannabis in adolescence (graph on the left) but not if they used it in adulthood (graph on the right). The Met/Met genotype showed no adverse effect of cannabis use in adolescence.

These results show that genetic variations can predispose people to show adverse effects from environmental experiences and that the

experiences may have age-related effects. Presumably these effects relate to the fact that the brain is undergoing significant development during adolescence.

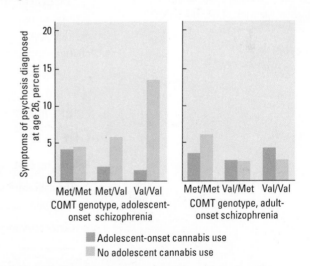

Adapted from "Moderation of the Effect of Adolescent-Onset Cannabis Use on Adult Psychosis by a Functional Polymorphism in the Catechol-O-Methyltransferase Gene: Longitudinal Evidence of a Gene X Environment Interaction," by A. Caspi, T. E. Moffitt, M. Cannon, J. McClay, R. Murray, et al., 2005, *Biological Psychiatry, 57,* 1117–1122.

his colleagues (2005) stands as a powerful example of gene–experience interactions. The investigators examined epigenetic patterns in 40 pairs of identical twins by measuring two molecular markers related to gene expression.

Although twins' patterns of gene expression were virtually identical when measured in childhood, 50-year-old twins exhibited differences so remarkable as to make them as different epigenetically as young non-twin siblings! The specific cause(s) of such differences is unknown but thought to be related to lifestyle factors, such as smoking and exercise habits, diet, stressors, drug use, and education, and to social experiences, such as marriage and child rearing, among others. The "epigenetic drift" in the twins supports the findings of less than 100 percent concordance for diseases in identical twins.

The role of epigenetic differences can also be seen across populations. Moshe Szyf, Michael Meaney, and their colleagues (e.g., 2008) have shown, for example, that the amount of maternal attention given by mother rats to their newborn infants alters the expression of certain genes in the adult hippocampus. These genes are related to the infants' stress response when they are adults. (Maternal attention is measured as the amount and type of mother–infant contact, and there can be a difference of up to 6 hours per day between attentive and inattentive mothers.)

A subsequent study by the same group (McGowan et al., 2009) examined epigenetic differences in hippocampal tissue obtained from two groups of humans: (1) suicides with histories of childhood abuse and (2) either suicides with no childhood abuse or controls who died from other causes. The epigenetic changes found in the abused suicide victims parallel those found in the rats with inattentive mothers, again suggesting that early experiences can alter hippocampal organization and function via changes in gene expression.

Experience-dependent changes in gene expression are found not only in the hippocampus but probably throughout the brain as well. For example, Richelle Mychasiuk and colleagues (2011) found that stressing pregnant rat dams led to large changes in gene expression in their offspring, in both the frontal cortex and the hippocampus. However, the investigators found virtually no overlap in the altered genes in the two brain regions: the same experience changed different brain regions differently.

Epigenetic studies promise to revolutionize our understanding of gene–brain interactions in normal brain development and brain function. They will also help researchers develop new treatments for neurological disorders. For example, specific epigenetic changes appear to be related to the presence or absence of functional recovery after stroke.

Consult the Index entry "Epigenetics" to locate coverage throughout the book.

REVIEW 7-5

Chemical and Genetic Measures of Brain and Behavior

Before you continue, check your understanding.

1. The concentrations of different chemicals in the brain can be measured in postmortem tissue using a(n) _____ assay or in vivo using _____ or _____.

2. Gene–environment interactions can be investigated in human populations by comparing _____ of behavioral traits in identical twins and adopted children.

3. The study of genes and behavior focuses on individual differences in _____, whereas the study of epigenetics and behavior examines differences in _____.

4. Describe briefly how epigenetic studies have led to the recognition that life experiences and the environment can alter brain functions.

Answers appear at the back of the book.

7-6 Comparing Neuroscience Research Methods

We have considered a wide range of research methods. **Table 7-1** summarizes them by group, including the goals and examples of each method. How do researchers choose among all these methods? Their main consideration is the research question being asked. Ultimately, that question is behavioral, but many steps lie along the route to understanding behavior.

Some researchers focus on morphology (structure) in postmortem tissue. This approach allows detailed analysis of both macro- and microstructure, depending on the method chosen. Identifying brain pathology, as in Parkinson's disease, can lead to insights about the causes and nature of a disorder.

Other investigators focus more on the ways neurons generate electrical activity in relation to behavior or on dynamic changes in brain activity during specific types of cognitive processing. Both approaches are legitimate: the goal is an understanding of brain–behavior relationships.

But investigators must consider practical issues, too. Temporal resolution (how quickly the measurement or image is obtained); spatial resolution (how accurate localization is in the brain); and the degree of invasiveness all are pertinent. It is impractical to consider MRI-based methods for studies of children, for example, because although the images are highly accurate, the participants must remain absolutely still for long periods.

Similarly, studies of brain-injured patients must take into account factors such as the subject's ability to maintain attention for long periods—during neuropsychological testing or imaging studies, for example. And practical problems such as motor or language impairment may limit the types of methods that researchers can use.

Of course, cost is an ever-present practical consideration. Studying brain and behavior linkages by perturbing the brain are generally less costly than some of the imaging methods, many of which require expensive machinery. EEG, ERP, and fNRIS are noninvasive and relatively inexpensive to set up (less than $100,000). MRI-based methods, MEG, and PET are very expensive (more than $2,000,000) and therefore typically located only in large research centers or hospitals. Similarly, epigenetic studies can be very expensive if investigators consider the entire genome in a large number of biological samples.

TABLE 7-1 Research Methods Used in Behavioral Neuroscience

Method	Goal	Examples	Method	Goal	Examples
Linking Brain and Behavior			**Brain Imaging**		
Tissue analysis (histology)	Identify cell types and connections; identify disease states	Stains	Static images	Noninvasive examination of brain structure	CT; MRI, DTI
Behavioral analysis	Generate tests to allow people and neuropsychological lab animals to demonstrate behavioral capacities	Tests; mazes	Dynamic images	Measure brain activity during specific behaviors	fMRI, PET, fNIRS
Brain perturbation	Modify brain activity to observe behavioral changes	Lesions; stimulation	**Chemical and Genetic Measures**		
			Postmortem chemistry	Measure distribution of transmitters	HPLC
Optogenetics	Use light to activate specific ion channels and relate to behavior	Insert specific light-sensitive proteins	In vivo chemistry	Relate fluctuations in transmitter release to behavior	Microdialysis; voltametry
Measuring Electrical Activity			Genetics	Relate genes to brain and behavior	DNA analysis
Recording graded potentials	Measure coordinated activity of thousands of neurons	EEG; ERP; MEG	Epigenetics	Identify effects of experience on gene expression, brain and behavior	Gene expression analysis
Recording action potentials	Record action potentials of individual neurons	Single-cell recording; multicell recording			

REVIEW 7-6

Comparing Neuroscience Research Methods

Before you continue, check your understanding.

1. Neuroscience measurements and imaging vary along the dimensions of _____, _____, and _____.

2. Relative to the expense of fMRI and PET imaging, noninvasively perturbing the brain using methods such as _____ or administering neuropsychological testing are _____.

3. The main consideration for choosing a research method is the question being asked. What is the fundamental goal of neuroscience research?

Answers appear at the back of the book.

7-7 Using Animals in Brain–Behavior Research

One major problem in understanding brain–behavior relationships in humans is that, aside from applying certain postmortem procedures and noninvasive imaging techniques, there are real constraints in using humans in neuroscience research. In addition, treatments for human neurological or psychiatric disorders, like most new treatments in medicine, are developed in nonhuman species before they are tested on humans.

Although obvious differences exist between the human and the nonhuman brain with respect to language, the general organization of the brain across mammalian species is similar, and the functioning of basic neural circuits in nonhuman mammals appears to generalize to humans. Thus, neuroscientists have been using animal models to make inferences about human brain function for well over a century.

Two important issues surface in developing treatments for brain and behavioral disorders with animal models. The first is whether the animals actually contract the same neurological diseases that humans do. The second surrounds the ethics of using animals in research. We consider each problem separately.

Benefits of Creating Animal Models of Disease

Some disorders—stroke, for example—seem relatively easy to model in laboratory animals because it is possible to interrupt blood supply to the brain and induce cortical injury and subsequent behavioral change. Obviously, it is much more difficult to determine whether behavioral disorders can actually be induced in laboratory animals. Consider *attention-deficit/hyperactivity disorder* (ADHD), a developmental disorder characterized by core behavioral symptoms of impulsivity, hyperactivity, and/or inattention. The most common issue for children with ADHD is that they have problems in school. Lab animals such as rats and mice do not go to school, so diagnosis is challenging.

ADHD has proved difficult to treat in children, and interest in developing an animal model is high. One way to proceed is to take advantage of the normal variance in the performance of rats on various tests of working memory and cognitive functioning. The idea is that we can think of ADHD in people or in rats as one extreme on a spectrum of behaviors that are part of a normal distribution in the general population. Many studies have now shown that treating rats with methylphendiate (Ritalin), a common treatment for children diagnosed with ADHD, actually improves the performance of rats that do poorly on tests of attentional processes.

One rat strain, the Kyoto SHR rat, has been widely used as an especially good model for ADHD. The strain presents known abnormalities in prefrontal dopaminergic

Focus 6-1 reports on the illicit use of prescription ADHD medications to boost performance at school and at work. Section 15-4 explores the nature of attention and disorders that result in deficits of attention.

innervation that correlate with behavioral abnormalities such as hyperactivity. Dopaminergic abnormalities are believed to be one underlying symptom of ADHD in children, as explained in Research Focus 7-4, "Attention-Deficit/Hyperactivity Disorder." Dopamine agonists such as methylphenidate can reverse behavioral abnormalities, both in children with ADHD and in the SHR rats.

Other models of ADHD focus on manipulating the animal's prefrontal development by perinatal anoxia (oxygen deprivation). This treatment leads to prefrontal abnormalities lateralized to the right hemisphere, which is also seen in humans with ADHD. There is no consistent evidence that perinatal anoxia is related to ADHD in human children, however.

Animal Welfare and Scientific Experimentation

Using nonhuman animals in scientific research has a long history, but only in the past half-century have ethical issues surrounding animal research gained considerable attention. Just as the scientific community has established ethical standards for research on human subjects, it has also developed regulations governing experimentation on animals.

RESEARCH FOCUS ✤ 7-4

Attention-Deficit/Hyperactivity Disorder

Together, attention-deficit/hyperactivity disorder (ADHD) and attention-deficit disorder (ADD) are probably the most common disorders of brain and behavior in children, with an incidence of from 4 percent to 10 percent of school-aged children. Although it is often not recognized, an estimated 50 percent of children with ADHD still show symptoms in adulthood, where its behaviors are associated with family breakups, substance abuse, and driving accidents.

The neurobiological basis of ADHD and ADD is generally believed to be a dysfunction in the noradrenergic or dopaminergic activating systems, especially in the frontal basal ganglia circuitry. Psychomotor stimulants such as methylphendiate (Ritalin) and Adderall (mainly dextroamphetamine) act to increase brain levels of noradrenaline and dopamine and are widely used for treating ADHD. About 70 percent of children show improvement of attention and hyperactivity symptoms with treatment, but there is little evidence that drugs directly improve academic achievement. This is important because about 40 percent of children with ADHD fail to get a high-school diploma, even though many receive special education for their condition.

A common view that ADHD is a cultural phenomenon reflecting the tolerance of parents and teachers to children's behavior has been challenged by a scholarly review by Faraone and coworkers (2003). These investigators conclude that the prevalence of ADHD worldwide is remarkably similar when the same rating criteria are used. Little is known about incidence in developing countries, however. It is entirely possible that the incidence may actually be higher in developing countries given that the learning environment for children is likely to be less structured than it is in developed nations.

The cause of ADHD is unknown but probably involves dopamine receptors in the forebrain. The most likely areas are the frontal lobe and

In this mainstream first-grade classroom, a special education student with ADHD uses the "turtle technique" to cope with frustration and stress.

subcortical basal ganglia. Evidence of reduced brain volumes in these regions in ADHD patients is growing, as is evidence of an increase in the dopamine transporter protein. The dopamine transporter increase would mean that dopamine reuptake into the presynaptic neuron occurs faster than it does in the brains of people without ADHD. The result is a relative decrease in dopamine. Ritalin would then be effective because it blocks dopamine reuptake.

ADHD is believed to be highly heritable, a conclusion supported by twin studies showing a concordance of about 75 percent in identical twins. Molecular genetic studies have identified at least seven candidate genes, and several of them are related to the dopamine synapse, in particular to the D_4 receptor gene.

Some experiments presented in this book predate ethical standards established for research on humans—for example, Bartholow's brain stimulation described in Section 4-1 and the inmate volunteers in Experiment 6-1—and on nonhuman animals, for example, Magendie's studies with puppies in Focus 2-4.

The governments of most developed nations regulate the use of animals in research; most states and provinces have additional legislation. Universities and other organizations engaged in research have their own rules governing animal use, as do professional societies of scientists and the journals in which they publish.

In Canada, the 20 organizations listed in **Table 7-2** make up the Canadian Council on Animal Care. The council is dedicated to enhanced animal care and use through education, voluntary compliance, and codes of ethics. The council is organized to respond flexibly to the concerns of both the scientific community and the general public, through rapid and frequent amendments to its guidelines.

The Canadian Council on Animal Care endorses four principles as guidelines for reviewing protocols for experiments that will use animals:

1. The use of animals in research, teaching, and testing is acceptable only if it promises to contribute to the understanding of environmental principles or issues, fundamental biological principles, or development of knowledge that can reasonably be expected to benefit humans, animals, or the environment.

2. Optimal standards for animal health and care result in enhanced credibility and reproducibility of experimental results.

3. Acceptance of animal use in science critically depends on maintaining public confidence in the mechanisms and processes used to ensure necessary, humane, and justified animal use.

4. Animals are used only if the researcher's best efforts to find an alternative have failed. Researchers who use animals employ the most humane methods on the smallest number of appropriate animals required to obtain valid information.

Legislation concerning the care and use of laboratory animals in the United States is set forth in the Animal Welfare Act, which includes laws passed by Congress in 1966, 1970, 1976, and 1985. Legislation in other countries is similar and in some European countries much more strict. The U.S. act covers mammals, including rats, mice, cats, and monkeys, and birds, but it excludes farm animals that are not used in research. It is administered by the U.S. Department of Agriculture (USDA) through inspectors in the Animal and Plant Health Inspection Service, Animal Care.

In addition, the Office of Human Research Protections of the National Institutes of Health (NIH) administers the Health Research Extension Act (passed in 1986). The act covers all animal uses conducted or supported by the U.S. Public Health Service and applies to any live vertebrate animal used in research, training, or testing. The act requires that each institution provide acceptable assurance that it meets all minimum regulations and conforms with *The Guide for the Care and Use of Laboratory Animals* (National Research Council, 2011) before conducting any activity that includes animals. The typical method for demonstrating conformance with the Guide is to seek voluntary accreditation from the Association for Assessment and Accreditation of Laboratory Animal Care International.

All accredited U.S. universities that receive government grant support are required to provide adequate treatment for all vertebrate animals. Reviews and specific protocols for fish, reptiles, mice, dogs, and monkeys to be used in research, teaching, or testing are administered through the same process. Anyone using animals in a U.S. university submits a protocol to the university's institutional animal care and use committee, composed of researchers, veterinarians, people who have some knowledge of science, and laypeople from the university and the community.

Companies that use animals for research are not required to follow this process. In effect, however, if they do not, they will be unable to publish the results of their research, because journals require that research conform to national guidelines on animal care. In

TABLE 7-2 Member Organizations in the Canadian Council on Animal Care

Agriculture Canada

Association of Canadian Faculties of Dentistry

Association of Canadian Medical Colleges

Association of Universities and Colleges of Canada

Canadian Association for Laboratory Animal Medicine

Canadian Association for Laboratory Animal Science

Canadian Federation of Humane Societies

Canadian Institute for Health Research

Canadian Society of Zoologists

Committee of Chairpersons for Departments of Psychology

Confederation of Canadian Faculties of Agriculture and Veterinary Medicine

Department of National Defense

Environment Canada

Fisheries and Oceans Canada

Health and Welfare Canada

Heart and Stroke Foundation of Canada

National Cancer Institute

National Research Council

Natural Sciences and Engineering Research Council

Canada's Research-Based Pharmaceutical Companies (Rx&D)

addition, discoveries made using animals are not recognized by government agencies that approve drugs for clinical trials with humans if they do not follow the prescribed process. Companies therefore use standards described as Good Laboratory Practice (GLP) that are as rigorous as those used by government agencies.

U.S. regulations specify that researchers consider alternatives to procedures that may cause more than momentary or slight pain or distress to animals. Most of the attention on alternatives has focused on the use of animals in testing and stems from high public awareness of some tests for pharmacological compounds, especially toxic compounds. Testing of such compounds is now regulated by the National Institute of Environmental Health Sciences.

In spite of the legislation related to animal use, considerable controversy remains over using animals in scientific research. At the extremes, people on one side approve and people on the other side disapprove of using animals for any form of research. Others fall somewhere in between. The debate centers on issues of law, morals, custom, and biology.

Because researchers in many branches of science experiment with animals to understand the functions of the human body, brain, and behavior, the issues in this debate are important to them. Because many people benefit from this research, including those who have diseases of the nervous system or nervous system damage, this debate is important to them. Because many people are philosophically opposed to using animals for work or food, this debate is important to them. And, because you, as a student, encounter many experiments on animals in this book, these issues are important to you as well.

REVIEW 7-7
Using Animals in Brain–Behavior Research

Before you continue, check your understanding.

1. Laboratory animals can model such human dysfunctions as _____ and _____.

2. One difficulty in using lab animals as models of human disease is determining _____.

3. Animal models provide a way to investigate both proposed _____ and _____ for behavioral disorders.

4. Outline the controversies that surround the use of animals in scientific research on brain and behavioral relationships.

Answers appear at the back of the book.

SUMMARY

7-1 Measuring Brain and Behavior

The brain's primary function is to produce behavior, so the fundamental technique of research in behavioral neuroscience is to study the direct relationship between brain and behavior. Investigators study healthy humans and other animals as well as human patients and laboratory animals with neurological problems.

Initially, scientists simply observed behavior, but they later developed neuropsychological testing measures designed to study specific functions such as fine movements, memory, and emotion. Today, researchers correlate these behavioral outcomes with anatomical, physiological, chemical, genetic, and other molecular measures of brain organization.

Brain and behavioral relations can be manipulated by altering brain function, either permanently or temporarily. Permanent changes involve damaging the brain directly by ablation or neurotoxins that remove or destroy brain tissue. Transient changes in brain activity can be induced either by using a mild electrical or magnetic current, as in DBS or TMS, or by administering drugs. Optogenetic stimulation is a transgenic technique that combines genetics and light to excite or inhibit targeted cells in living tissue.

7-2 Measuring the Brain's Electrical Activity

Electroencephalographic or magnetoencephalographic recordings measure electrical activity, either from thousands of neurons at once or from small numbers of or even individual neurons.

EEG can reveal a gross relationship between brain and behavior, as when a person is alert and displays the beta-wave pattern versus when the person is resting or sleeping, as indicated by the slower alpha-wave patterns. Event-related potentials tell us, on the other hand, that even though the entire brain is active during waking, certain parts are momentarily much more active than others. ERP records how the location of increased activity changes as information moves from one brain area to another.

EEG and ERP are noninvasive methods that record from electrodes on the scalp or, in the case of MEG, from magnetic detectors above the head. Electrocorticography, by contrast, records by attaching electrodes directly on the cortex. ECoG and single-cell recording are invasive.

Recording from single or multiple cells shows that neurons employ a code and that cortical neurons are organized into functional groups that work as a coordinated network. Neurons in sensory areas respond to specific characteristics of stimuli, such as color or pitch. Neurons in other regions can code for more complex information such as location of an object in space.

7-3 Static Imaging Techniques: CT and MRI

Computer tomography and magnetic resonance imaging methods are sensitive to the density of different brain structures, ventricles, nuclei, and pathways. CT is a form of three-dimensional X-ray, whereas MRI works on the principle that hydrogen atoms behave like spinning bar magnets in the presence of a magnetic field.

Although CT scans are less expensive and can be done quickly, MRI provides an exceptionally clear image both of nuclei and of fiber pathways in the brain and indicates that, structurally, different people's brains can be quite different. Both CT and MRI can be used to assess brain damage from neurological disease or injury, but MRI is more useful as a research tool.

Diffusion tensor imaging is a form of MRI that makes it possible to identify normal or abnormal fiber tracts and myelin in the brain. Magnetic resonance spectroscopy is another form of MRI that permits practitioners to detect brain metabolites, such as those produced following concussion.

7-4 Dynamic Brain Imaging

Metabolic imaging methods show that any behavior requires the collaboration of widespread neural circuits. Positron emission tomography records blood flow and other metabolic changes in periods of time measured in minutes and requires complex subtraction procedures and the averaging of responses across a number of subjects. Records of blood flow obtained

by using functional magnetic resonance imaging can be combined with static MRI images to identify the location of changes in the individual brain and to complement ERP results. Resting-state fMRI allows investigators to measure connectivity across brain regions.

Functional near-infrared spectroscopy is the form of optical tomography normally used for dynamic brain imaging studies. It works on the principle that an object, including brain tissue, can be reconstructed by gathering light transmitted through the object. fNIRS is much simpler to use than PET or fMRI, but because light does not penetrate very far into the brain, it can be used only to study cortical function.

7-5 Chemical and Genetic Measures of Brain and Behavior

Analysis of changes in both genes and neurochemicals provides insight into the molecular correlates of behavior. Although genes code all the information needed to construct and regulate cells, epigenetic research reveals that the environment and life experience can modify gene expression. Even identical twins, who have identical genomes at birth, have quite different patterns of gene expression and very different brains in adulthood.

7-6 Comparing Neuroscience Research Methods

The main consideration in neuroscience research is the question. Whatever the approach, the goal is to understand brain–behavior relationships. Table 7-1 on page 239 summarizes the research methods reviewed in the chapter. Among all the practical issues of measurement resolution and invasiveness, cost may prove the ultimate consideration.

7-7 Using Animals in Brain–Behavior Research

Understanding brain function, in both the normal and the abnormal brain, often benefits from the development of animal models. Animal models allow investigators to manipulate the brain to determine the effect of both experiential factors and neurological treatments on brain function.

Because animal subjects cannot protect themselves from abuse, governments and researchers have cooperated to develop ethical guidelines for the use of laboratory animals. These guidelines are designed to ensure that discomfort is minimized, as is the number of animals used for invasive procedures.

KEY TERMS

akinesia, p. 220

alpha rhythms, p. 224

behavioral neuroscience, p. 215

cerebral voltammetry, p. 236

computerized tomography (CT), p. 229

deep-brain stimulation (DBS), p. 223

diffusion tensor imaging (DTI), p. 231

electrocorticography (ECoG), p. 224

event-related potentials (ERPs), p. 224

functional magnetic resonance imaging (fMRI), p. 231

functional near-infrared spectroscopy (fNIRS), p. 212

magnetic resonance imaging (MRI), p. 229

magnetic resonance spectroscopy (MRS), p. 231

magnetoencephalogram (MEG), p. 227

microdialysis, p. 236

neuropsychology, p. 212

optogenetics, p. 223

place cells, p. 227

positron emission tomography (PET), p. 232

resting-state fMRI (rs-fMRI), p. 232

stereotaxic apparatus, p. 220

striatum, p. 236

transcranial magnetic stimulation (TMS), p. 223

 Please refer to the Companion Web Site at www.worthpublishers.com/kolbintro4e *for Interactive Exercises and Quizzes.*

CHAPTER

8

How Does the Nervous System Develop and Adapt?

RESEARCH FOCUS 8-1 LINKING SEROTONIN TO SIDS

8-1 THREE PERSPECTIVES ON BRAIN DEVELOPMENT

PREDICTING BEHAVIOR FROM BRAIN STRUCTURE

CORRELATING BRAIN STRUCTURE AND BEHAVIOR

INFLUENCES ON BRAIN AND BEHAVIOR

8-2 NEUROBIOLOGY OF DEVELOPMENT

GROSS DEVELOPMENT OF THE HUMAN NERVOUS SYSTEM

ORIGINS OF NEURONS AND GLIA

GROWTH AND DEVELOPMENT OF NEURONS

CLINICAL FOCUS 8-2 AUTISM SPECTRUM DISORDER

UNIQUE ASPECTS OF FRONTAL-LOBE DEVELOPMENT

GLIAL DEVELOPMENT

8-3 CORRELATING BEHAVIOR WITH NERVOUS-SYSTEM DEVELOPMENT

MOTOR BEHAVIORS

LANGUAGE DEVELOPMENT

DEVELOPMENT OF PROBLEM-SOLVING ABILITY

CAUTION ABOUT LINKING CORRELATION TO CAUSATION

8-4 BRAIN DEVELOPMENT AND THE ENVIRONMENT

EXPERIENCE AND CORTICAL ORGANIZATION

RESEARCH FOCUS 8-3 INCREASED CORTICAL ACTIVATION FOR SECOND LANGUAGES

EXPERIENCE AND NEURAL CONNECTIVITY

CRITICAL PERIODS FOR EXPERIENCE AND BRAIN DEVELOPMENT

CLINICAL FOCUS 8-4 ROMANIAN ORPHANS

ABNORMAL EXPERIENCE AND BRAIN DEVELOPMENT

HORMONES AND BRAIN DEVELOPMENT

INJURY AND BRAIN DEVELOPMENT

DRUGS AND BRAIN DEVELOPMENT

OTHER KINDS OF ABNORMAL BRAIN DEVELOPMENT

DEVELOPMENTAL DISABILITY

CLINICAL FOCUS 8-5 SCHIZOPHRENIA

8-5 HOW DOES ANY OF US DEVELOP A NORMAL BRAIN?

Linking Serotonin to SIDS

Sudden infant death syndrome (SIDS), the unexplained death while asleep of a seemingly healthy infant less than 1 year old, kills about 2500 babies yearly in the United States alone. Autopsies have historically failed to identify a clear cause of death from this tragic and troubling disorder.

More-recent postmortem studies reveal that SIDS victims are more likely than typical babies to have a particular gene variation that makes the serotonin (5-HT) transporter unusually efficient. Normally, the serotoninergic system helps to stimulate a mechanism that responds to high carbon dioxide levels in the blood and acts to expel the gas. In babies who have succumbed to SIDS, serotonin is cleared from the synapse more rapidly than normal.

This action makes serotonin less effective in regulating life-threatening events such as carbon dioxide buildup during sleep. Babies can breathe excessive levels of carbon dioxide that is trapped in their bedding, for example, and suffocate.

In addition to the serotonin-transporter abnormality, David Paterson and his colleagues (2006) found an abnormally low occurrence of 5-HT_{1A} receptors in SIDS victims' brains. This could reduce the serotonergic system's effectiveness in regulating behavior. The researchers found that boys have significantly fewer 5-HT_{1A} receptors than do females, a result consistent with higher SIDS mortality in boys.

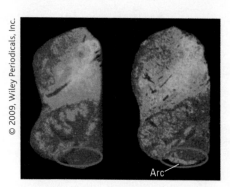

© 2009, Wiley Periodicals, Inc.

Serotonin in the arcuate nucleus (arc) of the midbrain is reduced in infants who have succumbed to SIDS (left) relative to controls (right). **From "Brainstem Mechanisms Underlying the Sudden Infant Death Syndome: Evidence from Human Pathologic Studies," by H.C. Kinney, 2009, *Developmental Psychobiology*, 51, pp. 223–233.**

Genetic manipulations of 5-HT receptors in mice provide further indirect evidence of a SIDS–serotonin connection. Enrica Audero and colleagues (2008) engineered mice with high levels of 5-HT_{1A} autoreceptors in the brainstem and unexpectedly found that 70% of the mice died before they were 120 days old.

These mice showed various abnormalities in the control of temperature and heartbeat, events that can also be seen in babies monitored during SIDS episodes. The serotonin autoreceptors are part of a feedback loop that turns off the serotonergic cells. This would act to reduce 5-HT transmission elsewhere in the brain.

A review by Hannah Kinney (2009), evaluating the SIDS–serotonin hypothesis, concludes that although other dysfunctional neurotransmitter systems in the brainstem may also be involved, evidence for the 5-HT defects is the most robust. She speculates that the primary defect is increased number of 5-HT cells, potentially arising during fetal development and owing to unknown causes, but augmented by adverse prenatal exposure to alcohol, nicotine, and/or other factors. This defect leads to the changes in 5-HT_{1A} receptors.

Although the SIDS–serotonin connection does not yet have a proactive treatment, it does provide evidence that SIDS is a real nervous-system disorder. The findings also offer parents some reassurance that they could have neither predicted nor prevented the deaths of their infants.

To understand how scientists go about studying the interconnected processes of brain and behavioral development, think about all the architectural parallels between how the brain is constructed and how a house is built. House plans are drawn as blueprints; the plans for a brain are encoded in genes. Architects do not specify every detail in a blueprint, nor do genes include every instruction for brain assembly and wiring.

The brain is just too complex to be encoded entirely and precisely in genes. For this reason, the fate of billions of brain cells is left partly undecided. This fact is true especially in regard to the massive undertaking of forming appropriate connections between cells.

If the structure and fate of each brain cell are not specified in advance, what controls brain development? Many factors are at work and, like house building, brain development is influenced by the environment in the course of the construction phase and by the quality of the materials used.

We can shed light on nervous-system development by viewing its architecture from different vantage points—structural, functional, and environmental. In this chapter, we consider the neurobiology of development first, explore the behavioral

sudden infant death syndrome (SIDS)
Unexplained death while asleep of a seemingly healthy infant less than 1 year old.

correlates of developing brain functions next, and then explore how experiences and environments influence neuroplasticity over the lifespan.

8-1 Three Perspectives on Brain Development

Brain and behavior develop apace. Scientists thus reason that both lines of development are closely linked. Events that alter behavioral development should similarly alter the brain's structural development and vice versa. As the brain develops, neurons become more and more intricately connected, and these increasingly complex interconnections underlie increasingly complex behavior. These observations enable neuroscientists to study the relation between brain and behavioral development from three different perspectives:

1. Structural development can be studied and correlated with the emergence of behavior.

2. Behavioral development can be analyzed and predictions made about what underlying circuitry must be emerging.

3. Factors that influence both brain structure and behavioral development, such as language or injury, can be studied.

Predicting Behavior from Brain Structure

We can look at the structural development of the nervous system and correlate it with the emergence of specific behaviors. For example, we can link the development of certain brain structures to the motor development of, say, grasping or crawling in infants. As brain structures mature, their functions emerge and develop, manifested in behaviors that we can observe.

Neural structures that develop quickly—the visual system, for instance—exhibit their functions sooner than do structures that develop more slowly, such as those for speech. Because the human brain continues to develop well into adulthood, you should not be surprised that some abilities emerge or mature rather late. Some cognitive behaviors controlled by the frontal lobes are among the last to develop. One such behavior, the ability to plan efficiently, is a skill important to many complexities of life, including organizing daily activities or making travel plans.

The Tower of Hanoi test, illustrated in **Figure 8-1,** shows how planning skills can be measured in the laboratory. The task is to mentally plan how to move colored discs one by one, in the minimum number of moves, from one configuration to another. Although 10-year-olds can solve simple configurations, more difficult versions of the task, such as that shown in Figure 8-1, cannot be performed efficiently until about ages 15 to 17. It should thus come as no surprise that adolescents can often appear disorganized and lack the ability to plan their activities in the way that fully mature adults can.

Adults with acquired frontal-lobe injuries also fail to perform well on the Tower of Hanoi test. Such evidence reinforces the idea that children are not miniature adults who simply need to learn the "rules" of adult behavior. The brain of a child is very different from that of an adult, and the brains of children at different ages are not really comparable either.

Correlating Brain Structure and Behavior

We can turn our sequence of observations around, scrutinizing behavior for the emergence of new abilities and then inferring underlying neural maturation. For example, as language emerges in the young child, we expect to find corresponding changes in neural structures that control language. In fact, neuroscientists do find such changes.

GOAL

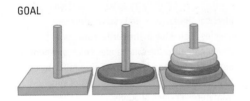

Move discs on towers below one by one to match goal above.

FIGURE 8-1 Testing Cognitive Development The Tower of Hanoi is a mathematical puzzle consisting of three rods and several different-sized discs. The task is to match the goal in as few moves as possible, obeying two rules: (1) only one disc may be moved at a time; (2) no disc may be placed on top of a smaller disc.

neural plate Thickened region of the ectodermal layer that gives rise to the neural tube.

neural tube Structure in the early stage of brain development from which the brain and spinal cord develop.

At birth, children do not speak, and even extensive speech training would not enable them to do so. The neural structures that control speech are not yet mature enough. As language emerges, the speech-related structures in the brain thus are undergoing the necessary maturation.

The same reasoning can be applied to frontal-lobe development. As frontal-lobe structures mature through adolescence and into early adulthood, we look for related changes in behavior. We can also do the reverse: because we observe new abilities emerging in the teenage years and even later, we infer that they must be controlled by late-maturing neural structures and connections.

Influences on Brain and Behavior

The third way to study interrelations between brain and behavioral development is to identify and study factors that influence both. From this perspective, the mere emergence of a certain fully developed brain structure is not enough. We must also know the events that shape how that structure functions and produces certain behaviors. Some events that influence brain function are sensory experience, injuries, and the actions of hormones and genes.

Logically, if one of these factors influences behavior, then structures in the brain that are changed by that factor are responsible for the behavioral outcomes. For example, we might study how the abnormal secretion of a hormone affects both a certain brain structure and a certain behavior. We can then infer that, because the observed behavioral abnormality results from the abnormally functioning brain structure, the structure must normally play some role in controlling the behavior.

REVIEW 8-1
Three Perspectives on Brain Development

Before you continue, check your understanding.

1. Structural brain development is correlated with the emergence of _____.

2. Behavioral development predicts the maturation of _____.

3. Three events that influence brain function are _____, _____, and _____.

4. What important constraint determines when behaviors emerge?

Answers appear at the back of the book.

8-2 Neurobiology of Development

Some 2000 years ago, the Roman philosopher Seneca proposed that a human embryo is an adult in miniature, and thus the task of development is simply to grow bigger. This idea of *preformation* was so appealing that it was widely believed for centuries. Even with the development of the microscope, the appeal of preformation proved so strong that biologists claimed to see microscopic horses in horse semen.

By the mid-1800s, the idea of preformation began to wane as people realized that embryos look nothing like the adults they become. In fact, it was obvious that the embryos of different species more closely resemble one another than their respective parents. The top row of **Figure 8-2** shows the striking similarity in the embryos of species as diverse as salamanders, chickens, and humans, shown in fetal form in the bottom row.

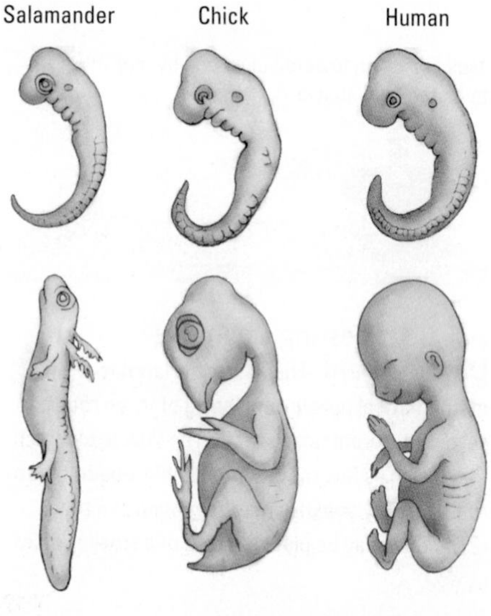

Early in development, all vertebrate species have a similar-looking primitive head, a region with bumps or folds, and all possess a tail. Only as an embryo develops does it acquire

FIGURE 8-2 Embryos and Evolution The physical similarity of embryos of different species is striking in the earliest stages of development, as the salamander, chick, and human embryos in the top row show. This similarity led to the conclusion that embryos are not simply miniature versions of adults.

Salamander   Chick   Human

the distinctive characteristics of its species. The similarity of young embryos is so great that many nineteenth-century biologists saw it as evidence for Darwin's view that all vertebrates arose from a common ancestor millions of years ago.

The embryonic nervous systems of vertebrates are as similar structurally as their bodies are. **Figure 8-3** details the three-chambered brain of a young vertebrate embryo: forebrain, midbrain, and hindbrain. The remaining neural tube forms the spinal cord. How do these three regions develop? We can trace the events as the embryo matures.

Gross Development of the Human Nervous System

When a sperm fertilizes an egg, the resulting human zygote consists of just a single cell. But this cell soon begins to divide and, by the 15th day, the emerging embryo resembles a fried egg **(Figure 8-4)**. This structure is formed by several sheets of cells with a raised area in the middle called the *embryonic disc*—essentially the primitive body.

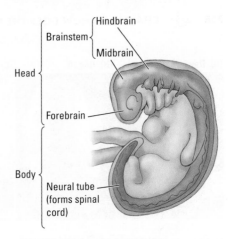

FIGURE 8-3 Embryonic Vertebrate Nervous System Forebrain, midbrain, and hindbrain are visible in the human embryo at about 28 days, as is the remaining neural tube, which will form the spinal cord.

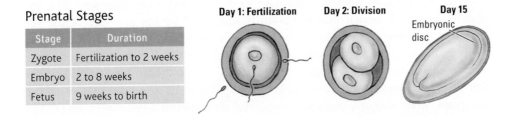

Prenatal Stages

Stage	Duration
Zygote	Fertilization to 2 weeks
Embryo	2 to 8 weeks
Fetus	9 weeks to birth

FIGURE 8-4 From Fertilization to Embryo Development begins at fertilization (day 1), with the formation of the zygote. On day 2, the zygote begins to divide. On day 15, the raised embryonic disk begins to form. Adapted from *The Developing Human: Clinically Oriented Embryology* (4th ed., p. 61), by K. L. Moore, 1988, Philadelphia: Saunders.

By day 21, 3 weeks after conception, primitive neural tissue, the **neural plate,** occupies part of the outermost layer of embryonic cells. The neural plate first folds to form the *neural groove,* detailed in **Figure 8-5**. The neural groove then curls to form the **neural tube,** much as a flat sheet of paper can be curled to make a cylinder.

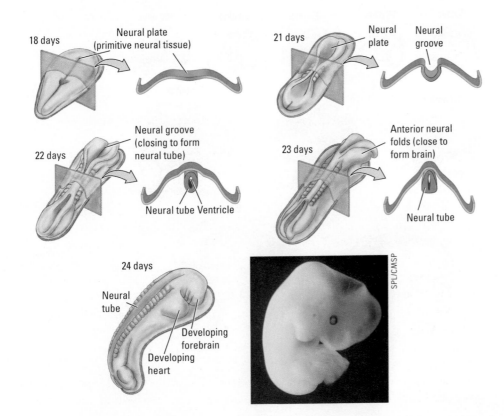

FIGURE 8-5 Formation of the Neural Tube A long depression, the neural groove, first forms in the neural plate. By day 21, the primitive brain and neural groove are visible. On day 23, the neural tube is forming as the neural plate collapses inward along the length of the dorsal surface of the embryo. The embryo is shown in a photograph at 24 days.

(A) Day 9   **(B) Day 10**   **(C) Day 11**

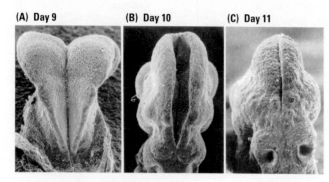

FIGURE 8-6 Neural-Tube Development
Scanning electron micrographs show the neural tube closing in a mouse embryo. Reproduced with the permission of Dr. R. E. Poelman, Laboratory of Anatomy, University of Leyden

Sections 6-5 and 12-5 detail the actions of testosterone and other gonadal hormones.

The cells that form the neural tube can be thought of as the nursery for the rest of the central nervous system. The open region in the center of the tube remains open and matures into the brain's ventricles and the spinal canal. Micrographs of the neural tube closing in a mouse embryo can be seen in **Figure 8-6**.

The human body and nervous system change rapidly in the next 3 weeks. By 7 weeks (49 days), the embryo begins to resemble a miniature person. **Figure 8-7** shows that the brain looks distinctly human by about 100 days after conception, but it does not begin to form gyri and sulci until about 7 months. By the end of the 9th month, the fetal brain has the gross appearance of the adult human brain, even though its cellular structure is different.

Another developmental process, shown in **Figure 8-8**, is sexual differentiation. Although the genitals begin to form in the seventh week after conception, they appear identical (indifferent) in the two sexes at this early stage. There is not yet any *sexual dimorphism,* or structural difference, between the sexes. Then, about 60 days after conception, male and female genitals start to become distinguishable.

What does sexual differentiation have to do with brain development? The answer is hormonal. Sexual differentiation is stimulated by the presence of the sex hormone **testosterone** in male embryos and by its absence from female embryos. Testosterone, secreted by the testes and responsible for the distinguishing characteristics of the male, changes the genetic activity of certain cells, most obviously those that form the genitals, but neural cells also respond to it, so certain regions of the embryonic brain also may begin to show sexual dimorphism beginning about 60 days after conception.

Prenatal exposure to *gonadal (sex) hormones* acts to shape male and female brains differently because these hormones activate different genes in the neurons of the two sexes. As described in Section 8-4, experience affects male and female brains differently; therefore genes and experience begin to shape the brain very early in life.

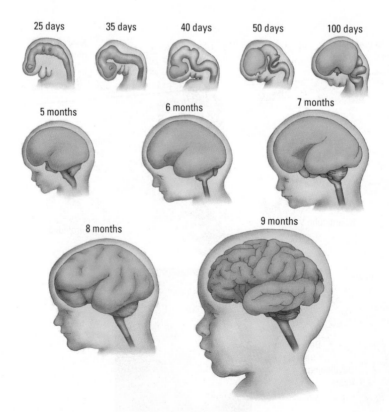

FIGURE 8-7 Prenatal Brain Development
The developing human brain undergoes a series of embryonic and fetal stages. You can identify the forebrain, midbrain, and hindbrain by color (review Figure 8-3) as they develop in the course of gestation. At 6 months, the developing forebrain has enveloped the midbrain structures. Adapted from "The Development of the Brain," by W. M. Cowan, 1979. *Scientific American,* 241(3), p. 116.

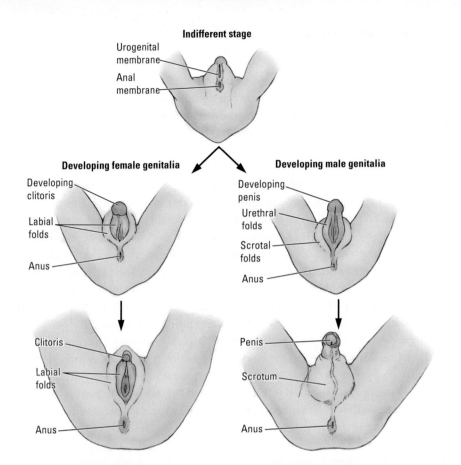

Indifferent stage

Urogenital membrane
Anal membrane

Developing female genitalia

Developing clitoris
Labial folds
Anus

Developing male genitalia

Developing penis
Urethral folds
Scrotal folds
Anus

Clitoris
Labial folds
Anus

Penis
Scrotum
Anus

FIGURE 8-8 Sexual Differentiation in the Human Infant Early in development (indifferent stage), male and female human embryos are identical. (*Left*) In the absence of testosterone, the female structure emerges. (*Right*) In response to testosterone, the genitalia begin to develop into the male structure at about 60 days. Parallel changes take place in the embryonic brain in response to the absence or presence of testosterone.

Origins of Neurons and Glia

The neural tube is the brain's nursery. The **neural stem cells** lining the neural tube have an extensive capacity for self-renewal. When a stem cell divides, it produces two stem cells; one dies and the other lives to divide again. This process repeats again and again throughout a person's lifetime. In an adult, neural stem cells line the ventricles, forming the **subventricular zone.**

If lining the ventricles were all that stem cells did throughout the decades of a human life, they would seem like odd kinds of cells to possess. But stem cells have a function beyond self-renewal: they give rise to so-called **progenitor cells** (*precursor cells*). These progenitor cells also can divide and, as shown in **Figure 8-9,** they eventually produce nondividing cells known as **neuroblasts** and **glioblasts.** In turn, neuroblasts and glioblasts mature into neurons and glia. Neural stem cells, then, are multipotent: they give rise to all the many specialized cell types in the CNS.

Adult stem cells line the subventricular zone and are also located in the spinal cord and in the retina of the eye.

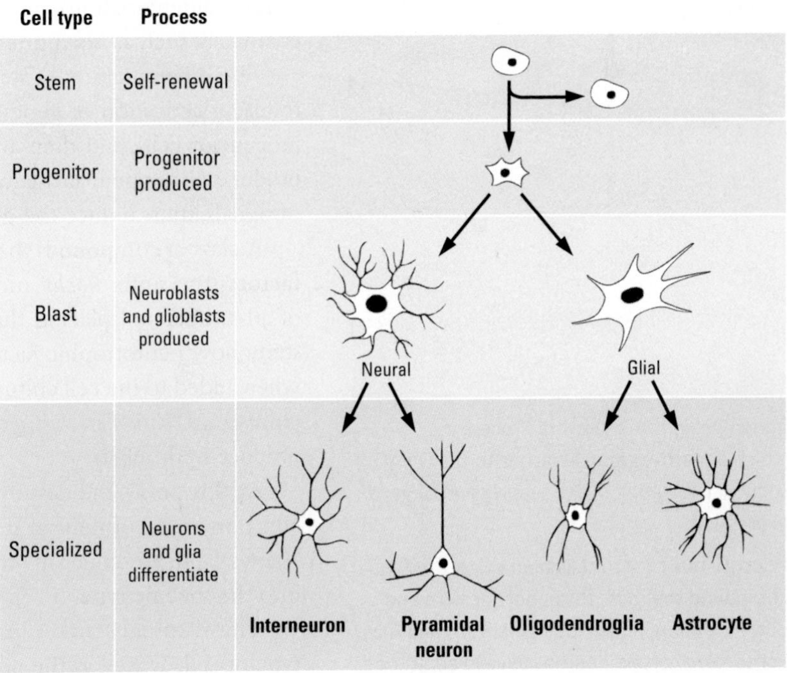

FIGURE 8-9 Origin of Brain Cells Cells in the brain begin as multipotential stem cells, develop into precursor cells, then produce blasts that finally develop into specialized neurons or glia.

Cell type	Process
Stem	Self-renewal
Progenitor	Progenitor produced
Blast	Neuroblasts and glioblasts produced
Specialized	Neurons and glia differentiate

Neural
Glial

Interneuron
Pyramidal neuron
Oligodendroglia
Astrocyte

Sam Weiss and his colleagues (1996) discovered that stem cells remain capable of producing neurons and glia not just into early adulthood but even in an aging brain. This important discovery implies that neurons that die in an adult brain should be replaceable. But neuroscientists do not yet know how to instruct stem cells to carry out this replacement process.

One possibility is to make use of signals that the brain normally uses to control stem-cell production in the adult brain. For example, when female mice are pregnant, the level of the neuropeptide *prolactin* increases, and this increase stimulates the brain to produce more neurons (Shingo et al., 2003). These naturally occurring hormonal signals have been shown to replace lost neurons in laboratory animals that have brain injuries.

How does a stem cell "know" to become a neuron rather than a skin cell? In each cell, certain genes are expressed (turned on) by a signal, and those genes then produce a particular cell type. *Gene expression* means that a formerly dormant gene becomes activated, resulting in the cell making a specific protein. You can easily imagine that certain proteins produce skin cells, whereas other proteins produce neurons.

The specific signals for gene expression are largely unknown, but these signals are probably chemical, and they form the basis of epigenetics. A common epigenetic mechanism that suppresses gene expression during development is **gene methylation,** or *DNA methylation*. Here, a methyl group (CH_3) attaches to the nucleotide base cytosine lying next to guanine on the DNA sequence. It is relatively simple to quantify the amount of gene methylation in different phenotypes, reflecting either an increase or decrease in overall gene expression.

Methylation alters gene expression dramatically during development. Prenatal stress can reduce gene methylation by 10 percent. This means that, relative to unstressed controls, more than 2000 genes (of the more than 20,000 in the human genome) now are expressed (Mychasiuk et al., 2011). Gene expression can be regulated by other epigenetic mechanisms, such as histone modification and mRNA modification, but these mechanisms are more difficult to quantify.

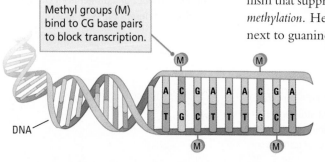

Gene Methylation Section 3-3 describes how researchers are applying the epigenetic code. Figure 3-25 diagrams the mechanisms of histone and mRNA methylation.

Thus, the chemical environment of a cell in the brain is different from that of a cell that forms skin, so different genes in these cells are activated, producing different proteins and different cell types. The different chemical environments needed to trigger this cellular differentiation could be caused by the activity of other neighboring cells or by chemicals, such as hormones, that are transported in the bloodstream.

The differentiation of stem cells into neurons must require a series of signals and the resulting activation of genes. A chemical signal must induce the stem cells to produce progenitor cells, and then another chemical signal must induce the progenitor cells to produce either neuroblasts or glioblasts. Finally, a chemical signal, or perhaps even a set of signals, must induce the genes to make a particular type of neuron.

A class of compounds that signal cells to develop in particular ways are **neurotrophic factors** (the suffix *trophic* means "nourishing"). By removing stem cells from the brain of an animal and placing those cells in solutions that keep them alive, researchers can study how neurotrophic factors function. One compound, *epidermal growth factor* (EGF), when added to the cell culture, stimulates stem cells to produce progenitor cells. Another compound, *basic fibroblast growth factor* (bFGF or FGF-2), stimulates progenitor cells to produce neuroblasts.

At this point, the destiny of a given neuroblast is undetermined. The blast can become any type of neuron if it receives the right chemical signal. The body relies on a "general-purpose neuron" that, when exposed to certain neurotrophic factors, matures into the specific type of cell that the nervous system requires in a particular location.

This flexibility makes brain development simpler than it would be if each different type of cell, as well as the number of cells of each type, had to be specified precisely in

gene (DNA) methylation Process in which a methyl group attaches to the DNA sequence resulting in the suppression of gene expression.

neurotrophic factor A chemical compound that acts to support growth and differentiation in developing neurons and may act to keep certain neurons alive in adulthood.

an organism's genes. In the same way, building a house from "all purpose" two-by-fours that can be cut to any length as needed is easier than specifying in a blueprint a precise number of precut pieces of lumber that can be used only in a certain location.

Growth and Development of Neurons

Human brains require approximately 10 billion (10^{10}) cells to form just the cortex that blankets a single hemisphere. To produce such a large number of cells, about 250,000 neurons must be born per minute at the peak of prenatal brain development. But as **Table 8-1** shows, this rapid formation of neurons (neurogenesis) and glia (gliogenesis) is just the first step in the growth of a brain. These cells must travel to their correct locations (a process called *migration*), they must differentiate into the right type of neuron or glial cell, and the neurons must grow dendrites and axons and subsequently form synapses.

The brain must also prune unnecessary cells and connections, sculpting itself according to the experiences and needs of the particular person. We consider each of these stages in brain development next, focusing on the development of the cerebral cortex, because more is known about cortical development than about the development of any other area of the human brain. However, the principles derived from our examination of the cortex apply to neural growth and development in other brain regions as well.

Neuronal Generation, Migration, and Differentiation

Figure 8-10 shows that neurogenesis is largely complete after about 5 months of gestation. (An important exception is the hippocampus, which continues to develop new neurons throughout life.) Until after full-term birth, however, the fetal brain is especially delicate and extremely vulnerable to injury, *teratogens* (chemicals that cause malformations), and trauma.

	TABLE 8-1 Stages of Brain Development

1	Cell birth (neurogenesis; gliogenesis)
2	Cell migration
3	Cell differentiation
4	Cell maturation (dendrite and axon growth)
5	Synaptogenesis (formation of synapses)
6	Cell death and synaptic pruning
7	Myelogenesis (formation of myelin)

Focus 11-2 describes outcomes that result from cerebral palsy, which results from brain trauma acquired perinatally (at or near birth).

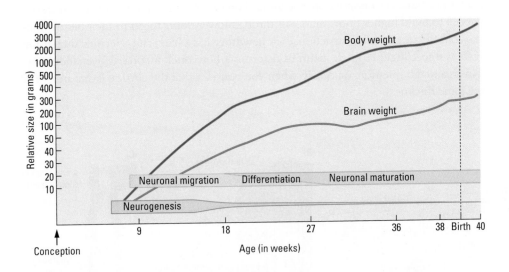

FIGURE 8-10 Prenatal Development of the Human Cerebral Cortex Brain weight and body weight increase rapidly and in parallel. The cortex begins to form about 6 weeks after conception, with neurogenesis largely complete by 20 weeks. Neural migration and cell differentiation begin at about 8 weeks and are largely complete by about 29 weeks. Neuron maturation, including axon and dendrite growth, begins at about 20 weeks and continues until well after birth. Adapted from "Pathogenesis of Late-Acquired Leptomeningeal Heterotopias and Secondary Cortical Alterations: A Golgi Study," by M. Marin-Padilla, 1993, in *Dyslexia and Development: Neurobiological Aspects of Extraordinary Brains* (p. 66), edited by A. M. Galaburda, Cambridge, MA: Harvard University Press.

Apparently, the developing brain can more easily cope with injury earlier, during neurogenesis, than it can during the final stages of cell migration or cell differentiation, when cell maturation begins. One reason may be that, once neurogenesis has slowed, it is very hard to start it up again. If neurogenesis is still progressing at a high rate, more neurons can be made to replace injured ones or perhaps existing neurons can be allocated differently.

Cell migration begins shortly after the first neurons are generated and continues for about 6 weeks in the cerebral cortex and even longer in the hippocampus. At this point, the process of cell differentiation, in which neuroblasts become specific types of neurons,

The hippocampus in the limbic system is critical to memory (Section 14-3) and vulnerable to stress (Section 6-5).

radial glial cell Path-making cell that a migrating neuron follows to its appropriate destination.

begins. Cell differentiation is essentially complete at birth, although neuron maturation, which includes the growth of dendrites, axons, and synapses, goes on for years and, in some parts of the brain, may continue throughout adulthood.

The cortex is organized into layers that are distinctly different from one another in their cellular makeup. How is this arrangement of differentiated areas created during development? Pasko Rakic and his colleagues have been finding answers to this question for more than three decades. Apparently, the subventricular zone contains a primitive map of the cortex that predisposes cells formed in a certain ventricular region to migrate to a certain cortical location. One region of the subventricular zone may produce cells destined to migrate to the visual cortex, whereas another region produces cells destined to migrate to the frontal lobes, for example.

But how do the cells know where these different parts of the cortex are located? They follow a path made by **radial glial cells.** A fiber from each of these path-making cells extends from the subventricular zone to the surface of the cortex, as illustrated in **Figure 8-11**A. The close-up views in Figure 8-11B and C show that neural cells from a given region of the subventricular zone need only follow the glial road and they will end up in the correct location.

As the brain grows, the glial fibers stretch, but they still go to the same place. Figure 8-11B also shows a nonradially migrating cell moving perpendicularly to the radial glial fibers. Although most cortical neurons follow the radial glial fibers, a small number appear to migrate by seeking some type of chemical signal. Researchers do not yet know why these cells function differently.

Figure 2-22 contrasts the six distinct layers of the sensory and motor cortices and their functions.

Cortical layers develop from the inside out, much like adding layers to a tennis ball. The neurons of innermost layer VI migrate to their locations first, followed by those destined for layer V, and so on, as successive waves of neurons pass earlier-arriving neurons to assume progressively more exterior positions in the cortex. The formation of the cortex is a bit like building a house from the ground up until you reach the roof. The materials needed to build higher floors must pass through lower floors to get to their destinations.

To facilitate house construction, each new story has a blueprint-specified dimension, such as 8 feet high. How do neurons determine how thick a cortical layer should be? This is a tough question, especially when you consider that the cortical layers are not all the same thickness.

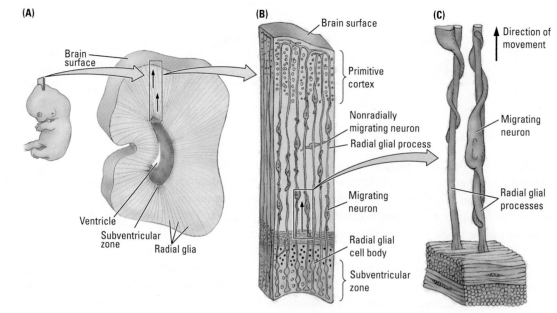

FIGURE 8-11 Neuronal Migration **(A)** Neuroscientists hypothesize that the map for the cortex is represented in the subventricular zone. **(B)** Radial glial fibers extend from the subventricular zone to the cortical surface. **(C)** Neurons migrate along the radial glial fibers, which take them from the protomap in the subventricular zone to the corresponding region in the cortex. Adapted from "Neurons in Rhesus Monkey Cerebral Cortex: Systematic Relation Between Time of Origin and Eventual Disposition," by P. Rakic, 1974, *Science, 183,* p. 425.

Local environmental signals—chemicals produced by other cells—likely influence the way cells form layers in the cortex. These intercellular signals progressively restrict the choice of traits a cell can express, as illustrated in **Figure 8-12.** Thus, the emergence of distinct cell types in the brain result not from the unfolding of a specific genetic program but rather from the interaction of genetic instructions, timing, and signals from other cells in the local environment.

Neuronal Maturation

After neurons migrate to their final destinations and differentiate into specific neuron types, they begin to mature in two ways. Maturing neurons (1) grow dendrites to provide surface area for synapses with other cells and (2) extend their axons to appropriate targets to initiate synapse formation.

Two events take place in the development of a dendrite: dendritic arborization (branching) and the growth of dendritic spines. As illustrated in **Figure 8-13,** dendrites in newborn babies begin as individual processes protruding from the cell body. In the first 2 years of life, dendrites develop increasingly complex extensions that look much like leafless branches of trees visible in winter: they undergo *arborization.* The dendritic branches then begin to form spines, where most synapses on dendrites are located.

Although dendritic development begins prenatally in humans, it continues for a long time after birth, as Figure 8-13 shows. Dendritic growth proceeds at a slow rate, on the order of micrometers per day. Contrast this with the development of axons, which grow on the order of a millimeter per day, about a thousand times as fast.

The disparate developmental rates of axons and dendrites are important because the faster-growing axon can contact its target cell before the cell's dendrites are completely formed. Thus the axon may play a role in dendritic differentiation and ultimately in neuron function—for example, as part of the visual, motor, or language circuitry of the brain. Abnormalities in the rate of maturation can produce abnormalities in patterns of neural connectivity, as explained in Clinical Focus 8-2, "Autism Spectrum Disorder" on page 256.

Axon-appropriate connections may be millimeters or even centimeters away in the developing brain, and the axon must find its way through complex cellular terrain to make them. Axon connections present a significant engineering problem for the

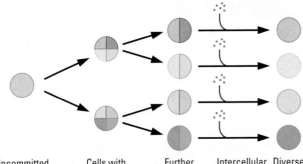

Uncommitted precursor | Cells with some segregation of determinants | Further segregation of determinants | Intercellular environment | Diverse cells

FIGURE 8-12 Cellular Commitment As diagrammed in Figure 8-9, precursor cells have an unlimited cell-fate potential but, as they develop, the interaction of genes, maturation, and environmental influences increasingly steer them toward a particular cell type.

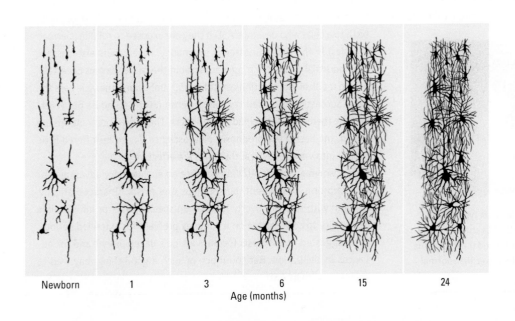

Newborn   1   3   6   15   24
Age (months)

FIGURE 8-13 Neuronal Maturation in Cortical Language Areas In postnatal differentiation of the human cerebral cortex—shown here around Broca's area, which controls speaking—neurons begin with simple dendritic fields that become progressively more complex until a child reaches about 2 years of age. Thus, brain maturation parallels the development of a behavior: the emergence of language. Adapted from *Biological Foundations of Language* (pp. 160–161), by E. Lenneberg, 1967, New York: Wiley.

Autism Spectrum Disorder

Leo Kanner and Hans Asperger first used the term *autism* (from the Greek *autos*, meaning "self") in the 1940s to describe children who seem to live in their own self-created worlds. Some were classified as intellectually disabled; others seemed to retain their intellectual functioning.

The contemporary term, **autism spectrum disorder** (ASD), accommodates this behavioral range to include children with either mild or severe symptoms. Severe symptoms include greatly impaired social interaction, a bizarre and narrow range of interests, marked abnormalities in language and communication, and fixed, repetitive movements.

The autism spectrum includes classic autism and related disorders. *Asperger's syndrome,* for example, is distinguished by an obsessive interest in a single topic or object to the exclusion of nearly any other. Children with Asperger's are socially awkward and also usually have delayed motor-skill development. Children with *Rett syndrome,* almost exclusively girls, are characterized by poor expressive language and clumsy hand use.

The rate of ASD has been rising over the past four decades, from fewer than 1 in 2000 people in 1980 to the 2011 estimate of the Centers for Disease Control that as many as 1 in 88 children has some form of autism. The cause of this increased incidence is uncertain. Suggestions include changes in diagnostic criteria, diagnosis of children at a younger age, and epigenetic influences. Although it knows neither racial nor ethnic nor social boundaries, ASD is four times as prevalent in boys as in girls.

The behavior of many children with ASD is noticeable from birth. To avoid physical contact, these babies arch their backs and pull away from caregivers or grow limp when held. But approximately one-third of children develop normally until, somewhere between 1 and 3 years of age, autism symptoms emerge.

Perhaps the most recognized characteristic of ASD is failure to interact socially. Some children do not relate to other people on any level. The attachments they do form are to inanimate objects, not to people. Some children who develop autism are severely impaired, others learn

to function quite well, and still others on the spectrum have exceptional abilities in music, art, or mathematics.

Another common characteristic of ASD is an extreme insistence on sameness. Children with autism vehemently resist even small modifications to their surroundings or their routines. Objects must always be placed in exactly the same locations, and tasks must always be carried out in precisely the same ways.

Children with ASD also have marked impairments in language development. Many do not speak at all; others repeat words aimlessly with little attempt to communicate or convey meaning. They may exhibit what seem like endlessly repetitive body movements, such as rocking, spinning, or flapping the hands. Some may engage as well in aggressive or self-injurious behaviors.

The brains of children diagnosed with ASD look remarkably normal. One emerging view is that their brains are characterized by unusual neuronal maturation rates. MRI studies show that, at about 6 months of age, the autistic brain's growth rate accelerates to the point that its total volume is 6 percent to 10 percent larger than the brains of typical children.

Abnormal brain volume is especially clear in the amygdala (Nordahl et al., 2012) and in the temporal and frontal lobes, the latter showing greater gray matter volume (see the review by Chen and coworkers, 2011). The subcortical amygdala plays an important role in generating fear, and the social withdrawal component of ASD may be related to the enlarged amygdala.

Accelerated brain growth associated with enlarged regions suggests that connections between cerebral regions are abnormal, which would in turn produce abnormal functioning. What leads to this abnormal brain development? More than 100 genetic differences have been described in children with ASD, so it is clear that no "autism gene" is at work.

The mechanism that translates genetic abnormalities into the autistic brain is unknown but is likely to include epigenetic factors that could be prenatal, postnatal, or both. Women have an increased risk of giving birth to a child who develops ASD if they are exposed to rubella (German measles) in the first trimester of pregnancy. Researchers also suspect that industrial toxins can trigger autism, but the cause remains uncertain.

No medical interventions exist for ASD. Behavioral therapies are the most successful, provided they are intense (20 to 40 hours per week) and the therapists are trained practitioners. The earlier interventions are begun, the better the prognosis. Neuroscience has so far offered little insight into why behavioral therapies are effective.

Autism may appear puzzling because no evolutionary advantage for its symptoms is apparent, but perhaps one exists. Characteristically, children with ASD are overly focused on specific tasks or information. The ability to concentrate on a complex problem for extended periods, it is suggested, is the basis for humankind's development and for advances in civilization. But too much of such a good thing may lead to problems such as ASD.

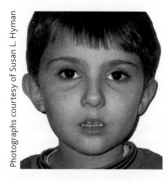
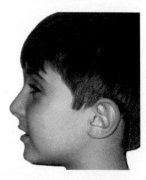

Photographs courtesy of Susan L. Hyman

Children with autism spectrum disorder often look typical, but some physical anomalies do characterize ASD. The corners of the mouth may be low compared with the upper lip, and the tops of the ears may flop over (*left*). The ears may be a bit lower than normal and have an almost square shape (*right*).

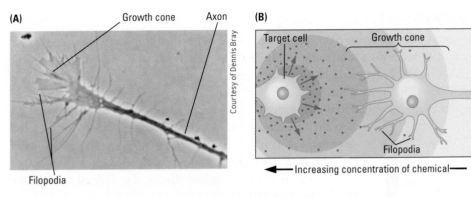

FIGURE 8-14 Seeking a Path **(A)** At the tip of this axon, nurtured in a culture, a growth cone sends out filopodia seeking specific molecules to guide the axon's growth direction. **(B)** Filopodia guide the growth cone toward a target cell that is releasing cell-adhesion or tropic molecules, represented in the drawing by red dots.

developing brain. Such a task could not possibly be specified in a rigid genetic program. Rather, genetic–environmental interaction is at work again as the formation of axonic connections is guided by various molecules that attract or repel the approaching axon tip.

Santiago Ramón y Cajal was the first scientist to describe this developmental process a century ago. He called the growing tips of axons **growth cones. Figure 8-14A** shows that as these growth cones extend, they send out shoots, analogous to fingers reaching out to find a pen on a cluttered desk. When one shoot, a **filopod** (plural *filopodia*), reaches an appropriate target, the others follow.

Growth cones are responsive to two types of cues (Figure 8-14B):

1. **Cell-adhesion molecules** (CAMs) are cell-manufactured molecules that either lie on the target cell's surface or are secreted into the intercellular space. Some CAMs provide a surface to which growth cones can adhere, hence their name; others serve to attract or repel growth cones.

2. **Tropic molecules,** to which growth cones respond, are produced by the targets being sought by the axons (*tropic* means "to move toward"). Tropic molecules essentially tell growth cones to "come over here." It is likely that they also tell other growth cones seeking different targets to "keep away."

Although Cajal predicted tropic molecules more than 100 years ago, the molecules have proved difficult to find. Only one group, **netrins** (from Sanskrit for "to guide"), has been identified in the brain so far. Given the enormous number of connections in the brain and the great complexity in wiring them, many other types of tropic molecules undoubtedly will be found.

Synaptic Development

The number of synapses in the human cerebral cortex is staggering, on the order of 10^{14}, or 100,000 trillion. This huge number could not possibly be determined by a genetic program that assigns each synapse a specific location. Like all stages of brain development, only the general outlines of neural connections in the brain are likely to be genetically predetermined. The vast array of specific synaptic contacts is then guided into place by a variety of environmental cues and signals.

A human fetus displays simple synaptic contacts in the fifth gestational month. By the seventh gestational month, synaptic development on the deepest cortical neurons is extensive. After birth, the number of synapses increases rapidly. In the visual cortex, synaptic density almost doubles between ages 2 months and 4 months and then continues to increase until age 1 year.

autism spectrum disorder (ASD) Range of cognitive symptoms, from mild to severe, that characterize autism; severe symptoms include greatly impaired social interaction, a bizarre and narrow range of interests, marked abnormalities in language and communication, and fixed, repetitive movements.

growth cone Growing tip of an axon.

filopod (pl. filopodia) Process at the end of a developing axon that reaches out to search for a potential target or to sample the intercellular environment.

cell-adhesion molecule (CAM) A chemical molecule to which specific cells can adhere, thus aiding in migration.

tropic molecule Signaling molecule that attracts or repels growth cones.

netrin Member of the only class of tropic molecules yet isolated.

Do not confuse *tropic* (guiding) molecules with the *trophic* (nourishing) molecules, discussed earlier, that support the growth of neurons and their processes.

Cell Death and Synaptic Pruning

Sculptors begin to create their statues with blocks of stone and chisel the unwanted pieces away. The brain does something similar by using cell death and synaptic pruning. The "chisel" in the brain could be a genetic signal, experience, reproductive hormones, and even stress. The effect of the "chisels" in the brain can be seen in changes in cortical thickness over time, as illustrated in **Figure 8-15,** a brain-imaging atlas. The cortex actually becomes measurably thinner in a caudal–rostral (back-to-front) gradient, a process that is probably mostly due to synaptic pruning.

The graph in **Figure 8-16** plots this rise and fall in synaptic density. Pasko Rakic estimated that, at the peak of synapse loss in humans, as many as 100,000 synapses may be lost per second. Synapse elimination is extensive. Peter Huttenlocher (1994) estimated it to be 42 percent of all synapses in the human cortex. We can only wonder what the behavioral consequence of this rapid synaptic loss might be. It is probably no coincidence that children, especially toddlers and adolescents, seem to change moods and behaviors quickly.

How does the brain eliminate excess neurons? The simplest explanation is competition, sometimes referred to as **neural Darwinism.** Charles Darwin believed that one key to evolution is the variation it produces in the traits possessed by a species. The environment then can select certain traits as favorable in aiding survival. From a Darwinian perspective, then, more animals are born than can survive to adulthood, and environmental pressures "weed out" the less-fit ones. Similar pressures cause neural Darwinism.

What exactly causes this cellular weeding out in the brain? It turns out that, when neurons form synapses, they become somewhat dependent on their targets for survival. In fact, deprived of synaptic targets, they eventually die. Neurons die because target cells produce neurotrophic factors that are absorbed by the axon terminals and function to regulate neuronal survival. *Nerve growth factor* (NGF), for example, is made by cortical cells and absorbed by cholinergic neurons in the basal forebrain.

If many neurons are competing for a limited amount of a neurotrophic factor, only some of those neurons can survive. The death of neurons deprived of a neurotrophic factor is different from the cell death caused by injury or disease. When neurons are deprived of a neurotrophic factor, certain genes seem to be expressed, resulting in a message for the cell to die. This programmed process is called **apoptosis.**

Apoptosis accounts for the death of overabundant neurons, but it does not account for the synaptic pruning from cells that survive. In 1976, French neurobiologist Jean-Pierre Changeux proposed a theory for synapse loss that also is based on competition. According to Changeux, synapses persist into adulthood only if they have become members of functional neural networks. If not, they are eventually eliminated from the brain. We can speculate that environmental factors such as hormones, drugs, and experience would influence the formation of active neural circuits and thus influence the processes of synapse stabilization and pruning.

In addition to outright errors in synapse formation that give rise to synaptic pruning, subtler changes in neural circuits may trigger the same process. One such change accounts for the findings of Janet Werker and Richard Tees (1992), who studied the ability of infants to discriminate speech sounds taken from widely disparate languages, such as English, Hindi (from India), and Salish (a Native American language). Their results show that young infants can discriminate speech sounds of different languages without

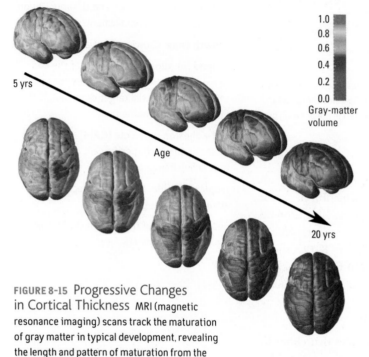

1.0
0.8
0.6
0.4
0.2
0.0
Gray-matter volume

5 yrs

Age

20 yrs

FIGURE 8-15 Progressive Changes in Cortical Thickness MRI (magnetic resonance imaging) scans track the maturation of gray matter in typical development, revealing the length and pattern of maturation from the back of the cortex to the front. Courtesy of Paul Thompson, Kiralee Hayashi, and Arthur Toga, University of California, Los Angeles; and Nitin Gogtay, Jay Gledd, and Judy Rappoport, National Institute of Mental Health.

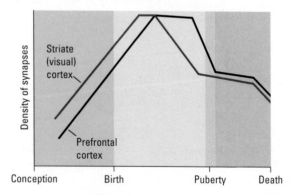

Density of synapses

Striate (visual) cortex

Prefrontal cortex

Conception   Birth   Puberty   Death

FIGURE 8-16 Synapse Formation and Pruning Changes in the relative density of synapses in the human visual cortex and prefrontal cortex (the frontmost part of the frontal lobe) as a function of age. Adapted from "Synaptogenesis in the Neocortex of the Newborn: The Ultimate Frontier for Individuation?" by J.-P. Bourgeois, 2001, in *Handbook of Developmental Cognitive Neuroscience,* edited by C. A. Nelson and M. Luciana, Cambridge, MA: MIT Press.

previous experience, but their ability to do so declines in the first year of life. An explanation for this declining ability is that synapses encoding speech sounds not normally encountered in an infant's daily environment are not active simultaneously with other speech-related synapses. As a result, they are eliminated.

Synaptic pruning may also allow the brain to adapt more flexibly to environmental demands. Human cultures are probably the most diverse and complex environments with which any animal must cope. Perhaps the flexibility in cortical organization achieved by the mechanism of selective synaptic pruning is a necessary precondition for successful development in a cultural environment.

Synaptic pruning may also be a precursor related to different perceptions that people develop about the world. Consider, for example, the obvious differences in "Eastern" and "Western" philosophies about life, religion, and culture. Given the obvious differences to which people in the East and West are exposed as their brains develop, imagine how differently their individual perceptions and cognitions may be. Considered together as a species, however, we humans are far more alike than we are different.

An important and unique characteristic common to all humans is language. As illustrated in Figure 8-15, the cortex generally thins from age 5 to age 20. But there is one exception: the major language regions of the cortex actually show an *increase* in gray matter. **Figure 8-17** contrasts the thinning of other cortical regions with the thickening of language-related regions (O'Hare & Sowell, 2008). Finding a different pattern of cerebral development for brain regions critical in language processing may not be surprising, given the unique role of language in cognition as well as the protracted nature of the language-learning process.

neural Darwinism Hypothesis that the processes of cell death and synaptic pruning are, like natural selection in species, the outcome of competition among neurons for connections and metabolic resources in a neural environment.

apoptosis Cell death that is genetically programmed.

Focus 2-1 considers the relevance of language to increases in human brain size.

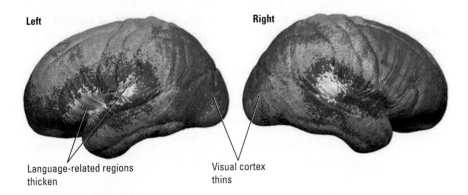

Language-related regions thicken

Visual cortex thins

FIGURE 8-17 Gray-Matter Thickness Brain maps showing the statistical significance of yearly change in cortical-thickness measures taken from MRIs. Color coding represents increasing (blue, white) or decreasing (yellow, green, red) cortical thickness. Red and white areas show statistically significant changes. Adapted from "Mapping Changes in the Human Cortex Throughout the Span of Life," by E. R. Sowell, P. M. Thompson, and A. W. Toga, 2004, *The Neuroscientist 10*, 372–392.

Unique Aspects of Frontal-Lobe Development

The brain-imaging atlas in Figure 8-15 confirms that the frontal lobe is the last brain region to mature. Since it was compiled, neuroscientists have confirmed that frontal-lobe maturation extends far beyond age 20, the boundary used for that atlas. Zdravko Petanjek and colleagues (2011) analyzed synaptic spine density in the dorsolateral area of the frontal lobe in a large sample of human brains. The brains in their sample range in age from deceased newborns to age 91 years at death.

The analysis confirms that dendritic spine density, which is a good measure of the number of excitatory synapses, is two to three times greater in children than in adults and that spine density begins to decrease during puberty. The analysis also shows that the elimination of dendritic spines continues well beyond age 20, stabilizing at the adult level around age 30. Two important correlates attend slow frontal-lobe development:

1. *The frontal lobe is especially sensitive to epigenetic influences* (Kolb et al., 2012). In a study of over 170,000 people, Robert Anda and colleagues (Anda et al., 2006) show that

3-D atlases guide researchers to the precise locations of brain regions (Section 7-1). The dorsolateral prefrontal cortex controls how we select appropriate movements (Sections 12-4 and 15-3).

such aversive childhood experiences (ACEs) as verbal or physical abuse, a family member's addiction, or loss of a parent are predictive of physical and mental health in middle age. People with two or more ACEs, for example, are 50 times more likely to acquire addictions or attempt suicide. Women with two or more ACEs are five times more likely to have suffered a sexual assault by age 50. We hypothesize that early aversive experiences promote these ACE-related susceptibilities by compromising frontal-lobe development. Consider, for example, that most victims of sexual assault know their attacker. Abnormal frontal-lobe development would make a person less likely to judge a situation as dangerous.

2. *The trajectory of frontal lobe development correlates with adult intelligence.* Philip Shaw and his colleagues (2006) used a longitudinal design, administering multiple structural MRIs to participants over time. The results show that it is not the thickness of the frontal cortex in adulthood that predicts IQ score but rather the change in trajectory of cortical thickness **(Figure 8-18)**. The children who score highest in intelligence show the greatest plastic changes in the frontal lobe over time. These changes are likely to reflect strong epigenetic influences.

Glial Development

Astrocytes nourish and support neurons, and oligodendroglia form the myelin that surrounds axons in the spinal cord and brain. See Table 3-1.

Astrocytes and oligodendrocytes begin to develop after most neurogenesis is complete, and continue to develop throughout life. Although CNS axons can function before they are myelinated, normal adult function is attained only after myelination is complete. Consequently, myelination is a useful rough index of cerebral maturation.

In the early 1920s, Paul Flechsig noticed that myelination of the cortex begins just after birth and continues until at least 18 years of age. He also noticed that some cortical regions were myelinated by age 3 to 4 years, whereas others showed virtually no myelination at that time. **Figure 8-19** shows one of Flechsig's cortical maps with areas shaded according to earlier or later myelination.

Flechsig hypothesized that the earliest-myelinating areas control simple movements or sensory analyses, whereas the latest-myelinating areas control the highest mental functions. MRI analyses of myelin development show that the thickness of white matter in the cortex largely does correspond to the progress of myelination, confirming Flechsig's

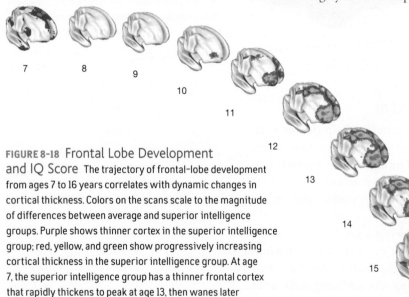

7   8   9   10   11   12   13   14   15   16

FIGURE 8-18 Frontal Lobe Development and IQ Score The trajectory of frontal-lobe development from ages 7 to 16 years correlates with dynamic changes in cortical thickness. Colors on the scans scale to the magnitude of differences between average and superior intelligence groups. Purple shows thinner cortex in the superior intelligence group; red, yellow, and green show progressively increasing cortical thickness in the superior intelligence group. At age 7, the superior intelligence group has a thinner frontal cortex that rapidly thickens to peak at age 13, then wanes later in adolescence. Adapted from "Intellectual Ability and Cortical Development in Children and Adolescents," by P. Shaw, D. Greenstein, J. Lerch, L. Clasen, R. Lenroot, et al., 2006, *Nature, 440,* pp. 666–679.

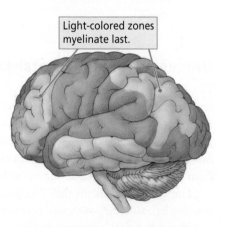

Light-colored zones myelinate last.

FIGURE 8-19 Progress of Myelination The fact that the light-colored zones are very late to myelinate led Flechsig to propose that they are qualitatively different in function from those that mature earlier.

FIGURE 8-20 Sex Differences in Brain Development Mean brain volume by age in years for males (pink) and females (purple). Females show more rapid growth than males, indicated by arrows, reaching maximum overall volume **(A)** and gray-matter volume **(B)** sooner. The decreasing gray matter corresponds to cell and synaptic loss. Increasing white-matter volume **(C)** largely corresponds to myelin development. Adapted from "Sexual Dimorphism of Brain Development Trajectories During Childhood and Adolescence," by R. K. Lenroot, N. Gogtay, D. K. Greenstein, et al., 2007, *NeuroImage 36*, 1065–1073.

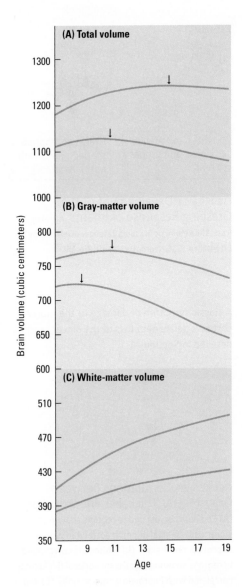

ideas. Myelination continues until at least 20 years of age, as illustrated in **Figure 8-20**, which graphs total brain volume, gray-matter volume, and white-matter volume during brain development.

REVIEW 8-2
Neurobiology of Development

Before you continue, check your understanding.

1. The central nervous system begins as a sheet of cells that folds into the _____.

2. The process of growing neurons is referred to as _____, whereas the process of forming glial cells is known as _____.

3. Growth cones are responsive to two types of cues: _____ and _____.

4. The adolescent period is characterized by two ongoing processes of brain maturation: _____ and _____.

5. What is the functional significance of the prolonged development of the frontal lobe?

Answers appear at the back of the book.

8-3 Correlating Behavior with Nervous-System Development

As particular brain areas mature, a person exhibits behaviors corresponding to the functions of the maturing areas. Stated differently, behaviors cannot emerge until the requisite neural machinery has developed. When that machinery is in place, however, related behaviors develop quickly through stages and are shaped significantly by epigenetic factors.

Researchers have studied these interacting changes in the brain and behavior, especially in regard to the emergence of motor skills, language, and problem solving in children. We now explore development in all three areas.

Motor Behaviors

The development of locomotion skills is easy to observe in human infants. At first, babies are unable to move about independently, but, eventually, they learn to crawl and then to walk.

Other motor skills develop in less obvious but no less systematic ways. Shortly after birth, infants are capable of flexing the joints of an arm in such a way that they can scoop something toward their bodies, and they can direct a hand, such as occurs toward a breast when suckling. Between 1 and 3 months of age, babies also begin to make spontaneous hand and digit movements that consist of almost all the skilled finger movements that they might make as an adult, a kind of "motor babbling."

2 months

| Orients hand toward an object and gropes to hold it. |

4 months

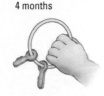

| Grasps appropriately shaped object with entire hand. |

10 months

| Uses pincer grasp with thumb and index finger opposed. |

FIGURE 8-21 Development of the Grasping Response of Infants Adapted from "The Automatic Grasping Response of Infants," by T. E. Twitchell, 1965, *Neuropsychologia, 3,* p. 251.

A classic symptom of damage to the motor cortex is permanent loss of the pincer grasp, detailed in Section 11-1.

FIGURE 8-22 Correlations Between Gray-Matter Thickness and Behavior
(A) Red dots correspond to regions showing significant cortical thinning correlated with improved motor skills. **(B)** White dots correspond to regions showing significant cortical thickening correlated with improved language skills. **(C)** Red dots show regions of decreased cortical thickness correlated with improved vocabulary scores.
(A) and (B) Adapted from "Normal Developmental Changes in Inferior Frontal Gray Matter Are Associated with Improvement in Phonological Processing: A Longitudinal MRI Analysis," by L. H. Lu, C. M. Leonard, P. M. Thompson, E. Kan, J. Jolley, et al., 2007, *Cerebral Cortex, 17,* pp. 1092–1099. (C) Adapted from "Longitudinal Mapping of Cortical Thickness and Brain Growth in Normal Children," by E. R. Sowell, P. M. Thompson, C. M. Leonard, S. E. Welcome, E. Kan, and A. W. Toga, 2004, *Journal of Neuroscience, 24,* pp. 8233–8223.

These movements are then directed toward handling parts of their bodies and their clothes (Wallace & Whishaw, 2003). Only then are reaching movements directed toward objects in space. For example, Tom Twitchell (1965) studied and described how the ability to reach for objects and grasp them progresses in a series of stages, illustrated in **Figure 8-21.**

Between 8 and 11 months, infants' grasping becomes more sophisticated as the "pincer grasp," employing the index finger and the thumb, develops. The pincer grasp is a significant development because it allows babies to make the very precise finger movements needed to manipulate small objects. What we see, then, is a sequence in the development of grasping: first scooping, then grasping with all the fingers, and then grasping by using independent finger movements.

If the development of increasingly well-coordinated grasping depends on the emergence of certain neural machinery, anatomical changes in the brain should accompany the emergence of these motor behaviors. Such changes do take place, especially in the development of dendritic arborizations and in connections between the neocortex and the spinal cord. And a correlation between myelin formation and the ability to grasp has been found (Yakovlev & Lecours, 1967).

In particular, a group of axons from motor-cortex neurons myelinate at about the same time that reaching and grasping with the whole hand develop. Another group of motor-cortex neurons that are known to control finger movements myelinate at about the time that the pincer grasp develops. MRI studies of changes in cortical thickness show that increased motor dexterity is associated with a decrease in cortical thickness in the hand region of the left motor cortex of right-handers (**Figure 8-22**A).

We can now make a simple prediction. If specific motor-cortex neurons are essential for adultlike grasping movements to emerge, the removal of those neurons should make an adult's grasping ability similar to that of a young infant, which is in fact what happens.

Language Development

The acquisition of speech follows a gradual series of developments that has usually progressed significantly by the age of 3 or 4. According to Eric Lenneberg (1967), children reach certain important speech milestones in a fixed sequence and at constant chronological ages. Children start to form a vocabulary by 12 months, and this 5-to-10-word repertoire typically doubles over the next six months. By 2 years, the vocabulary will range from 200 to 300 words that include mostly everyday objects. In another year, the vocabulary approaches 1000 words and begins to include simple sentences. At 6 years, children have a vocabulary of about 2500 words and can understand more than 20,000 words en route to an adult vocabulary of more than 50,000 words.

(A)

(B)

(C)

Although language skills and motor skills generally develop in parallel, the capacity for language depends on more than just the ability to make controlled movements of the mouth, lips, and tongue. Precise movements of the muscles controlling these body parts develop well before children can speak. Furthermore, even when children have sufficient motor skill to articulate most words, their vocabularies do not rocket ahead but rather progress gradually.

A small proportion of children (about 1 percent) have normal intelligence and normal motor-skill development, yet their speech acquisition is markedly delayed. Such children may not begin to speak in phrases until after age 4, despite an apparently normal environment and the absence of any obvious neurological signs of brain damage. Because the timing of speech onset appears universal in the remaining 99 percent of children across all cultures, something different has likely taken place in the brain maturation of a child with late language acquisition. But it is hard to specify what that difference is.

Because the age of language onset is usually between 1 and 2 and language acquisition is largely complete by age 12, the best strategy is to consider how the cortex is different before and after these two milestones. By age 2, cell division and migration are complete in the language zones of the cerebral cortex. The major changes that take place between the ages of 2 and 12 are in neuronal connectivity and the myelination of the speech zones.

Changes in dendritic complexity in these areas are among the most impressive in the brain. Recall from Figure 8-13 that the axons and dendrites of the speech zone called Broca's area are simple at birth but grow dramatically more dense between 15 and 24 months of age. This neuronal development correlates with an equally dramatic change in language ability, given that a baby's vocabulary starts to expand rapidly at about age 2.

We can therefore infer that language development may be constrained, at least in part, by the maturation of language areas in the cortex. Individual differences in the speed of language acquisition may be accounted for by differences in this neural development. Children with early language abilities may have early maturation of the speech zones, whereas children with delayed language onset may have later speech-zone maturation.

Results of MRI studies of the language cortex show that, in contrast with the thinning of the motor cortex associated with enhanced dexterity shown in Figure 8-22A, a *thickening* of the left inferior frontal cortex (roughly Broca's area) is associated with enhanced phonological processing (understanding speech sounds), as shown in Figure 8-22B. The unique association between cortical thickening and phonological processing is not due to a general relation between all language functions and cortical thickening, however. Figure 8-22C shows significant thinning of diffuse cortical regions associated with better vocabulary—regions outside the language areas—and vocabulary is one of the best predictors of general intelligence.

Development of Problem-Solving Ability

The first person to try to identify discrete stages of cognitive development was Swiss psychologist Jean Piaget (1952). He realized that he could infer children's understanding of the world by observing their behavior. For example, a baby who lifts a cloth to retrieve a hidden toy shows an understanding that objects continue to exist even when out of sight. This understanding of *object permanence,* is revealed by the behavior of the infant in the upper row of photographs in **Figure 8-23**.

An absence of understanding also can be seen in children's behavior, as shown by the actions of the 5-year-old girl in the lower row of photographs in Figure 8-23. She was shown two identical beakers with identical volumes of liquid in each and then watched as one beaker's liquid was poured into a taller, narrower beaker. When asked which beaker contained more liquid, she pointed to the taller beaker, not understanding that the amount of liquid remains constant despite the difference in appearance. Children display an understanding of this principle, the *conservation of liquid volume,* at about age 7.

Focus 7-1 describes research on the reactions of newborns to language.

FIGURE 8-23 Two Stages of Cognitive Development (*Top*) The infant shows that she understands object permanence—that things continue to exist when they are out of sight. (*Bottom*) This girl does not yet understand the principle of conservation of liquid volume. Beakers with identical volumes but different shapes seem to her to hold different amounts of liquid.

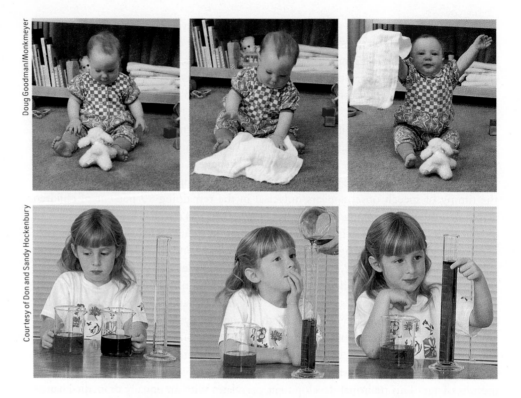

By studying children engaged in such tasks, Piaget concluded that cognitive development is a continuous process. Children's strategies for exploring the world and their understanding of it are constantly changing. These changes are not simply the result of acquiring specific pieces of new knowledge. Rather, at certain points in development, fundamental changes take place in the organization of a child's strategies for learning about the world and for solving problems. With these developing strategies comes new understanding.

Piaget identified four major stages of cognitive development, which are summarized in **Table 8-2**:

- Stage I is the *sensorimotor* period, from birth to about 18 to 24 months of age. During this time, babies learn to differentiate themselves from the external world, come to realize that objects exist even when out of sight, and gain some understanding of cause-and-effect relations.

- Stage II, the *preoperational* period, extends from 2 to 6 years of age. Children gain the ability to form mental representations of things in their world and to represent those things in words and drawings.

- Stage III is the period of *concrete operations*, typically from 7 to 11 years of age. Children are able to mentally manipulate ideas about material (concrete) things such as volumes of liquid, dimensions of objects, and arithmetic problems.

- Stage IV, the period of *formal operations*, is attained sometime after age 11. Children are now able to reason in the abstract, not just in concrete terms.

If we take Piaget's stages as rough approximations of qualitative changes that take place in children's thinking as they grow older, we can ask what neural changes might underlie them. One place to look for brain changes is in the relative rate of brain growth.

After birth, brain and body do not grow uniformly but rather tend to increase in mass during irregularly occurring periods commonly called **growth spurts**. In his

growth spurt Sporadic period of sudden growth that lasts for a finite time.

TABLE 8-2 Piaget's Stages of Cognitive Development

Typical age range	Description of stage	Developmental phenomena
Birth to 18–24 months	*Stage I: Sensorimotor* Experiences the world through senses and actions (looking, touching, mouthing)	Object permanence Stranger anxiety
About 2–6 years	*Stage II: Preoperational* Represents things with words and images but lacks logical reasoning	Pretend play Egocentrism Language development
About 7–11 years	*Stage III: Concrete operational* Thinks logically about concrete events; grasps concrete analogies and performs arithmetical operations	Conservation Mathematical transformations
About 12+ years	*Stage IV: Formal operational* Reasons abstractly	Abstract logic Potential for mature moral reasoning

analysis of brain-weight-to-body-weight ratios, Herman Epstein (1979) found consistent spurts in brain growth between 3 and 10 months (accounting for an increase of 30 percent in brain weight by the age of 1½ years) as well as from the ages of 2 to 4, 6 to 8, 10 to 12, and 14 to 16+ years. The increments in brain weight were from about 5 to 10 percent in each of these 2-year periods.

Brain growth takes place without a concurrent increase in the number of neurons, and so it is most likely due to the growth of glial cells, blood vessels, myelin, and synapses. Although synapses themselves would be unlikely to add much weight to the brain, their growth is accompanied by increased metabolic demands that cause neurons to become larger, new blood vessels to form, and new astrocytes to be produced for neuronal support and nourishment.

We would expect such an increase in the complexity of the cortex to generate more-complex behaviors, and so we might predict significant, perhaps qualitative, changes in cognitive function during each growth spurt. The first four brain-growth spurts identified by Epstein coincide nicely with the four main stages of cognitive development described by Piaget. Such correspondence suggests significant alterations in neural functioning with the onset of each cognitive stage.

At the same time, differences in the rate of brain development, or perhaps in the rate at which specific groups of neurons mature, may account for individual differences in the age at which the various cognitive advances identified by Piaget emerge. Although Piaget did not identify a fifth stage of cognitive development in later adolescence, the presence of a growth spurt then implies one.

A difficulty in linking brain-growth spurts to cognitive development is that growth spurts are superficial measures of changes taking place in the brain. We need to know at a deeper level what neural events are contributing to brain growth and just where they are taking place. A way to find out is to observe children's attempts to solve specific problems that are diagnostic of damage to discrete brain regions in adults. If children perform a particular task poorly, then whatever brain region regulates the performance of that task in adults must not yet be mature in children. Similarly, if children can perform one task but not another, the tasks apparently require different brain structures, and these structures mature at different rates.

William Overman and Jocelyne Bachevalier (Overman et al., 1992) used this logic to study the development of forebrain structures required for learning and memory in

EXPERIMENT 8-1

Question: In what sequence do the forebrain structures required for learning and memory mature?

Procedure

I. Displacement task

II. Nonmatching-to-sample learning task

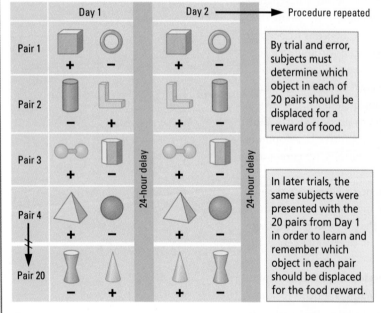

Subject is shown object that can be displaced for a food reward (+).

15 seconds

Preceding object and new object are presented.

Displacement of new object is rewarded with food.

III. Concurrent-discrimination learning task

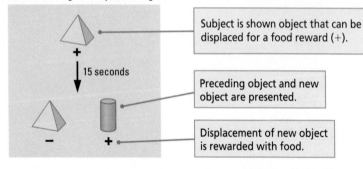

	Day 1		Day 2		→ Procedure repeated
Pair 1	+	−	+	−	By trial and error, subjects must determine which object in each of 20 pairs should be displaced for a reward of food.
Pair 2	−	+	+	−	
Pair 3	+	−	+	−	
Pair 4	+	−	+	−	
Pair 20	−	+	+	−	

24-hour delay

24-hour delay

In later trials, the same subjects were presented with the 20 pairs from Day 1 in order to learn and remember which object in each pair should be displaced for the food reward.

Results

Both humans and monkey infants learn the concurrent-discrimination task at a younger age than the nonmatching-to-sample task.

Conclusion: Neural structures underlying the concurrent-discrimination task mature sooner than those underlying the nonmatching-to-sample task.

Adapted from "Object Recognition Versus Object Discrimination: Comparison Between Human Infants and Infant Monkeys," by W. H. Overman, J. Bachevalier, M. Turner, and A. Peuster, 1992, *Behavioral Neuroscience, 106,* p. 18.

young children and in monkeys. The Procedure section of **Experiment 8-1** shows the three intelligence-test items presented to their participants. The first task was simply to learn to displace an object to obtain a food reward. When participants had learned this *displacement* task, they were trained in two more tasks believed to measure the functioning of the temporal lobes and the basal ganglia, respectively.

In the *nonmatching-to-sample* task, participants were shown an object that they could displace to receive a food reward. After a brief (15-second) delay, two objects were presented: the first object and a novel object. The participants then had to displace the novel object to obtain the food reward. Nonmatching to sample is thought to measure object recognition, which is a function of the temporal lobes. The participant can find the food only by recognizing the original object and not choosing it.

In the third task, *concurrent discrimination,* participants were presented with a pair of objects and had to learn that one object in that pair was always associated with a food reward, whereas the other object was never rewarded. The task was made more difficult by sequentially giving participants 20 different object pairs. Each day, they were presented with one trial per pair. Concurrent discrimination is thought to measure trial-and-error learning of specific object information, which is a function of the basal ganglia.

Adults easily solve both the nonmatching and the concurrent tasks but report that the concurrent task is more difficult because it requires remembering far more information. The key question developmentally is whether there is a difference in the ages at which children (or monkeys) can solve these two tasks.

It turns out that children can solve the concurrent task by about 12 months of age, but not until about 18 months of age can they solve what most adults believe to be the easier nonmatching task. These results imply that the basal ganglia, the critical area for the concurrent-discrimination task, mature more quickly than the temporal lobe, the critical region for the nonmatching-to-sample task.

Caution about Linking Correlation to Causation

Throughout this section we have described research that implies that changes in the brain cause changes in behavior. Neuroscientists assert that, by looking at behavioral development and brain development in parallel, they can make some inferences regarding the causes of behavior. Bear in mind, however, that just because two things correlate (take place together) does not prove that one of them causes the other.

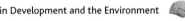

The correlation–causation problem raises red flags in studies of the brain and behavior because research in behavioral neuroscience, by its very nature, is often based on such correlations. Nevertheless, correlational studies, especially in the developmental area, have proved a powerful source of insight into fundamental principles of brain and behavior.

REVIEW 8-3

Correlating Behavior with Nervous-System Development

Before you continue, check your understanding.

1. The last stage in motor development in infants is the ability to make _____.

2. Development of language is correlated with cortical thinning related to _____ and cortical thickening related to _____.

3. Brain growth spurts correlate with _____.

4. Nonmatching-to-sample is believed to measure the function of the _____; concurrent discrimination learning is believed to measure the function of the _____.

5. What is a major challenge in relating changes in brain development to the emergence of behaviors?

Answers appear at the back of the book.

8-4 Brain Development and the Environment

Developing behaviors are shaped not only by the emergence of brain structures but also by each person's environments and experiences. Neuroplasticity suggests that the brain is pliable and can be molded into different forms, at least at the microscopic level. Brains exposed to different environmental experiences are molded in different ways. Culture is an important aspect of the human environment, so culture must help to mold the human brain. We would therefore expect people raised in widely different cultures to acquire differences in brain structure that have lifelong effects on their behavior.

The brain is plastic in response not only to external events but also to events within a person's body, including the effects of hormones, injury, and genetic mutations. The developing brain early in life is especially responsive to these internal factors, which in turn alter the way that the brain reacts to external experiences. In this section, we explore a whole range of external and internal environmental influences on brain development. We start with the question of exactly how experience alters brain structure.

Experience and Cortical Organization

Researchers can study the effects of experience on the brain and behavior by placing laboratory animals in different environments and observing the results. In one of the earliest such studies, Donald Hebb (1947) took a group of young laboratory rats home and let them grow up in his kitchen. A control group grew up in standard laboratory cages at McGill University.

The "home rats" had many experiences that the caged rats did not, including being chased with a broom by Hebb's less-than-enthusiastic wife. Subsequently, Hebb gave both groups a rat-specific "intelligence test" that consisted of learning to solve a series of mazes, collectively known as Hebb–Williams mazes. A sample maze is shown in **Figure 8-24**. The home rats performed far better on these tasks than the caged rats did. Hebb therefore concluded that experience must influence intelligence.

Section 5-4 describes the Hebb synapse, named for his predictions about synaptic plasticity. Section 14-4 elaborates on his contributions to learning theory.

FIGURE 8-24 Hebb–Williams Maze In this version of the maze, a rat is placed in the start box (S) and must learn to find the food in the goal box (G). Investigators can reconfigure the walls of the maze to create new problems. Rats raised in complex environments solve such mazes much quicker than do rats raised in standard laboratory cages.

Focus 5-5 describes some structural changes that neurons undergo as a result of learning.

(A)　　　　**(B)**

Laboratory housed　　　Complex-environment housed

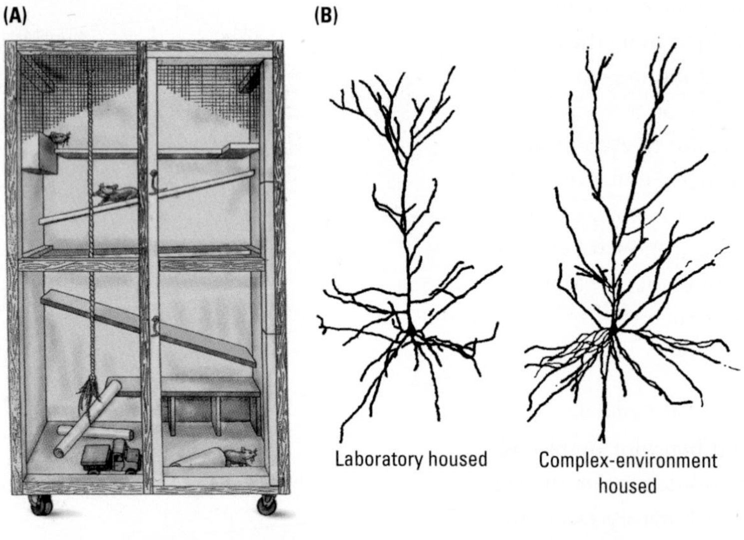

FIGURE 8-25 Enriched Environment, Enhanced Development **(A)** A complex environment for a group of about six rats allows the animals to move about and to interact with one another and with toys that are changed weekly. **(B)** Representative neurons from the parietal cortex of a laboratory-housed rat and a complex-environment-housed rat, which is more complex and has about 25 percent more dendritic space for synapses.

Figure 15-11 shows enhanced nerve-tract connectivity in people with perfect pitch.

On the basis of his research, Hebb reasoned that people reared in "stimulating" environments will maximize their intellectual development, whereas people raised in "impoverished" environments will not reach their intellectual potential. Hebb's reasoning may seem logical, but how do we define the ways that environments may be stimulating or impoverished?

People living in slums, for example, typically have few formal educational resources—decidedly not an "enriched" setting—but that does not mean that the environment offers no cognitive stimulation or challenge. On the contrary, people raised in slums are better adapted for survival in a slum than are people raised in upper-class homes. Does this adaptability make them more intelligent in a certain way?

In contrast, slum dwellers are not likely to be well adapted for college life. This is probably closer to what Hebb had in mind when he referred to a slum environment as limiting intellectual potential. Indeed, Hebb's logic led to the development of preschool television programs, such as *Sesame Street,* that offer enrichment for children who would otherwise have little preschool exposure to reading.

Hebb's studies used complex stimulating environments, but much simpler stimulation can also influence brain development. Tactile stimulation of human infants may be important not only for bonding with caregivers but also for stimulating brain development. For example, tactile stimulation of premature infants in incubators speeds their growth and allows for quicker release from hospital. Laboratory studies show that brushing infant rats for 15 min 3 times per day for the first three weeks of life also speeds up growth and development. The animals show enhanced motor and cognitive skills in adulthood as well. Tactile stimulation also dramatically improves recovery from brain injury incurred early in development.

The idea that early experience can change later behavior seems sensible enough, but we are left to question why experience should make such a difference. One reason is that experience changes the structure of neurons, which is especially evident in the cortex. Neurons in the brains of animals raised in complex environments, such as that shown in **Figure 8-25**A, are larger and have more synapses than do those of animals reared in barren cages (Figure 8-25B). Similarly, three weeks of tactile stimulation increases synapse numbers all over the cortex in adulthood.

Presumably, the increased number of synapses results from increased sensory processing in a complex and stimulating environment. The brains of animals raised in complex settings also display more (and larger) astrocytes. Although complex-rearing studies do not address the effects of human culture directly, predictions about human development are easily made on the basis of their findings. We know that experience can modify the brain, so we can predict that different experiences might modify the brain differently. This effect seems to be the case in language development, as explained in Research Focus 8-3, "Increased Cortical Activation for Second Languages."

Like exposure to language during development, early exposure to music also alters the brain. Perfect (absolute) pitch, the ability to recreate a musical note without external reference, for example, is believed to require early musical training during a period when brain development is most sensitive to this experience. Similarly, adults exposed only to Western music since childhood usually find Eastern music peculiar, even nonmusical, on first encountering it. Both examples demonstrate that neurons in the auditory system are altered by early exposure to music.

Increased Cortical Activation for Second Languages

Most of the world's population is bilingual, but people rarely learn their second language as early as their first. Denise Klein and her colleagues (2006) used both PET and fMRI to determine whether native and second languages differ in cortical activation.

Both languages overlap greatly in neural representation, but when participants are asked to repeat words, the second language shows greater activation in motor regions such as the striatum and cerebellum as well as in the frontal and temporal language regions. The investigators speculate that the second language has greater articulatory demands. These demands correspond to the increased neural involvement in motor as well as language areas, as shown in the adjoining illustration.

Further support for the conclusions from these imaging studies comes from a review of cortical-mapping studies of bilingual patients undergoing neurosurgery. Carlo Giussani and his colleagues (2007) conclude that, although all studies show that language representation is grossly located in the same cortical regions, distinct language-specific areas exist in the language regions of the frontal and temporoparietal regions.

Blue regions show increased activation in motor structures when speaking a second language. **From "Word and Nonword Repetition in Bilingual Subjects: A PET Study," by D. Klein, K. E. Watkins, R. J. Zatorre, and B. Milner, 2006, *Human Brain Mapping, 27*, pp. 153–161.**

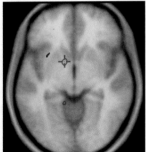
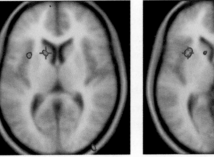
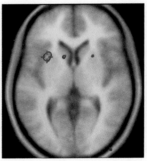
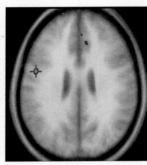

| Ventral striatum | Caudate nucleus | Anterior insula | Ventral premotor |

4.00
3.63
3.25
2.88
2.50

Such loss of plasticity does not mean that the adult human brain grows fixed and unchangeable. Adults' brains are influenced by exposure to new environments and experiences, although more slowly and less extensively than children's brains are. In fact, evidence reveals that experience affects the brain well into old age, which is good news for those of us who are no longer children.

Experience and Neural Connectivity

If experience can influence the structure of the cerebral cortex after a person is born, can it also sculpt the brain prenatally? It can. This prenatal influence of experience is very clearly illustrated in studies of the developing visual system.

Consider the problem of connecting the eyes to the rest of the visual system in development. A simple analogy will help. Imagine that students in a large lecture hall are each viewing the front of the room (the visual field) through a small cardboard tube, such as an empty paper-towel roll. If each student looks directly ahead, he or she will see only a small bit of the total visual field.

Essentially, this is how the photoreceptor cells in the eyes act. Each cell sees only a small bit of the visual field. The problem is to put all the bits together to form a complete picture. To do so, analogously to students sitting side by side, receptors that see adjacent views must send their information to adjacent regions in the various parts of the brain's visual system, such as the midbrain. How do they accomplish this feat?

Roger Sperry (1963) suggested the **chemoaffinity hypothesis,** the idea that specific molecules in different cells in the various regions of the midbrain give each cell a distinctive chemical identity. Each cell, in other words, has an identifiable biochemical label.

chemoaffinity hypothesis Proposal that neurons or their axons and dendrites are drawn toward a signaling chemical that indicates the correct pathway.

Section 9-2 describes the anatomy of the visual system. Midbrain structures are detailed in Figure 2-18.

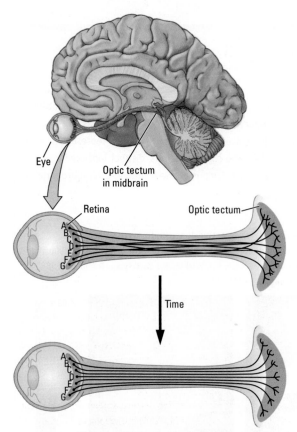

FIGURE 8-26 Chemoaffinity in the Visual System
Neurons A through G project from the retina to the tectum in the midbrain. The activities of adjacent neurons (C and D, say) are more likely to coincide than are the activities of widely separated neurons such as A and G. As a result, adjacent retinal neurons are more likely to establish permanent synapses on the same tectal neurons. By using chemical signals, axons grow to the approximate location in the tectum (*top*). The connections become more precise with the passage of time by the correlated activity (*bottom*).

amblyopia Condition in which vision in one eye is reduced as a result of disuse; usually caused by a failure of the two eyes to point in the same direction.

critical period Developmental "window" during which some event has a long-lasting influence on the brain; often referred to as a *sensitive period*.

imprinting Process that predisposes an animal to form an attachment to objects or animals at a critical period in development.

Presumably, incoming axons seek out a specific chemical, such as the tropic factors discussed earlier, and consequently land in the correct general region of the midbrain.

Many experiments have shown this process to take place prenatally as the eye and brain are developing. But the problem is that chemical affinity "directs" incoming axons only to a general location. To return to our two adjacent retinal cells, how do they now place themselves in the precisely correct position?

Here is where postnatal experience comes in: fine-tuning of neural placement is believed to be activity dependent. Because adjacent receptors tend to be activated at the same time, they tend to form synapses on the same neurons in the midbrain after chemoaffinity has drawn them to a general midbrain region. **Figure 8-26** illustrates this process. Neurons A and G are unlikely to be activated by the same stimulus, so they seldom fire synchronously. Neurons A and B, in contrast, are apt to be activated by the same stimuli, as are B and C. Through this simultaneous activity and with the passage of time, cells eventually line up correctly in the connections that they form.

Now consider what happens to axons coming from different eyes. Although the neural inputs from the two eyes may be active simultaneously, cells in the same eye are more likely to be active together than are cells in different eyes. The net effect is that inputs from the two eyes tend to organize themselves into neural bands, called *columns,* that represent the same region of space in each eye, as shown in **Figure 8-27**. The formation of these segregated cortical columns therefore depends on the patterns of coinciding electrical activity on the incoming axons.

If experience is abnormal—if one eye were covered during a crucial time in development, for example—then the neural connections will not be guided appropriately by experience. As shown at the right in Figure 8-27, the effect of suturing one eye closed has the most disruptive effect on cortical organization in kittens between 30 and 60 days after birth. In a child who has a "lazy eye," visual input from that eye does not contribute to fine-tuning the neural connections as it should. So the details of those connections do not develop normally, much as if the eye had been covered. The resulting loss of sharpness in vision is **amblyopia.**

To summarize, experience modifies the details of neural connections. An organism's genetic blueprint is vague in regard to exactly which connections in the brain go to exactly which neurons. Experience fine-tunes neural connectivity.

Critical Periods for Experience and Brain Development

The preceding examples of perfect pitch and visual connectivity show that for development to be normal, specific sensory experiences occurring at particular times are especially important. A time span during which brain development is most sensitive to a specific experience is called either a **critical period** or a *sensitive period.*

The absence of appropriate sensory experience during a critical period may result in abnormal brain development, leading to abnormal behavior that endures even into adulthood. Our colleague Richard Tees offered an analogy to help explain the concept of critical periods. He pictured the developing animal as a little train traveling past an environmental setting, perhaps the Rocky Mountains. All the windows are closed at the beginning of the journey (prenatal development), but, at particular stages of the trip, the windows in certain cars open, exposing the occupants (different parts of the brain) to the outside world. Some windows open to expose the brain to specific sounds, others to certain smells, others to particular sights, and so on.

This exposure affects the brain's development, and the absence of any exposure through an "open window," severely disturbs that development. As the journey continues, the windows become harder to open until finally they close permanently. This closure does not mean that the brain can no longer change, but changes become much harder to induce.

Now imagine two different trains, one headed through the Rocky Mountains and another, the Orient Express, traveling across Eastern Europe. The "views" from the windows are very different, and the effects on the brain are correspondingly different. In other words, not only is the brain altered by the experiences it has during a critical period, but the particular kinds of experiences encountered matter too.

An extensively studied, related behavior is **imprinting,** a critical period during which an animal learns to restrict its social preferences to a specific class of objects, usually the members of its own species. In birds, such as chickens or waterfowl, the critical period for imprinting is often shortly after hatching. Normally, the first moving object a young hatchling sees is a parent or sibling, so the hatchling's brain appropriately imprints to its own species.

Appropriate imprinting is not inevitable. Konrad Lorenz (1970) demonstrated that, if the first animal or object that baby goslings encounter is a person, the goslings imprint to that person as though he or she were their mother. **Figure 8-28** shows a flock of goslings that imprinted to Lorenz and followed him wherever he went. Incorrect imprinting has long-term consequences for the hatchlings. They often direct their subsequent sexual behavior toward humans. A Barbary dove that had become imprinted to Lorenz directed its courtship toward his hand and even tried to copulate with the hand if it was held in a certain orientation.

Birds can imprint not just to humans but also to inanimate objects, especially moving objects. Chickens have been induced to imprint to a milk bottle sitting on the back of a toy train moving around a track. But the brain is not entirely clueless when it comes to selecting an imprinting target. Given a choice, young chicks will imprint on a real chicken over any other stimulus.

The brain's rapid acquisition and permanent behavioral consequences suggest that, during imprinting, the brain makes a rapid change of some kind, probably a structural change, given the permanence of the new behavior. Gabriel Horn and his colleagues at Cambridge University (1985) tried to identify the changes in the brains of chicks during imprinting. The results of Horn's electron microscopic studies show that the synapses in a specific region of the forebrain enlarge with imprinting. Thus, imprinting seems to be a good model for studying brain plasticity during development, in part because the changes are rapid, related to specific experience, and localized in the brain.

Abnormal Experience and Brain Development

If complex or enriched experiences can stimulate brain growth and influence later behavior, severely restricted experiences seem likely to retard both brain growth and behavior. To study the effects of such restrictions, Donald Hebb and his colleagues (Clarke et al., 1951) placed young Scottish terriers in the dark with as little stimulation as possible and compared their behavior to that of dogs raised in a normal environment.

When the dogs raised in the barren environment were later removed from it, their behavior was very unusual. They showed virtually no reaction to people or other dogs, and they appeared to have lost the sensation of pain. Even sticking pins in them produced no response. When given a dog version of the Hebb–Williams intelligence test for rats, these dogs performed very poorly and were unable to learn some tasks that dogs raised in more stimulating settings were able learn easily.

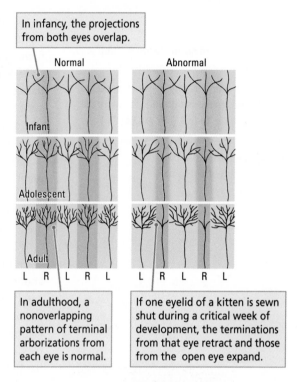

In infancy, the projections from both eyes overlap.

In adulthood, a nonoverlapping pattern of terminal arborizations from each eye is normal.

If one eyelid of a kitten is sewn shut during a critical week of development, the terminations from that eye retract and those from the open eye expand.

FIGURE 8-27 Ocular-Dominance Columns In the normal postnatal development of the cat brain, axons from each eye enter the cortex, where they grow large terminal arborizations. (L, left eye; R, right eye).

Thomas D. McAvoy/*Time* Magazine

FIGURE 8-28 Strength of Imprinting Ethologist Konrad Lorenz followed by goslings that imprinted on him. He was the first object that the geese encountered after hatching, so he became their "mother."

androgen Class of hormones that stimulates or controls masculine characteristics.

masculinization Process by which exposure to androgens (male sex hormones) alters the brain, rendering it identifiably male.

estrogens Variety of sex hormones responsible for the distinguishing characteristics of the female.

Section 6-5 explains the neurobiology of the stress response. Section 16-4 connects mood and reactivity to stress.

Results of subsequent studies show that depriving young animals specifically of visual input or of maternal contact has devastating consequences for their behavioral development and presumably for their brain development. For instance, Austin Riesen (1982) and his colleagues extensively studied animals raised in the dark. They found that even though the animals' eyes still work, they may be functionally blind after early visual deprivation. An absence of visual stimulation results in the atrophy of dendrites on cortical neurons, which is essentially the opposite of the results observed in the brains of animals raised in complex and stimulating environments.

Not only does the absence of specific sensory inputs adversely affect brain development, so do more complex abnormal experiences. In the 1950s, Harry Harlow began the first systematic laboratory studies of analogous deprivation in laboratory animals. Harlow showed that infant monkeys raised without maternal (or paternal) contact develop grossly abnormal intellectual and social behaviors in adulthood.

Harlow separated baby monkeys from their mothers shortly after birth and raised them in individual cages. Perhaps the most stunning effect was that, in adulthood, these animals were totally unable to establish normal relations with other animals. Unfortunately, Harlow did not analyze the brains of the deprived monkeys. We would predict atrophy of cortical neurons, especially in the frontal-lobe regions related to normal social behavior. Harlow's student Stephen Suomi continues to study early experiences in monkeys at the U.S. National Institute of Child Health and Human Development. He has found a wide variety of hormonal and neurological abnormalities among motherless monkeys (see the review by Suomi, 2011).

Children exposed to barren environments or to abuse or neglect are at a serious disadvantage later in life. Proof is the hampered intellectual and motor development displayed by children raised in dreadful circumstances and described in Clinical Focus 8-4, "Romanian Orphans." Although some argue that children can succeed in school and in life if they really want to, abnormal developmental experiences can clearly alter the brain irrevocably. As a society, we cannot be complacent about the environments to which our children are exposed.

Early exposure to stress, including prenatally, also has a major effect on a child's later behavior. Stress can alter the expression of certain genes, such as those related to serotonin reuptake (see Research Focus 8-1). Early alteration in serotonin activity can severely alter how the brain responds to stressful experiences later in life.

Stress early in life may predispose people to develop such behavioral disorders as depression (Sodhi & Sanders-Bush, 2004). Early stress can also leave a lasting imprint on brain structure: the amygdala is enlarged and the hippocampus is reduced in size (Salm et al., 2004). Changes in frontal-lobe anatomy have been associated with the development of depressive and anxiety disorders and may be linked to the epigenetic effects described in Section 8-2.

Hormones and Brain Development

The determination of sex is largely genetic. In mammals, the Y chromosome present in males controls the process by which an undifferentiated primitive gonad develops into testes, as illustrated in Figure 8-8. The testes subsequently secrete testosterone, which stimulates the development of male reproductive organs and, in puberty, the appearance of male secondary sexual characteristics such as facial hair and the deepening of the voice.

Gonadal Hormones and Brain Development

Gonadal hormones also influence neuronal development. Testosterone, the best-known **androgen** (the class of hormones that stimulates or controls masculine characteristics), is released during a brief period in the course of prenatal brain development and subsequently acts to alter the brain much as it alters the sex organs. This process is **masculinization**.

Romanian Orphans

In the 1970s, Romania's Communist regime outlawed all forms of birth control and abortion. The natural result was more than 100,000 unwanted children placed in state-run orphanages. The conditions were appalling.

The children were housed and clothed but given virtually no environmental stimulation. Mostly they were confined to cots with few, if any, playthings and virtually no personal interaction with overworked caregivers who looked after 20 to 25 children at once. Bathing often consisted of being hosed down with cold water.

After the Communist government fell, the outside world intervened. Hundreds of these children were placed in adoptive homes throughout the world, especially in the United States, Canada, and the United Kingdom. Studies of these severely deprived children on arrival in their new homes document malnourishment, chronic respiratory and intestinal infections, and severe developmental impairments.

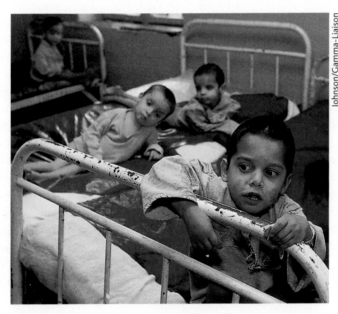

Romanian orphans warehoused in the 1970s and 1980s endured the conditions shown in this photograph. The utter absence of stimulation hampered their normal brain development.

A British study by Michael Rutter (1998) and his colleagues assessed the orphans at two standard deviations below age-matched children for weight, height, and head circumference (taken as a very rough measure of brain size.) Scales of motor and cognitive development assess most of the children in the retarded range.

The improvement these children showed in the first 2 years after placement in their adoptive homes was nothing short of spectacular. Average height and weight advanced to nearly normal, although head circumference remained below normal. Many tested in the normal range of motor and cognitive development. But a significant number were still considered intellectually impaired. What caused these individual differences in recovery from the past deprivation?

The key factor was age at adoption. Children adopted before 6 months of age did significantly better than those adopted later. In a Canadian study by Elenor Ames (1997), Romanian orphans who were adopted before 4 months of age and then tested at age 4½ had an average Stanford–Binet IQ score of 98. Age-matched Canadian controls had an average score of 109. Brain-imaging studies showed that children adopted at an older age had smaller-than-normal brains.

Charles Nelson and his colleagues (Nelson et al., 2007; Marshall et al., 2008; Smyke et al., 2012) analyzed cognitive and social development as well as event-related potential (ERP) measures in a group of children who had remained in Romania. Whether the children had moved to foster homes or remained in institutions, the studies reveal severe abnormalites at about 4 years of age. The age at adoption was again important, but in the Nelson studies the critical age appears to be before 24 months rather than 6 months as in the earlier studies.

The inescapable conclusion is that the human brain may be able to recover from a brief period of extreme deprivation in early infancy, but periods longer than 6 months produce significant developmental abnormalities that cannot be overcome completely. The studies of Romanian orphans make clear that the developing brain requires stimulation for normal development. Although the brain may be able to catch up after a brief deprivation, severe deprivation of more than a few months results in a smaller-than-normal brain and associated behavioral abnormalities, especially in cognitive and social skills.

Testosterone does not affect all body organs or all regions of the brain, but it does affect many brain regions in many different ways. It affects the number of neurons formed in certain brain areas, reduces the number of neurons that die, increases cell growth, increases or reduces dendritic branching, increases or reduces synaptic growth, and regulates the activity of synapses, among other effects.

Estrogens, a variety of sex hormones responsible for the distinguishing characteristics of the female, also probably influence postnatal brain development. Jill Goldstein and her colleagues found sex differences in the volume of cortical regions that are known to have differential levels of receptors for testosterone (androgen receptors) and estrogen,

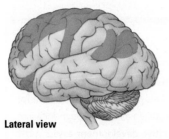
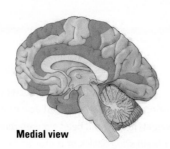

Lateral view · · · **Medial view**

FIGURE 8-29 Sex Differences in Brain Volume Cerebral areas related to sex differences in the distribution of estrogen (purple) and androgen (pink) receptors in the developing brain correspond to areas of relatively larger cerebral volumes in adult women and men. Adapted from "Normal Sexual Dimorphism of the Adult Human Brain Assessed by In Vivo Magnetic Resonance Imaging," by J. M. Goldstein, L. J. Seidman, N. J. Horton, N. Makris, D. N. Kennedy et al., 2001, *Cerebral Cortex, 11*, 490–497

respectively, as diagrammed in **Figure 8-29** (Goldstein et al., 2001). Clearly, hormones alter brain development: a male brain and a female brain are not the same.

Testosterone's effects on brain development were once believed to be unimportant because this hormone was thought to primarily influence brain regions related to sexual behavior but not regions of "higher" functions. This belief is false. Testosterone changes the structure of cells in many regions of the cortex, with diverse behavioral consequences that include influences on cognitive processes. Clear sex differences appear in the rate of brain development (see Figure 8-20).

Reconsider Experiment 8-1, described in Section 8-3. Jocelyne Bachevalier trained infant male and female monkeys in the concurrent-discrimination task, in which the animal has to learn which of two objects in a series of object pairs conceals a food reward. Bachevalier also trained the animals in another task, *object-reversal learning*. The task is to learn that one particular object always conceals a food reward, whereas another object never does. After the animal learns this pattern, the reward contingencies are reversed so that the particular object that has always been rewarded is now never rewarded, and the formerly unrewarded object now conceals the reward. When the animal learns this new pattern, the contingencies are reversed again, and so on, for five reversals.

Bachevalier found that 2½-month-old male monkeys were superior to female monkeys on the object-reversal task, but females did better on the concurrent task. Apparently, the different brain areas required for these two tasks mature at different rates in male and female monkeys. Bachevalier later tested additional male monkeys whose testes had been removed at birth and so were no longer exposed to testosterone. These animals performed like females on the tasks, implying that testosterone was influencing the rate of brain development in areas related to certain cognitive behaviors.

Bachevalier and her colleague William Overman (Overman et al., 1996) then repeated the experiment, this time using as their participants children from 15 to 30 months old. The results were the same: boys were superior at the object-reversal task and girls were superior at the concurrent task. The investigators found no such male–female differences in performance among older children (32–55 months of age). Presumably, by this older age, the brain regions required for each task had matured in both boys and girls. At the earlier age, however, gonadal hormones seemed to be influencing the rate of maturation in certain regions of the brain, just as they had in the baby monkeys.

Lifelong Effects of Gonadal Hormones

Although the biggest effects of gonadal hormones may come during early development, their role is by no means finished in infancy. Gonadal hormones (including both testosterone and estrogen, which is produced in large quantities by the ovaries in females) continue to influence brain structure throughout an animal's life. In fact, removal of the ovaries in middle-aged laboratory rats leads to marked growth of dendrites and the production of more glial cells in the cortex. This finding of widespread neural change in the cortex associated with estrogen loss has implications for treating postmenopausal women with hormone-replacement therapy.

Gonadal hormones also affect how the brain responds to environmental events. For instance, among rats housed in complex environments, males show more dendritic growth in neurons of the visual cortex than do females (Juraska, 1990). In contrast, females housed in this setting show more dendritic growth in the hippocampus than do males. Apparently, the same experience can affect the male and female brain differently owing to the mediating influence of gonadal hormones.

As females and males develop, then, their brains continue to diverge more and more from each other, much like following different forks in a road. After choosing one path, your direction is forever changed as the roads increasingly course farther apart.

To summarize, gonadal hormones alter the basic development of neurons, shape the nature of experience-dependent changes in the brain, and influence the structure of neurons throughout our lifetimes. Those who believe that behavioral differences between males and females are solely the result of environmental experiences must consider these neural effects of sex hormones.

In part, it is true that environmental factors exert a major influence. But one reason they do so may be that male and female brains are different to start with. Even the same events experienced by structurally different brains may lead to different effects on those brains. Evidence now shows that significant experiences, such as prenatal stress, produce markedly different changes in gene expression in the frontal cortex of male and female rats (Mychasiuk et al., 2011).

Another key question related to hormonal influences on brain development is whether any sex differences in brain organization might be independent of hormonal action. In other words, are differences in the action of sex-chromosome genes unrelated to sex hormones? Although little is known about such genetic effects in humans, studies of birds clearly show that genetic effects on brain cells may indeed contribute to sex differentiation.

Songbirds have an especially interesting brain dimorphism: in most species, males sing and females do not. This behavioral difference between the sexes is directly related to a neural birdsong circuit that is present in males but not in females. Robert Agate and his colleagues (2003) studied the brain of a rare strain of zebra finch, a *gynandromorph* that exhibits physical characteristics of both sexes, as shown in **Figure 8-30.**

Genetic analysis shows that cells on one half of the brain and body are genetically female and genetically male on the other half. Both sides of the gynandromorph's body and brain were exposed to the same hormones in the bloodstream during prenatal development. Thus, the effect of male and female genes on the birdsong circuit can be examined to determine how the genes and hormones might interact.

If the sex difference in the birdsong circuit were totally related to the presence of hormones prenatally, then both sides of the brain should be equally masculine or feminine. Agate's results confirm the opposite: the neural song circuit is masculine on the male side of the brain. Only a genetic difference in the brain that was at least partly independent of the effects of the hormones could explain such a structural difference.

Adolescent Onset of Mental Disorders

Adolescence is a time of rapid brain change related both to pubertal hormones and to stressful psychosocial factors. Relationships with parents and peers are among the prime stressors, as is school. Add to this the finding, charted in **Figure 8-31,** that the peak age of onset for any mental disorder is estimated at 14 years (Paus et al., 2008).

Figure 8-31 reveals that age differences in onset exist across disorders. However, anxiety disorders, psychoses (including schizophrenia), bipolar disorder, depression, eating disorders, and substance abuse most commonly emerge by or during adolescence. From an evolutionary perspective, the neurobiological and associated behavioral changes linked with the period we define as adolescence are designed to optimize the brain for challenges that lie ahead in adulthood. But the brain's plasticity in adolescence can also make it vulnerable to psychopathologies that can last for the rest of the individual's life.

Injury and Brain Development

Dating back to the late 1800s, infants and children were generally believed to show better recovery from brain injury than adults. In the 1930s, Donald Hebb studied children with major birth-related injuries to the frontal lobes and found them to have severe and

FIGURE 8-30 Gynandromorph This rare zebra finch has dull female plumage on one side of the body and bright male plumage on the other side.

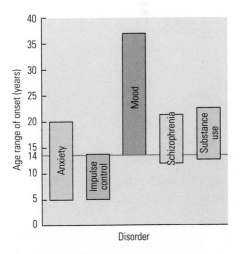

FIGURE 8-31 Emergence of Mental Disorders in Adolescence Adapted from "Why Do So Many Psychiatric Disorders Emerge During Adolescence?" by T. Paus, M. Keshavan, and J. N. Giedd, 2008, *Nature Reviews Neuroscience, 9*, pp. 947–957.

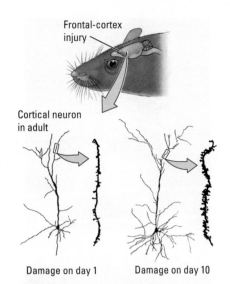

FIGURE 8-32 Time-Dependent Effects
Damage to the rat's frontal cortex on the day of birth leads to cortical neurons with simple dendritic fields and sparse growth of spines in the adult (*left*). In contrast, damage to the frontal cortex at 10 days of age leads to cortical neurons with expanded dendritic fields and denser spines than normal in adults (*right*). Adapted from "Possible Anatomical Basis of Recovery of Function After Neonatal Frontal Lesions in Rats," by B. Kolb and R. Gibb, 1993, *Behavioral Neuroscience, 107,* p. 808.

Focus 6-2 describes the lasting effects of an expectant mother's alcohol consumption on the developing brain of her child.

Focus 7-4 details ADHD and Focus 14-1 details dyslexia.

permanent behavioral abnormalities in adulthood. He concluded that brain damage early in life can alter the subsequent development of the rest of the brain and may be worse than injury later in life.

To what extent have other studies confirmed Hebb's conclusion? Few anatomical studies of humans with early brain injuries exist, but we can make some general predictions from studying laboratory animals. In general, early brain injuries do produce abnormal brains, especially at certain critical periods in development.

For humans, the worst time appears to be in the last half of the intrauterine period and the first couple of months after birth. Rats and cats that suffer injuries at a comparable time have significantly smaller brains than normal, and their cortical neurons show a generalized atrophy relative to normal brains, as illustrated in **Figure 8-32.** Behaviorally, these animals appear cognitively deficient in a wide range of skills.

The effect of injury to the developing brain is not always devastating. For example, researchers have known for more than 100 years that children with brain injuries in the first couple of years after birth almost never have the severe language disturbances common to adults with equivalent injuries. Animal studies help explain why.

Whereas damage to the rat brain in the developmental period comparable to the last few months of gestation in humans produces widespread cortical atrophy, damage at a time in the development of the rat brain roughly comparable to age 6 months to 2 years in humans actually produces more dendritic development in rats, as seen on the right in Figure 8-32. Furthermore, these animals show dramatic recovery of functions, implying that the brain has a capacity during development to compensate for injury. Parallel studies in cats have shown extensive reorganization of cortical-to-cortical connections after early injury to the visual cortex (see the review by Payne and Lomber, 2001).

Drugs and Brain Development

The U.S. National Institute on Drug Abuse (NIDA; 1998) estimates that 25 percent of all live births in the United States today are exposed to nicotine in utero. Similar statistics on alcohol consumption by pregnant mothers are not available, but the effects in the etiology of fetal alcohol effects are well documented. Even low doses of commonly prescribed drugs, including antidepressants and antipsychotics, appear to alter prenatal neuron development in the prefrontal cortex. It manifests after birth in abnormalities in behaviors controlled by the affected regions (see the review by Halliwell and coworkers, 2009).

NIDA also estimates that 5.5 percent of all expectant mothers, approximately 221,000 pregnant women each year in the United States, use an illicit drug at least once in the course of their pregnancies. Among pregnant teenagers aged 15 to 17, that statistic climbs to 16%, or about 14,000 women. And what about caffeine? More than likely most children were exposed to caffeine (from coffee, tea, cola and energy drinks, and chocolate) in utero.

The precise effects of prenatal drug intake on brain development are poorly understood, but the overall conclusion from current knowledge is that children with prenatal exposure to a variety of psychoactive drugs have an increased likelihood of later drug use (Malanga and Kosofsky, 2003). Many experts suggest that, although, again, poorly studied, childhood disorders—learning disabilities and attention-deficit-hyperactivity disorder (ADHD) are examples—may be related to prenatal exposure to drugs such as nicotine or caffeine or both. Carl Malanga and Barry Kosofsky note poignantly that "society at large does not yet fully appreciate the impact that prenatal drug exposure can have on the lives of its children."

Other Kinds of Abnormal Brain Development

The nervous system need not be damaged by external forces to develop abnormally. For instance, many genetic abnormalities are believed to result in abnormalities in the development and, ultimately, the structure, of the brain. *Spina bifida,* a condition in which the genetic blueprint goes awry and the neural tube does not close completely, leads to an incompletely formed spinal cord. After birth, children with spina bifida usually have serious motor problems.

Imagine what would happen if some genetic abnormality caused the front end of the neural tube not to close properly. Because the front end of the neural tube forms the brain (see Figure 8-5), this failure would result in gross abnormalities in brain development. Such a condition exists and is known as **anencephaly.** Infants affected by anencephaly die soon after birth.

Abnormal brain development can be much subtler than anencephaly. For example, if cells do not migrate to their correct locations, and if these mispositioned cells do not subsequently die, they can disrupt brain function and may lead to disorders ranging from seizures to schizophrenia (see review by Guerrini et al., 2007). In a variety of conditions, neurons fail to differentiate normally. In certain cases, the neurons fail to produce long dendrites or spines. As a result, connectivity in the brain is abnormal, leading to developmental disabilities.

The opposite condition also is possible: neurons continue to make dendrites and form connections with other cells to the point at which these neurons become extraordinarily large. The functional consequences of all the newly formed connections can be devastating. Excitatory synapses in the wrong location effectively short-circuit a neuron's function.

A curious consequence of abnormal brain development is that behavioral effects may emerge only as the brain matures and the maturing regions begin to play a greater role in behavior. This consequence is especially true of frontal-lobe injuries. The frontal lobes continue to develop into early adulthood (see Figure 8-18), and often not until adolescence do the effects of frontal-lobe abnormalities begin to be noticed.

Schizophrenia is a disease characterized by its slow development, usually not becoming obvious until late adolescence. Clinical Focus 8-5, "Schizophrenia" on page 278 relates the progress and possible origin of the disease.

Developmental Disability

Impaired cognitive functioning accompanies abnormal brain development. Impairment may range from mild, allowing an almost normal lifestyle, to severe, requiring constant care. As summarized in **Table 8-3,** such developmental disability can result from chronic malnutrition, genetic abnormalities such as Down syndrome, hormonal abnormalities, brain injury, or neurological disease. Different causes produce different abnormalities in brain organization, but the critical similarity across all types of developmental disability is that the brain is not normal.

anencephaly Failure of the forebrain to develop.

Section 16-4 describes many abnormalities of the schizophrenic brain and Section 5-3 the possible relation between schizophrenia and excessive dopamine or serotonin activity.

Figure 3-22 illustrates trisomy, the chromosomal abnormality that causes Down syndrome.

TABLE 8-3 Causes of Developmental Disability

Cause	Example mechanism	Example condition
Genetic abnormality	Error of metabolism Chromosomal abnormality	Phenylketonuria (PKU) Down syndrome
Abnormal embryonic development	Exposure to a toxin	Fetal alcohol syndrome (FAS)
Prenatal disease	Infection	Rubella (German measles) Retardation
Birth trauma	Anoxia (oxygen deprivation)	Cerebral palsy
Malnutrition	Abnormal brain development	Kwashiorkor
Environmental abnormality	Sensory deprivation	Children in Romanian orphanages

Schizophrenia

When Mrs. T. was 16 years old, she began to experience her first symptom of schizophrenia: a profound feeling that people were staring at her. These bouts of self-consciousness soon forced her to end her public piano performances. Her self-consciousness led to withdrawal, then to fearful delusions that others were speaking about her behind her back, and finally to suspicions that they were plotting to harm her.

At first Mrs. T.'s illness was intermittent, and the return of her intelligence, warmth, and ambition between episodes allowed her to complete several years of college, to marry, and to rear three children. She had to enter a hospital for the first time at age 28, after the birth of her third child, when she began to hallucinate.

Now, at 45, Mrs. T. is never entirely well. She has seen dinosaurs on the street and live animals in her refrigerator. While hallucinating, she speaks and writes in an incoherent, but almost poetic way. At other times, she is more lucid, but even then the voices she hears sometimes lead her to do dangerous things, such as driving very fast down the highway in the middle of the night, dressed only in a nightgown. . . . At other times and without any apparent stimulus, Mrs. T. has bizarre visual hallucinations. For example, she saw cherubs in the grocery store. These experiences leave her preoccupied, confused, and frightened, unable to perform such everyday tasks as cooking or playing the piano. (Gershon & Rieder, 1992, p. 127)

It has always been easier to identify schizophrenic behavior than to define what schizophrenia is. Perhaps the one universally accepted criterion for its diagnosis is the absence of other neurological disturbances or affective (mood) disorders that could cause a person to lose touch with reality—a definition by default.

Symptoms of schizophrenia are heterogeneous, suggesting that biological abnormalities vary from person to person. Most patients appear to stay at a fairly stable level after the first few years of displaying schizophrenic symptoms, with little evidence of a decline in neuropsychological functioning. Symptoms come and go, much as for Mrs. T., but the severity is relatively constant after the first few episodes.

Numerous studies have investigated the brains of schizophrenia patients, both in autopsies and in MRI and CT scans. Although the results vary, most neuroscientists agree that the brains of people who develop

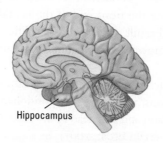

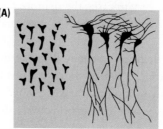

Pyramidal-cell orientation in the hippocampus of **(A)** a normal brain and **(B)** a schizophrenic brain. Adapted from "A Neurohistologic Correlate of Schizophrenia," by J. A. Kovelman and A. B. Scheibel, 1984, *Biological Psychiatry, 19*, p. 1613.

Hippocampus

(A) Organized (normal) pyramidal neurons   **(B)** Disorganized (schizophrenic) pyramidal neurons

schizophrenia weigh less than normal and have enlarged ventricles. Research findings also suggest that schizophrenic brains have smaller frontal lobes (or at least a reduction in the number of neurons in the prefrontal cortex) and thinner parahippocampal gyri.

Joyce Kovelman and Arnold Scheibel (1984) found abnormalities in the orientation of hippocampal neurons in schizophrenics. Rather than the consistently parallel orientation of neurons in this region characteristic of normal brains, schizophrenic brains have a more haphazard organization, as shown in the accompanying drawings.

Evidence is increasing that the abnormalities observed in schizophrenic brains are associated with disturbances of brain development. William Bunney and his colleagues (1997) suggested that at least a subgroup of schizophrenia sufferers experience either environmental insults or some type of abnormal gene activity in the fourth to sixth month of fetal development.

These events are thought to result in abnormal cortical development, particularly in the frontal lobes. Later in adolescence, as the frontal lobes approach maturity, the person begins to experience symptoms deriving from this abnormal prenatal development.

Dominique Purpura (1974) conducted one of the few systematic investigations of developmentally disabled children's brains. Purpura used Golgi stain to examine the neurons of children who had died from accident or disease unrelated to the nervous system. When he examined the brains of children with various forms of intellectual disability, he found that dendrite growth was stunted and the spines were very sparse relative to dendrites from children of typical intelligence, as illustrated in **Figure 8-33**.

The simpler structure of these neurons is probably indicative of a marked reduction in the number of connections in the brain, which presumably caused the developmental disability. Variation in both the nature and the extent of neuronal abnormality in different children would lead to different behavioral syndromes.

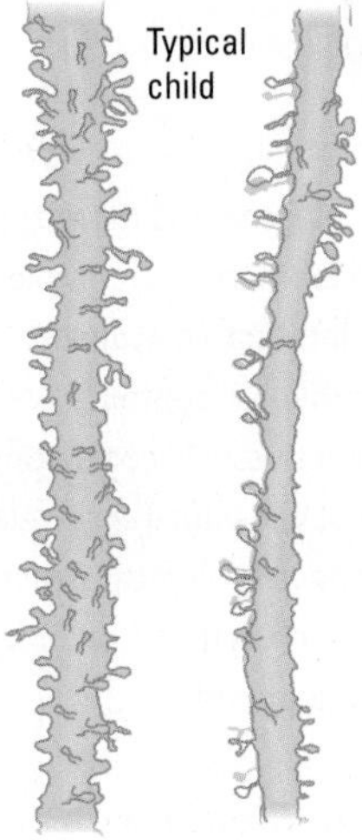

Typical child Developmentally disabled child

REVIEW 8-4

Brain Development and the Environment

Before you continue, check your understanding.

1. The idea that specific molecules exist in different cells in the various regions of the midbrain, giving each cell a distinctive chemical identity, is known as the _____.

2. Abnormal visual stimulation to one eye during early development can lead to a loss of acuity, known as _____.

3. The hormone _____ masculinizes the brain during development.

4. The brain's sensitivity to experience is highest during _____.

5. Why do so many mental disorders appear during adolescence?

Answers appear at the back of the book.

FIGURE 8-33 Neuronal Contrast
Representative dendritic branches from cortical neurons in a child of typical intelligence (*left*) and a developmentally disabled child (*right*), whose neurons are thinner and have far fewer spines.
Adapted from "Dendritic Spine 'Dysgenesis' and Mental Retardation," by D. P. Purpura, 1974, *Science, 186,* p. 1127.

8-5 How Does Any of Us Develop a Normal Brain?

When we consider the brain's complexity, the less-than-precise process of brain development, and the myriad factors that can influence development, we are left to marvel at how so many of us end up with brains that pass for normal. We all must have had neurons that migrated to wrong locations, made incorrect connections, were exposed to viruses or other harmful substances. If the brain were as fragile as it might seem, to end up with a normal brain would be almost impossible.

Apparently, animals have evolved a substantial capacity to repair minor abnormalities in brain development. Most people have developed in the range that we call "normal" because the human brain's plasticity and regenerative powers overcome minor developmental deviations. By initially overproducing neurons and synapses, the brain has the capacity to correct any errors that might have arisen accidentally.

These same plastic properties later allow us to cope with the ravages of aging. Neurons are dying throughout our lifetimes. By age 60, we ought to be able to see significant effects of all this cell loss, especially considering the cumulative results of exposure to environmental toxins, drugs, traumatic brain injuries, and other neural insults. But this is not what happens.

Although some teenagers may not believe it, relatively few 60-year-olds are demented. By most criteria, the 60-year-old who has been intellectually active throughout adulthood is likely to be much wiser than the 18-year-old whose brain has lost relatively few neurons. A 60-year old chess player will have a record of many more chess matches from which to draw game strategies than does an 18-year old, for example.

Clearly, some mechanism must enable us to compensate for loss and minor injury to our brain cells. This capacity for plasticity and change, for learning and adapting, is a most important characteristic of the human brain during development and throughout life.

We return to learning, memory, and neuroplasticity in Chapter 14.

SUMMARY

8-1 Three Perspectives on Brain Development

Nervous-system development is more than the simple unfolding of a genetic blueprint. Development is a complex dance of genetic and environmental events that interact to sculpt the brain to fit within a particular cultural and environmental context. We can approach this dance from three different perspectives: (1) correlating nervous-system development with the emergence of behavior, (2) correlating the emergence of behavior and the likely maturing neurological structures, and (3) observing the relations among factors, such as music or injury, that influence both behavioral and neurological development.

8-2 Neurobiology of Development

Human brain maturation is a long process, lasting as late as age 30. Neurons, the units of brain function, develop a phenotype, migrate, and, as their processes elaborate, establish connections with other neurons even before birth. The developing brain produces many more neurons and connections than it needs and then prunes back in toddlerhood and again in adolescence and early adulthood to a stable level maintained by some neurogenesis throughout the life span. Experiences throughout development can trigger epigenetic mechanisms, such as gene methylation, that alter gene expression.

8-3 Correlating Behavior with Nervous-System Development

Throughout the world, across the cultural spectrum, from newborn to adult, we all develop through similar behavioral stages. As infants develop physically, motor behaviors emerge in a predictable sequence from gross, poorly directed movements toward objects to controlled pincer grasps to pick up objects as small as pencils by about 11 months. Cognitive behaviors also develop through a series of testable stages of logic and problem solving. Beginning with Jean Piaget, researchers have identified and characterized four or more distinct stages of cognitive development, each of which can be identified by specific behavioral tests.

Behaviors emerge as the neural systems that produce them develop. The hierarchical relation between brain structure and brain function can be inferred by matching the median timetables of neurodevelopment with observed behavior. Motor behaviors emerge in synchrony with the maturation of motor circuits in the cerebral cortex, basal ganglia, and cerebellum, as well as in the connections from these areas to the spinal cord. Similar correlations between emerging behaviors and neuronal development can be seen in the maturation of cognitive behavior, as circuits in the frontal and temporal lobes mature in early adulthood.

8-4 Brain Development and the Environment

The brain is most plastic during its development, and the structure of neurons and their connections can be molded by various factors throughout development. The brain's sensitivity to factors such as external events, quality of environment, tactile stimulation, drugs, gonadal hormones, stress, and injury varies over time. At critical periods in the course of development, different brain regions are particularly sensitive to different events.

Perturbations of the brain in the course of development from, say, anoxia, trauma, or toxins can alter brain development significantly; result in severe behavioral abnormalities, including intellectual disability; and may be related to such disorders as ASD. Many other behavioral disorders emerge in adolescence, a time of prolonged frontal lobe change.

8-5 How Does Any of Us Develop a Normal Brain?

The brain has a substantial capacity to repair or correct minor abnormalities, allowing most people to develop normal behavioral repertoires and to maintain brain function throughout life.

KEY TERMS

amblyopia, p. 270

androgen, p. 272

anencephaly, p. 277

apoptosis, p. 259

autism spectrum disorder (ASD), p. 257

cell-adhesion molecule (CAM), p. 257

chemoaffinity hypothesis, p. 269

critical period, p. 270

estrogens, p. 272

filopod (pl. filopodia), p. 257

gene (DNA) methylation, p. 252

glioblast, p. 251

growth cone, p. 257

growth spurt, p. 264

imprinting, p. 270

masculinization, p. 272

netrin, p. 257

neural Darwinism, p. 259

neural plate, p. 248

neural stem cell, p. 251

neural tube, p. 248

neuroblast, p. 251

neurotrophic factor, p. 252

progenitor (precursor) cell, p. 251

radial glial cell, p. 254

sudden infant death syndrome (SIDS), p. 246

subventricular zone, p. 251

testosterone, p. 251

tropic molecule, p. 257

 Please refer to the Companion Web Site at www.worthpublishers.com/kolbintro4e *for Interactive Exercises and Quizzes.*

CLINICAL FOCUS 9-1 MIGRAINES AND A CASE OF BLINDSIGHT

9-1 NATURE OF SENSATION AND PERCEPTION

SENSORY RECEPTORS

NEURAL RELAYS

SENSORY CODING AND REPRESENTATION

PERCEPTION

9-2 FUNCTIONAL ANATOMY OF THE VISUAL SYSTEM

STRUCTURE OF THE RETINA

THE BASICS VISIBLE LIGHT AND THE STRUCTURE OF THE EYE

PHOTORECEPTORS

CLINICAL FOCUS 9-2 VISUAL ILLUMINANCE

RETINAL-NEURON TYPES

VISUAL PATHWAYS

DORSAL AND VENTRAL VISUAL STREAMS

9-3 LOCATION IN THE VISUAL WORLD

CODING LOCATION IN THE RETINA

LOCATION IN THE LATERAL GENICULATE NUCLEUS AND REGION V1

VISUAL CORPUS CALLOSUM

9-4 NEURONAL ACTIVITY

SEEING SHAPE

SEEING COLOR

RESEARCH FOCUS 9-3 COLOR-DEFICIENT VISION

NEURONAL ACTIVITY IN THE DORSAL STREAM

9-5 THE VISUAL BRAIN IN ACTION

INJURY TO THE VISUAL PATHWAY LEADING TO THE CORTEX

INJURY TO THE "WHAT" PATHWAY

CLINICAL FOCUS 9-4 CARBON MONOXIDE POISONING

INJURY TO THE "HOW" PATHWAY

CHAPTER

9

How Do We Sense, Perceive, and See the World?

281

Migraines and a Case of Blindsight

D. B.'s recurring headaches began at about age 14. A visual aura warned of a headache's approach: an oval-shaped area of flashing (scintillating) light appeared just to the left of center in his field of vision. Over the next few minutes, the oval enlarged. After about 15 minutes, the flashing light vanished, and D. B. was blind in the region of the oval.

D. B. described the oval as an opaque white area surrounded by a rim of color. A headache on the right side of his head followed and could persist for as long as 48 hours. D. B. usually fell asleep before that much time elapsed. When he awakened, the headache was gone and his vision was normal again.

D. B. suffered from severe *migraine,* a recurrent headache usually localized to one side of the head. Migraines vary in severity, frequency, and duration and are often accompanied by nausea and vomiting. Migraine is perhaps the most common of all neurological disorders, afflicting some 5 to 20 percent of the population at some time in their lives.

Auras may be auditory or tactile as well as visual and may result in an inability to move or to talk. After an aura passes, most people suffer a severe headache caused by a dilation of cerebral blood vessels. The headache is usually localized to one side of the head, just as the aura is on one side of the visual field. Left untreated, migraines may last for hours or even days.

D. B.'s attacks continued at intervals of about 6 weeks for 10 years. After one attack, he was left with a small blind spot, or *scotoma,* illustrated in the accompanying photographs. When D. B. was 26 years old, a neurologist found that a collection of abnormal blood vessels at the back of his right occipital lobe was causing the migraine attacks—a most unusual cause.

By the time D. B. was 30, the migraines began to interfere with his family life, social life, and job. No drug treatment was effective, so D. B. had the malformed blood vessels surgically removed. The operation relieved his pain and generally improved his life, but a part of his right occipital lobe, deprived of blood, had died. D. B. was blind in the left half of his visual field: as he looks at the world through either eye, he is unable to see anything to the left of the midline.

Lawrence Weizkrantz (1986) made a remarkable discovery about D. B.'s blindness. D. B. could not identify objects in his blind area but could very accurately "guess" if a light had blinked on there and even where the light was located. Apparently, D. B.'s brain knew when a light blinked and where it appeared. This phenomenon is called *blindsight.* D. B.'s brain knew more than he was consciously aware of.

X = Fixation point

As a migraine scotoma develops, a person looking at the small white "✕" in the photograph at the far left would first see a small patch of lines. This striped area continues growing outward, leaving an opaque area (scotoma) where the stripes had been, almost completely blocking the visual field within 15 to 20 minutes. Normal vision returns shortly thereafter.

What applies to D. B. applies to everyone. You are consciously aware of only part of the visual information that your brain is processing. This selective awareness is an important working principle behind human sensation and perception. Weizkrantz, a world-renowned visual neuroscientist at Oxford University, detected it in the visual system only because of D. B.'s injury.

We are also unaware of much sensory processing that takes place in the pathways for hearing, touch, taste, and smell. All our senses function to convert energy into neural activity that has meaning for us. We begin this chapter with a general summary of sensation and perception that explores how this energy conversion takes place.

Vision is the main topic of this chapter, and hearing of Chapter 10. Section 11-4 covers the body senses and balance. Section 12-2 explains smell and taste.

- For taste and olfaction, various chemical molecules carried by the air or contained in food fit themselves into receptors of various shapes to activate action potentials.

Were our visual receptors somewhat different, we would be able to see in the ultraviolet as well as the visible parts of the electromagnetic spectrum, as honeybees and butterflies can. The receptors of the human ear respond to a wide range of sound waves, but elephants and bats can hear and produce sounds far below and above the range in which humans hear. In fact, in comparison with those of other animals, human sensory abilities are rather average.

Even our pet dogs have "superhuman" powers: they can detect odors, hear the low-range sounds of elephants, and see in the dark. We can hold up only our superior color vision. Thus, for each species and its individual members, sensory systems filter the sensory world to produce an idiosyncratic representation of reality.

Receptive Fields

Every sensory-receptor organ and cell has a **receptive field,** a specific part of the world to which it responds. If you fix your eyes on a point directly in front of you, for example, what you see of the world is the scope of your eyes' receptive field. If you close one eye, the visual world shrinks, and what the remaining eye sees is the receptive field for that eye.

Each photoreceptor cell within the eye points in a slightly different direction and so has a unique receptive field. You can grasp the conceptual utility of the receptive field by considering that the brain uses information from the receptive field of each sensory receptor not only to identify sensory information but also to contrast the information that each receptor field is providing.

Receptive fields not only sample sensory information but also help locate sensory events in space. Because the receptive fields of adjacent sensory receptors may overlap, their relatively different responses to events help us localize sensations. The spatial dimensions of sensory information produce cortical patterns and maps of the sensory world that form, for each of us, our sensory reality.

Our sensory systems are organized to tell us both what is happening in the world around us and what we ourselves are doing. When you move, you change the perceived properties of objects in the world, and you experience sensations that have little to do with the external world. When we run, visual stimuli appear to stream by us, a stimulus configuration called **optic flow.** When you move past a sound source, you hear an **auditory flow,** changes in the intensity of the sound that take place because of your changing location. Optic flow and auditory flow are useful in telling us how fast we are going, whether we are going in a straight line or up or down, and whether it is we who are moving or an object in the world that is moving.

Try this experiment. Slowly move your hand back and forth before your eyes and gradually increase the speed of the movement. Your hand will eventually get a little blurry because your eye movements are not quick enough to follow its movement. Now keep your hand still and move your head back and forth. The image of the hand remains clear. When receptors in the inner ear inform your visual system that your head is moving, the visual system compensates for the head movements, and you observe the hand as a stationary image.

Receptor Density and Sensitivity

Receptor density is particularly important in determining the sensitivity of a sensory system. For example, the tactile receptors on the fingers are numerous compared with those on the arm. This difference explains why the fingers can discriminate touch remarkably well and the arm cannot do so as well.

Our sensory systems use different receptors to enhance sensitivity under different conditions. For example, the visual system uses different sets of receptors to respond to

---

An animal's perception of the world depends on the complexity and organization of its nervous system.

receptive field Region of the visual world that stimulates a receptor cell or neuron.

optic flow Streaming of visual stimuli that accompanies an observer's forward movement through space.

auditory flow Change in sound heard as a person moves past a sound source or as a sound source moves past a person.

The ability to lose conscious visual perception while retaining unconscious vision, as D.B. did, leads us to the chapter's central question: How do we "see" the world? We begin by overviewing the visual system's anatomy. Next we consider the connections between the eyes and the sections of the brain that process visual information.

Turning to the perceptual experience of sight, we focus on how neurons respond to visual input and enable the brain to perceive different features, such as color, shape, and movement. At the chapter's end, we explore vision's culmination: understanding what we see. How do we infuse light energy with meaning, to grasp the meaning of written words or to see the beauty in a painting?

9-1 Nature of Sensation and Perception

We may believe that we see, hear, touch, smell, and taste real things in a real world. In fact, the only input our brains receive from the "real" world is a series of action potentials passed along the neurons of our various sensory pathways. Although we experience visual and body sensations as being fundamentally different from one another, the nerve impulses coursing in the neurons of these two sensory systems are very similar, as are the neurons themselves.

Neuroscientists understand how nerves can turn energy, such as light waves, into nerve impulses. They also know the pathways those nerve impulses take to reach the brain. But they do not know how we end up perceiving one set of nerve impulses as what the world looks like and another set as what makes us move.

How much of what you know comes through your senses? Taken at face value, this question seems reasonable. At the same time, we realize that our senses can deceive us—that two people can look at the same optical illusion and see very different images, that a person dreaming does not normally think that the dream images are real, that you often do not think that a picture of you looks like you. Many scientists think that much of what we know comes to us through our senses, but they also think that our brains actively transform sensory information into forms that help us to adapt and are thus behaviorally useful.

Our sensory systems appear to be extremely diverse. At first glance, vision, audition, body senses, taste, and olfaction appear to have little in common. Although our perceptions and behaviors in relation to them are very different, each sensory system is organized on a similar hierarchical plan. We now consider the features common to the sensory systems—receptors, neural relays between receptor and neocortex, sensory coding and representation, and perception.

Sensory Receptors

Sensory receptors are specialized cells that transduce (convert) sensory energy—light, for example—into neural activity. If we put flour into a sieve and shake it, the more finely milled particles will fall through the holes, whereas the coarser particles and lumps will not. Sensory receptors are designed to respond only to a narrow band of energy—analogous to particles of certain sizes—within each modality's energy spectrum.

Each sensory system's receptors are specialized to filter a different form of energy:

- For vision, light energy is converted into chemical energy in the photoreceptors of the retina, and the chemical energy is in turn converted into action potentials.

- In the auditory system, air-pressure waves are converted first into mechanical energy, which activates the auditory receptors that produce action potentials.

- In the somatosensory system, mechanical energy activates receptor cells that are sensitive to touch, pressure, or pain. Somatosensory receptors in turn generate action potentials.

Vision begins in the photoreceptor cells, the rods and cones shown here. Section 9-2 details how they work.

SPL/Photo Researchers

light and color. Color photoreceptors are small and densely packed to make sensitive color discriminations in bright light. The receptors for black–white vision are larger and more scattered, but their sensitivity to light—say, a lighted match at a distance of 2 miles on a dark night—is truly remarkable.

Differences in the density of sensory receptors determine the special abilities of many animals—the excellent olfactory ability in dogs and the excellent tactile ability in the digits of raccoons. Variations in receptor density in the human auditory-receptor organ may explain such abilities as perfect pitch displayed by some musicians.

Neural Relays

Inasmuch as receptors are common to each sensory system, all receptors connect to the cortex through a sequence of three or four intervening neurons. The visual and somatosensory systems have three, for example, and the auditory system has four. Information can be modified at different stages in the relay, allowing the sensory system to mediate different responses.

Neural relays also allow sensory systems to interact. There is no straight-through, point-to-point correspondence between one neural relay and the next; rather, there is a recoding of activity in each successive relay. Sensory neural relays are central to the hierarchy of motor responses in the brain.

Some of the three to four relays in each sensory system are in the spinal cord, others are in the brainstem, and still others are in the neocortex. At each level, the relay allows a sensory system to produce relevant actions that define the hierarchy of our motor behavior. For example, the first relay for pain receptors in the spinal cord is related to reflexes that produce withdrawal movements of a body part from a painful stimulus. Thus, even after section of the spinal cord from the brain, a limb still withdraws from a painful stimulus.

A dramatic effect of sensory interaction is the visual modification of sound. If a speech syllable such as "ba" is played by a recorder to a listener who at the same time is observing someone whose lips are articulating the syllable "ga," the listener hears not the actual sound "ba" but the articulated sound "da." The viewed lip movements modify the auditory perception of the listener.

The potency of this interaction effect highlights the fact that our perception of speech sounds is influenced by the facial gestures of a speaker. As described by Roy Hamilton and his coworkers (2006), the synchrony of gestures and sounds is an important aspect of our acquisition of language. A difficulty for people learning a foreign language can be related to the difficulty they have in blending a speaker's articulation movements with the sounds the speaker produces.

Sensory Coding and Representation

After it has been transduced, all sensory information from all sensory systems is encoded by action potentials that travel along peripheral nerves in the somatic nervous system until they enter the spinal cord or brain and, from there, on nerve tracts within the central nervous system. Every bundle carries the same kind of signal. How do action potentials encode different sensations (how does vision differ from touch), and how do they encode the features of particular sensations (how does purple differ from blue)?

Parts of these questions seem easy to answer; others pose a fundamental challenge to neuroscience. The presence of a stimulus can be encoded by an increase or decrease in the discharge rate of a neuron, and the amount of increase or decrease can encode stimulus intensity. As detailed in Section 9-4, qualitative visual changes, such as from red to green, can be encoded by activity in different neurons or even by different levels of discharge in the same neuron (for example, more activity might signify redder and less activity greener).

Section 10-4 explains how we perceive music.

Recall the principle from Section 2-6: brain systems are organized both hierarchically and in parallel.

Recall the principle from Section 2-6: the nervous system works by juxtaposing excitation and inhibition.

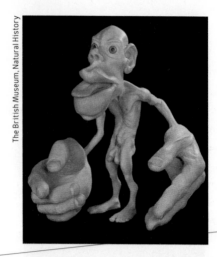

This curious figure reflects the topographic map in the sensorimotor cortex. Relatively larger areas control body parts we use to make the most skilled movements. See Sections 11-2 and 11-6.

(A)

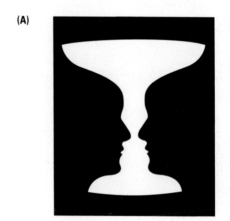

(B)

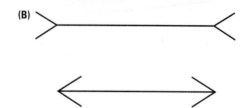

FIGURE 9-1 Perceptual Illusions **(A)** Edgar Rubin's ambiguous reversible image can be perceived as a vase or as two faces. **(B)** The top line of the Müller–Lyer illusion appears longer than the bottom line because of the contextual cues provided by the arrowheads. Both lines are equal in length.

What is less clear is how we perceive such sensations as touch, sound, and smell as different from one another. Part of the explanation is that different sensations are processed in distinct regions of the cortex. Also, we learn through experience to distinguish them. Third, each sensory system has a preferential link with certain kinds of reflex movements, constituting a distinct wiring that helps keep each system distinct at all levels of neural organization. For example, pain stimuli produce withdrawal responses, and fine-touch and pressure stimuli produce approach responses.

The distinctions among the sensory systems, however, are not always clear: some people hear in color or identify smells by how the smells sound to them. This mixing of the senses is called *synesthesia.* Anyone who has shivered when hearing a piece of music or at the noise that chalk or fingernails can make on a blackboard has "felt" sound.

In most mammals, the neocortex represents the sensory field of each modality—vision, hearing, touch, smell, or taste—as a spatially organized neural representation of the external world. This **topographic map** is a neural–spatial representation of the body or of the areas of the sensory world perceived by a sensory organ. All mammals have at least one primary cortical area for each sensory system. Additional areas are usually referred to as *secondary* because most of the information that reaches these areas is relayed through the primary area. Each additional representation is probably dedicated to encoding one specific aspect of the sensory modality. For vision, different additional representational areas may take part in perceiving color, movement, and form.

Perception

There is far more to **sensation** than the simple registration of physical stimuli from the environment by the sensory organs. Compared with the richness of actual sensation, our description of sensory neuroanatomy and function is bound to seem sterile. Part of the reason for the disparity is that our sensory impressions are affected by the contexts in which they take place, by our emotional states, and by our past experiences. All these factors contribute to **perception,** the subjective experience of sensation—how we interpret what we sense. Perception, rather than sensation, is of most interest to neuropsychologists.

Clear proof that perception is more than sensation lies in the fact that different people transform the same sensory stimulation into totally different perceptions. The classic demonstration is an ambiguous image such as the well-known Rubin's vase shown in **Figure 9-1A.** This image may be perceived either as a vase or as two faces. If you fix your eyes on the center of the picture, the two perceptions will alternate, even though the sensory stimulation remains constant.

The Müller–Lyer illusion in Figure 9-1B demonstrates the influence of context on perception. We perceive the top line as longer than the bottom line, even though both are exactly the same length. The contextual cues (the arrowheads) alter the perception of each line's length. Such ambiguous images and illusions demonstrate the workings of complex perceptual phenomena and are an enlightening source of insight into our cognitive processes.

REVIEW 9-1
Nature of Sensation and Perception

Before you continue, check your understanding.

1. _____ are energy filters that transduce incoming physical energy into neural activity.

2. _____ fields locate sensory events. Receptor _____ determines sensitivity to sensory stimulation.

3. We distinguish one sensory modality from another by its _____.

4. Sensation registers physical stimuli from the environment by the sensory organs. Perception is the _____.

5. How is the anatomical organization similar for each sense?

Answers appear at the back of the book.

9-2 Functional Anatomy of the Visual System

Vision is our primary sensory experience. Far more of the human brain is dedicated to vision than to any other sense. Understanding the visual system's organization is therefore key to understanding human brain function. To build this understanding, we begin by following the routes that visual information takes to the brain and within it. This exercise is a bit like traveling a road to discover where it goes.

Structure of the Retina

Light energy travels from the outside world, through the pupil, and into the eye, where it strikes a light-sensitive surface, the **retina,** at the back of the eye **(Figure 9-2).** From this stimulation of **photoreceptor** cells on the retina, we begin to create a visual world. If you are familiar with the properties of the electromagnetic spectrum and with the structure of the eye, read on. To refresh your knowledge of these topics, read "The Basics: Visible Light and the Structure of the Eye" on pages 288–289 before you continue.

Figure 9-2 includes a photograph of the retina, which is composed of photoreceptors beneath a layer of neurons connected to them. The neurons lie in front of the photoreceptor cells, but they do not prevent incoming light from being absorbed by those receptors because the neurons are transparent and the photoreceptors are extremely sensitive to light.

Together, the photoreceptor cells and the neurons of the retina perform some amazing functions. They translate light into action potentials, discriminate wavelengths so that we can distinguish colors, and work in a range of light intensities from very bright to very dim. These cells afford visual precision sufficient for us to see a human hair lying on the page of this book from a distance of 18 inches.

As in a camera, the image of objects projected onto the retina is upside down and backward. This flip-flopped orientation poses no problem for the brain. Remember that the brain is creating the outside world, so it does not really care how the image is oriented initially. In fact, the brain can make adjustments regardless of the orientation of the images that it receives.

If you were to put on glasses that invert visual images and keep the glasses on for several days, the world would first appear upside down but then would suddenly appear right side up again because your brain would correct the distortion (Held, 1968). Curiously, when you removed the glasses, the world would temporarily seem upside down once more, because your brain at first would be unaware that you had tricked it one more time. Eventually, though, your brain would solve this puzzle too, and the world would flip back into the right orientation.

FIGURE 9-2 Central Focus This cross section through the retina shows the depression at the fovea—also shown in the scanning electron micrograph at bottom left—where photoreceptors are packed most densely and where our vision is clearest.

topographic map Spatially organized neural representation of the external world.

sensation Registration of physical stimuli from the environment by the sensory organs.

perception Subjective interpretation of sensations by the brain.

retina Light-sensitive surface at the back of the eye consisting of neurons and photoreceptor cells.

photoreceptor Specialized type of retinal cell that transduces light into neural activity

Virtually all neurons in the retina are insensitive to light and so are unaffected by light passing through them.

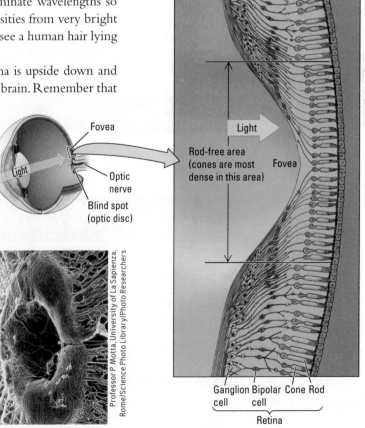

Professor P. Motta, University of La Sapienza, Rome/Science Photo Library/Photo Researchers

Visible Light and the Structure of the Eye

The brain's visual system analyzes visible light—the part of the electromagnetic (EM) spectrum that the human eye evolved to capture and focus.

Light: The Stimulus for Vision

Light can enter the eye directly from a source that produces it—a lamp, for example, or the sun—or indirectly after reflecting off a surface—the pages of a book, for example, or the surface of water. Not all light waves are the same length, and only a sliver of the EM spectrum is visible to us, as illustrated in "Electromagnetic Spectrum." If our photoreceptors could detect light in the shorter ultraviolet or longer infrared range of wavelengths, we would see additional colors.

Structure of the Eye

The range of light visible to humans is constrained not by the properties of light waves but rather by the properties of our visual receptors. How do photoreceptor cells in the retina absorb light energy and initiate the processes leading to vision? "How the Eye Works" illustrates the structure of the eye and shows how its design captures and focuses light.

Optical Errors of Refraction

A web of muscles adjusts the shape of the eye's lens to bend light to greater or lesser degrees, which allows near or far images to be focused on the retina. When images are not properly focused, we require corrective lenses.

The eye, like a camera, works correctly only when sufficient light passes through the lens and is focused on the receptor surface—the retina in the eye or the light-sensitive material in the camera. If the focal point of the light falls slightly in front of the receptor surface or slightly behind it, a refractive error causes objects to appear blurry. Refractive errors in the eye are of two basic types, diagrammed in "Refractive Errors."

Myopia (nearsightedness) afflicts about 50 percent of young people in the developed world. Hyperopia (farsightedness) is a less-common refractive error, but as people age, the lens loses its elasticity and consequently becomes unable to refract light from nearby objects correctly. This form of hyperopia, called presbyopia ("old sightedness"), is so common that you rarely find people older than 50 who do not need glasses to see up close, especially for reading.

It is also common to see young children wearing corrective lenses. The incidence of myopia in the United States has doubled in the past 40 years to about 42 percent. It is even higher in Northern Europe (50 percent) and Asia (50 percent to 80 percent). Two factors probably account for the increase.

First, more young people are attending school longer and thus are doing more "close work," especially reading. Close work strains the eye muscles. Second, people are spending less and less time outdoors in bright light. Bright light makes the pupil contract, which improves visual

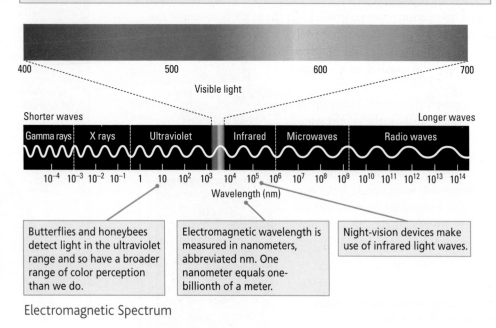

Electromagnetic Spectrum

How the Eye Works

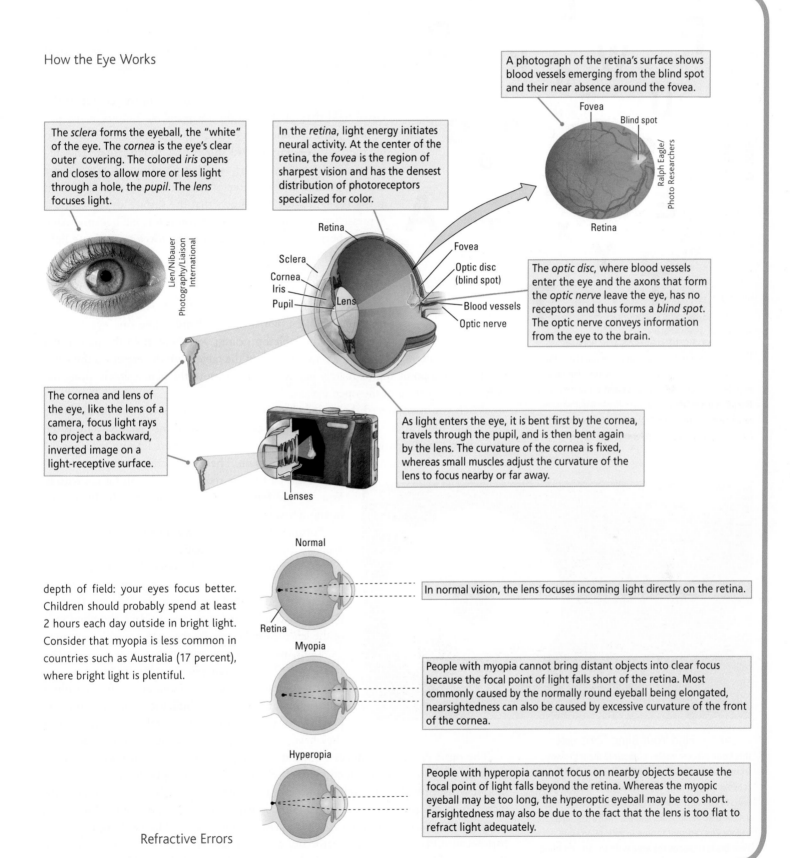

A photograph of the retina's surface shows blood vessels emerging from the blind spot and their near absence around the fovea.

Fovea

Blind spot

Ralph Eagle/ Photo Researchers

Retina

The *sclera* forms the eyeball, the "white" of the eye. The *cornea* is the eye's clear outer covering. The colored *iris* opens and closes to allow more or less light through a hole, the *pupil*. The *lens* focuses light.

In the *retina*, light energy initiates neural activity. At the center of the retina, the *fovea* is the region of sharpest vision and has the densest distribution of photoreceptors specialized for color.

Lien/Nibauer Photography/Liaison International

Retina
Sclera
Cornea
Iris
Pupil
Lens

Fovea
Optic disc (blind spot)
Blood vessels
Optic nerve

The *optic disc*, where blood vessels enter the eye and the axons that form the *optic nerve* leave the eye, has no receptors and thus forms a *blind spot*. The optic nerve conveys information from the eye to the brain.

The cornea and lens of the eye, like the lens of a camera, focus light rays to project a backward, inverted image on a light-receptive surface.

Lenses

As light enters the eye, it is bent first by the cornea, travels through the pupil, and is then bent again by the lens. The curvature of the cornea is fixed, whereas small muscles adjust the curvature of the lens to focus nearby or far away.

depth of field: your eyes focus better. Children should probably spend at least 2 hours each day outside in bright light. Consider that myopia is less common in countries such as Australia (17 percent), where bright light is plentiful.

Normal

Retina

In normal vision, the lens focuses incoming light directly on the retina.

Myopia

People with myopia cannot bring distant objects into clear focus because the focal point of light falls short of the retina. Most commonly caused by the normally round eyeball being elongated, nearsightedness can also be caused by excessive curvature of the front of the cornea.

Hyperopia

People with hyperopia cannot focus on nearby objects because the focal point of light falls beyond the retina. Whereas the myopic eyeball may be too long, the hyperoptic eyeball may be too short. Farsightedness may also be due to the fact that the lens is too flat to refract light adequately.

Refractive Errors

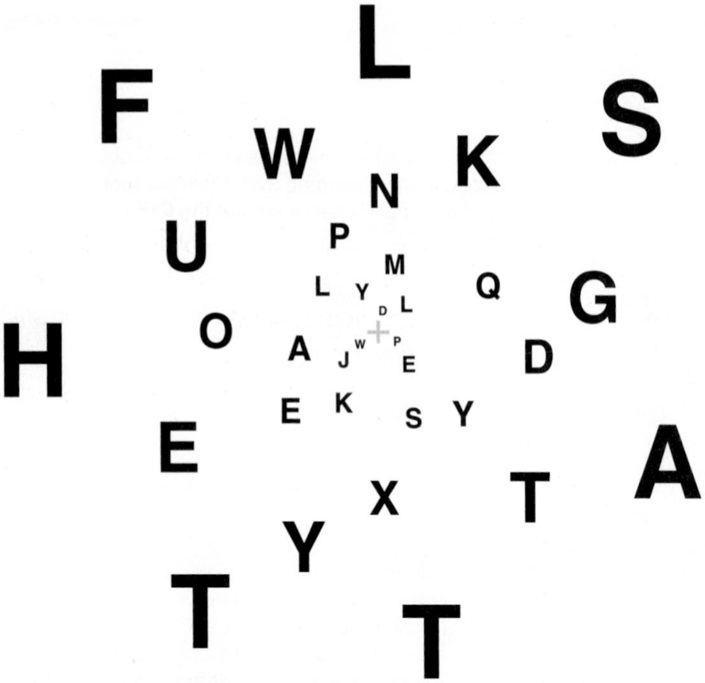

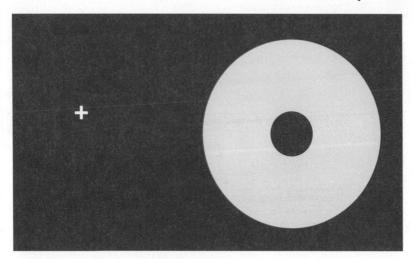

FIGURE 9-3 Acuity Across the Visual Field Focus on the plus sign in the middle of the chart to demonstrate the relative sizes of letters legible in the central field of vision compared with the peripheral field. From *Basic Vision: An Introduction to Visual Perception*, by R. Snowden, P. Thompson, and T. Troscianko, 2008, Oxford: Oxford University Press.

Fovea

Try this experiment. Focus on the print at the left edge of this page. The words will be clearly legible. Now, while holding your eyes still, try to read the words on the right side of the page. It will be very difficult, even impossible, even though you can see that words are there.

The lesson is that our vision is better in the center of the visual field than at the margins, or *periphery*. Letters at the periphery must be much larger than those in the center for us to see them as well. **Figure 9-3** shows how much larger. The difference is due partly to the fact that photoreceptors are more densely packed at the center of the retina, in a region known as the **fovea**. Figure 9-2 shows that the surface of the retina is depressed at the fovea. This depression is formed because many of the fibers of the optic nerve skirt the fovea to facilitate light access to its receptors.

Blind Spot

Now try another experiment. Stand with your head over a tabletop and hold a pencil in your hand. Close one eye. Stare at the edge of the tabletop nearest you. Now hold the pencil in a horizontal position and move it along the edge of the table, with the eraser on the table. Beginning at a point approximately below your nose, move the pencil slowly along the table in the direction of the open eye.

When you have moved the pencil about 6 inches, the eraser will vanish. You have found your **blind spot,** a small area of the retina also known as the *optic disc*. This is the area where blood vessels enter and exit the eye and where fibers leading from retinal neurons form the optic nerve that goes to the brain. There are therefore no photoreceptors in this part of the retina, and you cannot see with it. You can use **Figure 9-4** to demonstrate the blind spot in another way.

Fortunately, your visual system solves the blind-spot problem by locating the optic disc in a different location in each of your eyes. The optic disc is lateral to the fovea in each eye, which means that it is left of the fovea in the left eye and right of the fovea in the right eye. Because the visual world of the two eyes overlaps, the blind spot of the left eye can be seen by the right eye and vice versa.

Thus, using both eyes together, you can see the whole visual world. For people who are blind in one eye, the sightless eye cannot compensate for the blind spot in the functioning eye. Still, the visual system compensates for the blind spot in several other ways, and so they have no sense of a hole in their field of vision.

FIGURE 9-4 Find Your Blind Spot Hold this book 30 centimeters (about 12 inches) away from your face. Shut your left eye and look at the cross with your right eye. Slowly bring the page toward you until the red spot in the center of the yellow disc disappears and the entire disc appears yellow. The red spot is now in your blind spot. Your brain replaces the area with the surrounding yellow to fill in the image. Turn the book upside down to test your left eye.

The optic disc that produces a blind spot is of particular importance in neurology. It allows neurologists to indirectly view the condition of the optic nerve that lies behind it while providing a window onto events within the brain. If intracranial pressure increases, as occurs with a tumor or brain abscess (an infection), the optic disc swells, leading to a condition known as *papilloedema* (swollen disc). The swelling occurs in part because, like all neural tissue, the optic nerve is surrounded by cerebrospinal fluid. Pressure inside the cranium can displace this fluid around the optic nerve, causing swelling at the optic disc.

Another reason for papilloedema is inflammation of the optic nerve itself, a condition known as *optic neuritis.* Whatever the cause, a person with a swollen optic disc usually loses vision owing to pressure on the optic nerve. If the swelling is due to optic neuritis, probably the most common neurological visual disorder, the prognosis for recovery is good.

Photoreceptors

The retina's photoreceptor cells convert light energy first into chemical energy and then into neural activity. When light strikes a photoreceptor, it triggers a series of chemical reactions that lead to a change in membrane potential (electrical charge) that in turn leads to a change in the release of neurotransmitter onto nearby neurons.

Rods and cones, the two types of photoreceptors shown in **Figure 9-5,** differ in many ways. They are structurally different. Rods are longer than cones and cylindrically shaped at one end, whereas cones have a tapered end. **Rods** are more numerous than cones, are sensitive to low levels of brightness (luminance), especially in dim light, and function mainly for night vision (see Clinical Focus 9-2, "Visual Illuminance" on page 292). **Cones** do not respond to dim light, but they are highly responsive in bright light. Cones mediate both color vision and our ability to see fine detail (visual acuity).

fovea Region at the center of the retina that is specialized for high acuity; its receptive fields are at the center of the eye's visual field.

blind spot Region of the retina where axons forming the optic nerve leave the eye and where blood vessels enter and leave; has no photoreceptors and is thus "blind."

rod Photoreceptor specialized for functioning at low light levels.

cone Photoreceptor specialized for color and high visual acuity.

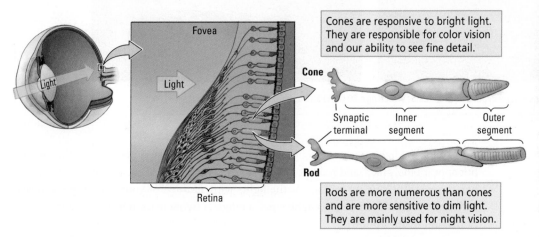

Cones are responsive to bright light. They are responsible for color vision and our ability to see fine detail.

Rods are more numerous than cones and are more sensitive to dim light. They are mainly used for night vision.

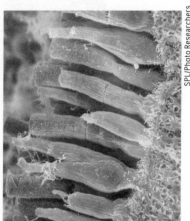

SPL/Photo Researchers

FIGURE 9-5 Photoreceptor Cells Rods and cones are tubelike structures, as the scanning electron micrograph at right shows. They differ, especially in the outer segment that contains the light-absorbing visual pigment. Rods are especially sensitive to broad-spectrum luminance, and cones to particular wavelengths of light.

Rods and cones are not evenly distributed over the retina. The fovea has only cones, but their density drops dramatically beyond the fovea. For this reason, our vision is not so sharp at the edges of the visual field, as demonstrated in Figure 9-3.

A final difference between rods and cones is in their light-absorbing pigments. All rods have the same pigment. Cones have three different pigments; any given cone has one of the three. These four different pigments, one in the rods and three in the cones, form the basis of our vision.

As shown on the spectrum in **Figure 9-6,** the three cone pigments absorb light across a range of visible frequencies, but each is most responsive to a small range of wavelengths— short (bluish light), medium (greenish light), and long (reddish light). As you can see on the background spectrum in Figure 9-6, however, if you

FIGURE 9-6 Range and Peak Sensitivity Our color perception corresponds to the summed activity of the three cone types: S cones, M cones, and L cones (for short, medium, and long wavelengths). Each type is most sensitive to a narrow range of the visible spectrum. Rods (white curve) prefer a range of wavelengths centered on 496 nm but do not contribute to our color perception. Rod activity is not summed with the cones in the color-vision system.

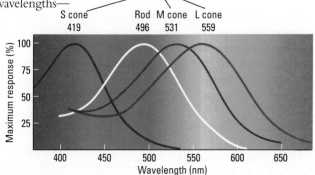

Visual Illuminance

The eye, like a camera, works correctly only when sufficient light passes through the lens and is focused on the receptor surface—the retina of the eye or the light-sensitive surface in the camera. Too little light entering the eye or the camera produces a problem of *visual illuminance*: it is hard to see any image at all.

Visual illuminance is typically a complication of the aging eye that cannot be cured by corrective lenses. As we age, the eye's lens and cornea allow less light through, and so less light strikes the retina. Don Kline

(1994) estimated that between ages 20 and 40, people's ability to see in dim light drops by 50 percent and by a further 50 percent over every 20 additional years. As a result, the ability to see in dim light becomes increasingly difficult, especially at night.

The only solution to compensate for visual illuminance is to increase lighting. Night vision is especially problematic. Not surprisingly, statistics show a marked drop in the number of people driving at night in each successive decade after age 40.

These photographs represent the drop in luminance between age 20 (*left*) and age 60 (*right*).

A nanometer is one-billionth of a meter.

were to look at lights with wavelengths of 419, 531, and 559 nanometers (nm), they would not appear blue, green, and red but rather blue-green, yellow-green, and orange. Remember, though, that you are looking at the lights with all three of your cone types and that each cone pigment responds to light across a range of frequencies, not just to its frequency of maximum absorption.

Both the presence of three different cone-receptor types and their relative numbers and distribution across the retina contribute to our perception of color. As **Figure 9-7** shows, the three cone types are distributed more or less randomly across the retina, making our ability to perceive different colors fairly constant across the visual field. The numbers of red and green cones are approximately equal, but blue cones are fewer in number. As a result, we are not as sensitive to wavelengths in the blue part of the visible spectrum as we are to red and green wavelengths.

FIGURE 9-7 Retinal Receptors The retinal mosaic of rods and three cone types. This diagram represents the distribution near the fovea, where cones outnumber rods. There are fewer blue cones than red and green cones.

Other species that have color vision similar to that of humans also have three types of cones with three color pigments. Because of slight variations in these pigments, the exact frequencies of maximum absorption differ among species. For humans, the exact frequencies are not identical with the numbers given earlier, which are an average across mammals. They are actually 426 and 530 nanometers for the blue and green cones, respectively, and 552 or 557 nanometers for the red cone. The two peak sensitivity levels given for red represent the two variants of the red cone that humans have evolved. The difference in these two red cones appears minuscule, but it does make a functional difference in some females' color perception.

The gene for the red cone is carried on the X chromosome. Males have only one X chromosome, so they have only one of these genes and only one type of red cone. The situation is more complicated for females, who possess two X chromosomes. Although most women have only one type of red cone, those who have both are more sensitive than the rest of us to color differences at the red end of the spectrum. We could say that women who have both types of red cone have a slightly rosier view of the world: their color receptors create a world with a richer range of red experiences. But they also have to contend with seemingly peculiar color coordination by others.

Retinal-Neuron Types

Photoreceptors are connected to two layers of retinal neurons. In the procession from the rods and cones toward the brain shown in **Figure 9-8**, the first layer contains three types of cells: *bipolar cells, horizontal cells,* and *amacrine cells.* Horizontal cells link photoreceptors with bipolar cells, whereas amacrine cells link bipolar cells with the cells of the second neural layer, the **retinal ganglion cells** (RGCs). RGC axons collect in a bundle at the optic disc and leave the eye to form the optic nerve.

retinal ganglion cell (RGC) One of a group of retinal neurons with axons that give rise to the optic nerve.

magnocellular (M) cell Large-celled visual-system neuron that is sensitive to moving stimuli.

parvocellular (P) cell Small-celled visual-system neuron that is sensitive to form and color differences.

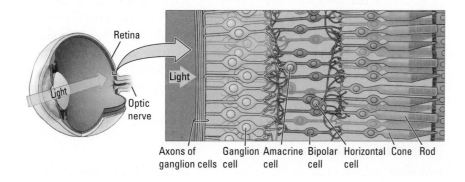

Retina / Light / Light / Optic nerve

Axons of ganglion cells Ganglion cell Amacrine cell Bipolar cell Horizontal cell Cone Rod

FIGURE 9-8 Retinal Cells The neurons in the retina—bipolar, horizontal, amacrine, and ganglion cells—form two layers moving outward from the rods and cones at the retinal surface. Light must pass through both transparent neuron layers to reach the photoreceptors.

Retinal ganglion cells fall into two major categories that, in the primate retina, are called M and P cells. The designations derive from the distinctly different populations of cells in the visual thalamus to which these two classes of RGCs send their axons. As shown in **Figure 9-9**, one population consists of **magnocellular cells** (hence M); the other consists of **parvocellular cells** (hence P). The larger M cells receive their input primarily from rods and so are sensitive to light but not to color. The smaller P cells, receive their input primarily from cones and so are sensitive to color.

In Latin, *magno* means "large" and *parvo* means "small."

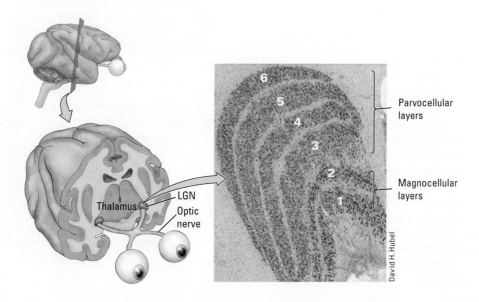

6 / 5 / 4 / 3 / 2 / 1

Thalamus LGN Optic nerve

Parvocellular layers

Magnocellular layers

David H. Hubel

FIGURE 9-9 Visual Thalamus The optic nerves connect with the lateral geniculate nucleus of the thalamus. The LGN has six layers: two magnocellular layers that receive input mainly from rods and four parvocellular layers that receive input mainly from cones.

M cells are found throughout the retina, including the periphery, where we are sensitive to movement but not to color or fine details. P cells are found largely in the region of the fovea, where we are sensitive to color and fine details. A distinction between these two categories of RGCs is maintained throughout the visual pathways, as you will see next, as we follow the ganglion cell axons into the brain.

Visual Pathways

The retinal ganglion cells form the optic nerve, the road into the brain. This road forks off to several places. The destinations of these branches give us clues to what the brain is doing with visual input and how the brain creates our visual world.

Crossing the Optic Chiasm

The optic chiasm gets its name from the shape of the Greek letter χ (pronounced "ki"—long "i").

We begin with the optic nerves, one exiting from each eye. Just before entering the brain, the optic nerves partly cross, forming the **optic chiasm.**

About half the fibers from each eye cross in such a way that the left half of each optic nerve goes to the left side of the brain, and the right half goes to the brain's right side, as diagrammed in **Figure 9-10.** The medial path of each retina, the *nasal retina,* crosses to the opposite side. The lateral path, the *temporal retina,* goes straight back on the same side. Because light that falls on the right half of each retina actually comes from the left side of the visual field, information from the left visual field goes to the brain's right hemisphere, and information from the right visual field goes to the left hemisphere. Thus, half of each retina's visual field is represented on each side of the brain.

Our visual system represents the world seen through two eyes as a single perception by connecting both eyes with both hemispheres.

Three Routes to the Visual Brain

Two main pathways lead to the visual cortex in the occipital lobe. Another, smaller pathway tracks into the hypothalamus.

GENICULOSTRIATE SYSTEM On entering the brain, the RGC axons separate, forming the two distinct pathways charted in **Figure 9-11.** All the axons of the P ganglion cells and some of the M ganglion cells form a pathway called the **geniculostriate system.**

optic chiasm Junction of the optic nerves, one from each eye, at which the axons from the nasal (inside—nearer the nose) halves of the retinas cross to the opposite side of the brain.

geniculostriate system Projections from the retina to the lateral geniculate nucleus to the visual cortex.

striate cortex Primary visual cortex (V1) in the occipital lobe; its striped appearance when stained gives it this name.

tectopulvinar system Projections from the retina to the superior colliculus to the pulvinar (thalamus) to the parietal and temporal visual areas.

retinohypothalamic tract Neural route formed by axons of photosensitive retinal ganglion cells from the retina to the suprachiasmatic nucleus; allows light to entrain the rhythmic activity of the SCN.

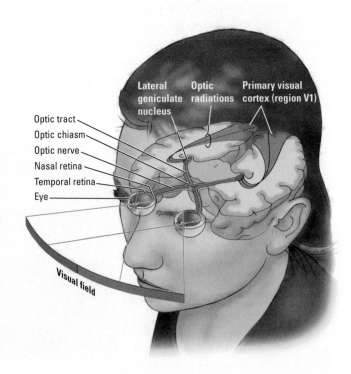

FIGURE 9-10 Crossing the Optic Chiasm This dorsal view shows the visual pathway from each eye to the primary visual cortex of each hemisphere. Information from the right side of the visual field (blue) moves from the two left halves of the retinas, ending in the left hemisphere. Information from the left side of the visual field (red) hits the right halves of the retinas and travels to the right side of the brain.

Labels in figure:
Lateral geniculate nucleus
Optic radiations
Primary visual cortex (region V1)
Optic tract
Optic chiasm
Optic nerve
Nasal retina
Temporal retina
Eye
Visual field

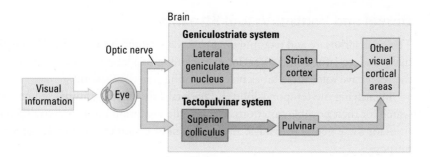

FIGURE 9-11 Main Visual Pathways into the Brain The optic nerve follows (1) the geniculostriate path to the primary visual cortex and (2) the tectopulvinar path to the temporal and parietal lobes. (The LGN of the thalamus is part of the diencephalon, shown in Figure 2-19; the superior colliculus in the tectum is part of the midbrain, shown in Figure 2-18.)

This pathway goes from the retina to the lateral geniculate nucleus (LGN) of the thalamus and then to layer IV of the primary visual cortex in the occipital lobe.

Figure 9-12 shows that, when stained, the primary visual cortex appears to have a broad stripe across it in layer IV and so is known as **striate** (striped) **cortex.** The geniculostriate system therefore bridges the thalamus (geniculate) and the striate cortex. From the striate cortex, the axon pathway now divides. One route goes to vision-related regions of the parietal lobe and another route goes to vision-related regions of the temporal lobe.

TECTOPULVINAR SYSTEM The second pathway leading from the eye is formed by the axons of the remaining M ganglion cells. These cells send their axons to the midbrain's superior colliculus, which sends connections to the pulvinar region of the thalamus. This pathway is therefore known as the **tectopulvinar system** because it runs from the eye through the midbrain tectum to the pulvinar (see Figure 9-11). The pulvinar then sends connections to the parietal and temporal lobes.

RETINOHYPOTHALAMIC TRACT Between 1 percent and 3 percent of retinal ganglion cells are unique in that they are *photosensitive:* they act as photoreceptors. These *pRGCs* contain the photosensitive pigment melanopsin and absorb blue light at a wavelength (between 460 and 480 nm) different from the wavelengths of rods or cones (see Figure 9-6). Axons of pRGCs form a small, third visual pathway, the **retinohypothalamic tract.**

The retinohypothalamic tract synapses in the tiny suprachiasmatic nucleus (SCN) in the hypothalamus, next to the optic chiasm. Photosensitive retinal ganglion cells play roles both in regulating circadian rhythms and in the pupillary reflex that expands and contracts the pupil in response to the amount of light falling on the retina. Farhan Zaidi and colleagues (2007) studied two profoundly blind subjects who lack functional rods and cones. The researchers found that stimulation with 480 nm (blue) light increases alertness and appears to play some rudimentary role in visual awareness.

Figure 13-6 maps the retinohypothalamic tract into the SCN.

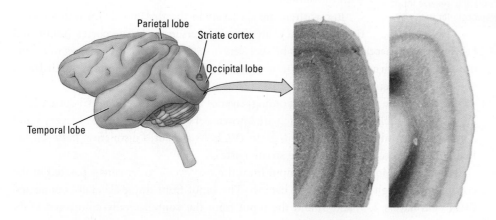

FIGURE 9-12 Striate Cortex Area V1 is also called the *striate cortex* because sections appear striated (striped) when stained with either a cell-body stain (*left*) or a myelin stain (*right*). The sections shown here come from a rhesus monkey's brain.

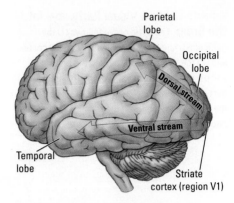

FIGURE 9-13 Visual Streaming Information travels from the occipital visual areas to the parietal and temporal lobes, forming the dorsal and ventral streams, respectively.

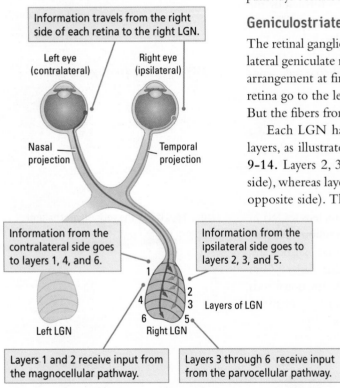

FIGURE 9-14 Geniculostriate Pathway

Figure 2-22 maps layers I through VI in the motor and sensory cortices and labels their functions.

Dorsal and Ventral Visual Streams

The geniculostriate and tectopulvinar pathways extend into the visual brain. Each eventually leads to either the parietal lobe or the temporal lobe. Our next task is to determine the respective roles these lobes play in creating our visual world.

Identifying the temporal-lobe and parietal-lobe visual pathways led researchers on a search for their possible functions. Why would evolution produce two different destinations for these neural pathways? Each route must create visual knowledge for a different purpose.

David Milner and Mel Goodale (2006) proposed that these two purposes are to identify what a stimulus is (the "what" function) and to use visual information to control movement (the "how" function). This "what–how" distinction came from an analysis of where visual information goes when it leaves the striate cortex. **Figure 9-13** shows the two distinct visual pathways that originate in the striate cortex, one progressing to the temporal lobe and the other to the parietal lobe. The pathway to the temporal lobe is known as the ventral stream, whereas the pathway to the parietal lobe is the dorsal stream.

To understand how the two streams function, we return to the details of how visual input from the eyes contributes to them. Both the geniculostriate and the tectopulvinar pathways contribute to the dorsal and ventral streams.

Geniculostriate Pathway

The retinal ganglion-cell fibers from the two eyes distribute their connections to the two lateral geniculate nuclei (left and right) of the thalamus in what appears to be an unusual arrangement at first glance. As seen in Figure 9-10, the fibers from the left half of each retina go to the left LGN; those from the right half of each retina go to the right LGN. But the fibers from each eye do not go to exactly the same LGN location.

Each LGN has six layers, and the projections from the two eyes go to different layers, as illustrated in anatomical context in Figure 9-9 and diagrammed in **Figure 9-14.** Layers 2, 3, and 5 receive fibers from the ipsilateral eye (the eye on the same side), whereas layers 1, 4, and 6 receive fibers from the contralateral eye (the eye on the opposite side). This arrangement provides both for combining the information from the two eyes and for segregating the information from the P and M ganglion cells.

Axons from the P cells go only to layers 3 through 6 (the parvocellular layers). Axons from the M cells go only to layers 1 and 2 (the magnocellular layers). Because the P cells are responsive to color and fine detail, LGN layers 3 through 6 must be processing information about color and form. In contrast, the M cells mostly process information about movement, so layers 1 and 2 must deal with movement.

Just as there are six layers in the thalamic LGN (numbered 1 through 6), there are also six layers in the striate cortex (numbered I through VI). That there happen to be six layers in each location is an accident of evolution found in all primate brains. Let us now see where these LGN cells from the thalamus send their connections within the visual cortex.

Layer IV is the main afferent (incoming) layer of the cortex. In the visual cortex, layer IV has several sublayers, two of which are known as IVCa and IVCb. LGN layers 1 and 2 go to IVCa, and layers 3 through 6 go to IVCb. As a result, a distinction between the P and M functions continues in the striate cortex.

As illustrated in **Figure 9-15,** input from the two eyes also remains separated in the cortex but through a different mechanism. The input from the ipsilaterally connected LGN cells (layers 2, 3, and 5) and the input from the contralaterally connected LGN

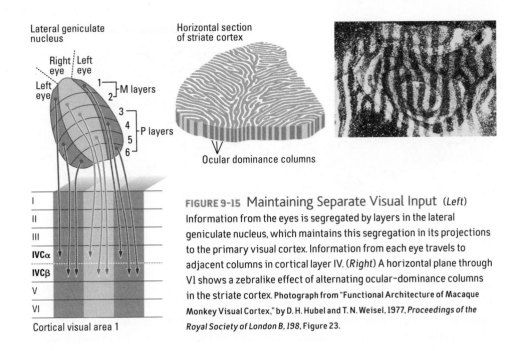

Lateral geniculate nucleus

Horizontal section of striate cortex

Ocular dominance columns

Cortical visual area 1

FIGURE 9-15 Maintaining Separate Visual Input (*Left*) Information from the eyes is segregated by layers in the lateral geniculate nucleus, which maintains this segregation in its projections to the primary visual cortex. Information from each eye travels to adjacent columns in cortical layer IV. (*Right*) A horizontal plane through V1 shows a zebralike effect of alternating ocular-dominance columns in the striate cortex. Photograph from "Functional Architecture of Macaque Monkey Visual Cortex," by D. H. Hubel and T. N. Weisel, 1977, *Proceedings of the Royal Society of London B, 198*, Figure 23.

cells (layers 1, 4, and 6) go to adjacent strips of cortex. These strips, which are about 0.5 millimeter across, are known as **cortical columns.**

In summary, the P and M ganglion cells of the retina send separate pathways to the thalamus, and this segregation continues in the striate cortex. The left and right eyes also send separate pathways to the thalamus, and these pathways, too, remain segregated in the striate cortex.

Tectopulvinar Pathway

To review, magnocellular RGCs found throughout the retina receive their input primarily from the rods, so they are sensitive to light but not to color. M cells in the periphery of the retina are sensitive to movement but not to color or fine details. In the brain, some M cells join P cells to form the geniculostriate pathway. The tectopulvinar pathway is formed by the axons of the remaining M cells.

These M cells send their axons to the superior colliculus in the midbrain's tectum. One function of the tectum is to produce orienting movements—to detect the location of stimuli and shift the eyes toward stimuli. The superior colliculus sends connections to the region of the thalamus known as the pulvinar.

The pulvinar has two main divisions. The medial pulvinar sends connections to the parietal lobe, and the lateral pulvinar sends connections to the temporal lobe. One type of information that these connections are conveying is related to "where," which is important in both the "what" and "how" visual streams.

The "where" function of the tectopulvinar system is useful in understanding D. B.'s blindsight, described in Clinical Focus 9-1. His geniculostriate system was disrupted by surgery, but his tectopulvinar system was not, thus allowing him to identify the location of stimuli (where) that he could not identify (what).

Occipital Cortex

Our route down the visual pathways has led us from the retina all the way back to the occipital lobe and into the parietal and temporal lobes. Now we explore how visual information proceeds from the striate cortex through the rest of the occipital lobe to the dorsal and ventral streams.

cortical column Cortical organization that represents a functional unit six cortical layers deep and approximately 0.5 millimeter square and that is perpendicular to the cortical surface.

Many textbooks emphasize the "how" pathway as a "where" function. We use Milner and Goodale's "what–how" distinction because "where" is both a property of "what" a stimulus is and a cue for "how" to control movement to a place.

FIGURE 9-16 Visual Regions of the Occipital Lobe

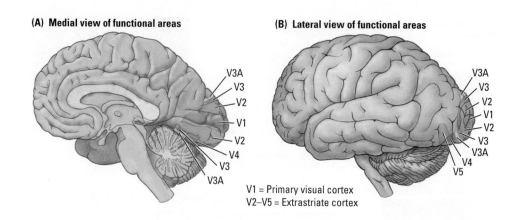

(A) Medial view of functional areas

V3A
V3
V2
V1
V2
V4
V3
V3A

(B) Lateral view of functional areas

V3A
V3
V2
V1
V2
V3
V3A
V4
V5

V1 = Primary visual cortex
V2–V5 = Extrastriate cortex

All mammals have at least one primary cortical area for each sensory system. The primary area relays most information that reaches secondary areas.

primary visual cortex (V1) Striate cortex that receives input from the lateral geniculate nucleus.

extrastriate (secondary visual) cortex Visual cortical areas outside the striate cortex.

blob Region in the visual cortex that contains color-sensitive neurons, as revealed by staining for cytochrome oxidase.

As shown in **Figure 9-16,** the occipital lobe is composed of at least six different visual regions: V1, V2, V3, V3A, V4, and V5. The striate cortex is region **V1,** the **primary visual cortex.** The remaining visual areas of the occipital lobe form the **extrastriate cortex** or *secondary visual cortex.* Because each occipital region has a unique cytoarchitecture (cellular structure) and unique inputs and outputs, we can infer that each must be doing something different from the others.

As shown in Figures 9-12 and 9-15, a remarkable feature of region V1 is its striations—its distinctly visible layers. When Margaret Wong-Riley and her colleagues (1993) stained region VI for the enzyme cytochrome oxidase, which has a role in cell metabolism, they found an unexpected heterogeneity. So they sectioned the V1 layers in such a way that each cortical layer was in one plane of section, much like peeling off the layers of an onion and laying them flat on a table. The surface of each flattened layer can then be viewed from above.

The heterogeneous cytochrome staining now appeared as random blobs in the layers of V1, as diagrammed in **Figure 9-17.** These darkened regions have in fact become known as **blobs,** the less-dark regions separating them as *interblobs.* Blobs and interblobs serve different functions. Neurons in the blobs take part in color perception; neurons in the interblobs participate in form and motion perception. Within region V1, then, input that arrives from the P-cell and M-cell pathways of the geniculostriate system is segregated into three separate types of information: color, form, and motion.

Color, form, and motion information from region V1 moves next to region V2, adjoining region V1. Here, the color, form, and motion inputs remain segregated, again seen through the pattern of cytochrome oxidase staining. But as Figure 9-17 shows, the staining pattern in region V2 is different from that in region V1. Region V2 has a pattern of thick and thin stripes intermixed with pale zones. The thick stripes receive input from the movement-sensitive neurons in region V1; the thin stripes receive input from V1's color-sensitive neurons; and the pale zones receive input from V1's form-sensitive neurons.

FIGURE 9-17 Heterogeneous Layering Blobs in region VI and stripes in region V2 are illustrated in this drawing of a flattened section through the visual cortex of a monkey. The blobs and stripes are revealed by a special stain for cytochrome oxidase, a marker for mitochondria, the organelles in cells that gather, store, and release energy.

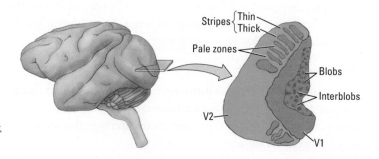

Stripes { Thin
Thick
Pale zones
Blobs
Interblobs
V2
V1

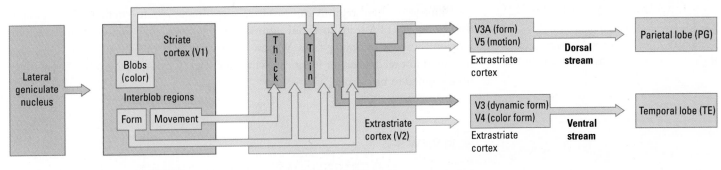

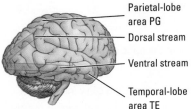

Parietal-lobe area PG
Dorsal stream
Ventral stream
Temporal-lobe area TE

FIGURE 9-18 Charting the Visual Streams The dorsal stream that controls visual action (*top*) begins in region V1 and flows through V2 to the other occipital areas and finally to the parietal cortex, ending in area PG. The ventral stream that controls object recognition (*bottom*) begins in region V1 and flows through V2 to the other occipital areas and finally to the temporal cortex, ending in area TE. Information from the blobs and interblobs in V1 flows to the thick, thin, and pale zones of V2 and then to regions V3 and V4 to form the ventral stream. Information in the thick and pale zones goes to regions V3A and V5 to form the dorsal stream.

As charted in **Figure 9-18,** the visual pathways proceed from region V2 to the other occipital regions and then to the parietal and temporal lobes, forming the dorsal and ventral streams. Although many parietal and temporal regions take part, the major regions are region G in the parietal lobe (thus called region PG) and region E in the temporal lobe (thus called region TE).

Within the dorsal and ventral streams, the simple records of color, form, and motion information from the occipital regions is assembled to produce a rich, unified visual world of complex objects, such as faces and paintings, and complex skills, such as bike riding and ball catching. We can think of the complex representations of the dorsal and ventral streams as consisting of "how" functions and "what" functions. "How" is action to be visually guided toward the "what" that identifies an object.

Vision Beyond the Occipital Cortex

Visual processing does not end in occipital cortex. It continues via the ventral and dorsal streams into the temporal and parietal cortex, respectively. Each region has multiple areas specialized for specific visual functions. For example, **Figure 9-19A** shows two regions on the ventral surface of the temporal lobes. One is specialized for recognizing faces (fusiform face area, FFA), the other for analyzing landmarks such as buildings or trees (parahippocampal place area, PPA). Figure 9-19B shows two regions in the parietal lobe related to eye movements (lateral intraparietal area, LIP) and visual control of grasping (anterior intraparietal area, AIP).

(A)

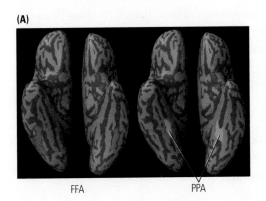

FFA   PPA

(B)

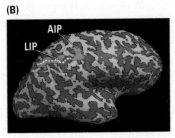

AIP
LIP

FIGURE 9-19 Vision Beyond the Occipital Cortex **(A)** In the temporal lobe, the fusiform face area (FFA) processes faces, and the parahippocampal place area (PPA) processes scenes. **(B)** In the parietal lobe, the lateral intraparietal area (LIP) contributes to eye movements; the anterior intraparietal area (AIP) is involved in visual control of grasping. **(A)** Adapted with permission from "Intersubject Synchronization of Cortical Activity During Natural Vision," by U. Hasson, Y. Nir, I. Levy, G. Fuhrmann, and R. Malach, 2004, *Science, 303,* pp. 1634–1640. **(B)** Modified from *The Visual Brain in Action* (2nd ed.), by A. D. Milner and M.A. Goodale, 2006, New York: Oxford University Press.

Agnosia literally means "not knowing." Section 15-7 ties conditions like agnosia to the search for a neural basis of consciousness.

Damage to these regions can produce surprisingly specific deficits. For example, damage to the FFA leads to **facial agnosia,** or *prosopagnosia,* a condition in which an individual cannot recognize faces. We saw one patient with prosopagnosia so severe that she could not recognize her identical twin sister's face. Curiously, her other visual functions seemed to be normal.

REVIEW 9-2

Functional Anatomy of the Visual System

Before you continue, check your understanding.

1. Neurons that project into the brain from the retina and form the optic nerve are called

   _____.

2. _____ retinal ganglion cells receive input mostly from cones and carry information about color and fine detail, whereas _____ retinal ganglion cells receive input mostly from rods and carry information about light but not color.

3. The two major pathways from the retina into the brain are _____ and _____.

4. Damage to the fusiform face area in the temporal lobe can produce _____.

5. Contrast the paths and functions of the dorsal and ventral streams.

Answers appear at the back of the book.

9-3 Location in the Visual World

As we move about, going from place to place, we encounter objects in specific locations. If we had no awareness of location, the world would be a bewildering mass of visual information. The next leg of our journey down the neural roads traces how the brain constructs a spatial map.

Neural coding of location begins in the retina and is maintained throughout all visual pathways. To understand how this spatial coding is accomplished, imagine your visual world as seen by your two eyes. Imagine the large red and blue rectangles in **Figure 9-20** as a wall. Focus your gaze on the black cross in the middle of the wall.

The part of the wall that you can see without moving your head is your **visual field.** It can be divided into two halves, the left and right visual fields, by drawing a vertical line through the middle of the black cross. Now recall from Figure 9-10 that the left half of each retina looks at the right side of the visual field, whereas the right half of each retina looks at the visual field's left side. Thus, input from the right visual field goes to the left hemisphere, and input from the left visual field goes to the right hemisphere.

Therefore, the brain can easily determine whether visual information is located to the left or right of center. If input goes to the left hemisphere, the source must be in the right visual field; if input goes to the right hemisphere, the source must be in the left visual field. This arrangement tells you nothing about the precise location of an object in the left or right side of the visual field, however. To understand how precise spatial localization is accomplished, we must return to the retinal ganglion cells.

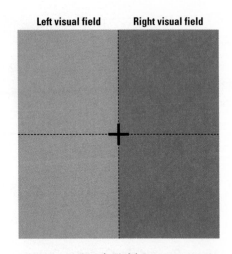

Left visual field   **Right visual field**

FIGURE 9-20 Visual-Field Demonstration
As you focus on the cross at center, information at the left of this focal point forms the left visual field (red) and travels to the right hemisphere. Information to the right of the focal point forms the right visual field (blue) and travels to the left hemisphere. The visual field can be split horizontally as well: information above the focal point is in the upper visual field and that below the focal point is in the lower visual field.

Coding Location in the Retina

Look again at Figure 9-8 and you can see that each RGC receives input through bipolar cells from several photoreceptors. In the 1950s, Stephen Kuffler, a pioneer in studying visual-system physiology, made an important discovery about how photoreceptors and retinal ganglion cells are linked (Kuffler, 1952). By shining small spots of light on the receptors, he found that each ganglion cell responds to stimulation on just a small circular patch of the retina, which is the ganglion cell's receptive field.

A ganglion cell's receptive field is therefore the region of the retina on which it is possible to influence that cell's firing. Stated differently, the receptive field represents the outer world as seen by a single cell. Each RGC sees only a small bit of the world, much as you would if you looked through a narrow cardboard tube. The visual field is composed of thousands of such receptive fields.

Now let us consider how receptive fields enable the visual system to interpret the location of objects. Imagine that the retina is flattened like a piece of paper. When a tiny light is shone on different parts of the retina, different ganglion cells respond. For example, when a light is shone on the top-left corner of the flattened retina, a particular RGC responds because that light is in its receptive field. Similarly, when a light is shone on the top-right corner, a different RGC responds.

By using this information, we can identify the location of a light on the retina by knowing which ganglion cell it activates. We can also interpret the light's location in the outside world because we know where the light must come from to hit a particular place on the retina. Light coming from above hits the bottom of the retina after passing through the eye's lens, for example, and light from below hits the top of the retina. Information at the top of the visual field stimulates ganglion cells on the bottom of the retina; information at the bottom of the field stimulates ganglion cells on the top of the retina.

Location in the Lateral Geniculate Nucleus and Region V1

Now consider the connection from the ganglion cells to the lateral geniculate nucleus. In contrast with the retina, the LGN is not a thin sheet; it is shaped more like a sausage. We can compare it to a stack of cards, with each card representing a layer of cells.

Figure 9-21 shows how the connections from the retina to the LGN can represent location. A retinal ganglion cell that responds to light in the top-left corner of the retina connects to the left side of the first card. A retinal ganglion cell that responds to light in the bottom-right corner of the retina connects to the right side of the last card. In this way, the location of left–right and top–bottom information is maintained in the LGN.

Like the ganglion cells, each LGN cell has a receptive field—the region of the retina that influences its activity. If two adjacent retinal ganglion cells synapse on a single LGN cell, the receptive field of that LGN cell will be the sum of the two ganglion cells' receptive fields. As a result, the receptive fields of LGN cells are bigger than those of RGCs.

The LGN projection to the striate cortex (region V1) also maintains spatial information. As each LGN cell, representing a particular place, projects to region V1, a spatially organized neural representation—a topographic map—is produced in the cortex. As illustrated in **Figure 9-22,** this representation is essentially a map of the visual world.

The central part of the visual field is represented at the back of the brain, whereas the periphery is represented more anteriorly. The upper part of the visual field is represented at the bottom of region V1 and the lower part at the top of V1. The other regions of the visual cortex (such as V3, V4, and V5) have topographical maps similar to that of V1. Thus the V1 neurons must project to the other regions in an orderly manner, just as the LGN neurons project to region V1 in an orderly way.

Within each visual cortical area, each neuron's receptive field corresponds to the part of the retina to which the neuron is connected. As a rule of thumb, the cells in the cortex have much larger receptive fields than those of retinal ganglion cells. This increase in

facial agnosia Face blindness—the inability to recognize faces; also called *prosopagnosia*.

visual field Region of the visual world that is seen by the eyes.

Like a camera lens, the lens in the eye focuses light rays to project a backward, inverted image on a light-receptive surface (see the illustration "How the Eye Works" in The Basics).

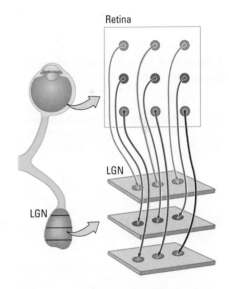

FIGURE 9-21 Receptive-Field Projection Information from a receptive field in the retina retains its spatial relation when sent to the lateral geniculate nucleus. Information at the top of the visual field goes to the top of the LGN, information from the bottom of the visual field goes to the bottom of the LGN, and information from the left or right goes to the left or right of the LGN, respectively.

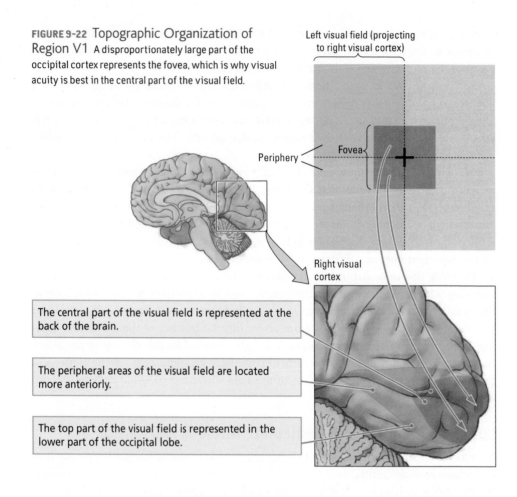

FIGURE 9-22 Topographic Organization of Region V1 A disproportionately large part of the occipital cortex represents the fovea, which is why visual acuity is best in the central part of the visual field.

Left visual field (projecting to right visual cortex)

Periphery

Fovea

Right visual cortex

The central part of the visual field is represented at the back of the brain.

The peripheral areas of the visual field are located more anteriorly.

The top part of the visual field is represented in the lower part of the occipital lobe.

receptive-field size means that the receptive field of a cortical neuron must be composed of the receptive fields of many RGCs, as illustrated in **Figure 9-23**.

There is one additional wrinkle to the organization of topographic maps. Jerison's principle of proper mass states that the amount of neural tissue responsible for a particular function is equivalent to the amount of neural processing required for that function. Jerison's principle extends to regions within the brain as well. The visual cortex provides some good examples.

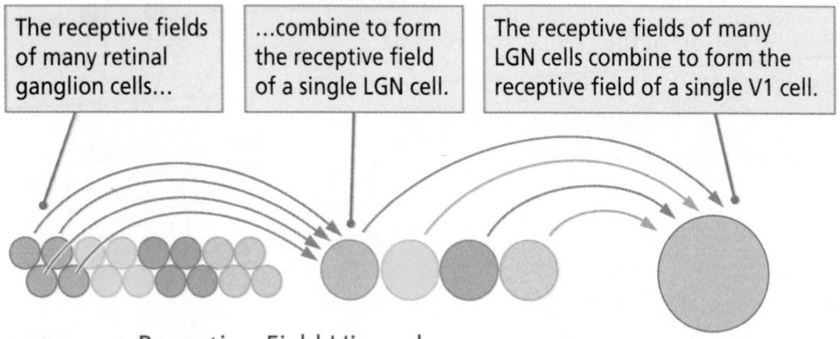

You can see in Figure 9-22 that not all parts of the visual field are equally represented in region V1. The small, central part of the visual field seen by the fovea is represented by a larger area in the cortex than the visual field's periphery, even though the periphery is a much larger part of the visual field. In accord with Jerison's principle, we would predict more processing of foveal information than of peripheral information in region V1. This prediction makes intuitive sense because we can see more clearly in the center of the visual field than at the periphery (see Figure 9-3). In other words, sensory areas that have more cortical representation provide a more-detailed creation of the external world.

In Figure 1-14, we apply Jerison's principle to overall brain size.

The receptive fields of many retinal ganglion cells...

...combine to form the receptive field of a single LGN cell.

The receptive fields of many LGN cells combine to form the receptive field of a single V1 cell.

FIGURE 9-23 Receptive-Field Hierarchy

Visual Corpus Callosum

Creating topographic maps based on neuronal receptive fields is an effective way for the brain to code object location. But if the left visual field is represented in the right cerebral hemisphere and the right visual field is represented in the left cerebral hemisphere, how are the two halves of the visual field ultimately bound together in a unified representation? After all, we have the subjective impression not of two independent visual fields but rather of a single, continuous field of vision.

The answer to how this unity is accomplished lies in the corpus callosum that binds the two sides of the visual field together at the midline. Until the 1950s, its function was largely a mystery. Physicians had occasionally cut the corpus callosum to control severe epilepsy or to reach a very deep tumor, but patients did not appear much affected by this surgery. The corpus callosum clearly linked the two hemispheres of the brain, but exactly which parts were connected was not yet known.

We now realize that the corpus callosum connects only certain brain structures. As shown in **Figure 9-24,** much of the frontal lobes have callosal connections, but the occipital lobes have almost none. If you think about it, there is no reason for a neuron in the visual cortex that is "looking at" one place in the visual field to be concerned with what another neuron in the opposite hemisphere is "looking at" in another part of the visual field.

Cells that lie along the midline of the visual field are an exception, however. These cells "look at" adjacent places in the visual field, one slightly to the left of center and one slightly to the right. Callosal connections between such cells zip the two visual fields together by combining their receptive fields to overlap at the midline. The two fields thus become one.

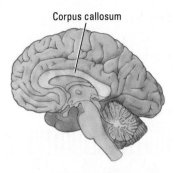

Corpus callosum

Section 15-4 describes the revelations learned from so-called "split-brain" patients whose corpus callosa have been severed.

REVIEW 9-3
Location in the Visual World

Before you continue, check your understanding.

1. The characteristic of visual receptive fields that allow us to detect exactly where a light source is coming from is their _____.

2. List four types of cells that have visual receptive fields: _____, _____, _____, and _____.

3. Inputs to different parts of cortical region V1 from different parts of the retina essentially form a _____ that represents the visual world within the brain.

4. The two sides of the visual world are bound together as one perception by the _____.

5. How does Jerison's principle of proper mass apply to the visual system?

Answers appear at the back of the book.

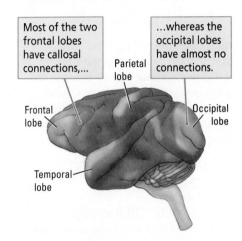

Most of the two frontal lobes have callosal connections,...

...whereas the occipital lobes have almost no connections.

Parietal lobe

Frontal lobe

Occipital lobe

Temporal lobe

FIGURE 9-24 Callosal Connections Darker areas show regions of the rhesus monkey cortex that receive projections from the opposite hemisphere through the corpus callosum.

9-4 Neuronal Activity

Individual neurons make up the visual system pathways. By studying how these cells behave when their receptive fields are stimulated, we can begin to understand how the brain processes different features of the visual world beyond the location of a light. We first examine how neurons in the ventral stream respond to shapes and colors and then briefly consider how neurons in the dorsal stream direct vision for action.

Seeing Shape

Imagine that a microelectrode placed near a neuron somewhere in the visual pathway from retina to cortex is recording changes in the neuron's firing rate. This neuron occasionally fires spontaneously, producing action potentials with each discharge. Assume that the neuron discharges, on average, once every 0.08 second. Each action potential is brief, on the order of 1 millisecond.

Figure 4-6 diagrams how microelectrodes work.

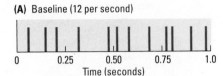

(A) Baseline (12 per second)

(B) Excitation

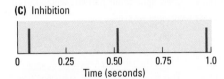

(C) Inhibition

FIGURE 9-25 Recording Neuronal Stimulation Each action potential is represented by a spike. **(A)** In a 1-second period at this neuron's baseline firing rate, 12 spikes were recorded. **(B)** An increase in firing rate over baseline signals excitation. **(C)** A decrease in firing rate under baseline signals inhibition.

FIGURE 9-26 On–Off Receptivity (A) In the receptive field of an RGC with an on-center and off-surround, a spot of light shining on the center excites the neuron, a spot of light in the surround inhibits it. When the light in the surround is turned off, firing rate increases briefly—an "offset" response. A light shining in both the center and the surround would produce a weak increase in firing. **(B)** In the receptive field of an RGC with an off-center and on-surround, light in the center produces inhibition, light on the surround produces excitation, and light across the entire field produces weak inhibition.

If we plot action potentials spanning a second, we see only spikes in the record because the action potentials are so brief. **Figure 9-25A** is a single-cell recording of 12 spikes in the span of 1 second. If the firing rate of this cell increases, we see more spikes (Figure 9-25B). If the firing rate decreases, we see fewer spikes (Figure 9-25C). The increase in firing represents excitation of the cell, whereas the decrease represents inhibition. Excitation and inhibition, as you know, are the principal mechanisms of information transfer in the nervous system.

Now suppose we present a stimulus to the neuron by illuminating its receptive field in the retina, perhaps by shining a light on a blank screen within the cell's visual field. We might place before the eye a straight line positioned at a 45° angle. The cell could respond to this stimulus either by increasing or decreasing its firing rate. In either case, we would conclude that the cell is creating information about the line.

Note that the same cell could show excitation to one stimulus, inhibition to another stimulus, and no reaction at all. The cell could be excited by lines oriented 45° to the left and inhibited by lines oriented 45° to the right. Similarly, the cell could be excited by stimulation in one part of its receptive field (such as the center) and inhibited by stimulation in another part (such as the periphery).

Finally, we might find that the cell's response to a particular stimulus is selective. Such a cell would be telling us about the importance of the stimulus to the animal. For instance, the cell might be excited when a stimulus is presented with food but inhibited when the same stimulus is presented alone. In each case, the cell is selectively sensitive to characteristics in the visual world.

Neurons at each level of the visual system have distinctly different characteristics and functions. Our goal is not to look at each neuron type but rather to consider generally how some typical neurons at each level differ from one another in their contributions to processing shape. We focus on neurons in three areas: the ganglion-cell layer of the retina, the primary visual cortex, and the temporal cortex.

Processing in Retinal Ganglion Cells

Neurons in the retina do not detect shape, because their receptive fields are miniscule dots. Each retinal ganglion cell responds only to the presence or absence of light in its receptive field, not to shape. Shape is constructed by processes in the cortex from the information that those ganglion cells pass on about events in their receptive fields.

The receptive field of a ganglion cell has a concentric circle arrangement, as illustrated in **Figure 9-26A**. A spot of light falling in the central circle of the receptive field

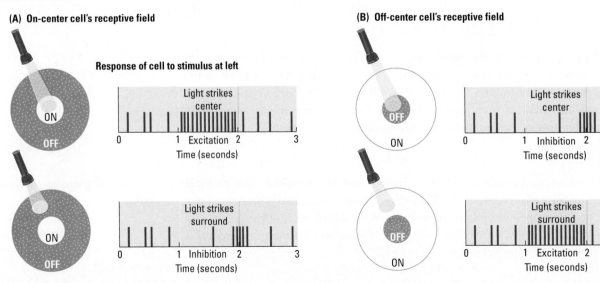

(A) On-center cell's receptive field

Response of cell to stimulus at left

Light strikes center

0   1   Excitation   2   3
Time (seconds)

Light strikes surround

0   1   Inhibition   2   3
Time (seconds)

(B) Off-center cell's receptive field

Light strikes center

0   1   Inhibition   2   3
Time (seconds)

Light strikes surround

0   1   Excitation   2   3
Time (seconds)

excites some of these cells, whereas a spot of light falling in the receptive field's surround (periphery) inhibits the cell. A spot of light falling across the entire receptive field weakly increases in the cell's firing rate.

This type of neuron is called an *on-center cell*. Other RGCs, called *off-center cells,* have the opposite arrangement, with light in the center of the receptive field inhibiting, light in the surround exciting, and light across the entire field producing weak inhibition (Figure 9-26B). The on–off arrangement of RGC receptive fields makes these cells especially responsive to very small spots of light.

This description of ganglion-cell receptive fields might mislead you into thinking that they form a mosaic of discrete little circles on the retina that do not overlap. In fact, neighboring retinal ganglion cells receive their inputs from an overlapping set of photoreceptors. As a result, their receptive fields overlap, as illustrated in **Figure 9-27**. In this way, a small spot of light shining on the retina is likely to produce activity in both on-center and off-center ganglion cells.

How can on-center and off-center ganglion cells tell the brain anything about shape? The answer is that a ganglion cell is able to tell the brain about the amount of light hitting a certain spot on the retina compared with the average amount of light falling on the surrounding retinal region. This comparison is known as **luminance contrast.**

To understand how luminance contrast tells the brain about shape, consider the hypothetical population of on-center ganglion cells represented in **Figure 9-28.** Their receptive fields are distributed across the retinal image of a light–dark edge. Some of the ganglion cells' receptive fields are in the dark area, others are in the light area, and still others' fields straddle the edge of the light.

The receptive fields of retinal ganglion cells overlap extensively,…	…so any two adjacent fields look at almost the same part of the world.

Receptive fields of neighboring ganglion cells

Two overlapping receptive fields

FIGURE 9-27 Overlapping Receptive Fields

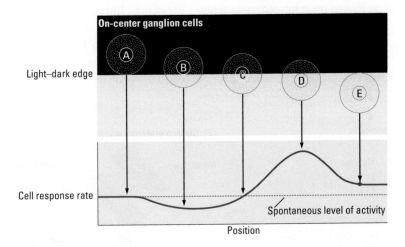

FIGURE 9-28 Activity at the Margins
Responses of a hypothetical population of on-center ganglion cells whose receptive fields (A–E) are distributed across a light–dark edge. The activity of the cells along the edge is most affected relative to those away from the edge. Adapted from *Neuroscience* (p. 195), edited by D. Purves, G. J. Augustine, D. Fitzpatrick, L. C. Katz, A.-S. LaMantia, and J. O. McNamara, 1997, Sunderland, MA: Sinauer.

The ganglion cells with receptive fields in the dark or light areas are least affected because they experience either no stimulation or stimulation of both the excitatory and the inhibitory regions of their receptive fields. The ganglion cells most affected by the stimulus are those lying along the edge. Ganglion cell B is inhibited because the light falls mostly on its inhibitory surround, and ganglion cell D is excited because its entire excitatory center is stimulated but only part of its inhibitory surround is.

Consequently, information transmitted from retinal ganglion cells to the visual areas in the brain does not give equal weight to all regions of the visual field. Rather, it emphasizes regions containing differences in luminance—areas along the edges. So RGCs are really sending signals about edges, and edges form shapes.

luminance contrast The amount of light reflected by an object relative to its surroundings.

ocular-dominance column Functional column in the visual cortex maximally responsive to information coming from one eye.

Processing Shape in the Primary Visual Cortex

Now consider cells in region V1 that receive their visual inputs from LGN cells, which in turn receive theirs from retinal ganglion cells. Because each V1 cell receives input from multiple RGCs, the receptive fields of the V1 neurons are much larger than those of retinal neurons. Consequently, V1 cells respond to stimuli more complex than simply "light on" or "light off." In particular, these cells are maximally excited by bars of light oriented in a particular direction rather than by spots of light. These V1 cells are therefore called *orientation detectors.*

Like the ganglion cells, some orientation detectors have an on–off receptive-field arrangement, but the arrangement is rectangular rather than circular. Visual cortex cells with this property are known as *simple cells.* Typical receptive fields for simple cells in the primary visual cortex are shown in **Figure 9-29.**

Simple cells are not the only kind of orientation detector in the primary visual cortex; several functionally distinct types of neurons populate region V1. For instance, the receptive fields of *complex cells,* such as those in **Figure 9-30,** are maximally excited by bars of light moving in a particular direction through the visual field. A *hypercomplex cell,* like a complex cell, is maximally responsive to moving bars but also has a strong inhibitory area at one end of its receptive field. As illustrated in **Figure 9-31,** a bar of light landing on the right side of the hypercomplex cell's receptive field excites the cell; but, if, for example, the bar lands mainly on the inhibitory area to the left, the cell's firing is inhibited.

Note that each class of V1 neurons responds to bars of light in some way, yet this response results from input originating in retinal ganglion cells that respond maximally not to bars but to spots of light. How does this conversion from responding to spots to responding to bars take place? An example will help explain the process.

A thin bar of light falls on the retinal photoreceptors, striking the receptive fields of perhaps dozens of retinal ganglion cells. The input to a V1 neuron comes from a group of ganglion cells that happen to be aligned in a row, as in **Figure 9-32.** That V1 neuron is activated (or inhibited) only when a bar of light hitting the retina strikes that particular row of ganglion cells. If the bar of light shines at a slightly different angle, only some of the retinal ganglion cells in the row are activated, so the V1 neuron is excited only weakly.

Figure 9-32 illustrates the connection between light striking the retina in a certain pattern and the activation of a simple cell in the primary visual cortex, one that responds

FIGURE 9-29 Typical Receptive Fields for Simple Visual-Cortex Cells Simple cells respond to a bar of light in a particular orientation, such as **(A)** horizontal or **(B)** oblique. The position of the bar in the visual field is important, because the cell either responds (ON) or does not respond (OFF) to light in adjacent regions of the visual field.

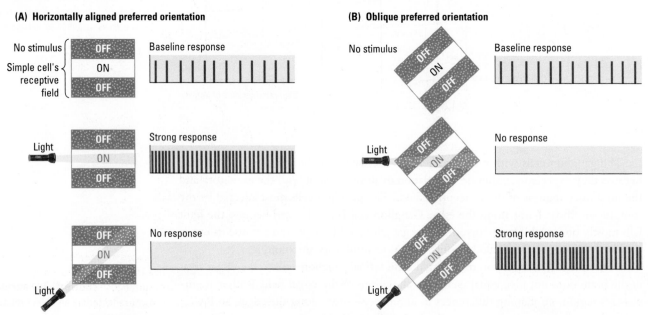

FIGURE 9-30 **Receptive Field of a Complex Cell** Unlike a simple cell's on-off response pattern, a complex cell in the visual cortex shows the same response throughout its circular receptive field, responding best to bars of light moving at a particular angle. The response is reduced or absent with the bar of light at other orientations.

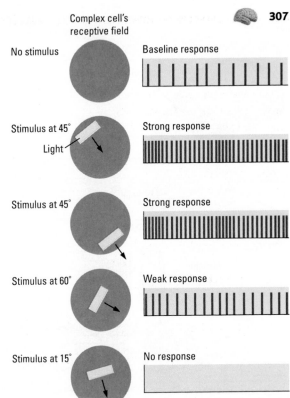

Complex cell's receptive field

No stimulus — Baseline response

Stimulus at 45° — Light — Strong response

Stimulus at 45° — Strong response

Stimulus at 60° — Weak response

Stimulus at 15° — No response

to a bar of light in a particular orientation. Using the same logic, we can also diagram the retinal receptive fields of complex or hypercomplex V1 neurons. Try it as an exercise yourself by adapting the format in Figure 9-32.

A characteristic of cortical structure is that the neurons are organized into functional columns. The pattern of connectivity in a column is vertical: inputs arrive in layer IV and then connect with cells in the other layers. **Figure 9-33** shows such a column, a 0.5-millimeter-diameter strip of cortex that includes representative neurons and their connections.

The neurons within a column have similar functions. For example, **Figure 9-34A** shows that neurons within the same column respond to lines oriented in the same direction. Adjacent columns house cells that are responsive to different line orientations. Figure 9-34B shows the columns of input coming from each eye, discussed earlier, called **ocular-dominance columns**. So the visual cortex has both orientation columns housing neurons of similar sensitivity and ocular-dominance columns with input from one eye or the other.

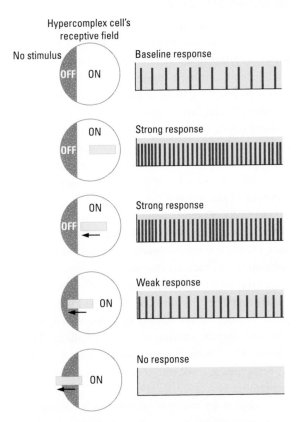

Hypercomplex cell's receptive field

No stimulus — OFF ON — Baseline response

ON OFF — Strong response

ON OFF — Strong response

ON — Weak response

ON — No response

FIGURE 9-31 **Receptive Field of a Hypercomplex Cell** A hypercomplex cell in the visual cortex responds to a moving bar of light in a particular orientation (horizontal, e.g.) anywhere in the excitatory (ON) part of its receptive field. If most of the bar extends into the inhibitory area (OFF), however, the response is inhibited.

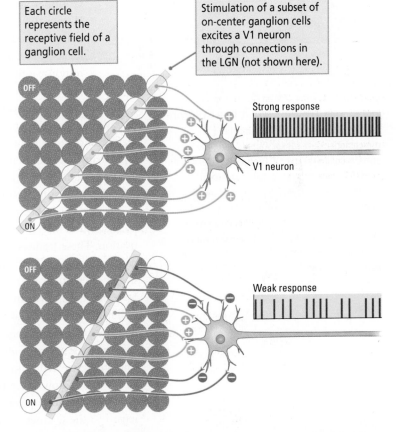

Each circle represents the receptive field of a ganglion cell.

Stimulation of a subset of on-center ganglion cells excites a V1 neuron through connections in the LGN (not shown here).

Strong response

V1 neuron

Weak response

FIGURE 9-32 **V1 Receptivity** A V1 cell responds to a row of ganglion cells in a particular orientation on the retina. The bar of light strongly activates a row of ganglion cells, each connected through the LGN to a V1 neuron. The activity of this V1 neuron is most affected by a bar of light at a 45° angle.

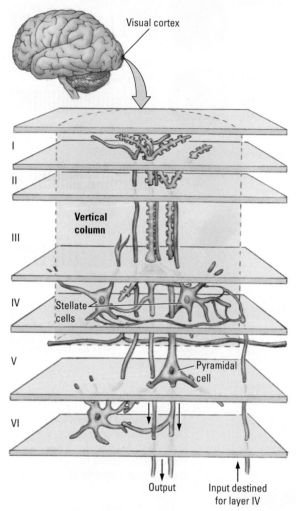

FIGURE 9-33 Neural Circuit in a Column in the Visual Cortex In this three-dimensional view, sensory inputs enter the cortical column at layer VI (bottom) and terminate on stellate cells in layer IV that synapse with pyramidal cells in layers III and V. The flow of information is vertical. Axons of the pyramidal cells leave the column to join other columns or structures. Adapted from "The 'Module-Concept' in Cerebral Architecture," by J. Szentagothai, 1975, *Brain Research*, 95, p. 490.

trichromatic theory Explanation of color vision based on the coding of three primary colors: red, green, and blue.

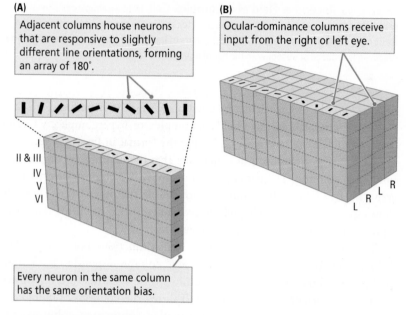

FIGURE 9-34 Organization of Functional Columns in V1

Processing Shape in the Temporal Cortex

Consider neurons along the ventral stream in temporal-lobe region TE. Rather than being responsive to spots or bars of light, TE neurons are maximally excited by complex visual stimuli, such as faces (see Figure 9-19A) or hands, and can be remarkably specific in their responsiveness. They may be responsive to particular faces seen head-on, to faces viewed in profile, to the posture of the head, or even to particular facial expressions.

How far does this specialized responsiveness extend? Would it be practical to have visual neurons in the temporal cortex specialized to respond to every conceivable feature of objects? Keiji Tanaka (1993) approached this question by presenting monkeys with many three-dimensional representations of animals and plants to find stimuli that are effective in activating particular neurons of the inferior temporal cortex.

Having identified stimuli that were especially effective, such as faces or hands, he then wondered which specific features of those stimuli are critical to stimulating the neurons. Tanaka found that most neurons in area TE require rather complex features for their activation. These features include a combination of characteristics such as orientation, size, color, and texture. Furthermore, neurons with similar, although slightly different, responsiveness to particular features tend to cluster together in columns, as shown in **Figure 9-35**.

Apparently, then, an object is represented not by the activity of a single neuron but rather by the activity of many neurons with slightly varying stimulus specificity. These neurons are grouped together in a column. This finding is important because it provides an explanation for *stimulus equivalence,* recognizing an object as remaining the same despite being viewed from different orientations.

Think of how the representation of objects by multiple neurons in a column can produce stimulus equivalence. If each neuron in the column module varies slightly in regard to the features to which it responds but the effective stimuli largely overlap, the effect of small changes in incoming visual images will be minimized, and we will continue to perceive an object as itself.

The stimulus specificity of neurons in the inferior temporal cortex in monkeys shows remarkable neuroplasticity. If monkeys are trained to discriminate particular shapes to obtain a food reward, not only do they improve their discriminatory ability but neurons in the temporal lobe also modify their preferred stimuli to fire maximally to some of the stimuli used in training. This result shows that the temporal lobe's role in visual processing is not determined genetically but is instead subject to experience, even in adults.

We can speculate that this neuroplastic characteristic evolved because it allows the visual system to adapt to different demands in a changing visual environment. Think of how different the demands on your visual recognition abilities are when you move from a dense forest to a treeless plain to a city street. The visual neurons of your temporal cortex can adapt to these differences (Tanaka, 1993). Experience-dependent visual neurons ensure in addition that people can identify visual stimuli that were never encountered as the human brain evolved.

Note that the preferred stimuli of neurons in the primary visual cortex are not modified by experience. This implies that the stimulus preferences of V1 neurons are genetically programmed. Regardless, the functions of the V1 neurons provide the building blocks for the more-complex and flexible characteristics of the inferior temporal cortex neurons.

Seeing Color

Scientists have long wondered why—and how—we see a world so rich in color. One hypothesis about "why" is that color vision evolved first in the great apes, specifically, in apes that eat fruit. Chimpanzees and humans are among this group. Over their evolution, both species have faced plentiful competition for ripe fruits—from other animals, insects, and each other. Scientists suspect that color vision gave them an evolutionary advantage.

An explanation of "how" we see color has its roots in the Renaissance that began 600 years ago in Italy. Painters of the time discovered that they could obtain the entire range of colors in the visual world by mixing only three colors of paint (red, blue, and yellow). This is the process of *subtractive color mixing* shown in **Figure 9-36**A.

We now know that such trichromatic color mixing is a property of the cones in the retina. Subtractive color mixing works by removing light from the mix. This is why black surfaces reflect no light: the darker the color, the less light it contains.

Conversely, the process of *additive color mixing* increases light to create color (Figure 9-36B). The lighter the color, the more light it contains, which is why a white surface reflects the entire visible spectrum of light. The primary colors of light are red, blue, and (unlike the primary colors of pigments used by painters) green. Light of different wavelengths stimulates the three different cone receptor types in different ways. It is the ratio of the activity of these three receptor types that creates our impression of different colors.

Trichromatic Theory

According to the **trichromatic theory,** the color we see—say, blue at short 400-nanometer wavelengths, green at medium 500 nm and red at long 600 nm—is determined by the relative responses of the different cone types (see Figure 9-6). If all three cone types are equally active, we see white.

The trichromatic theory predicts that, if we lack one type of cone receptor, we cannot process as many colors as we could with all three. This is exactly what happens when a person is born with only two cone types. The colors the person is unable to perceive

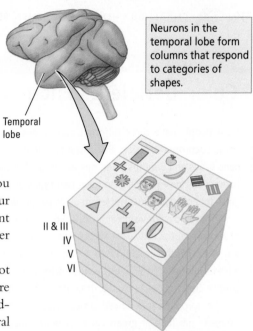

Neurons in the temporal lobe form columns that respond to categories of shapes.

Temporal lobe

I
II & III
IV
V
VI

FIGURE 9-35 Columnar Organization of Area TE Neurons with similar but slightly different pattern selectivity cluster in vertical columns, perpendicular to the cortical surface.

Section 1-4 recounts several ideas on how the primate lifestyle, including diet, encouraged evolution of their complex nervous systems.

(A)

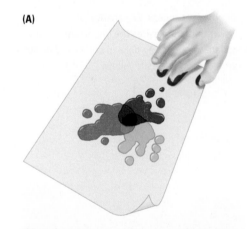

(B)

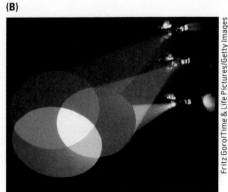

Fritz Goro/Time & Life Pictures/Getty Images

FIGURE 9-36 Color Mixing **(A)** Subtractive color mixing absorbs light waves that we see as red, blue, or yellow. When all visible wavelengths are absorbed, we see black. **(B)** Additive color mixing reflects light waves that we see as red, blue, and green. When all visible wavelengths are reflected, we see white.

Color-Deficient Vision

Most people have three different cone types in the retina and thus enjoy trichromatic vision. But some people are missing one or more cone types and are thus often mistakenly said to be "color-blind." Mistakenly, because people who have two types of cones still can distinguish lots of colors, just not as many as people with three cones can.

To have no color vision at all, one would have to have only one type of photoreceptor, rods. This occurrence is rare, but the authors do have a friend with the condition: he has no concept of color. It has led to a lifetime of practical jokes because others (especially his wife) must choose clothing colors that coordinate for him to wear.

The complete lack of red cones leads to a condition called *protanopia*; the lack of green cones is *deuteranopia*; the lack of blue cones is *tritanopia*. The frequency of each condition is about 1.00 percent in men and 0.01 percent in women. Having only a partial lack of one of the cones, most commonly green cones, also is possible. This condition afflicts about 5.0 percent of men and 0.4 percent of women.

Robert Snowden, Peter Thompson, and Tom Troscianko (2008) asked what it would look like to see with one cone missing. The adjoining illustration provides a simple approximation. People with protanopia, deuteranopia, and tritanopia still see plenty of color, but it is largely different from the color trichomats see.

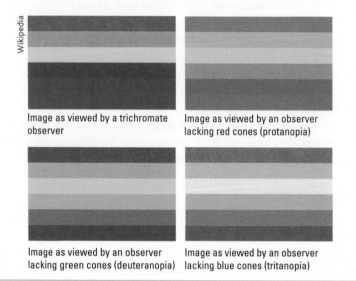

Image as viewed by a trichromate observer

Image as viewed by an observer lacking red cones (protanopia)

Image as viewed by an observer lacking green cones (deuteranopia)

Image as viewed by an observer lacking blue cones (tritanopia)

opponent process Explanation of color vision that emphasizes the importance of the apparently opposing pairs of colors: red versus green and blue versus yellow.

depend on which receptor type is missing, as illustrated in Research Focus 9-3, "Color-Deficient Vision."

The mere presence of cones in an animal's retina does not mean that the animal has color vision. It simply means that the animal has photoreceptors that are particularly sensitive to light. Many animals lack color vision as we know it, but the only animal with eyes known to have no cones at all is a fish, the skate.

Opponent Processes

Although the beginning of color perception in the cones follows the trichromatic model, succeeding levels of color processing use a different strategy. Try staring at the red and blue box in **Figure 9-37** for about 30 seconds and then at the white box next to it. When you shift your gaze to the white surface, you will experience a color afterimage in the colors opposite to red and blue—green and yellow. Conversely, if you stare at a green and yellow box and then shift to white, you will see a red and blue afterimage.

FIGURE 9-37 Demonstrating Opposing Color Pairs Stare at the rectangle on the left for about 30 seconds. Then stare at the white box on the right. You will experience an afterimage of green on the red side and of yellow on the blue side.

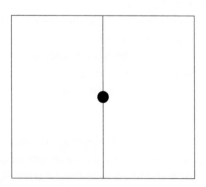

Such afterimages lead to the sense that there are actually four "basic" colors (red, green, yellow, and blue).

A characteristic of RGCs explains the presence of two opposing pairs of four basic colors. Remember that RGCs have an on–off and center–surround organization. Stimulation to the center of the cell's receptive field is either excitatory (in some cells) or inhibitory (in other cells), whereas stimulation to the periphery of the receptive field has the opposite effect (see Figure 9-26).

This arrangement can be adapted to create color-opponent cells. If one wavelength of light produced excitation and another inhibition, cells would evolve that are excited by red and inhibited by green (or vice versa), as would cells that are excited by blue and inhibited by yellow (or vice versa). Red–green and blue–yellow would therefore be linked to each other as color opposites, or "opponents."

In fact, about 60 percent of human retinal ganglion cells are color-sensitive in this way, with the center responsive to one wavelength and the surround to another. The most common **opponent-process** pairing, shown in **Figure 9-38,** is medium-wavelength (green) versus long-wavelength (red), but blue versus yellow RGCs also exist. Most likely, opponent-process cells evolved to enhance the relatively small differences in spectral absorption among the three types of cones.

Cortical neurons in region V1 also respond to color in an opponent-process manner reminiscent of retinal ganglion cells. Recall that color inputs in the primary visual cortex go to the blobs that appear in sections stained for cytochrome oxidase (see Figure 9-17). These blobs are where the color-sensitive cells are found.

Figure 9-39 models how the color-sensitive cells in the blobs are inserted amid the orientation-sensitive and ocular-dominance columns. Thus the primary visual cortex appears to be organized into modules that include ocular-dominance and orientation-sensitive columns as well as blobs. Think of V1 as composed of several thousand modules, each analyzing color and contour for a particular region of the visual world. This organization allows the primary visual cortex to perform several functions concurrently.

How do neurons in the visual system beyond region V1 process color? You have already learned that cells in region V4 respond to color, but, in contrast with the cells

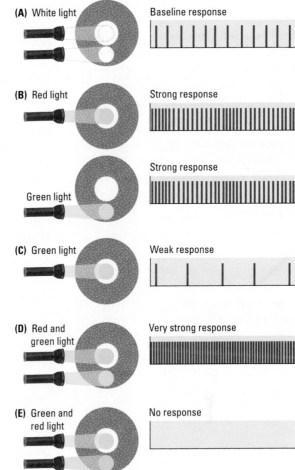

FIGURE 9-38 Opponent-Color-Contrast Response **(A)** A red–green color-sensitive RGC responds weakly to white-light illumination on its center and surround because red and green cones absorb white light to similar extents. Their inputs cancel out. **(B)** The cell responds strongly to a spot of red light in its center as well as to red's paired wavelength, green, in the surround. **(C)** It is strongly inhibited by a small spot of green in its center. **(D)** The RGC responds very strongly to simultaneous illumination of the center with red and the surround with green, and **(E)** it is completely inhibited by the simultaneous illumination of the center with green and the surround with red.

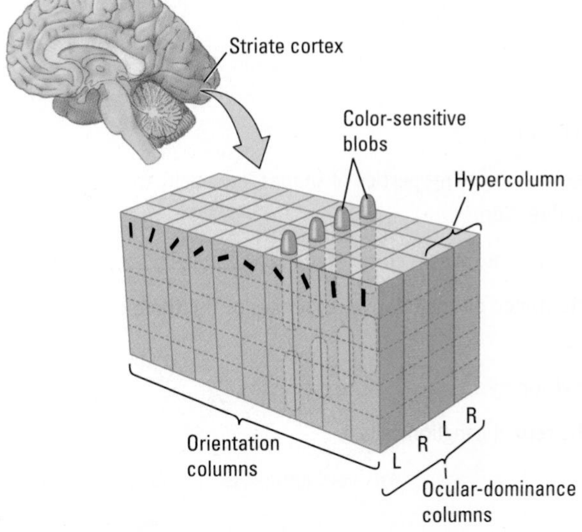

FIGURE 9-39 V1 Modules A model of striate cortex showing the orientation-sensitive columns, ocular-dominance columns, and color-sensitive blobs as composed of two hypercolumns. Each consists of a full set (red and blue) of orientation-sensitive columns spanning 180° of preferred angle as well as a pair of blobs. All cells in the hypercolumn share the same receptive field.

Striate cortex

Color-sensitive blobs

Hypercolumn

Orientation columns

Ocular-dominance columns

color constancy Phenomenon whereby the perceived color of an object tends to remain constant relative to other colors, regardless of changes in illumination.

homonymous hemianopia Blindness of an entire left or right visual field.

quadrantanopia Blindness of one quadrant of the visual field.

scotoma Small blind spot in the visual field caused by migraine or by a small lesion of the visual cortex.

in region V1, these V4 cells do not respond to particular wavelengths. Rather, they are responsive to different perceived colors, with the center of the field being excited by a certain color and the surround being inhibited.

Speculation swirls about the function of these V4 cells. One idea is that they are important for **color constancy,** the property of color perception whereby colors appear to remain the same relative to one another despite changes in light. For instance, if you were to look at a bowl of fruit through light-green glasses, the fruit would take on a greenish tinge, but bananas would still look yellow relative to red apples. If you removed all the fruit except the bananas and looked at them through the tinted glasses, the bananas would appear green because the color you perceive would not be relative to any other. Monkeys with V4 lesions lose color constancy, though they can discriminate different color wavelengths.

Neuronal Activity in the Dorsal Stream

A striking characteristic of many cells in the visual areas of the parietal cortex is that they are virtually silent to visual stimulation when a person is under anesthesia. This characteristic is true of neurons in the posterior parietal regions of the dorsal stream. In contrast, cells in the temporal cortex do respond to visual stimulation even when a person is anesthetized.

The silence on the part of neurons in the posterior parietal cortex under anesthesia makes sense if their role is to process visual information for action. In the absence of action when a person is unconscious, there is no need for processing. Hence, the cells are quiescent.

Cells in the dorsal stream are of many types, their details varying with the nature of the movement in which a particular cell is taking part. One interesting category of cells processes the visual appearance of an object to be grasped. For instance, if a monkey is going to pick up an apple, these cells respond even when the monkey is only looking at the apple. The cells do not respond when the monkey encounters the same apple in a situation where no movement is to be made.

Curiously, these cells respond if the monkey merely watches another monkey making movements to pick up the apple. Apparently, the cells have some sort of "understanding" of what is happening in the external world. But that understanding is always related to action performed with respect to visually perceived objects. These cells are what led David Milner and Mel Goodale (2006) to conclude that the dorsal stream is a "how" visual system.

REVIEW 9-4

Neuronal Activity

Before you continue, check your understanding.

1. Neurons in the primary visual cortex respond to properties of shapes, especially to _____ oriented in a certain direction.

2. Recognition of complex visual stimuli such as faces is completed in the _____ lobe.

3. The idea that the color we see is determined by the relative responses of the three cone types in the retina is called _____.

4. Retinal ganglion cells mediate color vision by _____ processes.

5. Describe the opponent process in the retinal ganglion cells.

Answers appear at the back of the book.

9-5 The Visual Brain in Action

Anatomical and physiological studies of brain systems leave one key question unanswered: How do all the cells in these systems act together to produce a particular function? One way to answer this question is to evaluate what happens when parts of the visual system are dysfunctional. Then we can see how these parts contribute to the workings of the whole. We use this strategy to examine the neuropsychology of vision—the study of the visual brain in action.

Injury to the Visual Pathway Leading to the Cortex

What happens when various parts of the visual pathway leading from the eye to the cortex are injured? For instance, destruction of the retina or optic nerve of one eye produces *monocular blindness,* the loss of sight in that eye. Partial destruction of the retina or optic nerve produces a partial loss of sight in one eye, with the loss restricted to the region of the visual field that has severed connections to the brain.

Injuries to the visual pathway beyond the eye also produce blindness. For example, complete cuts of the optic tract, the LGN, or region V1 of the cortex result in **homonymous hemianopia,** blindness of one entire side of the visual field, as shown in **Figure 9-40A.** We encountered this syndrome in Focus 9-1, the story of D. B.'s lesion in region V1. Should a lesion in one of these areas be partial, as is often the case, the result is **quadrantanopia,** destruction of only a part of the visual field, illustrated in Figure 9-40B.

Figure 9-40C shows that small lesions in V1 often produce small blind spots, or **scotomas,** in the visual field. Focus 9-1 observes that, for migraine sufferers, scotomas are a warning symptom. But brain-injured people are often totally unaware of them. One reason is that the eyes are usually moving.

We make tiny, involuntary eye movements almost constantly. Because of this usually constant eye motion, called *nystagmus,* a scotoma moves about the visual field, allowing the intact regions of the brain to perceive all the information in that field. If the eyes are temporarily held still, the visual system actually compensates for a scotoma through pattern completion—filling in the hole so to speak—so that the people and objects in the visual world are perceived as whole. The result is a seemingly normal set of perceptions.

The visual system may cover up a scotoma so successfully that its presence can be demonstrated to the patient only by "tricking" the visual system. The trick is to place an object entirely within the scotoma and, without allowing the patient to shift gaze, asking what the object is. If the patient reports seeing nothing, the examiner moves the object out of the scotoma so that it suddenly "appears" in the intact region of the visual field to confirm the existence of a blind area.

This technique is similar to demonstrating the presence of the blind spot that is due to the optic disc (as in Figure 9-4). When a person is looking at an object with only one eye, the brain compensates for the scotoma in the same way as it does for the optic-disc blind spot. As a result, the person does not notice the scotoma.

FIGURE 9-40 Consequences of Lesions in Region V1 The shaded areas indicate regions of visual loss. **(A)** A complete lesion of V1 in the left hemisphere results in hemianopia affecting the right visual field. **(B)** A large lesion of the lower lip of the calcarine fissure produces quadrantanopia that affects most of the upper-right visual quadrant. **(C)** A smaller lesion of the lower lip of the calcarine fissure results in a smaller scotoma.

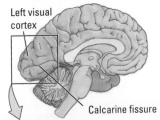

Left visual cortex

Calcarine fissure

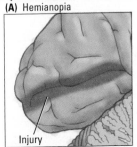

(A) Hemianopia

Injury

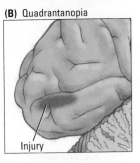

(B) Quadrantanopia

Injury

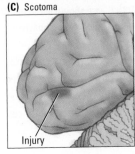

(C) Scotoma

Injury

Left visual field   Right visual field

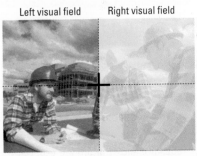

Photographs by Jim Pickerell/Stock Connection/PictureQuest

visual-form agnosia Inability to recognize objects or drawings of objects.

Thus the type of blindness that a person suffers gives clues about where in the visual pathway the cause of the problem lies. If the loss of vision is in one eye only, the problem must be in that eye or its optic nerve; if the loss of vision affects both eyes, the problem is most likely in the brain. Many people have difficulty understanding why a person with damage to the visual cortex has difficulty with both eyes. They fail to realize that it is the visual field, not the eye, that is represented in the brain.

Beyond region V1, the nature of visual loss caused by injury is considerably more complex. It is also very different in the ventral and dorsal streams. We therefore look at each pathway separately.

Injury to the "What" Pathway

While taking a shower, D. F., a 35-year-old woman, suffered carbon monoxide poisoning from a faulty gas-fueled water heater. The length of her exposure to the carbon monoxide is unclear, but when her roommate found her, the shower water was cold. Although carbon monoxide poisoning can cause several kinds of neurological damage, as discussed in Clinical Focus 9-4, "Carbon Monoxide Poisoning," the result in D. F. was an extensive lesion of the lateral occipital region, including cortical tissue in the ventral visual pathway.

Section 9-2 describes damage to the temporal lobe area that casues facial agnosia.

D. F.'s principal deficit was **visual-form agnosia,** an inability to recognize objects, real or drawn (see Farah, 1990). Not only was D. F. unable to recognize objects, especially

Section 9-2 describes damage to the temporal lobe area that casues facial agnosia.

CLINICAL FOCUS ⚛ 9-4

Carbon Monoxide Poisoning

Brain damage from carbon monoxide (CO) poisoning is usually caused either by a faulty furnace or by motor-vehicle exhaust fumes. The blood absorbs carbon monoxide gas, resulting in swelling and bleeding of the lungs and anoxia (loss of oxygen) in the brain. The cerebral cortex, hippocampus, cerebellum, and striatum are especially sensitive to CO-induced anoxia.

Only a small proportion of people who succumb to carbon monoxide poisoning have permanent neurological symptoms. Among those who do, the symptoms are highly variable. The most common symptoms are cortical blindness and various forms of agnosia, as seen in D. F. In addition, many victims suffer language difficulties.

The peculiarities of the language difficulties are shown clearly in a young woman whose case was described by Norman Geschwind (1972). Geschwind studied this patient for 9 years after her accidental poisoning. She required complete nursing care during this time, never uttered spontaneous speech, and did not comprehend spoken language. Nonetheless, she could repeat with perfect accuracy sentences that had just been said to her.

Areas damaged by carbon monoxide poisoning are shown in red in this postmortem diagram of Geschwind's patient's brain.

She could also complete certain well-known phrases. For example, if she heard "Roses are red," she would say, "Roses are red, violets are blue, sugar is sweet, and so are you." She could also learn new songs. She did not appear to understand the content of the songs, yet with only a few repetitions, she began to sing along. Eventually, she could sing the song spontaneously, making no errors in either words or melody.

Postmortem examination of this woman's brain found that, although she had extensive damage to the parietal and temporal lobes, as shown in the accompanying diagram, her speech areas were intact. Geschwind proposed that she could not comprehend speech because the words that she heard did not arouse associations in other parts of her cortex.

She could, however, repeat sentences because the internal connections of the speech regions were undamaged. Geschwind did not comment on whether this woman suffered from agnosia, but it is likely that she did. The difficulty would be in diagnosing agnosia in a person who is unable to communicate.

line drawings of objects, she could neither estimate their size and their orientation nor copy drawings of objects. Yet, interestingly, as **Figure 9-41** illustrates, although she did not recognize what she was drawing, D. F. could draw reasonable facsimiles of objects from memory. D. F. clearly had a lesion that interfered with her ventral-stream "what" pathway.

Remarkably, despite her inability to identify objects or to estimate their size and orientation, D. F. still retained the capacity, illustrated in **Figure 9-42,** to appropriately shape her hand when reaching out to grasp something. Goodale, Milner, and their research colleagues (1991) studied D. F. extensively for years, and they have devised a way to demonstrate D. F.'s skill at reaching for objects.

The middle column in **Figure 9-43** shows the grasp patterns of a control participant (S. H.) when she picks up something irregularly shaped. S. H. grasps the object along one of two different axes that makes it easiest to pick up. When D. F. is presented with the same task, shown in the left-hand column, she is as good as S. H. at placing her index finger and thumb on appropriately opposed "grasp" points.

Clearly, D. F. remains able to use the structural features of objects to control her visually guided grasping movements, even though she is unable to "perceive" these same features. This result demonstrates once more that we are consciously aware of only a small part of the sensory processing that goes on in the brain. Furthermore, D. F.'s ability to use structural features of objects for guiding movement but not for perceiving shapes again shows us that the brain has separate systems for these two types of visual operations.

D. F.'s lesion is located quite far back in the ventral visual pathway. Lesions located more anteriorly produce other types of deficits, depending on the exact location. For example, J. I., described by Oliver Sacks and Robert Wasserman (1987), was an artist who became color-blind owing to a cortical lesion presumed to be in region V4. His principal symptom was achromatopsia, or color agnosia. Despite his inability to distinguish any colors whatsoever, J. I.'s vision appeared otherwise unaffected.

Similarly, L. M., a woman described by Josef Zihl and his colleagues (1983), lost her ability to detect movement after suffering a lesion presumed to be in region V5. In her case, objects either vanished when they moved or appeared frozen despite their

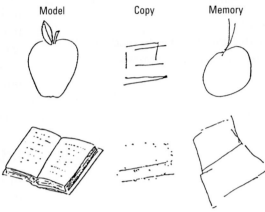

FIGURE 9-41 Injury to the Ventral Stream D. F. recognized neither of the two drawings on the left. Nor was she able to make recognizable copies of the drawings (center column). She was able to draw reasonable renditions from memory (right), but later, when D. F. was shown her drawings, she had no idea what they were. Adapted from *The Visual Brain in Action* (p. 127), by A. D. Milner and M. A. Goodale, 1995, Oxford: Oxford University Press.

FIGURE 9-42 Visual Guidance You may consciously reach for an object such as a pen or a mug, but your hand forms the appropriate posture automatically, without your conscious awareness. Figure 11-1 details this type of sequentially organized movement.

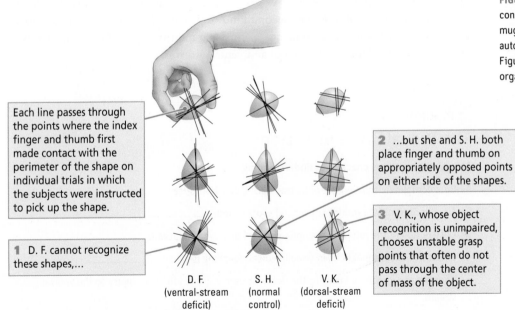

Each line passes through the points where the index finger and thumb first made contact with the perimeter of the shape on individual trials in which the subjects were instructed to pick up the shape.

1 D. F. cannot recognize these shapes,…

2 …but she and S. H. both place finger and thumb on appropriately opposed points on either side of the shapes.

3 V. K., whose object recognition is unimpaired, chooses unstable grasp points that often do not pass through the center of mass of the object.

D. F.
(ventral-stream deficit)

S. H.
(normal control)

V. K.
(dorsal-stream deficit)

FIGURE 9-43 Grasp Patterns The Milner–Goodale experiments confirm that the brain has different systems for visual object recognition and visual guidance of movement. Adapted with permission from *The Visual Brain in Action*, 2nd ed. (p. 132), by A. D. Milner and M. A. Goodale, 2006, Oxford: Oxford University Press.

optic ataxia Deficit in the visual control of reaching and other movements.

Section 11-5 details the somatosenses, including proprioception, or body awareness.

movement. L. M. had particular difficulty pouring tea into a cup, because the fluid appeared to be frozen in midair. Yet she could read, write, and recognize objects, and she appeared to have normal form vision—until objects moved.

These varied cases demonstrate that cortical injuries in the ventral stream all somehow interfere with determining "what" things are or are like. In each case, the symptoms are somewhat different, however, which is thought to be indicative of damage to different subregions or substreams of the ventral visual pathway.

Injury to the "How" Pathway

In 1909, R. Balint described a rather peculiar set of visual symptoms associated with a bilateral parietal lesion. The patient had full visual fields and could recognize, use, and name objects, pictures, and colors normally. But he had a severe deficit in visually guided reaching, even though he could still make accurate movements directed toward his own body (presumably guided by tactile or proprioceptive feedback from his joints). Balint called this syndrome **optic ataxia.**

Since Balint's time, many descriptions of optic ataxia associated with parietal injury have been recorded. Goodale has studied several such patients, one of whom is a woman identified as R. V. (Milner & Goodale, 2006). In contrast with patient D. F.'s visual-form agnosia, R. V.'s perception of drawings and objects was normal, but she could not guide her hand to reach for objects.

The rightmost column in Figure 9-43 shows that, when asked to pick up the same irregularly shaped objects that D. F. could grasp normally, R. V. often failed to place her fingers on the appropriate grasp points, even though she could distinguish the objects easily. In other words, although R. V.'s perception of the features of an object was normal for the task of describing that object, her perception was not normal for the task of visually guiding her hand to reach for the object.

To summarize, people with damage to the parietal cortex in the dorsal visual stream can "see" perfectly well, yet they cannot accurately guide their movements on the basis of visual information. Guidance of movement is the function of the dorsal stream. In contrast, people with damage to the ventral stream cannot "see" objects, because perception of objects is a ventral-stream function. Yet these same people can guide their movements to objects on the basis of visual information.

The first kind of patient, like R. V., has an intact ventral stream that analyzes the visual characteristics of objects. The second kind of patient, like D. F., has an intact dorsal stream that visually directs movements. Comparing the two types of cases enables us to infer the visual functions of the dorsal and ventral streams.

REVIEW 9-5

The Visual Brain in Action

Before you continue, check your understanding.

1. Cuts completely through the optic tract, LGN, or V1 produce _____.

2. Small lesions of V1 produce small blind spots called _____.

3. Destruction of the retina or the optic nerve of one eye produces _____.

4. The effect of severe deficits in visually guided reaching is called _____.

5. Contrast the effects of injury to the dorsal stream and the effects of injury to the ventral stream.

Answers appear at the back of the book.

SUMMARY

9-1 Nature of Sensation and Perception

Sensory systems allow animals, including ourselves, to adapt. Animals adapted to different environments vary widely in their sensory abilities. What is distinctive about humans is the extent to which we can transform sensations into perceptual information to mediate many aspects of language, music, and culture. For each sense, mammals represent the world in topographic maps that form neural–spatial representations in the cortex.

9-2 Functional Anatomy of the Visual System

Like all sensory systems, vision begins with receptor cells. The visual photoreceptors (rods and cones), located in the retina at the back of the eye, transduce the physical energy of light waves into neural activity.

Rods are sensitive to dim light. Cones are sensitive to bright light and mediate color vision. Each of the three cone types is maximally sensitive to a different wavelength—short, medium, or long. We see these wavelengths, respectively, as the colors blue, green, or red, and thus the cone receptor types often are referred to as blue, green, or red.

Retinal ganglion cells receive input from photoreceptors through bipolar cells and send their axons out from the eyes' retinas to form the optic nerve. P ganglion cells receive input mostly from cones and convey information about color and fine detail. M cells receive input from rods and convey information about luminance and movement but not color.

The optic nerve forms two distinct major routes into the brain. The geniculostriate pathway synapses first in the lateral geniculate nucleus of the thalamus and then in the primary visual cortex. The tectopulvinar pathway synapses first in the superior colliculus of the midbrain's tectum, then in the pulvinar of the thalamus, and finally in the visual cortex of the temporal and parietal lobes. A few optic-nerve fibers also form the retinohypothalamic tract that functions in part to control circadian rhythms.

Among the visual regions in the occipital cortex, V1 and V2 carry out multiple functions; the remaining regions (V3, V3A, V4, and V5) are more specialized. Visual information flows from the thalamus to V1 and V2 and then divides to form the two distinctly different visual streams. The unconscious dorsal stream aids in the visual guidance of movements, whereas the conscious ventral stream aids in object perception.

9-3 Location in the Visual World

At each step along the visual pathways, neuronal activities are distinctly different, yet the sum of the neural activity in all regions is our visual experience. Like all cortical regions, each functional column in the visual regions is about 0.5 millimeter in diameter and extends to the depth of the cortex. Columns in the visual system are specialized for processes such as analyzing lines of a particular orientation or comparing similar shapes, such as faces.

9-4 Neuronal Activity

Neurons in the ventral stream are selective for the different characteristics of shape. In the visual cortex, cells are maximally responsive to lines of different orientations. Cells in the inferior temporal cortex are responsive to different shapes, which, in some cases, appear to be abstract and, in other cases, have concrete forms such as hands or faces.

Cones in the retina are maximally responsive to different wavelengths of light, roughly corresponding to the perception of green, blue, and red. RCGs have a center–surround organization that facilitates their are opponent-process function—the cells are excited by one hue and inhibited by another (e.g., red versus green; blue versus yellow).

Color-sensitive cells in V1, the primary visual cortex, are located in the blobs and also have opponent-process properties. Cells in region V4 respond to the colors that we perceive rather than to particular wavelengths of visible light. Perceived color is influenced both by luminance and by the color of nearby objects.

9-5 The Visual Brain in Action

When visual information enters the brain, information from the left and right visual fields goes to the right and left sides of the brain, respectively. As a result, damage to the visual areas on one side of the brain results in visual disturbance in both eyes.

Specific visual functions are localized to different brain regions, so localized damage to a particular region results in the loss of a particular function. Damage to region V4 produces a loss of color constancy, for example; damage to regions in the parietal cortex inhibits the ability to shape the hand appropriately to grasp objects.

As summarized in the illustration, the visual streams perform two distinctly different functions: object recognition (the what) in the ventral stream and visual action (the how) in the dorsal stream. We are largely unconscious of ongoing "online analysis" in the dorsal stream that allows us to make accurate movements related to objects.

The ventral stream begins in V1 and flows through V2 to V3 and V4 into the temporal visual areas. The dorsal stream begins in V1 and flows through V5 and V3A to the posterior parietal visual areas. The double-headed arrows show that information flows back and forth between the dorsal and ventral streams—between recognition and action

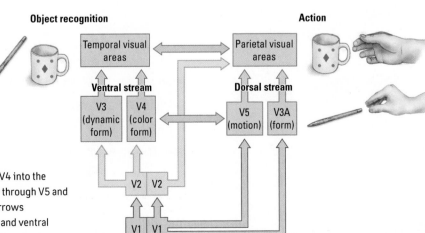

KEY TERMS

auditory flow, p. 284

blind spot, p. 291

blob, p. 298

color constancy, p. 312

cone, p. 291

cortical column, p. 297

extrastriate (secondary visual) cortex, p. 298

facial agnosia, p. 301

fovea, p. 291

geniculostriate system, p. 294

homonymous hemianopia, p. 312

luminance contrast, p. 305

magnocellular (M) cell, p. 293

ocular-dominance column, p. 306

opponent process, p. 310

optic ataxia, p. 316

optic chiasm, p. 294

optic flow, p. 284

parvocellular (P) cell, p. 293

perception, p. 287

photoreceptor, p. 287

primary visual cortex (V1), p. 298

quadrantanopia, p. 312

receptive field, p. 284

retina, p. 287

retinal ganglion cell (RGC), p. 293

retinohypothalamic tract, p. 294

rod, p. 291

scotoma, p. 312

sensation, p. 287

striate cortex, p. 294

tectopulvinar system, p. 294

topographic map, p. 287

trichromatic theory, p. 308

visual field, p. 301

visual-form agnosia, p. 314

Please refer to the Companion Web Site at www.worthpublishers.com/kolbintro4e *for Interactive Exercises and Quizzes.*

CHAPTER

10

How Do We Hear, Speak, and Make Music?

RESEARCH FOCUS 10-1 EVOLUTION OF LANGUAGE AND MUSIC

10-1 SOUND WAVES: STIMULUS FOR AUDITION

PHYSICAL PROPERTIES OF SOUND WAVES

PERCEPTION OF SOUND

PROPERTIES OF LANGUAGE AND MUSIC AS SOUNDS

10-2 FUNCTIONAL ANATOMY OF THE AUDITORY SYSTEM

STRUCTURE OF THE EAR

AUDITORY RECEPTORS

PATHWAYS TO THE AUDITORY CORTEX

AUDITORY CORTEX

RESEARCH FOCUS 10-2 SEEING WITH SOUND

10-3 NEURAL ACTIVITY AND HEARING

HEARING PITCH

DETECTING LOUDNESS

DETECTING LOCATION

DETECTING PATTERNS IN SOUND

10-4 ANATOMY OF LANGUAGE AND MUSIC

PROCESSING LANGUAGE

CLINICAL FOCUS 10-3 LEFT-HEMISPHERE DYSFUNCTION

CLINICAL FOCUS 10-4 ARTERIOVENOUS MALFORMATIONS

PROCESSING MUSIC

CLINICAL FOCUS 10-5 CEREBRAL ANEURYSMS

RESEARCH FOCUS 10-6 THE BRAIN'S MUSIC SYSTEM

10-5 AUDITORY COMMUNICATION IN NONHUMAN SPECIES

BIRDSONG

ECHOLOCATION IN BATS

Rock band

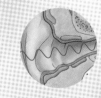

319

Evolution of Language and Music

The finding that modern humans (*Homo sapiens*) made music early on implies that music has been important in our evolution. Behavioral scientists have shown that music plays as central a role in our social and emotional lives as language does.

Thomas Geissmann (2001) noted that, among most of the 26 species of singing primates, males and females sing duets. All singing primates are monogamous, suggesting that singing may somehow relate to sexual behaviors. Music may also play a role in primates' parenting behaviors.

The modern human brain is specialized for analyzing certain aspects of music in the right temporal lobe, which is complemented by specialization for analyzing aspects of speech in the left temporal lobe. Did music and language evolve simultaneously in our species? Possibly.

Neanderthals (*Homo neanderthalensis*) have long fascinated researchers. The species originated about 300,000 years ago and disappeared about 30,000 years ago. During that time, it is likely that they coexisted in Europe and the Middle East with *Homo sapiens*, whom they resembled in many ways. In some locations, the two species may have even shared resources and tools.

Neanderthals had brains as large as or larger than the brains of *Homo sapiens* and appear to have shared many cultural similarities. Until recently, researchers hypothesized that Neanderthal culture was significantly less developed than that of early modern humans. Neanderthals buried their dead with artifacts, which implies that they held spiritual beliefs, but no conclusive evidence reveals that they created visual art. In contrast, *Homo sapiens* began painting on cave walls some 30,000 years ago, near the end of the Neanderthal era.

Anatomically, some skeletal analyses of the larynx suggest that Neanderthals' had poorer language ability than their *Homo sapiens* contemporaries. What about music? It appears that Neanderthals did have some form of music.

Shown in the accompanying photo is the bone flute found in 1995 by Ivan Turk, a paleontologist at the Slovenian Academy of Sciences in Ljubljana. Turk was excavating a cave in northern Slovenia used by Neanderthals long ago as a hunting camp. Buried in the cave among a cache of stone tools was the leg bone of a young bear that looked as if it had been fashioned into a flute.

The bone had holes aligned along one side that could not have been made by gnawing animals. Rather, the hole-spacing resembles the positions found on a modern flute. But the bone flute is at least 43,000 years old—perhaps as old as 82,000 years. All the evidence suggests that Neanderthals, not modern humans, made the instrument.

Bob Fink, a musicologist, analyzed the flute's musical qualities. He found that an eight-note scale similar to a do-re-mi scale could be played on the flute, but, compared with the scale most familiar in European music, one note was slightly off. That "blue note," a staple of jazz, is found as well in musical scales throughout Africa and India today.

The similarity between Neanderthal and contemporary musical scales encourages us to speculate about the brain that made this ancient flute. Like modern humans, Neanderthals probably had complementary hemispheric specialization for language and music. This may have played a role in the cohabitation of the two species and the interbreeding that led to 4 percent of Caucasian genes being of Neanderthal origin.

Ancient Bone Flute The hole alignment in this piece of bear femur, found in a cave in northern Slovenia, suggests that Neanderthals made a flute from the bone and made music with the flute. **Courtesy of Ivan Turk/Institut 2A Archeologijo, ZRC-Sazu, Slovenia**

Language and music are universal among humans. The oral language of every known culture follows similar basic structural rules, and people in all cultures create and enjoy music. Music and language allow us both to organize and to interact socially. Like music, language probably improves parenting. People who can communicate their intentions to one another and to their children presumably are better parents.

Humans' capacities for language and music are linked conceptually because both are based on sound. Understanding how and why we engage in speech and music is the goal of this chapter. We first examine the physical energy that we perceive as sound and then how the human ear and nervous system detect and interpret sound. We next examine the complementary neuroanatomy of human language and music processing. Finally, we investigate how two other species, birds and bats, interpret and utilize auditory stimuli.

sound wave Undulating displacement of molecules caused by changing pressure.

frequency Number of cycles that a wave completes in a given amount of time.

hertz (Hz) Measure of frequency (repetition rate) of a sound wave; 1 hertz is equal to 1 cycle per second.

10-1 Sound Waves: Stimulus for Audition

When you strike a tuning fork, the energy of its vibrating prongs displaces adjacent air molecules. **Figure 10-1** shows how, as one prong moves to the left, air molecules to the left compress (grow more dense) and air molecules to the right become more rarefied (grow less dense). The opposite happens when the prong moves to the right. The undulating energy generated by this displacement of molecules causes waves of changing air pressure—**sound waves**—to emanate from the fork. Sound waves can move through water, as well, and even through the ground.

What we experience as sound is a creation of the brain, as is what we see. Without a brain, sound and sight do not exist. Does a tree that falls in the forest make a sound if no one is there to hear it? The answer is no. A falling tree hitting the ground makes sound waves but not sound.

The top graph in **Figure 10-2** represents waves of changing air pressure emanating from a falling tree or a tuning fork by plotting air-molecule density against time at a single point. The bottom graph shows how the energy from the right-hand prong of the tuning fork moves to create the air-pressure changes associated with a single cycle. A *cycle* is one complete peak and valley on the graph—that is, the change from one maximum or minimum air-pressure level of the sound wave to the next maximum or minimum level, respectively.

Recall the principle from Section 2-6. The nervous system produces movement within a perceptual world the brain creates.

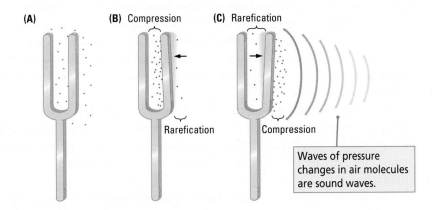

(A) **(B)** Compression **(C)** Rarefication

Rarefication Compression

Waves of pressure changes in air molecules are sound waves.

FIGURE 10-1 How a Tuning Fork Produces Sound Waves **(A)** The fork is still, and air molecules are distributed randomly. **(B)** When struck, the right arm of the fork moves to the left, air on the leading edge compresses and air on the trailing edge rarefies. **(C)** As the right arm moves to the right, air to the right compresses and air to the left rarefies.

Physical Properties of Sound Waves

Light is electromagnetic energy that we see; sound is mechanical energy that we hear. Sound-wave energy has three physical attributes—*frequency, amplitude,* and *complexity*—produced by the displacement of air molecules and summarized in **Figure 10-3**. The auditory system analyzes each property separately, just as the visual system analyzes color and form separately.

Section 9-4 explains how we see shapes and colors.

Sound-Wave Frequency

Although sound waves travel at a fixed speed of 1100 feet per second, sound energy varies in wavelength. **Frequency** is the number of cycles a wave completes in a given amount of time. Sound-wave frequencies are measured in cycles per second, called **hertz** (Hz), named after the German physicist Heinrich Rudolph Hertz.

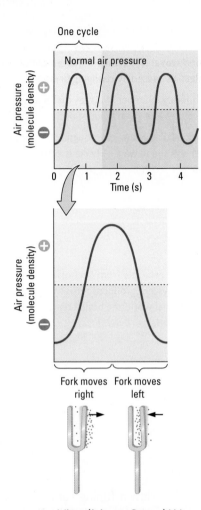

FIGURE 10-2 Visualizing a Sound Wave
Air-molecule density plotted against time at a particular point relative to the right prong of the tuning fork. Physicists call the resulting cyclical waves *sine waves*.

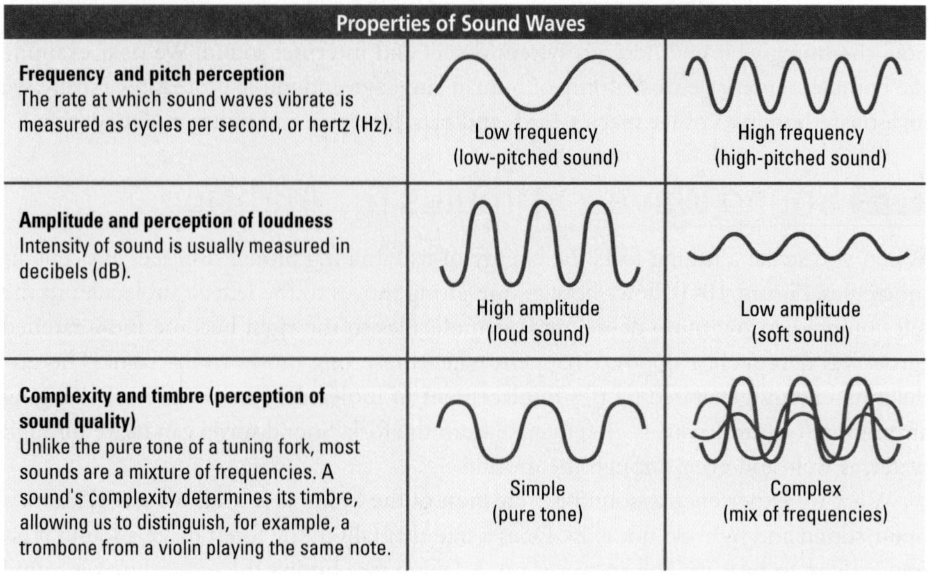

Properties of Sound Waves		
Frequency and pitch perception The rate at which sound waves vibrate is measured as cycles per second, or hertz (Hz).	Low frequency (low-pitched sound)	High frequency (high-pitched sound)
Amplitude and perception of loudness Intensity of sound is usually measured in decibels (dB).	High amplitude (loud sound)	Low amplitude (soft sound)
Complexity and timbre (perception of sound quality) Unlike the pure tone of a tuning fork, most sounds are a mixture of frequencies. A sound's complexity determines its timbre, allowing us to distinguish, for example, a trombone from a violin playing the same note.	Simple (pure tone)	Complex (mix of frequencies)

FIGURE 10-3 Physical Dimensions of Sound Waves The frequency, amplitude, and complexity of sound-wave sensations correspond to the perceptual dimensions of pitch, loudness, and timbre.

One hertz is 1 cycle per second, 50 hertz is 50 cycles per second, 6000 hertz is 6000 cycles per second, and so on. Sounds that we perceive as low in pitch have slower wave frequencies (fewer cycles per second), whereas sounds that we perceive as high pitched have faster wave frequencies (many cycles per second), as shown in the top panel of Figure 10-3.

Just as we can perceive light only at visible wavelengths, we can perceive sound waves only in the limited range of frequencies plotted in **Figure 10-4**. Humans' hearing range is from about 20 to 20,000 hertz. Many animals communicate with sound: their auditory systems are designed to interpret their species-typical sounds. After all, there is no point in making complicated songs or calls if other members of your species cannot hear and interpret them.

The range of sound-wave frequencies heard by different species varies extensively. Some species (such as frogs and birds) have rather narrow hearing ranges; others (such as dogs, whales, and humans) have broad ranges. Some species use extremely high frequencies (bats are off the scale in Figure 10-4); others use the low range (fish, for example).

The auditory systems of whales and dolphins are responsive to a remarkably wide range of sound waves. The characteristics at the extremes of these frequencies allow marine mammals to use them in different ways. Very-low-frequency sound waves travel long distances in water. Whales produce them for underwater communication over miles of distance. High-frequency sound waves create echoes and form the basis of sonar. Dolphins produce them in bursts, listening for the echoes that bounce back from objects and help the dolphins to navigate and locate prey.

Differences in the frequency of sound waves become differences in pitch when heard. Each note in a musical scale must have a different frequency because each has a different pitch. Middle C on the piano, for instance, has a frequency of 264 hertz.

Most people can discriminate between one musical note and another, but some can actually name any note they hear (A, B flat, C sharp, and so forth). This *perfect* (or *absolute*) *pitch* runs in families, suggesting a genetic influence. On the side of experience, most people who develop perfect pitch also receive musical training in matching pitch to note from an early age.

Section 8-4 suggests a critical period in brain development most sensitive to early musical training. Figure 15-11 shows enhanced neural connectivity in people with perfect pitch.

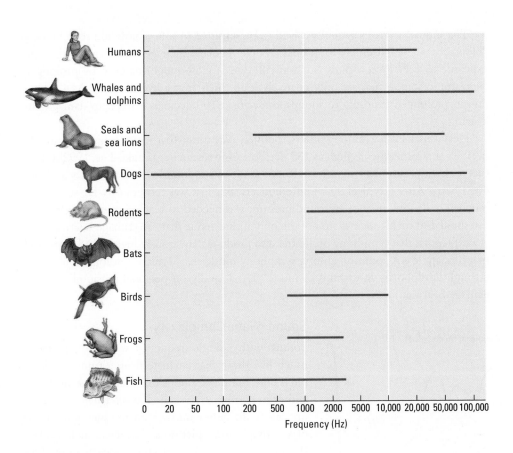

FIGURE 10-4 **Hearing Ranges among Animals** Frogs and birds hear a relatively narrow range of frequencies; whales' and dolphins' ranges are extensive, as are dogs'. Humans' hearing range is broad, yet we do not perceive many sound frequencies that other animals can both make and hear.

Sound-Wave Amplitude

Sound waves vary not only in frequency that causes differences in perceived pitch but also in **amplitude** (strength) that causes differences in perceived *intensity,* or *loudness.* If you hit a tuning fork lightly, it produces a tone with a frequency of, say, 264 hertz (middle C). If you hit it harder, the frequency remains 264 hertz but you also transfer more energy into the vibrating prong, increasing its amplitude.

The fork now moves farther left and right but at the same frequency. Increased compression of air molecules intensifies the energy in a sound wave, which "amps" the sound—makes it louder. Differences in amplitude are graphed by increasing the height of a sound wave, as shown in the middle panel of Figure 10-3.

Sound-wave amplitude is usually measured in **decibels** (dB), the strength of a sound relative to the threshold of human hearing as a standard, pegged at 0 decibels (**Figure 10-5**). Normal speech sounds, for example, measure about 40 decibels. Sounds that register more than about 70 dB we perceive as loud; those of less than about 20 dB we perceive as soft, or quiet.

amplitude Intensity of a stimulus; in audition, roughly equivalent to loudness, graphed by increasing the height of a sound wave.

decibel (dB) Unit for measuring the relative physical intensity of sounds.

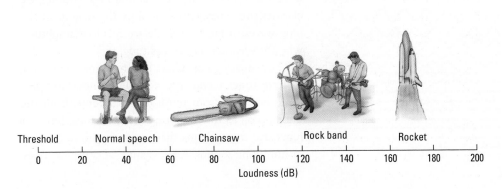

FIGURE 10-5 Sound Intensity

The human nervous system evolved to be sensitive to soft sounds and so is literally "blown away" by extremely loud ones. People regularly damage their hearing through exposure to very loud sounds (such as rifle fire at close range) or even by prolonged exposure to sounds that are only relatively loud (such as at a live concert). As a rule of thumb, prolonged exposure to sounds louder than 100 decibels is likely to damage our hearing.

Heavy-metal bands, among others, routinely play music that registers higher than 120 decibels and sometimes as high as 135 decibels. One researcher (Drake-Lee, 1992) found that rock musicians had a significant loss of sensitivity to sound waves, especially at about 6000 hertz. After a typical 90-minute concert, this loss was temporarily far worse—as much as a 40-fold increase in sound pressure was needed to reach a musician's hearing threshold. But rock concerts are not the only music venue that can damage hearing. Teie (1998) reports that symphony orchestras also produce dangerously high sound levels and that hearing loss is common among symphony musicians. Similarly, prolonged listening through headphones or earbuds to music played loudly on personal music players can damage hearing.

Sound-Wave Complexity

Sounds with a single frequency wave are *pure tones,* much like those that emanate from a tuning fork or pitch pipe, but most sounds mix wave frequencies together in combinations called *complex tones* (see Figure 10-3). To better understand the blended nature of a complex tone, picture a clarinetist, such as Don Byron in **Figure 10-6,** playing a steady note. The upper graph in Figure 10-6 represents the sound wave a clarinet produces.

The waveform pattern is more complex than the simple, regular waves visualized in Figures 10-2 or 10-3. Even when a musician plays a single note, the instrument is making a complex, not a pure, tone. Using a mathematical technique known as Fourier analysis, we can break this complex tone into its many component pure tones, the numbered waves traced at the bottom of Figure 10-6.

The *fundamental frequency* (wave 1) is the rate at which the complex waveform pattern repeats. Waves 2 through 20 are *overtones,* a set of higher-frequency sound waves that vibrate at whole-number (integer) multiples of the fundamental frequency. Different musical instruments sound unique because they produce overtones of different amplitudes. Among the clarinet overtones, represented by the heights of the blue waves in Figure 10-6, wave 5 is low amplitude, whereas wave 2 is high amplitude.

As primary colors blend into near-infinite variety, so pure tones blend into complex tones. Complex tones emanate from musical instruments, from the human voice, from birdsong, and from machines or repetitive mechanisms that make rhythmic buzzing or humming sounds. A key feature of complex tones,

FIGURE 10-6 Breaking Down a Complex Tone The waveform of a single note (*top*) from Don Byron's clarinet and the simple sound waves—the fundamental frequency (*middle*) and overtones (*bottom*)—that make up the complex tone. From *Stereo Review,* copyright 1977 by Diamandis Communications Inc.

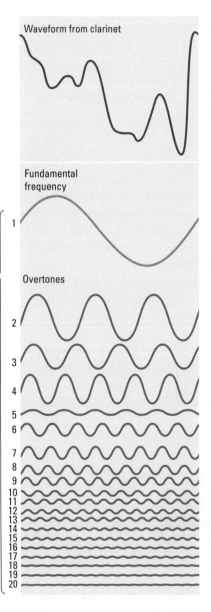

Waveform from clarinet

Fundamental frequency

Simple waves that make up sound of clarinet

Overtones

1
2
3
4
5
6
7
8
9
10
11
12
13
14
15
16
17
18
19
20

besides being made up of two or more pure tones, is some sort of periodicity: the fundamental frequency repeats at regular intervals. Sounds that are aperiodic, or random, we call *noise*.

Perception of Sound

Visualize what happens when you toss a pebble into a pond. Waves of water emanate from the point where the pebble enters the water. These waves produce no audible sound. But if your skin were able to convert the energy of the water waves (sensation) into neural activity that stimulated your auditory system, you would "hear" the waves when you placed your hand into the rippling water (perception). When you removed your hand, the "sound" would stop.

The pebble hitting the water is much like a tree falling to the ground, and the waves that emanate from the pebble's point of entry are like the air-pressure waves that emanate from the place where the tree strikes the ground. The frequency of the waves determines the pitch of the sound heard by the brain, whereas the height (amplitude) of the waves determines the sound's loudness.

Our sensitivity to sound waves is extraordinary. At the threshold of human hearing, we can detect the displacement of air molecules of about 10 picometers. We are rarely in an environment where we can detect such a small air-pressure change: there is usually too much background noise. A quiet, rural setting is probably as close as we ever get to an environment suitable for testing the acuteness of our hearing. The next time you visit the countryside, take note of the sounds you can hear. If there is no sound competition, you can often hear a single car engine miles away.

In addition to detecting very small changes in air pressure, the auditory system is also adept at simultaneously perceiving different sounds. As you sit reading this chapter, you are able to differentiate all sorts of sounds around you—traffic on the street, people talking next door, your air conditioner humming, footsteps in the hall. If you are listening to music, you detect the sounds of different instruments and voices.

You can perceive different sounds simultaneously because the different frequencies of air-pressure change associated with each sound wave stimulate different neurons in your auditory system. The perception of sounds is only the beginning of your auditory experience. Your brain interprets sounds to obtain information about events in your environment, and it analyzes a sound's meaning. Your use of sound to communicate with other people through both language and music clearly illustrate these processes.

Properties of Language and Music As Sounds

Language and music differ from other auditory sensations in fundamental ways. Both convey meaning and evoke emotion. The analysis of meaning in sound is a considerably more complex behavior than simply detecting a sound and identifying it. The brain has evolved systems that analyze sounds for meaning, speech in the left temporal lobe and music in the right.

Infants are receptive to speech and musical cues before they have any obvious utility, suggesting both the innate presence of these skills and the effects of prenatal experiences. Humans have an amazing capacity for learning and remembering linguistic and musical information. We are capable of learning a vocabulary of tens of thousands of words, often in many languages, and we have a capacity for recognizing thousands of songs.

Language facilitates communication. We can organize our complex perceptual worlds by categorizing information with words. We can tell others what we think and know and imagine. Imagine the efficiency that gestures and spoken language added to the cooperative food hunting and gathering behaviors of early humans.

1 picometer = one-trillionth of a meter

Focus 7-1 offers evidence that the brain is tuned in prenatally to the language into which it will be born.

Elvis Presley's memory for lyrics suits his legend: while serving in the U.S. Army, he wagered all comers that he could sing any song they named. Elvis never lost a bet.

Auditory constancy is reminiscent of the visual system's capacity for object constancy described in Section 9-4.

All these benefits of oral language seem obvious, but the benefits of music may seem less straightforward. In fact, music helps us to regulate our own emotions and to affect the emotions of others. After all, when do people most commonly make music? We sing and play music to communicate with infants and put children to sleep. We play music to enhance social interactions and gatherings and romance. We use music to bolster group identification—school songs and national anthems are examples. Music as we know it is unique to humans. Studies of nonhumans provide little evidence for preferences for human music over other sounds.

Another characteristic that distinguishes speech and musical sounds from other auditory inputs is their delivery speed. Nonspeech and nonmusical noise produced at a rate of about 5 segments per second is perceived as a buzz. (A sound segment is a distinct unit of sound.) Normal speed for speech is on the order of 8 to 10 segments per second, and we are capable of understanding speech at nearly 30 segments per second. Speech perception at these higher rates is truly amazing, because the speed of input far exceeds the auditory system's ability to transmit all the speech segments as separate pieces of information.

Properties of Language

Experience listening to a particular language helps the brain to analyze rapid speech, which is one reason why people who are speaking languages unfamiliar to you often seem to be talking incredibly fast. Your brain does not know where the foreign words end and begin, making them seem to run together in a rapid-fire stream.

A unique characteristic of our perception of speech sounds is our tendency to hear variations of a sound as if they were identical, even though the sound varies considerably from one context to another. For instance, the English letter "d" is pronounced differently in the words "deep," "deck," and "duke," yet a listener perceives the pronunciations to be the same "d" sound.

The auditory system must therefore have a mechanism for categorizing sounds as being the same despite small differences in pronunciation. Experience must affect this mechanism, because different languages categorize speech sounds differently. A major obstacle to mastering a foreign language after the age of 10 is the difficulty of learning the categories of sound that are treated as equivalent.

Properties of Music

Like other sounds, the subjective properties that people perceive in musical sounds differ from one another. One subjective property is *loudness,* the magnitude of the sound as judged by a person. Loudness is related to the amplitude of a sound wave and is measured in decibels, but loudness is also subjective. What is "very loud" music for one person may be only "moderately loud" for another, whereas music that seems "soft" to one listener may not seem at all soft to someone else.

Another subjective property of musical sounds is *pitch,* the position of each tone on a musical scale as judged by the listener. Although pitch is clearly related to sound-wave frequency, there is more to it than that. Consider the note middle C as played on a piano. This note can be described as a pattern of sound frequencies, as is the clarinet note in Figure 10-6.

Like the note played on the piano, any musical note is defined by its fundamental frequency. This is the lowest frequency of the sound-wave pattern, or the rate at which the overall pattern is repeated. For middle C, the fundamental frequency is 264 Hertz and the sound waves for notes C, E, and G, as measured by a spectrograph, are shown in **Figure 10-7.** Notice that, by convention, sound-wave spectrographs are measured in kilohertz (kHz), units of thousands of hertz. Thus, if we look at the fundamental frequency for middle C, it is the first large wave on the left, which is at 0.264 kilohertz. The fundamental frequencies for E and G are 0.330 and 0.392 kilohertz, respectively.

prosody Melodical tone of the spoken voice.

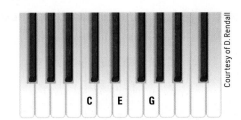

An important feature of the human brain's analysis of music is that middle C is perceived as being the same note regardless of whether it is played on a piano or on a guitar, even though the sounds made by these instruments are very different. The right temporal lobe has a special function in extracting pitch from sound, whether the sound is speech or music. In speech, pitch contributes to the perceived melodical tone of a voice, or **prosody.**

A final property of musical sound is *quality,* or *timbre,* the perceived characteristics that distinguish a particular sound from all others of similar pitch and loudness. We can easily distinguish the sound of a violin from that of a trombone even though both instruments are playing the same note at the same loudness. The quality of their sounds differs.

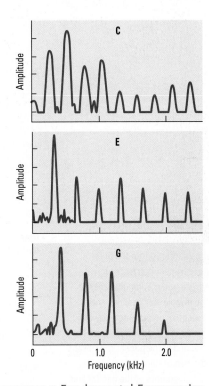

REVIEW 10-1
Sound Waves: Stimulus for Audition

Before you continue, check your understanding.

1. Sound-wave energy, the physical stimulus for the auditory system, is produced by changes in _____, a form of mechanical energy that is converted into neural activity in the ear.

2. Sound waves have three physical attributes: _____, _____, and _____.

3. Four properties of musical sounds are _____, _____, _____, and _____.

4. Sound is processed in the _____ lobes.

5. What distinguishes speech and musical sounds from other auditory inputs?

Answers appear at the back of the book.

FIGURE 10-7 Fundamental Frequencies of Piano Notes Waveforms of the notes C, E, and G as played on a piano and recorded on a spectrograph. The first wave in each graph is the fundamental frequency; the secondary waves are the overtones.

10-2 Functional Anatomy of the Auditory System

To understand how the nervous system analyzes sound waves, we begin by tracing the pathway taken by sound energy to and through the brain. The ear collects sound waves from the surrounding air and converts their mechanical energy into electrochemical neural energy that begins a long route through the brainstem to the auditory cortex.

Before we can trace the journey from ear to cortex, we need to ask what the auditory system is designed to do. Because sound waves have the properties of frequency, amplitude, and complexity, we can predict that the auditory system is structured to decode these properties. Most animals can tell where a sound comes from, so some mechanism must locate sound waves in space. Finally, many animals, including humans, not only analyze sounds for meaning but also make sounds. Because the sounds they produce are often the same as the ones they hear, we can infer that the neural systems for sound production and analysis must be closely related.

In humans, the evolution of sound-processing systems for both language and music led to the enhancement of specialized cortical regions, especially in the temporal lobes. In fact, a major difference between the human and the monkey cortex is a marked expansion of auditory areas in humans.

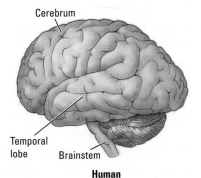

Structure of the Ear

The ear is a biological masterpiece in three acts: the outer ear, middle ear, and inner ear, all illustrated in **Figure 10-8.**

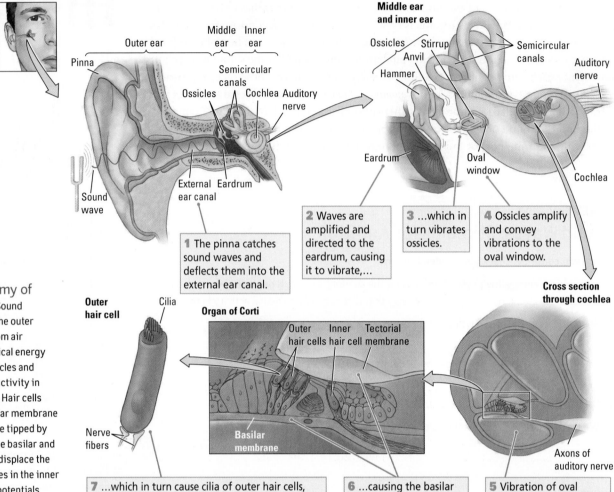

FIGURE 10-8 Anatomy of the Human Ear Sound waves gathered into the outer ear are transduced from air pressure into mechanical energy in the middle-ear ossicles and into electrochemical activity in the inner-ear cochlea. Hair cells embedded in the basilar membrane (the organ of Corti) are tipped by cilia. Movements of the basilar and tectorial membranes displace the cilia, leading to changes in the inner hair cells' membrane potentials and resultant activity of auditory bipolar neurons.

1 The pinna catches sound waves and deflects them into the external ear canal.

2 Waves are amplified and directed to the eardrum, causing it to vibrate,...

3 ...which in turn vibrates ossicles.

4 Ossicles amplify and convey vibrations to the oval window.

5 Vibration of oval window sends waves through cochlear fluid,...

6 ...causing the basilar and tectorial membranes to bend,...

7 ...which in turn cause cilia of outer hair cells, embedded in the tectorial membrane, to bend. This bending generates neural activity in hair cells.

ossicles Bones of the middle ear: malleus (hammer), incus (anvil), and stapes (stirrup).

cochlea Inner-ear structure that contains the auditory receptor cells.

basilar membrane Receptor surface in the cochlea that transduces sound waves into neural activity.

hair cell Sensory neurons in the cochlea tipped by cilia; when stimulated by waves in the cochlear fluid, outer hair cells generate graded potentials in inner hair cells, which act as the auditory receptor cells.

Processing Sound Waves

Both the *pinna,* the funnel-like external structure of the outer ear, and the external ear canal, which extends a short distance from the pinna inside the head, are made of cartilage and flesh. The pinna is designed to catch sound waves in the surrounding environment and deflect them into the external ear canal.

Because it narrows from the pinna, the external canal amplifies sound waves somewhat and directs them to the *eardrum* at its inner end. When sound waves strike the eardrum, it vibrates, the rate of vibration varying with the frequency of the waves. On the inner side of the eardrum, as depicted in Figure 10-8, is the middle ear, an air-filled chamber that contains the three smallest bones in the human body, connected in a series.

These three **ossicles** are called the *hammer,* the *anvil,* and the *stirrup* because of their distinctive shapes. The ossicles attach the eardrum to the *oval window,* an opening in the bony casing of the **cochlea,** the inner-ear structure that contains the auditory receptor cells. These receptor cells and the cells that support them are collectively called the *organ of Corti,* shown in detail in Figure 10-8.

When sound waves vibrate the eardrum, the vibrations are transmitted to the ossicles. The leverlike action of the ossicles conveys and amplifies the vibrations onto the membrane that covers the cochlea's oval window. As Figure 10-8 shows, the cochlea coils around itself and looks a bit like a snail shell. Inside its bony exterior, the cochlea is hollow, as the cross-sectional drawing reveals.

The hollow cochlear compartments are filled with a lymphatic fluid, and floating in its midst is the thin **basilar membrane.** Embedded in a part of the basilar membrane are outer and inner **hair cells.** At the tip of each hair cell are several filaments called *cilia*, and the cilia of the outer hair cells are embedded in an overlying membrane. The inner hair cells loosely contact this *tectorial membrane.*

Pressure from the stirrup on the oval window makes the cochlear fluid move because a second membranous window in the cochlea (the *round window*) bulges outward as the stirrup presses inward on the oval window. In a chain reaction, the waves traveling through the cochlear fluid bend the basilar and tectorial membranes, and the bending membranes stimulate the cilia at the tips of the outer hair cells. This stimulation generates graded potentials in the inner hair cells that act as the auditory receptor cells. The change in the membrane potential of the inner hair cells varies the amount of neurotransmitter that they release onto auditory neurons that go to the brain.

> The name *cochlea* actually means "snail shell" In Latin.

> Graded potentials are integral to the cell membrane's electrical activity, as Section 4-2 explains.

Transducing Sound Waves into Neural Impulses

How does the conversion of sound waves into neural activity code the various properties of sound that we perceive? In the late 1800s, the German physiologist Hermann von Helmholtz proposed that sound waves of different frequencies cause different parts of the basilar membrane to resonate. Von Helmholtz was partly correct. Actually, all parts of the basilar membrane bend in response to incoming waves of any frequency. The key is where on the basilar membrane the peak displacement takes place.

This solution to the coding puzzle was determined in 1960, when George von Békésy observed the basilar membrane directly. He saw a traveling wave moving along the membrane all the way from the oval window to the membrane's apex. The structure and function of the basilar membrane are easier to visualize if the cochlea is uncoiled and laid flat, as in **Figure 10-9.**

The coiled cochlea in Figure 10-9A maps the frequencies to which each part of the basilar membrane is most responsive. When the oval window vibrates in response to the vibrations of the ossicles, shown beside the uncoiled membrane in Figure 10-9B, it generates waves that travel through the cochlear fluid. Békésy placed little grains of silver along the basilar membrane and watched them jump in different places to different frequencies of incoming waves. Faster wave frequencies caused maximum peaks of displacement near the base of the basilar membrane; slower wave frequencies caused maximum displacement peaks near the membrane's apex.

As a rough analogy, consider what happens when you shake a rope. If you shake it very quickly, the rope waves are very small and short and remain close to the base—the hand holding the rope. But if you shake the rope slowly, with a broader movement, the

FIGURE 10-9 Anatomy of the Cochlea **(A)** The basilar membrane is maximally responsive to frequencies that are mapped as the cochlea uncoils. **(B)** Sound waves of different frequencies produce maximal displacement of the basilar membrane (shown uncoiled) at different locations.

(A) Uncoiling of cochlea

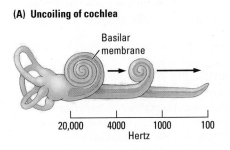

Basilar membrane

20,000 4000 1000 100
Hertz

(B) Uncoiled cochlea

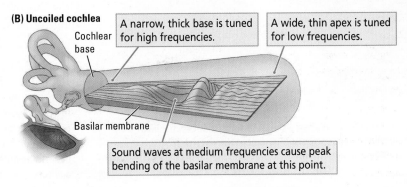

Cochlear base

Basilar membrane

A narrow, thick base is tuned for high frequencies.

A wide, thin apex is tuned for low frequencies.

Sound waves at medium frequencies cause peak bending of the basilar membrane at this point.

longer waves reach their peak farther along the rope—toward the apex. The key point is that, although both rapid and slow shakes produce movement along the rope's entire length, the maximum displacement is found at one end of the rope or the other, depending on whether the wave movements are rapid or slow.

This same response pattern holds for the basilar membrane and sound-wave frequency. All sound waves cause some displacement along the entire length of the membrane, but the amount of displacement at any point varies with the frequency of the sound wave. In the human cochlea, shown uncoiling in Figure 10-9A, the basilar membrane near the oval window is maximally affected by frequencies as high as about 20,000 hertz, the upper limit of our hearing range. The most effective frequencies at the membrane's apex register less than 100 hertz, closer to our lower limit of about 20 Hz (see Figure 10-4).

Intermediate frequencies maximally displace points on the basilar membrane between its two ends, as shown in Figure 10-9B. When a wave of a certain frequency travels down the basilar membrane, hair cells at the point of peak displacement are stimulated, resulting in a maximal neural response in those cells. An incoming signal composed of many frequencies causes several different points along the basilar membrane to vibrate and excites hair cells at all these points.

Not surprisingly, the basilar membrane is much more sensitive to changes in frequency than the rope in our analogy is because the basilar membrane varies in thickness along its entire length. It is narrow and thick at the base, near the oval window, and wider and thinner at its tightly coiled apex. The combination of varying width and thickness enhances the effect of small differences in frequency on the basilar membrane. As a result, the cochlear receptors can code small differences in sound-wave frequency as neural impulses.

Auditory Receptors

Hair cells ultimately transform sound waves into neural activity. Figure 10-8 (bottom left) shows the anatomy of the hair cells; **Figure 10-10** illustrates how sound waves stimulate them. Each human cochlea has two sets of hair cells: 3500 inner hair cells and 12,000 outer hair cells. Only the inner hair cells are the auditory receptors. This total number of receptor cells is small, considering the number of different sounds we can hear.

As diagrammed in Figure 10-10, the hair cells are anchored in the basilar membrane. The tips of the cilia of outer hair cells are attached to the overlying tectorial membrane, but the cilia of the inner hair cells do not touch that membrane. Nevertheless, the movement of the basilar and tectorial membranes causes the cochlear fluid to flow past the cilia of the inner hair cells, bending them back and forth. Animals with intact outer hair cells but no inner hair cells are effectively deaf.

The outer hair cells function simply to sharpen the cochlea's resolving power by contracting or relaxing and thereby changing the stiffness of the tectorial membrane. How this outer hair-cell function is controlled is puzzling. What stimulates these cells to contract or relax?

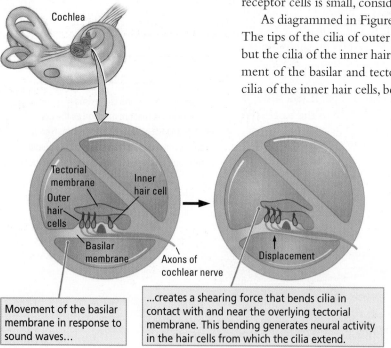

Cochlea

Tectorial membrane

Outer hair cells

Inner hair cell

Basilar membrane

Axons of cochlear nerve

Displacement

Movement of the basilar membrane in response to sound waves...

...creates a shearing force that bends cilia in contact with and near the overlying tectorial membrane. This bending generates neural activity in the hair cells from which the cilia extend.

FIGURE 10-10 Transducing Waves into Neural Activity Movement of the basilar membrane creates a shearing force in the cochlear fluid that bends the cilia, leading to the opening or closing of calcium channels in the outer hair cells. An influx of calcium ions leads the inner hair cells to release neurotransmitter that stimulates increased action potentials in auditory neurons.

The answer seems to be that, through connections with axons in the auditory nerve, the outer hair cells send a message to the brainstem auditory areas and receive a message back that causes the cells to alter tension on the tectorial membrane. In this way, the brain helps the hair cells to create an auditory world. The outer cells are also part of a mechanism that modulates auditory nerve firing, especially in response to very loud sounds.

A final question remains: How does movement of the cilia alter neural activity? The neurons of the auditory nerve have a spontaneous baseline rate of firing action potentials, and this rate is changed by how much neurotransmitter the hair cells release. It turns out that movement of the cilia changes the hair cell's polarization and its rate of neurotransmitter release. Look at Figure 10-8 again and you'll notice that the hair-cell cilia differ in height.

Movement of the cilia toward the tallest results in depolarization: calcium channels open and release neurotransmitter onto the dendrites of the cells that form the auditory nerve, generating more nerve impulses. Movement toward the shortest cilia hyperpolarizes the cell membrane and transmitter release decreases, thus decreasing activity in auditory neurons.

Hair cells are amazingly sensitive to the movement of their cilia. A movement sufficient to allow sound-wave detection is only about 0.3 nm, about the diameter of a large atom! Now you can understand why our hearing is so incredibly sensitive.

Pathways to the Auditory Cortex

The inner hair cells in the organ of Corti synapse with neighboring bipolar cells, the axons that form the auditory (cochlear) nerve. The auditory nerve in turn forms part of the eighth cranial nerve, the auditory vestibular nerve that governs hearing and balance. Whereas ganglion cells in the eye receive inputs from many receptor cells, bipolar cells in the ear receive input from only a single inner hair-cell receptor.

The cochlear-nerve axons enter the brainstem at the level of the medulla and synapse in the cochlear nucleus, which has ventral and dorsal subdivisions. Two nearby structures in the hindbrain (brainstem), the superior olive (a nucleus in the olivary complex) and the trapezoid body, each receive connections from the cochlear nucleus, as charted in **Figure 10-11.** The projections from the cochlear nucleus connect with cells on the same side of the brain as well as with cells on the opposite side. This arrangement mixes the inputs from the two ears to form the perception of a single sound.

Both the cochlear nucleus and the superior olive send projections to the inferior colliculus in the dorsal midbrain. Two distinct pathways emerge from the inferior colliculus, coursing to the **medial geniculate nucleus,** which lies in the thalamus. The ventral region of the medial geniculate nucleus projects to the **primary auditory cortex (area A1),** whereas the dorsal region projects to the auditory cortical regions adjacent to area A1.

Analogous to the two distinct visual pathways—the ventral stream for object recognition and the dorsal stream for the visual control of movement—a similar distinction exists in the auditory cortex (Romanski et al., 1999). Just as we can identify objects by their sound characteristics, we can direct our movements by the sound we hear.

medial geniculate nucleus Major thalamic region concerned with audition.

primary auditory cortex (area A1) Asymmetrical structures, found within Heschl's gyrus in the temporal lobes, that receive input from the ventral region of the medial geniculate nucleus.

Section 4-2 reviews the phases of the action potential and its propagation as a nerve impulse.

Figure 2-27 lists and locates all the cranial nerves, and the caption offers a mnemonic device for remembering them in order.

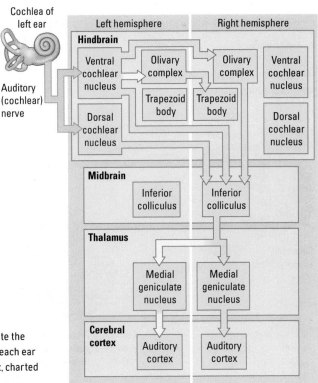

FIGURE 10-11 Auditory Pathways Auditory inputs cross to the hemisphere opposite the ear in the hindbrain and midbrain then recross in the thalamus so that information from each ear reaches both hemispheres. Multiple nuclei process inputs en route to the auditory cortex, charted here for the left ear.

The role of sound in guiding movement is less familiar to sight-dominated people than it is to the blind. Nevertheless, the ability exists in us all. Imagine waking up in the dark and reaching to pick up a ringing telephone or to turn off an alarm clock. Your hand automatically forms the appropriate shape in reaction to just the sound you have heard. That sound is guiding your movements much as a visual image guides them.

Relatively little is known about the what–how auditory pathways in the cortex. One appears to continue through the temporal lobe, much like the ventral visual pathway, and plays a role in identifying auditory stimuli. A second auditory pathway apparently goes to the posterior parietal region, where it forms a type of dorsal route for the auditory control of movement. It appears as well that auditory information can gain access to visual cortex, as illustrated in Research Focus 10-2, "Seeing with Sound."

Figure 9-13 maps the visual pathways through the cortex.

Auditory Cortex

In humans, the primary auditory cortex (A1) lies within Heschl's gyrus, surrounded by secondary cortical areas (A2), as shown in **Figure 10-12A**. The secondary cortex lying behind Heschl's gyrus is called the *planum temporale* ("temporal plane").

In right-handed people, the planum temporale is larger on the left side of the brain than it is on the right, whereas Heschl's gyrus is larger on the right side than on the left. The cortex of the left planum forms a speech zone known as **Wernicke's area** (the posterior speech zone), whereas the cortex of the larger, right-hemisphere Heschl's gyrus has a special role in analyzing music.

Recall the principle from Section 2-6: the brain is both symmetrical and asymmetrical.

These hemispheric differences mean that the auditory cortex is anatomically and functionally asymmetrical. Although cerebral asymmetry is not unique to the auditory system, it is most obvious here because auditory analysis of speech takes place only in the left hemisphere of right-handed people. About 70 percent of left-handed people have the same anatomical asymmetries as right-handers, an indication that speech organization is not related to hand preference. Language, including speech and other functions such as reading and writing, also is asymmetrical, although the right hemisphere also contributes to these broader functions.

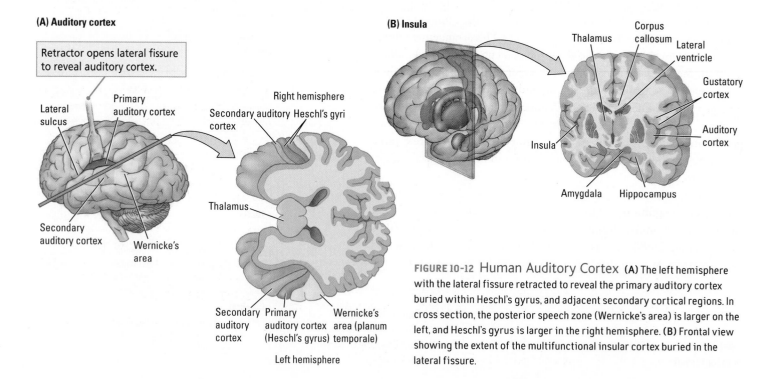

(A) Auditory cortex

Retractor opens lateral fissure to reveal auditory cortex.

Lateral sulcus
Primary auditory cortex
Secondary auditory cortex
Wernicke's area

Right hemisphere
Secondary auditory cortex
Heschl's gyri
Thalamus
Secondary auditory cortex
Primary auditory cortex (Heschl's gyrus)
Wernicke's area (planum temporale)

Left hemisphere

(B) Insula

Thalamus
Corpus callosum
Lateral ventricle
Gustatory cortex
Auditory cortex
Insula
Amygdala
Hippocampus

FIGURE 10-12 Human Auditory Cortex (A) The left hemisphere with the lateral fissure retracted to reveal the primary auditory cortex buried within Heschl's gyrus, and adjacent secondary cortical regions. In cross section, the posterior speech zone (Wernicke's area) is larger on the left, and Heschl's gyrus is larger in the right hemisphere. **(B)** Frontal view showing the extent of the multifunctional insular cortex buried in the lateral fissure.

Seeing with Sound

As detailed in Section 10-5, the ability to locate objects in space by echolocation has been extensively studied in species such as bats and dolphins. But it was reported 50 years ago that some blind people also echolocate.

More recently, anecdotal reports have surfaced of blind people who navigate around the world by using clicks made with their tongues and mouths and then listening to the returning echoes. In fact, videos show congenitally blind people riding bicycles down a street with silent obstacles such as parked cars. But how do they do this, and what part of the brain enables it?

Behavioral studies of blind people reveal that echolocators make short, spectrally broad clicks by moving the tongue backwards and downwards from the roof of the mouth directly behind the teeth. Skilled echolocators can identify properties of objects that include their position, distance, size, shape, and texture (Teng & Whitney, 2011).

Thaler and colleagues (2011) investigated the neural basis of this ability using fMRI. They studied two blind echolocation experts and compared brain activity for sounds that contain both clicks and returning echoes with brain activity for control sounds that did not contain the echoes. The participants use echolocation to localize objects in the environment, but more important, they also perceive object shape, motion, and even identity!

When the blind participants listened to recordings of their echolocation clicks and echoes, compared to silence, both the auditory cortex and the primary visual cortex showed activity. Sighted controls showed activation only in the auditory cortex. Remarkably, when the investigators compared the controls' brain activity to recordings that contained echoes versus those that did not, the auditory activity disappeared. By contrast, as illustrated in the figure, the blind echolocators showed activity only in the visual cortex when sounds with and without echoes were compared. Sighted controls (not shown) showed no activity in either the visual or auditory cortex in this comparison.

These results suggest that blind echolocation experts process click–echo information using brain regions normally devoted to vision. Thaler and his colleagues propose that the primary visual cortex is performing a spatial computation using information from the auditory cortex. Perhaps future research will determine how this process might work.

More immediately, the study suggests that echolocation could be taught to blind and visually impaired people to provide them increased independence in their daily life.

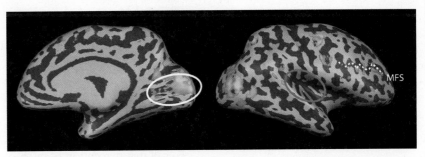

Seeing with Sound When cortical activation for sound with and without echoes is imaged in a blind echolocator, only the visual cortex shows activation (*left*) relative to the auditory cortex (*right*). Adapted from "Neural Correlates of Natural Humans Echolocation in Early and Late Blind Echolocation Experts," by L. Thaler, S. R. Arnott, and M. A. Goodale, 2011, *PLoS ONE, 6,* e20162.

The remaining 30 percent of left-handers fall into two distinct groups. The organization in about half of these people is opposite that of right-handers. The other half has some idiosyncratic bilateral speech representation. That is, about 15 percent of all left-handed people have some speech functions in one hemisphere and some in the other hemisphere.

The localization of language on one side of the brain is an example of **lateralization.** Note here simply that, as a rule of thumb in neuroanatomy, if one hemisphere is specialized for one type of analysis—as, for example, the left hemisphere is for language—the other hemisphere has a complementary function—the right hemisphere appears to be lateralized for music.

The temporal-lobe sulci enfold a large volume of cortical tissue far more extensive than the auditory cortex (Figure 10-12B). Buried in the lateral fissure, cortical tissue called the **insula** contains not only lateralized regions related to language but also areas controlling the perception of taste (the gustatory cortex) and areas linked to the neural structures underlying social cognition. As you might expect, injury to the insula can produce such diverse deficits as disturbance of both language and taste.

Wernicke's area Secondary auditory cortex (planum temporale) lying behind Heschl's gyrus at the rear of the left temporal lobe that regulates language comprehension; also called posterior speech zone.

lateralization Process whereby functions become localized primarily on one side of the brain.

insula Located within the lateral fissure, multifunctional cortical tissue that contains regions related to language, to the perception of taste, and to the neural structures underlying social cognition.

We consider gustation in Section 12-2 and social cognition in Section 15-6.

REVIEW 10-2

Functional Anatomy of the Auditory System

Before you continue, check your understanding.

1. Incoming sound-wave energy vibrates the eardrum, which in turn vibrates the _____.

2. The auditory receptors, known as _____, are found in the _____.

3. The motion of the cochlear fluid causes displacement of the _____ and _____ membranes.

4. The axons of bipolar cells from the cochlea form the _____, nerve, which is part of the _____ cranial nerve.

5. The auditory nerve originating in the cochlea goes to various nuclei in the brainstem and then projects to the _____ in the midbrain and the _____ in the thalamus.

6. Describe the asymmetrical structure and functions of the auditory cortex.

Answers appear at the back of the book

10-3 Neural Activity and Hearing

We now turn to the activities of neurons in the auditory system that create our perception of sound. Neurons at different levels in this system serve different functions. To get an idea of what individual hair cells and cortical neurons do, we consider how the auditory system codes sound-wave energy so that we perceive pitch, loudness, location, and pattern.

Hearing Pitch

Recall that perception of pitch corresponds to the frequency (repetition rate) of sound waves measured in hertz (cycles per second). Hair cells in the cochlea code frequency as a function of their location on the basilar membrane. In this **tonotopic representation,** hair-cell cilia at the base of the cochlea are maximally displaced by high-frequency waves that we hear as high-pitched sounds, and those at the apex are displaced the most by low-frequency waves that we hear as low-pitched sounds. Because bipolar-cell axons that form the cochlear nerve are each connected to only one hair cell, they convey information about the spot on the basilar membrane, from apex to base, that is being stimulated.

Recordings from single fibers in the cochlear nerve reveal that, although each axon transmits information about only a small part of the auditory spectrum, each cell does respond to a range of sound-wave frequencies—if the wave is sufficiently loud. That is, each hair cell is maximally responsive to a particular frequency and also responds to nearby frequencies, but the sound wave's amplitude must be greater (louder) for those nearby frequencies to excite the receptor's membrane potential.

This range of hair-cell responses to different frequencies at different amplitudes can be plotted to form a tuning curve. As graphed in **Figure 10-13,** each hair-cell receptor

Tonotopic literally means "tone place."

The hair cell's frequency range is analogous to a photoreceptor's response to a range of light wavelengths. See Figure 9-6 .

FIGURE 10-13 Tuning Curves Graphs plotted by the frequency and amplitude of sound-wave energy required to increase the firing rate of two different axons in the cochlear nerve. The lowest point on each tuning curve is the frequency to which that hair cell is most sensitive. The upper curve is centered on a frequency of 1000 hertz, in the midrange of human hearing; the lower curve is centered on a frequency of 10,000 hertz, in the high range.

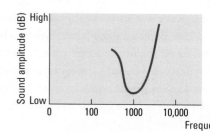
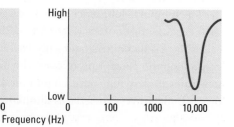

is maximally sensitive to a particular wavelength but still responds somewhat to nearby wavelengths.

The axons of the bipolar cells in the cochlea project to the cochlear nucleus in an orderly manner (see Figure 10-11). Axons entering from the base of the cochlea connect with one location, those entering from the middle connect to another location, and those entering from the apex connect to yet another. Thus the tonotopic representation of the basilar membrane is reproduced in the cochlear nucleus.

This systematic representation is maintained throughout the auditory pathways and into the primary auditory cortex. **Figure 10-14** shows the distribution of projections from the base and apex of the cochlea across area A1. Similar tonotopic maps can be constructed for each level of the auditory system.

tonotopic representation Property of audition in which sound waves are processed in a systematic fashion from lower to higher frequencies.

cochlear implant Electronic device implanted surgically into the inner ear to transduce sound waves into neural activity and allow a deaf person to hear.

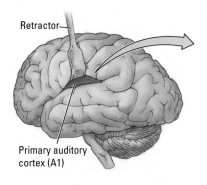
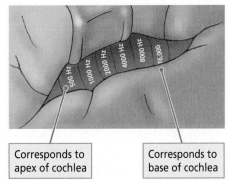

Retractor

Primary auditory cortex (A1)

500 Hz · 1000 Hz · 2000 Hz · 4000 Hz · 8000 Hz · 16,000

Corresponds to apex of cochlea

Corresponds to base of cochlea

FIGURE 10-14 Tonotopic Representation of Area A1 A retractor holds the lateral fissure open to reveal the underlying primary auditory cortex. The anterior end of area A1 corresponds to the apex of the cochlea and hence, low frequencies. The posterior end corresponds to the base of the cochlea and hence, high frequencies.

This systematic organization has enabled the development of **cochlear implants**—electronic devices surgically inserted in the inner ear to allow deaf people to hear (see Loeb, 1990). A miniature microphonelike processor detects the component frequencies of incoming sound waves and sends them to the appropriate place on the basilar membrane through tiny wires. The nervous system does not distinguish between stimulation coming from this artificial device, shown in **Figure 10-15**, and stimulation coming through the middle ear.

As long as appropriate signals go to the correct locations on the basilar membrane, the brain will "hear." Cochlear implants work very well, allowing the deaf to detect even the fluctuating pitches of speech. Their success corroborates the tonotopic representation of pitch in the basilar membrane.

Even so, the quality of sound stimulated by cochlear implants is degraded relative to natural hearing. Many people with implants find music unpleasant and difficult to listen to. Researchers are trying to increase the number of electrodes that stimulate the basilar membrane with the hope of enabling more users to perceive and enjoy music.

One minor difficulty with the tonotopic theory of frequency detection is that the cochlea does not use this mechanism at the very apex of the basilar membrane, where hair cells, as well as the bipolar cells to which they are connected, respond to frequencies below about 200 hertz (see Figure 10-9B). At this location, all the cells respond to movement of the basilar membrane, but they do so in proportion to the frequency of the incoming wave. Higher rates of bipolar cell firing signal a relatively higher frequency, whereas lower rates of firing signal a lower frequency.

Why the cochlea uses a different system to differentiate pitch within this range of very-low-frequency sound waves is not clear. It probably has to do with the physical limitations of the basilar membrane. Discriminating among low-frequency sound waves is not important to humans, but other animals such as elephants and whales depend on these frequencies to communicate. These species most likely have more neurons at the apex of the basilar membrane than we humans do.

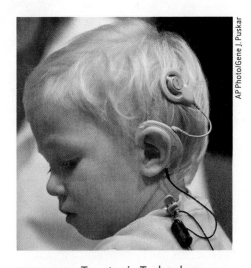

AP Photo/Gene J. Puskar

FIGURE 10-15 Tonotopic Technology Cochlear implants electronically process incoming sound-wave stimulation directly to the correct locations on the basilar membrane via a microphone linked to a small speech-processing computer worn behind the ear.

Detecting Loudness

The simplest way for cochlear (bipolar) cells to indicate sound-wave intensity is to fire at a higher rate when amplitude is greater, which is exactly what happens. More intense air-pressure changes produce more intense vibrations of the basilar membrane and therefore greater shearing of the cilia. Increased shearing leads to more neurotransmitter released onto bipolar cells. As a result, the bipolar axons fire more frequently, telling the auditory system that the sound is getting louder.

Detecting Location

The fact that each cochlear nerve synapses on both sides of the brain provides mechanisms for locating the source of a sound. In one mechanism, neurons in the brainstem compute the difference in a sound wave's arrival time at each ear. Such differences in arrival time need not be large to be detected. If two sounds presented through earphones are separated in time by as little as 10 microseconds, the listener will perceive that a single sound came from the leading ear.

This computation of left-ear–right-ear arrival times is carried out in the medial part of the superior olivary complex (see Figure 10-11). Because these hindbrain cells receive inputs from each ear, they can compare exactly when the signal from each ear reaches them.

Figure 10-16 shows how sound waves originating on the left reach the left ear slightly before they reach the right ear. As the sound source moves from the side of the head toward the middle, a person has greater and greater difficulty locating it: the difference in arrival time becomes smaller and smaller until there is no difference at all.

When we detect no difference, we infer that the sound is either directly in front of us or directly behind us. To locate it, we move our heads, making the sound waves strike one ear sooner. We have a similar problem distinguishing between sounds directly above and below us. Again, we solve the problem by tilting our heads, thus causing the sound waves to strike one ear before the other.

Another mechanism used by the auditory system to detect the source of a sound is the sound's relative loudness on the left and the right. The head acts as an obstacle to higher-frequency sound waves that do not easily bend around the head. As a result, higher-frequency waves on one side of the head are louder than on the other.

The lateral part of the superior olive and the trapezoid body detect this difference. Again, sound waves coming from directly in front or behind or from directly above or below require the same solution of tilting or turning the head.

Head tilting and turning take time, which usually is not important for humans. Time is important for other animals, such as owls, that hunt using sound. Owls need to know the location of a sound simultaneously in at least two directions—left and below, for example, or right and above.

Owls, like humans, can orient in the horizontal plane to sound waves by using the different times at which sound waves reach the two ears. Additionally, the owl's ears have evolved to detect the relative loudness of sound waves in the vertical plane. As diagrammed in **Figure 10-17**, owls' ears are slightly displaced in the vertical direction. This solution allows owls to hunt entirely by sound in the dark. Bad news for mice.

Detecting Patterns in Sound

Music and language are perhaps the primary sound-wave patterns that humans recognize. Perceiving sound-wave patterns as meaningful units thus is fundamental to our auditory analysis. Because music perception and language perception are lateralized in the right and left temporal lobes, respectively, we can guess that neurons in the right and left temporal cortex take part in pattern recognition and analysis of both auditory experiences. Studying the activities of auditory neurons in humans is not easy, however.

Extra distance that sound must travel to reach the right ear.

FIGURE 10-16 Locating a Sound Sound waves originating on the left side of the body reach the left ear slightly before the right, but the difference in arrival time is subtle, and the auditory system fuses the dual stimuli so that we perceive a single, clear sound coming from the left.

Horizontal orienting is *azimuth detection;* vertical orienting is *elevation detection.*

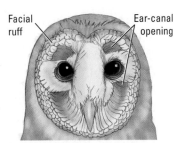

Facial ruff   Ear-canal opening

FIGURE 10-17 Hunting by Ear (*Left*) This barn owl has aligned its talons with the body axis of the mouse that it is about to catch in the dark. (*Right*) The owl's facial ruff collects and funnels sound waves into ear-canal openings through tightly feathered troughs formed by the ruff above and below the eyes. The owl's left ear is more sensitive to sound waves from the left and below because the ear canal is higher on the left side and the trough is tilted down. The right-side ear canal is lower and the trough is tilted up, making the right ear more sensitive to sound waves from the right and above. Drawing adapted from "The Hearing of the Barn Owl," by E. I. Knudsen, 1981, *Scientific American, 245*(6), p. 115.

Most of the knowledge that neuroscientists have comes from studies of how individual neurons respond in nonhuman primates. Both human and nonhuman primates have a ventral and dorsal cortical pathway for audition. Neurons in the ventral pathway decode spectrally complex sounds—referred to by some investigators as auditory object recognition—including the meaning of speech sounds for people and species-typical vocalizations in monkeys (for a review, see Rauschecker, 2012). Less is known about the properties of neurons in the dorsal auditory stream, but this path clearly has a role in integrating auditory and somatosensory information to control speech production. We could call it *audition for action*.

Audition for action parallels unconscious control of visually guided movements, shown in Figure 9-42, by the dorsal stream.

REVIEW 10-3
Neural Activity and Hearing

Before you continue, check your understanding.

1. Bipolar neurons in the cochlea form _____ maps that code sound-wave frequencies.

2. Loudness is decoded by the firing rate of cells in the _____.

3. Detecting the location of a sound is a function of neurons in the _____ and _____ of the brainstem.

4. The function of the dorsal auditory pathway can be described as _____.

5. Explain how the brain detects a sound's location.

Answers appear at the back of the book.

10-4 Anatomy of Language and Music

This chapter began with the discovery of the Neanderthal flute and its evolutionary implications (see Focus 10-1). The fact that Neanderthals made flutes implies not only that they processed musical sound-wave patterns but also that they made music. In our brains, musical ability is generally a right-hemisphere specialization complementary to language ability, localized in the left hemisphere in most people.

Section 7-4 surveys dynamic brain imaging methods, and Section 7-2 reviews methods for measuring the brain's electrical activity.

No one knows whether these complementary systems evolved together in the hominid brain, but it is highly likely. Language and music abilities are highly developed in the modern human brain. Although little is known about how each is processed at the cellular level, electrical stimulation and recording and blood-flow imaging studies yield important insights into the cortical regions that process them. We investigate such studies next, focusing first on how the brain processes language.

Processing Language

More than 4000 human languages are spoken in the world today, and probably many more have gone extinct in past millennia. Researchers have wondered whether the brain has a single system for understanding and producing any language, regardless of its structure, or whether very different languages, such as English and Japanese, are processed in different ways. To answer this question, it helps to analyze languages to determine just how fundamentally similar they are, despite their obvious differences.

Uniformity of Language Structure

Foreign languages often seem impossibly complex to nonspeakers. Their sounds alone may seem odd and difficult to make. If you are a native speaker of English, for instance, Asian languages, such as Japanese, probably sound peculiarly melodic and almost without obvious consonants to you, whereas European languages, such as German or Dutch, may sound heavily guttural.

Even within such related languages as Spanish, Italian, and French, marked differences can make learning one of them challenging, even if the student already knows another. Yet as real as all these linguistic differences may be, they are superficial. The similarities among human languages, although not immediately apparent, are actually far more fundamental than their differences.

Noam Chomsky is usually credited with being the first linguist to stress similarities over differences in human language structure. In a series of books and papers written over the past half-century, Chomsky has made a sweeping claim, as have researchers such as Steven Pinker (1997) more recently. They argue that all languages have common structural characteristics stemming from a genetically determined constraint. Humans, apparently, have a built-in capacity for creating and using language.

Chomsky was greeted with deep skepticism when he first proposed this idea in the 1960s, but it has since become clear that human language is indeed genetically based. An obvious piece of evidence: language is universal in human populations. All people everywhere use language.

Complexity of a language is not related to the technological complexity of a culture. The languages of technologically primitive peoples are every bit as complex and elegant as the languages of postindustrial cultures. Nor is the English of Shakespeare's time inferior or superior to today's English; it is just different.

A year-old baby's 5- to 10-word vocabulary doubles in the next 6 months and by 36 months mushrooms into a 1000-word repertoire. See Section 8-3.

Another piece of evidence that Chomsky adherents cite for the genetic basis of human language is that humans learn language early in life and seemingly without effort. By about 12 months of age, children everywhere have started to speak words. By 18 months, they are combining words, and by age 3 years, they have a rich language capability.

Broca's area Anterior speech area in the left hemisphere that functions with the motor cortex to produce the movements needed for speaking.

Perhaps the most amazing thing about language development is that children are not formally taught the structure of their language. As toddlers, they are not painstakingly instructed in the rules of grammar. In fact, their early errors—sentences such as "I goed to the zoo"—are seldom even corrected by adults. Yet children master language rapidly. They also acquire language through a series of stages that are remarkably similar across cultures. Indeed, the process of language acquisition plays an important role in Chomsky's theory of its innateness—which is not to say that language development is not influenced by experience.

At the most basic level, for example, children learn the language they hear spoken. In an English household, they learn English; in a Japanese home, they learn Japanese. They also pick up the language structure—the vocabulary and grammar—of the people around them, even though that structure can vary from one speaker to another. Children go through a sensitive period for language acquisition, probably from about 1 to 6 years of age. If they are not exposed to language throughout this critical period, their language skills are severely compromised. If children learn two languages simultaneously, both share the same part of Broca's area. In fact, their neural representations overlap (Kim et al., 1997).

Both its universality and natural acquisition favor the theory for a genetic basis of human language. A third piece of evidence is the many basic structural elements that all languages have in common. Granted, every language has its own particular rules that specify exactly how various parts of speech are positioned in a sentence (syntax), how words are inflected to convey different meanings, and so forth. But overarching rules also apply to all human languages.

For instance, all languages employ grammar, the parts of speech that we call subjects, verbs, and direct objects. Consider the sentence "Jane ate the apple." "Jane" is the subject, "ate" is the verb, and "apple" is the direct object. Syntax is not specified by any universal rule but rather is a characteristic of the particular language. In English, syntactical order is subject, verb, object; in Japanese, the order is subject, object, verb; in Gaelic, the order is verb, subject, object. Nonetheless, all have both syntax and grammar.

The existence of these two structural pillars in all human languages is seen in the phenomenon of *creolization*—the development of a new language from what was formerly a rudimentary language, or *pidgin*. Creolization took place in the seventeenth-century Americas when slave traders and colonial plantation owners brought together, from various parts of West Africa, people who lacked a common language. The newly enslaved needed to communicate, and they quickly created a pidgin based on whatever language the plantation owners spoke—English, French, Spanish, or Portuguese.

The pidgin had a crude syntax (word order) but lacked a real grammatical structure. The children of the slaves who invented this pidgin grew up with caretakers who spoke only pidgin to them. Yet within a generation, these children had created their own creole, a language complete with a genuine syntax and grammar.

Clearly, the pidgin invented of necessity by adults was not a learnable language for children. Their innate biology shaped a new language similar in basic structure to all other human languages. All creolized languages seem to evolve in a similar way, even though the base languages are unrelated. This phenomenon could happen only if there were an innate, biological component to language development.

Localizing Language in the Brain

Finding a universal basic language structure set researchers on the search for an innate brain system that underlies language use. By the late 1800s, it had become clear that language functions were at least partly localized—not just within the left hemisphere but to specific areas there. Clues that led to this conclusion began to emerge early in the nineteenth century, when neurologists observed patients with frontal-lobe injuries who suffered language difficulties.

Then, in 1861, the French physician Paul Broca confirmed that certain language functions are localized in the left hemisphere. Broca concluded, on the basis of several postmortem examinations, that language is localized in the left frontal lobe, in a region just anterior to the central fissure. A person with damage in this area is unable to speak despite both an intact vocal apparatus and normal language comprehension. The confirmation of **Broca's area** was significant because it triggered the idea that the left and right hemispheres might have different functions.

Focus 8-3 describes how cortical activation differs for second languages learned later in life.

Section 7-1 links Broca's observations and contributions to the emergence of neuropsychology.

(A)

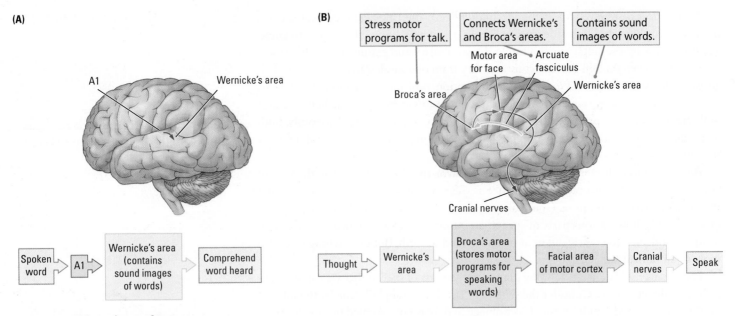

(B)

FIGURE 10-18 Neurology of Language (A) In Wernicke's model of speech recognition, stored sound images are matched to spoken words in the left posterior temporal cortex, shown in yellow. (B) Speech is produced through the connection that the arcuate fasciculus makes between Wernicke's area and Broca's area.

aphasia Inability to speak or comprehend language despite the presence of normal comprehension and intact vocal mechanisms. Broca's aphasia is the inability to speak fluently despite the presence of normal comprehension and intact vocal mechanisms. Wernicke's aphasia is the inability to understand or to produce meaningful language even though the production of words is still intact.

Other neurologists of the time believed that Broca's area might be only one of several left-hemisphere regions that control language. In particular, they suspected a relation between hearing and speech. Proving this suspicion correct, Karl Wernicke later described patients who had difficulty comprehending language after injury to the posterior region of the left temporal lobe, identified as Wernicke's area in **Figure 10-18.**

In Section 10-2 we identified Wernicke's area as a speech zone (see Figure 10-12A). Damage to any speech area produces some form of **aphasia,** the general term for any inability to comprehend or produce language despite the presence of normal comprehension and intact vocal mechanisms. At one extreme, people who suffer *Wernicke's aphasia* can speak fluently, but their language is confused and makes little sense, as if they have no idea what they are saying. At the other extreme, a person with *Broca's aphasia* cannot speak despite normal comprehension and intact physiology.

Wernicke went on to propose a model, diagrammed in Figure 10-18A, for how the two language areas of the left hemisphere interact to produce speech. He theorized that images of words are encoded by their sounds and stored in the left posterior temporal cortex. When we hear a word that matches one of those sound images, we recognize it, which is how Wernicke's area contributes to speech comprehension.

To *speak* words, Broca's area in the left frontal lobe must come into play, because the motor program to produce each word is stored in this area. Messages travel to Broca's area from Wernicke's area through the *arcuate fasciculus,* a pathway that connects the two regions. Broca's area in turn controls the articulation of words by the vocal apparatus, as diagrammed in Figure 10-18B.

Wernicke's model provided a simple explanation both for the existence of two major language areas in the brain and for the contribution each area makes to the control of language. But the model was based on postmortem examinations of patients with brain lesions that were often extensive. Not until neurosurgeon Wilder Penfield's pioneering studies, begun in the 1930s, were the language areas of the left hemisphere clearly and accurately mapped.

Auditory and Speech Zones Mapped by Brain Stimulation

It turns out that, among Penfield's discoveries, neither is Broca's area the independent site of speech production nor is Wernicke's area the independent site of language comprehension. Electrical stimulation of either region disrupts both processes.

(A)

Montreal Neurological Institute

(B)

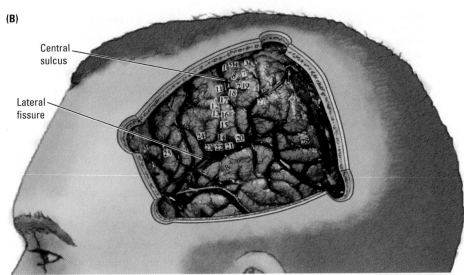

Central sulcus

Lateral fissure

FIGURE 10-19 Mapping Cortical Functions **(A)** During neurosurgery for intractable epilepsy, the patient is fully conscious, lying on his right side, and kept comfortable with local anesthesia. Wilder Penfield stimulates discrete cortical areas in the patient's exposed left hemisphere. In the background, a neurologist monitors an EEG recorded from each stimulated area to help identify the eleptogenic focus. The anesthetist (seated) observes the patient's responses to the cortical stimulation. **(B)** A drawing overlies a photograph of the patient's exposed brain. The numbered tickets identify points Penfield stimulated to map the cortex in this patient's brain. At points 26, 27, and 28, a stimulating electrode produced speech disruption. Point 26 presumably is in Broca's area, 27 (not shown) is the motor cortex facial-control area, and 28 is in Wernicke's area.

Penfield took advantage of the chance to map the brain's auditory and language areas when he operated on patients undergoing elective surgery to treat intractable epilepsy. The goal of this surgery is to remove tissues where the abnormal discharges are localized. A major challenge is preventing injury to critical neural regions that serve important functions. To determine the location of these critical regions, Penfield used a weak electrical current to stimulate the brain surface. By monitoring the patient's response to stimulation in different locations, Penfield could map brain functions along the cortex.

Typically, two neurosurgeons perform the operation (Penfield is shown operating in **Figure 10-19**A), and a neurologist analyzes the electroencephalogram in an adjacent room. Because patients are awake, they can contribute during the procedure, and the effects of brain stimulation in specific regions can be determined in detail and mapped. Penfield placed little numbered tickets on different parts of the brain's surface where the patient noted that stimulation had produced some noticeable sensation or effect, producing the cortical map shown in Figure 10-19B.

When Penfield stimulated the auditory cortex, patients often reported hearing various sounds, a ringing that sounded like a doorbell, a buzzing noise, or a sound like birds chirping. This result is consistent with later studies of single-cell recordings from the auditory cortex in nonhuman primates. Findings in these later studies showed that the auditory cortex has a role in pattern recognition.

Penfield also found that stimulation in area A1 seemed to produce simple tones—ringing sounds, and so forth—whereas stimulation in the adjacent auditory cortex (Wernicke's area) was more apt to cause some interpretation of a sound—ascription of a buzzing sound to a familiar source such as a cricket, for instance. There was no difference in the effects of stimulation of the left or right auditory cortex, and the patients heard no words when the brain was stimulated.

Sometimes, however, stimulation of the auditory cortex produced effects other than the perception of sounds. Stimulation of one area, for example, might cause a patient

Section 7-1 describes an array of brain-stimulation techniques used in neuroscience research.

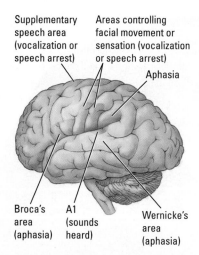

Supplementary speech area (vocalization or speech arrest)

Areas controlling facial movement or sensation (vocalization or speech arrest)

Aphasia

Broca's area (aphasia)

A1 (sounds heard)

Wernicke's area (aphasia)

FIGURE 10-20 Cortical Regions That Control Language This map, based on Penfield's extensive study, summarizes areas in the left hemisphere where direct stimulation may disrupt speech or elicit vocalization. Adapted from *Speech and Brain Mechanisms* (p. 201), by W. Penfield and L. Roberts, 1956. London: Oxford University Press.

to experience a sense of deafness, whereas stimulation of another area might produce a distortion of sounds actually being heard. As one patient exclaimed after a certain region had been stimulated, "Everything you said was mixed up!"

Penfield was most interested in the effects of brain stimulation not on simple sound-wave processing but on language. He and later researchers used electrical stimulation to identify four important cortical regions that control language. The two classic regions—Broca's area and Wernicke's area—are left-hemisphere regions. Located on both sides of the brain are the other two major regions of language use: the dorsal area of the frontal lobes and the areas of the motor and somatosensory cortex that control facial, tongue, and throat muscles and sensations. Although the effects on speech vary depending on the region, stimulating any of them disrupts speech in some way.

Clearly, much of the left hemisphere takes part in audition. **Figure 10-20** shows the left-hemisphere areas that Penfield found engaged in some way in processing language. In fact, Penfield mapped cortical language areas in two ways, first by disrupting speech and then by eliciting speech. Not surprisingly, damage to any speech area produces some form of aphasia.

DISRUPTING SPEECH Penfield expected that electrical current might disrupt ongoing speech by effectively "short-circuiting" the brain. He stimulated different regions of the cortex while the patient was speaking. In fact, the speech disruptions took several forms, including slurring, word confusion, and difficulty in finding the right word. Such aphasias are detailed in Clinical Focus 10-3, "Left-Hemisphere Dysfunction."

Electrical stimulation of the **supplementary speech area** on the dorsal surface of the frontal lobes (shown in Figure 10-20) can even stop ongoing speech completely, a reaction that Penfield called *speech arrest*. Stimulation of other cortical regions far removed from the temporal and frontal speech areas has no effect on ongoing speech, with the

---

CLINICAL FOCUS ✛ 10-3

Left-Hemisphere Dysfunction

Susan S., a 25-year-old college graduate and mother of two, suffered from epilepsy. When she had a seizure, which was almost every day, she lost consciousness for a short period during which she often engaged in repetitive behaviors, such as rocking back and forth.

Medication can usually control such psychomotor seizures, but the drugs were ineffective for Susan. The attacks were very disruptive to her life: they prevented her from driving and restricted the types of jobs she could hold. So Susan decided to undergo neurosurgery to remove the region of abnormal brain tissue that was causing the seizures.

The procedure has a high success rate. Susan's surgery entailed the removal of a part of the left temporal lobe, including most of the cortex in front of the auditory areas. Although it may seem a substantial amount of the brain to cut away, the excised tissue is usually abnormal, so any negative consequences are typically minor.

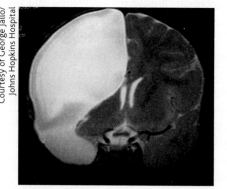

Courtesy of George Jallo/Johns Hopkins Hospital

Postoperative MRI of a patient who has lost most of the left hemisphere.

After the surgery, Susan did well for a few days, but then she suffered unexpected and unusual complications. As a result, she lost the remainder of her left temporal lobe, including the auditory cortex and Wernicke's area. The extent of lost brain tissue resembles that shown in the accompanying MRI.

Susan no longer understood language, except to respond to the sound of her name and to speak just one phrase: "I love you." Susan was also unable to read, showing no sign that she could even recognize her own name in writing.

To find ways to communicate with Susan, one of us (BK) tried humming nursery rhymes to her. She immediately recognized them and could say the words. We also discovered that her singing skill was well within the normal range and she had a considerable repertoire of songs.

Susan did not seem able to learn new songs, however, and she did not understand "messages" that were sung to her. Apparently, Susan's musical repertoire was stored and controlled independently of her language system.

exception of regions of the motor cortex that control movements of the face, shown in Figure 10-20. This exception makes sense because talking requires movement of facial, tongue, and throat muscles.

ELICITING SPEECH The second way Penfield mapped language areas was to stimulate the cortex when a patient was not speaking to see if he could cause the person to utter a speech sound. Penfield did not expect to trigger coherent speech; cortical stimulation is not physiologically normal and so probably would not produce actual words or word combinations. His expectation was borne out.

Stimulation of regions on both sides of the brain—for example, the supplementary speech areas—produces a sustained vowel cry, such as "Oooh" or "Eee." Stimulation of the facial areas in the motor cortex and the somatosensory cortex produces some vocalization related to movements of the mouth and tongue. Stimulation outside these speech-related zones produces no such effects.

Auditory Cortex Mapped by Positron Emission Tomography

Today, researchers use PET, a brain-imaging technique that detects changes in blood flow by measuring changes in the uptake of compounds such as oxygen or glucose, to study the metabolic activity of brain cells engaged in processing language. PET imaging is based on a surprisingly old idea.

In the late 1800s, Angelo Mosso was fascinated by the observation that pulsations in the living brain keep pace with the heartbeat. Mosso believed that the pulsations were related to changes in blood flow in the brain. He later noticed that the pulsations appeared to be linked to mental activity. For example, when a subject was asked to perform a simple calculation, the increase in brain pulsations and presumably in blood flow was immediate.

But to demonstrate a relation between mental activity and blood flow within the brain requires a more quantifiable measure than visual observation. Various procedures for measuring blood flow in the brain were devised in the twentieth century. (One is described in Clinical Focus 10-4, "Arteriovenous Malformations" on page 344.) Not until the development of PET in the 1970s, however, could blood flow in the brain of a living human be measured safely and precisely (Posner & Raichle, 1997).

What happens when PET is used while people listen to sounds? Although there are many PET studies of auditory stimulation, a series conducted by Robert Zatorre and his colleagues (1992, 1995) serves as a good example. These researchers hypothesized that simple auditory stimulation, such as bursts of noise, are analyzed by area A1, whereas more complex auditory stimulation, such as speech syllables, are analyzed in adjacent secondary auditory areas.

The researchers also hypothesized that performing a speech-sound-discrimination task would selectively activate left-hemisphere regions. This selective activation is exactly what they found. **Figure 10-21**A shows increased activity in the primary auditory cortex

supplementary speech area Speech-production region on the dorsal surface of the left frontal lobe.

Section 7-4 details the procedures used to obtain a PET scan.

FIGURE 10-21 Cortical Activation in Language-Related Tasks **(A)** Passively listening to noise bursts activates the primary auditory cortex. **(B)** Listening to words activates the posterior speech area, including Wernicke's area. **(C)** Making a phonetic discrimination activates the frontal region, including Broca's area.

(A) Listening to bursts of noise

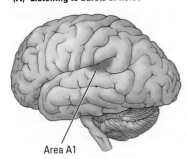

Area A1

(B) Listening to words

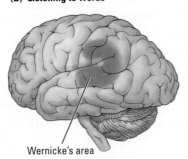

Wernicke's area

(C) Discriminating speech sounds

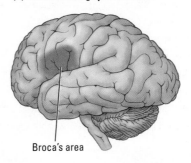

Broca's area

Arteriovenous Malformations

An arteriovenous malformation (also called an AV malformation or angioma) is a mass of enlarged and tortuous cortical blood vessels that form congenitally. AV malformations are quite common, accounting for as many as 5 percent of all cases of cerebrovascular disease.

Although angiomas may be benign, they often interfere with underlying brain functioning and can produce epileptic seizures. The only treatment is to remove the malformation. This procedure carries significant risk, however, because the brain may be injured in the process.

Walter K. was diagnosed with an AV malformation when he was 26 years old. He had consulted a physician because of increasingly severe headaches, and a neurological examination revealed an angioma over his occipital lobe. A surgeon attempted to remove the malformation, but the surgery did not go well. Walter was left with a defect in the bone overlying his visual cortex. This bone defect made it possible to listen to the blood flow through the malformation.

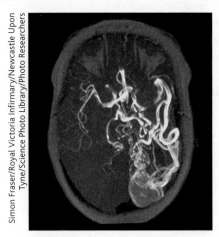

Simon Fraser/Royal Victoria Infirmary/Newcastle Upon Tyne/Science Photo Library/Photo Researchers

MRI, dorsal view, shows the brain surface of an 18-year-old girl with an angioma. The abnormal cerebral blood vessels (in white) form a balloonlike structure (the blue area at lower right) that caused the death of brain tissue around it in the right occipital cortex.

Dr. John Fulton noticed that when Walter suddenly began to use his eyes after being in the dark, there was a prompt increase in the noise (known as a *bruit*, French for "noise") associated with blood flow. Fulton documented his observations by recording the sound waves of the bruit while Walter performed visual experiments.

For example, if Walter had his eyes closed and then opened them to read a newspaper, there was a noticeable increase in blood flow through the occipital lobe. If the lights went out, the noise of the blood flow subsided. Merely shining light into Walter's eyes had no effect, nor was there an effect when he inhaled the scent of vanilla or strained to listen to faint sounds.

Apparently, the bruit and its associated blood flow were triggered by mental effort related to vision. To reach this conclusion was remarkable, given that Fulton used only a stethoscope and a simple recording device for his study. Modern instrumentation, such as that of positron emission tomography, has shown that Fulton's conclusion was correct.

in response to bursts of noise, whereas secondary auditory areas are activated by speech syllables (Figure 10-21B and C).

Both types of stimuli produced responses in both hemispheres, but there was greater activation in the left hemisphere for the speech syllables. These results imply that area A1 analyzes all incoming auditory signals, speech and nonspeech, whereas the secondary auditory areas are responsible for some higher-order signal processing required for analyzing language sound patterns.

As Figure 10-21C shows, the speech-sound-discrimination task yielded an intriguing additional result: Broca's area in the left hemisphere was activated as well. The involvement of this frontal-lobe region during auditory analysis may seem surprising. In Wernicke's model, Broca's area is considered the place where the motor programs needed to produce words are stored. It is not normally a region thought of as the site of speech-sound discrimination.

A possible explanation is that, to determine that the "g" in "bag" and "pig" is the same speech sound, the auditory stimulus must be related to how the sound is actually articulated. That is, the speech-sound perception requires a match with the motor behaviors associated with making the sound.

This role for Broca's area in speech analysis is confirmed further when investigators ask people to determine whether a stimulus is a word or a nonword (e.g., "tid" versus "tin" or "gan" versus "tan"). In this type of study, information about how the words are articulated is irrelevant, and Broca's area would not need to be recruited. Imaging reveals that it is not.

Processing Music

Although Penfield did not study the effect of brain stimulation on musical analysis, many researchers study musical processing in brain-damaged patients. Clinical Focus 10-5, "Cerebral Aneurysms," describes one such case. Collectively, the results of these studies confirm that musical processing is in fact largely a right-hemisphere specialization, just as language processing is largely a left-hemisphere one.

Localizing Music in the Brain

An excellent example of right-hemisphere predominance for music processing is seen in a famous patient—the French composer Maurice Ravel (1875–1937). "Bolero" is perhaps his best-known work. At the peak of his career, Ravel suffered a left-hemisphere stroke and developed aphasia. Yet many of Ravel's musical skills remained intact after the stroke because they were localized to the right hemisphere. He could still recognize melodies, pick up tiny mistakes in music he heard being played, and even judge the tuning of pianos. His music perception was largely intact.

Interestingly, however, skills that had to do with producing music were among those destroyed. Ravel could no longer recognize written music, play the piano, or compose. This dissociation of music perception and music production is curious. Apparently, the left hemisphere plays at least some role in certain aspects of music processing, especially those that have to do with making music.

To find out more about how the brain carries out the perceptual side of music processing, Zatorre and his colleagues (1994) conducted PET studies. When participants

---

CLINICAL FOCUS ✛ 10-5

Cerebral Aneurysms

C. N. was a 35-year-old nurse described by Isabelle Peretz and her colleagues (1994). In December 1986, C. N. suddenly developed severe neck pain and headache. A neurological examination revealed an aneurysm in the middle cerebral artery on the right side of her brain.

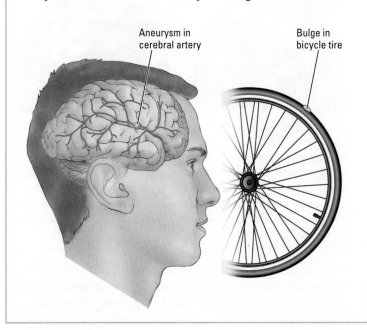

Aneurysm in cerebral artery

Bulge in bicycle tire

An *aneurysm* is a bulge in a blood-vessel wall caused by weakening of the tissue, much like the bulge that appears in a bicycle tire at a weakened spot. Aneurysms in a cerebral artery are dangerous: if they burst, severe bleeding and subsequent brain damage result.

In February 1987, C. N.'s aneurysm was surgically repaired, and she appeared to suffer few adverse effects. However, postoperative brain imaging revealed that a new aneurysm had formed in the same location but on the opposite side of the brain. This second aneurysm was repaired 2 weeks later.

After her surgery, C. N. had temporary difficulty finding the right word when she spoke, but, more important, her perception of music was deranged. She could no longer sing, nor could she recognize familiar tunes. In fact, singers sounded to her as if they were talking instead of singing. But C. N. could still dance to music.

A brain scan revealed damage along the lateral fissure in both temporal lobes. The damage did not include the primary auditory cortex, nor did it include any part of the posterior speech zone. For these reasons, C. N. could still recognize nonmusical sound patterns and showed no evidence of language disturbance. This finding reinforces the hypothesis that nonmusical sounds and speech sounds are analyzed in parts of the brain separate from those that process music.

(A) Listening to bursts of noise

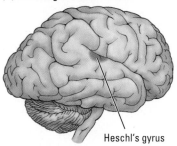

Heschl's gyrus

(B) Listening to melodies

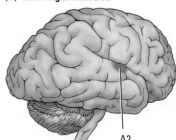

A2

(C) Comparing pitches

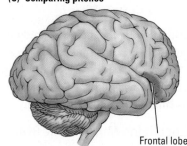

Frontal lobe

FIGURE 10-22 Cortical Activation in Music-Related Tasks **(A)** Passively listening to noise bursts activates Heschl's gyrus. **(B)** Listening to melodies activates the secondary auditory cortex. **(C)** Making relative pitch judgments about two notes of each melody activates a right-frontal-lobe area.

Focus 14-5 describes how playing a musical instrument can affect sensorimotor maps in the cortex.

listened simply to bursts of noise, Heschl's gyrus became activated (**Figure 10-22A**), but this was not the case when they listened to melodies. As shown in Figure 10-22B, the perception of melody triggers major activation in the right-hemisphere auditory cortex lying in front of Heschl's gyrus, as well as minor activation in the same region of the left hemisphere (not shown).

In another test, participants listened to the same melodies but this time were asked to indicate whether the pitch of the second note was higher or lower than that of the first note. During this task, which requires short-term memory of what has just been heard, blood flow in the right frontal lobe increased (Figure 10-22C). As with language, then, the frontal lobe plays a role in auditory analysis when short-term memory is required. People with enhanced or impaired musical abilities show differences in frontal-lobe organization, as demonstrated in Research Focus 10-6, "The Brain's Music System."

RESEARCH FOCUS ⁘ 10-6

The Brain's Music System

Nonmusicians have musical ability and enjoy music. Musicians show an enormous range of ability: some have perfect pitch and some do not, for example. About 4 percent of the population is tone deaf. Their difficulties with music, characterized as **amusia**—an inability to distinguish between musical notes—are lifelong.

Robert Zatorre and his colleagues (Bermudez et al., 2009; Hyde et al., 2007) have used MRI to look at differences among the brains of musicians, nonmusicians, and amusics. MRIs of the left and right hemispheres show that, compared to nonmusicians, cortical thickness is greater in musicians' dorsolateral frontal and superior temporal regions. Curiously, musicians with perfect pitch have thinner cortex in the posterior part of the dorsolateral frontal lobe. Thinner appears to be better for some music skills.

Compared to nonmusicians, then, musicians with thicker-than-normal cortex must have enhanced neural networks in the right-hemisphere frontal–temporal system linked to performing musical tasks. But thicker-than-normal cortex can bestow both advantage and impairment.

Analysis of music participants' brains showed thicker cortex in the right frontal area and in the right auditory cortex regions. Some

Right hemisphere   **Left hemisphere**

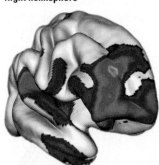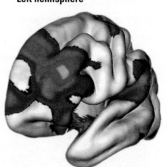

MRIs showing that, compared to nonmusicians, musicians have thicker cortex in frontal and temporal areas that contribute to performing musical tasks. The differences are greatest in the green, yellow, and red regions.
Adapted from "Neuroanatomical Correlates of Musicianship as Revealed by Cortical Thickness and Voxel-Based Morphometry," by P. Bermudez, J. P. Lerch, A. C. Evans, and R. J. Zatorre, 2009, *Cerebral Cortex*.

abnormality in neuronal migration during brain development is likely to have led to an excess of neurons in the right frontal–temporal music pathway of the amusics. Their impaired music cognition is the result.

As noted earlier, the capacity for language appears to be innate. Sandra Trehub and her colleagues (1999) showed that music may be innate as well, as we hypothesized at the beginning of the chapter. Trehub found that infants show learning preferences for musical scales versus random notes. Like adults, children are very sensitive to musical errors, presumably because they are biased for perceiving regularity in rhythms. Thus, it appears that the brain is prepared at birth for hearing both music and language and, presumably, selectively attends to these auditory signals.

While our musical capacity seems to be innate, it may not be expressed except under extraordinary circumstances. In a fascinating book, *Musicophilia,* neurologist Oliver Sacks (2007) describes the case of a surgeon who was struck by lightning. Dr. Anthony Cicoria had some initial cognitive problems, but they cleared over the weeks following the incident. Then, unexpectedly and suddenly, a desire to hear piano music overwhelmed him.

As the doctor's condition developed, piano music consumed his life: he was compelled to learn to play and to write piano music. He continued to work as a surgeon, but his life revolved around piano music. Neither a divorce nor serious head injuries sustained in a motorcycle accident had any effect on his passion for playing and writing music. The cause of Dr. Cicoria's sudden obsession with music is unknown, but presumably the lightning strike somehow changed circuits in his brain and released his musical passion.

Music as Therapy

The power of music to engage cerebral regions has led to its use as a therapeutic tool for brain dysfunctions. Music is being used as a treatment for mood disorders such as depression, for example, but the best evidence of its effectiveness lies in studies of motor disorders such as stroke and Parkinson's disease (Johansson, 2012).

Listening to rhythm activates the motor and premotor cortex and can improve gait and arm training after stroke. Musical experience reportedly also enhances the ability to discriminate speech sounds and to distinguish speech from background noise in patients with aphasia. Parkinson's patients who step to the beat of music can improve their gait length and walking speed. With all these applications, perhaps researchers will decide to use noninvasive imaging to determine which brain areas are recruited with music therapy.

Music therapy also appears to be a useful complement to more traditional therapies, especially when there are problems with mood, such as in depression or brain injury. This may turn out to be very important in the treatment of stroke and traumatic brain injury where depression is a common complication in recovery.

amusia Tone deafness—an inability to distinguish between musical notes.

Links to music appear in Focus features 1-1 and 5-2 and in the dance class for Parkinson patients pictured on page 161. Sections 16-2 and 16-3 revisit music therapy.

REVIEW 10-4
Anatomy of Language and Music

Before you continue, check your understanding.

1. The human auditory system has complementary specialization for the perception of sounds: left for _____ and right for _____.

2. The three frontal-lobe regions that play a role in producing language are _____, _____, and _____.

3. _____ area identifies speech syllables and words and stores their representations in that location.

4. _____ area matches speech sounds to the motor programs necessary to articulate them.

5. At one end of the spectrum for musical ability are people with _____ and at the other are people who are _____.

6. What evidence supports the idea that language is innate?

Answers appear at the back of the book.

10-5 Auditory Communication in Nonhuman Species

Sound has survival value. You will appreciate this if you've ever narrowly escaped becoming an accident statistic by crossing a busy intersection on foot while listening to a music player or talking on a cell phone. Audition is as important a sense to many animals as vision is to humans. Many animals also communicate with other members of their species by using sound, as humans do.

Here we consider just two types of auditory communication in nonhumans: birdsong and echolocation. Each provides a model for understanding different aspects of brain–behavior relations in which the auditory system plays a role.

Birdsong

Of about 8500 living species of birds, about half are considered songbirds. Birdsong has many functions, including attracting mates (usually employed by males), demarcating territories, and announcing location or even mere presence. Although all birds of the same species have a similar song, the song's details vary markedly from region to region, much as dialects of the same human language vary.

Parallels Between Birdsong and Language

Figure 10-23 includes sound-wave spectrograms for the songs of male white-crowned sparrows that live in three different localities near San Francisco. These songs are quite different from region to region. The differences stem from the fact that song development in young birds is influenced not just by genes but also by early experience and learning. In fact, young birds can acquire more elaborate songs than can other members of their species if the young birds have a good tutor (Marler, 1991).

These gene–experience interactions are epigenetic mechanisms. For example, brain areas that control singing in adult song sparrows show altered gene expression in spring as the breeding—and singing—season begins (Thompson et al., 2012). Such studies have not yet targeted young birds, but it is safe to predict that researchers will find parallel changes. Birdsong and human language have broad similarities beyond regional variation. Both appear to be innate yet are sculpted by experience. Both are diverse and can vary in

Figure 3-25 shows how the epigenetic mechanism of methylation can affect gene expression.

White-crowned sparrow

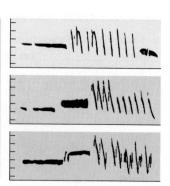

FIGURE 10-23 Birdsong Dialects The songs of male white-crowned sparrows recorded in three locales around San Francisco Bay are very similar, but sound-wave spectrograms reveal that the dialects differ. Like humans, birds acquire regional dialects. Adapted from "The Instinct to Learn," by P. Marler, 1991, in *The Epigenesis of Mind: Essays on Biology and Cognition* (p. 39), edited by S. Carey and R. German, Hillsdale, NJ: Erlbaum.

complexity. Humans seem to have a basic template for language that is programmed into the brain, and experience adds a variety of specific structural forms to this template.

If a young bird is not exposed to song until it is a juvenile and then listens to recordings of birdsongs of different species, the young bird shows a general preference for its own species' song. This preference must mean that a species-specific song template exists in the brain of each bird species. As for language, the details of this birdsong template are modified by experience.

Another broad similarity between birdsong and human language is their great diversity. Among birds, diversity is apparent in the sheer number of songs that a species possesses. Species such as the white-crowned sparrow have but a single song; the marsh wren has as many as 150.

The number of syllables in birdsong also varies greatly, ranging from 30 for the canary to about 2000 for the brown thrasher. Similarly, even though all modern human languages are equally complex, they vary significantly in the type and number of elements they employ. The number of meaningful speech-sound patterns in human languages ranges from about 15 (for some Polynesian languages) to about 100 (for some dialects spoken in the Caucasus Mountains). English has 24.

A final broad similarity between birdsong and human language lies in how they develop. In many bird species, song development is heavily influenced by experience during a critical period, just as language development is in humans. Birds also go through stages in song development, just as humans go through stages in language development. Hatchlings make noises that attract their parents' attention, usually for feeding, and human babies, too, emit cries to signal hunger, among other things.

The fledgling begins to make noises that Charles Darwin compared to the prespeech babbling of human infants. These noises, called *subsong,* are variable in structure, low in volume, and often produced as the bird appears to doze. Presumably, subsong, like human babbling, is a sort of practice for the later development of adult communication after the bird has left the nest.

As a young bird matures, it starts to produce sound-wave patterns that contain recognizable bits of the adult song. Finally, the adult song emerges. In most species, the adult song remains remarkably stable, although a few species, such as canaries, can develop a new song every year that replaces the previous year's song.

Neurobiology of Birdsong

The neurobiology of birdsong has been a topic of intense research partly because it provides an excellent model of brain changes that accompany learning and partly because it offers insight into how sex hormones influence behavior. Fernando Nottebohm and his colleagues first identified the major structures controlling birdsong in the late 1970s (Nottebohm & Arnold, 1976). These structures are illustrated in **Figure 10-24.** The largest are the higher vocal control center (HVC) and the nucleus *robustus archistriatalis* (RA). The axons of the HVC connect to the RA, which in turn sends axons to the 12th cranial nerve. This nerve controls the muscles of the *syrinx,* the structure that actually produces the song.

The HVC and RA have several important and some familiar characteristics:

- They are asymmetrical in some bird species, with the structures in the left hemisphere larger than those in the right hemisphere. In many cases, this asymmetry is similar to the lateralized control of language in humans: if the left-hemisphere pathways are damaged, the birds stop singing, but similar injury in the right hemisphere has no effect on song.

FIGURE 10-24 Avian Neuroanatomy Lateral view of the canary brain shows several left-hemisphere nuclei that control song learning. Two critical nuclei, necessary both for adult singing and for learning the song, are the higher vocal control center (HVC) and the nucleus robustus archistriatalis (RA). Other regions necessary for learning the song during development but not required for adult singing include the dorsal archistriatum (Ad), the lateral magnocellular nucleus of the anterior neostriatum (IMAN), area X of the avian striatum, and the medial dorsolateral nucleus of the thalamus (DLM).

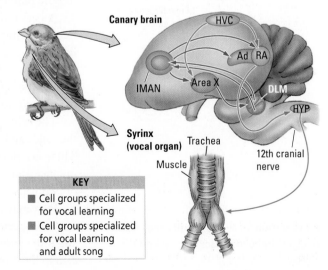

Canary brain

HVC
Ad RA
IMAN
Area X
DLM
HYP

Syrinx (vocal organ) Trachea
Muscle
12th cranial nerve

KEY
- Cell groups specialized for vocal learning
- Cell groups specialized for vocal learning and adult song

A rare strain of zebra finch, shown in Figure 8-30, exhibits physical characteristics of both sexes.

- Birdsong structures are sexually dimorphic. That is, they are much larger in males than in females. In canaries, they are five times as large in the male bird. This sex difference is due to the hormone testosterone in males. Injection of testosterone into female birds causes the song-controlling nuclei to increase in size.

- The size of the birdsong-controlling nuclei is related to singing skill. For instance, unusually talented singers among male canaries tend to have larger HVCs and RAs than do less-gifted singers.

- The HVC and RA contain not only cells that produce birdsong but also cells responsive to hearing song, especially the song of a bird's own species.

The same structures therefore play a role in both song production and song perception. This avian neural anatomy is comparable to the overlapping roles of Broca's and Wernicke's areas in language perception and production in humans.

Echolocation in Bats

Next to rodents, bats are the most numerous order of mammals. The two general groups, or suborders, of bats are the smaller echolocating bats (Microchiroptera) and the larger fruit-eating and flower-visiting bats (Megachiroptera), sometimes called flying foxes. The echolocating bats interest us here because they use sound waves to navigate, to hunt, and to communicate.

Dolphins use an auditory strategy similar to bats, but in water. Focus 10-2 profiles human echolocators.

Most of the 680 species of echolocating bats feed on insects. Some others live on blood (vampire bats), and some catch frogs, lizards, fishes, birds, and small mammals. Bats' auditory system is specialized to use echolocation not only to locate targets in the dark but also to analyze the features of targets, as well as features of the environment in general. Through echolocation, a bat identifies prey, navigates through the leaves of trees, and locates surfaces suitable to land on. Perhaps a term analogous to visualization, such as "audification," would be more appropriate.

Microchiroptera
(an echolocating bat)

Megachiroptera
(a fruit-eating bat)

Echolocation works rather like sonar. The larynx of a bat emits bursts of sound waves at ultrasonic frequencies that bounce off objects and return to the bat's ears, allowing the animal to identify what is in the surrounding environment. The bat, in other words, navigates by the echoes it hears, differentiating among the various characteristics of the echoes.

Objects that are moving (such as insects) have a moving echo, smooth objects give a different echo from rough objects, and so on. A key component of this echolocation system is the analysis of differences in the return times of echoes. Close objects return echoes sooner than more distant objects do, and the textures of various objects' surfaces impose minute differences in return times.

A bat's cries are of short duration (ranging from 0.3 to 200 milliseconds) and high frequency (from 12,000 to 200,000 hertz, charted in Figure 10-4). Most of this range lies at too high a frequency for the human ear to detect. Different bat species produce sound waves of different frequencies that depend on the animal's ecology. Bats that catch prey in the open use different frequencies from those used by bats that catch insects in foliage and from those used by bats that hunt prey on the ground.

The echolocation abilities of bats are phenomenal, as shown in **Figure 10-25.** Bats in the wild can be trained to catch small food particles thrown up into the air in the dark. These echolocating skills make the bat a most efficient hunter. The little brown bat, for instance, can capture very small flying insects, such as mosquitoes, at the remarkable rate of two per second.

Researchers have considerable interest in the neural mechanisms of bat echolocation. Each bat species emits sound waves in a relatively narrow range of frequencies, and a bat's auditory pathway has cells specifically tuned to echoes in its species' frequency range. For example, the mustached bat sends out sound waves ranging from 60,000 to 62,000 hertz,

echolocation Ability to identify and locate an object by bouncing sound waves off the object.

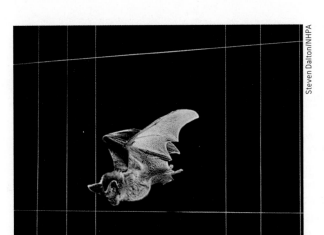

FIGURE 10-25 Born with Sonar Based entirely on auditory information, a bat with a 40-centimeter wingspan can navigate through openings in a 14-by-14-centimeter mesh made of 80-millimeter nylon thread while flying in total darkness.

Steven Dalton/NHPA

and its auditory system has a cochlear fovea (a maximally sensitive area in the organ of Corti) that corresponds to that frequency range.

In this way, more neurons are dedicated to the frequency range used for echolocation than to any other range of frequencies. Analogously, our visual system dedicates more neurons to the retina's fovea, the area responsible for our most detailed vision. In the cortex of the bat's brain, several distinct areas process complex echo-related inputs. One area computes the distance of given targets from the animal, for instance, whereas another area computes the velocity of a moving target. This neural system makes the bat exquisitely adapted for nighttime navigation.

Figure 9-2 details the fovea's structure.

REVIEW 10-5
Auditory Communication in Nonhuman Species

Before you continue, check your understanding.

1. Song development in young birds is influenced by both genes and early experience and learning, interactions indicative of _____.

2. In many bird species the control of song in the brain is lateralized to the _____ hemisphere.

3. Bats use _____ to locate prey in the dark. This system is much like the _____ ships use to locate underwater objects.

4. What does the presence of dialects in birdsong in the same species demonstrate?

Answers appear at the back of the book.

SUMMARY

Although we take language and music for granted, both play central roles in our mental lives and in our social lives. Language and music provide us ways to communicate with other people—and with ourselves. They facilitate social identification, parenting, and cultural transmission.

fundamental physical qualities of sound-wave energy: frequency (repetition rate), amplitude (size), and complexity. Perceptually, neural networks in the brain then translate these energies into the pitch, loudness, and timbre of the sounds that we hear.

10-1 Sound Waves: The Stimulus for Audition

The stimulus for the auditory system is the mechanical energy of sound waves that results from changes in air pressure. The ear transduces three

10-2 Functional Anatomy of the Auditory System

Beginning in the ear, mechanical and electrochemical systems combine to transform sound waves into auditory perceptions—what we hear. Changes in air pressure are conveyed in a mechanical chain reaction from the eardrum to the bones of the middle ear to the oval window of the cochlea and the cochlear fluid that lies behind it in the inner ear. Movements of the cochlear fluid produce movements in specific regions of the basilar membrane, leading to changes in the electrochemical activity of the auditory receptors, the inner hair cells on the basilar membrane that send neural impulses through the auditory nerve into the brain.

10-3 Neural Activity and Hearing

The basilar membrane has a tonotopic organization. High-frequency sound waves maximally stimulate hair cells at the base, whereas low-frequency sound waves maximally stimulate hair cells at the apex, enabling cochlear neurons to code various sound frequencies.

Tonotopic organization analyzes sound waves at all levels of the auditory system, and the system also detects both amplitude and location. The firing rate of cochlear neurons codes sound amplitude, with louder sounds producing higher firing rates than softer sounds do. Location is detected by structures in the brainstem that compute differences in the arrival times and the loudness of a sound in the two ears.

The hair cells of the cochlea synapse with bipolar neurons that form the cochlear nerve, which in turn forms part of the eighth cranial nerve. The cochlear nerve takes auditory information to three structures in the hindbrain: the cochlear nucleus, the superior olive, and the trapezoid body. Cells in these areas are sensitive to differences in both sound-wave intensity and arrival times at the two ears. In this way, they enable the brain to locate a sound.

The auditory pathway continues from the hindbrain areas to the inferior colliculus of the midbrain, then to the medial geniculate nucleus in the thalamus, and finally to the auditory cortex. As for vision, dorsal and ventral pathways exist in the auditory cortex, one for pattern recognition and the other for controlling movements in auditory space. Cells in the cortex are responsive to specific categories of sounds, such as species-specific communication.

10-4 Anatomy of Language and Music

Despite differences in speech-sound patterns and structures, all human languages have the same basic foundation of a syntax and a grammar.

This fundamental similarity implies an innate template for creating language. The auditory areas of the cortex in the left hemisphere play a special role in analyzing language-related information, whereas those in the right hemisphere play a special role in analyzing music-related information. The right temporal lobe also analyzes prosody, the melodical qualities of speech.

Among several language-processing areas in the left hemisphere, Wernicke's area identifies speech syllables and words and so is critically engaged in speech comprehension. Broca's area matches speech-sound patterns to the motor behaviors necessary to make them and so plays a major role in speech production. Broca's area also discriminates between closely related speech sounds. Aphasias result from an inability to speak (Broca's aphasia) or comprehend language (Wernicke's aphasia) despite the presence of normal comprehension and intact vocal mechanisms.

Auditory analysis of music draws more on right hemisphere activity than on the left. Nor is music production localized to the right hemisphere: it recruits left hemisphere involvement as well. The perception of music engages both the right temporal and frontal regions. Music's power to engage both right and left hemisphere activity makes it a powerful tool to engage the injured or dysfunctioning brain. Music therapy likely will play an increasingly important role in coming years.

10-5 Auditory Communication in Nonhuman Species

Nonhuman animals have evolved specialized auditory structures and behaviors. Regions of songbirds' brains are specialized for producing and comprehending song. In many species, these regions are lateralized to the left hemisphere, analogous in a way to how language areas are lateralized to the left hemisphere in most humans. The similarities between the development of song in birds and the development of language in humans, as well as similarities in the neural mechanisms underlying both the production and the perception of birdsong and language, are striking.

Both owls and bats can fly and catch prey at night using only auditory information to guide their movement. Echolocating bats evolved a biological sonar that allows them to map the objects in their auditory world, as humans map their visual worlds. Although some blind humans employ this strategy, the mainly auditory reality of bats, dolphins, and other echolocators is one most humans can only try to imagine.

KEY TERMS

aphasia, p. 340

amplitude, p. 323

amusia, p. 347

basilar membrane, p. 328

Broca's area, p. 338

cochlea, p. 328

cochlear implant, p. 335

decibel (dB), p. 323

echolocation, p. 350

frequency, p. 321

hair cell, p. 328

hertz (Hz), p. 321

insula, p. 333

lateralization, p. 333

medial geniculate nucleus, p. 331

ossicles, p. 328

primary auditory cortex (area A1), p. 331

prosody, p. 326

sound wave, p. 321

supplementary speech area, p. 343

tonotopic representation, p. 335

Wernicke's area, p. 333

CHAPTER

11

How Does the Nervous System Respond to Stimulation and Produce Movement?

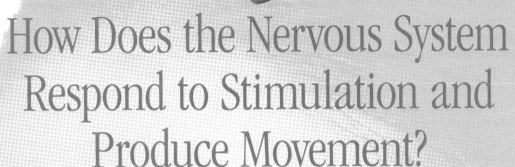

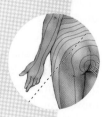

RESEARCH FOCUS 11-1 NEUROPROSTHETICS

11-1 HIERARCHY OF MOVEMENT CONTROL

THE BASICS RELATING THE SOMATOSENSORY AND MOTOR SYSTEMS

FOREBRAIN AND INITIATION OF MOVEMENT

BRAINSTEM AND SPECIES-TYPICAL MOVEMENT

CLINICAL FOCUS 11-2 CEREBRAL PALSY

SPINAL CORD AND EXECUTION OF MOVEMENT

CLINICAL FOCUS 11-3 SPINAL-CORD INJURY

11-2 MOTOR SYSTEM ORGANIZATION

MOTOR CORTEX

MOTOR CORTEX AND SKILLED MOVEMENT

PLASTICITY IN THE MOTOR CORTEX

CORTICOSPINAL TRACTS

MOTOR NEURONS

CONTROL OF MUSCLES

11-3 BASAL GANGLIA, CEREBELLUM, AND MOVEMENT

BASAL GANGLIA AND MOVEMENT FORCE

CLINICAL FOCUS 11-4 TOURETTE'S SYNDROME

CEREBELLUM AND MOVEMENT SKILL

11-4 ORGANIZATION OF THE SOMATOSENSORY SYSTEM

SOMATOSENSORY RECEPTORS AND PERCEPTION

DORSAL-ROOT GANGLION NEURONS

SOMATOSENSORY PATHWAYS TO THE BRAIN

SPINAL REFLEXES

FEELING AND TREATING PAIN

RESEARCH FOCUS 11-5 PHANTOM-LIMB PAIN

VESTIBULAR SYSTEM AND BALANCE

11-5 EXPLORING THE SOMATOSENSORY CORTEX

SOMATOSENSORY HOMUNCULUS

RESEARCH FOCUS 11-6 TICKLING

EFFECTS OF DAMAGE TO THE SOMATOSENSORY CORTEX

SOMATOSENSORY CORTEX AND COMPLEX MOVEMENT

Neuroprosthetics

Most of us seamlessly control the approximately 650 muscles that move our bodies. But if the motor neurons that control the muscles no longer connect to them, as happens in amyotrophic lateral sclerosis (ALS, or Lou Gehrig's disease) then movement and eventually breathing become impossible.

ALS hit Scott Mackler, a neuroscientist and marathon runner, when he was about 40 years old. Placed on a respirator that enabled him to breathe, he developed *locked-in syndrome:* Mackler lost virtually all ability to communicate.

ALS has no cure, and death often occurs within 5 years of diagnosis. Yet Scott Mackler is no longer locked in. He's back at work, in touch with family and friends, and he even gave an interview on CBS's *60 Minutes* in 2008.

Mackler has learned to translate his mental activity into movement. His method: a *brain–computer interface* (BCI). BCIs employ electrical signals from the brain to direct computer-controlled devices. BCI is part of the field of **neuroprosthetics,** in which computer-assisted devices replace lost biological function.

A *computer–brain interface* (CBI) employs electrical signals from a computer to instruct the brain. Cochlear implants that deliver sound-related signals to the inner ear to allow hearing are one type of CBI. *Brain–computer–brain interfaces* (BCBI) combine the BCI and CBI approaches. Using BCBIs under development, the brain will be able to command robotic devices that provide sensory feedback to the brain.

Right now, these interfaces are slow. Mackler may take up to 20 seconds to make a single command. Devices currently in development will enhance speed. Future devices will also greatly increase signal precision by using electrodes placed directly adjacent to brain cells.

Electrode arrays that interface with about 100 brain cells will be replaced with arrays that interface with thousands of brain cells. Experimental approaches will use optogenetics, incorporating light-sensitive channels into motor cortex and sensory cortex neurons. Light signals are faster than electrical signals and produce less tissue damage.

BCBI interfaces will command robotic hands to grasp objects while tactile receptors on the robot deliver touch and other sensory information to the user at the same time. BCBIs will also control exoskeletal devices that produce reaching and walking movements and return touch, body position, and balance information to guide movement (Lebedev et al., 2011).

Scott Mackler in his office checking e-mail with the assistance of his BCI.

Section 1-1 offers a simple definition of behavior as any kind of movement in a living organism.

Movement is a defining feature of animals, and this chapter explores how the nervous system produces movement. The body senses and movement interrelate at all levels of the nervous system. Somatosensation is more closely related to movement than are the other senses.

At the level of the spinal cord, somatosensory information contributes to motor reflexes. In the brainstem, it contributes to movement timing and control. In the cerebrum, somatosensory information contributes to complex voluntary movements. Indeed, for many functions, the other senses work through the somatosensory system to produce movement. If the motor system is a vehicle and the somatosensory system is the driver, the other sensory systems act like backseat drivers.

neuroprosthetics Field that develops computer-assisted devices to replace lost biological function.

In this chapter, we consider how movement is organized in the central nervous system before we turn to the role of the body senses and balance and how these somatosenses contribute to movement. If you want to review the relation of the motor system and somatosensation before you read on, turn to "The Basics: Relating the Somatosensory and Motor Systems" on pages 356–357.

11-1 Hierarchy of Movement Control

When we move, our behavior is complex. We decide on a goal, then we choose how to achieve the goal, and then we move. We are conscious of the choices we make but frequently unaware of how our motor system produces the appropriate movement sequence.

The major components of our motor system are the cerebrum (forebrain), the brainstem, and the spinal cord. The cerebrum contributes to our conscious control of movement while the brainstem and spinal cord perform our more automatic movements. In the face of impaired brainstem or spinal-cord function, the forebrain can imagine movements but can no longer produce them. Neuroprosthetic devices like those described in Research Focus 11-1 replace the automatic control of movement provided by the brainstem and spinal cord and restore control to the forebrain.

Figure 11-1 shows the stepwise sequence your CNS performs in directing your hand to pick up a mug. You visually inspect the cup to determine what part of it to grasp. The visual cortex relays this information through somatosensory regions of the neocortex to the motor regions of the neocortex that plan and initiate the movement. Only then does the brain send instructions to the part of the spinal cord that controls the muscles of your arm and hand.

As you grasp the mug's handle, information from sensory receptors in your fingers travels to the spinal cord and from there back to the somatosensory cortex to confirm that the cup has been grasped. The somatosensory cortex, in turn, informs the motor cortex that the cup is now being held. Other regions of the brain also participate in controlling the movement. The subcortical *basal ganglia* help to produce the appropriate amount of force for grasping the cup's handle, while in the brainstem, the cerebellum helps to regulate the timing and accuracy of the movement.

Here the journey begins with movement. In Section 4-4 we begin the journey with sensation: how a sensory stimulus initiates a nerve impulse that ends in movement, a muscular contraction.

The important concept to remember for now is the hierarchical organization of the entire sensoritmotor system. When you reach the end of the chapter, review Figure 11-1 to reinforce what you've learned.

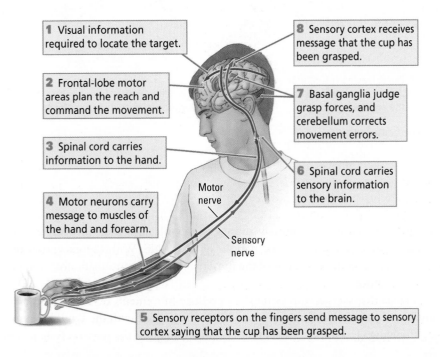

1 Visual information required to locate the target.

2 Frontal-lobe motor areas plan the reach and command the movement.

3 Spinal cord carries information to the hand.

4 Motor neurons carry message to muscles of the hand and forearm.

Motor nerve

Sensory nerve

8 Sensory cortex receives message that the cup has been grasped.

7 Basal ganglia judge grasp forces, and cerebellum corrects movement errors.

6 Spinal cord carries sensory information to the brain.

5 Sensory receptors on the fingers send message to sensory cortex saying that the cup has been grasped.

FIGURE 11-1 Sequentially Organized Movement

⊕ THE BASICS

Relating the Somatosensory and Motor Systems

The intimate relationship between the motor system that allows us to move and the somatosensory system that guides our movements is apparent in their close anatomical relationships. *Afferent* somatosensory information travels from the body inward via the somatic nervous system.

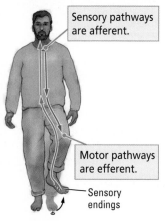

Sensory pathways are afferent.

Motor pathways are efferent.

Sensory endings

Information Flow

Movement information travels out of the central nervous system via a parallel, *efferent* motor system.

As diagrammed in "Information Flow," when you step on a tack, the sensory signals transmitted by the SNS from the body through the spinal cord and into the brain are afferent. Efferent signals from the CNS trigger a motor response: you lift your foot.

The spinal cord connects the somatosensory and motor systems throughout the CNS. "Spinal-Nerve

Connections" shows the spinal cord in cross section. In the outer part, which consists of white matter, dorsal tracts are sensory and ventral tracts are motor, with some exceptions. The inner part of the cord is gray matter composed largely of cell bodies and shaped like a butterfly.

SNS nerves entering the spinal cord's dorsal side carry information inward from the body's sensory receptors and collect together into a *dorsal root* as the fibers enter a spinal-cord segment of the CNS. Fibers leaving the spinal cord's ventral side carry information out from the spinal cord to the muscles. They, too, bundle together as the fibers exit the spinal cord, forming a *ventral root*. (Bundles of nerve fibers within the CNS are called *tracts;* outside the CNS they are called *nerves.*)

The spinal cord lies within a series of small bones called *vertebrae* categorized into the five anatomical regions diagrammed on the left in "Spinal Segments and Dermatomes." Each spinal segment corresponds to a region of body surface called a *dermatome* (literally, a "skin cut"), shown on the right. From top to bottom, the cervical, thoracic, lumbar, sacral, and coccygeal regions are identified by spinal segment number: C5 (cervical segment 5) at the base of the neck, for example, and L2 in the lower back.

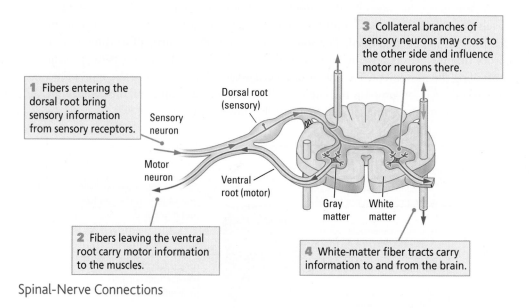

1 Fibers entering the dorsal root bring sensory information from sensory receptors.

3 Collateral branches of sensory neurons may cross to the other side and influence motor neurons there.

Dorsal root (sensory)

Sensory neuron

Motor neuron

Ventral root (motor)

Gray matter

White matter

2 Fibers leaving the ventral root carry motor information to the muscles.

4 White-matter fiber tracts carry information to and from the brain.

Spinal-Nerve Connections

Clearly, an action as seemingly simple as picking up a mug involves widespread regions of your CNS and specific regions of your SNS. With the exception of your decision to pick up the cup, most of the movement happens automatically, without your conscious control.

The idea that the nervous system can produce movements by recruiting many different neural structures organized hierarchically originated with the English neurologist John Hughlings-Jackson over a century ago. He thought of the nervous system as being

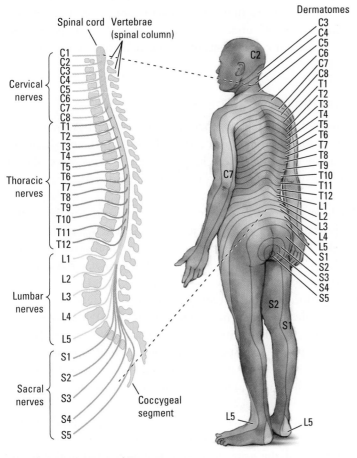

Spinal Segments and Dermatomes

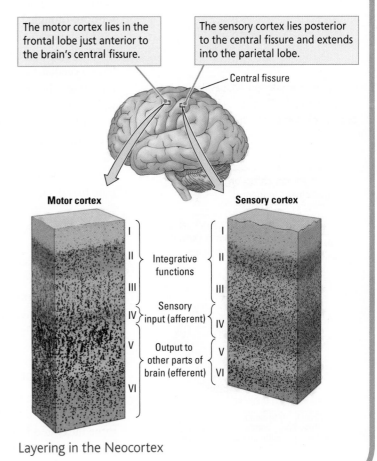

characteristics, and functions. Layer IV is relatively thick in the sensory cortex but relatively thin in the motor cortex, for example. Layer V is relatively thick in the motor cortex and relatively thin in the sensory cortex. Cortical layer IV is afferent, layer V is efferent, and that makes sense: sensory regions have a large input layer and motor regions a large output layer.

Layering in the Neocortex

The segmentation of the body and the nervous system has a long evolutionary history that can be seen in worms as well as in vertebrates. The cervical and lumbar dermatomes represent the human forelimbs and hind limbs. Their arrangement is sequential if you imagine a human in an "all-fours" posture.

"Layering in the Neocortex" plumbs the depths of the primary motor cortex (shown in blue) and adjacent sensory (red) cortical regions. Viewed through a microscope, the six cortical layers differ in appearance,

organized in a number of levels. Successively higher levels control more complex aspects of behavior by acting through the lower levels. The three major levels in Hughlings-Jackson's model are those just mentioned: forebrain, brainstem, and spinal cord. He also proposed that further levels of organization exist within these divisions, an idea we will pursue when we describe the control of movement by the motor cortex in Section 11-2.

Hughlings-Jackson adopted the concept of hierarchical organization from Darwin's evolutionary theory. He knew that the chordate nervous system had evolved gradually:

Recall Principle 4 from Section 2-6: The CNS functions on multiple levels.

To review this evolution, see "The Basics: Classification of Life" in Section 1-3.

motor sequence Movement modules preprogrammed by the brain and produced as a unit.

mirror neuron Cell in the primate premotor cortex that fires when an individual observes a specific action taken by another individual.

As described in Section 7-1, Lashley experimented for three decades to find the location of memory in the brain. He failed.

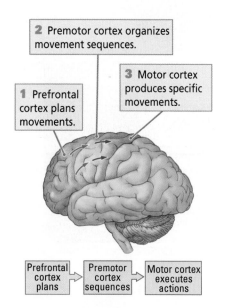

2 Premotor cortex organizes movement sequences.

1 Prefrontal cortex plans movements.

3 Motor cortex produces specific movements.

Prefrontal cortex plans → Premotor cortex sequences → Motor cortex executes actions

FIGURE 11-2 Initiating a Motor Sequence

Sections 12-4 and 15-2 explore the cognitive deficits created by frontal-lobe injury.

the spinal cord first developed in worms; the brainstem in fish, amphibians, and reptiles; and the forebrain in birds and mammals. Rather than replacing lower levels, Hughlings-Jackson reasoned, the higher levels must act by controlling the lower levels.

In describing how we grasp a mug, we observed how a hierarchically organized structure, the mammalian nervous system, functions. Higher regions work through and influence the actions of lower areas. To understand how all these CNS regions work together, we now consider the major components of the hierarchy one by one, starting at the top with the forebrain.

Forebrain and Initiation of Movement

Complex movements consist of many acts. Consider playing basketball. At every moment, players must make decisions and perform actions. Dribble, pass, and shoot are different categories of movement, and each can be performed in many ways. Skilled players choose among the categories effortlessly and blend them together seemingly without thought.

One explanation for how we control such complex movements, popular in the 1930s, centers on feedback: after we act, we wait for feedback about how well the action has succeeded; then we make the next movement accordingly. The pioneering neuroscience researcher Karl Lashley (1951), in an article titled "The Problem of Serial Order in Behavior," found fault with this explanation.

Lashley argued that we perform skilled movements too quickly to rely on feedback about one movement shaping the next movement. The time required to receive feedback about the first movement combined with the time needed to develop a plan for the subsequent movement and send a corresponding message to muscles is simply too long for effective action. Lashley suggested that movements must be performed as **motor sequences,** with one sequence held in readiness while an ongoing sequence is being completed.

According to this view, all complex behaviors, including speaking, playing the piano, and playing basketball, require selecting and executing multiple movement sequences. As one sequence is being executed, the next sequence is being prepared so that it can follow the first smoothly. The act of speaking illustrates Lashley's view. When people use complex rather than simple word sequences, they are more likely to pause and make "umm" and "ahh" sounds, suggesting that it is taking them more time than usual to organize their word sequences.

Initiating a Motor Sequence

The frontal lobe of each hemisphere is responsible for planning and initiating motor sequences. The frontal lobe is divided into a number of different regions, including the three illustrated in **Figure 11-2.** From front to back, they are the prefrontal cortex, the premotor cortex, and the primary motor cortex. These frontal-lobe motor areas are arranged hierarchically.

PREFRONTAL CORTEX Atop the hierarchy, the *prefrontal cortex* (PFC) plans complex behavior. Deciding to get up at a certain hour to arrive at work on time, to stop at the library to return a book that is due, even whether a behavior is right or wrong and whether it should be performed at all are examples. Humans with prefrontal cortex injury often break social and legal rules not because they do not know the rules or the consequences of breaking them but because their decision making is faulty. The prefrontal cortex does not specify the precise movements to be made. It simply specifies the goal.

PREMOTOR CORTEX To bring a plan to completion, the prefrontal cortex sends instructions to the premotor cortex, which produces the appropriate complex movement sequences. If the premotor cortex is damaged, the sequences cannot be coordinated and

Normal animal

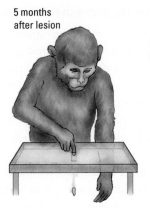

5 months after lesion

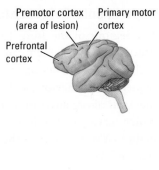

Premotor cortex (area of lesion)   Primary motor cortex

Prefrontal cortex

FIGURE 11-3 Premotor Control On a task requiring both hands, the normal monkey can push the peanut out of a hole with one hand and catch it in the other, but 5 months after lesioning of the premotor cortex, the brain-injured monkey cannot. Adapted from "Supplementary Motor Area of the Monkey's Cerebral Cortex: Short- and Long-Term Effects after Unilateral Ablation and the Effects of Subsequent Callosal Section," by C. Brinkman, 1984, *Journal of Neuroscience*, 4, p. 925.

the goal cannot be accomplished. For example, the monkey at the right in **Figure 11-3** has a lesion in the dorsal premotor cortex. The monkey has been given the task of extracting a piece of food wedged in a hole in a table (Brinkman, 1984).

If the monkey simply pushes the food with a finger, the food will drop to the floor and be lost. The monkey has to catch the food by holding a palm beneath the hole as the food is being pushed out. But this brain-injured animal is unable to make the two complementary movements together. It can push the food with a finger and extend an open palm, but it cannot coordinate these actions of its two hands, as the normal monkey on the left can.

One way that the premotor cortex organizes movement sequences involves specialized **mirror neurons** that discharge when we perform an action such as reaching for food. But mirror neurons have additional abilities: they discharge when we observe another individual performing a movement, even though we may be making no movement at all. They can also discharge if we simply think about performing the movement.

Mirror neurons allow us to produce movement sequences and also to observe, understand, and copy the movement sequences of others (Casile et al., 2011). The motor system's ability not only to produce movements but also to recognize the movements of others and to imagine movements is enabled in part by the mirror neurons in the premotor cortex. Mirror neurons also populate the parietal cortex, primary motor cortex, and even the cerebellum.

PRIMARY MOTOR CORTEX The premotor cortex organizes movement sequences but does not specify how each movement is to be carried out. Those details are the responsibility of the primary motor cortex. To understand its role, consider the rich array of movements we can use to grasp objects.

In using the precision pincer grip (**Figure 11-4A**), we hold an object between the thumb and index finger. We can perform many precision grips using the thumb and other fingers in opposition. The pincer grip not only allows us to pick up small objects easily but also allows us to use whatever is held with considerable skill. In contrast, using a power grip (Figure 11-4B), we hold an object much less dexterously but with far more power. This whole-hand grasp simply involves closing the thumb and one or more fingers around the object.

Clearly, a precision grip is the more demanding movement because the two fingers must be placed precisely on the object. People with damage to the primary motor cortex have difficulty shaping their fingers correctly to perform precision grips, although they may perform whole hand grasps. They also have difficulty in performing many skilled movements of the hands, arms, and trunk (Lang & Schieber, 2004).

Recall Principle 6 from Section 2-6: Brain circuits process information both hierarchically and in parallel.

Section 15-2 describes the evolutionary advantages of mirror neurons. Focus 15-2 reports on mirror neuron dysfunction in autism spectrum disorder.

(A) Pincer grip

(B) Power grip

The Photo Works

FIGURE 11-4 Getting a Grip

Figure 8-21 shows that grasping responses in infants progress from the whole-hand to the pincer grip.

Experimental Evidence for the Movement Hierarchy

The regions of the frontal lobe in each hemisphere that plan, coordinate, and execute precise movements are hierarchically related. After the prefrontal cortex has formulated a plan of action, it instructs the premotor cortex to organize the appropriate sequence of behaviors, and the primary motor cortex executes the movements. This hierarchical organization is supported by findings from studies of cerebral blood flow, which serves as an indicator of neural activity. **Figure 11-5** shows the regions of the brain that were active as the participants in one such study performed different tasks (Roland, 1993).

As the participants use a finger to push a lever, increased blood flow is limited to the primary somatosensory and primary motor cortex (Figure 11-5A). As the participants execute a sequence of finger movements, blood flow also increases in the premotor cortex (Figure 11-5B). And as the participants use a finger to trace their way through a maze, a task that requires coordinated movements in relation to a goal, blood flow increases in the prefrontal cortex as well (Figure 11-5C). As the participants were performing these tasks, notice that relative blood flow increased only in the regions taking part in the required movements rather than throughout the frontal lobe.

Section 7-4 describes dynamic imaging methods that record and measure blood flow in the brain.

FIGURE 11-5 **Hierarchical Control of Movement in the Brain** Adapted from *Brain Activation* (p. 63), by P. E. Roland, 1993, New York: Wiley-Liss.

(A) Simple movement

Blood flow increases in the hand area of the primary somatosensory and primary motor cortex when subjects use a finger to push a lever.

Motor cortex | Sensory cortex

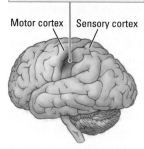

(B) Movement sequence

Blood flow increases in the premotor cortex when subjects perform a sequence of movements.

Dorsal premotor cortex

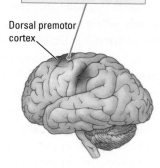

(C) Complex movement

Blood flow also increases in the prefrontal, temporal, and parietal cortex when subjects use a finger to find a route through a maze.

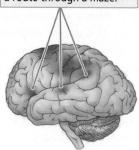

Brainstem and Species-Typical Movement

In a series of studies, the Swiss neuroscientist Walter Hess (1957) found that the brainstem controls *species-typical behaviors,* actions displayed by every member of a species—the pecking of a robin, the hissing of a cat, or the breaching of a whale. Hess developed the technique of implanting electrodes into the brains of cats and other animals and cementing them in place. These electrodes could then be attached to stimulating leads in the freely moving animal without causing it much discomfort.

By stimulating the brainstem, Hess was able to elicit the innate movements that the animal might be expected to make. A resting cat could be induced to suddenly leap up with an arched back and erect hair as though frightened by an approaching dog, for example. The elicited movements began abruptly when the stimulating current was turned on and ended equally abruptly when the stimulating current was turned off. An animal performed such species-typical behaviors in a subdued manner when the stimulating current was low and displayed increased vigor as the stimulating current was turned up.

Section 1-5 introduces species-typical behavior, noting that evolutionary principles apply *across* species but not to individuals *within* a species.

locked-in syndrome Condition in which a patient is aware and awake but cannot move or communicate verbally due to complete paralysis of nearly all voluntary muscles except the eyes.

The actions varied, depending on the brainstem site that was stimulated. Stimulating some sites produced head turning, others produced walking or running, and still others elicited displays of aggression or fear. The animal's reaction toward a particular stimulus could be modified accordingly. For instance, when shown a stuffed toy, a cat responded to electrical stimulation of some brainstem sites by stalking the toy and to stimulation of other sites with a fearful response and withdrawal.

Hess's experiments have been confirmed and expanded by other researchers using many different animal species. **Experiment 11-1** shows the effects of brainstem stimulation on a chicken under various conditions (von Holst, 1973). Notice the effect of context: how the neural site stimulated interacts both with the object presented and with the stimulation's duration.

With stimulation of a certain site alone, the chicken displays only restless behavior. When a fist is displayed, the same stimulation elicits slightly threatening behavior. When the object displayed is then switched from a fist to a stuffed polecat, the chicken responds with vigorous threats. Finally, with continued stimulation in the presence of the polecat, the chicken flees, screeching.

Such experiments show that an important brainstem function is to produce complex patterns of adaptive behavior. These patterns include movements used in eating and drinking and in sexual behavior. Animals can be induced to display these survival-related behaviors when certain areas of the brainstem are stimulated. An animal can even be induced to eat nonfood objects, such as chips of wood, if the part of the brainstem that triggers chewing is stimulated.

Grooming illustrates how complex patterns of action are coordinated by the brainstem (Kalueff et al., 2007). A grooming rat sits back on its haunches, licks its paws, wipes its nose with its paws, then wipes its paws across its face, and finally turns to lick the fur on its body. These movements are always performed in the same order, from the face to the shoulders and then toward the rear of the body. The next time you dry off after a shower or swimming, note the "grooming sequence" you use. Humans' grooming sequence is very similar to the one rats use.

The brainstem is also important for maintaining posture, standing upright, coordinating movements of the limbs, swimming and walking, grooming the fur, and making nests. The effects of damage to regions of the brainstem that organize many adaptive movements can be seen in the effects of **locked-in syndrome,** which Scott Mackler experienced in connection with Lou Gehrig's disease (see Research Focus 11-1). The patient with locked-in syndrome is aware and awake but cannot

Focus 4-4 has more detail on ALS and Lou Gehrig.

EXPERIMENT 11-1

Question: What are the effects of brainstem stimulation under different conditions?

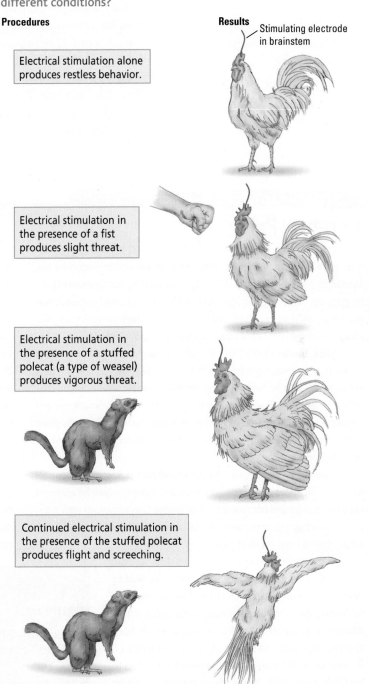

Procedures

Electrical stimulation alone produces restless behavior.

Electrical stimulation in the presence of a fist produces slight threat.

Electrical stimulation in the presence of a stuffed polecat (a type of weasel) produces vigorous threat.

Continued electrical stimulation in the presence of the stuffed polecat produces flight and screeching.

Results

Stimulating electrode in brainstem

Conclusion: Stimulation of some brainstem sites produces behavior that depends on context, suggesting that an important function of the brainstem is to produce appropriate species-typical behavior.

Adapted from *The Collected Papers of Erich von Holst* (p. 121), translated by R. Martin, 1973, Coral Gables, FL: University of Miami Press.

move or communicate verbally due to complete paralysis of nearly all voluntary muscles except the muscles of the eyes.

The effects of brainstem damage on behavior also can be seen in **cerebral palsy** (CP), a disorder primarily of motor function: voluntary movements become difficult to make whereas conscious behavior controlled by the cortex may remain intact. Cerebral palsy is caused by brainstem trauma before or shortly after birth. As described in Clinical Focus 11-2, "Cerebral Palsy," trauma leading to cerebral palsy can sometimes happen in early infancy as well.

Clearly, a brain injury that causes cerebral palsy can be extremely damaging to movement while leaving sensory abilities and cognitive capacities unimpaired. Advances in neuroprosthetic technology (see Research Focus 11-1), touch-screen technology, and

Section 8-4 surveys the effects of injuries on the developing brain.

CLINICAL FOCUS ⊕ 11-2

Cerebral Palsy

E. S. suffered a cold and infection when he was about 6 months old. Subsequently, he had great difficulty coordinating his movements. As he grew up, his hands and legs were almost useless and his speech was extremely difficult to understand. E. S. was considered intellectually disabled and spent most of his childhood in a custodial school.

When E. S. was 13 years old, the school bought a computer, and one of the teachers attempted to teach E. S. to use it by pushing the keys with a pencil that he held in his mouth. Within a few weeks, the teacher realized that E. S. was extremely intelligent and could communicate and complete school assignments on the computer. He was eventually given a motorized wheelchair that he could control with finger movements of his right hand.

Assisted by the computer and the wheelchair, E. S. soon became almost self-sufficient and eventually attended college, where he achieved excellent grades and became a student leader. On graduation with a degree in psychology, he became a social worker and worked with children who suffered from cerebral palsy.

William Little, an English physician, first noticed in 1853 that difficult or abnormal births could lead to later motor difficulties in children. The disorder that Little described was cerebral palsy (also called *Little's disease*), a group of brain disorders that result from brain damage acquired perinatally (at or near birth). CP is common worldwide, with an incidence estimated to be 1.5 in every 1000 births. Among surviving babies who weigh less than 2.5 kilograms at birth, the incidence is much higher—about 10 in 1000.

The most common cause of cerebral palsy is birth injury, especially due to *anoxia*, a lack of oxygen. Anoxia may result from a defect in the placenta, the organ that allows oxygen and nutrients to pass from mother to child in utero, or it may be caused by a tangled umbilical cord during birth that reduces the oxygen supply to the infant. Other causes include

infections, hydrocephalus, seizures, and prematurity, and infection. All produce a defect in the immature brain before, during, or just after birth.

Most children with cerebral palsy appear normal in the first few months of life, but as the nervous system develops, the motor disturbances become progressively more noticeable. Common symptoms include spasticity, an exaggerated contraction of muscles when they are stretched; dyskinesia, involuntary extraneous movements such as tremors and uncontrollable jerky twists (athetoid movements); and rigidity, or resistance to passive movement. Everyday movements are abnormal or the affected person may be confined to a wheelchair.

As a means for investigating the relationship between brain development and susceptibility to brain injury, Tymofiyeva and colleagues (2012) have used MRI to develop a "baby connectome," a map of the connections in the brain at each developmental age. Investigators can derive abnormalities in the development of brain connections from the connectome.

Many people with cerebral palsy have successful professional careers. Dr. John Melville, second from right, monitors a CT scan and discusses the patient's progress with a colleague. Cerebral palsy has not impeded—and may have inspired—his medical career.

remotely controlled computers all signal important markers toward allowing disabled people to become more independent.

Spinal Cord and Execution of Movement

The late Christopher Reeve, a well-known actor who portrayed Superman in three films, was thrown from a horse during a riding competition in 1995. Reeve's spinal cord was severed near its upper end, at the C1–C2 level. The injury left Reeve's brain intact and functioning and his remaining spinal cord intact and functioning too, but his brain and spinal cord were no longer connected.

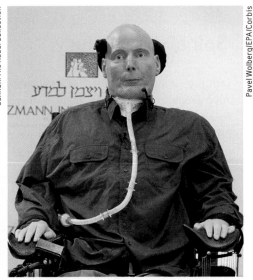

Christopher Reeve (*left*) portraying Superman in 1984 and (*right*) in 2004, nine years after his spinal-cord injury.

As a result, other than movements of his head and slight movement in his shoulders, Reeve's body was completely paralyzed. He was even unable to breathe without assistance. A century ago such a severe injury would have been fatal, but modern and timely medical treatment allowed Reeve to survive for nearly a decade.

A cut high on the spinal cord, such as Christopher Reeve survived, entails paralysis and loss of sensation in the arms and legs, a condition called **quadriplegia** (see "Spinal Segments and Dermatomes" on page 357). If the cut is low, **paraplegia** results: paralysis and loss of sensation are confined to the legs and lower body, as described in Clinical Focus 11-3, "Spinal-Cord Injury" on page 364. Christopher Reeve and his late wife Dana founded the Christopher and Dana Reeve Foundation for spinal-cord research. It is dedicated to improving the life and function of spinal-cord-injured people and also to searching for cures for spinal-cord injury.

Far from a mere relay between the body and brain, the spinal cord contains complex motor programs. A spinal-cord patient can walk on a conveyor belt if the body is supported. Indeed, Christopher Reeve was able to "walk" in a swimming pool where his body was supported by water.

When the leg of a spinal-cord patient is moved backward on a conveyor belt, causing the foot to lose support, the limb reflexively lifts off the belt and swings forward underneath the body. As the foot then touches the surface of the belt again, tactile receptors initiate the reflex that causes the foot to push against the surface and support the body's weight. In this way, several spinal reflexes work together to facilitate the complex movement of walking. This reflexive organization of walking can even be obtained in a premature or newborn baby: when held in the correct position the baby will perform stepping movements.

Scratch reflex

Among the complex reflexes that can be observed in other vertebrates is the **scratch reflex.** Here, an animal reflexively scratches a part of its body in response to a stimulus from the surface of the body. The complexity of the scratch reflex is revealed in the accuracy of the movement. Without direction from the brain, the tip of a limb, usually a hind limb, can be correctly directed to the part of the body that is irritated. Typically, itching is the sensation that elicits scratching; it is likely that the sensory receptors on the surface of the skin that produce the itch evolved for detecting parasites and other foreign objects on the skin surface. We will return to itching sensations in Section 11-4.

In humans and other animals with a severed spinal cord, spinal reflexes still function, even though the spinal cord is cut off from communication with the brain. As a result, the paralyzed limbs may display spontaneous movements or spasms. But the brain can no longer guide the timing of these automatic movements. Consequently, reflexes related to bladder and bowel control may need to be artificially stimulated by caregivers.

cerebral palsy Group of brain disorders that result from brain damage acquired perinatally (at or near birth).

quadriplegia Paralysis of the legs and arms due to spinal-cord injury.

paraplegia Paralysis of the legs due to spinal-cord injury.

scratch reflex Automatic response in which an animal's hind limb reaches to remove a stimulus from the surface of the body.

Spinal-Cord Injury

Each year, on average, about 11,000 people in the United States and 1000 people in Canada suffer spinal-cord injury (as reported by the Foundation for Spinal Cord Injury). Nearly 40 percent of these injuries occur in traffic accidents and another 40 percent occur as a result of falls. Often the spinal cord is completely severed, leaving the victim with no sensation or movement from the site of the cut downward.

Although 12,000 annual spinal-cord injuries may seem like a large number, it is small relative to the number of people in the United States and Canada who suffer other kinds of nervous-system damage each year. To increase public awareness of their condition and promote research into possible treatments, some, like Christopher Reeve and, pictured here, Canadian Rick Hansen, have been especially active.

Hansen's paraplegia resulted from a lower thoracic spinal injury in 1975. Twelve years later, to raise public awareness of the potential of people with disabilities, he wheeled himself 40,000 kilometers around the world to raise funds for the Man in Motion Legacy Trust Fund. The fund contributes to rehabilitation, wheelchair sports, and public-awareness programs. In 2008 it sponsored the Blusson Spinal Cord Centre in Vancouver, Canada, the largest institution dedicated to spinal-cord research in the world, housing over 300 investigators.

Spinal-cord injury is usually due to trauma to the cord that then results in a number of secondary degenerative processes that contribute significantly to the size of the lesion. Thereafter, the formation of

Rick Hansen on the Man in Motion Tour in 1987.

scar tissue, a cavity, and cysts block communication between the two severed sides. Research on spinal-cord injury is directed at minimizing the acute changes that take place after the insult, devising ways to facilitate neural communication across the injury, and improving mobility and home care.

Nanotechnology, the science of creating molecular-sized tools that can serve biological functions in the body, holds future promise for both decreasing the acute effects of injury and bridging the two sides of the injury. Nanotechnology works with substances between 1 and 100 nanometers (nm) in size. (A nanometer is one billionth, or 10^{-9}, of a meter.)

Nanotubes or nanovesicles can be engineered to transport drugs, RNA, or new stem cells into the area of injury where they can arrest degenerative changes and help form neural bridges across the injury (Sharma & Sharma, 2012). Nanotubes that can carry chemicals or conduct electrical impulses can be threaded into the injury through blood vessels and then into the very small capillaries within the spinal cord. Or they can be injected as molecules that self-assemble into scaffolding or tubes when they reach a target area.

Nanoscaffolding, introduced into the injury to form a bridge, can aid the regrowth of axons across the injury (Cho et al., 2012). Nanoaxons can be introduced into the region of injury to synapse with neurons on both sides of the injury and to carry messages across the injury. Because they are small, nanomedicinal substances can interface with spinal cord cells on both sides of an injury.

REVIEW 11-1
Hierarchy of Movement Control

Before you continue, check your understanding.

1. The motor system is organized as a _____.

2. The _____ cortex plans movements, the _____ cortex organizes movement sequences to carry out the plan, and the _____ cortex executes the precise movements.

3. The _____ is responsible for species-typical movements, for survival-related actions, and for posture and walking.

4. In addition to serving as a pathway between the brain and the rest of the body, the _____ independently produces reflexive movements.

5. Explain what happens when the brain is disconnected from the spinal cord and why.

Answers appear at the back of the book.

11-2 Motor System Organization

Although we humans tend to rely primarily on our hands for manipulating objects, we can still learn to handle things with other body parts, such as the mouth or a foot, if we have to. Some people without arms become proficient at using a foot for writing or painting or even for driving. What properties of the motor system allow such versatility in carrying out such skilled movements? You will find the answer to this question in this section as we examine first the organization of the motor cortex, then the descending pathways from the motor cortex to the brainstem and spinal cord and the motor neurons that in turn connect with the muscles of the body.

Motor Cortex

In 1870, two Prussian physicians, Gustav Fritsch and Eduard Hitzig, discovered that they could electrically stimulate the neocortex of an anesthetized dog to produce movements of the mouth, limbs, and paws on the opposite side of the dog's body. They provided the first direct evidence that the neocortex controls movement. Later researchers confirmed the finding by experimenting with a variety of animals as subjects, including rats, monkeys, and apes.

Based on this research background, beginning in the 1930s Wilder Penfield used electrical stimulation to map the cortices of conscious human patients who were about to undergo neurosurgery. Penfield's aim was to use the results to assist in surgery. He and his colleagues confirmed that movements in humans are triggered mainly in response to stimulation of the primary motor cortex.

Mapping the Motor Cortex

Penfield summarized his results by drawing cartoons of body parts to represent the areas of the motor cortex that produce movement in those parts. The result was a **homunculus** (little person) that could be spread out across the motor cortex, as illustrated in **Figure 11-6**. Because the body is symmetrical, an equivalent motor homunculus is represented in the primary motor cortex of each hemisphere, and each motor cortex mainly controls movement in the opposite side of the body. Penfield also identified another, smaller motor homunculus in the dorsal premotor area of each frontal lobe, a region sometimes referred to as the *supplementary motor cortex*.

The most striking feature of the motor homunculus shown in **Figure 11-7** is the disproportionate relative sizes of its body parts compared with the relative sizes of actual parts of the human body. The homunculus has huge hands with an especially large thumb. Its lips and tongue are also prominent. In contrast, the trunk, arms, and legs that constitute most of the area of a real body are smaller in relative size.

These distortions illustrate the fact that extensive areas of the motor cortex allow precise regulation of the hands, fingers, lips, and tongue (see Figure 11-6). Areas of the body over which we have much less motor control have a much smaller representation in the motor cortex.

homunculus Representation of the human body in the sensory or motor cortex; also any topographical representation of the body by a neural area.

Section 4-1 describes the milestones that led to understanding how the nervous system uses electrical charge to convey information.

Figure 10-19 shows Penfield using brain stimulation to map the cortex.

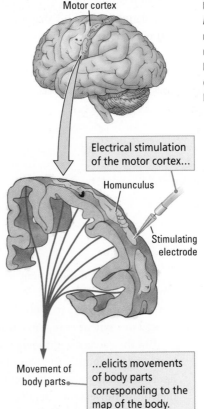

Motor cortex

Electrical stimulation of the motor cortex...

Homunculus

Stimulating electrode

Movement of body parts

...elicits movements of body parts corresponding to the map of the body.

FIGURE 11-6 Penfield's Homunculus Movements are topographically organized in the motor cortex. Stimulation of the dorsal medial regions of the cortex produces movements in the lower limbs. Stimulation in ventral regions of the cortex produces movements in the upper body, hands, and face.

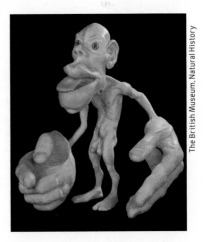

The British Museum, Natural History

FIGURE 11-7 Homuncular Man An artist's representation illustrates the disproportionate areas of the sensory and motor cortices that control different parts of the body.

topographic organization Neural spatial representation of the body or areas of the sensory world perceived by a sensory organ.

Another curious feature of the homunculus as laid out across the motor cortex is that the body parts are arranged differently from those of an actual body. The area of the cortex that produces eye movements is located in front of the homunculus head on the motor cortex (see the top drawing in Figure 11-6), and the head is oriented with the chin up and the forehead down (bottom drawing). The tongue is located below the forehead.

Modeling Movement

The motor homunculus shows at a glance that relatively larger areas of the brain control the parts of the body that we use to make the most skilled movements. This makes it useful for understanding the **topographic organization** (functional layout) of the primary motor cortex. Debate over how the motor areas represented by Penfield's homunculus might produce movement has been considerable.

An early idea is that each part of the homunculus controls muscles in that part of the body. Information from other cortical regions could be sent to the motor homunculus, and neurons in the appropriate part of the homunculus could then activate body muscles required for producing the movement. If you need to pick up a coin, for example, messages from the finger area of the motor cortex would instruct the fingers to so move.

In contrast to this idea, more recent experiments suggest that the motor cortex represents not muscles but rather a repertoire of fundamental movement categories. Penfield had used very brief electrical pulses, and the movements he elicited were little more than muscle twitches. More recent studies have used half-second-long trains of electrical stimulation that allow time for movements to occur (Grazaino, 2006).

The drawings in **Figure 11-8** illustrate several movement categories elicited by long-duration stimulation: (A) defensive facial postures, (B) movements of the hand to mouth, (C) manipulation and shaping of the hand and digits (fingers) in central body space, (D) outward reach with the hand, and (E) climbing and leaping postures.

(A) Defensive facial posture   **(B) Hand to mouth**   **(C) Central body space**   **(D) Outward reach**   **(E) Climbing/reaching posture**

   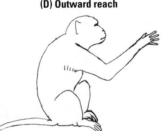 

FIGURE 11-8 Motor Cortex Movement Categories Five categories of movements evoked by electrical stimulation of the motor cortex in the monkey. Adapted from "The Organization of Behavioural Repertoire in Motor Cortex," by M. Graziano, 2006, *Annual Reviews of Neuroscience, 29*, pp. 105–134.

Each observed movement has the same end regardless of the location of a monkey's limb or its other ongoing behavior. Electrical stimulation that results in the hand coming to the mouth always recruits the hand. If a weight is attached to the monkey's arm, the evoked movement compensates for the added load.

But the categorized movement is inflexible: when an obstacle is placed between the hand and the mouth, the hand hits the obstacle. If stimulation continues after the hand has reached the mouth, the hand remains there for the duration of the stimulation. Further, broad movement categories—for example, reaching—cluster together on the motor cortex, but reaching directed to different parts of space is elicited from slightly different cortical points in the topographic "reaching map."

Studies on human subjects using MRI suggests that the human motor cortex, like the monkey motor cortex, is organized in terms of functional movement categories (Meier et al., 2008). The motor cortex maps appear to represent basic "types" of movement that learning and practice can modify. In other words, the motor cortex encodes not muscle twitches but a "lexicon," or dictionary, of movements that is not large. If you think about it, you will find that these few movements used in different combinations can produce all the movements that you can produce, even in activities as complex as playing basketball.

So although Penfield's map is accurate in revealing a relationship between different parts of the motor cortex and the body, more recent research suggests that the motor cortex really represents the repertoire of movements that each species of animal can make.

Motor Cortex and Skilled Movement

In a study designed to investigate how neurons in the motor cortex control movement, Edward Evarts (1968) used the simple procedure illustrated in **Experiment 11-2.** He trained a monkey to flex its wrist to move a bar. Different weights could be attached to the bar. An electrode implanted in the wrist region of the monkey's motor cortex recorded the activity of neurons there.

Evarts discovered that the neurons began to discharge even before the monkey flexed its wrist, as shown in the Results section of Experiment 11-2. Thus, they take part in planning the movement as well as initiating it. The neurons continued to discharge as the wrist moved, confirming that they play a role in producing the movement. Finally, the neurons discharged at a higher rate when the bar was loaded with a weight. This finding shows that motor-cortex neurons increase the force of a movement by increasing their rate of firing and its duration, as stated in the Conclusion of the experiment.

Evarts's findings also reveal that the motor cortex has a role in specifying the direction of a movement. Motor-cortex neurons in the wrist might discharge when the monkey flexed its wrist inward but not when the wrist was extended back to its starting position. These on–off neuronal responses are a simple way of coding the direction in which the wrist is moving. Evart's finding that motor cortex neurons are involved in planning movements is confirmed in much less formal situations. For example, recordings of a monkey's bicep muscle that produces arm flexion and of motor cortex neurons that, when stimulated, produce arm flexion show correlated activity during a monkey's spontaneous behaviors.

Nevertheless, studies using human participants reveal a number of situations in which motor-cortex neurons are active at the same time that no movement occurs (Schieber, 2011). These situations include planning a movement, withholding a movement on instruction, and mental imagery. So the versatility of the motor cortex extends past the range of movements that it can produce. These flexible properties of motor neurons probably underlie our ability to imagine movements and also allow them to control brain–computer interfaces (see Section 11-1).

EXPERIMENT 11-2

Question: How does the motor cortex take part in the control of movement?

Procedure

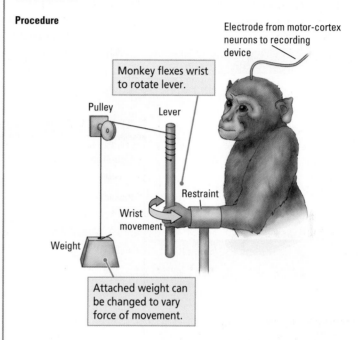

Electrode from motor-cortex neurons to recording device

Monkey flexes wrist to rotate lever.

Pulley

Lever

Restraint

Wrist movement

Weight

Attached weight can be changed to vary force of movement.

Results

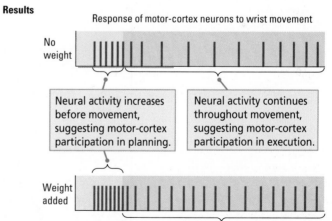

Response of motor-cortex neurons to wrist movement

No weight

Neural activity increases before movement, suggesting motor-cortex participation in planning.

Neural activity continues throughout movement, suggesting motor-cortex participation in execution.

Weight added

Movement begins

Neural activity increases over no-weight condition, suggesting that motor-cortex neurons code force of movement.

Conclusion: The motor cortex takes part in planning movement, executing movement, and adjusting the force and duration of a movement.

Adapted from "Relation of Pyramidal Tract Activity to Force Exerted During Voluntary Movement," by E. V. Evarts, 1968, *Journal of Neurophysiology, 31*, p. 15.

Plasticity in the Motor Cortex

An intimate relationship exists between the activity of neurons in the motor cortex and movement in the body, and the studies described above show that flexibility is part of the relationship. This flexibility underlies another property of the motor cortex: its plasticity, which underlies motor learning and also contributes to recovery after the motor cortex is damaged, as the following example explains.

A study by Randy Nudo and his coworkers (1996), summarized in the Procedure section of **Experiment 11-3,** illustrates change in a map of the motor cortex due to cortical damage. These researchers mapped the motor cortices of monkeys to identify the hand and digit areas. They then surgically removed a small part of the cortex that represents the digit area. After undergoing this electrolytic lesion, the monkeys used the affected hand much less, relying mainly on the good hand.

When the researchers examined the monkeys three months later, the animals were unable to produce many movements of the lower arm, including the wrist, the hand, and

Section 14-4 explores how motor maps change in response to learning.

Section 7-1 describes several ablation techniques used by neuroscience researchers to manipulate the brain.

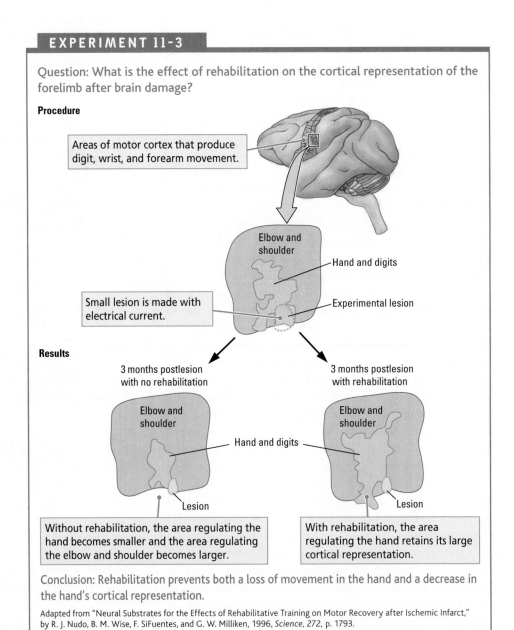

EXPERIMENT 11-3

Question: What is the effect of rehabilitation on the cortical representation of the forelimb after brain damage?

Procedure

Areas of motor cortex that produce digit, wrist, and forearm movement.

Elbow and shoulder

Hand and digits

Small lesion is made with electrical current.

Experimental lesion

Results

3 months postlesion with no rehabilitation

3 months postlesion with rehabilitation

Elbow and shoulder

Elbow and shoulder

Hand and digits

Lesion

Lesion

Without rehabilitation, the area regulating the hand becomes smaller and the area regulating the elbow and shoulder becomes larger.

With rehabilitation, the area regulating the hand retains its large cortical representation.

Conclusion: Rehabilitation prevents both a loss of movement in the hand and a decrease in the hand's cortical representation.

Adapted from "Neural Substrates for the Effects of Rehabilitative Training on Motor Recovery after Ischemic Infarct," by R. J. Nudo, B. M. Wise, F. SiFuentes, and G. W. Milliken, 1996, *Science, 272,* p. 1793.

the digits surrounding the area with the lesion. Much of the area representing the hand and lower arm had disappeared from the animals' cortical maps. The shoulder, upper arm, and elbow areas had spread out to take up what had formerly been space representing the hand and digits. The Results section of Experiment 11–3 shows this topographic change.

The experimenters wondered whether the change could have been prevented had they forced the monkeys to use the affected arm. To find out, they used the same procedure on other monkeys, except that during the postsurgery period they forced the animals to rely on the bad arm by binding the good arm in a sling.

Three months later, when the experimenters reexamined these monkeys' motor maps, they found that the hand and digit area retained its large size. Even though no neural activity occurred in the spot with the lesion, the monkeys had gained some function in the digits that had been connected to the damaged spot. Apparently, the remaining digit area of the cortex was now controlling the movement of these fingers.

Most likely, plasticity is promoted in the formation of new connections and the strengthening of existing connections among different parts of the motor homunculus. Humans who suffer a stroke to the motor cortex also display plasticity-mediated recovery. They may at first be completely unable to use their contralateral forelimb, but with time and practice they may recover a great deal of movement.

One way to enhance recovery is to restrain the good limb. **Restraint-induced therapy** forces the person to use the affected limb and is a major therapy for stroke-induced limb paralysis. Its effectiveness depends on frustration of the good limb, a concerted effort in using the bad limb, and neural plasticity.

restraint-induced therapy Procedure in which restraint of a healthy limb forces a patient to use an impaired limb to enhance recovery of function.

corticospinal tract Bundle of nerve fibers directly connecting the cerebral cortex to the spinal cord, branching at the brainstem into an opposite-side lateral tract that informs movement of limbs and digits and a same-side ventral tract that informs movement of the trunk; also called *pyramidal tract*.

Focus 2-3 describes disruptions stroke causes; Section 16-3 reviews stroke treatments.

Corticospinal Tracts

The main efferent pathways from the motor cortex to the brainstem to the spinal cord are the **corticospinal tracts.** The axons from these tracts originate mainly in layer V pyramidal cells of the motor cortex but also extend from the premotor cortex and the sensory cortex (see "Layering in the Neocortex" on page 357). The axons descend into the brainstem, sending collaterals to a few brainstem nuclei, and eventually emerge on the brainstem's ventral surface where they form a large bump on each side. These bumps, or "pyramids," give the corticospinal tracts their alternate name, the *pyramidal tracts*.

At this point, some of the axons descending from the left hemisphere cross over to the right side of the brainstem. Likewise, some of the axons descending from the right hemisphere cross over to the left side of the brainstem. The remaining axons stay on their original sides. This division produces two corticospinal tracts, one crossed and the other uncrossed, entering each side of the spinal cord. **Figure 11-9** illustrates the division of tracts originating in the left-hemisphere cortex. The dual tracts on each side of the brainstem then descend into the spinal cord, forming the two spinal-cord tracts.

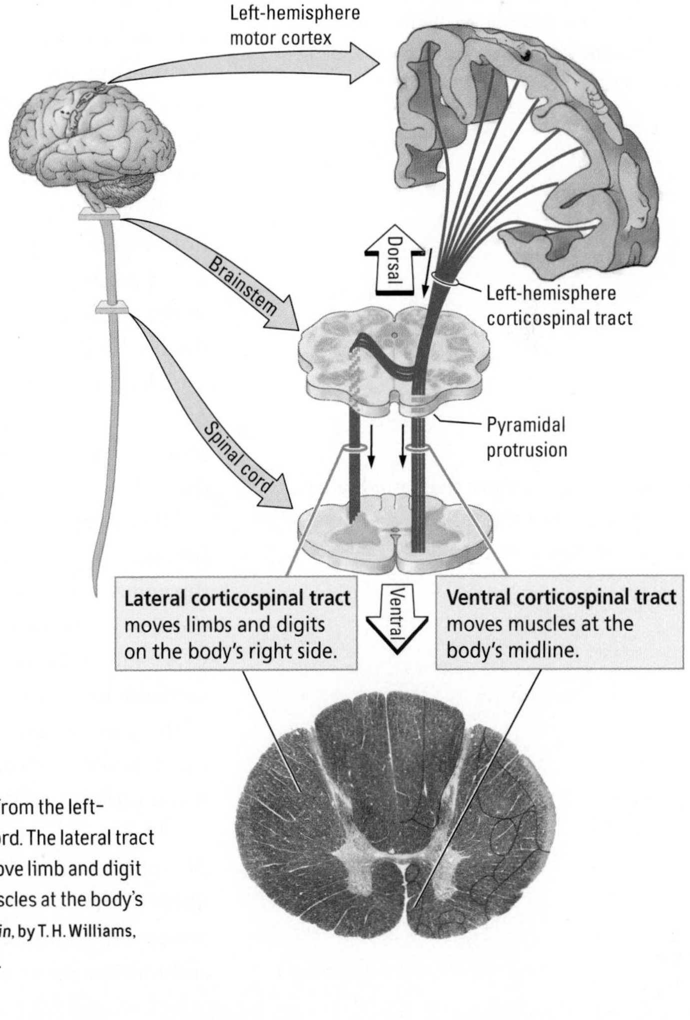

Left-hemisphere motor cortex

Dorsal

Brainstem

Left-hemisphere corticospinal tract

Spinal cord

Pyramidal protrusion

Ventral

| **Lateral corticospinal tract** moves limbs and digits on the body's right side. | **Ventral corticospinal tract** moves muscles at the body's midline. |

FIGURE 11-9 Left-Hemisphere Corticospinal Tract Nerve fibers descend from the left-hemisphere motor cortex to the brainstem, where the tract branches into the spinal cord. The lateral tract crosses the brainstem's midline, descending into the right side of the spinal cord to move limb and digit muscles on the body's right side. The ventral tract remains on the left side to move muscles at the body's midline. Photograph of spinal cord reproduced from *The Human Brain: Dissections of the Real Brain*, by T. H. Williams, N. Gluhbegovic, and J. Jew, on CD-ROM. Published by Brain University, brain-university.com 2000.

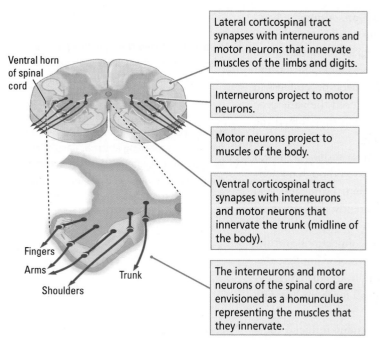

Ventral horn of spinal cord

Lateral corticospinal tract synapses with interneurons and motor neurons that innervate muscles of the limbs and digits.

Interneurons project to motor neurons.

Motor neurons project to muscles of the body.

Ventral corticospinal tract synapses with interneurons and motor neurons that innervate the trunk (midline of the body).

Fingers

Arms

Trunk

Shoulders

The interneurons and motor neurons of the spinal cord are envisioned as a homunculus representing the muscles that they innervate.

FIGURE 11-10 Motor-Tract Organization The interneurons and motor neurons in the left and right tracts of the ventral spinal cord are topographically arranged: the more lateral neurons innervate more distal parts of the limbs (those farther from the midline), and the more medial neurons innervate more proximal muscles of the body (those closer to the midline).

By the way, the neurons that your brain is using to carry out this task are the same neurons that you are tracing.

The spinal-cord cross section in **Figure 11-10** shows the location of the two tracts, on the left and right sides. The fibers that cross to the opposite side of the brainstem descend the spinal cord in a lateral (side) position, to form the *lateral corticospinal tract*. The fibers that remain on their original side continue from the brainstem down the spinal cord in a ventral (front) position, to form the *ventral corticospinal tract*.

To retrace the pathway, the coriticospinal tracts originate in the neocortex and terminate in the spinal cord. Within the spinal cord, corticospinal fibers make synaptic connections with both interneurons and motor neurons, but the motor neurons carry all nervous system commands out to the muscles.

Motor Neurons

The spinal-cord motor neurons that connect to muscles are located in the ventrolateral part of the spinal cord and jut out to form the spinal column's ventral horns, which contain two kinds of neurons. Interneurons lie just medial to the motor neurons and project onto them. The motor neurons send their axons to the muscles of the body. The fibers from the corticospinal tracts make synaptic connections with both the interneurons and the motor neurons, but all nervous system commands to the muscles are carried by the motor neurons.

Figure 11-10 shows that a homunculus of the body is represented again in the spinal cord. The more laterally located motor neurons project to muscles that control the fingers and hands, whereas intermediately located motor neurons project to muscles that control the arms and shoulders. The most medially located motor neurons project to muscles that control the trunk of the body. Axons of the lateral corticospinal tract connect mainly with the lateral interneurons and motor neurons, and axons of the ventral corticospinal tract connect mainly to the medial interneurons and motor neurons.

To visualize how the motor homunculus in the cortex relates to the motor-neuron homunculus in the spinal cord, look at Figure 11-9. Place your finger on the index-finger region of the motor homunculus on the left side of the brain. Tracing the axons of the cortical neurons downward, your route takes you through the brainstem, across its midline, and down the right lateral corticospinal tract.

The journey ends at the interneurons and motor neurons in the most lateral region of the spinal cord's right ventral horn—the horn on the opposite side of the nervous system from which you began. Following the axons of these motor neurons, you find that they synapse on muscles that move the index finger on the right-hand side of the body.

If you repeat the procedure by tracing the pathway from the trunk area of the motor homunculus, located near the top on the left side of the brain, you follow the same route through the upper part of the brainstem. However, you do not cross over to the opposite side of the brainstem. Instead, you descend into the spinal cord on the left side, the same side of the nervous system on which you began, eventually ending up in the most medially located interneurons and motor neurons of the left side's ventral horn. (Some of these axons also cross over to the other side of the spinal cord.) Thus, if you follow these motor-neuron axons, you end up at their synapses with the muscles that move the trunk on both sides of the body.

This visualization should help you to remember the routes taken by the axons of the motor system. The limb regions of the motor homunculus contribute most of their fibers to the lateral corticospinal tract. Because these fibers have crossed over to the opposite side of the brainstem, they activate motor neurons that move the arm, hand, leg, and foot on the opposite side of the body.

In contrast, the trunk regions of the motor homunculus contribute their fibers to the ventral corticospinal tract. These fibers do not cross over at the brainstem, although some do cross over in the spinal cord. In short, the neurons of the motor homunculus in the left-hemisphere cortex control the trunk on both sides of the body and the limbs on the body's right side. Similarly, neurons of the motor homunculus in the right-hemisphere cortex control the trunk on both sides of the body and the limbs on the body's left side.

Thus, one hemisphere of the cortex controls the hands and fingers of the opposite side of the body and the trunk on both sides of the body. The motor cortex is organized in terms of a number of functional movement categories, such as reaching to a point in space (see Figure 11-8D), and it follows that the spinal cord must have a similar organization. A similar template in the spinal cord ensures that instructions from the motor cortex concerning, for example, reaching to a part of space, are reproduced faithfully.

In addition to the corticospinal pathways, about 24 other pathways from the brainstem to the spinal cord carry instructions, such as information related to posture and balance (see Section 11-4), and they control the autonomic nervous system (ANS). For all these functions, remember that the motor neurons are the final common path.

Figure 2-30 diagrams the pathways from CNS to ANS.

Control of Muscles

The spinal-cord motor neurons synapse on the muscles that control body movements. For example, the biceps and triceps of the upper arm control movement of the lower arm. Limb muscles are arranged in pairs, as shown in **Figure 11-11**. One member of a pair, the *extensor*, moves (extends) the limb away from the trunk. The other member of the pair, the *flexor*, moves (flexes) the limb in toward the trunk. Experiment 11-2 on page 368 demonstrates the on–off responses of cortical motor neurons, depending on whether the flexor or extensor muscle is being used.

Connections between the interneurons and motor neurons of the spinal cord ensure that the muscles work together so that, when one muscle contracts, the other relaxes. Thus, the interneurons and motor neurons of the spinal cord not only relay instructions from the brain but also, through their connections, cooperatively organize the movement of many muscles. As you know, the neurotransmitter at the motor-neuron–muscle junction is acetylcholine.

Figure 4-26 illustrates ACh action at a motor-neuron–muscle junction.

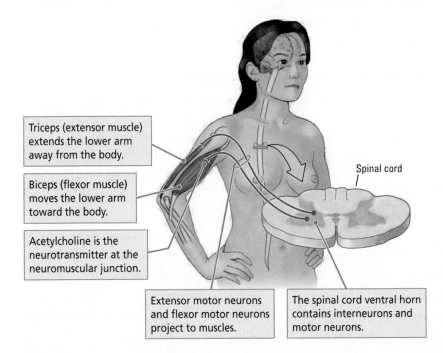

Triceps (extensor muscle) extends the lower arm away from the body.

Biceps (flexor muscle) moves the lower arm toward the body.

Acetylcholine is the neurotransmitter at the neuromuscular junction.

Spinal cord

Extensor motor neurons and flexor motor neurons project to muscles.

The spinal cord ventral horn contains interneurons and motor neurons.

FIGURE 11-11 Coordinating Muscle Movement

hyperkinetic symptom Symptom of brain damage that results in excessive involuntary movements, as seen in Tourette's syndrome.

hypokinetic symptom Symptom of brain damage that results in a paucity of movement, as seen in Parkinson's disease.

REVIEW 11-2
Motor System Organization

Before you continue, check your understanding.

1. The _____ organization of the motor cortex is represented by a _____, in which parts of the body that are capable of the most skilled movements (especially the mouth, fingers, and thumbs) are regulated by _____ cortical regions.

2. Change can take place in the cortical _____ to aid in recovery of function after injury to the motor cortex.

3. Instructions regarding movement travel out from the motor cortex through the _____ tracts to terminate on interneurons that project to motor neurons in the ventral horn of the spinal cord. Many corticospinal-tract fibers cross to the opposite side of the spinal cord to form the _____ tracts; some stay on the same side to form the _____ tracts.

4. The ventral corticospinal tracts carry instructions for _____ movements, whereas the lateral corticospinal tracts carry instructions for _____ and _____ movements.

5. The axons of motor neurons in the spinal cord carry instructions to _____ that are arranged in pairs. One _____ a limb while the other _____ the limb.

6. Some neurons in the premotor and primary motor cortex respond to imagined movements and recognize the movements of others. Name this type of neuron and describe the technology that takes advantage of its attributes.

Answers appear at the back of the book.

FIGURE 11-12 Basal Ganglia Connections
The caudate putamen in the basal ganglia connects to the amygdala through the tail of the caudate nucleus. The lateral see-through view shows the basal ganglia relative to surrounding structures, including the substantia nigra, with which it shares reciprocal connections. The basal ganglia receive input from most regions of the cortex and send input into the frontal lobes through the subthalamic nucleus.

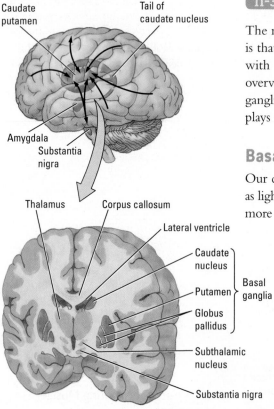

11-3 Basal Ganglia, Cerebellum, and Movement

The main evidence that the basal ganglia and the cerebellum perform motor functions is that damage to either structure impairs movement. Both have extensive connections with the motor cortex, further suggesting their participation in movement. After an overview of each structure's anatomy, we look at some symptoms that arise after the basal ganglia or the cerebellum is damaged. Then we consider the roles that each structure plays in controlling movement.

Basal Ganglia and Movement Force

Our control over the force of our movements is remarkable. We can manipulate objects as light as a needle for sewing or swing objects as heavy as a baseball bat for driving a ball more than 100 yards. The brain areas that allow us to adjust the force of our movements in these ways include the subcortical basal ganglia within the forebrain, shown in **Figure 11-12.** Its collection of nuclei, shown in the cross section, connect with the motor cortex and with the midbrain.

Anatomy of the Basal Ganglia

As shown in Figure 11-12, a prominent structure in the basal ganglia is the caudate putamen, a large cluster of nuclei that extends as a "tail" into the temporal lobe, ending in the amygdala. The basal ganglia receive inputs from two main sources:

1. All areas of the neocortex and limbic cortex, including the motor cortex, project to the basal ganglia.

2. The nigrostriatial dopaminergic activating system projects to the basal ganglia from the substantia nigra, a cluster of darkly pigmented cells in the midbrain.

Basal ganglia nuclei project to both the motor cortex and the substantia nigra. It is not surprising, given the widespread connections of the basal ganglia with subcortical structures and with the neocortex, that they have been implicated in a wide range of functions, including association or habit learning, motivation, emotion, and motor control. The behavioral deficits that follow damage to the basal ganglia provide the best insights into their motor functions.

How the Basal Ganglia Control Movement Force

Two different—in many ways opposite—kinds of movement disorders result from damage to the basal ganglia. If cells of the caudate putamen are damaged, unwanted writhing and twitching movements called *dyskinesias* result. For example, Huntington's disease, in which cells of the caudate putamen are destroyed, is characterized by involuntary and exaggerated movements. Other examples of involuntary movements related to caudate putamen damage are the unwanted tics and vocalizations peculiar to Tourette's syndrome, the topic of Clinical Focus 11-4, "Tourette's Syndrome" on page 374.

In addition to causing involuntary movements, or **hyperkinetic symptoms,** damage to the basal ganglia can result in a loss of motor ability, or **hypokinetic symptoms** that lead to rigidity and difficulty initiating and producing movement. Parkinson's disease is caused by the loss of dopamine cells in the substantia nigra that project into forebrain basal ganglia nuclei, and the disease is characterized by hypokinetic symptoms.

That two different and seemingly opposing kinds of symptoms—hyperkinetic and hypokinetic—arise subsequent to basal ganglia damage suggests that one function of these nuclei is in regulating movement force. The idea is that hyperkinetic disorders such as Huntington's disease result from errors of too much force and so result in excessive movement. Hypokinetic disorders such as Parkinson's disease result from errors of too little force and so result in insufficient movement.

To confirm these ideas, Moisello and colleagues (2011) used a reaching task in which they examined the velocity and force of the movements. Huntington's disease subjects reached using too much force, thus seemingly flinging a limb. Parkinson's disease patients reached with too little force, thus producing a slowed movement.

Kurniawan and his coworkers (2010) used MRI to examine the basal ganglia activity of participants who, for a small monitory reward, considered how much force to apply in a gripping task. The imaging showed more basal ganglia activity when participants contemplated using a more forceful grip and less activity when contemplating a less forceful grip. The researchers suggest that the basal ganglia may play a role not just in producing force but also in computing the effortful costs of making movements.

What neural pathways enable the basal ganglia to select movements or modulate the force of movements? One theory holds that two pathways from the basal ganglia affect motor cortex activity: an inhibitory pathway and an excitatory pathway (Kreitzer & Malenka, 2008). These pathways are illustrated **Figure 11-13.**

The regions colored in red in Figure 11-13 have an inhibitory action on structures to which they project, and the regions colored in green have an excitatory action on structures to which they project. Notice that the pathway from the thalamus to the cortex and from there to the brainstem and spinal cord is green. It is over this pathway that movement is produced.

Figure 5-17 traces the nigrostriatial pathways in the dopaminergic activating system and highlights their importance in maintaining normal motor behavior.

Focus 3-4 describes the genetic basis of Huntington's disease.

Detailed coverage of Parkinson's disease and its treatment appears in Chapters 5, 7, and 16.

FIGURE 11-13 Regulating Movement Force Two pathways in the basal ganglia modulate movements produced in the cortex. Green pathways are excitatory, red are inhibitory. The indirect pathway excites the GP$_i$ whereas the direct pathway has an inhibitory effect. If activity in the indirect pathway dominates, the thalamus shuts down, and the cortex is unable to produce movement. If direct-pathway activity dominates, the thalamus can become overactive, amplifying movement. Adapted from "Functional Architecture of Basal Ganglia Circuits: Neural Substrates of Parallel Processing," by R. E. Alexander and M. D. Crutcher, 1990, *Trends in Neuroscience, 13,* p. 269.

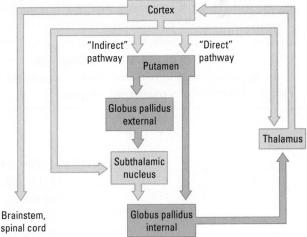

Remember that behaviors (movements) are lost when a disorder prevents excitatory instructions and released when a disorder prevents inhibitory instructions.

Tourette's Syndrome

The neurological disorder Tourette's syndrome (TS) was first described in 1885 by Georges Gilles de la Tourette, a French neurologist, who described the symptoms as they appeared in Madame de D., one of his patients:

> Madame de D., presently age 26, at the age of 7 was afflicted by convulsive movements of the hands and arms. These abnormal movements occurred above all when the child tried to write, causing her to crudely reproduce the letters she was trying to trace. After each spasm, the movements of the hand became more regular and better controlled until another convulsive movement would again interrupt her work. She was felt to be suffering from over-excitement and mischief, and because the movements became more and more frequent, she was subject to reprimand and punishment. Soon it became clear that these movements were indeed involuntary and convulsive in nature. The movements involved the shoulders, the neck, and the face, and resulted in contortions and extraordinary grimaces. As the disease progressed, and the spasms spread to involve her voice and speech, the young lady made strange screams and said words that made no sense. (Friedhoff & Chase, 1982)

The statistical incidence of Tourette's syndrome is about 1 in 1000 people. TS affects all racial groups and seems to be hereditary. The age range of onset is between 2 and 25 years.

The most frequent symptoms of Tourette's syndrome are involuntary tics and complex movements, such as hitting, lunging, or jumping. People with TS may also suddenly emit cries and other vocalizations or inexplicably utter words that do not make sense in the context, including scatology and swearing.

Tourette's syndrome is thought to reflect an abnormality of the basal ganglia, especially in the right hemisphere, because its symptoms can be controlled with haloperidol, an antipsychotic drug that blocks dopamine synapses in the basal ganglia. Using fMRI to correlate resting activity in different regions of the brain, Church and colleagues (2009) documented cortical changes associated with TS. These changes include increases in connectivity, mainly in the parietal cortex of Tourette's patients, and decreases in connectivity between the parietal and frontal cortex. This finding, diagrammed in the illustration, suggests alterations in the function of neural circuits that connect the posterior sensory regions of the cortex to its anterior motor regions.

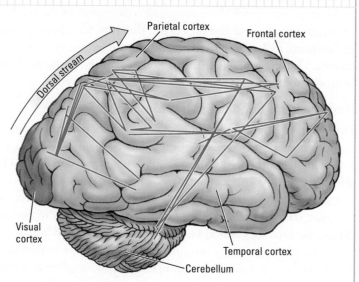

Areas of the brain that show enhanced connectivity (green) or decreased connectivity (red) in fMRI analysis of young adults with Tourette's syndrome suggest abnormalities in dorsal-stream structures linking the parietal cortex to the frontal cortex. **Adapted from "Control Networks in Paediatric Tourette Syndrome Show Immature and Anomalous Patterns of Functional Connectivity," by J. A. Church, D. A. Fair, N. U. Dosenbach, et al., 2009, _Brain, 32,_ pp. 225–238.**

In an attempt to understand the urge to make involuntary movements and vocalizations, Jackson and his coworkers (2011) proposed that many behaviors that people perform, such as yawning, can be characterized as having an "urge-to-action." Using fMRI they found that both involuntary movements and yawning are related to activity in the frontal cortex and in the insula, an area of the temporal cortex located within the brain's lateral fissure.

The investigators suggested that there is an urge-to-action system in the brain. Why particular movements and vocalizations that characterize individuals with Tourette's syndrome should be caught up by this urge system remains a mystery.

The cortex can influence the basal ganglia via the two pathways shown in Figure 11-13. If the direct pathway is activated, the globus pallidus internal (GP$_i$) is inhibited, and the pathway is freed to produce movement. If the indirect pathway is activated, the GP$_i$ is activated and inhibits the thalamus, thus blocking movement.

The GP$_i$ thus acts like a volume control. If it is turned up, movement is blocked; if it is turned down, movement is allowed. This model proposes that diseases of the basal ganglia in one way or the other affect this volume control function, resulting in impairments of excessive or slowed movement.

The idea that the GP$_i$ acts like a volume control over movement is the basis for a number of treatments for Parkinson's disease. If the GP$_i$ is surgically destroyed in

Parkinson patients—the equivalent of activating the green pathway—muscular rigidity is reduced and the ability to make normal movements is improved. Similarly, *deep brain stimulation* (DBS) of the GP$_i$ inactivates it, freeing movement. Consistent with this "volume hypothesis," recordings made from cells of the globus pallidus show that they are excessively active in people with Parkinson's disease.

Figure 1-6 shows electrodes implanted in the brain for DBS.

Cerebellum and Movement Skill

Musicians have a saying: "Miss a day of practice and you're OK, miss two days and you notice, miss three days and the world notices." Apparently, some change must take place in the brain when we neglect to practice a motor skill. The cerebellum may be the motor system component that is affected. Whether the skill is playing a musical instrument, pitching a baseball, or texting, the cerebellum is critical for acquiring and maintaining motor skills.

Anatomy of the Cerebellum

The cerebellum, a large and conspicuous part of the motor system, sits atop the brainstem, clearly visible just behind the cerebrum, and like the cerebrum, it is divided into two hemispheres **(Figure 11-14)**. A small lobe, the *flocculus,* projects from its ventral, or inferior, surface. Although smaller than the cerebrum, the cerebellum has many more gyri and sulci, and it contains about one-half of all the neurons in the entire nervous system.

As Figure 11-14 shows, the cerebellum can be divided into several regions, each specialized for a different aspect of motor control. At its base, the flocculus receives projections from the middle-ear vestibular system, described in Section 11-4, and takes part in controlling balance. Many of its projections go to the spinal cord and to the motor nuclei that control eye movements.

Just as the motor cortex has a homuncular organization and a number of homunculi, the hemispheres of the cerebellum have at least two, as shown in Figure 11-14. The most medial part of each homunculus controls the face and the midline of the body. The more lateral parts connect to areas of the motor cortex and are associated with movements of

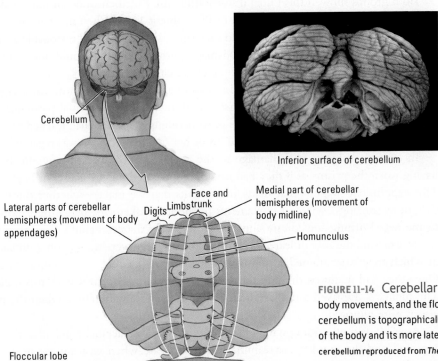

Cerebellum

Inferior surface of cerebellum

Lateral parts of cerebellar hemispheres (movement of body appendages)

Digits Limbs Face and trunk

Medial part of cerebellar hemispheres (movement of body midline)

Homunculus

Floccular lobe (eye movements and balance)

FIGURE 11-14 Cerebellar Homunculus The cerebellar hemispheres control body movements, and the flocculus controls eye movements and balance. The cerebellum is topographically organized: its more medial parts represent the midline of the body and its more lateral parts represent the limbs and digits. Photograph of cerebellum reproduced from *The Human Brain: Dissections of the Real Brain*, by T. H. Williams, N. Gluhbegovic, and J. Jew, on CD-ROM. Published by Brain University. brain-university.com 2000.

Ramón y Cajal's drawing of a Purkinje cell, circa 1900.

the limbs, hands, feet, and digits. The pathways from the cerebellar hemispheres project to nuclei at the interface of the cerebellum and spinal cord, which in turn project to other brain regions, including the motor cortex.

To summarize the cerebellum's topographic organization, the midline of the homunculus is represented in its central part; the limbs and digits are represented in the lateral parts. Tumors or damage to midline areas of the cerebellum disrupt balance, eye movement, upright posture, and walking but do not substantially disrupt other movements such as reaching, grasping, and using the fingers. For example, a person with medial damage to the cerebellum may, when lying down, show few symptoms. Damage to lateral parts of the cerebellum disrupts arm, hand, and finger movements much more than movements of the body's trunk.

The arrangement and connections of the cerebellum are built to a common plan. The cerebellar cortex consists of three layers of cells with the very distinctive Purkinje cells forming the second layer. The Purkinje cells are the output cells of the cerebellum. This common plan suggests that the cerebellum has a common function with respect to the control that it has over other regions of the motor system.

How the Cerebellum Improves Movement Control

Attempts to understand how the cerebellum controls movement have centered on two major ideas. Damage to the cerebellum (1) does not abolish any movement but (2) does disrupt the timing and execution of movement. Thus, the cerebellum must regulate the timing of movements and adjust the flow of movement as required in different situations.

Tom Thach (2007), in an intriguing experiment, illustrates how the cerebellum helps make the adjustments needed to keep movements accurate. The experiment involved having control participants and subjects with cerebellar damage throw darts at a target, as shown in the Procedure section of **Experiment 11-4.** After a number of throws that allowed them to become reasonably accurate, both groups donned glasses containing wedge-shaped prisms that displaced the apparent location of the target to the left. Now when they threw a dart, it landed to the left of the intended target.

Both groups showed this initial distortion in aim. But then came an important difference, graphed in the Results section of Experiment 11-4. When normal participants saw the dart miss the mark, they adjusted each successive throw until reasonable accuracy was restored. In contrast, subjects with damage to the cerebellum could not correct for this error. Time after time, they missed the target far to the left.

Next, the controls removed the prism glasses and threw a few more darts. Again, a significant difference emerged. The first dart thrown by each normal participant was much too far to the right (owing to the previous adjustment they had learned to make), but soon each adjusted once again until his or her former accuracy was regained.

In contrast, subjects with damage to the cerebellum showed no aftereffects from having worn the prisms, as if they had never compensated for the glasses to begin with. This experiment suggests that many movements that we make—whether throwing a dart, or organizing a series of movements to shoot a basketball—depend on moment-to-moment learning and adjustments that are made by the cerebellum.

To examine the role of learning in this task, monkeys were trained on a similar task in which they were trained to point with a finger to a target on a computer screen. Once they had mastered the task they were required to perform it with prism glasses. The monkeys displayed a displacement of pointing and an aftereffect when the prism glasses were removed.

After 30 days of training with and without prisms, displacement and aftereffect disappeared, and the monkeys were immediately accurate when wearing the prisms and after they were removed. When the arm region of the cerebellum homunculus was then

EXPERIMENT 11-4

Question: Does the cerebellum help to make adjustments required to keep movements accurate?

Procedure

Prism glasses

Subject throws dart at target

Subject wears prisms that divert gaze

Prisms removed, subject adapts

Results

Normal participant

Initial throws | With prisms | Prisms removed

Distance from target (to the right) / (to the left)

Trials

A normal participant adapts when wearing the prisms and shows aftereffects when the prisms are removed.

Patient with damage to cerebellum

Initial throws | With prisms | Prisms removed

Trials

A patient with damage to the cerebellum fails to correct throws while wearing the prisms and shows no aftereffects when the prisms are removed.

Conclusion: Many movements we make depend on moment-to-moment learning and adjustments made by the cerebellum.

Adapted from "The Cerebellum and the Adaptive Coordination of Movement," by W. T. Thach, H. P. Goodkin, and J. G. Keating, 1992, *Annual Review of Neuroscience, 15,* p. 429.

anesthetized with the local anesthetic agent lidocaine, pointing without prisms was normal, but the learned adaptation to prisms was abolished. This experiment illustrates that the cerebellum is responsible not only for online adjustments of movement but also for learning relatively permanent movement skills (Norris et al., 2011).

To better understand how the cerebellum improves motor skills by adjusting movements, imagine throwing a dart yourself. Suppose you aim at the bull's eye, throw the dart, and find that it misses the board completely. You then aim again, this time adjusting your throw to correct for the original error. Notice that there are actually two versions of your action: (1) the movement that you intended to make and (2) the actual movement as recorded by sensory receptors in your arm and shoulder.

If you carry out the intended movement successfully, you need make no correction on your next try. But if you miss, an adjustment is called for. One way in which the adjustment might be made is through the feedback circuit shown in **Figure 11-15.**

The cortex sends instructions to the spinal cord to throw a dart at the target. A copy of the same instructions is sent to the cerebellum through the inferior olive, a nucleus in

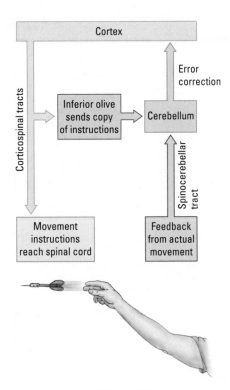

FIGURE 11-15 **Intention, Action, and Feedback** By comparing the message for the intended movement with the movement that was actually performed, the cerebellum sends an error message to the cortex to improve the accuracy of a subsequent movement.

the brainstem that projects to the cerebellum. When you next throw the dart, the sensory receptors in your arm and shoulder code the actual movement that you make and send a message about it back to the cerebellum through the spinocerebellar tract. The cerebellum now has information about both versions of the movement—what you intended to do and what you actually did—and can calculate the error and tell the cortex how to correct the movement. When you next throw a dart, you incorporate the correction into your throw.

REVIEW 11-3
The Basal Ganglia, the Cerebellum, and Movement

Before you continue, check your understanding.

1. The _____ contribute to motor control by adjusting the _____ associated with each movement.

2. Damage to the basal ganglia results either in unwanted, involuntary _____ movements (too much force is being exerted) or in such _____ rigidity that movements are difficult to perform (too little force is being exerted).

3. The cerebellum contributes to motor control by improving movement _____ and the learning of motor _____.

4. Describe how the cerebellum improves motor-skill accuracy.

Answers appear at the back of the book.

11-4 Organization of the Somatosensory System

The motor system produces movements, but without sensation movement would quickly become impaired. The somatosensory system tells us what's going on in the environment and also what we are doing. It allows us to distinguish what is done to us from what we do. When someone pushes you sideways, for example, the somatosensory system tells you that you have been pushed. If you lunge to the side yourself, the somatosensory system tells you that you have moved yourself.

In considering the motor system, we started at the cortex and followed the motor pathways out to the spinal cord (review Figures 11-9 and 11-10). This efferent route makes sense because it follows the outward flow of instructions regarding movement. As we explore the somatosensory system, we will proceed in the opposite direction because afferent sensory information flows inward, from sensory receptors in the body through sensory pathways in the spinal cord to the cortex.

Somatosensation is unique among sensory systems. It is not localized in the head, as are vision, hearing, taste, and smell but rather is distributed throughout the body. Somatosensory receptors are found in all parts of the body, and neurons from these receptors carry information in to the spinal cord.

Within the spinal cord, two somatosensory pathways project to the brain and eventually to the somatosensory cortex. One part of the system, however, is confined to a single organ, the inner ear, which houses the vestibular system that contributes to our sense of balance and head movement. Before we detail its workings, we investigate the anatomy of the somatosensory system and how it contributes to movement.

Somatosensory Receptors and Perception

Our bodies are covered with sensory receptors. Sensory receptors include our body hair and many types of receptors that are embedded in both surface layers and deeper layers of the skin and in muscles, tendons, and joints. Some receptors consist simply of the surface of a sensory neuron dendrite. Others include a dendrite and other tissue, such as the dendrite attached to a hair or covered by a special capsule or attached by a sheath of connective tissue to adjacent tissue.

The density of sensory receptors in the skin, muscles, tendons, and joints varies greatly in different parts of the body. The variation in density is one reason that different parts of the body are more or less sensitive to stimulation. Body parts that are very sensitive to touch—including the hands, feet, lips, and eyes—have many more sensory receptors than other body parts do. Sensitivity to different somatosensory stimuli is also a function of the receptors in a particular region.

Humans have two kinds of skin, hairy skin and **glabrous skin.** Glabrous skin, which includes the skin on the palms of the hands and feet, the lips, and the tongue, is hairless and exquisitely sensitive to a wide range of stimuli. It covers the body parts that we use to explore objects—hence, its heightened sensitivity.

The touch sensitivity of skin is often measured with a two-point sensitivity test. By touching the skin with two sharp points simultaneously, we can observe how close together the points can be placed while still being detected by the participant as two points rather than one. On glabrous skin, we can detect the two points when they are as close as 3 millimeters apart.

On hairy skin, two-point sensitivity is weaker by about a factor of 10. The two points seem to merge into one below a separation distance ranging from 2 to 5 cm, depending on exactly which part of the body is tested. You can confirm these differences in sensitivity on your own body by touching two sharp pencil points to a palm and to a forearm, varying the distances that you hold the points apart. Be sure not to look as you touch each surface.

Classifying Somatosensory Receptors

The varied types of somatosensory receptors in the human body may total as many as 20 or more, but they can all be classified into the three groupings illustrated in **Figure 11-16,** nociception, hapsis, and proprioception.

glabrous skin Skin that does not have hair follicles but contains larger numbers of sensory receptors than do other skin areas.

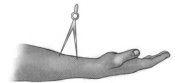
Two-point sensitivity test

FIGURE 11-16 Somatosensory Receptors Perceptions derived from the body senses of nocioception, hapsis, and proprioception depend on different receptors located in different parts of the skin, muscles, joints, and tendons.

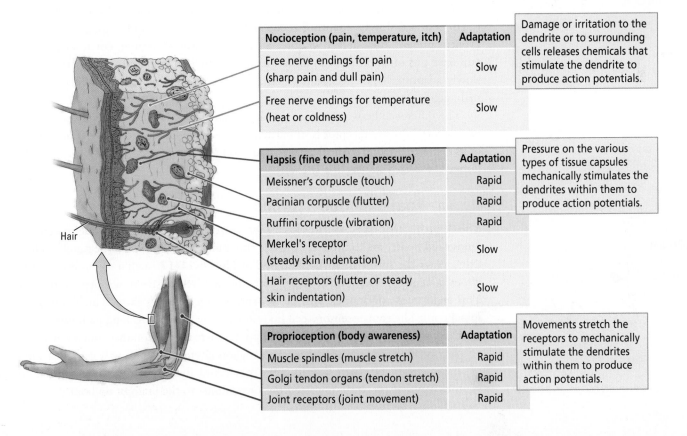

Nocioception (pain, temperature, itch)	Adaptation	
Free nerve endings for pain (sharp pain and dull pain)	Slow	Damage or irritation to the dendrite or to surrounding cells releases chemicals that stimulate the dendrite to produce action potentials.
Free nerve endings for temperature (heat or coldness)	Slow	

Hapsis (fine touch and pressure)	Adaptation	
Meissner's corpuscle (touch)	Rapid	Pressure on the various types of tissue capsules mechanically stimulates the dendrites within them to produce action potentials.
Pacinian corpuscle (flutter)	Rapid	
Ruffini corpuscle (vibration)	Rapid	
Merkel's receptor (steady skin indentation)	Slow	
Hair receptors (flutter or steady skin indentation)	Slow	

Proprioception (body awareness)	Adaptation	
Muscle spindles (muscle stretch)	Rapid	Movements stretch the receptors to mechanically stimulate the dendrites within them to produce action potentials.
Golgi tendon organs (tendon stretch)	Rapid	
Joint receptors (joint movement)	Rapid	

Hair

IRRITATION **Nocioception** is the perception of pain, temperature, and itch. Most nocio-ceptors consist of free nerve endings, as diagrammed at the top of Figure 11-16. When damaged or irritated, these endings secrete chemicals, usually peptides, that stimulate the nerve to produce an action potential. The action potential then conveys a message about pain, temperature, or itch to the central nervous system.

PRESSURE **Hapsis** (from the Greek for "touch") is the ability to discriminate objects on the basis of touch. Haptic receptors enable us to perceive fine touch and pressure, and to identify objects that we touch and grasp. Haptic receptors occupy both superficial layers and deep layers of the skin and are attached to body hairs as well.

Figure 4-25 illustrates the cellular processes at work in the dendrite of a sensory neuron when a touch receptor is activated.

As diagrammed in the center of Figure 11-16, haptic receptors consist of a dendrite attached to a hair or to connective tissue, or a dendrite encased within a capsule of tissue. Mechanical stimulation of the hair, tissue, or capsule activates special channels on the dendrite, which in turn initiate an action potential. Differences in the tissue forming the capsule determine the kinds of mechanical energy conducted through the haptic recep-tor to the nerve. For example, pressure that squeezes the capsule of a Pacinian corpuscle is the necessary stimulus for initiating an action potential.

MOVEMENT **Proprioception** is the perception of body location and movement. Proprio-ceptors are encapsulated nerve endings that are sensitive to the stretch of muscles and tendons and the movement of joints. In the Golgi tendon organ shown at the bottom of Figure 11-16, for instance, an action potential is triggered when the tendon moves, stretching the receptor attached to it.

Duration of Receptor Response

Somatosensory receptors are specialized to tell us when a sensory event occurs and whether it is still occurring. Information about when a stimulus occurs is handled by **rapidly adapting receptors** that respond to the beginning and the end of a stimulus and produce only brief bursts of action potentials. As shown in Figure 11-16, haptic receptors that respond to touch (Meissner's corpuscles), to fluttering sensations (Pacin-ian corpuscles), and to vibration (Ruffini corpuscles) all are rapidly adapting receptors.

In contrast, **slowly adapting receptors** continue to respond as long as a sensory event is present: they detect whether a stimulus is still occurring. For instance, after you have put on an article of clothing and become accustomed to how it feels, only slowly adapting haptic receptors (such as Merkel's receptors and hair receptors) re-main active.

The difference between a rapidly adapting and a slowly adapting receptor rests on two factors: how the receptor is stimulated and how the ion channels in the membrane of its dendrite respond to mechanical stimulation. The stimulation may be sharp or cold, fluttery or deep, a stretch or a swerve.

Dorsal-Root Ganglion Neurons

The dendrites that carry somatosensory information into the CNS belong to neurons whose cell bodies are located just outside the spinal cord in dorsal-root ganglia. Their axons enter the spinal cord. As illustrated in **Figure 11-17,** such a *dorsal-root ganglion neuron* contains a single long dendrite. Only the tip is responsive to sensory stimulation. This dendrite is continuous with the somatosensory neuron's axon, which enters the spinal cord. The somatosensory cell body sits to one side of this long pathway.

Every spinal-cord segment is flanked by a dorsal-root ganglion that contains neurons of many types. Each type of dorsal-root ganglion neuron responds to a particular kind of somatosensory information. Within the spinal cord, the axons of dorsal-root ganglion neurons may synapse with other neurons or continue to the brain or do both.

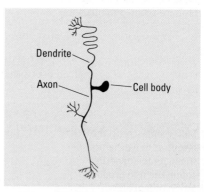

Somatosensory neuron

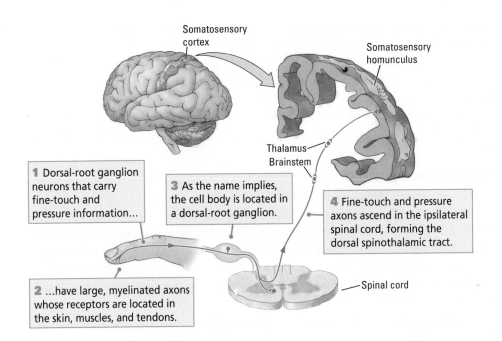

Somatosensory cortex

Somatosensory homunculus

Thalamus
Brainstem

1 Dorsal-root ganglion neurons that carry fine-touch and pressure information...

3 As the name implies, the cell body is located in a dorsal-root ganglion.

4 Fine-touch and pressure axons ascend in the ipsilateral spinal cord, forming the dorsal spinothalamic tract.

2 ...have large, myelinated axons whose receptors are located in the skin, muscles, and tendons.

Spinal cord

FIGURE 11-17 Haptic Dorsal-Root Ganglion Neuron The dendrite and axon of this dorsal-root ganglion neuron are contiguous and carry sensory information from the skin to the central nervous system. The large, myelinated dorsal-root axons travel up the spinal cord to the brain in the dorsal column, whereas the small axons synapse with neurons whose axons cross the spinal cord and ascend on the other side (shown in Figure 11-19).

The axons of dorsal-root ganglion neurons vary in diameter and myelination. These structural features are related to the kind of information the neurons carry. Proprioceptive information (location and movement) and haptic information (touch and pressure) are carried by dorsal-root ganglion neurons that have large, well-myelinated axons. Nociceptive information (pain, temperature, itch) is carried by dorsal-root ganglion neurons that have smaller axons with little or no myelin.

Because of their size and myelination, the larger neurons carry information faster than the smaller neurons do. One explanation for why proprioceptive and haptic neurons are designed to carry messages quickly is that their information requires rapid response. Imagine that you've touched a hot stove. A myelinated pain fiber activates and instructs the hand to withdraw quickly. A nonmyelinated pain fiber will let you know for some time afterwards that you burned your fingers.

Disruption of Dorsal-Root Ganglion Function

We can support the claim that sensory information is essential for movement by describing what happens when dorsal-root ganglion cells do not function. A clue comes from a visit to the dentist. If you have ever had a tooth "frozen" for dental work, you have experienced the very strange effect of losing sensation on one side of your face. Not only do you lose pain perception but you also seem to lose the ability to move your facial muscles properly, making it awkward to talk, eat, and smile. So even though the anesthetic is blocking only sensory nerves, your movement ability is affected as well.

In much the same way, damage to sensory nerves affects both sensory perceptions and motor abilities. John Rothwell and his coworkers (1982) described a patient, G. O., who was **deafferentated** (had lost afferent sensory fibers) by a disease that destroyed somatosensory dorsal-root ganglion neurons. G. O. had no sensory input from his hands. He could not, for example, feel when his hand was holding something.

However, G. O. could still accurately produce a range of finger movements, and he could outline figures in the air even with his eyes closed. He could also move his thumb accurately through different distances and at different speeds, judge weights, and match forces Nevertheless, his hands were relatively useless to him in daily life. Although G. O.

As explained in Section 3-1, myelin is the coating around axons, formed by glial cells, that speeds neurotransmission.

nociception Perception of pain, temperature, and itch.

hapsis Perceptual ability to discriminate objects on the basis of touch.

proprioception Perception of the position and movement of the body, limbs, and head.

rapidly adapting receptor Body sensory receptor that responds briefly to the onset of a stimulus on the body.

slowly adapting receptor Body sensory receptor that responds as long as a sensory stimulus is on the body.

deafferentation Loss of incoming sensory input usually due to damage to sensory fibers; also loss of any afferent input to a structure.

could drive his old car, he was unable to learn to drive a new one. He was also unable to write, to fasten shirt buttons, and to hold a cup.

G. O. began movements quite normally, but as he proceeded, the movement patterns gradually fell apart, ending in failure. Part of G. O.'s difficulties lay in maintaining muscle force for any length of time. When he tried to carry a suitcase, he would quickly drop it unless he continually looked down to confirm that he was carrying it. Clearly, although G. O. had damage only to his sensory neurons, he suffered severe motor disability as well, including the inability to learn new motor skills.

Disruption of Body Awareness

Movement abnormalities also result from more selective damage to neurons that carry proprioceptive information about body location and movement. Neurologist Oliver Sacks (1998) gives a dramatic example in his description of a patient, Christina, who suffered damage to proprioceptive sensory fibers throughout her body after taking mega-doses of vitamin B_6. Christina was left with very little ability to control her movements and spent most of each day lying prone. Here is how she describes what a loss of proprioception means:

> "What I must do then," she said slowly, "is use vision, use my eyes, in every situation where I used—what do you call it?—proprioception before. I've already noticed," she added, musingly, "that I may lose my arms. I think they are in one place, and I find they're in another. This proprioception is like the eyes of the body, the way the body sees itself. And if it goes, as it's gone with me, it's like the body's blind. My body can't see itself if it's lost its eyes, right? So I have to watch it—be its eyes." (Sacks, 1998, p. 46)

Clearly, Christina's motor system is intact, but without a sense of where her body is in space and what her body is doing, she is almost completely immobilized. Jonathan Cole (1995) has described the case of Ian Waterman, who lost proprioception after a presumed viral infection at age 19. He is the only person reported to have learned how to move again and this relearning took a period of years. He was even able to drive. All this regained movement was mediated by vision, however, without which he was as helpless as Christina.

Somatosensory Pathways to the Brain

As the axons of somatosensory neurons enter the CNS in the spinal cord, they divide, forming two pathways to the brain. The haptic-proprioceptive axons for touch and body awareness ascend the spinal cord ipsilaterally, whereas nociceptive (pain, temperature, itch) nerve fibers synapse with neurons whose axons cross to the contralateral side of the spinal cord before ascending to the brain. **Figure 11-18** shows these two routes through the spinal cord. The dorsal haptic-proprioceptive pathway is shown as a solid red line, the ventral nociceptive pathway as a dashed red line.

The Dorsal Spinothalamic Tract

Haptic-proprioceptive axons are located in the dorsal portion of the spinal cord and form the **dorsal spinothalamic tract.** These axons for fine touch and pressure synapse in the dorsal-column nuclei located at the base of the brain. As shown in Figure 11-18, axons of neurons in the dorsal-column nuclei then cross over to the other side of the brainstem and ascend through the brainstem as part of a pathway called the *medial lemniscus.*

These dorsal-column axons synapse in the **ventrolateral thalamus,** the part of the thalamus that carries afferent information about body senses to the somatosensory cortex. The thalamic neurons send most of their axons to the somatosensory cortex; some go to the motor cortex. Thus, three relay neurons are required to carry haptic-proprioceptive information to the brain: dorsal-root ganglia neurons, dorsal-column nuclei neurons, and thalamic neurons.

Oliver Sacks's research has informed scientific understanding of conditions as diverse as Parkinson's disease (Focus 5-3 and Section 16-3), injury to the ventral visual stream (Section 9-4), an unexpected obsession with music (Section 10-4), and the thought process (Section 15-1).

Ipsilateral connections lie on the same side of the body on which they enter, contralateral connections lie on the opposite side.

dorsal spinothalamic tract Pathway that carries fine-touch and pressure fibers.

ventrolateral thalamus Part of the thalamus that carries information about body senses to the somatosensory cortex.

ventral spinothalamic tract Pathway from the spinal cord to the thalamus that carries information about pain and temperature.

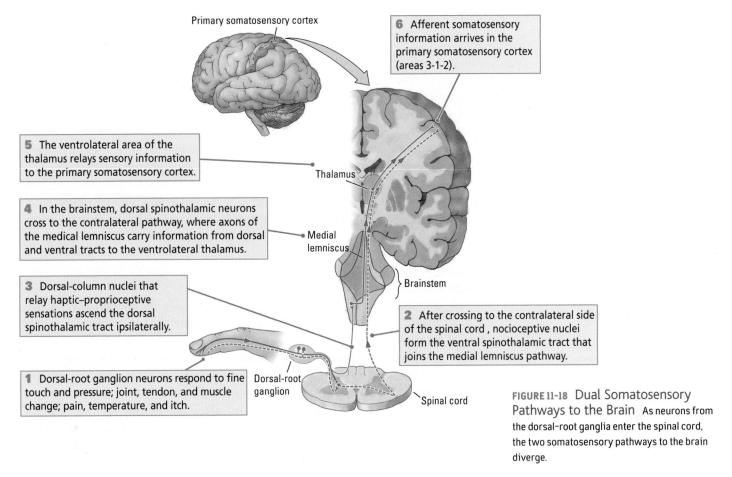

Primary somatosensory cortex

6 Afferent somatosensory information arrives in the primary somatosensory cortex (areas 3-1-2).

5 The ventrolateral area of the thalamus relays sensory information to the primary somatosensory cortex.

Thalamus

4 In the brainstem, dorsal spinothalamic neurons cross to the contralateral pathway, where axons of the medical lemniscus carry information from dorsal and ventral tracts to the ventrolateral thalamus.

Medial lemniscus

3 Dorsal-column nuclei that relay haptic–proprioceptive sensations ascend the dorsal spinothalamic tract ipsilaterally.

Brainstem

2 After crossing to the contralateral side of the spinal cord , nocioceptive nuclei form the ventral spinothalamic tract that joins the medial lemniscus pathway.

1 Dorsal-root ganglion neurons respond to fine touch and pressure; joint, tendon, and muscle change; pain, temperature, and itch.

Dorsal-root ganglion

Spinal cord

FIGURE 11-18 Dual Somatosensory Pathways to the Brain As neurons from the dorsal-root ganglia enter the spinal cord, the two somatosensory pathways to the brain diverge.

The Ventral Spinothalamic Tract

Nocioceptive axons take a different route to the brain. As shown in Figure 11-18, they first synapse with neurons in the dorsal part of the spinal cord's gray matter, and these neurons, in turn, send their axons across to the other side of the spinal cord. There, on the ventral side, they form the **ventral spinothalamic tract** that carries afferent information about pain, temperature, and itch to the thalamus.

The ventral tract joins the medial lemniscus in the brainstem to continue on to the ventrolateral thalamus. Some thalamic neurons receiving input from axons of the ventral spinothalamic tract send their axons to the somatosensory cortex. So again, three groups of neurons are required to convey nocioceptive information to the brain: dorsal-root neurons, spinal-cord gray-matter neurons, and ventrolateral thalamic neurons.

Effects of Unilateral Spinal-Cord Damage

The two separate somatosensory pathways in the spinal cord—haptic-proprioceptive and nocioceptive—enter the spinal cord together, separate in the spinal cord, and join up again in the brainstem. Because of this arrangement, damage to one side of the spinal cord results in distinctive sensory losses to both sides of the body below the site of injury.

As is illustrated in **Figure 11-19,** loss of hapsis and proprioception occurs unilaterally, on the side of the body where the damage occurred, and loss of nocioception occurs contralaterally, on the opposite side of the body. Unilateral damage at the points where the pathways come together; that is, to the dorsal roots or in the brainstem or the thalamus, affects hapsis, proprioception, and nocioception equally because these parts of the pathways lie in close proximity.

FIGURE 11-19 Effects of Unilateral Spinal-Cord Injury

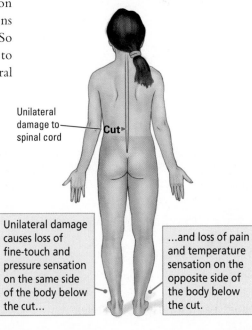

Unilateral damage to spinal cord

Cut

Unilateral damage causes loss of fine-touch and pressure sensation on the same side of the body below the cut...

...and loss of pain and temperature sensation on the opposite side of the body below the cut.

monosynaptic reflex Reflex requiring one synapse between sensory input and movement.

Spinal Reflexes

Somatosensory nerve fibers not only convey information to the cortex, they also participate in behaviors mediated by the spinal cord and brainstem. Spinal-cord somatosensory axons, even those ascending the dorsal columns, give off axon collaterals that synapse with interneurons and motor neurons on both sides of the spinal cord. The circuits made between sensory receptors and muscles through these connections mediate spinal reflexes.

The simplest spinal reflex is formed by a single synapse between a sensory neuron and a motor neuron. **Figure 11-20** illustrates such a **monosynaptic reflex,** the knee jerk that affects the quadriceps muscle of the thigh, which is anchored to the leg bone by the patellar tendon. When the lower leg hangs free and this tendon is tapped with a small hammer, the quadriceps muscle is stretched, activating the stretch-sensitive sensory receptors embedded in it.

The sensory receptors then send a signal to the spinal cord through sensory neurons that synapse with motor neurons projecting back to the same thigh muscle. The discharge from the motor neurons stimulates the muscle, causing it to contract to resist the stretch. Because the tap is brief, the stimulation is over before the motor message arrives, and the muscle contracts even though it is no longer stretched. This contraction pulls the leg up, producing the reflexive knee jerk.

This simplest of reflexes entails monosynaptic connections between single sensory neurons and single motor neurons. Somatosensory axons from other receptors, especially those in the skin, make much more complex connections with both interneurons and motor neurons. These multisynaptic connections are responsible for more complex spinal reflexes, such as those involved in standing and walking, actions that include many muscles on both sides of the body.

Feeling and Treating Pain

As many as 30 percent of visits to physicians are for pain symptoms, as are 50 percent of emergency room visits. People suffer pain as a result of acute injuries such as broken bones, chronic conditions such as cancer and arthritis, and intermediate conditions such

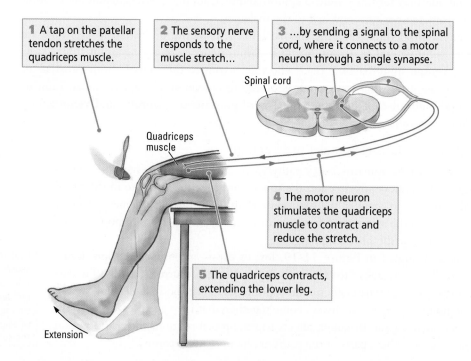

1 A tap on the patellar tendon stretches the quadriceps muscle.

2 The sensory nerve responds to the muscle stretch...

3 ...by sending a signal to the spinal cord, where it connects to a motor neuron through a single synapse.

Spinal cord

Quadriceps muscle

4 The motor neuron stimulates the quadriceps muscle to contract and reduce the stretch.

5 The quadriceps contracts, extending the lower leg.

Extension

FIGURE 11-20 Monosynaptic Reflex

as stiff muscles from exercising. Women experience pain during menstruation, pregnancy, and childbirth. The incidence of people living with pain increases as people age, and for many people, pain is a constant companion.

Perceiving Pain

People can experience "central pain" in a part of the body that is not obviously injured. One type of central pain is "phantom-limb pain," which, as described in Research Focus 11-5, seems to occur in a limb that has been lost.

People suffering pain would happily dispense with it. But pain is necessary: the occasional person born without pain receptors experiences body deformities through failure to adjust posture and acute injuries through failure to avoid harmful situations.

Pain perception results from synthesizing a plethora of sensory information. There may be as many as eight different kinds of pain fibers, judging from the peptides and other chemicals released by these nerves when irritated or damaged. Some of these chemicals irritate surrounding tissue, stimulating it to release other chemicals to stimulate blood flow and to stimulate the pain fibers themselves. These reactions contribute to pain, redness, and swelling at the site of an injury.

Consider itch as an example. We may feel itchy and consequently scratch when a foreign object is on our body. We also frequently feel itch in the absence of an obvious

---

RESEARCH FOCUS ⊕ 11-5

Phantom-Limb Pain

Up to 80 percent of people who have had a limb amputated also endure phantom-limb phenomena, including pain and other sensations and motor phantoms such as phantom movement and cramps (Kern et al., 2009). Phantom sensations and movements are illusions that indicate that their source is the brain.

Various techniques have been used to minimize phantom-limb pain, including drug-based pain management with opioids and the injection of pain medications into the spinal cord. An innovative method devised by V. S. Ramachandran (1996) assists the patient in creating a counterillusion that the limb is intact and is providing normal sensory input to the brain.

Ramachandran devised a mirror box into which an amputee who has lost an arm inserts the intact arm and then observes a reflection of it in the mirror. The mirror-image reflection suggests that the missing arm is present and can be controlled, as shown in the illustration. The perception of the limb as intact counteracts phantom pain and cramps.

Inspired by Ramachandran's mirror, researchers have developed other illusions to suggest that the missing arm is present and can be controlled. One method uses virtual-reality goggles to suggest that the limb is present. Another method uses the so-called "rubber limb" phenomenon to create the illusion that a missing limb is present. To induce this phenomenon, the stump of the amputated limb is stimulated tactually while the subject observes a prosthetic limb being touched. All the illusions lessen phantom-limb pain and cramps by suggesting that a normal limb is present.

Alessandria and colleagues (2011) asked whether phantom limbs and their associated sensations also occur during dreaming. They awoke sleeping participants with amputations during rapid eye movement, or

REM, sleep and asked them to recount their dreams. In none of the dreams did the participants remember having an amputated limb or phantom-limb sensations.

This finding suggests that the integrity of brain regions associated with a limb is preserved even though a limb is absent. When awake, the loss of moment-to-moment sensory activity from the limb creates the opportunity for phantom perceptions.

pain gate Hypothetical neural circuit in which activity in fine-touch and pressure pathways diminishes the activity in pain and temperature pathways.

periaqueductal gray matter (PAG) Nuclei in the midbrain that surround the cerebral aqueduct joining the third and fourth ventricles; PAG neurons contain circuits for species-typical behaviors (e.g., female sexual behavior) and play an important role in the modulation of pain.

Section 6-2 notes the similarities between endogenous peptide hormones in the brain and the opioid analgesics, including opium and its synthetic derivatives such as morphine.

FIGURE 11-21 Pain Gate An interneuron in the spinal cord receives excitatory input (plus sign) from the fine-touch and pressure pathway and inhibitory input (minus sign) from the pain and temperature pathway. The relative activity of the interneuron then determines whether pain and temperature information is sent to the brain. Adapted from *The Puzzle of Pain* (p. 154), by R. Melzack, 1973, New York: Basic Books.

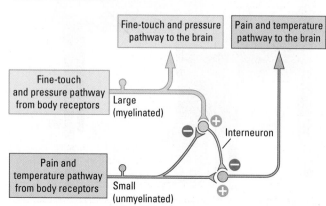

stimulus, and some drugs, including opioids, enhance the itch sensation in the absence of a physical stimulus at the itchy part of the body.

Haptic information also contributes to the perception of pain. For example, people can accurately report the location and characteristics of various kinds of pain, but in the absence of fine-touch and pressure information, pain is more difficult to identify and localize.

The ventral spinothalamic tract illustrated in Figure 11-18 is the main pain pathway to the brain, but as many as four other pathways may carry pain information from the spinal cord to the brain. These pathways are both crossed and uncrossed and project to the reticular formation of the midbrain, where they produce arousal; to the amygdala, where they produce emotional responses typically associated with pain; and to the hypothalamus, where they activate hormonal and cardiovascular responses.

The existence of multiple pain pathways in the spinal cord makes it difficult to treat chronic pain by selectively cutting the ventrospinothalamic tract—one radical procedure used to control chronic pain. It is likely that each pain pathway has its own function, whether for sensation, arousal, emotional responses, or other physiological responses.

Responding to Pain

Neuronal circuits in the spinal cord allow haptic–proprioceptive and nociceptive pathways to interact. Such interactions may be responsible for our puzzling and variable responses to pain. For example, people who are engaged in combat or intense athletic competition may receive a serious injury to the body but start to feel the pain only much later.

A friend of ours was attacked by a grizzly bear while hiking and received 200 stitches to bind his wounds. When friends asked if it hurt to be bitten by a grizzly bear, he surprisingly answered no, explaining, "I had read the week before about someone who was killed and eaten by a grizzly bear. So I was thinking that this bear was going to eat me unless I got away. I did not have time for pain. I was fighting for my life. It was not until the next day that I started feeling pain—and fear."

The primacy of our friend's fear over his pain is related to the stress he was under. Failure to experience pain in a fight-or-flight-situation is obviously adaptive, as is illustrated by this story, and may be related to the activation of endorphins, our endogeneous opioids. Treatments for pain include opioid drugs (such as morphine), acupuncture (which entails the rapid vibration of needles embedded in the skin), and simply rubbing the area surrounding the injury. Psychological factors interact with pain treatments because most studies on pain management find that placebo effects can be as effective as actual treatments. Pain is puzzling in the variety of ways in which it can be lessened.

To explain both the perception of pain and how it can be suppressed in so many different ways, Ronald Melzack and Patrick Wall (1965) proposed a "gate" theory of pain. The essence of gate theory as applied to pain perception is that activities in different sensory pathways play off against each other and so determine whether and how much pain is perceived as a result of an injury. Melzack and Wall propose that haptic–proprioceptive stimulation can reduce pain perception, whereas the absence of such stimulation can increase pain perception through interactions at a pain gate.

A model of the **pain gate**, illustrated in **Figure 11-21**, consists of a haptic–proprioceptive fiber that conveys fine touch and pressure information and a nociceptive fiber that conducts pain information. Each fiber synapses with the same interneuron. Collaterals from the haptic–proprioceptive pathway excite the interneuron, whereas collaterals from the nociceptive pathway inhibit it. The interneuron, in turn, inhibits a neuron that relays pain information from the spinal cord to the brain. Consequently, when the haptic–proprioceptive pathway is active, the interneuron is stimulated, it inhibits the secondary pain neuron, and the interneuron acts as a gate, reducing the sensation of pain.

Treating Pain

Gate theory helps to explain how different pain treatments work. When you stub your toe, for instance, you feel pain because the pain pathway to the brain is open. Rubbing the toe activates the haptic–proprioceptive pathway and reduces the flow of information in the pain pathway because the pain gate partly closes, relieving the pain sensation.

Similarly, a variety of treatments for pain, including massage, immersion in warm water, and acupuncture, may produce pain-relieving effects by selectively activating haptic and proprioceptive fibers relative to pain fibers, thus closing the pain gate. For acupuncture, vibrating needles on different body points presumably activate fine-touch and pressure fibers. The pain gate model may also explain why opioid drugs influence pain. The interneuron that is the gate uses an endogenous opiate as an inhibitory neurotransmitter. Thus, opioids have one of their effects in relieving pain by mimicking the actions of the endogeneous opioid neurotransmitter of the interneuron.

One of the most successful treatments for pain is the injection of small amounts of morphine under the dura mater, the outer layer of the meninges that protects the spinal cord. This epidural anesthesia is mediated by the action of morphine on interneurons in the spinal cord. Although morphine is a very useful treatment for pain, its effects lessen with continued use. This form of habituation may be related to changes that take place on the postsynaptic receptors of pain neurons in the spinal cord and brain.

Gate theory suggests an explanation for the "pins and needles" that we feel after sitting too long in one position. Loss of oxygen from reduced blood flow first deactivates the large myelinated axons that carry touch and pressure information, leaving the small unmyelinated fibers that carry pain and temperature messages unaffected. As a result, "ungated" sensory information flows in the pain and temperature pathway, leading to the pins-and-needles sensation.

Melzack and Wall propose that pain gates may be located in the brainstem and cortex as well as the spinal cord. These additional gates help to explain how other approaches to pain relief work. For example, researchers have found that feelings of severe pain can be lessened when people have a chance to shift their attention from the pain to other stimuli. Dentists have long used this technique by giving their patients something soothing to watch or listen to while undergoing procedures.

The influence of attention on pain sensations may work through a cortical pain gate. Electrical stimulation at a number of sites in the brainstem can reduce pain, perhaps by closing brainstem pain gates. Another way in which pain perceptions might be lessened is through descending pathways from the forebrain and the brainstem to the spinal-cord pain gate.

The presence in the spinal cord of relatively complex neural circuits, such as the pain gate, is related both to the variable nature of pain and to some successful treatments and some problems in treating pain. In response to noxious stimulation, pain neurons in the spinal cord can undergo sensitization: successive pain experiences can produce an escalating response to a similar noxious stimulus. Spinal-cord neurons thus learn to produce a larger pain signal.

The brain can also influence the pain signal it receives from the spinal cord. The cell bodies of **periaqueductal gray matter** (PAG) neurons surround the cerebral aqueduct connecting the third and fourth ventricles. Electrical stimulation of the PAG is effective in suppressing pain.

Neurons in the PAG produce their pain-suppressing effect by exciting pathways (including serotonergic and noradrenergic pathways) in the brainstem that project to the spinal

Figure 2-4 diagrams the triple-layered meninges that encases the brain and spinal cord.

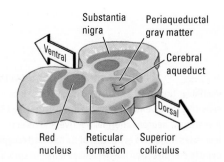

The PAG is one nucleus within the midbrain's tegmentum (floor), shown here in cross section.

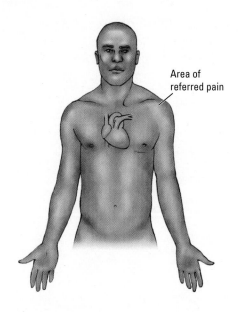

FIGURE 11-22 Referred Pain During a heart attack, pain from receptors in the heart is felt in the left shoulder and upper arm.

cord where they inhibit neurons that form the ascending pain pathways. Activation in these inhibitory circuits explain in part why the sensation and perception of pain is lessened during sleep. Deep brain stimulation of the PAG by implanted microelectrodes is one way of treating pain that proves resistant to all other therapies, including treatment with opioid drugs.

Many internal organs, including the heart, the kidneys, and the blood vessels, have pain receptors, but the ganglion neurons carrying information from these receptors do not have their own pathway to the brain. Instead, they synapse with spinal-cord neurons that receive nocioceptive information from the body's surface. Consequently, the neurons in the spinal cord that relay pain and temperature messages to the brain receive two sets of signals: one from the body's surface and the other from the internal organs.

These spinal-cord neurons cannot distinguish between the two sets of signals—nor can we. As a result, pain in body organs is felt as **referred pain** coming from the body surface. For example, pain in the heart associated with a heart attack is felt as pain in the left shoulder and upper arm **(Figure 11-22)**. Pain in the stomach is felt as pain in the midline of the trunk; pain in the kidneys is felt as pain in the lower back. Pain in blood vessels in the head is felt as diffuse pain that we call a headache (remember that the brain has no pain receptors).

Vestibular System and Balance

The only localized part of the somatosensory system, the **vestibular system,** consists of two organs, one located in each inner ear. As **Figure 11-23**A shows, each vestibular organ is made up of two groups of receptors: the three *semicircular canals* and the *otolith organs,* the *utricle* and the *saccule.* These vestibular receptors do two jobs: (1) they tell us the position of the body in relation to gravity and (2) they signal changes in the direction and the speed of head movements.

Note in Figure 11-23A that the semicircular canals are oriented in three different planes that correspond to the three dimensions in which we move through space. Each

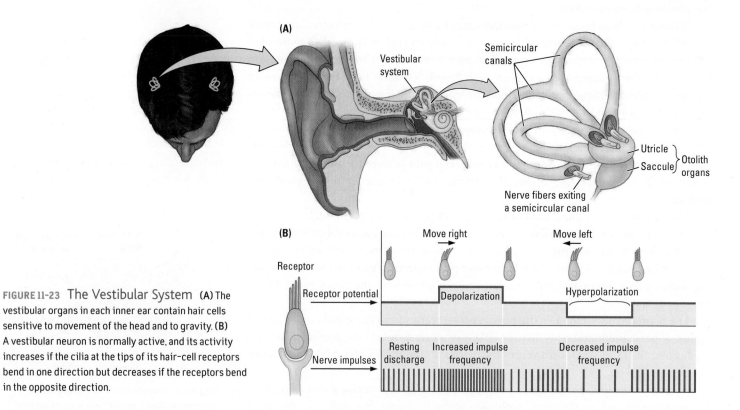

FIGURE 11-23 The Vestibular System (A) The vestibular organs in each inner ear contain hair cells sensitive to movement of the head and to gravity. (B) A vestibular neuron is normally active, and its activity increases if the cilia at the tips of its hair-cell receptors bend in one direction but decreases if the receptors bend in the opposite direction.

canal furnishes information about movement in its particular plane. The semicircular canals are filled with a fluid called *endolymph.* Immersed in the endolymph is a set of hair cells.

When the head moves, the endolymph also moves, pushing against the hair cells and bending the cilia at their tips. The force of the bending is converted into receptor potentials in the hair cells that send action potentials over vestibular nerve axons to the brain. These axons are normally quite active: bending the cilia in one direction increases receptor potentials, consequently increasing vestibular nerve axon activity; bending them in the other direction decreases vestibular afferent axon activity. These responses are diagrammed in Figure 11-23B. Typically, when the head turns in one direction, the receptor message on that side of the body increases neuronal firing. The message on the body's opposite side leads to a decrease in firing.

The utricle and saccule lie stacked just beneath the semicircular canals, as shown Figure 11-23A. They also contain hair cells, but these receptors are embedded in a gelatinlike substance that contains small crystals of the salt calcium carbonate called *otoconia.* When you tilt your head, the gelatin and otoconia press against the hair cells, bending them. The mechanical action of the hair bending modulates the rate of action potentials in vestibular afferent axons that convey messages about the position of the head in three-dimensional space.

The receptors in the vestibular system tell us about our location relative to gravity, about acceleration and deceleration of our movements, and about changes in movement direction. They also allow us to ignore the otherwise very destabilizing influence that our movements might have on us. When you are standing on a moving bus, for example, even slight movements of the vehicle could potentially throw you off balance, but they do not. Similarly, when you make movements yourself, you easily avoid tipping over, despite the constant shifting of your body weight. Your vestibular system enables your stability.

To demonstrate the role of vestibular receptors in helping you to compensate for your own movements, try this experiment. Hold your hand in front of you and shake it. Your hand appears blurry. Now shake your head instead of your hand, and the hand remains in focus. Compensatory signals from your vestibular system allow you to see the hand as stable even though you are moving around.

Vertigo (from the Latin for "spinning"), a sensation of dizziness when one is not moving, is a dysfunction of the inner ear and can be accompanied by nausea as well as difficulty maintaining balance while walking. A common way to induce vertigo is to spin, as children do when playing. Vertigo can also occur from looking down from a height or looking up at a tall object and as one is simply standing up or sitting down. One intoxicating effect of alcohol is vertigo. **Ménière's disease,** named after a French physician, is a disorder of the middle ear resulting in vertigo and loss of balance.

Vestibular hair cells work on the same principles as the cochlear hair cells that mediate hearing, described in Section 10-2.

Cranial nerve 8, the auditory vestibular nerve, is one of 12 pairs diagrammed in Figure 2-27 that link the brain to the head and neck and to various internal organs.

referred pain Pain felt on the surface of the body that is actually due to pain in one of the internal organs of the body.

vestibular system Somatosensory system that comprises a set of receptors in each inner ear that respond to body position and to movement of the head

Ménière's disease Disorder of the middle ear resulting in vertigo and loss of balance.

REVIEW 11-4
Organization of the Somatosensory System

Before you continue, check your understanding.

1. Body senses contribute to the perception of _____ (touch and pressure), _____ (location and movement), and _____ (temperature, pain, itch).

2. Haptic-proprioceptive information is carried into the CNS by the _____ spinothalamic tract; nociceptive information is carried in by the _____ spinothalamic tract.

3. The two tracts interact in the spinal cord to regulate pain perception via a _____.

4. In the midbrain, the _____ effectively suppresses pain by activating neuromodulatory circuits that inhibit pain pathways.

5. The only localized somatosensory system is the _____ system, which helps us to maintain _____ by signaling information about the head's position and our movement through space.

6. Explain how proprioception acts as the "eyes of the body."

Answers appear at the back of the book.

11-5 Exploring the Somatosensory Cortex

Somatosensory neurons do more than convey sensation to the brain: they enable us to perceive things that we describe as pleasant or unpleasant, the shape and texture of objects, the effort required to complete tasks, and even our spatial environment. **Figure 11-24** illustrates the two main somatosensory areas in the cortex.

FIGURE 11-24 Somatosensory Cortex Stimulation of the primary somatosensory cortex in the parietal lobe produces sensations that are referred to appropriate body parts. Information from the primary somatosensory cortex travels to the secondary somatosensory cortex for further perceptual analysis and to contribute to movement sequences mediated in the frontal lobes.

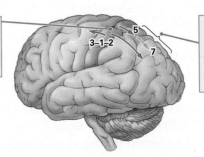

Primary somatosensory cortex receives sensory information from the body.

Secondary somatosensory cortex receives sensory information from the primary somatosensory cortex.

Korbinian Brodmann numbered these areas more than a century ago on his map of the cortex (see Figure 2-23).

The *primary somatosensory cortex* receives projections from the thalamus and consists of Brodmann's areas 3-1-2 (all appear red in Figure 11-24). The primary somatosensory cortex begins the process of constructing perceptions from somatosensory information. It mainly consists of the postcentral gyrus just behind the central fissure in the parietal lobe. Thus, the primary somatosensory cortex lies adjacent to the primary motor cortex, on the other side of the central fissure in the frontal lobe.

The *secondary somatosensory cortex* (Brodmann's areas 5 and 7, shaded orange and yellow in Figure 11-24) is located in the parietal lobe just behind the primary somatosensory cortex. The secondary somatosensory cortex refines perceptual constructions and sends information to the frontal cortex.

Somatosensory Homunculus

In his studies of human patients undergoing brain surgery, Wilder Penfield electrically stimulated the somatosensory cortex and recorded the patients' responses. Stimulation at some sites elicited sensations in the foot; stimulation of other sites produced sensations in a hand, the trunk, or the face. By mapping these responses, Penfield was able to construct a somatosensory homunculus in the cortex, shown in **Figure 11-25A**. The sensory homunculus looks very similar to the motor homunculus shown in Figure 11-6 in that the most sensitive areas of the body are accorded a relatively larger cortical area.

Using smaller electrodes and more precise recording techniques in monkeys, Jon Kaas (1987) found that the primary somatosensory cortex does not consist of a single homunculus, as Penfield's original model proposed. When Kaas stimulated sensory receptors on the body and recorded the activity of cells in the sensory cortex, he found that the somatosensory cortex comprises four representations of the body. Each is associated with a class of sensory receptors.

The progression of these representations across the primary somatosensory cortex from front to back is shown in Figure 11-25B. Area 3a cells are responsive to muscle receptors; area 3b cells are responsive to slow-responding skin receptors. Area 1 cells are responsive to rapidly adapting skin receptors, and area 2 cells are responsive to deep tissue pressure and joint receptors. In other studies, Hiroshi Asanuma (1989) and his coworkers found still another sensory representation in the motor cortex (area 4) in which cells respond to muscle and joint receptors.

Perceptions constructed from elementary sensations depend on combining the elementary sensations. This combining takes place as areas 3a and 3b project onto area 1, which in turn projects onto area 2. Whereas a cell in area 3a or 3b may respond to activity in only a certain area on a certain finger, for example, cells in area 1 may respond to similar information from a number of different fingers.

At the next level of synthesis, cells in area 2 respond to stimulation in a number of different locations on a number of different fingers as well as to stimulation from different kinds of somatosensory receptors. Thus, area 2 contains *multimodal neurons* that are responsive to movement force, orientation, and direction. We perceive all these properties when we hold an object in our hands and manipulate it.

With each successive information relay, both the size of the pertinent receptive fields and the synthesis of somatosensory modalities increase. One reason that sensory information remains segregated at the level of the cortex could be that we often need to distinguish among different kinds of sensory stimuli coming from different sources. For example, we need to be able to tell the difference between tactile stimulation on the surface of the skin, which is usually produced by some external agent, and stimulation coming from muscles, tendons, and joints, which is usually produced by our own movements.

At the same time, we need to know about the combined sensory properties of a stimulus. For instance, when we manipulate an object, it is useful to "know" the object both by its sensory properties, such as temperature and texture, and by the movements we make as we handle it. For this reason, the cortex provides for somatosensory synthesis too. The tickle sensation seems rooted in an "other versus us" somatosensory distinction, as described in Research Focus 11-6, "Tickling" on page 392.

Research by Vernon Mountcastle (1978) shows that cells in the somatosensory cortex are arranged in functional columns running from layer I to layer VI, similar to the functional columns found in the visual cortex. Every cell in a functional cortical column responds to a single class of receptors. Some columns are activated by rapidly adapting skin receptors, others by slowly adapting skin receptors, still others by pressure receptors, and so forth. All neurons in a functional column receive information from the same local area of skin. In this way, neurons lying within a column seem to be an elementary functional unit of the somatosensory cortex.

(A) Penfield's single-homunculus model

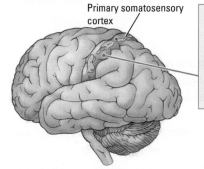

Primary somatosensory cortex

The primary somatosensory cortex is organized as a single homunculus with large areas representing body parts that are very sensitive to sensory stimulation.

(B) Four-homunculus model

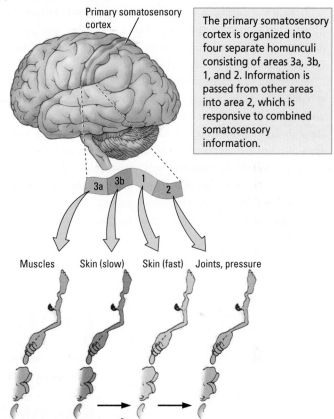

Primary somatosensory cortex

The primary somatosensory cortex is organized into four separate homunculi consisting of areas 3a, 3b, 1, and 2. Information is passed from other areas into area 2, which is responsive to combined somatosensory information.

3a 3b 1 2

Muscles   Skin (slow)   Skin (fast)   Joints, pressure

FIGURE 11-25 Two Models of the Somatosensory Cortex

Figure 9-33 shows the organization of these functional columns in the primary visual cortex (V1).

Tickling

Everyone knows the effects and consequences of tickling. The perception is a curious mixture of pleasant and unpleasant sensations. There are two kinds of tickling. The sensation from a light caress is *kinismesis;* the pleasurable effect of hard rhythmic probing is *gargalesis.*

The tickle sensation is experienced not only by humans but also by other primates and by cats, rats, and probably most mammals. Play in rats is associated with 50-kilohertz vocalizations, and tickling body regions that are targets of the rats' own play also elicits 50-kilohertz vocalizations (Panksepp, 2007).

Tickling is rewarding in that people and animals solicit tickles from others. They even enjoy observing others being tickled. Using a robot and brain-imaging techniques, Sarah Blakemore and her colleagues (1998) explained why we cannot tickle ourselves.

Blakemore had participants deliver two kinds of identical tactile stimuli to the palms of their hand. In one condition, the stimulus was predictable and in the other a robot introduced an unpredictable delay in the stimulus. Only the unpredictable stimulus was perceived as a tickle. Thus, it is not the stimulation itself but its unpredictability that accounts for the tickle perception. This is why we cannot tickle ourselves.

One interesting feature of tickling is its accompanying, distinctive laughter. This laughter can be identified by sonograms (sound analysis), and people can distinguish tickle-related laughter from other forms of laughter.

Intrigued by findings that all apes appear to laugh during tickling, Ross and coworkers (2009) compared tickle-related laughter in apes and found that human laughter is more similar to chimpanzee laughter than to the laughter of gorillas and other apes. We humans thus have inherited not only susceptibility to tickling from our ape ancestors but laughter as well.

LWA-Dann Tardif/Corbis

Effects of Damage to the Somatosensory Cortex

Damage to the primary somatosensory cortex impairs the ability to make even simple sensory discriminations and movements. Suzanne Corkin and her coworkers (1970) demonstrated this effect by examining patients with cortical lesions that included most of areas 3-1-2 in one hemisphere. The researchers mapped the primary sensory cortices of these patients before they underwent elective surgery for removal of a carefully defined piece of that cortex, including the hand area. The patients' sensory and motor skills in both hands were tested on three different occasions: before the surgery, shortly after the surgery, and almost a year afterward.

The tests included pressure sensitivity, two-point touch discrimination, position sense (reporting the direction in which a finger was being moved), and haptic sense (using touch to identify objects, such as a pencil, a coin, eyeglasses, and so forth). For all the sensory abilities tested, the surgical lesions produced a severe and seemingly permanent deficit in the contralateral hand. Sensory thresholds, proprioception, and hapsis all were greatly impaired.

The results of other studies of both humans and animals have shown that damage to the somatosensory cortex also impairs simple movements. For example, limb use in reaching for an object is impaired, as is the ability to shape the hand to hold an object (Leonard et al., 1991). Nevertheless, like the motor cortex, the somatosensory cortex is plastic. Plasticity is illustrated by the reorganization of somatosensory cortex after deafferentation.

(A) Control monkey

This area of the somatosensory cortex represents the arm and face.

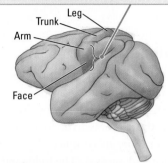

Leg
Trunk
Arm
Face

This normal pattern is illustrated by a normal face.

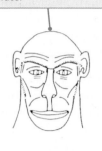

FIGURE 11-26 **Somatosensory Plasticity**
Adapted from "Massive Cortical Reorganization after Sensory Deafferentation in Adult Macaques," by T. P. Pons, P. E. Garraghty, A. K. Ommaya, J. H. Kaas, and M. Mishkin, 1991, *Science,* 252, p. 1858.

(B) Deafferented monkey

The area of the somatosensory cortex that formerly represented the arm has been taken over by expansion of the face area.

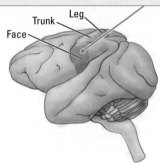

Leg
Trunk
Face

This expansion is illustrated by an elongated face.

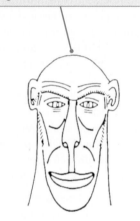

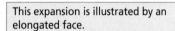

In 1991, Tim Pons and his coworkers reported a dramatic change in the somatosensory maps of monkeys in which the ganglion cells for one arm had been deafferentated a number of years earlier. The researchers had wanted to develop an animal model of damage to sensory nerves that could be a source of insight into human injuries, but they were interrupted by a legal dispute with an animal advocacy group. Years later, as the health of the animals declined, a court injunction allowed the mapping experiment to be conducted.

Pons and his coworkers discovered that the area of the somatosensory cortex that had formerly represented the arm no longer did so. Light touches on the lower face of a monkey now activated cells in what had formerly been the cortical arm region. As illustrated in **Figure 11-26,** the facial area in the cortex had expanded by as much as 10 to 14 millimeters, virtually doubling its original size by entering the arm area.

This massive change was completely unexpected. The stimulus–response patterns associated with the new expanded facial area of the cortex appeared indistinguishable from those associated with the original facial area. Furthermore, the trunk area, which bounded the other side of the cortical arm area, did not expand into the vacated arm area.

What could account for this expansion of the face area into the arm area? There is evidence for preexisting axon collaterals that are not normally active, but these collaterals would probably not be able to extend far enough to account for all the cortical reorganization. Another possibility is that, within the thalamic relay nuclei, facial-area neurons project collaterals to arm-area neurons. These neurons are close together, so the collaterals need travel only a millimeter or so. Whatever the

Section 7-7 covers some debates over using animals in brain–behavior research.

We return to this story in Section 14-4, where we look at how the brain changes in response to experience.

mechanism, the very dramatic cortical reorganizations observed in this study are helpful in understanding other remarkable phenomena, including phantom-limb sensations. In humans who have lost a forelimb, touches to the face can be felt as touches to the missing forearm.

Somatosensory Cortex and Complex Movement

How are our abilities to move and to interpret stimulation on our body related? The somatosensory cortex plays an important role in confirming that movements have taken place. Damage to the secondary somatosensory cortex does not disrupt the plans for making movements, but it does disrupt how the movements are performed, leaving their execution fragmented and confused. This inability to complete a plan of action accurately—to make a voluntary movement—is called **apraxia.** The following case highlights its symptoms.

The word *apraxia* derives from the Greek words for "no" and "action."

> A woman with a biparietal lesion [damage on the left- and right-hemisphere secondary somatosensory cortex] had worked for years as a fish-filleter. With the development of her symptoms, she began to experience difficulty in carrying on with her job. She did not seem to know what to do with her knife. She would stick the point in the head of a fish, start the first stroke, and then come to a stop. In her own mind she knew how to fillet fish, but yet she could not execute the maneuver. The foreman accused her of being drunk and sent her home for mutilating fish.
>
> The same patient also showed another unusual phenomenon that might possibly be apraxic in nature. She could never finish an undertaking. She would begin a job, drop it, start another, abandon that one, and within a short while would have four or five uncompleted tasks on her hands. This would cause her to do such inappropriate actions as putting the sugar bowl in the refrigerator, and the coffeepot inside the oven. (Critchley, 1953, pp. 158–159)

The somatosensory cortex contributes to movement by participating in both the dorsal and the ventral visual streams. The dorsal (how) stream, working without conscious awareness, provides vision for action, as when we automatically shape a hand as we reach to grasp a cup (recall Figure 11-1). The ventral (what) stream, in contrast, works with conscious awareness to identify the object as a cup.

As **Figure 11-27** illustrates, the dorsal visual stream projects to the secondary somatosensory cortex and then to the frontal cortex (Kaas et al., 2012). In this way, visual information is integrated with somatosensory information to produce unconscious movements appropriately shaped and directed to their targets. The secondary somatosensory area contributes perceptual information to the ventral stream by providing conscious haptic information about the identity of objects and completed movements. From this information the frontal cortex can select the appropriate actions that should follow from those that are already complete.

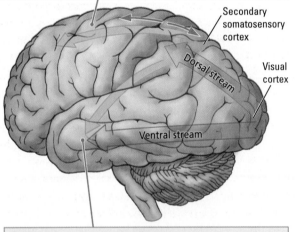

Information from the secondary somatosensory cortex contributes to the dorsal stream by specifying the movement used for grasping a target.

Secondary somatosensory cortex

Dorsal stream

Visual cortex

Ventral stream

Information from the secondary somatosensory cortex contributes to the ventral stream by providing information about object size, shape, and texture.

FIGURE 11-27 Visual Aid Section 9-4 explains how visual information from the dorsal and ventral streams contributes to movement

REVIEW 11-5

Exploring the Somatosensory Cortex

Before you continue, check your understanding.

1. The _____ somatosensory cortex, arranged as a series of homunculi, feeds information to the _____ somatosensory cortex which is responsible for somatosensory perception.

2. Damage to the secondary somatosensory cortex produces _____, an inability to complete a series of movements.

apraxia Inability to make voluntary movements in the absence of paralysis or other motor or sensory impairment, especially an inability to make proper use of an object.

3. The somatosensory cortex provides information to the _____ stream to produce unconscious movements and also provides information to the _____ stream for conscious recognition of objects.

4. Explain briefly what phantom-limb pain tells about the brain.

Answers appear at the back of the book.

SUMMARY

The somatosensory system and the motor system are interrelated at all levels of the nervous system. At the level of the spinal cord, sensory information contributes to motor reflexes; in the brainstem, sensory information contributes to complex regulatory movements. At the level of the neocortex, sensory information is used to record just-completed movements as well as to represent the sizes, shapes, and position of objects.

11-1 Hierarchy of Movement Control

Movement is organized hierarchically, using the entire nervous system (review Figure 11-1). The forebrain plans, organizes, and initiates movements, whereas the brainstem coordinates regulatory functions, such as eating and drinking, and controls neural mechanisms that maintain posture and produce locomotion. Many reflexes are organized at the level of the spinal cord and occur without the brain's involvement.

11-2 Motor System Organization

Maps produced by stimulating the motor cortex show that it is organized topographically as a homunculus, with the parts of the body capable of fine movements occupying large regions of motor cortex.

Motor-cortex neurons initiate movement, produce movement, control the force of movement, and indicate movement direction. Disuse of a limb, as might result from a motor-cortex injury, results in shrinkage of the limb's representation in the motor cortex. This shrinkage can be prevented, however, if the limb can be somehow forced into use, as in restraint therapy.

Two corticospinal pathways emerge from the motor cortex to the spinal cord. The lateral corticospinal tract consists of axons from the digit, hand, and arm regions of the motor cortex. The tract synapses with spinal interneurons and motor neurons located laterally in the spinal cord, on the side of the cord opposite the side of the brain on which the corticospinal tract started. The ventral corticospinal tract consists of axons from the trunk region of the motor cortex. This tract synapses with interneurons and motor neurons located medially in the spinal cord, on the same side of the cord as the side of the brain on which the corticospinal tract started.

Interneurons and motor neurons of the spinal cord also are topographically organized: more laterally located motor neurons project to digit, hand, and arm muscles, and more medially located motor neurons project to trunk muscles.

11-3 Basal Ganglia, Cerebellum, and Movement

Damage to the basal ganglia or to the cerebellum results in movement abnormalities. This tells us that both brain structures participate in movement control. The basal ganglia regulate the force of movements, whereas the cerebellum plays a role in learning and in maintaining movement accuracy.

11-4 Organization of the Somatosensory System

The somatosensory system is distributed throughout the entire body and consists of more than 20 types of specialized receptors, each sensitive to a particular form of mechanical energy. Each somatosensory receptor projecting from skin, muscles, tendons, or joints is associated with a dorsal-root ganglion neuron that carries the sensory information from the spinal cord into the brain.

Fibers carrying proprioceptive (location and movement) information and haptic (touch and pressure) information ascend the spinal cord as the dorsal spinothalamic tract. These fibers synapse in the dorsal-column nuclei at the base of the brain, at which point axons cross over to the other side of the brainstem to form the medial lemniscus, which ascends to the ventrolateral thalamus. Most of the ventrolateral thalamus cells project to the somatosensory cortex.

Nociceptive (pain, temperature, and itch) dorsal-root ganglion neurons synapse on entering the spinal cord. Their relay neurons cross the spinal cord to ascend to the thalamus as the ventral spinothalamic tract.

Because the two somatosensory pathways take somewhat different routes, unilateral spinal-cord damage impairs proprioception and hapsis ipsilaterally below the site of injury and nociception contralaterally below the site.

11-5 Exploring the Somatosensory Cortex

The somatosensory system is represented topographically as a homunculus in the primary somatosensory region of the parietal cortex (areas 3-1-2). The most sensitive parts of the body are accorded the largest regions of somatosensory neocortex, as the body parts most capable of fine movements are accorded the largest regions in the motor system homunculus.

A number of homunculi represent different sensory modalities, and these regions are hierarchically organized. If sensory input from a part of the body is cut off from the cortex by damage to sensory fibers, adjacent functional regions of the sensory cortex can expand into the now-unoccupied region.

The somatosensory cortex contributes to the dorsal visual stream to direct hand movements to targets. The somatosensory cortex also contributes to the ventral visual stream to create representations of external objects.

KEY TERMS

apraxia, p. 394

cerebral palsy, p. 363

corticospinal tract, p. 369

deafferentation, p. 381

dorsal spinothalamic tract, p. 382

glabrous skin, p. 379

hapsis, p. 381

homunculus, p. 365

hyperkinetic symptom, p. 372

hypokinetic symptom, p. 372

locked-in syndrome, p. 360

Ménière's disease, p. 389

mirror neuron, p. 358

monosynaptic reflex, p. 384

motor sequence, p. 358

nocioception, p. 381

neuroprosthetics, p. 354

pain gate, p. 386

paraplegia, p. 363

periaqueductal gray matter (PAG), p. 386

proprioception, p. 381

quadriplegia, p. 363

rapidly adapting receptor, p. 381

referred pain, p. 389

restraint-induced therapy, p. 369

scratch reflex, p. 363

slowly adapting receptor, p. 381

topographic organization, p. 366

ventral spinothalamic tract, p. 382

ventrolateral thalamus, p. 382

vestibular system, p. 389

Please refer to the Companion Web Site at www.worthpublishers.com/kolbintro4e for Interactive Exercises and Quizzes.

What Causes Emotional and Motivated Behavior?

RESEARCH FOCUS 12-1 PAIN OF REJECTION

12-1 IDENTIFYING THE CAUSES OF BEHAVIOR

BEHAVIOR FOR BRAIN MAINTENANCE

NEURAL CIRCUITS AND BEHAVIOR

12-2 CHEMICAL SENSES

OLFACTION

GUSTATION

12-3 EVOLUTION, ENVIRONMENT, AND BEHAVIOR

EVOLUTIONARY INFLUENCES ON BEHAVIOR

ENVIRONMENTAL INFLUENCES ON BEHAVIOR

INFERRING PURPOSE IN BEHAVIOR: TO KNOW A FLY

12-4 NEUROANATOMY OF MOTIVATED AND EMOTIONAL BEHAVIOR

REGULATORY AND NONREGULATORY BEHAVIOR

REGULATORY FUNCTION OF THE HYPOTHALAMIC CIRCUIT

ORGANIZING FUNCTION OF THE LIMBIC CIRCUIT

EXECUTIVE FUNCTION OF THE FRONTAL LOBES

CLINICAL FOCUS 12-2 AGENESIS OF THE FRONTAL LOBES

STIMULATING AND EXPRESSING EMOTION

AMYGDALA AND EMOTIONAL BEHAVIOR

PREFRONTAL CORTEX AND EMOTIONAL BEHAVIOR

EMOTIONAL DISORDERS

CLINICAL FOCUS 12-3 ANXIETY DISORDERS

12-5 CONTROL OF REGULATORY AND NONREGULATORY BEHAVIOR

CONTROLLING EATING

CLINICAL FOCUS 12-4 WEIGHT-LOSS STRATEGIES

CONTROLLING DRINKING

CONTROLLING SEXUAL BEHAVIOR

CLINICAL FOCUS 12-5 ANDROGEN-INSENSITIVITY SYNDROME AND THE ANDROGENITAL SYNDROME

SEXUAL ORIENTATION, SEXUAL IDENTITY, AND BRAIN ORGANIZATION

COGNITIVE INFLUENCES ON SEXUAL BEHAVIOR

12-6 REWARD

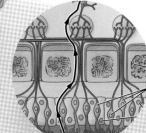

Pain of Rejection

We use words like *sorrow, grief,* and *heartbreak* to describe a loss. Loss evokes painful feelings, and the loss or absence of contact that comes with social rejection leads to "hurt" feelings. Several investigators have attempted to discover whether physically painful and emotionally hurtful feelings are manifested in the same neural regions.

Physical pain (e.g., hot or cold stimuli) is easy to inflict, but inducing equivalently severe emotional, or affective, pain is more difficult. Ethan Kross and colleagues (2011) performed an experiment that may have succeeded in balancing the degree of emotional and physical pain participants experienced.

Using fMRI, they scanned 40 participants who had recently experienced an unwanted breakup. The participants viewed two photographs in order to compare brain activation: a photograph of their ex-partner, to evoke negative emotion, or the photo of a same-gender friend with whom they shared a positive experience at about the time of the breakup. The picture order was randomized across subjects.

Emotional cue phrases associated with each photograph directed the participants to focus on a specific experience they shared with each person. The physical pain employed by the researchers, which was administered in a separate session, was either painfully hot or nonpainfully warm stimulations on participants' forearms.

The research question is, Does the experience and regulation of both physical and social pain have a common neuroanatomical basis? The results, shown in the illustration, reveal that four regions respond to both types of pain: the insula, dorsal anterior cingulate (limbic) cortex, somatosensory thalamus, and secondary somatosensory cortex. The conclusion: Social rejection hurts in the same way physical pain hurts.

Previous studies (see, for example, Eisenberger and colleagues, 2003; 2006) had shown anterior cingulate activity during the experience of both physical and emotional pain. But few results also showed

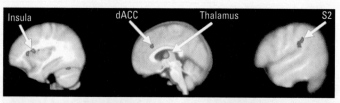
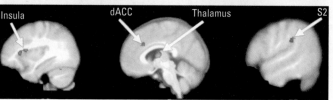

Social Rejection/Physical Pain Overlap These fMRIs result from averaging scans from 40 participants to image the brain's response to physical or emotional pain. We see activation in the insula, dorsal anterior cingulate cortex (dACC), somatosensory thalamus, or secondary somatosensory cortex (S2). Adapted from "Social Rejection Shares Somatosensory Representations with Physical Pain," by E. Kross, M. G. Berman, W. Mischel, E. E. Smith, and T. D. Water, 2011, *Proceedings of the National Academy of Sciences (USA), 108,* 6270–6275.

activity in cortical somatosensory regions related to physical pain. The results imaged by Kross and colleagues suggest that the brain systems underlying emotional reactions to social rejection may have developed by co-opting brain circuits that support the affective component of physical pain.

Another insight from this study is that normalizing the activity of these brain regions probably provides a basis for both physical and mental restorative processes. Seeing the similarity in brain activation during both social and physical pain helps us understand why social support can reduce physical pain, much as it soothes emotional pain.

Knowing that the brain makes emotional experience real—more than mere metaphors of "hurt" or "pain"—how do we incorporate our thoughts and reasons for behaving as we do? Clearly, our subjective feelings and thoughts influence our actions. The cognitive interpretations of subjective feelings are **emotions**—anger, fear, sadness, jealousy, embarrassment, joy. These feelings can operate outside our immediate awareness as well.

This chapter begins by exploring the causes of behavior. Sensory stimulation, neural circuits, hormones, and reward are of primary importance in explaining behavior. We focus both on emotion and on the underlying reasons for **motivation**—behavior that seems purposeful and goal-directed. Like emotion, motivated behavior is both inferred and subjective and can occur without awareness or intent. It includes both regulatory behaviors, such as eating, which are essential for survival, and nonregulatory behaviors, such as curiosity, which are not required to meet the basic needs of an animal.

Research on the neuroanatomy responsible for emotional and motivated behavior focuses on a neural circuit formed by the hypothalamus, the limbic system, and the frontal lobes. But behavior is influenced as much by the interaction of our social and natural environments and by evolution as it is by biology. To explain all this interaction in regard to how the brain controls behavior, we concentrate on the specific examples of feeding and sexual activity. Our exploration leads us to revisit the topic of reward, which plays a key role in explaining emotional and motivated behaviors.

12-1 Identifying the Causes of Behavior

emotion Cognitive interpretation of subjective feelings.

motivation Behavior that seems purposeful and goal-directed.

We may think that the most obvious explanation for why we behave as we do is simply that that we act in a state of free will: we do what we want to and we always have a choice. But free will is not a likely cause of behavior.

Consider Roger. We first met 25-year-old Roger in the admissions ward of a large mental hospital when Roger approached us and asked if we had any snacks. We had chewing gum, which he accepted eagerly. We thought little about this encounter until 10 minutes later when we noticed Roger eating the flowers from the stems in a vase on a table. A nurse took the vase away but said little to Roger.

Later, as we wandered about the ward, we encountered a worker replacing linoleum floor tiles. Roger was watching the worker, and as he did, he dipped his finger into the pot of gluing compound and licked the glue from his finger, as if he were sampling honey from a jar. When we asked Roger what he was doing, he said that he was really hungry and that this stuff was not too bad. It reminded him of peanut butter.

One of us tasted the glue and concluded not only that it did not taste like peanut butter but that it tasted awful. Roger was undeterred. We alerted a nurse, who quickly removed him from the glue. Later, we saw him eating flowers from another bouquet.

Neurological testing revealed that a tumor had invaded Roger's hypothalamus at the base of his brain. He was indeed hungry all the time and in all likelihood could consume more than 20,000 calories a day if allowed to do so.

Would you say that Roger had free will regarding his appetite and food preferences? Probably not. Roger seemed compelled to eat whatever he could find, driven by a ravenous hunger. The nervous system, not an act of free will, produced this behavior and undoubtedly produces many others.

If free will does not adequately explain why we act was we do, what does? One obvious answer is that we do things that are rewarding. Experiences that are rewarding must activate brain circuits that make us feel good. Consider the example of prey killing by domestic cats.

One frustrating thing about being a cat owner is that even well-fed cats kill birds—often lots of birds. Most people are not much bothered when their cats kill mice: they view mice as a nuisance. But birds are different. People enjoy watching birds in their yards and gardens. Many cat owners wonder why their pets keep killing birds.

To provide an answer, we can look to the activities of neural circuits. Cats must have a brain circuit that controls prey killing. When this circuit is active, a cat makes an appropriate kill. Viewed in an evolutionary context, it makes sense for cats to have such a circuit because, in the days when cats were not owned by doting humans, they did not have food dishes that were regularly filled.

Why does this prey-killing circuit become active when a cat does not need food? One explanation is that, to secure survival, the activity of circuits like the prey-killing circuit are in some way rewarding—they make the cat "feel good." As a result, the cat will engage in the pleasure-producing behavior often. This helps to guarantee that it will usually not go hungry.

Watching and sharing cute cat videos is a rewarding behavior for many people: it makes them feel good and they do it often.

In the wild, after all, a cat that did not like killing would probably be a dead cat. In the early 1960s, Steve Glickman and Bernard Schiff first proposed the idea that behaviors such as prey killing are rewarding. We return to reward in Section 12-6 because it is important to understanding the causes of behavior.

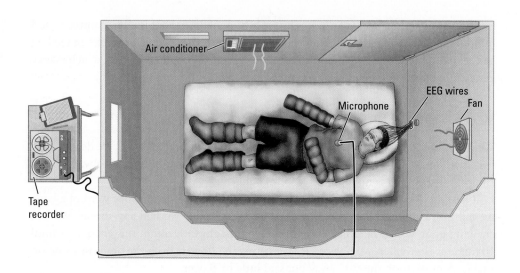

FIGURE 12-1 Sensory Deprivation
Experimenters record the EEG of a participant lying on a bed in a dimly lit environmental cubicle 24 hours a day, with time out only for meals and bathroom breaks. A translucent plastic visor restricts the participant's vision; a U-shaped pillow and the noise of a fan and air conditioner limit hearing. The sense of touch is restricted by cotton gloves and long cardboard cuffs. Adapted from "The Pathology of Boredom," by W. Heron, 1957, *Scientific American, 197*(4), p. 52.

FIGURE 12-2 Brain Maintenance Monkeys quickly learn to solve puzzles or perform other tricks to gain access to a door that looks out from their dimly lit quarters into an adjacent room. A toy train is a strong visual incentive for the monkey peeking through the door; a bowl of fruit is less rewarding. From "Persistence of Visual Exploration in Monkeys," by R. A. Butler and H. F. Harlow, 1954, *Journal of Comparative and Physiological Psychology, 47,* p. 260.

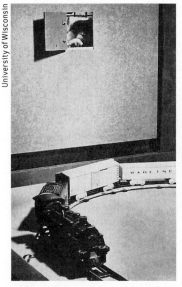

Behavior for Brain Maintenance

Some experiences are rewarding; others are aversive, again because of brain circuits. In this case, brain circuits are activated to produce behaviors that will reduce the aversive experience. One example is the brain's inherent need for stimulation.

Donald Hebb and his coworkers (Heron, 1957) studied the effects of **sensory deprivation,** depriving people of nearly all sensory input. They wanted to see how well-fed, physically comfortable college students who were paid handsomely for their time would react if they did nothing, saw nothing, and heard or touched very little 24 hours per day. **Figure 12-1** shows the setting for this experiment.

Each student lay on a bed in a small, soundproofed room with ears enveloped by a hollowed-out pillow that muffled the monotonous hums of a nearby fan and air conditioner. Cardboard tubes covered the hands and arms, cutting off the sense of touch, and a translucent visor covered the eyes, blurring the visual world. The participants were given food on request and access to bathroom facilities. Otherwise, they were asked simply to enjoy the peace and quiet. For doing so, they would receive $20 per day, which was about four times what a student could earn 60 years ago, even for a day's hard labor.

Wouldn't you think the participants would be quite happy to contribute to scientific knowledge in such a painless way? In fact, they were far from happy. Most were content for perhaps 4 to 8 hours; then they became increasingly distressed. They craved stimulation of almost any kind. In one version of the experiment, the participants could listen, on request, to a talk for 6-year-old children on the dangers of alcohol. Some of them asked to hear it 20 times a day. Few lasted more than 24 hours in these conditions.

What caused their distress? Why did they find sensory deprivation so aversive? The answer, Hebb and colleagues concluded, must be that the brain has an inherent need for stimulation.

Psychologists Robert Butler and Harry Harlow (1954) came to a similar conclusion through a series of experiments they conducted at about the same time Hebb conducted his sensory-deprivation studies. Butler placed rhesus monkeys in a dimly lit room with a small door that could be opened to view an adjoining room. As shown in **Figure 12-2,** the researchers could vary the stimuli in the adjoining room so that the monkeys could view different objects or animals each time they opened the door.

Monkeys in these conditions spent a lot of time opening the door and viewing whatever was on display, such as toy trains circling a track. The monkeys were even willing to perform various tasks just for an opportunity to look through the door. The longer they were deprived of a chance to look, the more time they spent looking when finally given the opportunity.

The Butler and Harlow experiments, together with Hebb and colleagues' research on sensory deprivation, show that in the absence of stimulation, the brain will seek it out.

Neural Circuits and Behavior

Researchers have identified brain circuits for reward and discovered that these circuits can modulate to increase or decrease activity. Researchers studying the rewarding properties of sexual activity in males, for example, found that a man's frequency of copulation correlates with his levels of **androgens** (male hormones). Unusually high androgen levels are related to very high sexual interest; abnormally low androgen levels are linked to low sexual interest or perhaps no interest at all. The brain circuits are still present but are more difficult to activate in the absence of androgens.

Another way to modulate reward circuits comes via our chemical senses, smell and taste. The odor of a mouse can stimulate hunting in cats, whereas the odor of a cat will move mice into hiding. Similarly, the smell from a bakery can make us hungry, whereas foul odors can reduce the rewarding value of our favorite foods. Although we tend to view the chemical senses as relatively minor in our daily lives, they play a central role in motivated and emotional behavior, as discussed next in Section 12-2.

The idea of a neural basis for understanding motivated behavior has wide application. For instance, we can say that Roger had a voracious and indiscriminate appetite either because the brain circuits that initiate eating were excessively active or because the circuits that terminate eating were inactive. Similarly, we can say that Hebb's participants were highly upset by sensory deprivation because the neural circuits that respond to sensory inputs were forced into abnormal underactivity. So the main reason why a particular thought, feeling, or action occurs lies in what is going on in brain circuits.

sensory deprivation Experimental setup in which a subject is allowed only restricted sensory input; subjects generally have a low tolerance for deprivation and may even display hallucinations.

androgen Male hormone related to level of sexual interest.

REVIEW 12-1
Identifying the Causes of Behavior

Before you continue, check your understanding.

1. One reason that cats kill when they may not be hungry is that the killing behavior is
   _____.

2. A reason animals get bored and seek new things to do is to maintain a(n) _____.

3. Neural circuits are strongly modulated by which senses?

4. Why is free will inadequate to explain why we do the things we do?

Answers appear at the back of the book.

12-2 Chemical Senses

Chemical reactions play a central role in nervous system activity, and *chemosignals* (chemical signals) play a central role in motivated and emotional behavior. Mammals identify group members by odor, mark their territories with urine and other odorants, identify favorite and forbidden foods by taste, and form associations among odors, tastes, and emotional events.

Olfaction

Olfaction is the most puzzling sensory system. We can discriminate thousands of odors, yet we have great difficulty finding words to describe what we smell. We may like or dislike smells or compare one smell to another, but we lack a vocabulary for olfactory perceptions.

Wine experts rely on olfaction to tell them about wines, but they must *learn* to use smell to do so. Training courses in wine sniffing typically run one full day per week for a year, and most course takers still have great difficulty passing the final test. The degree of difficulty contrasts with that of vision and audition, senses designed to analyze the specific qualities of the sensory input, such as pitch in audition or color in vision. In contrast, olfaction seems designed to discriminate whether information is safe or familiar—is the smell from an edible food? from a friend or from a stranger?—or identifies a signal, perhaps a receptive mate.

Receptors for Smell

Conceptually, identifying chemosignals is similar to identifying other sensory stimuli (light, sound, touch). But instead of converting physical energy such as light or sound waves into receptor potentials, scent interacts with chemical receptors. This constant chemical interaction appears to be tough on the receptors, so in contrast with the receptors for light, sound, and touch, chemical receptors are constantly being replaced. The life of an olfactory receptor is about 60 days.

The receptor surface for olfaction is the olfactory epithelium, which lies in the nasal cavity, as illustrated in **Figure 12-3.** The epithelium is composed of receptor cells and support cells. Each receptor cell sends a process that ends in 10 to 20 cilia into a mucous layer, the *olfactory mucosa*. Chemicals in the air we breathe dissolve in the mucosa to interact with the cilia. If the receptors are affected by an olfactory chemosignal, metabotropic activation of a specific G protein leads to an opening of sodium channels and a change in membrane potential.

The receptor surface of the epithelium varies widely across species. In humans, the area is estimated to range from 2 to 4 square centimeters, in dogs the area is about 18 square centimeters, and in cats about 21 square centimeters. No wonder our sensitivity to odors is less acute than that of dogs and cats: they have 10 times as much receptor area

Figure 5-15A illustrates the activity of a metabotropic receptor coupled to an ion channel.

FIGURE 12-3 Olfactory Epithelium

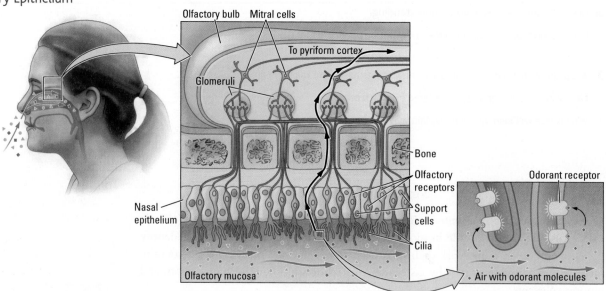

as humans have! Roughly analogous to the tuning characteristics of cells in the auditory system, olfactory receptor neurons in vertebrates do not respond to specific odors but rather to a range of odors.

How does a limited number of receptor types allow us to smell many different odors? The simplest explanation is that any given odorant stimulates a unique pattern of receptors, and the summed activity, or pattern of activity, produces our perception of a particular odor. Analogously, the visual system enables us to identify many different colors with only three receptor types in the retina: the summed activity of the three cones leads to our rich color life.

A fundamental difference, however, is that there are far more receptors in the olfactory system than in the visual system. Richard Axel and Linda Buck won the Nobel Prize in medicine in 2004 for their discovery that a novel gene family (about 350 genes in humans) encodes a huge and diverse set of olfactory receptors. The combination of these receptors allows us to discriminate about 10,000 different smells.

Olfactory Pathways

Olfactory receptor cells project to the olfactory bulb, ending in ball-like tufts of dendrites called glomeruli, shown in Figure 12-3. There they form synapses with the dendrites of mitral cells. The mitral cells send their axons from the olfactory bulb to the broad range of forebrain areas summarized in **Figure 12-4.** Many olfactory targets, such as the amygdala and pyriform cortex, have no connection *through* the thalamus, as do other sensory systems. However, a thalamic connection (to the dorsomedial nucleus) does project to the **orbitofrontal cortex** (OFC), the area of the prefrontal cortex located behind the eye sockets (the *orbits*) that receives projections from the dorsomedial nucleus of the thalamus. The OFC plays a central role in a variety of emotional and social behaviors as well as in eating.

Accessory Olfactory System

A unique class of odorants is **pheromones,** biochemicals released by one animal that act as chemosignals and can affect the physiology or behavior of another animal. For example, Karen Stern and Martha McClintock (1998) found that when women reside together, their estrous cycles begin to synchronize. Furthermore, the researchers found that the synchronization of menstrual cycles is conveyed by odors.

Pheromones appear able to affect more than sex-related behavior. A human chemosignal, androstadienone, has been shown to alter glucose utilization in the neocortex—that is, how the brain uses energy (Jacob et al., 2001). Thus, a chemosignal appears to affect cortical processes even though the signal was not actually detected consciously. The puzzle is why we would evolve such a mechanism and how it might actually affect cerebral functioning.

Pheromones are unique odors because they are detected by a special olfactory receptor system known as the *vomeronasal organ,* which is made up of a small group of sensory receptors connected by a duct to the nasal passage. The receptor cells in the vomeronasal organ send their axons to the accessory olfactory bulb, which lies adjacent to the main olfactory bulb. The vomeronasal organ connects primarily with the amygdala and hypothalamus by which it probably plays a role in reproductive and social behavior.

It is likely that the vomeronasal organ participates not in general olfactory behavior but rather in the analysis of pheromones such as those in urine. You may have seen bulls or cats engage in a behavior known as *flehmen,* illustrated in **Figure 12-5.** When exposed to novel urine from a cat or human, cats raise their upper lip to close off the nasal passages and suck air into the mouth. The air flows through the duct on the roof of the mouth en route to the vomeronasal organ.

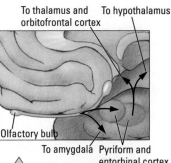
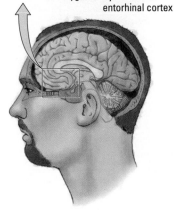

FIGURE 12-4 Olfactory Pathways

The phenomenon of synchronizing menstrual cycles is called the *Whitten effect.*

orbitofrontal cortex (OFC) Prefrontal cortex located behind the eye sockets (the *orbits*) that receives projections from the dorsomedial nucleus of the thalamus; plays a central role in a variety of emotional and social behaviors as well as in eating; also called *orbital frontal cortex.*

pheromone Odorant biochemical released by one animal that acts as a chemosignal and can affect the physiology or behavior of another animal.

FIGURE 12-5 Response to Pheromones
(*Left*) A cat sniffs a urine-soaked cotton ball, (*center*) raises its upper lip to close off the nasal passages, and (*right*) follows with the full gape response characteristic of flehmen, a behavior mediated by the accessory olfactory system.

Courtesy of Arthur Nonneman and Bryan Kolb

Human Olfactory Processing

One common misperception is that people have a miserable sense of smell relative to other mammals. Our threshold for detecting many smells is certainly inferior to that of our pet dogs, cats, and horses, but humans have a surprisingly acute sensitivity to smells that are behaviorally relevant.

Several studies show convincingly that people can identify their own odor, the odor of kin versus non-kin, and the odor of friends versus strangers with accuracy well above chance (e.g., Olsson et al., 2006). Johan Lundstrom and colleagues (2008) used PET scans to identify the neural networks that process human body odors. They made two surprising findings.

First, the brain analyzes common odors and body odors differently. Although both activate primary olfactory regions, body odors also activate structures that were not previously believed to be involved in olfactory processing, including the posterior cingulate cortex, occipital cortex, and anterior cingulate cortex—regions that are also activated by visually emotional stimuli. Considered in an evolutionary perspective, the ability to identify human odors probably is uniquely important to safety.

Second, smelling a stranger's odor activates the amygdala and insular cortex, similar to activation observed for fearful visual stimuli such as masked or fearful faces. The investigators also asked participants to rate the intensity and pleasantness of odors and found that the odors of strangers rated as stronger and less pleasant.

Lundstrom and colleagues conclude that processing body odors is mostly unconscious and represents an automatic process that matches odors to a learned "library" of smells. Similar unconscious processes seem to occur during visual and auditory information processing and to play an important role in our emotional reactions.

The multifunctional insula contains regions related to language, taste perception, and social cognition (see Figures 12-7 and 10-12B).

Gustation

Research reveals significant differences in taste preferences both between and within species. Humans and rats like sucrose and saccharin solutions, but dogs reject saccharin and cats are indifferent to both, inasmuch as they do not detect sweetness at all. The failure of cats to taste sweet may not be surprising: they are pure carnivores, and nothing that they normally eat is sweet.

Within the human species, clear differences in taste thresholds and preferences are obvious. An example is the preference for or dislike of bitter tastes—the flavor of brussels sprouts, for instance. People tend to either love them or hate them. Linda Bartoshuk (2000) showed absolute differences among adults: some perceive certain tastes as very bitter, whereas others are indifferent to them. Presumably, the latter group is more tolerant of brussels sprouts.

The sensitivity to bitterness is related to genetic differences in the ability to detect a specific bitter chemical (6-n-propylthiouracil, or PROP). PROP bitterness associates with allelic variation in the taste receptor gene, *TAS2R38*. People who are able to detect minute quantities find the taste extremely bitter and are sometimes referred to as *supertasters.* Those who do not taste PROP as very bitter are referred to as *nontasters.* The advantage of being a supertaster is that bitter foods are often poisonous. The disadvantage is that supertasters avoid many nutritious fruits and vegetables that they find bitter.

Valerie Duffy and her colleagues (2010) investigated sensitivity to quinine (normally perceived as bitter) in participants who were assessed for the *TAS2R38* genotype; the taste bud density of the participants was estimated by counting the number of papillae (the little bumps on the tongue). Quinine was reported as more bitter to those who tasted PROP as very bitter or to those who had more taste buds. Thus, bitterness is related both to *TAS2R38* and to tongue anatomy.

Nontasters, by either genotype or phenotype (for few taste buds), reported greater consumption of vegetables, regardless of whether they are typically thought to be bitter. Nontasters with higher numbers of taste buds reported eating about 25% more vegetables than the other groups. These data suggest that genetic variation in taste can explain differences in overall consumption of all types of vegetables. It also suggests that convincing people to eat more healthy foods—that is, vegetables—may prove difficult if they are supertasters.

Differences in taste thresholds also emerge as we age. Children are much more responsive than adults to taste and are often intolerant of spicy foods because they have more taste receptors than adults have. By age 20, humans have lost at least an estimated 50 percent of their taste receptors. No wonder children and adults have different food preferences.

Receptors for Taste

Taste receptors are found within taste buds located on the tongue, under the tongue, on the soft palate on the roof of the mouth, on the sides of the mouth, and at the back of the mouth on the nasopharynx. Each of the five different taste-receptor types responds to a different chemical component of food. The four most familiar are sweet, sour, salty, and bitter. The fifth type, called the *umami* receptor, is specially sensitive to glutamate, a neurotransmitter molecule, and perhaps to protein.

Taste receptors are grouped into taste buds, each containing several receptor types, as illustrated in **Figure 12-6.** Gustatory stimuli interact with the receptor tips, the *microvilli,* to open ion channels, leading to changes in membrane potential. The base of the taste bud is contacted by the branches of afferent nerves that come from cranial nerves 7 (facial nerve), 9 (glossopharyngeal nerve), or 10 (vagus nerve).

Gustatory Pathways

Cranial nerves 7, 9, and 10 form the main gustatory nerve, the *solitary tract.* On entering the brainstem, the tract divides in two, as illustrated in **Figure 12-7.** One route (traced in red in Figure 12-7) travels through the posterior medulla to the ventroposterior medial nucleus of the thalamus. This nucleus in turn sends out two pathways, one to the primary somatosensory cortex and the other to a region just rostral to the secondary somatosensory cortex in the gustatory cortex of the insula.

The gustatory region in the insula is dedicated to taste, whereas the primary somatosensory region is also responsive to tactile information and is probably responsible for localizing tastes on the tongue and for our reactions to a food's texture. The gustatory cortex sends a projection to the orbital cortex in a region near the input from the olfactory cortex. It is likely that the mixture of olfactory and gustatory input in the orbital cortex gives rise to our perception of flavor.

A meta-analysis of noninvasive imaging studies that demonstrate taste-responsive brain regions, conducted by Maria Veldhuizen and colleagues (2011), also shows a brain asymmetry. The investigators conclude that areas in the right orbital cortex mediate the pleasantness of tastes, whereas the same region in the left hemisphere mediates the

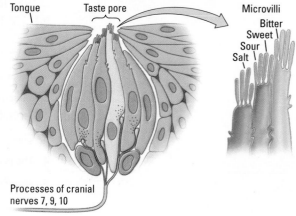

FIGURE 12-6 Anatomy of a Taste Bud
Adapted from "Chemical Senses: Taste and Olfaction," by D. V. Smith and G. M. Shepherd, 2003, in *Fundamental Neuroscience* (2nd ed., pp. 631–667), ed. L. R. Squire, F. E. Bloom, S. K. McConnell, J. L. Roberts, N. C. Spitzer, and M. J. Zigmond, New York: Academic Press.

Glutatmatelike substances described in Section 6-4 include the food additive monosodium glutamate (MSG) and domoic acid, a neurotoxin.

Figure 2-27 illustrates the cranial nerves.

FIGURE 12-7 Gustatory Pathways

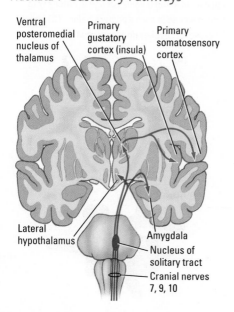

unpleasantness of tastes. Whereas the insula identifies the nature and intensity of flavors, the OFC evaluates the affective properties of tastes.

The second pathway from the gustatory nerve (shown in blue in Figure 12-7) projects through the pons to the hypothalamus and amygdala. Researchers hypothesize that these inputs play some role in feeding behavior, possibly evaluating the pleasantness and salience of flavors.

REVIEW 12-2
The Chemical Senses

Before you continue, check your understanding.

1. The receptor surface for olfaction is the _____.

2. Olfactory and gustatory pathways eventually merge in the orbitofrontal cortex, leading to the perception of _____.

3. Chemosignals that convey information about the sender are called _____.

4. The perception of bitter is related to both the _____ and the _____.

5. How does a limited number of receptor types allow us to smell many different odors?

Answers appear at the back of the book.

12-3 Evolution, Environment, and Behavior

Odor and taste play a fundamental role in the biology of emotional and motivated behavior. Why does the sight or smell of a bird or a mouse trigger stalking and killing in a cat? Why does the human body stimulate sexual interest? We can address such questions by investigating the evolutionary and environmental influences on brain-circuit activity that contribute to behavior.

Evolutionary Influences on Behavior

The evolutionary explanation hinges on the concept of **innate releasing mechanisms** (IRMs), activators for inborn, adaptive responses that aid an animal's survival. IRMs help an animal to successfully feed, reproduce, and escape predators. The concept is best understood by analyzing its parts.

IRMs are present from birth rather than acquired through experience, as the term *innate* implies. These mechanisms have proved adaptive and therefore have been maintained in the genome of the species. The term *releasing* indicates that IRMs act as triggers for behaviors set in motion by internal programs.

Let us return to prey killing by cats. The cat's brain must have a built-in mechanism that triggers appropriate stalking and killing in response to such stimuli as a bird or a mouse. A cat must also have a built-in mechanism that triggers mating behavior in the presence of a suitable cat of the opposite sex. Not all cat behaviors are due to IRMs, but you probably can think of other innate releasing mechanisms that cats possess—arching and hissing on encountering a threat, for example. For all these IRMs, the animal's brain must have a set of norms against which it can match stimuli to trigger an appropriate response.

The following experiment suggests the existence of such innate, internalized norms. One of us (B. K.) and Arthur Nonneman allowed a litter of 6-week-old kittens to play in a room and become familiar with it. After this adjustment period, we introduced a two-dimensional image of an adult cat in a "Halloween" posture, as shown in **Figure 12-8A**.

FIGURE 12-8 Innate Releasing Mechanism in Cats **(A)** Displaying the "Halloween" cat stimulates cats to respond defensively, with raised fur, arched backs, and bared teeth. This behavior appears at about 6 weeks of age in kittens who have never seen such a posture before. **(B)** The "Picasso" cat evokes no response at all.

(A)

(B)

FIGURE 12-9 Innate Releasing Mechanism in Humans Facial expressions made by young infants in response to expressions made by the experimenter. From "Discrimination and Imitation of Facial Expression by Neonates," by T. M. Field, R. Woodson, R. Greenberg, and D. Cohen, 1982, *Science, 218,* p. 180.

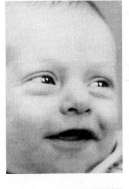

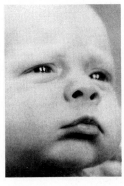

Courtesy of Dr. Tiffany M. Field

The kittens responded with raised fur, arched backs, and bared teeth, all signs of being threatened by the image of the adult. Some even hissed at the model. These kittens had no experience with any adult cat except their mother, and there was no reason to believe that she had ever shown them this behavior. Some sort of template of this posture must be prewired in the kitten brain. Seeing the model that matched this preexisting template automatically triggered a threat response. This innate trigger is an IRM.

The IRM concept also applies to humans. In one study, Tiffany Field and her colleagues (1982) had an adult display to young infants various exaggerated facial expressions, such as happiness, sadness, and surprise. As **Figure 12-9** shows, the babies responded with very much the same expressions the adults displayed. These newborns were too young to be imitating the adult faces intentionally. Rather, babies must innately match these facial expressions to internal templates, in turn triggering some prewired program to reproduce the expressions in their own faces. Such an IRM would have adaptive value if these facial expressions serve as important social signals for humans.

Evidence for a prewired motor program related to facial expressions also comes from the study of congenitally blind children, who spontaneously produce the very same facial expressions that sighted people do, even though they have never seen them in others. IRMs are prewired into the brain, but they can be modified by experience. Our cat Hunter's stalking skills were not inherited fully developed at birth but rather matured functionally as she grew older. The same is true of many human IRMs, such as those for responding to sexually arousing stimuli.

Different cultures may emphasize different stimuli as arousing. Even within a single culture, there is variation in what different people find sexually stimulating. Nonetheless, some human attributes are universally sexually arousing. For most human males, an example is the hip-to-waist ratio of human females. This ratio is probably part of an IRM.

The IRM concept can be related to the Darwinian view of nervous system evolution. Natural selection favors behaviors that prove adaptive for an organism, and these behaviors are passed on to future generations. Because behavior patterns are produced by the activity of neurons in the brain, the natural selection of specific behaviors is really the selection of particular brain circuits.

Animals that survive long enough to reproduce and have healthy offspring are more likely to pass on their brain circuit genes than are animals with traits that make them less likely to survive and successfully reproduce. Thus, cats with brain circuits that make them adept at stalking prey or responding fiercely to threats are more likely to survive and produce many offspring, passing on their adaptive brain circuits and behaviors to their young. In this way, the behaviors become widespread in the species over time.

Although the Darwinian view seems straightforward when considering how cats evolved brain circuits for stalking prey or responding to threats, it is less obvious when applied to many complex human behaviors. Why, for instance, have humans evolved the behavior of killing other humans? At first glance, it seems counterproductive to the survival of the human species. Why has it endured? For an answer, we turn to the field of **evolutionary psychology,** which applies principles of natural selection to explanations of human behavior.

Section 1-2 reviews Darwin's theory, the philosophy of materialism, and contemporary perspectives.

innate releasing mechanism (IRM) Hypothetical mechanism that detects specific sensory stimuli and directs an organism to take a particular action.

evolutionary psychology Discipline that seeks to apply principles of natural selection to understand the causes of human behavior.

Section 15-5 posits a genetically related explanation for the evolution of sex-related cognitive differences.

Evolutionary psychologists assume that any behavior, including homicide, exists because natural selection has favored the neural circuits that produce it. When two men fight a duel, one commonsense explanation might be that they are fighting over grievances. But evolutionary psychologists would instead ask, Why is a behavior pattern that risks people's lives sustained in a population? Their answer: Fights are about social status.

Men who fought and won duels passed on their genes to future generations. Through time, therefore, the traits associated with successful dueling—strength, aggression, agility—became more prevalent among humans, and so, too, did dueling. Martin Daly and Margot Wilson (1988) extended this evolutionary analysis to further account for homicide. In their view, homicide may endure in our society despite its severe punishment because it is related to behaviors that were adaptive in the human past.

Suppose that natural selection favored sexually jealous males who, to guarantee that they had fathered all offspring born of their mates, effectively intimidated their rivals and bullied their mates. Male jealousy would become a prevalent motive for interpersonal violence, including homicide. Homicide itself does not help a man produce more children. But men who are apt to commit homicide are more likely to engage in other behaviors (bullying and intimidation) that improve their social status and therefore their reproductive fitness. Homicide therefore is related to adaptive traits that have been selected through millennia.

Evolutionary theory cannot account for all human behavior, perhaps not even homicide. But evolutionary psychologists can generate intriguing hypotheses about how natural selection might have shaped the brain and behavior and provide an evolutionary perspective on the neurological bases of behavior.

Environmental Influences on Behavior

Many psychologists have emphasized learning as a cause of behavior. No one would question that we modify our behavior as we learn, but the behaviorist B. F. Skinner went much further. He believed that behaviors are selected by environmental factors.

Skinner's argument is simple. Certain events function as rewards, or **reinforcers,** and when a reinforcing event follows a particular response, similar responses are more likely to occur again. Skinner argued that reinforcement can be manipulated to encourage the display of complex behaviors.

The power of experience to shape behavior by pairing stimuli and rewards is typified by one of Skinner's experiments. A pigeon is placed in a box that has a small disc on one wall (the stimulus). If the pigeon pecks at the disc (the response), a food tray opens and the pigeon can feed (the reinforcement or reward). The pigeon quickly learns the association between the stimulus and the response, especially if the disc has a small spot on it. It pecks at the spot and, within minutes, it has mastered the response needed to receive a reward.

Now the response requirement can be made more complex. The pigeon might be required to turn 360 degrees before pecking the disc to gain the reward. The pigeon can learn this response too. Other contingencies might then be added, making the response requirements even more complex. For instance, the pigeon might be trained to turn in a clockwise circle if the disc is green, to turn in a counterclockwise circle if the disc is red, and to scratch at the floor if the disc is yellow.

If you suddenly came across this complex behavior in a pigeon, you would probably be astounded. But if you understood the experience that had shaped the bird's behavior, you would understand its cause. The rewards offered to the pigeon altered its behavior: its responses were controlled by the color of the disc on the wall.

Skinner extended behavioral analysis to include actions of all sorts—behaviors that at first do not appear easily explained. For instance, he argued that various phobias could be accounted for by understanding a person's reinforcement history. Someone who

Skinner box

once was terrified by a turbulent plane ride thereafter avoids air travel and manifests a phobia of flying. The avoidance of flying is rewarding because it lowers the person's anxiety level, so the phobic behavior is maintained.

Skinner also argued against the commonly held view that much of human behavior is under our own control. From Skinner's perspective, free will is only an illusion, because behavior is controlled not by the organism but rather by the environment, through experience. But what is the experience actually doing? Increasing evidence suggests that epigenetic changes regulate changes in memory circuits. Skinner was not studying the brain directly, but it is becoming clear that epigenetics supports his perspective. We learn many complex behaviors through changes in memory-related genes that act to modify neural circuits (see the review by Day and Sweatt, 2011).

The environment does not always effect change in the brain. A case in point can be seen again in pigeons. A pigeon in a Skinner box can quickly learn to peck a disc to receive a bit of food, but it cannot learn to peck a disc to escape from a mild electric shock to its feet. Why not? Although the same simple pecking behavior is being rewarded, apparently the pigeon's brain is not prewired for this second kind of association. The bird is prepared genetically to make the first association, for food, but not prepared for the second, which makes adaptive sense. Typically, it flies away from noxious situations.

The specific nature of the behavior–consequence associations that animals are able to learn was first shown in 1966 by psychologist John Garcia. He observed that farmers in the western United States are constantly shooting at coyotes for attacking lambs, yet, despite the painful consequences, the coyotes never seem to learn to stop killing lambs in favor of safer prey. The reason, Garcia speculated, is that a coyote's brain is not prewired to make this kind of association.

So Garcia proposed an alternative to deter coyotes from killing lambs—one that uses an association that a coyote's brain is prepared to make: the connection between eating something that makes one sick and avoiding that food in the future. Garcia gave the coyotes a poisoned lamb carcass, which made them sick but did not kill them. With only one pairing of lamb and illness, most coyotes learned not to eat sheep for the rest of their lives.

Many humans have similarly acquired food aversions because the taste of a certain food—especially a novel one—was subsequently paired with illness. This **learned taste aversion** is acquired even when the food that was eaten is in fact unrelated to the later illness. As long as the taste and the nausea are paired in time, the brain is prewired to make a connection between them.

One of us ate his first Caesar salad the night before coming down with a stomach flu. A year later, he was offered another Caesar salad and, to his amazement, felt ill just at the smell of it. Even though his earlier illness had not been due to the salad, he had formed an association between the novel flavor and the illness. This strong and rapid *associative learning* makes adaptive sense. Having a brain that is prepared to make a connection between a novel taste and subsequent illness helps an animal avoid poisonous foods and so aids in its survival. A curious aspect of taste-aversion learning is that we are not even aware of having formed the association until we encounter the taste and/or smell again.

The fact that the nervous system is often prewired to make certain associations but not to make others has led to the concept of **preparedness** in learning theories. Preparedness can help account for some complex behaviors. For example, if two rats are paired in a small box and exposed to a mild electric shock, they will immediately fight with one another, even though neither was responsible for the shock. Apparently, the rat brain is prepared to associate injury with nearby objects or other animals. The extent to which we might extend this idea to explain such human behaviors as bigotry and racism is an interesting topic to ponder.

Section 14-4 reports that epigenetic mechanisms mediate synaptic plasticity, especially in learning and memory.

Our brains are wired to link unrelated stimuli together, as explained in Section 14-4.

reinforcer In operant conditioning, any event that strengthens the behavior it follows.

learned taste aversion Acquired association between a specific taste or odor and illness; leads to an aversion to foods that have the taste or odor.

preparedness Predisposition to respond to certain stimuli differently from other stimuli.

Inferring Purpose in Behavior: To Know a Fly

A pitfall in studying the causes of behavior is the tendency to assume that behavior is intentional. The problems in inferring purpose from an organism's actions are illustrated in a wonderful little book, *To Know a Fly,* by Vincent Dethier.

When a fly lands on a kitchen table, it wanders about, occasionally stomping its feet. Eventually, it finds a bit of food and sticks its proboscis (a trunklike extension) into the food and eats. The fly may then walk to a nearby place and begin to groom by rubbing its legs together quickly. Finally, it spends a long period motionless.

If you observed a fly engaged in these behaviors, it might appear to have been initially searching for food because it was hungry. When it found food, you might assume that it gorged itself until it was satisfied, and then it cleaned up and rested. In short, the fly's behavior might seem to you to have some purpose or intention.

Dethier studied flies for years to understand what a fly is actually doing when it engages in these behaviors. His findings have little to do with purpose or intention. When a fly wanders about a table, it is not deliberately searching. It is tasting what it walks on.

As **Figure 12-10** shows, a fly's taste receptors are on its feet. Tasting is automatic when a fly walks. An adult fly's nervous system has a built-in preference for sweet tastes and aversions to sour, salty, or bitter flavors. Therefore, when a fly encounters something sweet, it automatically lowers its proboscis and eats—or drinks if the sweet is liquid. The sweeter the food, the more a fly will consume. (Sweet foods attract us humans too, as Focus 12-4 reports.)

Why does a fly stop eating? A logical possibility is that its blood-sugar level rises to some threshold. If this were correct, injecting glucose into the circulatory system of a fly would prevent the fly from eating. But that does not happen. Blood-glucose level has no effect on a fly's feeding. Furthermore, injecting food into the animal's stomach or intestine has no effect either. So what is left?

Flies have a nerve (the recurrent nerve) that extends from the neck to the brain and carries information about whether any food is present in the esophagus. If the recurrent nerve is cut, the fly is chronically "hungry" and never stops eating. Such flies become so full and fat that their feet no longer reach the ground, and they become so heavy that they cannot fly.

Even though a fly appears to act with a purpose in mind, a series of very simple mechanisms actually control its behavior—mechanisms not remotely related to our concept of thought or intent. Hunger is simply the activity of the nerve. Clearly, we should not assume simply from appearances that a behavior carries intent. Behavior can have very subtle causes that do not include conscious purpose. How do we know that any behavior is purposeful? That question turns out to be difficult to answer.

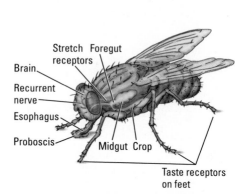

FIGURE 12-10 Feeding System of the Fly After sampling by taste buds on the fly's feet, food is taken in through the proboscis and passes through the esophagus to the gut. Stretch receptors at the entrance to the gut determine when the esophagus is full. The recurrent nerve alerts the brain to signal cessation of eating.

REVIEW 12-3
Evolution, Environment, and Behavior

Before you continue, check your understanding.

1. Skinner argued that behaviors could be shaped by _____ in the environment.

2. John Garcia used the phenomenon of _____ to discourage coyotes from killing lambs.

3. The brain of a species is prewired to produce _____ to specific sensory stimuli selected by evolution to prompt associations between certain environmental events.

4. When a fly wanders around on a table, it is not exploring so much as _____.

5. Explain briefly how the concept of preparedness accounts for puzzling human behaviors.

Answers appear at the back of the book.

12-4 Neuroanatomy of Motivated and Emotional Behavior

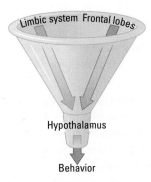

The neural circuits that control behavior encompass regions at all levels of the brain, but the critical neural structures in emotional and motivated behavior are the hypothalamus and associated pituitary gland, the limbic system, and the frontal lobes. The expression of emotions includes physiological changes, in heart rate, blood pressure, and hormone secretions. It also includes motor responses, especially movements of the facial muscles that produce facial expressions (see Figure 12-9). So much of human life revolves around emotions, that understanding them is central to understanding our humanness.

But emotions are not restricted to people. A horse that is expecting to get alfalfa for dinner will turn its nose up at grass hay and may stomp its front feet and toss its head around. Two dogs that are in competition for attention may snap at one another. Charles Darwin interpreted these types of behaviors as emotions in his classic book *The Expression of the Emotions in Man and Animals,* published in 1872. We now know that the expression of emotion in all mammals is related to activity in the limbic system and frontal lobes.

Although the hypothalamus plays a central role in controlling motivated behavior, it takes its instructions from the limbic system and the frontal lobes. The limbic and frontal regions project to the hypothalamus, which houses many basic neural circuits for controlling behavior and for autonomic processes that maintain critical body functions within a narrow, fixed range—that is, **homeostatic mechanisms.** In **Figure 12-11,** the hypothalamus is represented by the neck of a funnel, and the limbic system and the frontal lobes form the funnel's rim. To produce behavior, the hypothalamus sends axons to other brainstem circuits.

FIGURE 12-11 Funneling Signals In this model, inputs from the frontal lobes and limbic system funnel through the hypothalamus, which sends its axons to control brainstem circuits that produce motivated behaviors.

Section 6-5 explores homeostatic mechanisms and the hormones that regulate them.

Regulatory and Nonregulatory Behavior

We seek mates, food, or sensory stimulation because of brain activity, but it is convenient to talk about such behavior as being "motivated." Motivated behaviors are not something that we can point to in the brain, however. Rather, motivations are inferences that we make about why someone, ourselves included, engages in a particular behavior. The two general classes of motivated behaviors are regulatory and nonregulatory. In this section we explore both categories before exploring the neuroanatomy of motivation and emotion.

Regulatory Behaviors

Regulatory behaviors—behaviors motivated by an organism's survival—are controlled by homeostatic mechanisms. By analogy, consider a house where the thermostat is set at 18 degrees Celsius. When the temperature falls below a certain tolerable range (say, to 16 degrees Celsius), the thermostat turns the furnace on. When the temperature rises above a certain tolerable level (say, 20 degrees Celsius), the thermostat turns on the air conditioner.

Human body temperature is controlled in a somewhat similar manner by a "thermostat" in the hypothalamus that holds internal temperature at about 37 degrees Celsius, a temperature referred to as *setpoint.* Even slight variations cause us to engage in various behaviors to regain the setpoint. For example, when body temperature drops slightly, neural circuits that increase body temperature turn on. These neural circuits might induce an involuntary response such as shivering or a seemingly voluntary behavior such as moving closer to a heat source. Conversely, if body temperature rises slightly, we sweat or move to a cooler place.

homeostatic mechanism Process that maintains critical body functions within a narrow, fixed range.

regulatory behavior Behavior motivated to meet the survival needs of the animal.

nonregulatory behavior Behavior unnecessay to the basic survival needs of the animal.

pituitary gland Endocrine gland attached to the bottom of the hypothalamus; its secretions control the activities of many other endocrine glands; known to be associated with biological rhythms.

medial forebrain bundle (MFB) Tract that connects structures in the brainstem with various parts of the limbic system; forms the activating projections that run from the brainstem to the basal ganglia and frontal cortex.

Categories of Motivated Behavior

Some Regulatory Behaviors
Internal body temperature
Eating and drinking
Salt consumption
Waste elimination
Some Nonregulatory Behaviors
Sex
Parenting
Aggression
Food preference
Curiosity
Reading

Figure 2-30 diagrams ANS pathways and connections.

Similar mechanisms control many other homeostatic processes, including the amount of water in the body, the balance of dietary nutrients, and the level of blood sugar. Control of many of these homeostatic systems is quite complex, requiring both neural and hormonal mechanisms. However, in some way, all the body's homeostatic systems include the activity of the hypothalamus.

Imagine that specific cells are especially sensitive to temperature. When they are cool, they become very active; when they are warm, they become less active. These cells could function as a thermostat, telling the body when it is too cool or too warm. A similar set of cells could serve as a "glucostat," controlling the level of sugar in the blood, or as a "waterstat," controlling the amount of H_2O in the body. In fact, the body's real homeostatic mechanisms are slightly more complex than this imagined one, but they work on the same general principle.

Mechanisms to hold conditions such as temperature constant have evolved because the body, including the brain, is a chemical "soup" in which thousands of reactions are taking place all the time. Maintaining constant temperature becomes critical. When temperature changes, even by 2 degrees Celsius, the rates at which chemical reactions take place change.

Such changes might be tolerable, within certain limits, if all the reaction times changed to the same extent. But they do not. Consequently, an increase of 2 degrees might increase one reaction by 10 percent and another by only 2 percent. Such uneven changes would wreak havoc with finely tuned body processes such as metabolism and the workings of neurons.

A similar logic applies to maintaining homeostasis in other body systems. For instance, cells require certain concentrations of water, salt, and glucose to function properly. If the concentrations were to fluctuate wildly, they would cause a gross disturbance of metabolic balance and a subsequent biological disaster.

Nonregulatory Behaviors

In contrast with regulatory behaviors, such as eating or drinking, **nonregulatory behaviors** are neither required to meet the basic survival needs of an animal nor controlled by homeostatic mechanisms. Thus, nonregulatory behaviors include everything else we do—from sexual intercourse to parenting to such curiosity-driven activities as conducting psychology experiments.

Some nonregulatory behaviors, such as sexual intercourse, entail the hypothalamus, but most of them probably do not. Rather, such behaviors entail a variety of forebrain structures, especially the frontal lobes. Presumably, as the forebrain evolved and enlarged, so did our range of nonregulatory behaviors.

Most nonregulatory behaviors are strongly influenced by external stimuli. As a result, sensory systems must play some role in controlling them. For example, the sexual behavior of most male mammals is strongly influenced by the pheromone emitted by receptive females. If the olfactory system is not functioning properly, we can expect abnormalities in sexual behavior. We will return to the topic of sexual behavior in Section 12-5, where we investigate it as an example of how a nonregulatory behavior is controlled. But first we explore the brain structures that take part in motivated behaviors—both nonregulatory and regulatory.

Regulatory Function of the Hypothalamic Circuit

The hypothalamus maintains homeostasis by acting on both the endocrine system and the autonomic nervous system (ANS) to regulate our internal environment. The hypothalamus also influences the behaviors selected by the rest of the brain, especially by the limbic system. Although it constitutes less than 1 percent of the human brain's volume, the hypothalamus controls an amazing variety of motivated behaviors, ranging from heart rate to feeding and sexual activity.

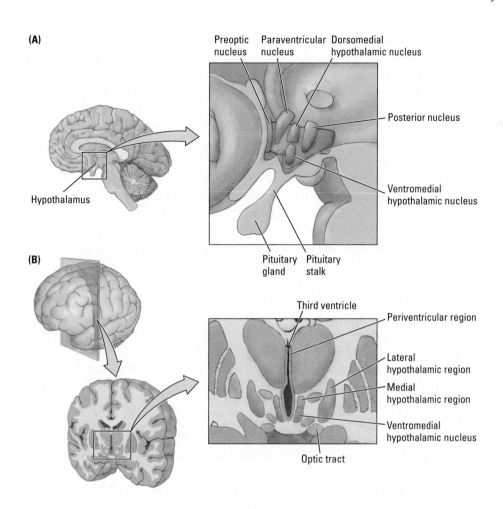

(A)

Preoptic nucleus
Paraventricular nucleus
Dorsomedial hypothalamic nucleus
Posterior nucleus
Ventromedial hypothalamic nucleus
Hypothalamus
Pituitary gland
Pituitary stalk

(B)

Third ventricle
Periventricular region
Lateral hypothalamic region
Medial hypothalamic region
Ventromedial hypothalamic nucleus
Optic tract

FIGURE 12-12 Nuclei and Regions of the Hypothalamus (**A**) Medial view shows the relation between the hypothalamic nuclei and the rest of the brain. (**B**) Frontal view shows the relative positions of the hypothalamus, thalamus, and in the midline between the left and right hemispheres, the third ventricle. Note the three principal hypothalmic regions: periventricular, lateral, and medial.

Hypothalamic Involvement in Hormone Secretions

A principal function of the hypothalamus is to control the **pituitary gland,** which is attached to it by a stalk (**Figure 12-12A**). Figure 12-12B diagrams the anatomic location of the hypothalamus in each hemisphere, with the thalamus above and the optic tracts just lateral.

The hypothalamus can be divided into three regions, lateral, medial, and periventricular, illustrated in frontal view in Figure 12-12B. The lateral hypothalamus is composed both of nuclei and of nerve tracts running up and down the brain, connecting the lower brainstem to the forebrain. The principal tract, shown in **Figure 12-13,** is the **medial forebrain bundle** (MFB).

The MFB connects structures in the brainstem with various parts of the limbic system and forms the activating projections that run from the brainstem to the basal ganglia and frontal cortex. Fibers that ascend from the dopamine- and noradrenaline-containing cells of the lower brainstem form a significant part of the MFB. The dopamine-containing fibers of the MFB contribute to the control of many motivated behaviors, including eating and sex. They also contribute to pathological behaviors, such as addiction and impulsivity.

Each hypothalamic nucleus is anatomically distinct, but most have multiple functions, in part because the cells in different nuclei contain various peptide neurotransmitters. Each peptide plays a role in different behaviors. For instance, transmitters in the cells in

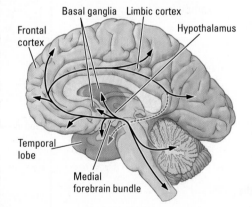

Basal ganglia
Limbic cortex
Frontal cortex
Hypothalamus
Temporal lobe
Medial forebrain bundle

FIGURE 12-13 Medial Forebrain Bundle The activating projections that run from the brainstem to the basal ganglia and frontal cortex are major components of the MFB, a primary pathway for fibers connecting various parts of the limbic system with the brainstem.

Section 6-4 elaborates on dopamine's importance in rewarding experiences related to drug use.

Section 5-2 reviews the structure and functions of peptide neurotransmitters.

releasing hormones Peptides that are released by the hypothalamus and act to increase or decrease the release of hormones from the anterior pituitary.

the paraventricular nucleus may be vasopressin, oxytocin, or various combinations of other peptides (such as enkephalin and neurotensin). When peptide neurotransmitters act, we may experience a range of feelings such as well-being (endorphins) or attachment (oxytocin and vasopressin). For example, oxytocin is released during intimate moments such as hugging or sex.

The production of various neuropeptides hints at the special relation between the hypothalamus and the pituitary. The pituitary consists of distinct anterior and posterior glands, as shown in **Figure 12-14.** The posterior pituitary is composed of neural tissue and is essentially a continuation of the hypothalamus.

Neurons in the hypothalamus make peptides (e.g., oxytocin and vasopressin) that are transported down their axons to terminals lying in the posterior pituitary. If these neurons become active, they send action potentials to the terminals, causing them to release the peptides stored there. But rather than affecting another neuron, as occurs at most synapses, these peptides are picked up by capillaries (tiny blood vessels) in the posterior pituitary's rich vascular bed.

The peptides then enter the body's bloodstream. The blood carries them to distant targets, where they exert their effects. Vasopressin, for example, affects water resorption by the kidneys, and oxytocin controls both uterine contractions and the ejection of milk by mammary glands in the breasts. Peptides can have multiple functions, depending on where their receptors are located. Thus, oxytocin not only controls milk ejection in females but also plays a more general role in several forms of affiliative behavior, including parental care, grooming, and sexual behavior in both men and women (Insel & Fernald, 2004).

The glandular tissue of the anterior pituitary synthesizes various hormones. The major hormones and their functions are listed in **Table 12-1.** The hypothalamus controls the release of these anterior pituitary hormones by producing chemicals known as **releasing hormones.** Produced by hypothalamic cell bodies, releasing hormones

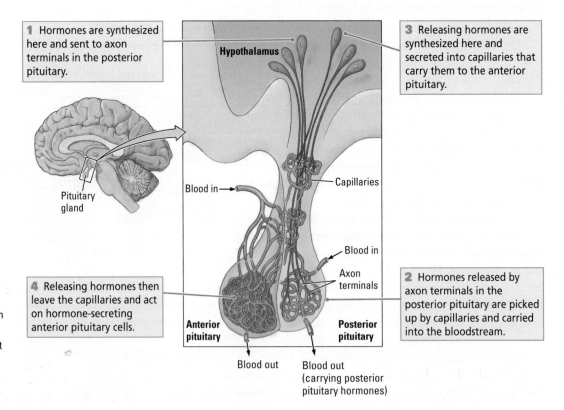

FIGURE 12-14 Hypothalamus and Pituitary Gland The anterior pituitary is connected to the hypothalamus by a system of blood vessels that carry hormones from the hypothalamus to the pituitary. The posterior pituitary receives input from axons of hypothalamic neurons. Both regions respond to hypothalamic input by producing hormones that travel in the bloodstream to stimulate target organs.

TABLE 12-1 Major Hormones Produced by the Anterior Pituitary

Hormone	Function
Adrenocorticotrophic hormone (ACTH)	Controls secretions of the adrenal cortex
Thyroid-stimulating hormone (TSH)	Controls secretions of the thyroid gland
Follicle-stimulating hormone (FSH)	Controls secretions of the gonads
Luteinizing hormone (LH)	Controls secretions of the gonads
Prolactin	Controls secretions of the mammary glands
Growth hormone (GH)	Promotes growth throughout the body

are secreted into capillaries that transport them to the anterior pituitary, as Figure 12-14 shows.

A releasing hormone can either stimulate or inhibit the release of an anterior pituitary hormone. For example, the hormone prolactin is produced by the anterior pituitary, but its release is controlled by a prolactin-releasing factor and a prolactin release–inhibiting factor, both synthesized in the hypothalamus. The release of hormones by the anterior pituitary in turn provides a means by which the brain can control what is taking place in many other parts of the body. Three factors control hypothalamic hormone-related activity: feedback loops, neural regulation, and responses based on experience.

FEEDBACK LOOPS When the level of, say, thyroid hormone is low, the hypothalamus releases thyroid-stimulating hormone–releasing hormone (TSH–releasing hormone) that stimulates the anterior pituitary to release TSH. TSH then acts on the thyroid gland to secrete more thyroid hormone.

There must be some control over how much hormone is secreted, and the hypothalamus has receptors to detect the level of thyroid hormone. When that level rises, the hypothalamus lessens its secretion of TSH–releasing hormone. This type of system is essentially a form of homeostatic control that works as a feedback mechanism, a system in which a neural or hormonal loop regulates the activity of neurons, initiating the neural activity or hormone release, as illustrated in **Figure 12-15**A.

The hypothalamus initiates a cascade of events that result in the secretion of hormones, but it pays attention to how much hormone is released. When a certain level is

Thyroid gland

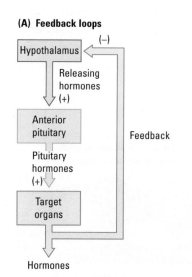

(A) Feedback loops

(B) Milk-letdown response

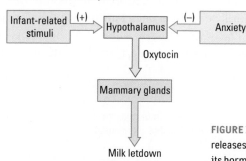

Milk letdown

FIGURE 12-15 Hypothalamic Controls **(A)** The hypothalamus releases hormones that stimulate the anterior pituitary to release its hormones that stimulate target organs such as the thyroid and adrenal gland to release their hormones. Those hormones act, in turn, to influence the hypothalamus to decrease its secretion of the releasing hormone. **(B)** Oxytocin release from the hypothalamus, which stimulates the mammary glands to release milk, is enhanced by infant-related stimuli and inhibited by maternal anxiety.

reached, it stops its hormone-stimulating signals. Thus, the feedback mechanism in the hypothalamus maintains a fairly constant circulating level of certain hormones.

NEURAL CONTROL A second control over hormone-related activities of the hypothalamus requires regulation by other brain structures, such as the limbic system and the frontal lobes. Figure 12-15B diagrams this type of control in relation to the effects of oxytocin released from the paraventricular nucleus of the hypothalamus. As stated earlier, one function of oxytocin is to stimulate cells of the mammary glands to release milk. As an infant suckles the breast, the tactile stimulation causes hypothalamic cells to release oxytocin, which stimulates milk letdown. In this way, the oxytocin cells participate in a fairly simple reflex that is both neural and hormonal.

Other stimuli also can influence oxytocin release, however, which is where control by other brain structures comes in. For example, the sight, sound, or even thought of her baby can trigger a lactating mother to eject milk. Conversely, as diagrammed in Figure 12-15B, feelings of anxiety in a lactating woman can inhibit milk ejection. These excitatory and inhibitory influences exerted by cognitive activity imply that the cortex can influence neurons in the paraventricular region. It is likely that projections from the frontal lobes to the hypothalamus perform this role.

EXPERIENTIAL RESPONSES A third control on the hormone-related activities of the hypothalamus is the brain's responses to experience. In response to experience, neurons in the hypothalamus undergo structural and biochemical changes just as cells in other brain regions do. In other words, hypothalamic neurons are like neurons elsewhere in the brain in that they can be changed by prolonged demands placed on them.

Such changes in hypothalamic neurons can affect the output of hormones. For instance, when a woman is lactating, the cells producing oxytocin increase in size to promote oxytocin release to meet the increasing demands of a growing infant for more milk. Through this control, which is mediated by experience, a mother provides her baby with sufficient milk over time.

Hypothalamic Involvement in Generating Behavior

Not only does the hypothalamus control hormone systems but it also has a central role in generating behavior. This function was first demonstrated by studies in which stimulating electrodes were placed in the hypothalami of various animals, ranging from chickens to rats and cats. When a small electric current was delivered through a wire electrode, an animal suddenly engaged in some complex behavior. The behaviors included eating and drinking, digging, and displaying fear, attack, predatory, or reproductive behavior. The particular behavior depended on which of many sites in the hypothalamus was stimulated. All the behaviors were smooth, well integrated, and indistinguishable from normally occurring ones. Furthermore, all were goal directed.

The onset and termination of the behaviors depended entirely on the hypothalamic stimulation. For example, if an electrode in a certain location elicited feeding behavior,

Recall the principle of neuroplasticity from Section 2-6: the details of nervous system functioning are constantly changing.

FIGURE 12-16 Generating Behavior
When rats receive electrical stimulation to the hypothalamus, they produce goal-directed behaviors. This rat is stimulated to dig when and only when the electricity is turned on. Note also that, if the sawdust is removed (not shown in the drawing at right), there is no digging.

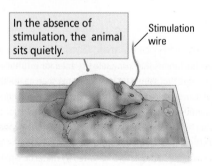

In the absence of stimulation, the animal sits quietly.

Stimulation wire

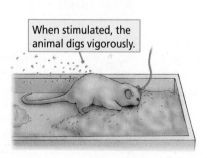

When stimulated, the animal digs vigorously.

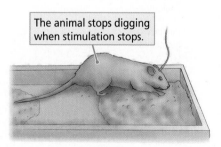

The animal stops digging when stimulation stops.

the animal ate as soon as the stimulation was turned on and continued to eat until the stimulation was turned off. If the food was removed, however, the animal would neither eat nor engage in other behaviors such as drinking. Recall that Roger, profiled in Section 12-1, ate continuously if foodlike materials were present, corresponding to the continuous hypothalamic activity caused by a tumor.

Figure 12-16 illustrates the effect of stimulation at a site that elicits digging. When no current is delivered, the animal sits quietly. When the current is turned on, the animal digs into the sawdust vigorously; when the current is turned off, the animal stops digging. If the sawdust is removed, there also is no digging.

Two more important characteristics of the behaviors generated by hypothalamic stimulation are related to (1) survival and (2) reward. Animals apparently find the stimulation of these behaviors pleasant, as suggested by the fact that they willingly expend effort, such as pressing a bar, to trigger the stimulation. Recall that cats kill birds and mice because the act of stalking and killing prey is rewarding to them. Similarly, we can hypothesize that animals eat because eating is rewarding, drink because drinking is rewarding, and mate because mating is rewarding.

Organizing Function of the Limbic Circuit

We now turn our attention to parts of the brain that interact with the hypothalamus in generating motivated and emotional behaviors. These brain structures evolved as a ring around the brainstem in early amphibians and reptiles. Nearly 150 years ago, Paul Broca was impressed by this evolutionary development and called these structures the "limbic lobe."

Known collectively as the *limbic system* today, these structures are actually a primitive cortex. In mammals, the limbic cortex encompasses the cingulate gyrus and the hippocampal formation, as shown in Figure 12-17. The hippocampal formation includes the **hippocampus**—a cortical structure that plays a role in species-specific behaviors, memory, and spatial navigation and is vulnerable to the effects of stress—and the *parahippocampal cortex* adjacent to it.

Organization of the Limbic Circuit

As anatomists began to study the limbic-lobe structures, connections to the hypothalamus became evident. It also became apparent that the limbic lobe has a role in emotion. For instance, in the 1930s, James Papez observed that people with rabies display radically abnormal emotional behavior, and postmortems showed that the rabies had selectively attacked the hippocampus.

hippocampus Distinctive, three-layered subcortical structure of the limbic system lying in the medial region of the temporal lobe; plays a role in species-specific behaviors, memory, and spatial navigation and is vulnerable to the effects of stress; named for the Greek word for seahorse.

The limbic cortex derives its name from the Latin word *limbus*, meaning "border" or "hem."

Figure 6-23 diagrams the vicious circle created when increased stress hormone levels destroy hippocampal neurons.

The definitive proof of rabies is still postmortem examination of the hippocampus.

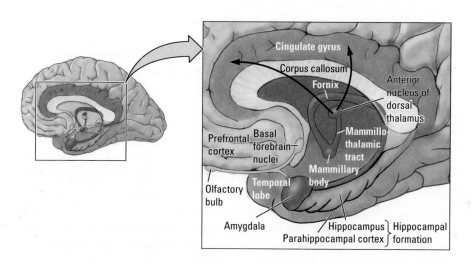

FIGURE 12-17 Limbic Lobe Encircling the brainstem, the limbic lobe as described by Broca consists of the cingulate gyrus and hippocampal formation (the hippocampus and parahippocampal cortex), the amygdala, the mammillothalamic tract, and the anterior thalamus.

(A)

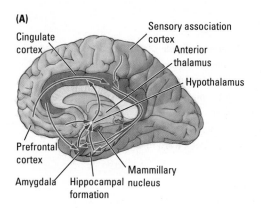

Cingulate cortex

Sensory association cortex

Anterior thalamus

Hypothalamus

Prefrontal cortex

Amygdala

Hippocampal formation

Mammillary nucleus

(C)

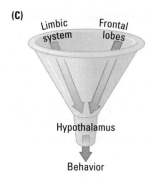

Limbic system

Frontal lobes

Hypothalamus

Behavior

FIGURE 12-18 Limbic System **(A)** In this contemporary conception of the limbic system, an interconnected network of structures, the Papez circuit, controls emotional expression. **(B)** A schematic representation, coded to brain areas shown in part A by color, charts the major connections of the limbic system. **(C)** A reminder that parts A and B can be conceptualized as a funnel of outputs through the hypothalamus

(B)

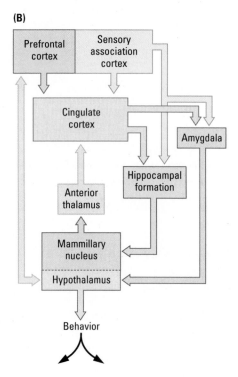

Prefrontal cortex

Sensory association cortex

Cingulate cortex

Amygdala

Hippocampal formation

Anterior thalamus

Mammillary nucleus

Hypothalamus

Behavior

Section 15-3 elaborates on multisensory integration and the binding problem.

Papez concluded from his observations that the limbic lobe and associated subcortical structures provide the neural basis of emotion. He proposed a circuit, traced in **Figure 12-18A**, now known as the *Papez circuit,* whereby emotion could reach consciousness, presumed at that time to reside in the cerebral cortex. In 1949, Paul MacLean expanded Papez's limbic-circuit concept to include the amygdala and prefrontal cortex. Figures 12-17 and 12-18A show the amygdala lying adjacent to the hippocampus in the temporal lobe, with the prefrontal cortex lying just anterior.

Figure 12-18B charts the limbic circuit schematically. The hippocampus, amygdala, and prefrontal cortex all connect with the hypothalamus. The mammillary nucleus of the hypothalamus connects to the anterior thalamus, which in turn connects with the cingulate cortex that then completes the circuit by connecting with the hippocampal formation, amygdala, and prefrontal cortex. This anatomical arrangement can be compared to the funnel in Figure 12-18C, which shows the hypothalamus as the spout leading to motivated and emotional behavior.

There is now little doubt that most structures of the limbic system, especially the amygdala and hypothalamus, take part in emotional behaviors, as detailed later in this section. But most limbic structures are now known to play an important role in various motivated behaviors as well, especially in motivating species-typical behaviors such as feeding and sexual activity. The critical structures for such motivated behaviors, as well as for emotion, are the hypothalamus, which we have already considered, and the amygdala, to which we turn now.

Amygdala

Named for the Greek word for "almond" because of its shape, the **amygdala** consists of three principal subdivisions: the corticomedial area, the basolateral area, and the central area. Like the hypothalamus, the amygdala receives inputs from all sensory systems. But, in contrast with the neurons of the hypothalamus, those of the amygdala require more complex stimuli to be excited.

In addition, many amygdala neurons are *multimodal:* they respond to more than one sensory modality. In fact, some neurons in the amygdala respond to the entire sensory array: sight, sound, touch, taste, and smell stimuli. These cells must create a rather complex image of the sensory world.

The amygdala sends connections primarily to the hypothalamus and the brainstem, where it influences neural activity associated with emotions and species-typical behavior. For example, when the amygdalae of epileptic patients are electrically stimulated before brain surgery, the patients become fearful and anxious. We observed a woman who responded with increased respiration and heart rate, saying that she felt as if something bad was going to happen, although she could not specify what.

Amygdala stimulation can also induce eating and drinking. We observed a man who drank water every time the stimulation was turned on. (There happened to be a pitcher of water on the table next to him.) Within 20 minutes, he had consumed about 2 liters of water. When asked if he was thirsty, he said, "No, not really. I just feel like drinking."

The amygdala's role in eating can be seen in patients with lesions in the amygdala. These patients, like Roger as a result of his tumor, are often much less discriminating in their food choices, eating foods that were formerly unpalatable to them. Lesions of the amygdala may also give rise to hypersexuality.

Executive Function of the Frontal Lobes

The amygdala is intimately connected with the functioning of the frontal lobes that constitute all cortical tissue anterior to the central sulcus. This large area is made up of several functionally distinct regions mapped in **Figure 12-19A**.

The motor cortex controls fine movements, especially of the fingers, hands, toes, feet, tongue, and face. The premotor cortex participates in the selection of appropriate movement sequences. For instance, a resting dog may get up in response to its owner's call, which serves as an environmental cue for a series of movements processed by one region of the premotor cortex. Or a dog may get up for no apparent reason and wander about the yard, a sequence of actions in response to an internal cue, this time processed by a different region of the premotor cortex.

Prefrontal Anatomy and Connections

As shown in Figure 12-19A, the **prefrontal cortex** (PFC) is anterior to the premotor cortex. The PFC plays a key role in controlling executive functions such as planning movements. Its four primary areas are the dorsolateral region; the orbitofrontal cortex, also shown from a ventral aspect in Figure 12-19B along with the ventromedial PFC; and the anterior cingulate cortex (ACC), shown in Figure 12-19C along with the PFC's medial regions.

The prefrontal cortex plays a role in specifying the goals toward which movement should be directed. It controls the processes by which we select movements that are appropriate for the particular time and context. This selection may be cued by internal information (such as memory and emotion) or it may be made in response to context (environmental information).

Like the amygdala, the frontal lobes receive highly processed information from all sensory areas, and many neurons in the prefrontal cortex, like those in the amygdala, are

amygdala Almond-shaped collection of nuclei located within the limbic system; plays a role in emotional and species-typical behaviors.

prefrontal cortex (PFC) The large frontal-lobe area anterior to the motor and premotor cortex; plays a key role in controlling executive functions such as planning.

Figure 11-2 charts the hierarchy of the frontal-lobe regions with regard to movement.

Prefrontal literally means "in front of the front."

Our focus here is the PFC's role in motivation and emotion. In Section 15-2 we turn to the PFC and cognition.

FIGURE 12-19 Gross Subdivisions of the Frontal Lobe and Prefrontal Cortex

(A) Lateral view

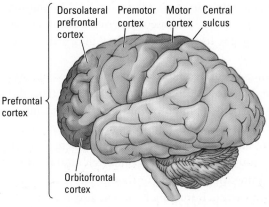

Dorsolateral prefrontal cortex • Premotor cortex • Motor cortex • Central sulcus

Prefrontal cortex

Orbitofrontal cortex

(B) Ventral view

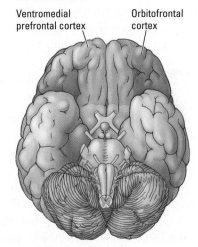

Ventromedial prefrontal cortex • Orbitofrontal cortex

(C) Medial view

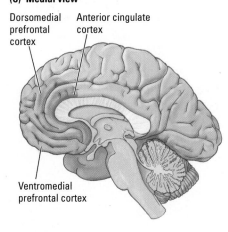

Dorsomedial prefrontal cortex • Anterior cingulate cortex

Ventromedial prefrontal cortex

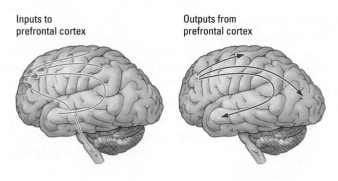

Inputs to prefrontal cortex

Outputs from prefrontal cortex

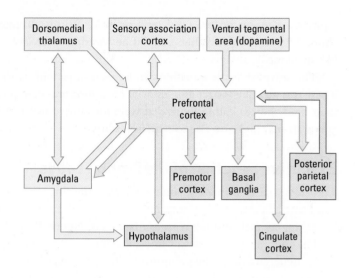

FIGURE 12-20 Prefrontal Connections The prefrontal cortex receives inputs from all sensory systems, the amygdala, the dorsal medial thalamus, the somatosensory cortex, and the dopamine-rich cells of the ventral tegmentum. The prefrontal cortex sends connections to the amygdala, premotor cortex, basal ganglia, posterior parietal cortex, hypothalamus, and cingulate cortex.

Section 16-4 elaborates on the causes of and treatments for schizophrenia.

These connections influence movement (Section 11-1), memory (Section 14-3) , and cognition (Section 15-2).

multimodal. As shown in **Figure 12-20,** the prefrontal cortex receives connections from the amygdala, the dorsomedial thalamus, the sensory association cortex, the posterior parietal cortex, and the dopaminergic cells of the ventral tegmental area.

The dopaminergic input plays an important role in regulating how prefrontal neurons react to stimuli, including emotional ones. Abnormalities in this dopaminergic projection may account for some disorders, including schizophrenia, in which people evince little emotional reaction to normally arousing stimuli.

Figure 12-20 also shows the areas to which the prefrontal cortex sends connections. The inferior prefrontal region projects axons to the amygdala and the hypothalamus in particular. These axons provide a route for influencing the ANS, which controls changes in blood pressure, respiration, and other internal processes. The dorsolateral prefrontal region sends its connections primarily to the sensory association cortex, the posterior parietal cortex, the cingulate cortex, the basal ganglia, and the premotor cortex.

Prefrontal Functions

The prefrontal cortex takes part in selecting behaviors appropriate to the particular time and place. Selection may be cued by internal information or made in response to the environmental context. Disruption to this selection function can be seen in people with injury to the dorsolateral frontal lobe. They become overly dependent on environmental cues to determine their behavior.

Like small children, they can be easily distracted by what they see or hear. We have all experienced this kind of loss of concentration to some extent, but for a frontal-lobe patient, the problem is exaggerated and persistent. Because the person becomes so absorbed in irrelevant stimuli, he or she is unable to act on internalized information most of the time.

A good example is J. C., whose bilateral damage to the dorsolateral prefrontal cortex resulted from having a tumor removed. J. C. would lie in bed most of the day fixated on television programs. He was aware of his wife's opinion of this behavior, but only the opening of the garage door when she returned home from work in the evening would stimulate him into action. Getting out of bed was controlled by this specific environmental cue; without it, he seemed to lack motivation. Television completely distracted him from acting on internal knowledge of things that he could or should do.

Adapting behavior appropriately to the environmental context also is a function of the prefrontal cortex. Most people readily change their behavior to match the situation at hand. We behave one way with our parents, another with our friends, another with our children, and yet another with our coworkers. Each set of people creates a different context, and we shift our behaviors accordingly. Our tone of voice, our use of slang or profanity, and the content of our conversations are quite different in different contexts.

Even among our peers we act differently, depending on who is present. We may be relaxed in the presence of some people and ill at ease with others. It is therefore no accident that the size of the frontal lobes is related to a species' sociability. Social behavior is extremely rich in contextual information, and humans are highly social.

Controlling behavior in context requires detailed sensory information, which is conveyed from all the sensory regions to the frontal lobes. This sensory input includes not only information from the external world but also internal information from the ANS. People with damage to the orbital prefrontal cortex, which is common in traumatic brain injuries, have difficulty adapting their behavior to the context, especially the social context. Consequently, they often make social gaffes.

In summary, the role of the frontal lobes in selecting behaviors is important in considering what causes behavior. The frontal lobes act much like a composer, but instead of selecting notes and instruments, they select our actions. Not surprisingly, the frontal lobes are sometimes described as housing the brain's executive functions. To grasp the full extent of frontal-lobe control of behavior, see Clinical Focus 12-2, "Agenesis of the Frontal Lobe."

Focus 1-1 and Section 1-2 recount some behavioral effects of brain trauma, Section 14-5 details recovery from TBI, and Section 16-3 explores its symptoms and treatments.

Section 15-2 considers the role of the frontal lobe in the executive function of planning.

CLINICAL FOCUS ✛ 12-2

Agenesis of the Frontal Lobes

The role of the frontal lobes in motivated behavior is perhaps best understood by looking at J. P.'s case, described in detail by Stafford Ackerly (1964). J. P., who was born in December 1914, was a problem child. Early on, he developed the habit of wandering. Policemen would find him miles from home, as he had no fear of being lost. Severe whippings by his father did not deter him.

J. P.'s behavioral problems continued and expanded as he grew older, and by adolescence, he was constantly in trouble. Yet J. P. also had a good side. When he started school, his first-grade teacher was so impressed with his polite manners that she began writing a letter to his parents to compliment them on having such a well-mannered child who was such a good influence in the class.

As she composed the letter, she looked up to find J. P. exposing himself to the class and masturbating. This juxtaposition of polite manners and odd behavior characterized J. P.'s conduct throughout his life. At one moment he was charming; at the next he was engaged in socially unacceptable behavior.

J. P. developed no close friendships with people of either sex, in large part because of his repeated incidents of public masturbation, stealing, excessive boastfulness, and wandering. He was a person of normal intelligence who seemed unaffected by the consequences of his behavior. Police officers, teachers, and neighbors all felt that he was willfully behaving in an asocial manner and blamed his parents for not enforcing strict enough discipline.

Perhaps as a result, not until he was 19 years old was J. P.'s true condition detected. To prevent him from serving a prison term for repeated automobile theft, a lawyer suggested that J. P. undergo psychiatric evaluation. He was examined by a psychiatrist, who ordered an X-ray (the only brain scan available at the time). The image revealed that J. P. lacked a right frontal lobe. Furthermore, his left frontal lobe was about 50 percent of normal size. It is almost certain that he simply never developed frontal lobes.

The failure of a structure to develop is known as *agenesis;* J. P.'s condition was agenesis of the frontal lobes. His case offers an unusual opportunity to study the role of the frontal lobes in motivated behavior.

Clearly, J. P. lacked the "bag of mental tricks" that most people use to come to terms with the world. Normally, behavior is affected both by its past consequences and by current environmental input. J. P. did not seem much influenced by either factor. As a result, the world was simply too much for him. He always acted childlike and was unable to formulate plans for the future or to inhibit many of his behaviors. He acted on impulse. At home, he was prone to aggressive outbursts about small matters, especially with regard to his mother.

Curiously, J. P. seemed completely unaware of his life situation. Even though the rest of his brain was working fairly well—his IQ was normal and his language skills were very good—the functional parts of his brain were unable to compensate for the absence of the frontal lobes.

somatic marker hypothesis Posits that "marker" signals arising from emotions and feelings act to guide behavior and decision making, usually in an unconscious process.

Klüver-Bucy syndrome Behavioral syndrome, characterized especially by hypersexuality, that results from bilateral injury to the temporal lobe.

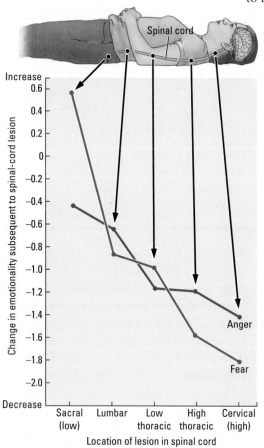

FIGURE 12-21 Losing Emotion Spinal-cord injury blunts the experience of emotion. Loss of emotionality is greatest when the lesion is high on the spine. Adapted from *Principles of Behavioral Neuroscience* (p. 339), by J. Beatty, 1995, Dubuque, IA: Brown & Benchmark.

Christopher Reeve's spinal cord was severed at the cervical level (high), as described in Section 11-1. Although Reeve's emotions may have been blunted, his motivation clearly remained intact.

Stimulating and Expressing Emotion

Emotion, like motivation, is intangible: it is an inferred state. But the importance of emotion to our everyday lives is hard to underestimate. Emotion can motivate us. It inspires artistic expression, for example, from poetry to filmmaking to painting. Many people enjoy the arts simply because they evoke emotions. And while people find certain emotions pleasant, severe and prolonged negative emotions, especially anxiety and depression, can cause clinical disorders.

To explore the neural control of emotions, we must first specify the types of behavior we want to explain. Think of any significant emotional experience you've had recently. Perhaps you had a serious disagreement with a close friend. Maybe you just got engaged to be married.

A common characteristic of such experiences includes autonomic responses such as rapid breathing, sweating, and dry mouth. Emotions may also entail strong subjective feelings that we often label as anger, fear, or love. Finally, emotions typically entail thoughts or plans related to the experience itself and may take the form of replaying conversations and events in your mind, anticipating what you might say or do under similar circumstances in the future or planning your married life.

These three forms of emotional experience suggest the influence of different neural systems. The autonomic component must include the hypothalamus and associated structures. The components of the feelings are more difficult to localize but clearly include the amygdala and probably parts of the frontal lobes. And the thoughts are likely to be cortical.

What is the relation between our cognitive experience of an emotion and the physiological changes associated with it? One view is that the physiological changes (such as trembling and rapid heartbeat) come first, and the brain then interprets these changes as an emotion of some kind. This perspective implies that the brain (most likely the cortex) creates a cognitive response to autonomic information.

That response varies with the context in which the autonomic arousal occurs. If we are frightened by a movie, we experience a weaker, more short-lived emotion than if we are frightened by a real-life encounter with a gang of muggers. Variations of this perspective have gone by many names, beginning with the *James-Lange theory*, named for its originators, but all assume that the brain concocts a story to explain bodily reactions.

Two lines of evidence support the James-Lange theory and similar points of view. One is that the same autonomic responses can accompany different emotions. That is, particular emotions are not tied to their own unique autonomic changes. This line of evidence leaves room for interpreting what a particular pattern of arousal means, even though particular physiological changes may suggest only a limited range of possibilities. The physiological changes experienced during fear and happiness are unlikely to be confused with one another.

The second line of evidence supporting the view that physiological changes are the starting point for emotions comes from people with reduced information about their own autonomic arousal, owing to spinal-cord injury, for example. These people suffer a decrease in perceived emotion, and its severity depends on how much sensory input they have lost. **Figure 12-21** illustrates this relation. People with the greatest loss of sensory input, which occurs with injuries at the uppermost end of the spinal cord, also have the greatest loss of emotional intensity. In contrast, people with low spinal injuries retain most of their visceral input and have essentially normal emotional reactions.

Antonio Damasio (1999) emphasized an important additional aspect of the link between emotional and cognitive factors in his **somatic marker hypothesis.** When Damasio studied patients with frontal-lobe injuries, he was struck by how they could be highly rational in analyzing the world yet still make decidedly irrational social and personal decisions. The explanation, he argued, is that the neural machinery that underlies emotion no longer affects the reasoning of people with frontal-lobe injury, either consciously or unconsciously. Cut off from critical emotional input, many social and personal decisions suffer.

To account for these observations, Damasio proposed that emotions are responses induced by either internal or external stimuli not normally attended to consciously. For example, if you encounter a bear as you walk down the street (presuming that you live in a place where this event could take place) the stimulus is processed rapidly without conscious appraisal. In other words, a sensory representation of the bear in the visual cortex is transmitted directly to brain structures, such as the amygdala, that initiate an emotional response.

This emotional response includes actions on structures in the forebrain and brainstem and ultimately on the ANS. The amygdala has connections to the frontal lobes, so the emotional response can influence the frontal lobes' appraisal. But if the frontal lobes are injured, the emotional information is excluded from cognitive processing, so the quality of emotion-related appraisals suffers, and the response to the bear might be inappropriate.

To summarize, Damasio's somatic marker hypothesis proposes how emotions are normally linked to a person's thoughts, decisions, and actions. In a typical emotional state, certain regions of the brain send messages to many other brain areas and to most of the rest of the body through hormones and the ANS. These messages produce a global change in the organism's state, and the altered state influences behavior, often unconsciously.

The account of a grizzly bear attack in Section 11-4 confirms the primacy of emotion—fear in this case—over other factors, including pain.

Amygdala and Emotional Behavior

In addition to controlling certain species-typical behaviors described earlier, the amygdala influences emotion (Davis et al., 2003). Its role can be seen most clearly in monkeys whose amygdalae have been removed. In 1939, Heinrich Klüver and Paul Bucy reported an extraordinary result, now known as the **Klüver-Bucy syndrome,** that followed the removal of the amygdalae and anterior temporal cortices of monkeys. The principal symptoms included the following:

1. Tameness and loss of fear

2. Indiscriminate dietary behavior (eating many types of formerly rejected foods)

3. Greatly increased autoerotic, homosexual, and heterosexual activity with inappropriate object choice (e.g., the sexual mounting of chairs)

4. Tendency to attend to and react to every visual stimulus

5. Tendency to examine all objects by mouth

6. Visual agnosia, an inability to recognize objects or drawings of objects

Visual agnosia results from damage to the ventral visual stream in the temporal lobe, but the other symptoms are related to the amygdala damage. Tameness and loss of fear after amygdalectomy is especially striking. Monkeys that normally show a strong aversion to stimuli such as snakes show no fear of them whatsoever. In fact, amygdalectomized monkeys may pick up live snakes and even put them in their mouths.

Although the Klüver-Bucy syndrome is not common in humans—because bilateral temporal lobectomies are rare—its symptoms can be seen in people with certain forms of encephalitis, a brain infection. In some cases, encephalitis centered on the base of the brain can damage both temporal lobes and produce many Klüver-Bucy symptoms, including especially indiscriminate sexual behavior and the tendency to examine objects by mouth.

Section 9-4 describes several varieties of visual form agnosia, including a range of case studies.

Focus 2-2 examines some causes and symptoms of encephalitis.

Exploiting fear has proved an especially effective technique for controlling group behavior throughout human history.

The amygdala's role in Klüver-Bucy syndrome points to its central role in emotion. So does its electrical stimulation, which produces an autonomic response (such as increased blood pressure and arousal) as well as a feeling of fear. Fear produced by the brain in the absence of an obvious threat may seem odd, but fear is basic to a species' survival. To improve their chances of surviving, most organisms using fear as a stimulus minimize their contact with dangerous animals, objects, and places and maximize their contact with safe things.

Awareness of danger and of safety has both an innate and a learned component, as Joe LeDoux (1996) emphasized. The innate component, much as in the IRMs described in Section 12-3, is the automatic processing of species-relevant sensory information—inputs from the visual, auditory, and olfactory systems. The importance of olfactory inputs is not obvious to us humans. Our senses are dominated by vision. But olfactory information connects directly to the amygdala in the human brain (see Figure 12-17). For other animals, olfactory cues often predominate.

A rat that has never encountered a ferret thus shows an immediate fear response to the odor of ferret. Other novel odors (such as peppermint or coffee) do not produce an innate fear reaction. The innate response triggers in the rat an autonomic activation that stimulates conscious awareness of danger.

In contrast, the learned component of fear consists of the avoidance of specific animals, places, and objects that the organism has come to associate with danger. The organism is not born with this avoidance behavior prewired. In a similar way, animals learn to increase contact with environmental stimuli that they associate with positive outcomes, such as food or sexual activity or, in the laboratory, drugs. Damage to the amygdala interferes with all these behaviors. The animal loses not only its innate fears but also its acquired fears and preferences for certain environmental stimuli.

To summarize, a species' survival requires the amygdala. It influences autonomic and hormonal responses through its connections to the hypothalamus. It influences our conscious awareness of the positive and negative consequences of events and objects through its connections to the prefrontal cortex.

Prefrontal Cortex and Emotional Behavior

At about the same time that Klüver and Bucy began studying their monkeys, Carlyle Jacobsen was studying the effects of frontal lobotomy on the cognitive capacities of two chimpanzees. A frontal lobotomy destroys a substantial amount of brain tissue as the result of inserting a sharp instrument into the frontal lobes and moving it back and forth.

In 1936, Jacobsen reported that one of the chimps that had been particularly neurotic before being subjected to this procedure became more relaxed after it. Incredibly, a leading Portuguese neurologist of the time, Egas Moniz, seized on this observation as a treatment for behavioral disorders in humans, and the frontal lobotomy, illustrated in **Figure 12-22,** was initiated as the first technique of **psychosurgery,** that is, neurosurgery intended to alter behavior.

The use of psychosurgery grew rapidly in the 1950s. In North America alone, nearly 40,000 people received frontal lobotomies as a treatment for psychiatric disorders. Not until the 1960s was any systematic research conducted into the effects of frontal lesions on social and emotional behavior. By this time, the frontal lobotomy had virtually vanished as a "treatment." We now know that prefrontal lesions in various species, including humans, severely affect social and emotional behavior.

Agnes is a case in point. We met Agnes at the psychiatric hospital where we met Roger. (We described Roger's indiscriminate eating in Section 12-1.) At the time, Agnes, a 57-year-old woman, was visiting one of the nurses. She had, however, once been a patient.

The first thing we noticed about Agnes was that she exhibited no outward sign of emotion. She showed virtually no facial expression. Agnes had been subjected to a

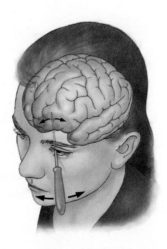

FIGURE 12-22 Transorbital Leukotomy In this procedure, a leukotome is inserted through the bone of the eye socket and the orbitofrontal cortex is disconnected from the rest of the brain.

procedure known as a *frontal leukotomy* because her husband, an oil tycoon, felt that she was too gregarious. Evidently, he felt that her "loose lips" were a detriment to his business dealings. He convinced two psychiatrists that she would benefit from psychosurgery, and her life was changed forever.

To perform a leukotomy, as illustrated in Figure 12-22, a surgeon uses a special knife called a leukotome to sever the connections of a region of the orbitofrontal cortex (see Figure 12-19). In our conversations with Agnes, we quickly discovered that she had considerable insight into the changes brought about by the leukotomy. In particular, she indicated that she no longer had any feelings about things or most people, although, curiously, she was attached to her dog. She said that she often just felt empty and much like a zombie.

Agnes's only moment of real happiness in the 30 years since her operation was the sudden death of her husband, whom she blamed for ruining her life. Unfortunately, Agnes had squandered her dead husband's considerable wealth as a consequence of her inability to plan or organize. This inability, we have seen, is another symptom of prefrontal injury.

The orbitofrontal area has direct connections with the amygdala and hypothalamus. Its stimulation can produce autonomic responses, and, as we saw in Agnes, damage to the orbital region can produce severe personality change characterized by apathy and loss of initiative or drive. The orbital cortex is probably responsible for the conscious awareness of emotional states produced by the rest of the limbic system, especially the amygdala.

Agnes's loss of facial expression is also typical of frontal-lobe damage. In fact, people with frontal-lobe injuries and people who suffer from schizophrenia or autism spectrum disorder are usually impaired at both producing and perceiving facial expressions, including a wide range of expressions found in all human cultures—happiness, sadness, fear, anger, disgust, and surprise. As with J. P.'s frontal-lobe agenesis, described in Clinical Focus 12-2, it is difficult to imagine how such people can function effectively in our highly social world without being able to emote or to recognize the emotions of others.

Although facial expression is a key to recognizing emotion, so is tone of voice, or *prosody.* Frontal-lobe patients are devoid of prosody, both in their own conversations and in understanding the prosody of others. The lost ability to comprehend or produce emotional expression in both faces and language partly explains the apathy of frontal-lobe patients. In some ways, they are similar to spinal-cord patients who have lost autonomic feedback and so can no longer feel the arousal associated with emotion. Frontal-lobe patients can no longer either read emotion in other people's faces and voices or experience it in their own.

Some psychologists have proposed that our own facial expressions provide us with important clues to the emotions we are feeling. This idea has been demonstrated in experiments reviewed by Pamela Adelmann and Robert Zajonc (1989). In one such study, people were required to contract their facial muscles by following instructions about which parts of the face to move. Unbeknown to the participants, the movements produced happy and angry expressions. Afterward, they viewed a series of slides and reported how the slides made them feel.

They said that they felt happier when they were inadvertently making a happy face and angrier when making the angry face. Frontal-lobe patients presumably have no such feedback from their own facial expressions, which could contribute to their emotional experiences being dampened.

Emotional Disorders

Major depression, a highly disruptive emotional disorder, is characterized by prolonged feelings of worthlessness and guilt, the disruption of normal eating habits, sleep disturbances, a general slowing of behavior, and frequent thoughts of suicide. A depressed person feels

psychosurgery Any neurosurgical technique intended to alter behavior.

ASD is the topic of Focus 8-2, and schizophrenia of Focus 8-5.

Of all psychological disorders, major depression, detailed in Focus 6-3, is one of the most treatable. As Section 16-4 notes, cognitive and intrapersonal therapies are as effective as drugs.

severely despondent for a prolonged time. Major depression is common in our modern world, with a prevalence of about 6 percent of the population at any given time.

Depression has a genetic component. It not only runs in families but also frequently tends to be found in both members of a pair of identical twins. The genetic component in depression implies a biological abnormality, but the cause remains

---

CLINICAL FOCUS ✛ 12-3

Anxiety Disorders

Animals normally become anxious at times, especially when they are in obvious danger. But anxiety disorders are different. They are characterized by intense feelings of fear or anxiety that are inappropriate for the circumstances.

People with anxiety disorders have persistent and unrealistic worries about impending misfortune. They also tend to suffer multiple physical symptoms attributable to hyperactivity of the sympathetic nervous system.

G. B.'s case is a good example. He was a 36-year-old man with two college degrees who began to experience severe spells initially diagnosed as some type of heart condition. He would begin to breathe heavily, sweat, experience heart palpitations, and sometimes suffer pains in his chest and arms. During these attacks, he was unable to communicate coherently and would lie helplessly on the floor until an ambulance arrived to take him to an emergency room.

Extensive medical testing and multiple attacks in a period of about 2 years eventually led to the diagnosis of **generalized anxiety disorder.** Like most of the 5 percent of the U.S. population who suffer an anxiety disorder at some point in their lives, G. B. was unaware that he was overly anxious.

The cause of generalized anxiety is difficult to determine, but one likely explanation is related to the cumulative effect of general stress. Although G. B. appeared outwardly calm most of the time, he had been a prodemocracy activist in communist Poland, a dangerous position to adopt.

Because of the dangers, he and his family eventually had to escape from Poland to Turkey, and from there they went to Canada. G. B. may have had continuing worries about the repercussions of his political activities—worries (and stress) that eventually found expression in generalized anxiety attacks.

The most common and least disabling type of anxiety disorders are **phobias.** A phobia pertains to a clearly defined dreaded object (such as spiders or snakes) or some greatly feared situation (such as enclosed spaces or crowds). Most people have mild aversions to some types of stimuli. This kind of aversion becomes a phobia only when a person's feelings toward a disliked stimulus lead to overwhelming fear and anxiety.

The incidence of disabling phobias is surprisingly high: phobias are estimated to affect at least one in ten people. For most people with a phobia, the emotional reaction can be controlled by avoiding what they

Up to 90 percent of people with animal phobias overcome their fears in a single exposure therapy session that lasts 2 or 3 hours.

dread. Others face their fears in controlled settings, with the goal of overcoming them.

Panic disorder has an estimated incidence on the order of 3 percent of the population. The symptoms of panic disorder include recurrent attacks of intense terror that come on without warning and without any apparent relation to external circumstances. Panic attacks usually last only a few minutes, but the experience is always terrifying. Sudden activation of the sympathetic nervous system leads to sweating, a wildly beating heart, and trembling.

Although panic attacks may occur only occasionally, the victim's dread of another episode may be continual. Consequently, many people with panic disorders also experience *agoraphobia,* a fear of public places or situations in which help might not be available. This phobia makes some sense because a person with a panic disorder may feel particularly vulnerable to having an attack in a public place.

Freud believed that anxiety disorders are psychological in origin and treatable with talking therapies in which people confront their fears. Today, cognitive-behavioral therapies are used for this purpose, as shown in the accompanying photo, but anxiety disorders are known to have a clear biological link.

Pharmacologically, anxiety disorders are most effectively treated with benzodiazepines, of which diazepam (Valium) is the best known. Alprazolam (Xanax) is the most commonly prescribed drug for panic attacks. Benzodiazepines act by augmenting GABA's inhibitory effect and are believed to exert a major influence on neurons in the amygdala.

Whether treatments are behavioral, pharmacological, or both, the general goal is to normalize brain activity in the limbic system.

unknown. However, neuroscience researchers' interest in the role of epigenetic changes in depression is increasing. One hypothesis is that early life stress may produce epigenetic changes in the prefrontal cortex (see the review by Schroeder and coworkers, 2010).

Excessive anxiety is an even more common emotional problem than depression. Anxiety disorders, including posttraumatic stress disorder (PTSD), phobias, generalized anxiety disorders, panic disorders, and obsessive-compulsive disorder (OCD), are estimated to affect from 15 to 35 percent of the population. As described in Clinical Focus 12-3, "Anxiety Disorders," symptoms include persistent fears and worries in the absence of any direct threat, usually accompanied by various physiological stress reactions, such as rapid heartbeat, nausea, and breathing difficulty.

As with depression, the root cause of anxiety disorders is not known, but the effectiveness of the drug treatments described in Clinical Focus 12-3 implies a biological basis. The most widely prescribed anxiolytic (antianxiety) drugs are the benzodiazepines, such as Valium, Librium, and Xanax.

Why would the brain have a mechanism for benzodiazepine action? It certainly did not evolve to allow us to take Valium. Probably this mechanism is part of a system that both increases and reduces anxiety levels. The mechanism for raising anxiety seems to entail a compound known as diazepam-binding inhibitor that appears to bind antagonistically with the $GABA_A$ receptor, resulting in greater anxiety.

An increase in anxiety can be beneficial, especially if we are drowsy and need to be alert to deal with some kind of crisis. Impairment of this mechanism or the one that reduces anxiety can cause serious emotional problems, even anxiety disorders.

generalized anxiety disorder Persistently high levels of anxiety often accompanied by maladaptive behaviors to reduce anxiety; the disorder is thought to be caused by chronic stress.

phobia Fear of a clearly defined object or situation.

panic disorder Recurrent attacks of intense terror that come on without warning and without any apparent relation to external circumstances.

Section 6-2 explains the actions of antianxiety agents; Figure 6-7 illustrates their action at the $GABA_A$ receptor.

Section 16-4 further explores causes for anxiety disorders and reviews treatments.

REVIEW 12-4
Neuroanatomy of Motivated and Emotional Behavior

Before you continue, check your understanding.

1. The two different types of motivated behaviors are _____ behaviors, which maintain homeostasis, and _____ behaviors, encompassing basically all other behaviors.

2. The brain's homeostat for many functions is found in the _____.

3. The three brain structures housing the major behavioral circuitry involved in motivation and emotion are _____, _____, and _____.

4. The prefrontal cortex has three main subdivisions: _____, _____, and _____.

5. Damage to the _____ is the primary cause of the Klüver-Bucy syndrome.

6. The anterior pituitary gland produces _____.

7. Contrast the functions of the limbic system and the frontal lobes.

Answers appear at the back of the book.

12-5 Control of Regulatory and Nonregulatory Behavior

The two distinctly different types of motivated behaviors described in Section 12-4 are *regulatory* behaviors that maintain vital body-system balance, or homeostasis, and *nonregulatory* behaviors not controlled by a homeostatic mechanism—basically

obesity Excessive accumulation of body fat.

anorexia nervosa Exaggerated concern with being overweight that leads to inadequate food intake and often excessive exercising; can lead to severe weight loss and even starvation.

all other behaviors. In this section, we focus first on the control of two regulatory behaviors in humans—eating and fluid intake. Then we explore the control of human sexual behavior. While sexual behavior is nonregulatory; that is, not essential for an individual organism's survival, it is of enormous psychological significance to humans.

Controlling Eating

There is more to feeding behavior than sustenance alone. We must eat and drink to live, but we also derive great pleasure from these acts. For many people, eating is a focus of daily life, if not for survival, for its centrality to social activities, from get-togethers with family and friends to business meetings and even to group identification. Are you a gourmet, a vegetarian, or a snack-food junkie? Do you diet?

Control over eating is a source of frustration and even grief for many people in the developed world. In 2000, the World Health Organization identified **obesity,** the excessive accumulation of body fat, as a worldwide epidemic. The United States is a case in point. From 1990 to 2010, the percentage of overweight people increased from about 50 percent to 65 percent of the population. The proportion of people considered obese increased from about 12 percent in 1990 to 33 percent in 2012.

The increasing numbers of overweight and obese children and adults persist despite a substantial decrease in fat intake in American diets. What behaviors might cause persistent weight gain? One key to understanding weight gain in the developed world is evolutionary. Even 40 years ago, much of our food was only seasonally available. In a world with uncertain food availability, it makes sense to store excess body calories in the form of fat to be used later when food is scarce. Down through history and in many cultures today, plumpness was and is desirable as a standard of beauty and a sign of health and wealth.

In postindustrial societies, where food is continuously and easily available, being overweight may not be the healthiest strategy. People eat as though food will be scarce and fail to burn off the extra calories by exercising, and the result is apparent. About half of the U.S. population has dieted at some point in their lives. At any given time, at least 25 percent report that they are currently on a diet. For a comparison of how some well-known dieting programs perform, see Clinical Focus 12-4, "Weight-Loss Strategies."

Most Americans are overweight but live in a culture obsessed with slimness. The human control system for feeding has multiple neurobiological inputs, including cognitive factors such as thinking about food and the association between environmental cues (e.g., watching television or studying) and the act of eating. The constant pairing of such cues with eating can result in the cues alone becoming a motivation, or incentive, to eat. We return to this phenomenon in the discussion of rewards and addictions in Section 12-6.

Eating disorders entail being either underweight or overweight. **Anorexia nervosa** is an eating disorder with a huge cognitive component—namely, self-image. A person's body image is highly distorted in anorexia. This misperception leads to an exaggerated concern with being overweight spiraling to excessive dieting, compulsive exercising, and severe, potentially life-threatening weight loss. Anorexia is especially identified with adolescent girls.

The neurobiological control of feeding behavior in humans is not as simple as it is in the fly described in Section 12-3. The multiple inputs to the human control system for feeding come from three major sources: the cognitive factors already introduced, the hypothalamus, and the digestive system.

Weight-Loss Strategies

Among the wide range of diets and weight-loss strategies on the market, none has stopped the obesity epidemic facing the developed world. Diets range widely in their recommended allowed proportions and types of fats, carbohydrates, and proteins.

A study by Iris Shai and colleagues (2008) compared a low-fat diet, a low-carbohydrate diet, and the Mediterranean diet—one high in fruits, vegetables, legumes, and whole grains and including fish, nuts, and low-fat dairy products (Panagiotakos et al., 2004). The Shai study lasted two years and had relatively low dropout rates (95.4 percent remained on their diets at 1 year and 84.6 percent at 2 years).

Figure A shows that all three diets led to weight loss. The low-carbohydrate diet produced the largest acute loss in weight. Over the 2-year period subjects gained back some weight, and in the second year the low-carbohydrate and Mediterranean dieters had similar weight loss.

The low-carbohydrate diet had more favorable effects on lipid levels, whereas the Mediterranean diet provided better control of glucose and insulin levels. Those who followed the low-fat diet lost weight but fared less well than those on the other regimes.

The question of which foods are most likely to lead to weight gain was studied in a 20-year prospective study of three different cohorts totaling about 120,000 U.S. women and men (Mozaffarian et al., 2011). None were obese when the study began. Participants gained an average of 3.4 pounds during every 4-year period.

Shown graphically in **Figure B,** weight gain was most strongly related to the intake of potato chips, potatoes, sugar-sweetened beverages, and red meat. Weight loss was related to the intake of vegetables, whole grains, fruits, nuts, and yogurt. Other lifestyle factors related to weight change included the amount of television watching (gain) and physical activity (loss).

Teresa Fung and her colleagues (2010) conducted a prospective study of mortality over a 26-year period in 85,000 women and 45,000 men. They found that a low-carbohydrate diet based on animal food sources was associated with higher mortality of all causes, whereas a vegetable-based, low-carbohydrate diet was associated with lower mortality of all causes.

Shai and her colleagues concluded that health-care professionals might suggest more than one dietary approach based on individual preferences and metabolic needs, as long as the effort is sustained. At present, the only certain solution to weight loss appears to be a permanent switch to a diet reduced in calories and fat combined with increased physical activity. The results of the Mozaffarian study suggest cutting back on foods and beverages with high sugar content as well.

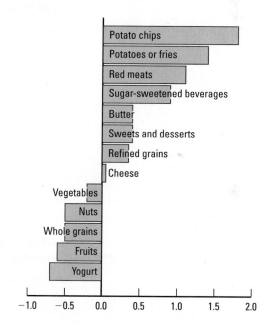

Average weight increase or decrease, in pounds per 4-year period, with each increase in daily serving

FIGURE B Foods Most Likely to Lead to Weight Gain or Loss Weight changes recorded for each increase in daily serving of a food per 4-year period, based on the diets and weights of nearly 125,000 people who were followed over 20 years. **Adapted from "Changes in Diet and Lifestyle and Long-Term Weight Gain in Women and Men," by D. Mozaffarinan, M. P. H. Tao Hao, E. B. Rimm, W. C. Willett, and F. B. Hu (2011), New England Journal of Medicine, 3654, 2392–2404.**

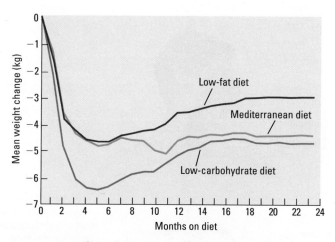

Months on diet

FIGURE A Benefits of Dieting Changes in body weight over a 2-year period on three different diets. Low-carbohydrate and Mediterranean diets were equivalent after 1 year, and both led to more weight loss than did the low-fat diet. **Adapted from "Weight Loss with a Low-Carbohydrate, Mediterranean, or Low-Fat Diet," by I. Shai, D. Schwarzfuchs, Y. Henkin, et al., 2008, New England Journal of Medicine, 13, 229–241.**

aphagia Failure to eat; may be due to an unwillingness to eat or to motor difficulties, especially with swallowing.

hyperphagia Disorder in which an animal overeats, leading to significant weight gain.

Section 6-5 reviews the general categories of hormones and how they work.

Digestive System and Control of Eating

The digestive tract, illustrated in **Figure 12-23**, begins in the mouth and ends at the anus. As food travels through the tract, the digestive system extracts three types of nutrients: lipids (fats), amino acids (the building blocks of proteins), and glucose (sugar). Each nutrient is a specialized energy reserve. Because we require varying amounts of these reserves depending on what we are doing, the body has detector cells to keep track of the level of each nutrient in the bloodstream.

Glucose is the body's primary fuel and is virtually the only energy source for the brain. Because the brain requires glucose even when the digestive tract is empty, the liver acts as a short-term reservoir of glycogen, a starch that acts as an inert form of glucose. When blood-sugar levels fall, as when we are sleeping, detector cells tell the liver to release glucose by converting glycogen into glucose.

Thus the digestive system functions mainly to break down food, and the body needs to be apprised of how well this breakdown is proceeding. Feedback mechanisms provide such information. When food reaches the intestines, it interacts with receptors there to trigger the release of at least 10 different peptide hormones, including one known as cholecystokinin (CCK).

The released peptides inform the brain (and perhaps other organs, in the digestive system) about the nature and quality of the food in the gastrointestinal tract. The level of CCK appears to play a role in *satiety,* the feeling of having eaten enough. For example, if CCK is infused into the hypothalamus of an animal, the animal's appetite diminishes.

Hypothalamus and Control of Eating

The hypothalamus, which controls hormone systems, is the key brain structure in feeding. Feeding behavior is influenced by hormones, including insulin, growth hormone, and sex steroids that stimulate and inhibit feeding and aid in converting nutrients into fat and fat into glucose.

Investigation into how the hypothalamus controls feeding began in the early 1950s, when researchers discovered that damage to the lateral hypothalamus in rats caused the

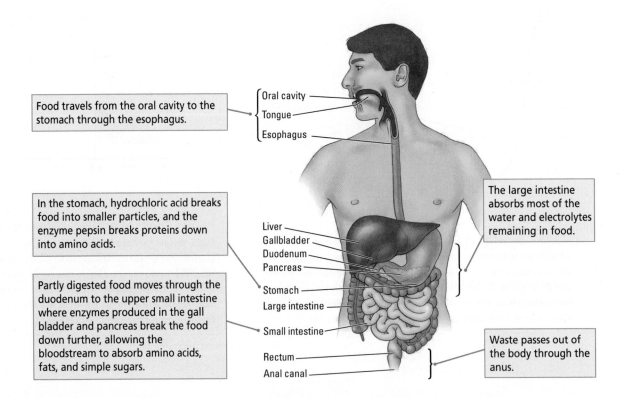

Food travels from the oral cavity to the stomach through the esophagus.

Oral cavity
Tongue
Esophagus

The large intestine absorbs most of the water and electrolytes remaining in food.

In the stomach, hydrochloric acid breaks food into smaller particles, and the enzyme pepsin breaks proteins down into amino acids.

Liver
Gallbladder
Duodenum
Pancreas

Partly digested food moves through the duodenum to the upper small intestine where enzymes produced in the gall bladder and pancreas break the food down further, allowing the bloodstream to absorb amino acids, fats, and simple sugars.

Stomach
Large intestine
Small intestine
Rectum
Anal canal

Waste passes out of the body through the anus.

FIGURE 12-23 The Digestive System

animals to stop eating, a symptom known as **aphagia** (in Greek, *phagein* means "to eat"). In contrast, damage to the ventromedial hypothalamus (VMH) caused the animals to overeat, a symptom known as **hyperphagia**. A VMH-lesioned rat that overate to the point of obesity is shown in the Procedure section of **Experiment 12-1**. The Results section reveals that the VMH-lesioned rat weighed more than a kilogram, three times the weight of her normal sister, which was 340 grams.

At about the same time, researchers also found that electrical stimulation of the lateral hypothalamus elicits feeding, whereas stimulation of the ventromedial hypothalamus inhibits feeding. The opposing effects of injury and stimulation to these two regions led to the idea that the lateral hypothalamus signals "eating on," whereas the VMH signals "eating off." This model quickly proved too simple.

Not only does the lateral hypothalamus contain cell bodies but fiber bundles also pass through it. Damage to either structure can produce aphagia. Similarly, damage to fibers passing through the VMH often causes injury as well to the paraventricular nucleus of the hypothalamus. And damage to the paraventricular nucleus alone is now known to produce hyperphagia. Clearly, the role of the hypothalamus in the control of feeding involves more than the activities of its lateral and ventromedial structures alone.

In the half-century since those first studies, researchers have learned that damage to the lateral and ventromedial hypothalamus and to the paraventricular nucleus has multiple effects. They include changes in hormone levels (especially insulin), in sensory reactivity (the taste and attractiveness of food is altered), in glucose and lipid levels in the blood, and in metabolic rate. The general role of the hypothalamus is to act as a sensor for the levels of lipids, glucose, hormones, and various peptides. Groups of hypothalamic neurons, for example, sense the level of glucose (glucostatic neurons) as well as the level of lipids (lipostatic neurons).

The sum of the activity of all such hypothalamic neurons creates a very complex homeostat that controls feeding. **Figure 12-24** shows that this homeostat receives inputs from three sources: the digestive system (such as information about blood-glucose levels), hormone systems (such as information about the level of CCK), and parts of the brain that process cognitive factors. We turn to these cognitive factors next.

Cognitive Factors and Control of Eating

Pleasure and its absence are cognitive factors in controlling eating. Just thinking about a favorite food can make any of us feel hungry. The cognitive aspect to feeding includes not only the images of food that we pull from memory but also external sensations, especially food-related sights and smells. Learned associations, such as the learned taste aversions discussed in Section 12-3, are also related to feeding.

Neural control of the cognitive factors important for controlling eating in humans probably originates in multiple brain regions. Two structures are clearly important: the amygdala and the orbital prefrontal cortex. Damage to the amygdala alters food preferences and abolishes taste-aversion learning. These effects are probably related to the amygdala's efferent connections to the hypothalamus.

EXPERIMENT 12-1

Question: Does the hypothalamus play a role in eating?

Procedure

The ventromedial hypothalamus (VMH) of the rat on the right was damaged, and her body weight was monitored for a year. Her sister on the left is normal.

Intact brain of sister rat   Rat brain with lesion

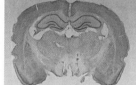

Results

The VMH-lesioned rat showed a dramatic increase in food intake and body weight.

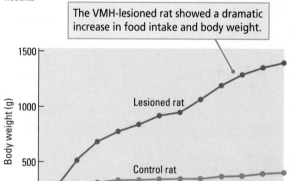

Conclusion: The VMH plays a role in controlling the cessation of eating. Damage to the VMH results in prolonged and dramatic weight gain.

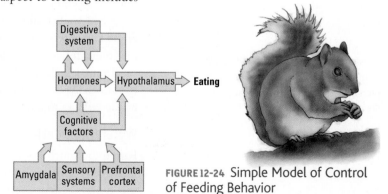

FIGURE 12-24 Simple Model of Control of Feeding Behavior

osmotic thirst Thirst that results from an increased concentration of dissolved chemicals, or *solutes*, in body fluids.

hypovolumic thirst Thirst that is produced by a loss of overall fluid volume from the body.

The amygdala's role in regulating species-typical behaviors is well established, but the role of the orbital prefrontal cortex is more difficult to pin down. Rats and monkeys with damage to the orbital cortex lose weight, in part because they eat less. Humans with orbital injuries are invariably slim, but we know of no formal studies on their eating habits. The orbital prefrontal cortex receives projections from the olfactory bulb, and cells in this region do respond to smells. Because odors influence the taste of foods, it is likely that damage to the orbital prefrontal cortex decreases eating because of diminished sensory responses to food odor and perhaps to taste.

An additional cognitive factor in controlling eating is the pleasure we derive from it, especially from eating foods with certain tastes. Think chocolate. What pleasure is and how the brain produces it are topics discussed in Section 12-6 in the context of reward.

Randy Seeley and Stephen Woods (2003) have noted that, in spite of the problem people now appear to have with weight gain, adult mammals do a masterful job of matching their caloric intake to caloric expenditure. Consider that a typical man eats 900,000 calories per year. To gain just one extra pound requires him to eat 4000 calories more than are burned in that year. This increase amounts to only 11 calories per day, equivalent to a single potato chip. But people rarely eat just one chip. As Figure B in Clinical Focus 12-4 illustrates, potato chips top the list of major food sources linked to weight gain over time.

Controlling Drinking

About 70 percent of the human body is composed of water that contains a range of chemicals that participate in the hundreds of chemical reactions involved in bodily functions. Essential homeostatic mechanisms control water levels (and hence chemical concentrations) within rather narrow limits. The rate of a chemical reaction is partly determined by how concentrated the supplies of participating chemicals are.

As with eating, we drink for many reasons. We consume some beverages, such as coffee, wine, beer, and juice, for an energy boost or to relax, as part of social activities, or just because they taste good. We drink water for its health benefits, to help wash down a meal or to intensify the flavor of dry foods. On a hot day, we drink water because we are thirsty, presumably because we have lost significant moisture through sweating and evaporation.

These examples illustrate the two kinds of thirst. **Osmotic thirst** results from an increase in the concentrations of dissolved chemicals, known as *solutes*, in the body fluids. **Hypovolemic thirst** results from a loss of overall fluid volume from the body.

Osmotic Thirst

The solutes found inside and outside cells in the body are ideally concentrated for the body's chemical reactions. Maintaining this concentration requires a kind of homeostat, much like the mechanism that controls body temperature. Deviations from the ideal solute concentration activate systems to reestablish it.

When we eat salty foods, such as potato chips, the salt (NaCl) spreads through the blood and enters the extracellular fluid that fills the spaces between our cells. This shifts the solute concentration away from the ideal. Receptors in the hypothalamus along the third ventricle detect the altered solute concentration and relay the message "too salty" to various hypothalamic areas that, in turn, stimulate us to drink. Other messages are sent to the kidneys to reduce water excretion.

Turning to sugar-sweetened beverages to quench thirst from eating salty foods increases the chance of weight gain (see Focus 12-4).

Water Intoxication

Eating too much leads to obesity. What happens when we drink too much water? Our kidneys are efficient at processing water, but if we drink it in large volumes all at once, the kidneys cannot keep up.

The result is a condition called "water intoxication." Body tissues swell with the excess fluid, essentially drowning the cells in fresh water. At the same time, the relative concentration of sodium drops, leading to an imbalance in electrolytes.

Water intoxication can produce a wide range of symptoms, from irregular heartbeat to headache, and in severe cases, people may act as though they are intoxicated from alcohol. The most likely way for an adult to develop water intoxication is to sweat heavily, by running a marathon in hot weather, for example, and then drink too much water without added electrolytes.

Hypovolemic Thirst

Unlike osmotic thirst, hypovolemic thirst arises when the total volume of body fluids declines, motivating us to drink more and replenish them. In contrast with osmotic thirst, however, hypovolemic thirst encourages us to choose something other than water, because water would dilute the solute concentration in the blood. Rather, we prefer to drink flavored beverages that contain salts and other nutrients.

Hypovolemic thirst and its satiation are controlled by a different hypothalamic circuit from the one that controls osmotic thirst. When fluid volume drops, the kidneys send a hormone signal (angiotensin) that stimulates midline hypothalamic neurons. These neurons, in turn, stimulate drinking.

Controlling Sexual Behavior

Individuals must feed and drink repeatedly to survive. This is the essence of regulatory behavior. But notwithstanding procreation, which is essential to the survival of the species, sexual behavior is nonregulatory: it is not essential for the individual organism's survival. That fact does nothing to convince most of us that sex is unimportant.

In Sigmund Freud's psychodynamic theory, sexual drives are central to human behavior. Sexual themes repeatedly appear in our art, literature, and films. They bombard us via advertising and other sales pitches. Such significance makes it all the more important to understand the control of human sexual behavior in both gonadal hormones and brain circuits.

Effects of Sex Hormones on the Brain

During the fetal stage of prenatal development, a male's Y chromosome controls the differentiation of embryonic gonad tissue into testes, which in turn secrete testosterone. This process is an organizing effect of gonadal hormones. Testosterone masculinizes both the sex organs and the brain during development. A major organizing effect that gonadal hormones have on the brain is in the hypothalamus, especially the preoptic area of the medial hypothalamus. Organizing effects also operate in other nervous system regions, notably the amygdala, the prefrontal cortex, and the spinal cord.

Sex-related differences in the nervous system make sense behaviorally. After all, animal courtship rituals differ between the sexes, as do copulatory behaviors, with females engaging in sexually receptive responses and males in mounting ones. The production of these sex differences in behaviors depends on the action of gonadal hormones on the brain during both development and adulthood.

The actions of hormones on the adult brain are referred to as *activating effects,* in contrast with the developmental *organizing effects.* Here we consider both, separately.

ORGANIZING EFFECTS OF SEX HORMONES During fetal development, a male's testes produce male hormones, the androgens. In the developing rat, androgens are produced during the last week of fetal development and the first week after birth. The androgens produced at this time greatly alter both neural structures and later behavior. For example, the hypothalamus and prefrontal cortex of a male rat differ structurally from both those of female rats and those of males that were not exposed to androgens during their development.

Section 8-4 explains the organizing influences of gonadal hormones and critical periods on the developing brain. The organizing effects of gonadal hormones have been studied most extensively in rats.

sexual dimorphism Differential development of brain areas in the two sexes.

Testosterone

Estradiol

Section 6-5 details the overall activating effects of sex hormones on male and female behavior.

In adulthood, males with little exposure to the androgen testosterone during development behave like genetically female rats. If given estrogen and progesterone, they become sexually "receptive" and display typical female behaviors when mounted by males. Male rats that are castrated in adulthood do not act in this way.

Sexual dimorphism, the differential development of brain areas in the two sexes, arises from a complex series of steps. Cells in the brain produce aromatase, an enzyme that converts testosterone into estradiol, one of the class of female sex hormones called *estrogens*. Thus, when males produce testosterone, it gets converted in the brain into an estrogen. Therefore a female hormone, estradiol, actually masculinizes a male brain.

Females are not masculinized by the presence of estrogens because the fetuses of both sexes produce a liver enzyme (*alpha fetoprotein*) that binds to estrogen, rendering it incapable of entering neurons. Testosterone is unaffected by alpha fetoprotein, so it enters neurons and is converted into estradiol.

The organizing effects of testosterone are clearly illustrated in the preoptic area of the hypothalamus (see Figure 12-12A), which plays a critical role in the copulatory behavior of male rats. Comparing this area in males and females, Roger Gorski and his colleagues found a nucleus about five times as large in the males as in the females (Gorski, 1984). Significantly, the sexual dimorphism of the preoptic area can be altered by manipulating gonadal hormones during development. Castrating male rats at birth leads to a smaller preoptic area, whereas treating infant females with testosterone increases its size.

The organizing effects of gonadal hormones are more difficult to study in humans. However, John Money and Anke Ehrhardt (1972) revealed an important role of these hormones in human development. Clinical Focus 12-5, "Androgen-Insensitivity Syndrome and the Androgenital Syndrome," describes this role.

ACTIVATING EFFECTS OF SEX HORMONES The sexual behavior of both males and females also depends on the actions that gonadal hormones have on the adult brain. In most vertebrate species, female sexual behavior varies in the course of an estrous cycle during which the levels of hormones that the ovaries produce fluctuate. The rat's estrous cycle is about 4 days long, with sexual receptivity occurring only in the few hours during which the production of the ovarian hormones estrogen and progesterone peaks. These ovarian hormones alter brain activity, which in turn alters behavior. Furthermore, in female rats, various chemicals are released after mating, and these chemicals inhibit further mating behavior.

The activating effect of ovarian hormones can be seen clearly in cells of the hippocampus. **Figure 12-25** compares hippocampal pyramidal neurons taken from female rats at two points in the estrous cycle: one when estrogen levels are high and the other when they are low. When estrogen levels are high, more dendritic spines and presumably more synapses emerge. These neural differences during the estrous cycle are all

FIGURE 12-25 Hormonal Effects A comparison of the dendrites of hippocampal pyramidal neurons at high and low levels of estrogen in the rat's (4-day) estrous cycle reveals far fewer dendritic spines in the low period. Adapted from "Naturally Occurring Fluctuation in Dendritic Spine Density on Adult Hippocampal Pyramidal Neurons," by C. S. Woolley, E. Gould, M. Frankfurt, and B. McEwen, 1990, *Journal of Neuroscience, 10*, p. 1289.

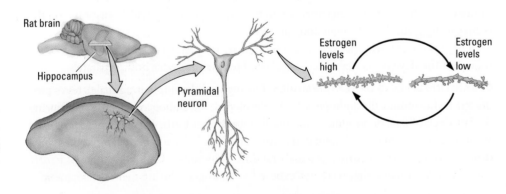

CLINICAL FOCUS ✛ 12-5

Androgen-Insensitivity Syndrome and the Androgenital Syndrome

After the testes have formed in a male fetus, sexual development depends on the actions of testicular hormones. Studying people with *androgen-insensitivity syndrome* makes this dependence crystal clear. In this syndrome, an XY (genetic male) fetus produces androgens, but the body cannot to respond to them.

Because androgen-insensitivity syndrome does not affect estrogen receptors, these people are still responsive to estrogen produced by both the adrenal gland and the testes. As a result, they develop female secondary sexual characteristics during puberty, even without additional hormone treatment. A person with androgen-insensitivity syndrome is therefore a genetic male who develops a female phenotype, that is, appears to be female, as shown in the photograph on the left.

If no Y chromosome is present to induce the growth of testes, an XX (genetic female) fetus develops ovaries and becomes a female. If the adrenal glands of either the mother or the infant produce an excessive amount of androgens, however, the female fetus is exposed to androgens, producing the *androgenital syndrome* (*congenital adrenal hyperplasia*).

The effects vary, depending on when the androgens are produced and on the level of exposure. In extreme cases, an enlarged clitoris develops that can be mistaken for a small penis, as shown in the photograph on the right.

In less severe cases, no gross abnormality in genital structure develops, but there is a behavioral effect: these girls show a high degree of tomboyishness. In early childhood, they identify with boys and prefer boys' clothes, toys, and games. One explanation for this behavioral effect is that the developing brain is masculinized, thus changing later behavior.

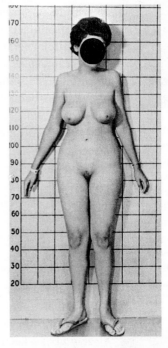
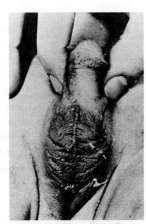

(*Left*) In androgen-insensitivity syndrome, a genetic male (XY) is insensitive to gonadally produced androgens but remains sensitive to estrogens, leading to the development of a female phenotype. (*Right*) In congenital adrenal hyperplasia, a genetic female (XX) is exposed to androgens produced by the adrenal gland embryonically, leading to the partial development of male external genitalia.
Reprinted from *Man and Woman, Boy and Girl*, by John Money and Anke A. Ehrhardt, 1972 (p. 116), Baltimore: Johns Hopkins University Press.

the more remarkable when we consider that cells in the female hippocampus are continually changing their connections to other cells every 4 days throughout the animal's adulthood.

In males, testosterone activates sexual behavior in two distinctly different ways. First, the actions of testosterone on the amygdala are related to the motivation to seek sexual activity. Second, the actions of testosterone on the hypothalamus are needed to produce copulatory behavior. We look at both processes next.

Hypothalamus, Amygdala, and Sexual Behavior

The hypothalamus is the critical structure controlling copulatory behaviors in both male and female mammals. The ventromedial hypothalamus controls the female mating posture, which in quadrapedal animals is called *lordosis:* arching the back and elevating the rump while the female otherwise remains quite still. Damage to the VMH abolishes lordosis. The role of the VMH is probably twofold: it controls the neural circuit that produces lordosis, and it influences hormonal changes in the female during coitus.

In males, neural control of sexual behavior is somewhat more complex. The medial preoptic area, which is larger in males than in females, controls copulation. Damage to

Typical posture

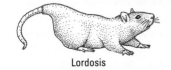
Lordosis

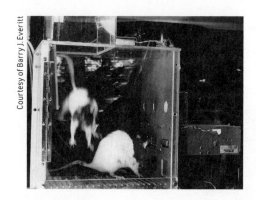

FIGURE 12-26 Studying Sexual Motivation and Mating In this experiment, a male rat is required to press the bar 10 times to gain access to a receptive female who "drops in" through a trap door. The copulatory behavior of the male rat illustrates mating behavior, whereas the bar pressing for access to a female rat illustrates sexual motivation. Adapted from "Sexual Motivation: A Neural and Behavioral Analysis of the Mechanisms Underlying Appetitive and Copulatory Responses of Male Rats," by B. J. Everitt, 1990, *Neuroscience and Biobehavioral Reviews, 14,* p. 227.

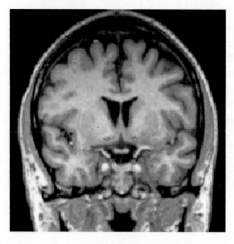

FIGURE 12-27 Hypothalamus and Sex in Males Red regions on the left and right represent bilateral activity in the hypothalamus of men viewing erotic film clips. This activity is absent as the men watched sports clips. From "The Hypothalamus, Sexual Arousal and Psychosexual Identity in Human Males: A Functional Magnetic Resonance Imaging Study," by M. Brunetti, C. Babiloni, A. Ferretti, C. Del Gratta, A. Merla, et al., 2008, *European Journal of Neuroscience, 27,* 2922–2927.

Figure 3-25 illustrates two aspects of methylation: histone and DNA modification.

the medial preoptic area greatly disrupts mating performance, whereas electrical stimulation of this area activates mating, provided that testosterone is circulating in the bloodstream. Curiously, although destruction of the medial preoptic area stops male mammals from mating, they continue to show interest in receptive females. For instance, monkeys with lesions in the medial preoptic area will not mate with receptive females, but they will masturbate while watching them from across the room.

Barry Everitt (1990) designed an ingenious apparatus that allows male rats to press a bar to deliver receptive females. After males were trained to use this apparatus, shown in **Figure 12-26,** lesions were made in their medial preoptic areas. Immediately, their sexual behavior changed. They would still press the bar to obtain access to females but would no longer mate with them.

Apparently, the medial preoptic area controls mating, but it does not control sexual motivation. The brain structure responsible for motivation appears to be the amygdala. When Everitt trained male rats in the apparatus and then lesioned their amygdalae, they would no longer press the bar to gain access to receptive females, but they would mate with receptive females that were provided to them.

It is not practical to discriminate small hypothalamic nuclei in fMRI studies of humans. Studies have shown a bilateral increase in activity in the hypothalamus when men view erotic video clips but not when they view sports video clips **(Figure 12-27).** Further, the degree of sexual arousal is related to the increase in hypothalamic activity (e.g., Brunetti et al., 2008).

In summary, the hypothalamus controls copulatory behavior in both male and female mammals. In males, the amygdala influences sexual motivation, and it probably plays a key role in female sexual motivation as well, especially among females of species, such as humans, in which sexual activity is not tied to fluctuations in ovarian hormones.

Sexual Orientation, Sexual Identity, and Brain Organization

Does **sexual orientation**—a person's sexual attraction to the opposite sex or to the same sex or to both sexes—have a neural basis? Sexual orientation appears to be determined during early development and is influenced by genetics and by epigenetic factors during prenatal brain development. No solid evidence points to any postnatal experience directing sexual orientation.

Indeed, it appears virtually impossible to change a person's sexual orientation. For example, children raised by lesbian couples are heterosexually oriented, and evidence is lacking to support the ideas that homosexuality is a lifestyle choice or an effect of social learning (Bao & Swaab, 2011). The place to look for differences, therefore, is in the brains of people who identify themselves as heterosexual and homosexual.

Like rats, humans have sex-related differences in the structure of the hypothalamus and amygdala. Several hypothalamic nuclei are two to three times larger in males (for a review, see Becker and colleagues, 2008). Sexual differentiation of the brain results from the effect of testosterone and is complete by birth.

But sex differences in the brain are not simply a matter of hormones. Epigenetics plays a role too, beginning early in development. For example, in females one of the two X chromosomes is largely silenced, but not all its genes are silenced, thus providing a basis for sex differences. Furthermore, emerging evidence suggests that sex differences in the hypothalamus result from differences in gene methylation (for a review, see McCarthy and coworkers, 2009).

Variations in epigenetic effects could lead to differences in the architecture and function of the hypothalamus in homosexuals. Differences in the hypothalami of heterosexual and homosexual men suggest that homosexual men form, in effect, a "third sex" because their hypothalami differ from those of both females and heterosexual males.

Differences in the hypothalamus may form a basis for **gender identity**—a person's feeling of being male or female. One atypical form of gender identity is **transsexuality**, the strong belief of having been born the wrong sex. Transsexuals' desire to live as the opposite sex can be so strong that they undergo sex-change surgery.

Several factors appear to influence the likelihood of transsexuality—chromosomal abnormalities, polymorphisms of the genes for the estrogen and androgen receptors, abnormal gonadal hormone levels, prenatal exposure to certain anticonvulsants, and immune system activity directed toward the Y chromosome. These factors are hypothesized to lead to changes in the architecture and function of brain structures, especially the hypothalamus, which matches transsexuals' gender identities while their sex organs do not (Bao & Swaab, 2011).

In summary, differences in sexual orientation and gender identity appear to result from prenatal events that influence the organization and function of the brain, not from postnatal social or environmental experiences.

Cognitive Influences on Sexual Behavior

People think about sex, dream about sex, make plans about sex. These behaviors may include activity in the amygdala or the hypothalamus, but they must certainly also include the cortex. This is not to say that the cortex is essential for sexual motivation and copulation.

In studies of rats whose entire cortices have been removed, both males and females still engage in sexual activity, although the males are somewhat clumsy. Nevertheless, the cortex must play a role in certain aspects of sexual behavior. For instance, imagery about sexual activity must include activity in the ventral visual pathway of the cortex. And thinking about sexual activity and planning for it must require the participation of the frontal lobes.

As you might expect, these aspects of sexual behavior are not easily studied in rats, and they remain uncharted waters in research on humans. However, changes in the sexual behavior of people with frontal-lobe injury are well documented. And recall J. P.'s case, described in Clinical Focus 12-2.

Although J. P. lost his inhibition about sexual behavior, frontal-lobe damage is just as likely to produce a loss of libido (sexual interest). The wife of a man who, 5 years earlier, had a small tumor removed from the medial frontal region, complained that she and her husband had since had no sexual contact whatever. He was simply not interested, even though they were both still in their twenties.

The husband said that he no longer had sexual fantasies or sexual dreams and, although he still loved his wife, he did not have any sexual urges toward her or anyone else. Such cases clearly indicate that the human cortex has an important role in controlling sexual behaviors. The exact nature of the role remains poorly understood.

 REVIEW 12-5

Control of Regulatory and Nonregulatory Behavior

Before you continue, check your understanding.

1. The three main hypothalamic regions that control feeding are _____, _____, and _____.

2. The key structures in the control of sexual behavior are the _____ and the _____.

3. The two types of effects that hormones exert on the brain are _____ and _____.

sexual orientation A person's sexual attraction to the opposite sex or to the same sex or to both sexes.

gender identity A person's feeling of being either male or female.

transsexuality Gender-identity disorder involving the strong belief of having been born the wrong sex.

4. _____ thirst results from an increase in the concentration of dissolved chemicals; _____ thirst results from a decline in the total volume of body fluids.

5. Discuss: Sex differences in the brain are not simply a matter of hormones.

Answers appear at the back of the book.

12-6 Reward

Throughout this chapter we have concluded repeatedly that animals engage in a wide range of voluntary behaviors because the behaviors are rewarding. They are rewarding because they increase the activity of neural circuits that function to maintain an animal's contact with certain environmental stimuli, either in the present or in the future. Presumably, an animal perceives the activity of these circuits as pleasant. This would explain why reward can help maintain not only adaptive behaviors such as feeding and sexual activity but also potentially nonadaptive behaviors such as drug addiction. After all, evolution would not have prepared the brain specifically for the eventual development of psychoactive drugs.

Section 7-1 describes how electrical stimulation is used as a treatment as well as a research tool.

The first clue to the presence of a reward system in the brain came with an accidental discovery by James Olds and Peter Milner in 1954. They found that rats would perform behaviors such as pressing a bar to administer a brief burst of electrical stimulation to specific sites in their brains. This phenomenon is called *intracranial self-stimulation* or *brain-stimulation reward*.

Typically, rats will press a lever hundreds or even thousands of times per hour to obtain this brain stimulation, stopping only when they are exhausted. Why would animals engage in such a behavior when it has absolutely no value to their survival or to that of their species? The simplest explanation is that the brain stimulation is activating the system underlying reward (Wise, 1996).

After more than a half-century of research on brain-stimulation reward, investigators now know that dozens of sites in the brain maintain self-stimulation. Some especially effective regions are the lateral hypothalamus and medial forebrain bundle (see Figures 12-12 and 12-13). Stimulation along the MFB tract activates fibers that form the ascending pathways from dopamine-producing cells of the midbrain tegmentum, shown in **Figure 12-28**. This mesolimbic dopamine pathway sends terminals to various sites, including especially the nucleus accumbens in the basal ganglia and the prefrontal cortex.

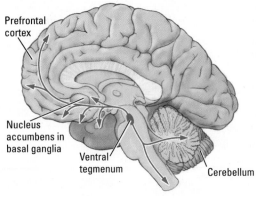

FIGURE 12-28 Mesolimbic Dopamine System Axons emanating from the ventral tegmentum (blue arrows) project diffusely through the brain. Dopamine release in these mesolimbic pathways has a role in feelings of reward and pleasure. The nucleus accumbens is a critical structure in this "reward system."

Neuroscientists believe that the mesolimbic dopamine system is central to circuits mediating reward for several reasons:

1. Dopamine release shows a marked increase when animals are engaged in intracranial self-stimulation.

2. Drugs that enhance dopamine release increase self-stimulation, whereas drugs that decrease dopamine release decrease self-stimulation. It seems that the amount of dopamine released somehow determines how rewarding an event is.

3. When animals engage in behaviors such as feeding or sexual activity, dopamine release rapidly increases in locations such as the nucleus accumbens.

4. Highly addictive drugs such as nicotine and cocaine increase the level of dopamine in the nucleus accumbens.

Even opiates appear to affect at least some of an animal's actions through the dopamine system. Animals quickly learn to press a bar to obtain an injection of opiates directly into the midbrain tegmentum or the nucleus accumbens. The same animals do not work to obtain the opiates if the dopaminergic neurons of the mesolimbic system are inactivated. Apparently, then, animals engage in behaviors that increase dopamine release.

Note, however, that dopamine is not the only rewarding compound in the brain. For example, opioid transmitters such as encephalin and dynorphin are also rewarding, as are benzodiazepines.

Robinson and Berridge (2008) propose that reward contains separable psychological components, corresponding roughly to "wanting," which is often called "incentive," and "liking," which is equivalent to an evaluation of pleasure. This idea can be applied to discovering why we increase contact with a stimulus such as chocolate.

Two independent factors are at work: our desire to have the chocolate (wanting) and the pleasurable effect of eating the chocolate (liking). This distinction is important. If we maintain contact with a certain stimulus because dopamine is released, the question becomes whether the dopamine plays a role in the wanting or the liking aspect of the behavior. Robinson and Berridge propose that wanting and liking processes are mediated by separable neural systems and that dopamine is the transmitter in wanting. Liking, they hypothesize, entails opioid and benzodiazepine–GABA systems.

According to Robinson and Berridge, wanting and liking are normally two aspects of the same process, so rewards are usually wanted and liked to the same degree. However, it is possible, under certain circumstances, for wanting and liking to change independently.

Consider rats with lesions of the ascending dopaminergic pathway to the forebrain. These rats do not eat. Is it simply that they do not desire to eat (a loss of wanting) or has food become aversive to them (a loss of liking)? To find out which factor is at work, the animals' facial expressions and body movements in response to food can be observed to see how liking is affected. After all, when animals are given various foods to taste, they produce different facial and body reactions, depending on whether they perceive the food as pleasant or aversive.

Among humans, typically when a person tastes something sweet, he or she responds by licking the fingers or the lips, as shown at the top of **Figure 12-29**. If the taste is unpleasantly salty, say, as shown in the bottom panel, the reaction is often spitting, grimacing, or wiping the mouth with the back of the hand. Rats, too, show distinctive positive and negative responses to pleasant and unpleasant tastes.

By watching the responses when food is squirted into the mouth of a rat that otherwise refuses to eat, we can tell to what extent a loss of liking for food is a factor in the

Section 6-4 details Robinson and Berridge's wanting-and-liking theory of addiction, including the idea that reward has multiple parts.

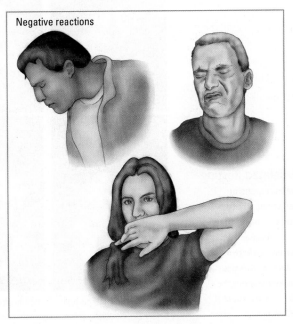

FIGURE 12-29 **Human Reactions to Taste** Sucrose and other palatable tastes elicit positive (hedonic) reactions, including licking the fingers and the lips. Quinine and other nonpalatable tastes elicit negative (aversive) reactions, including spitting, expressing distaste, and wiping the mouth with the back of the hand. Adapted from "Food Reward: Brain Substrates of Wanting and Liking," by K. C. Berridge, 1996, *Neuroscience and Biobehavioral Reviews, 20,* p. 6.

animal's food rejection. Interestingly, rats that do not eat after receiving lesions to the dopamine pathway act as though they still like food.

Now consider a rat with a self-stimulation electrode in the lateral hypothalamus. This rat will often eat heartily while the stimulation is on. The obvious inference is that the food must taste good—presumably even better than it does usually. But what happens if we squirt food into the rat's mouth and observe its behavior when the stimulation is on versus when it is off?

If the brain stimulation primes eating by evoking pleasurable sensations, we would expect that the animal would be more positive in its facial and body reactions toward foods when the stimulation is turned on. In fact, the opposite is found. During stimulation, rats react more aversively to tastes such as sugar and salt than when stimulation is off. Apparently, the stimulation increases wanting but not liking.

Experiments of this sort show that what appears to be a single event—reward—is actually composed of at least two independent processes. Just as our visual system independently processes "what" and "how" information in two separate streams, our reward system appears to process wanting and liking independently. Reward is not a single phenomenon any more than perception or memory are.

Like the networks underlying perception and memory, the prefrontal and limbic networks underlying reward are diffuse. Amy Janes and her colleagues (2012) used resting-state fMRI to identify a limbic network and three discrete prefrontal networks believed to support reward, illustrating the breadth of the reward system **(Figure 12-30)**. They then compared these networks in nicotine addicts and nonaddicts. The results show greater connectivity in the prefrontal–striatial connections and in the limbic pathway in smokers.

The general conclusions from this study, and from related studies of gamblers, are that the reward pathways are diffuse and that the size and activity of these pathways are related to the intensity of reward. The hope is that resting-state fMRI can be used as a biomarker, for both addiction severity and treatment efficacy.

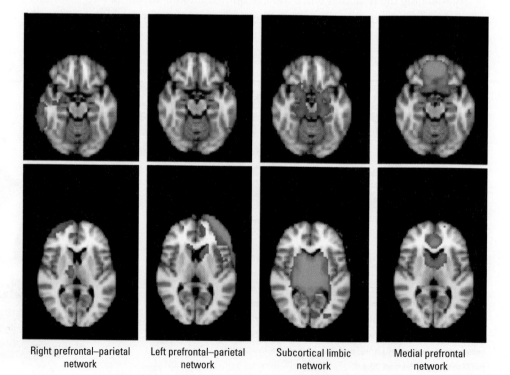

FIGURE 12-30 Addiction-Related Resting-State Networks The green regions in the images indicate enhanced connectivity in prefrontal and limbic circuits in the brains of nicotine-dependent smokers. Adapted from "Prefrontal and Limbic Resting State Brain Network Functional Connectivity Differs Between Nicotine-Dependent Smokers and Non-Smoking Controls," by A. C. Janes, L. D. Nickerson, B. deB. Frederick, and M. H. Kaufman, 2012, *Drug and Alcohol Dependence*, in press

Right prefrontal–parietal network   Left prefrontal–parietal network   Subcortical limbic network   Medial prefrontal network

REVIEW 12-6
Reward

Before you continue, check your understanding.

1. Animals engage in voluntary behaviors because the behaviors are _____.

2. Neural circuits maintain contact with rewarding environmental stimuli in the present or in the future through _____ and _____ subsystems.

3. The neurotransmitter systems hypothesized to be basic to reward are _____, _____, and _____ systems.

4. What is intracranial self-stimulation, and why is it rewarding?

Answers appear at the back of the book.

SUMMARY

12-1 Identifying the Causes of Behavior

Our inner, subjective feelings (emotions) and goal-directed thoughts (motivations) influence how we behave and adapt as individuals and as a species. Emotion and motivation are inferred states that can escape conscious awareness or intent and make the case for free will difficult to argue.

Biologically, reward motivates animals to engage in behavior, whereas aversive circumstances prompt brain circuits to produce behaviors that will reduce them. Such are the effects of sensory deprivation. The brain inherently needs stimulation. In its absence, the brain will seek it out.

Sensory stimulation leads to hormone activity and to dopamine activity in the brainstem. Neural circuits organized in the brainstem control species-typical behaviors such as mouse killing by cats and singing by birds.

These brainstem circuits manifest their evolutionary advantage: they are rewarding. Rewarding behavior motivates living beings. When animals disengage from behaviors that motivate their species, they go extinct.

12-2 The Chemical Senses

In the olfactory and gustatory senses, chemical neuroreceptors in the nose and tongue interact with chemosignals, leading to neural activity in cranial nerve 1 for olfaction and cranial nerves 7, 9, and 10 for taste. The cranial nerves enter the brainstem, and through a series of synapses, pass into the forebrain. Smell and taste input merges in the orbitofrontal cortex to produce our perception of flavor.

12-3 Evolution, Environment, and Behavior

Behavior is controlled by its consequences as well as by its biology. Consequences may affect the evolution of the species or the behavior of an individual. Behaviors selected by evolution are often triggered by innate releasing mechanisms. Behaviors selected only in an individual animal are shaped by that animal's environment and are learned.

12-4 Neuroanatomy of Motivated and Emotional Behavior

The neural structures that initiate emotional and motivated behaviors are the hypothalamus, the pituitary gland, the amygdala, the dopaminergic and noradrenergic activating pathways from nuclei in the lower brainstem, and the frontal lobes.

The experience of both emotion and motivation is controlled by activity in the ANS, hypothalamus, and forebrain, especially the amygdala and frontal cortex. Emotional and motivated behavior may be unconscious responses to internal or external stimuli controlled either by the activity of innate releasing mechanisms or by cognitive responses to events or thoughts.

12-5 Control of Regulatory and Nonregulatory Behavior

The two distinctly different types of motivated behaviors are (1) regulatory (homeostatic) behaviors that maintain vital body-system balance and (2) nonregulatory behaviors, basically consisting of all other behaviors that are not controlled by a homeostatic mechanism nor are reflexive. Feeding is a regulatory behavior controlled by the interaction of the digestive and hormonal systems and the hypothalamic and cortical circuits. Sexual activity is a nonregulatory behavior motivated by the amygdala. Copulatory behavior is controlled by the hypothalamus (the ventromedial hypothalamus in females and the preoptic area in males). Sexual orientation (a person's attraction to the opposite or same sex) and gender identity (a person's feeling of being male or female) are related to the organization of the hypothalamus. Differences in hypothalamic organization are likely related to epigenetic effects in early development.

12-6 Reward

Survival depends on maximizing contact with some environmental stimuli and minimizing contact with others. Reward is a mechanism for controlling this differential. Two independent features of reward are wanting and liking. The wanting component is thought to be controlled by dopaminergic activating systems, whereas the liking component is thought to be controlled by opiate and GABA–benzodiazepine systems.

KEY TERMS

amygdala, p. 419

androgen, p. 401

anorexia nervosa, p. 428

aphagia, p. 430

emotion, p. 399

evolutionary psychology, p. 407

gender identity, p. 437

generalized anxiety disorder, p. 427

hippocampus, p. 417

homeostatic mechanism, p. 411

hyperphagia, p. 430

hypovolumic thirst, p. 432

innate releasing mechanism (IRM), p. 407

Klüver-Bucy syndrome, p. 422

learned taste aversion, p. 409

medial forebrain bundle (MFB), p. 412

motivation, p. 399

nonregulatory behavior, p. 412

obesity, p. 428

orbitofrontal cortex (OFC), p. 403

osmotic thirst, p. 432

panic disorder, p. 427

pheromone, p. 403

phobia, p. 427

pituitary gland, p. 412

prefrontal cortex (PFC), p. 419

preparedness, p. 409

psychosurgery, p. 425

regulatory behavior, p. 411

reinforcer, p. 409

releasing hormone, p. 414

sensory deprivation, p. 401

sexual dimorphism, p. 434

sexual orientation, p. 437

somatic marker hypothesis, p. 422

transsexuality, p. 437

Please refer to the Companion Web Site at www.worthpublishers.com/kolbintro4e for Interactive Exercises and Quizzes.

CHAPTER

13

Why Do We Sleep and Dream?

CLINICAL FOCUS 13-1 DOING THE RIGHT THING AT THE RIGHT TIME

13-1 A CLOCK FOR ALL SEASONS

ORIGINS OF BIOLOGICAL RHYTHMS

BIOLOGICAL CLOCKS

BIOLOGICAL RHYTHMS

FREE-RUNNING RHYTHMS

ZEITGEBERS

CLINICAL FOCUS 13-2 SEASONAL AFFECTIVE DISORDER

13-2 NEURAL BASIS OF THE BIOLOGICAL CLOCK

SUPRACHIASMATIC RHYTHMS

KEEPING TIME

RESEARCH FOCUS 13-3 SYNCHRONIZING BIORHYTHMS AT THE MOLECULAR LEVEL

PACEMAKING CIRCADIAN RHYTHMS

PACEMAKING CIRCANNUAL RHYTHMS

COGNITIVE AND EMOTIONAL RHYTHMS

13-3 SLEEP STAGES AND DREAMING

MEASURING HOW LONG WE SLEEP

MEASURING SLEEP IN THE LABORATORY

STAGES OF WAKING AND SLEEPING

A TYPICAL NIGHT'S SLEEP

CONTRASTING NREM SLEEP AND REM SLEEP

CLINICAL FOCUS 13-4 RESTLESS LEGS SYNDROME

DREAMING

WHAT WE DREAM ABOUT

13-4 WHAT DOES SLEEP ACCOMPLISH?

SLEEP AS A BIOLOGICAL ADAPTATION

SLEEP AS A RESTORATIVE PROCESS

SLEEP AND MEMORY STORAGE

13-5 NEURAL BASES OF SLEEP

RETICULAR ACTIVATING SYSTEM AND SLEEP

NEURAL BASIS OF THE EEG CHANGES ASSOCIATED WITH WAKING

NEURAL BASIS OF REM SLEEP

13-6 SLEEP DISORDERS

DISORDERS OF NON-REM SLEEP

DISORDERS OF REM SLEEP

CLINICAL FOCUS 13-5 SLEEP APNEA

13-7 WHAT DOES SLEEP TELL US ABOUT CONSCIOUSNESS?

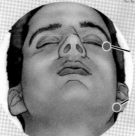

Doing the Right Thing at the Right Time

We have all heard this advice: Get eight hours of sleep every night. Eat three balanced meals each day. Scientific evidence supports this good advice. Regular sleeping and eating habits contribute to good health.

We humans are **diurnal animals** (from the Latin *dies*, meaning "day"): we are active during daylight, and we sleep when it is dark. This **circadian rhythm,** the day–night rhythm found in most animals, evolved to maximize food acquisition during the day, when we see best, and to minimize expending our energy stores by sleeping during the night when do not see well.

Today we live in environments that allow us to intrude on our circadian rhythm in two ways. Artificial lighting allows us to extend our waking hours well into the night and into sleep time. Handy food sources—often easily metabolized, high-calorie foods—allow us to enjoy mealtime anytime.

Together, these intrusions into our natural circadian rhythm contribute to a combination of medical disorders known as **metabolic syndrome** that, collectively, increase the risk of developing cardiovascular disease and diabetes (Buxton et al., 2012; Delezie & Challet, 2011).

The roots of metabolic syndrome lie in disruptions of our **biological clock,** the neural system that times behavior. One clock controls sleep–waking. Another controls the functioning of body organs, such as the liver, pancreas, and gut, related to feeding. The clock that controls sleep–waking responds to light, whereas the clock that controls feeding responds to eating. Their activity is reciprocal: disrupting one disrupts the other.

Irregular sleep and meal schedules thus change the synchrony of biological clocks. Metabolic rate, plasma glucose and pancreatic insulin secretion can slow down or speed up at inappropriate times, contributing to obesity and diabetes. Prevention and treatment for obesity and diabetes can include doing right thing at the right time when it comes to sleep schedules and mealtimes.

© AntonVengo/SuperStock.

Humans and other animals perform a remarkable number of behaviors to adapt to daily and seasonal cycles. Our daily rhythms of sleeping and waking, feeding, exercising, and social interaction vary through the year. Other animals share these daily activities and also migrate, hibernate, and shed or grow feathers or hair as the seasons change. In this chapter, we answer questions related to these daily and seasonal rhythms:

• How does the brain produce biological rhythms?

• Why has sleep evolved?

• What neural mechanisms regulate sleeping and waking?

• How do disorders of natural sleep rhythms occur?

13-1 A Clock for All Seasons

We first consider evidence that proved the existence of a biological clock, how the clock keeps time, and how it regulates our behavior. Because environmental cues themselves are not always consistent, we examine the role of biological clocks in helping us interpret environmental cues in an intelligent way.

Origins of Biological Rhythms

Biorhythms, the inherent timing mechanisms that control or initiate various biological processes, are linked to the cycles of days and seasons produced by Earth's rotation on its axis and by its progression in orbit around the sun **(Figure 13-1).** Earth rotates on its axis once every 24 hours, producing a 24-hour cycle of day and night. The day–night cycle changes across the seasons, however.

Earth's axis is tilted slightly, so as Earth orbits the sun once each year, the North and South Poles incline slightly toward the sun for part of the year and slightly away from it for the rest of the year. When inclined toward the sun, the Southern Hemisphere experiences summer: it gets more direct sunshine for more hours each day, and the climate is warmer. At the same time, the Northern Hemisphere, inclined away from the sun, experiences winter: it receives less direct sunlight, making the days shorter and the climate colder. Tropical regions near the equator undergo little seasonal or day-length change as Earth progresses around the sun.

Daily and seasonal changes have combined effects on organisms, inasmuch as the onset and duration of daily change depend on the season and latitude. Animals living in polar regions have to cope with greater fluctuations in daily temperature, light, and food availability than do animals living near the equator.

We humans evolved as equatorial animals, and our behavior is dominated by a circadian rhythm of daylight activity and nocturnal sleep. Not only does human waking and sleep behavior cycle daily, so also do pulse rate, blood pressure, body temperature, rate of cell division, blood-cell count, alertness, urine composition, metabolic rate, sexual drive, feeding behavior, and responsiveness to medications. The activity of nearly every cell in our bodies, including gene expression, also has a daily rhythm.

Biorhythms are not unique to animals. Plants display rhythmic behavior, exemplified by species in which leaves or flowers open during the day and close at night. Even unicellular algae and fungi display rhythmic behaviors related to the passage of the day. Some animals, including lizards and crabs, change color in a rhythmic pattern. The Florida chameleon, for example, turns green at night, whereas its coloration matches its environment during the day. In short, almost every living organism and every living cell displays rhythms of some sort that are related to daily changes.

diurnal animal Organism that is active chiefly during daylight.

circadian rhythm Day–night rhythm.

metabolic syndrome Combination of medical disorders, including obesity and insulin abnormalities, that collectively increase the risk of developing cardiovascular disease and diabetes.

biological clock Neural system that times behavior.

biorhythm Inherent timing mechanism that controls or initiates various biological processes.

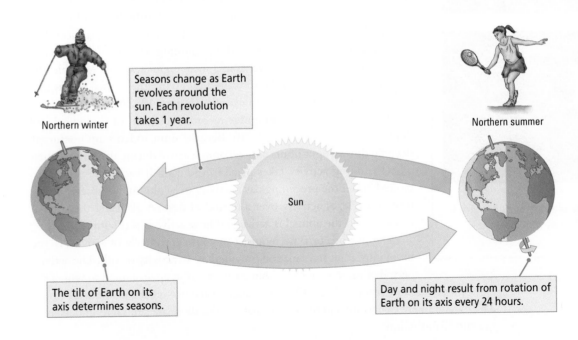

Northern winter

Seasons change as Earth revolves around the sun. Each revolution takes 1 year.

Sun

Northern summer

The tilt of Earth on its axis determines seasons.

Day and night result from rotation of Earth on its axis every 24 hours.

FIGURE 13-1 Origins of Biorhythms Each point on Earth faces the sun for part of its daily rotation cycle (daytime) and faces away from the sun for the other part (nighttime). Seasonal changes in temperature and in the amount of daylight result from the annual revolution of Earth around the sun and the tilt of its axis.

EXPERIMENT 13-1

Question: Is plant movement exogenous or endogenous?

Procedure

The movements of the plant's leaves are recorded in constant dim light.

A pen attached to a leaf is moved when the leaf moves,...

Revolving drum

Pen

Results

...producing a record of the movement.

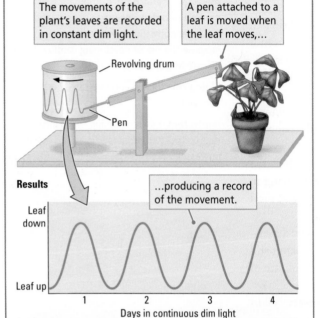

Leaf down

Leaf up

| 1 | 2 | 3 | 4 |

Days in continuous dim light

Leaf up

Leaf down

Conclusion: Movement of the plant is endogenous. It is caused by an internal clock that matches the temporal passage of a real day.

Biological Clocks

If animal behavior were affected only by daily changes in external cues, the neural mechanisms that account for changes in behavior would be simple to study. An external cue could be isolated and the neural processes that respond to the cue identified.

That behavior is not driven simply by external cues was first recognized in 1729 by the French geologist Jean Jacques d'Ortous de Mairan (see Raven et al., 1992). In an experiment similar to the one illustrated in the Procedure section of **Experiment 13-1,** de Mairan isolated a plant from daily light, dark, and temperature cues. He noted that the rhythmic movements of its leaves, seen over a light–dark cycle, continued when it was isolated, as graphed in the Results section of the experiment.

What concerned investigators who came after de Mairan was the possibility that some undetected external cue stimulates the rhythmic behavior of the plant. Such cues could include changes in gravity, in electromagnetic fields, and even in the intensity of cosmic rays from outer space. But further experiments showed that daily fluctuations are endogenous—they come from within the plant. Thus, the plant must have an endogenous biological clock.

Subsequent experiments do indeed show that most organisms have biological clocks that synchronize their behavior to the temporal passage of a real day. A biological clock signals that if daylight lasts for a given time today, it will last for about the same time tomorrow. A biological clock allows an animal to anticipate events in advance and prepare for them both physiologically and cognitively. And unless external factors get in the way, a biological clock regulates feeding times, sleeping times, and metabolic activity so that they are appropriate to day–night cycles. Biological clocks also regulate gene expression in every cell in the body so that cells function in harmony.

Biological Rhythms

Although the existence of endogenous biological clocks was demonstrated nearly 300 years ago, detailed study of biorhythms had to await the development of electrical- and computer-based timing devices. Behavioral analysis requires a method for counting behavioral events and a method for displaying the events in a meaningful way. For example, the behavior of a rodent can be measured by giving the animal access to a running wheel in which it can exercise **(Figure 13-2A).**

A chart recorder or computer records each turn of the wheel and displays the result (Figure 13-2B). Because most rodents are nocturnal, sleeping during light hours and becoming active during dark hours, their wheel-running activity takes place in the dark. If each day's activity is plotted under the preceding day's activity in a column, we observe a pattern, or cycle, of activity over a period of time. A glance at the pattern reveals when the animal is active and how active it is (Figure 13-2C).

The **period,** or time required to complete a cycle of activity, is one bit of important information provided by an activity record. The activity period of most animals is about 24 hours in an environment in which the lights go on and off with regularity. Our own sleep–wake period also is about 24 hours. The measurement of periods and the events that control them are central to understanding circadian rhythms.

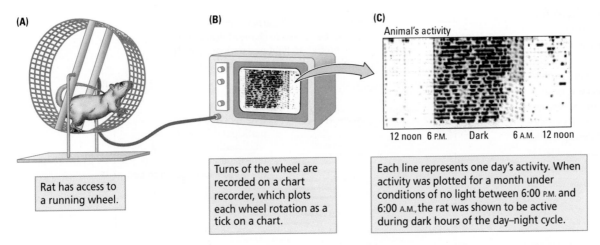

(A) Rat has access to a running wheel.

(B) Turns of the wheel are recorded on a chart recorder, which plots each wheel rotation as a tick on a chart.

(C) Animal's activity

12 noon 6 P.M. Dark 6 A.M. 12 noon

Each line represents one day's activity. When activity was plotted for a month under conditions of no light between 6:00 P.M. and 6:00 A.M., the rat was shown to be active during dark hours of the day–night cycle.

FIGURE 13-2 Recording the Daily Activity Cycle of a Rat Adapted from *Biological Clocks in Medicine and Psychiatry* (pp. 12–15), by C. P. Richter, 1965, Springfield, IL: Charles C Thomas.

In Latin, *circa* means "about," *annum* means "year," and *dies* means "day."

Many behaviors have periods that are longer or shorter than this 24-hour circadian rhythm. *Circannual rhythms* last about a year. The yearly migratory and mating cycles of animals are circannual. Other biorhythms have monthly or seasonal periods greater than a day but less than a year. These are *infradian rhythms.* The menstrual cycle of female humans, with an average period of about 28 days, is an infradian biorhythm linked to the cycle of the moon and thus also referred to as a *circalunar cycle.*

Ultradian rhythms have a period of less than one day. Our eating behavior, which takes place about every 90 minutes to 2 hours, including snacks, is one ultradian rhythm. Rodents, although active throughout the night, display an ultradian rhythm in being most active at the beginning and end of the dark period.

Biological rhythm	Time frame	Example
Circannual	Yearly	Migratory cycles of birds
Circadian	Daily	Human sleep–wake cycle
Ultradian	Less than a day	Human eating cycles
Infradian	More than a day	Human menstrual cycle

The fact that a behavior appears to be rhythmic does not mean that it is ruled by a biological clock. Animals may postpone migrations as long as food supplies last. They adjust their circadian activities in response to food availability, the presence of predators, and competition from other members of their own species. We humans obviously change our daily activities in response to seasonal changes, work schedules, and play opportunities. Therefore, whether a rhythmic behavior is produced by a biological clock and the extent to which it is controlled by a clock must be demonstrated experimentally.

Free-Running Rhythms

To determine whether a rhythm is produced by a biological clock, researchers have designed three types of tests in which they remove relevant cues. A test can be given (1) in continuous light or (2) continuous darkness, or (3) the selection of light or darkness can be left to the participant. Each treatment gives a slightly different insight into the periods of biological clocks.

Jurgen Aschoff and Rutger Weber first demonstrated that the human sleep–waking rhythm is governed by a biological clock (see Kleitman, 1965). They allowed participants to select their light–dark cycle and studied them in an underground bunker where no cues signaled when day began or ended. The participants selected the periods when their

period Time required to complete a cycle of activity.

FIGURE 13-3 Free-Running Rhythm in a Human Subject The record for days 1 through 3 shows the daily sleep period under normal day–night conditions. The record for days 4 through 20 shows the free-running rhythm that developed while the subject was isolated in a bunker and allowed to control day and night length. The daily activity period shifts from 24 hours to 25.9 hours. On days 21 through 25, the period returns to 24 hours when the subject is again exposed to a normal light-and-dark cycle. Adapted from *Sleep* (p. 33), by J. A. Hobson, 1989, New York: Scientific American Library.

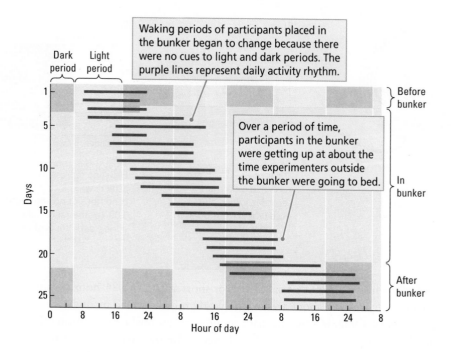

lights were on or off, when they were active, and when they slept. In short, they selected the length of their own day and night.

Measures of ongoing behavior and recording of sleeping periods with sensors on the beds revealed that the participants continued to show daily sleep–activity rhythms. This finding demonstrated that humans have an endogenous biological clock that governs sleep–waking behavior. **Figure 13-3** shows, however, that the biorhythm was different when compared with biorhythms before and after isolation. Although the period of the participants' sleep–wake cycles approximated 24 hours before and after the test, during the test they progressively deviated from clock time. Rather than being 24 hours, the period in the bunker ranged from about 25 to 27 hours, depending on the participant.

The participants were choosing to go to bed from 1 to 2 hours later every "night." Soon they were getting up at about the time the experimenters outside the bunker were going to bed. Clearly, the participants were displaying their own personal cycles. Such a **free-running rhythm** runs at a frequency of the body's own devising when environmental cues are absent. Humans' free-running rhythm is slightly longer than 24 hours.

The period of free-running rhythms also depends on the way in which external cues are removed. When hamsters, a nocturnal species, are tested in constant darkness, their free-running periods are a little shorter than 24 hours; when they are tested in constant light, their free-running periods are a little longer than 24 hours. This test dependency is typical of nocturnal animals.

As **Figure 13-4** shows, the opposite free-running periods are typical of diurnal animals (Binkley, 1990). When sparrows, which are diurnal birds, are tested in constant darkness, their free-running periods are a little longer than 24 hours; when they are tested in constant light, their free-running periods are a little shorter than 24 hours.

A rule of thumb to explain the period of free-running rhythms in light or dark is that animals expand and contract their sleep periods as the sleep-related period—light for hamsters and dark for sparrows—expands or contracts. Understanding this point enables us to predict how excess artificial lighting, which expands the light portion of

free-running rhythm Rhythm of the body's own devising in the absence of all external cues.

Zeitgeber Environmental event that entrains biological rhythms: a "time giver."

entrain Determine or modify the period of a biorhythm.

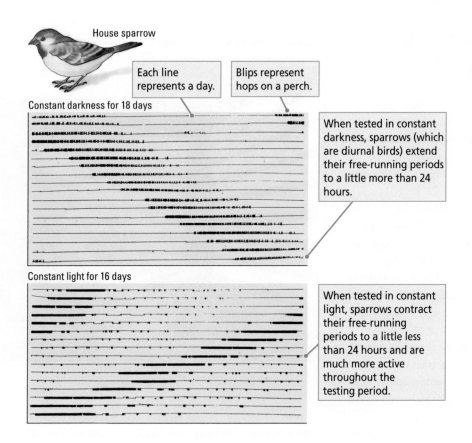

House sparrow

Each line represents a day.

Blips represent hops on a perch.

Constant darkness for 18 days

When tested in constant darkness, sparrows (which are diurnal birds) extend their free-running periods to a little more than 24 hours.

Constant light for 16 days

When tested in constant light, sparrows contract their free-running periods to a little less than 24 hours and are much more active throughout the testing period.

FIGURE 13-4 Free-Running Rhythms of a Diurnal Animal *Adapted from The Clockwork Sparrow* (p. 16), by S. Binkley, 1990, Englewood Cliffs, NJ: Prentice Hall.

our days, influences our circadian periods. Our sleep periods contract and we get less sleep each night.

Zeitgebers

Endogenous rhythmicity is not the only factor that contributes to circadian periods. There must be a mechanism for setting rhythms to correspond to environmental events. To be useful, the biological clock must keep to a time that predicts actual changes in the day–night cycle. If a biological clock is like a slightly defective wristwatch, it will eventually provide times that are inaccurate by hours and so become useless.

If we reset an errant wristwatch each day, however—say, when we awaken—it will then provide useful information even though it is not perfectly accurate. There is an equivalent way of resetting a free-running biological clock. Sunrise and sunset, eating times, and many other activities all influence the period of the circadian clock.

Aschoff and Weber called a clock-setting cue a **Zeitgeber** ("time giver" in German). When a clock is reset by a Zeitgeber, it is said to be **entrained.** Normally, light is the most potent entraining stimulus. Clinical Focus 13-2, "Seasonal Affective Disorder" on page 450, explains the importance of light in entraining circadian rhythms.

The property that allows the biological clock to be entrained allows the circadian rhythms to synchronize with seasonal changes in day–night duration. North and south of the equator, you recall, the time of onset and the length of day and night change as the seasons progress. At extreme latitudes, daylight begins very early in the morning in summer and very late in the morning in winter. An entrained biological clock allows an animal to synchronize its activity with seasonal changes.

A biological clock that is reset each day tells an animal that daylight will begin to-morrow at approximately the same time that it began today and that tomorrow will last

Seasonal Affective Disorder

In *seasonal affective disorder* (SAD), a form of depression associated with winter, sunlight does not entrain the circadian rhythm. The perception of longer nights by the circadian pacemaker stimulates pressure for more sleep. If not satisfied, cumulative sleep deprivation can result. Consequently, a person's biorhythm becomes a free-running rhythm.

Because people vary in the duration of their free-running rhythms, the lack of entrainment affects individuals differently. Some are "phase-retarded," with desired sleep time coming earlier each day; some are "phase-delayed," with desired sleep time coming later each day.

The cumulative changes associated with altered circadian rhythms can result in depression. The finding that the incidence of symptoms of depression increases as a function of the latitude at which a person lives supports this idea.

Researchers report that light can ameliorate the depression of SAD, and one treatment, *phototherapy*, uses light to entrain the circadian rhythm. The idea is to increase the short winter photoperiod by exposing a person to artificial bright light in the morning or both morning and evening. Typical room lighting is not bright enough.

Recent findings show that a class of retinal ganglion cells that express a photosensitive pigment called melanopsin are responsive to blue light (see Section 13-2). Meesters and colleagues (2011) compared standard bright-light treatment (10,000 lux) to a blue-enriched white light

A participant in research on SAD designed to investigate how differing types of lighting affect mood.

of lower intensity (750 lux). They found both light regimes equally effective in treating the symptoms of SAD and propose that blue-enriched white lighting could be used in the home and workplace in northern or southern latitudes to prevent SAD from developing.

If a hamster happens to blink during this Zeitgeber, the light will still penetrate its closed eyelids and entrain its biological clock.

approximately as long as today did. Current research finds that light Zeitgebers are effective at both sunrise and sunset. Schwartz and coworkers (2011) suggest that clock time can be adjusted twice each day; morning light sets the biological clock by advancing it, and evening darkness sets the clock by retarding it.

The very potent entraining effect of light Zeitgebers is illustrated by laboratory studies of Syrian hamsters, perhaps one of the most compulsive animal timekeepers. When given access to running wheels, the hamsters exercise during the night segment of the laboratory day–night cycle. A single brief flash of light is an effective Zeitgeber for entraining their biological clocks.

Considering the less compulsive behavior that most of us display, we should shudder at the way we entrain our own clocks when we stay up late in artificial light, sleep late some days, and get up early by using an alarm clock on other days. **Light pollution,** the extent to which we are exposed to artificial lighting, disrupts circadian rhythms and accounts for a great deal of inconsistent behavior associated with accidents, daytime fatigue, alterations in emotional states, and obesity and diabetes.

Entrainment works best if the adjustment to the biological clock is not too large. People who work shifts are often subject to huge adjustments, especially when they work the graveyard shift (11:00 P.M. to 7:00 A.M.), the period when they would normally sleep. The results of studies show that adapting to such a change is difficult and stressful. Metabolic syndrome, described in Clinical Focus 13-1, is higher in people who work shifts compared with people who have daytime work schedules (Pietroiusti et al., 2010).

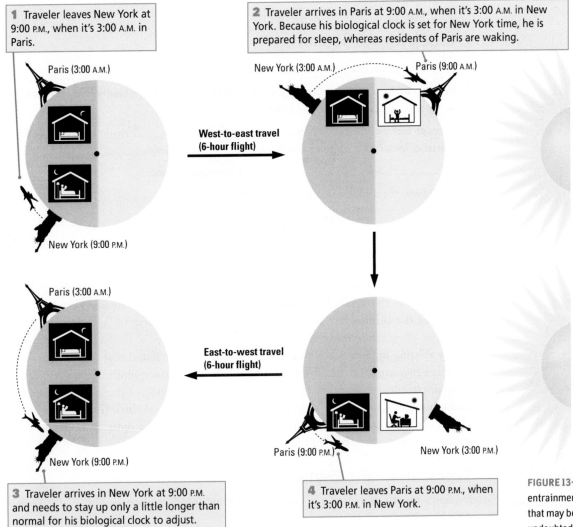

1 Traveler leaves New York at 9:00 P.M., when it's 3:00 A.M. in Paris.

Paris (3:00 A.M.)

New York (9:00 P.M.)

West-to-east travel (6-hour flight)

2 Traveler arrives in Paris at 9:00 A.M., when it's 3:00 A.M. in New York. Because his biological clock is set for New York time, he is prepared for sleep, whereas residents of Paris are waking.

New York (3:00 A.M.)   Paris (9:00 A.M.)

Paris (3:00 A.M.)

New York (9:00 P.M.)

East-to-west travel (6-hour flight)

Paris (9:00 P.M.)   New York (3:00 P.M.)

3 Traveler arrives in New York at 9:00 P.M. and needs to stay up only a little longer than normal for his biological clock to adjust.

4 Traveler leaves Paris at 9:00 P.M., when it's 3:00 P.M. in New York.

FIGURE 13-5 Jet Lag Disruption in the entrainment of a person's biological clock that may be brought on by jet travel is undoubtedly more pronounced in west-to-east travel because the disruption in the person's circadian rhythm is dramatic. On the return journey, the traveler's biological clock has a much easier adjustment to make.

Thus, shift workers should be vigilant in maintaining good sleep habits and diet and in exercising to minimize other risk factors for metabolic syndrome. Adaptations to shift work are better if people first work the swing shift (3:00 P.M. to 11:00 P.M.) for a time before beginning the graveyard shift.

Long-distance air travel—say, from North America to Europe or Asia—also demands a large and difficult time adjustment. For example, travelers flying east from New York to Paris will begin their first European day just when their biological clocks are signaling that it is time for sleep **(Figure 13-5)**. The difference between a person's circadian rhythm and the daylight cycle in a new environment can produce the feeling of disorientation and fatigue called **jet lag.**

The west-to-east traveler generally has a more difficult adjustment than does the east-to-west traveler, who needs to stay up only a little longer than normal. The occasional traveler may cope with jet lag quite well, but frequent travelers, such as airline personnel, face a substantial adaptive challenge. Jet lag for the occasional traveler can be managed with sleep on arrival or shortly after. The brain's biological clock resets in a day, and other body organs follow after about a week. For frequent travelers and flight crews, resetting is not so easy. Persistent asynchronous rhythms generated by jet lag are associated with altered sleep and temperature rhythms, fatigue, and stress. (Ariznavarreta et al., 2002)

light pollution Exposure to artifical light that changes activity patterns and so distrupts circadian rhythms.

jet lag Fatigue and disorientation resulting from rapid travel through time zones and exposure to a changed light–dark cycle.

A Clock for All Seasons

Before you continue, check your understanding.

1. Many behaviors occur in a rhythmic pattern in relation to time. These biorhythms may display a yearly, or _____, cycle or a daily _____ cycle.

2. Although biological clocks keep fairly good time, their _____ rhythms may be slightly shorter or longer than 24-hours unless they are reset each day by _____.

3. _____ and _____ can disrupt circadian rhythms.

4. Explain why the circadian rhythm is important.

Answers appear at the back of the book.

13-2 Neural Basis of the Biological Clock

Curt Richter (1965) was the first researcher who attempted to locate biological clocks in the brain. In the 1930s, he captured wild rats and tested them in activity wheels. He found that the animals ran, ate, and drank when the lights were off and were relatively quiescent when the lights were on.

By ablating brain tissue with electric current, Richter found that animals lost their circadian rhythms after damage to the hypothalamus. Subsequently, by making much more discrete lesions, experimenters have shown that a region of the hypothalamus, the **suprachiasmatic nucleus** (SCN), acts as the master biological clock (Ralph & Lehman, 1991). The SCN is named for its location just above (supra) the optic chiasm, where the optic tracts cross at the base of the hypothalamus.

Suprachiasmatic Rhythms

Evidence for the SCN's role in circadian rhythms comes from several lines of evidence (Schibler, 2009):

1. If the suprachiasmatic nuclei are selectively damaged, animals still eat, drink, exercise, and sleep, but they do these activities at haphazard times.

2. If a form of glucose that is taken up by metabolically active cells but is not used by them and cannot escape from them is tagged with a radioactive label, cells that are more active will subsequently emit more radioactivity. When this tracer is injected into rodents, more tracer is found in the SCN after injections given in the light period of the light–dark cycle than after injections given in the dark period. This experiment demonstrates that suprachiasmatic cells are more active during the light period.

3. Recording electrodes placed in the SCN confirm that neurons in this region are more active during the light period of the cycle than during the dark period.

4. If all the pathways into and out of the suprachiasmatic nucleus are cut, SCN neurons maintain their rhythmic electrical activity.

5. SCN cells removed from the brain and cultured in a dish retain a periodic rhythm.

Clearly, suprachiasmatic neurons have an intrinsically rhythmic pattern of activity, which has earned the SCN the designation "master biological clock."

Although the SCN is the master clock, it is not the sole biological clock. Two other neural structures, the intergeniculate leaflet and the pineal gland, also display clocklike activity. Further, nearly every cell in the body has its own clock.

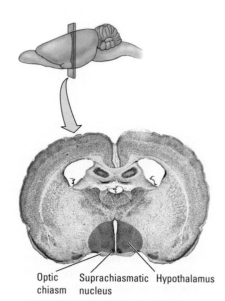

Optic   Suprachiasmatic   Hypothalamus
chiasm   nucleus

Suprachiasmatic Nucleus in a Rat Brain

suprachiasmatic nucleus (SCN) Master biological clock, located in the hypothalmus just above the optic chiasm.

retinohypothalamic tract Neural route formed by axons of photosensitive retinal ganglion cells from the retina to the suprachiasmatic nucleus; allows light to entrain the rhythmic activity of the SCN.

After the SCN is destroyed, some behaviors retain a timed occurrence. One is feeding. Animals without an SCN can still display anticipatory behavior—becoming active in relation to scheduled meal times—and can organize related behaviors, including memory for food locations in relation to meal times (Antle & Silver, 2009). How this anticipatory behavior is timed is not known, but whatever the mechanism, it can also act as a Zeitgeber for the main SCN clock. That is, a regular feeding schedule can entrain the SCN clock and many other organs and cells of the body.

Keeping Time

If SCN neurons are isolated from one another, each remains rhythmic, but the rhythmicity of some cells is different from that of other cells. Thus, rhythmic activity is a property of SCN cells, but the timing of the rhythm must be set so that the cells can both synchronize their activity in relation to each other and in relation to Zeitgebers. In the brain, SCN cells connect one to another through inhibitory GABA synapses, and these connections allow them to act in synchrony. Their entrainment depends upon external inputs.

The SCN receives information about light through the **retinohypothalamic tract (Figure 13-6).** This pathway begins with specialized retinal ganglion cells (RGCs) that contain the photosensitive pigment melanopsin. These melanopsin-containing, or *photosensitive RGCs,* receive light-related signals from the rods and cones and send that information to visual centers in the brain. However, pRGCs also can be activated directly by certain wavelengths of blue light in the absence of rods and cones.

Photosensitive retinal ganglion cells are distributed across the retina and, in humans, make up between 1 and 3 percent of all RGCs. Their axons project to various regions in the brain, including the SCN, which they innervate bilaterally. Melanopsin-containing ganglion cells use glutamate as their primary neurotransmitter but also contain two cotransmitters, substance P and pituitary adenylate cyclase-activating polypeptide (PACAP).

When stimulated by light, pRGCs are excited and in turn excite cells in the SCN. The existence of light-sensitive retinal ganglion cells that are involved in entraining the circadian rhythm explains the continued presence of an entrained rhythm in people who are blind due to retinal degeneration that destroys the rods and cones (Zaidi et al., 2007). Even so, pRGCs do receive inputs from cones and rods. Cones can influence their activity in bright daylight, and rods can influence their activity in dim light.

As illustrated in Figure 13-6, the SCN consists of two parts, a more ventrally located *core* and a more dorsally located *shell.* The retinohypothalamic tract activates the core cells. Core neurons are not rhythmic, but they entrain the shell neurons, which are rhythmic.

In addition to the retinohypothalamic input, the SCN receives projections from a number of brain regions. These include the intergeniculate leaflet in the thalamus and the raphé nucleus, the nonspecific serotonergic activating system of the brainstem. The terminal region of these inputs onto the SCN display regional variations, suggesting that various portions of the shell and the core have somewhat different functions.

The SCN's circadian rhythm is usually entrained by morning and evening light, but it can also be entrained or disrupted by sudden changes in lighting, by arousal, by

GABA is main inhibitory neurotransmitter in the CNS.

Section 9-2 traces three main routes from the retina to the visual brain. Figure 9-8 diagrams the retina's cellular structure.

Glutamate is main excitatory neurotransmitter in the CNS.

FIGURE 13-6 The Retinohypothalamic Tract and the SCN

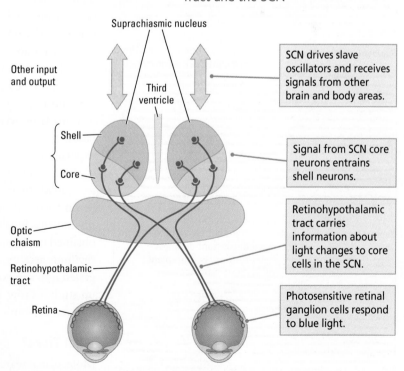

Suprachiasmic nucleus

Other input and output

Third ventricle

Shell

Core

Optic chaism

Retinohypothalamic tract

Retina

SCN drives slave oscillators and receives signals from other brain and body areas.

Signal from SCN core neurons entrains shell neurons.

Retinohypothalamic tract carries information about light changes to core cells in the SCN.

Photosensitive retinal ganglion cells respond to blue light.

moving about, and by feeding. These influences differ from the light entrainment provided over the retinohypothalamic tract.

The intergeniculate leaflet and the raphé nucleus are pathways through which other photic and nonphotic events influence the SCN rhythm (Cain et al., 2007). The neural structures mediating food-related timing are not known, but these other entraining pathways explain why being subjected to bright light during the dark portion of the circadian cycle and being aroused or eating during the sleep portion of the circadian cycle disrupt the cycle (Mistlberger & Antle, 2011).

It is likely that the regional variation in the anatomy of SCN shell area provides the substrate for different rhythms within the circadian cycle. Findings from studies on the genes that control rhythms in fruit flies suggest the existence of two separate groups of circadian neurons. *M cells* control morning activity and need morning light for entrainment; *E cells* control evening activity and need darkness onset for entrainment (Stoleru et al., 2007).

Some people are early to bed and early to rise and are energetic in the morning. Other people are late to rise and late to bed and are energetic in the evening. It seems likely that the individual differences between these "lark" and "owl" **chronotypes** are due to differences in SCN shell neurons. The differences may be the equivalent of fruit fly M cells and E cells that, when expressed differently in people, account for differences in the amplitude of phases in the circadian period (Brown et al., 2008).

In hamsters and mice, mutant gene variations produce strains of chronotypes with circadian periods as varied as 24, 20 or 17 hours (Monecke et al., 2011). As genetic analysis becomes less expensive, the study of the genes underlying human chronotypes will provide insights into individual differences in our biological clocks.

Immortal Time

How do suprachiasmatic cells develop their rhythmic activity? The endogenous rhythm is not learned. When animals are raised in constant darkness, their behavior still becomes rhythmic. In experiments in which animals have been maintained without entraining cues for a number of generations, each generation continues to display rhythmic behavior. Even if the mother has received a lesion of the SCN so that her behavior is not rhythmic, the behavior of her offspring is rhythmic.

A line of evidence supporting the idea that suprachiasmatic cells are genetically programmed for rhythmicity comes from studies performed in Canada by Martin Ralph and his coworkers with the use of transplantation techniques (Ralph & Lehman, 1991). **Figure 13-7** illustrates the experiment.

First, hamsters are tested in constant dim light or in constant darkness to establish their free-running rhythm. They then receive a suprachiasmatic lesion followed by another test to show that the lesion has abolished their rhythmicity. Finally, the hamsters receive transplants of suprachiasmatic cells obtained from hamster embryos. About 60 days later, the hamsters again show rhythmic activity, demonstrating that the transplanted cells have become integrated into the host brain and have reestablished rhythmic behavior. Follow-up studies show that the rhythms of many but not all body organs show restored rhythmic activity.

What Ticks?

Molecular research is directed toward determining what genes control the ticking of the circadian clock. The timing device is in each SCN neuron and in most other cells of the body as well.

FIGURE 13-7 Circadian Rhythms Restored by Neural Transplantation

Adapted from "Transplantation: A New Tool in the Analysis of the Mammalian Hypothalamic Circadian Pacemaker," by M. R. Ralph and M. N. Lehman, 1991, *Trends in Neurosciences, 14,* p. 363.

Normal

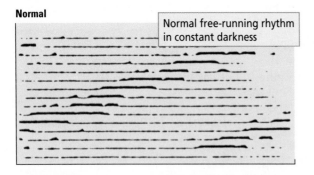

Normal free-running rhythm in constant darkness

Suprachiasmatic lesion

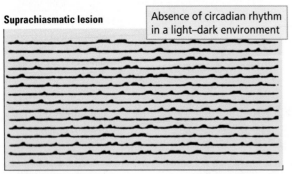

Absence of circadian rhythm in a light–dark environment

Suprachiasmatic transplant

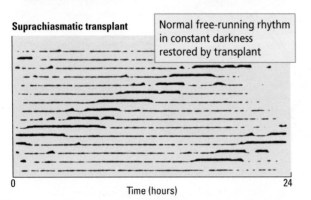

Normal free-running rhythm in constant darkness restored by transplant

0   Time (hours)   24

RESEARCH FOCUS ⟷ 13-3

Synchronizing Biorhythms at the Molecular Level

The retinohypothalamic tract carries information about light changes from the melanopsin-containing photosensitive retinal ganglion cells to the core of the SCN, the brain's main biological timekeeper (see Figure 13-6). Core cells synapse with SCN neurons in the shell and entrain the shell cells to the light signal's 24-hour rhythm.

The timing mechanism in the shell cells is a pair of interlocking feedback loops that pace the SCN's function over a 24-hour period. The complete clockwork in the two feedback loops involves as many as ten genes and their protein products. The secondary loop regulates functions for the main clockwork loop.

The main clock mechanism is the *transcription-translation-inhibition-feedback* loop. Its sequence, which is mapped in the illustration, follows:

STEP 1: TRANSCRIPTION In the cell nucleus, three *Period* genes (*Per1*,* *Per2*, *Per3*) and two *Cryptochrome* genes (*Cry1*, *Cry2*) are transcribed into *Per1*, *Per2*, and *Per3* messenger RNA and *Cry1* and *Cry2* mRNA.

STEP 2: TRANSLATION In the endoplasmic reticulum, ribosomes translate these mRNAs into the proteins PER1,[†] PER2, PER3 and CRY1, CRY2. In the intracellular fluid, the proteins then form various **dimers,** or two-protein combinations, such as PERCRY.

STEP 3: INHIBITION PERCRY dimers enter the cell nucleus, where they bind to and inhibit the CB dimer (formed by the CLOCK and BMAL proteins). The CB dimer turns on the Enhancer box (Ebox), a part of the DNA that activates transcription of the *Period* and *Cryptochrome* genes. So when the CB dimer is inhibited, the *Per* and *Cry* genes are no longer expressed.

Gene and *mRNA* names are italicized.
[†]PROTEIN names are capitalized.

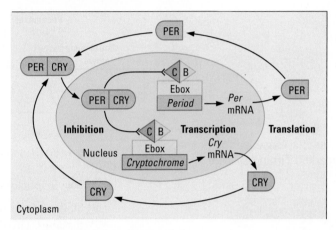

Main SCN Clock Mechanism

STEP 4: DECAY After they play their inhibitory role, the PERCRY proteins decay. Then the CB dimer resumes its activity, the *Per* and *Cry* genes resume expression, and the 24-hour cycle begins anew.

This sequence of gene turn-on followed by gene turn-off occurs in an inexorable, daily loop.

Mutations in any circadian gene can lead to circadian alterations, including absence of a biorhythm or an altered biorhythm. For example, alleles of *Period 1* and *Period 2* genes determine chronotype—whether an individual will be "early to bed and early to rise" or "late to bed and late to rise." The search for therapies for clock disorders includes looking for small molecules that can serve as drugs to reset biorhythms disrupted by jet lag, shift work, and epigenetic and inherited gene irregularities.

The circadian rhythm involves a feedback loop in which proteins are first made and then combine. The combined protein, called a *dimer* for "two proteins," inhibits the production of its component proteins. Then the dimer degrades and the process begins anew. Research Focus 13-3, "Synchronizing Biorhythms at the Molecular Level," describes the main feedback loop in mammals. Just as the back-and-forth swing of a pendulum makes a grandfather clock tick, the increase and decrease in protein synthesis once each day produces the cellular rhythm.

Figure 3-13 diagrams the process of protein synthesis.

Pacemaking Circadian Rhythms

The SCN itself is not responsible for directly producing behavior. After it has been damaged, drinking and eating, sleeping and wakefulness still occur. They no longer occur at appropriate times, however.

An explanation for how the SCN controls biological rhythms is illustrated in **Figure 13-8.** In this model, light entrains the SCN, and the pacemaker in turn drives a number of "slave" oscillators. Each slave oscillator is responsible for the rhythmic occurrence of one activity. In other words, drinking, eating, body temperature, and sleeping are each produced by a separate slave oscillator.

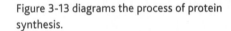

chronotype Individual differences in circadian activity.

dimer Two proteins combined into one.

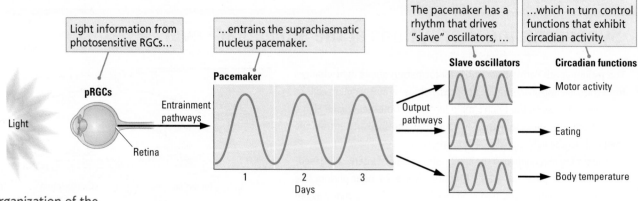

Light information from photosensitive RGCs...

...entrains the suprachiasmatic nucleus pacemaker.

The pacemaker has a rhythm that drives "slave" oscillators, ...

...which in turn control functions that exhibit circadian activity.

Slave oscillators   Circadian functions

Pacemaker

pRGCs

Light

Entrainment pathways

Retina

Output pathways

Motor activity

Eating

Body temperature

Days

FIGURE 13-8 Organization of the Circadian Timing System

Figure 12-14 diagrams how the pituitary gland works.

Figure 2-30 diagrams the arousing and calming divisions of the autonomic nervous system. Section 6-5 describes how glucocorticoids affect the body and brain.

melatonin Hormone secreted by the pineal gland during the dark phase of the day–night cycle; influences daily and seasonal biorhythms.

The SCN clock entrains slave oscillators in at least three ways:

1. SCN neurons send axonal connections to nuclei close by in the hypothalamus and thalamus. These nuclei in turn have extensive connections with other brain and body structures to which they pass on the entraining signal.

2. The SCN connects with pituitary endocrine neurons to control the release of a wide range of hormones. These hormones circulate through the body to entrain many body tissues and organs.

3. The SCN also sends indirect messages to autonomic neurons in the spinal cord to inhibit the pineal gland from producing the hormone **melatonin,** which influences daily and seasonal biorhythms.

4. SCN cells themselves release hormones. Silver and colleagues (1996) used a transplantation technique in which encapsulated SCN cells were transplanted into hamsters that had received SCN lesions. Even through the transplanted cells did not make axonal connections, they restored many circadian behaviors, indicating that some SCN signals must be hormonal and travel through the bloodstream.

An illustration of the widespread effects of the SCN is seen in its control of two hormones, melatonin and glucocorticoids. The SCN controls the release of melatonin from the pineal gland so that it circulates during the dark phase of the circadian cycle, and it controls the release of glucocorticoids from the adrenal gland so that they circulate during the light phase of the circadian cycle.

Melatonin has sleep-promoting actions and influences the parasympathetic rest and digest system. Glucocorticoids mobilize glucose for cellular activity and can support arousal responses in the sympathetic system. These two hormones will entrain any body organ that has receptors for them, and most organs do. Thus, melatonin promotes rest activities and glucocorticoids promote arousal activities during the dark and the light portions of the circadian cycle, respectively. Their actions explain in part why it is difficult to sleep during the day and stay awake to work at night.

Pacemaking Circannual Rhythms

The suprachiasmatic nucleus not only controls daily rhythms, it also controls circannual rhythms. Russel Reiter (1980) illustrates this form of pacemaking in hamsters. Hamsters are summertime (long-day) breeders. As the days lengthen in springtime, the gonads of male hamsters grow and release hormones that stimulate sexual behavior. As the days shorten in the winter, the gonads shrink, the amount of the hormones produced by the gonads decreases, and the males lose interest in sex.

During the dark phase of the day–night cycle, the pineal gland secretes melatonin and during the light phase it does not. **Figure 13-9** shows that, when a hamster's

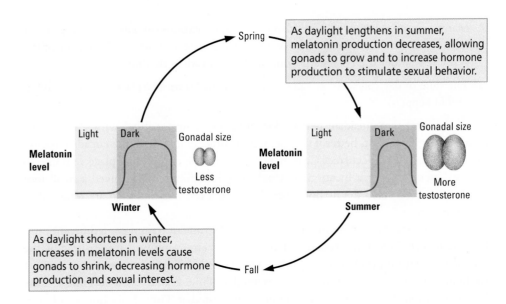

Spring

As daylight lengthens in summer, melatonin production decreases, allowing gonads to grow and to increase hormone production to stimulate sexual behavior.

Light Dark Gonadal size

Melatonin level

Less testosterone

Winter

Light Dark Gonadal size

Melatonin level

More testosterone

Summer

As daylight shortens in winter, increases in melatonin levels cause gonads to shrink, decreasing hormone production and sexual interest.

Fall

FIGURE 13-9 Hamster's Circannual Pacemaker Adapted from "The Pineal and Its Hormones in the Control of Reproduction in Mammals," by R. J. Reiter, 1980, *Endocrinology Review, I,* p. 120.

melatonin level is low, the gonads enlarge and when it is high, the gonads shrink. The control that the pineal gland exerts over the gonads is in turn controlled by the suprachiasmatic nucleus. Through connections in the autonomic nervous system, the SCN drives the pineal gland as a slave oscillator.

During the daylight period of the circadian cycle, the SCN inhibits melatonin secretion by the pineal gland. Thus, as the days become shorter, the period of inhibition becomes shorter and the period in which melatonin is released becomes longer. When the daylight period is shorter than 12 hours, melatonin release becomes sufficiently long to inhibit the hamster's gonads, and they shrink.

Melatonin also influences the testes of animals that are short-day breeders, such as sheep and deer, which mate in the fall and early winter. The effect of melatonin on reproductive behavior in these species is the reverse of that in the hamster: their reproductive activities begin as melatonin release increases.

In his classic book *Biological Clocks in Medicine and Psychiatry,* Curt Richter (1965) hypothesized that many physical and behavioral disorders might be caused by "shocks," either physical or environmental, that upset the timing of biological clocks. For example, the record of psychotic attacks suffered by the English writer Mary Lamb, illustrated in **Figure 13-10,** is one of many rhythmic records that Richter thought represented the action of an abnormally functioning biological clock.

FIGURE 13-10 Dysfunctional Clock? Attacks of mental illness displayed by the English writer Mary Lamb through her adult life appear to have had a cyclical component. Such observations would be difficult to obtain today because the drugs used to treat psychiatric disorders can mask abnormal biorhythms. Adapted from *Biological Clocks in Medicine and Psychiatry* (p. 92), by C. P. Richter, 1965, Springfield, IL: Charles C Thomas.

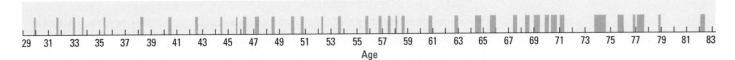

| 29 | 31 | 33 | 35 | 37 | 39 | 41 | 43 | 45 | 47 | 49 | 51 | 53 | 55 | 57 | 59 | 61 | 63 | 65 | 67 | 69 | 71 | 73 | 75 | 77 | 79 | 81 | 83 |

Age

Cognitive and Emotional Rhythms

Because circadian rhythms are synchronized in many parts of the body, they influence cognitive and emotional functions, including emotional experience, learning and retention, decision making, and motivation, in a phasic manner. Among the reasons to postulate that cognitive and emotional function is least in part influenced by circadian rhythms are the following:

1. The extent to which a process depends on the expenditure of metabolic energy dictates optimal times for its activity during the circadian cycle.

2. Many cognitive events require changes in gene expression, either turning on or turning off genes. Studies of gene expression suggest as much as 10 percent of the genome is under the epigenetic control of the circadian rhythm.

3. The time of day can serve as a good index for the "time and place" at which things should happen.

Studies find that circadian period does influence emotional behavior. Animals remember previous events better if they are tested at the same time of the circadian period that the initial event occurred. Cain et al. (2008) used a conditional place response to test rats' memories for a location at which they had previously received a small foot shock. The rats receive the noxious stimulation in a test box and then are returned to their home cage.

The following day, groups of rats are tested at the time of training or a different time. Rats trained at Time 1 and tested at Time 1 show better retention as measured by the time they spent immobile (freezing) than rats trained at Time 1 and tested at Time 2, six hours later in the day. Similarly, rats trained at Time 2 and tested at Time 2 show better retention than rats trained at Time 2 and tested at Time 1, six hours earlier in the day. The results show that the circadian cycle puts a "time stamp" on the memory that allows it to be better recalled at the same time of the period that the initial event occurred.

Cognitive behaviors also synchronize with circadian periods. Animals have a variety of needs, for food, water, and sleep, and it is adaptive for them to know when these resources become available. If food is available on a schedule, animals display anticipatory behaviors, such as becoming active, as the feeding time arrives. Other anticipatory events include salivation, intestinal activity, and sensations of hunger. Anticipatory activity might also include recalling events related to the timing of feeding and the location of food. Although many such cognitive activities can occur in the absence the SCN, it is adaptive for them to occur at the right time and place, and the SCN enables them to do so (Antle & Silver, 2009).

REVIEW 13-2
Neural Basis of the Biological Clock

Before you continue, check your understanding.

1. Biological rhythms are timed by internal biological clocks. The master clock is the

   _____.

2. Light cues entrain the suprachiasmatic nucleus to control daily rhythms via the _____ tract, which receives information via _____ cells.

3. Pacemaking produced by the SCN is a product of its _____ cells, which activate slave oscillators via both _____ signals and _____ connections.

4. Why should studying for an exam and taking the exam occur at the same time of day?

Answers appear at the back of the book.

13-3 Sleep Stages and Dreaming

Waking behavior encompasses periods when we are physically inactive and periods when we are physically active, and so does sleep. Our sleeping behavior consists of periods of resting, napping, long bouts of sleep, and various sleep-related events including snoring, dreaming, thrashing about, and even sleepwalking. In this section, we describe some sleeping and dreaming behaviors and the neural processes that underlie them.

Measuring How Long We Sleep

A crude measure of sleeping and waking behavior is the *self-report:* people record in a diary when they wake and when they retire to sleep. The diaries show considerable variation in sleep–waking behavior. People sleep more when they are young than when they are old. Most people sleep about 7 to 8 hours per night, but some people sleep much more or much less than that, even as little as 1 hour each day.

Some people nap for a brief period in the daytime; others never nap. Benjamin Franklin is credited with the aphorism "Early to bed and early to rise makes a man healthy, wealthy, and wise," but measures of sleep behavior indicate that the correlation Franklin proposed does not actually exist. Variations in sleeping times are normal and napping can be good.

Measuring Sleep in the Laboratory

Laboratory sleep studies allow researchers to measure sleep accurately and to record physiological changes associated with sleep. Measuring sleep requires recording at least three electrical body signals, and the signals are obtained with a polygraph. **Figure 13-11** illustrates a typical polygraph setup in a sleep laboratory and some commonly used measures that define sleep.

Electrodes pasted onto standard locations on the skull's surface yield an electroencephalogram (EEG), a record of brain-wave activity. Electrodes placed on neck muscles provide an electromyogram (EMG), a record of muscle activity. Electrodes located near the eyes provide an electrooculogram (EOG), a record of eye movements. Functions including body temperature, circulating hormones, and blood glucose levels can also be measured.

Section 7-1 reviews several methods for measuring the brain's electrical activity.

Stages of Waking and Sleeping

Biological measurements show that sleep, like waking, is not a unitary state but consists of a number of different stages. For example, the EEG recording shows distinct patterns of brain-wave activity as the neocortex generates distinct rhythmic patterns from states categorized as awake, drowsy, sleeping, and dreaming.

Figure 7-9 diagrams EEG brain-wave patterns that reflect a range of conscious states in humans.

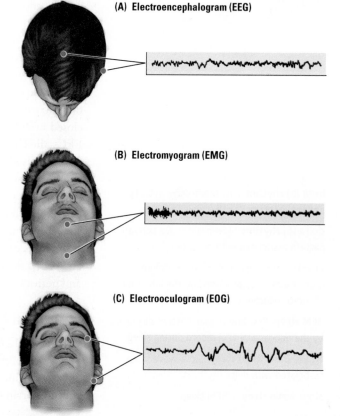

(A) Electroencephalogram (EEG)

(B) Electromyogram (EMG)

(C) Electrooculogram (EOG)

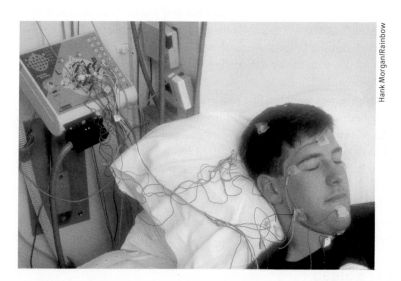

Hank Morgan/Rainbow

FIGURE 13-11 Sleep-Laboratory Setup Electronic equipment records readouts from the electrodes attached to the sleeping subject. **(A)** Electroencephalogram made from a point on the skull relative to a neutral point on the ear. **(B)** Electromyogram made between two muscles, such as those on the chin and throat. **(C)** Electrooculogram made between the eye and a neutral point on the ear.

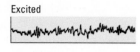

Excited

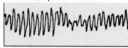

Relaxed, eyes closed

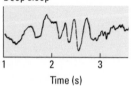

Deep sleep

1   2   3
Time (s)

Waking State

When a person is awake, the EEG pattern consists of small-amplitude (height) waves with a fast frequency (repetition period). This pattern, the **beta (β) rhythm,** is also called fast-wave activity, activated EEG, or waking EEG. Beta-rhythm waves have a frequency ranging from 15 to 30 Hz (times per second). Also associated with waking, the EMG is active, and the EOG indicates that the eyes move.

Drowsy State

When a person becomes drowsy, the EEG indicates that beta-wave activity in the neocortex disappears. The amplitude of the EEG waves increases, and their frequency becomes slower. Concurrently, the EMG remains active, as the muscles have tone, and the EOG indicates that the eyes are not moving.

When participants relax and close their eyes, they may produce the *alpha (α) rhythm*—large, extremely regular brain waves with a frequency ranging from 7 to 11 Hz. Humans generate alpha rhythms in the region of the visual cortex at the back of the brain, and the rhythms abruptly stop if a relaxed person is disturbed or opens his or her eyes. Not everyone displays alpha rhythms, and some people display them much better than others.

Sleeping State

As participants enter deeper sleep, they produce yet slower, larger EEG waves called **delta (δ) rhythms,** also known as slow-wave activity or resting activity. Delta-rhythm waves have a frequency of 1 to 3 Hz. The slowing of brain-wave activity is associated with the loss of consciousness that characterizes sleep. Still, the EMG indicates muscle activity, signifying that the muscles retain tone, although the EOG indicates that the eyes do not move.

REM and Non-REM Sleep Phases

Sleep consists of periods when a sleeper is relatively still and periods when the mouth, fingers, and toes twitch. This behavior is readily observed in household pets and bed partners. In 1955, Eugene Aserinsky and Nathaniel Kleitman (Lamberg, 2003), working at the University of Chicago, observed that the twitching periods are also associated with rapid eye movements (REM). Other than twitches and eye movements, the EMG indicates that muscles are inactive, a condition termed **atonia** ("without tone").

REM coincides with distinct brain-wave patterns recorded on the EEG that suggest that the participant is awake, with the eyes flickering back and forth behind the sleeper's closed eyelids (see Dement, 1972). By accumulating and analyzing REM recorded on EEGs, the Chicago investigators were the first to identify **REM sleep,** the fast-wave pattern displayed by the neocortical EEG record. This discovery let to the contemporary naming of two states of sleep.

The delta-rhythm sleep period during which the EEG pattern is slow and large and the EOG is inactive is called **NREM** (for non-REM) **sleep** to distinguish it from REM sleep. It is sometimes called **slow-wave sleep.** We use the terms REM and NREM sleep. Both states can be further subdivided—NREM sleep on the basis of the slowing EEG and REM sleep on the bases of periods when twitches and rapid eye movements occur and periods in which these movements do not occur.

A Typical Night's Sleep

Figure 13–12A displays the EEG patterns associated with waking, NREM sleep and REM sleep. NREM sleep is divided into four stages on the basis of EEG records. Notice that the main change characterizing these stages is the increase in size and slowing in speed of brain waves in the progression from stage 1 sleep through stage 4 sleep.

beta (β) rhythm Fast brain-wave activity pattern associated with a waking EEG.

delta (δ) rhythm Slow brain-wave activity pattern associated with deep sleep.

atonia No tone; condition of complete muscle inactivity produced by the inhibition of motor neurons.

REM sleep Fast brain-wave pattern displayed by the neocortical EEG record during sleep.

NREM (non-REM) sleep Slow-wave sleep associated with delta rhythms.

slow-wave sleep NREM sleep.

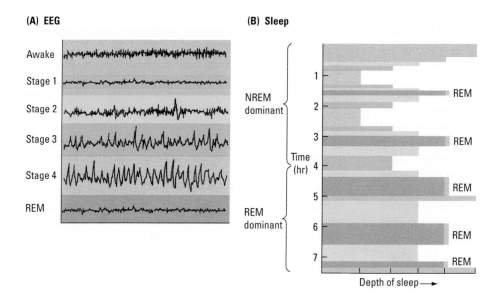

(A) EEG **(B) Sleep**

FIGURE 13-12 Sleep Recording and Revelations **(A)** Electroencephalograph patterns associated with waking, with the four NREM sleep stages, and with REM sleep. **(B)** In a typical night's sleep, a person undergoes a number of sleep-state changes, in roughly 90-minute periods. NREM sleep dominates the early sleep periods, and REM sleep dominates later sleep. The duration of each sleep stage is reflected in the thickness of each bar, which is color-coded to the corresponding stage in part A. The depth of each stage is graphed as the relative length of the bar. Adapted from "Sleep and Dreaming," by D. D. Kelley, in E. R. Kandel, J. H. Schwartz, and T. M. Jessell (Eds.), *Principles of Neuroscience*, 1991, New York: Elsevier, p. 794.

The designation of these stages assumes that the sleeper moves from relatively shallow sleep in stage 1 to deep sleep in stage 4. Self-reports from participants who are awakened from sleep at different times suggest that stage 4 is the deepest sleep, and participants act groggy when awakened from it. As described earlier, REM sleep is distinctive because, although a participant is asleep, EEG activity shows a waking pattern.

Figure 13-12B graphs one participant's sleep stages in the course of a night's sleep. Notice that the depth of sleep changes several times. The participant cycles through the four stages of NREM sleep and then enters REM sleep. This NREM–REM sequence lasts approximately 90 minutes and occurs five times in the course of the sleep period.

The labels indicating REM sleep in Figure 13-12B tell us that the durations of the different sleep stages roughly divide the sleep period into two parts, the first dominated by NREM sleep, the second dominated by REM sleep. Separate measures of body temperature record that it is lowest (about 1.5 degrees below a normal temperature of 37.7 degrees Celsius) during the earlier NREM-dominated part of the sleep period and rises during the later REM-dominated part.

Findings from sleep-laboratory studies confirm that the sleep patterns of individual people follow this pattern, with some variability. Studies also confirm that REM sleep takes up a substantial proportion of sleep time: adults who typically sleep about 8 hours spend about 2 of those hours in REM. A person's REM sleep durations also vary at different times of life. Periods of REM sleep increase during growth spurts, in conjunction with physical exertion, and for women during pregnancy.

The time spent in REM sleep also changes dramatically over the life span. **Figure 13-13** shows that most people sleep less as they grow older. In the first 2 years of life, however, REM sleep makes up nearly half of sleep time, but it declines proportionately until, in middle age, it constitutes little more than 10 percent of sleep time.

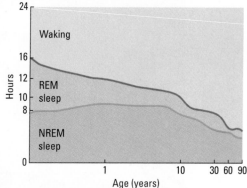

FIGURE 13-13 Sleeping and Waking over the Life Span The amount of time that humans spend sleeping decreases with age. The proportion of REM sleep is especially high in the first few years of life. Adapted from "Ontogenetic Development of the Human Sleep-Dream Cycle," by H. P. Roffward, J. Muzio, and W. C. Dement (1966), *Science, 152*, 604–619.

Contrasting NREM Sleep and REM Sleep

Although it seems an inactive period, a remarkable range of activities takes place during sleep. During NREM sleep, body temperature declines, heart rate decreases, blood flow decreases, body weight decreases because of water loss in perspiration, and levels of growth hormone increase.

Restless Legs Syndrome

I've always been a fairly untalented sleeper. Even as a child, it would take me some time to fall asleep, and I would often roll around searching for a comfortable position before going under. But my real difficulties with sleeping did not manifest themselves until early adulthood. . . .

Initially, my symptoms consisted of a mild tingling in my legs. It caused me to be fidgety and made it hard to fall asleep. Eventually, I went through a number of days without much sleep and reached a point where I simply could not function. I went to a doctor who prescribed a small course of sleeping medication (a benzodiazepine). I was able to get good sleep and my sleep cycle seemed to get back on track. Over the next decade I had periodic bouts of tingling in my legs which caused me to be fidgety and interfered with sleep. As time passed, the bouts occurred with increasing frequency and the symptoms became more noticeable and uncomfortable. I would simply suffer through these bouts, sleeping poorly and paying the consequences. . . .

Being a student, I did not have a regular doctor. Unfortunately, most physicians I met did not know about restless legs syndrome, or RLS, and thought I was "drug seeking" or merely stressed out. I received a variety of patronizing responses and found these experiences insulting and demeaning. . . .

When I took my current position, I started seeing a doctor on a regular basis, and experimented with better medications. By this time, my sleep was

being seriously affected by RLS. The sensations in my legs were something like a combination of an ache in my muscles (much like one gets after exercising) and an electrical, tingling sensation. They would be briefly relieved with movement, such as stretching, rubbing, contracting my muscles, or changing position, but would return within seconds. In fact, my wife says my cycle is about 13 to 15 seconds between movements. I do this either when awake or during sleep. Trying not to move greatly increases the discomfort—much like trying to not scratch a very bad itch. The symptoms get worse in the evening and at night. Most nights, I have trouble falling asleep. Other nights, I wake up after an hour or so and then have trouble going back under. . . .

I am very up front about the fact that I have RLS. In fact, whenever I teach the topic of sleep and sleep disorders in my brain and behavior classes, I always make some time to talk about my experiences with RLS. Occasionally, students approach me with their own difficulties, and I try to provide them with information and resources. (Stuart Hall, Ph.D., University of Montana)

Restless legs syndrome (RLS) is a sleep disorder in which a person experiences unpleasant sensations in the legs described as creeping, crawling, tingling, pulling, or pain. The sensations are usually in the calf area but may be felt anywhere from the thigh to the ankle. One or both legs may be affected; some people also feel the sensations in their arms.

Many people with RLS have a related sleep disorder called *periodic limb movement in sleep*. PLMS is characterized by involuntary jerking or bending leg movements that typically occur every 10 to 60 seconds during the sleep period. Some people experience hundreds of such movements per night that can wake them, disturb their sleep, and annoy bed partners.

RLS may affect as much as 10 percent of the population and is more common in women than in men. In mild cases, massage, exercise, stretching, and hot baths may be helpful. For more severe cases, patients can restrict their intake of caffeine and take benzodiazepines to help them get to sleep.

L-dopa, a drug that is also used to treat Parkinson's disease, is an effective treatment, but there is no evidence that RLS is more frequent in Parkinson's patients, whose condition is related to low dopamine function. Nor is RLS associated with gross changes in dopamine cells in the brain (Thomas & Watson, 2008). RLS has been associated with poor iron uptake, especially in the substantia nigra, and some people have been helped by iron supplements. One focus of research into RSL is to improve iron absorption by the brain.

NREM sleep is also the time when we toss and turn in bed, pull on the covers, and engage in other movements (for an extreme example, see Clinical Focus 13-4, "Restless Legs Syndrome"). If we talk in our sleep, we do so during NREM sleep. If we make flailing movements, such as banging an arm or kicking a foot, we usually do so in NREM sleep. Some people even get up and walk while asleep, and this "sleepwalking" takes place during NREM sleep. All this activity during our so-called resting state is remarkable!

REM sleep is no less exciting and remarkable than NREM sleep. During REM sleep, our eyes move, our toes, fingers, and mouths twitch, and males have penile erections. Still, we are paralyzed, as indicated by atonia. This absence of muscle tone is due to inhibition of motor neurons by sleep regions of our brainstem. In the sleep lab, atonia is recorded on an electromyogram as the absence of muscle activity.

Posture also differs during NREM and REM sleep. You can get an idea of postural differences by observing a cat or dog sleeping. During NREM they may be lying down but still have some posture, with the head partly supported in a partially upright position. At the onset of REM sleep, the animal usually subsides into a sprawled posture as the paralysis of its muscles sets in. **Figure 13-14** illustrates the sleep postures of a horse. Horses can sleep while standing up by locking their knee joints, and they can sleep while lying down with their heads held slightly up. At these times, they are in NREM sleep. When they are completely sprawled out, they are in REM sleep.

During REM sleep, mammals' limbs twitch visibly, and if you look carefully at the face of a dog or cat, you will also see the skin of the snout twitch and the eyes move behind the eyelids. It might seem strange that an animal that is paralyzed can make small twitching movements, but the neural pathways that mediate these twitches obviously are spared the paralysis. One explanation for the twitching of eyes, face, and distal parts of the limbs is that such movements may help to maintain blood flow in those parts of the body.

An additional change resulting from atonia during REM sleep is that mechanisms that regulate body temperature stop working, and body temperature moves toward room temperature. You may wake up from REM sleep feeling cold or hot, depending on the temperature of the room, because your body has drifted toward room temperature during a REM period.

Dreaming

The most remarkable aspect of REM sleep—dreaming—was discovered by William Dement and Nathaniel Kleitman in 1957 (Dement, 1972). When participants were awakened from REM sleep, they reported that they had been having vivid dreams. In contrast, participants aroused from NREM sleep were much less likely to report that they had been dreaming, and the dreams they did report were much less vivid. Children may experience brief, very frightening dreams called *night terrors* in NREM sleep. Night terrors can be so vivid that the child may continue to experience the dream and the fear after awaking.

The technique of electrical recording from a sleeping participant in a sleep laboratory made it possible to subject dreams to experimental analysis. Such studies provided objective answers to a number of interesting questions concerning dreaming.

How often do people dream? Reports by people on their dreaming behavior had previously suggested that dreaming was quite variable: some reported that they dreamed frequently and others that they never dreamed. Waking participants up during periods of REM showed that everyone dreams, that they dream a number of times each night, and that dreams last longer as a sleep session progresses. Those who claimed not to dream presumably forgot their dreams. Perhaps people forget their dreams because they do not wake up in the course of a dream or immediately afterward, thus allowing subsequent NREM sleep activity to erase the memory of the dream.

How long do dreams last? Common wisdom suggests that dreams last but an instant. By waking people up at different intervals after the onset of REM sleep and matching the reported dream content to the previous duration of REM sleep, however, researchers

FIGURE 13-14 Nap Time Horses usually seek an open, sunny area for brief periods of sleep. I. Q. W.'s horse, Lady Jones, illustrates three sleep postures. (*Top*) She displays NREM sleep, standing with legs locked and head down. (*Middle*) She displays NREM sleep, lying down with head up. (*Bottom*) She is in REM sleep, in which all postural and muscle tone is lost.

demonstrated that dreams appear to take place in real time. An action that a person performed in a dream lasts about as long as it would take to perform while awake. It is likely that time shrinking is a product of remembering a dream just as time shrinking is a feature of our recall of other memories.

What We Dream About

The study of dreaming in sleep laboratories allows researchers to study other questions that have always intrigued people: Why do we dream? What do we dream about? What do dreams mean?

Past explanations of dreaming have ranged from messages from the gods to indigestion. The first modern treatment of dreams was described by the founder of psychoanalysis, Sigmund Freud, in *The Interpretation of Dreams,* published in 1900. Freud reviewed the early literature on dreams, described a methodology for studying them, and provided a theory to explain their meaning. We briefly consider Freud's theory because it remains popular in psychoanalysis and in the arts, and it is representative of other psychoanalytical theories of dreams.

Freud suggested that the function of dreams is the symbolic fulfillment of unconscious wishes. His theory of personality is that people have both a conscious and an unconscious. Freud proposed that the unconscious contains unacknowledged desires and wishes, which are sexual. He further proposed that dreams have two levels of meaning. The *manifest content* of a dream is a series of often bizarre, loosely connected images and actions. The *latent content* of the dream contains its true meaning. As interpreted by a psychoanalyst, the symbolic events of a dream provide a coherent account of the dreamer's unconscious wishes—for Freud, the wishes related to sex.

Freud provided a method for interpreting manifest symbols and reconstructing the latent content of dreams. For example, he pointed out that a dream usually begins with an incident from the previous day, incorporates childhood experiences, and includes ongoing unfulfilled wishes. He also identified several types of dreams, such as those that deal with childhood events, anxiety, and wish fulfillment. The content of the dream was important to Freud and other psychoanalysts in clinical practice because, when interpreted, dreams offer insight into a patient's problems.

Other psychoanalysts, unhappy with Freud's emphasis on sex, developed their own methods of interpretation. The psychoanalyst Carl Jung, a contemporary of Freud, proposed that dream symbolism signifies distant human memories encoded in the brain but long since lost to conscious awareness. Jung proposed that dreams allow the dreamer to relive the history of the human race, our "collective unconscious." As more theories of dream interpretation developed, their common central weakness became apparent: it was impossible to know which interpretation was correct.

The dream research of Freud and his contemporaries was impeded by their reliance on a subject's memory of a dream and by the fact that many subjects were patients. This situation unquestionably resulted in the selection of the unusual by both the patient and the analyst. Now researchers study dreams more objectively, by waking participants and questioning them.

Experimental analysis indicates that most dreams are related to events that happened quite recently and concern ongoing problems. Colors of objects, symbols, and emotional content most often relate to events taking place in a person's recent waking period. Calvin Hall and his colleagues (1982) documented more than 10,000 dreams of normal people and found that more than 64 percent are associated with sadness, anxiety, or anger. Only about 18 percent are happy. Hostile acts against the dreamer outnumber friendly acts by more than two to one. Surprisingly, in regard to Freud's theory, only about 1 percent of dreams include sexual feelings or acts.

Figure 16-1 presents a contemporary take on Freud's model of the mind.

Contemporary researchers continue to attempt to interpret dream content. The two hypotheses that follow are polar opposites: one sees no meaning in dreams, and the other sees the content of dreams as reflecting biologically adaptive coping mechanisms.

Dreams as Meaningless Brain Activity

J. Allan Hobson (2004) proposed in the *activation–synthesis hypothesis* that, during a dream, the cortex is bombarded by signals from the brainstem, and these signals produce the pattern of waking (or activated) EEG. The cortex, in response to this excitation, generates images, actions, and emotion from personal memory stores. In the absence of external verification, these dream events are fragmented and bizarre and reveal nothing more than that the cortex has been activated.

Furthermore, Hobson proposed, on the basis of PET-imaging results, that part of the frontal cortex is less active in dreaming than in waking. The frontal cortex controls working memory, memory for events that have just happened, and attention. The dreamer cannot remember and link dream events as they take place because monitoring by the frontal cortex is required for these functions. On waking, the dreamer may attempt to create a story line for these fragmented, meaningless images.

Chapter 14 describes the extensive involvement of the frontal cortex in memory. Attention is the topic of Section 15-2.

In Hobson's hypothesis, dreams are personal in that memories and experiences are activated, but they have no meaning. So the following dream, for example, with its bizarre, delusional, and fragmented elements, represents images synthesized to accompany brain activation. Any meaning that the dream might seem to have is created after the fact by the middle-aged dreamer recounting it:

> I found myself walking in a jungle. Everything was green and fresh and I felt refreshed and content. After some time I encountered a girl whom I did not know. The most remarkable thing about her was her eyes, which had an almost gold color. I was really struck by her eyes not only because of their unique color but also because of their expression. I tried to make out other details of her face and body but her eyes were so dominating that was all I could see. Eventually, however, I noticed that she was dressed in a white robe and was standing very still with her hands at her side. I then noticed that she was in a compound with wire around it. I became concerned that she was a prisoner. Soon, I noticed other people dressed in white robes and they were also standing still or walking slowly without swinging their arms. It was really apparent that they were all prisoners. At this time I was standing by the fence that enclosed them, and I was starting to feel more concerned. Suddenly it dawned on me that I was in the compound and when I looked down at myself I found that I was dressed in a white robe as well. I remember that I suddenly became quite frightened and woke up when I realized that I was exactly like everyone else. The reason that I remembered this dream is the very striking way in which my emotions seemed to be going from contentment, to concern, to fear as the dream progressed. I think that this dream reflected my desire in the 1970s to maintain my individuality. (Recounted by A. W.)

Dreams as a Coping Strategy

Anttio Revonsuo of Finland uses content analysis to argue that dreams are biologically adaptive in that they lead to enhanced coping strategies in dealing with threatening life events (Valli & Revonsuo, 2009). The evolutionary aspect of this "coping hypothesis" is that enhanced performance is especially important for people whose environment typically includes dangerous events that constitute extreme threats to reproductive success. Revonsuo notes that dreams are highly organized and significantly biased toward threatening images, as was A. W.'s dream. People seldom dream about reading, writing, and calculating, even if these behaviors occupy much of their day.

Dream threats are the same events that are threatening in waking life **(Figure 13-15)**. For example, animals and strange men who could be characterized as "enemies" figure prominently in dreams. Dream content incorporates the current emotional problems of the dreamer and leads to improvements in and adjustments to life problems.

FIGURE 13-15 Dream Content The terrifying visions that may persist even after awakening from a frightening dream are represented in *The Night*, by Swiss painter Ferdinand Hodler. *The Night, by Ferdinand Hodler (1853–1918), oil on canvas, 116 3 299 cm, Kunstmuseum, Berne, Switzerland.*

In contrast with the "threat" interpretation of dreams and from their own analysis of dream content, Malcolm-Smith and her coworkers (2012) found that approach behavior occurs more frequently in dreams than does avoidance behavior. They suggest, therefore, that reward-seeking behavior is as likely to represent the latent content of a dream as is avoidance behavior.

As you can see from these marked differences in interpretation, dreams remain a puzzle and a challenge just waiting for clever experimentation to reveal their secrets. In this respect, it is interesting to contrast night dreams and daydreams. Eric Klinger (1990) suggests that daydreams are ordinary and often fun, with little violence or sex, and so are just the opposite of night dreams.

REVIEW 13-3
Sleep Stages and Dreaming

Before you continue, check your understanding.

1. Sleep consists of two phases, _____, which stands for _____, and _____, which stands for _____.

2. REM sleep is characterized by eye movement recorded by the _____, atonia recorded by the _____, and waking activity recorded by the _____.

3. There are about _____ REM sleep periods each night, with each period _____ as sleep progresses.

4. Evidence from sleep lab analysis suggests that _____ dreams and that dreams take place in _____.

5. What major factor makes interpreting dreams difficult?

Answers appear at the back of the book.

13-4 What Does Sleep Accomplish?

Sleep is not a passive process that takes place as a result of a decrease in sensory stimulation. Findings from sensory-deprivation research reveal that, when participants are isolated in quiet bedrooms, they spend less time asleep, not more. These results do not support the idea that sleep sets in because there is nothing else to do. Here we consider three contemporary explanations for sleep: sleep as adaptive, restorative, and supportive of memory.

basic rest–activity cycle (BRAC) Recurring cycle of temporal packets, about 90-minute periods in humans, during which an animal's level of arousal waxes and wanes.

Sleep As a Biological Adaptation

Many lines of evidence argue that sleep is a biologically adaptive behavior influenced by the ways a species has evolved to interact with its environment:

- Sleep serves as an energy-conserving strategy to cope with times when food is scarce. Each animal species gathers food at optimal times and conserves energy the rest of the time. If the food that a species eats has a high nutrient value, the species can spend less time foraging and more time sleeping.

- Whether a species is predator or prey influences its sleep behavior. A predator can sleep at its ease; the prey's sleep time is reduced because it must remain alert and ready to fight or flee at unpredictable times **(Figure 13-16)**.

- Strictly nocturnal or diurnal animals are likely to sleep when they cannot travel easily. Colloquially, Dement proposes: "We sleep to keep from bumping into things in the dark."

Much about animals' sleep patterns are consistent with the adaptive explanation. **Figure 13-17** charts the average sleep time of some common mammals. Herbivores, including donkeys, horses, and cows, spend a long time collecting enough food to sustain themselves. This reduces their sleep time. Because they are also prey, their sleep time is further reduced as they watch for predators. Carnivores, including domestic cats and dogs, eat nutrient-rich foods and usually consume most of a day's or even a week's food at a single meal. Because they do not need to eat constantly and because by resting they can conserve energy, carnivores spend a great deal of time each day sleeping.

The behavior of some animals does appear odd, however, so understanding any animal's sleep behavior requires understanding its natural history. Opossums, which spend much of their time asleep, may have specialized in energy conservation as a survival strategy. We humans are average among species in our sleep time. As omnivores not subjected to overwhelming predation, our sleep is intermediate between that of herbivores and carnivores.

Sleep can contribute to energy conservation. During sleep, energy is not being expended in moving the body or supporting its posture. The brain is a major energy user, so switching off the brain during sleep, especially NREM sleep, conserves energy. The drop in body temperature that typically accompanies sleep slows metabolic activity, so it too contributes to energy conservation.

A good explanation of sleep must account not only for sleep but also for NREM and REM sleep. Before the discovery of REM sleep, Kleitman suggested that animals have a **basic rest–activity cycle** (BRAC) that, for humans, has a period of about 90 minutes (see Dement, 1972). Kleitman based his hypothesis on the observation that human infants have frequent feeding periods between which they sleep.

As illustrated in **Figure 13-18**, the behavior of adult humans does suggest that activity and rest are organized into 90-minute temporal packets. School classes, work periods, exercise sessions, meal times, coffee breaks, and snack times appear to be divided into intervals of 90 minutes or so. The later discovery that REM sleep occurs at intervals of about 90 minutes added support to Kleitman's hypothesis, because the REM periods could be considered a continuation into sleep of the 90-minute BRAC cycle. The hypothesis assumes that periods of eating are periods of high brain activity, just as are periods of REM.

FIGURE 13-16 Do Not Disturb Biological theories of sleep suggest that sleep is an energy-conserving strategy and serves other functions as well, such as staying safe during the night.

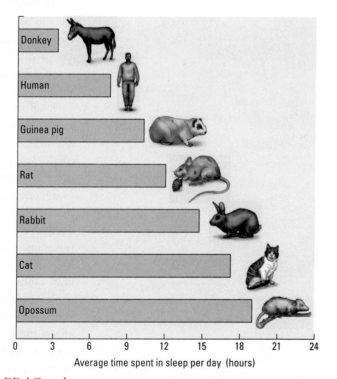

Average time spent in sleep per day (hours)

FIGURE 13-17 Average Sleep Time Sleep time is affected both by the amount of time required to obtain food and by the risk of predation.

FIGURE 13-18 **Behavioral Rhythms** Our behavior is dominated by a basic rest–activity cycle (red) through which our activity levels change in the course of the day and by an NREM–REM sleep cycle (purple) during the night.

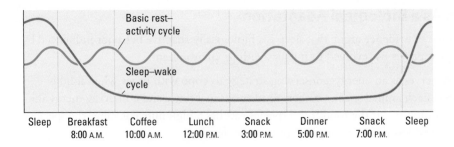

Basic rest– activity cycle

Sleep–wake cycle

| Sleep | Breakfast 8:00 A.M. | Coffee 10:00 A.M. | Lunch 12:00 P.M. | Snack 3:00 P.M. | Dinner 5:00 P.M. | Snack 7:00 P.M. | Sleep |

Kleitman proposed that the BRAC rhythm is so fundamental that it cannot be turned off. Accordingly, for a night's sleep to be uninterrupted by periodic waking (and perhaps snacking), the body is paralyzed and only the brain is active. To use an analogy, rather than turning off your car's engine when you're stopped at a red light, you apply the brakes to keep the idling car from moving. For REM sleep, the atonia that paralyzes movement is the brake.

Sleep As a Restorative Process

The idea that sleep has restorative properties is illustrated by Shakespeare in Macbeth's description of sleep:

Sleep that knits up the ravell'd sleave of care,
The death of each day's life, sore labour's bath,
Balm of hurt minds, great nature's second course,
Chief nourisher in life's feast.

Sleep-Deprivation Studies

We can understand the idea of sleep as a restorative from our personal perspectives. Toward the end of the day, we grow tired, and when we awaken from sleep, we are refreshed. If we do not get enough sleep, we become irritable. One hypothesis of sleep as restorative proposes that the chemical events that provide energy to cells are reduced during waking and replenished during sleep.

Even so, fatigue and alertness may simply be aspects of the circadian rhythm and have nothing at all to do with wear and tear on the body or depletion of essential bodily resources. To evaluate whether sleep is essential for bodily processes, investigators have conducted sleep-deprivation studies. These studies have not clearly identified any function for which sleep is essential.

Dement participated as an observer in one case study on sleep deprivation that illustrates this point. In 1965, as part of a science-fair project, a student named Randy Gardner planned to break the world record of 260 hours (almost 11 days) of consecutive wakefulness with the help of two classmates, who would keep him awake. Gardner did break the record, then slept for 14 hours and reported no ill effects. The world record now stands at a little more than 18 days.

It is important to note that one of Gardner's observers reported that he experienced hallucinations and cognitive and memory lapses during deprivation. These negative effects did not last. Reviews of sleep-deprivation research are consistent in concluding that, at least for these limited periods of sleep deprivation, no marked physiological alterations ensue.

Sleep deprivation does not seem to have adverse physiological consequences, but it is associated with poor cognitive performance. Performance on tasks that require attention declines as a function of hours of sleep deprivation. Irregular sleep can be associated with metabolic syndrome, described in Clinical Focus 13-1. Finally, sleep deprivation contributes to accidents at work and on the road. The sleep-deprivation deficit does not manifest itself in an inability to do a task, because sleep-deprived participants can perform even very complex tasks; rather, the deficit is in sustained attention.

A confounding factor in evaluating sleep-deprived participants is that they take **microsleeps,** brief sleep periods lasting up to a few seconds. During microsleep, participants may remain sitting or standing, but their eyelids droop briefly and they become less responsive to external stimuli. If you have driven a car while tired, you may have experienced a microsleep and awakened just in time to prevent yourself from driving off the road or worse.

REM-Sleep Deprivation

Some studies have focused on the selective benefits of REM sleep. To deprive a participant of REM sleep, researchers allow participants to sleep but awaken them as they start to go into REM sleep. REM-sleep deprivation has two effects:

1. Participants show an increased tendency to go into REM sleep in subsequent sleep sessions, so awakenings must become more and more frequent.

2. After REM deprivation, participants experience "REM rebound," showing more than the usual amount of REM sleep in the first available sleep session.

Some early reports from REM-deprivation studies stated that participants could begin to hallucinate and display other abnormalities in behavior, but these reports have not been confirmed.

Two kinds of observations, however, argue against effects from prolonged or even complete deprivation of REM sleep. Virtually all antidepressant drugs, including MAO inhibitors, tricyclic antidepressants, and SSRIs, suppress REM sleep either partly or completely. The clinical effectiveness of these drugs may in fact derive from their REM-suppressant effects (Wilson & Argyropoulos, 2005). No studies report adverse consequences from prolonged REM deprivation as a consequence of treatment with antidepressants. It is possible that these drugs do something usually subserved by REM sleep, so their effect is substitutive rather than inhibitory.

In a number of reported cases, lower-brainstem damage has resulted in a complete loss of REM sleep. For example, patients with brainstem lesions reportedly remained ambulatory and verbally communicative, but their REM was abolished. They were reported to live quite satisfactorily without REM sleep (Osorio & Daroff, 1980).

It is important to acknowledge that because most sleep studies are relatively brief, they are unable to capture the many potential long-term changes that can take place in the body during or after sleep deprivation. The consequences of sleep deprivation may become apparent only some time after an experimental manipulation has taken place. Of further importance is recognizing that sleep disruptions, as we discussed in Section 13-3 in relation to circadian rhythms, do have health consequences.

Sleep and Memory Storage

The suggestion that sleep plays a role in memory dates back over a century, and in the interval, a lot has changed with respect to our understanding of memory and of sleep. We now know of two general categories of memory. *Episodic memory* includes conscious information, such as our autobiographical memories and knowledge of facts. *Implicit memory* includes unconscious processes such as motor-skills learning.

We now also know that it takes time to store memories. Part of this process of **consolidation,** the process of stabilizing a memory trace after learning, may involve a memory "moving" from an initial coding site in one part of the brain to a permanent location in another part—from the temporal lobes to the frontal lobes, for example. We know as well that we frequently recall memories, rehash them, and integrate them

microsleep Brief period of sleep lasting a second or so.

consolidation The process of stabilizing a memory trace after learning.

Section 6-2 reviews the full spectrum of antidepressant drugs.

Section 14-1 expands on the workings of the explicit and implicit memory systems.

with new related events, a process called **reconsolidation,** the process of restabilizing a memory trace after the memory is revisited.

Does sleep enable these different aspects of memory? One research approach associates episodic memory consolidation with NREM sleep, and another research approach associates implicit memory consolidation with REM sleep. A third approach proposes that the process of shifting a memory to a new location during consolidation occurs during sleep. A fourth approach to memory and sleep suggests that reconsolidation is ongoing during both waking and sleeping and involves both daydreaming and night dreaming.

Other approaches also include the idea that memory is selectively processed in only certain parts of sleep. For example, stage 2 NREM sleep features short, rapid discharges of electrical activity called **sleep spindles** and slower, large-amplitude waves referred to as **K-complexes.** Both have been proposed to play a role in memory. We can recall memories in seconds, whereas the actual events may have taken place over hours. Thus, sleep spindles' brief, bursting appearance has been proposed to represent the instantaneous recall of a memory, akin to replaying a video at 20 times normal speed.

Understanding the possible role of sleep in memory formation also requires some understanding of the many events associated with sleep. For example, REM sleep takes up much of sleep time in infancy, a life stage when learning takes up a good part of our waking activity. Is REM related to general maturation of brain components and connections, to memory, or to both processes?

When we are older, our sleep schedules become less regular. Is good memory storage associated with good sleep and poor memory storage with sleep deprivation (Dudai, 2012)? The remainder of this section describes three of the many current approaches to studying the role of sleep in memory.

NREM and Explicit Memory

Gerrard and colleagues (2008) make use of the finding that many hippocampal cells fire when a rat is in a certain location in an environment. These **place cells** are relatively inactive until the rat passes through that place again. Recordings made from as many as 100 cells at the same time in three conditions—during NREM sleep, during a food-search task, and during NREM sleep after a food-search session—show that during the food-search task, the activity of some place cells becomes correlated. During the subsequent periods of the rats' NREM sleep, these correlations recur **(Figure 13-19).**

This result suggests that the memory of the previous food-searching experience is being replayed and thus stored during NREM sleep. The Gerrard team's experiment illustrates the role of NREM sleep in consolidating an explicit memory—the memory relating a particular environment and the food reward found in that environment.

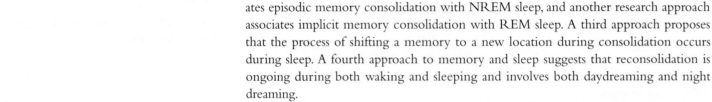

K-complex

Sleep spindle

Figure 7-12 shows variation in place-cell firing rates among young, old, and transgenic mice.

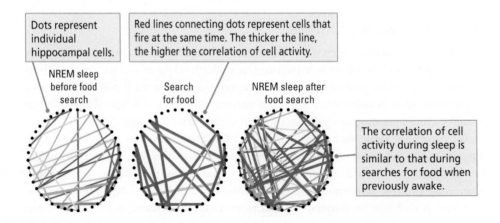

FIGURE 13-19 Neural Replay? The activity of hippocampal cells suggests that rats dream about previous experiences. The dots on the periphery of the circles represent the activity of 42 hippocampal cells recorded at the same time during (*left*) NREM sleep before a food-searching task, (*middle*) the food-searching task, and (*right*) NREM sleep after the task. No strong correlations between cells emerged during the NREM sleep that preceded the food-searching task, but correlations between cells during the food search and during the subsequent slow-wave sleep were strong. Adapted from "Reactivation of Hippocampal Ensemble Memories During Sleep," by M. A. Wilson and B. L. McNaughton, 1994, *Science, 165,* p. 678.

Dots represent individual hippocampal cells.

Red lines connecting dots represent cells that fire at the same time. The thicker the line, the higher the correlation of cell activity.

NREM sleep before food search

Search for food

NREM sleep after food search

The correlation of cell activity during sleep is similar to that during searches for food when previously awake.

REM and Implicit Memory

To determine whether humans' dreams are related to memory, Pierre Maquet and his coworkers in Belgium (2000) trained participants on a serial reaction task and observed regional blood flow in the brain with PET scans during training and during REM sleep on the subsequent night. The participants faced a computer screen displaying six positional markers. They were to push one of six keys when a corresponding positional marker was illuminated. They did not know that the sequence in which the positional markers were illuminated was predetermined. This is an example of an implicit memory task, one in which a motor skill is mastered.

Consequently, as training progressed, the participants indicated that they were learning because their reaction time improved in trials in which one positional marker was correlated with a preceding marker. On the PET-scan measures of brain activation, a similar pattern of neocortical activation was found during task acquisition and during REM sleep **(Figure 13-20)**. On the basis of this result, Maquet and coworkers suggest first that the participants were dreaming about their learning experience and second that the replay during REM strengthened the memory of the task. This research finding thus suggests that REM sleep is associated with the replay of implicit memory and may be related to the storage or consolidation of that memory.

Reaction-time task

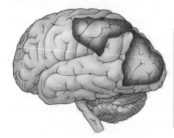

Participants are trained on a reaction-time task, and brain activity is recorded with PET.

REM sleep that night

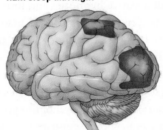

Participants display a similar pattern of brain activity during subsequent REM sleep.

FIGURE 13-20 Do We Store Implicit Memories During REM Sleep? Adapted from "Experience-Dependent Changes in Cerebral Activation During Human REM Sleep," by P. Maquet, S. Laureys, P. Peigneux, S. Fuchs, C. Petiau, et al., 1998, *Nature Neuroscience, 3*, p. 832.

Storing Memories During Sleep

In an interesting approach to sleep and memory formation—one that controls for many of the difficulties in devising appropriate control conditions in sleep research—Nelini and coworkers (2012) studied spatial memory formation in chicks, a species in which such memories are stored mainly in the right hemisphere. Chickens, like many bird species, alternate sleep in each hemisphere, and the researchers were able to show that after a learning experience, the right hemisphere displayed more sleep than did the left hemisphere. The selectivity in hemispheric sleep opens up the possibility of comparing plastic changes in the two hemispheres as a way of understanding genetic, biochemical, and plastic changes associated with memory formation.

Memory-storing explanations of sleep are extremely interesting, and the technology required for pursuing the question is quite new. Thus, debate concerning the fate of memories during sleep continues (Brankačk et al., 2009). Among the more interesting speculations is a proposal that elaborate memories are formed during sleep and then pruned to more useful dimensions during waking. Another suggestion is that only certain events are likely to be stored during sleep and that they may become associated with unrelated events, thus risking creating a false or distorted memory of an event. Nevertheless, as we have described, there are large variations in how long people sleep, and there are conditions in which people have little or no REM sleep. As yet, little evidence shows that either condition is deleterious to memory formation.

reconsolidation The process of restabilizing a memory trace after the memory is revisited.

sleep spindle Brief burst of EEG activity typically occurring during NREM sleep.

K-complex Sharp, high-amplitude EEG wave occurring during NREM sleep.

place cell Hippocampal neurons maximally responsive to specific locations in the world.

REVIEW 13-4

What Does Sleep Accomplish?

Before you continue, check your understanding.

1. Sleep is proposed to occur as a _____ adaption, as a _____ process, or as an aid in storing _____.

2. _____ memory is associated with NREM sleep, and _____ memory is associated with REM sleep.

3. In rats performing a spatial task, correlations develop between _____ firing in the hippocampus that is then replayed in _____ sleep.

4. When you are sleep-deprived, you are more likely to slip into a _____ for a few seconds.

5. Describe a difficulty in relating memory formation to sleep.

Answers appear at the back of the book.

13-5 Neural Bases of Sleep

The idea that the brain contains a sleep-inducing substance has long been popular and is reinforced by the fact that a variety of chemical agents induce sleep. Such substances include sedative hypnotics and morphine. Our understanding of circadian rhythms suggests, however, that changes in many neurochemicals and hormones and the metabolic activity of most of the body's cells produce our sleep–waking cycles.

The hormone melatonin, secreted from the pineal gland during the dark phase of the light–dark cycle, causes sleepiness, and a synthetic form can be taken as an aid for sleep, so melatonin might be thought to be the sleep-producing substance. Sleep, however, survives the removal of the pineal gland. Thus, melatonin and many other chemical substances may only contribute to sleep, not cause it (see Research Focus 13-3).

Some observations suggest that sleep is not produced by a compound circulating in the bloodstream. When dolphins and birds sleep, only one brain hemisphere sleeps at a time. This ability presumably allows an animal's other hemisphere to remain behaviorally alert and suggests that sleep is produced by the action of some region within each hemisphere.

In this section, we consider two points about the neural basis of sleep. First, we examine evidence that sleep is produced by the activity of a slave oscillator of the suprachiasmatic nucleus (see Figure 13-8). Second, we look at evidence that the various events associated with sleep, including events associated with REM and NREM sleep, are controlled by a number of different brainstem nuclei.

Reticular Activating System and Sleep

A dramatic experiment by Giuseppe Moruzzi and Horace Magoun (1949) began to answer the question of which brain areas regulate sleep. Moruzzi and Magoun were recording the cortical EEG from anesthetized cats while electrically stimulating the cats' brainstems. They discovered that in response to the electrical stimulation, the large, slow delta EEG typical of anesthesia was dramatically replaced by the low-voltage, fast-wave beta EEG typical of waking.

The beta-EEG activity outlasted the period of stimulation, demonstrating that the pattern was produced by the activity of neurons in the region of the stimulating electrode. During the "waking period," the cat did not become behaviorally aroused because it was anesthetized, but its cortical EEG appeared to indicate that it was awake.

Subsequent experiments show that a waking EEG can be induced from a large area running through the center of the brainstem. Anatomically, this area is composed of

reticular activating system (RAS) Large reticulum (mixture of cell nuclei and nerve fibers) that runs through the center of the brainstem; associated with sleep–wake behavior and behavioral arousal; often called the *reticular formation*.

coma Prolonged state of deep unconsciousness resembling sleep.

a mixture of cell nuclei and nerve fibers that form a *reticulum*. Moruzzi and Magoun named this brainstem area the **reticular activating system** (RAS) and proposed that it is responsible for sleep–waking behavior. **Figure 13-21** diagrams the location of the RAS.

If someone disturbs you when you are asleep, you usually wake up. To explain how sensory stimulation and the RAS are related, Moruzzi and Magoun proposed that sensory pathways entering the brainstem have collateral axons that synapse with neurons in the RAS. They proposed that sensory stimulation is conveyed to RAS neurons by these collaterals then RAS neurons produce the desynchronized EEG via axons that project to the cortex.

Because Moruzzi and Magoun could possibly have stimulated various sensory pathways passing through the brainstem, it was necessary to demonstrate that brainstem neurons, not sensory-pathway stimulation, produced the waking EEG. After cuts to the brainstem just behind the RAS severed incoming sensory pathways, RAS stimulation still produced a desynchronized EEG.

The idea that the brainstem plays a role in waking behavior helps to explain why brainstem damage can result in **coma,** a prolonged state of deep unconsciousness resembling sleep. In a well-publicized case, after taking a minor tranquilizer and having a few drinks at a birthday party, a 21-year-old woman named Karen Ann Quinlan sustained RAS damage that put her in a coma (Quinlan & Quinlan, 1977). She was hospitalized, placed on a respirator to support breathing, and fed by tubes. Her family fought a protracted legal battle to have her removed from life support, which they finally won before the Supreme Court of New Jersey. Even after having been removed from life support, however, Quinlan lived for 10 more years in a perpetual coma.

Neural Basis of EEG Changes Associated with Waking

Building on the pioneering studies on the RAS, research has since revealed a number of neural systems in the brainstem that play a role in sleeping and waking behavior. Case Vanderwolf and his coworkers (Vanderwolf, 2002) showed that two different systems in the brainstem influence waking EEG. **Figure 13-22** illustrates the locations of these structures. Both send neural pathways into the neocortex, where they make diffuse connections with cortical neurons.

The basal forebrain contains large cholinergic cells. These neurons secrete acetylcholine (ACh) from their terminals onto neocortical neurons to stimulate a waking EEG (beta rhythm). The midbrain structure, the median raphé, contains serotonin (5-HT) neurons whose axons also project diffusely to the neocortex, where they also stimulate neocortical cells to produce a beta rhythm, recorded as a waking EEG.

Although both pathways produce a very similar pattern of waking EEG activity, the relations of the two types of waking EEG to behavior are different. If the activity of the cholinergic projection is blocked by drugs or by lesions to the cells of the basal forebrain, the waking EEG normally recorded from an immobile rat is replaced by EEG activity resembling that of NREM sleep. Only if the rat walks or is otherwise active is a waking EEG obtained from the neocortex. These findings, graphed in Figure 13-22, suggest that the cholinergic EEG is responsible for the waking associated with being still yet alert, whereas the serotonergic activation is additionally responsible for the waking EEG associated with movement.

Note that neither the basal forebrain system nor the median raphé system is responsible for behavior. In fact, if both structures are pharmacologically or surgically destroyed, a rat can still stand and walk around. Its neocortical EEG, however, permanently resembles that of a sleeping animal.

As long as one of the activating systems is producing a waking EEG, rats can learn simple tasks. If both systems are destroyed, however, an animal, although still able to walk around, is no longer able to learn or display

A *reticulum*, derived from the Latin word *rete*, meaning "net," appears as a mottled mixture of gray matter and white matter.

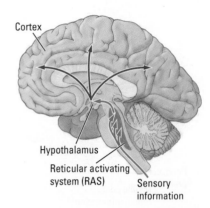

FIGURE 13-21 **Sleep–Wake Controller** The reticular activating system, a region in the middle of the brainstem, is characterized by a mixture of cell bodies and fiber pathways. Stimulation of the RAS produces a waking EEG, whereas damage to it produces a slow-wave, sleeplike EEG.

Figure 5-17 summarizes the major neural activating systems and their functions.

FIGURE 13-22 **Brain Activators** In the rat, basal forebrain ACh neurons produce an activated EEG pattern when a rat is alert but immobile. The 5-HT raphé neurons of the midbrain produce an activated EEG pattern when the rat moves.

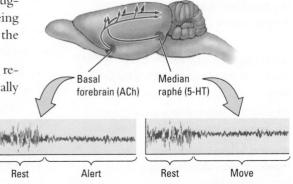

intelligent behavior. In a sense, the cortex is like a house in which the lights are powered by two separate sources: both must fail for the house to be left in darkness, but, if at least one source is operating, the lights stay on.

It is likely that the basal forebrain and median raphé produce the same two desynchronized EEG patterns in humans that they produce in rats. Consequently, when we are alert and still, the cholinergic neurons are active, and when we move, the serotonin neurons additionally are active.

You may have had the experience, when you felt sleepy in a class or behind the wheel of a car, of being able to wake yourself up by moving—shaking your head or stretching. Presumably, your arousal level decreased as your cholinergic neurons became inactive. When you moved, your serotonergic neurons became active and restored your level of arousal. When we enter sleep, both cholinergic and serotonergic neurons become less active, allowing slow waves to emanate from the cortex.

Neural Basis of REM Sleep

Barbara Jones (1993) and her colleagues described a group of cholinergic neurons, known as the **peribrachial area,** which are responsible for REM sleep. This area is located in the dorsal part of the brainstem just anterior to the cerebellum (**Figure 13-23**). Jones selectively destroyed these cells by spraying them with the neurotoxin kainic acid. She found that REM sleep was drastically reduced. This result suggests that the peribrachial area is the region responsible for producing REM sleep and REM-related behaviors.

The peribrachial area extends into a more ventrally located nucleus called the **medial pontine reticular formation** (MPRF). Lesions of the MPRF also abolish REM sleep, and injections of cholinergic agonists (drugs that act like ACh) into the MPRF induce REM sleep. Thus, both the peribrachial area and the MPRF, illustrated in Figure 13-23, take part in producing REM sleep.

If these two brain areas are responsible for producing REM sleep, how do other events related to REM sleep take place? As you know, such events include the following:

- EEG pattern in the neocortex similar to waking EEG

- Rapid eye movements (REM)

- Atonia, the absence of muscle tone

Figure 13-24 charts an explanation showing how other REM-related activities are induced.

- The peribrachial area initiates REM sleep and activates the medial pontine reticular formation.

- The MPRF sends projections to activate basal forebrain cholinergic neurons, resulting in an activated EEG recorded from the cortex.

- The MPRF excites brainstem motor nuclei to produce rapid eye movements and other twitches.

- The atonia of REM sleep is produced by the MPRF through a pathway that sends input to the subcoerulear nucleus, located just behind it.

- The subcoerulear nucleus excites the magnocellular nucleus of the medulla, which sends projections to the spinal motor neurons to inhibit them so that paralysis is achieved during the REM-sleep period.

In support of such a neural arrangement, French researcher Michael Jouvet (1972) observed that cats with lesions in the subcoerulear nucleus display a remarkable behavior when they enter REM sleep. Rather than stretching out in the atonia that typically accompanies REM sleep, the cats he was studying stood up, looked around, and made movements

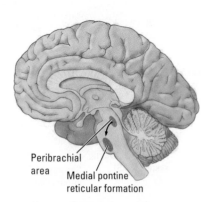

FIGURE 13-23 Brainstem Nuclei Responsible for REM Damage to either the peribrachial area or the medial pontine formation reduces or abolishes REM sleep.

Peribrachial area

Medial pontine reticular formation

Figure 6-5 shows agonist and antagonist action at the ACh synapse. Table 6-2 lists some natural neurotoxins, their sources, and their actions.

peribrachial area Cholinergic nucleus in the dorsal brainstem having a role in REM sleep behaviors; projects to medial pontine reticulum.

medial pontine reticular formation (MPRF) Nucleus in the pons participating in REM sleep.

insomnia Disorder of slow-wave sleep resulting in prolonged inability to sleep.

narcolepsy Slow-wave sleep disorder in which a person uncontrollably falls asleep at inappropriate times.

of catching an imaginary mouse or running from an imaginary threat. Apparently, if cats with damage to this brain region dream about catching mice or escaping from a threat, they act out their dreams.

REVIEW 13-5
Neural Bases of Sleep

Before you continue, check your understanding.

1. The _____ in the central region of the brainstem is responsible for producing _____ sleep.

2. Loss of the RAS produces _____.

3. The peribrachial area and the MPRF, through activating pathways to the neocortex and spinal cord, are responsible for producing events associated with _____.

4. Cats with lesions to the _____ nucleus act out their dreams.

5. If you nod off to sleep at an inconvenient time, why does moving awaken you?

Answers appear at the back of the book.

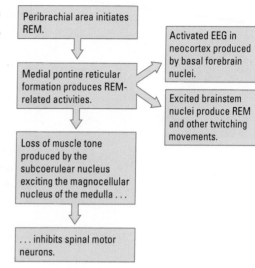

FIGURE 13-24 Neural Control of REM Sleep

13-6 Sleep Disorders

Occasional sleep disturbances are annoying and may result in impaired performance the following day. About 15 percent of people complain of ongoing sleep problems; an additional 20 percent complain of occasional sleep problems. As people age, the incidence of complaints increases.

In the extreme, a rare genetic condition, *fatal famial insomnia,* causes individuals to stop sleeping altogether. Their insomnia may contribute to death after a number of months without sleep (Synofzik et al., 2009). In this section, we consider more common abnormalities of NREM sleep and REM sleep.

Disorders of Non-REM Sleep

The two most common sleep disorders are **insomnia,** prolonged inability to sleep, and **narcolepsy,** uncontrollably falling asleep at inconvenient times. Both are disorders of NREM sleep. Insomnia and narcolepsy are related, as anyone who has stayed up late at night can confirm: a night without sleep is often accompanied by a tendency to fall asleep at inconvenient times the next day.

Narcolepsy derives from the Greek words meaning "numbness" and "to be seized."

Insomnia

Our understanding of insomnia is complicated by a large variation in how much time people spend asleep. Some short sleepers may think that they should sleep more, and some long sleepers may think that they should sleep less. Yet for each, the sleeping pattern may be appropriate.

People's sleep is disrupted by lifestyle choices, such as those described in Clinical Focus 13-1. Staying up late, for example, may set a person's circadian rhythm forward, encouraging a cascade of late sleep followed by staying up still later. Indoors and outdoors, light pollution contributes to sleep disorders by disrupting circadian rhythms. Some sleep problems are brought on by shift work or by jet lag, as described in Section 13-1. Other common causes of sleep disorders are stress, long hours of work, and irregular lifestyles. Just worrying about insomnia is estimated to play a major role in 15 percent of cases.

People who are depressed may sleep too much or too little. Anxiety and depression account for about 35 percent of insomnias. Quantitative differences also exist in the sleep of depressed patients because they enter REM sleep very quickly. Entering REM sleep quickly, however, is also related to sleep deprivation rather than directly to depression, because people who are sleep-deprived also enter REM very quickly.

Insomnia is brought on by sedative-hypnotic drugs, including seconal, sodium amytal, and many minor tranquilizers. These "sleeping pills" do help people get to sleep, but they are likely to feel groggy and tired the next day, which defeats the purpose of taking the drug. Although sleeping pills promote NREM sleep, they deprive the user of REM sleep.

In addition, people develop tolerance to these medications, become dependent on them, and display rebound insomnia when they stop taking them. A person then increases the dose each time the drug fails to produce the desired effect. The syndrome in which patients unsuccessfully attempt to sleep by increasing their drug dosage is called **drug-dependency insomnia.**

Narcolepsy

Like many people, you may suddenly have been overcome by an urge to sleep at an inconvenient time, perhaps while attending a lecture. For some people, such experiences with narcolepsy are common and disruptive. J. S., a junior in college, sat in the front row of the classroom for his course on the brain. Within a few minutes after each class began, he dropped off to sleep. The instructor became concerned and asked J. S. to stay after class to discuss his sleeping behavior.

J. S. reported that sleeping in classes was a chronic problem. Not only did he sleep in class, he fell asleep whenever he tried to study. He even fell asleep at the dinner table and in other inappropriate locations. His sleeping problem had made getting through high school a challenge and was making it difficult for J. S. to pass his college courses.

About 1 percent of people suffer from narcolepsy, which takes a surprising number of forms. J. S. had a form of narcolepsy that caused him to fall asleep while sitting still, and his sleeping bouts consisted of brief spurts of NREM sleep lasting from 5 to 10 minutes. This pattern is very similar to napping and to dropping off to sleep in class after a late night, but it is distinguishable as narcolepsy by its frequency and by the disruptive effect. J. S. eventually discussed his problem with his physician and received a prescription for Ritalin, an amphetaminelike drug that stimulates dopamine transmission. The treatment proved helpful.

Some people who suffer from daytime sleepiness attend sleep clinics. Studies of narcoleptic people in sleep clinics resulted in a surprising discovery concerning one cause of narcolepsy: **sleep apnea,** an inability to breathe during sleep. Clinical Focus 13-5, "Sleep Apnea," describes a person who spent all night, every night waking up to breathe. This nighttime behavior left him extremely tired and caused him to nod off in the daytime.

Disorders of REM Sleep

REM sleep is associated with muscular atonia and dreaming, but remarkably, both events can occur when a person is awake. This kind of REM happened to L. M., a college senior who recounted the following experience.

The student had just gone to sleep when her roommate came into their room. The student woke up and intended to ask her roommate if she wanted to go skating the next morning but found herself unable to speak. She tried to turn her head to follow her roommate's movements across the room but found that she was paralyzed. She had the terrifying feeling that some creature was hiding in the bathroom waiting for her roommate. She tried to cry out but produced only harsh, gurgling noises. In response to these peculiar noises, the roommate knocked her out of her paralysis by hitting her with a pillow.

This form of narcolepsy, called **sleep paralysis,** is common. In informal class surveys, almost a third of students report having had such an experience, as do some war veterans during group therapy sessions. The atonia is typically accompanied by a feeling of dread or fear. It seems likely that in sleep paralysis, a person has entered REM sleep, is dreaming, and atonia has occurred, but the person remains "awake."

Section 6-4 explains theories of drug tolerance and dependence.

Focus 6-1 explores amphetamine use for cognitive enhancement. Focus 7-4 describes how Ritalin mitigates symptoms of ADHD.

Apnea comes from the Latin words *a*, "not," and *pnea*, "breathing."

drug-dependency insomnia Condition resulting from continuous use of "sleeping pills"; drug tolerance also results in deprivation of either REM or NREM sleep, leading the user to increase the drug dosage.

sleep apnea Inability to breathe during sleep; person has to wake up to breathe.

sleep paralysis Inability to move during deep sleep owing to the brain's inhibition of motor neurons.

cataplexy Form of narcolepsy linked to strong emotional stimulation in which an animal loses all muscle activity or tone, as if in REM sleep, while awake.

hypnogogic hallucination Dreamlike event at the beginning of sleep or while a person is in a state of cataplexy.

Sleep Apnea

The first time I went to a doctor for my insomnia, I was twenty-five—that was about thirty years ago. I explained to the doctor that I couldn't sleep; I had trouble falling asleep, I woke up many, many times during the night, and I was tired and sleepy all day long. As I explained my problem to him, he smiled and nodded. Inwardly, this attitude infuriated me—he couldn't possibly understand what I was going through. He asked me one or two questions: Had any close friend or relative died recently? Was I having any trouble in my job or at home? When I answered no, he shrugged his shoulders and reached for his prescription pad. Since that first occasion I have seen I don't know how many doctors, but none could help me. I've been given hundreds of different pills—to put me to sleep at night, to keep me awake in the daytime, to calm me down, to pep me up—have even been psychoanalyzed. But still I cannot sleep at night. (In Dement, 1972, p. 73)

This patient went to the Stanford University Sleep Disorders Clinic in 1972. Recording electrodes monitored his brain, muscle, eye, and breathing activity while he slept (see Figure 13-11). The experimenters were amazed to find that he had to wake up to breathe. They observed that he would go for more than a minute without breathing, wake up and gasp for breath, then return to sleep. Then the sequence began again.

Sleep apnea may be produced by a central problem, such as a weak command to the respiratory muscles, or it may be obstructive, caused by collapse of the upper airway. When people who suffer from sleep apnea stop breathing, they either wake up completely and have difficulty getting back to sleep or they have repeated partial awakenings throughout the night to gasp for breath.

Sleep apnea affects all ages and both sexes, and 30 percent of people older than 65 years of age may have some form of it. Sleep apnea can

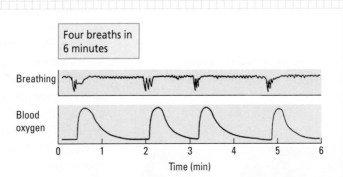

Four breaths in 6 minutes

Breathing rate and blood-oxygen level recorded from a person with sleep apnea during REM sleep. Blood oxygen increased after each breath, then continued to fall until another breath was taken. This person inhaled only 4 times in the 6-minute period; a normal sleeper would breathe more than 60 times in the same interval.

even occur in children and may be related to some cases of *sudden infant death syndrome* (SIDS), or crib death, in which otherwise healthy infants inexplicably die in their sleep. Sleep apnea is thought to be more common among people who are overweight and who snore, conditions in which air flow is restricted.

Treatments for sleep apnea include surgery or appliances that expand the upper airway, weight loss, and face masks that deliver negative pressure to open the airway. Untreated, sleep apnea can cause high blood pressure and other cardiovascular disease, memory problems, weight gain, impotence, headaches, and brain damage due to oxygen insufficiency (Wenner et al., 2009).

The atonia of REM sleep can occur when a person is awake and active; this form is called **cataplexy**. In cataplexy, a person loses muscle tone gradually or even quickly falls to the floor, atonic. The collapse can be so sudden that there is a real risk of injury. Cataplexy can be triggered by excitement or laughing. While in an atonic condition, the person sees imaginary creatures or hears imaginary voices. People who fall into a state of cataplexy with these **hypnogogic hallucinations** give every appearance of having fallen into REM sleep while remaining "awake."

Cataplexy can have a genetic basis. In 1970, William Dement was given a litter of Doberman pinscher dogs and later a litter of Labrador retrievers. These dogs displayed cataplexy. The disease is transmitted as a recessive trait: to develop it, a dog must inherit the gene from both its mother and its father. The descendants of those dogs continue to provide animal models for investigating the neural basis of the disease as well as its treatment.

When such a dog is excited—when it is running for a piece of food, for example—it may suddenly collapse, as illustrated at the left in **Figure 13-25.** Jerome Siegel (2004) investigated the cause of narcolepsy in dogs. He found that neurons in the subcoerulear nucleus become inactive and neurons in the magnocellular nucleus of the medulla become active during attacks of cataplexy, just as they do during REM sleep.

On the basis of anatomical examinations of the brains of narcoleptic dogs, Siegel suggested that the death of neurons in the amygdala and adjacent forebrain areas is a

The word *cataplexy* comes from the Greek word *kataplessein*, meaning "to strike down." *Hypnogogic* comes from the Greek *hypnos*, "sleep," and *agogos*, "leading into."

Figure 3-21 explains inheritance patterns for genetic disorders.

Section 3-3 investigates knockout technology and other genetic engineering techniques.

one-time event that occurs just before the onset of the disease, early in life. A subset of these neurons produces a peptide called *orexin* (also called *hypocretin*) that serves as a signaling molecule to maintain wakefulness (Clark et al., 2009).

Orexin cells are also located in the hypothalamus and send projections to many other brain regions, suggesting that they play a role maintaining our normal waking state. To test the idea that orexin loss is related to cataplexy, the investigators bred knockout mice that had no orexin. When these mice became active, such as at feeding time, they collapsed into cataplexy, supporting the idea that an orexin system contributes to a normal waking state.

Recall Jouvet's experiment: he reported that cats with lesions to the subcoerulear region of the brainstem entered REM sleep without accompanying atonia and so apparently acted out their dreams. A similar condition has been reported in people and may either have a genetic basis or be caused by brain damage. The condition has been named *REM without atonia*. The behavior of people who have REM without atonia suggests that they are acting out their dreams. Following is the account of a 67-year-old patient (Schenck et al., 1986):

> I had a dream where someone was shooting at me with a rifle and it was in a field that had ridges in it, so I decided to crawl behind a ridge—and I then had a gun too—and I look over the ridge so when he showed up I would shoot back at him and when I came to [i.e., awakened] I was kneeling alongside the bed with my arms extended like I was holding the rifle up and ready to shoot.

In the dream, the patient saw vivid images, but he heard nothing and felt afraid. Although many patients who have had such experiences have been described, most are elderly and suffer from brain injury or other brain-related disorders. REM without atonia can be treated with benzodiazepines, which block REM sleep.

REVIEW 13-6
Sleep Disorders

Before you continue, check your understanding.

1. Disorders of NREM sleep include _____, in which a person has difficulty falling asleep at night, and _____, in which a person falls asleep involuntarily in the daytime.

2. Treating insomnia with sleeping pills, usually sedative hypnotics, may cause _____: progressively higher doses must be taken to achieve sleep.

3. Disorders of REM sleep include _____, in which a person awakes but is paralyzed and experiences fear, and _____, in which a person may lose muscle tone and collapse in the daytime.

4. People who act out their dreams, a condition termed _____, may have damage to the _____ nucleus.

5. Is orexin the substance that produces waking?

Answers appear at the back of the book.

FIGURE 13-25 Cataplexy In both dog and human, an attack of catalepsy causes the head to droop and the back and legs to sag and can progress to a complete loss of muscle tone while the person or dog is awake and conscious. Cataplexy is distinct from narcolepsy in that people hear and remember what is said around them, and dogs can track a moving object with their eyes. (*Left*) James Arnovsky and (*right*) Joel Deutsch, Slim Films

13-7 What Does Sleep Tell Us about Consciousness?

René Descartes conceived his idea of a mind through a dream. He dreamed that he was interpreting the dream as it occurred, a behavior called *lucid dreaming*. Later, when awake, he reasoned that, if he could think and analyze a dream while asleep, his mind must be able to function during both waking and sleeping. He proposed therefore that the mind must be independent of the body that undergoes sleeping and waking transitions. Contemporary sleep research tells us that our sleep–waking states are still more complex.

As described in preceding sections, what we colloquially refer to as "waking" comprises at least two different states: alert consciousness, mediated by the cholinergic system, and consciousness with movement, mediated by the serotonergic system. Similarly, sleep consists of NREM and REM phases. NREM sleep in turn consists of four different substages, as indicated by the EEG (see Figure 13-12A). REM sleep periods of twitching and periods of non-twitching can occur both during sleep and during waking. People who are sleeping may awake and find themselves in a condition of sleep paralysis in which they experience the hallucinations and fear common in dreams. People who are awake may fall into a state of cataplexy in which they are conscious of being awake while experiencing the visual and emotional features of dreams.

Sleep researcher Allan Hobson reported the peculiar symptoms he suffered after a brainstem stroke (Hobson, 2002). For the first 10 days after the lesion, he suffered from complete insomnia and experienced neither REM nor NREM sleep. Whenever he closed his eyes, however, he did experience sudden visual hallucinations that had a dreamlike quality. This experience suggested that eye closure is sufficient to produce the visual components of REM sleep but with neither loss of consciousness nor atonia. Hobson eventually recovered normal sleeping patterns, and the hallucinations stopped.

Beyond being a source of insight into the neural basis of consciousness, the study of sleep states and events may help to explain some psychiatric and drug-induced conditions. For example, among the symptoms of schizophrenia are visual and auditory hallucinations. Are these hallucinations dream events that occur unexpectedly during waking? Many people who take hallucinogenic drugs such as LSD report visual hallucinations. Does the drug initiate the visual features of dreams? People who have panic attacks suffer from very real fright that has no obvious cause. Are they experiencing the fear attacks that commonly occur during sleep paralysis and cataplexy?

What the study of sleep tells us about consciousness is that many conscious states exist. Some are associated with waking and some with sleeping and the two can mix together to produce a variety of odd conditions. When it comes to consciousness, there is far more to sleeping and waking than just sleeping and waking.

Section 1-2 recounts how, in the mid-1600s, Descartes chose the pineal gland as the seat of the mind.

Section 15-7 explores the neural basis of consciousness and ideas about why humans are conscious.

SUMMARY

13-1 A Clock for All Seasons

Biorhythms are cyclic behavior patterns of varying length displayed by animals, plants, and even single-celled organisms. Mammals display a number of biorhythms, including circadian (daily) rhythms and circannual (yearly) rhythms. In the absence of environmental cues, circadian rhythms are free running, lasting a little more or a little less than their usual period of about 24 hours, depending on the individual organism or the environmental conditions. Cues that reset a biological clock to a 24-hour rhythm are called Zeitgebers.

13-2 Neural Basis of the Biological Clock

A biological clock is a neural structure responsible for producing rhythmic behavior. Our master biological clock is the suprachiasmatic nucleus. The SCN is responsible for circadian rhythms, and it has its own free-running rhythm with a period that is a little more or a little less than 24 hours. Stimuli from the environment, such as sunrise and sunset, entrain the free-running rhythm so that its period is 24 hours.

Neurons of the suprachiasmatic nucleus are active in the daytime and inactive at night. These neurons display their rhythmicity when disconnected from other brain structures, when removed from the brain and cultured in a dish, and after having been cultured in a dish for a number of generations. When reimplanted into a brain without an SCN, they restore the animal's circadian rhythms. The different aspects of neuronal circadian rhythms, including their period, are under genetic control.

13-3 Sleep Stages and Dreaming

Sleep events are measured by recording the brain's activity to produce an electroencephalogram (EEG), muscular activity to produce an electromyogram (EMG), and eye movements to produce an electrooculogram (EOG).

A typical night's sleep, as indicated by physiological measures, consists of stages that take place in a number of cycles over the course of the night. During REM sleep, the EEG has a waking pattern, and the sleeper displays rapid eye movements. Stages of sleep in which the EEG has a slower rhythm are called non-REM (NREM) sleep.

Intervals of NREM sleep and REM sleep alternate four or five times each night. The duration of NREM sleep is longer earlier in the sleep period, whereas the duration of REM sleep is longer in the later part of the sleep period. These intervals also vary with age.

A sleeper in slow-wave sleep has muscle tone, may toss and turn, and has dreams that are not especially vivid. A sleeper in REM sleep has no muscle tone and so is paralyzed and has vivid dreams in real time. Dream duration coincides with the duration of the REM period.

The activation–synthesis hypothesis proposes that dreams are not meaningful and are only a by-product of the brain's state of excitation during REM. The coping hypothesis suggests that dreaming evolved as a mechanism to cope with real threats and fears posed by the environment.

13-4 What Does Sleep Accomplish?

Several theories of sleep have been advanced, including the propositions that sleep is a biological adaptation that conserves energy resources and that it is a restorative process that fixes wear and tear in the brain and body. Sleep may also organize and store memories.

13-5 Neural Bases of Sleep

Separate neural regions of the brain are responsible for NREM and REM sleep. The reticular activating system located in the central area of the brainstem is responsible for NREM sleep. If the RAS is stimulated, a sleeper awakes; if it is damaged, a person may enter a coma.

The peribrachial area and the medial pontine reticular formation of the brainstem are responsible for REM sleep. If these areas are damaged, REM sleep may no longer occur. Pathways from these areas project to the cortex to produce the cortical activation of REM and to the brainstem to produce the muscular paralysis of REM.

13-6 Sleep Disorders

Disorders of NREM sleep include insomnia, the inability to sleep at night, and narcolepsy, inconveniently falling asleep in the daytime. Sedative hypnotics used to induce sleep may induce drug-dependency insomnia, a sleep disorder in which progressively larger doses of the drug are required to produce sleep.

Disorders of REM sleep include sleep paralysis, in which a dreaming person awakens but remains unable to move and sometimes feels fear and dread. Cataplexy is a disorder in which an awake person collapses into a state of paralysis. At the same time, the person may remain awake and have hypnogogic hallucinations similar to dreaming.

13-7 What Does Sleep Tell Us about Consciousness?

Sleep research provides insight into consciousness by revealing that many kinds of waking and sleeping exist. Just as the events of wakefulness intrude into sleep, the events of sleep can intrude into wakefulness. The array of conditions thus produced demonstrates that consciousness is not a unitary state.

KEY TERMS

atonia, p. 460

basic rest–activity cycle (BRAC), p. 466

beta (β) rhythm, p. 460

biological clock, p. 445

biorhythm, p. 445

cataplexy, p. 476

chronotype, p. 455

circadian rhythm, p. 445

coma, p. 472

consolidation, p. 469

delta (δ) rhythm, p. 460

dimer, p. 455

diurnal animal, p. 445

drug-dependency insomnia, p. 476

entrain, p. 448

free-running rhythm, p. 448

hypnogogic hallucination, p. 476

insomnia, p. 474

jet lag, p. 451

K-complex, p. 471

light pollution, p. 451

medial pontine reticular formation (MPRF), p. 474

melatonin, p. 456

metabolic syndrome, p. 445

microsleep, p. 469

narcolepsy, p. 474

NREM (non-REM) sleep, p. 460

peribrachial area, p. 474

period, p. 447

place cell, p. 471

reconsolidation, p. 471

REM sleep, p. 460

reticular activating system (RAS), p. 472

retinohypothalamic tract, p. 452

sleep apnea, p. 476

sleep paralysis, p. 476

sleep spindle, p. 471

slow-wave sleep, p. 460

suprachiasmatic nucleus (SCN), p. 452

Zeitgeber, p. 448

CHAPTER 14

How Do We Learn and Remember?

CLINICAL FOCUS 14-1 REMEDIATING DYSLEXIA

14-1 CONNECTING LEARNING AND MEMORY

STUDYING LEARNING AND MEMORY IN THE LABORATORY

TWO CATEGORIES OF MEMORY

WHAT MAKES EXPLICIT AND IMPLICIT MEMORY DIFFERENT?

WHAT IS SPECIAL ABOUT PERSONAL MEMORIES?

14-2 DISSOCIATING MEMORY CIRCUITS

DISCONNECTING EXPLICIT MEMORY

DISCONNECTING IMPLICIT MEMORY

CLINICAL FOCUS 14-2 PATIENT BOSWELL'S AMNESIA

14-3 NEURAL SYSTEMS UNDERLYING EXPLICIT AND IMPLICIT MEMORIES

NEURAL CIRCUIT FOR EXPLICIT MEMORIES

CLINICAL FOCUS 14-3 ALZHEIMER'S DISEASE

CLINICAL FOCUS 14-4 KORSAKOFF'S SYNDROME

CONSOLIDATION OF EXPLICIT MEMORIES

NEURAL CIRCUIT FOR IMPLICIT MEMORIES

NEURAL CIRCUIT FOR EMOTIONAL MEMORIES

14-4 STRUCTURAL BASIS OF BRAIN PLASTICITY

LONG-TERM POTENTIATION

MEASURING SYNAPTIC CHANGE

ENRICHED EXPERIENCE AND PLASTICITY

SENSORY OR MOTOR TRAINING AND PLASTICITY

RESEARCH FOCUS 14-5 MOVEMENT, LEARNING, AND NEUROPLASTICITY

EXPERIENCE-DEPENDENT CHANGE IN THE HUMAN BRAIN

EPIGENETICS OF MEMORY

PLASTICITY, HORMONES, TROPHIC FACTORS, AND DRUGS

SOME GUIDING PRINCIPLES OF BRAIN PLASTICITY

14-5 RECOVERY FROM BRAIN INJURY

DONNA'S EXPERIENCE WITH TRAUMATIC BRAIN INJURY

THREE-LEGGED CAT SOLUTION

NEW-CIRCUIT SOLUTION

LOST-NEURON-REPLACEMENT SOLUTION

Remediating Dyslexia

As children absorb their society's culture, acquiring language skills seems virtually automatic. Yet some people have lifelong difficulties in mastering language-related tasks. Educators classify these difficulties under the umbrella of *learning disabilities*.

Dyslexia, impairment in learning to read, may be the most common learning disability. Children with dyslexia (from Greek words suggesting "bad" and "reading") have difficulty learning to write as well as to read.

In 1895, James Hinshelwood, an eye surgeon, examined some schoolchildren who were having reading problems, but he could find nothing wrong with their vision. Hinshelwood was the first to suggest that children with reading problems were impaired in brain areas associated with language use. Norman Geshwind and Albert Galaburda (1985) proposed how such impairment might come about.

Struck by the finding that dyslexia is far more common in boys than in girls, they reasoned that hormonal influences in early development influence brain development too. They examined postmortem the brains of a small sample of people who had experienced dyslexia and found abnormal collections of neurons, or "warts," in and around the language areas of the brain.

This relation between structural abnormalities in the brain and learning disabilities is further evidence that an intact brain is necessary for typical human functioning. Geshwind and Galaburda also found abnormalities in the auditory thalamus, suggesting a deficit in auditory processing. More recently, noninvasive brain imaging has determined that, relative to the brains of normal participants, activity is reduced in the left temporoparietal cortex of people with dyslexia.

Michael Merzenich and his colleagues designed a remedial treatment program based on the assumption that the fundamental problem in learning disabilities lies in auditory processing, specifically of language sounds (e.g., Temple et al., 2003). Remediation involves learning to make increasingly difficult sound discriminations, for example, discriminating "ba" and "da."

When the sounds are spoken slowly discriminating between them is easy, but as they grow briefer and occur faster, discrimination becomes more difficult. Previous studies using rats and monkeys had shown that discrimination training stimulates neural plasticity in the auditory system, making it capable of discriminating sounds that previously were not possible.

The representative fMRIs shown here reveal decreased activation in many brain regions in untreated dyslexic children compared with typical children. With training, dyslexic readers can normalize their brain activity and, presumably, its connectivity.

The extent of increased brain activation in the language-related regions (circled in the images) correlates to the amount of increased brain activation overall. The results suggest that the remedial treatment both improves brain function in regions associated with phonological processing and produces compensatory activation in related brain regions.

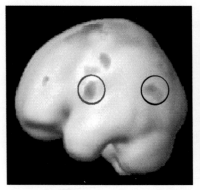
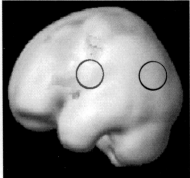

Typical-reading children while rhyming

Dyslexic-reading children while rhyming (before remediation)

Regions of the frontal and temporoparietal cortex that showed decreased activation in children with untreated dyslexia. **Adapted from "Neural Deficits in Children with Dyslexia Ameliorated by Behavioral Remediation: Evidence from Functional MRI," by E. Temple, G. K. Deutsch, R. A. Poldrack, S. L. Miller, P. Tallal, M. M. Merzenich, and J. D. E. Gabrieli, 2003,** *Proceedings of the National Academy of Sciences (USA) 100,* pp. 2860–2865.

The brain is plastic. It changes throughout life, allowing us to modify our behavior to adapt and learn and to remember. If we reflect on our own lives, we can easily compile a list of experiences that must change the brain:

Neuroplasticity, the hallmark of nervous-system function, is the nervous system's potential for physical or chemical change that enhances its adaptability.

- Profound changes during development

- Acquisition of culture

- Preferences among foods and beverages, art and music, and for other experiences

- Ability to cope with neurodegeneration in the aging process and, for many, to accommodate neurological injury or disease

Learning is common to all these experiences. Understanding how the brain supports learning is a fundamental question in neuroscience. At the level of the neuron, synapses change with experience—learning new information, for example. Such changes can take place anywhere in the brain.

We can investigate neuronal changes that support learning specific types of information by describing the changes in cells exposed to specific sensory experiences. Or we can look at the neural changes that mediate brain plasticity—recovery from brain injury, addiction to drugs, or conquering a learning disability. The goal of this chapter is to move beyond the general concept of neuroplasticity to an understanding of what stimulates plastic change in the brain. We inspect changes related to environment and experience, learning and memory, electrical stimulation, chemical influences, and brain injury.

14-1 Connecting Learning and Memory

Learning is a relatively permanent change in an organism's behavior as a result of experience. **Memory** is the ability to recall or recognize previous experience. Memory thus implies a mental representation of the previous experience, sometimes referred to as a *memory trace*. Neuroscientists presume that this hypothetical memory trace corresponds to some physical change in the brain, most likely involving synapses.

At the macro level, we infer what we know about learning and memory formation from changes in behavior, not by observing the brain directly. Studying learning and memory therefore requires behavioral measures that evaluate how these changes come about. We begin here by reviewing how learning and memory researchers study animals in the laboratory. The results obtained in this research suggest, in a general way, how the brain organizes its learning and memory systems.

Studying Learning and Memory in the Laboratory

A challenge for psychologists studying memory in laboratory animals (or people) is to get the subjects to reveal what they can remember. Because laboratory animals do not talk, investigators must devise ways for a subject to show its knowledge. Different species can "talk" to us in different ways, so the choice of test must be matched to the capabilities of the species.

Mazes or swimming pools are typically used to study rats because rats live in tunnels and near water. Studies of monkeys take advantage of their sharp vision and avid curiosity by requiring them to look under objects for food or at television monitors. When birds are the subjects, natural behaviors such as singing are used. And for human participants, investigators tend to use paper-and-pencil tests.

Psychologists have devised hundreds of different tests over the past century, and the test results reveal that many types of learning and memory exist. Each appears to have its own neural circuitry. Two classic traditions for training animals to "talk" to investigators emerged a century ago. These very different approaches to studying learning and memory are based on the work of Edward Thorndike in the United States and on experiments conducted by Ivan Pavlov in Russia.

Pavlovian Conditioning

In the early years of the twentieth century, Ivan Pavlov, a Russian physiologist, discovered that when a food reward accompanies some stimulus, such as a tone, dogs learn to associate the stimulus with the food. Then whenever they hear the tone, they salivate even though no food is present. This type of learning has many names, including **Pavlovian conditioning,** *respondent conditioning,* and *classical conditioning,* and its characteristics have been documented in many studies.

A key feature of Pavlovian conditioning is that animals learn to associate two stimuli (such as the presentation of the food and the tone) and to communicate to us that they have learned it by giving the same response (such as salivation) to both stimuli. Pet owners are familiar with this type of learning: to a cat or dog, the sound of a can being

The memory trace is one among many *psychological constructs,* abstract mental processes that can be inferred only from behavior (see Section 15-1). Others include learning, language, emotion, motivation, and thinking.

Figure 7-4 illustrates swimming-pool tests for rats, Experiment 15-1 monkeys' perceptual threshold for apparent motion, and Focus 15-4 tests that reveal the effects of brain injuries on cognitive performance in humans.

dyslexia Impairment in learning to read and write; probably the most common learning disability.

learning Relatively permanent change in an organism's behavior as a result of experience.

memory Ability to recall or recognize previous experience.

Pavlovian conditioning Learning procedure whereby a neutral stimulus (such as a tone) comes to elicit a response because of its repeated pairing with some event (such as the delivery of food); also called *classical conditioning* or *respondent conditioning.*

FIGURE 14-1 Eye-Blink Conditioning
Neural circuits in the cerebellum mediate this form of stimulus–response learning.

1 Headgear is arranged for eye-blink conditioning.

Electrodes

2 Puff of air to eye causes eye to blink.

Air jet tube

Audio speaker

3 After pairing air puff with tone, tone alone comes to elicit a blink.

opened is a clear stimulus for food. Two forms of Pavlovian conditioning are commonly used in experiments today: eye-blink conditioning and fear conditioning. Each is associated with neural circuits in discrete brain regions; thus both have proved especially useful.

Eye-blink conditioning has been used to study Pavlovian learning in rabbits and people **(Figure 14-1).** In these studies, a tone (or some other stimulus) is associated with a painless puff of air to the participant's eye. The tone is the **conditioned stimulus** (CS) that comes to elicit a blink produced initially by the air puff. The air puff is the **unconditioned stimulus** (UCS), because blinking is the normal reaction—the **unconditioned response** (UCR)—to a puff of air. The participant communicates that it has learned that the signal stimulus predicts the puff by blinking in response to the signal (the CS) alone—a **conditioned response** (CR).

Circuits in the cerebellum mediate Pavlovian learning. The cerebellum does not have special circuits just for eye-blink conditioning, which is an artificial situation. Rather, the cerebellum has circuits designed to pair various motor responses with environmental events. Eye-blink conditioning experiments simply take advantage of this biological predisposition.

In the cerebellum, the flocculus controls eye movements. You can examine the cerebellar homunculus in Figure 11-14.

In **fear conditioning,** a noxious stimulus is used to elicit fear, an emotional response. A rat or other animal is placed in a box. A mild but noxious electric current can be passed through the grid floor. As shown in **Experiment 14-1,** a tone (the CS) is presented just before a brief, unexpected, mild electric shock.

When the tone is presented later without the shock, the animal acts afraid. It may become motionless and may urinate in anticipation of the shock. Presentation of a novel stimulus, perhaps a light, in the same environment has little effect on the animal. Thus, the animal communicates to us that it has learned the association between the tone and the shock.

The shock is roughly equivalent to a jolt of static electricity you might get when you rub your feet on a carpet and then touch a metal object or another person.

Because the CR is emotional, circuits of the amygdala rather than the cerebellum mediate fear conditioning. Although both eye-blink and fear conditioning are Pavlovian, different parts of the brain mediate the learning.

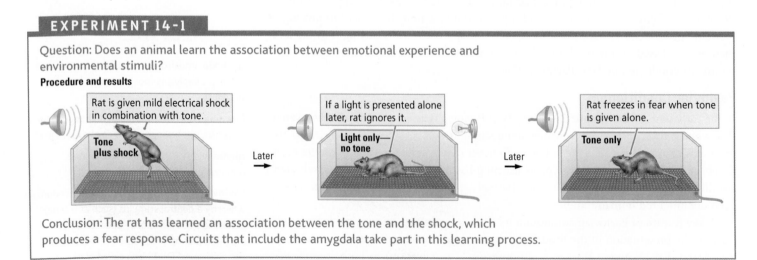

EXPERIMENT 14-1

Question: Does an animal learn the association between emotional experience and environmental stimuli?

Procedure and results

Rat is given mild electrical shock in combination with tone.

Tone plus shock

Later

If a light is presented alone later, rat ignores it.

Light only—no tone

Later

Rat freezes in fear when tone is given alone.

Tone only

Conclusion: The rat has learned an association between the tone and the shock, which produces a fear response. Circuits that include the amygdala take part in this learning process.

Operant Conditioning

In the United States, Edward Thorndike (1898) began a second tradition of studying learning and memory. Thorndike was interested in how animals solve problems. In one series of experiments, he placed cats in a box with a plate of fish outside it **(Figure 14-2).** The only way for a hungry cat to get to the fish was to figure out how to get out of the box.

The solution was to press on a lever, which activated a system of pulleys that opened the box door. The cat gradually learned that its actions had consequences: on the initial trial, the cat touched the releasing mechanism only by chance as it restlessly paced inside the box. The cat apparently learned that something it had done opened the door, and it tended to repeat the behaviors that had occurred just before the door opening. After a few trials, the cat took just seconds to get the door open so that it could devour the fish.

Later studies by B. F. Skinner (e.g., 1938) used a similar strategy of *reinforcement* to train rats to press bars or pigeons to peck keys to obtain food. Just as Thorndike's cats learned to escape his puzzle boxes, many animals learn to bar press or key peck simply if they are placed in the apparatus and allowed to discover the response that obtains the reward. This type of learning is **operant conditioning,** or *instrumental conditioning,* as Thorndike called it. The animal demonstrates that it has learned the association between its actions and the consequences by performing the task faster.

The variety of operant associations is staggering: we are constantly learning to associate our behavior with its consequences. It should be no surprise, then, that operant learning is not localized to any particular circuit in the brain. The circuits needed vary with the requirements of the task. For example, olfactory tasks involve olfactory-related structures like the orbitofrontal cortex and the amygdala, spatial tasks recruit the hippocampus, and motor tasks require the basal ganglia.

Two Categories of Memory

Humans present a distinct challenge to studying memory because so much of our learning is verbal. Psychologists have studied human memory since the mid-1800s. More recently, cognitive psychologists have developed sophisticated measures of learning and memory for neuropsychological investigations. Two such measures help to distinguish between two categories of memory in humans.

In one kind of task, a group of participants reads a list of words, such as *spring, winter, car,* and *boat.* Another group reads a list consisting of *trip, tumble, run,* and *sun.* All the participants are then asked to define a series of words. One is *fall.*

Section 12-3 describes how Skinner used reinforcers to shape behavior.

eye-blink conditioning Commonly used experimental technique in which subjects learn to pair a formerly neutral stimulus with a defensive blinking response.

conditioned stimulus (CS) In Pavlovian conditioning, an originally neutral stimulus that, after association with an unconditioned stimulus (UCS), triggers a conditioned response.

unconditioned stimulus (UCS) A stimulus that unconditionally—naturally and automatically—triggers a response.

unconditioned response (UCR) In classical conditioning, the unlearned, naturally occurring response to the unconditioned stimulus, such as salivation when food is in the mouth.

conditioned response (CR) In Pavlovian conditioning, the learned response to a formerly neutral conditioned stimulus (CS).

fear conditioning Learned association, a conditioned emotional response, between a neutral stimulus and a noxious event such as a shock.

operant conditioning Learning procedure in which the consequences (such as obtaining a reward) of a particular behavior (such as pressing a bar) increase or decrease the probability of the behavior occurring again; also called *instrumental* conditioning.

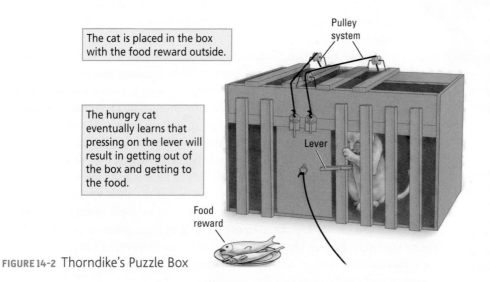

The cat is placed in the box with the food reward outside.

Pulley system

The hungry cat eventually learns that pressing on the lever will result in getting out of the box and getting to the food.

Lever

Food reward

FIGURE 14-2 Thorndike's Puzzle Box

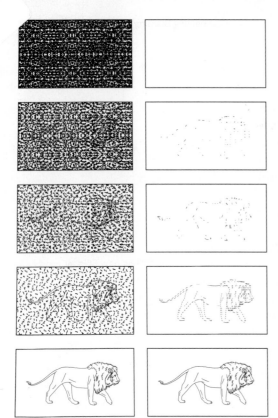

FIGURE 14-3 Gollin Figure Test Participants are shown a series of drawings in sequence, from least to most clear, and asked to identify the image. Most people must see several panels before they can identify it. On a retention test some time later, however, participants identify the image sooner than they did on the first test, indicating some form of memory for the image. Amnesic subjects also show improvement on this test, even though they do not recall taking it.

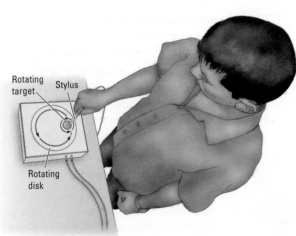

Rotating target
Stylus
Rotating disk

FIGURE 14-4 Pursuit-Rotor Task The participant must keep the stylus in contact with the metal disc that is moving in a circular pattern on a turntable, which also is rotating in a circular pattern. Although the task is difficult, most people show significant improvement after a brief training period. Given a second test at some later time, both normal participants and amnesics show retention of the task, but the amnesics typically do not recall learning it before.

The word *fall* has multiple meanings, including the season and a tumble. People who have just read the word list containing names of seasons are likely to give the "season" meaning; those who have read the second list, containing action words, typically give the "tumble" meaning. Some form of unconscious (and unintentional) learning takes place as the participants read the word lists.

This task measures **implicit memory:** participants demonstrate knowledge—a skill, conditioned response, or recalling events on prompting—but cannot explicitly retrieve the information. People with **amnesia,** a partial or total loss of memory, perform normally on tests of implicit memory. The amnesic person has no recollection of having read the word list yet acts as though some neural circuit has been influenced by it. In amnesia, a *dissociation*—a disconnect—occurs between the memory of the unconscious (or implicit) learning and **explicit memory,** the conscious recollection of training. Nonamesic people can retrieve an explicit memory and indicate that they know the retrieved item is correct.

This implicit–explicit distinction is not restricted to verbal learning; it is true of visual learning and motor learning tasks as well. For example, when people are shown the top panel of the Gollin figure test in **Figure 14-3** and asked what it shows, they are unlikely to be able to identify an image. They are then presented with a succession of more nearly complete sketches until they can identify the picture. When control participants and amnesics are later shown the same sketch, both groups identify the figure sooner than they could the first time. Even though the amnesic subjects may not recall seeing the sketches before, they behave as though they had.

To measure implicit motor-skills learning, a person is taught a skill, such as the pursuit-rotor task shown in **Figure 14-4.** A small metal disc moves in a circular pattern on a turntable that also is moving. The task is to hold a stylus on the small disc as it spins. This task is not as easy as it looks, especially when the turntable is moving quickly.

Nonetheless, with an hour's practice most people become reasonably proficient. Presented with the same task a week later, both normal participants and amnesics take less time to perform it. Here, too, the amnesics fail to recall performing the task before.

The distinction between tests of implicit and explicit memory is consistent and therefore must offer a key to how the brain stores information. Some theorists make subtle distinctions between the implicit–explicit dichotomy we use for categorizing unconscious and conscious memory and other terminologies. Many researchers prefer to distinguish between **declarative memory,** the specific contents of specific experiences that can be verbally recalled (times, places, or circumstances), and **procedural memory,** the ability to perform a task. As applied to humans there is little practical difference.

Table 14-1 lists commonly used dichotomies, the general distinction being that one memory category requires recalling specific information, whereas the other refers to knowledge of which we are not consciously aware. We can include Pavlovian conditioning and Thorndike's and Skinner's operant learning in this analysis too: all are forms of implicit learning.

Nonspeaking animals can display explicit memory. One of us owned a cat that loved to play with a little ball. One day, as the cat watched, the ball was temporarily put on a high shelf to keep it away from an inquisitive toddler. For weeks afterward, the cat sat and stared at the shelf where the ball had been placed, even though the ball was not visible—an example of explicit memory.

Animals also display explicit memory when they learn psychological tasks. Rats can be trained to find highly palatable food in a new location in a large compound each day. The task is to go to the most recent location. This piece of information is explicit and can be demonstrably forgotten.

Suppose a well-trained rat is given one trial with the food at a new location for several trials and then retested an hour, a day, 3 days, or a week later. The rat has no difficulty with a delay of an hour or perhaps even a day. Some rats are flawless at 3 days, but most have forgotten the location by the time a week has elapsed. Instead, they wander around looking for the food. This behavior illustrates their implicit memory of the **learning set,** the "rules of the game"—an implicit understanding of how a problem can be solved with a rule that can be applied in many different situations—namely, here, that a desired food can be found with a certain type of search strategy.

What Makes Explicit and Implicit Memory Different?

One reason explicit and implicit memories differ is that each is housed in a different set of neural structures. Another reason they differ is that explicit and implicit information are processed differently.

Encoding Memories

Implicit information is encoded in very much the same way as it is perceived and can be described as data-driven, or bottom-up, processing. The idea is that information enters the brain through the sensory receptors and is then processed in a series of subcortical and cortical regions. For example, visual information about an object goes from the visual receptors (the "bottom") to the thalamus, the occipital cortex, and then through the ventral stream to the temporal lobe (the "top"), where the object is recognized.

Explicit memory, in contrast, depends on conceptually driven, or top-down, processing: the person reorganizes the data. For example, if you were searching for a particular object, such as your keys, you would ignore other objects. This is a top-down process because circuits in the temporal lobe (the "top") form an image that influences how incoming visual information (the "bottom") is processed, which in turn greatly influences information recall later.

Because a person has a relatively passive role in encoding implicit memory, he or she has difficulty recalling the memory spontaneously but recalls it more easily when there is **priming** by the original stimulus or some feature of it. Because a person plays an active role in processing information explicitly, the internal cues that were used in processing can also be used to initiate spontaneous recall.

Findings from studies of eyewitness testimony demonstrate the active nature of explicit-memory recall—and its potential fallibility (e.g., Loftus, 1997). In a typical experiment, people are shown a video clip of an accident in which a car collides with another car stopped at an intersection. One group of participants is asked to estimate how fast the moving car was going when it "smashed" into the other car. A second group is asked how fast the car was going when it "bumped" into the other car.

Later questioning indicates that the memory of how fast the moving car was going is biased by the instruction: participants looking at "smashing" cars estimate faster speeds than the speeds estimated by participants looking at "bumping" cars. The instruction actually causes the information to be processed differently. In both cases, the participants were certain that their memories were accurate.

Other experiments show that implicit memory also is fallible. For example, participants are read the following list of words: *sweet, chocolate, shoe, table, candy, horse, car, cake, coffee, wall, book, cookie, hat.* After a few minutes' delay, the participants hear another list of words that includes some from the first list and some that are new. Participants are asked to identify which words were present on the first list and to indicate how certain they are of the identification.

TABLE 14-1 Differentiating the Two Categories of Memory

Term for conscious memory	Term for unconscious memory
Explicit	Implicit
Declarative	Nondeclarative
Fact	Skill
Memory	Habit
Knowing that	Knowing how
Locale	Taxon
Conscious recollection	Skills
Elaboration	Integration
Memory with record	Memory without record
Autobiographical	Perceptual
Representational	Dispositional
Episodic	Procedural
Semantic	Nonassociative
Working	Reference

Note: This list of paired terms used by various theorists to differentiate conscious from unconscious forms of memory is intended to help you relate other discussions of memory to the one in this book, which favors the explicit–implicit distinction.

implicit memory Unconscious memory: subjects can demonstrate knowledge, such as a skill, conditioned response, or recalling events on prompting, but cannot explicitly retrieve the information.

amnesia Partial or total loss of memory.

explicit memory Conscious memory: subjects can retrieve an item and indicate that they know that the retrieved item is the correct item.

declarative memory Ability to recount what one knows, to detail the time, place, and circumstances of events; often lost in amnesia.

procedural memory Ability to recall a movement sequence or how to perform some act or behavior.

learning set The "rules of the game;" implicit understanding of how a problem can be solved with a rule that can be applied in many different situations.

priming Using a stimulus to sensitize the nervous system to a later presentation of the same or a similar stimulus.

episodic memory Autobiographical memory for events pegged to specific place and time contexts.

One of the words on the second list is *sugar*. Most subjects indicate not just that *sugar* was on the first list but that they are certain it was. Although other sweet things were, *sugar* was not. This demonstration is intriguing, because it shows the ease with which we can form "false memories" and defend their veracity with certainty.

Processing Memories

Although we can distinguish memories generally as implicit or explicit, the brain does not process all implicit or all explicit memories in the same way. Memories can be divided according to categories that differ from those listed in Table 14-1. For example, we can make a distinction between memories for different types of sensory information.

Visual and auditory information is processed by different neural areas, so it is reasonable to assume that auditory memories are stored in different brain regions from the regions that store visual memories. We can also make a distinction between information stored in so-called *short-term memory* and information held for a longer time in *long-term memory*. In short-term memory, information—the final score of a playoff game or the combination of your friend's bike lock, for instance—is held in memory only briefly, for a few minutes at most, and then discarded. In long-term memory, information—such as a close friend's name—is held in memory indefinitely, perhaps for a lifetime.

The frontal lobes play an important role in short-term memory, whereas the temporal lobe plays a central role in long-term storage of verbal information. The crucial point is that no single place in the nervous system can be identified as the location for memory or learning. Virtually the entire nervous system can be changed by experience, but different parts of an experience change different parts of the nervous system. One challenge for the experimenter is to devise ways of manipulating experience to demonstrate change in different parts of the brain.

Storing Memories

Understanding that every part of the brain can learn influences how we view the neural circuits that mediate memory. We could expect that areas that process information also house the memory of that information. Areas that process visual information, for example, probably house visual memory. Since the temporal lobe has specialized regions for processing color, shape, and other visual characteristics of an object, we can predict that the memory for various visual attributes of objects is stored separately.

A series of PET studies by Alex Martin and colleagues (1995) at the U.S. National Institute of Mental Health confirmed this prediction. In one study, participants were shown black-and-white line drawings of objects and asked to generate words denoting either colors or actions of the objects. The idea is that processing color and motion are carried out in different locations in the temporal lobe, and thus the activity linked with the memories of color and motion also might be dissociated.

Just such a dissociation was demonstrated. **Figure 14-5** shows that recall of colors activates a region in the ventral temporal lobe, just anterior to the area controlling color perception, whereas recall of action words activates a region in the middle temporal gyrus, just anterior to the area controlling motion perception. This distribution of neural activation shows not only that object memory is at least partly located in the temporal lobes but also that it is found in regions associated with the original perception of the objects.

What Is Special about Personal Memories?

One aspect of memory unique to each of us is our personal, or *autobiographical*, memory. This **episodic memory** includes not only a record of events (episodes) that took place but also a record of our presence and role in the events.

FIGURE 14-5 Memory Distribution Blood flow in left-hemisphere regions increases when participants generate color words (red) and action words (blue) to describe static, black-and-white drawings of objects. Purple areas indicate overlap. The red region extends into the ventral temporal lobe, suggesting that object memory is organized as a distributed system. Object attributes are stored close to the cortical regions that mediate their perceptions. Parietal lobe activation likely is related to movements associated with action words and frontal lobe activation to the spontaneously generated behavior. Adapted from "Discrete Cortical Regions Associated with Knowledge of Color and Knowledge of Action," by A. Martin, J. V. Haxby, F. M. Lalonde, C. L. Wiggs, and L. G. Ungerleider, 1995, *Science, 270*, p. 104.

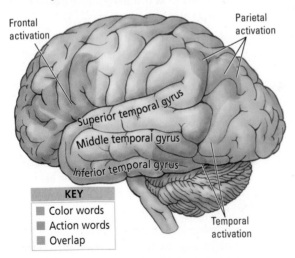

Our personal experiences form the basis of who we are and the rules by which we live. That is, we have memories not only for events but also for their context at a particular time in a particular place. We thus gain a concept of time and a sense of our personal role in a changing world.

Imagine what would happen if we lost our personal memories. We would still recall events that took place but would be unable to see our role in them. People with frontal-lobe injuries sometimes exhibit such symptoms, as illustrated in a case described by Endel Tulving (2002).

K. C. suffered a serious traumatic brain injury in a motorcycle accident that produced multiple cortical and subcortical lesions. What is remarkable about K. C. is that his cognitive abilities are intact and indistinguishable from those of most normal healthy adults. He can still play chess and the organ, and his short-term memory is intact. He knows who he is, when his birthday is, the names of schools he attended, the location of the family cottage, and so on.

What K. C. cannot do is recall any personally experienced events. This episodic amnesia covers his entire life, from birth until the present. He knows facts about himself but has no memory for events that included him personally. For example, K. C. cannot describe an event that took place in school that specifically included him, while at the same time recalling going to school and the knowledge he gained there.

Findings from neuroimaging studies of people with episodic amnesia suggest that they consistently have frontal-lobe injuries (Lepage et al., 2001), but exactly why these lesions produce episodic amnesia remains unclear. Nonetheless, Tulving made the interesting proposal that episodic memory is a marvel of nature: it transforms the brain into a kind of time machine that allows us to dwell on the past and make plans for the future. He goes further, suggesting that this ability may be unique to humans and is presumably due to some novel evolutionary development of the frontal lobe.

Not all people with episodic amnesia have brain injury, however. Many case reports describe patients with massive memory disturbances resulting from some sort of "psychiatric" or "psychogenic" disorder. Such cases have been fodder for numerous movie plots.

Hans Markowitsch (2003) noted that the amnesia reported in some of these cases is remarkably similar to episodic amnesia seen in neurological patients. Neuroimaging of patients with *psychogenic amnesia* shows a massive reduction in brain activity in frontal regions that is remarkably similar to that seen in neurological patients with episodic amnesia **(Figure 14-6).** Therefore we can assume that patients with psychogenic amnesias have a dysfunction of frontal-brain activity that acts to block the retrieval of autobiographical memory.

Just as there are people with very poor autobiographical memory, there is also a rare group of people with "superior" autobiographical memory. Elizabeth Parker and her colleagues (2006) described A. J., a woman whose personal memory dominates her life. Seemingly without effort A. J., can tell you what she was doing on any specific date and even what day of the week it was. Several other cases have now been documented, and the investigators propose calling this rare ability *hyperthymestic syndrome* (from the Greek *thymesis,* meaning "remembering").

FIGURE 14-6 **Lost Episodes** Horizontal sections from two patients with selective retrograde amnesia for autobiographical information. The section on the left is from an amnesic patient who had a brain infection (herpes simplex encephalitis). On the right is the brain of a patient with psychogenic amnesia. In each, the right frontal and temporal lobes are dark (white arrows), owing to a metabolic reduction in the right temporofrontal region. From "Functional Neuroimaging Correlates of Functional Amnesia," by H. J. Markowitsch, 1999, *Memory, 7,* Plate 2. Reprinted by permission of Psychology Press Ltd., Hove.

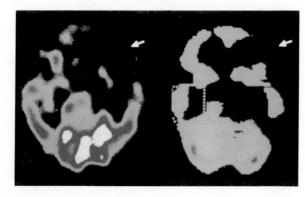

REVIEW 14-1
Connecting Learning and Memory

Before you continue, check your understanding.

1. An organism learns that some stimulus is paired with a reward. This is _____ conditioning.

2. After learning that consequences follow its behavior, an organism modifies its behavior. This is _____ conditioning.

3. Memory of information that is unconsciously learned is _____ memory, whereas memory of specific, factual information is _____ memory.

4. Memory that is autobiographical and unique to each person is _____ memory.

5. Where is memory stored in the brain?

Answers appear at the back of the book.

14-2 Dissociating Memory Circuits

Beginning in the 1920s and continuing until the early 1950s, American psychologist Karl Lashley searched in vain for the neural circuits underlying memories for the solutions to mazes learned by laboratory rats and monkeys. Lashley's working hypothesis was that memories must be represented in the perceptual and motor circuitry used to learn solutions to problems. He believed that, if he removed bits of this circuitry or disconnected it, amnesia should result.

In fact, neither ablation procedure produced amnesia. Lashley found instead that the severity of the memory disturbance was related to the size of the injury rather than to its location. In 1951, after 30 years of searching, Lashley concluded that he had failed to find the location of the memory trace, although he believed that he knew where it was *not* located (Lashley, 1960).

Just two years later William Scoville, a neurosurgeon, made a serendipitous discovery that Lashley's studies had not predicted. Scoville was attempting to rid people of seizures by removing the abnormal brain tissue that caused them. On August 23, 1953, Scoville performed a bilateral medial-temporal-lobe resection on a young man, Henry Molaison (H. M.), whose severe epilepsy was not controlled by medication.

H. M.'s seizures originated in the region that includes the amygdala, hippocampal formation, and associated subcortical structures, so Scoville removed them bilaterally, leaving the more lateral temporal-lobe tissue intact. As shown in **Figure 14-7,** the removal specifically included the anterior part of the hippocampus, the amygdala, and the adjacent cortex.

Disconnecting Explicit Memory

The behavioral symptoms Scoville noted after the surgery were completely unexpected. He invited Brenda Milner to study H. M. Milner had been studying memory difficulties in patients with unilateral temporal-lobe removals for the treatment of epilepsy. She and her colleagues worked with H. M. for more than 50 years, making him the most studied case in neuroscience (e.g., Corkin, 2002). H. M. died in 2008.

Section 7-1 describes brain lesions as an aspect of ablation, the first and simplest technique used to manipulate the brain.

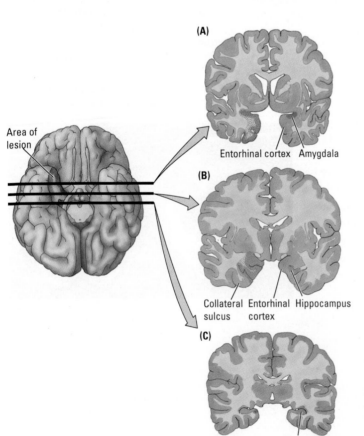

(A)

Entorhinal cortex   Amygdala

(B)

Area of lesion

Collateral Entorhinal Hippocampus
sulcus   cortex

(C)

Hippocampus

FIGURE 14-7 Extent of H. M.'s Surgery H. M.'s right-hemisphere lesion is highlighted in the brain viewed ventrally. The lesion runs along the wall of the medial temporal lobe. The left side of the brain has been left intact to show the relative location of the medial temporal structures. Parts A, B, and C, based on MRI scans, depict a series of coronal sections of H. M.'s brain. Adapted from "H. M.'s Medial Temporal Lobe Lesion: Findings from Magnetic Resonance Imaging," by S. Corkin, D. G. Amaral, R. G. Gonzalez, K. A. Johnson, and B. T. Hyman, 1997, *Journal of Neuroscience, 17,* p. 3966.

H. M.'s most remarkable symptom was severe amnesia: he was unable to recall anything that had happened since his surgery in 1953. H. M. retained an above-average I.Q. score (118 on the Wechsler Adult Intelligence Scale; 100 is average), and he performed normally on perceptual tests. His recall of events from his childhood and school days was intact. Socially, H. M. was well mannered, and he engaged in sophisticated conversations. However, he had no recall for recent events. H. M. lacked any explicit memory.

In one study by Suzanne Corkin (2002), H. M. was given a tray of hospital food, which he ate. A few minutes later, he was given another tray. He did not recall having eaten the first meal and proceeded to eat another. A third tray was brought, and this time he ate only the dessert, complaining that he did not seem to be very hungry.

To understand the implications and severity of H. M.'s condition, one need only consider a few events in his postsurgical life. His father died, but H. M. continued to ask where his father was, only to experience anew the grief of learning that his father had passed away. (Eventually H. M. stopped asking about his father, suggesting that some type of learning had taken place.)

Similarly, when in the hospital, he typically asked, with many apologies, the nurses to tell him where he was and how he came to be there. He remarked on one occasion, "Every day is alone in itself, whatever enjoyment I've had and whatever sorrow I've had." His experience was that of a person who perceives his surroundings but cannot comprehend the situation he is in because he does not remember what has gone before.

Formal tests of H. M.'s memory showed what one would expect: he had no recall for specific information just presented. In contrast, his implicit-memory performance was nearly intact. He performed normally on tests such as the incomplete-figure or pursuit-rotor tasks illustrated in Figures 14-3 and 14-4. Whatever systems are required for implicit memory must therefore have been intact, but the systems crucial to explicit memory were missing or dysfunctional. Clinical Focus 14-2, "Patient Boswell's Amnesia" on page 492, describes a case similar to that of H. M.

Curiously, H. M. recognized faces, including his own, and he recognized that he aged. Face recognition depends on the parahippocampal gyrus, which was partly intact on H. M.'s right side.

Section 9-2 explains prosopagnosia, an inability to recognize faces; Section 9-5 describes other visual-form agnosias.

Disconnecting Implicit Memory

Among the reasons Lashley's research did not find a syndrome like that shown by H. M., the two most important are that Lashley did not damage the medial temporal regions. Nor did he use tests of explicit memory, so his animal subjects would not have shown H. M.'s deficits. Rather, Lashley's tests were mostly measures of implicit memory, with which H. M. had no problems.

The following case illustrates that Lashley probably should have been looking in the basal ganglia for the deficits that his tests of implicit memory revealed. The basal ganglia play a central role in motor control. Among the compelling examples of implicit memory are examples of motor learning—driving and playing musical instruments or online games, to name a few.

J. K. was above average in intelligence and worked as a petroleum engineer for 45 years. In his mid-70s, he began to show symptoms of Parkinson's disease, in which the projections from the dopaminergic cells of the brainstem to the basal ganglia die. At about age 78, J. K. started to have memory difficulties.

Focus features 5-2, 5-3, and 5-4 and Section 7-1 detail aspects of Parkinson's disease. Section 16-3 reviews treatments.

Curiously, J. K.'s memory disturbance was related to tasks that he had done all his life. On one occasion, he stood at the door of his bedroom frustrated by his inability to recall how to turn on the lights. "I must be crazy," he remarked. "I've done this all my life and now I can't remember how to do it!" On another occasion, he was seen trying to turn the radio off with the television remote control. This time he explained, "I don't recall how to turn off the radio so I thought I would try this thing!"

Patient Boswell's Amnesia

At the age of 48, Boswell developed herpes simplex encephalitis, a brain infection. Boswell had completed 13 years of schooling and had worked for nearly 30 years in the newspaper advertising business. By all accounts a normal, well-adjusted person, he was successful in his profession.

Boswell recovered from the acute symptoms of the disease, which included seizures and a 3-day coma. His postdisease intelligence was low average, probably owing to the neurological damage caused by the infection. Nonetheless, his speech and language remained normal in every respect, and he suffered no defects of sensory perception or of movement.

But Boswell was left with a severe amnesic syndrome. If he hears a short paragraph and is asked to describe its main points, he routinely gets scores of zero. He can only guess the day's date and is unable even to guess what year it is. When asked what city he is in, he simply guesses.

Boswell does know his place of birth, and he can correctly recall his birth date about half the time. In sum, Boswell has severe amnesia for events both before and since his encephalitis. Like H. M., he does show implicit memory on tests such as the pursuit-rotor task.

Antonio Damasio and his colleagues (1989) have investigated Boswell's amnesia extensively, and his brain pathology is now well documented. The critical damage, diagrammed in the adjoining illustration, is bilateral destruction of the medial temporal regions and a loss of the basal forebrain and the posterior part of the orbitofrontal cortex. In addition, Boswell has lost the insular cortex, which is found in the lateral fissure and not visible in the illustration.

Boswell's sensory and motor cortices are intact, as are his basal ganglia, but Boswell's injury is more extensive than H. M.'s. Like H. M., he

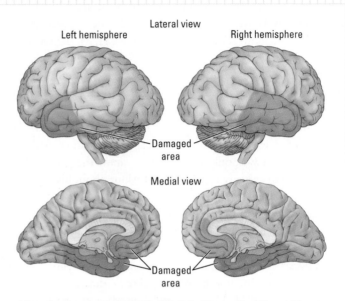

After a herpes simplex encephalitis infection, patient Boswell has great difficulty remembering events before and after his illness. Areas of damage in the medial temporal region, the basal forebrain, and the posterior orbitofrontal cortex are highlighted in red. Compare Figure 14-6.

has a loss of new memories. Unlike H. M., he also has a severe loss of access to old information, probably because of his insular and prefrontal injuries. Nonetheless, again like H. M., Boswell's procedural memory is intact, a fact that illustrates the dissociation between neural circuits underlying explicit and implicit forms of memory.

J. K.'s clear deficit in implicit memory contrasts sharply with his awareness of daily events. He could recall explicit events as well as most men his age and speak intelligently on issues of the day that he had just read about. Once when we visited him, one of us entered the room first and he immediately asked where the other was, even though it had been 2 weeks since we told him that we would be coming to visit.

This intact long-term memory is very different from the situation of H. M., who would not have remembered that anybody was coming even 5 minutes after being told. Because Parkinson's disease primarily affects the basal ganglia, J. K.'s deficit in implicit memory was probably related to his basal ganglia dysfunction.

REVIEW 14-2

Dissociating Memory Circuits

Before you continue, check your understanding.

1. Based on the case of H. M., we can conclude that the structures involved in explicit memory include the _____, the _____, and adjacent cortex.

2. Implicit memory deficits in patients with Parkinson's disease demonstrate that a major structure in implicit memory is the _____.

3. What is the main difference between the Lashley and Milner studies?

Answers appear at the back of the book.

14-3 Neural Systems Underlying Explicit and Implicit Memories

Findings from laboratory studies, largely on rats and monkeys, have reproduced the symptoms of patients such as H. M. and J. K. by injuring the animals' medial temporal regions and basal ganglia, respectively. Other structures, most notably in the frontal and temporal lobes, also play roles in certain types of explicit memory. We now consider the systems for explicit and implicit memory separately.

Neural Circuit for Explicit Memories

The dramatic amnesic syndrome discovered in H. M. in the 1950s led investigators to focus on the hippocampus, at the time regarded as a large brain structure in search of a function. But H. M. had other damaged structures, too, and the initial focus on the hippocampus as the location of explicit-memory processing turned out to be misguided.

After decades of anatomical and behavioral studies sorted out the complexities, consensus on the anatomy of explicit memory had coalesced by the mid-1990s. The prime structures for explicit memory include the medial temporal region and the frontal cortex and structures closely related to them.

Before considering the model, we must first revisit the anatomy of the medial temporal region. Review Figure 14-5, which summarizes findings from the studies by Martin and colleagues showing that memories of the color and motion characteristics of objects reside in separate locations in the temporal lobe. The medial temporal region thus receives multiple sensory inputs.

The macaque monkey's medial temporal region shares many anatomical similarities with the human brain, and this monkey has been the principal subject for anatomical study of the region. In addition to the subcortical hippocampus and amygdala, three areas in the medial temporal cortex take part in explicit memory. As illustrated in **Figure 14-8,** these regions, lying adjacent to the hippocampus, are the **entorhinal cortex,** the **parahippocampal cortex,** and the **perirhinal cortex.** A sequential arrangement of two-way connections, charted in **Figure 14-9,** project from the major cortical regions

entorhinal cortex Located on the medial surface of the temporal lobe; provides a major route for neocortical input to the hippocampal formation; often degenerates in Alzheimer's disease.

parahippocampal cortex Cortex located along the dorsal medial surface of the temporal lobe.

perirhinal cortex Cortex lying next to the rhinal fissure on the base of the brain.

We now know that the hippocampus participates in species-specific behaviors and spatial navigation, and memory and is vulnerable to stress.

If you consult books or reviews published before 1995, you may find explanations for memory quite different from those in this chapter (see Gazzaniga, 2000).

In Greek, *para* means "beside," *rhino* means "nose." The perirhinal cortex lies beside the rhinal sulcus on the bottom of the brain.

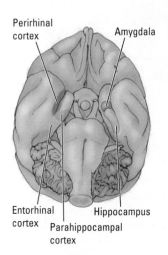

FIGURE 14-8 Medial Temporal Cortex and Subcortical Structures Ventral view of the rhesus macaque monkey brain, showing the medial temporal regions on the left. Each plays a distinct role in processing sensory information for memory storage. The hippocampus and amygdala are not directly visible from the surface of the brain because they lie within the cortical regions illustrated on the left. All these cortical and subcortical structures are present on both sides of the brain.

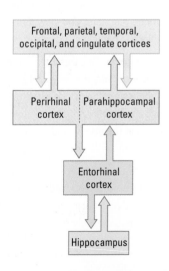

FIGURE 14-9 Reciprocal Medial Temporal Connections Input from the sensory cortices flows to the parahippocampal and perirhinal regions, then to the entorhinal cortex, and finally to the hippocampus, which then feeds back to the medial temporal regions and on back to the sensory regions in the neocortex.

into the perirhinal and parahippocampal cortices, which in turn project to the entorhinal cortex and then to the hippocampus.

The prominent input from the neocortex to the perirhinal region is from the visual regions of the ventral stream coursing through the temporal lobe. The perirhinal region is thus a prime candidate for visual object memory. Similarly, the parahippocampal cortex has a strong input from regions of the parietal cortex believed to take part in visuospatial processing. Thus, the parahippocampal region likely has a role in **visuospatial memory,** that is, using visual information to recall an object's location in space.

Section 9-2 traces these visual pathways in detail.

Because both the perirhinal and the parahippocampal regions project to the entorhinal cortex, it is likely that this region participates in more integrative forms of memory. The entorhinal cortex is in fact the first area to show cell death in Alzheimer's disease, a form of dementia characterized by severe deficits in explicit memory (see Clinical Focus 14-3, "Alzheimer's Disease").

Section 16-3 elaborates on Alzheimer's and other dementias.

CLINICAL FOCUS ✛ 14-3

Alzheimer's Disease

In the 1880s it was noted that the brain undergoes atrophy with aging, but the reason was not really understood until the German physician Alois Alzheimer published a landmark study in 1906. Alzheimer described a set of behavioral symptoms and associated neuropathology in a 51-year-old woman who was demented. The cellular structure of her cerebral and limbic cortices showed various abnormalities.

An estimated 5.4 million people are now afflicted with Alzheimer's disease in the United States, although the only certain diagnostic test remains postmortem examination of cerebral tissue. The disease progresses slowly, and many people with Alzheimer's disease probably die from other causes before the cognitive symptoms incapacitate them.

We knew of a physics professor who continued to work until he was nearly 80 years old, at which time he succumbed to a heart attack. Postmortem examination of his brain revealed significant Alzheimer's pathology. His slipping memory had been attributed by his colleagues to "old-timer's disease."

The cause of Alzheimer's disease remains unknown, although it has been variously attributed to genetic predisposition, abnormal levels of trace elements (e.g., aluminum), immune reactions, and slow viruses. Two principal neuronal changes take place in Alzheimer's disease:

1. *Loss of cholinergic cells in the basal forebrain.* One treatment for Alzheimer's disease, therefore, is medication that increases acetylcholine levels in the forebrain. An example is Exelon, which is the trade name for rivastigmine, a cholinergic agonist that appears to provide temporary relief from the progression of the disease and is available both orally and as a skin patch.

2. *Development of neuritic plaques in the cerebral cortex.* A **neuritic plaque** consists of a central core of homogeneous protein material (*amyloid*) surrounded by degenerative cellular fragments. The plaques, illustrated here, are not distributed evenly throughout the

cortex but are concentrated especially in the temporal-lobe areas related to memory. Neuritic plaques are often associated with another abnormality, neurofibrillary tangles, which are paired helical filaments found in both the cerebral cortex and the hippocampus.

Cortical neurons begin to deteriorate as the cholinergic loss, plaques, and tangles develop. The first cells to die are in the entorhinal cortex (see Figure 14-8). Significant memory disturbance ensues.

An idea emerging from stroke neurologists is that dementia may reflect a chronic cerebrovascular condition, marginal high blood pressure. Marginal elevations in blood pressure can lead to cerebral microbleeds, especially in white matter. The cumulative effect of years or even decades of tiny bleeds would eventually lead to increasingly disturbed cognition. This may first appear as a "mild cognitive impairment" that slowly progresses with cumulative microbleeds.

Neuritic plaque, as is often found in the cerebral cortices of Alzheimer patients. The amyloid is the dark spot in the center of the image, which is surrounded by the residue of degenerated cells.

Cecil Fox/Science Source/Photo Researchers

The Hippocampus and Spatial Memory

We are left with a conundrum. If the hippocampus is not the key structure in explicit memory yet is the recipient of the entorhinal connections, what does it do? The hippocampus is probably engaged in visuospatial memory processes required for places, such as recalling the location of an object. John O'Keefe and Lynn Nadel were the first to advance this idea, in 1978.

Certainly both laboratory animals and human patients with selective hippocampal injury have severe deficits in various forms of spatial memory. Similarly, monkeys with hippocampal lesions have difficulty learning the location of objects (*visuospatial learning*), as can be demonstrated in tasks such as the ones illustrated in **Figure 14-10.**

Monkeys are trained to displace objects to obtain a food reward (Figure 14-10A) and then are given one of two tasks. In the *visual-recognition task* shown in Figure 14-10B, the animal displaces a sample object to obtain a food reward. After a short delay, the animal is presented with two objects. One is novel. The task is to learn that the novel object must be displaced to obtain a food reward. This task is a test of explicit visual object memory. Monkeys with perirhinal lesions are impaired at the task.

In the *object-position task* in Figure 14-10C, the monkey is shown one object to be displaced for a food reward. Then the monkey is shown the same object along with a second, identical one. The task is to learn to displace the object that is in the same position as it was in the initial presentation. Monkeys with hippocampal lesions are selectively impaired at this task.

From the results of these studies on the hippocampus, we would predict that animals with especially good spatial memories should have bigger hippocampi than do species with poorer spatial memories. David Sherry and his colleagues (1992) tested this hypothesis in birds.

Many birds are cachers: they harvest sunflower seeds and other favored foods and hide (cache) them to eat later. Some birds can find hundreds of items that they have cached. To evaluate whether the hippocampus plays a role in this activity, Sherry and his coworkers measured hippocampal size in closely related bird species, only one of

visuospatial learning Using visual information to recall an object's location in space.

neuritic plaque Area of incomplete necrosis (dead tissue) consisting of a central protein core (amyloid) surrounded by degenerative cellular fragments; often seen in the cortex of people with senile dementias such as Alzheimer's disease.

Section 1-4 explains the encephalization quotient (EQ), an index of brain-to-body-size ratios that allows comparisons of different species' relative brain sizes.

FIGURE 14-10 Two Memory Tasks for Monkeys **(A)** In "basic training," a monkey learns to displace an object to obtain a food reward. In **(B)** and **(C)** the plus and minus signs indicate whether the object is (1) or is not (2) associated with food.

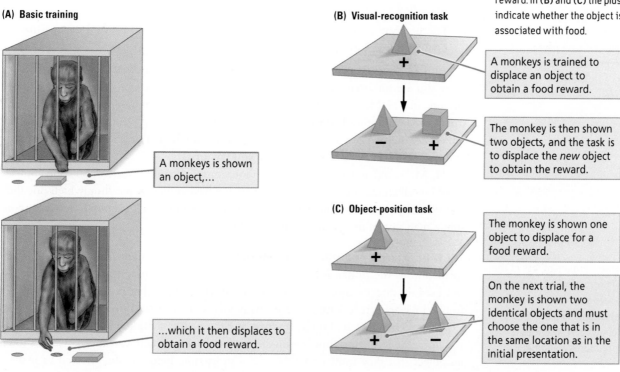

(A) Basic training

A monkeys is shown an object,…

…which it then displaces to obtain a food reward.

(B) Visual-recognition task

A monkeys is trained to displace an object to obtain a food reward.

The monkey is then shown two objects, and the task is to displace the *new* object to obtain the reward.

(C) Object-position task

The monkey is shown one object to displace for a food reward.

On the next trial, the monkey is shown two identical objects and must choose the one that is in the same location as in the initial presentation.

FIGURE 14-11 Inferring Spatial Memory
This graph relates hippocampal volume to forebrain volume in 3 food-storing (*left*) and 10 non-food-storing (*right*) families of songbirds. The hippocampi of birds that cache food, such as the black-capped chickadee, are about twice as large as the hippocampi of birds that do not, such as the sparrow. Data from "Spatial Memory and Adaptive Specialization of the Hippocampus," by D. F. Sherry, L. F. Jacobs, and S. J. C. Gaulin, 1992, *Trends in Neuroscience*, 15, pp. 298–303.

Chickadee

Common sparrow

Relative volumetric ratio of hippocampus to forebrain

Food-storing   Non-food-storing

which is a food cacher. As shown in **Figure 14-11**, the hippocampal formation is larger in birds that cache food than in birds that do not. In fact, the hippocampi of food-storing birds are more than twice as large as expected for birds of their brain size and body weight.

Sherry found a similar relation when he compared different species of food-storing rodents. Merriam's kangaroo rats, rodents that store food in various places throughout their territory, have larger hippocampi than bannertail kangaroo rats that store food only in their burrows. Hippocampal size in both birds and mammals appears to be related to the cognitive demands of two highly spatial activities, foraging for and storing food.

One prediction that we might make from the Sherry experiments is that people who have jobs with high spatial demands have large hippocampi. Taxi drivers in London fit this category. Successful candidates for a cab driver's license in London must demonstrate that they know the location of every street in that huge and ancient city. Using MRI, Eleanor Maguire and her colleagues (2000) found the posterior region of the hippocampus in London taxi drivers to be significantly larger than the same region in the control participants. This finding presumably explains why a select few pass a spatial-memory test that most of us would fail miserably.

Reciprocal Connections for Explicit Memory

The temporal pathway of explicit memory is reciprocal: connections from the neocortex run to the entorhinal cortex and then back to the neocortex (see Figure 14-9). Reciprocal connections have two benefits:

1. Signals from the medial temporal regions back to the cortical sensory regions keep the sensory experience alive in the brain: the neural record of an experience outlasts the actual experience.

2. The pathway back to the neocortex means that it is kept apprised of information being processed in the medial temporal regions.

Although we have focused on the role of the medial temporal regions, other structures also are important in explicit memory. People with frontal-lobe injuries are not amnesic like H. M. or J. K., but they do have difficulties with memory for the temporal (time) order of events. Imagine that you are shown a series of photographs and asked to remember them. A few minutes later, you are asked whether you recognize two photographs and, if so, to indicate which one you saw first.

H. M. would not remember the photographs. People with frontal-lobe injuries would recall seeing the photographs but would have difficulty recalling which one they had seen most recently. The role of the frontal lobe in explicit memory clearly is more subtle than that of the medial temporal lobe.

THE FRONTAL LOBE AND SHORT-TERM MEMORY All sensory systems in the brain send information to the frontal lobe, as do the medial temporal regions. This information is not used for direct sensory analysis, so it must have some other purpose. In general, the frontal lobe appears to have a role in many forms of short-term memory.

Joaquin Fuster (e.g., Fuster, Bodner, & Kroger, 2000) studied single-cell activity in the frontal lobe during short-term-memory tasks. For example, if monkeys are shown an object that they must remember for a short time before being allowed to make a

Feedback to the cortex is not part of the basal ganglia systems that take part in implicit memory, which helps to explain its unconscious nature.

Korsakoff's syndrome Permanent loss of the ability to learn new information (anterograde amnesia) and to retrieve old information (retrograde amnesia) caused by diencephalic damage resulting from chronic alcoholism or malnutrition that produces a vitamin B_1 deficiency.

response, neurons in the prefrontal cortex show a sustained firing during the delay. Consider the tests illustrated in **Figure 14-12**:

- In the general design for each test, a monkey is shown a light, which is the cue, and after a delay must then make a response to get a reward.

- In the *delayed-response task,* the monkey is shown two lights in the choice test and must choose the one that is in the same location as the cue.

- In the *delayed-alternation task,* the monkey is again shown two lights in the choice tests but now must choose the light that is *not* in the same location as the cue.

- In the *delayed-matching-to-sample task,* the monkey is shown, say, a red light and then, after a delay, is shown a red and a green light. The task is to choose the red light, regardless of its new location.

Fuster found that in each task certain cells in the prefrontal cortex fire throughout the delay. Animals that have not learned the task show no such cell activity. Curiously, if a trained animal makes an error, its cellular activity corresponds: the cells stop responding before the error occurs. They have "forgotten" the cue.

TRACING THE EXPLICIT MEMORY CIRCUIT People who have chronically abused alcohol can develop an explicit-memory disturbance known as **Korsakoff's syndrome.** In some cases, severe deficits in explicit memory extend to implicit memory as well. Korsakoff's syndrome is caused by a thiamine (vitamin B$_1$) deficiency that kills cells in the medial part of the diencephalon—the "between brain" at the top of the brainstem—including the medial thalamus and mammillary bodies in the hypothalamus of Korsakoff patients. The frontal lobes of 80 percent show atrophy (loss of cells). The memory disturbance is probably so severe in many Korsakoff patients because the damage includes not only forebrain but also brainstem structures (see Clinical Focus 14-4, "Korsakoff's Syndrome" on page 498).

Mortimer Mishkin and his colleagues (Mishkin, 1982; Murray, 2000) at the U.S. National Institute of Mental Health proposed a neural circuit for explicit memory. It incorporates the evidence from both humans and laboratory animals with injuries to the temporal and frontal lobes. **Figure 14-13** presents a modified version of the Mishkin model that includes not only the frontal and temporal lobes but also the medial thalamus,

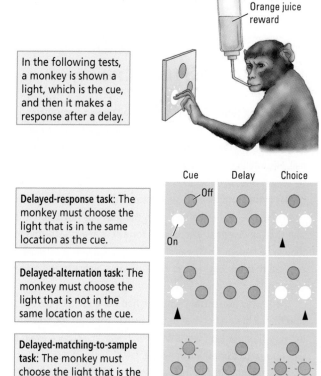

Orange juice reward

In the following tests, a monkey is shown a light, which is the cue, and then it makes a response after a delay.

Delayed-response task: The monkey must choose the light that is in the same location as the cue.

Delayed-alternation task: The monkey must choose the light that is not in the same location as the cue.

Delayed-matching-to-sample task: The monkey must choose the light that is the same color as the cue.

Cue Delay Choice

Time →

FIGURE 14-12 Testing Short-Term Memory A monkey performing a short-term memory task responds by pressing the disc to get a fruit juice reward (*top*). The correct disc varies, depending on the requirements of the task (*bottom*). (For each task, an arrowhead indicates the correct choice.) Adapted from *Memory in the Cerebral Cortex* (p. 178), by J. Fuster, 1995, Cambridge, MA: MIT Press.

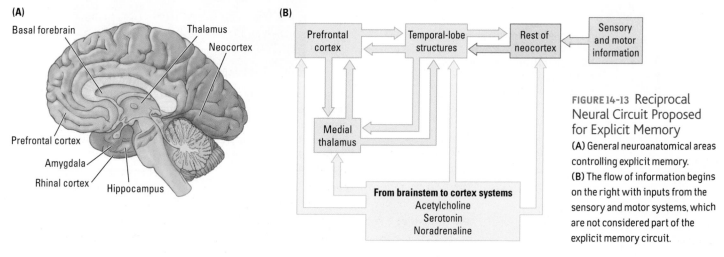

(A)

Basal forebrain
Thalamus
Neocortex
Prefrontal cortex
Amygdala
Rhinal cortex
Hippocampus

(B)

Prefrontal cortex → Temporal-lobe structures ← Rest of neocortex ← Sensory and motor information

Medial thalamus

From brainstem to cortex systems
Acetylcholine
Serotonin
Noradrenaline

FIGURE 14-13 Reciprocal Neural Circuit Proposed for Explicit Memory
(A) General neuroanatomical areas controlling explicit memory.
(B) The flow of information begins on the right with inputs from the sensory and motor systems, which are not considered part of the explicit memory circuit.

Korsakoff's Syndrome

Over the long term, alcoholism, especially when accompanied by malnutrition, obliterates memory. When Joe R., a 62-year-old man, was hospitalized, his family complained that his memory had become abysmal. His intelligence was in the average range, and he had no obvious sensory or motor difficulties. Nevertheless, he was unable to say why he was in the hospital and usually stated that he was actually in a hotel.

When asked what he had done the previous night, Joe R. typically said that he "went to the Legion for a few beers with the boys." Although he had, in fact, been in the hospital, it was a sensible response because that is what he had done on most nights in the preceding 30 years.

Joe R. was not certain what he had done for a living but believed that he had been a butcher. In fact, he had been a truck driver for a local delivery firm. His son was a butcher, however, so once again his story was related to something in his life.

Joe's memory for immediate events was little better. On one occasion, we asked him to remember having met us, and then we left the room. On our return 2 or 3 minutes later, he had no recollection of ever having met us or of having taken psychological tests that we administered.

Joe R. had Korsakoff's syndrome, a condition named after Sergei Korsakoff, a Russian physician who in the 1880s first called attention to a syndrome that accompanies chronic alcoholism. The most obvious symptom is severe loss of memory, including amnesia for both information learned in the past (**retrograde amnesia**) and information learned since the onset of the memory disturbance (**anterograde amnesia**).

One unique characteristic of the amnesic syndrome in Korsakoff patients is that they tend to make up stories about past events rather than admit that they do not remember. These stories, like those of Joe R., are generally plausible because they are based on actual experiences.

Curiously, Korsakoff patients have little insight into their memory disturbance and are generally indifferent to suggestions that they have a memory problem. Such patients are generally apathetic to things going on around them too. Joe R. was often seen watching television when the set was turned off.

The cause of Korsakoff's syndrome is a thiamine (vitamin B_1) deficiency resulting from prolonged intake of large quantities of alcohol. (In addition to a "few beers with the boys," Joe R. had a long history of drinking a 26-ounce bottle of rum every day.) The thiamine deficiency results in the death of cells in the midline diencephalon, including especially the medial regions of the thalamus and the mammillary bodies of the hypothalamus.

Most Korsakoff patients also show cortical atrophy, especially in the frontal lobe. With the appearance of the Korsakoff symptoms, which can happen quite suddenly, prognosis is poor. Only about 20 percent of patients show much recovery after a year on a vitamin B_1–enriched diet. Joe R. has shown no recovery after several years and will spend the rest of his life in a hospital setting.

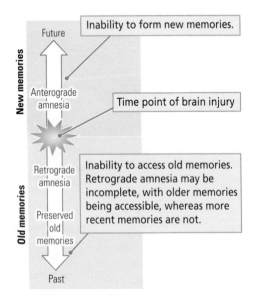

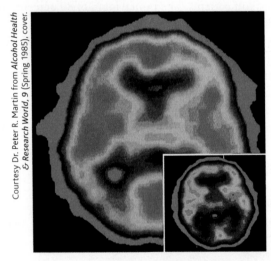

Courtesy Dr. Peter R. Martin from *Alcohol Health & Research World*, 9 (Spring 1985), cover.

These PET scans, from a normal patient (larger image) and a Korsakoff patient (inset), demonstrate reduced activity in the frontal lobe of the diseased brain. (The frontal lobes are at the bottom center of each scan.) Red and yellow represent areas of high metabolic activity versus the lower level of activity in the darker areas.

which is implicated in Korsakoff's syndrome, and the basal forebrain-activating systems that are implicated in Alzheimer's disease:

- The sensory neocortical areas send their connections to the medial temporal regions, which are in turn connected to the medial thalamus and prefrontal cortex.

- The basal forebrain structures are hypothesized to play a role in maintaining appropriate levels of activity in the forebrain structures so that they can process information.

- The temporal-lobe structures are hypothesized to be central to the formation of long-term explicit memories.

- The prefrontal cortex is central to the maintenance of temporary (short-term) explicit memories as well as memory for the recency (chronological order) of explicit events.

Consolidation of Explicit Memories

Amnesia often appears to be time-dependent. H.M., for example, was unable to form new explicit memories, but he appeared to have good recall of facts and events from time periods remote from his surgery, such as his childhood. Such findings led to the idea that the medial temporal region could *not* be the ultimate storage site for long-term memories; it was more likely to be the neocortex (for a review, see Squire and Wixted, 2011).

This idea led to the hypothesis that the hippocampus *consolidates* new memories, a process that makes them permanent. In **consolidation,** or stabilizing a memory trace after learning, memories move from the hippocampus to diffuse regions in the neocortex. Once they move, hippocampal involvement is no longer needed.

It is not clear how memories are moved, however, or how long it takes. Robert Sutherland and his colleagues (2010) propose a model of consolidation, the *distributed reinstatement theory.* In their model, a learning episode rapidly creates a stored memory representation that is strong in the hippocampus but weak elsewhere. The memory is replayed on the time scale of hours or days after the learning, leading to enhanced representations outside the hippocampus. With each repetition of the learning—that is, practice—the subsequent replay progressively enhances the nonhippocampal memory representation. If the hippocampus is extensively damaged, the memory remains.

Memory is not constant over time. When people get misleading information about events they have experienced, for example, their later recall of the event often is modified. Indeed, this contributes to the notorious unreliability of eyewitness testimony in legal cases. (Do you remember the "smashing" and "bumping" car collisions from Section 14-1) The fact that memories appear changeable seems to fly in the face of the concept of consolidation, which presumes that, once consolidated, memories are fixed.

One solution to this conundrum is to suggest that whenever a memory is replayed in the mind, it is open to further consolidation, a phenomenon known as **reconsolidation,** the process of restabilizing a memory trace after the memory is revisited. Interest in this idea has been intense over the past few years, and although the final story is yet to be written, it does appear that reconsolidation does occur, at least for some types of memories.

One way to think of this process is to see consolidation of memories as never-ending: new information is constantly being integrated into existing memory networks (for a review, see McKenzie and Eichenbaum, 2011). After all, we frequently recall memories, rehash them, and integrate them with new events. Our memories are not laid on a tabula rasa but must be interwoven into a lifetime of memories.

One implication of reconsolidation is that it ought to be possible to erase negative memories by using amnesic agents when the memory is revisited. This idea has important implications for reducing or eliminating the effects of strong emotional experiences, such as those seen in posttraumatic stress disorders.

retrograde amnesia Inability to remember events that took place before the onset of amnesia.

anterograde amnesia Inability to remember events subsequent to a disturbance of the brain such as head trauma, electroconvulsive shock, or certain neurodegenerative diseases.

consolidation Process of stabilizing a memory trace after learning.

reconsolidation Process of restabilizing a memory trace after the memory is revisited.

Section 5-4 links sensitization to PTSD. Sections 6-5 and 12-4 document how stress fosters and prolongs its effects, and Focus 16-1 covers treatment strategies.

(A)

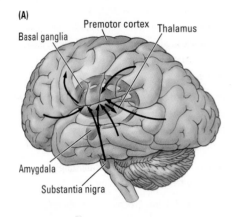

(B)

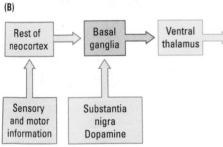

FIGURE 14-14 Unidirectional Neural Circuit Proposed for Implicit Memory **(A)** General anatomical areas controlling implicit memory. **(B)** Circuit diagram showing the one-way flow of information, beginning with inputs from the sensory and motor systems, which are not considered part of the memory circuit.

Section 12-4 details the amygdala's influence on emotional behavior. Focus 12-3 describes panic and other anxiety disorders.

The ANS, diagrammed in Figure 2-30, monitors and controls life-support functions. Section 11-4 reports on the PAG's role in pain perception.

Neural Circuit for Implicit Memories

Hypothesizing that the basal ganglia are central to implicit memory, Mishkin and his colleagues also proposed a neural circuit for implicit memories (Mishkin, 1982; Mishkin et al., 1997). As **Figure 14-14** shows, the basal ganglia receive input from the entire neocortex and send projections to the ventral thalamus and then to the premotor cortex. The basal ganglia also receive widely and densely distributed projections from dopamine-producing cells in the substantia nigra. Dopamine appears necessary for circuits in the basal ganglia to function and may indirectly participate in implicit-memory formation.

The connection from the cortex to the basal ganglia in the implicit-memory system flows only in one direction. Most of the neocortex receives no direct information regarding the activities of the basal ganglia. Mishkin believes that this unidirectional flow accounts for the unconscious nature of implicit memories. For memories to be conscious, the neocortical regions involved must receive feedback, as they do in the explicit-memory system (see Figure 14-13).

Mishkin's models show why people with basal ganglia dysfunction, as occurs in Parkinson's disease, have deficits in implicit memory, whereas people with injuries to the frontal or temporal lobes have relatively good implicit memories, even though they may have profound disturbances of explicit memory. Some people with Alzheimer's disease are able to play games expertly, even though they have no recollection of having played them before.

Daniel Schacter (1983) wrote of a golfer with Alzheimer's disease. The golfer's medial temporal system was severely compromised by the disease, but his basal ganglia were unaffected. Despite impairment of his explicit knowledge, as indexed by his inability to find shots or to remember his strokes on each hole, the man retained his ability to play the game.

Neural Circuit for Emotional Memories

Whether **emotional memory** for the affective properties of stimuli or events is implicit or explicit is not altogether clear. It could be both. Certainly people can react with fear to specific stimuli that they can identify, and we have seen that they can also fear situations for which they do not seem to have specific memories.

Panic disorder is a common pathology of emotional memory. People show marked anxiety but cannot identify a specific cause. Emotional memory has a unique anatomical component—the amygdala, mentioned in Section 14-1 in regard to fear conditioning. The amygdala seems to evoke our feelings of anxiety toward stimuli that by themselves would not normally produce fear.

Emotional memory has been studied most thoroughly in fear conditioning by pairing noxious stimuli, such as foot shock, with a tone (see Experiment 14-1). Michael Davis (1992) and Joseph LeDoux (1995) used fear conditioning to demonstrate that the amygdala is critical to emotional memory. Damage to the amygdala abolishes emotional memory but has little effect on implicit or explicit memory.

The amygdala has close connections with the medial temporal cortical structures as well as with the rest of the cortex. It also sends projections to the brainstem structures that control autonomic responses such as blood pressure and heart rate, to the hypothalamus that controls hormonal systems, and to the periaqueductal gray matter (PAG) that affects the perception of pain **(Figure 14-15)**. The amygdala hooks in to the implicit-memory system through its connections with the basal ganglia.

Fear is not the only aspect of emotional memory coded by the amygdala. A study of severely demented patients by Bob Sainsbury and Marjorie Coristine (1986) nicely

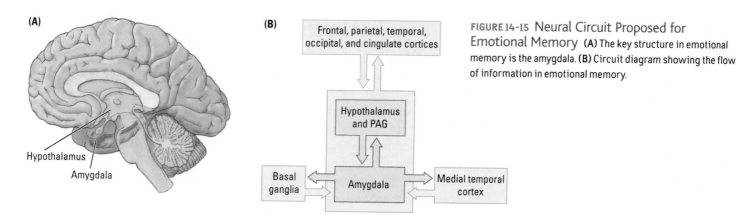

(A)

Hypothalamus
Amygdala

(B)

Frontal, parietal, temporal, occipital, and cingulate cortices

Hypothalamus and PAG

Basal ganglia — Amygdala — Medial temporal cortex

FIGURE 14-15 Neural Circuit Proposed for Emotional Memory **(A)** The key structure in emotional memory is the amygdala. **(B)** Circuit diagram showing the flow of information in emotional memory.

illustrates this point. The patients were believed to have severe cortical abnormalities but intact amygdalar functioning.

The researchers first established that the patients' ability to recognize photographs of close relatives was severely impaired. The patients were then shown four photographs, one of which depicted a relative (either a sibling or a child) who had visited in the past 2 weeks. The task was to identify the person whom they liked better than the other three. Although the subjects were unaware that they knew anyone depicted in the photographs, they consistently preferred the photographs of their relatives. This result suggests that, although the explicit and probably the implicit memory of the relative was gone, each patient's emotional memory guided his or her preference.

Emotionally arousing experiences tend to be vividly remembered, a fact confirmed by findings from both animal and human studies. James McGaugh (2004) concluded that emotionally significant experiences, both pleasant and unpleasant, must activate hormonal and brain systems that act to "stamp in" these vivid memories.

McGaugh noted that many neural systems probably take part, but the basolateral part of the amygdala is critical. The general idea is that emotionally driven hormonal and neurochemical activating systems (probably cholinergic and noradrenergic) stimulate the amygdala. The amygdala in turn modulates the laying down of emotional memory circuits in the rest of the brain, especially in the medial temporal and prefrontal regions and the basal ganglia. We would not expect people with amygdala damage to have enhanced memory for emotion-laden events, and they do not (Cahill et al., 1995).

emotional memory Memory for the affective properties of stimuli or events.

Figure 5-17 traces the connections of the neural activating systems. Section 6-5 explains how hormones work.

REVIEW 14-3
Neural Systems Underlying Explicit and Implicit Memories

Before you continue, check your understanding.

1. The two key structures for explicit memory are _____ and _____.

2. A system consisting of the basal ganglia and neocortex forms the neural basis of the _____ memory system.

3. The _____ and associated structures form the neural basis for emotional memory.

4. The progressive stabilization of memories is known as _____.

5. Why do we remember emotionally arousing experiences so vividly?

Answers appear at the back of the book.

associative learning Linkage of two or more unrelated stimuli to elicit a behavioral response.

long-term potentiation (LTP) Long-lasting increase in synaptic effectiveness after high frequency stimulation.

long-term depression (LTD) Long-lasting decrease in synaptic effectiveness after low-frequency electrical stimulation.

Cajal's neuron theory—that neurons are the functional units of the nervous system—is now universally accepted. It includes the idea that interactions between these discrete cells enable behavior.

Experiment 5-2 explains habituation at the neuronal level, and Experiment 5-3 explains sensitization.

Experiment 4-1 illustrates how excitatory postsynaptic potentials (EPSPs) increase probability that a neuron will produce an action potential.

14-4 Structural Basis of Brain Plasticity

We have encountered three different categories of memory—explicit, implicit, and emotional—and the different brain circuits that underlie each. Our next task is to consider how the neurons in these circuits change to allow us to consolidate and store memories. The consensus among neuroscientists is that the changes take place at the synapse, in part simply because that is where neurons influence one another.

This idea dates back to 1928 and the Spanish anatomist Santiago Ramón y Cajal. He suggested that learning might produce prolonged morphological (structural) changes in the efficiency of the synapses activated in the learning process. This idea turned out to be easier to propose than to study.

Researchers still encounter a major challenge as they investigate Cajal's suggestion because it is unclear where in the brain to look for synaptic changes that might correlate with memory for a specific stimulus. This task is formidable. Imagine trying to find the exact location of the neurons responsible for storing your grandmother's name. You would face a similar challenge in trying to find the neurons responsible for the memory of an object in a monkey's brain as the monkey performs the visual-recognition task illustrated in Figure 14-10B.

One approach to finding the neuronal correlates of memory aims, first, to determine that synaptic changes are correlated with memory in the mammalian brain; second, to localize the synaptic changes to specific neural pathways; and, third, to analyze the nature of the synaptic changes themselves. This section reviews studies that have begun to show how experience is correlated with synaptic changes related to memory.

We first consider a strategy based on neuronal physiology and experience. We then look at the gross neural changes correlated with different forms of experience, ranging from potentially good—living in enriched environments and learning specific tasks—to probably bad—chronic administration of trophic factors, hormones, and addictive drugs. Each of these diverse experiences modifies the general synaptic organization of the brain in a strikingly similar manner that we summarize at the end the section.

Long-Term Potentiation

Findings from studies of behavioral *habituation* (a weakened response to a stimulus) and *sensitization* (a strengthened response) in the sea snail *Aplysia* show that physical changes in synapses do underlie learning. Adaptive synapses in the mammalian brain participate in **associative learning,** a response elicited by linking unrelated stimuli together—by learning that A goes with B.

Learned associations are a common type of explicit memory. Associating a face with a person, an odor with a food, or a sound with a musical instrument are everyday examples. Learning that learning takes place at synapses is another. The phenomenon underlying associative learning entails an enduring neural change in a postsynaptic cell after an excitatory signal, or EPSP, from the presynaptic cell crosses the synaptic gap.

Both the relatively simple circuitry of the hippocampus and the ease of recording postsynaptic potentials there make it ideal for studying the neural basis of associative learning. In 1973, Timothy Bliss and Terje Lømo demonstrated that repeated electrical stimulation of the pathway entering the hippocampus produces a progressive increase in EPSP size recorded from hippocampal cells. The enhancement in the size of these "field potentials" lasts for a number of hours to weeks or even longer. Bliss and Lømo called it **long-term potentiation** (LTP), a long-lasting increase in synaptic effectiveness after high-frequency stimulation.

Figure 14-16A illustrates the experimental procedure for obtaining LTP. The presynaptic neuron is stimulated electrically while the electrical activity produced by the

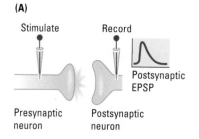

(A)

Stimulate Record

Presynaptic Postsynaptic
neuron neuron

Postsynaptic
EPSP

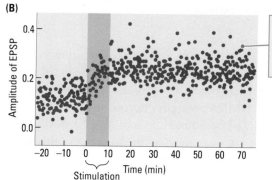

(B)

Each dot represents
the amplitude of the
EPSP in response to one
weak test stimulation.

Stimulation Time (min)

FIGURE 14-16 Recording Long-Term
Potentiation **(A)** In this experimental setup,
the presynaptic neuron is stimulated with a
test pulse and the EPSP is recorded from the
postsynaptic neuron. **(B)** After a period of intense
stimulation, the amplitude of the EPSP produced
by the test pulse increases: LTP has taken place.

stimulation is recorded from the postsynaptic neuron. The readout in Figure 14-16A
shows the EPSP produced by a single pulse of electrical stimulation.

In a typical experiment, a number of test stimuli are given to estimate the size of the
induced EPSP. Then a strong burst of stimulation, consisting of a few hundred pulses
of electrical current per second, is administered (Figure 14-16B). The test pulse is then
given again. The increased amplitude of the EPSP endures for as long as 90 minutes after
the high-frequency burst: LTP has taken place. For the EPSP to increase in size, more
neurotransmitter must be released from the presynaptic membrane or the postsynaptic
membrane must become more sensitive to the same amount of transmitter or both
changes must take place.

The discovery of LTP led to a revolution in thinking about how memories are stored.
As investigators varied the stimulation that produced LTP, they discovered its opposite.
Instead of using high-frequency stimulation (e.g., 100 Hz), they used low-frequency
stimulation (e.g., 5 Hz) and recorded a *decrease* in EPSP size, termed **long-term depression** (LTD). If LTP is a mechanism for creating memories, perhaps LTD is a mechanism
for clearing out old memories.

If LTP and LTD form a basis for understanding synaptic changes underlying memory,
two predictions follow. First, when animals learn problems, we should see enhanced LTP
in the recruited pathways and, second, LTP should produce enduring changes in synaptic
morphology that resemble those seen in memory. Both appear to be true.

The original studies of LTP concentrated on excitatory glutamate synapses. Glutamate
is released from the presynaptic neuron and acts on two different types of receptors on the
postsynaptic membrane, the NMDA and AMPA receptors, as shown in **Figure 14-17A**.
AMPA receptors ordinarily mediate the responses produced when glutamate is released

Section 16-1 reports that deep brain
stimulation used for severely depressed
subjects induces a change, similar to LTP, that
appears to make the brain more plastic and
receptive to other treatments.

FIGURE 14-17 Lasting Effects of
Glutamate Enhanced glutamate prompts a
neurochemical cascade that underlies synaptic
change and LTP.

(A) Weak electrical stimulation

Because the NMDA receptor pore is
blocked by a magnesium ion, release of
glutamate by a weak electrical
stimulation activates only the AMPA
receptor.

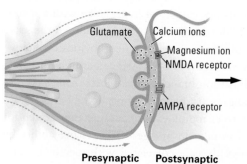

Glutamate Calcium ions

Magnesium ion
NMDA receptor

AMPA receptor

**Presynaptic Postsynaptic
neuron neuron**

(B) Strong electrical stimulation (depolarizing EPSP)

A strong electrical stimulation can
depolarize the postsynaptic membrane
sufficiently that the magnesium ion is
removed from the NMDA receptor pore.

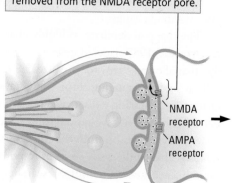

NMDA
receptor

AMPA
receptor

(C) Weak electrical stimulation

Now glutamate, released by weak
stimulation, can activate the NMDA
receptor to allow Ca^{2+} influx, which,
through a second messenger, increases
the function or number of AMPA
receptors or both.

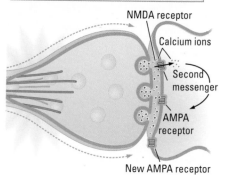

NMDA receptor

Calcium ions

Second
messenger

AMPA
receptor

New AMPA receptor

NMDA is shorthand for *N*-methyl-D-aspartate, and AMPA stands for alpha-amino-3-hydroxy-5-methylisoazole-4-proprionic acid.

from a presynaptic membrane. They allow Na$^+$ ions to enter, depolarizing and thus exciting the postsynaptic membrane. The initial amplitude of the EPSP in Figure 14-16A is produced by this action of the AMPA receptor.

NMDA receptors do not usually respond to glutamate, because their pores are blocked by magnesium ions (Mg^{2+}). NMDA receptors are doubly gated ion channels that can open to allow the passage of calcium ions if two events take place at approximately the same time:

1. The postsynaptic membrane is depolarized, displacing the magnesium ion from the NMDA pore (Figure 14-17B). The strong electrical stimulation delivered by the experimenter serves as a way of displacing magnesium.

2. NMDA receptors are activated by glutamate from the presynaptic membrane (Figure 14-17C).

With the doubly gated NMDA channels open, calcium ions enter the postsynaptic neuron and act through second messengers to initiate the cascade of events associated with LTP. These events include an increased responsiveness of AMPA receptors to glutamate, the formation of new AMPA receptors, and even retrograde messages to the presynaptic terminal to enhance the release of glutamate. The final amplitude of the EPSP in Figure 14-16B is produced by one or more of these actions.

Although the studies generated by the Bliss and Lømø discoveries have focused on excitatory synapses, experiments on inhibitory GABA interneurons demonstrate phenomena similar to LTP and LTD labeled *LTPi* and *LTDi* (for a review, see Maffei, 2011). This discovery was a surprise. At the time, it was generally believed that inhibitory neurons were not plastic, but they definitely are. It appears that plasticity of GABAergic (inhibitory) synapses plays some fundamental role in modulating networks of excitatory neurons.

Measuring Synaptic Change

In principle, experience could cause the brain to change in either of two ways: by modifying existing circuitry or by creating novel circuitry. In actuality, the plastic brain uses both strategies.

Modifying Existing Circuits

The simplest way to find synaptic change is to look for gross changes in the morphology of dendrites. Essentially, dendritic spines are extensions of the neuron membrane that allow more space for synapses. Cells that have few or no dendrites have limited space for inputs, whereas cells with complex dendritic protrusions may have space for tens of thousands of inputs.

Figure 3-5 shows how dendrites branch from sensory neurons, motor neurons, and interneurons.

So more dendrites mean more connections. Change in dendritic structure, therefore, implies change in synaptic organization. In complex neurons, such as pyramidal cells, 95 percent of synapses are on the dendrites. Measuring the extent of dendritic changes can infer synaptic change.

The shape of dendrites is highly changeable. Dale Purves and his colleagues (Purves & Voyvodic, 1987) labeled cells in the dorsal-root ganglia of living mice with a special dye that allowed them to visualize the cells' dendrites. When they examined the same cells at intervals ranging from a few days to weeks, they identified obvious qualitative changes in dendritic extent, as represented in **Figure 14-18**. We can assume that new dendritic branches have new synapses and that lost branches mean lost synapses.

An obvious lesson from the Purves studies is that neuronal morphology is not static: neurons change their structure in response to changing experiences. As they search for neural correlates of memory, researchers can take advantage of this changeability by studying variations in dendritic morphology that are correlated with specific experiences, such as learning some task.

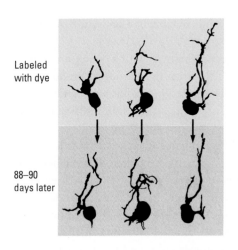

FIGURE 14-18 Dendritic Plasticity
Reconstructions of parts of the dendrites of three mouse superior cervical ganglion cells observed at an interval of 3 months evince changes in both the extension and the retraction of particular dendritic branches. Adapted from "Imaging Mammalian Nerve Cells and Their Connections over Time in Living Animals," by D. Purves and J. T. Voyvodic, 1987, *Trends in Neuroscience, 10*, p. 400.

Labeled with dye

88–90 days later

What do changes in dendritic morphology reveal? Let us consider a given neuron that generates more synaptic space. The new synapses can be either additional contacts between neurons that were already connected with the neuron in question or contacts between neurons that were not formerly connected. Examples of these distinctly different synapse types are illustrated in **Figure 14-19.**

New synapses can result either from the growth of new axon terminals or from the formation of synapses along axons as they pass by dendrites (Figure 14-19A and B). In both cases, new synapses correspond to changes in the local circuitry of a region and not to the development of new connections between distant parts of the brain. Forming new connections between widely separated brain regions would be very difficult in a fully grown brain because of the dense plexus of cells, fibers, and blood vessels that blocks the way.

Thus, the growth of new synapses indicates modifications to basic circuits already in the brain. This strategy has an important implication for the location of synaptic changes underlying memory. During development, the brain forms circuits to process sensory information and to produce movement (behavior). These circuits are the most likely to be modified to form memories (see Figure 14-5).

Focus 5-5 explains how dendritic spines form and diagrams why they provide the structural basis for behavior.

Creating Novel Circuits

Twenty years ago, the general assumption was that the mammalian brain did not make new neurons in adulthood. The unexpected discovery in the 1970s that the brains of songbirds such as canaries grow new neurons to produce songs in the mating season led

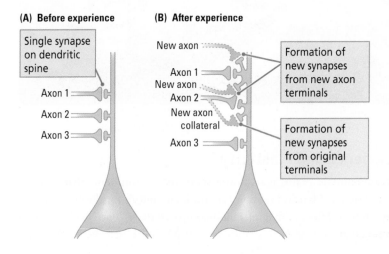

(C) Various observed shapes of new dendritic spines

FIGURE 14-19 Effects of Experience on Dendrites **(A)** Three inputs to a pyramidal cell dendrite. Each axon forms a synapse with a different dendritic spine. **(B)** In forming multiple spine heads, either the original axons can divide and innervate two spine heads or new axons or axon collaterals (dotted outlines) can innervate the new spine heads. **(C)** Single dendritic spines may sprout multiple synapses.

researchers to reconsider this assumption. The adult mammalian brain, too, might be capable of generating new neurons.

This possibility can be tested directly by injecting animals with a compound—bromode-oxyuridine (BrdU)—that is taken up by cells when they divide to produce new cells, including neurons. When the compound is injected into adult rats, dividing cells incorporate it into their DNA. In later analysis, a specific stain can be used to identify the new neurons.

The BrdU technique has yielded considerable evidence that the mammalian brain, including the primate brain, can generate neurons destined for the olfactory bulb, the hippocampal formation, and possibly even the neocortex of the frontal and temporal lobes (Eriksson et al., 1998; Gould et al., 1999). The reason is not yet clear, but adult neurogenesis may enhance brain plasticity, particularly with respect to processes underlying learning and memory. Elizabeth Gould and her colleagues (1999) showed, for example, that generation of new neurons in the hippocampus is enhanced when animals learn explicit-memory tasks.

Experience appears to increase the generation of these new neurons. Perhaps the most interesting demonstration of experience driving neurogenesis comes from a study by Katherine Woollett and Eleanor Maguire (2011). We noted in Section 14-3 that London taxi drivers, who must learn the locations and pass an exam on central London's roughly 25,000 irregular streets, have larger-than-normal volumes in the posterior regions of their hippocampi. The investigators asked whether the increase resulted from taking a 4-year course to pass the exam or was already present when the candidate drivers started the course.

Woollett and Maguire recorded structural MRIs from the would-be taxi drivers before and after training then compared those trainees who qualified ($n = 39$) to those who failed ($n = 20$). The images in **Figure 14-20** show that hippocampal volume increased in those who qualified. Those who failed showed no changes. The average age of the trainees was about 40 years, leading Woollett and Maguire to conclude that the capacity for memory improvement and correlated structural changes in the hippocampus extends well into adulthood.

Experiment 7-1 confirms the hypothesis that hippocampal neurons contribute to memory formation.

FIGURE 14-20 Learning Effects on Hippocampal Volume After a 4-year course devoted to learning London's street layout, trainees who qualified as licensed taxi drivers show increased gray-matter volume in the most posterior part of the hippocampus (orange and yellow areas). From "Acquiring 'the Knowledge' of London's Layout Drives Structural Brain Changes," by K. Woollett and E. A. Maguire, 2011, *Current Biology, 21,* pp. 2109–2114.

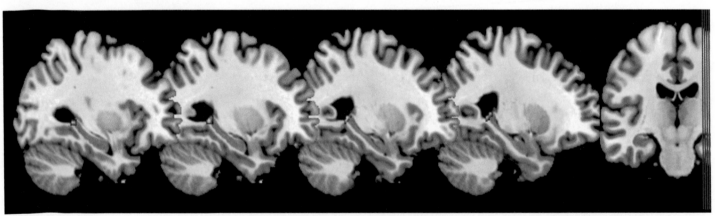

Enriched Experience and Plasticity

Section 8-4 details Hebb's research and his wife's reaction to that first enrichment exercise.

One way to stimulate animals' brains is to house them in environments that provide sensory or motor experience. Donald Hebb (1947) took laboratory rats home and gave them the run of his kitchen. After an interval, Hebb compared these "enriched" rats with a group that had remained in cages in his laboratory at McGill University by training both groups to solve various mazes. The enriched animals performed better, and Hebb concluded that one effect of the enriched experience is to enhance later learning. This important conclusion laid the foundation for the U.S. Head Start programs that provide academic experiences for disadvantaged preschool-aged children.

Subsequent investigators have opted for a more constrained type of "enriched enclosure." For example, in our own studies, we place groups of six rats in enclosures. The enclosures give animals a rich social experience as well as extensive sensory and motor experience. The most obvious consequence is an increase in brain weight that may be on the order of 10 percent relative to cage-reared animals, even though the enriched rats typically weigh less, in part because they get more exercise.

The key question is What is responsible for the increased brain weight? A comprehensive series of studies by Anita Sirevaag and William Greenough (1988) used light- and electron-microscopic techniques to analyze 36 different aspects of cortical synaptic, cellular, and vascular morphology in rats raised either in cages or in complex environments. The simple conclusion: a coordinated change occurs not only in the extent of dendrites but also in glial, vascular, and metabolic processes in response to differential experiences **(Figure 14-21)**.

Animals with enriched experience have more synapses per neuron and also more astrocytes, more blood capillaries, and higher mitochondrial volumes. Clearly, when the brain changes in response to experience, the expected neural changes take place, and adjustments in the metabolic requirements of the now larger neurons take place as well.

Gerd Kempermann and his colleagues (1998) sought to determine whether experience actually alters the number of neurons in the brain. To test this idea, they compared the generation of neurons in the hippocampi of mice housed in complex environments with that of mice reared in laboratory cages. They located the number of new neurons by injecting the animals with BrdU several times in the course of their complex-housing experience.

The new neurons generated in the brain during the experiment incorporated the BrdU. When the researchers later looked at the hippocampi, they found more new neurons in the complex-housed rats than in the cage-housed rats. Although the investigators did not look in other parts of the brain, such as the olfactory bulb, we can reasonably expect that similar changes took place in other neural structures. This result is exciting because it implies that experience not only can alter existing circuitry but also can influence the generation of new neurons and thus new circuitry.

Sensory or Motor Training and Plasticity

Studies showing neuronal change in animals housed in complex environments demonstrate that large areas of the brain can change with such experience. This finding leads us to ask whether specific experiences produce synaptic changes in localized cerebral regions. One way to approach this question is to give animals specific experiences and then see how their brains have changed. Another way is to look at the brains of people who have had a lifetime of some particular experience. We consider each research strategy separately.

Manipulating Experience Experimentally

Fen-Lei Chang and William Greenough (1982) conducted perhaps the most convincing manipulated-experience study. They took advantage of the fact that the visual pathways of the laboratory rat are about 90 percent crossed. That is, about 90 percent of the connections from the left eye to the cortex project through the right thalamus to the right hemisphere and vice versa for the right eye.

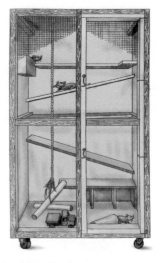

Enriched Rat Enclosure

Mitochondria gather, store, and release energy within the cell. Figure 3-10 diagrams the organelles and other internal components of a neuron.

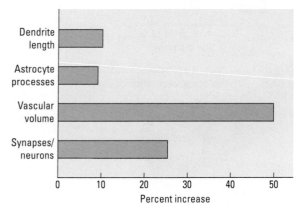

FIGURE 14-21 Consequences of Enrichment Cortical changes that take place in response to experience are found not only in neurons but also in astrocytes and vasculature. Based on data from "Differential Rearing Effects on Rat Visual Cortex Synapses. I. Synaptic and Neuronal Density and Synapses per Neuron," by A. Turner and W. T. Greenough, 1985, *Brain Research, 329*, pp. 195–203; "Differential Rearing Effects on Rat Visual Cortex Synapses. III. Neuronal and Glial Nuclei," by A. M. Sirevaag and W. T. Greenough, 1987, Brain Research, 424, pp. 320–332; and "Experience-Dependent Changes in Dendritic Arbor and Spine Density in Neocortex Vary with Age and Sex," by B. Kolb, R. Gibb, and G. Gorny, 2003, *Neurobiology of Learning and Memory, 79*, pp. 1–10.

In humans, only about half the optic fibers cross. Figure 9-10 diagrams the pathways.

EXPERIMENT 14-2

Question: Does the learning of a fine motor skill alter the cortical motor map?

Procedures

Difficult task

One group of monkeys was trained to retrieve food from a small well.

Simple task

Another group of monkeys was trained to retrieve food from a large well.

Both groups were allowed 12,000 finger flexions. The small-well task was more difficult and required the learning of a fine motor skill in order to match performance of the simpler task.

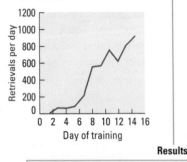

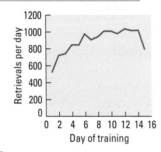

Results

The motor representation of digit, wrist, and arm was mapped.

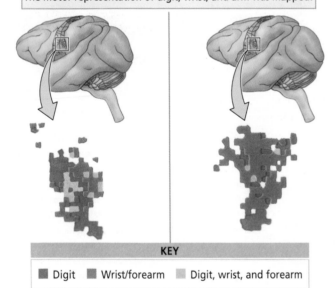

KEY

■ Digit   ■ Wrist/forearm   ■ Digit, wrist, and forearm

Conclusion: The digit representation in the brain of the animal with the more difficult task is larger, corresponding to the neuronal changes necessary for the acquired skill.

Adapted from "Adaptive Plasticity in Primate Motor cortex as a Consequence of Behavioral Experience and Neuronal Injury," by R. J. Nudo, E. J. Plautz, and G. W. Miliken, 1997, *Seminars in Neuroscience,* 9, p. 20.

Chang and Greenough placed a patch over one eye of each rat and then trained the animals in a maze. The visual cortex of only one eye received input about the maze, but the auditory, olfactory, tactile, and motor regions of both hemispheres were equally active as the animals explored. A comparison of the neurons in the two hemispheres revealed that those in the visual cortex of the trained hemisphere had more extensive dendrites. The researchers concluded that some feature associated with the encoding, processing, or storage of visual input from training was responsible for forming new synapses because the hemispheres did not differ in other respects.

Complementary studies that mapped the motor cortex of monkeys were conducted by Randy Nudo and his colleagues (1997). They noted striking individual differences in topography. The investigators speculated that the variability among individuals might be due to each monkey's experiences up to the time at which the cortical map was derived. To test this idea directly, Nudo and colleagues trained two groups of squirrel monkeys to retrieve banana-flavored food pellets from either a small or a large food well. A monkey was able to insert its entire hand into the large well but only one or two fingers into the small well, as illustrated in the Procedures section of **Experiment 14-2.**

Monkeys in the two groups were matched for number of finger flexions, which totaled about 12,000 for the entire study. The monkeys trained on the small well improved with practice, making fewer finger flexions per food retrieval as training proceeded. Maps of forelimb movements were produced by microelectrode stimulation of the cortex.

The maps showed systematic changes in the animals trained on the small but not on the large well. Presumably, these changes are due to the more demanding motor requirements of the small-well condition. The results of this experiment demonstrate that the functional topography of the motor cortex is shaped by learning new motor skills, not simply by repetitive motor use.

Most studies demonstrating plasticity in the motor cortex have been performed with laboratory animals in which the cortex was mapped by microelectrode stimulation. Today, imaging techniques such as transcranial magnetic stimulation (TMS) and functional magnetic resonance imaging (fMRI) make it possible to show parallel results in humans who have special motor skills. For example, right-handed musicians who play stringed instruments show an increased cortical representation of the fingers of the left hand and Braille readers an increased cortical representation of the reading finger.

Thus, the functional organization of the motor cortex is altered by skilled use in humans. It can also be altered by chronic injury in humans and laboratory animals. Jon Kaas (2000) showed that when the sensory nerves in one limb are severed in monkeys, large-scale changes in the somatosensory maps ensue. In particular, in the absence of input, the relevant part of the cortex no longer responds to stimulation of the limb, which is not surprising. But this cortex does not remain inactive. Rather, the deafferentated cortex begins to respond to input from other body parts. The region that formerly responded to the stimulation of the hand now responds to stimulation on the face, a cortical area normally adjacent to the hand area.

Similar results can be found in the cortical maps of people whose limbs have been amputated. For example, Vilayanur Ramachandran (1993) found that when the face of a hand amputee is brushed lightly with a cotton swab, the person has a sensation of the amputated hand being touched. **Figure 14-22** illustrates the rough map of the hand that Ramachandran was actually able to chart on the face.

Figure 11-26 diagrams the extent of cortical reorganization: the monkey's facial area virtually doubled after deafferentation of the arm.

(A)

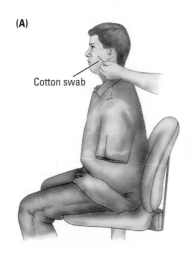

Cotton swab

(B)

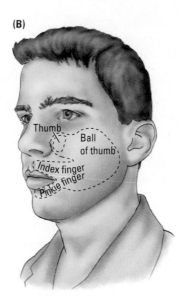

Thumb
Ball of thumb
Index finger
Pinkie finger

FIGURE 14-22 Cortical Reorganization
When a hand amputee's face is stroked lightly with a cotton swab **(A)**, the person experiences the touch as the missing hand being lightly touched **(B)** as well as experiencing touch to the face. The deafferentated cortex forms a representation of the amputated hand on the face. As in the normal somatosensory homunculus, the thumb is disproportionately large. Adapted from "Behavioral and Magnetoencephalographic Correlates of Plasticity in the Adult Human Brain," by V. S. Ramachandran, 1993, *Proceedings of the National Academy of Sciences (USA)*, *90*, p. 10418.

The likely explanation is that the face area in the motor cortex has expanded to occupy the deafferentated limb cortex, but the brain circuitry still responds to the activity of this cortex as representing input from the limb. This response may explain the phantom-limb pain often experienced by amputees.

The idea that experience can alter cortical maps can be demonstrated with other types of experience. For example, if animals are trained to make certain digit movements over and over again, the cortical representation of those digits expands at the expense of the remaining motor areas. Similarly, if animals are trained extensively to discriminate among different sensory stimuli such as tones, the auditory cortical areas responding to those stimuli increase in size.

As described in Research Focus 14-5, "Movement, Learning, and Neuroplasticity" on page 510 one effect of musical training is to alter the motor representations of the digits used to play different instruments. We can speculate that musical training probably alters the auditory representations of specific sound frequencies as well. Both changes are essentially forms of memory, and it is likely that the underlying synaptic changes take place on the appropriate sensory or motor cortical maps.

Focus 11-5 recounts Ramachandran's therapy for minimizing phantom-limb pain.

Sections 10-3 and 15-3 feature discussions of music's benefits for the the brain.

RESEARCH FOCUS ⚙ 14-5

Movement, Learning, and Neuroplasticity

Many lines of research show that practicing a motor skill—playing a musical instrument, for instance—induces changes in the cortical somatosensory and motor maps. The mental maps generally become larger, at least the finger and hand representations.

Presumably, musical skill improves with practice, but are other abilities enhanced too? Patrick Ragert and colleagues (2003) showed that professional pianists have not only better motor skills in their fingers but enhanced somatosensory perception as well.

When the researchers measured the ability to detect subtle sensory stimulation of the fingertips, they found that the pianists were more sensitive than controls. They also found that the enhancement in tactile sensitivity was related to the hours per day that the musicians spent practicing.

The investigators then asked whether the enhanced perceptual ability precluded further improvement in the musicians. Surprisingly, when both the musicians and controls were given a 3-hour training session designed to improve tactile sensitivity, the musicians showed more improvement than did the controls. Again the extent of improvement correlated with daily practice time.

This result implies that well-practiced musicians not only learn to play music but also develop a greater capacity for learning. Rather than using up all the available synapses, they develop the capacity to make even more.

Not all motor learning is good, however. Many musicians develop *focal hand dystonia*—abnormal finger and hand positions, cramps, and

LaBrecht Music & Arts/Corbis

difficulty in coordinating hand and finger movements. Dystonia can be so disabling that some musicians must give up their occupation.

Typically, dystonia afflicts musicians who practice trying to make perfect finger movements on their instruments. Musicians at high risk include string players, who receive vibratory stimulation at their fingertips. The constant practice has been suggested to lead not only to improved musical ability but also to distortion or disordering of the motor maps in the cortex. Synchronous activation of the digits by the vibration leads to this unwanted side effect.

Victor Candia and colleagues (2003) reasoned that musicians' dystonia was probably an example of disordered learning and could be treated by retuning the motor map. The investigators used magnetoencephalography (MEG) to measure changes in sensory-evoked magnetic fields in the cortex.

At the beginning of the study, the musicians with dystonia had a disordered motor map: the finger areas overlapped one another. In training, each subject used a hand splint tailored to his or her hand. The splint allowed for the immobilization of different fingers while the subjects made independent movements of the others.

After 8 days of training for about 2 hours per day, the subjects showed marked alleviation in the dystonic symptoms, and the neuroimaging showed a normalization of the cortical map with distinct finger areas. Thus, training reversed the learned changes in the motor map and treated the dystonia. The musicians had actually "learned" a disorder, and they were able to "unlearn" it.

Experience-Dependent Change in the Human Brain

According to Ramachandran's amputee study, the human brain appears to change with altered experience. But this study did not examine neuronal change directly; neuronal change was inferred from behavior. The only way to examine synaptic change directly is to look directly at brain tissue. In living humans, this is not an option, but the brains of people who died from nonneurological causes can be examined and the structure of their cortical neurons can be related to their experiences.

One way to approach this idea is to look for a relation between neuronal structure and education. Arnold Scheibel and his colleagues conducted many such studies in the 1990s (e.g., Jacobs and Scheibel, 1993; Jacobs, Scholl, and Scheibel, 1993). In one, they found a relation between the size of the dendrites in Wernicke's area and the amount of education. In the brains of deceased people with a college education, the cortical

Wernicke's area, diagrammed in Figure 10-18, contributes to speech and to language comprehension.

neurons from this language area had more dendritic branches than did those from people with a high-school education, which, in turn, had more dendritic material than did those from people with less education. People who have more dendrites may be more likely to go to college, but that possibility is not easy to test.

Another way to look at the relation between neurons in Wernicke's area and behavior is to take advantage of the now well-documented observation that on average the verbal abilities of females are superior to those of males. When Scheibel and his colleagues examined the structure of neurons in Wernicke's area, they found that females do have more extensive dendritic branching there than males do.

Finally, these investigators approached the link between experience and neuronal morphology in a slightly different way. They began with two hypotheses. First, they suggested a relation between the complexity of dendritic branching and the nature of the computational tasks performed by a brain area.

To test this hypothesis, they examined the dendritic structure of neurons in different cortical regions that handle different computational tasks. For example, when they compared the structure of neurons corresponding to the somatosensory representation of the trunk with those for the fingers, they found the latter to have more complex cells **(Figure 14-23)**. They reasoned that the somatosensory inputs from receptive fields on the chest wall would constitute less of a computational challenge to cortical neurons than would those from the fingers and that the neurons representing the chest would therefore be less complex.

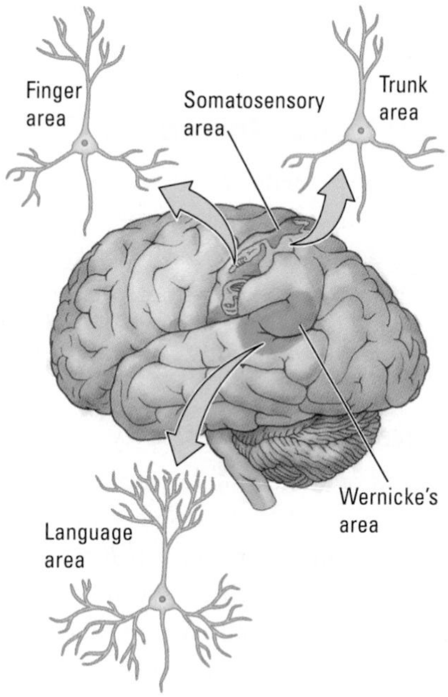

The group's second hypothesis was that dendritic branching in all regions is subject to experience-dependent change. The researchers hypothesized that predominant life experience (e.g., occupation) should, as a result, alter dendritic structure. Although they did not test this hypothesis directly, they did make an interesting observation. In their study comparing cells in the trunk area, in the finger area, and in the supramarginal gyrus—a region of the parietal lobe associated with higher cognitive processes (thinking)—they found curious individual differences.

For example, especially large differences in trunk and finger neurons were found in the brains of people who had a high level of finger dexterity maintained over long periods of time (for example, career word processors). In contrast, no difference between trunk and finger neurons was found in sales representatives. Remember, Scheibel and colleagues conducted their research 20 years ago, in the days before portable electronic devices entered the workplace. We would not expect a good deal of specialized finger use among sales reps of that time and thus less-complex demands on their finger neurons.

In summary, although the studies showing a relation between experience and neuronal structure in humans depend on correlations rather than actual experiments, the findings are consistent with those observed in experimental studies of other species. We are thus led to the general conclusion that specific experiences can produce localized changes in the synaptic organization of the brain and that such changes form the structural basis of memory.

Figure 15-16 diagrams some tasks that consistently show, on average, that females' verbal fluency is better than males' and that males do better than females on spatial reasoning tasks.

FIGURE 14-23 Experience and Neuronal Complexity Confirmation of Scheibel's hypothesis that cell complexity is related to the computational demands required of the cell. Neurons that represent the trunk area of the body have relatively less computational demand than do cells representing the finger region. In turn, cells engaged in more-cognitive functions (such as language, as in Wernicke's area) have greater computational demand than do those engaged in finger functions.

Epigenetics of Memory

An enigma in the search for neural mechanisms underlying memory is the fact that whereas memories remain stable over time, all cells are constantly undergoing molecular turnover. The simplest explanation for this is epigenetic: specific sites in the DNA of neurons involved in specific memories might exist in either methylated on nonmethylated states.

Courtney Miller and David Sweatt (2011) tested this idea directly by measuring methylation in the hippocampi of rats that underwent contextual fear conditioning (see Experiment 14-1). They showed that fear conditioning is associated with rapid methylation, but if they blocked methylation, there was no memory. The investigators conclude that epigenetic mechanisms mediate synaptic plasticity broadly, but especially in learning and memory. One implication of these results is that cognitive disorders, including memory defects, could result from aberrant epigenetic modifications (for a review, see Day and Sweatt, 2011).

Figure 3-25 illustrates two aspects of methylation: histone and DNA modification.

Plasticity, Hormones, Trophic Factors, and Drugs

The news media often report that psychoactive drugs can damage your brain. Some drugs certainly do act as toxins and can selectively kill brain regions, but a more realistic mode of drug action is to *change* the brain. Although not many studies have looked at drug-induced morphological changes, there is evidence that some compounds can greatly change the synaptic organization of the brain. These compounds include hormones, neurotrophic factors, and psychoactive drugs. We briefly consider each category.

Hormones and Plasticity

Levels of circulating hormones play a critical role both in determining the structure of the brain and in eliciting certain behaviors in adulthood. Although the structural effects of hormones were once believed only to be expressed in the course of development, current belief is that adult neurons also can respond to hormonal manipulations with dramatic structural changes. We consider the actions of gonadal hormones and stress hormones in this section.

Section 6-5 explains the classes, functions, and control exerted by hormones; Section 8-4 their organizing effects during development; and Section 12-5 the activating effects in adulthood.

Research findings have established that the structural differences in cortical neurons of male and female rats depend on gonadal hormones. What is more surprising, perhaps, is that gonadal hormones continue to influence cell structure and behavior in adulthood. Elizabeth Hampson and Doreen Kimura (1988) showed that women's performance on various cognitive tasks changes throughout the menstrual cycle as their estrogen levels fluctuate.

Changes in estrogen level appear to alter the structure of neurons and astrocytes in the neocortex and hippocampus, which probably accounts for at least part of the performance fluctuation. **Figure 14-24** illustrates changes in dendritic spines in the hippocampal cells of female rats at different phases of their 4-day estrous cycle. As the estrogen level rises, the number of synapses rises; as the estrogen level drops, the number of synapses declines.

Curiously, estrogen's influence on cell structure may be different in the hippocampus and neocortex. Jane Stewart found, for example, that when the ovaries of middle-aged female rats are removed, estrogen levels drop sharply, producing an increase in the number of spines on pyramidal cells throughout the neocortex but a decrease in spine density in the hippocampus (Stewart & Kolb, 1994). How these synaptic changes might influence processes such as memory is not immediately obvious, but the question is reasonable, especially because menopausal women also experience sharp drops in estrogen levels and a corresponding decline in verbal memory ability.

FIGURE 14-24 Hormones and Neuroplasticity Sections of dendrites from hippocampal cells during times of high and low levels of estrogen during the rat's 4-day estrous cycle reveal many more dendritic spines during the period when estrogen levels are high. Adapted from "Naturally Occurring Fluctuation in Dendritic Spine Density on Adult Hippocampal Pyramidal Neurons," by C. S. Woolley, E. Gould, M. Frankfurt, and B. S. McEwen, 1990, *Journal of Neuroscience, 10*, p. 4038.

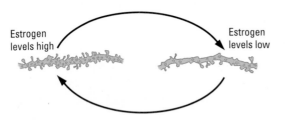

Estrogen levels high

Estrogen levels low

This question is also relevant to middle-aged men, who show a slow decline in testosterone levels that correlates with a drop in spatial ability. Rats that are gonadectomized in adulthood show an increase in cortical spine density, much like the ovariectomized females, although we do not know how this change relates to spatial behavior. Nonetheless, a reasonable supposition is that testosterone levels might influence spatial memory throughout life.

When the body is stressed, the pituitary gland produces adrenocorticotrophic hormone (ACTH), which stimulates the adrenal cortex to produce steroid hormones known as *glucocorticoids.* Important in protein and carbohydrate metabolism, controlling sugar levels in the blood, and the absorption of sugar by cells, glucocorticoids have many actions on the body, including the brain. Robert Sapolsky (1992) proposed that glucocorticoids can sometimes be neurotoxic.

In particular, he found that, with prolonged stress, glucocorticoids appear to kill cells in the hippocampus. Elizabeth Gould and her colleagues (1998) showed that even brief periods of stress can reduce the number of new granule cells produced in the hippocampi of monkeys, presumably through the actions of stress hormones. Evidence of neuron death and reduced neuron generation in the hippocampus has obvious implications for the behavior of animals, especially for processes such as spatial memory.

In sum, hormones can alter the brain's synaptic organization and even the number of neurons in the brain. Little is known today about the behavioral consequences of such changes, but it is likely that hormones can alter the course of plastic changes in the brain.

Neurotrophic Factors and Plasticity

Neurotrophic factors, chemical compounds that signal stem cells to develop into neurons or glia (listed in **Table 14-2**), also act to reorganize neural circuits. The first neurotrophic factor, **nerve growth factor** (NGF), was discovered in the peripheral nervous system more than a generation ago. NGF is trophic (nourishing) in the sense that it stimulates neurons to grow dendrites and synapses, and in some cases, it promotes the survival of neurons.

Trophic factors produced in the brain by neurons and glia can affect neurons both through cell-membrane receptors and by actually entering the neuron to act internally on its operation. For example, trophic factors may be released postsynaptically to act as signals that can influence the presynaptic cell. Experience stimulates their production, so neurotrophic factors have been proposed as agents of synaptic change. For example, brain-derived neurotrophic factor (BDNF) increases when animals solve specific problems such as mazes. This finding has led to speculation that BDNF release may enhance plastic changes, such as the growth of dendrites and synapses.

Unfortunately, although many researchers would like to conclude that BDNF has a role in learning, this conclusion does not necessarily follow. When animals solve mazes, their behavior differs from their behavior when they remain in cages. So we must first demonstrate that changes in BDNF, NGF, or any other trophic factor are actually related to forming new synapses. Nevertheless, if we assume that trophic factors do act as agents of synaptic change, then we should be able to use increased trophic factor activity during learning as a marker for where to look for changed synapses associated with learning and memory.

Psychoactive Drugs and Plasticity

Many people regularly use stimulants such as caffeine, and some use more psychoactively stimulating drugs such as nicotine, amphetamine, or cocaine. The long-term consequences of abusing psychoactive drugs

nerve growth factor (NGF) Neurotrophic factor that stimulates neurons to grow dendrites and synapses and, in some cases, promotes the survival of neurons.

Figure 6-23 illustrates the body's response to stress.

Section 8-2 explains how neurotrophic factors send these signals.

Section 5-4 notes that the Hebb synapse—one that changes with use so that learning takes place—is hypothesized to employ just such a mechanism.

TABLE 14-2 Molecules Exhibiting Neurotrophic Activities

Proteins initially characterized as neurotrophic factors
Nerve growth factor (NGF)
Brain-derived neurotrophic factor (BDNF)
Neurotrophin 3 (NT-3)
Ciliary neurotrophic factor (CNTF)
Growth factors with neurotrophic activity
Fibroblast growth factor, acidic (aFGF or FGF-1)
Fibroblast growth factor, basic (bFGF or FGF-2)
Epidermal growth factor (EGF)
Insulinlike growth factor (ILGF)
Transforming growth factor (TGF)
Lymphokines (interleukin 1, 3, 6 or IL-1, IL-3, IL-6)
Protease nexin I, II
Cholinergic neuronal differentiation factor

behavioral sensitization Escalating behavioral response to the repeated administration of a psychomotor stimulant such as amphetamine, cocaine, or nicotine; also called *drug-induced* behavioral sensitization.

metaplasticity Interaction among different plastic changes in the brain.

are now well documented, but the question of why these drugs cause problems remains to be answered. One explanation for the behavioral changes associated with chronic psychoactive drug abuse is that the drugs change the brain.

One experimental demonstration of these changes is *drug-induced behavioral sensitization,* often referred to simply as **behavioral sensitization,** the progressive increase in behavioral actions in response to repeated administration of a drug. Behaviors increase even when the amount given in each dose does not change. Behavioral sensitization occurs with most psychoactive drugs, including amphetamine, cocaine, morphine, and nicotine.

Aplysia becomes more sensitive to a stimulus after repeated exposure. Psychoactive drugs appear to have a parallel action: they lead to increased behavioral sensitivity to their actions. For example, a rat given a small dose of amphetamine may show an increase in activity. When the rat is given the same dose of amphetamine on subsequent occasions, the increase in activity is progressively larger. If no drug is given for weeks or even months, and then the amphetamine is given in the same dose as before, behavioral sensitization picks up where it left off and continues. Some long-lasting change must have taken place in the brain in response to the drug. Drug-induced behavioral sensitization can therefore be viewed as a memory for a particular drug.

The parallel between drug-induced behavioral sensitization and other forms of memory leads us to ask if the changes in the brain after behavioral sensitization are similar to those found after other forms of learning. They are. For example, there is evidence of increased numbers of receptors at synapses and of more synapses in sensitized animals.

In a series of studies, Terry Robinson and his colleagues found a dramatic increase in dendritic growth and spine density in rats that were sensitized to amphetamine, cocaine, or nicotine relative to rats that received injections of a saline solution (Robinson & Kolb, 2004). **Experiment 14-3** compares the effects of amphetamine and saline treatments on cells in the nucleus accumbens, a structure in the basal ganglia. Neurons in the amphetamine-treated brains have more dendritic branches and increased spine density. Repeated exposure to psychoactive stimulants thus alters the structure of cells in the brain. These changes in turn may be related to "learned addictions."

These plastic changes were not found throughout the brain. Rather, they were localized to such regions as the prefrontal cortex and nucleus accumbens that receive a large dopamine projection. Dopamine is believed to play a significant role in the rewarding properties of drugs (Wise, 2004). Other psychoactive drugs also appear to alter neuronal structure. Marijuana, morphine, and certain antidepressants change dendritic length and spine density, although in somewhat different ways from those of stimulants. Morphine, for example, produces a decrease in dendritic length and spine density in the nucleus accumbens and prefrontal cortex (Robinson & Kolb, 2004).

What do drug-induced changes in synaptic organization mean for later experience-dependent plasticity? If rats are given amphetamine, cocaine, or nicotine for 2 weeks before being placed in complex environments, the expected increases in dendritic length and spine density in the cortex do not happen (Kolb et al., 2003). This is not because the brain can no longer change: giving the animals additional drug doses can still produce change.

EXPERIMENT 14-3

Question: What effect do repeated doses of amphetamine, a psychomotor stimulant, have on neurons?

Procedure

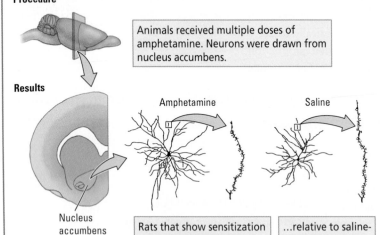

Animals received multiple doses of amphetamine. Neurons were drawn from nucleus accumbens.

Results

Amphetamine   Saline

Nucleus accumbens

Rats that show sensitization to amphetamine have increased dendritic growth and spine density...

...relative to saline-treated rats that served as controls.

Conclusion: The sensitization induced by repeated exposure to amphetamine changes the structure of neurons in certain brain areas.

Adapted from "Persistent Structural Adaptations in Nucleus Accumbens and Prefrontal Cortex Neurons Produced by Prior Experience with Amphetamine," by T. E. Robinson and B. Kolb, 1997, *Journal of Neuroscience, 17,* p. 8496.

Rather, something about prior drug exposure alters the way in which the brain later responds to experience.

Why prior drug exposure has this effect is not yet known, but obviously drug taking can have long-term effects on brain plasticity. One possible explanation is epigenetic. Giving animals repeated doses of amphetamine or nicotine increases methylation, both in the prefrontal cortex and nucleus accumbens. Methylation may render the synapses less able to change in response to later experiences (Mychasiuk et al., 2012).

Some Guiding Principles of Brain Plasticity

Brain plasticity will continue to be a fundamental concept underlying research into brain–behavior relationships through the coming decade. Some basic rules have emerged to guide this research (see Kolb and Gibb, 2008, for more details). Here we list seven.

1. Behavioral Change Reflects Change in the Brain

The primary function of the brain is to produce behavior, but behavior is not static. We learn and remember, we think new thoughts or visualize new images, and we change throughout our lifetimes. All these processes require changes in neural networks. Whenever neural networks change, behavior, including mental behavior, also changes. A corollary of this principle is especially important as neuroscientists search for treatments for brain injuries or behavioral disorders: *To change behavior, we must change the brain.*

2. All Nervous Systems Are Plastic in the Same General Way

Even the simplest animals, such as the roundworm *C. elegans,* can show simple learning that correlates with neuronal plasticity. The molecular details may differ between simple and complex systems, but the general principles of neuroplasticity appear to be conserved across both simple and complex animals. This conservation allows more studies of neural plasticity among a wider range of animal species than in most areas of neuroscience.

3. Plastic Changes Are Age-Specific

The brain responds to the same experiences differently at different ages—and especially during development. The prefrontal cortex is late to mature, for example, so the same experience affects this region differently in infancy than it does in adolescence and on throughout life.

4. Prenatal Events Can Influence Brain Plasticity Throughout Life

Even prenatal experiences alter brain organization. Potentially negative experiences, such as prenatal exposure to recreational or prescription drugs, as well as positive experiences, such as tactile stimulation of the mother's skin, may alter gene expression or produce other epigenetic effects and produce enduring effects on brain organization.

5. Plastic Changes Are Brain-Region Dependent

Although we are tempted to expect plastic changes in neuronal networks to be fairly general, it is becoming clear that many experience-dependent changes are highly specific. We saw this specificity in the effects of psychoactive drugs on the prefrontal cortex but not on other cortical regions. Not only do drugs selectively change the prefrontal cortex but the dorsolateral and orbital prefrontal areas also show opposite changes—the precise changes varying with the particular drug. For example, stimulants such as amphetamine increase the spine density in the dorsolateral region but decrease it in the orbital region.

6. Experience-Dependent Changes Interact

Metaplasticity is a property of a lifetime's interaction among different plastic changes in the brain. As an animal travels through life, an infinity of experiences can alter its brain organization. A lifetime's experiences might interact. Housing animals in complex

Figure 12-30 shows enhanced connectivity in prefrontal and limbic circuits in the brains of nicotine-dependent smokers.

Investigators currently study neuroplasticity in species ranging from worms and insects to fish, birds, and mammals.

Section 8-4 traces how brain organization details change rapidly—and sometimes critically—during development.

Focus 6-2 reports on the tragedy of fetal alcohol syndrome, Section 8-4 on the benefits of tactile stimulation, and Focus 8-2 on epigenetic factors in the autism spectrum.

Figure 12-19 diagrams these prefrontal regions.

environments produces profound changes in their neural network organization, but prior exposure to psychoactive drugs completely blocks the enrichment effect. Conversely, although complex housing does not block the effects of drugs, the effects are markedly attenuated. Prenatal events can affect later drug effects: prenatal tactile stimulation of the mother, for example, reduces the later effects of psychoactive drugs on the child.

7. Plasticity Has Pros and Cons

We have mainly emphasized the neuroplastic changes that can support improved motor and cognitive function. But as noted for the effects of psychoactive drugs, plastic changes in neural networks can also interfere with behavior. Drug addicts whose prefrontal cortex has been altered are prone to poor judgment in their personal lives. People who have posttraumatic stress disorders show altered blood flow in the amygdala and cingulate cortex. That's the bad news.

Encouraging plastic changes that reverse these prefrontal alterations is the good news, and is associated with a loss of the prefrontal disorder. Age-related dementia is related to synaptic loss that various forms of cognitive therapy can reverse (e.g., Mahncke, Bron-stone, & Merzenich, 2006).

REVIEW 14-4

Structural Basis of Brain Plasticity

Before you continue, check your understanding.

1. Repeated high-frequency stimulation of excitatory neurons leads to the phenomenon of _____, whereas repeated low-frequency stimulation leads to _____.

2. LTPi and LTDi are found in _____ neurons.

3. Structural changes underlying memory include changes in both _____ and _____.

4. Learning complex spatial information has been linked to increased gray matter in the _____.

5. The progressive increase in behavioral actions in response to repeated administration of a drug is called _____.

6. How can plastic changes in the brain produce adverse effects?

Answers appear at the back of the book.

14-5 Recovery from Brain Injury

The nervous system appears to be conservative in its use of the mechanisms related to behavioral change. If neuroscientists wish to change the brain, as after injury or disease, then they should look for treatments that will produce plastic changes related to learning, memory, and other behaviors.

Recall that H. M. failed to recover his lost memory capacities, even after 55 years of practice in trying to remember information. Relearning simply was not possible for H. M. He had lost the requisite neural structures. But other people do show some recovery.

An average person would probably say that the process of recovery after brain trauma requires the injured person to relearn lost skills, whether walking, talking, or using the fingers. But what exactly does recovery entail? Partial recovery of function is common after brain injury, but a person with brain trauma or brain disease has lost neurons. The brain may be missing structures critical for relearning or remembering.

Donna's Experience with Traumatic Brain Injury

Donna started dancing when she was 4 years old, and she was a "natural." By the time she finished high school she had the training and skill necessary to apprentice with and later join a major dance company. Donna remembers vividly the day she was chosen to dance a leading role in *The Nutcracker*. She had marveled at the costumes as she watched the popular Christmas ballet as a child, and now she would dance in those costumes!

The births of two children interrupted her career as a dancer, but Donna never lost the interest. In 1968, when both her children were in school, she began dancing again with a local company. To her amazement, she could still perform most of the movements, although she was rusty on the choreography of the classical dances that she had once memorized so meticulously. Nonetheless, she quickly relearned. In retrospect, she should not have been so surprised, because she had always had an excellent memory.

One evening in 1990, while on a bicycle ride, Donna was struck by a drunk driver. Although she was wearing a helmet, she suffered damage to the brain that results from a blow to the head—a **traumatic brain injury,** or TBI. She was comatose for several weeks. As she regained consciousness, she was confused and had difficulty talking to and understanding others. Her memory was very poor, spatial disorientation meant she often got lost, she had various motor disturbances, and she had difficulty recognizing anyone but her family and closest friends.

Over the ensuing 10 months, Donna regained most of her motor abilities and language skills, and her spatial abilities improved significantly. Nonetheless, she was short-tempered and easily frustrated by the slowness of her recovery, symptoms typical of people with brain trauma. She suffered periods of depression.

Donna also found herself prone to inexplicable surges of panic when doing simple things. On one occasion early in her rehabilitation, she was shopping in a large supermarket and became overwhelmed by the number of salad dressing choices. She ran from the store, and only after she sat outside and calmed herself could she go back inside to continue shopping.

Two years later, Donna was dancing once again, but she now found it very difficult to learn and remember new steps. Her emotions were still unstable, which was a strain on her family, but her episodes of frustration and temper outbursts became much less frequent. A year later, they were gone and her life was not obviously different from that of other middle-aged women.

Even so, some cognitive changes persisted. Donna seemed unable to remember the names or faces of new people she met. She lost concentration if background distractions such as a television or a radio playing intruded. She could not dance as she had before her injury, but she did work at it diligently. Her balance on sudden turns gave her the most difficulty. Rather than risk falling, she retired from her life's first love.

Donna's case demonstrates the human brain's capacity for continuously changing its structure and ultimately its function throughout a lifetime. From what we have learned in this chapter, we can identify three different ways in which Donna could recover from her brain injury: she could learn new ways to solve problems, she could reorganize the brain to do more with less, and she could generate new neurons to produce new neural circuits. We briefly examine these three possibilities.

Three-Legged Cat Solution

Cats that lose a leg quickly learn to compensate for the missing limb and once again become mobile. They show recovery of function: the limb is gone, but behavior has changed to compensate. This simplest solution to recovery from TBI we call the "three-legged cat solution."

traumatic brain injury (TBI) Damage to the brain that results from a blow to the head.

Focus 1-1 and Section 1-2 introduce some consequences of and treatments for TBI, which are elaborated here and in Section 16-3.

The brain changes in response to these dancers' new experiences and new abilities. After her accident, Donna's brain had to change to allow her to regain her lost abilities, but she never recovered the ability these young women have to learn new dances.

Digital Stock

A similar explanation can account for many instances of apparent recovery of function after TBI. Imagine that a right-handed person has a stroke that leads her to lose the use of her right hand and arm. Unable to write with the affected limb, she switches to her left hand. Behavioral compensation like this—a person learns to use the opposite hand to write—presupposes that some changes in the nervous system underlie this new skill.

New-Circuit Solution

A second way to recover from brain damage is for the brain to form new connections that allow it to "do more with less." This change is most easily accomplished by processes similar to those we considered for other forms of plasticity. The brain changes its neural connections to overcome the loss.

Without some form of intervention, recovery from most instances of brain injury is relatively modest. Recovery can be increased significantly if the person engages in behavioral, pharmacological, or brain-stimulation therapy that encourages the brain to make new connections.

Behavioral therapy, such as speech therapy, physiotherapy, and music therapy presumably increases brain activity, which facilitates the neural changes. In a pharmacological intervention, the patient takes a drug, such as nerve growth factor, known to influence brain plasticity. When NGF is given to animals with strokes that damaged the motor cortex, their motor functions improve **(Experiment 14-4)**. The behavioral changes are correlated with a dramatic increase in dendritic branching and spine density in the remaining, intact motor regions. The morphological changes are correlated with improved motor functions, such as reaching with the forelimb to obtain food, as illustrated in Experiment 14-2 (Kolb et al., 1997). But because brain tissue is still missing, recovery is by no means complete.

In principle, we might expect that any drug that stimulates the growth of new connections would help people recover from brain injury. However, the neural growth must occur in regions of the brain that can influence a particular lost function. A drug that stimulates the growth of synapses on cells in the visual cortex, for example, would not enhance recovery of hand use. The visual neurons play no direct role in moving the hand.

A third strategy to generate new neural circuits uses either deep brain stimulation (DBS) or direct electrical stimulation of perilesional regions. The goal of electrical stimulation is to directly increase activity in remaining parts of specific, damaged neural networks. In DBS, it is to put the brain into a more plastic ("trainable") state so that rehabilitation therapies work better. Both strategies are currently in preliminary clinical trials.

Lost-Neuron-Replacement Solution

The third strategy a patient like Donna could pursue is to generate new neurons to produce new neural circuits. The idea that brain tissue could be transplanted from one animal to another goes back a century. The evidence is good that tissue from fetal brains can be transplanted and will grow and form some connections in the new brain.

Unfortunately, in contrast with transplanted hearts or livers, transplanted brain tissue functions poorly. The procedure seems most suited to conditions in which a small number of functional cells are required, as in the replacement of dopamine-producing cells in Parkinson's disease or in the replacement of suprachiasmatic cells to restore circadian rhythms.

By 2004, dopamine-producing cells had been surgically transplanted into the striata of many Parkinson patients. Although the disease has not been reversed, some patients, especially the younger ones, have shown functional gains that justify the procedure. Nonetheless, ethical issues remain because the tissue is taken from aborted human fetuses.

Focus 5-4 recounts a successful case of fetal stem cell transplantation, and Section 13-2 describes SCN cell replacement.

The striatum, a region in the basal ganglia, includes the caudate nucleus and putamen.

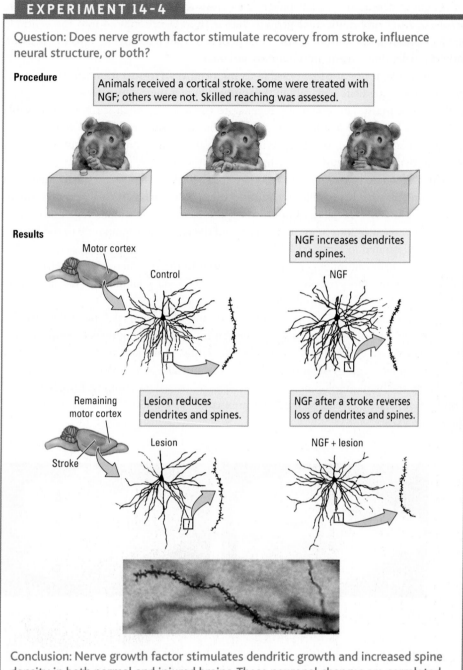

EXPERIMENT 14-4

Question: Does nerve growth factor stimulate recovery from stroke, influence neural structure, or both?

Procedure

Animals received a cortical stroke. Some were treated with NGF; others were not. Skilled reaching was assessed.

Results

Motor cortex

Control

NGF

NGF increases dendrites and spines.

Remaining motor cortex

Stroke

Lesion

Lesion reduces dendrites and spines.

NGF + lesion

NGF after a stroke reverses loss of dendrites and spines.

Conclusion: Nerve growth factor stimulates dendritic growth and increased spine density in both normal and injured brains. These neuronal changes are correlated with improved motor function after stroke.

Adapted from "Nerve Growth Factor Treatment Prevents Dendritic Atrophy and Promotes Recovery of Function after Cortical Injury," by B. Kolb, S. Cote, A. Ribeiro-da-Silva, and A. C. Cuello, 1997, *Neuroscience, 76*, p. 1146.

Adult stem cells are a second way to replace lost neurons. Investigators know that the brain is capable of making neurons in adulthood. The challenge is to get the brain to do so after an injury. The first breakthrough in this research was made by Brent Reynolds and Sam Weiss (1992).

Cells lining the ventricles of adult mice were removed and placed in a culture medium. The researchers demonstrated that if the correct trophic factors are added, the

Even in adults, neural stem cells line the subventricular zone, diagrammed in Figure 8-11A.

epidermal growth factor (EGF)
Neurotrophic factor that stimulates the subventricular zone to generate cells that migrate into the striatum and eventually differentiate into neurons and glia.

cells begin to divide and can produce new neurons and glia. Furthermore, if the trophic factors—particularly **epidermal growth factor** (EGF)—are infused into the ventricle of a living animal, the subventricular zone generates cells that migrate into the striatum and eventually differentiate into neurons and glia.

In principle, it ought to be possible to use trophic factors to stimulate the subventricular zone to generate new cells in the injured brain. If these new cells were to migrate to the site of injury and essentially regenerate the lost area, then it might be possible to restore at least some lost function. All lost behaviors could not be restored, however, because the new neurons would have to establish the same connections with the rest of the brain that the lost neurons once had.

This task would be daunting, because the connections would have to be formed in an adult brain that already has billions of connections. Nonetheless, such a treatment might someday be feasible. Cocktails of trophic factors are effective in stimulating neurogenesis in the subventricular zone after brain injury, and the new cells can migrate to the injured region, as illustrated in **Figure 14-25.**

The new cells can influence behavior and lead to improvement. The mechanism of influence is poorly understood (e.g., Kolb et al., 2007), yet preliminary clinical trials are under way and so far show no ill effects in volunteers.

The hippocampus and olfactory bulb are among the regions that normally produce new neurons in adulthood, and experience can influence the numbers of neurons produced. It is possible, therefore, that researchers could stimulate the generation of new neurons in intact regions of an injured brain. It is possible as well that the new neurons could help the brain develop new circuits to restore partial functioning. Experience and trophic factors are likely to be used in combination in studies of recovery from TBI in the coming years.

FIGURE 14-25 Stem Cells Do the Trick
After cortical stroke (*left*), infusion of epidermal growth factor into the lateral ventricle of a rat induced neurogenesis in the subventricular zone. The stem cells migrated to the site of injury and filled in the damaged area (*right*).

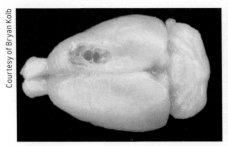 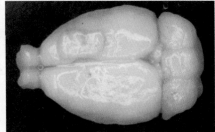

Courtesy of Bryan Kolb

REVIEW 14-5

Recovery from Brain Injury

Before you continue, check your understanding.

1. Three ways to compensate for the loss of neurons are (1) _____, (2) _____, and (3) _____.

2. Two ways of using electrical stimulation to enhance postinjury recovery are _____ and _____.

3. Endogenous stem cells can be recruited to enhance functional improvement by using _____ factors.

4. What is the lesson of the three-legged cat?

Answers appear at the back of the book.

SUMMARY

14-1 Connecting Learning and Memory

Learning is a change in an organism's behavior as a result of experience. Memory is the ability to recall or recognize previous experience. For more than a century, laboratory studies using animals have uncovered two fundamentally different types of learning: Pavlovian and operant (or instrumental).

The two basic types of memory are implicit (unconscious) and explicit (conscious). Episodic memory includes not only a record of events (episodes) that occurred but also our presence there and our role in the events. It is likely that the frontal lobe plays a unique role in this autobiographical memory.

14-2 Dissociating Memory Circuits

Multiple subsystems control different aspects of memory within the explicit and implicit systems. People with damaged explicit memory circuits have impaired recall for facts and events. People with damaged implicit memory circuits are impaired in their recall of skills and habits.

14-3 Neural Systems Underlying Explicit and Implicit Memories

The neural circuits underlying implicit and explicit memory are distinctly different: the reciprocal system for explicit memory includes medial temporal structures; the unidirectional system for implicit memory includes the basal ganglia. Emotional memory has characteristics of both implicit and explicit memory. The neural circuits for emotional memory are unique in that they include the amygdala. The chart below summarizes broad categories within these multiple memory systems.

For memories to become established in the brain—the process of consolidation—experiences must change neural connections, and these changes must become relatively permanent. When memories are revisited, neural connectivity can become less fixed, allowing the

neural networks and thus the memory to be modified, a process called reconsolidation.

14-4 Structural Basis of Brain Plasticity

The brain has the capacity for structural change, and structural change presumably underlies functional change. The brain changes structure in two fundamental ways in response to experience.

First, existing neural circuits change largely by modifying synaptic connections. One proposed mechanism of synaptic change is long-term potentiation (LTP), which is reflected in the modification of EPSPs following learning.

Second, novel neural circuits are formed both by new connections among existing neurons and by generating new neurons. One mechanism for establishing novel circuits is generating new neurons in the hippocampus. A likely mechanism for maintaining synaptic changes is epigenetic: specific sites in the DNA of neurons in modified circuits can exist in either methylated or unmethylated states.

Neuronal activity is the key to brain plasticity; through it, synapses form and change. Neuronal activity can be induced by general or specific experience as well as by electrical or chemical stimulation. Chemical stimulation may range from hormones and neurotrophic compounds to psychoactive drugs.

Much of the brain is capable of plastic change with experience. Different experiences lead to changes in different neural systems. The table summarizes seven basic principles that guide research about brain plasticity and behavior.

Some Guiding Principles of Brain Plasticity

1. Behavioral change reflects change in the brain.
2. All nervous systems are plastic in the same general way.
3. Plastic changes are age-specific.
4. Prenatal events can influence brain plasticity throughout life.
5. Plastic changes are brain-region dependent.
6. Experience-dependent changes interact.
7. Plasticity has pros and cons.

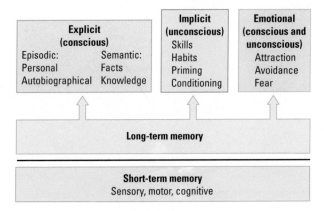

Multiple Memory Systems

14-5 Recovery from Brain Injury

Plastic changes after brain injury parallel those seen when the brain changes with experience. Changes related to recovery do not always occur spontaneously, however, and must be stimulated by behavioral training, by the effects of psychoactive drugs or neurotrophic factors, or by electrical brain stimulation. The key to stimulating recovery from brain injury is to increase the plastic changes underlying the recovery.

KEY TERMS

amnesia, p. 487

anterograde amnesia, p. 499

associative learning, p. 502

behavioral sensitization, p. 514

conditioned response (CR), p. 485

conditioned stimulus (CS), p. 485

consolidation, p. 499

declarative memory, p. 487

dyslexia, p. 483

emotional memory, p. 501

entorhinal cortex, p. 493

epidermal growth factor (EGF), p. 520

episodic memory, p. 488

explicit memory, p. 487

eye-blink conditioning, p. 485

fear conditioning, p. 485

implicit memory, p. 487

Korsakoff's syndrome, p. 496

learning, p. 483

learning set, p. 487

long-term depression (LTD), p. 502

long-term potentiation (LTP), p. 502

memory, p. 483

metaplasticity, p. 514

nerve growth factor (NGF), p. 513

neuritic plaque, p. 495

operant conditioning, p. 485

parahippocampal cortex, p. 493

Pavlovian conditioning, p. 483

perirhinal cortex, p. 493

priming, p. 487

procedural memory, p. 487

reconsolidation, p. 499

retrograde amnesia, p. 499

traumatic bain injury (TBI), p. 517

unconditioned response (UCR), p. 485

unconditioned stimulus (UCS), p. 485

visuospatial learning, p. 495

Please refer to the Companion Web Site at www.worthpublishers.com/kolbintro4e *for Interactive Exercises and Quizzes.*

How Does the Brain Think?

RESEARCH FOCUS 15-1 SPLIT BRAIN

15-1 NATURE OF THOUGHT

CHARACTERISTICS OF HUMAN THOUGHT

NEURAL UNIT OF THOUGHT

COMPARATIVE FOCUS 15-2 ANIMAL INTELLIGENCE

15-2 COGNITION AND THE ASSOCIATION CORTEX

KNOWLEDGE ABOUT OBJECTS

MULTISENSORY INTEGRATION

SPATIAL COGNITION

ATTENTION

PLANNING

IMITATION AND UNDERSTANDING

RESEARCH FOCUS 15-3 CONSEQUENCES OF MIRROR-NEURON DYSFUNCTION

15-3 EXPANDING FRONTIERS OF COGNITIVE NEUROSCIENCE

MAPPING THE BRAIN

CLINICAL FOCUS 15-4 NEUROPSYCHOLOGICAL ASSESSMENT

SOCIAL NEUROSCIENCE

NEUROECONOMICS

15-4 CEREBRAL ASYMMETRY IN THINKING

ANATOMICAL ASYMMETRY

FUNCTIONAL ASYMMETRY IN NEUROLOGICAL PATIENTS

FUNCTIONAL ASYMMETRY IN THE NORMAL BRAIN

FUNCTIONAL ASYMMETRY IN THE SPLIT BRAIN

EXPLAINING CEREBRAL ASYMMETRY

LEFT HEMISPHERE, LANGUAGE, AND THOUGHT

15-5 VARIATIONS IN COGNITIVE ORGANIZATION

SEX DIFFERENCES IN COGNITIVE ORGANIZATION

HANDEDNESS AND COGNITIVE ORGANIZATION

CLINICAL FOCUS 15-5 SODIUM AMOBARBITAL TEST

SYNESTHESIA

CLINICAL FOCUS 15-6 A CASE OF SYNESTHESIA

15-6 INTELLIGENCE

CONCEPT OF GENERAL INTELLIGENCE

MULTIPLE INTELLIGENCES

DIVERGENT AND CONVERGENT INTELLIGENCE

INTELLIGENCE, HEREDITY, EPIGENETICS, AND THE SYNAPSE

15-7 CONSCIOUSNESS

WHY ARE WE CONSCIOUS?

WHAT IS THE NEURAL BASIS OF CONSCIOUSNESS?

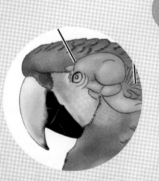

Split Brain

Epileptic seizures may begin in a restricted region of one brain hemisphere and then spread through the fibers of the corpus callosum to the corresponding location in the opposite hemisphere. To prevent the spread of seizures that cannot be controlled through medication, neurosurgeons sometimes sever the 200 million nerve fibers of the corpus callosum.

The procedure is medically beneficial for many epilepsy patients, leaving them virtually seizure-free with only minimal effects on their everyday behavior. In special circumstances, however, the aftereffects of a severed corpus callosum become more readily apparent, as extensive psychological testing by Roger Sperry, Michael Gazzaniga, and their colleagues (Sperry, 1968; Gazzaniga, 1970) has demonstrated.

On close inspection, such **split-brain** patients reveal a unique behavioral syndrome that offers insight into the nature of cerebral asymmetry. Cortical asymmetry is essential for such integrative tasks as language and body control.

One split-brain subject was presented with several blocks. Each block had two red sides, two white sides, and two half-red and half-white sides, as illustrated. The task was to arrange the blocks to form patterns identical with those shown on cards.

When the subject used his right hand to perform the task, he had great difficulty. His movements were slow and hesitant. In contrast, when he performed the task with his left hand, his solutions were not only accurate but also quick and decisive.

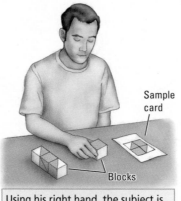

Using his right hand, the subject is unable to duplicate the pattern...

...but with his left hand, the split-brain patient performs the task correctly.

In this experiment, a split-brain patient's task is to arrange a set of blocks to match the pattern shown on a card. Adapted from Cognitive Neuroscience: *The Biology of the Mind* (p. 323), by M. S. Gazzaniga, R. B. Ivry, and G. R. Mangun, 1999, New York: Norton.

Findings from studies of other split-brain patients have shown that, as tasks of this sort become more difficult, left-hand superiority increases. Normal participants perform equally well with either hand, indicating the intact connection between the two hemispheres. But in split-brain subjects, each hemisphere works on its own.

Apparently, the right hemisphere, which controls the left hand, has visuospatial capabilities that the left hemisphere does not.

split brain Surgical disconnection of the two hemispheres in which the corpus callosum is cut.

psychological construct Idea, resulting from a set of impressions, that some mental ability exists as an entity; examples include memory, language, and emotion.

cognition Act or process of knowing or coming to know; in psychology, used to refer to the processes of thought.

syntax Ways in which words are put together to form phrases, clauses, or sentences; proposed to be a unique characteristic of human language.

Studies of split-brain patients reveal that the left and right cerebral hemispheres engage in fundamentally different types of thinking. Yet typically, we are unaware of these brain asymmetries. In this chapter, we examine the neural systems and subsystems that control thinking. In the mammalian brain, these systems are in the cortex.

Our first task is to define the mental processes we wish to study—to ask, What is the nature of thought? Then we consider the cortical regions—for vision, audition, movement and associative function—that play major roles in thinking. We examine how these cortical connections are organized into such systems and subsystems as the dorsal and ventral visual streams, and how neuroscientists study them.

Next we explore the brain's asymmetrical organization and delve deeper into split-brain phenomena. Another distinguishing feature of human thought is the different ways that individual people think. We consider several sources of these differences, including those related to sex and to what we call intelligence. Finally, we address consciousness and how it may relate to the neural control of thought.

15-1 Nature of Thought

Studying abstract mental processes such as thought, language, memory, emotion, and motivation is tricky. They cannot be seen but can only be inferred from behavior and are best thought of as **psychological constructs,** ideas that result from a set of impressions. The mind constructs the idea as being real, even though it is not tangible.

We run into trouble when we try to locate constructs such as thought or memory in the brain. That we have words for these constructs does not mean that the brain is organized around them. Indeed, it is not. For instance, although people talk about memory as a unitary thing, the brain neither treats memory as unitary nor localizes it in one particular place. The many forms of memory are each treated differently by widely distributed brain circuits. The psychological construct of memory that we think of as being a single thing turns out not to be unitary at all.

Assuming a neurological basis for psychological constructs such as memory and thought is risky, but we certainly should not give up searching for where and how the brain produces them. After all, thought, memory, emotion, motivation, and other such constructs are the most interesting activities the brain performs.

Psychologists typically use the term **cognition** (knowing) to describe thought processes, that is, how we come to know about the world. For behavioral neuroscientists, cognition usually entails the ability to pay attention to stimuli, whether external or internal, to identify stimuli, and to plan meaningful responses to them. External stimuli cue neural activity in our sensory receptors. Internal stimuli can spring from the autonomic nervous system (ANS) as well as from neural processes—from constructs such as memory and motivation.

> Section 14-1 details the varied forms of memory. Section 14-4 explains how neuroplasticity contributes to memory processing and storage.

Characteristics of Human Thought

Human cognition is widely believed to have unique characteristics. One is that human thought is verbal, whereas the thought of other animals is nonverbal. Language is presumed to give humans an edge in thinking, and in some ways it does:

- Language provides the brain a means of categorizing information. It allows us easily to group together objects, actions, and events that have common factors.

- Language provides a means of organizing time, especially future time. It enables us to plan our behavior around time (Monday at 3:00 P.M.) in ways that nonverbal animals cannot.

- Perhaps most important, human language has **syntax**—sets of rules about putting words together to create meaningful utterances.

> The appearance of human language correlates with the dramatic increase in human brain size described in Focus 2-1. Section 10-4 explains the foundations underlying all languages.

Linguists argue that, although other animals, such as chimpanzees, can use and recognize large numbers of vocalizations (about three dozen for chimps), they do not rearrange these sounds to produce new meanings. This lack of syntax, linguists maintain, makes chimpanzee language literal and inflexible. Human language, in contrast, has enormous flexibility that enables us to talk about virtually any topic, even highly abstract ones like psychological constructs. In this way, our thinking is carried beyond a rigid here and now.

> Before you accept the linguists' position, review Focus 1-2, featuring the chimp Kanzi.

Neurologist Oliver Sacks illustrates the importance of syntax to human thinking in his description of Joseph, an 11-year-old deaf boy who was raised without sign language for his first 10 years and so was never exposed to syntax. According to Sacks:

> Joseph saw, distinguished, used; he had no problems with perceptual categorization or generalization, but he could not, it seemed, go much beyond this, hold abstract ideas in mind, reflect, play, plan. He seemed completely literal—unable to juggle images or hypotheses

cell assembly Hypothetical group of neurons that become functionally connected because they receive the same sensory inputs. Hebb proposed that cell assemblies were the basis of perception, memory, and thought.

Figure 11-2 diagrams the frontal-lobe hierarchy that initiates a motor sequence.

or possibilities, unable to enter an imaginative or figurative realm. . . . He seemed, like an animal, or an infant, to be stuck in the present, to be confined to literal and immediate perception. (Sacks, 1989, p. 40)

Language, including syntax, develops innately in children because the human brain is programmed to use words in a form of universal grammar. However, in the absence of words—either spoken or signed—no grammar can develop. Without the linguistic flexibility that grammar allows, no "higher-level" thought can emerge. Without syntactical language, thought is stuck in the world of concrete, here-and-now perceptions. Syntax, in other words, influences the very nature of our thinking.

In addition to arranging words in syntactical patterns, the human brain has a passion for stringing together events, movements, and thoughts. We combine musical notes into melodies, movements into dance, images into videos. We design elaborate rules for games and governments. To conclude that the human brain is organized to chain together events, movements, and thoughts seems reasonable. Syntax is merely one example of this innate human way of thinking about the world.

We do not know how this propensity to string things together evolved, but one possibility involves natural selection. Stringing movements together into sequences can be highly adaptive. It would allow for building houses or weaving threads into cloth, for instance.

William Calvin (1996) proposed that the motor sequences most important to ancient humans were those used in hunting. Throwing a rock or a spear at a moving target is a complex act that requires much planning. Sudden ballistic movements, such as throwing, last less than an eighth of a second and cannot be corrected by feedback. The brain has to plan every detail of these movements and then spit them out as a smooth-flowing sequence.

Today, a football quarterback does just this when he throws a football to a receiver running a zigzag pattern to elude a defender. A skilled quarterback can hit the target on virtually every throw, stringing his movements together rapidly in a continuous sequence with no pauses or gaps. This skill is unique to humans. Chimpanzees can throw objects, but their throws are not accurate. No chimpanzee could learn to throw a ball to hit a moving target.

The human predisposition to sequence movements may have encouraged language development. Spoken language, after all, is a sequence of movements involving the tongue and mouth. Viewed in this way, language is the by-product of a brain that was already predisposed to operate by stringing movements, events, or even ideas, together.

A critical characteristic of human motor sequencing is our ability to create novel sequences with ease. We constantly produce new sentences. Composers and choreographers earn their livings creating new music and dance sequences. Creating novel sequences of movements or thoughts is a function of the frontal lobes.

People with frontal-lobe damage have difficulty generating novel solutions to problems. They are described as lacking imagination. The frontal lobes are critical not only to organizing behavior but also to organizing thinking. One major difference between the human brain and other primates' brains is the size of the frontal lobes.

Neural Unit of Thought

What exactly goes on within the brain to produce what we call thinking? Is thought an attribute exclusive to humans? Before you answer, consider the mental feats of Alex the parrot, profiled in Comparative Focus 15-2, "Animal Intelligence."

Alex's cognitive abilities are unexpected in a bird. In the past 40 years, the intellectual capacities of chimpanzees and dolphins have provoked great interest, but Alex's mental life appears to have been just as rich as the mental life of those two large-brained mammals.

The fact that birds are capable of thought is a clue to the neural basis of thought. A logical presumption may be that thinking, which humans are so good at, must be due neuron. If at some point the random activity of the dots increases to a level that obscures movement in a neuron's preferred direction, the neuron will stop responding because it does not detect any consistent pattern.

So the question becomes How does the activity of any given neuron correlate with the perceptual threshold for apparent motion? On the one hand, if our perception of apparent motion results from the summed activity of many dozens or even thousands of neurons, little correlation would exist between the activity of any one neuron and the perception. On the other hand, if our perception of apparent motion is influenced by individual neurons, then a strong correlation would exist between the activity of a single cell and the perception.

The results of Experiment 15-1 are unequivocal: the sensitivity of individual neurons is very similar to the perceptual sensitivity of the monkeys to apparent motion. As shown in the Results section, if individual neurons failed to respond to the stimulus, the monkeys behaved as if they did not perceive any apparent motion.

This finding is curious. Given the large number of V5 neurons, it seems logical that perceptual decisions are based on the responses of a large pool of neurons. But Newsome's results show that the activity of individual cortical neurons is correlated with perception rather than perception being the property of a particular brain region.

Still, Hebb's idea of a cell assembly—an ensemble of neurons that represents a complex concept—suggests some way of converging the inputs of individual neurons to arrive at a consensus. Here, the neuronal ensemble represents a sensory event (apparent motion) that the activity of the ensemble detects. Cell assemblies could be distributed over fairly large regions of the brain or they could be confined to smaller areas, such as cortical columns.

Figures 9-33 through 9-35 diagram functional columns in the visual and temporal cortices.

Via computer modeling, cognitive scientists have demonstrated the capacity of cell assembly circuits to perform sophisticated statistical computations. Other complex tasks, such as Alex the parrot's detecting an object's color, also are believed to entail neuronal ensembles. Cell assemblies provide the basis for cognition. Different ensembles come together, much like words in language, to produce coherent thoughts.

What do individual neurons contribute to a cell assembly? Each acts as a computational unit. As Experiment 15-1 shows, even one solitary neuron can decide on its own when to fire if its summed inputs indicate that movement is taking place. Neurons are the only elements in the brain that combine evidence and make decisions. Neurons are the foundation of cognitive processes and of thought. The combination of individual neurons into novel neural networks produces complex mental representations—ideas, for instance.

Section 4-3 describes how individual neurons integrate information by summing up their inputs.

REVIEW 15-1
Nature of Thought

Before you continue, check your understanding.

1. Cognition, or thought, entails the abilities of _____, _____, and _____ stimuli.

2. Unlike thought in other animals, humans have the added advantage of _____, which adds _____ to thought.

3. The _____ is the basic unit of thought.

4. Describe the most important way in which human thought differs from thinking in other animals.

Answers appear at the back of the book.

15-2 Cognition and the Association Cortex

All together, the primary sensory and motor cortical regions occupy about a third of the neocortex **(Figure 15-1)**. The remaining two thirds, located in the frontal, temporal, parietal, and occipital lobes, is referred to generally as the **association cortex.** Its function is to produce cognition.

A fundamental difference between the association cortex and the primary sensory and motor areas is that the association cortex has a distinctive pattern of connections. A major source of input to all cortical areas is the thalamus, which rests atop the brainstem. The primary sensory cortex receives inputs from thalamic nuclei that receive information from the body's sense organs. But inputs to the association cortex come from thalamic areas that receive their inputs from other regions in the cortex.

As a result, inputs to the association cortex are already highly processed. So this information must be fundamentally different from the raw information reaching the primary sensory and motor cortex. The association regions contain knowledge either about our external or internal world or about movements.

Owing to its close relationship to the visual and auditory sensory regions, the temporal association regions tend to produce cognition related to visual and auditory processing. Similarly, the parietal cortex is closely related to somatosensation and movement control. In contrast, the frontal cortex coordinates information coming from the parietal and temporal association regions with information coming from subcortical regions.

As diagrammed in **Figure 15-2,** the multiple subdivisions of the *prefrontal cortex* (PFC) encompass its dorsal, lateral, orbital frontal, and medial regions. Activity in each of these prefrontal regions is associated with the different types of cognitive processing that we describe throughout this chapter. An additional frontal-lobe region shown in Figure 15-2 is the anterior cingulate cortex (ACC). Once believed only to play a role in

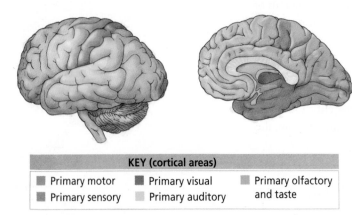

KEY (cortical areas)

■ Primary motor	■ Primary visual	■ Primary olfactory and taste
■ Primary sensory	■ Primary auditory	

FIGURE 15-1 Cortical Functions Lateral view of the left and medial view of the right hemisphere, showing the primary motor and sensory areas. All remaining cortical areas are collectively referred to as the association cortex, which functions in thinking.

FIGURE 15-2 Prefrontal Association Cortex **(A)** Lateral view of the left hemisphere, **(B)** medial view of the right hemisphere, and **(C)** ventral view of the brain represent regions of the prefrontal cortex (PFC) in relation to the associated anterior cingulate (limbic) cortex, shown in the medial view. Figure 12-19 illustrates how these areas function in producing emotional behavior.

(A) Lateral view

Dorsolateral prefrontal cortex

Orbital prefrontal cortex

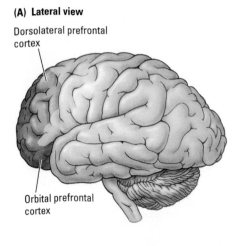

(B) Medial view

Dorsomedial prefrontal cortex

Anterior cingulate cortex

Ventromedial prefrontal cortex

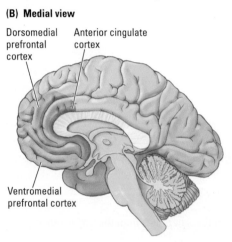

(C) Ventral view

Ventromedial prefrontal cortex

Orbital prefrontal cortex

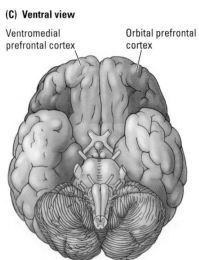

emotion, it is becoming clear that the ACC functions as an interface between emotion and cognition.

To understand the types of knowledge that the association areas contain, we next consider different forms of cognitive behavior and then trace these behaviors to different parts of the association cortex.

Knowledge about Objects

Visualize a milk carton sitting on a counter directly in front of you. What do you see? Now imagine moving the carton a few inches off to one side as you continue to stare directly ahead. What do you see now? Next, imagine that you tilt the carton toward you at a 45-degree angle. Again, what do you see? Probably you answered that you saw the same thing in each situation: a rectangular box with lettering on it.

Intuitively, you feel that the brain must "see" the object much as you have perceived it. The brain's "seeing," however, is more compartmentalized than are your perceptions. Compartmentalization is revealed in people who suffer damage to different regions of the occipital cortex and often lose one particular aspect of visual perception. For instance, those with damage to visual area V4 can no longer perceive color, whereas those with damage to area V5 can no longer see movement (when the milk carton moves, it becomes invisible to them).

Moreover, your perception of the consistently rectangular shape of the milk carton does not always match the forms that your visual system is processing. When you tip the carton toward you, you still perceive it as rectangular, even though it is no longer presenting a rectangular shape to your eyes. Your brain has somehow ignored the change in information about shape that your retinas have sent it and concluded that this shape is still the same milk carton.

There is more to your conception of the milk carton than merely perceiving and processing its physical characteristics. You also know what a milk carton is, what it contains, and where you can get one. The knowledge about milk cartons that you have acquired is represented in the temporal association cortex that forms the ventral stream of visual processing. If the temporal association regions are destroyed, a person loses visual knowledge not only about milk cartons but also about all other objects. The person becomes *agnosic* (unknowing).

Knowledge about objects includes even more than how they look and what they are used for. It depends on what will be done with the information—how to pick up the milk carton, for example. Knowledge of *what* things are is temporal; knowledge of *how* to grasp the object is parietal **(Figure 15-3)**.

Multisensory Integration

Our knowledge about information in the world comes through multisensory channels. We see and hear a barking dog, and the visual information and auditory information fit together seamlessly. How do all our different neural systems and functional levels combine to afford us a unified conscious experience?

Philosophers, impressed with this integrative capacity, identified the **binding problem,** which asks how the brain ties its single and varied sensory and motor events together into a unified perception or behavior. Gradually, it is becoming clear how the brain binds up our perceptions and how this ability is gradually acquired in postnatal life (see the review by Stein and Rowland, 2011).

One solution to the sensory integration aspect of the binding problem lies in regions of the association cortex that are *multimodal,* that is, populated by neurons that respond to information from more than one sensory modality, as illustrated in **Figure 15-4.** Investigators presume that multimodal regions combine characteristics of stimuli across

association cortex Neocortex outside the primary sensory and motor cortices that functions to produce cognition.

binding problem Philosophical question focused on how the brain ties single and varied sensory and motor events together into a unified perception or behavior.

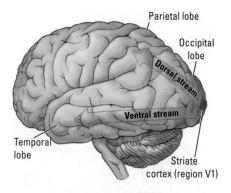

FIGURE 15-3 Streaming Visual Information The dorsal visual stream mediates vision for action. The ventral stream mediates vision for object recognition.

Section 9-4 recounts cases illustrative of various visual-form agnosias.

FIGURE 15-4 Multisensory Areas in the Monkey Cortex Colored areas represent regions where anatomical data and/or electrical stimulation demonstrate multisensory interactions. Dashed lines represent multimodal areas revealed when sulci are opened. Adapted from "Is Neocortex Essentially Multisensory?" by A. A. Ghanzanfar and C. E. Schroeder, 2006, in *Trends in Cognitive Science*, 10, pp. 278–285.

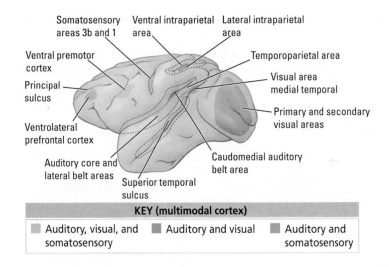

Section 12-2 explains how the senses of smell and taste combine to create the experience of flavor.

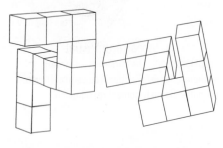

FIGURE 15-5 Spatial Cognition These two figures are the same, but they are oriented differently in space. Researchers test spatial cognition by giving subjects pairs of stimuli like this and asking if the shapes are the same or different.

The Basics in Section 1-3 traces the evolution of the nervous system across the animal kingdom.

Alex the parrot manipulated objects with his beak.

different senses when we encounter them, separately or together. For example, the fact that we can visually identify objects that we have only touched implies a common perceptual system linking the visual and somatic circuits. Clinical Focus 15-6 profiles a man with a rare capacity: stimulation in one sensory modality (in this case, taste) concurrently induces experience in a different modality (in this case, touch).

Spatial Cognition

The location of objects is just one aspect of what we know about space. *Spatial cognition* refers to a whole range of mental functions that vary from navigational ability (getting from point A to point B) to the mental manipulation of complex visual arrays like those shown in **Figure 15-5**.

Imagine going for a walk in an unfamiliar park. You do not go around and around in circles. Rather, you proceed in an organized, systematic way. You also need to find your way back. These abilities require a representation of the physical environment in your mind's eye.

At some time during the walk, let's assume that you are uncertain of where you are—a common problem. One solution is to create a mental image of your route, complete with various landmarks and turns. It is a small step from mentally manipulating these kinds of navigational landmarks and movements to manipulating other kinds of images in your mind. Thus the ability to mentally manipulate visual images seems likely to have arisen in parallel with the ability to navigate in space.

The evolution of skill in mental manipulation is also closely tied to the evolution of physical movements. It is likely that animals first moved by using whole-body movements (the swimming motion of a fish), then developed coordinated limb movements (quadrupedal walking), and finally mastered discrete limb movements, such as the reaching movement of a human arm. As the guidance strategies for controlling movements became more sophisticated, cognitive abilities evolved to support the guidance systems.

It seems unlikely that more sophisticated cognitive abilities evolved on their own. Why would a fish, say, be able to manipulate an object in its mind that it could not manipulate in the real world? But a human, who *can* manipulate objects by hand, should be able to imagine such manipulations. After all, we are constantly observing our hands manipulating things: we must have many mental representations of such activities.

Once the brain can manipulate objects that are physically present, it seems a small step to manipulating objects that are only imagined—to solve problems like the one depicted in **Figure 15-6.** The ability to manipulate an object in the mind's eye probably flows from the ability to manipulate tangible objects with the hands.

Research findings provide clues to the brain regions participating in various aspects of spatial cognition. For instance, the dorsal stream in the parietal lobes plays a central role in controlling vision for action. Discrete limb movements are made to points in space, so a reasonable supposition is that the evolutionary development of the dorsal stream provided a neural basis for such spatial cognitive skills as the mental rotation of objects. In fact, people with damaged parietal association regions, especially in the right hemisphere, have deficits in processing complex spatial information, both in the real world and in their imaginations.

If we trace the evolutionary development of the human brain, we find that the parietal association regions expanded considerably more in humans than in other primates. This expanded brain region functions, in part, to perform the complex spatial operations just discussed. Humans have a capacity for building that far exceeds that of our nearest relative, the chimpanzee. A long leap of logic may be required in making the assertion, but perhaps our increased capacity for building and manipulating objects played an important role in developing our cognitive spatial abilities.

Attention

Imagine that you're meeting some friends at a football game. You search for them as you meander through the crowd in the stadium. Suddenly, you hear the distinctive laugh of one friend, and you turn to scan in that direction. You see your group and rush to join them.

This everyday experience demonstrates the nature of **attention,** selective narrowing or focusing of awareness to part of the sensory environment or to a class of stimuli. Even as you are bombarded by sounds, smells, feelings, and sights, you can still detect a familiar laugh or spot a familiar face: you can direct your attention.

More than 100 years ago, William James (1890) defined attention: "It is the taking possession by the mind in clear and vivid form of one out of what seem several simultaneous objects or trains of thought." James's definition goes beyond our example of locating friends in a crowd, inasmuch as he notes that we can attend selectively to thoughts as well as to sensory stimuli. Who hasn't at some time been so preoccupied with a thought as to exclude all else from mind? So attention can be directed inward as well as outward.

Selective Attention

Like many other inferred mental processes, the neural basis of attention is particularly difficult to study. However, research with monkeys has identified neurons in the cortex and midbrain that show enhanced firing rates to particular locations or visual stimuli to which the animals have been trained to attend. Significantly, the same stimulus can activate a neuron at one time but not at another, depending on the monkey's learned focus of attention.

In the study shown in **Experiment 15-2** on page 534, James Moran and Robert Desimone (1985) trained monkeys to hold a bar while gazing at a fixation point on a screen. A sample stimulus (e.g., a vertical red bar) appeared briefly at one location in the visual field, followed about 500 milliseconds later by a test stimulus at the same location. When the test stimulus was identical with the initial sample stimulus, an animal was rewarded if it immediately released the bar.

Each animal was trained to attend to stimuli presented in one particular area of the visual field and to ignore stimuli in any other area. In this way, the same visual stimulus could be presented to different regions of a neuron's receptive field to test whether the cell's response varied with stimulus location.

As the animals performed the task, the researchers recorded neurons firing in visual area V4. Neurons in V4 are sensitive to color and form, and different neurons respond to different combinations of these two variables (e.g, a red vertical bar or a

attention Selective narrowing or focusing of awareness to part of the sensory environment or to a class of stimuli.

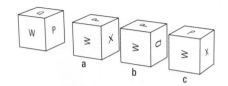

FIGURE 15-6 Mental Manipulation Try this sample test item used to measure spatial orientation. Compare the three cubes on the right with the one on the left. No letter appears on more than one face of a given cube. Which cube—a, b, or c—could be a different view of the cube on the left? The correct answer is at the bottom of the page.

The answer to the mental manipulation in Figure 15-6 is a.

Question: Can neurons learn to respond selectively to stimuli?

Procedure

Monkeys were trained to release a bar when a certain stimulus was presented in a certain location. The monkeys learned to ignore stimuli in all other locations.

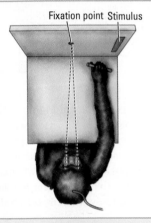

Fixation point Stimulus

Results

During performance of this task, researchers recorded the firing of neurons in visual area V4, which are sensitive to color and form. Stimuli were presented in either rewarded or unrewarded locations.

Pretraining recordings:

Rewarded location

Strong response

Unrewarded location

Strong response

Before training, neurons responded to stimuli in all locations.

Posttraining recordings:

Rewarded location

Strong response

Unrewarded location

Baseline response

After training, neurons responded only when the visual stimuli were in the rewarded location.

Conclusion: Neurons can learn to respond selectively to information in their receptive field.

contralateral neglect Ignoring a part of the body or world on the side opposite (contralateral to) that of a brain injury.

extinction In neurology, neglect of information on one side of the body when it is presented simultaneously with similar information on the other side of the body.

green horizontal bar). Visual stimuli were presented either in the correct location for a reward or in an incorrect location for no reward.

As diagrammed in the Results section of Experiment 15-2, neurons responded only when a visual stimulus was in the correct location, even though the same stimulus was presented in the incorrect location. Before training, the neurons responded to all stimuli in both locations. This finding tells us that the ability to attend to specific parts of the sensory world is a property of single neurons. Once again, we see that the neuron is the computational unit of cognition.

Deficits of Attention

Attention is probably a property of neurons throughout the brain, with some regions playing a more central role than others. The frontal lobes, for instance, play a very important part in attention. People with frontal-lobe injuries tend to become overly focused on environmental stimuli. They seem to selectively direct attention to an excessive degree or to have difficulty shifting attention. The results of studying these people suggest that the frontal association cortex plays a critical role in the ability to flexibly direct attention where it is needed. Indeed, planning, a key frontal-lobe function, requires this ability.

The parietal association cortex plays a key role in other aspects of attention. This role is perhaps best illustrated by studying the attention deficit referred to as *neglect*. Neglect occurs when a brain-injured person ignores sensory information that should be considered important. Usually the condition affects only one side of the body, in which case it is called **contralateral neglect. Figure 15-7** shows contralateral neglect in a dog that would eat food from only the right side of its dish. Neglect is a fascinating symptom because it often entails no damage to sensory pathways. Rather, neglect is a failure of attention.

People with damage to the parietal association cortex of the right hemisphere may have particularly severe neglect of objects or events in the left side of their world. For example, one man dressed only the right side of his body, shaved only the right side of his face, and read only the right side of a page (if you can call that reading). He could move his left limbs spontaneously, but when asked to raise both arms, he would raise only the right. When pressed, he could be induced to raise the left arm, but he quickly dropped it to his side again.

As people with contralateral neglect begin to recover, they show another interesting symptom. They neglect information on one side of the body when it is presented simultaneously with similar information on the other side of the body. **Figure 15-8** shows a common clinical test for this symptom, called **extinction.**

In an extinction test, the patient is asked to keep his or her eyes fixed on the examiner's face and to report objects presented in one or both sides of the visual field. When presented with a single object (a fork) to one side or the other, the patient orients himself or herself toward the appropriate side of the visual field, so we know that he or she cannot be blind on either side. But now suppose that two forks are presented, one on the left and one on the right. Curiously, the patient ignores the fork on the left and reports that there is one on the right. When asked about the left side, the patient is quite certain that nothing appeared there and that only one fork was presented, on the right.

Perhaps the most curious aspect of neglect is that people who have it fail to pay attention not only to one side of the physical world around them but also to one side of the world represented in their minds. We studied one woman who had complete neglect for everything on her left side. She complained that she could not use her kitchen because she could never remember the location of anything on her left.

We asked her to imagine standing at the kitchen door and to describe what was in the various drawers on her right and left. She could not recall anything on her left. We then asked her to imagine walking to the end of the kitchen and turning around. We again asked her what was on her right, the side of the kitchen that had previously been on her left. She broke into a big smile and tears ran down her face as she realized that she now knew what was on that side of the room. All she had to do was reorient her body in her mind's eye. She later wrote and thanked us for changing her life, because she was now able to cook again. Clearly, neglect can exist in the mind as well as in the physical world.

Although complete contralateral neglect is normally associated with parietal-lobe injury, specific forms of neglect can arise from other injuries. Ralph Adolphs and his colleagues (2005) describe the case of S. M., a woman with bilateral amygdala damage who could not recognize fear in faces. On further study, the reason was discovered: S. M. failed to look at the eyes when she looked at faces; instead, she looked at other facial features such as the nose. Because fear is most clearly identified in the eyes, not the nose, she did not identify the emotion. When she was specifically instructed to look at the eyes, her recognition of fear became entirely normal. Thus, the amygdala plays a role in directing attention to the eyes to identify facial expressions.

Planning

At noon on a Friday, a friend proposes that she and you go to a nearby city for the weekend to attend a concert. She will pick you up at 6:00 P.M. and you will drive there together.

Because you are completely unprepared for this invitation and because you are going to be busy until 4:00, you must rush home and get organized. En route you stop at a fast food restaurant so that you won't be hungry on the 2-hour drive. You also need cash, so you head to the nearest ATM. When you get home, you grab various pieces of clothing appropriate for the concert and the trip. You also pack your toiletries. You somehow manage to get ready by 6:00, when your friend arrives.

Although the task of getting ready in a hurry may make us a bit harried, most of us can manage it. People with frontal-lobe injury cannot. To learn why, let's consider what the task requires.

1. To plan your behavior, you must select from many options. What do you need to take with you? Cash? Then which ATM is closest and what is the quickest route to it? Are you hungry? Then what is the fastest way to get food on a Friday afternoon?

2. In view of your time constraint, you have to ignore irrelevant stimuli. If you pass a sign advertising a sale in your favorite store, for instance, you have to ignore it and persist with the task at hand.

3. You have to keep track of what you have done already, a requirement especially important while you are packing. You do not want to forget items or pack duplicates. You do not want to take four pairs of shoes but no toothbrush.

The general requirements of this task can be described as the temporal (time) organization of behavior. You are planning what you need to do and when you need to do it. This kind of planning is the general function of the frontal lobes, especially the prefrontal cortex.

FIGURE 15-7 **Contralateral Neglect in a Dog** This dog had a right hemisphere brain tumor and would eat the food on the right side of its dish but ignore food on the left side.

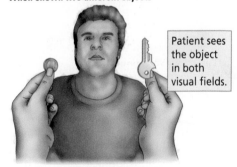

When shown two identical objects

Patient's right visual field

Patient's left visual field

Patient sees only the object in his right visual field.

When shown two different objects

Patient sees the object in both visual fields.

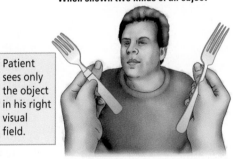

When shown two kinds of an object

Patient sees only the object in his right visual field.

FIGURE 15-8 **Testing for Extinction** A stroke patient who shows neglect for information presented to his left responds differently depending on whether objects in the left and right visual fields are similar or different.

FIGURE 15-9 Wisconsin Card Sorting Test The subject's task is to place each card in the pile under the appropriate card in the row, sorting by one of three possible categories. Subjects are never explicitly told what the correct sorting category is—color, number, or form; they are told only whether their responses are correct or incorrect. After subjects have begun sorting by one category, the tester unexpectedly changes to another category.

But to perform the task of planning, you also need to recognize objects (an occipital- and temporal-lobe function) and to make appropriate movements with respect to them (a parietal-lobe function). You can therefore think of the frontal lobes as acting like an orchestra conductor. The frontal lobes make and read some sort of motor plan to organize behavior in space and time—a kind of "motor score," analogous to the musical score a conductor uses. People with frontal-lobe injuries are simply unable to organize their behavior.

Performance on the Wisconsin Card Sorting Test exemplifies the kinds of deficits frontal-lobe injury creates. **Figure 15-9** shows the testing materials. The subject is presented with the four stimulus cards arrayed at the top. These cards bear designs that differ in color, form, and number of elements, thus creating three possible sorting categories to be used in the task. The subject must sort a deck of cards into piles in front of the various stimulus cards, depending on the sorting category called for. But the correct sorting category is never stated. The subject is simply told after placing each card whether the choice is correct or incorrect.

In one trial, for example, the first correct sorting category is color. After the subject has sorted a number of cards by color, the correct solution switches, without warning, to form. When the subject has started to sort by form, the correct solution again changes unexpectedly, this time to the number of items on each card. The sorting rule later becomes color again, and so on, with each change in rule coming unannounced.

Shifting response strategies is particularly difficult for people with frontal-lobe lesions: they may continue responding to the original stimulus (color) for as many as 100 cards until the test ends. This pattern, known as **perseveration,** is the tendency to emit repeatedly the same verbal or motor response to varied stimuli.

Frontal-lobe subjects may even comment that they know that color is no longer the correct category, but they continue to sort by color. One stated: "Form is probably the correct solution now so this [sorting by color] will be wrong, and this will be wrong, and wrong again." Despite knowing the correct sorting category, the frontal-lobe patient cannot shift behavior in response to the new external information.

Imitation and Understanding

In all communication—both verbal and nonverbal—the sender and receiver must have a common understanding of what counts. If a person speaks a word or makes a gesture, another person will understand only if he or she interprets it correctly. To accomplish this coordination in communication, the processes of producing and perceiving a message must share a common representation in the brains of the sender and the receiver.

How do both the sender and the receiver of a potentially ambiguous gesture, such as a raised hand or a faint smile, achieve a common understanding of what the gesture means? Giacomo Rizzolatti and his colleagues (Rizzolatti, 2007; Rizzolatti & Craighero, 2004) proposed an answer to these questions. In the frontal lobes of monkeys, they identified neurons that discharge during active movements of the hand or mouth or both. These neural discharges do not precede the movements but instead occur in synchrony with them. But it takes time for a neural message to go from a frontal lobe to a hand, so we would predict that, if these cells are controlling the movements, they will discharge *before* the movements take place. The cells must therefore be recording a movement that is taking place.

In the course of his studies, Rizzolatti also found that many "movement" neurons located in the inferior frontal and posterior parietal cortex discharge when a monkey sees other monkeys make the same movements. They also discharge when the monkey sees the experimenter make the movements. Rizzolatti called them **mirror neurons.** The researchers proposed that mirror neurons represent actions, one's own or those of others. Such neural representations could be used both for imitating others' actions and for understanding their meaning, thus enabling appropriate responses. Mirror neurons therefore provide the link between the sender and the receiver of a communication.

Mirror neurons, introduced in Section 11-1 as part of the frontal-lobe system that initiates movement, respond not only to objects but also to specific observed actions.

Rizzolatti and his colleagues used PET to look for these same neuron populations in humans. Participants were asked to watch a movement, to make the same movement, or to imagine the movement. In each case, a region of the lateral frontal lobe in the left hemisphere, including Broca's area, was activated. Taken together with the results of the monkey studies, this finding suggests that primates have a fundamental mechanism for recognizing action. People apparently recognize actions made by others because the neural patterns produced when they observe those actions are similar to those produced when they themselves make those same actions.

According to Rizzolatti, the human capacity to communicate with words may have resulted from a progressive evolution of the mirror-neuron system observed in the monkey brain. After all, mimicking behaviors, such as dancing and singing, is central to human culture. Evolving the capacity for mimicry was perhaps the precursor to the evolution of language. For language, the same neurons would recognize words spoken by others and produce the same words in speech. Research Focus 15-3, "Consequences of Mirror-Neuron Dysfunction," elaborates on the implications of mirror neurons.

Section 7-4 reviews how positron emission tomography works and how neuroscientists use it in research.

perseveration Tendency to emit repeatedly the same verbal or motor response to varied stimuli.

mirror neuron Cell in the primate premotor cortex that fires when an individual observes a specific action taken by another individual.

RESEARCH FOCUS ✦ 15-3

Consequences of Mirror-Neuron Dysfunction

Mirror neurons, located in the inferior frontal cortex and the posterior parietal cortex, are believed to form a neural system for imitation. The original studies by Giacomo Rizzolatti and his colleagues demonstrate that mirror neurons fire in response to seeing specific movements, or, if the intention of the movement is clear—to retrieve food, for example— even in response to seeing only the beginning of the movement (Rizzolatti & Craighero, 2004).

Investigators also wondered whether mirror neurons might be important in social cognition. People tend to imitate one another when interacting socially. This imitation may form a basis for developing empathy.

Catherine Carr and her colleagues (2003) showed that the mirror-neuron system is activated when people observe or imitate emotional faces. The investigators conclude that mirror-neuron activation allows us to construct an "inner imitation" of others' actions, such as facial expressions and body postures. Essentially, mirror neurons allow us to understand the mental states other people experience.

Marco Iacoboni and Mirella Dapretto (2006) hypothesized that, given the key role of imitation and mirror neurons in social cognition, a dysfunctional mirror-neuron network could lead to deficits in social behavior. A prime example is autism spectrum disorder (ASD).

Iacoboni and Dapretto predicted that the cortical regions containing mirror neurons would show delayed development in individuals with

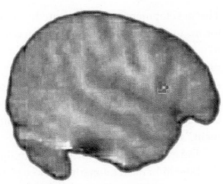

Mirror-neuron-system activity imaged during testing of observation and imitation of emotional expressions. Typically developing children show greater right frontal activity (red and yellow area) than do those with ASD. **From "Understanding Emotions in Others: Mirror Neuron Dysfunction in Children with Autism. Spectrum Disorders," by M. Dapretto et al., 2006, in *Nature Neuroscience, 9*, pp. 28–30.**

ASD and might show reduced or absent activity during imitation. Dapretto and her colleagues tested the hypothesis by measuring fMRI activity in children as they imitated facial expressions.

As shown in the figure, individuals with ASD had markedly reduced right frontal activity during imitation that correlates with the severity of the disorder: the greater the ASD symptoms, the lower the mirror-neuron system activity. The ASD–mirror-neuron relationship has been called the "broken mirror" hypothesis of autism (Ramachandran & Oberman, 2006).

Rizzolatti and Fabbri-Destro (2010) summarized more recent studies on the broken mirror hypothesis. They conclude that, in autism, the mirror system is silent during the observation of actions. Those with ASD therefore cannot grasp others' intentions directly.

This conclusion can be appreciated in a study by Sonia Boria and colleagues (2009). They showed typically developing children and children with ASD photographs of goal-directed motor acts (e.g., grasping a mug; picking up a pair of scissors). Then they asked the children to report both *what* the actor was doing and *why* he was doing it.

Both groups of children recognized what the actor was doing, but children with autism were impaired at understanding the why. Regardless of the actor's hand shape, these children inferred that the why was the typical use for the object. For example, the sight of a hand picking up a mug triggered the response, "for drinking," even when the observed handgrip made this action implausible.

cognitive neuroscience Study of the neural bases of cognition.

brain connectome Map of the complete structural and functional neural connections of the human brain in vivo.

A major difference between humans and monkeys is that the mirror neurons are localized in the human left hemisphere. This unilateral representation may be significant for understanding how language is organized in the brain. If the abilities to mimic and to understand gestures were present before language developed and if the neural circuits for these abilities became lateralized, then language would also have become lateralized: the system on which it is based already existed in the left hemisphere.

REVIEW 15-2
Cognition and the Association Cortex

Before you continue, check your understanding.

1. The association cortex contains _____ and functions to produce _____.

2. As a general rule, the _____ lobes generate knowledge about objects, whereas the _____ lobes produce various forms of spatial cognition.

3. The frontal lobes function not only to make movements but also to _____ and to _____.

4. _____ neurons in the frontal and parietal lobes represent actions, one's own or those of others.

5. Describe the function of multimodal cortex.

Answers appear at the back of the book.

15-3 Expanding Frontiers of Cognitive Neuroscience

The development of sophisticated, noninvasive stimulation and recording techniques for measuring the brain's electrical activity and noninvasive brain imaging methods led to a major shift in the study of brain and behavior: **cognitive neuroscience,** the field that studies the neural bases of cognition. Cognitive neuroscience focuses on high-tech research methods but continues to rely on the decidedly low-tech tools of neuropsychological assessment—behavioral tests that compare the effects that injuries to different brain regions have on particular tasks. Clinical Focus 15-4, "Neuropsychological Assessment," illustrates its benefits.

Today, sophisticated imaging techniques are helping cognitive neuroscientists map the human brain. The methods of cognitive neuroscience are assisting social psychologists discover how the brain mediates social interactions and helping economists discover how the brain makes decisions.

Mapping the Brain

Figure 7-16 shows virtual nerve tracts imaged by DTI, and Section 7-4 reviews dynamic imaging, including fMRI-based methods.

An important step in identifying the neural bases of cognition is to describe the connections in the cerebral cortex. Two promising imaging tools for mapping the human brain's connectivity are diffusion tensor imaging (DTI) and functional connectivity magnetic resonance imaging (fcMRI). While each has strengths in different applications, both techniques are allowing researchers to develop a comprehensive map, sometimes referred to as the **brain connectome,** of functional connections in the living brain.

Using fMRI in Brain Mapping

The fcMRI technique uses resting-state fMRI (rs-fMRI) to measure functional correlations between brain regions. Pooling rs-fMRI data across thousands of healthy young adults makes it possible to identify consistent patterns of connectivity, or nerve tracts, in

CLINICAL FOCUS ⊕ 15-4

Neuropsychological Assessment

In an age of "high-tech" procedures such as PET, fMRI, and ERP, "low-tech" behavioral assessment continues to be one of the best, simplest, and most economical ways to measure cognitive function.

To illustrate the nature and power of neuropsychological assessment, we compare the test performance of three patients on an array of tests from among those used in a complete neuropsychological assessment. The five tests presented here measure verbal and visual memory, verbal fluency, abstract reasoning, and reading. Performance was compared with that of a normal control participant.

In the delayed memory tests—one verbal, the other visual—patients were read a list of words and two short stories. They were also shown a series of simple drawings. Their task was to repeat the words and stories immediately after hearing them and to draw the simple figures.

Half an hour later, without warning, they were asked to perform the task again. Their performances on the delayed tests yielded the delayed verbal and visual memory scores listed in the table.

Subjects' Scores

Test	Control	J. N.	E. B.	J. W.
Delayed verbal memory	17	9*	16	16
Delayed visual memory	12	14	8*	12
Verbal fluency	62	62	66	35*
Card-sorting errors	9	10	12	56*
Reading	15	21	22	17

*Abnormally poor score.

In the verbal fluency test, patients had 5 minutes to write down as many words as they could think of that start with the letter *s,* excluding numbers and people's names. Then came the Wisconsin Card Sorting Test, which assesses abstract reasoning (see Figure 15-9). Finally, the patients were given a reading test.

The first patient, J. N., was a 28-year-old man who had developed a tumor in the anterior and medial part of the left temporal lobe. Preoperative psychological tests showed this man to be of superior intelligence. His only significant deficits appeared on tests of verbal memory.

When we saw J. N. a year after surgery that successfully removed the tumor, he had returned to his job as a personnel manager. His intelligence was still superior, but as the accompanying score summary shows, he was still impaired on the delayed verbal memory test, recalling only about 50 percent as much as the control participants and other subjects did.

The second patient, E. B., was a college senior majoring in psychology. An aneurysm in her right temporal lobe had burst, and the anterior part of that lobe had been removed. E. B. was of above-average intelligence and completed her bachelor of arts degree with good grades. Her score on the delayed visual memory test, where she recalled just over half of what the other subjects did, clearly showed her residual deficit.

The third patient, also of above-average intelligence, was J. W., a 42-year-old police detective who had earned a college degree. A benign tumor had been removed from his left frontal lobe.

We saw J. W. 10 years after his surgery. He was still on the police force but at a desk job. His verbal fluency was markedly reduced, as was his ability to solve the card-sorting task. His reading skill, however, was unimpaired. This was also true of the other patients.

Two general principles emerge from the results of these three neuropsychological assessments:

1. *Brain functions are localized to different cerebral regions.* Thus, damage to different parts of the brain produces different symptoms.

2. *Brain organization is asymmetrical.* Left-hemisphere damage preferentially affects verbal functions, whereas right-hemisphere damage preferentially affects nonverbal functions.

---

the brain. Utilizing rs-fMRI data from 1000 participants, Thomas Yeo and colleagues (2011) parcellated the human cerebral cortex into the 17 networks illustrated in **Figure 15-10.**

The cerebral cortex is made up of primary sensory and motor networks as well as the multiple large-scale networks that form the association cortex. The sensory and motor networks are largely local: adjacent areas tend to show strong functional coupling with one another. In Figure 15-10, the turquoise and blue/gray regions in somatosensory and motor cortex and the purple region in visual cortex illustrate these couplings.

In contrast, the association networks include areas distributed throughout the prefrontal, parietal,

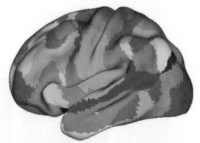 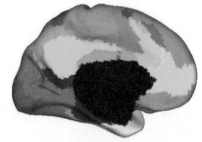

Left hemisphere, lateral view   Left hemisphere, Medial view

FIGURE 15-10 Parcellation of Cerebral Cortical Networks An estimate of 17 cortical networks based on fcMRI data from 1000 participants. Each color represents a network. Some, such as the blue auditory areas in temporal lobe, are localized; others are widely distributed, such as the yellow regions, which reveal prefrontal–posterior parietal connectivity. Image courtesy of Thomas Yeo et al., 2011, *Journal of Neurophysiology, 106,* pp. 1125–1165.

anterior temporal, and midline regions. In Figure 15-10, the distributed yellow regions show prefrontal–posterior parietal connectivity. Some distributed networks, shown in light red, include temporal, posterior parietal, and prefrontal regions.

Unlike DTI, fcMRI does not measure static anatomical connectivity but rather uses temporal (time-based) correlations between neurophysiological activity in different regions to infer functional connectivity. One obvious direction of investigations based on fcMRI is in searching for phenotypic variations among individuals. Such mapping will allow the examination of specific traits (e.g., musical ability) or psychiatric diagnoses (see the review by Kelly and colleagues, 2012).

Tractography Using Diffusion Tensor Imaging

DTI studies provide results, often called *tractography*, that complement the networks mapped by fcMRI. Tractography measures actual neuroanatomical pathways that can be related to specific traits. Traditional postmortem tract tracing was performed on single brains. Today, tractography can be done quickly on many living brains, and measurements can be made simultaneously in the entire brain. This advance allows researchers to correlate specific behavioral traits with specific patterns of connectivity.

Psyche Loui and her colleagues (2011) were interested in the neural basis of perfect (or absolute) pitch, the ability not only to discriminate among musical notes but also to name any note heard. Perfect pitch is rare among humans but shared by people with remarkable musical talents. Think Mozart. Development of perfect pitch is sensitive to early experiences, including musical training and exposure to tonal languages such as Japanese.

Loui's team studied musicians with and without absolute pitch who were matched in gender, age, handedness, ethnicity, IQ score, and years of musical training. Participants were given a test of pitch-labeling that allowed the researchers to place the musicians with absolute pitch into two categories: more accurate and less accurate. The less accurate group was superior to participants without absolute pitch who could not accurately identify the note.

The investigators hypothesized that absolute pitch could be related to increased connectivity in brain regions that process sounds. They used DTI to reconstruct white-matter tracts connecting two regions of the temporal cortex involved in auditory processing, the superior and middle temporal gyri, as well as regions not involved in auditory processing (corticospinal tract). The results, reproduced in **Figure 15-11,** show that people with absolute pitch have greater connectivity in temporal-lobe regions responsible for pitch

FIGURE 15-11 Tractography of Temporal Lobe Auditory Circuits Color overlaid on DTI images demonstrate tracts between the superior temporal gyrus and middle temporal gyrus in three individuals. Cases A and B had more accurate and less accurate perfect pitch, respectively; Case C did not. The tracts that connect the regions in the left hemisphere (colored orange) are larger in Cases A and B than in C. Tracts colored purple connect the regions in right hemisphere. Images © 2011, Massachusetts Institute of Technology; from P. Loui, H. C. C. Li, A. Hohmann, and G. Schlaug, 2011, *Journal of Cognitive Neuroscience, 23,* pp. 1015–1026.

(A) More accurate perfect pitch

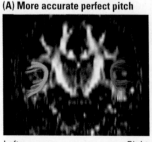

Left hemisphere | Right hemisphere

(B) Less accurate perfect pitch

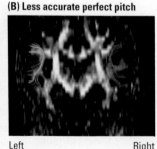

Left hemisphere | Right hemisphere

(C) Lacking perfect pitch

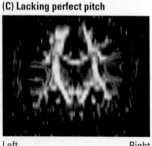

Left hemisphere | Right hemisphere

perception than do people with non-absolute pitch. The effect is largest in people in the more accurate group.

Although enhanced connectivity was present in both hemispheres of the musicians, when the investigators correlated performance on a test of absolute pitch with tract volume, only the tract volume in the left hemisphere predicted performance. It appears that having more local connections, or **hyperconnectivity,** in the left hemisphere is responsible for absolute pitch.

It is tempting to speculate that other exceptional talents, such as creativity, might be related to hyperconnectivity in cerebral regions. Similarly, we can speculate that reduced structural and functional connectivity is related to cognitive impairments after acquired brain injuries and/or neurodevelopmental and psychiatric disorders.

Social Neuroscience

By combining the tools of cognitive neuroscience, especially dynamic neuroimaging, with more abstract concepts from social psychology, **social neuroscience** seeks to understand how the brain mediates social interactions. Matthew Lieberman (2007) identified broad themes that attempt to encompass all cognitive processes involved in understanding and interacting with others as well as understanding ourselves.

Understanding Others

One difference between perceiving objects and animals is that animals have minds and experiences that are not open to direct inspection. We infer the minds of other animals in part by observing their behaviors and, in the case of people, by listening to their words. In doing so, we may develop a **theory of mind,** the attribution of mental states to others.

Theory of mind includes an understanding that others may have feelings and beliefs that are different from our own. This broader understanding has led some investigators to conclude that theory of mind may be uniquely human. Many researchers who study apes strongly believe that apes, too, possess a theory of mind.

Many fMRI studies over the past decade suggest that the brain region believed most closely associated with theory of mind is the dorsolateral prefrontal cortex (see Figure 15-2 on page 530). The prefrontal regions of humans are disproportionately large when we correct for brain size, but other apes also have large prefrontal regions, which supports the likelihood that they also have a theory of mind.

The understanding of others can also be inferred from the presence of empathy. For example, when participants watch videos of others smelling disgusting odors they report a mutual feeling of disgust. Lieberman and his colleagues (Rameson, Morelli, & Lieberman, 2012) used fMRI to assess the neural correlates of empathy by asking participants to emphasize with sad images. Empathy was correlated with increased activity in the medial prefrontal region, suggesting that it is critical for instantiating empathetic experience.

Understanding Oneself

Not only are we humans aware of the intentions of others, we also have a sense of self. Humans and apes have a unique ability to recognize themselves in a mirror, an ability that human infants can demonstrate by about 21 months of age. Studies using fMRI have shown that, when we recognize our own faces versus the faces of familiar others, brain activity increases in right lateral prefrontal cortex and in the lateral parietal cortex. The parietal cortex activation is thought to reflect the body's recognition of what "itself" feels like.

hyperconnectivity Increased local connections between two related brain regions.

social neuroscience Interdisciplinary field that seeks to understand how the brain mediates social interactions.

theory of mind Ability to attribute mental states to others.

Looking at its reflection and pointing to a dot that has been placed on its forehead, this chimpanzee displays self-recognition, a cognitive ability possessed by higher primates.

Courtesy of Cognitive Evolution Group, University of Louisiana at Lafayette, New Iberia Research

But there is more to understanding oneself than self-recognition. People also have a self-concept that includes beliefs about their own personal traits (e.g., "kind," "intelligent"). When participants are asked to determine whether trait words or sentences are self-descriptive, brain activity in medial prefrontal regions increases.

Self-Regulation

Self-regulation is our ability to control our emotions and impulses as a means for achieving long-term goals. We may wish to yell at the professor because an exam was unfair but most of us recognize that this course will not be productive. Dynamic imaging studies again reveal that the prefrontal regions are critical in social cognition, in this case in self-regulation.

Section 8-2 describes unique aspects of frontal-lobe development that extend beyond childhood—up to age 30.

Children are often poor at self-regulation, which probably reflects the slow development of prefrontal regions responsible for impulse control. A uniquely human ability is to self-regulate by putting feelings into words, a strategy that allows us to control emotional outbursts. Curiously, such verbal labeling is associated with increased activity in the right lateral prefrontal regions but *not* in the left.

Humans not only can control their emotions, they also have expectations about how a stimulus might feel (e.g., an injection by syringe). Our expectations can alter the actual feeling when we experience an event. It is common for people to say "ouch" when they do something like stub their toe, even if they actually experience no pain. Nobukatsu Sawamoto and colleagues (2000) found that when participants expect pain, activity increases in the anterior cingulate cortex (see Figure 15-2), a region associated with pain perception, even if the stimulus turns out not to be painful.

See Focus 12-1. Feeling and treating pain is a topic in Section 11-4.

Living in a Social World

Much of our waking time is spent interacting with others socially. In a sense, our understanding of our "self" and our social interactions link together as a single mental action. One important aspect of this behavior includes the formation of attitudes and beliefs about ourselves and about others. When we express attitudes (including prejudices) toward ideas or human groups, brain imaging shows activation in prefrontal, anterior cingulate, and lateral parietal regions.

Samuel McLure and colleagues (2004) took advantage of the fact that many people have strong attitudes towards cola-flavored sodas. Coca-Cola and Pepsi Cola are nearly identical in chemical composition, yet subjectively, people routinely prefer one over the other. The researchers ran blind taste tests among people who stated a cola preference.

As a group, participants failed to discriminate the drinks accurately when they were presented in a blind taste test. Brain activity was also equivalent for each cola in the blind condition. However, when participants believed that they were drinking Coca-Cola, significant changes in brain activity were recorded in many regions, including the hippocampus and dorsolateral prefrontal cortex. The investigators concluded that cultural information biases brain systems, which in turn biases attitudes.

Social Cognition and Brain Activity

Social cognitions, ranging from understanding ourselves to understanding others, clearly are associated with activation of specific brain regions, especially prefrontal regions. The obvious conclusion is that the activity of prefrontal regions produces our social cognitions, just as activity in visual regions produces our visual perceptions. But this conclusion has proven controversial.

Ed Vul and colleagues (2009) go so far as to suggest that "correlations in social neuroscience are voodoo." Their assertion has led to strong disputations (e.g., Lieberman, Berkman, & Wager, 2012). The arguments are complex, focus on the nature of the analysis of fMRI data, and will certainly continue. In our view, the debate does not impugn the general conclusion that brain states produce behavioral states.

neuroeconomics Interdisciplinary field that seeks to understand how the brain makes decisions.

Neuroeconomics

Historically, economics was based on the "rational actor," the belief that people make rational decisions. In the real world, however, people often make decisions based on assumption or intuition, as is common in gambling. (For many everyday examples, see a wonderful book by Leonard Mlodinow, 2009.) Why don't people always make rational decisions?

The cerebral processes underlying human decision making are not easily inferred from behavioral studies. But investigators in the field of **neuroeconomics,** which combines ideas from economics, psychology, and neuroscience, are attempting to explain those processes by studying patterns of brain activity as people make decisions.

The general assumption is that two different neural decision pathways influence our choices. One pathway is deliberate, slow, rule-driven, and emotionally neutral, and it acts as a *reflective system*. The other pathway—fast, automatic, emotionally biased—forms a *reflexive system*.

If people must make quick decisions that they believe will provide immediate gain, there is widespread activity in the dopaminergic reward system, including the ventromedial prefrontal cortex and ventral striatum (nucleus accumbens), that maps the reflexive pathway. If slower, deliberative decisions are possible, activity is greater in the lateral prefrontal, medial temporal, and posterior parietal cortex, the areas that form the reflective pathway.

Neuroeconomists are working to identify different patterns of neural activity in everyday decision making that may help account for how people make decisions about their finances, social relations, and other personal choices. Although most neuroeconomic studies to date have used fMRI, in principle these studies could also use other forms of noninvasive imaging. Further, epigenetic factors probably play an important role in developing the balance between the reflective and reflexive systems in individuals. Epigenetic studies therefore may help explain why many people make decisions that are not in their long-term best interest.

Based on predictions that all mammals will have similar reflective and reflexive decision systems, and because all animals make decisions, the neural bases of decision processes in nonhumans undoubtedly will receive more study in the future.

Figure 6-17 maps the dopaminergic pathways associated with reward.

REVIEW 15-3
Expanding Frontiers of Cognitive Neuroscience

Before you continue, check your understanding.

1. The development of noninvasive imaging techniques enabled cognitive psychologists to investigate the neural bases of thought in the "normal" brain, leading to the field called

   _____ .

2. Imaging methods such as DTI and fcMRI are allowing researchers to develop a _____ , a map of the complete structural and functional neural connections in the living human brain.

3. Social neuroscience is an interdisciplinary field that seeks to understand how the brain mediates

   _____ .

4. Our attribution of mental states to others is known as _____ .

5. Neuroeconomics seeks to understand the neural bases of _____ .

6. List four general themes of social neuroscience research.

Answers appear at the back of the book.

15-4 Cerebral Asymmetry in Thinking

A fundamental discovery in behavioral neuroscience was the finding by Paul Broca and his contemporaries in the mid-1800s that language is lateralized to the brain's left hemisphere. But the implications of lateralized brain functions were not really understood until the 1960s, when Roger Sperry (1968) and his colleagues began to study people who, as described in Research Focus 15-1, had undergone surgical separation of the two hemispheres as a treatment for intractable epilepsy.

It soon became apparent that the two cerebral hemispheres are more specialized in their functions than researchers had previously realized. Before considering how the two sides of the brain cooperate in generating cognitive activity, we look at the anatomical differences between the left and right hemispheres.

Anatomical Asymmetry

The media seized on Sperry's findings in the 1980s with an avalanche of self-help books about "left-brained" and "right-brained" people. The novelty has worn off, but the concept of cerebral asymmetry remains important to understanding how the human brain thinks.

Building on Broca's findings, investigators have learned how the language- and music-related areas of the left and right temporal lobes differ anatomically. In particular, the primary auditory area is larger on the right, whereas the secondary auditory areas are larger on the left in most people. Other brain regions also are asymmetrical.

Figure 15-12 shows that the lateral fissure, which partly separates the temporal and parietal lobes, has a sharper upward course in the right hemisphere relative to the left. As a result, the posterior part of the right temporal lobe is larger than the same region on the left side of the brain, as is the left parietal lobe relative to the right.

Among the anatomical asymmetries in the frontal lobes, the region of sensory–motor cortex representing the face is larger in the left hemisphere than in the right, a difference that presumably corresponds to the special role of the left hemisphere in talking. Broca's area is organized differently on the left and the right. The area visible on the surface of the brain is about one-third larger on the right than on the left, whereas the area of cortex buried in the sulci of Broca's area is greater on the left than on the right.

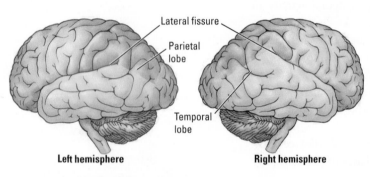

Lateral fissure

Parietal lobe

Temporal lobe

Left hemisphere

Right hemisphere

FIGURE 15-12 Cerebral Asymmetry The lateral fissure takes a flatter course in the left hemisphere than in the right. As a result, the posterior part of the right temporal lobe is larger than the same region on the left side, and the inferior parietal region is larger on the left than on the right.

Not only do these gross anatomical differences exist but so too do hemispheric differences in the details of their cellular and neurochemical structures. For example, the neurons in Broca's area on the left have larger dendritic fields than do the corresponding neurons on the right. The discovery of structural asymmetries tells us little about why such differences exist. Ongoing research is revealing that they are due to underlying differences in cognitive processing by the two sides of the brain.

Although many anatomical asymmetries in the human brain are related to language, brain asymmetries are not unique to humans. Most if not all mammals have asymmetries, as do many species of birds. Cerebral asymmetry therefore cannot simply be present for language processing. Rather, it is likely that human language evolved after the brain had become asymmetrical. Language simply took advantage of processes, including the development of mirror neurons, that had already been lateralized by natural selection in earlier members of the human lineage.

Functional Asymmetry in Neurological Patients

The specialized functions of the two cerebral hemispheres become obvious in people with damage to the left side or right side of the brain. To see these functional differences clearly, compare the cases of G. H. and M. M.

When G. H. was 5 years old, he went on a hike with his family and was hit on the head by a large rock that rolled off an embankment. He was unconscious for a few minutes and had a severe headache for a few days but quickly recovered. Around age 18, however, he started having seizures.

Neurosurgical investigation revealed that G. H. had suffered a right posterior parietal injury from the rock accident. **Figure 15-13A** shows the area affected. After surgery to remove this area, G. H. had weakness on the left side of his body and showed contralateral neglect. But these symptoms lessened fairly quickly, and a month after the surgery, they had completely cleared.

Nevertheless, G. H. suffered chronic difficulties in copying drawings; 4 years later, he still performed this task at about the level of a 6-year-old. He also had trouble assembling puzzles, which he found disappointing because he had enjoyed doing puzzles before his surgery. When asked to perform mental manipulations like the one in Figure 15-6, he became very frustrated and refused to continue.

G. H. also had difficulty finding his way around familiar places. The landmarks he had used to guide his travels before the surgery no longer seemed to work for him. G. H. now has to learn street names and use a verbal strategy to go from one place to another.

Left Parietal Damage

M. M.'s difficulties were quite different. A meningioma had placed considerable pressure on the left parietal region. The tumor was surgically removed when M. M. was 16 years old. It had damaged the area shown in Figure 15-13B.

After the surgery, M. M. experienced a variety of problems, including *aphasia,* impairment in the use of language. The condition lessened over time: a year after the surgery, M. M. spoke fluently. Unfortunately, other difficulties persisted.

In solving arithmetic problems, in reading, and even in simply calling objects or animals by name, M. M. performed at about the level of a 6-year-old. She had no difficulty making movements spontaneously but when asked to copy a series of arm movements, such as those diagrammed in **Figure 15-14,** she had great difficulty. She seemed unable to figure out how to make her arm move to match the example. A general impairment in making voluntary movements in the absence of paralysis or a muscular disorder is a symptom of *apraxia,* the inability to complete a plan of action accurately.

(A) Case G. H.

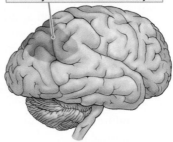

Injury to this area of the right hemisphere caused difficulties in copying drawings, assembling puzzles, and finding the way around a familiar city.

(B) Case M. M.

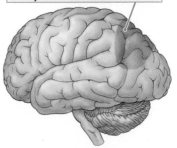

Injury to this area of the left hemisphere caused difficulties in language, copying movements, reading, and generating names of objects or animals.

FIGURE 15-13 Contrasting Parietal-Lobe Injuries

Meningioma, a tumor of the brain's protective coverings (meninges), is imaged in Focus 3-3.

Section 10-4 describes how left-hemisphere damage causes aphasias.

Section 11-5 explains how damage to the somatosensory cortex contributes to apraxia.

Series 1

Series 2

FIGURE 15-14 Two Arm-Movement Series Subjects observe the tester perform each sequence and then copy it as accurately as they can. People with left-hemisphere injury, especially in the posterior parietal region, are impaired at copying such movements.

dichotic listening Experimental procedure for simultaneously presenting a different auditory input to each ear through stereophonic earphones.

Lessons from G. H. and M. M.

What can we learn about brain function by comparing the patients G. H. and M. M.? Their lesions were in approximately the same location but in opposite hemispheres, and their symptoms were very different.

Judging from G. H.'s difficulties, the right hemisphere plays a role in controlling spatial skills, such as drawing, assembling puzzles, and navigating in space. In contrast, M. M.'s condition reveals that the left hemisphere seems to play some role in controlling language functions and in various cognitive tasks related to schoolwork—namely, reading and arithmetic. In addition, the left hemisphere plays a role in controlling sequences of voluntary movement that differs from the right hemisphere's role.

To some extent then, the left and right hemispheres think about different types of information. The question is whether these functional differences can be observed in a normal brain.

Functional Asymmetry in the Normal Brain

In the course of studying the auditory capacities of people with temporal-lobe lesions, Doreen Kimura (1967) came upon an unexpected finding. She presented her normal-control participants with two strings of digits, one played into each ear, a procedure known as **dichotic listening.** The task was to recall as many digits as possible.

Kimura found that the normal controls recalled more digits presented to the right ear than to the left. This result is surprising because the auditory system is repeatedly crossed, beginning in the midbrain. Nonetheless, information coming from the right ear seems to have preferential access to the left (speaking) hemisphere.

In a later study, Kimura (1973) played two pieces of music for participants, one to each ear. She then gave the participants a multiple-choice test, playing four bits from musical selections and asking the participants to pick out the bits that they had heard before. In this test, she found that participants were more likely to recall the music played to the left ear than to the right. This result implies that the left ear has preferential access to the right (musical) hemisphere.

The demonstration of this functional asymmetry in the normal brain provoked much interest in the 1970s, leading to demonstrations of functional asymmetries in the visual and tactile systems as well. Consider the visual system. If we fixate on a target, such as a dot positioned straight ahead, all the information to the left of the dot goes to the right hemisphere and all the information to the right of the dot goes to the left hemisphere, as shown in **Figure 15-15.**

If information is presented for a relatively long time—say, 1 second—we can easily report what was in each visual field. If, however, the presentation is brief—say, only 40 milliseconds—then the task is considerably harder. This situation reveals a brain asymmetry.

Words presented briefly to the right visual field and hence sent to the left hemisphere are more easily reported than are words presented briefly to the left visual field. Similarly, if complex geometric patterns or faces are shown briefly, those presented to the left visual field and hence sent to the right hemisphere are more accurately reported than are those presented to the right visual field.

Apparently, the two hemispheres not only think about different types of information, they also process information differently. The left hemisphere seems biased toward processing language-related information, whereas the right hemisphere seems biased toward processing nonverbal, especially spatial, information.

A word of caution: Although asymmetry studies are fascinating, what they tell us about the differences between the two hemispheres is not entirely clear. They tell us that *something* is different, but it is a long leap to conclude that the two hemispheres house entirely different kinds of skills.

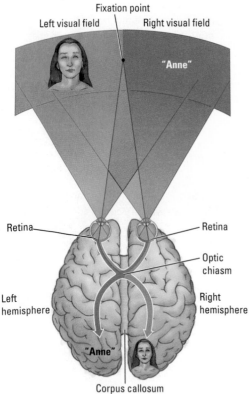

FIGURE 15-15 Visual Pathways to the Two Hemispheres When fixating at a point, each eye sees both visual fields but sends information about the right visual field only to the left hemisphere and information about the left visual field only to the right hemisphere. In normal participants given short exposures to stimuli (well under 1 second), the left hemisphere is more accurate at perceiving words, whereas the right hemisphere is more accurate at perceiving objects, such as faces.

The two hemispheres have many functions in common, such as controlling movement in the contralateral hand and processing sensory information through the thalamus. Still, differences in the cognitive operations of the two hemispheres do exist. These differences can be better understood by studying split-brain patients, whose cerebral hemispheres have been surgically separated for medical treatment.

Functional Asymmetry in the Split Brain

Before considering the details of split-brain studies, let us review what we already know about cerebral asymmetry. First, the left hemisphere has language; the right hemisphere does not. Second, as demonstrated in Research Focus 15-1, on page 524 the right hemisphere performs better at certain nonverbal tasks, especially those that engage visuospatial skills.

But how does a severed corpus callosum affect how the brain thinks? After the corpus callosum has been cut, the two hemispheres have no way of communicating with one another. The left and right hemispheres are therefore free to think about different things. In a sense, a split-brain patient has two brains.

One way to test the different cognitive functions of the two hemispheres in a split-brain patient takes advantage of the fact that information in the left visual field goes to the right hemisphere and information in the right field goes to the left hemisphere (see Figure 15-15). With the corpus callosum cut, information presented to one side of the brain has no way of traveling to the other side. It can be processed only in the hemisphere that receives it.

Experiments 15-3 and **15-4** show some basic testing procedures that use this approach. The split-brain subject fixates on the dot in the center of the screen while information is presented to the left or right visual field. The person must respond with the left hand (controlled by the right hemisphere), with the right hand (controlled by the left hemisphere), or verbally (also a left-hemisphere function). In this way, researchers can observe what each hemisphere knows and what it is capable of doing.

As illustrated in Experiment 15-3, for instance, a picture—say, of a spoon—might be flashed and the subject asked to state what he or she sees. If the picture is presented to the right visual field, the person will answer, "Spoon." If the picture is presented to the left visual field, however, the person will say, "I see nothing." The subject responds in this way for two reasons:

1. The right hemisphere (that receives the visual input) does not talk, so it cannot respond verbally, even though it sees the spoon in the left visual field.

2. The left hemisphere does talk, but it does not see the spoon, so it answers—quite correctly, from its own perspective—that nothing was presented.

Now suppose that the task changes. In Experiment 15-4A, the picture of a spoon is still presented to the left visual field, but the subject is asked to use the left hand to pick out the object shown on the screen. In this case, the left hand, controlled by the right hemisphere, which sees the spoon, readily picks out the correct object.

Can the right hand also choose correctly? No, because it is controlled by the left hemisphere that cannot see a spoon. If the person in

Those popular "left-brain–right-brain" media accounts of the 1980s ignored the fact that the two hemispheres have many functions in common.

EXPERIMENT 15-3

Question: Will severing the corpus callosum affect the way in which the brain responds?

Procedure

The split-brain subject fixates on the dot in the center of the screen while an image is projected to the left or right visual field. He is asked to identify verbally what he sees.

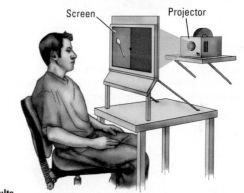

Screen   Projector

Results

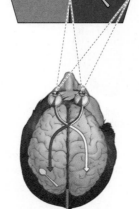

| If the spoon is presented to the right visual field, the subject verbally answers, "Spoon." | If the spoon is presented to the left visual field, the subject verbally answers, "I see nothing." |

Left visual field   Right visual field   Left visual field   Right visual field

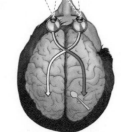

Severed corpus callosum

Conclusion: When the left hemisphere, which can speak, sees the spoon in the right visual field, the subject responds correctly. When the right hemisphere, which cannot speak, sees the spoon in the left visual field, the subject does not respond.

(A) Question: How can the right hemisphere of a split-brain subject show that it knows information?

Procedure

The split-brain subject is asked to use his left hand to pick out the object shown on the screen to the left visual field (right hemisphere).

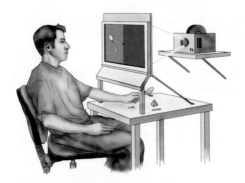

Results

The subject chooses the spoon with his left hand because the right hemisphere sees the spoon and controls the left hand. If the right hand is forced to choose, it will do so by chance because no stimulus is shown to the left hemisphere.

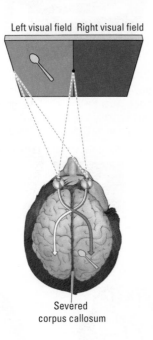

Left visual field Right visual field

Severed
corpus callosum

(B) Question: What happens if both hemispheres are asked to respond to competing information?

Procedure

Each visual field is shown a different object—a spoon to the left and a pencil to the right. The split-brain subject is asked to use both hands to pick up the object seen.

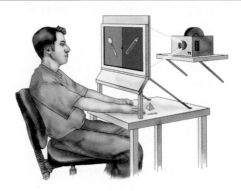

Results

In this case, the right and left hands do not agree. They may each pick up a different object, or the right hand may prevent the left hand from performing the task.

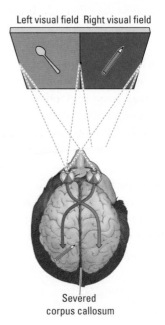

Left visual field Right visual field

Severed
corpus callosum

Conclusion: Each hemisphere is capable of responding independently. The left hemisphere may dominate in a competition, even if the response is not verbal.

this situation is forced to select an object with the right hand, the left hemisphere does so at random.

Now let's consider an interesting twist. In the Procedure for Experiment 15-4B, each hemisphere is shown a different object—say, a spoon to the right hemisphere and a pencil to the left. The subject is asked to use both hands to pick out the object seen.

The problem here is that the right hand and left hand do not agree. While the left hand tries to pick up the spoon, the right hand tries to pick up the pencil or tries to prevent the left hand from performing the task.

This conflict between the hemispheres can be seen in the everyday behavior of some split-brain subjects. One woman, P. O. V., reported frequent interhemispheric competition for at least 3 years after her surgery. "I open the closet door. I know what I want to wear. But as I reach for something with my right hand, my left comes up and takes something different. I can't put it down if it's in my left hand. I have to call my daughter."

We know from Experiment 15-3 that the left hemisphere is capable of using language, and Research Focus 15-1 reveals that the right hemisphere has visuospatial capabilities that the left hemisphere does not. Although findings from nearly half a century of studies of split-brain patients show that the two hemispheres process information differently, another word of caution is needed. There is more overlap in function between the hemispheres than was at first suspected. The right hemisphere, for instance, does have some language functions, and the left hemisphere does have some spatial abilities. Nonetheless, the two hemispheres undoubtedly are different.

Explaining Cerebral Asymmetry

Various hypotheses propose to explain hemispheric differences. One idea, that the left hemisphere plays an important role in controlling fine movements, dates back a century. Recall M. M., the meningioma patient with left-parietal-lobe damage who suffered apraxia (see Figure 15-13B). Although that condition subsided, she was left with chronic trouble in copying movements.

Perhaps one reason that the left hemisphere has a role in language is that speaking requires fine motor movements of the mouth and tongue. Significantly, damage to the language-related areas of the left hemisphere almost always interferes with both language and movement, regardless of whether the person speaks or signs. Reading Braille, however, may not be so affected by left-hemisphere lesions. Most people prefer to use the left hand to read Braille, which essentially consists of spatial patterns, so processes related to reading Braille may reside in the right hemisphere.

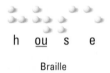

h <u>ou</u> s e

Braille

That said, another clue that the left hemisphere's specialization for language may be related to its special role in controlling fine movements comes from investigating where certain parts of speech are processed in the brain. Recall that cognitive systems for representing abstract concepts are likely to be related to systems that produce more-concrete behaviors. Consequently, we might expect that the left hemisphere would participate in forming concepts related to fine movements.

Concepts that describe movements are the parts of speech we call verbs. A fundamental difference between left- and right-hemisphere language abilities is that verbs seem to be processed only in the left hemisphere, whereas nouns are processed in both hemispheres. In other words, not only does the left hemisphere have a special role in producing actions, it also produces the mental representations of actions in the form of words.

If the left hemisphere excels at language because it is better at controlling fine movements, what is the basis of the right hemisphere's abilities? One idea is that the right hemisphere has a special role in controlling movements in space. In a sense, this role is an elaboration of the functions of the dorsal visual stream (diagrammed in Figure 15-3).

Once again, we can propose a link between movement at a concrete level and movement at a more abstract level. If the right hemisphere is producing movements in space, then it is also likely to produce mental images of such movements. We would therefore predict that right-hemisphere patients would be impaired both at making spatially guided movements and at thinking about such movements. And they are.

Recall Principle 1 from Section 2-6: The nervous system produces movement within a perceptual world that the brain creates.

Bear in mind that theories about the reasons for hemispheric asymmetry are highly speculative. The brain has evolved to produce movement and to create a sensory reality, so the observed asymmetry must somehow be related to these overriding functions. That is, more recently evolved functions, such as language, likely are extensions of preexisting functions.

In other words, the fact that language is represented asymmetrically does not mean that the brain is asymmetrical because of language. After all, other species that do not talk have asymmetrically organized brains. Once again, we see evidence that more recent adaptations, such as mirror neurons, probably play an important role in the emergence of unique human functions.

Left Hemisphere, Language, and Thought

We end our examination of brain asymmetry by considering one more provocative idea. Michael Gazzaniga (1992) proposed that the left hemisphere's superior language skills are important in understanding the differences in thinking between humans and other animals. He called the speaking hemisphere the "interpreter." What he meant is illustrated in the following experiment, using split-brain patients as subjects.

Each hemisphere is shown the same two pictures—a picture of a match followed by a picture of a piece of wood, for example. Another set of pictures is then shown. The task is to pick from this set a third picture that has an inferred relation to the other two. In this example, the third related picture might be a bonfire. The right hemisphere is incapable of making the inference that a match struck and held to a piece of wood could create a bonfire, whereas the left hemisphere can easily arrive at this interpretation.

An analogous task uses words. One or the other hemisphere might be shown the words *pin* and *finger* and then be asked to pick out a third word that is related to the other two. In this case, the correct answer might be *bleed*.

The right hemisphere is not able to make this connection. Although it has enough language ability to pick out close synonyms for *pin* and *finger* (*needle* and *thumb*, respectively), it cannot make the inference that pricking a finger with a needle will result in bleeding.

Again, the left hemisphere has no difficulty with this task. Apparently, the left hemisphere's language capability gives it a capacity for interpretation that the right hemisphere lacks. One reason may be that language serves to label and express the computations of other cognitive systems.

Gazzaniga goes even further. He suggests that the evolution of the language abilities possessed by the left hemisphere makes humans a "believing" species: humans can make inferences and have beliefs about sensory events. By contrast, Alex, the gray parrot profiled in Comparative Focus 15-2, would not have been able to make inferences or hold beliefs because he did not have a system analogous to our left-hemisphere language system. Alex could use language but could not make inferences about sensory events with language.

Gazzaniga's idea is certainly intriguing. It implies a fundamental difference in the nature of cerebral asymmetry—and therefore in the nature of cognition—between humans and other animals that exists because of the nature of human language. We return to this idea in Section 15-7.

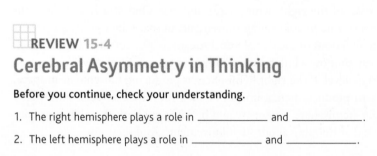

REVIEW 15-4

Cerebral Asymmetry in Thinking

Before you continue, check your understanding.

1. The right hemisphere plays a role in _____ and _____.

2. The left hemisphere plays a role in _____ and _____.

3. The split-brain results from cutting apart the _____.

4. Why does it matter that the two cerebral hemispheres process information differently?

Answers appear at the back of the book.

15-5 Variations in Cognitive Organization

No two brains are identical. Some differences are genetically determined; others result from plastic changes such as those created by experience and learning or epigenetic factors. Some brain differences are idiosyncratic (unique to a particular person), whereas many others are systematic and common to whole categories of people. In this section, we consider two systematic variations in brain organization, those related to sex and handedness, and one idiosyncratic variation, the fascinating sensory ability of *synesthesia*.

Sex Differences in Cognitive Organization

The idea that men and women think differently probably originated with the first men and women. Science backs it up. Books, including one by Doreen Kimura (1999), have compiled considerable evidence for marked sex differences in the way men and women perform on many cognitive tests. As illustrated in **Figure 15-16,** paper-and-pencil tests

(A) Spatial relation–type task

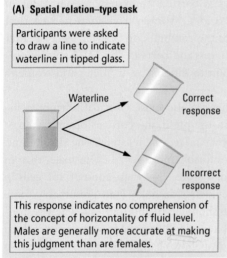

Participants were asked to draw a line to indicate waterline in tipped glass.

Waterline — Correct response

Incorrect response

This response indicates no comprehension of the concept of horizontality of fluid level. Males are generally more accurate at making this judgment than are females.

(B) Mental rotation–type task

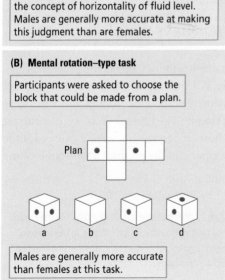

Participants were asked to choose the block that could be made from a plan.

Plan

a   b   c   d

Males are generally more accurate than females at this task.

(C) Short-term-memory–type task

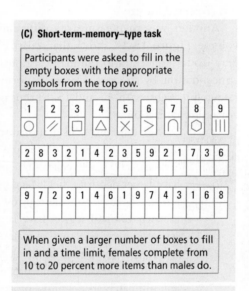

Participants were asked to fill in the empty boxes with the appropriate symbols from the top row.

1	2	3	4	5	6	7	8	9

| 2 | 8 | 3 | 2 | 1 | 4 | 2 | 3 | 5 | 9 | 2 | 1 | 7 | 3 | 6 |

| 9 | 7 | 2 | 3 | 1 | 4 | 6 | 1 | 9 | 7 | 4 | 3 | 1 | 6 | 8 |

When given a larger number of boxes to fill in and a time limit, females complete from 10 to 20 percent more items than males do.

(D) Verbal fluency–type task

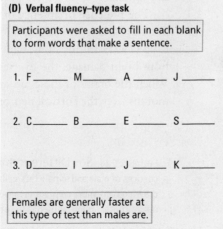

Participants were asked to fill in each blank to form words that make a sentence.

1. F_____ M_____ A _____ J _____

2. C_____ B _____ E _____ S _____

3. D_____ I _____ J _____ K _____

Females are generally faster at this type of test than males are.

FIGURE 15-16 Tasks That Reliably Show Sex-Related Cognitive Differences

FIGURE 15-17 Sex Differences in Brain Volume Women's brain volume (purple) in prefrontal and medial paralimbic regions is significantly higher than men's. Men have larger relative volumes (pink) in the medial and orbital frontal cortex and the angular gyrus. Purple areas correspond to regions that have high levels of estrogen receptors during development, pink to regions high in androgen receptors during development. From "Normal Sexual Dimorphism of the Adult Human Brain Assessed by In Vivo Magnetic Resonance Imaging," by J. M. Goldstein, L. J. Seidman, N. J. Horton, N. Makris, D. N. Kennedy, et al., 2001, Cerebral Cortex 11, pp. 490–497.

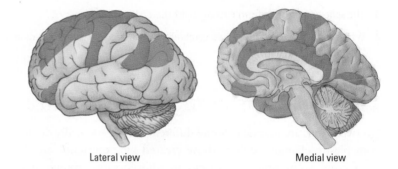

Lateral view   Medial view

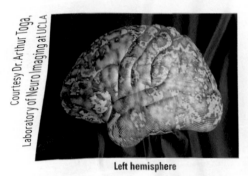

Courtesy Dr. Arthur Toga, Laboratory of Neuro Imaging at UCLA

Left hemisphere

FIGURE 15-18 Sex Differences in Gray-Matter Concentration Women show increased gray-matter concentration in the cortical regions shown in color on this MRI. Gray-shaded regions are not statistically different in males and females.

consistently show that, on average, females have better verbal fluency than males do, whereas males do better on tests of spatial reasoning. Our focus here is on how such differences relate to the brain.

Neural Basis of Sex Differences

Considerable evidence points to sex differences, both in the brain's gross cerebral structure and at a neuronal level. Jill Goldstein and her colleagues (2001) conducted a large MRI study of sexual dimorphism in the human brain. They found that women have larger volumes of dorsal prefrontal and associated paralimbic regions, whereas men have larger volumes of more ventral prefrontal regions **(Figure 15-17).** (Brain size is related to body size, and on average, male brains are bigger than female brains, so the investigators corrected for size.)

Another way to measure sex differences is cortical thickness, independent of volume. **Figure 15-18** shows that, relative to men, women have increased cortical gray-matter concentration in many regions of the cerebral cortex. Men's gray-matter concentration, by contrast, is more uniform across the cortex. The MRI studies represented in Figures 15-17 and 15-18 thus point to differences in men's and women's cortical organization.

Sex differences in neuronal structure also exist. Gonadal hormones influence the structures of neurons in the prefrontal cortices of rats (Kolb & Stewart, 1991). The cells in one prefrontal region, located along the midline, have larger dendritic fields (and presumably more synapses) in males than in females, as shown in the top row of **Figure 15-19.** In contrast, the cells in the orbitofrontal region have larger dendritic fields (and presumably more synapses) in females than in males, as shown in the bottom row. These sex differences are not found in rats that have had their gonads or ovaries removed at birth. Presumably, sex hormones somehow change the brain's organization and ultimately its cognitive processing.

The presence or absence of gonadal hormones affects the brain in adulthood as well as in early development, a finding in a study by Stewart and Kolb (1994). In this study, which focused on how hormones affect recovery from brain damage, the ovaries of middle-aged female rats were removed. When the brains of these rats and those of control rats were examined some months later, the cortical neurons—especially the prefrontal neurons—of the

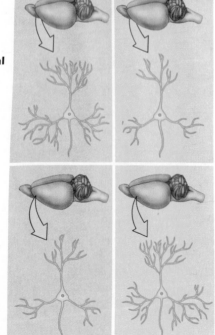

Male rat   Female rat

Cells from medial frontal cortex

Cells from orbitofrontal region

FIGURE 15-19 Sex Differences in the Architecture of Neurons In the frontal cortices of male and female rats, cells in the midline frontal region (shown in the top two drawings) are more complex in males than in females, whereas the opposite is true of the orbitofrontal region (shown in the bottom two drawings).

rats whose ovaries had been removed had undergone structural changes. Specifically, the cells had grown 30 percent more dendrites, and their spine density increased compared with the cells in control rats. Clearly, gonadal hormones can affect the neuronal structure of the brain at any point in an animal's life.

An additional way to consider the neural basis of sex differences is to look at the effects of cortical injury in men and women. If sex differences exist in the neural organization of cognitive processing, there ought to be differences in the effects of cortical injury in the two sexes. Doreen Kimura (1999) conducted this kind of study and showed that the pattern of cerebral organization within each hemisphere may in fact differ between the sexes.

Investigating people who had sustained cortical strokes in adulthood, Kimura tried to match the location and extent of injury in her male and female subjects. She found that men and women were almost equally likely to be aphasic subsequent to left-hemisphere lesions of some kind. But men were more likely to be aphasic and apraxic after damage to the left posterior cortex, whereas women were far more likely to be aphasic and apraxic after lesions to the left frontal cortex. These results, summarized in **Figure 15-20,** suggest a difference in intrahemispheric organization between the sexes.

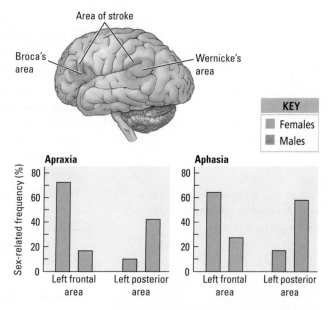

FIGURE 15-20 Evidence for Sex Differences in Cortical Organization Apraxia and aphasia are associated with frontal damage to the left hemisphere in women and with posterior damage in men. Adapted from *Sex and Cognition*, by D. Kimura, 1999, Cambridge, MA: MIT Press.

Evolution of Sex-Related Cognitive Differences

Although gonadal hormones have taken center stage in explaining sex differences in cognitive function, we are still left to question how these differences arose in the first place. To answer this question, we must look back at human evolution. Ultimately, males and females of a species have virtually all their genes in common. Mothers pass their genes to both sons and daughters, and fathers do the same.

The only way a gene can affect one sex preferentially is for the gene's activities to be influenced by the animal's gonadal hormones, which in turn are determined by the presence or absence of the Y chromosome. The Y chromosome carries a gene called the *testes-determining factor* (TDF). This gene stimulates the body to produce testes, which then manufacture androgens, which subsequently influence the activities of other genes.

Like other body organs, the brain is a potential target of natural selection. We should therefore expect to find sex-related differences in the brain whenever the two sexes differ in the adaptive problems they have faced in the evolutionary history of the species. The degree of aggressive behavior produced by the brain is a good example.

Males are more physically aggressive than females in most mammalian species. This trait presumably improved males' reproductive success, reinforcing natural selection for greater aggressiveness in males. Producing higher levels of aggression entails male hormones. We know from studies of nonhuman species that aggression is related directly to the presence of androgens and to their effects on gene expression both during brain development and later in life. In this case, therefore, natural selection has worked on gonadal hormone levels to favor aggressiveness in males.

Explaining sex-related differences in cognitive processes, such as language or spatial skills, is more speculative than explaining sex-related differences in aggressive behavior. Nevertheless, some hypotheses come to mind. We can imagine, for instance, that in the history of mammalian evolution, males have tended to range over larger territories than have females. This behavior requires spatial abilities, so these skills would have been favored in males.

Section 6-5 outlines activating effects of gonadal and synthetic steroid hormones, Section 8-4 the organizing effects of gonadal hormones during brain development, and Section 12-5 the effects of gonadal hormones in controlling sexual behavior.

Support for this hypothesis comes from comparing spatial problem-solving abilities in males of closely related mammalian species—species in which the males range over large territories versus species in which the males do not have such extensive ranges. Pine voles, for example, have restricted ranges and no sex-related difference in range, whereas meadow voles have ranges about 20 times as large as those of pine voles, and the males range more widely than the females.

When the spatial skills of pine voles and meadow voles are compared, meadow voles are far superior. A sex difference in spatial ability among meadow voles favors males, but no such sex difference exists among pine voles. The hippocampus is implicated in spatial navigation skills. Significantly, the hippocampus is larger in meadow voles than in pine voles, and it is larger in male meadow voles than in females (Gaulin, 1992). A similar logic could help explain sex-related differences in spatial abilities between human males and females (see Figure 15-16).

Explaining sex-related differences in language skills also is speculative. One hypothesis holds that, if males were hunters and often away from home, the females left behind in social groups would be selectively favored to develop tools for social interaction, one of which is language. We might also argue that females were selected for fine motor skills (such as foraging for food and making clothing and baskets). Because of the relation between language and fine motor skills, enhanced language capacities also might have evolved in females.

Although such speculations are interesting, they are not testable. We will probably never know with certainty why sex-related differences in brain organization developed.

Handedness and Cognitive Organization

Nearly everyone prefers one hand over the other for writing or throwing a ball. Most people prefer the right hand. In fact, left-handedness has historically been viewed as odd. But it is not rare. An estimated 10 percent of the human population worldwide is left-handed. This proportion represents the number of people who write with the left hand. When other criteria are used to determine left-handedness, estimates range from 10 percent to 30 percent of the population.

Because the left hemisphere controls the right hand, the general assumption is that right-handedness is somehow related to the presence of speech in the left hemisphere. If this were so, then language would be located in the right hemispheres of left-handed people. This hypothesis is easily tested, and it turns out to be false.

In the course of preparing epileptic patients for surgery to remove the abnormal tissue causing their seizures, Ted Rasmussen and Brenda Milner (1977) injected the left or right hemisphere with sodium amobarbital (see Clinical Focus 15-5, "Sodium Amobarbital Test"). This drug produces a short-acting anesthesia of the entire hemisphere, making it possible to determine where speech is located. For instance, if a person becomes aphasic when the drug is injected into the left hemisphere but not when the drug is injected into the right, then speech must reside in that person's left hemisphere.

Rasmussen and Milner found that virtually all right-handed people had speech in the left hemisphere, but the reverse was not true for left-handed people. About 70 percent of left-handers also had speech in the left hemisphere. Of the remaining 30 percent, about half had speech in the right hemisphere and half had speech in both hemispheres.

Findings from neuroanatomical studies have subsequently shown that left-handers with speech in the left hemisphere have asymmetries similar to those of right-handers. By contrast, in left-handers with speech located in the right hemisphere or in both hemispheres—known as **anomalous speech representation**—the anatomical symmetry is reversed or absent.

Section 14-3 explains how visuospatial learning and visuospatial memory both recruit the hippocampus.

anomalous speech representation
Condition in which a person's speech zones are located in the right hemisphere or in both hemispheres.

Sodium Amobarbital Test

Guy, a 32-year-old lawyer, had a vascular malformation over the region corresponding to the posterior speech zone. The malformation was beginning to cause neurological symptoms, including epilepsy. The ideal surgical treatment was removal of the abnormal vessels.

The complication with this surgery is that removing vessels sitting over the posterior speech zone poses a serious risk of permanent aphasia. Because Guy was left-handed, his speech areas could be in the right hemisphere. If so, the surgical risk would be much lower.

To achieve certainty in such doubtful cases, Jun Wada and Ted Rasmussen (1960) pioneered the technique of injecting sodium amobarbital, a barbiturate, into the carotid artery to produce a brief period of anesthesia of the ipsilateral hemisphere. (Injections are now normally made through a catheter inserted into the femoral artery.) This procedure enables an unequivocal localization of speech because injection into the speech hemisphere results in an arrest of speech lasting as long as several minutes. As speech returns, it is characterized by aphasic errors.

Injection into the nonspeaking hemisphere may produce no or only brief speech arrest. The amobarbital procedure has the advantage of allowing each hemisphere to be studied separately in the functional absence of the other (anesthetized) hemisphere. Because the period of anesthesia lasts several minutes, a variety of functions, including memory and movement, can be studied to determine a hemisphere's capabilities.

The sodium amobarbital test is always performed bilaterally, with the second cerebral hemisphere being injected several days after the

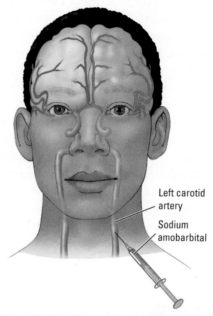

Left carotid artery

Sodium amobarbital

To avoid damaging speech zones in patients about to undergo brain surgery, surgeons inject sodium amobarbital into the carotid artery. The drug anesthetizes the hemisphere where it is injected (here, the left hemisphere), allowing the surgeon to determine if that hemisphere is dominant for speech.

first one to make sure that there is no residual drug effect. In the brief period of drug action, the patient is given a series of simple tasks requiring the use of language, memory, and object recognition. Speech is tested by asking the patient to name some common objects presented in quick succession, to count and to recite the days of the week forward and backward, and to spell simple words.

If the injected hemisphere is nondominant for speech, the patient may continue to carry out the verbal tasks, although there is often a period as long as 30 seconds during which he or she appears confused and is silent but can resume speech with urging. When the injected hemisphere is dominant for speech, the patient typically stops talking and remains completely aphasic until recovery from the anesthesia is well along, somewhere in the range of 4 to 10 minutes.

Guy was found to have speech in the left hemisphere. During the test of his left hemisphere, he could not talk. Later, he said that, when he was asked about a particular object, he wondered just what the question meant. When he finally had some vague idea, he had no idea of what the answer was or how to say anything. By then he realized that he had been asked all sorts of other questions to which he had also not responded.

When asked which objects he had been shown, he said he had no idea. However, when given an array of objects and asked to choose with his left hand, he was able to identify the objects by pointing because his nonspeaking right hemisphere controlled that hand. In contrast, his speaking left hemisphere had no memory of the objects because it had been asleep.

Sandra Witelson and Charlie Goldsmith (1991) asked whether any other gross differences might exist in the brain structure of right- and left-handers. One possibility is that the connectivity of the cerebral hemispheres may differ. To test this idea, the investigators studied the hand preference of terminally ill subjects on a variety of one-handed tasks. They later performed postmortem studies of these patients' brains, paying particular attention to the size of the corpus callosum. They found that the callosal cross-sectional area was 11 percent greater in left-handed and ambidextrous (no hand preference) people than in right-handed people.

Whether this enlarged callosum is due to a greater number of fibers, to thicker fibers, or to more myelin remains to be seen. If the larger corpus callosum is due to more fibers, the difference would be on the order of 25 million more fibers. Presumably, such a difference would have major implications for the organization of cognitive processing in left- and right-handers.

Synesthesia

Some variations in brain organization are idiosyncratic rather than systematic. **Synesthesia** is an individual's capacity to join sensory experiences across sensory modalities, as discussed in Clinical Focus 15-6, "A Case of Synesthesia." Examples include the ability to hear colors or taste shapes. Edward Hubbard (2007) estimated the incidence of synesthesia at about 1 in every 23 people, although for most it is likely to be limited in scope.

Synesthesia runs in families; an example is the family of Russian novelist Vladimir Nabokov. As a toddler, Nabokov complained to his mother that the letter colors on his wooden alphabet blocks were "all wrong." His mother understood what he meant, because she too perceived letters and words in particular colors. Nabokov's son is synesthetic in the same way.

If you recall shivering on hearing a particular piece of music or the noise of fingernails scratching across a chalkboard, you have "felt" sound. Even so, other sensory blendings may be difficult to imagine. How can sounds or letters possibly produce colors? Studies of synesthetes show that the same stimuli always elicit the same experiences for them.

The most common form of synesthesia is colored hearing. For many synesthetes, this means hearing both speech and music in color—experiencing a visual mélange of colored shapes, movement, and scintillation. The fact that colored hearing is more common than other types of synesthesia is curious.

Musician and composer Stevie Wonder is a synesthete, as were music legends Duke Ellington and Franz Liszt and Nobel Prize-winning physicist Richard Feynman.

CLINICAL FOCUS ⊕ 15-6

A Case of Synesthesia

Michael Watson tastes shapes. He first came to the attention of the neurologist Richard Cytowic over dinner. After tasting a sauce that he was making for roast chicken, Watson blurted out, "There aren't enough points on the chicken."

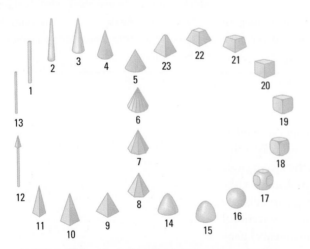

Neurologist Richard Cytowic devised this set of figures to help Michael Watson communicate the shapes he senses when he tastes food.

When Cytowic quizzed him about this strange remark, Watson said that all flavors had shape for him. "I wanted the taste of this chicken to be a pointed shape, but it came out all round. Well, I mean it's nearly spherical. I can't serve this if it doesn't have points" (Cytowic, 1998, p. 4).

Watson has synesthesia, which literally means "feeling together." All his life Watson has experienced the feeling of shape when he tastes or smells food. When he tastes intense flavors, he reports an experience of shape that sweeps down his arms to his fingertips. He experiences the feeling of weight, texture, warmth or cold, and shape, just as though he were grasping something.

The feelings are not confined to his hands, however. Watson experiences some taste shapes, such as points, over his whole body. He experiences others only on the face, back, or shoulders. These impressions are not metaphors, as other people might use when they say that a cheese is "sharp" or that a wine is "textured." Such descriptions make no sense to Watson. He actually feels the shapes.

Cytowic systematically studied Watson to determine whether his feelings of shape were always associated with particular flavors and found that they were. Cytowic devised the set of geometric figures shown here to allow Watson to communicate which shapes he associated with various flavors.

The five primary senses (vision, hearing, touch, taste, and smell) all generate synesthetic pairings. Most, however, are in one direction. For instance, whereas synesthetes may see colors when they hear, they do not hear sounds when they look at colors. Furthermore, some sensory combinations occur rarely, if at all. In particular, taste or smell rarely triggers a synesthetic response.

Because each case is idiosyncratic, the neurological basis of synesthesia is difficult to study. Few studies have related it directly to brain function or brain organization, and different people may experience it for different reasons. Various hypotheses have been advanced to account for synesthesia:

- Extraordinary neural connections between the different sensory regions are related in a particular synesthete.

- Activity is increased in multimodal areas of the frontal lobes that receive inputs from more than one sensory area.

- Particular sensory inputs elicit unusual patterns of cerebral activation.

Whatever the explanation, the brains of synesthetes clearly think differently from the brains of other people when it comes to certain types of sensory inputs.

synesthesia Ability to perceive a stimulus of one sense as the sensation of a different sense, as when sound produces a sensation of color; literally, "feeling together."

REVIEW 15-5
Variations in Cognitive Organization

Before you continue, check your understanding.

1. The two major contributors to organizational differences in individual brains are _____ and _____.

2. Differences in the cerebral organization of thinking are probably related to differences in the _____ that underlie different types of cognitive processing.

3. People who automatically experience sensations in more than one sensory modality are said to have _____.

4. What roles do gonadal hormones play in brain organization and function?

Answers appear at the back of the book.

15-6 Intelligence

Intelligence is a major influence on anyone's thinking ability. It is easy to identify in people and even easy to observe in other animals. Yet intelligence is not at all easy to define. Despite years of study, researchers have not yet reached agreement on what intelligence entails. We therefore begin this section by reviewing some hypotheses of intelligence.

Concept of General Intelligence

In the 1920s, Charles Spearman proposed that although different kinds of intelligence may exist, there is also an underlying general intelligence, which he called the "g" factor. Consider for a moment what a general intelligence factor might mean for the brain. Presumably, brains with high or low "g" would have some general difference in brain architecture—in gyral patterns, cytoarchitectonics, vascular patterns, or neurochemistry, for example.

This difference could not be something as simple as size, because human brain size (which varies from about 1000 to 2000 grams) correlates poorly with intelligence. Another possibility is that "g" is related to some special characteristic of cerebral

Cytoarchitectonics refers to the organization, structure, and distribution of brain cells.

Section 1-5 reveals the fallacies inherent in attempting to correlate human brain size with intelligence.

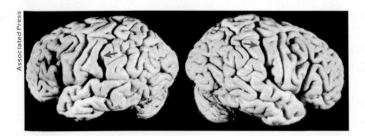

FIGURE 15-21 Einstein's Brain The lateral fissure (at arrows) takes an exaggerated upward course relative to its course in typical brains, essentially fusing the posterior temporal regions with the inferior parietal regions. Reprinted with the permission of S. Witelson, D. Kigar, T. Harvey, and *The Lancet*, June 19, 1999, p. 2151.

connectivity or even to the ratio of neurons to glia. Still another possibility is that "g" is related to the activation of specific brain regions, possibly in the frontal lobe (Duncan et al., 2000; Gray & Thompson, 2004).

The results of preliminary studies of Albert Einstein's brain imply that cerebral connectivity and glia-to-neuron ratio may play important roles. Sandra Witelson and her colleagues (Witelson, Kigar, & Harvey, 1999) found that, although Einstein's brain is the same size and weight as the average male brain, its lateral fissure is short, and both the left and the right lateral fissures take a particularly striking upward deflection (**Figure 15-21**, and compare Figure 15-12). This arrangement essentially fuses the inferior parietal area with the posterior temporal area.

The inferior parietal cortex has a role in mathematical reasoning, so it is tempting to speculate that Einstein's mathematical abilities were related to neural rearrangements in this area. But another important difference may distinguish Einstein's brain. Marion Diamond and her colleagues (1985) looked at its glia-to-neuron ratio versus the mean for a control population. They found that Einstein's inferior parietal cortex had a higher glia-to-neuron ratio than average: each neuron in this region had an unusually high number of glial cells supporting it.

The glia-to-neuron ratio was not unusually high in any other cortical areas of Einstein's brain measured by these researchers. Possibly, then, certain types of intelligence could be related to differences in cell structure in localized regions of the brain. But even if this hypothesis proves correct, it still offers little neural evidence in favor of a general intelligence factor.

A neuropsychological possibility is that the "g" factor is related to language processes in the brain. Recall that language ability qualitatively changes the nature of cognitive processing in humans. So perhaps people with very good language skills also have an advantage in general thinking ability.

Multiple Intelligences

Many other hypotheses of intelligence have been set forth since Spearman's, but few have considered the brain directly. One exception, proposed by Howard Gardner (1983), a neuropsychologist at Harvard, considers the effects of neurological injury on people's behavior. Gardner concludes that seven distinctly different forms of intelligence exist and that brain injury can selectively damage each form. The idea of multiple human intelligences should not be surprising given the many different types of cognitive operations that the human brain can perform.

Gardner's seven categories of intelligence are linguistic, musical, logical-mathematical, spatial, bodily-kinesthetic, intrapersonal, and interpersonal. Linguistic and musical intelligence are straightforward concepts, as is logical-mathematical intelligence. Spatial intelligence refers to the abilities discussed in this chapter, especially navigating in space, and to the ability to draw and paint. Bodily-kinesthetic intelligence refers to superior motor abilities, such as those exemplified by skilled athletes and dancers.

The two types of "personal" intelligence are less obvious. They refer to operations of the frontal and temporal lobes required for success in a highly social environment. The intrapersonal aspect is an awareness of one's own feelings, whereas the interpersonal aspect is the ability to recognize the feelings of others and to respond appropriately. Gardner's definition of intelligence has the advantage not only of being inclusive but also of acknowledging forms of intelligence not typically recognized by standard intelligence tests, abilities such as theory of mind, described in Section 15-3.

One prediction stemming from Gardner's analysis of intelligence is that brains ought to differ in some way when people have more of one form of intelligence and less of another. Logically, we could imagine that, if a person were higher in musical intelligence and lower in interpersonal intelligence, then the regions of the brain for music (especially the temporal lobe) would differ in some fundamental way from the "less efficient" regions for interpersonal intelligence. One way to examine such differences is to use fcMRI or DTI to identify differences in pathways, as in the example of absolute pitch (see Figure 15-11).

Divergent and Convergent Intelligence

One clear difference between lesions in the parietal and temporal lobes and lesions in the frontal lobes is in the way they affect performance on standardized intelligence tests. Posterior lesions produce reliable and often large decreases in intelligence test scores, whereas frontal lesions do not. This is puzzling. If frontal-lobe damage does not diminish a person's intelligence test score, why do people with this kind of damage often do "stupid" things? The answer lies in the difference between two kinds of intelligence referred to as divergent and convergent.

According to J. P. Guilford (1967), traditional intelligence tests measure what is called **convergent thinking**—applying knowledge and reasoning skills to narrow the range of possible solutions to a problem, then zeroing in on one correct answer. Typical intelligence test items using vocabulary words, arithmetic problems, puzzles, block designs, and so forth, all require convergent thinking. They demand a single correct answer that can be easily scored.

In contrast, **divergent thinking** reaches outward from conventional knowledge and reasoning skills to explore new, more unconventional solutions to problems. Divergent thinking assumes a variety of possible approaches and answers to a question rather than a single "correct" solution. A task that requires divergent thinking is to list all the possible uses you can imagine for a coat hanger. Clearly, a person who is very good at divergent thinking might not necessarily be good at convergent thinking, and vice versa.

The distinction between divergent and convergent intelligence is useful because it helps us to understand the effects of brain injury on thought. Frontal-lobe injury is believed to interfere with divergent thinking. The convergent thinking measured by standardized IQ tests is often impaired in people with damage to the temporal and parietal lobes.

Injury to the left parietal lobe, in particular, causes devastating impairment in the ability to perform cognitive processes related to academic work. People with this kind of injury may be aphasic, alexic, and apraxic. They often have severe deficits in arithmetic ability. All such impairments would interfere with school performance and performance at most jobs.

Patient M. M., discussed in Section 15-4, had left-parietal-lobe injury and was unable to return to school. In contrast with people like M. M., people with frontal-lobe injuries seldom have deficits in reading, writing, or arithmetic. And they show no decrement in standardized IQ tests. C. C.'s case provides a good example.

C. C. had a meningioma along the midline between the frontal lobes. Extracting it required removing brain tissue from both hemispheres. C. C. had been a prominent lawyer before his surgery. Afterward, although he still had a superior IQ and superior memory, he was unable to work, in part because he no longer had any imagination. He could not generate the novel solutions to legal problems that had characterized his career before the surgery. Thus, both M. M. and C. C. suffered problems that prevented them from working, but their problems differed because different kinds of thinking were affected by their injuries.

convergent thinking Form of thinking that searches for a single answer to a question (such as 2 + 2 = ?); contrasts with divergent thinking.

divergent thinking Form of thinking that searches for multiple solutions to a problem (such as How many different ways can a pen be used?); contrasts with convergent thinking.

Intelligence, Heredity, Epigenetics, and the Synapse

Another way to categorize human intelligence was proposed by Donald Hebb. Like Guilford, Hebb thought of people as having two forms of intelligence, which he called intelligence A and intelligence B. Unlike Guilford's convergent-divergent dichotomy, Hebb's **intelligence A** refers to innate intellectual potential, which is highly heritable: it has a strong genetic component. **Intelligence B** is observed intelligence, which is influenced by experience as well as other factors, such as disease, injury, or exposure to environmental toxins, especially during development.

Hebb (1980) understood that experience can influence the structure of brain cells significantly. In his view, experiences influence brain development, and thus observed intelligence because they alter the brain's synaptic organization. It follows that people with lower-than-average intelligence A can raise their intelligence B by appropriate postnatal experiences, whereas people with higher-than-average intelligence A can be negatively affected by a poor environment. The task is to identify a "good" and a "bad" environment in which to stimulate people to reach their highest potential intelligence.

One implication of Hebb's view of intelligence is that the brain's synaptic organization plays a key role. Synaptic organization is partly directed by a person's genes, but it is also epigenetic. Variations in the kinds of experiences to which people are exposed, coupled with variations in genetic patterns, undoubtedly contribute to the individual differences in intelligence that we observe—both quantitative differences (as measured by IQ tests) and qualitative differences (as in Gardner's view).

Hikaru Takeuchi and colleagues (2012) used fMRI to characterize brain activation while participants performed working memory tasks that varied in complexity. Performance on IQ tests and memory are highly correlated, so the investigators reasoned that brain activity during memory tests might reflect brain differences related to IQ score.

Performance speed correlated with increased activation in the right dorsolateral prefrontal cortex as well as an increase in the interaction between the prefrontal cortex and right posterior parietal cortex. Gray-matter volume in the right dorsolateral prefrontal region correlated with the participant's accuracy in working memory tasks, which in turn correlated with psychometric measures of intelligence.

Parallel results, obtained using event-related potentials (ERP), have been found by others (e.g., Langer et al., 2009). The general conclusion from these types of studies is that general intelligence is related to the efficiency of cortical networks linking prefrontal and parietal regions. We can speculate that the efficiency of different neural networks, as might be seen in fcMRI studies, will underlie the variation in each of Gardner's seven different forms of intelligence.

We recount Hebb's pioneering contributions to the development of enriched environments and their importance in early childhood education in Section 8-4.

intelligence A Hebb's term for innate intellectual potential, which is highly heritable and cannot be measured directly.

intelligence B Hebb's term for observed intelligence, which is influenced by experience as well as other factors in the course of development and is measured by intelligence tests.

consciousness The mind's level of responsiveness to impressions made by the senses.

Section 7-2 describes the use of ERP in mapping brain function.

REVIEW 15-6
Intelligence

Before you continue, check your understanding.

1. Different concepts of intelligence include Spearman's _____, Gardner's _____, Guilford's concepts of _____ and _____ thinking, and Hebb's _____ and _____.

2. Each form of intelligence that humans possess is probably related to the brain's _____ organization as well as to its _____ efficiency.

3. No two brains are alike. They differ, for example, in _____, _____, and _____.

4. Evidence that Hebb's intelligence A and intelligence B can be altered by experience is evidence of _____ influences on brain organization.

5. How might intelligence be related to brain activity?

Answers appear at the back of the book.

15-7 Consciousness

Our conscious experience is familiar and intimate yet remains largely a mysterious product of the brain. Everyone has an idea of what it means to be conscious, but like thinking and intelligence, consciousness is easier to identify than to define. Definitions range from a mere manifestation of complex thought processes to slipperier notions that see it as the subjective experience of awareness or the "inner self."

Despite the difficulty of defining consciousness, scientists generally agree that it is a process, not a thing. And consciousness is probably not a single process but a collection of several processes, such as those associated with seeing, talking, thinking, emotion, and so on.

Consciousness is not unitary but can take various forms. A person is not necessarily equally conscious at all stages of life. We don't think of a newborn baby as being conscious in the same way that a healthy older child or adult is. Indeed, we might say that part of the process of maturation is becoming fully conscious. The level of consciousness even changes across the span of a day as we pass through various states of drowsiness, sleep, and waking. One trait that characterizes consciousness, then, is its constant variability.

Why Are We Conscious?

Countless people, including neuroscience researchers, have wondered why we have the experience that we call **consciousness,** the mind's level of responsiveness to impressions made by the senses. The simplest explanation is that consciousness provides an adaptive advantage. Either our creation of the sensory world or our selection of behavior is enhanced by being conscious. Consider visual consciousness.

According to Francis Crick and Christof Koch (1998), an animal such as a frog acts a bit like a zombie when it responds to visual input. Frogs respond to small, preylike objects by snapping and to large, looming objects by jumping. These responses are controlled by different visual systems and are best thought of as reflexive rather than conscious. These visual systems work well for the frog. So why do humans need to add consciousness?

Crick and Koch suggest that reflexive systems are fine when their number is limited, but as their numbers grow, reflexive arrangements become inefficient, especially when two or more systems are in conflict. As the amount of information about an event increases, it becomes advantageous to produce a single, complex representation and make it available for a sufficient time to the parts of the brain—such as the frontal lobes—that choose among many possible plans of action. This sustained, complex representation is consciousness.

Of course, to survive we must retain the ability to respond quickly and unconsciously when we need to. This ability exists alongside our ability to process information consciously. The ventral visual stream is conscious, but the dorsal stream, which acts more rapidly, is not. The action of the unconscious, online dorsal stream can be seen in athletes. To hit a baseball or tennis ball traveling at more than 100 miles per hour requires athletes to swing before they are consciously aware of actually seeing the ball. Conscious awareness of the ball comes just after hitting it.

In a series of experiments, Marc Jeannerod and his colleagues (Castiello, Paulignan, & Jeannerod, 1991) found a similar dissociation between behavior and awareness in

Section 1-2 describes why the Glasgow Coma Scale, which objectively scores degrees of consciousness in people with TBI, is more useful than subjective statements that consciousness has "improved."

Section 13-3 explores various stages of sleeping and dream states.

Section 12-1 describes sensory-deprivation experiments and the conclusion investigators drew from them: the brain has an inherent need for stimulation.

Section 11-1 notes that movement sequences often happen automatically, without conscious control, as when you reach for your coffee mug.

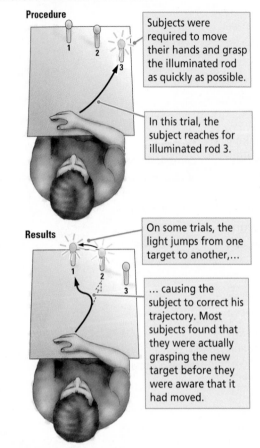

Question: Can people alter their movements without conscious awareness?

Procedure

Subjects were required to move their hands and grasp the illuminated rod as quickly as possible.

In this trial, the subject reaches for illuminated rod 3.

Results

On some trials, the light jumps from one target to another,…

… causing the subject to correct his trajectory. Most subjects found that they were actually grasping the new target before they were aware that it had moved.

Conclusion: It is possible to dissociate behavior and conscious awareness.

Adapted from "The Neural Correlations of Conscious Experience," by C. Frith, R. Perry, and E. Lumer, 1999, *Trends in Cognitive Sciences, 3,* pp. 105–114.

Focus 9-1 describes blindsight and Section 9-4 visual form agnosias. Section 14-2 reviews implicit learning in amnesia and Section 16-4 OCD.

Focus 11-5 outlines techniques that minimize phantom limb pain and Figure 14-20 how limb amputation remaps the cortex. Focus 8-5 describes brain abnormalities accompanying schizophrenia and Section 16-4 its diagnosis and treatment.

normal volunteers as they make grasping movements. **Experiment 15-5** illustrates the results of a representative experiment. Participants were required to grasp one of three rods as quickly as possible. The correct target rod on any given trial was indicated by a light on that rod.

On some trials, unknown to the participants, the light jumped from one target to another. Participants were asked to report whether such a jump had occurred. As shown in the Results section of the experiment, although participants were able to make the trajectory correction, they were sometimes actually grasping the correct target before they were aware that it had changed.

On some trials, the extent of dissociation between motor and vocal responses was so great that, to their surprise, participants had grasped the target some 300 milliseconds before they emitted the vocal response. Like baseball players, they experienced conscious awareness of the stimulus event only after their movements had taken place. No thought was required to make the movement, just as frogs catch flies without having to think about it.

Such movements are different from those consciously directed toward a specific object, as when we reach into a bowl of jellybeans to select a candy of a certain color. In this case, we must be aware of all the different colors surrounding the color we want. Here the conscious ventral stream is needed to discriminate among particular stimuli and respond differentially to them. Consciousness, then, allows us to select behaviors that correspond to an understanding of the nuances of sensory inputs.

What Is the Neural Basis of Consciousness?

Consciousness must be related in some way to neural-system activity in the brain, particularly in the forebrain. One way to investigate these systems is to contrast two kinds of neurological conditions.

In the first condition, a person lacks conscious awareness about some subset of information, even though he or she processes that information unconsciously. Examples include blindsight, visual form agnosia, implicit learning in amnesia, and visual neglect (discussed in Section 15-2). Another example is obsessive-compulsive disorder, in which people persist in some checking behavior—to see that the stove is off or the door is locked, even though they have already checked many times.

All these phenomena show that stimuli can be highly processed by the brain without entering conscious awareness. This is quite different from the second type of neurological condition, in which people are consciously aware of stimuli that are not actually there. Examples include phantom limbs and the hallucinations of schizophrenia. In both, consciousness of specific events, such as pain in a missing limb or hearing voices, exists even though these events clearly are not "real."

We can draw two conclusions from these contrasting conditions. First, the representation of a visual object or event is likely to be distributed over many parts of the visual system and probably over parts of the frontal lobes as well. Damage to different areas not only produces different specific symptoms, such as agnosia or neglect, but can also produce a specific loss of visual consciousness. Disordered functioning can induce faulty consciousness, such as hallucinations. Second, because visual consciousness can be lost, it follows that parts of the neural circuit must produce this awareness.

In Section 15-1, we appointed the neuron the unit of thinking. It is unlikely, however, that the neuron can be the unit of conscious experience. Instead, consciousness presumably is a process that emerges from neural circuits, with greater degrees of consciousness associated with increasingly complex circuitry.

For this reason, humans, with their more complex brain circuits, are often credited with a greater degree of consciousness than other animals have. Simple animals such as worms are assumed to have less consciousness (if any) than dogs, which in turn are assumed to have less consciousness than humans. Brain injury may alter self-awareness in humans, as in contralateral neglect, but unless a person is in a coma, he or she still retains some conscious experience.

Some people argue that language fundamentally changes the nature of consciousness. Recall Gazzaniga's belief that the left hemisphere, with its language capabilities, acts as an interpreter of stimuli (see Section 15-4). He maintains that this ability is an important difference between the functions of the two hemispheres.

Yet people who are aphasic have not lost consciousness. Although language may alter the nature of our conscious experience, equating any one brain structure with consciousness seems an unlikely hypothesis. Rather, viewing consciousness as a product of all cortical areas, their connections, and their cognitive operations holds more promise.

We end our discussion of thinking on an interesting, if speculative, note. David Chalmers (1995) proposes that consciousness includes not only the information the brain experiences through its sensory systems but also the information the brain has stored and presumably the information the brain can imagine. In his view, consciousness is the end product of all the brain's cognitive processes.

An interesting implication of Chalmers's notion is that, as the brain changes with experience, so does the state of consciousness. As our sensory experiences become richer and our store of information greater, our consciousness may become more complex. From this perspective, there may indeed be some advantage to growing old.

REVIEW 15-7
Consciousness

Before you continue, check your understanding.

1. Over the course of human evolution, one characteristic of sensory processing is that it has become more _____.

2. _____ is the mind's level of responsiveness to impressions made by the senses.

3. As relative human brain size and complexity have increased, so too has our degree of _____.

4. Not all behavior is under conscious control. What types of behaviors are not conscious?

Answers appear at the back of the book.

SUMMARY

15-1 Nature of Thought

One product of both human and nonhuman brain activity is the complex processes we call thinking, or cognition. We use such words as *language* and *memory* to describe various cognitive operations. These concepts are not physical things but rather psychological constructs. They are merely inferred and are not found in discrete places in the brain.

The brain carries out multiple cognitive operations. Perception, action for perception, imagery, planning, spatial cognition, and attention—each requires the widespread activity of many cortical areas. The unit of cognition, however, is the neuron.

15-2 Cognition and the Association Cortex

The brain's association cortex includes medial, dorsal and orbital subdivisions of the prefrontal cortex, the posterior parietal cortex, and anterior regions of the temporal lobe. Cell assemblies in the association cortex specifically take part in most forms of cognition.

The frontal lobes not only plan, organize, and initiate movements; they also organize our behavior over time. As a general rule, the temporal lobes generate knowledge about objects, whereas the parietal lobes produce various forms of spatial cognition. Neurons in both the temporal and the parietal lobes seem to contribute to our ability to selectively attend to particular sensory information.

Regions in the frontal and parietal lobes contain mirror neurons that represent actions, one's own or those of others. Such neural representations could be used both for imitating others' actions and for understanding their meanings. A significant area of the cortex is multisensory, which allows the brain to combine characteristics of stimuli across different sensory modalities, whether we encounter them together or separately.

15·3 Expanding Frontiers of Cognitive Neuroscience

Neuropsychological studies that began in the late 1800s to examine the behavioral capacities of people and laboratory animals with localized brain injuries do not allow investigators to study "normal" brains. But the development of noninvasive brain-recording systems and imaging techniques has led to the field of cognitive neuroscience, which studies the neural basis of cognition by measuring brain activity while normal participants are engaged in various cognitive tasks.

An important step in identifying the neural bases of cognition is to map the connections of the cerebral cortex, known as the brain connectome. Two promising imaging tools for mapping the connectome are functional connectivity magnetic resonance imaging (fcMRI) and tractography using diffusion tensor imaging.

One focus of social neuroscience, a field that combines cognitive neuroscience with social psychology, is exploring how we understand the intentions of others by creating a theory of mind. Social neuroscience also investigates how we develop a sense of self and attitudes and beliefs. Using noninvasive imaging techniques such as fMRI, researchers have shown that social cognition primarily involves activity in the prefrontal cortex.

Neuroeconomics is a new field that combines psychology, neuroscience, and economics and seeks to understand human decision making. fMRI studies have shown two decision-making pathways. One is reflective, involving diffuse regions of association cortex. The other is reflexive, involving the dopaminergic reward system.

15·4 Cerebral Asymmetry in Thinking

Cognitive operations are organized asymmetrically in the left and right cerebral hemispheres: each carries out complementary functions. Various syndromes result from association-cortex injury, among them agnosia, apraxia, aphasia, and amnesia. Each includes the loss or disturbance of a form of cognition. The most obvious functional difference is language, which normally is housed in the left hemisphere.

Cerebral asymmetry, manifested in anatomical differences between the two hemispheres, can be inferred from the differential effects of injury to opposite sides of the brain. Asymmetry can also be seen in the normal brain and in the brain that is surgically split for the relief of intractable epilepsy.

15·5 Variations in Cognitive Organization

Unique brains produce unique thought patterns. Marked variations in brain organization that exist among individuals are exhibited in idiosyncratic capacities such as synesthesia. Systematic differences in cognition exist as well, manifested in the performance of females and males on various cognitive tests, especially on tests of spatial and verbal behavior.

Sex differences in cognition result from the action of gonadal hormones on cortical organization, possibly on the architecture of cortical neurons and ultimately on neural networks. Marked differences are observed in the anatomical organization of the female and male cerebral hemispheres.

Differences in hemispheric organization also appear in right- and left-handers. Rather than being a single group, left-handers constitute at least three different groups. One appears to have speech in the left hemisphere, as right-handers do, and two have anomalous speech representation, either in the right hemisphere or in both hemispheres. The reasons for these organizational differences remain unknown.

15·6 Intelligence

Intelligence is easy to spot but difficult to define. Obvious differences in intelligence exist across species, as well as within a species, and we find varied forms of intelligence among humans within our own culture and in other cultures. Intelligence is not related to differences in brain size within a species or to any obvious gross structural differences among different members of the species. It may be related to differences in synaptic organization and processing efficiency.

15·7 Consciousness

The larger a species' brain is relative to its body size, the more knowledge the brain creates. Consciousness, the mind's level of responsiveness to impressions made by the senses, is a property that emerges from the complexity of the nervous system.

KEY TERMS

anomalous speech representation, p. 554

association cortex, p. 531

attention, p. 533

binding problem, p. 531

brain connectome, p. 538

cell assembly, p. 526

cognition, p. 524

cognitive neuroscience, p. 538

consciousness, p. 560

contralateral neglect, p. 534

convergent thinking, p. 559

dichotic listening, p. 546

divergent thinking, p. 559

extinction, p. 534

hyperconnectivity, p. 541

intelligence A, p. 560

intelligence B, p. 560

mirror neuron, p. 537

neuroeconomics, p. 542

perseveration, p. 537

psychological construct, p. 524

social neuroscience, p. 541

split brain, p. 524

synesthesia, p. 557

syntax, p. 524

theory of mind, p. 541

Please refer to the Companion Web Site at www.worthpublishers.com/kolbintro4e *for Interactive Exercises and Quizzes.*

CHAPTER

16

What Happens When the Brain Misbehaves?

16-1 MULTIDISCIPLINARY RESEARCH ON BRAIN AND BEHAVIORAL DISORDERS

RESEARCH FOCUS 16-1 POSTTRAUMATIC STRESS DISORDER

CAUSES OF ABNORMAL BEHAVIOR

INVESTIGATING THE NEUROBIOLOGY OF BEHAVIORAL DISORDERS

16-2 CLASSIFYING AND TREATING BRAIN AND BEHAVIORAL DISORDERS

IDENTIFYING AND CLASSIFYING BEHAVIORAL DISORDERS

TREATMENTS FOR DISORDERS

RESEARCH FOCUS 16-2 TREATING BEHAVIORAL DISORDERS WITH TRANSCRANIAL MAGNETIC STIMULATION

16-3 UNDERSTANDING AND TREATING NEUROLOGICAL DISORDERS

TRAUMATIC BRAIN INJURY

CLINICAL FOCUS 16-3 CONCUSSION

STROKE

EPILEPSY

MULTIPLE SCLEROSIS

NEURODEGENERATIVE DISORDERS

ARE PARKINSON'S AND ALZHEIMER'S ASPECTS OF ONE DISEASE?

AGE-RELATED COGNITIVE LOSS

16-4 UNDERSTANDING AND TREATING BEHAVIORAL DISORDERS

PSYCHOTIC DISORDERS

MOOD DISORDERS

RESEARCH FOCUS 16-4 ANTIDEPRESSANT ACTION AND BRAIN REPAIR

ANXIETY DISORDERS

16-5 IS MISBEHAVIOR ALWAYS BAD?

Posttraumatic Stress Disorder

Life is filled with stress. Routinely, we cope. But some events are so physically threatening and emotionally shattering that people who endure them experience long-term consequences. Flashbacks and nightmares persist long after any physical danger has passed. These symptoms can lead to emotional numbness and a diagnosis of **posttraumatic stress disorder** (PTSD).

Traumatic events that may trigger PTSD include violent personal assault, natural or human-caused disaster, accident, and war. An estimated 1 in 6 veterans of the conflicts in Iraq and Afghanistan, including many not directly exposed to combat, developed symptoms of depression, anxiety, and PTSD.

Understanding the neural basis and identifying new PTSD treatments has spurred intense interest. Treatment is often difficult.

Patients with PTSD have significant reductions bilaterally, both in the volume of the hippocampus and amygdala of the temporal lobes and of the frontal cortex. Reduced cortical thickness, shown in the accompanying cortical-imaging maps from Geuze and colleagues (2008), is associated with reduced cerebral blood flow and deficits in performance on neuropsychological tests of frontal- and temporal-lobe function.

Predisposing factors may contribute to PTSD, and these include reduced volume in brain areas such as the hippocampal and amygdala, a susceptibility to stress and a life background characterized by poor social support (Geuze et al., 2012).

As recently as 1980, when posttraumatic stress disorder acquired its name, PTSD was labeled a "psychological" problem related to individuals who were trying to repress unpleasant experiences. Treatment for PTSD was psychotherapy. Patients were encouraged to imagine and talk about the stressful experiences they endured.

But people enduring PTSD are trying to forget rather than relive. Today, treatment based on virtual-reality (VR) simulations is improving PTSD treatment outcomes for war veterans.

In **virtual-reality (VR) exposure therapy,** a controlled, virtual-immersion environment combines realistic street scenes, sounds, and odors that allows war veterans to relive traumatic events. The Virtual Iraq program, for example, can be customized to start with benign events—such as children playing—and gradually add increasingly stressful components, culminating in such traumatic events as a roadside bomb exploding in the virtual space around an armored personnel carrier.

To make Virtual Iraq realistic, the system pumps in smells, stepping up from the scent of bread baking to body odor to the reek of gunpowder and burning rubber. Speakers provide sounds while off-the-shelf subwoofers mounted under the subject's chair recreate movements.

The first study on virtual-reality exposure therapy reported positive results (Reger et al., 2011). VR exposure therapy is now used prior to combat exposure as a means of preventing PTSD from developing (Rizzo et al., 2012).

Cortical thickness (mm)

5.00

0.00

Veterans with PTSD

Veterans without PTSD

Cortical Thickness and PTSD Blue shading reflects reduced cortical thickness in group-averaged brains of veterans with PTSD (*top*) compared with veterans without PTSD (*bottom*). **Adapted from "Thinner Prefrontal Cortex in Veterans with Posttraumatic Stress Disorder," by E. Geuze, H. G. M. Westenberg, A. Heinecke, C. S. de Kloet, R. Goebel, & E. Vermetten, 2008, *NeuroImage, 41,* 675–681 (Figure 2, p. 678).**

The search for neural markers and effective treatments for PTSD underscores a shift in thinking about most "psychological disorders" as brain disorders. Just as the brain is the source of normal behavior, it is also the source of abnormal behavior. Progress toward identifying the neural bases of brain disorders is hampered by the same challenges that hamper understanding the typical brain.

The brain is complex. The functions of all its parts are not yet understood, nor is it yet clear how the brain produces mind, a sense of well-being, and a sense of self. Still,

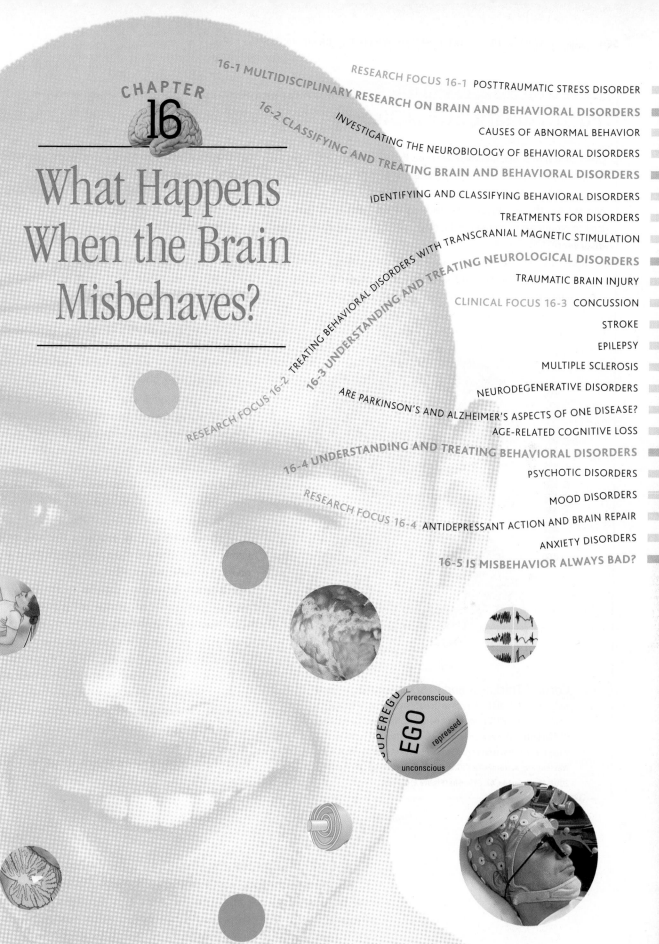

CHAPTER

16

What Happens When the Brain Misbehaves?

RESEARCH FOCUS 16-1 POSTTRAUMATIC STRESS DISORDER

16-1 MULTIDISCIPLINARY RESEARCH ON BRAIN AND BEHAVIORAL DISORDERS

CAUSES OF ABNORMAL BEHAVIOR

INVESTIGATING THE NEUROBIOLOGY OF BEHAVIORAL DISORDERS

16-2 CLASSIFYING AND TREATING BRAIN AND BEHAVIORAL DISORDERS

IDENTIFYING AND CLASSIFYING BEHAVIORAL DISORDERS

TREATMENTS FOR DISORDERS

RESEARCH FOCUS 16-2 TREATING BEHAVIORAL DISORDERS WITH TRANSCRANIAL MAGNETIC STIMULATION

16-3 UNDERSTANDING AND TREATING NEUROLOGICAL DISORDERS

TRAUMATIC BRAIN INJURY

CLINICAL FOCUS 16-3 CONCUSSION

STROKE

EPILEPSY

MULTIPLE SCLEROSIS

NEURODEGENERATIVE DISORDERS

ARE PARKINSON'S AND ALZHEIMER'S ASPECTS OF ONE DISEASE?

AGE-RELATED COGNITIVE LOSS

16-4 UNDERSTANDING AND TREATING BEHAVIORAL DISORDERS

PSYCHOTIC DISORDERS

MOOD DISORDERS

RESEARCH FOCUS 16-4 ANTIDEPRESSANT ACTION AND BRAIN REPAIR

ANXIETY DISORDERS

16-5 IS MISBEHAVIOR ALWAYS BAD?

Posttraumatic Stress Disorder

Life is filled with stress. Routinely, we cope. But some events are so physically threatening and emotionally shattering that people who endure them experience long-term consequences. Flashbacks and nightmares persist long after any physical danger has passed. These symptoms can lead to emotional numbness and a diagnosis of **posttraumatic stress disorder** (PTSD).

Traumatic events that may trigger PTSD include violent personal assault, natural or human-caused disaster, accident, and war. An estimated 1 in 6 veterans of the conflicts in Iraq and Afghanistan, including many not directly exposed to combat, developed symptoms of depression, anxiety, and PTSD.

Understanding the neural basis and identifying new PTSD treatments has spurred intense interest. Treatment is often difficult.

Patients with PTSD have significant reductions bilaterally, both in the volume of the hippocampus and amygdala of the temporal lobes and of the frontal cortex. Reduced cortical thickness, shown in the accompanying cortical-imaging maps from Geuze and colleagues (2008), is associated with reduced cerebral blood flow and deficits in performance on neuropsychological tests of frontal- and temporal-lobe function.

Predisposing factors may contribute to PTSD, and these include reduced volume in brain areas such as the hippocampal and amygdala, a susceptibility to stress and a life background characterized by poor social support (Geuze et al., 2012).

As recently as 1980, when posttraumatic stress disorder acquired its name, PTSD was labeled a "psychological" problem related to individuals who were trying to repress unpleasant experiences. Treatment for PTSD was psychotherapy. Patients were encouraged to imagine and talk about the stressful experiences they endured.

But people enduring PTSD are trying to forget rather than relive. Today, treatment based on virtual-reality (VR) simulations is improving PTSD treatment outcomes for war veterans.

In **virtual-reality (VR) exposure therapy,** a controlled, virtual-immersion environment combines realistic street scenes, sounds, and odors that allows war veterans to relive traumatic events. The Virtual Iraq program, for example, can be customized to start with benign events—such as children playing—and gradually add increasingly stressful components, culminating in such traumatic events as a roadside bomb exploding in the virtual space around an armored personnel carrier.

To make Virtual Iraq realistic, the system pumps in smells, stepping up from the scent of bread baking to body odor to the reek of gunpowder and burning rubber. Speakers provide sounds while off-the-shelf subwoofers mounted under the subject's chair recreate movements.

The first study on virtual-reality exposure therapy reported positive results (Reger et al., 2011). VR exposure therapy is now used prior to combat exposure as a means of preventing PTSD from developing (Rizzo et al., 2012).

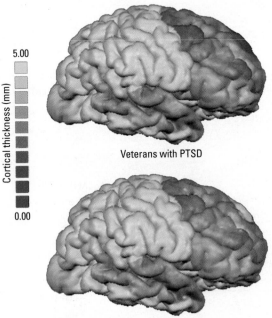

Veterans with PTSD

Veterans without PTSD

Cortical Thickness and PTSD Blue shading reflects reduced cortical thickness in group-averaged brains of veterans with PTSD (*top*) compared with veterans without PTSD (*bottom*). **Adapted from "Thinner Prefrontal Cortex in Veterans with Posttraumatic Stress Disorder," by E. Geuze, H. G. M. Westenberg, A. Heinecke, C. S. de Kloet, R. Goebel, & E. Vermetten, 2008, *NeuroImage, 41*, 675–681 (Figure 2, p. 678).**

The search for neural markers and effective treatments for PTSD underscores a shift in thinking about most "psychological disorders" as brain disorders. Just as the brain is the source of normal behavior, it is also the source of abnormal behavior. Progress toward identifying the neural bases of brain disorders is hampered by the same challenges that hamper understanding the typical brain.

The brain is complex. The functions of all its parts are not yet understood, nor is it yet clear how the brain produces mind, a sense of well-being, and a sense of self. Still, significant advances in understanding have led to the realization that under some circumstances, our brains are competent in coping with life's challenges while under other circumstances, its "misbehavior" produces PTSD and other disorders.

To illustrate progress in studying brain and behavior over the past century, contrast the theories of Sigmund Freud with present-day views of behavior and the brain. Freud's theories were based on his observations of his patients without the help of the anatomical or imaging data available today. The underlying tenet of Freud's theory is that our motivations are largely hidden in our unconscious minds.

Freud posited that a mysterious, repressive force actively withholds our sexual and aggressive motivation from conscious awareness. He believed that mental illness results from the failure of the repressive processes. Freud proposed the three components of mind illustrated in **Figure 16-1A**:

1. Primitive functions, including the "instinctual drives" of sex and aggression, are located in the *id,* the part of the mind that Freud thought operated on an unconscious level.

2. The rational part of the mind he called the *ego.* Much of the ego's activity Freud also believed to be unconscious, although experience (to him, our perceptions of the world) is conscious.

3. The *superego* aspect of mind acts to repress the id and to mediate ongoing interactions between the ego and the id.

For Freudians, abnormal behaviors result from the emergence of unconscious drives into voluntary, conscious behavior. The aim of *psychoanalysis,* the original talking therapy, is to trace symptoms back to their unconscious roots and thus expose them to rational judgment.

By the 1970s, scientific studies of the brain were making the whole notion of id-ego-superego seem antiquated. Nevertheless, some resemblance between Freud's theory and brain theory is apparent (Figure 16-1B). The limbic system and brainstem have properties akin to the id: they produce emotional and motivated behaviors, including the will to survive and to reproduce. The neocortex has properties akin to the ego; it allows us to learn and to solve everyday problems. The frontal cortex has properties akin to the superego: it enables us to be aware of others and learn to follow social norms.

Furthermore, many processes underlying these functions are unconscious: they operate outside our awareness. What is different between Freud's view and present-day neuroscience is the knowledge that the brain is composed of hundreds of interacting structures, not just three.

Investigating the origins and treatment of abnormal behavior is perhaps the most fascinating pursuit in the study of the brain and behavior. Today, neurologists treat *organic*

posttraumatic stress disorder (PTSD) Syndrome characterized by physiological arousal symptoms brought on by recurring memories and dreams related to a traumatic event for months or years after the event.

virtual-reality (VR) exposure therapy Controlled, virtual-immersion environment that, by allowing individuals to relive traumatic events, gradually desensitizes them to stress.

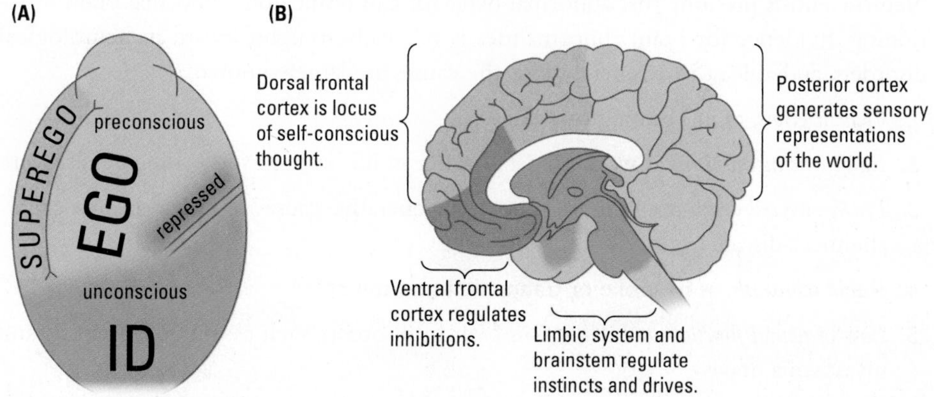

(A)

SUPEREGO
preconscious
EGO
repressed
unconscious
ID

(B)

Dorsal frontal cortex is locus of self-conscious thought.

Posterior cortex generates sensory representations of the world.

Ventral frontal cortex regulates inhibitions.

Limbic system and brainstem regulate instincts and drives.

FIGURE 16-1 Mind Models **(A)** Freud based his model of the mind, drawn in 1933, solely on clinical observations (color added). **(B)** In a contemporary brain-imaging and lesioning studies map, the brainstem and limbic system correlate with Freud's depiction of the id, the ventral frontal and posterior cortex with the ego, and the dorsal frontal cortex with the superego.
Part A from A. W. Freud by arrangement with Paterson Marsh Ltd, London; coloring added by Oliver Turnbull. Part B adapted from a drawing by Oliver Turnbull.

disorders of the nervous system—conditions such as Parkinson's disease and stroke—medically. Psychiatrists treat *behavioral disorders* such as schizophrenia and PTSD with pharmacological and other medical treatments in combination with behavioral treatments. Increasingly, these practitioners are synthesizing insights from both disciplines into a unified understanding of mind and brain—a **neuropyschoanalysis.**

Psychologists, social workers, and other health-care professionals treat and assist clients with many everyday complexities of living such as those posed by PTSD. Family members also participate because they are most affected when loved ones develop brain disorders, and they are usually the primary caregivers. Clearly, it is beneficial when all these groups share a common view—that the brain is ultimately the source of behavior—and of misbehavior.

With the organic–neurological and behavioral–psychiatric distinction in mind, we first survey how researchers investigate the neurobiology of organic and behavioral disorders. We then examine how disorders are classified, treated, and distributed in the population. In the last half of the chapter we review established and emerging treatments for disorders.

16-1 Multidisciplinary Research on Brain and Behavioral Disorders

Chapter 7 surveys a full range of research methods, from single-cell recordings to dynamic brain imaging.

Brain research is multidisciplinary. One way to summarize the methods of studying the link between brain and behavior is to consider them from the macro level of the whole organism down to the molecular level of neuronal excitation. Behavioral studies by their very nature investigate the whole organism, but understanding the whole organism requires understanding its parts—its cells, its chemistry, and its genes.

Section 3-3 reviews genetics, genetic engineering, and epigenetic mechanisms. Section 7-5 explores techniques for measuring genetic and epigenetic influences on brain and behavior.

The emergence of molecular biology offers neuroscientists varied approaches to studying behavior. Scientists can breed strains of animals, usually mice, with either a gene "knocked out" (deleted or inactivated) or a gene inserted. Knockout technology is used both to create animal models of human disorders and to generate treatments for neurobehavioral disorders.

Improvements in brain-imaging techniques enable researchers and physicians to measure changes in anatomy and brain activity without direct access to the brain. They can describe structures and pathways in an individual brain and the changes that the brain undergoes during development, learning, and after damage. The effects of treatments for brain disorders can be observed both in behavior and in ongoing brain activity. Hand in hand with these improvements, our understanding of brain behavior and of brain misbehavior are undergoing constant revision.

Causes of Abnormal Behavior

Neuroscientists presume that abnormal behavior can result from abnormal brain functioning. Evidence for brain abnormalities is relatively straightforward in neurological disorders, and at least in a general sense, the causes are largely known:

1. *Genetic errors,* as in Huntington's disease

2. *Epigenetic mechanisms* at work prenatally, later in life, even in succeeding generations

Focus 3-4 describes the genetic basis of Huntington's disease, Focus 5-2 neural degeneration in Parkinson's disease, Focus 2-3 symptoms and aftereffects of stroke, and Focus 3-3 the neural breakdown attendant to MS.

3. *Progressive cell death* resulting from neurodegenerative causes, as in Parkinson's or Alzheimer's disease

4. *Rapid cell death,* as in stroke or traumatic brain injury

5. *Loss of neural function and connections* seen in disorders such as multiple sclerosis and myasthenia gravis

In contrast to these organic–neurological disorders, far less is known about the neurobiological causes of behavioral–psychiatric disorders. To date, no large-scale neurobiological studies have been conducted of either postmortem pathology or biochemical pathology in the population at large. Still, clues to possible causes of psychiatric behaviors have been uncovered. In each case, some abnormality of the brain's structure or activity must be implicated. The questions asked by behavioral neuroscientists are What is that particular brain abnormality? What is its cause?

Table 16-1 lists the most likely categories of causes underlying behavioral disorders, micro to macro. At the microscopic level is genetic error, such as that responsible for Tay-Sachs disease and Huntington's disease. Genetic error is probably linked to some other proposed causes, such as hormonal or developmental anomalies, as well.

Genes may be the source not only of anatomical, chemical, or physiological defects but also of susceptibility to other factors that may cause behavioral problems. A person's genetic vulnerability to stress, infection, or pollution may be the immediate cause of some abnormal conditions listed in Table 16-1. In other cases, no direct genetic predisposition is needed: abnormal behavior arises strictly from epigenetic factors that influence gene expression and function.

The triggering environmental factor that produces epigenetic change might be poor nutrition or exposure to toxins, including naturally occurring toxins, manufactured chemicals, and infectious agents. Other disorders are undoubtedly related to negative experiences ranging from developmental deprivation to extreme psychosocial neglect and traumas or chronic stress in later life.

Investigating the Neurobiology of Behavioral Disorders

A single brain abnormality can cause a behavioral disorder, explaining everything about the disorder and its treatment. **Phenylketonuria** (PKU) is such a disorder. PKU results from a defect in the gene for phenylalanine hydroxylase, an enzyme that breaks down the amino acid phenylalanine. Babies with PKU have elevated levels of phenylalanine in their blood.

Left untreated, PKU causes severe mental retardation, but PKU can easily be treated just by restricting the dietary intake of phenylalanine—foods high in protein, including beef, fish, cheese, and soy. Expectant mothers who have had PKU might provide an in utero environment with high phenylalanine levels, but that, too, can be controlled if the mother restricts her dietary intake of phenylalanine.

If other behavioral disorders were as simple and well understood as PKU, research in neuroscience could quickly yield cures for them. Many disorders do not result from a single genetic abnormality, however, and the causes of most disorders remain largely conjectural. The major problem is that psychiatric diagnosis of a disability is based mainly on behavioral symptoms, and behavioral symptoms give few clues to specific neurochemical or neurostructural causes.

This problem also can be seen in treating PKU. **Table 16-2** lists what is known about PKU at different levels of analysis: genetic, biochemical, histological, neurological, behavioral, and social. The underlying problem becomes less apparent with the procession of entries in the table. In fact, it is not possible to predict the specific biochemical abnormality from information at the neurological, behavioral, or social levels. But the primary information available *is* at the neurological, behavioral, and social levels.

For most psychiatric diseases, the underlying pathology is unknown. For PKU, an elevated phenylpyruric acid level in the urine of a single patient was the organic clue

TABLE 16-1 Causes of Certain Behavioral Disorders

Cause	Disorder
Genetic error	Tay-Sachs disease
Hormonal anomaly	Androgenital syndrome
Developmental anomaly	Schizophrenia
Infection	Encephalitis
Injury	Traumatic brain injury
Toxins	MPTP poisoning
Poor nutrition	Korsakoff's syndrome
Stress	Anxiety disorders
Negative experience	Developmental delays among Romanian orphans

The Index of Disorders inside the book's front cover lists where each disorder in Table 16-1 is discussed.

neuropsychoanalysis Movement within neuroscience and psychoanalysis to combine the insights of both to yield a unified understanding of mind and brain.

phenylketonuria (PKU) Behavioral disorder caused by elevated levels of the amino acid phenylalanine in the blood and resulting from a defect in the gene for the enzyme phenylalanine hydroxylase; the major symptom is severe mental retardation.

TABLE 16-2 Phenylketonuria: Known Neurobiological Pathogenesis of a Behavioral Disorder

Level of analysis	Information known
Genetic	Inborn error of metabolism; autosomal recessive defective gene
Biochemical pathogenesis	Impairment in the hydroxylation of phenylalanine to tyrosine, causing elevated blood levels of phenylalanine and its metabolites
Histological abnormality	Decreased neuron size and dendritic length, and lowered spine density; abnormal cortical lamination
Neurological findings	Severe mental retardation, slow growth, abnormal EEG
Behavioral symptoms	For 95 percent of patients, IQ below 50
Social disability	Loss of meaningful, productive life; significant social and economic cost
Treatment	Restrict dietary intake of phenylalanine

Source: Adapted from "Special Challenges in the Investigation of the Neurobiology of Mental Illness," by G. R. Heninger, 1999, in *The Neurobiology of Mental Illness* (p. 90), edited by D. S. Charney, E. J. Nestler, and B. S. Bunney. New York: Oxford University Press.

needed to understand and treat the behavioral disorder. The task for future study and treatment of most behavioral disorders is to identify the biological markers that will lead to similar understandings.

Challenges to Diagnosis

Knowledge about behavioral disorders is hampered by its subjective nature. Most diagnostic information gathered concerns a patient's behavior, which comes from both patients and their families. Unfortunately, people seldom are objective observers of their own behavior or that of a loved one. We tend to be selective in noticing and reporting symptoms. If we believe that someone has a memory problem, for example, we often notice memory lapses that we might ordinarily ignore.

Nor are we often specific in identifying symptoms. Simply identifying a memory problem is not really helpful. Treatment requires knowing exactly what type of memory deficit underlies the problem. The losses of memory for words, places, and habits have very different underlying pathologies and brain systems.

Just as patients and their loved ones make diagnosis difficult, so too do those who perform the diagnosing. Behavioral information about patients may be interpreted by general physicians, psychiatrists, neurologists, psychologists, or social workers, as well as others. Evaluators with different conceptual biases shape and filter the questions they ask and the information they gather differently.

One evaluator believes that most behavioral disorders are genetic in origin, another believes that most result from a virus, and a third believes that many can be traced to repressed sexual experiences during childhood. Each makes different types of observations and administers different kinds of diagnostic tests. In contrast, diagnosing organic disorders is less dependent on subjective observations than on objective experimental methods, but these, too, have limitations.

Research Challenges

Even if the problems of diagnosis were solved, major obstacles to investigating behavioral disorders would still exist. Following is a partial list.

ORGANIZATIONAL COMPLEXITY The nervous system far outstrips other body systems in complexity. The brain has a wider variety of cell types than does any other organ, and the cells and their connections are plastic: they change with experience. These features of neurons add a whole new dimension to understanding normal and abnormal functioning.

Sections 14-2 and 14-3 explain a range of memory deficits.

SYSTEMIC COMPLEXITY As our understanding of brain and behavior has progressed, it has become apparent that multiple receptor systems serve many different functions. For example, the activating systems for acetylcholine, dopamine, norepinephrine, serotonin, and GABA are diffuse, with little specificity between biochemistry and behavior. As George Heninger (1999) pointed out, no clear demonstration of a single receptor system with a specific relation to a specific behavior has as yet been made. The neurotransmitter GABA affects some 30 percent of the synapses in the brain. When people ingest GABA agonists, such as benzodiazepines, multiple effects on behavior become apparent. It is difficult to administer enough of a benzodiazepine to reduce anxiety to a "normal" level without producing sedative side effects.

NEURONAL PLASTICITY Even the nigrostriatal dopaminergic system's close relation to Parkinson's disease is enigmatic. It is impossible to tie dopamine depletion to a consistent behavioral syndrome. Two people with Parkinson's disease can have vastly different symptoms, even though the common basis of the disease is a loss of neurons from the substantia nigra. Furthermore, only when the loss of dopamine neurons exceeds about 60 to 80 percent do investigators see clinical signs of Parkinson's disease. Are all those cells not needed? That is unlikely, but the result shows that the brain's compensatory plasticity is considerable. When diseases progress slowly, the brain has a remarkable capacity for adapting.

COMPENSATORY PLASTICITY Even the best technology produces uncertain relationships. Magnetic resonance imaging may show that a person with multiple sclerosis has many nervous system lesions, yet the person displays very few outward symptoms. Just as brain lesions do not always produce behavioral symptoms, behavioral symptoms are not always linked to obvious neuropathology. Clearly, people display compensatory plasticity: they can change their behavior to adapt to neural change, and they can display abnormal behavior without obvious brain pathology.

TECHNOLOGICAL RESOLUTION Some people have notable behavioral problems after suffering brain trauma, yet no obvious signs of brain damage appear on an MRI. Infants may seem to have a typical brain only to display severe cerebral palsy later. The resolution of technology may always lack the detail to detect subtle neuronal change, such as a drop in dendritic-spine density or injury so diffuse that it is hard to identify. Given the current diagnostic methods for both behavioral disorders and neuropathology, identifying disorders and their causes is seldom easy.

MODELING SIMPLICITY A major avenue for investigating the causes of disorders is to develop and study animal models. Rats with specific lesions of the nigrostriatal dopamine system are used to model Parkinson's disease. Models lead to significant advances in understanding of neural conditions and their treatments. One problem with animal models is that the view they provide of the neurobiology behind behavioral abnormalities is oversimplified.

The fact that a drug reduces symptoms does not necessarily mean that it is acting on a key biochemical aspect of the pathology. Aspirin can get rid of a headache, but that does not mean that the headache is caused by the receptors on which aspirin acts. Similarly, antipsychotic drugs block D_2 receptors, but that does not mean that schizophrenia is caused by abnormal D_2 receptors. Schizophrenia quite possibly results from a disturbance in glutamatergic systems, and for some reason dopamine antagonists are effective in rectifying the abnormality.

MODELING LIMITATIONS Modeling human disorders is a complex task, so use caution when you encounter news stories about studies using animal models that point toward possible cures for human behavioral diseases. Caution applies especially to psychiatric

Figure 5-17 summarizes the major neural activating systems. Figure 6-7 explains how psychoactive drugs affect the $GABA_A$ receptor site.

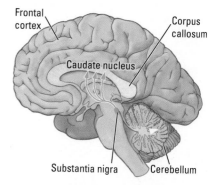

Nigrostriatal Dopamine Pathways
Axons of neurons in the midbrain substantia nigra project to the basal ganglia, supplying dopamine to maintain normal motor behavior. Dopamine loss is related to muscle rigidity and dyskinesia seen in Parkinson's disease.

Section 7-7 reviews the benefits of creating animal models of disorders.

disorders in which causes are still unknown. Further, many symptoms of disorders such as schizophrenia and anxiety are largely cognitive. Objectively identifying any cognitive processes mimicked by a laboratory model is difficult.

REVIEW 16-1
Multidisciplinary Research on Brain and Behavioral Disorders

Before you continue, check your understanding.

1. Neural correlates of Freud's id, ego, and super ego could be, respectively, the _____, _____, and _____.

2. Causes of abnormal behavior include _____, _____, _____, _____, and _____.

3. For most psychiatric disorders, the causes are unknown, but _____ is an exception.

4. A major challenge in diagnosing disorders is that diagnosis tends to be _____.

5. Describe a research challenge for understanding brain injury.

Answers appear at the back of the book.

16-2 Classifying and Treating Brain and Behavioral Disorders

Behavioral disorders afflict millions every year. The National Institute for Mental Disorders estimates that about one in four people in the United States suffers a diagnosable mental disorder in a given year and nearly one-half of the population does over their lifetime. Of these people, only a minority receives treatment of any kind, and even fewer receive treatment from a mental-health specialist. Large-scale surveys of neurological disorders show a pattern of prevalence similar to behavioral disorders. Together, behavioral and neurological disorders are the leading cause of disability after age 15.

Figure 8-31 pegs the peak age of onset for mental disorders at 14 years.

Identifying and Classifying Behavioral Disorders

Epidemiology is the study of the distribution and causes of diseases in human populations. A major contribution of epidemiological studies has been to help define and assess behavioral disorders, including especially those labeled psychiatric. We can categorize psychiatric disorders by three general types—disorders of psychoses, mood, and affect.

The first set of criteria for diagnoses in psychiatry was developed in 1972. Since that time, two parallel sets of criteria have gained prominence and new versions of both appear periodically. One is the World Health Organization's *International Classification of Diseases* (ICD-10); the other is the most recent edition of the American Psychiatric Association's *Diagnostic and Statistical Manual of Mental Disorders*, or **DSM.**

ICD-11 is slated for publication in 2015, DSM-5 in 2013.

Table 16-3 summarizes the classification scheme used in the DSM. Like any classification of psychiatric disorders, the DSM is to some extent arbitrary and unavoidably depends on prevailing cultural views. A good example is the social definition of abnormal sexual behavior. At its inception, the DSM listed homosexual behavior as pathological. Since 1980, however, the manual has omitted homosexual behavior from its list of disorders.

Section 12-5 explores the relationship of sexual orientation and sexual identity to brain organization.

TABLE 16-3 Summary of DSM Classification of Abnormal Behaviors

Diagnostic category	Core features and examples of specific disorders
Disorders usually first diagnosed in infancy, childhood, and adolescence	Tend to emerge and sometimes dissipate before adult life: pervasive developmental disorders (such as ASD), learning disorders, attention-deficit hyperactivity disorder, conduct disorder, separation-anxiety disorder
Delirium, dementia, amnesia, and other cognitive disorders	Dominated by impairment in cognitive functioning: Alzheimer's disease, Huntington's disease
Mental disorders due to a general medical condition	Caused primarily by a general medical disorder: mood disorder due to a general medical condition
Substance-related disorders	Brought about by the use of substances that affect the central nervous system: alcohol-use disorders, opioid-use disorders, amphetamine-use disorders, cocaine-use disorders, hallucinogen-use disorders
Schizophrenia and other psychotic disorders	Functioning deteriorates toward a state of psychosis, or loss of contact with reality
Mood disorders	Severe disturbances of mood resulting in extreme and inappropriate sadness or elation for extended periods of time: major depressive disorder, bipolar disorder
Anxiety disorders	Anxiety: generalized anxiety disorder, phobias, panic disorder, obsessive-compulsive disorder, acute stress disorder, posttraumatic stress disorder
Somatoform disorders	Physical symptoms that are apparently caused primarily by psychological rather than physiological factors: conversion disorder, somatization disorder, hypochondriasis
Fictitious disorders	Intentional production or feigning of physical or psychological symptoms
Dissociative disorders	Significant changes in consciousness, memory, identity, or perception, without a clear physical cause: dissociative amnesia, dissociative fugue, dissociative identity disorder (multiple personality disorder)
Eating disorders	Abnormal patterns of eating that significantly impair functioning: anorexia nervosa, bulimia nervosa
Sexual disorders and sexual-identity disorder	Chronic disruption in sexual functioning, behavior, or preferences: sexual dysfunctions, paraphilias, sexual-identity disorder
Sleep disorders	Chronic sleep problems: primary insomnia, primary hypersomnia, sleep-terror disorder, sleepwalking disorder
Impulse-control disorders	Chronic inability to resist impulses, drives, or temptations to perform certain acts that are harmful to the self or others: pathological gambling, kleptomania, pyromania, intermittent explosive disorder
Adjustment disorder	Maladaptive reaction to a clear stressor, such as divorce or business difficulties, that first occurs within 3 months after the onset of the stressor
Other conditions that may be a focus of clinical attention	Conditions or problems that are worth noting because they cause significant impairment, such as relational problems, problems related to abuse or neglect, medication-induced movement disorders, and psychophysiological disorders

Source: Adapted from *Diagnostic and Statistical Manual of Mental Disorders* (4th ed.), 1994. Washington, DC: American Psychiatric Association.

Each revision responds to new classification-related information. For example, the DSM-5 classifies all forms of autism as autism spectrum disorder (ASD). Classifying a broad range of conditions as one simplifies diagnosis but draws criticism for its associated loss of descriptive power and, in the case of ASD, a perceived difficulty in obtaining remedial help for children at the severe end of the spectrum.

One continually emerging means of looking for indicators of behavioral disorders is brain imaging, including MRI and PET. These tools are not currently used clinically, but they may soon be used both to classify disorders and to monitor the effectiveness of treatment. To be useful, imaging tests must be sensitive enough to detect unique features of brain disorders and specific enough to rule out similar conditions. This feature is problematic because many behavioral disorders display similar abnormalities. Enlarged ventricles indicate a loss of brain cells and may appear in schizophrenia, Alzheimer's disease, alcoholism, or head trauma, for example.

Nonetheless, imaging technology is shedding new light on behavioral disturbances. In an impressive imaging example, Judith Rapoport and coworkers (2012) examined changes in the brains of subjects with childhood-onset schizophrenia. They propose that,

Focus 8-2 describes the autism spectrum.

DSM Abbreviation of *Diagnostic and Statistical Manual of Mental Disorders,* the American Psychiatric Association's classification system for psychiatric disorders.

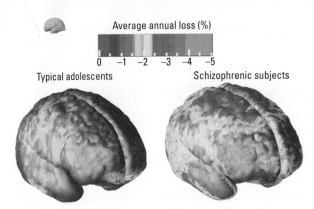

Average annual loss (%)

0 −1 −2 −3 −4 −5

Typical adolescents   Schizophrenic subjects

FIGURE 16-2 Early-Onset Schizophrenia
Comparison of three-dimensional maps derived from MRI scans reveals that, compared with healthy teenagers aged 13 to 18 (*left*), patients with childhood-onset schizophrenia (*right*) have widespread loss of gray matter across the cerebral hemispheres. *Left:* Courtesy of Paul Thompson and Arthur W. Toga, University of California. *Right:* Laboratory of Neuro Imaging, Los Angeles, and Judith L. Rapoport, National Institute of Mental Health.

FIGURE 16-3 Adult-Onset Schizophrenia
Note the abnormally low blood flow in the prefrontal cortex at the top of the PET scan in the brain of (*left*) an adult schizophrenia patient compared with (*right*) an adult who does not have schizophrenia.

The Index of Disorders inside the book's front cover lists the disorders discussed in this book—a mere fraction of the total.

in part, the condition begins in utero and is characterized by an excessive pruning of short-distance cortical connections. **Figure 16-2** shows that between the ages of 13 and 18, children who developed schizophrenia showed a remarkable loss of gray matter in the cerebral cortex.

An earlier study by the same group found a delayed growth rate in white matter—on the order of 2 percent per year—in children with schizophrenia compared with healthy children (Gogtay et al., 2008). The abnormality was found throughout the brain but was greater in the frontal lobes, especially on the right. This loss in cerebral gray and white matter correlates with the onset of a variety of behavioral disturbances characteristic of schizophrenia.

Not all disorders show such obvious loss of tissue, but they may show abnormal blood flow or metabolism that can be detected by either fMRI or PET. The PET images in **Figure 16-3** illustrate the metabolic changes in adult-onset schizophrenia. The scan on the left reveals an obvious abnormality in activity in the prefrontal cortex compared with the scan on the right from an adult who does not suffer from schizophrenia. Note that the prefrontal area does not show loss of gray matter in the early-onset-schizophrenia MRI study reproduced in Figure 16-2. Therefore, it is likely that the two diseases have different origins.

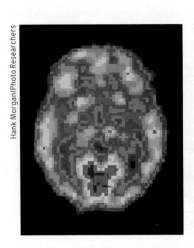
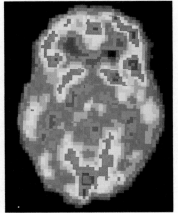

Hank Morgan/Photo Researchers

Combining behavioral diagnoses with neuroimaging will move practitioners beyond symptom checklists like those published in the DSM to more objective medical diagnoses. Imaging analyses will help target treatments to reduce the severity of such serious disorders as schizophrenia and Alzheimer's disease. Remember, however, that current imaging techniques do not detect all brain pathology. Part of the challenge for the future is to improve current techniques and to develop others that can identify more subtle molecular abnormalities in the nervous system.

Treatments for Disorders

An inclusive list of brain and behavioral disorders would consist of some 2000 entries. The long-term prospects for curing organic or behavioral disorders on the macro level depend on the capacity to treat structural and biochemical abnormalities at the micro level.

Organic abnormalities include genetic disorders (such as Huntington's disease), developmental disorders (such as ASD), infectious diseases (such as meningitis), nervous system injuries (such as brain or spinal-cord trauma), and degenerative dementias (such as Alzheimer's disease). Each may be associated with various structural changes, the congenital absence of neurons or glia, the presence of abnormal neurons or glia, the death of neurons or glia, and neurons or neural connections with unusual structures.

Abnormalities may also appear in any improper balances in the biochemical organization or operation of the nervous system. Biochemical abnormalities include disordered proteins in cell-membrane channels, low or high numbers of neuroreceptors, and low or high numbers of molecules, especially neurotransmitters or hormones.

The ultimate clinical problem for behavioral neuroscientists is applying their knowledge to generate treatments that can restore a disordered brain (and mind) to the range of normality. This challenge is daunting because the first task is so difficult: learning the cause of a particular behavioral disturbance. Few behavioral disorders have as simple a cause as PKU does. Most, like schizophrenia, are complex.

Still, several more or less effective treatments for a range of behavioral disorders have been developed. Treatments fall into four general categories:

1. *Neurosurgical* The skull is opened and some intervention is performed on the brain.

2. *Electrophysiological* Brain function is modified by stimulation through the skull.

3. *Pharmacological* A chemical that affects the brain is either ingested or injected.

4. *Behavioral* Treatment manipulates the body or the experience, which in turn influences the brain.

Neurosurgical Treatments

Neurosurgical manipulations of the nervous system with the goal of directly altering it have been largely reparative, as when tumors are removed or arteriovenous malformations are corrected. More recent neurosurgical approaches aim to alter brain activity to alleviate a neurological or behavioral disorder. The surgery either damages some dysfunctional area of the brain or stimulates dysfunctional areas with electrodes.

Treatment for Parkinson's disease can employ both neurosurgical approaches. In the first technique, an electrode is placed in the motor thalamus and an electric current is used to damage neurons that are responsible for producing the tremor characteristic of Parkinson's. In the second neurosurgical technique, **deep brain stimulation** (DBS), an electrode fixed in place in the globus pallidus or subthalamic nucleus is connected to an external electrical stimulator that the patient can activate **(Figure 16-4)**. The stimulation can inactivate the tremor-producing area of the brain or activate the brain in other ways and so restore more normal movement.

DBS has been used experimentally to treat traumatic brain injury (TBI) and behavioral dysfunctions such as obsessive-compulsive disorder (OCD) and major depression. Kennedy and colleagues (2011) have followed the effects of DBS in a group of people with severe and intractable depression. Even after three years of implantation, the DBS electrodes are well tolerated and the stimulation remains effective.

Following the onset of stimulation, an acute shift in brain activity induces a change in baseline neural activity similar to long-term potentiation. The change appears to make the brain more plastic and receptive to other treatments. Recovery is a prolonged process, however, that involves depressed subjects learning more effective patterns of thought and behavior. One difficulty: if DBS is turned off, patients tend to relapse, although it is not known why.

Another highly experimental neurosurgical strategy draws on the fixed sequence of prenatal brain development from cell division and cell differentiation to cell migration and synaptogenesis. If a brain region is functioning abnormally or if it is diseased or dead, as occurs in TBI or after stroke, it should be possible to return this region to the embryonic state and regrow a normal region. The use of so-called "induced neurogenesis" has

Brain tumors are the topic of Focus 3-2; arteriovenous malformations are the topic of Focus 10-4.

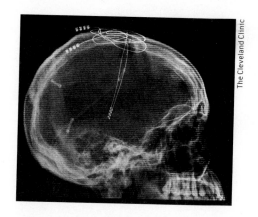

The Cleveland Clinic

FIGURE 16-4 Deep Brain Stimulation
X-ray of a human brain showing electrodes implanted in the thalamus for DBS.

Section 14-4 explains how LTP takes place.

deep brain stimulation (DBS) Neurosurgery in which electrodes implanted in the brain stimulate a targeted area with a low-voltage electrical current to facilitate behavior.

Figure 14-25 shows neurogenesis induced in a rat brain to repair a cortical stroke.

Stem cells are multipotent in that they have the potential to develop into any cell type. Figure 8-9 diagrams the origins of specialized brain cells from neural stem cells.

a science-fiction ring to it but may someday be feasible. In laboratory rats, for example, stem cells can be induced by *neurotrophic factors* to generate new cells that can migrate to the site of an injury.

Where would the stem cells come from? In the 1980s, neurosurgeons experimented with implanting fetal cells into adult brains, but this approach has had limited success. Another idea comes from the discovery that multipotent stem cells in other body regions, such as in bone marrow and skin, appear capable of manufacturing neural stem cells. Indeed, using appropriate manipulations, any cell can potentially be returned to a stem-cell state. The advantage is that these cells are not rejected by the immune system because they are the subject's own cells.

If people's own multipotent stem cells prove practical in generating neural stem cells, it should be possible to extract stem cells, place them in a special culture medium to generate thousands or millions of cells, and then place these stem cells in the damaged brain. The cells would then be instructed to differentiate appropriately and develop the correct connections. The challenge is formidable, but meeting it is well within the realm of possibility.

Stem-cell transplantation is taken seriously today as a treatment for disorders such as TBI and stroke. Douglas Kondziolka and his colleagues (2000) tried cell transplants with 12 stroke victims. They harvested progenitor cells from a rare tumor known as a teratocarcinoma and chemically altered them to develop a neuronal phenotype. Between 2 million and 6 million cells were transplanted into regions around the stroke.

The patients were followed for a year, and for 6 of them, PET scans showed increased metabolic activity in the areas that had received the transplanted cells. The transplants were having some effect on the host brain. Behavioral analyses also showed some modest improvement in these patients.

A follow-up study by the same group (Kondziolka et al., 2005) added behavioral therapy to the treatment but again showed only modest functional improvement. In an ongoing parallel study, patients are receiving growth factors to stimulate endogenous stem-cell production. Although the procedure appears safe, its efficacy is unclear. Nonetheless, these studies show that some form of stem-cell treatment for brain injury and stroke is feasible.

Electrophysiological Treatments

Treating the mind by treating the body is an ancient notion. In the 1930s, researchers used insulin to lower blood sugar and produce seizures as a treatment for depression. By the 1950s, insulin therapy had been replaced by electroconvulsive therapy (ECT), the first electrical brain-stimulation treatment.

ECT was developed as a treatment for otherwise untreatable depression, and although its mode of action was not understood, it did prove useful. Although rarely used today, ECT is still sometimes the only treatment that works for people with severe depression. One reason may be that it stimulates the production of a variety of neurotrophic factors, especially BDNF (brain-derived neurotrophic factor) that in turn restore inactive cells to a more active mode.

Neurotrophic factors are nourishing chemical compounds that support neuronal growth, development, and viability.

Problems with ECT include the massive convulsions caused by the electrical stimulation. Large doses of medications are normally required to prevent them. ECT also leads to memory loss, a symptom that can be troublesome with repeated treatments.

A newer, noninvasive technique, transcranial magnetic stimulation (TMS), uses magnetic rather than electrical stimulation and is an FDA-approved treatment for depression. Clinical applications for TMS, reviewed in Research Focus 16-2, "Treating Behavioral Disorders with Transcranial Magnetic Stimulation," are growing.

Figure 7-7 diagrams how TMS works.

Transcranial magnetic stimulation is a more focused treatment than ECT and will probably be far more widely used. TMS can be applied narrowly, to a focal area, rather than diffusely, as with ECT. And the prospective range of applications for TMS is broad.

RESEARCH FOCUS ✣ 16-2

Treating Behavioral Disorders with Transcranial Magnetic Stimulation

In transcranial magnetic stimulation (TMS), a magnetic coil is placed over the scalp to induce an electrical current in underlying brain regions. TMS can be applied to localized brain regions (focal areas) thought to be implicated in specific disorders. Manipulation of the magnetic field can stimulate an area of cortex as small as a quarter, only the cortical surface, or deeper layers of brain tissue.

The primary clinical use of TMS, which the U.S. Food and Drug Administration formally approved in 2008, is for depression. A number of studies report positive effects using TMS, but the required duration of treatment and the duration of beneficial effects is still under investigation.

The effects of brief pulses of TMS do not outlive the stimulation. *Repetitive TMS* (rTMS), however, which involves continuous stimulation for up to several minutes, produces more long-lasting effects. What is needed to fully evaluate the effects of TMS in alleviating depression is a double-blind study in which both the therapists and patients are unaware of whether real or sham stimulation is administered (Bersani et al., 2012).

In addition to treating depression, small but promising studies have extended the benefits of TMS to schizophrenic auditory hallucinations, anxiety disorders, neurodegenerative diseases, hemiparesis, and pain syndrome (Wassermann & Zimmerman, 2012).

Among the problems in all studies of TMS are questions related to the duration and intensity of stimulation and also to the area stimulated. Each person's brain is slightly different, so to ensure that appropriate structures are stimulated, MRI must be performed on each subject.

Does TMS stimulation make the brain more plastic? If so, can learning be enhanced? Rajji and colleagues (2011) stimulated the motor cortex of control participants and found that when a train of TMS produced a lasting change in the excitability of the motor cortex, subsequent learning of a motor task was facilitated. This finding implies that appropriately applied stimulation could improve the therapeutic effects of cognitive therapy.

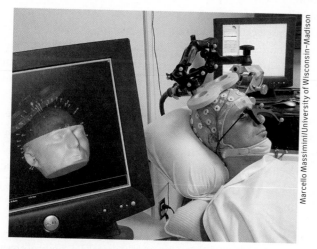

In clinical therapy for depression, TMS influences neural activity in a localized brain area.

Marcello Massimini/University of Wisconsin–Madison

In a study of 100 patients with depression, TMS was found to be well tolerated and effective (Connolly et al., 2012). TMS is also used to explore normal brain function and is well tolerated by participants.

Pharmacological Treatments

Several accidental discoveries, beginning in the 1950s, led to a pharmacological revolution in the treatment of behavioral disorders:

1. The development of phenothiazines (neuroleptics) as a treatment for schizophrenia stemmed from a drug used to premedicate surgical patients. In the following decades, neuroleptic drugs became increasingly more selective, and they remain effective.

2. A new class of antianxiety drugs was invented: the anxiolytics. Medications such as Valium quickly became—and remain—the most widely prescribed drugs in the United States.

3. L–Dopa provided the first drug treatment for serious motor dysfunction in Parkinson's disease. Once taken, L–dopa is converted into dopamine and replaces dopamine lost due to Parkinson's disease.

The power of psychoactive drugs to change disordered behavior revolutionized the pharmaceutical industry. L–Dopa's effectiveness led to optimism that drugs might be

tardive dyskinesia Inability to stop the tongue or other body parts from moving; motor side effect of neuroleptic drugs.

behavioral therapy Treatment that applies learning principles, such as conditioning, to eliminate unwanted behaviors.

cognitive therapy Psychotherapy based on the perspective that thoughts intervene between events and emotions, and thus the treatment of emotional disorders requires changing maladaptive patterns of thinking.

psychotherapy Talking therapy derived from Freudian psychoanalysis and other psychological interventions.

Section 6-2 recounts the classes of psychoactive drugs and notes that their therapeutic effects were originally discovered by accident.

Focus 12-3 recounts a case of generalized anxiety disorder.

developed as "magic bullets" to correct the chemical imbalances found in Alzheimer's disease and other disorders. Research was directed toward making drugs more selective in targeting specific disorders while producing fewer side effects. Both goals have been difficult to achieve.

Pharmacological treatments have significant downsides. Acute and chronic side effects top the list, and long-term effects may create new problems. Consider a person who receives antidepressant medication. The drug may ease the depression, but it may also produce unwanted side effects, including decreased sexual desire, fatigue, and sleep disturbance. These last two effects may also interfere with cognitive functioning.

Thus, although a medication may be useful for getting a person out of a depressed state, it may produce other symptoms that are themselves disturbing and may complicate recovery. Furthermore, in depression related to a person's life events, a drug does not provide the behavioral tools needed to cope with an adverse situation. Some psychologists say, "A pill is not a skill."

Negative side effects of drug treatments can be seen in many people being treated for schizophrenia with neuroleptics. Antipsychotic drugs act on the mesolimbic dopamine system that affects motivation, among other functions. The side effect emerges because the drugs also act on the nigrostriatal dopaminergic system that controls movement.

Patients who take neuroleptics eventually develop motor disturbances. **Tardive dyskinesia,** an inability to stop the tongue, hands, or other body parts from moving, is a motor symptom of long-term neuroleptic administration. Movement-disorder side effects can persist after the psychoactive medication has been stopped. Taking drugs for behavioral disorders, then, does carry risk. Rather than acting like "magic bullets," these medications often act like "magic shotguns."

Behavioral Treatments

Treatments for behavioral disorders need not be direct biological or medical interventions. Just as the brain can alter behavior, behavior can alter the brain. Behavioral treatments focus on key environmental factors that influence how a person acts. As behavior changes in response to treatment, the brain is affected as well.

An example is the treatment of *generalized anxiety disorders* attributed to chronic stress. People endure persistently high levels of anxiety and often engage in maladaptive behaviors to reduce it. They require immediate treatment with antianxiety medication, but long-term treatment entails changing their behavior. Generalized anxiety disorder is not simply a problem of abnormal brain activity. It is also a problem of experiential and social factors that fundamentally alter the person's perception of the world.

Perhaps you are thinking, Behavioral treatments may help somewhat in treating brain dysfunction, but the real solution must lie in altering brain activity. Remember, though, that every aspect of behavior is the product of brain activity: behavioral treatments *do* act by changing brain function. That is, not only does altering the brain change our behavior, altering our behavior changes the brain.

If people can change the way they think and feel about themselves or some aspect of their lives, this change has taken place because "talking about their problems" or "resolving a problem" alters how their brains function. In a sense, then, behavioral treatments are "biological interventions." They may sometimes be helped along by drug treatments that make the brain more receptive to change through behavioral therapies. In this way, drug treatments and behavioral treatments have synergistic effects, each helping the other to be more effective.

Your behavior is a product of all your learning and social experiences. An obvious approach to developing a treatment is to recreate a learning environment that replaces a maladaptive behavior with an adaptive behavior. Thus, the various approaches to behavioral treatment use principles derived from experimentally based learning theory. Following is a sampling of these approaches.

BEHAVIOR MODIFICATION **Behavioral therapies** apply well-established learning principles to eliminate unwanted behaviors. Therapists apply the principles developed in studying learning by reinforcement in laboratory settings, including operant and classical conditioning. For example, if a person is debilitated by a fear of insects, rather than looking for inner causes, the behavioral therapist tries to replace the maladaptive behaviors with more constructive ways of behaving. These might include training the patient to relax while systematically exposing him to unthreatening insects (butterflies) followed by gradual exposure to more threatening insects (bees). This form of habituation (adaption to a repeatedly presented stimulus) is called *systematic desensitization.*

Systematic desensitization for a phobia, the most common among anxiety disorders, as Focus 12-3 reports.

COGNITIVE THERAPY **Cognitive therapies** take the perspective that thoughts intervene between events and emotions. Consider responses to losing a job. One thought could be "I am a loser, life is hopeless." An alternate thought is "The job was a dead end for me and the boss did me a favor." You can imagine that the former cognition might lead to depression, whereas the latter would not. Cognitive therapies challenge a person's self-defeating attitudes and assumptions. Cognitive therapy is important for people with brain injuries, too, because it is easy for people to think that they are "crazy" or "stupid" after brain injury. Equally if not more powerful is *cognitive-behavioral therapy,* discussed in Section 16-4.

NEUROPSYCHOLOGICAL THERAPY If a relative or friend had a stroke and became aphasic, you would expect the person to receive speech therapy, which is a behavioral treatment for an injured brain. The logic in speech therapy is that by practicing (relearning) the basic components of speech and language, the patient should be able to regain at least some of the lost function. The same logic can be applied to other types of behavioral disorders, whether motor or cognitive.

Therapies for cognitive disorders resulting from brain trauma or dysfunction aim to retrain people in the fundamental cognitive processes they have lost. Although cognitive therapy seems as logical as speech therapy after a stroke, the difficulty is that cognitive therapy assumes that we know what fundamental elements of cognitive activity are meaningful to the brain. Cognitive scientists are far from understanding these elements well enough to generate optimal therapies. Still, neuropsychologists are developing neurocognitive programs that can improve functional outcomes following TBI and stroke (Mateer & Sira, 2006; Sohlberg & Mateer, 1989). Treatment effectiveness can be improved with computer-based tools, as can follow-up therapy.

EMOTIONAL THERAPY In the 1920s, Sigmund Freud developed the idea that talking about emotional problems enables people to gain insights into their causes and serves as treatment too. "Talking cures" and other forms of psychological intervention may be broadly categorized as **psychotherapies.**

Since Freud's time, many ideas have been put forth about the best type of therapy for emotional disorders. The key point here is that for many disorders, whether neurological or psychiatric, medical treatments are not effective unless patients also receive psychotherapy. Indeed, in many cases, the only effective treatment lies in addressing the unwanted behaviors directly—in acquiring the skill rather than taking the pill.

Consider a 25-year-old woman pursuing a promising career as a musician who suffered a traumatic brain injury in an automobile accident. After the accident, she found that she was unable to read music. Not surprisingly, she soon became depressed. Part of her therapy required her to confront her disabling cognitive loss by talking about it rather than by simply stewing about it. Only when she pursued psychotherapy did she begin to recover from her intense depression.

Exercise boosts your mood because it boosts your dopamine levels. Section 10-4 observes that practitioners have only begun to tap music's power as a therapeutic tool.

Focus 11-5 describes a low-tech strategy for controlling phantom-limb pain.

real-time fMRI (rt-fMRI) Behavior-modification technique in which individuals learn to change their behavior by controlling their own patterns of brain activation.

For many people with emotional impairments resulting from brain disease or trauma, the most effective treatment for depression or anxiety is helping them adjust by encouraging them to talk about their difficulties. Group therapy provides such encouragement and is standard treatment in brain-injury rehabilitation units.

PHYSICAL ACTIVITY AND MUSIC AS THERAPY Exercise and music have positive effects on peoples' attitudes, emotional well-being, and brain function. Music affects arousal and activates the motor and premotor cortex. Listening to music can improve gait in Parkinson's and stroke patients. Physical activity, including playing sports, combined with other therapies, improves well-being and counteracts the effects of depression.

REAL-TIME fMRI (rt-fMRI) Using this behavior-modification technique, individuals learn to change their behavior by controlling their own patterns of brain activation. **Real-time fMRI** was first used to treat intractable pain (deCharms, 2008). Pain produces a characteristic pattern of brain activity. The researchers proposed that if subjects could see their brain activity via fMRI in real time as they were feeling pain, they could be trained to reduce the neural activity and lessen their pain. Real-time fMRI uses a form of operant conditioning in which the gradual modification of a participant's behavior increases the probability of reward.

Think of rt-fMRI as a form of neural plasticity in which the individual learns new strategies guided by brain activation information. When subjects decrease brain activation in regions associated with pain, they report decreased pain perception. Conversely, through learning to increase brain activation in these regions, they would be able to increase their pain—although it seems unlikely that this ability would be much cultivated!

An actual potential application of rt-fMRI is in monitoring brain activation when treatment for disorders occurs in the context of behavioral therapy. Participants have been trained to control emotion while viewing real-time activation from areas involved in emotional processing, including the insular cortex and the amygdala.

VIRTUAL-REALITY THERAPY The general principle behind VR therapy is that patients are placed in a virtual world or interact with a virtual world displayed on a computer screen or worn goggles. One example is the armored personnel carrier described in Research Focus 16-1. The participant can experience sights, sounds, and even smells that mimic situations related to the behavioral disorder, PTSD in this case. Modified virtual-reality therapy involves having a patient interact with a virtual world, as if they were a character in a computer game. Winning the game involves making choices that are adaptive; poor choices result in losing the game (Opris et al., 2012).

REVIEW 16-2

Classifying and Treating Brain and Behavioral Disorders

Before you continue, check your understanding.

1. A system for the classification of behavioral disorders is _____.

2. Four types of treatment for behavioral disorders are _____, _____, _____, and _____.

3. A therapy in which an electrode delivers stimulation to the brain directly is called _____.

4. An effective replacement for electroconvulsive therapy (ECT) is _____.

5. What common factor underlies most types of behavioral treatments?

Answers appear at the back of the book.

16-3 Understanding and Treating Neurological Disorders

In each of our lifetimes, at least one close friend or relative will develop a neurological disorder, even if we ourselves escape them. Their causes are understood in a general sense, and for most, rehabilitative treatment is emerging. In this section we review some common neurological disorders: traumatic brain injury, stroke, epilepsy, multiple sclerosis, and neurodegenerative disorders.

Traumatic Brain Injury

Traumatic brain injury (TBI), a wound to the brain that results from a blow to the head, is the most common form of brain damage in people under age 40. TBI commonly results from the head making impact with other objects—as can occur in automobile and industrial accidents—and in sports injuries. TBI can also follow blows to the chest that result in a rapid increase in blood pressure that can damage the brain indirectly. The incidence of TBI is about eight times that of breast cancer, AIDS, spinal-cord injury, and multiple sclerosis combined.

The two most important factors in the incidence of head trauma are age and sex. Children and elderly people are more likely to suffer head injuries from falls than are others, and males between 15 and 30 incur brain injuries, especially from automobile and motorcycle accidents (**Figure 16-5**). A child's chance of suffering significant traumatic brain injury before he or she is old enough to drive is 1 in 30.

Section 14-5 details the dancer Donna's recovery from TBI.

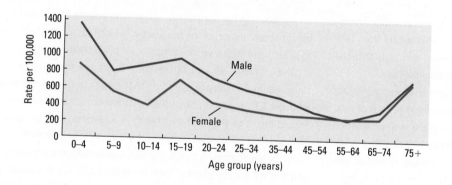

FIGURE 16-5 Incidence Rates of Head Trauma Based on combined reports of emergency-room visits, hospitalizations, and deaths, this chart graphs the estimated frequency of TBI in males and females across the life span. Adapted from the Centers for Disease Control report *TBI in the United States: Emergency Department Visits, Hospitalizations, and Deaths*, 2004.

Concussion is a critical concern for both professional and amateur athletes, especially those who play football, ice hockey, lacrosse, and soccer, and no less for those serving in the military, in light of the high incidence of roadside bomb blasts aimed at foreign troops in Iraq and Afghanistan. Sports account for about 20 percent of TBIs, and the U.S. Army Institute of Surgical Research reports that traumatic brain injury affects more than 1 in 5 U.S. soldiers wounded in war.

A large-scale longitudinal study is underway in cooperation with football and ice hockey players who have a history of concussion and have agreed to donate their brains for postmortem analysis. Preliminary examination of the brains of deceased professional football players who had a history of concussion and severe postconcussion symptoms, described in Clinical Focus 16-3, "Concussion" on page 582, reveal extensive, diffuse loss of cerebral tissue.

In longitudinal research, investigators repeatedly observe or examine subjects over time with respect to the study's variable(s).

Symptoms and Outcome of Brain Trauma

TBI can cause direct damage to the brain. Trauma can disrupt the brain's blood supply, induce bleeding (leading to increased intracranial pressure), cause swelling (leading to increased intracranial pressure), expose the brain to infection, and scar brain tissue (the

Concussion

Early in 2011, 50-year-old former Chicago Bears defensive back Dave Duerson shot himself in the chest and died. He left a note asking that his brain be studied. Duerson had played 11 years in the National Football League, won two Super Bowls, and received numerous awards.

As a pro player he endured at least 10 concussions, but they did not seem serious enough to cause him to leave the game. After retiring from football he went to Harvard and obtained a business degree. He pursued a successful business career until he began to experience problems in decision making and temper control.

Eventually, Duerson's business and marriage failed. After his suicide, the Center for the Study of Traumatic Encephalopathy in Boston did study his brain. The center is conducting postmortem anatomical analyses of the brains of former athletes as part of a long-term longitudinal study.

Duerson's diagnosis is **chronic traumatic encephalopathy** (CTE), a progressive degenerative disease found in individuals with a history of multiple concussion and other closed-head injuries. (A variant long associated with boxing is *dementia pugilistica,* or DP.)

Concussion, the common term for mild traumatic brain injury (MTBI), is common in sports, especially contact sports including American football, ice hockey, and rugby. Concussion also results from falls in many other sports and from vehicular accidents. It is likely that the incidence of concussion is higher than 6 per 1000 individuals.

Most concussions go unrecognized. For those that are diagnosed, little apparent pathology appears after relatively short periods of rest, the usual treatment. Nevertheless, the relationship is well established between concussion and a range of degenerative diseases that occur later in life—dementias, including Alzheimer's disease, as well as Parkinson's disease, motor neuron disease, and CTE.

The relationship between concussion in early life and later degenerative brain disease suggests that concussion can initiate a cascade of pathological events that, over years, develop into CTE (Gavett et al.,

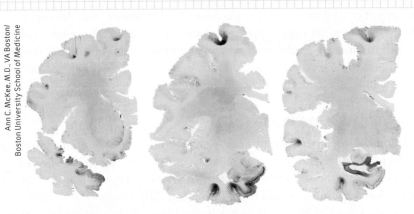

Ann C. McKee, M.D., VA Boston/ Boston University School of Medicine

Dave Duerson's Brain   Staining for tau protein highlights degenerating brain tissue (dark brown areas) in the frontal cortex on medial temporal lobe of these coronal sections through Duerson's anterior right hemisphere. Damage to the subcortical basal ganglia in the sections' central portions is sparse by comparison.

2010). CTE is characterized by neurofibrillary tangles, plaques, and neuronal death. Cerebral atropy and expanded ventricles due to cell loss are typical in advanced cases.

As shown in the illustration, researchers test for cell death by staining for accumulation of the tau protein, which is associated with neuronal death and so is a sensitive marker for brain trauma.

The unknowns about CTE are many. Is just one or are many concussions required to initiate a cascade that results in CTE? Are individuals who get CTE especially susceptible? What constitutes a concussion? Should blows to the head that result in no pronounced symptoms be distinguished from blows that result in loss of consciousness?

What we do know is that many well-known athletes, especially football and ice hockey players, have developed CTE. Clearly, much more care needs to be taken, beginning in childhood, to prevent concussion and to ensure that concussion is treated, even though what constitutes adequate treatment remains uncertain.

---

scarred tissue becomes a focus for later epileptic seizures). The disruption in blood supply tends to be brief, but a parallel disruption of energy production by neuronal mitochrondria can persist for weeks and is related to many postconcussion behavioral symptoms.

Traumatic brain injuries are commonly accompanied by a loss of consciousness that may be brief (minutes) or prolonged (coma). The duration of unconsciousness can serve as a measure of the severity of damage, because it correlates directly with mortality, intellectual impairment, and deficits in social skills. The longer the coma lasts, the greater the possibility of serious impairment and death.

Two kinds of behavioral effects result from TBI: (1) impairment of the specific functions mediated by the cortex at the coup (the site of impact) or countercoup (opposite side) lesion, as illustrated in **Figure 16-6,** and (2) more generalized impairments from

Section 1-2 presents a case study on recovering consciousness following TBI.

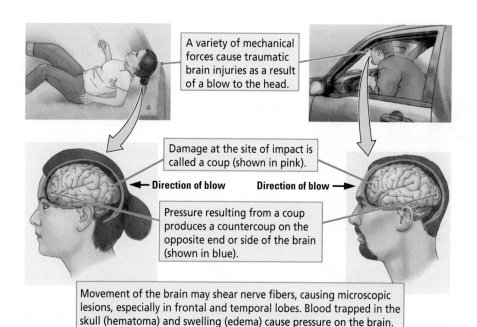

FIGURE 16-6 Mechanics of Traumatic Brain Injury Pink and blue shading mark brain regions most frequently damaged in closed-head injury. A blow can produce a contusion both at the site of impact and on the opposite side of the brain, owing to compression against the front or the back of the skull.

widespread trauma throughout the brain. Discrete impairment is most commonly associated with damage to the frontal and temporal lobes, the brain areas most susceptible to TBI (see tissue samples in Focus 16-3).

More generalized impairment results from minute lesions and lacerations scattered throughout the brain. Movement of the hemispheres in relation to one another causes tearing characterized by a loss of complex cognitive functions, including reductions in mental speed, concentration, and overall cognitive efficiency.

TBI patients generally complain of poor concentration or lack of ability. They fail to do things as well as they could before the injury, even though their intelligence is unimpaired. In fact, in our experience, people with high skill levels seem to be the most affected by TBI, in large part because they are acutely aware of loss of a skill that prevents them from returning to their former competence level.

Traumatic brain injuries that damage the frontal and temporal lobes also tend to significantly affect personality and social behavior. According to Muriel Lezak (2004), few victims of traffic accidents who have sustained severe head injuries ever resume their studies or return to gainful employment. If they do reenter the work force, they do so at a lower level than before their accidents.

One frustrating problem with traumatic brain injuries is misdiagnosis: the chronic effects of the injuries often are not accompanied by any obvious neurological signs or abnormalities in CT or MRI scans. Patients may therefore be referred for psychiatric or neuropsychological evaluation. An MRI-based imaging technique, **magnetic resonance spectroscopy** (MRS), holds promise for accurate diagnosis of TBI.

MRS, a modification of MRI, can identify changes in specific markers of neuronal function. One such marker is N-acetylaspartate (NAA), the second most abundant amino acid in the human brain (Tsai & Coyle, 1995). The level of NAA expression assesses the integrity of neurons, and deviations from normal levels (up or down) can be taken as a marker of abnormal brain function. People with traumatic brain injuries show a chronic decrease in NAA that correlates with the severity of the injury. Although not widely used clinically yet, MRS is a promising tool not only in identifying brain abnormalities but also in monitoring cellular response to therapeutic interventions (Hunter et al., 2012).

chronic traumatic encephalopathy (CTE) Progressive degenerative disease caused by multiple concussions and other closed-head injuries, characterized by neurofibrillary tangles, plaques, and cerebral atrophy and expanded ventricles due to cell loss.

magnetic resonance spectroscopy (MRS) Modification of MRI to identify changes in specific markers of neuronal function; promising for accurate diagnosis of traumatic brain injuries.

Recovery from Traumatic Brain Injury

Recovery from head trauma may continue for 2 to 3 years and longer, but most cognitive recovery occurs in the first 6 to 9 months. Recovery of memory functions appears to be slower than recovery of general intelligence, and the final level of memory performance is lower than for other cognitive functions. People with brainstem damage, as inferred from oculomotor disturbance, have a poorer cognitive outcome, and a poorer outcome is probably true of people with initial dysphasias or hemiparesis as well.

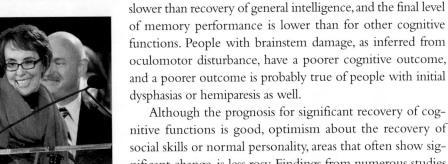

Gabrielle Giffords before (*left*) and a year after (*right*) she survived a gunshot to the head. She had regained limited speech, partly with the help of singing therapy, and the trauma to her brain's left hemisphere continued to limit mobility on her right side. Section 14-5 profiles recovery from TBI.

Although the prognosis for significant recovery of cognitive functions is good, optimism about the recovery of social skills or normal personality, areas that often show significant change, is less rosy. Findings from numerous studies support the conclusions that quality of life—in social interactions, perceived stress levels, and enjoyment of leisure—is significantly reduced after TBI and that this reduction is chronic (Murphy & Carmine, 2012). Attempts to develop tools to measure changes in psychosocial adjustment in brain-injured people are few, so we must rely largely on subjective descriptions and self-reports. Both provide little information about the specific causes of these problems.

Stroke

Focus 2-3 describes the symptoms and aftereffects of stroke.

Diagnosticians may be able to point to a specific immediate cause of *stroke,* an interruption of blood flow from either the blockage of a vessel or bleeding from a vessel. However, this initial event merely sets off a sequence of damage that progresses even if the blood flow is restored. Stroke results in a lack of blood, called **ischemia,** followed by a cascade of cellular events that wreak the real damage. Changes at the cellular level can seriously compromise not only the injured part of the brain but other brain regions as well.

Effects of Stroke

Consider what happens after a stroke that interrupts the blood supply to one of the cerebral arteries. In the first seconds to minutes after ischemia, as illustrated in **Figure 16-7,** changes begin in the ionic balance of the affected regions, including changes in pH and in the properties of the cell membrane. These ionic changes result in a number of pathological events.

Figure 5-4 details how calcium affects neurotransmitter release. Figure 5-15 diagrams how metabotropic receptors can activate second messengers.

1. Release of massive amounts of glutamate results in prolonged opening of calcium channels in cell membranes.

2. Open calcium channels in turn allow toxic levels of calcium to enter the cell, not only producing direct toxic effects but also instigating various second-messenger pathways that can harm neurons. In the ensuing minutes to hours, mRNA is stimulated, altering the production of proteins in the neurons and possibly proving toxic to the cells.

3. Brain tissues become inflamed and swollen, threatening the integrity of cells that may be far removed from the stroke site. As in TBI, an energy crisis ensues as mitochondria reduce their production of ATP to produce cerebral energy.

4. A form of neural shock occurs. During this **diaschisis,** areas distant from the damage are functionally depressed. Thus, not only are localized neural tissue and its function lost but areas related to the damaged region also suffer a sudden withdrawal of excitation or inhibition.

5. Stroke may also be followed by changes in the metabolism of the injured hemisphere, its glucose utilization, or both, which may persist for days. Like diaschisis,

these metabolic changes can have severe effects on the functioning of otherwise normal tissue. For example, after a cortical stroke, metabolic rate has been shown to decrease about 25 percent throughout the rest of the hemisphere.

Treatments for Stroke

The ideal treatment is to restore blood flow in blocked vessels before the cascade of nasty events begins. One clot-busting drug is tissue plasminogen activator (t-PA). The difficulty is that t-PA must be administered within 3 to 5 hours to be effective. Currently, only a small percentage of stroke patients arrive at the hospital soon enough, in large part because stroke is not quickly identified, transportation is slow, or the stroke is not considered an emergency.

Other drugs called **neuroprotectants** can be used to try to block the cascade of postinjury events, but to date truly effective drugs do not exist. Clinical trials based on animal studies have generally failed, in part because understanding what the appropriate brain targets should be is limited.

When the course of the stroke leads to dead brain tissue, the only treatments that can be beneficial are those that facilitate plastic changes in the remaining brain. Examples are speech therapy or physical therapy. Revolutionary approaches to stroke rehabilitation use virtual-reality treatments, computer games, and robotics to construct machines that can assist therapy.

Still, some simple treatments are surprisingly effective. One is *constraint-induced therapy*, pioneered by Edward Taub in the 1990s. Its logic confronts a problem in poststroke recovery related to learned nonuse. For example, stroke patients with motor deficits in a limb often compensate by overusing the intact limb, which in turn leads to increased loss of use in the impaired limb.

In constraint-induced therapy, the intact limb is held in a sling for several hours per day, forcing the patient to use the impaired limb. There is nothing magical about Taub's procedure: virtually any treatment that forces patients to practice behaviors extensively is successful. An important component of these treatments, however, is a posttreatment contract in which the patients continue to practice after the formal therapy is completed. If they fail to do so, the chances for learned nonuse and a return of symptoms are high.

Another common effect of stroke is loss of speech. Specific speech therapy programs can aid in the recovery of speech. Music and singing, mediated in part by the right hemisphere, can augment speech therapy after left-hemisphere stroke.

Therapies using pharmacological interventions (e.g., noradrenergic, dopaminergic, cholinergic agonists) combined with behavioral therapies provide equivocal gains in stroke patients. The bulk of evidence suggests that patients with small, gray-matter strokes are most likely to show benefits from these treatments, whereas those with large strokes that include white matter show little benefit.

Finally, there have been many attempts to use either direct cortical stimulation or TMS in combination with behavioral therapy as a stroke treatment. The idea is to induce plasticity in regions adjacent to the dead tissue with the goal of enhancing the efficiency of the residual parts of the neuronal networks. These treatments have proved beneficial in patients with good residual motor control, but again, those with larger injuries show much less benefit, presumably because a sufficient residual neuronal network is not available.

Epilepsy

In epilepsy, a person suffers recurrent seizures that register on an electroencephalogram (EEG) associated with disturbances of consciousness. The character of epileptic episodes can vary greatly, and seizures are common; 1 person in 20 experiences at least one seizure

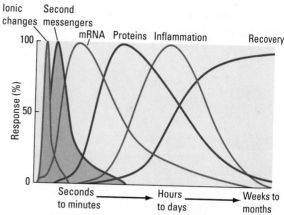

FIGURE 16-7 Results of Ischemia A cascade of events takes place after blood flow is blocked as a result of stroke. Within seconds, ionic changes at the cellular level spur changes in second-messenger molecules and RNA production. Changes in protein production and inflammation follow and resolve slowly, in hours to days. Recovery begins within hours to days and continues from weeks to months or years.

Experiment 11-3 shows the research with monkeys that contributed to developing constraint-induced therapy for people.

ischemia Lack of blood to the brain as a result of stroke.

diaschisis Neural shock that follows brain damage in which areas connected to the site of damage show a temporary arrest of function.

neuroprotectant Drug used to try to block the cascade of poststroke neural events.

Focus 4-1 describes a diagnosis of epilepsy and shows an EEG being recorded.

Section 3-3 contrasts Mendelian genetics and epigenetics.

TABLE 16-4 Factors That May Precipitate Seizures in Susceptible Persons

Drugs
Alcohol
Analeptics
Excessive anticonvulsants
Phenothiazines
Tricyclic antidepressants
Emotional stress
Fever
Hormonal changes
Adrenal steroids
Adrenocorticotrophic hormone (ACTH)
Menses
Puberty
Hyperventilation
Sensory stimuli
Flashing lights
Laughing
Reading, speaking, coughing
Sounds: music, bells
Sleep
Sleep deprivation
Trauma

Source: Adapted from *Behavioral Neurobiology* (p. 5), by J. H. Pincus and G. J. Tucker, 1974. New York: Oxford University Press.

in his or her lifetime. The prevalence of multiple seizures is much lower, however—about 1 in 200.

Epileptic seizures are classified as **symptomatic** if they can be identified with a specific cause, such as infection, trauma, tumor, vascular malformation, toxic chemicals, very high fever, or other neurological disorders. Seizures are **idiopathic** if they appear spontaneously and in the absence of other diseases of the central nervous system.

Table 16-4 summarizes the great variety of circumstances that appear to precipitate seizure. The range of circumstances is striking, but seizures do have a consistent feature: the brain is most epileptogenic when it is inactive and the patient is sitting still.

Although epilepsy has long been known to run in families, its incidence is lower than a one-gene, Mendelian model would predict. It is more likely that epigenetic factors are at work: certain genotypes carry a predisposition to seizure, given certain environmental circumstances. The most remarkable clinical feature of epileptic disorders is the widely varying intervals between attacks—from minutes to hours to weeks or even years. In fact, it is almost impossible to describe a basic set of symptoms to be expected in all or even most people with epilepsy. Nevertheless, three particular symptoms are found within the variety of epileptic episodes:

1. An *aura,* or warning, of impending seizure may take the form of a sensation—an odor or a noise—or may simply be a "feeling" that the seizure is going to occur.

2. *Loss of consciousness* ranges from complete collapse in some people to simply staring off into space in others. The period of lost consciousness is often accompanied by amnesia, including forgetting the seizure itself.

3. Seizures commonly have a *motor component,* but as noted, the movement characteristics vary considerably. Some people shake; others exhibit automatic movements, such as rubbing the hands or chewing.

A diagnosis of epilepsy usually is confirmed by EEG. But some seizures are difficult to document except under special circumstances (e.g., an EEG recorded during sleep). Moreover, not all persons with an EEG that suggests epilepsy actually have seizures. Some statistics estimate that as many as 4 people in 20 simply have abnormal EEG patterns, many more than the 1 in 200 thought to suffer from epilepsy. Among the many types of epileptic seizures, we compare only two here: focal and generalized seizures.

Focal Seizures

Focal seizures begin in one place in the brain and then spread out. John Hughlings-Jackson hypothesized in 1870 that focal seizures probably originate from the point (focus) in the neocortex representing the region of the body where the movement is first seen. He was later proved correct. In *Jacksonian focal seizures,* for example, the attack begins with jerking movements in one part of the body—a finger, a toe, or the mouth—and then spreads to adjacent parts. If the attack begins with a finger, the jerks might spread to other fingers, then the hand, the arm, and so on, producing so-called "Jacksonian marches."

Complex partial seizures, another focal type, originate most commonly in the temporal lobe and somewhat less frequently in the frontal lobe. Complex partial seizures are characterized by three common manifestations:

1. Subjective experiences—for example, forced, repetitive thoughts, alterations in mood, feelings of déjà vu, or hallucinations—before the attack

2. **Automatisms**—repetitive, stereotyped movements such as lip smacking or chewing or activities such as undoing buttons during the attack

3. Postural changes, such as when the person assumes a **catatonic** (frozen) **posture,** during the attack

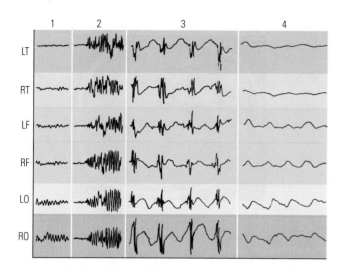

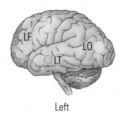

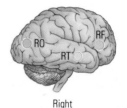

FIGURE 16-8 Grand Mal Seizure Patterns Examples of EEG patterns recorded during a grand mal seizure. Dots on the hemispheres at right indicate the approximate recording sites. Abbreviations: LT and RT, left and right temporal; LF and RF, left and right frontal; LO and RO, left and right occipital. Column numbers mark the seizure's stages: (1) normal record before the attack; (2) onset of the attack; (3) clonic phase, in which the person makes rhythmic movements in time with the large abnormal discharges; and (4) period of coma after the seizure ends.

Left   Right

Generalized Seizures

Generalized seizures lack focal onset and often occur on both sides of the body. The **grand mal** ("big bad") attack is characterized by loss of consciousness and stereotyped motor activity. Patients typically go through the four stages charted in **Figure 16-8:** (1) a tonic stage, in which the body stiffens and breathing stops, (2) a clonic stage, in which there is rhythmic shaking, (3) a postseizure **postictal depression** during which the patient is confused, and (4) a period of coma after the seizure ends. Aura precedes about half of grand mal seizures.

The **petit mal** ("little bad") attack involves a loss of awareness with no motor activity except for blinking, turning the head, or rolling the eyes. Petit mal attacks are of brief duration, seldom exceeding about 10 seconds. The typical EEG recording of a petit mal seizure has a 3-per-second spike-and-wave pattern.

Treatment of Epilepsy

The treatment of choice for epilepsy is anticonvulsant drugs, including diphenylhydantoin (DPH, Dilantin), phenobarbital, or one of about twenty other approved drugs (Rogawski & Loscher, 2004). These drugs are mainly sedative and anesthetic agents when given in low doses, so patients are advised not to drink alcohol. A major site of their action is the inhibitory $GABA_A$ receptor, which acts on a wide variety of neurons.

If medication fails to alleviate the seizure problem satisfactorily in patients with focal seizures, surgery can be performed to remove the focus of abnormal functioning. The abnormal tissue is localized by both EEG and cortical stimulation. It is then removed surgically with the goal of eliminating the cause of the seizures. Many patients show complete recovery and are seizure free, although some must remain on anticonvulsants after the surgery to ensure that the seizures do not return.

Multiple Sclerosis

In multiple sclerosis (MS), the myelin that encases axons is damaged and the functions of the neurons disrupted. MS is characterized by the loss of myelin in both motor and sensory tracts and nerves. The myelin sheath and in some cases the axons are destroyed. Brain imaging with MRI, as shown in **Figure 16-9,** allows areas of sclerosis to be identified in the brain and spinal cord.

symptomatic seizure Identified with a specific cause, such as infection, trauma, tumor, vascular malformation, toxic chemicals, very high fever, or other neurological disorders.

idiopathic seizure Appears spontaneously and in the absence of other diseases of the central nervous system.

focal seizure Seizure that begins locally (at a focus) and then spreads out to adjacent areas.

automatism Unconscious, repetitive, stereotyped movement characteristic of seizure.

catatonic posture Rigid or frozen pose resulting from a psychomotor disturbance.

grand mal seizure Seizure characterized by loss of consciousness and stereotyped motor activity.

postictal depression Postseizure state of confusion and reduced affect.

petit mal seizure Seizure of brief duration, characterized by loss of awareness with no motor activity except for blinking, turning the head, or rolling the eyes.

Figure 10-19 shows neurologists using brain stimulation to map an epilepsy patient's motor cortex.

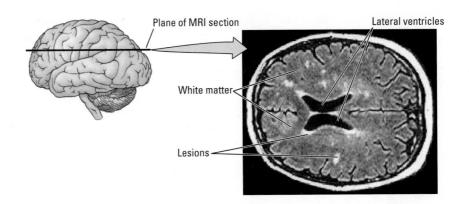

FIGURE 16-9 Diagnosing MS Imaged by MRI, discrete multiple sclerosis lesions appear as white patches all around the lateral ventricles and in the white matter of the brain. Adapted from "Disability and Lesion Load in MS: A Reassessment with MS Functional Composite Score and 3D Fast Flair," by P. A. Ciccarelli, A. J. Brex, A. J. Thompson, and D. H. Miller, 2000, *Journal of Neurology, 249*, pp. 18–24.

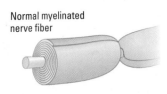

Normal myelinated nerve fiber

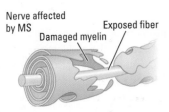

Nerve affected by MS

Damaged myelin

Exposed fiber

MS destroys the myelin sheath. *Sclerosis* comes from the Greek word for "hardness." Adapted from Mayo Foundation for Medical Education and Research.

Focus 4-3 describes another autoimmune disease, myasthenia gravis.

Remissions and relapses are a striking feature of MS: in many cases, early symptoms are initially followed by improvement. The course varies, running from a few years to as long as 50 years. Paraplegia, the classic feature of MS, may eventually confine the affected person to bed.

Worldwide, about 1 million people are afflicted with MS; women outnumber men about two to one. Multiple sclerosis is most prevalent in northern Europe and northern North America, and rare in Japan and in more southerly or tropical countries. Depending on region, the incidence of MS ranges from 2 to 150 per 100,000 people, making it one of the most common structural diseases of the nervous system.

The cause of MS is unknown. Proposed causes include bacterial infection, a virus, environmental factors including pesticides, and an immune response of the central nervous system. Often a number of cases are seen in a single family, and many genes have been associated with MS. Presently there is no clear evidence that MS is inherited or that it is transmitted from one person to another.

Research has focused on the relation of the immune system to MS and the possibility that MS is an **autoimmune disease.** The ability to discriminate between a foreign pathogen in the body and the body itself is a central feature of the immune system. If this discrimination fails, the immune system makes antibodies to a person's own body, in this case the myelin.

As the genomes of various organisms have been sequenced in recent years, it has become apparent that all biological organisms have many genes in common, and thus the proteins found in different organisms are surprisingly similar. And here is the problem for the human immune system: a foreign microbe may have proteins that are very similar to the body's own proteins. If the microbe and human have a common gene sequence, the immune system can mistakenly attack itself, a process known as *horror autotoxicus.* Many microbial protein sequences are homologous with structures found in myelin, which leads to an attack against the microbe and a person's own myelin.

The work showing the important role of the immune system in MS has led to intense research to develop new treatments (Steinman et al., 2002). One strategy is to build up tolerance in the immune system by injecting DNA-encoding myelin antigens as well as DNA-encoding specific molecules in the cascade of steps that leads to the death of myelin cells.

That MS is more common in extreme northern and southern latitudes has raised the possibility that inadequate direct sunlight, which is necessary for the body to synthesize vitamin D, is a factor in precipitating the condition (Niino, 2010). Understanding a possible relationship is complicated by the lack of understanding whether it is an acute or a long-standing deficiency that is relevant. It is also possible that a deficiency of vitamin D interacts with other factors to in some way increase susceptibility to MS.

A recent study suggests that MS might originate from insufficient blood drainage from the brain and that it can be improved or alleviated by cleaning or expanding veins from the brain, including the jugular, to improve drainage. More than 20 papers have been published on what has been called "liberation therapy" for this condition of *chronic cerebrospinal venous insufficiency (CCSVI).*

No clinical trials have been performed and no evidence-based research indicates that liberation therapy is effective. Nevertheless, active Internet discussion worldwide has led to Web sites advertising treatment using vascular surgery for patients with MS (Fragoso, 2011). Numerous studies are examining a plethora of questions related to CCSVI. Primary among them: Is venous flow restricted in people with MS? Optimism about an outcome has far outpaced the science, but it is a hot topic, and ongoing studies should soon resolve the question.

Neurodegenerative Disorders

The demographics now developing in North America and Europe have never been experienced by human societies. Since 1900, the percentage of older people has increased steadily. In 1900, about 4 percent of the population had attained 65 years of age. By 2030, about 20 percent of the population will be older than 65—about 50 million in the United States alone.

Dementias affect from 1 to 6 percent of the population older than age 65 and from 10 to 20 percent older than age 80. For every person diagnosed with dementia, it has been estimated that several others suffer undiagnosed cognitive impairments that affect their quality of life (Larrabee & Crook, 1994). Currently, more than 6 million people in the United States are diagnosed with dementia. This number is projected to rise to about 15 million by 2050, at which time there will be 1 million new cases per year. When this projection is extended across the rest of the developed world, the social and economic costs are truly staggering.

Types of Dementia

Dementia is an acquired and persistent syndrome of intellectual impairment. The two essential features of dementia are (1) loss of memory and other cognitive deficits and (2) impairment in social and occupational functioning. Dementia is not a singular disorder, but there is no clear agreement on how to split up subtypes. The DSM sees the memory symptom as central to amnesic dementia, which it contrasts to vascular dementia. In contrast, Daniel Kaufer and Steven DeKosky (1999) divide dementias into the broad categories of degenerative and nondegenerative **(Table 16-5).**

Nondegenerative dementias, a heterogeneous group of disorders with diverse etiologies, including diseases of the vascular or endocrine systems, inflammation, nutritional deficiency, and toxic conditions are summarized in the right column of Table 16-5. The most prevalent cause is vascular. The most significant risk factors for nondegenerative dementias are chronic hypertension, obesity, sedentary lifestyle, smoking, and diabetes. All are risk factors for cardiovascular disease as well. *Degenerative dementias,* listed in the left column of the table, presumably have a degree of genetic transmission.

We now review two degenerative dementias, Parkinson's disease and Alzeimer's disease. Both pathological processes are primarily intrinsic to the nervous system, and both tend to affect certain neural systems selectively.

Parkinson's Disease

Parkinson's disease is common; estimates of its incidence vary up to 1.0 percent of the population. Incidence rises sharply in old age. In view of the increasingly aging population in Western Europe and North America, the incidence

autoimmune disease Illness resulting from the loss of the immune system's ability to discriminate between foreign pathogens in the body and the body itself.

dementia Acquired and persistent syndrome of intellectual impairment characterized by memory and other cognitive deficits and impairment in social and occupational functioning.

TABLE 16-5 Degenerative and Nondegenerative Dementias

Degenerative	Nondegenerative
Alzheimer's disease	Vascular dementias (e.g., multi-infarct dementia)
Extrapyramidal syndromes (e.g., progressive supernuclear palsy)	Infectious dementia (e.g., AIDS dementia)
Wilson's disease	
Huntington's disease	Neurosyphillis
Parkinson's disease	Posttraumatic dementia
Frontal temporal dementia	Demyelinating dementia (e.g., multiple sclerosis)
Corticobasal degeneration	
Leukodystrophies (e.g., adrenoleukodystrophy)	Toxic or metabolic disorders (e.g., vitamin B_{12} and niacin deficiencies)
Prion-related dementias (e.g., Creutzfeld-Jakob disease)	Chronic alcohol or drug abuse (e.g., Korsakoff's syndrome)

Source: Adapted from "Diagnostic Classifications: Relationship to the Neurobiology of Dementia," by D. I. Kaufer and S. T. DeKosky, 1999, in *The Neurobiology of Mental Illness* (p. 642), edited by D. S. Charney, E. J. Nestler, and B. S. Bunney, New York: Oxford University Press.

Figure 7-2 diagrams degeneration in the substantia nigra associated with Parkinson's symptoms.

of Parkinson's disease is certain to rise in the coming decades, but Parkinsonism is also of interest for several other reasons:

- Parkinson's disease seems related to degeneration of the substantia nigra and to the loss of the neurotransmitter dopamine produced there and released in the striatum. The disease is therefore the source of an important insight into the roles of the substantia nigra and dopamine in movement control.

- Although Parkinsonism is described as a disease entity, symptoms vary enormously among people, which illustrates the complexity in understanding a neurological disorder. A well-defined set of cells degenerates in Parkinson's disease, yet the symptoms are not the same in every sufferer.

- Many symptoms of Parkinson's disease strikingly resemble changes in motor activity that take place as a consequence of aging. Thus, the disease is a source of indirect insight into the more general problems of neural changes in aging.

The symptoms of Parkinson's disease begin insidiously, often with a tremor in one hand and slight stiffness in the distal parts of the limbs. Movements may then become slower, the face becoming masklike with loss of eye blinking and poverty of emotional expression. Thereafter the body may become stooped, and the gait becomes a shuffle with the arms hanging motionless at the sides. Speech may become slow and monotonous, and difficulty in swallowing may cause drooling.

Although the disease is progressive, the rate at which the symptoms worsen is variable, and only rarely is progression so rapid that a person becomes disabled within 5 years. Usually from 10 to 20 years elapse before symptoms cause incapacity. A curious aspect of Parkinson's disease is its on-again–off-again quality: symptoms may appear suddenly and disappear just as suddenly.

Partial remission may also occur in response to interesting or stimulating situations. Neurologist Oliver Sacks (1998) recounts an incident in which a stationary Parkinson patient leaped from his wheelchair at the seaside and rushed into the breakers to save a drowning man, only to fall back into his chair immediately afterward and become inactive again. Remission of some symptoms in activating situations is common but is not usually as dramatic as this case. Incredibly, simply playing familiar music can help an otherwise inactive patient get up and dance. Or a patient who has difficulty walking may ride a bicycle or skate effortlessly. These activities can be used as physical therapy and physical therapy is important because it may slow disease progression.

The four major symptoms of Parkinson's disease are tremor, rigidity, loss of spontaneous movement (*akinesia*), and disturbances of posture. Each symptom may be manifest in different body parts in different combinations. Because some symptoms entail the appearance of abnormal behaviors (positive symptoms) and others the loss of normal behaviors (negative symptoms), we consider the symptoms in both major categories.

Positive symptoms are behaviors not typically seen in people. *Negative symptoms* are absences of typical behaviors or the inability to engage in an activity.

POSITIVE SYMPTOMS Because positive symptoms are common in Parkinson's disease, they are thought to be inhibited, or held in check, in unaffected people but released from inhibition in the process of the disease. Following are the three most common positive symptoms:

1. *Tremor at rest.* Alternating movements of the limbs occur when they are at rest and stop during voluntary movements or sleep. Hand tremors often have a "pill rolling" quality, as if a pill were being rolled between the thumb and forefinger.

2. *Muscular rigidity.* Rigidity, or increased muscle tone simultaneously in both extensor and flexor muscles, is particularly evident when the limbs are moved passively at a joint. Movement is resisted, but with sufficient force, the muscles yield for a short distance and then resist movement again. Thus, complete passive flexion or extension of a joint occurs in a series of steps, giving rise to the term *cogwheel rigidity*. The

rigidity may be severe enough to make all movements difficult, like moving in slow motion and being unable to speed up the process.

3. *Involuntary movements.* Small movements or changes in posture, sometimes referred to as **akathesia** or "cruel restlessness," may be concurrent with general inactivity to relieve tremor and sometimes to relieve stiffness but often occur for no apparent reason. Other involuntary movements are distortions of posture, such as occur during *oculogyric crisis* (involuntary turns of the head and eyes to one side), which last for periods of minutes to hours.

NEGATIVE SYMPTOMS After detailed analysis of negative symptoms, Jean Prudin Martin (1967) divided patients severely affected with Parkinson's disease into five groups:

1. *Disorders of posture.* A *disorder of fixation* presents as an inability or difficulty in maintaining a part of the body in its normal position in relation to other parts. A person's head may droop forward or a standing person may gradually bend forward, ending up on the knees. *Disorders of equilibrium* create difficulties in standing or even sitting unsupported. In less severe cases, people may have difficulty standing on one leg, or if pushed lightly on the shoulders, they may fall passively without taking corrective steps or attempting to catch themselves.

2. *Disorders of righting.* A person has difficulty in achieving a standing position from a supine position. Many advanced patients have difficulty even in rolling over.

3. *Disorders of locomotion.* Normal locomotion requires support of the body against gravity, stepping, balancing while the weight of the body is transferred from one leg to the other, and pushing forward. Parkinson patients have difficulty initiating stepping. When they do walk, they shuffle with short footsteps on a fairly wide base of support because they have trouble maintaining equilibrium when shifting weight from one leg to the other. On beginning to walk, Parkinson patients often demonstrate **festination**: they take faster and faster steps and end up running forward.

4. *Speech disturbances.* One symptom most noticeable to relatives is the almost complete absence of prosody (rhythm and pitch) in the speaker's voice.

5. *Akinesia.* Poverty or slowness of movement may also manifest itself in a blankness of facial expression, a lack of blinking or swinging the arms when walking, a lack of spontaneous speech, or an absence of normal fidgeting. Akinesia is also manifested in difficulty making repetitive movements, such as tapping, even in the absence of rigidity. People who sit motionless for hours show akinesia in its most striking manifestation.

COGNITIVE SYMPTOMS Although Parkinson's disease is usually thought of as a motor disorder, changes in cognition occur as well. Psychological symptoms in Parkinson patients are as variable as the motor symptoms. Nonetheless, a significant percentage of patients show cognitive symptoms that mirror their motor symptoms.

Oliver Sacks (1998), for example, reports impoverishment of feeling, libido, motive, and attention; people may sit for hours, apparently lacking the will to begin or continue any activity. In our experience, thinking seems generally to be slowed and is easily confused with dementia because patients do not appear to be processing the content of conversations. In fact, they are simply processing very slowly.

CAUSES OF PARKINSONISM The ultimate cause of Parkinson's disease—loss of cells in the substantia nigra—may be due to disease, such as encephalitis or syphilis, to drugs such as MPTP, or to unknown causes. *Idiopathic* causes—those related to the individual—may include environmental pollutants, insecticides, and herbicides. Demographic studies of patient admission in the cities of Vancouver, Canada, and Helsinki, Finland, show an increased incidence of patients contracting the disease at ages younger than 40. This

akathesia Small, involuntary movements or changes in posture; motor restlessness.

festination Tendency to engage in a behavior, such as walking, at faster and faster speeds.

Cognitive slowing in Parkinson's patients has some parallels to Alzheimer's disease.

Actor Michael J. Fox, pictured in Focus 5-2, was diagnosed with young-onset Parkinson's disease at age 30.

finding has prompted the suggestion that water and air might contain environmental toxins that work in a fashion similar to MPTP (1-methyl-4-phenylpyridinium), a modified form of heroin that causes Parkinson's disease.

For Parkinson's patients, rhythmic movement is a helpful addition to treatments directed toward replacing depleted dopamine and restoring the balance between neural excitation and inhibition—between the loss and the release of behavior.

Figure 11-13 charts how the GP$_i$, a structure in the basal ganglia, regulates movement force.

TREATING PARKINSON'S DISEASE The cure for Parkinson's disease is either to stop degeneration in the substantia nigra or replace it. Neither goal is achievable at present. Thus, current treatment is pharmacological and directed toward support and comfort.

Psychological factors influence Parkinsonism's major symptoms: a person's outcome is affected by how well he or she copes with the disability. Consequently, patients should seek behaviorally oriented treatment early—counseling on the meaning of symptoms, the nature of the disease, and the potential for most patients to lead long and productive lives. Physical therapy should consist of simple measures, such as heat and massage to alleviate painful muscle cramps and training and exercise to cope with the debilitating changes in movement. Used therapeutically for Parkinson's patients, music and exercise can improve other aspects of behavior, including balance and walking, and may actually slow the course of the disease.

Pharmacological treatment has two main objectives:

1. To increase the activity in whatever dopamine synapses remain

2. To suppress the activity in structures that show heightened activity in the absence of adequate dopamine action

L-Dopa is converted into dopamine in the brain and enhances effective dopamine transmission, as do drugs such as amantadine, amphetamine, monoamine oxidase inhibitors, and tricyclic antidepressants. Naturally occurring anticholinergic drugs, such as atropine and scopolamine, and synthetic anticholinergics, such as benztropine (Cogentin) and trihexyphenidyl (Artane), block the cholinergic systems of the brain that seem to show heightened activity in the absence of adequate dopamine activity.

A drawback of drug therapies is that, as the disease progresses, they become less effective, and the incidence of side effects increases. Some drug treatments that stimulate dopamine receptors directly have been reported to result in increased sexuality and an increased incidence of compulsive gambling.

Several treatments for Parkinson's disease focus on treating its positive symptoms. Two surgical treatments described in Section 16-2 are based on the idea that increased activity of globus pallidus neurons inhibits motor function. A lesion of the internal part of the globus pallidus (GP$_i$) can reduce rigidity and tremor. Hyperactivity of GP$_i$ neurons can also be reduced neurosurgically by electrically stimulating the neurons via deep brain stimulation (see Figure 16-4). A stimulating electrode is permanently implanted in the GP$_i$ or an adjacent brain area, the subthalamic nucleus. Patients carry a small electrical stimulator that they can turn on to produce DBS and so reduce the symptoms of rigidity and tremor. These two treatments may be used sequentially: when DBS becomes less effective as the disease progresses, a GP$_i$ lesion may be produced.

A promising prospective treatment involves increasing the number of dopamine-producing cells. The simplest way to do so is to transplant embryonic dopamine cells into the basal ganglia. In the 1980s and 1990s, this treatment was used with varying degrees of success marked by many poorly conducted studies with inadequate preassessment and postassessment procedures.

A newer course of treatment proposes to increase the number of dopamine cells either by transplanting stem cells that could then be induced to take a dopaminergic phenotype or by stimulating the production of endogenous stem cells and their migration to the basal ganglia. The advantage of stem cells is that they do not have to be derived from embryonic tissue but can come from a variety of sources, including the person's own body.

All these treatments are highly experimental (Politis & Lindvall, 2012). Before cell replacement will become a useful therapy, many experimental questions need to be resolved, including which cell source is best, where grafts should be located in the brain, and how new cells can be integrated into existing brain circuits. Stem cells are not a quick fix for Parkinson's disease, but the pioneering work on this disease will be instrumental in applying this technology to other diseases.

Anatomical Correlates of Alzheimer's Disease

Alzheimer's disease accounts for about 65 percent of all dementias. Given the increasing population of elderly people and thus of Alzheimer's disease, research is being directed toward potential causes, including personal lifestyle, environmental toxins, high levels of trace elements such as aluminum in the blood, an autoimmune response, a slow-acting virus, reduced blood flow to the cerebral hemispheres, and genetic predisposition.

In some families, the incidence of Alzheimer's disease can be especially high, making genetic causes especially pertinent in understanding disease progression. Risk factors include the presence of the *Apoe4* gene, below-average IQ, poor education, and TBI. Presumably, better educated and/or more intelligent people and those who carry the *Apoe2* gene are better able to compensate for the cell death in degenerative dementia.

A decade ago, the only way to identify and study Alzheimer's disease was to study postmortem pathology. This approach was less than ideal because determining which brain changes came early in the disease and which followed as a result of those early changes was impossible. Nonetheless, it became clear that widespread changes take place in the neocortex and limbic cortex and associated changes take place in a number of neurotransmitter systems. Most of the brainstem, cerebellum, and spinal cord are relatively spared its major ravages.

The principal neuroanatomical change in Alzheimer's disease is the emergence of amyloid plaques (clumps of protein from dead neurons and astrocytes), chiefly in the limbic cortex and neocortex. Increased plaque concentration in the cortex has been correlated with the magnitude of cognitive deterioration. Plaques are generally considered nonspecific phenomena in that they can be found in non-Alzheimer patients and in dementias caused by other known events.

Focus 14-3, on the etiology of Alzheimer's disease, includes a micrograph showing an amyloid plaque.

Another anatomical correlate of Alzheimer's disease is neurofibrillary tangles (accumulations of microtubules from dead cells) found in both the neocortex and the limbic cortex, where the posterior half of the hippocampus is affected more severely than the anterior half. Neurofibrillary tangles have been described mainly in human tissue and have also been observed in patients with Down syndrome and Parkinson's disease and other dementias.

Neurofilaments are one type of *tubules* that reinforce the cell's structure, aid in its movement, and transport proteins to their destinations.

Finally, neocortical changes that correlate with Alzheimer's disease are not uniform. As **Figure 16-10** shows plainly, the cortex atrophies (shrinks) and can lose as much as

(A) Healthy Brain

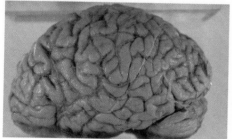

(B) Brain with Alzheimer's

Courtesy of the Nun Study, University of Minnesota

FIGURE 16-10 Cortical Degeneration in Alzheimer's Disease Brain of **(A)** a healthy elderly adult contrasted with **(B)** an elderly adult's brain that shows shriveling due to cell shrinkage characteristic of Alzheimer's disease.

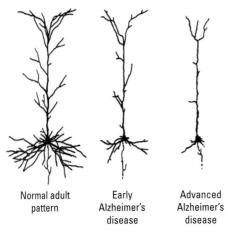

FIGURE 16-11 Stripped Branches As their dendritic trees degenerate and neurons atrophy, patients with Alzheimer's disease experience worsening symptoms, including memory loss and personality changes. Neurons drawn from Golgi-stained sections in "Age-Related Changes in the Human Forebrain," by A. Scheibel, 1982, *Neuroscience Research Program Bulletin, 20,* pp. 577–583.

Figure 14-17 shows how glutamate can affect NMDA and AMPA receptors at the synapse to promote learning by association.

FIGURE 16-12 Midbrain Lewy Body Lewy bodies (*arrow*) characteristic of Parkinson's disease are found in the brains of patients with other disorders as well.

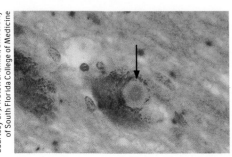

one-third of its volume as the disease progresses. But cellular analyses at the microscopic level reveal that some areas, including the primary sensory and motor areas, especially the visual cortex and the sensory–motor cortex, are relatively spared. The frontal lobes are less affected than is the posterior cortex.

Areas of most extensive change are the association areas of the neocortex and the limbic cortex. The entorhinal cortex is affected earliest and most severely. The entorhinal cortex is the major relay for information from the neocortex traveling to the hippocampus and related structures, then back to the neocortex. Entorhinal damage is associated with memory loss. Given that memory loss is an early and enduring symptom of Alzheimer's disease, it is most likely caused by the degenerative changes that take place in this area.

Many studies describe loss of cells in the cortices of Alzheimer patients, but this finding is disputed. There seems to be a substantial reduction in large neurons, but these cells may shrink rather than disappear. The more widespread cause of cortical atrophy appears to be a loss of dendritic arborization **(Figure 16-11)**.

In addition cell loss and shrinkage, there are changes in the neurotransmitters of the remaining cells. In the 1970s, researchers believed that a treatment for Alzheimer's disease could be found to parallel the L-dopa treatment of Parkinson's disease. The prime candidate neurotransmitter was acetylcholine. One treatment developed for Alzheimer's disease is medication that increases acetylcholine levels in the forebrain. An example, available both orally and as a skin patch, is Exelon, the trade name for rivastigmine, a cholinergic agonist that appears to provide temporary relief from disease progression. Unfortunately, Alzheimer's has proved far more complex because other transmitters clearly are changed as well. Noradrenaline, dopamine, and serotonin are reduced, as are the NMDA and AMPA receptors for glutamate.

Are Parkinson's and Alzheimer's Aspects of One Disease?

Neither Parkinson's disease nor Alzheimer's disease is related to a single brain structure or region, although dopamine in the case of the former and acetylcholine in the case of the latter seem more affected. Other similarities in their pathologies suggest some common neurodegenerative processes (Calne & Mizuno, 2004), such as a loss of cells from the substantia nigra.

The best-studied similarity in the two diseases is the **Lewy body (Figure 16-12)**, a circular, fibrous structure that forms within the cytoplasm of neurons and is thought to correspond to abnormal neurofilament metabolism. Until recently, the Lewy body was believed to be a hallmark of Parkinson's disease. It was most often found in the brainstem in the region of the substantia nigra. It is now clear that Lewy bodies are found in several neurodegenerative disorders, including Alzheimer's disease. There are even reports of people with Alzheimer's-like dementias who do not have plaques and tangles but do have extensive Lewy bodies in the cortex.

Donald Calne noted that, when investigators went to Guam at the end of World War II to investigate a report of widespread dementia described as similar to Alzheimer's disease, they did indeed report a high incidence of Alzheimer's. Many years later, Calne and his colleagues, also experts in Parkinson's disease, examined the same general group of people and found that they had Parkinson's disease. Calne noted that, if you look for Alzheimer symptoms in these people, you find them and miss the Parkinson symptoms. And vice versa.

Indeed, as we age, we all show a loss of cells in the substantia nigra, but only after we have lost about 60 percent of them will we start to show Parkinsonian symptoms. From this perspective, we begin to understand that the dementias of aging share many anatomical similarities.

Age-Related Cognitive Loss

Most people who grow old do not become demented, but virtually everyone shows an age-related cognitive loss, even while living active, healthy, productive lives. Aging is associated with declines in perceptual functions, especially vision, hearing, and olfaction, and declining motor, cognitive, and executive (planning) functions as well. Older people tend to learn at a slower pace and typically do not attain the same mastery of new skills as do younger adults.

Noninvasive imaging studies reveal that aging is correlated with a decrease in white-matter volume probably related to myelin loss. This condition is reparable. There is little evidence of neuronal loss in normal aging, although there is a reduction in neurogenesis in the hippocampus.

Compared with younger people, older people tend to activate larger regions of their attentional and executive networks (parietal and prefrontal cortex) when they perform complex cognitive and executive tasks. This increased activation correlates with reduced performance on tests of working memory as well as attentional and executive tasks.

Two lines of evidence suggest that that age-related declines in function can be slowed. The first is aerobic exercise to enhance general health. The goal is to improve the brain's plasticity via increased neurogenesis, gliogenesis, and trophic factor support. The second treatment is exercise for the brain, employing training strategies that enhance neural plasticity and reverse learned nonuse.

Most of us have experienced the frustration of losing a skill (whether it's trigonometry or tennis) after having not done it for some time, perhaps not for decades. Loss of skill does not reflect dementia but simply a "use it or lose it" scenario. Training programs designed to stimulate plasticity in the appropriate cerebral circuitry include motor-, auditory-, or visual-system–based cognitive and/or attentional training. Brain training is designed to stimulate plasticity rather than to rehabilitate specific losses.

William Milberg and his colleagues (e.g., Kuo et al., 2005; Leritz et al., 2011) have been exploring the interesting idea that aging changes in the brain take place in the context of the entire body's aging. They asked, for example, whether changes in the body such as weight gain, high cholesterol, or hypertension might be related to cognitive change and brain structure in people who show no dementia. Both hypertension and obesity were related to decreased cortical thickness in the frontal lobe, which was correlated with decreased performance on tests of frontal lobe function. In contrast, high cholesterol was associated with temporal lobe cortical thickening and impaired temporal lobe functioning.

An idea emerging from stroke neurologists is that dementia may reflect a chronic cerebrovascular condition, marginal high blood pressure. Marginal elevations in blood pressure can lead to cerebral microbleeds, especially in white matter. The cumulative effect of years or even decades of tiny bleeds would eventually lead to increasingly disturbed cognition. This may first appear as a *mild cognitive impairment (MCI)* that slowly progresses, with cumulative microbleeds, toward dementia.

Sandy Huffaker/The New York Times/Redux

Brain Workout Engaging in cognitively stimulating activities can keep neural networks and general cognitive function from declining with age. From "At the Bridge Table: Clues to a Lucid Old Age," *New York Times,* May 22, 2009 (pp. A1, A18).

Lewy body Circular fibrous structure found in several neurodegenerative disorders; forms within the cytoplasm of neurons and is thought to result from abnormal neurofilament metabolism.

REVIEW 16-3

Understanding and Treating Neurological Disorders

Before you continue, check your understanding.

1. As the world's population ages, _____ disease will become more common than _____ disease.

2. Even imaging techniques may miss pathology produced by _____.

3. Interruption of blood to the brain is called _____ and if prolonged can result in _____ .

4. Although superficially appearing to be very different diseases, Parkinson's and Alzheimer's share similarities such as _____ and _____ that suggest that they may be part of a common disease spectrum.

5. Describe two strategies that can reduce or reverse neurological and cognitive decline with aging.

Answers appear at the back of the book.

16-4 Understanding and Treating Behavioral Disorders

The DSM summarizes a wide range of psychiatric disorders. **Figure 16-13** summarizes the prevalence of specific disorders and lists common symptoms. Added together, nearly 50 percent of the world's population is experiencing a behavioral disorder right now. The statistic holds even when we account for those who acquire more than one disorder.

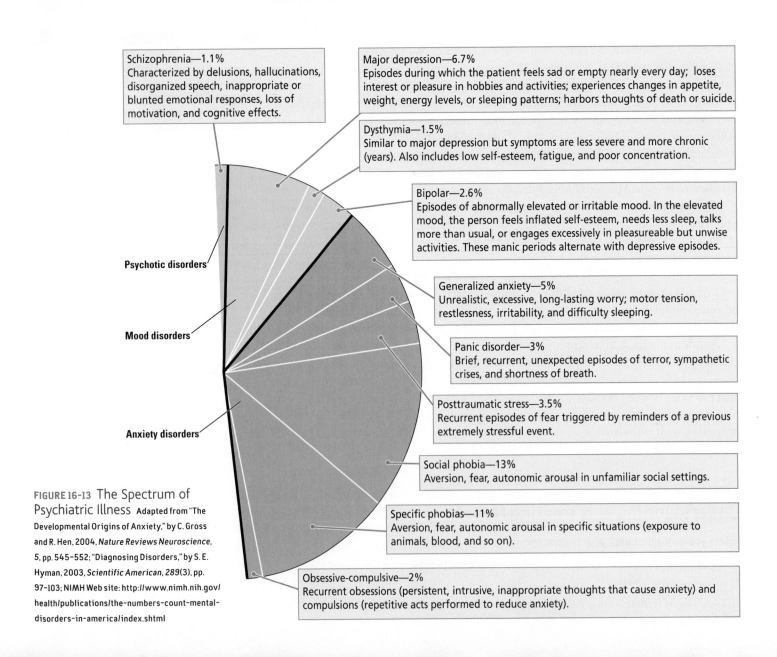

Schizophrenia—1.1%
Characterized by delusions, hallucinations, disorganized speech, inappropriate or blunted emotional responses, loss of motivation, and cognitive effects.

Major depression—6.7%
Episodes during which the patient feels sad or empty nearly every day; loses interest or pleasure in hobbies and activities; experiences changes in appetite, weight, energy levels, or sleeping patterns; harbors thoughts of death or suicide.

Dysthymia—1.5%
Similar to major depression but symptoms are less severe and more chronic (years). Also includes low self-esteem, fatigue, and poor concentration.

Bipolar—2.6%
Episodes of abnormally elevated or irritable mood. In the elevated mood, the person feels inflated self-esteem, needs less sleep, talks more than usual, or engages excessively in pleasureable but unwise activities. These manic periods alternate with depressive episodes.

Generalized anxiety—5%
Unrealistic, excessive, long-lasting worry; motor tension, restlessness, irritability, and difficulty sleeping.

Panic disorder—3%
Brief, recurrent, unexpected episodes of terror, sympathetic crises, and shortness of breath.

Posttraumatic stress—3.5%
Recurrent episodes of fear triggered by reminders of a previous extremely stressful event.

Social phobia—13%
Aversion, fear, autonomic arousal in unfamiliar social settings.

Specific phobias—11%
Aversion, fear, autonomic arousal in specific situations (exposure to animals, blood, and so on).

Obsessive-compulsive—2%
Recurrent obsessions (persistent, intrusive, inappropriate thoughts that cause anxiety) and compulsions (repetitive acts performed to reduce anxiety).

Psychotic disorders

Mood disorders

Anxiety disorders

FIGURE 16-13 The Spectrum of Psychiatric Illness Adapted from "The Developmental Origins of Anxiety," by C. Gross and R. Hen, 2004, *Nature Reviews Neuroscience*, 5, pp. 545–552; "Diagnosing Disorders," by S. E. Hyman, 2003, *Scientific American*, 289(3), pp. 97–103; NIMH Web site: http://www.nimh.nih.gov/health/publications/the-numbers-count-mental-disorders-in-america/index.shtml

Personal, family, and social costs are not reflected in these statistics. Schizophrenia, bipolar disorder, and major depression affect a relatively smaller number of people, for example, but their costs, in loss of social relationships, productivity, and medical care, are disproportionate (McEvoy, 2007). We focus on the three general behavioral categories—psychoses, mood disorders, and anxiety disorders—that are the best studied and understood.

Psychotic Disorders

Psychoses are psychological disorders in which a person loses contact with reality, experiencing irrational ideas and distorted perceptions. Although many psychotic disorders exist, schizophrenia, schizoaffective disorder, and schizophreniform disorder among them, schizophrenia is the most common and best understood. The complexity of the behavioral and neurobiological factors that characterize schizophrenia makes it especially difficult to diagnose and classify. Understanding schizophrenia is an evolving process and far from complete.

Diagnosing Schizophrenia

The DSM lists six diagnostic symptoms of schizophrenia:

1. Delusions—beliefs that distort reality

2. Hallucinations—distorted perceptions—such as hearing voices

3. Disorganized speech, such as incoherent statements or senselessly rhyming talk

4. Disorganized behavior or excessive agitation

5. The opposite extreme: catatonic behavior

6. Negative symptoms, such as blunted emotions or loss of interest and drive, all characterized by the absence of some normal response

The DSM criteria are subjective. They are more helpful in clinical diagnoses than in relating schizophrenia to objective, measurable brain abnormalities.

Classifying Schizophrenia

Timothy Crow (1980, 1990) addressed the classifying difficulties by looking for a relation between brain abnormalities and specific schizophrenia symptoms. He proposed two distinct syndromes, which he called type I and type II.

- **Type I schizophrenia** is characterized predominantly by positive symptoms that manifest behavioral excesses, such as hallucinations and agitated movements. Type I schizophrenia is also associated with acute onset, good prognosis, and a favorable response to neuroleptics.

- **Type II schizophrenia,** by contrast, is characterized by negative symptoms that entail behavioral deficits. Type II schizophrenia is associated with chronic affliction, poor prognosis, poor response to neuroleptics, cognitive impairments, enlarged ventricles, and cortical atrophy, particularly in the frontal cortex (see Figure 16-3).

Crow's analysis had a major effect on clinical thinking about schizophrenia. However, between 20 and 30 percent of patients show a pattern of mixed type I and type II symptoms. So the two types may actually represent points along a continuum of biological and behavioral manifestations of schizophrenia.

Neurobiology of Schizophrenia

Although schizophrenia may present over a wide range of symptoms, individual differences in the behavioral effects exist as well. In recent years, an explosion of research related to the origins and causes of schizophrenia has emerged, but the three lines of research summarized in this section have transformed ideas about the condition—its genetics, its development, and its associated brain correlates.

type I schizophrenia Disorder characterized predominantly by positive symptoms (e.g., behavioral excesses such as hallucinations and agitated movements) likely due to a dopaminergic dysfunction and associated with acute onset, good prognosis, and a favorable response to neuroleptics.

type II schizophrenia Disorder characterized by negative symptoms (behavioral deficits) and associated with chronic affliction, poor prognosis, poor response to neuroleptics, cognitive impairments, enlarged ventricles, and cortical atrophy, particularly in the frontal cortex.

Neuroleptics are antipsychotic drugs described in Section 6-2.

mania Disordered mental state of extreme excitement.

bipolar disorder Mood disorder characterized by periods of depression alternating with normal periods and periods of intense excitation, or *mania*.

GENETICS It has long been recognized that genetics plays an important role in schizophrenia, as does environment. The concordance of schizophrenia in identical twins is high, up to about 70 percent. It is unlikely that a single gene or a mutation on that gene accounts for the condition, however.

More likely, mutations on a number of chromosomes (chromosomes 1, 6, 8, 13, and 22 are candidates) predispose an individual to schizophrenia. It is also likely that many different mutations in candidate genes on those chromosomes are involved. Genetic studies suggest that schizophrenia is probably a family of disorders and that the family may include other conditions, such as major depression and bipolar disorder (Claes et al., 2012). What may be common to the genes involved is that they contribute to brain development and so potentially contribute to various abnormalities in brain development.

DEVELOPMENT Typically, schizophrenia is diagnosed in young adulthood, but a body of evidence suggests that its origins occur much earlier in development, even prenatally. Its expression in adulthood must therefore await the conclusion of a host of developmental processes that ultimately shape the adult human brain. That schizophrenia has developmental origins has many implications.

First, environmental factors acting through epigenetic mechanisms are likely to influence brain development such that a subset of at-risk individuals develops adult schizophrenia. Second, identifying developmental factors that contribute to schizophrenia provides the best opportunity for intervention to reduce the risks of developing it. Any potential "cure" for schizophrenia will depend on early detection and remediation through epigenetic mechanisms (Gebicke-Haerter, 2012).

BRAIN CORRELATES Many studies show anatomical changes in the brain associated with schizophrenia, especially in the temporal and frontal lobes.

1. Suggesting cell loss in these areas, the schizophrenic brain generally has large ventricles and thinner cortex in the medial temporal regions and frontal cortex.

2. Some aspects of the composition of neurons and fibers of the temporal lobes and the frontal lobes are changed, as indicated by changes in their density imaged by MRI (Kong et al., 2012).

3. Alterations in neuronal structure show abnormal dendritic fields in cells in the dorsal prefrontal regions; as shown in **Figure 16-14,** in the hippocampus (Cho et al., 2004); and in the entorhinal cortex (Arnold et al., 1997).

FIGURE 16-14 Organic Dysfunction **(A)** Rather than the consistently parallel orientation of hippocampal neurons typical in normal brains, **(B)** the orientation of hippocampal neurons in the schizophrenic brain is haphazard. Adapted from "A Neurohistologic Correlate of Schizophrenia," by J. A. Kovelman and A. B. Scheibel, 1984, *Biological Psychiatry, 19,* p. 1613.

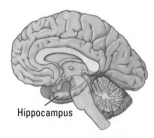

Hippocampus

(A)

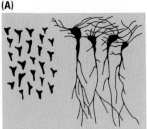

Organized (normal) pyramidal neurons

(B)

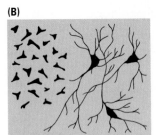

Disorganized (schizophrenic) pyramidal neurons

In sum, this evidence seems almost definitive in identifying schizophrenia as a developmental brain disorder associated in the main with alteration in the temporal and frontal cortex. These are brain regions associated with memory, language, and decision making.

Neurochemical Correlates of Schizophrenia

Neuroscientists also consider the neurochemical correlates of brain–behavior relations in schizophrenia. Dopamine abnormalities were the first to be linked to schizophrenia. First, that most neuroleptic drugs act on the dopamine synapse was taken as evidence that schizophrenia is a disease of heightened activity in the ventral tegmental dopamine system. Second, drugs that enhance dopaminergic activity, such as amphetamine, can produce psychotic symptoms reminiscent of schizophrenia.

This dopamine theory of schizophrenia appears too simple, however, because many other neurochemical abnormalities, summarized in **Table 16-6,** are also associated with schizophrenia—in particular, abnormalities in dopamine and dopamine receptors and in GABA and GABA-binding sites. Recent evidence also suggests changes in glutamate receptors in schizophrenia, a finding that is leading to the development of new classes of drug therapy (Pannese et al., 2012). Considerable variability exists

GABA is the main inhibitory neurotransmitter and Glu the main excitatory neurotransmitter in the brain.

among patients in the extent of each abnormality, however. How these neurochemical variations might relate to the presence or absence of specific symptoms is not yet known.

To summarize, schizophrenia is a complex disorder associated with both positive and negative symptoms, with abnormalities in brain structure and metabolism (especially in the prefrontal and temporal cortex), and with neurochemical abnormalities in dopamine, glutamate, and GABA. Given the complexity of all these behavioral and neurobiological factors, it comes as no surprise that schizophrenia is so difficult to characterize and to treat.

Mood Disorders

The DSM identifies a continuum of mood disorders, but the ones of principal interest here—depression and mania—represent the extremes of affect. The main symptoms of *major depression* are prolonged feelings of worthlessness and guilt, disruption of normal eating habits, sleep disturbances, a general slowing of behavior, and frequent thoughts of suicide.

Mania, the opposite affective extreme from depression, is characterized by excessive euphoria, which the subject perceives as normal. The affected person often formulates grandiose plans and is uncontrollably hyperactive. Periods of mania often change, sometimes abruptly, into states of depression and back again to mania, a condition designated **bipolar disorder.**

Neurobiology of Depression

The neurobiology of depression is complex in that both the brain and the environment contribute. Predisposing factors related to brain anatomy and chemistry thus may contribute more to affective changes in some people, whereas mainly life experiences contribute to affective changes in others. As a result, a bewildering number of life, health, and brain factors have been related to depression. These factors include economic or social failure, circadian rhythm disruption, vitamin D and other nutrient deficiency, pregnancy, brain injury, diabetes, cardiovascular events, and childhood abuse, among many others.

A major approach in neurobiological studies of depression is to ask whether a common brain substrate exists for depression. We know that antidepressant drugs acutely increase the synaptic levels of norepinephrine and serotonin. This finding led to the idea that depression results from a decrease in the availability of one or both neurotransmitters. Lowering their levels in normal participants does not produce depression, however. And while antidepressant medications increase the level of norepinephrine and serotonin within days, it takes weeks for drugs to start relieving depression.

Various explanations for these confounding results have been suggested. None is completely satisfactory. Ronald Duman (2004) reviewed evidence to suggest that antidepressants act, at least in part, on signaling pathways, such as on cAMP, in the postsynaptic cell. Neurotrophic factors appear to affect the action of antidepressants and furthermore may underlie the neurobiology of depression. Investigators know, for example, that brain-derived neurotrophic factor (BDNF) is down-regulated by stress and up-regulated by antidepressant medication (Wang et al., 2012).

Given that BDNF acts to enhance the growth and survival of cortical neurons and synapses, BDNF dysfunction may adversely affect norepinephrine and serotonin systems through the loss of either neurons or synapses. Antidepressant medication may increase the release of BDNF through its actions on cAMP. The key point here is that the cause is probably not just a simple decrease in transmitter levels. Many brain changes are related to depression.

TABLE 16-6 Biochemical Changes Associated with Schizophrenia

Decreased dopamine metabolites in cerebrospinal fluid
Increased striatal D_2 receptors
Decreased expression of D_3 and D_4 mRNA in specific cortical regions
Decreased cortical glutamate
Increased cortical glutamate receptors
Decreased glutamate uptake sites in cingulate cortex
Decreased mRNA for the synthesis of GABA in prefrontal cortex
Increased $GABA_A$-binding sites in cingulate cortex

Source: Adapted from "The Neurochemistry of Schizophrenia," by W. Byne, E. Kemegther, L. Jones, V. Harouthunian, and K. L. Davis, 1999, in *The Neurobiology of Mental Illness* (p. 242), edited by D. S. Charney, E. J. Nestler, and B. S. Bunney. New York: Oxford University Press.

Focus 6-3 explains the threat of suicide attendant to major depression that is left untreated.

Section 14-4 explores the relations of hormones, trophic factors, and psychoactive drugs to neuroplasticity.

Section 6-5 explains the neurobiology of the stress response—how it begins and ends.

Mood and Reactivity to Stress

A significant psychological factor in understanding depression is reactivity to stress. Monoamines—the noradrenergic and serotoninergic activating systems diagrammed in **Figure 16-15A**—modulate hormone secretion by the hypothalamic-pituitary-adrenal system—the **HPA axis**—illustrated in Figure 16-15B. When we are stressed, the HPA axis is stimulated to secrete corticotropin-releasing hormone, which stimulates the pituitary to produce adrenocorticotropic hormone (ACTH). ACTH circulates through the blood and stimulates the adrenal medulla to produce cortisol. Normally, cortisol helps us deal with stress. If we cannot cope, or if stress is intense, excessive cortisol can wield a negative influence on the brain, damaging the feedback loops the brain uses to turn off the stress response.

Excessive stress in early life may be especially detrimental (Gould et al., 2012). During critical periods in early childhood, abuse or other severe environmental stress can permanently disrupt the reactivity of the HPA axis so that it is constantly overactive. HPA-axis overactivity results in oversecretion of cortisol, an imbalance associated with depression in adulthood. Gould and coworkers find, for example, that 45 percent of adults with depression lasting 2 years or more experienced abuse, neglect, or parental loss as children.

Patrick McGowan and colleagues (2010) wondered if early experiences could alter gene expression related to the activity of cortisol in the HPA axis. They compared, postmortem, hippocampi obtained from suicide victims with a history of childhood abuse and hippocampi from other suicide victims with no childhood abuse or from controls. Abused suicide victims showed decreased gene expression for cortisol receptors relative to the controls. These results, derived from epigenetics, confirm that early neglect or abuse alters the HPA axis.

To summarize, the diffuse distribution of the noreprinephrine- and serotonin-activating systems makes relating depression to a single brain structure impossible. Findings

This research confirms studies on the effects of stress on hippocampal function reported in Sections 6-5, 7-5, and 8-4.

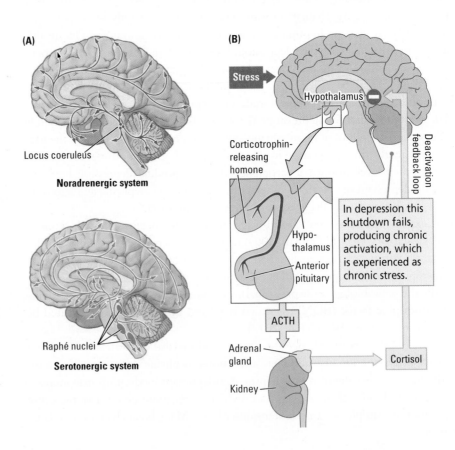

FIGURE 16-15 Stress-Activating System
(A) Medial view showing that (*top*) cell bodies of noradrenergic (norepinephrine) neurons emanate from the locus coeruleus and (*bottom*) cell bodies of the serotonergic activating system emanate from the Raphé nuclei. **(B)** When activated, the HPA axis affects mood, thinking, and indirectly, cortisol secretion by the adrenal glands. HPA deactivation begins when cortisol binds to hypothalamic receptors.

from neuroimaging studies show that depression is accompanied by increased blood flow and glucose metabolism in the orbitofrontal cortex, the anterior cingulate cortex, and the amygdala. Blood flow drops as the symptoms of depression remit when a patient takes antidepressant medication (Drevets, Kishore, & Krishman, 2004). Antidepressants effectively increase the amount of serotonin in the cortex and may also stimulate brain repair. For example, fluoxetine (Prozac) stimulates both BDNF production and neurogenesis in the hippocampus, resulting in a net increase in the number of granule cells (see Research Focus 16-4, "Antidepressant Action and Brain Repair").

Although we have emphasized biological correlates of depression, the best treatment need not be a direct biological intervention. **Cognitive-behavioral therapy** (CBT) is an excellent—arguably the best—therapy for depression. It focuses on challenging the reality of the patient's beliefs and perceptions. The objective is to identify dysfunctional thoughts and beliefs that accompany negative emotions and replace them with more realistic ones.

Simply pointing out to a person that the person's beliefs are faulty is not likely to be effective, however, because it probably took months or years to develop those beliefs. The

HPA axis Hypothalamic-pituitary-adrenal circuit that controls the production and release of hormones related to stress.

cognitive-behavioral therapy (CBT) Problem-focused, action-oriented, structured, treatment for eliminating dysfunctional thoughts and maladaptive behaviors.

RESEARCH FOCUS ✛ 16-4

Antidepressant Action and Brain Repair

Excessive stress, poor coping skills, or both contribute to excessive cortisol levels in the brain and to depression and anxiety. Cells in the hippocampus that are especially sensitive to cortisol can die when cortisol levels either are reduced, as shown in the figure, or increase excessively.

Hippocampal cells may play a role in regulating cortisol levels and "switching off" the brain's stress response. Their absence leaves the HPA axis unchecked. Then, increased cortisol levels contribute to more neuron death and to depression and anxiety. It's a vicious circle.

A remarkable finding emerged from studying the effects on the hippocampus of antidepressants such as fluoxetine (Prozac) and other SSRIs. The drugs increased neurogenesis in the subventricular zone, cell migration into the hippocampus, and the number of functioning cells in the hippocampus (Malberg et al., 2000).

That antidepressants take some weeks to produce their positive effects on behavior and new neurons take some weeks to migrate and incorporate into the hippocampus suggest that the antidepressants' mechanism of action is to enhance hippocampal function—in essence, to replace the neural switch that turns off the stress response.

Luca Santarelli and colleagues (2003) used knockout mice that lacked a specific serotonin receptor (5-HT-1A) that is sensitive to fluoxetine.

They exposed these mice and mice with intact 5-HT-1A receptors to chronic, unpredictable stress. Both groups developed a general deterioration in the state of their fur coats.

Chronic antidepressant treatment affected only the mice with intact serotonin receptors, and only these mice displayed enhanced neurogenesis. Thus, because antidepressants activate serotonin receptors, serotonin must play a role in enhancing neurogenesis.

The literature on neurogenesis in the hippocampus is the most prolific in neuroscience, represented by over 3000 papers using mainly animal models. To summarize their findings, phenomena we perceive as bad (infant deprivation, conflict, stress, poor diet, most drugs) depress neurogenesis, and phenomena we perceive as good (exercise, enriched environments, sufficient sleep, good diet) enhance neurogenesis.

Against this background, the effects of antidepressants are instructive. But do antidepressants do more than enhance neurogenesis? Kobayashi and colleagues (2010) find that fluoxetine reverses neuronal maturation of hippocampal cells making the cells appear younger. Should this prove to be a general effect of antidepressants, their effectiveness against depression possibly is produced by inducing a more juvenile brain state—a state that is more plastic and adaptable.

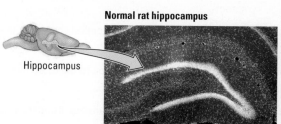

Normal rat hippocampus

Hippocampus

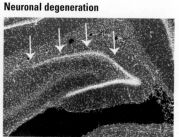

Neuronal degeneration

neural circuits underlying the beliefs must be changed, just as the strategies for developing new ones must change. In a real sense then, cognitive-behavioral therapy is effective if it induces neural plasticity and changes brain activity.

Anxiety Disorders

We all experience anxiety, usually acutely as a response to stress or, less commonly, as a chronic reactivity—an increased anxiety response—even to seemingly minor stressors. Anxiety reactions certainly are not pathological, and it is likely that they are an evolutionary adaptation for coping with adverse conditions. But anxiety can become pathological to the point of making life miserable.

Anxiety disorders are among the most common psychiatric conditions. The DSM-IV lists six classes of anxiety disorders that together affect an estimated 4 of 10 people at some point in their lifetimes (see Figure 16-13). Those six classes expand to ten in the DSM-5.

Imaging studies of people with anxiety disorders record increased baseline activity in the cingulate cortex and parahippocampal gyrus and an enhanced response to anxiety-provoking stimuli in the amygdala and prefrontal cortex. This finding suggests excessive excitatory neurotransmission in a circuit involving anterior cingulate cortex, prefrontal cortex, amygdala, and parahippocampal region. Researchers hypothesize that, because drugs that enhance the inhibitory transmitter GABA are particularly effective in reducing anxiety, excessive excitatory neurotransmission in this circuit is "anxiety." But what causes it?

Considerable interest has developed in investigating why some people show a pathological level of anxiety to stimuli to which others have a milder response. One hypothesis, just covered in the section on depression, is that stressful experiences early in life increase a person's susceptibility to a variety of behavioral abnormalities, especially anxiety disorders.

Although anxiety disorders used to be treated primarily with benzodiazepines such as Valium, now they are also treated with SSRIs such as Prozac, Paxil, Celexa, and Zoloft. Antidepressant drugs do not act immediately, however, suggesting that the treatments must stimulate some gradual change in brain structure, much as these drugs act in treating depression.

Cognitive-behavioral therapy is as effective as drugs in treating anxiety. The most effective behavioral therapies expose and reexpose patients to their fears. For example, treating a phobic fear of germs requires exposing the patient repeatedly to potentially germy environments, such as public washrooms, until the discomfort abates (Abramowitz, 1998).

Focus 12-3 describes symptoms of anxiety disorders, Section 6-5 how stress-induced damage contributes to PTSD, and Section 5-3 increased serotonergic activity in OCD.

Figure 12-18 diagrams these limbic system structures and charts their major connections.

One more time: a pill is not a skill.

REVIEW 16-4

Understanding and Treating Behavioral Disorders

Before you continue, check your understanding.

1. Type I schizophrenia features _____ symptoms whereas Type II schizophrenia features _____ symptoms.

2. Schizophrenia is associated with pronounced anatomical changes in the _____ and _____ cortices.

3. The monoamine activating systems that have received the most investigation related to understanding depression are _____ and _____.

4. The most effective treatment for depression and anxiety disorders is _____.

5. Describe the main difficulty in linking genes to schizophrenia.

Answers appear at the back of the book.

 16-5 Is Misbehavior Always Bad?

You know this movie plot: a person sustains some sort of blow to the head and becomes a better person. You might wonder whether pathological changes in the brain and behavior sometimes lead to behavioral improvement in real life. A report by Jim Giles (2004) on Tommy McHugh's case is thought provoking.

McHugh, a heroin addict, had committed multiple serious crimes and had spent a great deal of time in jail. He suffered a cerebral hemorrhage (bleeding into the brain) from an aneurysm. The bleeding was repaired surgically by placing a metal clip on the leaking artery. After he recovered from the injury, McHugh showed a dramatic change in personality, took up painting, which he had never done before, and became a successful artist.

McHugh's injury-induced brain changes appear to have been beneficial. The exact nature of McHugh's brain injury has not been identified because the metal clip in his brain precludes the use of MRI. Aspects of his cognitive behavior suggest that he may have frontal-lobe damage.

The phenomenon in which an individual acquires a new skill after an injury is called *acquired savant syndrome*. There are other reports of people who have developed new musical or artistic talents after their injuries (Miller et al., 2000). Corrigan and colleagues (2011) propose that allied savant skills can be acquired by depressing inhibitory systems in the brain so that new skill strategies can be activated. In experiments, for example, depressing participants' left hemispheres with TMS briefly improved mathematical skills, which are subserved by the right hemisphere.

The general idea of manipulating the brain for the better is controversial (Heinz et al., 2012). Influencing brain function more scientifically through a strategy loosely described as **cognitive enhancement** enlists current knowledge of pharmacology, brain plasticity, brain stimulation, neurogenetics, and so on, to boost brain functioning. Of course, people already use drugs to alter brain function, and the basis of many therapies is to enhance brain function. Psychosurgical techniques such as frontal lobotomies were based on the general idea that brain function could be improved. As yet, however, evidence is lacking that cognitive enhancement for the average person is better than old-fashioned but readily available methods: learning, practice, and a healthy lifestyle.

cognitive enhancement Brain-function enhancement by pharmacological, physiological, or surgical manipulation.

Section 10-4 describes a surgeon obsessed with playing and writing piano music after he was struck by lightning.

Focus 6-1 offers a brief history of cognitive enhancers, as context for the current trend among some students of procuring, as a study aid, prescription medication used to treat ADHD.

SUMMARY

Contemporary understanding of brain and behavior is providing new insights, explanations, and treatments for brain misbehavior. Neurologists, who treat organic disorders, and psychiatrists, who treat behavioral disorders, are forging a unified understanding of mind and brain: neuropsychoanalysis.

16-1 Multidisciplinary Research on Brain and Behavioral Disorders

Most behavioral disorders have multiple causes—genetic, biochemical, anatomical, and social–environmental variables—all of them interacting. Research methods directed toward these causes include family studies designed to find a genetic abnormality that might be corrected, biochemical anomalies that might be reversed by drug or hormone therapy, anatomical pathologies that might account for behavioral changes, and social–environmental variables.

Investigators rely increasingly on neuroimaging (fMRI, PET, TMS, ERP) to examine brain–behavior relations in vivo in normal participants as well as in people experiencing disorders. Interest in more refined behavioral measurements is growing, especially for cognitive behavior, the better to understand behavioral symptoms.

16-2 Classifying and Treating Brain and Behavioral Disorders

Disorders can be classified according to presumed etiology (cause), symptomatology, or pathology. The primary etiological classification, neurological versus psychiatric, is artificial because it presupposes that two categories encompass all types of disorders. As more is learned about etiology, more disorders fall into the neurological category.

Symptomatological classification requires a checklist, such as the DSM. The problem is that symptoms of psychiatric disorders overlap. The checklist of likely symptoms for disorders is thus open to interpretation. Symptoms may appear more or less prominent, depending on subjective perceptions.

Pathological classification of some behavioral disorders may be possible with MRI or other imaging techniques but often requires postmortem examination. What is becoming clear is that disorders overlap more in pathology than previously recognized.

The table summarizes the range of available treatments for brain and behavioral disorders, from highly invasive neurosurgery to noninvasive electrophysiology and from moderately invasive pharmacology to indirect behavioral treatments.

General Treatment Categories

Neurosurgical	Behavioral
Direct intervention	*Manipulation of experience*
DBS	Behavior modification
Stem-cell transplantation	Cognitive/cognitive-behavioral therapy
Tissue removal or repair	Neuropsychological
Electrophysiological	Emotional therapy/psychotherapy
Noninvasive manipulation	Physical activity/music
ECT	rt-fMRI
TMS, rTMS	Virtual-reality and other computer-based simulations
Pharmacological	
Chemical administration	
Antibiotics or antivirals	
Psychoactive drugs	
Neurotrophic factors	
Nutritional	

16-3 Understanding and Treating Neurological Disorders

If a disorder, such as depression, is presumably caused primarily by a biochemical imbalance, treatment is likely to be pharmacological. If the disorder has a suspected anatomical cause, treatment may include the removal of pathological tissue (as in epilepsy) or implanted electrodes to activate underactive regions (as in Parkinson's disease and stroke). Brain activation with TMS is promising and noninvasive. Many disorders, however, require medical treatment concurrent with behavioral therapy, including physiotherapy or cognitive rehabilitation for stroke or TBI and cognitive-behavioral therapies for depression and anxiety disorders.

16-4 Understanding and Treating Psychiatric Disorders

The number of people who endure hidden diseases of behavior, especially neurodegenerative disorders and stroke, is increasing as the population of the Western world ages. Like other plagues in human history, dementias affect not only the person with the disease but also the caregivers. About half of the caregivers for people with disorders linked to aging seek psychiatric care themselves.

16-5 Is Misbehavior Always Bad?

In rare cases, people who experience disordered behaviors may inadvertently benefit from neurological disease, suggesting that some brain misbehavior can prove beneficial. The logic of cognitive enhancement is that, by employing genetic manipulations, transplants, and brain stimulation to alter brain organization, enhancement might alter improve behavior more generally.

KEY TERMS

akathesia, p. 591

autoimmune disease, p. 589

automatism, p. 587

behavioral therapy, p. 578

bipolar disorder, p. 598

catatonic posture, p. 587

chronic traumatic encephalopathy (CTE), p. 583

cognitive-behavioral therapy (CBT), p. 601

cognitive enhancement, p. 603

cognitive therapy, p. 578

deep brain stimulation (DBS), p. 575

dementia, p. 589

diaschisis, p. 585

DSM, p. 573

festination, p. 591

focal seizure, p. 587

grand mal seizure, p. 587

HPA axis, p. 601

idiopathic seizure, p. 587

ischemia, p. 585

Lewy body, p. 595

magnetic resonance spectroscopy (MRS), p. 583

mania, p. 598

neuroprotectant, p. 585

neuropsychoanalysis, p. 569

petit mal seizure, p. 587

phenylketonuria (PKU), p. 569

postictal depression, p. 587

posttraumatic stress disorder (PTSD), p. 567

psychotherapy, p. 578

real-time fMRI (rt-fMRI), p. 580

symptomatic seizure, p. 587

tardive dyskinesia, p. 578

type I schizophrenia, p. 597

type II schizophrenia, p. 597

virtual-reality (VR) exposure therapy, p. 567

Chapter 1
What Are the Origins of Brain and Behavior?

Review 1-1
Neuroscience in the Twenty-First Century

1. cerebrum, forebrain; hemispheres; brainstem
2. central nervous system (CNS); peripheral nervous system (PNS)
3. inherited; learning
4. Research on embodied language proposes that we understand each other not only by listening to words but also by observing gestures and other body language and that we think not only with silent language but also with overt gestures and body language.

Review 1-2
Perspectives on Brain and Behavior

1. mentalism; dualism; Materialism
2. natural selection; Charles Darwin
3. traumatic brain injury (TBI); minimally conscious state (MCS); persistent vegetative state (PVS)
4. In formulating the theory of natural selection, Darwin relied on observation to conclude that living organisms are related and pass traits from parents to offspring. Mendel used experimentation to show that heritable factors underlie phenotypic variation among species.

Review 1-3
Evolution of Brains and of Behavior

1. nervous systems
2. nerve net; bilaterally symmetrical; ganglia; chordate
3. Humans possess the largest brain of all animals relative to body size.

Review 1-4
Evolution of the Human Brain and Behavior

1. common ancestor; chimpanzee
2. hominid
3. *in any order:* changes in climate, changes in lifestyle skills, including preferred food resources, changes in skull anatomy, neoteny

4. Changes in climate may have driven many physical changes in hominids, including the nearly threefold increase in brain size from apes to modern humans. Evidence suggests that each new hominid species appeared after climate changes devastated old environments and produced new environments. Eventually, modern humans evolved adaptability sufficient to allow us to populate almost every climatic region on earth.

Review 1-5
Modern Human Brain Size and Intelligence

1. species-typical behavior
2. culture
3. In comparing different species, a larger brain correlates with more complex behavior. In comparing individual members within a species, brain size and intelligence are not particularly related; rather, the complexity of different brain regions is related to behavioral abilities. Humans, for example, vary widely in body size and in brain size as well as in having different kinds of intelligence. All these factors make any simple comparison of individuals' brain size and general intelligence impossible.

Chapter 2
How Does the Nervous System Function?

Review 2-1
Overview of Brain Function and Structure

1. behavior; brain
2. *in any order:* frontal, temporal, parietal, occipital
3. neuroplasticity
4. white matter; gray matter
5. tracts; nerves
6. Compare your diagram with Figure 2-2B.

Review 2-2
Evolutionary Development of the Nervous System

1. forebrain; midbrain; hindbrain
2. behavior

3. The forebrain region has grown dramatically over the course of vertebrate evolution. But as more complex nervous systems have emerged, more primitive forms have not been discarded and replaced. They have been added to. So the forebrain's growth represents the elaboration of functions already present in the other brain regions and leads to its functioning on multiple levels.

Review 2-3
Central Nervous System: Mediating Behavior

1. levels of function
2. spinal cord
3. hindbrain; midbrain; diencephalon
4. basal ganglia; limbic system
5. The forebrain regulates cognitive activity, including thought and memory, and holds ultimate control over movement (behavior).

Review 2-4
Somatic Nervous System: Transmitting Information

1. cranial nerves; spinal (peripheral) nerves
2. same
3. head; internal body organs and glands
4. The law of Bell and Magendie, which states that sensory (afferent) spinal nerve fibers are located dorsally and motor (efferent) spinal fibers are located ventrally, is important because it allows neurologists to predict the location of spinal-cord damage accurately, based on changes in sensation or movement experienced or reported by patients.

Review 2-5
Autonomic Nervous System: Balancing Internal Functions

1. ganglia
2. sympathetic; parasympathetic; in opposition
3. The autonomic system operates largely outside our conscious awareness, whether we are awake or asleep, to regulate the vegetative functions essential to life.

Review 2-6
Ten Principles of Nervous-System Function

1. olfactory system; somatic nervous system
2. multiple levels of functioning
3. excitation; inhibition
4. Any individual's perceived reality is only a rough approximation of what is actually present. An animal's representation of the world depends on the nature of the information sent to the animal's brain.

Chapter 3
What Are the Functional Units of the Nervous System?

Review 3-1
Cells of the Nervous System

1. *in either order:* neurons; glia
2. *in either order:* excite; inhibit
3. *in any order:* sensory neurons, interneurons, motor neurons
4. *in any order:* ependymal cells, astrocytes, microglia, oligodendroglia, Schwann cells; nourishing, removing waste, insulating, supporting, repairing
5. The main obstacle is duplicating the complexity of a mammalian brain—the sheer number of neurons and of the connections they make with each other, and their ability to change. Advances in computing and miniaturization technologies are bringing the idea of complex robots within the realm of the possible.

Review 3-2
Internal Structure of a Cell

1. *in any order:* cell membrane, nucleus, endoplasmic reticulum, Golgi bodies, microtubules (*or* tubules), vesicles
2. proteins; *in any order:* channels, gates, pumps
3. DNA, RNA, protein
4. endoplasmic reticulum; Golgi bodies; microtubules
5. By using most of the proteins that it makes, a cell enables itself to interact with other cells and to modify their behavior. The collective action of cells then mediates behavior.

Review 3-3
Genes, Cells, and Behavior

1. 23; protein
2. alleles; proteins
3. mutation; Down syndrome
4. recessive; dominant
5. Selective breeding; Cloning; transgenic animals
6. Mendelian genetics concentrates on inheritance patterns—on which genes parents pass to their offspring and offspring pass to succeeding generations. Epigenetics studies how the environment and experience can affect the inherited genome.

Chapter 4
How Do Neurons Use Electrical Signals to Transmit Information?

Review 4-1
Searching for Electrical Activity in the Nervous System

1. stimulation; recording
2. *in any order:* recording from the giant axons of the North Atlantic squid; using an oscilloscope to measure small changes in voltage; crafting microelectrodes small enough to place on or into an axon
3. *in either order:* concentration gradient, from an area of relatively high concentration to an area of lower concentration; voltage gradient, from an area of relatively high charge to an area of lower charge
4. Ion channels in cell membranes may open to facilitate ion movement, close to impede ion movement, or pump ions across the membrane.

Review 4-2
Electrical Activity of a Membrane

1. resting potential; ions
2. semipermeable; negative
3. hyperpolarization; depolarization
4. action potential; nerve impulse
5. Nerve impulses travel more rapidly on myelinated axons because of saltatory conduction: action potentials jump at the speed of light between the nodes separating the glial cells that form the axon's myelin sheath.

Review 4-3
How Neurons Integrate Information

1. excitatory postsynaptic potentials (EPSPs); inhibitory postsynaptic potentials (IPSPs)
2. time; space; integrates
3. cell body; axon hillock; axon
4. Some neurons have voltage-sensitive channels on their dendrites that allow the reverse movement of an action potential into the neurons' dendritic fields.

Review 4-4
Into the Nervous System and Back Out

1. sensory system *or* sense
2. sensory receptor cell; voltage-sensitive
3. motor; muscle
4. The varieties of membrane channels explain a wide range of neural events. Channels generate the transmembrane charge, mediate graded potentials, and trigger the action potential.

Chapter 5
How Do Neurons Communicate and Adapt?

Review 5-1
A Chemical Message

1. chemical synapses; gap junction
2. presynaptic cell membrane; postsynaptic cell membrane; quanta
3. axodendritic; axosomatic; axomuscular; axoaxonic; axosynaptic; axoextracellular; axosecretory; dendrodendritic
4. dendrite; cell body
5. When an action potential is propagated on an axon terminal, (1) a chemical transmitter that has been synthesized and stored in the axon terminal (2) is released from the presynaptic membrane into the synaptic cleft. The transmitter (3) diffuses across the cleft and binds to receptors on the postsynaptic membrane. (4) Then the transmitter is deactivated.

Review 5-2
Varieties of Neurotransmitters

1. synthesis, release, receptor action, inactivation
2. *in any order:* small-molecule transmitters, peptide transmitters, transmitter gases; ionotropic, metabotropic
3. An ionotropic receptor contains a pore or channel that can be opened or closed to regulate the flow-through of ions, directly bringing about rapid and usually excitatory voltage changes on the cell membrane. Metabotropic receptors are generally inhibitory, are slow acting, and activate second

messengers to indirectly produce changes in the function and structure of the cell.

Review 5-3
Neurotransmitter Systems and Behavior

1. neurotransmitter; neurotransmitter
2. acetylcholine; acetylcholine; acetylcholine; norepinephrine
3. *in any order:* cholinergic, dopaminergic, noradrenergic, serotonergic
4. This idea has been attractive for a long time because there is a clear relationship between DA loss and Parkinson's disease, and acetylcholine and norepinephrine are clearly related to somatic and autonomic behaviors. But for other neurotransmitter systems in the brain, establishing clear one-to-one relationships has proved difficult.

Review 5-4
Adaptive Role of Synapses in Learning and Memory

1. synapse; learning
2. *in either order:* habituation; sensitization
3. presynaptic axon terminal; sensory; calcium; less
4. interneurons; potassium; calcium ions, *or* Ca^{2+}
5. posttraumatic stress disorder, or PTSD; sensitization
6. Permanent responses to frequently occurring stimuli are biologically (*or* behaviorally *and/or* metabolically) efficient, but if stimuli change suddenly, a lack of flexibility becomes maladaptive.

Chapter 6
How Do Drugs and Hormones Influence the Brain and Behavior?

Review 6-1
Principles of Psychopharmacology

1. psychoactive drugs; psychopharmacology
2. blood–brain barrier; brain
3. synapses; agonists; antagonists
4. tolerance; sensitization
5. *in any order:* feces; urine; sweat; breath; breast milk
6. (a) Drug use at home is unlikely to condition drug-taking behavior to familiar home cues, so tolerance is likely to occur. (b) Novel cues in a work setting may enhance conditioning and so sensitize the occasional drug user.

Review 6-2
Grouping Psychoactive Drugs

1. *in either order:* behavioral; psychoactive
2. $GABA_A$; Cl^- *or* chloride ion
3. MAO inhibitors; SSRIs
4. endorphins
5. release; reuptake; D_2
6. Psychotropic drugs act on many neurotransmitters, including acetylcholine, anandamide, dopamine, epinephrine, glutamate, norepinephrine, and serotonin.

Review 6-3
Factors Influencing Individual Responses to Drugs

1. disinhibition; learning; alcohol myopia theory
2. Substance abuse; addiction *or* substance dependence
3. psychomotor activation; mesolimbic dopamine system
4. females; males
5. Alcohol myopia theory suggests that intoxicated individuals are unusually responsive to local and immediate cues, so the environment excessively influences their behavior while consequences go ignored.

Review 6-4
Explaining and Treating Drug Abuse

1. liking (pleasure); tolerance; wanting (craving); sensitization
2. frontal cortex; brainstem; mesolimbic dopamine system (pathways); dorsal striatum
3. inheritance; epigenetics
4. drugs; other life experiences
5. A reasonable approach to treatment views drug addiction in the same way as chronic behavioral addictions and medical problems are viewed: as a lifelong problem for most people.

Review 6-5
Hormones

1. neurohormones; pituitary gland; releasing hormones; brain
2. *in either order:* steroid hormones; peptide hormones
3. homeostatic; gonadal; glucocorticoids
4. anabolic (*or* anabolic–androgenic) steroids; muscle mass; masculinizing
5. epinephrine; cortisol
6. The hippocampus plays an important role in ending the stress response by reg-

ulating cortisol levels. If cortisol remains elevated by prolonged stress, eventually it damages the hippocampus.

Chapter 7
How Do We Study the Brain's Structure and Functions?

Review 7-1
Measuring Brain and Behavior

1. brain function; behavior; *any one from among:* block tapping, mirror drawing, recent memory; *any one from among:* place learning, matching to place, landmark learning
2. sectioning and staining; multiphoton microscope
3. *in any order:* brain lesions; brain stimulation; optogenetics
4. Brain-stimulation methods include using electrical pulses, as in DBS; magnetic fields, as in TMS; chemicals, by administering drugs; or light, as in the transgenic technique of optogenetics.

Review 7-2
Measuring the Brain's Electrical Activity

1. *in any order:* EEG; ERP; MEG; single-cell recording
2. graded potentials
3. action potentials
4. electrical activity of many neurons; three-dimensional localization of the cell groups generating the measured field
5. EEG measures the brain's electrical activity, and ERP allows scientists to determine which brain areas are processing various kinds of stimuli and in which order the areas come into play.

Review 7-3
Static Imaging Techniques: CT and MRI

1. *in either order:* computer tomography *or* CT scan; magnetic resonance imaging *or* MRI
2. neural connections *or* fiber pathways; concentrations of brain metabolites
3. brain injury *or* brain damage
4. CT creates X-ray images of one object from many different angles then uses scanning software to combine the images into a 3-D image of the brain.

Review 7-4
Dynamic Brain Imaging

1. *in any order:* fMRI; PET; optical tomography *or* fNIRS

2. radioactively labeled molecules; neurochemical

3. cerebral blood flow

4. Resting-state images in PET and rs-fMRI can identify abnormalities in brain function, and rs-fMRI can also identify functional connections in the resting brain.

Review 7-5
Chemical and Genetic Measures of Brain and Behavior

1. biochemical; *in either order:* microdialysis; voltammetry

2. concordance rates

3. DNA; gene expression

4. Epigenetic studies show that life experience can alter gene expression and that these changes are associated with changes in neuronal structure and connectivity. The altered neuronal organization is, in turn, associated with changes in behavior.

Review 7-6
Comparing Neuroscience Research Methods

1. *in any order:* temporal resolution; spatial resolution; degree of invasiveness

2. *any one or more:* EEG, ERP, *and/or* fNIRS; inexpensive

3. The fundamental goal of neuroscience research is an understanding of brain–behavior relationships.

Review 7-7
Using Animals in Brain–Behavior Research

1. *any two in any order:* stroke; ADHD; Parkinson's disease; schizophrenia

2. whether laboratory animals experience the same symptoms that humans do

3. neural bases; treatments

4. Using laboratory animals in research leads to concerns about animal welfare and raises ethical issues about whether animals should be used in research and, if so, in what types of research.

Chapter 8
How Does the Nervous System Develop and Adapt?

Review 8-1
Three Perspectives on Brain Development

1. behavior

2. neural circuits

3. *any 3 in any order:* hormones; sensory experience; injuries; genes

4. Behaviors cannot emerge until the requisite neural structures are sufficiently mature.

Review 8-2
Neurobiology of Development

1. neural tube

2. neurogenesis; gliogenesis

3. *in either order:* cell-adhesion molecules; tropic factors

4. *in either order:* myelination; synaptic pruning

5. Dynamic changes in frontal-lobe structure (morphology) are related to the development of intelligence.

Review 8-3
Correlating Behavior with Nervous-System Development

1. independent finger movements *or* the pincer grasp

2. vocabulary; phonological processing

3. Piaget's stages of cognitive development

4. temporal lobe; basal ganglia

5. Correlation does not prove causation.

Review 8-4
Brain Development and the Environment

1. chemoaffinity hypothesis

2. amblyopia

3. testosterone

4. critical periods

5. Adolescence is a time of rapid brain change related to both pubertal hormones and psychosocial stressors that make the brain vulnerable to disorders.

Chapter 9
How Do We Sense, Perceive, and See the World?

Review 9-1
Nature of Sensation and Perception

1. sensory receptors

2. receptive; density

3. target in the brain

4. subjective experience of sensation

5. Each modality has many receptors and sends information to the cortex to form topographic maps.

Review 9-2
Functional Anatomy of the Visual System

1. retinal ganglion cells *or* RGCs

2. P *or* parvocellular; M *or* magnocellular

3. geniculostriate; tectopulvinar

4. facial agnosia *or* prosopagnosia

5. The dorsal stream to the parietal lobe processes the visual guidance of movements (the how). The ventral stream to the temporal lobe processes the visual perception of objects (the what).

Review 9-3
Location in the Visual World

1. small size

2. photoreceptors; retinal ganglion cells, lateral geniculate neurons, cortical neurons

3. topographic map

4. corpus callosum

5. The fovea is represented by a larger area in the cortex than the visual field's periphery, and thus there is more processing of foveal information in region V1 than of peripheral information.

Review 9-4
Neuronal Activity

1. bars of light

2. temporal

3. trichromatic theory

4. opponent

5. RGCs are excited by one wavelength of light and inhibited by another, producing two pairs of what seem to be color opposites—red versus green and blue versus yellow.

Review 9-5
Visual Brain in Action

1. hemianopia

2. scotomas

3. monocular blindness

4. optic ataxia

5. Damage to the dorsal stream produces deficits in visually guided movements. Damage to the ventral stream produces deficits in object recognition.

Chapter 10
How Do We Hear, Speak, and Make Music?

Review 10-1
Sound Waves: Stimulus for Audition

1. air-pressure waves

2. *in any order:* frequency; amplitude; complexity

3. *in any order:* loudness; pitch; prosody, quality *or* timbre

4. temporal

5. Delivery speed, or the number of sound segments that can be analyzed per second, distinguishes speech and musical sounds from other auditory inputs. Nonlanguage sounds faster than 5 segments per second are heard as a buzz, yet we are capable of understanding speech delivered at nearly 30 segments per second.

Review 10-2
Functional Anatomy of the Auditory System

1. ossicles *or, in any order,* hammer, anvil, and stirrup
2. hair cells; cochlea
3. *in either order:* basilar; tectorial
4. auditory *or* cochlear; auditory vestibular *or* eighth
5. inferior colliculus; medial geniculate nucleus
6. The planum temporale is larger in the left hemisphere, and Heschl's gyrus is larger in the right. This anatomical asymmetry is correlated to a functional asymmetry: the left temporal cortex analyzes language-related sounds, whereas the right temporal cortex analyzes music-related ones.

Review 10-3
Neural Activity and Hearing

1. tonotopic
2. cochlea
3. superior olive; trapezoid body
4. action for audition
5. There are two mechanisms. Neurons in the brainstem (hindbrain) compute the time difference in a sound wave's arrival at each ear. Other neurons in the brainstem compute the difference in sound amplitude (loudness) in each ear.

Review 10-4
Anatomy of Language and Music

1. language; music
2. *in any order:* Broca's area; supplementary speech area; face area of motor cortex
3. Wernicke's
4. Broca's
5. *in either order:* perfect (*or* absolute) pitch; amusic *or* tone deaf
6. Three lines of evidence support the idea that language is innate: the universality of language, the natural acquisition of language by children, and the presence of syntax in all languages.

Review 10-5
Auditory Communication in Nonhuman Species

1. epigenetic mechanisms
2. left
3. echolocation; sonar
4. Birdsong dialects demonstrate that the songs young birds hear influence how they sing their songs.

Chapter 11
How Does the Nervous System Respond to Stimulation and Produce Movement?

Review 11-1
A Hierarchy of Movement Control

1. hierarchy
2. prefrontal; premotor; motor cortex
3. brainstem
4. spinal cord
5. Lower-level functions in the motor hierarchy can continue in the absence of higher-level ones, but the higher levels provide voluntary control over movements. When the brain is disconnected from the spinal cord then, movement can no longer be controlled at will.

Review 11-2
Motor System Organization

1. topographic; homunculus; larger
2. motor map
3. corticospinal; lateral corticospinal; ventral corticospinal
4. trunk; arm; finger
5. muscles; *in either order:* flexes, extends
6. Mirror neurons in the premotor and primary motor cortex can be active at the same time that no actual movement occurs, a property that allows them to control neuroprosthetic devices such as brain–computer interfaces.

Review 11-3
Basal Ganglia, Cerebellum, and Movement

1. basal ganglia; force
2. hyperkinetic; hypokinetic
3. accuracy; skills
4. The cerebellum compares an intended movement with the actual movement, calculates any necessary corrections, and informs the cortex to correct the movement.

Review 11-4
Organization of the Somatosensory System

1. hapsis; proprioception; nocioception
2. dorsal spinothalamic; ventral spinothalamic
3. pain gate
4. periaqueductal gray matter *or* PAG
5. vestibular system; balance
6. Without proprioception, sensory information about body location and movement is lost and can only be regained using vision.

Review 11-5
Exploring the Somatosensory Cortex

1. primary somatosensory; secondary somatosensory
2. apraxia
3. dorsal; ventral
4. Pain perception does not depend simply on pain sensations but is a creation of the brain.

Chapter 12
What Causes Emotional and Motivational Behavior?

Review 12-1
Identifying the Causes of Behavior

1. rewarding
2. minimum level of sensory stimulation
3. smell and taste *or* chemical senses
4. In general, behavior is controlled by neural circuits that are modulated by a wide range of factors.

Review 12-2
The Chemical Senses

1. olfactory epithelium
2. flavor
3. pheromones
4. allele of the taste receptor gene, TAS2R38; number of taste buds
5. Any given odorant stimulates a unique pattern of receptors, and the summed activity, or pattern of activity, produces our perception of a particular odor.

Review 12-3
Evolution, Environment, and Behavior

1. rewards *or* reinforcers
2. taste-aversion learning
3. innate releasing mechanisms *or* IRMs
4. tasting
5. When two unrelated events are experienced together they may become

inadvertently associated. For example, unexpected pain in the presence of a stranger may lead to a faulty association between the events.

Review 12-4
Neuroanatomy of Motivated and Emotional Behavior

1. regulatory; nonregulatory
2. hypothalamus
3. *in any order:* hypothalamus; limbic system; frontal lobes
4. *in any order:* dorsolateral; orbitofrontal; medial
5. amygdala
6. hormones
7. The limbic system stimulates emotional reactions and species-typical behaviors, whereas the frontal lobes generate the rationale for behavior at the right time and context, taking factors such as external events and internal information into account.

Review 12-5
Control of Regulatory and Nonregulatory Behavior

1. *in any order:* lateral hypothalamus; ventromedial hypothalamus; paraventricular nucleus
2. *in either order:* hypothalamus; amygdala
3. *in either order:* organizing; activating
4. osmotic; hypervolumic
5. Variations in epigenetic effects could lead to a difference in the architecture and function of the hypothalamus in homosexuals and transsexuals.

Review 12-6
Reward

1. rewarding
2. wanting; liking
3. *in any order:* dopamine; opioid; benzodiazepine–GABA systems.
4. Intracranial self-stimulation is a phenomenon whereby animals learn to turn on a stimulating electric current in their brains, presumably because it activates the neural system that underlies reward.

Chapter 13
Why Do We Sleep and Dream?

Review 13-1
A Clock for All Seasons

1. circannual; circadian
2. free-running; Zeitgebers

3. *any 2:* light pollution; jet lag; working swing shifts
4. Circadian rhythm allows us to synchronize our behavior with our bodies' metabolic processes—so that we are hungry at optimal times for eating, for example.

Review 13-2
Neural Basis of the Biological Clock

1. superchiasmatic nucleus *or* SCN
2. retinohypothalamic; melanopsin ganglion
3. shell; chemical; anatomical
4. Experimental evidence suggests that the circadian rhythm can put a time stamp on a behavioral event, rendering it easier to recall at the same time in the circadian cycle that it occurred in previously.

Review 13-3
Sleep Stages and Dreaming

1. *in either order:* REM, rapid eye movement; NREM, non-rapid eye movement
2. EOG *or* electrooculogram; EMG *or* electromyogram; EEG *or* electroencephalogram
3. 5; lengthening
4. everyone; real time
5. Interpreting dreams is difficult because it is always possible that the dream interpreter will impose his or her own interpretation on dreams.

Review 13-4
What Does Sleep Accomplish?

1. biological; restorative; memories
2. Explicit; implicit
3. place cell; NREM
4. microsleep
5. If a memory can be stored during waking, sleep may not be essential for its storage.

Review 13-5
Neural Bases of Sleep

1. reticular activating system *or* RAS; NREM
2. coma
3. REM sleep
4. subcoerulear
5. We have separate neural systems for keeping us awake while we are still (cholinergic) and awake when we move (serotinergic).

Review 13-6
Sleep Disorders

1. Insomnia; narcolepsy
2. drug-dependent insomnia
3. sleep paralysis; cataplexy
4. REM without atonia; subcoerulear
5. Orexin is probably only one of many factors that are related to waking behavior, as animals with narcolepsy can be awake but then collapse into sleep.

Chapter 14
How Do We Learn and Remember?

Review 14-1
Connecting Learning and Memory

1. Pavlovian
2. operant
3. implicit; explicit
4. episodic
5. Memory is not localized to any particular circuit in the brain. Rather, multiple memory circuits vary with the actual requirements of the memory task.

Review 14-2
Dissociating Memory Circuits

1. *in either order:* hippocampus; amygdala
2. basal ganglia
3. Lashley searched for explicit memory in the perceptual and motor systems of his animal subjects using tests designed mostly for implicit memory. Milner studied a patient with medial temporal removal and used tests of both explicit and implicit memory.

Review 14-3
Neural Systems Underlying Explicit and Implicit Memories

1. *in either order:* hippocampus; neocortex (*or* cortex)
2. implicit
3. amygdala
4. consolidation
5. Emotional experiences stimulate hormonal and neurochemical activating systems that stimulate the amygdala. The amygdala in turn modulates the laying down of memory circuits in the rest of the brain.

Review 14-4
Structural Basis of Brain Plasticity

1. long-term potentiation; long-term depression
2. GABAergic *or* inhibitory

3. *in either order:* synapse number; neuron number
4. hippocampus
5. behavioral sensitization
6. Behavioral disorders such as addiction and PTSD are examples of plastic changes that are not good.

Review 14-5
Recovery from Brain Injury

1. *in any order:* learn new ways to solve problems; reorganize the brain to do more with less; replace the lost neurons
2. *in either order:* direct cortical stimulation; deep brain stimulation *or* DBS
3. neurotrophic
4. Functional improvement after brain injury reflects compensation rather than recovery.

Chapter 15
How Does the Brain Think?

Review 15-1
Nature of Thought

1. attending to; identifying; making meaningful responses to
2. language; flexibility
3. neuron
4. Much of human thought is verbal. Language allows us to categorize information and provides a way to organize our behavior around time.

Review 15-2
Cognition and the Association Cortex

1. knowledge; cognition
2. temporal; parietal
3. *in either order:* plan movements; organize behavior over time
4. Mirror
5. Multimodal cortex allows the brain to combine characteristics of stimuli across different sensory modalities, whether we encounter them together or separately.

Review 15-3
Expanding Frontiers of Cognitive Neuroscience

1. cognitive neuroscience
2. brain connectome
3. social interactions
4. theory of mind
5. decision making
6. *in any order:* understanding others; understanding oneself; self-regulation; social living

Review 15-4
Cerebral Asymmetry in Thinking

1. *in either order:* spatial behavior; music
2. *in either order:* controlling voluntary movement sequences; language
3. corpus callosum
4. Because the hemispheres process information differently, they think differently. And the existence of language in the left hemisphere allows it to label computations and thus make inferences that the right cannot.

Review 15-5
Variations in Cognitive Organization

1. *in either order:* sex; handedness
2. neural circuits
3. synesthesia
4. Gonadal hormones influence brain development and shape neural circuits in adulthood.

Review 15-6
Intelligence

1. "g" factor *or* general intelligence; multiple intelligences; convergent and divergent; intelligence A; intelligence B
2. structural; functional
3. *any 3, in any order:* gyral patterns; cytoarchitectonics; vascular patterns; neurochemistry
4. epigenetic
5. Both fMRI and ERP studies show that the efficiency of prefrontal–parietal circuits is related to standard measures of general intelligence.

Review 15-7
Consciousness

1. complex
2. Consciousness
3. consciousness
4. Movements in which speed is critical, such as hitting a pitched ball, cannot be controlled consciously.

Chapter 16
What Happens When The Brain Misbehaves?

Review 16-1
Multidisciplinary Research on Brain and Behavioral Disorders

1. brainstem and limbic system; ventral frontal and posterior cortex; dorsal frontal cortex

2. *in any order:* genetic errors; epigenetic mechanisms; progressive cell death; rapid cell death; loss of neural connections
3. PKU *or* phenylketonuria
4. subjective
5. Brain pathology can exist without obvious clinical symptoms and clinical symptoms without obvious pathology.

Review 16-2
Classifying and Treating Brain and Behavioral Disorders

1. ICD (*International Classification of Diseases*) *or* DSM (*Diagnostic and Statistical Manual of Mental Disorders*)
2. *in any order:* neurosurgical, electrophysiological, pharmacological, behavioral
3. deep brain stimulation *or* DBS
4. transcranial magnetic stimulation *or* TSM
5. The principles underlying behavioral treatments are derived mainly from learning theory.

Review 16-3
Understanding and Treating Neurological Disorders

1. Alzheimer's; Parkinsons's
2. TBI *or* traumatic brain injury
3. ischemia; stroke
4. loss of cells from the substantia nigra; accumulation of Lewy bodies
5. Aerobic exercise and "brain training" are strategies for enhancing or stimulating neuroplasticity as we age.

Review 16-4
Understanding and Treating Behavioral Disorders

1. positive; negative
2. *in either order:* temporal; frontal
3. *in either order:* norepinephrine; serotonin
4. cognitive-behavioral therapy
5. Many genetic and epigenetic influences contribute to every behavior, including schizophrenia.

absolutely refractory Refers to the state of an axon in the repolarizing period during which a new action potential cannot be elicited (with some exceptions), because gate 2 of sodium channels, which is not voltage sensitive, is closed.

acetylcholine (ACh) First neurotransmitter discovered in the peripheral and central nervous systems; activates skeletal muscles in the somatic nervous system and may either excite or inhibit internal organs in the autonomic system.

action potential Large, brief reversal in the polarity of an axon.

activating system Neural pathways that coordinate brain activity through a single neurotransmitter; cell bodies are located in a nucleus in the brainstem and axons are distributed through a wide region of the brain.

addiction Desire for a drug manifested by frequent use of the drug, leading to the development of physical dependence in addition to abuse; often associated with tolerance and unpleasant, sometimes dangerous, withdrawal symptoms on cessation of drug use. Also called *substance dependence.*

afferent Conducting toward a central nervous system structure.

agonist Substance that enhances the function of a synapse.

akathesia Small, involuntary movements or changes in posture; motor restlessness.

akinesia Slowness or absence of movement.

alcohol myopia "Nearsighted" behavior displayed under the influence of alcohol: local and immediate cues become prominent, and remote cues and consequences are ignored.

allele Alternate form of a gene; a gene pair contains two alleles.

alpha rhythm Regular wave pattern in an electroencephalogram; found in most people when they are relaxed with closed eyes.

Alzheimer's disease Degenerative brain disorder related to aging that first appears as progressive memory loss and later develops into generalized dementia.

amblyopia Condition in which vision in one eye is reduced as a result of disuse; usually caused by a failure of the two eyes to point in the same direction.

amnesia Partial or total loss of memory.

amphetamine Drug that releases the neurotransmitter dopamine into its synapse and, like cocaine, blocks dopamine reuptake.

amplitude Intensity of a stimulus; in audition, roughly equivalent to loudness, graphed by increasing the height of a sound wave.

amusia Tone deafness—an inability to distinguish between musical notes.

amygdala Almond-shaped collection of nuclei located within the limbic system; plays a role in emotional and species-typical behaviors.

anabolic steroid Class of synthetic hormones related to testosterone that have both muscle-building (anabolic) and masculinizing (androgenic) effects; also called *anabolic–androgenic steroid.*

androgen Class of hormones that stimulates or controls masculine characteristics; male hormone related to level of sexual interest.

anencephaly Failure of the forebrain to develop.

anomalous speech representation Condition in which a person's speech zones are located in the right hemisphere or in both hemispheres.

anorexia nervosa Exaggerated concern with being overweight that leads to inadequate food intake and often excessive exercising; can lead to severe weight loss and even starvation.

antagonist Substance that blocks the function of a synapse.

anterograde amnesia Inability to remember events subsequent to a disturbance of the brain such as head trauma, electroconvulsive shock, or certain neurodegenerative diseases.

antianxiety agent Drug that reduces anxiety; examples are minor tranquillizers such as benzodiazepines and sedative-hypnotic agents.

aphagia Failure to eat; may be due to an unwillingness to eat or to motor difficulties, especially with swallowing.

aphasia Inability to speak or comprehend language despite the presence of normal comprehension and intact vocal mechanisms. Broca's aphasia is the inability to speak fluently despite the presence of normal comprehension and intact vocal mechanisms. Wernicke's aphasia is the inability to understand or to produce meaningful language even though the production of words is still intact.

apoptosis Cell death that is genetically programmed.

apraxia Inability to make voluntary movements in the absence of paralysis or other motor or sensory impairment, especially an inability to make proper use of an object.

association cortex Neocortex outside the primary sensory and motor cortices that functions to produce cognition.

associative learning Linkage of two or more unrelated stimuli to elicit a behavioral response.

astrocyte Star-shaped glial cell that provides structural support to neurons in the central nervous system and transports substances between neurons and blood vessels.

atonia No tone; condition of complete muscle inactivity produced by the inhibition of motor neurons.

attention Selective narrowing or focusing of awareness to part of the sensory environment or to a class of stimuli.

attention-deficit/hyperactivity disorder (ADHD) Developmental disorder characterized by core behavioral symptoms of impulsivity, hyperactivity, and/or inattention.

auditory flow Change in sound heard as a person moves past a sound source or as a sound source moves past a person.

autism spectrum disorder (ASD) Range of cognitive symptoms, from mild to severe, that characterize autism; severe symptoms include greatly impaired social interaction, a bizarre and narrow range of interests, marked abnormalities in language and communication, and fixed, repetitive movements.

autoimmune disease Illness resulting from the loss of the immune system's ability to discriminate between foreign pathogens in the body and the body itself.

automatism Unconscious, repetitive, stereotyped movement characteristic of seizure.

autonomic nervous system (ANS) Part of the PNS that regulates the functioning of internal organs and glands.

autoreceptor "Self-receptor" in a neural membrane that responds to the transmitter released by the neuron.

axon "Root," or single fiber, of a neuron that carries messages to other neurons.

axon collateral Branch of an axon.

axon hillock Juncture of soma and axon where the action potential begins.

back propagation Reverse movement of an action potential into the dendritic field of a neuron; postulated to play a role in plastic changes that underlie learning.

barbiturate Drug that produces sedation and sleep.

basal ganglia Subcortical forebrain nuclei that coordinate voluntary movements of the limbs and body; connected to the thalamus and to the midbrain.

basic rest–activity cycle (BRAC) Recurring cycle of temporal packets, about 90-minute periods in humans, during which an animal's level of arousal waxes and wanes.

basilar membrane Receptor surface in the cochlea that transduces sound waves into neural activity.

behavioral neuroscience Study of the biological bases of behavior.

behavioral sensitization Escalating behavioral response to the repeated administration of a psychomotor stimulant such as amphetamine, cocaine, or nicotine; also called *drug-induced behavioral sensitization.*

behavioral therapy Treatment that applies learning principles, such as conditioning, to eliminate unwanted behaviors.

beta (β) rhythm Fast brain-wave activity pattern associated with a waking EEG.

bilateral symmetry Body plan in which organs or parts present on both sides of the body are mirror images in appearance. For example, the hands are bilaterally symmetrical, whereas the heart is not.

binding problem Philosophical question focused on how the brain ties single and varied sensory and motor events together into a unified perception or behavior.

biological clock Neural system that times behavior.

biorhythm Inherent timing mechanism that controls or initiates various biological processes.

bipolar disorder Mood disorder characterized by periods of depression alternating with normal periods and periods of intense excitation, or *mania*.

bipolar neuron Sensory neuron with one axon and one dendrite.

blind spot Region of the retina where axons forming the optic nerve leave the eye and where blood vessels enter and leave; has no photoreceptors and is thus "blind."

blob Region in the visual cortex that contains color-sensitive neurons, as revealed by staining for cytochrome oxidase.

blood–brain barrier Tight junctions between the cells that compose blood vessels in the brain, providing a barrier to the entry of an array of substances, including toxins, into the brain.

brain connectome Map of the complete structural and functional neural connections of the human brain in vivo.

brainstem Central structures of the brain, including the hindbrain, midbrain, thalamus, and hypothalamus, that are responsible for most unconscious behavior.

Broca's area Anterior speech area in the left hemisphere that functions with the motor cortex to produce the movements needed for speaking.

carbon monoxide (CO) Gas that acts as a neurotransmitter in the activation of cellular metabolism.

cataplexy Form of narcolepsy linked to strong emotional stimulation in which an animal loses all muscle activity or tone, as if in REM sleep, while awake.

catatonic posture Rigid or frozen pose resulting from a psychomotor disturbance.

cell-adhesion molecule (CAM) A chemical molecule to which specific cells can adhere, thus aiding in migration.

cell assembly Hypothetical group of neurons that become functionally connected because they receive the same sensory inputs. Hebb proposed that cell assemblies were the basis of perception, memory, and thought.

cell body (soma) Core region of the cell containing the nucleus and other organelles for making proteins.

central nervous system (CNS) The brain and spinal cord that together mediate behavior.

cerebellum Major structure of the brainstem specialized for coordinating and learning skilled movements. In large-brained animals, the cerebellum may also have a role in coordinating other mental processes.

cerebral cortex Thin, heavily folded film of nerve tissue composed of neurons that is the outer layer of the forebrain. Also called *neocortex*.

cerebral palsy Group of brain disorders that result from brain damage acquired perinatally (at or near birth).

cerebral voltammetry Technique used to identify the concentration of specific chemicals in the brain as animals behave freely.

cerebrospinal fluid (CSF) Clear solution of sodium chloride and other salts that fills the ventricles inside the brain and circulates around the brain and spinal cord beneath the arachnoid layer in the subarachnoid space.

cerebrum Major structure of the forebrain that consists of two virtually identical hemispheres (left and right) and is responsible for most conscious behavior.

channel Opening in a protein embedded in the cell membrane that allows the passage of ions.

chemical synapse Junction at which messenger molecules are released when stimulated by an action potential.

chemoaffinity hypothesis Proposal that neurons or their axons and dendrites are drawn toward a signaling chemical that indicates the correct pathway.

cholinergic neuron Neuron that uses acetylcholine as its main neurotransmitter. The term *cholinergic* applies to any neuron that uses Ach as its main transmitter.

chordate Animal that has both a brain and a spinal cord.

chronic traumatic encephalopathy (CTE) Progressive degenerative disease caused by multiple concussions and other closed-head injuries, characterized by neurofibrillary tangles, plaques, and cerebral atrophy and expanded ventricles due to cell loss.

chronotype Individual differences in circadian activity.

circadian rhythm Day–night rhythm.

cladogram Phylogenetic tree that branches repeatedly, suggesting a taxonomy of organisms based on the time sequence in which evolutionary branches arise.

clinical trial Consensual experiment directed toward developing a treatment.

cochlea Inner-ear structure that contains the auditory receptor cells.

cochlear implant Electronic device implanted surgically into the inner ear to transduce sound waves into neural activity and allow a deaf person to hear.

cognition Act or process of knowing or coming to know; in psychology, used to refer to the processes of thought.

cognitive-behavioral therapy (CBT) Problem-focused, action-oriented, structured, treatment for eliminating dysfunctional thoughts and maladaptive behaviors.

cognitive enhancement Brain-function enhancement by pharmacological, physiological, or surgical manipulation.

cognitive neuroscience Study of the neural bases of cognition.

cognitive therapy Psychotherapy based on the perspective that thoughts intervene between events and emotions, and thus the treatment of emotional disorders requires changing maladaptive patterns of thinking.

color constancy Phenomenon whereby the perceived color of an object tends to remain constant relative to other colors, regardless of changes in illumination.

coma Prolonged state of deep unconsciousness resembling sleep.

common ancestor Forebearer from which two or more lineages or family groups arise and so is ancestral to both groups.

competitive inhibitor Drug such as nalorphine and naloxone that acts quickly to block the actions of opioids by competing with them for binding sites; used to treat opioid addiction.

computerized tomography (CT) X-ray technique that produces a static, three-dimensional image of the brain in cross section—a *CT scan*.

concentration gradient Differences in concentration of a substance among regions of a container that allow the substance to diffuse from an area of higher concentration to an area of lower concentration.

conditioned response (CR) In Pavlovian conditioning, the learned response to a formerly neutral conditioned stimulus (CS).

conditioned stimulus (CS) In Pavlovian conditioning, an originally neutral stimulus that, after association with an unconditioned stimulus (UCS), triggers a conditioned response.

cone Photoreceptor specialized for color and high visual acuity.

consciousness The mind's level of responsiveness to impressions made by the senses.

consolidation Process of stabilizing a memory trace after learning.

contralateral neglect Ignoring a part of the body or world on the side opposite (contralateral to) that of a brain injury.

convergent thinking Form of thinking that searches for a single answer to a question (such as $2 + 2 = ?$); contrasts with divergent thinking.

corpus callosum Band of white matter containing about 200 million nerve fibers that connects the two cerebral hemispheres to provide a route for direct communication between them.

cortical column Cortical organization that represents a functional unit six cortical layers deep and approximately 0.5 millimeter square and that is perpendicular to the cortical surface.

corticospinal tract Bundle of nerve fibers directly connecting the cerebral cortex to the spinal cord, branching at the brainstem into an opposite-side lateral tract that informs movement of limbs and digits and a same-side ventral tract that informs movement of the trunk; also called *pyramidal tract*.

cranial nerve One of a set of 12 nerve pairs that control sensory and motor functions of the head, neck, and internal organs.

critical period Developmental "window" during which some event has a long-lasting influence on the brain; often referred to as a *sensitive period*.

cross-tolerance Reduction of response to a novel drug because of tolerance developed in response to a chemically related drug.

culture Learned behaviors that are passed on from one generation to the next through teaching and experience.

cytoarchitectonic map Map of the neocortex based on the organization, structure, and distribution of the cells.

deafferentation Loss of incoming sensory input usually due to damage to sensory fibers; also loss of any afferent input to a structure.

decibel (dB) Unit for measuring the relative physical intensity of sounds.

declarative memory Ability to recount what one knows, to detail the time, place, and circumstances of events; often lost in amnesia.

deep-brain stimulation (DBS) Neurosurgery in which electrodes implanted in the brain stimulate a targeted area with a low-voltage electrical current to facilitate behavior.

delta (δ) rhythm Slow brain-wave activity pattern associated with deep sleep.

dementia Acquired and persistent syndrome of intellectual impairment characterized by memory and other cognitive deficits and impairment in social and occupational functioning.

dendrite Branching extension of a neuron's cell membrane that greatly increases the surface area of the cell and collects information from other cells.

dendritic spine Protrusion from a dendrite that greatly increases the dendrite's surface area and is the usual point of dendritic contact with the axons of other cells.

depolarization Decrease in electrical charge across a membrane, usually due to the inward flow of sodium ions.

dermatome Body segment corresponding to a segment of the spinal cord.

diaschisis Neural shock that follows brain damage in which areas connected to the site of damage show a temporary arrest of function.

dichotic listening Experimental procedure for simultaneously presenting a different auditory input to each ear through stereophonic earphones.

diencephalon The "between brain" that integrates sensory and motor information on its way to the cerebral cortex.

diffusion Movement of ions from an area of higher concentration to an area of lower concentration through random motion.

diffusion tensor imaging (DTI) Magnetic resonance imaging method that, by detecting the directional movements of water molecules, can image fiber pathways in the brain.

dimer Two proteins combined into one.

disinhibition theory Explanation holding that alcohol has a selective depressant effect on the cortex, the region of the brain that controls judgment, while sparing subcortical structures responsible for more primitive instincts, such as desire.

diurnal animal Organism that is active chiefly during daylight.

divergent thinking Form of thinking that searches for multiple solutions to a problem (such as How many different ways can a pen be used?); contrasts with convergent thinking.

dopamine (DA) Amine neurotransmitter that plays a role in coordinating movement, in attention and learning, and in behaviors that are reinforcing.

dopamine hypothesis of schizophrenia Idea that excess activity of the neurotransmitter dopamine causes symptoms of schizophrenia.

dorsal spinothalamic tract Pathway that carries fine-touch and pressure fibers.

Down syndrome Chromosomal abnormality resulting in mental retardation and other abnormalities, usually caused by an extra chromosome 21.

drug-dependency insomnia Condition resulting from continuous use of "sleeping pills"; drug tolerance also results in deprivation of either REM or NREM sleep, leading the user to increase the drug dosage.

DSM Abbreviation of *Diagnostic and Statistical Manual of Mental Disorders,* the American Psychiatric Association's classification system for psychiatric disorders.

dualism Philosophical position that holds that both a nonmaterial mind and a material body contribute to behavior.

dyslexia Impairment in learning to read and write; probably the most common learning disability.

echolocation Ability to identify and locate an object by bouncing sound waves off the object.

efferent Conducting away from a central nervous system structure.

electrical stimulation Passage of an electrical current from the uninsulated tip of an electrode through tissue, resulting in changes in the electrical activity of the tissue.

electrocorticography (ECoG) Graded potentials recorded with electrodes placed directly on the brain's surface.

electroencephalogram (EEG) Graph that records electrical activity through the skull or from the brain and represents graded potentials of many neurons.

embodied language Hypothesis that the movements we make and the movements we perceive in others are central to communication with others.

emotion Cognitive interpretation of subjective feelings.

emotional memory Memory for the affective properties of stimuli or events.

encephalization quotient (EQ) Jerison's quantitative measure of brain size obtained from the ratio of actual brain size to expected brain size, according to the principle of proper mass, for an animal of a particular body size.

endorphin Peptide hormone that acts as a neurotransmitter and may be associated with feelings of pain or pleasure; mimicked by opioid drugs such as morphine, heroin, opium, and codeine.

end plate On a muscle, the receptor–ion complex that is activated by the release of the neurotransmitter acetylcholine from the terminal of a motor neuron.

entorhinal cortex Located on the medial surface of the temporal lobe; provides a major route for neocortical input to the hippocampal formation; often degenerates in Alzheimer's disease.

entrain Determine or modify the period of a biorhythm.

ependymal cell Glial cell that makes and secretes cerebrospinal fluid; found on the walls of the ventricles in the brain.

epidermal growth factor (EGF) Neurotrophic factor that stimulates the subventricular zone to generate cells that migrate into the striatum and eventually differentiate into neurons and glia.

epigenetics Differences in gene expression related to environment and experience.

epinephrine (EP, *or* adrenaline) Chemical messenger that acts as a hormone to mobilize the body for fight or flight during times of stress and as a neurotransmitter in the central nervous system.

episodic memory Autobiographical memory for events pegged to specific place and time contexts.

estrogens Variety of sex hormones responsible for the distinguishing characteristics of the female.

event-related potentials (ERPs) Complex electroencephalographic waveforms related in time to a specific sensory event.

evolutionary psychology Discipline that seeks to apply principles of natural selection to understand the causes of human behavior.

excitation Increase in the activity of a neuron or brain area.

excitatory postsynaptic potential (EPSP) Brief depolarization of a neuron membrane in response to stimulation, making the neuron more likely to produce an action potential.

explicit memory Conscious memory: subjects can retrieve an item and indicate that they know that the retrieved item is the correct item.

extinction In neurology, neglect of information on one side of the body when it is presented simultaneously with similar information on the other side of the body.

extrastriate (secondary visual) cortex Visual cortical areas outside the striate cortex.

eye-blink conditioning Commonly used experimental technique in which subjects learn to pair a formerly neutral stimulus with a defensive blinking response.

facial agnosia Face blindness—the inability to recognize faces; also called *prosopagnosia*.

fear conditioning Learned association, a conditioned emotional response, between a neutral stimulus and a noxious event such as a shock.

festination Tendency to engage in a behavior, such as walking, at faster and faster speeds.

fetal alcohol spectrum disorder (FASD) Range of physical and intellectual impairments observed in some children born to alcoholic mothers.

filopod (pl. filopodia) Process at the end of a developing axon that reaches out to search for a potential target or to sample the intercellular environment.

focal seizure Seizure that begins locally (at a focus) and then spreads out to adjacent areas.

forebrain Evolutionarily the newest part of the brain; coordinates advanced cognitive functions such as thinking, planning, and language; contains the limbic system, basal ganglia, and the neocortex.

fovea Region at the center of the retina that is specialized for high acuity; its receptive fields are at the center of the eye's visual field.

free-running rhythm Rhythm of the body's own devising in the absence of all external cues.

frequency Number of cycles that a wave completes in a given amount of time.

frontal lobe Part of the cerebral cortex often generally characterized as performing the brain's "executive" functions, such as decision making; lies anterior to the central sulcus and beneath the frontal bone of the skull.

functional magnetic resonance imaging (fMRI) Magnetic resonance imaging in which changes in elements such as iron or oxygen are measured during the performance of a specific behavior; used to measure cerebral blood flow during behavior or resting.

functional near-infrared spectroscopy (fNIRS) Noninvasive technique that gathers light transmitted through cortical tissue to image blood-oxygen consumption; form of optical tomography.

gamma-aminobutyric acid (GABA) Amino acid neurotransmitter that inhibits neurons.

ganglia Collection of nerve cells that function somewhat like a brain.

gap junction (electrical synapse) Fused prejunction and postjunction cell membrane in which connected ion channels form a pore that allows ions to pass directly from one neuron to the next.

gate Protein embedded in a cell membrane that allows substances to pass through the membrane on some occasions but not on others.

gender identity A person's feeling of being either male or female.

gene DNA segment that encodes the synthesis of a particular protein.

gene (DNA) methylation Addition of a methyl group to cytosine lying next to guanine in DNA. Methylation acts to suppress gene expression.

generalized anxiety disorder Persistently high levels of anxiety often accompanied by maladaptive behaviors to reduce anxiety; the disorder is thought to be caused by chronic stress.

geniculostriate system Projections from the retina to the lateral geniculate nucleus to the visual cortex.

genotype Particular genetic makeup of an individual.

glabrous skin Skin that does not have hair follicles but contains larger numbers of sensory receptors than do other skin areas.

glial cell Nervous-system cell that provides insulation, nutrients, and support and that aids in repairing neurons and eliminating waste products.

glioblast Product of a progenitor cell that gives rise to different types of glial cells.

glucocorticoid One of a group of steroid hormones, such as cortisol, secreted in times of stress; important in protein and carbohydrate metabolism.

glutamate (Glu) Amino acid neurotransmitter that excites neurons.

gonadal (sex) hormone One of a group of hormones, such as testosterone, that control reproductive functions and bestow sexual appearance and identity as male or female.

G protein Guanyl-nucleotide-binding protein coupled to a metabotropic receptor that, when activated, binds to other proteins.

graded potential Small voltage fluctuation in the cell membrane restricted to the vicinity on the axon where ion concentrations change to cause a brief increase (hyperpolarization) or decrease (depolarization) in electrical charge across the cell membrane.

grand mal seizure Seizure characterized by loss of consciousness and stereotyped motor activity.

gray matter Areas of the nervous system composed predominantly of cell bodies and capillary blood vessels that function either to collect and modify information or to support this activity.

growth cone Growing tip of an axon.

growth spurt Sporadic period of sudden growth that lasts for a finite time.

gyrus (pl. gyri) A small protrusion or bump formed by the folding of the cerebral cortex.

habituation Learning behavior in which a response to a stimulus weakens with repeated stimulus presentations.

hair cell Sensory neurons in the cochlea tipped by cilia; when stimulated by waves in the cochlear fluid, outer hair cells generate graded potentials in inner hair cells, which act as the auditory receptor cells.

hapsis Perceptual ability to discriminate objects on the basis of touch.

hemisphere Literally, half a sphere, referring to one side of the cerebrum.

hertz (Hz) Measure of frequency (repetition rate) of a sound wave; 1 hertz is equal to 1 cycle per second.

heterozygous Having two different alleles for the same trait.

hindbrain Evolutionarily the oldest part of the brain; contains the pons, medulla, reticular formation, and cerebellum, structures that coordinate and control most voluntary and involuntary movements.

hippocampus Distinctive, three-layered subcortical structure of the limbic system lying in the medial region of the temporal lobe; plays a role in species-specific behaviors, memory, and spatial navigation and is vulnerable to the effects of stress; named for the Greek word for seahorse.

histamine (H) Neurotransmitter that controls arousal and waking; can cause the constriction of smooth muscles and so, when activated in allergic reactions, contributes to asthma, a constriction of the airways.

homeostatic hormone One of a group of hormones that maintain internal metabolic balance and regulate physiological systems in an organism.

homeostatic mechanism Process that maintains critical body functions within a narrow, fixed range.

hominid General term referring to primates that walk upright, including all forms of humans, living and extinct.

homonymous hemianopia Blindness of an entire left or right visual field.

homozygous Having two identical alleles for a trait.

homunculus Representation of the human body in the sensory or motor cortex; also, any topographical representation of the body by a neural area.

HPA axis Hypothalamic-pituitary-adrenal circuit that controls the production and release of hormones related to stress.

Huntington's disease Hereditary disease characterized by chorea (ceaseless, involuntary, jerky movements) and progressive dementia, ending in death.

hydrocephalus Buildup of pressure in the brain and, in infants, swelling of the head caused if the flow of cerebrospinal fluid is blocked; can result in retardation.

hyperconnectivity Increased local connections between two related brain regions.

hyperkinetic symptom Symptom of brain damage that results in excessive involuntary movements, as seen in Tourette's syndrome.

hyperphagia Disorder in which an animal overeats, leading to significant weight gain.

hyperpolarization Increase in electrical charge across a membrane, usually due to the inward flow of chloride or sodium ions or the outward flow of potassium ions.

hypnogogic hallucination Dreamlike event at the beginning of sleep or while a person is in a state of cataplexy.

hypokinetic symptom Symptom of brain damage that results in a paucity of movement, as seen in Parkinson's disease.

hypothalamus Diencephalon structure that contains many nuclei associated with temperature regulation, eating, drinking, and sexual behavior.

hypovolumic thirst Thirst that is produced by a loss of overall fluid volume from the body.

idiopathic seizure Seizure that appears spontaneously and in the absence of other diseases of the central nervous system.

implicit memory Unconscious memory: subjects can demonstrate knowledge, such as a skill, conditioned response, or recalling events on prompting, but cannot explicitly retrieve the information.

imprinting Process that predisposes an animal to form an attachment to objects or animals at a critical period in development.

inhibition Decrease in the activity of a neuron or brain area.

inhibitory postsynaptic potential (IPSP) Brief hyperpolarization of a neuron membrane in response to stimulation, making the neuron less likely to produce an action potential.

innate releasing mechanism (IRM) Hypothetical mechanism that detects specific sensory stimuli and directs an organism to take a particular action.

insomnia Disorder of slow-wave sleep resulting in prolonged inability to sleep.

insula Located within the lateral fissure, multifunctional cortical tissue that contains regions related to language, to the perception of taste, and to the neural structures underlying social cognition.

intelligence A Hebb's term for innate intellectual potential, which is highly heritable and cannot be measured directly.

intelligence B Hebb's term for observed intelligence, which is influenced by experience as well as other factors in the course of development and is measured by intelligence tests.

interneuron Association neuron interposed between a sensory neuron and a motor neuron; thus, in mammals, interneurons constitute most of the neurons of the brain.

ionotropic receptor Embedded membrane protein that acts as (1) a binding site for a neurotransmitter and (2) a pore that regulates ion flow to directly and rapidly change membrane voltage.

ischemia Lack of blood to the brain as a result of stroke.

jet lag Fatigue and disorientation resulting from rapid travel through time zones and exposure to a changed light–dark cycle.

K-complex Sharp, high-amplitude EEG wave occurring during NREM sleep.

Klüver-Bucy syndrome Behavioral syndrome, characterized especially by hypersexuality, that results from bilateral injury to the temporal lobe.

Korsakoff's syndrome Permanent loss of the ability to learn new information (anterograde amnesia) and to retrieve old information (retrograde amnesia) caused by diencephalic damage resulting from chronic alcoholism or malnutrition that produces a vitamin B_1 deficiency.

lateralization Process whereby functions become localized primarily on one side of the brain.

law of Bell and Magendie The general principle that sensory fibers are located dorsally and motor fibers are located ventrally.

learned taste aversion Acquired association between a specific taste or odor and illness; leads to an aversion to foods that have the taste or odor.

learning Relatively permanent change in behavior that results from experience.

learning set The "rules of the game;" implicit understanding of how a problem can be solved with a rule that can be applied in many different situations.

Lewy body Circular fibrous structure found in several neurodegenerative disorders; forms within the cytoplasm of neurons and is thought to result from abnormal neurofilament metabolism.

light pollution Exposure to artifical light that changes activity patterns and so distrupts circadian rhythms.

limbic system Disparate forebrain structures lying between the neocortex and the brainstem that form a functional system controlling affective and motivated behaviors and certain forms of memory; includes cingulate cortex, amygdala, and hippocampus, among other structures.

locked-in syndrome Condition in which a patient is aware and awake but cannot move or communicate verbally due to complete paralysis of nearly all voluntary muscles except the eyes.

long-term depression (LTD) Long-lasting decrease in synaptic effectiveness after low-frequency electrical stimulation.

long-term potentiation (LTP) Long-lasting increase in synaptic effectiveness after high frequency stimulation.

luminance contrast The amount of light reflected by an object relative to its surroundings.

magnetic resonance imaging (MRI) Technique that produces a static, three-dimensional brain image by passing a strong magnetic field through the brain, followed by a radio wave, then measuring the radiation emitted from hydrogen atoms.

magnetic resonance spectroscopy (MRS) Modification of MRI to identify changes in specific markers of neuronal function; promising for accurate diagnosis of traumatic brain injuries.

magnetoencephalogram (MEG) Magnetic potentials recorded from detectors placed outside the skull.

magnocellular (M) cell Large-celled visual-system neuron that is sensitive to moving stimuli.

major depression Mood disorder characterized by prolonged feelings of worthlessness and guilt, the disruption of normal eating habits, sleep disturbances, a general slowing of behavior, and frequent thoughts of suicide.

mania Disordered mental state of extreme excitement.

masculinization Process by which exposure to androgens (male sex hormones) alters the brain, rendering it identifiably male.

materialism Philosophical position that holds that behavior can be explained as a function of the brain and the rest of the nervous system without explanatory recourse to the mind.

medial forebrain bundle (MFB) Tract that connects structures in the brainstem with various parts of the limbic system; forms the activating projections that run from the brainstem to the basal ganglia and frontal cortex.

medial geniculate nucleus Major thalamic region concerned with audition.

medial pontine reticular formation (MPRF) Nucleus in the pons participating in REM sleep.

melatonin Hormone secreted by the pineal gland during the dark phase of the day–night cycle; influences daily and seasonal biorhythms.

memory Ability to recall or recognize previous experience.

Ménière's disease Disorder of the middle ear resulting in vertigo and loss of balance.

meninges Three layers of protective tissue—dura mater, arachnoid, and pia mater—that encase the brain and spinal cord.

mentalism Explanation of behavior as a function of the nonmaterial mind.

metabolic syndrome Combination of medical disorders, including obesity and insulin abnormalities, that collectively increase the risk of developing cardiovascular disease and diabetes.

metabotropic receptor Embedded membrane protein, with a binding site for a neurotransmitter but no pore, linked to a G protein that can affect other receptors or act with second messengers to affect other cellular processes.

metaplasticity Interaction among different plastic changes in the brain.

microdialysis Technique used to determine the chemical constituents of extracellular fluid.

microelectrode A microscopic insulated wire or a salt-water-filled glass tube of which the uninsulated tip is used to stimulate or record from neurons.

microglia Glial cells that originate in the blood, aid in cell repair, and scavenge debris in the nervous system.

microsleep Brief period of sleep lasting a second or so.

midbrain Central part of the brain that contains neural circuits for hearing and seeing as well as orienting movements.

mind Proposed nonmaterial entity responsible for intelligence, attention, awareness, and consciousness.

mind–body problem Quandary of explaining how a nonmaterial mind and a material body interact.

minimally conscious state (MCS) Condition in which a person can display some rudimentary behaviors, such as smiling or uttering a few words, but is otherwise not conscious.

mirror neuron Cell in the primate premotor cortex that fires when an individual observes a specific action taken by another individual.

monoamine oxidase (MAO) inhibitor Antidepressant drug that blocks the enzyme monoamine oxidase from degrading neurotransmitters such as dopamine, noradrenaline, and serotonin.

monosynaptic reflex Reflex requiring one synapse between sensory input and movement.

mood stabilizer Drug for treatment of bipolar disorder that mutes the intensity of one pole of the disorder, thus making the other pole less likely to recur.

motivation Behavior that seems purposeful and goal-directed.

motor neuron Neuron that carries information from the brain and spinal cord to make muscles contract.

motor sequence Movement modules preprogrammed by the brain and produced as a unit.

multiple sclerosis (MS) Nervous-system disorder that results from the loss of myelin (glial-cell covering) around neurons.

mutation Alteration of an allele that yields a different version of the allele.

myelin Glial coating that surrounds axons in the central and peripheral nervous systems; prevents adjacent neurons from short-circuiting.

narcolepsy Slow-wave sleep disorder in which a person uncontrollably falls asleep at inappropriate times.

natural selection Darwin's theory for explaining how new species evolve and how existing species change over time. Differential success in the reproduction of different characteristics (phenotypes) results from the interaction of organisms with their environment.

neocortex (cerebral cortex) Newest, outer layer ("new bark") of the forebrain, composed of about six layers of gray matter; creates our reality.

neoteny Process in which maturation is delayed and so an adult retains infant characteristics; idea derived from the observation that newly evolved species resemble the young of their common ancestors.

nerve Large collection of axons coursing together outside the central nervous system.

nerve growth factor (NGF) Neurotrophic factor that stimulates neurons to grow dendrites and synapses and, in some cases, promotes the survival of neurons.

nerve impulse Propagation of an action potential on the membrane of an axon.

nerve net Simple nervous system that has no brain or spinal cord but consists of neurons that receive sensory information and connect directly to other neurons that move muscles.

netrin Member of the only class of tropic molecules yet isolated.

neural Darwinism Hypothesis that the processes of cell death and synaptic pruning are, like natural selection in species, the outcome of competition among neurons for connections and metabolic resources in a neural environment.

neural plate Thickened region of the ectodermal layer that gives rise to the neural tube.

neural stem cell Self-renewing, multipotential cell that gives rise to any of the different types of neurons and glia in the nervous system.

neural tube Structure in the early stage of brain development from which the brain and spinal cord develop.

neuritic plaque Area of incomplete necrosis (dead tissue) consisting of a central protein core (amyloid) surrounded by degenerative cellular fragments; often seen in the cortex of people with senile dementias such as Alzheimer's disease.

neuroblast Product of a progenitor cell that gives rise to any of the different types of neurons.

neuroeconomics Interdisciplinary field that seeks to understand how the brain makes decisions.

neuron Specialized nerve cell engaged in information processing.

neuropeptide Multifunctional chain of amino acids that acts as a neurotransmitter; synthesized from mRNA on instructions from the cell's DNA. Peptide neurotransmitters can act as hormones and may contribute to learning.

neuroplasticity The nervous system's potential for physical or chemical change that enhances its adaptability to environmental change and its ability to compensate for injury.

neuroprosthetics Field that develops computer-assisted devices to replace lost biological function.

neuroprotectant Drug used to try to block the cascade of poststroke neural events.

neuropsychoanalysis Movement within neuroscience and psychoanalysis to combine the insights of both to yield a unified understanding of mind and brain.

neuropsychology Study of the relations between brain function and behavior.

neurotransmitter Chemical released by a neuron onto a target with an excitatory or inhibitory effect.

neurotrophic factor Chemical compound that acts to support growth and differentiation in developing neurons and may act to keep certain neurons alive in adulthood.

nitric oxide (NO) Gas that acts as a chemical neurotransmitter—for example, to dilate blood vessels, aid digestion, and activate cellular metabolism.

nocioception Perception of pain, temperature, and itch.

node of Ranvier Part of an axon that is not covered by myelin.

nonregulatory behavior Behavior unnecessary to the basic survival needs of the animal.

noradrenergic neuron From adrenaline, Latin for "epinephrine"; a neuron containing norepinephrine.

norepinephrine (NE, or noradrenaline) Neurotransmitter found in the brain and in the sympathetic division of the autonomic nervous system; accelerates heart rate in mammals.

NREM (non-REM) sleep Slow-wave sleep associated with delta rhythms.

nucleus (pl. nuclei) A group of cells forming a cluster that can be identified with special stains to form a functional grouping.

obesity Excessive accumulation of body fat.

obsessive-compulsive disorder (OCD) Behavior disorder characterized by compulsively repeated acts (such as hand washing) and repetitive, often unpleasant, thoughts (obsessions).

occipital lobe Part of the cerebral cortex where visual processing begins; lies at the back of the brain and beneath the occipital bone.

ocular-dominance column Functional column in the visual cortex maximally responsive to information coming from one eye.

oligodendroglia Glial cells in the central nervous system that myelinate axons.

operant conditioning Learning procedure in which the consequences (such as obtaining a reward) of a particular behavior (such as pressing a bar) increase or decrease the probability of the behavior occurring again; also called *instrumental conditioning.*

opioid analgesic Drug like morphine, with sleep-inducing (narcotic) and pain-relieving (analgesic) properties; originally *narcotic analgesic.*

opponent process Explanation of color vision that emphasizes the importance of the apparently opposing pairs of colors: red versus green and blue versus yellow.

optic ataxia Deficit in the visual control of reaching and other movements.

optic chiasm Junction of the optic nerves, one from each eye, at which the axons from the nasal (inside—nearer the nose) halves of the retinas cross to the opposite side of the brain.

optic flow Streaming of visual stimuli that accompanies an observer's forward movement through space.

optogenetics Transgenic technique that combines genetics and light to control targeted cells in living tissue.

orbitofrontal cortex (OFC) Prefrontal cortex located behind the eye sockets (the *orbits*) that receives projections from the dorsomedial nucleus of the thalamus; plays a central role in a variety of emotional and social behaviors as well as in eating; also called *orbital frontal cortex.*

organizational hypothesis Proposal that actions of hormones in development alter tissue differentiation; for example, testosterone masculinizes the brain.

orienting movement Movement related to sensory inputs, such as turning the head to see the source of a sound.

oscilloscope Device that serves as a sensitive voltmeter by registering the flow of electrons to measure voltage.

osmotic thirst Thirst that results from an increased concentration of dissolved chemicals, or *solutes,* in body fluids.

ossicles Bones of the middle ear: malleus (hammer), incus (anvil), and stapes (stirrup).

pain gate Hypothetical neural circuit in which activity in fine-touch and pressure pathways diminishes the activity in pain and temperature pathways.

panic disorder Recurrent attacks of intense terror that come on without warning and without any apparent relation to external circumstances.

parahippocampal cortex Cortex located along the dorsal medial surface of the temporal lobe.

paralysis Loss of sensation and movement due to nervous-system injury.

paraplegia Paralysis of the legs due to spinal-cord injury.

parasympathetic division Part of the autonomic nervous system; acts in opposition to the sympathetic division—for example, preparing the body to rest and digest by reversing the alarm response or stimulating digestion.

parietal lobe Part of the cerebral cortex that functions to direct movements toward a goal or to perform a task, such as grasping an object; lies posterior to the central sulcus and beneath the parietal bone at the top of the skull.

Parkinson's disease Disorder of the motor system correlated with a loss of dopamine in the brain and characterized by tremors, muscular rigidity, and a reduction in voluntary movement.

parvocellular (P) cell Small-celled visual-system neuron that is sensitive to form and color differences.

Pavlovian conditioning Learning procedure whereby a neutral stimulus (such as a tone) comes to elicit a response because of its repeated pairing with some event (such as the delivery of food); also called *classical conditioning* or *respondent conditioning.*

peptide hormone Chemical messenger synthesized by cellular DNA that acts to affect the target cell's physiology.

perception Subjective interpretation of sensations by the brain.

periaqueductal gray matter (PAG) Nuclei in the midbrain that surround the cerebral aqueduct joining the third and fourth ventricles; PAG neurons contain circuits for species-typical behaviors (e.g., female sexual behavior) and play an important role in the modulation of pain.

peribrachial area Cholinergic nucleus in the dorsal brainstem having a role in REM sleep behaviors; projects to medial pontine reticulum.

period Time required to complete a cycle of activity.

peripheral nervous system (PNS) All the neurons in the body located outside the brain and spinal cord; provides sensory and motor connections to and from the central nervous system.

perirhinal cortex Cortex lying next to the rhinal fissure on the base of the brain.

perseveration Tendency to emit repeatedly the same verbal or motor response to varied stimuli.

persistent vegetative state (PVS) Condition in which a person is alive but unable to communicate or to function independently at even the most basic level.

petit mal seizure Seizure of brief duration, characterized by loss of awareness with no motor activity except for blinking, turning the head, or rolling the eyes.

phenotype Individual characteristics that can be seen or measured.

phenotypic plasticity An individual's capacity to develop into more than one phenotype.

phenylketonuria (PKU) Behavioral disorder caused by elevated levels of the amino acid phenylalanine in the blood and resulting from a defect in the gene for the enzyme phenylalanine hydroxylase; the major symptom is severe mental retardation.

pheromone Odorant biochemical released by one animal that acts as a chemosignal and can affect the physiology or behavior of another animal.

phobia Fear of a clearly defined object or situation.

photoreceptor Specialized type of retinal cell that transduces light into neural activity

pituitary gland Endocrine gland attached to the bottom of the hypothalamus; its secretions control the activities of many other endocrine glands; known to be associated with biological rhythms.

place cells Hippocampal neurons maximally responsive to specific locations in the world.

positron emission tomography (PET) Imaging technique that detects changes in blood flow by measuring changes in the uptake of compounds such as oxygen or glucose; used to analyze the metabolic activity of neurons.

postictal depression Postseizure state of confusion and reduced affect.

postsynaptic membrane Membrane on the transmitter-input side of a synapse (dendritic spine).

posttraumatic stress disorder (PTSD) Syndrome characterized by physiological arousal symptoms brought on by recurring memories and dreams related to a traumatic event for months or years after the event.

prefrontal cortex (PFC) Large frontal-lobe area anterior to the motor and premotor cortex; plays a key role in controlling executive functions such as planning.

preparedness Predisposition to respond to certain stimuli differently from other stimuli.

presynaptic membrane Membrane on the transmitter-output side of a synapse (axon terminal).

primary auditory cortex (area A1) Asymmetrical structures, found within Heschl's gyrus in the temporal lobes, that receive input from the ventral region of the medial geniculate nucleus.

primary visual cortex (V1) Striate cortex that receives input from the lateral geniculate nucleus.

priming Using a stimulus to sensitize the nervous system to a later presentation of the same or a similar stimulus.

procedural memory Ability to recall a movement sequence or how to perform some act or behavior.

progenitor cell Precursor cell derived from a stem cell; it migrates and produces a neuron or a glial cell.

proprioception Perception of the position and movement of the body, limbs, and head.

prosody Melodical tone of the spoken voice.

protein Folded-up polypeptide chain.

psyche Synonym for *mind,* an entity once proposed to be the source of human behavior.

psychedelic drug Drug that can alter sensation and perception; examples are lysergic acid dielthylmide, mescaline, and psilocybin.

psychoactive drug Substance that acts to alter mood, thought, or behavior; is used to manage neuropsychological illness; or is abused.

psychological construct Idea, resulting from a set of impressions, that some mental ability exists as an entity; examples include memory, language, and emotion.

psychomotor activation Increased behavioral and cognitive activity; at certain levels of consumption, the drug user feels energetic and in control.

psychopharmacology Study of how drugs affect the nervous system and behavior.

psychosurgery Any neurosurgical technique intended to alter behavior.

psychotherapy Talking therapy derived from Freudian psychoanalysis and other psychological interventions.

pump Protein in the cell membrane that actively transports a substance across the membrane.

Purkinje cell Distinctive interneuron found in the cerebellum.

pyramidal cell Distinctive interneuron found in the cerebral cortex.

quadrantanopia Blindness of one quadrant of the visual field.

quadriplegia Paralysis of the legs and arms due to spinal-cord injury.

quantum (pl. quanta) Amount of neurotransmitter, equivalent to the contents of a single synaptic vesicle, that produces a just observable change in postsynaptic electric potential.

radial glial cell Path-making cell that a migrating neuron follows to its appropriate destination.

radiator hypothesis Idea that selection for improved brain cooling through increased blood circulation in the brains of early hominids enabled the brain to grow larger.

rapidly adapting receptor Body sensory receptor that responds briefly to the onset of a stimulus on the body.

rate-limiting factor Any enzyme that is in limited supply, thus restricting the pace at which a chemical can be synthesized.

real-time fMRI (rt-fMRI) Behavior-modification technique in which individuals learn to change their behavior by controlling their own patterns of brain activation.

receptive field Region of the visual world that stimulates a receptor cell or neuron.

reconsolidation The process of restabilizing a memory trace after the memory is revisited.

referred pain Pain felt on the surface of the body that is actually due to pain in one of the internal organs of the body.

regulatory behavior Behavior motivated to meet the survival needs of the animal.

reinforcer In operant conditioning, any event that strengthens the behavior it follows.

relatively refractory Refers to the state of an axon in the later phase of an action potential during which increased electrical current is required to produce another action potential; a phase during which potassium channels are still open.

releasing hormones Peptides that are released by the hypothalamus and act to increase or decrease the release of hormones from the anterior pituitary.

REM sleep Fast brain-wave pattern displayed by the neocortical EEG record during sleep.

resting potential Electrical charge across the cell membrane in the absence of stimulation; a store of potential energy produced by a greater negative charge on the intracellular side relative to the extracellular side.

resting-state fMRI (rs-fMRI) Magnetic resonance imaging method that measures changes in elements such as iron or oxygen when the individual is resting (not engaged in a specific task).

restraint-induced therapy Procedure in which restraint of a healthy limb forces a patient to use an impaired limb to enhance recovery of function.

reticular activating system (RAS) Large reticulum (mixture of cell nuclei and nerve fibers) that runs through the center of the brainstem; associated with sleep–wake behavior and behavioral arousal; often called the *reticular formation.*

reticular formation Midbrain area in which nuclei and fiber pathways are mixed, producing a netlike appearance; associated with sleep–wake behavior and behavioral arousal.

retina Light-sensitive surface at the back of the eye consisting of neurons and photoreceptor cells.

retinal ganglion cell (RGC) One of a group of retinal neurons with axons that give rise to the optic nerve.

retinohypothalamic tract Neural route formed by axons of photosensitive retinal ganglion cells from the retina to the suprachiasmatic nucleus; allows light to entrain the rhythmic activity of the SCN.

retrograde amnesia Inability to remember events that took place before the onset of amnesia.

reuptake Deactivation of a neurotransmitter when membrane transporter proteins bring the transmitter back into the presynaptic axon terminal for subsequent reuse.

rod Photoreceptor specialized for functioning at low light levels.

saltatory conduction Propagation of an action potential at successive nodes of Ranvier; saltatory means "jumping" or "dancing."

schizophrenia Behavioral disorder characterized by delusions, hallucinations, disorganized speech, blunted emotion, agitation or immobility, and a host of associated symptoms.

Schwann cell Glial cell in the peripheral nervous system that myelinates sensory and motor axons.

scotomas Small blind spot in the visual field caused by migraine or by a small lesion of the visual cortex.

scratch reflex Automatic response in which an animal's hind limb reaches to remove a stimulus from the surface of the body.

second-generation antidepressant Drug whose action is similar to that of tricyclics (first-generation antidepressants) but more selective in its action on the serotonin reuptake transporter proteins; also called *atypical antidepressant.*

second messenger Chemical that carries a message to initiate a biochemical process when activated by a neurotransmitter (the first messenger).

segmentation Division into a number of parts that are similar; refers to the idea that many animals, including vertebrates, are composed of similarly organized body segments.

selective serotonin reuptake inhibitor (SSRI) Tricyclic antidepressant drug that blocks the reuptake of serotonin into the presynaptic terminal.

sensation Registration of physical stimuli from the environment by the sensory organs.

sensitization Learning behavior in which the response to a stimulus strengthens with repeated presentations of that stimulus because the stimulus is novel or because the stimulus is stronger than normal—for example, after habituation has occurred.

sensory deprivation Experimental setup in which a subject is allowed only restricted sensory input; subjects generally have a low tolerance for deprivation and may even display hallucinations.

sensory neuron Neuron that carries incoming information from sensory receptors into the spinal cord and brain.

serotonin (5-HT) Amine neurotransmitter that plays a role in regulating mood and aggression, appetite and arousal, the perception of pain, and respiration.

sexual dimorphism Differential development of brain areas in the two sexes.

sexual orientation A person's sexual attraction to the opposite sex or to the same sex or to both sexes.

sleep apnea Inability to breathe during sleep; person has to wake up to breathe.

sleep paralysis Inability to move during deep sleep owing to the brain's inhibition of motor neurons.

sleep spindle Brief burst of EEG activity typically occurring during NREM sleep.

slowly adapting receptor Body sensory receptor that responds as long as a sensory stimulus is on the body.

slow-wave sleep NREM sleep.

small-molecule transmitter Quick-acting neurotransmitter synthesized in the axon terminal from products derived from the diet.

social neuroscience Interdisciplinary field that seeks to understand how the brain mediates social interactions.

somatic marker hypothesis Posits that "marker" signals arising from emotions and feelings act to guide behavior and decision making, usually in an unconscious process.

somatic nervous system (SNS) Part of the PNS that includes the cranial and spinal nerves to and from the muscles, joints, and skin that produce movement, transmit incoming sensory input, and inform the CNS about the position and movement of body parts.

somatosensory neuron Brain cell that brings sensory information from the body into the spinal cord.

sound wave Undulating displacement of molecules caused by changing pressure.

spatial summation Graded potentials that occur at approximately the same location and time on a membrane are summed.

species Group of organisms that can interbreed.

species-typical behavior Behavior that is characteristic of all members of a species.

spinal cord Part of the central nervous system encased within the vertebrae (spinal column) that provides most of the connections between the brain and the rest of the body.

split brain Surgical disconnection of the two hemispheres in which the corpus callosum is cut.

stereotaxic apparatus Surgical instrument that permits the researcher to target a specific part of the brain.

steroid hormone Fat-soluble chemical messenger synthesized from cholesterol.

storage granule Membranous compartment that holds several vesicles containing a neurotransmitter.

stretch-sensitive channel Ion channel on a tactile sensory neuron that activates in response to stretching of the membrane, initiating a nerve impulse.

striate cortex Primary visual cortex (V1) in the occipital lobe; its striped appearance when stained gives it this name.

striatum Caudate nucleus and putamen of the basal ganglia.

stroke Sudden appearance of neurological symptoms as a result of severely interrupted blood flow.

substance abuse Use of a drug for the psychological and behavioral changes it produces aside from its therapeutic effects.

subventricular zone Lining of neural stem cells surrounding the ventricles in adults.

sudden infant death syndrome (SIDS) Unexplained death while asleep of a seemingly healthy infant less than 1 year old.

sulcus (pl. sulci) A groove in brain matter, usually a groove found in the neocortex or cerebellum.

supplementary speech area Speech-production region on the dorsal surface of the left frontal lobe.

suprachiasmatic nucleus (SCN) Master biological clock, located in the hypothalamus just above the optic chiasm.

sympathetic division Part of the autonomic nervous system; arouses the body for action, such as mediating the involuntary fight-or-flight response to alarm by increasing heart rate and blood pressure.

symptomatic seizure Identified with a specific cause, such as infection, trauma, tumor, vascular malformation, toxic chemicals, very high fever, or other neurological disorders.

synapse Junction between one neuron and another that forms the information-transfer site between neurons.

synaptic cleft Gap that separates the presynaptic membrane from the postsynaptic membrane.

synaptic vesicle Organelle consisting of a membrane structure that encloses a quantum of neurotransmitter.

synesthesia Ability to perceive a stimulus of one sense as the sensation of a different sense, as when sound produces a sensation of color; literally, "feeling together."

syntax Ways in which words are put together to form phrases, clauses, or sentences; proposed to be a unique characteristic of human language.

tardive dyskinesia Inability to stop the tongue or other body parts from moving; motor side effect of neuroleptic drugs.

Tay-Sachs disease Inherited birth defect caused by the loss of genes that encode the enzyme necessary for breaking down certain fatty substances; appears 4 to 6 months after birth and results in retardation, physical changes, and death by about age 5.

tectopulvinar system Projections from the retina to the superior colliculus to the pulvinar (thalamus) to the parietal and temporal visual areas.

tectum Roof (area above the ventricle) of the midbrain; its functions are sensory processing, particularly visual and auditory, and the production of orienting movements.

tegmentum Floor (area below the ventricle) of the midbrain; a collection of nuclei with movement-related, species-specific, and pain-perception functions.

temporal lobe Part of the cerebral cortex that functions in connection with hearing, language, and musical abilities; lies below the lateral fissure, beneath the temporal bone at the side of the skull.

temporal summation Graded potentials that occur at approximately the same time on a membrane are summed.

terminal button (end foot) Knob at the tip of an axon that conveys information to other neurons.

testosterone Sex hormone secreted by the testes and responsible for the distinguishing characteristics of the male.

thalamus Diencephalon structure through which information from all sensory systems is integrated and projected into the appropriate region of the neocortex.

theory of mind Ability to attribute mental states to others.

threshold potential Voltage on a neural membrane at which an action potential is triggered by the opening of Na^+ and K^+ voltage-sensitive channels; about 250 millivolts relative to extracellular surround.

tolerance Decrease in response to a drug with the passage of time.

tonotopic representation Property of audition in which sound waves are processed in a systematic fashion from lower to higher frequencies.

topographic map Spatially organized neural representation of the external world.

topographic organization Neural spatial representation of the body or areas of the sensory world perceived by a sensory organ.

Tourette's syndrome Disorder of the basal ganglia characterized by tics, involuntary vocalizations (including curse words and animal sounds), and odd, involuntary movements of the body, especially of the face and head.

tract Large collection of axons coursing together within the central nervous system.

transcranial magnetic stimulation (TMS) Procedure in which a magnetic coil is placed over the skull to stimulate the underlying brain; used either to induce behavior or to disrupt ongoing behavior.

transgenic animal Product of technology in which number of genes or a single gene from one species is introduced into the genome of another species and passed along and expressed in subsequent generations.

transmitter-activated receptor Protein that has a binding site for a specific neurotransmitter and is embedded in the membrane of a cell.

transmitter-sensitive channel Receptor complex that has both a receptor site for a chemical and a pore through which ions can flow.

transporter Protein molecule that pumps substances across a membrane.

transsexuality Gender-identity disorder involving the strong belief of having been born the wrong sex.

traumatic brain injury (TBI) Damage to the brain that results from a blow to the head.

trichromatic theory Explanation of color vision based on the coding of three primary colors: red, green, and blue.

tricyclic antidepressant First-generation antidepressant drug with a chemical structure characterized by three rings that blocks serotonin reuptake transporter proteins.

tropic molecule Signaling molecule that attracts or repels growth cones.

tumor Mass of new tissue that grows uncontrolled and independent of surrounding structures.

type I schizophrenia Disorder characterized predominantly by positive symptoms (e.g., behavioral excesses such as hallucinations and agitated movements) likely due to a dopaminergic dysfunction and associated with acute onset, good prognosis, and a favorable response to neuroleptics.

type II schizophrenia Disorder characterized by negative symptoms (behavioral deficits) and associated with chronic affliction, poor prognosis, poor response to neuroleptics, cognitive impairments, enlarged ventricles, and cortical atrophy, particularly in the frontal cortex.

unconditioned response (UCR) In classical conditioning, the unlearned, naturally occurring response to the unconditioned stimulus, such as salivation when food is in the mouth.

unconditioned stimulus (UCS) A stimulus that unconditionally—naturally and automatically—triggers a response.

ventral spinothalamic tract Pathway from the spinal cord to the thalamus that carries information about pain and temperature.

ventricle One of four cavities in the brain that contain cerebrospinal fluid that cushions the brain and may play a role in maintaining brain metabolism.

ventrolateral thalamus Part of the thalamus that carries information about body senses to the somatosensory cortex.

vertebrae (sing. vertebra) The bones that form the spinal column.

vestibular system Somatosensory system that comprises a set of receptors in each inner ear that respond to body position and to movement of the head.

virtual-reality (VR) exposure therapy Controlled, virtual-immersion environment that, by allowing individuals to relive traumatic events, gradually desensitizes them to stress.

visual field Region of the visual world that is seen by the eyes.

visual-form agnosia Inability to recognize objects or drawings of objects.

visuospatial learning Using visual information to recall an object's location in space.

voltage gradient Difference in charge between two regions that allows a flow of current if the two regions are connected.

voltage-sensitive channel Gated protein channel that opens or closes only at specific membrane voltages.

voltmeter Device that measures the flow and the strength of electrical voltage by recording the difference in electrical potential between two bodies.

wanting-and-liking theory When a drug is associated with certain cues, the cues themselves elicit desire for the drug; also called *incentive-sensitization theory*.

Wernicke's area Secondary auditory cortex (planum temporale) lying behind Heschl's gyrus at the rear of the left temporal lobe that regulates language comprehension; also called posterior speech zone.

white matter Areas of the nervous system rich in fat-sheathed neural axons that form the connections between brain cells.

wild type Refers to a normal (most common in a population) phenotype or genotype.

withdrawal symptom Physical and psychological behavior displayed by an addict when drug use ends.

Zeitgeber Environmental event that entrains biological rhythms: a "time giver."

Chapter 1

Bird, C. D., & Emery, N. J. (2009). Insightful problem solving and creative tool modification by captive nontool-using rooks. *Proceedings of the National Academy of Sciences, 106,* 10370–10375.

Bronson, R. T. (1979). Brain weight–body weight scaling in dogs and cats. *Brain, Behavior and Evolution, 16,* 227–236.

Darwin, C. (1963). *On the origin of species by means of natural selection, or the preservation of favored races in the struggle for life.* New York: New American Library. (Original work published 1859.)

Darwin, C. (1965). *The expression of the emotions in man and animals.* Chicago: University of Chicago Press. (Original work published 1872.)

Deary, I. J. (2000). *Looking down on human intelligence: From psychometrics to the brain.* Oxford Psychology Series, No. 34. New York: Oxford University Press.

Dennett, D. (1978). *The Intentional Stance.* Cambridge, MA: MIT Press.

Descartes, R. (1972). *Treatise on man* (T. S. Hall, Trans.). Cambridge, MA: Harvard University Press. (Original work published 1664.)

Eibl-Eibesfeldt, I. (1970). *Ethology: The biology of behavior.* New York: Holt, Rinehart and Winston.

Dunbar, R. (1998). *Grooming, gossip, and the evolution of language.* Cambridge, MA: Harvard University Press.

Falk, D. (2004). Hominid brain evolution: New century, new directions. *Collegium antroplogicum, 228* (Suppl. 2), 59–64.

Gardner, H. (2006). *Multiple intelligences: New horizons.* New York: Basic Books.

Gardner R. A., & Gardner B. T. (1969). Teaching sign language to a chimpanzee. *Science 165,* 664–672.

Gladwell, M. (2000). *The tipping point: How little things can make a big difference.* Boston: Little Brown.

Goodall, J. (1986). *The chimpanzees of Gombe.* Cambridge, MA: Harvard University Press.

Gordon, A. D., Nevell, L., & Wood, B. (2008). The *Homo floresiensis* cranium (LB1): Size, scaling, and early *Homo* affinities. *Proceedings of the National Academy of Sciences of the United States of America, 105,* 4650–4655.

Gould, S. J. (1981). *The mismeasure of man.* New York: Norton.

Hebb, D. O. (1949). *The organization of behavior: A neuropsychological theory.* New York: Wiley.

Heron, W. (1957). The pathology of boredom. *Scientific American, 196*(1), 52–56.

Iwaniuk, A. N., Lefebvre, L., & Wylie, D. R. W. (2006). Comparative morphology of the avian cerebellum III: Correlations with tool use. *Brain, Behavior and Evolution, 68,* 113.

Jacobson, E. (1932). Electrophysiology of mental activities. *American Journal of Psychology, 44,* 677–694.

Jerison, H. J. (1973). *The evolution of the brain and intelligence.* New York: Academic Press.

Lewin, R. (1998). *The Origin of Modern Humans.* New York: Scientific American Library.

Linge, F. R. (1990). Faith, hope, and love: Nontraditional therapy in recovery from serious head injury, a personal account. *Canadian Journal of Psychology, 44,* 116–129.

McKinney, M. L. (1998). The juvenilized ape myth: Our "overdeveloped" brain. *Bioscience, 48,* 109–116.

Milton, K. (2003). The critical role played by animal source foods in human *(Homo)* evolution. *Journal of Nutrition, 133* (Suppl. 2), 3886S–3892S.

Murdock, G. P. (1965). *Culture and society: Twenty-four essays.* Pittsburgh: University of Pittsburgh Press.

Pickering, R., Dirks, P. H., Jinnah, Z., de Ruiter, D. J., Churchill, S. E., et al. (2011). *Australopithecus sediba* at 1.977 Ma and implications for the origins of the genus *Homo. Science, 333,* 1421–1423.

Potts, R., & Sloan, C. (2010). *What Does It Mean To Be Human?* Companion volume to the Smithsonian National Museum of Natural History's David H. Koch Hall of Human Origins. Washington, DC: National Geographic Society.

Prinz, J. (2008). Is consciousness embodied? In P. Robbins and. M. Aydede (Eds.), *Cambridge Handbook of Situated Cognition.* Cambridge: Cambridge University Press.

Savage-Rumbaugh, S. (1999). *Ape communication: Between a rock and a hard place in origins of language—What non-human primates can tell us.* School of American Research Press.

Schiff, N. D., & Fins, J. J. (2007). Deep brain stimulation and cognition: Moving from animal to patient. *Current Opinion in Neurology, 20,* 638–642.

Schuiling, G. A. (2005). On sexual behavior and sex-role reversal. *Journal of Psychosomatic and Obstetric Gynaecology, 26*(3), 217–223.

Stedman, H. H., Kozyak, B. W., Nelson, A., Thesier, D. M., Su, L. T., Low, D. W., Bridges, C. R., Shrager, J. B., Minugh-Purvis, N., & Mitchell, M. A. (2004). Myosin gene mutation correlates with anatomical changes in the human lineage. *Nature, 428,* 415–418.

Tagliatela, J. P., Russell, J. L., Schaeffer, J. A., & Hopkins, W. D. (2011). Chimpanzee vocal signaling points to a multimodal origin of human language. *PLoS One, 6,* e18852.

Terkel, J. (1995). Cultural transmission in the black rat: Pinecone feeding. *Advances in the Study of Behavior, 24,* 119–154.

Weiner, J. (1995). *The beak of the finch.* New York: Vintage.

Zhang, G., Pei, Z., Ball, E. V., Mort, M., Kehrer-Sawatzki, H., & Cooper, D. N. (2011). Cross-comparison of the genome sequences from human, chimpanzee, Neanderthal and a Denisovan hominin identifies novel potentially compensated mutations. *Human Genomics, 5,* 453–484.

Chapter 2

Brodmann, K. (1909). *Vergleichende Lokalisationlehr der Grosshirnrinde in ihren Prinzipien dargestellt auf Grund des Zellenbaues.* Leipzig: J. A. Barth.

Felleman, D. J., & van Essen, D. C. (1991). Distributed hierarchical processing in the primate cerebral cortex. *Cerebral Cortex, 1,* 1–47.

Fiorito, G., & Scotto, P. (1992). Observational learning in *Octopus vulgaris. Science, 256,* 545–547.

Fox, D. (2011). The limits of intelligence. *Scientific American 305*(1), 36–43.

Gilbert, S. F., & Epel, D. (2009). *Ecological developmental biology: Integrating epigenetics, medicine, and evolution.* New York: Sinauer.

Hatcher, M. A., & Starr, J. A. (2011). Role of tissue plasminogen activator in ischemic stroke. *Annals of Pharmacotherapy, 45,* 364–371.

Jerison, H. J. (1991). *Brain size and the evolution of mind.* New York: American Museum of Natural History.

Langhorne, P., Bernhardt, J., & Kwakkel, G. (2011). Stroke rehabilitation. *Lancet, 277,* 1693–1702.

Luria, A. R. (1973). *The working brain.* Harmondsworth, UK: Penguin.

Papez, J. W. (1937). A proposed mechanism of emotion. *Archives of Neurology and Psychiatry, 38,* 724–744.

Passingham, R. (2008). *What is special about the human brain?* Oxford: Oxford University Press.

Chapter 3

Adar, E., Nottebohm, E. F., & Barnea A. (2008). The relationship between nature of social change, age, and position of new neurons and their survival in adult zebra finch brain. *Journal of Neuroscience, 28,* 5394–5400.

Bainbridge, M. N., Wiszniewski, W., Murdock, D. R., Friedman, J., Gonzaga-Jauregui, C., et al. (2011). Whole-genome sequencing for optimized patient management. *Science Translational Medicine, 3,* 87re3.

Balaban, E. (2005). Brain switching: Studying evolutionary behavioral changes in the context of individual brain development. *International Journal of Developmental Biology 49,* 117–124.

Central Brain Tumor Registry of the United States (2011).

Charney, E. (2012). Behavior genetics and postgenomics. *Behavioral Brain Sciences,* in press.

Eisener-Dorman, A. F., Lawrence, D. A., & Bolivar, V. J. (2008). Cautionary insights on knockout mouse studies: The gene or not the gene? *Brain, Behavior, and Immunity,* September 12 e-publication.

Ghosh, K. K., Burns, L.D., Cocker, E. D., Nimmerjahn, A., Ziv, Y., Gamal, A. E., & Schnitzer, M. J. (2011). Miniaturized integration of a fluorescence microscope. *Nature Methods 8,* 871–878.

Gill, J. M., & Rego, A. C. (2009). The R6 lines of transgenic mice: A model for screening new therapies for Huntington's disease. *Brain Research Reviews, 59,* 410–431.

Kaati, G., Bygren, L. O., Pembrey, M., & Sjöström, M. (2007). Transgenerational response to nutrition, early life circumstances and longevity. *European Journal of Human Genetics, 15,* 784–790.

Karlsson, E. K., & Lindblad-Toh, K. (2008). Leader of the pack: Gene mapping in dogs and other model organisms. *National Review of Genetics, 9,* 713–725.

Kues, W. A., & Niemann, H. (2011). Advances in farm animal transgenesis. *Preventive Veterinary Medicine, 102,* 146–156.

Livet, J., Weissman, T. A., Kang, H., Draft, R. W., Lu, J., Bennis, R. A., Sanes, J. A., & Lichtman, J. W. (2007). Transgenic strategies for combinatorial expression of fluorescent proteins in the nervous system. *Nature, 450,* 56–62.

McDonald, C. L., Bandtlow, C., & Reindl, M. (2011). Targeting the Nogo receptor complex in diseases of the central nervous system. *Current Medicinal Chemistry, 18,* 234–244.

McFarland, K. N., & Cha, J. H. (2011). Molecular biology of Huntington's disease. *Handbook of Clinical Neurology, 100,* 25–81.

Naert, G., & Rivest, S. (2011). The role of microglial cell subsets in Alzheimer's disease. *Current Alzheimer Research, 8*(2), 151–155.

Ramón y Cajal, S. (1909–1911). *Histologie du système nerveux de l'homme et des vertébrés.* Paris: Maloine.

Reeve, R., van Schaik, A., Jin, C., Hamilton, T., Torben-Neilsen, B., & Webb, B. (2007). Directional hearing in a silicon cricket. *Biosystems, 87,* 307–313.

Riva, G., Gaggioli, A., & Mantovani, F. (2008). Are robots present? From motor simulation to "being there." *Cyberpsychology and Behaviour, 11,* 631–636.

Wahlsten, D., & Ozaki, H. S. (1994). Defects of the fetal forebrain in acallosal mice. In M. Lassonde & M. A. Jeeves (Eds.), *Callosal agenesis* (pp. 125–132). New York: Plenum.

Webb, B. (1996). A cricket robot. *Scientific American, 275*(6), 94–99.

Zamboni, P., Galeotti, R., Weinstock-Guttman, B., Kennedy, C., Salvi, F., & Zivadinov, R. (2012). Venous angioplasty in patients with multiple sclerosis: Results of a pilot study. *European Journal of Vascular and Endovascular Surgery, 43,* 116–122.

Chapter 4

Allen, M. J., Lacroix, J. J., Ramachandran, S., Capone, R., Whitlock, J. L., et al. (2011). Mutant SOD1 forms ion channel: Implications for ALS pathophysiology. *Neurobiology of Diseases, 45,* 831–838.

Bano, S., Yadav, S. N., Chaudhary, V., & Garga, U. C. (2011). Neuroimaging in epilepsy. *Journal of Pediatrics and Neuroscience, 6,* 19–26.

Bartholow, R. (1874). Experimental investigation into the functions of the human brain. *American Journal of Medical Sciences, 67,* 305–313.

Cao, Z. F., Burdakov, D., & Sarnyai, Z. (2011).Optogenetics: potentials for addiction research. *Addiction Biology, 16,* 519–531.

Debanne, D. (2011). The nodal origin of intrinsic bursting. *Neuron, 25,* 569–570.

Descartes, R. (1972). *Treatise on man* (T. S. Hall, Trans.). Cambridge, MA: Harvard University Press. (Original work published 1664.)

Eccles, J. (1965). The synapse. *Scientific American, 21*(1), 56–66.

Fenno, L., Yizhar, O., & Deisseroth, K. (2011). The development and application of optogenetics. *Annual Review of Neuroscience, 34,* 389–412.

Hodgkin, A. L., & Huxley, A. F. (1939). Action potentials recorded from inside nerve fiber. *Nature, 144,* 710–711.

Legenstein, R., & Maass, W. (2011). Branch-specific plasticity enables self-organization of nonlinear computation in single neurons. *Journal of Neuroscience, 31,* 10787–10802.

Liewald, J. F., Brauner, M., Stephens, G. J., Bouhours, M., Schultheis, C., Xhen, M., & Gottschalk, A. (2008). Optogenetic analysis of synaptic function. *Nature Methods, 10,* 895–902.

Rezania, K., Soliven, B., Baron, J., Lin, H., Penumalli, V., & van Besien, K. (2012). Myasthenia gravis, an autoimmune manifestation of lymphoma and lymphoproliferative disorders: Case reports and review of literature. *Leukemia and Lymphoma, 53,* 371–380.

Zhang, F., Aravanis, A. M., Adamantidis, A., de Lecea, L., & Deisseroth, K. (2007). Circuit-breakers: Optical technologies for probing neural signals and systems. *Nature Reviews Neuroscience, 8,* 577–581.

Chapter 5

Bailey, C. H., & Chen, M. (1989). Time course of structural changes at identified sensory neuron synapses during long-term sensitization in *Aplysia. Journal of Neuroscience, 9,* 1774–1780.

Ballard, A., Tetrud, J. W., & Langston, J. W. (1985). Permanent human Parkinsonism due to 1-methyl-4-phenyl-1,2,3,6-tetrahydropyridine (MPTP). *Neurology, 35,* 949–956.

Barbeau, A., Murphy, G. F., & Sourkes, T. L. (1961). Les catecholamines dans la maladie de Parkinson. Bel-Air Symposium on Monoamines and the Central Nervous System, Georg et Cie, Geneva.

Birkmayer, W., & Hornykiewicz, O. (1961). Der L-3,4-Dioxyphenylalanin (L-DOPA) Effekt bei der Parkinson A Kinase. *Wiener Klinische, 73,* 787–789.

Bosch, M., & Hayashi, Y. (2012). Structural plasticity of dendritic spines. *Current Opinion in Neurobiology, 22,* 383–388.

Debello, W.M. (2008) Micro-rewiring as a substrate for learning. *Trends in Neurosciences, 31,* 577–584.

Dere, E., & Zlomuzica, A. (2012). The role of gap junctions in the brain in health and disease. Neuroscience and Biobehavioral Reviews, 36, 206–217.

Ehringer, H., & Hornykiewicz, O. (1960/1974). Distribution of noradrenaline and dopamine (3-hydroxytyramine) in the human brain and their behavior in the presence of disease affecting the extrapyramidal system. In J. Marks (Ed.), *The treatment of Parkinsonism with L-dopa* (pp. 45–56). Lancaster, UK: MTP Medical and Technical Publishing.

Hamilton, T. J., Wheatley, B. M., Sinclair, D. B., Bachmann, M., Larkum, M. E., & Colmers, W. F. (2010). Dopamine modulates synaptic plasticity in dendrites of rat and human dentate granule cells. *Proceedings of the National Academy of Sciences USA, 107*(42): 18185–18190.

Hebb, D. O. (1949). *The organization of behavior.* New York: Wiley.

Kandel, E. R., Schwartz, J. H., & T. M. Jessell (Eds.). (2000). *Principles of neural science* (4th ed.). New York: McGraw-Hill.

Katz, B. (1970). On the quantal mechanism of neural transmitter release. In *Nobel Lectures, Physiology or Medicine 1963–1970* (pp. 485–492). Amsterdam; Elsevier.

Lane, E. L., Björklund, A., Dunnett, S. B., & Winkler C. (2010). Neural grafting in Parkinson's disease unraveling the mechanisms underlying graft-induced dyskinesia. *Progress in Brain Research, 184,* 295–309.

Langston, J. W. (2008). *The case of the frozen addicts.* Pantheon Books, 1995. Original from the University of Michigan, digitized August 5, 2008.

Miniaci, M. C., Kim, J. H., Puthanveettil, S. V., Si, K., Zhu, H., Kandel, E. R., & Bailey, C. H. (2008). Sustained CPEB-dependent local protein synthesis is required to stabilize synaptic growth for persistence of long-term facilitation in *Aplysia. Neuron, 59,* 1024–1036.

Parkinson, J. (1817/1989). An essay on the shaking palsy. In A. D. Morris & F. C. Rose (Eds.), *James Parkinson: His life and times* (pp. 151–175). Boston: Birkhauser.

Sacks, O. (1976). *Awakenings.* New York: Doubleday.

Tréatikoff, C. (1974). Thesis for doctorate in medicine, 1919. In J. Marks (Ed.), *The treatment of Parkinsonism with L-dopa* (pp. 29–38). Lancaster, UK: MTP Medical and Technical Publishing.

Ungerstedt, U. (1971). Adipsia and aphagia after 6-hydroxydopamine induced degeneration of the nigrostriatal dopamine system in the rat brain. *Acta Physiologica (Scandinavia), 82* (Suppl. 367), 95–122.

Widner, H., Tetrud, J., Rehngrona, S., Snow, B., Brundin, P., Gustavii, B., Bjorklund, A., Lindvall, O., & Langston, J. W. (1992). Bilateral fetal mesencephalic grafting in two patients with Parkinsonism induced by 1-methyl-4-phenyl-1,2,3,6-tetrahydropyridine (MPTP). *New England Journal of Medicine, 327,* 1556–1563.

Chapter 6

Aagaard, L., Hansen, E. H. (2011). The occurrence of adverse drug reactions reported for attention deficit hyperactivity disorder (ADHD) medications in the pediatric population: a qualitative review of empirical studies. *Neuropsychiatric Disorders and Treatments, 7,* 729–744.

Asarnow, J. R., Porta, G., Spirito, A., Emslie, G., Clarke, G., et al. (2011). Suicide attempts and nonsuicidal self-injury in the treatment of resistant depression in adolescents: findings from the TORDIA study. *Journal of the American Academy of Child and Adolescent Psychiatry, 50,* 772–781.

Barrós-Loscertales, A., Garavan, H., Bustamante, J. C., Ventura-Campos, N., Llopis, J. J., et al. (2011). Reduced striatal volume in cocaine-dependent patients. *Neuroimage, 56,* 1021–10266.

Becker, J. B., & Hu, M. (2008). Sex differences in drug abuse. *Frontiers in Neuroendocrinology, 29*(1), 36–47.

Becker, J. B., Breedlove, S. M., Crews, D., & McCarthy, M. M. (2002). *Behavioral endocrinology* (2nd ed.). Cambridge, MA: Bradford.

Büttner, A. (2011). Review: The neuropathology of drug abuse. *Neuropathology and Applied Neurobiology, 37,* 118–134.

Comer, R. J. (2011). *Fundamentals of abnormal psychology* (7th ed.). New York: Worth Publishers.

Cooper Center Longitudinal Study (2011). See **Hoang** et al., 2011.

Cowan, R. L., Roberts, D. M., & Joers, J. M. (2008). Neuroimaging in human MDMA (Ecstasy) users. *Annals of the New York Academy of Sciences, 1139,* 291–298.

DeCarolis, N. A., & Eisch, A. J. (2010). Hippocampal neurogenesis as a target for the treatment of mental illness: A critical evaluation. *Neuropharmacology, 58,* 884–893.

DeLisi, L. E. (2008). The effect of Cannabis on the brain: Can it cause brain anomalies that lead to increased risk for schizophrenia? *Current Opinion in Psychiatry, 21*(2), 140–50.

Durell, T. M., Kroutil, L. A., Crits-Christoph, P., Barchha, N., & Van Brunt, D. E. (2008). Prevalence of nonmedical methamphetamine use in the United States. Substance Abuse Treatment, Prevention, and Policy, 3, 19.

Everitt, B. J., Belin, D., Economidou, D., Pelloux, Y., Dalley, J. W., & Robbins, T. W. (2008). Review: Neural mechanisms underlying the vulnerability to develop compulsive drug-seeking habits and addiction. *Philosophical Transactions of the Royal Society of London, 363,* 3125–3135.

Fraioli, S., Crombag, H. S., Badiani, A., & Robinson, T. E. (1999). Susceptibility to amphetamine-induced locomotor sensitization is modulated by environmental stimuli. *Neuropsychopharmacology, 20,* 533–541.

Freud, S. (1974). *Cocaine papers* (R. Byck, Ed.). New York: Penguin.

Gilbertson, M. W., Shenton, M. E., Ciszewski, A., Kasai, K., Lasko, N. B., Orr, S. P., & Pitman, R. K. (2002). Smaller hippocampal volume predicts pathologic vulnerability to psychological trauma. *Nature Neuroscience, 5,* 1242–1247.

Greely, H., Sahakian, B., Harris, J., Kessler, R.C., Gazzaniga, M., et al. (2008). Towards responsible use of cognitive-enhancing drugs by the healthy. *Nature, 456,* 702–705.

Griffin, J. A., Umstattd, M. R., & Usdan, S. L. (2010). Alcohol use and high-risk sexual behavior among collegiate women: A review of research on alcohol myopia theory. *Journal of the American College of Health, 58,* 523–532.

Hampson, E., & Kimura, D. (2005). Sex differences and hormonal influences on cognitive function in humans. In J. B. Becker, S. M. Breedlove, & D. Crews (Eds.), *Behavioral endocrinology* (pp. 357–398). Cambridge, MA: MIT Press.

Henquet, C., Di Forti, M., Morrison, P., Kuepper, R., & Murray, R.M. (2008). Gene-environment interplay between cannabis and psychosis. *Schizophrenia Bulletin, 34,* 1111–1121.

Hoang, M. T., Defina, L. F., Willis, B. L., Leonard, D. S., Weiner, M. F., & Brown, E. S. (2011). Association between low serum 25-hydroxyvitamin D and depression in a large sample of healthy adults: Cooper Center Longitudinal Study. *Mayo Clinical Proceedings, 86,* 1050–1055.

Isbell, H., Fraser, H. F., Wikler, R. E., Belleville, R. E., & Eisenman, A. J. (1955). An experimental study of the etiology of "rum fits" and delirium tremens. *Quarterly Journal of Studies on Alcohol, 16,* 1–35.

Julien, R. M., Advokat, C. D., & Comaty, J. E. (2011). *A primer of drug action* (12th ed.). New York: Worth Publishers.

Kutcher, S., & Gardner, D. M. (2008). Use of selective serotonin reuptake inhibitors and youth suicide: Making sense from a confusing story. *Current Opinion in Psychiatry, 21,* 65–69.

Landré, L., Destrieux, C., Baudry, M., Barantin, L., Cottier, J. P., et al. (2010). Preserved subcortical volumes and cortical thickness in women with sexual abuse-related PTSD. *Psychiatry Research, 183,* 181–186.

MacAndrew, C., & Edgerton, R. B. (1969). *Drunken comportment: A social explanation.* Chicago: Aldine.

MacDonald, T. K., MacDonald, G., Zanna, M. P., & Fong, G. T. (2000). Alcohol, sexual arousal, and intentions to use condoms in young men: Applying alcohol myopia theory to risky sexual behavior. *Health Psychology, 19,* 290–298.

McCann, U. D., Lowe, K. A., & Ricaurte, G. A. (1997). Long-lasting effects of recreational drugs of abuse on the central nervous system. *The Neurologist, 3,* 399–411.

McGowan, P. O., Sasaki, A., D'Alessio, A. C., Dymov, S., Labonté, B., Szyf, M., Turecki, G., & Meaney, M. J. (2009). Epigenetic regulation of the glucocorticoid receptor in human brain associates with childhood abuse. *Nature Neurosciences, 12,* 342–348.

Milroy, C. M., & Parai, J. L. (2011). The histopathology of drugs of abuse. *Histopathology, 59,* 579–593.

Olney, J. W., Ho, O. L., & Rhee, V. (1971). Cytotoxic effects of acidic and sulphur-containing amino acids on the infant mouse central nervous system. *Experimental Brain Research, 14,* 61–67.

Popova, S., Lange, S., Bekmuradov, D., Mihic, A., & Rehm, J. (2011). Fetal alcohol spectrum disorder prevalence estimates in correctional systems: A systematic literature review. *Canadian Journal of Public Health, 102,* 336–340.

Quinn, P. D., & Fromme, K. (2012). Event-level associations between objective and subjective alcohol intoxication and driving after drinking across the college years. *Psychology of Addicted Behavior, 26,* 384–392.

Radjenović, J., Petrović, M., & Barceló, D. (2009). Fate and distribution of pharmaceuticals in wastewater and sewage sludge of the conventional activated sludge (CAS) and advanced membrane bioreactor (MBR) treatment. *Water Research, 43,* 831–841.

Robinson, T. E., & Becker, J. B. (1986). Enduring changes in brain and behavior produced by chronic amphetamine administration: A review and evaluation of animal models of amphetamine psychosis. *Brain Research Reviews, 11,* 157–198.

Robinson, T. E., & Berridge, K. C. (2008). Incentive sensitization theory. *Philosophical Transactions of the Royal Society of London, 363,* 3137–3146.

Robison, A. J., & Nestler, E. J. (2011). Transcriptional and epigenetic mechanisms of addiction. *Nature Reviews Neuroscience, 12,* 623–637.

Sapolsky, R. M. (1992). *Stress, the aging brain, and the mechanisms of neuron death.* Cambridge, MA: MIT Press.

Sapolsky, R. M. (2004). *Why zebras don't get ulcers* (3rd ed.). New York: Henry Holt and Company.

Sapolsky, R. M. (2005). The influence of social hierarchy on primate health. *Science, 308*(5722) 648–652.

Sarne, Y., Asaf, F., Fishbein, M., Gafni, M., & Keren, O. (2011). The dual neuroprotective-neurotoxic profile of cannabinoid drugs. *British Journal of Pharmacology, 163,* 1391–1401.

Severus, E., Schaaff, N., & Möller, H. J. (2012). State of the art: Treatment of bipolar disorders. *CNS Neuroscience and Therapeutics, 18,* 214–218.

Smith, A. D., Smith, S. M., de Jager, C. A., Whitbread, P., Johnston, C., et al. (2010). Homocysteine-lowering by B vitamins slows the rate of accelerated brain atrophy in mild cognitive impairment: a randomized controlled trial. *PLoS One, 8,* e12244.

Steen, E., Terry, B. M., Rivera, E. J., Cannon, J. L., Neely, T. R., Tavares, R., Xu, X. J., Wands, J. R., & de la Monte, S. M. (2005). Impaired insulin and insulin-like growth factor expression and signaling mechanisms in Alzheimer's disease: Is this type 3 diabetes? *Journal of Alzheimer's Disease, 7,* 63–80.

Vevelstad, M., Oiestad, E. L., Middelkoop, G., Hasvold, I., Lilleng, P., et al. (2012). The PMMA epidemic in Norway: Comparison of fatal and non-fatal intoxications. *Forensic Science International, 219,* 151–157.

Wenger, J. R., Tiffany, T. M., Bombardier, C., Nicholls, K., & Woods, S. C. (1981). Ethanol tolerance in the rat is learned. *Science, 213,* 575–577.

Whishaw, I. Q., Mittleman, G., & Evenden, J. L. (1989). Training-dependent decay in performance produced by the neuroleptic *cis*(Z)-flupentixol on spatial navigation by rats in a swimming pool. *Pharmacology, Biochemistry, and Behavior, 32,* 211–220.

Chapter 7

Bueller, J. A., Aftab, M., Sen, S., Gomez-Hassan, D. Burmeister, M., & Zubieta, J.-K. (2006). BDNF Val66Met allele is associated with reduced hippocampal volume in healthy subjects. *Biological Psychiatry, 59,* 812–815.

Cacucci, F., Yi, M., Wills, T. J., Chapmans, P., & O'Keefe, J. (2008). Place cell firing correlates with memory deficits and amyloid plaque burden in Tg2576 Alzheimer mouse model. *Proceedings of the National Academy of Sciences (USA), 105,* 7863–7868.

Caspi, A., Moffitt, T. E., Cannon, M., McClay, J., Murray, R., et al. (2005). Moderation of the effect of adolescent-onset cannabis use on adult psychosis by a functional polymorphism in the catechol-O-methyltransferase gene: longitudinal evidence of a gene X environment interaction. *Biological Psychiatry, 57,* 1117–1127.

Chamberlain, S. R., Robbins, T. W., & Sahakian, B. J. (2007). The neurobiology of attention-deficit/hyperactivity disorder. *Biological Psychiatry, 61,* 1317–1319.

Damasio, H., & Damasio, A. R. (1989). *Lesion analysis in neuropsychology.* New York: Oxford University Press.

Faraone, S. V., Segeant, J., Gillberg, C., & Biederman, J. (2003). The worldwide prevalence of ADHD: is it an American condition? *World Psychiatry, 2,* 104–113.

Fox, P. T., and Raichle, M. E. (1986). Focal physiological uncoupling of cerebral blood flow and oxidative metabolism during somatosensory stimulation in human subjects. *Proceedings of the National Academy of Sciences (USA), 83,* 1140–1144.

Fraga, M. F., Ballestar, E., Paz, M. F., Ropero, S., Setien, F., Ballestar, M. L., et al. (2005). Epigenetic differences arise during the lifetime of monozygotic twins. *Proceedings of the National Academy of Sciences (USA), 102,* 10604–10609.

Gonzalez, C. L. R., Gharbawie, O. A., & Kolb, B. (2006). Chronic low-dose administration of nicotine facilitates recovery and synaptic change after focal ischemia in rats. *Neuropharmacology, 50,* 777–787.

Hoshi, Y. (2007) Functional near-infrared spectroscopy: current status and future prospects. *Journal of Biomedical Optics, 12,* 062106-1–062106-9.

Kolb, B., & Walkey, J. (1987). Behavioural and anatomical studies of the posterior parietal cortex of the rat. *Behavioural Brain Research, 23,* 127–145.

Kwong, K. K., et al. (1992). Dynamic magnetic resonance imaging of human brain activity during primary sensory stimulation. *Proceedings of the National Academy of Sciences (USA), 89,* 5678.

May, L., Byers-Heinlein, K., Gervain, J., & Werker, J. F. (2011). Language and the newborn brain: Does prenatal language experience shape the neonate neural response to speech? *Frontiers in Psychology, 2,* 1–9.

McGowan, P. O., Sasaki, A., D'Alessio, A. C., Dymov, S., Labonté, B., Szyf, M., Turecki, G., & Meaney, M. J. (2009). Epigenetic regulation of the glucocorticoid receptor in human brain associates with childhood abuse. *Nature Neuroscience, 12,* 342–348.

Morris, R. G. M. (1981) Spatial localization does not require the presence of local cues. *Learning and Motivation, 12,* 239–260.

Mychsiuk, R., Schmold, N., Ilnytskyy, S., Kovalchuk, O., Kolb, B., & Gibb, R. (2011). Prenatal bystander stress alters brain, behavior, and the epigenome of developing rat offspring. *Developmental Neuroscience, 33,* 159–169.

National Research Council (1996). *The guide for the care and use of laboratory animals.* Washington, DC: National Academy Press.

Ogawa, S., Lee, T. M., Kay, A. R., & Tank, D. W. (1990). Brain magnetic resonance imaging with contrast dependent on blood oxygenation. *Proceedings of the National Academy of Sciences (USA), 87,* 9868–9872.

O'Keefe, J., & Dostrovksy, J. (1971). The hippocampus as a spatial map. *Brain Research, 34,* 171–175.

Penfield, W., & Jasper, H. H. (1954). *Epilepsy and the functional anatomy of the human brain.* Boston: Little, Brown.

Posner, M. I., & Raichle, M. E. (1997). *Images of mind.* New York: W. H. Freeman and Company.

Reza, M. F., Ikoma, K., Ito, T, Ogawa, T., & Mano, Y. (2007). N200 latency and P300 amplitude in depressed mood post-traumatic brain injury patients. *Neuropsychological Rehabilitation, 17,* 723–734.

Schallert, T., Whishaw, I. Q., Ramirez, V. D., & Teitelbaum, P. (1978). Compulsive, abnormal walking caused by anticholinergics in akinetic, 6-hydroxydopamine-treated rats. *Science, 199,* 1461–1463.

Scoville, W. B., & Milner, B. (1957). Loss of recent memory after bilateral hippocampal lesions. *Journal of Neurology, Neurosurgery, and Psychiatry, 20,* 11–21.

Spinney, L. (2005) Optical topography and the color of blood. *The Scientist, 19,* 25–27.

Sullivan, R. M., and Brake, W. G. (2003). What the rodent prefrontal cortex can tell us about attention deficit/hyperactive disorder: The critical role of early developmental events on prefrontal function. *Behavioural Brain Research, 146:* 43–55.

Szyf, M., McGowan, P., & Meaney, M. J. (2008). The social environment and the epigenome. *Environmental Molecular Mutagenetics, 49,* 46–60.

Taga, G., Asakawa, K., Maki, A., Konishi, Y., & Koizumi, H. (2003). Brain imaging in awake infants by near-infrared optical topography. *Proceedings of the National Academy of Sciences (USA), 100,* 10722–10777.

Tisdall, M. M., & Smith, M. (2006). Cerebral microdialysis: research technique or clinical tool. *British Journal of Anaesthesia, 97,* 18–25.

Van den Heuvel, M. P., & Hulshoff Pol, H. E. (2010). Exploring the brain network: A review on resting-state fMRI functional connectivity. *European Neuropsychopharmcology, 20,* 519–534.

Watanabe, H., Homae, F., Nakano, T., & Taga, G. (2008). Functional activation in diverse regions of the developing brain in infants. *NeuroImage, 43,* 346–357.

Whishaw, I. Q. (1989). Dissociating performance and learning deficits in spatial navigation tasks in rats subjected to cholinergic muscarinic blockade. *Brain Research Bulletin, 23,* 347–358.

Whishaw, I. Q., & Kolb, B. (2005). *The behavior of the laboratory rat.* New York: Oxford University Press.

Chapter 8

Anda, R. F., Felitti, V. J., Bremner, J. D., Walker, J. D., Whitfield, C., et al. (2006). The enduring effects of abuse and related adverse experiences in childhood: A convergence of evidence from neurobiology and epidemiology. *European Archives of Psychiatry and Clinical Neuroscience, 256,* 174–s186.

Agate, R. J., Grisham, W., Wade, J., Mann, S., Wingfield, J., Schanen, C., Palotie, A., & Arnold, A. P. (2003). Neural, but not gonadal, origin of brain sex differences in a gynandromorphic finch. *Proceedings of the National Academy of Sciences of the United States of America, 100,* 4873–4878.

Ames, E. W. (1997). The development of Romanian orphanage children adopted to Canada. Final report to Human Resources Development, Canada.

Audero, E., Coppi, E., Mlinar, B., Rossetti, T., Caprioli, A., Banchaabouchi, A., Corradetti, R., & Gross, C. (2008). Sporadic autonomic dysregulation and death associated with excessive serontonin autoinhibition. *Science, 321,* 130–133.

Bourgeois, J.-P. (2001). Synaptogenesis in the neocortex of the newborn: The ultimate frontier for individuation? In Nelson, C. A,. & Luciana, M., *Handbook of Developmental Cognitive Neuroscience.* Cambridge, MA: MIT Press.

Bunney, B. G., Potkin, S. G., & Bunney, W. E. (1997). Neuropathological studies of brain tissue in schizophrenia. *Journal of Psychiatric Research, 31,* 159–173.

Changeux, J.-P., & Danchin, A. (1976). Selective stabilization of developing synapses as a mechanism for the specification of neuronal networks. *Nature, 264,* 705–712.

Chen, R., Jiao, Y., & Herskovits, E. H. (2011). Structural MRI in autism spectrum disorder. *Pediatric Research, 69,* 63R–68R.

Clarke, R. S., Heron, W., Fetherstonhaugh, M. L., Forgays, D. G., & Hebb, D. O. (1951). Individual differences in dogs: Preliminary report on the effects of early experience. *Canadian Journal of Psychology, 5*(4), 150–156.

Cowan, W. M. (1979). The development of the brain. *Scientific American, 241*(3), 116.

Edelman, G. M. (1987). *Neural darwinism: The theory of neuronal group selection.* New York: Basic Books.

Epstein, H. T. (1979). Correlated brain and intelligence development in humans. In M. E. Hahn, C. Jensen, & B. C. Dudek (Eds.), *Development and evolution of brain size: Behavioral implications* (pp. 111–131). New York: Academic Press.

Gershon, E. S., & Rieder, R. O. (1992). Major disorders of mind and brain. *Scientific American, 267*(3), 126–133.

Giussani, C., Roux, F. E., Lubrano, V., Gaini, S. M., & Bello, L. (2007). Review of language organization in bilingual patients: What can we learn from direct brain mapping? *Acta Neurochir (Wien), 149,* 1109–1116.

Goldstein, J. M., Seidman, L. J., Horton, N. J., Makris, N., Kennedy, D. N., Caviness, V. S., Jr., Faraone, S. V., & Tsuang, M. T. (2001). Normal sexual dimorphism of the adult human brain assessed by in vivo magnetic resonance imaging. *Cerebral Cortex, 11,* 490–497.

Guerrini, R., Dobyns, W. B., & Barkovich, A. J. (2007). Abnormal dvelopment of the human cerebral cortex: Genetics, functional consequences and treatment options. *Trends in Neurosciences, 31,* 154–162.

Halliwell, C., Comeau, W., Gibb, R., Frost, D. O., & Kolb, B. (2009). Factors influencing frontal cortex development and recovery from early frontal injury. *Developmental Neurorehabilitation, 12,* 269–278.

Harlow, H. F. (1971). *Learning to love.* San Francisco: Albion.

Hebb, D. O. (1947). The effects of early experience on problem solving at maturity. *American Psychologist, 2,* 737–745.

Horn, G., Bradley, P., & McCabe, B. J. (1985). Changes in the structure of synapses associated with learning. *Journal of Neuroscience, 5,* 3161–3168.

Huttenlocher, P. R. (1994). Synaptogenesis in human cerebral cortex. In G. Dawson & K. W. Fischer (Eds.), *Human behavior and the developing brain* (pp. 137–152). New York: Guilford Press.

Juraska, J. M. (1990). The structure of the cerebral cortex: Effects of gender and the environment. In B. Kolb & R. Tees (Eds.), *The cerebral cortex of the rat* (pp. 483–506). Cambridge, MA: MIT Press.

Kinney, H. C. (2009). Brainstem mechanisms underlying the sudden infant death syndrome: Evidence from human pathologic studies. *Developmental Psychobiology, 51,* 223–233.

Klein, D., Watkins, K. E., Zatorre, R. J., & Milner, B. (2006). Word and nonword repetition in bilingual subjects: A PET study. *Human Brain Mapping, 27,* 153–161.

Kolb, B., & Gibb, R. (1993). Possible anatomical basis of recovery of function after neonatal frontal lesions in rats. *Behavioral Neuroscience, 107,* 799–811.

Kolb, B., Mychasiuk, R., Muhammad, A., Li, Y., Frost, D.O., & Gibb, R. (2012). Experience and the developing prefrontal cortex. *Proceedings of the National Academy of Sciences (USA).* doi:10.1073/pmas.1121251109.

Kovelman, J. A., & Scheibel, A. B. (1984). A neurohistologic correlate of schizophrenia. *Biological Psychiatry, 19,* 1601–1621.

Lenneberg, E. H. (1967). *Biological foundations of language.* New York: Wiley.

Lenroot, R. K., Gogtay, N., Greenstein, D. K., et al. (2007). Sexual dimorphism of brain development trajectories during childhood and adolescence. *NeuroImage, 36,* 1065–1073.

Lorenz, K. (1970). *Studies on animal and human behavior* (Vols. 1 and 2). Cambridge, MA: Harvard University Press.

Lu, L. H., Leonard, C. M., Thompson, P. M., Kan, E., Jolley, J., Welcome, S. E., Toga, A. W., & Sowell, E. R. (2007). Normal developmental changes in inferior frontal gray matter are associated with improvement in phonological processing: A longitudinal MRI analysis. *Cerebral Cortex 17,* 1092–1099.

Malanga, C. J., & Kosofsky, B. E. (2003). Does drug abuse beget drug abuse? Behavioral analysis of addiction liability in animal models of prenatal drug exposure. *Developmental Brain Research, 147,* 47–57.

Marin-Padilla, M. (1988). Early ontogenesis of the human cerebral cortex. In A. Peters & E. G. Jones (Eds.), *Cerebral cortex, Vol. 7: Development and maturation of the cerebral cortex* (pp.1–34). New York: Plenum.

Marin-Padilla, M. (1993). Pathogenesis of late-acquired leptomeningeal heterotopias and secondary cortical alterations: A Golgi study. In A. M. Galaburda (Ed.), *Dyslexia and development: Neurobiological aspects of extraordinary brains* (pp. 64–88). Cambridge, MA: Harvard University Press.

Marshall, P. J., Reeb, C., Fox, N. A., Nelson, C. A., & Zeanah, C. H. (2008). Effects of early intervention on EEG power and coherence in previously institutionalized children in Romania. *Developmental Psychopathology, 20,* 861–880.

Moore, K. L. (1988). *The developing human: Clinically oriented embryology* (4th ed.). Philadelphia: Saunders.

Mychasiuk, R., Gibb, R., & Kolb, B. (2011). Prenatal stress produces sexually dimorphic and regionally specific changes in gene expression in hippocampus and frontal cortex of developing rat offspring. *Developmental Neuroscience, 33,* 531–538.

Myers, D. G. (2010). *Psychology* (9th ed.). New York: Worth Publishers.

National Institute on Drug Abuse. (1999). Prenatal cocaine exposure costs at least $352 million per year. *NIDA Research Findings, 13*(5), 1–2.

Nelson, C. A., & Luciana, M. (Eds.) (2008). *Handbook of developmental cognitive neuroscience,* 2nd ed. Cambridge, MA: MIT Press.

Nelson, C. A., Zeanah, C. H., Fox, N. A., Marshall, P. J., Smyke, A. T., & Guthrie, D. (2007). Cognitive recovery in socially deprived young children: The Bucharest Early Intervention Project. *Science, 318,* 1937–1940.

Nordahl, C. W., Scholz, R., Yang, X., Buonocore, M. H., Simon, T., Rogers S., et al. (2012). Increased rate of amygdala growth in children aged 2 to 4 years with autism spectrum disorders: A longitudinal study. *Archives of General Psychiatry, 69,* 53–61.

O'Hare, E. D., & Sowell, E. R. (2008). Imaging developmental changes in gray and white matter in the human brain. In C. A. Nelson and M. Luciana (Eds.), *Handbook of developmental cognitive neuroscience* (pp. 23–38), Cambridge, MA: MIT Press.

Overman, W., Bachevalier, J., Schuhmann, E., & Ryan, P. (1996). Cognitive gender differences in very young children parallel biologically based cognitive gender differences in monkeys. *Behavioral Neuroscience, 110,* 673–684.

Overman, W., Bachevalier, J., Turner, M., & Peuster, A. (1992). Object recognition versus object discrimination: Comparison between human infants and infant monkeys. *Behavioral Neuroscience, 106,* 15–29.

Paterson, D. S., Trachtenberg, F. L., Thompson, E. G., Belliveau, R. A., Beggs, A. H., Dranall, R., Chadwick, A. E., Krous, H. F., & Kinney, H. C. (2006). Multiple serotonergic brainstem abnormalities in sudden infant death syndrome. *Journal of the American Medical Association, 296,* 2124–2132.

Paus, T., Keshavan, M., & Giedd, J. N. (2008). Why do so many psychiatric disorders emerge during adolescence? *Nature Reviews Neuroscience, 9,* 947–957.

Payne, B. R., & Lomber, S. G. (2001). Reconstructing functional systems after lesions of cerebral cortex. *Nature Reviews Neuroscience, 2,* 911–919.

Petanjek, Z., Judas, M., Simic, G., Rasin, M. R., Uylings, H. B. M., et al. (2011). Extraordinary neoteny of synaptic spines in the human prefrontal cortex. *Proceedings of the National Academy of Sciences (USA), 108,* 13281–13286.

Piaget, J. (1952). *The origins of intelligence in children.* New York: Norton.

Purpura, D. P. (1974). Dendritic spine "dysgenesis" and mental retardation. *Science, 186,* 1126–1127.

Rakic, P. (1974). Neurons in rhesus monkey cerebral cortex: Systematic relation between time of origin and eventual disposition. *Science, 183,* 425.

Rakic, P. (1995). Corticogenesis in human and nonhuman primates. In M. Gazzaniga (Ed.), *The cognitive neurosciences* (pp. 127–145). Cambridge, MA: MIT Press.

Riesen, A. H. (1982). Effects of environments on development in sensory systems. In W. D. Neff (Ed.), *Contributions to sensory physiology* (Vol. 6, pp. 45–77). New York: Academic Press.

Rutter, M. (1998). Developmental catch-up, and deficit, following adoption after severe global early privation. *Journal of Child Psychology and Psychiatry, 39,* 465–476.

Salm, A. K., Pavelko, M., Drouse, E. M., Webster, W., Kraszpulski, M., & Birkle, D. L. (2004). Lateral amygdala nucleus expansion in adult rats is associated with exposure to prenatal stress. *Developmental Brain Research, 148,* 159–167.

Shaw, P., Greenstein, D., Lerch, J., Clasen, L., Lenroot, R., et al. (2006). Intellectual ability and cortical development in children and adolescents. *Nature, 440,* 666–679.

Shingo, T., Gregg, C., Enwere, E., Fujikawa, H., Hassam, R., Geary, C., Cross, J. C., & Weiss, S. (2003). Pregnancy-stimulated neurogenesis in the adult female forebrain mediated by prolactin. *Science, 299,* 117–120.

Smyke, A. T., Zeanah, C. H., Gleason, M. M., Drury, S. S., Fox, N. A., et al. (2012). A randomized controlled trial comparing foster care and institutional care for children with signs of reactive attachment disorder. *American Journal of Psychiatry, 169,* 508–514.

Sodhi, M. S., & Sanders-Bush, E. (2004). Serotonin and brain development. *International Review of Neurobiology, 59,* 111–174.

Sowell, E. R., Thompson, P. M., Leonard, C. M., Welcome, S. E., Kan, E., & Toga, A. W. (2004). Longitudinal mapping of cortical thickness and brain growth in normal children. *Journal of Neuroscience, 24,* 8223–8231.

Sowell, E. R., Thompson, P. M., & Toga, A.W. (2004). Mapping changes in the human cortex throughout the span of life. *The Neuroscientist, 10,* 372–392.

Sperry, R. W. (1963). Chemoaffinity in the orderly growth of nerve fiber patterns and connections. *Proceedings of the National Academy of Sciences of the United States of America, 50,* 703–710.

Stevens, H., Leckman, J., Coplan, J., & Suomi, S. (2009). Risk and resilience: Early manipulation of macaque social experience and persistent behavioral and neurophysiological outcomes. *Journal of the American Academy of Child and Adolescent Psychiatry, 48,* 114–127.

Suomi, S. J. (2011). Risk, resilience, and gene-environment interplay in primates. *Journal of the Canadian Academy of Child Adolescent Psychiatry, 20,* 289–297.

Twitchell, T. E. (1965). The automatic grasping response of infants. *Neuropsychologia, 3,* 247–259.

Wallace, P. S., & Whishaw, I. Q. (2003). Independent digit movements and precision grip patterns in 1–5-month-old human infants: Hand-babbling, including vacuous than self-directed hand and digit movements, precedes targeted reaching. *Neuropsychologia, 41,* 1912–1918.

Weiss, S., Reynolds, B. A., Vescovi, A. L., Morshead, C., Craig, C. G., & van der Kooy, D. (1996). Is there a neural stem cell in the mammalian forebrain? *Trends in Neurosciences, 19*(9), 387–393.

Werker, J. F., & Tees, R. C. (1992). The organization and reorganization of human speech perception. *Annual Review of Neuroscience, 15,* 377–402.

Wiesel, T. N. (1982). Postnatal development of the visual cortex and the influence of environment. *Nature, 299,* 583–591.

Yakovlev, P. E., & Lecours, A.-R. (1967). The myelogenetic cycles of regional maturation of the brain. In A. Minkowski (Ed.), *Regional development of the brain in early life.* Oxford: Blackwell.

Chapter 9

Balint, R. (1909). Seelenlähmung des "Schauens," optische Ataxie, räumliche Störung der Aufmerksamkeit. *Monatschrift für Psychiatrie und Neurologie, 25,* 51–81.

Farah, M. J. (1990). *Visual agnosia.* Cambridge, MA: MIT Press.

Geschwind, N. (1972). Language and the brain. *Scientific American, 226*(4), 78–83.

Goodale, M. A., Milner, D. A., Jakobson, L. S., & Carey, J. D. P. (1991). A kinematic analysis of reaching and grasping movements in a patient recovering from optic ataxia. *Nature, 349,* 154–156.

Hamilton, R. H., J. T. Shenton, and H. B. Coslett. (2006). An acquired deficit of audiovisual speech processing. *Brain and Language, 98,* 66–73.

Hasson, U., Nir, Y., Levy, I., Fuhrmann, G., & Malach, R. (2004). Intersubject synchronization of cortical activity during natural vision. *Science, 303,* 1634–1640.

Held, R. (1968). Dissociation of visual function by deprivation and rearrangement. *Psychologische Forschung, 31,* 338–348.

Hubel, D. H. (1988). *Eye, brain, and vision.* New York: Scientific American Library.

Hubel, D. H., & Weisel, T. N. (1977). Functional architecture of macaque monkey visual cortex. *Proceedings of the Royal Society of London B, 198,* Figure 23.

Kline, D. W. (1994). Optimizing the visibility of displays for older observers. *Experimental Aging Research, 20,* 11–23.

Kuffler, S. W. (1952). Neurons in the retina: Organization, inhibition and excitatory problems. *Cold Spring Harbor Symposia on Quantitative Biology, 17,* 281–292.

Lashley, K. S. (1941). Patterns of cerebral integration indicated by the scotomas of migraine. *Archives of Neurology and Psychiatry, 46,* 331–339.

Milner, A. D., & Goodale, M. A. (2006). *The visual brain in action (2nd ed.).* Oxford: Oxford University Press.

Purves, D., Augustine, G. J., Fitzpatrick, D., Katz, L. C., LaMantia, A.-S., & McNamara, J. O. (Eds.). (2001) *Neuroscience.* Sunderland, MA: Sinauer.

Sacks, O., & Wasserman, R. (1987). The case of the colorblind painter. *The New York Review of Books, 34,* 25–33.

Snowden, R., Thompson, P., & Troscianko, T. (2008). *Basic vision: An introduction to visual perception.* Oxford: Oxford University Press.

Szentagothai, J. (1975). The "module-concept" in cerebral cortex architecture. *Brain Research, 95,* 475–496.

Tanaka, K. (1993). Neuronal mechanisms of object recognition. Science, 262, 685–688.

Tanaka, K. (1996). Inferotemporal cortex and object vision. *Annual Review of Neuroscience, 19,* 100–139.

Weizkrantz, L. (1986). *Blindsight: A case study and implications.* Oxford: Oxford University Press.

Wong-Riley, M. T. T., Hevner, R. F., Cutlan, R., Earnest, M., Egan, R., Frost, J., & Nguyen, T. (1993). Cytochrome oxidase in the human visual cortex: Distribution in the developing and the adult brain. *Visual Neuroscience, 10,* 41–58.

Zaidi, F. H., Hull, J. T., Peirson, S. N., Wulff, K., Aeschbach, D., et al. (2007). Short-wavelength light sensitivity of circadian, pupillary, and visual awareness in humans lacking an outer retina. *Current Biology, 17,* 2122–2128.

Zeki, S. (1997). *A vision of the brain.* Oxford: Blackwell Scientific.

Zihl, J., von Cramon, D., & Mai, N. (1983). Selective disturbance of movement vision after bilateral brain damage. *Brain, 106,* 313–340.

Chapter 10

Belin, P., Zatorre, R. J., Lafaille, P., & Pike, B. (2000). Voice-selective areas in the human auditory cortex. *Nature, 403,* 309–312.

Bermudez, P., Lerch, J. P., Evans, A. C., & Zatorre, R. J. (2009). Neuroanatomical correlates of musicianship as revealed by cortical thickness and voxel-based morphometry. *Cerebral Cortex, 19,* 1583–1596.

Chomsky, N. (1965). Aspects of the theory of syntax. Cambridge, MA: MIT Press.

Drake-Lee, A. B. (1992). Beyond music: Auditory temporary threshold shift in rock musicians after a heavy metal concert. *Journal of the Royal Society of Medicine, 85,* 617–619.

Fiez, J. A., Raichle, M. E., Balota, D. A., Tallal, P., & Petersen, S. E. (1996). PET activation of posterior temporal regions during auditory word presentation and verb generation. *Cerebral Cortex, 6,* 1–10.

Fiez, J. A., Raichle, M. E., Miezin, F. M., & Petersen, S. E. (1995). PET studies of auditory and phonological processing: Effects of stimulus characteristics and task demands. *Journal of Cognitive Neuroscience, 7,* 357–375.

Geissmann, T. (2001). Gibbon songs and human music from an evolutionary perspective. In N. L. Wallin, B. Merker, & S. Brown (Eds.), *The origins of music* (pp. 103–124). Cambridge, MA: MIT Press.

Hyde, K. L., Lerch, J. P., Zatorre, R. J., Griffiths, T. D., Evans, A. C., & Peretz, I. (2007). Cortical thickness in congenital amusia: When less is better than more. *Journal of Neuroscience, 27,* 13028–13032.

Johansson, B. B. (2012). Multisensory stimulation in stroke rehabilitation. *Frontiers in Human Neuroscience, 6,* 60.

Kim, K. H. S., Relkin, N. R., Young-Min Lee, K., and Hirsch, J. (1997). Distinct cortical areas associated with native and second languages. *Nature 388,* 171–174.

Knudsen, E. I. (1981). The hearing of the barn owl. *Scientific American, 245*(6), 113–125.

Loeb, G. E. (1990). Cochlear prosthetics. *Annual Review of Neuroscience, 13,* 357–371.

Marler, P. (1991). The instinct to learn. In S. Carey and R. German (Eds.), *The epigenesis of mind: Essays on biology and cognition* (p. 39). Hillsdale, NJ: Erlbaum.

Neuweiler, G. (1990). Auditory adaptations for prey capture in echolocating bats. *Physiological Reviews, 70,* 615–641.

Nottebohm, F., & Arnold, A. P. (1976). Sexual dimorphism in vocal control areas of the songbird brain. *Science, 194,* 211–213.

Penfield, W., & Roberts, L. (1956). *Speech and brain mechanisms.* Oxford: London: Oxford University Press.

Penfield, W., & Roberts, L. (1959). *Speech and brain mechanisms.* Princeton, NJ: Princeton University Press.

Peretz, I., Kolinsky, R., Tramo, M., Labrecque, R., Hublet, C., Demeurisse, G., & Belleville, S. (1994). Functional dissociations following bilateral lesions of auditory cortex. *Brain, 117,* 1283–1301.

Pinker, S. (1997). *How the mind works.* New York: Norton.

Posner, M. I., & Raichle, M. E. (1997). *Images of mind.* New York: W. H. Freeman and Company.

Romanski, L. M., Tian, B., Fritz, J., Mishkin, M., Goldman-Rakic, P. S., & Rauschecker, J. P. (1999). Dual streams of auditory afferents target multiple domains in the primate prefrontal cortex. *Nature Neuroscience, 2,* 1131–1136.

Rauschecker, J. P. (2012). Ventral and dorsal streams in the evolution of speech and language. *Frontiers in Evolutionary Neuroscience, 4*(7), 10.3389/fnevo.2012.00007.

Sacks, O. (2007). *Musicophilia: Tales of music and the brain.* New York: Knopf.

Teie, P. U. (1998). Noise-induced hearing loss and symphony orchestra musicians: risk factors, effects, and management. *Maryland Medical Journal, 47,* 13–18.

Teng, S., & Whitney, D. (2011). The acuity of echolocation: Spatial resolution in the sighted compared to expert performance. *Journal of Visual Impairment Blindness, 105,* 20–32.

Thaler, L., Arnott, S. R., & Goodale, M. A. (2011). Neural correlates of natural human echolocation in early and late blind echolocation experts. *PLoS ONE, 6,* e20162.

Thompson, C. K., Meitzen, J., Replogle, K., Drnevich, J., Lent, K. L., et al. (2012). Seasonal changes in patterns of gene expression in avian song control brain regions. *PLoS One, 7,* e35119.

Trehub, S., Schellenberg, E. G., & Ramenetsky, G. B. (1999). Infants' and adults' perception of scale structures. *Journal of Experimental Psychology: Human Perception and Performance, 25,* 965–975.

Zatorre, R. J., Evans, A. C., & Meyer, E. (1994). Neural mechanisms underlying melodic perception and memory for pitch. *Journal of Neuroscience, 14,* 1908–1919.

Zatorre, R. J., Evans, A. C., Meyer, E., & Gjedde, A. (1992). Lateralization of phonetic and pitch discrimination in speech processing. *Science, 256,* 846–849.

Zatorre, R. J., Meyer, E., Gjedde, A., & Evans, A. C. (1995). PET studies of phonetic processing of speech: Review, replication, and reanalysis. *Cerebral Cortex, 6,* 21–30.

Chapter 11

Alessandria, M., Vetrugno, R., Cortelli, P., & Montagna, P. (2011). Normal body schema and absent phantom limb experience in amputees while dreaming. *Consciousness and Cognition, 20,* 831–834.

Alexander, R. E., & Critcher, M. D. (1990). Functional architecture of basal ganglia circuits: Neural substrates of parallel processing. *Trends in Neuroscience, 13,* 269.

Asanuma, H. (1989). *The motor cortex.* New York: Raven Press.

Blakemore, S. J., Wolpert, D. M., & Frith, C. D. (1998). Central cancellation of self-produced tickle sensation. *Nature Neuroscience, 1,* 635–640.

Brinkman, C. (1984). Supplementary motor area of the monkey's cerebral cortex: Short- and long-term deficits after unilateral ablation and the effects of subsequent callosal section. *Journal of Neuroscience, 4,* 918–992.

Casile, A., Caggiano, V., & Ferrari, P. F. (2011). The mirror neuron system: A fresh view. *Neuroscientist, 17,* 524–538.

Cho, Y., & Borgens, R. B. (2012). Polymer and nano-technology applications for repair and reconstruction of the central nervous system. *Experimental Neurology, 233,* 126–144.

Church, J. A., Fair, D. A., Dosenbach, N. U., Cohen, A. L., Miezin, F. M., Petersen, S. E., & Schlaggar, B. L. (2009). Control networks in paediatric Tourette syndrome show immature and anomalous patterns of functional connectivity. *Brain, 32,* 225–238.

Cole, J. (1995). *Pride and a daily marathon.* Cambridge, MA: MIT Press.

Corkin, S., Milner, B., & Rasmussen, T. (1970). Somatosensory thresholds. *Archives of Neurology, 23,* 41–58.

Critchley, M. (1953). *The parietal lobes.* London: Arnold.

Evarts, E. V. (1968). Relation of pyramidal tract activity to force exerted during voluntary movement. *Journal of Neurophysiology, 31,* 14–27.

Friedhoff, A. J., & Chase, T. N. (1982). Gilles de la Tourette syndrome. *Advances in Neurology, 35,* 1–17.

Graziano, M. (2006). The organization of behavioural repertoire in motor cortex. *Annual Reviews of Neuroscience, 29,* 105–134.

Hess, W. R. (1957). *The functional organization of the diencephalon.* London: Grune & Stratton.

Hughlings-Jackson, J. (1931). *Selected writings of John Hughlings-Jackson* (Vols. 1 and 2, J. Taylor, Ed.). London: Hodder.

Jackson, S. R., Parkinson, A., Kim, S. Y., Schüermann, M., & Eickhoff, S. B. (2011). On the functional anatomy of the urge-for-action. *Cognition and Neuroscience, 2,* 227–243.

Kaas, J. H. (1987). The organization and evolution of neocortex. In S. P. Wise (Ed.), *Higher brain functions* (pp. 237–298). New York: Wiley.

Kaas, J. H., Stepniewska, I., & Gharbawie, O. (2012). Cortical networks subserving upper limb movements in primates. *European Journal of Physical and Rehabilitation Medicine, 48,* 299–306.

Kalueff, A. V., Aldridge, J. W., LaPorte, J. L., Murphy D. L., & Tuohimaa, P. (2007). Analyzing grooming microstructure in neurobehavioral experiments. *Nature Protocols, 25,* 38–44.

Kern, U., Busch, V., Rockland, M., Kohl, M., & Birklein, F. (2009). Prevalence and risk factors of phantom limb pain and phantom limb sensations in Germany: A nationwide field survey. *Schmerz, 23,* 479–488.

Kreitzer, A. C., & Malaenka, R. C. (2008) Striatal plasticity and basal ganglia circuit function. *Neuron, 60,* 543–554.

Lang, C. E., & Schieber, M. H. (2004). Reduced muscle selectivity during individuated finger movements in humans after damage to the motor cortex or corticospinal tract. *Journal of Neurophysiology, 91,* 1722–1733.

Lashley, K. S. (1951). The problem of serial order in behavior. In L. A. Jeffress (Ed.), *Cerebral mechanisms and behavior* (pp. 112–136). New York: Wiley.

Leonard, C. M., Glendinning, D. S., Wilfong, T., Cooper, B. Y., & Vierck, C. J., Jr. (1991). Alterations of natural hand movements after interruption of fasciculus cuneatus in the macaque. *Somatosensory and Motor Research, 9,* 61–75.

Melzack, R. (1973). *The puzzle of pain.* New York: Basic Books.

Melzack, R., & Wall, P. D. (1965). Pain mechanisms: A new theory. *Science, 150,* 971–979.

Mountcastle, V. B. (1978). An organizing principle for cerebral function: The unit module and the distributed system. In G. M. Edelman & V. B. Mountcastle (Eds.), *The mindful brain* (pp. 7–50). Cambridge, MA: MIT Press.

Nudo, R. J., Wise, B. M., SiFuentes, F., & Milliken, G. W. (1996). Neural substrates for the effects of rehabilitative training on motor recovery after ischemic infarct. *Science, 272,* 1791–1794.

Panksepp, J. (2007). Neuroevolutionary sources of laughter and social joy: modeling primal human laughter in laboratory rats. *Behavioural Brain Research, 4,* 231–244.

Penfield, W., & Boldrey, E. (1958). Somatic motor and sensory representation in the cerebral cortex as studied by electrical stimulation. *Brain, 60,* 389–443.

Pons, T. P., Garraghty, P. E., Ommaya, A. K., Kaas, J. H., Taum, E., & Mishkin, M. (1991). Massive cortical reorganization after sensory deafferentation in adult macaques. *Science, 252,* 1857–1860.

Roland, P. E. (1993). *Brain activation.* New York: Wiley-Liss.

Rothwell, J. C., Taube, M. M., Day, B. L., Obeso, J. A., Thomas, P. K., & Marsden, C. D. (1982). Manual motor performance in a deafferented man. *Brain, 105,* 515–542.

Sacks, O. W. (1998). *The man who mistook his wife for a hat: And other clinical tales.* New York: Touchstone Books.

Thach, W. T. (2007). On the mechanism of cerebellar contributions to cognition. *Cerebellum, 6,* 163–167.

von Holst, E. (1973). *The collected papers of Erich von Holst* (R. Martin, Trans.). Coral Gables, FL: University of Miami Press.

Williams, T. H., Gluhbegovic, N., & Jew, J. (2000). *The human brain: Dissections of the real brain,* on CD-ROM. Published by Brain University, brain-university.com 2000.

Chapter 12

Ackerly, S. S. (1964). A case of paranatal bilateral frontal lobe defect observed for thirty years. In J. M. Warren & K. Akert (Eds.), *The frontal granular cortex and behavior* (pp. 192–218). New York: McGraw-Hill.

Adelmann, P. K., & Zajonc, R. B. (1989). Facial efferents and the experience of emotion. *Annual Review of Psychology, 40,* 249–280.

Bartoshuk, L. M. (2000). Comparing sensory experiences across individuals: Recent psychophysical advances illuminate genetic variation in taste perception. *Chemical Senses, 25,* 447–460.

Beatty, J. (1995). *Principles of behavioral neuroscience* (p. 339). Dubuque, IA: Brown & Benchmark.

Bao, A.-M., & Swaab, D. F. (2011). Sexual differentiation of the human brain: Relation to gender identity, sexual orientation and neuropsychiatric disorders. *Frontiers in Neuroendocrinology, 32,* 214–226.

Becker, J. B., Berkley, K. J., Geary, N., Hampson, E., Herman, J. P., & Young, E. A. (2008). *Sex differences in the brain: From genes to behavior.* New York: Oxford University Press.

Berridge, K. C. (1996). Food reward: Brain substrates of wanting and liking. *Neuroscience and Biobehavioral Reviews, 20,* 6.

Brunetti, M., Babiloni, C., Ferretti, A., Del Gratta, C., Merla, A., et al. (2008). *European Journal of Neuroscience, 27,* 2922–2927.

Butler, R. A., & Harlow, H. F. (1954). Persistence of visual exploration in monkeys. *Journal of Comparative and Physiological Psychology, 47,* 257–263.

Daly, M., & Wilson, M. (1988). *Homicide.* New York: Aldine.

Damasio, A. R. (1999). *The feeling of what happens: Body and emotion in the making of consciousness.* New York: Harcourt Brace.

Davis, M., Walker, D. L., & Myers, K. M., (2003). Role of the amygdala in fear extinction measured with potentiated startle. *Annals of the New York Academy of Sciences, 985,* 218–232.

Day, J. J., & Sweatt, J. D. (2011). Cognitive neuroepigenetics: A role for epigenetic mechanisms in learning and memory. *Neurobiology of Learning and Memory, 96,* 2–12.

Dethier, V. G. (1962). *To know a fly.* San Francisco: Holden-Day.

Duffy, V. B., Hayes, J. E., Davidson, A. C., Kidd, J. R., Kidd, K. K., & Bartoshuk, L. M. (2010). Vegetable intake in college-aged adults is explained by oral sensory phenotypes and TAS2R38 genotype. *Chemosensory Perception, 3,* 137–148.

Eisenberger, N. I., Jarcho, J. M., Lieberman, M. D., & Naliboff, B. D. (2006). An experimental study of shared sensitivity to physical pain and social rejection. *Pain, 126,* 132–138.

Eisenberger, N. I., Lieberman, M. D., & Williams, K. D. (2003). Does rejection hurt? An fMRI study of social exclusion. *Science, 302,* 290–292.

Everitt, B. J. (1990). Sexual motivation: A neural and behavioral analysis of the mechanisms underlying appetitive and copulatory responses of male rats. *Neuroscience and Biobehavioral Reviews, 14,* 217–232.

Field, T. M., Woodson, R., Greenberg, R., & Cohen, D. (1982), Discrimination and imitation of facial expression by neonates. *Science, 218,* 179–181.

Fung, T. T., van Dam, R. M., Hankinson S. E., Stampfer, M., Willett, W. C., & Hu, F. B. (2010). Low-carbohydrate diets and all-cause and cause-specific mortality: two cohort studies. *Annals of Internal Medicine, 153,* 289–298.

Garcia, J., & Koelling, R. A. (1966). Relation of cue to consequences in avoidance learning. *Psychonomic Science, 4,* 123–124.

Glickman, S. E., & Schiff, B. B. (1967). A biological theory of reinforcement. *Psychological Review, 74,* 81–109.

Gorski, R. A. (1984). Critical role for the medial preoptic area in the sexual differentiation of the brain. *Progress in Brain Research, 61,* 129–146.

Heron, W. (1957). The pathology of boredom. *Scientific American, 196*(1), 52–56.

Insel, T. R., & Fernald, R. D. (2004). How the brain processes social information: Searching for the social brain. *Annual Review of Neuroscience, 27,* 697–722.

Jacob, S., Kinnunen, L. H., Metz, J., Cooper, M., & McClintock, M. K. (2001). Sustained human chemosignal unconsciously alters brain function. *Neuroreport, 12,* 2391–2394.

Jacobsen, C. F. (1936). Studies of cerebral function in primates. *Comparative Psychology Monographs, 13,* 1–68.

Janes, A. C., Nickerson, L. D., Frederick, B. deB., & Kaufman, M. H. (2012). Prefrontal and limbic resting state brain network functional connectivity differs between biotine-dependent smokers and non-smoking controls. *Drug and Alcohol Dependence, 125,* 252–259.

Klüver, H., & Bucy, P. C. (1939). Preliminary analysis of the temporal lobes in monkeys. *Archives of Neurology and Psychiatry, 42,* 979–1000.

Kross, E., Berman, M. G., Mischel, W., Smith, E. E., & Water, T. D. (2011). Social rejection shares somatosensory representations with physical pain. *Proceedings of the National Academy of Sciences (USA), 108,* 6270–6275.

LeDoux, J. (1996). *The emotional brain.* New York: Simon & Schuster.

Lundstrom, J. N., Boyle, J. A., Zatorre, R. J., & Jones-Gotman, M. (2008) Functional neuronal processing of body odors differs from that of similar common odors. *Cerebral Cortex, 18,* 1466–1474.

MacLean, P. D. (1949). Psychosomatic disease and the "visceral brain": Recent developments bearing on the Papez theory of emotion. *Psychosomatic Medicine, 11,* 338–353.

McCarthy, M. M., Auger, A. P., Bale, T. L., De Vries, G. J., Dunn, G. A., et al. (2009). The epigenetics of sex differences in the brain. *Journal of Neuroscience, 29,* 12815–12823.

Money, J., & Ehrhardt, A. A. (1972). *Man and woman, boy and girl.* Baltimore: Johns Hopkins University Press.

Mozaffarinan, D., Tao Hao, M. P. H., Rimm, E. B., Willett, W. C., & Hu, F. B. (2011). Changes in diet and lifestyle and long-term weight gain in women and men. *New England Journal of Medicine, 3654,* 2392–2404.

Olds, J., & Milner, P. (1954). Positive reinforcement produced by electrical stimulation of septal area and other regions of rat brain. *Journal of Comparative and Physiological Psychology, 47,* 419–427.

Olsson, S. B., Barnard, J., & Turri, L. (2006) Olfaction and identification of unrelated individuals: Examination of the mysteries of human odor recognition. *Journal of Chemical Ecology, 32,* 1635–1645.

Panagiotakos, D. B., Pitsavos, C., Polychronopoulos, E., Chrysohoou, C., Zampelas, A., & Trichopoulou, A. (2004). Can a Mediterranean diet moderate the development and clinical progression of coronary heart disease? A systematic review. *Medical Science Monitor, 10,* RA193–RA198.

Robinson, T. E., & Berridge, K. C. (2008). Incentive sensitization theory. *Philosophical Transactions of the Royal Society of London; 363,* 3137–3146.

Schroeder, M., Krebs, M. O., Bleich, S., & Frieling, H. (2010). Epigenetics and depression. *Current Opinion in Psychiatry, 23,* 588–592.

Seeley, R. J., & Woods, S. C. (2003). Monitoring of stored and available fuel by the CNS: Implications for obesity. *Nature Reviews Neuroscience, 4,* 885–909.

Shai, I., Schwarzfuchs, D., Henkin, Y., Shahar, D. R., Witkow, S., et al. (2008). Weight loss with a low-carbohydrate, Mediterranean, or low-fat diet. *New England Journal of Medicine, 13,* 229–241.

Skinner, B. F. (1938). *The behavior of organisms.* New York: Appleton-Century-Crofts.

Smith, D. V., & Shepherd, G. M. (2003). Chemical senses: Taste and olfaction. In L. R. Squaire, F. E. Bloom, S. K. McConnell, J. L. Roberts, N. C. Spitzer, & M. J. Zigmond (Eds.), *Fundamental Neuroscience* (2nd ed., pp. 631–667). New York: Academic Press.

Stern, K., & McClintock, M. K. (1998). Regulation of ovulation by human phermones. *Nature, 392,* 177–179.

Veldhuizen, M. G., Albrecht, J., Zelano, C., Boesveldt, S., Breeslin, P., & Lundstrom, J. N. (2011). Identification of human gustatory cortex by activation likelihood estimation. *Human Brain Mapping, 32,* 2256–2266.

Woolley, C. S., Gould, E., Frankfurt, M., & McEwen, B. (1990). Naturally occurring fluctuation in dendritic spine density on adult hippocampal pyramidal neurons. *Journal of Neuroscience, 10,* 4035–4039.

World Health Organization. (2000). *Obesity: Preventing and managing the global epidemic.* WHO Technical Report Series, No. 894, 1–253.

Chapter 13

Antle, M. C., & Silver, R. (2009). Neural basis of timing and anticipatory behaviors. *European Journal of Neuroscience, 30,* 1643–1649.

Ariznavarreta, C., Cardinali, D. P., Villanúa, M. A., Granados, B., Martín, M., et al. (2002). Circadian rhythms in airline pilots submitted to long-haul transmeridian flights. *Aviation, Space and Environmental Medicine, 73,* 445–455.

Binkley, S. (1990). *The clockwork sparrow.* Englewood Cliffs, NJ: Prentice Hall.

Brankačk, J., Platt B., & Riedel, G. (2009). Sleep and hippocampus: Do we search for the right things? *Progress in Neuro-Psychopharmacology and Biological Psychiatry, 33,* 806–812.

Brown, S. A., Kunz, D., Dumas, A., Westermark, P. O., Vanselow, K., et al. (2008). Molecular insights into human daily behavior. *Proceedings of the National Academy of Sciences (USA), 105,* 1602–1907.

Buxton, O. M., Cain, S. W., O'Connor, S. P., Porter, J. H., Duffy, J. F., et al. (2012). Adverse metabolic consequences in humans of prolonged sleep restriction combined with circadian disruption. *Science and Translation Medicine, 12,* 129–143.

Cain, S. W., McDonald, R. J., & Ralph, M. R. (2008). Time stamp in conditioned place avoidance can be set to different circadian phases. *Neurobiology of Learning and Memory, 89,* 591–5944.

Cain, S. W., Verwey, M., Szybowska, M., Ralph, M. R., & Yeomans, J. S. (2007). Carbachol injections into the intergeniculate leaflet induce nonphotic phase shifts. *Brain Research, 1177,* 59–65.

Clark, E. L., Baumann, C. R., Cano, G., Scammell, T. E., & Mochizuki, T. (2009). Feeding-elicited cataplexy in orexin knockout mice. *Neuroscience, 161,* 970–977.

Delezie, J., & Challet, E. (2011). Interactions between metabolism and circadian clocks: Reciprocal disturbances. *Annals of the New York Academy of Sciences, 1243,* 30–46.

Dement, W. C. (1972). *Some must watch while some must sleep.* Stanford, CA: Stanford Alumni Association.

Dudai, Y. (2012). The restless engram: Consolidations never ends. *Annual Review of Neuroscience, 35,* 227–247.

Freud, S. (1990). *The interpretation of dreams.* Leipzig and Vienna: Franz Deuticke.

Gerrard, J. L., Burke, S. N., McNaughton, B. L., & Barnes, C. A. (2008). Sequence reactivation in the hippocampus is impaired in aged rats. *Journal of Neuroscience 30,* 7883–7890.

Hall, C. S., Domhoff, G. W., Blick, K. A., & Weesner, K. E. (1982). The dreams of college men and women in 1950 and 1980: A comparison of dream contents and sex differences. *Sleep, 5,* 188–194.

Hobson, J. A. (2002). Sleep and dream suppression following a lateral medullary infarct: A first-person account. *Consciousness and Cognition, 11,* 377–390.

Hobson, J. A. (2004). *13 dreams Freud never had: A new mind science.* New York: Pi Press.

Jones, B. E. (1993). The organization of central cholinergic systems and their functional importance in sleep-waking states. *Progress in Brain Research, 98,* 61–71.

Jouvet, M. (1972). The role of monoamines and acetylcholine-containing neurons in the regulation of the sleep–waking cycle. *Ergebnisse der Physiologie, 64,* 166–307.

Klinger, E. (1990). *Daydreaming.* Los Angeles, CA: Tarcher (Putnam).

Lamberg, L. (2003). Scientists never dreamed finding would shape half-century of sleep research. *Journal of the American Medical Association, 290,* 2652–2654.

Malcolm-Smith, S., Koopowitz, S., Pantelis, E., & Solms, M. (2012). Approach/avoidance in dreams. *Consciousness and Cognition, 21,* 408–412.

Maquet, P., Laureys, S., Peigneux, P., Fuchs, S., Petiau, C., et al. (2000). Experience-dependent changes in cerebral activation during human REM sleep. *Nature Neuroscience, 3,* 831–836.

Meesters, Y., Dekker, V., Schlangen, L.J., Bos, E. H., & Ruiter, M. J. (2011). Low-intensity blue-enriched white light (750 lux) and standard bright light (10,000 lux) are equally effective in treating SAD: A randomized controlled study. *BMC Psychiatry, 28,* 11–17.

Mistlberger, R. E., & Antle, M.C. (2011). Entrainment of circadian clocks in mammals by arousal and food. *Essays in Biochemistry, 49,* 119–136.

Monecke, S., Brewer, J. M., Krug, S., & Bittman, E. L. (2011). Duper: A mutation that shortens hamster circadian period. *Biological Rhythms, 26,* 283–292.

Moruzzi, G., & Magoun, H .W. (1949). Brain stem reticular formation and activation of the EEG. *Electroencephalography and Clinical Neurophysiology, 1,* 455–473.

Nelini, C., Bobbo, D., & Mascetti, G. G. (2012). Monocular learning of a spatial task enhances sleep in the right hemisphere of domestic chicks (*Gallus gallus*). *Experimental Brain Research, 18,* 381–388.

Osorio, I., & Daroff, R. B. (1980). Absence of REM and altered NREM sleep in patients with spinocerebellar degeneration and slow saccades. *Annals of Neurology, 7,* 277–280.

Pietroiusti, A., Neri, A., Somma, G., Coppeta, L., Iavicoli, I., et al. (2010). Incidence of metabolic syndrome among night-shift healthcare workers. *Occupation and Environment Medicine, 67,* 54–57.

Quinlan, J., & Quinlan, J. (1977). *Karen Ann: The Quinlans tell their story.* Toronto: Doubleday.

Ralph, M. R., & Lehman, M. N. (1991). Transplantation: A new tool in the analysis of the mammalian hypothalamic circadian pacemaker. *Trends in Neurosciences, 14,* 363–366.

Raven, P. H., Evert, R. F., & Eichorn, S. E. (1992). *Biology of plants.* New York: Worth Publishers.

Reiter, R. J. (1980). The pineal and its hormones in the control of reproduction in Mammals. *Endocrinology Review, 1,* 120.

Richter, C. P. (1965). *Biological clocks in medicine and psychiatry.* Springfield, IL: Charles C Thomas.

Roffward, H. P., Muzio, J., & Dement, W. C. (1966). Ontogenetic development of the human sleep–dream cycle. *Science, 152,* 604–619.

Schenck, C. H., Bundlie, S. R., Ettinger, M. G., & Mahowald, M. W. (1986). Chronic behavioral disorders of human REM sleep: A new category of parasomnia. *Sleep, 25,* 293–308.

Schwartz, W. J., Tavakoli-Nezhad, M., Lambert, C. M., Weaver, D. R., & de la Iglesia, H. O. (2011). Distinct patterns of Period gene expression in the suprachiasmatic nucleus underlie circadian clock photoentrainment by advances or delays. *Proceedings of the National Academy of Sciences (USA), 108,* 17219–17224.

Siegel, J. (2004). Brain mechanisms that control sleep and waking. *Naturwissenschaften, 91,* 355–365.

Silver, R., LeSauter, J., Tresco, P. A., & Lehman, M. N. (1996). A diffusible coupling signal from the transplanted suprachiasmatic nucleus controlling circadian locomotor rhythms. *Nature, 382,* 810–813.

Stoleru, D., Nawathean, P., Fernández, M. P., Menet, J. S., Ceriani, M. F., & Rosbash, M. (2007). The *Drosophila* circadian network is a seasonal timer. *Cell, 6(129),* 207–219.

Synofzik, M., Bauer, P., & Schöls, L. (2009). Prion mutation D178N with highly variable disease onset and phenotype. *Journal of Neurology, Neurosurgery and Psychiatry, 80,* 345–346.

Thomas, K., & Watson, C. B. (2008). Restless legs syndrome in women: A review. *Journal of Women's Health (Larchmt), 17,* 859–868.

Valli, K., & Revonsuo, A. (2009). The threat simulation theory in light of recent empirical evidence: A review. *American Journal of Psychology, 122,* 17–38.

Vanderwolf, C. H. (2002). *An odyssey through the brain, behavior and the mind.* Amsterdam: Kluwer Academic.

Wenner, J. B., Cheema, R., & Ayas, N. T. (2009). *Journal of Cardiopulmonay Rehabilitation and Prevention, 29,* 76–83.

Wilson, M. A., & McNaughton, B. L. (1994). Reactivation of hippocampal ensemble memories during sleep. *Science, 265,* 676–679.

Wilson, S., & Argyropoulos, S. (2005). Antidepressants and sleep: A qualitative review of the literature. *Drugs, 65,* 927–947.

Zaidi, F. H., Hull, J. T., Peirson, S. N., Wulff, K., Aeschbach, D., et al. (2007). Short-wavelength light sensitivity of circadian, pupillary, and visual awareness in humans lacking an outer retina. *Current Biology, 17,* 2122–2128.

Chapter 14

Cahill, L., Babinsky, R., Markowitsch, H. J., & McGaugh, J. L. (1995). The amygdala and emotional memory. *Nature, 377*, 295–296.

Candia, V., Wienbruch, C., Elbert, T., Rockstroh, B., & Ray, W. (2003). Effective behavioral treatment of focal hand dystonia in musicians alters somatosensory cortical organization. *Proceedings of the National Academy of Sciences (USA), 100*, 7942–7946.

Chang, F.-L. F., & Greenough, W. T. (1982). Lateralized effects of monocular training on dendritic branching in adult split-brain rats. *Brain Research, 232*, 283–292.

Corkin, S. (2002). What's new with the amnesic patient H. M.? *Nature Reviews Neuroscience, 3*, 153–160.

Corkin, S., Amaral, D. G., Gonzalez, R. G., Johnson, K. A., & Hyman, B. T. (1997). H. M.'s medial temporal lobe lesion: Findings from magnetic resonance imaging. *Journal of Neuroscience, 17*, 3964–3979.

Damasio, A. R., Tranel, D., & Damasio, H. (1989). Amnesia caused by herpes simplex encephalitis, infarctions in basal forebrain, Alzheimer's disease and anoxia/ischemia. In F. Boller & J. Grafman (Eds.), *Handbook of neuropsychology* (Vol. 3, pp. 149–166). New York: Elsevier.

Davis, M. (1992). The role of the amygdala in fear and anxiety. *Annual Review of Neuroscience, 15*, 353–375.

Eriksson, P. S., Perfilieva, E., Bjork-Eriksson, T., Alborn, A. M., Nordborg, C., Peterson, D. A., & Gage, F. H. (1998). Neurogenesis in the adult human hippocampus. *Nature Medicine, 4*, 1313–1317.

Fuster, J. M. (1995). *Memory in the cerebral cortex.* Cambridge, MA: MIT Press.

Fuster, J. M., Bodner, M., & Kroger, J. K. (2000). Cross-modal and cross-temporal association in neurons of frontal cortex. *Nature, 405*, 347–351.

Gazzaniga, M. S. (Ed.). (2000). *The new cognitive neurosciences.* Cambridge, MA: MIT Press.

Geshwind, N., & Galaburda, A. M. (1985). *Cerebral lateralization.* Cambridge, MA: MIT Press.

Gould, E., Tanapat, P., Hastings, N. B., & Shors, T. J. (1999). Neurogenesis in adulthood: A possible role in learning. *Trends in Cognitive Sciences, 3*, 186–191.

Gould, E., Tanapat, P., McEwen, B. S., Flugge, G., & Fuchs, E. (1998). Proliferation of granule cell precursors in the dentate gyrus of adult monkeys is diminished by stress. *Proceedings of the National Academy of Sciences (USA), 95*, 3168–3171.

Hampson, E., & Kimura, D. (1988). Reciprocal effects of hormonal fluctuations on human motor and perceptual-spatial skills. *Behavioral Neuroscience, 102*, 456–459.

Hebb, D. O. (1947). The effects of early experience on problem solving at maturity. *American Psychologist, 2*, 737–745.

Hebb, D. O. (1949). *The organization of behavior: A neuropsychological theory.* New York: Wiley.

Jacobs, B., Schall, M., & Scheibel, A. B. (1993). A quantitative dendritic analysis of Wernicke's area in humans: II. Gender, hemispheric, and environmental factors. *Journal of Comparative Neurololgy, 327*, 97–111.

Jacobs, B., & Scheibel, A. B. (1993). A quantitative dendritic analysis of Wernicke's area in humans: I. Lifespan changes. *Journal of Comparative Neurololgy, 327*, 83–96.

Kaas, J. (2000). The reorganization of sensory and motor maps after injury in adult mammals. In M. S. Gazzaniga (Ed.), *The cognitive neurosciences* (pp. 223–236). Cambridge, MA: MIT Press.

Kempermann, G., Kuhn, H. G., & Gage, F. H. (1998). Experience-induced neurogenesis in the senescent dentate gyrus. *Journal of Neuroscience, 18*, 3206–3212.

Kolb, B. (1999). Towards an ecology of cortical organization: Experience and the changing brain. In J. Grafman & Y. Christen (Eds.), *Neuropsychology: From lab to the clinic* (pp. 17–34). Paris: Springer Verlag.

Kolb, B., Cote, S., Ribeiro-da-Silva, A., & Cuello, A. C. (1997). "Nerve growth factor treatment prevents dendritic atrophy and promotes recovery of function after cortical injury." *Neuroscience, 76*, 1146.

Kolb, B., & Gibb, R. (2008). Principles of brain plasticity and behavior. In D. Stuss, I. Robertson, & G. Winocur (Eds.), *Brain plasticity and rehabilitation.* New York: Oxford University Press.

Kolb, B., Gibb, R., & Gorny, G. (2003). Experience-dependent changes in dendritic arbor and spine density in neocortex vary with age and sex. *Neurobiology of Learning and Memory, 79*, 1–10.

Kolb, B., Gorny, G., Cote, S., Ribeiro-da-Silva, & Cuello, A. C. (1997). Nerve growth factor stimulates growth of cortical pyramidal neurons in young adult rats. *Brain Research, 751*, 289–294.

Kolb, B., Gorny, G., Li, Y., Samaha, A. N., & Robinson, T. E. (2003). Amphetamine or cocaine limits the ability of later experience to promote structural plasticity in the neocortex and nucleus accumbens. *Proceedings of the National Academy of Science (USA), 100*, 10523–10528.

Kolb, B., Morshead, C., Gonzalez, C., Kim, N., Shingo, T., & Weiss, S. (2007). Growth factor-stimulated generation of new cortical tissue and functional recovery after stroke damage to the motor cortex of rats. *Journal of Cerebral Blood Flow and Metabolism, 27*, 983–397.

Kolb, B., & Whishaw, I. Q. (1998). Brain plasticity and behavior. *Annual Review of Psychology, 49*, 43–64.

Lashley, K. S. (1960). In search of the engram. Symposium No. 4 of the Society of Experimental Biology. In F. A. Beach, D. O. Hebb, C. T. Morgan, & H. T. Nissen (Eds.), *The neuropsychology of Lashley* (pp. 478–505). New York: McGraw-Hill. (Reprinted from R. Sutton (Ed.), *Physiological mechanisms of animal behavior*, pp. 454–482, 1951. Cambridge: Cambridge University Press.)

LeDoux, J. E. (1995). In search of an emotional system in the brain: Leaping from fear to emotion and consciousness. In M. S. Gazzaniga (Ed.), *The cognitive neurosciences* (pp. 1047–1061). Cambridge, MA: MIT Press.

Lepage, M., McIntosh, A. R., & Tulving, E. (2001). Transperceptual encoding and retrieval process in memory: A PET study of visual and haptic objects. *Neuroimage, 14*, 572–584.

Loftus, E. F. (1997). Creating false memories. *Scientific American, 277*(3), 70–75.

Maguire, E. A., Gadian, D. G., Johnsrude, I. S., Good, C. D., Ashburner, J., Frackowiak, R. S., & Frith, C. D. (2000). Navigation-related structural change in the hippocampi of taxi drivers. *Proceedings of the National Academy of Sciences (USA), 97*, 4398–4403.

Mahncke, H. W., Bronstone, A., & Merzenich, M. M. (2006). Brain plasticity and functional losses in the aged: scientific bases for a novel intervention. *Progress in Brain Research, 157*, 81–109.

Markowitsch, H. J. (1999). Functional neuroimaging correlates of functional amnesia. *Memory 7*, Plate 2.

Markowitsch, H. J. (2003). Psychogenic amnesia. *Neuroimage, 20*(Suppl. 1), S132–S138.

Martin, A., Haxby, J. V., Lalonde, F. M., Wiggs, C. L., & Ungerleider, L. G. (1995). Discrete cortical regions associated with knowledge of color and knowledge of action. *Science, 270*, 102–105.

Martin, P. R. (1985). Cover, *Alcohol Health & Research World, 9* (Spring 1985).

McGaugh, J. (2004). The amygdala modulates the consolidation of memories of emotionally arousing experiences. *Annual Review of Neuroscience, 27*, 1–28.

Milner, B., Corkin, S., & Teuber, H. (1968). Further analysis of the hippocampal amnesic syndrome: 14 year follow-up study of HM. *Neuropsychologia, 6*, 215–234.

Mishkin, M. (1982). A memory system in the brain. *Philosophical Transactions of the Royal Society of London, Biological Sciences, 298*, 83–95.

Mishkin, M., Suzuki, W. A., Gadian, D. G., & Vargha-Khadem, F. (1997). Hierarchical organization of cognitive memory. *Philosophical Transactions of the Royal Society of London, Biological Sciences, 352*, 1461–1467.

Murray, E. (2000). Memory for objects in nonhuman primates. In M. S. Gazzaniga (Ed.), *The new cognitive neurosciences* (pp. 753–763). Cambridge, MA: MIT Press.

Nudo, R. J., Plautz, E. J., & Milliken, G. W. (1997). Adaptive plasticity in primate motor cortex as a consequence of behavioral experience and neuronal injury. *Seminars in Neuroscience, 9*, 13–23.

O'Keefe, J., & Nadel, L. (1978). *The hippocampus as a spatial map.* New York: Oxford University Press.

Purves, D., & Voyvodic, J. T. (1987). Imaging mammalian nerve cells and their connections over time in living animals. *Trends in Neurosciences, 10*, 398–404.

Ragert, P., Schmidt, A., Altenmuller, E., & Dinse, H. R. (2003). Superior tactile performance and learning in professional pianists: Evidence for meta-plasticity in musicians. *European Journal of Neuroscience, 19*, 473–478.

Ramachandran, V. S. (1993). Behavioral and magnetoencephalographic correlates of plasticity in the adult human brain. *Proceedings of the National Academy of Sciences (USA), 90*, 10413–10420.

Ramón y Cajal, S. (1928). *Degeneration and regeneration of the nervous system.* London: Oxford University Press.

Reynolds, B., & Weiss, S. (1992). Generation of neurons and astrocytes from isolated cells of the adult mammalian central nervous system. *Science, 255*, 1707–1710.

Robinson, T. E., & Kolb, B. (1997). Persistent structural adaptations in nucleus accumbens and prefrontal cortex neurons produced by prior experience with amphetamine. *Journal of Neuroscience, 17*, 8491–8498.

Robinson, T. E., & Kolb, B. (2004). Structural plasticity associated with drugs of abuse. *Neuropharmacology, 47*(Suppl. 1), 33–46.

Sainsbury, R. S., & Coristine, M. (1986). Affective discrimination in moderately to severely demented patients. *Canadian Journal on Aging, 5*, 99–104.

Sapolsky, R. M. (1992). *Stress, the aging brain, and the mechanisms of neuron death.* Cambridge, MA: MIT Press.

Schacter, D. L. (1983). Amnesia observed: Remembering and forgetting in a natural environment. *Journal of Abnormal Psychology, 92,* 236–242.

Sherry, D. F., Jacobs, L. F., & Gaulin, S. J. C. (1992). Spatial memory and adaptive specialization of the hippocampus. *Trends in Neuroscience, 15,* 298–303.

Sirevaag, A. M., & Greenough, W. T. (1987). Differential rearing effects on rat visual cortex synapses: III. Neuronal and glial nuclei, boutons, dendrites, and capillaries. *Brain Research, 424,* 320–332.

Sirevaag, A. M., & Greenough, W. T. (1988). A multivariate statistical summary of synaptic plasticity measures in rats exposed to complex, social and individual environments. *Brain Research, 441,* 386–392.

Skinner, B. F. (1938). *The behavior of organisms.* New York: Appleton-Century-Crofts.

Temple, E., Deutsch, G. K., Poldrack, R. A., Miller, S. L., Tallal, P., Merzenich, M. M., & Gabrieli, J. D. E. (2003). Neural deficits in children with dyslexia ameliorated by behavioral remediation: Evidence from functional MRI. *Proceedings of the National Academy of Sciences (USA), 100,* 2860–2865.

Thorndike, E. L. (1898). Animal intelligence: An experimental study of the associative processes in animals. *Psychological Review Monograph Supplements, 2,* 1–109.

Tulving, E. (2002). Episodic memory: From mind to brain. *Annual Review of Psychology, 53,* 1–25.

Turner, A., & Greenough, W. T. (1985). Differential rearing effects on rat visual cortex synapses: I. Synaptic and neuronal density and synapses per neuron. *Brain Research, 329,* 195–203.

Wise, R. A. (2004). Dopamine, learning and motivation. *Nature Neuroscience Reviews, 5,* 483–494.

Woolley, C. S., Gould, E., Frankfurt, M., & McEwen, B. S. (1990). Naturally occurring fluctuation in dendritic spine density on adult hippocampal pyramidal neurons. *Journal of Neuroscience, 10,* 4035–4039.

Chapter 15

Adolphs, R., Gjosselin, F., Buchanan, T. W., Tranel, D., Schyns, P., & Damasio, A. R. (2005). A mechanism for impaired fear recognition after amygdala damage. *Nature, 433,* 68–72.

Boria, S., Fabbri-Destro, M., Cattaneo, L., Sparaci, L., Sinigaglia, C., et al. (2009). Intention understanding in autism. *PLoS ONE, 4,* e5596.

Calvin, W. H. (1996). *How brains think.* New York: Basic Books.

Carr, L., Iacoboni, M., Dubeau, M. C., Mazziotta, J. C. & Lenzi, G. L. (2003). Neural mechanisms of empathy in humans: a relay from neural systems for imitation to limbic areas. *Proceedings of the National Academy of Sciences (USA), 100,* 5497–5502.

Castiello, U., Paulignan, Y., & Jeannerod, M. (1991). Temporal dissociation of motor responses and subjective awareness. *Brain, 114,* 2639–2655.

Chalmers, D. J. (1995). *The conscious mind: In search of a fundamental theory.* Oxford: Oxford University Press.

Crick, F., & Koch, C. (1998). Consciousness and neuroscience. *Cerebral Cortex, 8,* 97–107.

Cytowic, R. E. (1998). *The man who tasted shapes.* Cambridge, MA: MIT Press.

Diamond, M. C., Scheibel, A. B., Murphy, G. M., Jr., & Harvey, T. (1985). On the brain of a scientist: Albert Einstein. *Experimental Neurology, 88,* 198–204.

Duncan, J., Seitz, R. J., Kolodny, J., Bor, D., Herzog, H., Ahmed, A., Newell, F. N., & Emslie, H. (2000). A neural basis for general intelligence. *Science, 289,* 457–459.

Gardner, H. (1983). *Frames of mind.* New York: Basic Books.

Gaulin, S. J. (1992). Evolution of sex differences in spatial ability. *Yearbook of Physical Anthropology, 35,* 125–131.

Gazzaniga, M. S. (1970). *The bisected brain.* New York: Appleton-Century-Crofts.

Gazzaniga, M. S. (1992). *Nature's mind.* New York: Basic Books.

Gazzaniga, M. S., Ivry, R. B., & Mangun, G. R. (1999). *Cognitive science: The biology of the mind.* New York: Norton.

Ghanzanfar, A. A., & Schroeder, C. E. (2006). Is neocortex essentially multisensory? *Trends in Cognitive Science, 10,* 278–285.

Goldstein, J. M., Seidman, J. L., Horton, N. J., Makris, N., Kennedy, D. N., et al. (2001). Normal sexual dimorphism of the adult human brain assessed by in vivo magnetic resonance imaging. *Cerebral Cortex, 11,* 490–497.

Gray, J. R., & Thompson, P. M. (2004). Neurobiology of intelligence: Science and ethics. *Nature Reviews Neuroscience, 5,* 471–482.

Guilford, J. P. (1967). *The nature of human intelligence.* New York: McGraw-Hill.

Hebb, D. O. (1980). *Essay on mind.* Hillsdale, NJ: Lawrence Erlbaum.

Hubbard, E. W. (2007). Neurophysiology of synesthesia. *Current Psychiatry Reports, 9,* 193–199.

Iacoboni, M., & Dapretto, M. (2006). The mirror neuron sytem and the consequences of its dysfunction. *Nature Neuroscience Reviews, 7,* 942–951.

James, W. (1890). *Principles of psychology.* New York: Henry Holt.

Kelly, C., Biswal, B. B., Craddock, R. C., Castellanos, F. X., & Milham, M. P. (2012). Characterizing variation in the functional connectome: Promise and pitfalls. *Trends in Cognitive Science, 16,* 181–188.

Kimura, D. (1967). Functional asymmetry of the brain in dichotic listening. *Cortex, 3,* 163–178.

Kimura, D. (1973). The asymmetry of the human brain. *Scientific American, 228*(3), 70–78.

Kimura, D. (1999). *Sex and cognition.* Cambridge, MA: MIT Press.

Kolb, B., & Stewart, J. (1991). Sex-related differences in dendritic branching of cells in the prefrontal cortex of rats. *Journal of Neuroendocrinology, 3,* 95–99.

Langer, N., Pedroni, A., Gianotti, L. R., Hanggi, J., Knoch D., & Jancke, L. (2009). Functional brain network efficiency predicts intelligence. *Human Brain Mapping, 4,* 299–307.

Lieberman, M. D. (2007). Social cognitive neuroscience: A review of core processes. *Annual Review of Psychology, 58,* 259–289.

Lieberman, M. D., Berkman, E. T., & Wager, T. D. (2012). Correlations in social neuroscience aren't voodoo. *Perspectives on Psychological Science, 33,* 1393–1406.

Loui, P., Li, H. C. Hohmann, A., & Schlaug, G. (2011). Enhanced cortical connectivity in absolute pitch musicians: A model for local hyperconnectivity. *Journal of Cognitive Neuroscience, 23,* 1015–1026.

McClure, S. M., Li, J., Tomlin, D., Cypert, K.S., Montague, L.M., & Montague, P. R. (2004). Neural correlates of behavioral preferences for culturally familiar drinks. *Neuron, 44,* 379–387.

Mlodinow, L. (2009). *The drunkard's walk: How randomness rules our lives.* New York: Vintage Books.

Moran, J., & Desimone, R. (1985). Selective attention gates visual processing in the extrastriate cortex. *Science, 229,* 782–784.

Mukerjee, M. (1996). Interview with a parrot [field note]. *Scientific American, 274*(4), 24.

Newsome, W. T., Shadlen, M. N., Zohary, E., Britten, K. H., & Movshon, J. A. (1995). Visual motion: Linking neuronal activity to psychophysical performance. In M. Gazzaniga (Ed.), *The cognitive neurosciences* (pp. 401–414). Cambridge, MA: MIT Press.

Pepperberg, I. M. (1990). Some cognitive capacities of an African grey parrot (*Psittacus erithacus*). In P. J. B. Slater, J. S. Rosenblatt, & C. Beer (Eds.), *Advances in the study of behavior* (Vol. 19, pp. 357–409). New York: Academic Press.

Pepperberg, I. M. (1999). *The Alex studies.* Cambridge, MA: Harvard University Press.

Pepperberg, I. M. (2006). Ordinality and inferential ability of a grey parrot (*Psittacus erithacus*). *Journal of Comparative Psychology, 120,* 205–216.

Ramachandran, V. S., & Oberman, L. M. (2006). Broken mirrors: A theory of autism. *Scientific American, 295,* 62–69.

Rameson, L. T., Morellis, S. A., & Lieberman, M. D. (2012). The neural correlates of empathy: Experience, automaticity, and prosocial behavior. *Journal of Cognitive Neuroscience, 24,* 235–245.

Rasmussen, T., & Milner, B. (1977). The role of early left brain injury in determining lateralization of cerebral speech functions. *Annals of the New York Academy of Sciences, 299,* 355–369.

Rizzolatti, G. (2007). *Mirrors on the mind.* New York: Oxford University Press.

Rizzolatti, G., & Craighero, L. (2004). The mirror-neuron system. *Annual Review of Neuroscience, 27,* 169–192.

Rizzolatti, G., & Fabbri-Destro, M. (2010). Mirror neurons: From discovery to autism. *Experimental Brain Research, 200,* 223–237.

Sacks, O. (1989). *Seeing voices.* Los Angeles: University of California Press.

Sawamoto, N., Honda., Okada, T., Hanakawa, T., Kanda, M., et al. (2000). Expectation of pain enhances responses to nonpainful somatosensory stimulation in the anterior cingulate cortex and parietal operculum/posterior insula: An event-related functional magnetic resonance imaging study. *Journal of Neuroscience, 20,* 7438–7445.

Sperry, R. (1968). Mental unity following surgical disconnection of the cerebral hemispheres. *Harvey Lectures, 62,* 293–323.

Stein, B. E., & Rowland, B. A. (2011). Organization and plasticity in multisensory integration: Early and late experience affects its governing principles. *Progress in Brain Research, 191,* 145–163.

Stewart, J., & Kolb, B. (1994). Dendritic branching in cortical pyramidal cells in response to ovariectomy in adult female rats: Suppression by neonatal exposure to testosterone. *Brain Research, 654,* 149–154.

Takeuchi, H., Sugiura, M., Sassa, Y., Sekiguchi, A., Yomogida, Y., et al. (2012). Neural correlates of the differences between working memory speed and simple sensorimotor speed: An fMRI study. *PLoS One, 7,* e30579.

Vul, E., Harris, C., Winkielman, P., & Pashler, H. (2009) Puzzlingly high correlations in fMRI studies of emotion, personality, and social cognition. *Perspectives on Psychological Science, 4,* 274–290.

Wada, J., & Rasmussen, T. (1960). Intracarotid injection of sodium amytal for the lateralization of cerebral speech dominance: Experimental and clinical observations. *Journal of Neurosurgery, 17,* 266–282.

Witelson, S. F., & Goldsmith, C. H. (1991). The relationship of hand preference to anatomy of the corpus callosum in men. *Brain Research, 545,* 175–182.

Witelson, S. F., Kigar, D. L., & Harvey, T. (1999). The exceptional brain of Albert Einstein. *Lancet, 353,* 2149–2153.

Yeo, B. T. T., Fienen, F. M., Sepulcre, J., Sabuncu, M. R., Lashkari, D., et al. (2011). The organization of the human cerebral cortex estimated by intrinsic functional connectivity. *Journal of Neurophysiology, 106,* 1125–1165.

Chapter 16

Abramowitz, J. S. (1998). Does cognitive-behavior therapy cure obsessive-compulsive disorder? A meta-analytic evaluation of clinical significance. *Behavior Therapy, 29,* 339–355.

American Psychiatric Association. (1994). *Diagnostic and statistical manual of mental disorders* (4th ed.). Washington, DC: American Psychiatric Association.

Arnold, S. E., Rushinsky, D. D., & Han, L. Y. (1997). Further evidence of abnormal cytoarchitecture in the entorhinal cortex in schizophrenia using spatial point analyses. *Biological Psychiatry, 142,* 639–647.

Bersani, F. S., Minichino, A., Enticott, P. G., Mazzarini, L., Khan, N., et al. (2012, May 3). Deep transcranial magnetic stimulation as a treatment for psychiatric disorders: A comprehensive review. *European Journal of Psychiatry.* [Epub ahead of print]

Calne, D. B., & Mizuno, Y. (2004). The neuromythology of Parkinson's disease. *Parkinsonism and Related Disorders, 10,* 319–322.

Caplan, A. L. (2003). Is better best? *Scientific American, 289*(3), 104–105.

Charney, D. S., Nestler, E. J., & Bunney, B. S. (Eds.). (1999). *The neurobiology of mental illness.* New York: Oxford University Press.

Cho, R. Y., Gilbert, H., & Lewis, D. A. (2004). The neurobiology of schizophrenia. In D. S. Charney & F. J. Nestler (Eds.), *The neurobiology of mental illness* (2nd ed., pp. 299–310). New York: Oxford University Press.

Ciccarelli, P. A., Brex, A. J., Thompson, A. J., & Miller, D. H. (2000). Disability and lesion load in MS: A reassessment with MS functional composite score and 3D fast flair. *Journal of Neurology, 249,* 18–24.

Claes, S., Tang, Y. L., Gillespie, C. F., & Cubells, J. F. (2012). Human genetics of schizophrenia. *Handbook of Clinical Neurology, 106,* 37–52.

Connolly, R. K., Helmer, A., Cristancho, M. A., Cristancho, P., & O'Reardon, J. P. (2012). Effectiveness of transcranial magnetic stimulation in clinical practice post-FDA approval in the United States: Results observed with the first 100 consecutive cases of depression at an academic medical center. *Journal of Clinical Psychiatry, 74,* 567–573.

Corrigan, N. M., Richards, T. L., Treffert, D. A., & Dager, S. R. (2012). Toward a better understanding of the savant brain. *Comprehensive Psychiatry, 53,* 706–717.

Crow, T. J. (1980). Molecular pathology of schizophrenia: More than one disease process? *British Medical Journal, 280,* 66–68.

Crow, T. J. (1990). Nature of the genetic contribution to psychotic illness: A continuum viewpoint. *Acta Psychiatrica Scandinavia, 81,* 401–408.

deCharms, R. C. (2008). Applications of real-time fMRI. *Nature Reviews Neuroscience 9,* 720–729.

Drevets, W. C., Kishore, M. G., & Krishman, K. R. R. (2004). Neuroimaging studies of mood disorders. In D. S. Charney & E. J. Nestler (Eds.), *The neurobiology of mental illness* (2nd ed., pp. 461–490). New York: Oxford University Press.

Duman, R. S. (2004). The neurochemistry of depressive disorders. In D. S. Charney & E. J. Nestler (Eds.), *The neurobiology of mental illness* (2nd ed., pp. 421–439). New York: Oxford University Press.

Fragoso, Y. D. (2011). The internet racing ahead of the scientific evidence: The case of "liberation treatment" for multiple sclerosis. *Arquives Neuropsiquiatria, 69,* 525–527.

Gavett, B. E., Stern, R. A., Cantu, R.C., Nowinski, C. J., & McKee, A. C. (2010). Mild traumatic brain injury: A risk factor for neurodegeneration. *Alzheimer's Research and Therapy, 2,* 18.

Gebicke-Haerter, P. J. (2012). Epigenetics of schizophrenia. *Pharmacopsychiatry, 45,* Suppl. 1, S42–S48.

Geuze, E., van Wingen, G. A., van Zuiden, M., Rademaker, A. R., Vermetten, E., et al. (2012). Glucocorticoid receptor number predicts increase in amygdala activity after severe stress. *Psychoneuroendocrinology, 375,* 1837–1844.

Geuze, E., Westenber, H. G. M., Heinecke, A., de Kloet, C. S., Goebel, & Vermetten, E. (2008). Thinner prefrontal cortex in veterans with posttraumatic stress disorder. *NeuroImage, 41,* 675–681.

Gogtay, N., Lu, A., Leow, A. D., Klunder, A. D., Lee, A. D., Chavez, A., Greenstein, D., Giedd, J. N., Toga, A. W., Rapoport, J. L., & Thompson, P. M. (2008). Three-dimensional brain growth abnormalities in childhood-onset schizophrenia visualized by using tensor-based morphometry. *Proceedings of the National Academy of Sciences (USA), 105,* 15979–15984.

Giles J. (2004). Neuroscience: Change of mind. *Nature, 430,* 14.

Gross, C., & Hen, R. (2004). The developmental origins of anxiety. *Nature Reviews Neuroscience, 5,* 545–552.

Gould, F., Clarke, J., Heim, C., Harvey, P. D., Majer, M., & Nemeroff, C. B. (2012). The effects of child abuse and neglect on cognitive functioning in adulthood. *Journal of Psychiatric Research, 46,* 500–506.

Heinz, A., Kipke, R., Heimann, H., & Wiesing, U. (2012). Cognitive neuroenhancement: False assumptions in the ethical debate. *Journal of Medical Ethics, 38,* 372–375.

Heninger, G. R. (1999). Special challenges in the investigation of the neurobiology of mental illness. In D. S. Charney, E. J. Nestler, & B. S. Bunney (Eds.), *The neurobiology of mental illness* (pp. 89–98). New York: Oxford University Press.

Hunter, J. V., Wilde, E. A., Tong, K. A., & Holshouser, B. A. (2012). Emerging imaging tools for use with traumatic brain injury research. *Journal of Neurotrauma, 29,* 654–671.

Hyman, S. E. (2003). Diagnosing disorders. *Scientific American, 289*(3), 97–103.

Kaufer, D. I., & DeKosky, S. T. (1999). Diagnostic classifications: Relationship to the neurobiology of dementia. In D. S. Charney, E. J. Nestler, & B. S. Bunney (Eds.), *The neurobiology of mental illness* (p. 642). New York: Oxford University Press.

Kennedy, S. H., Giacobbe, P., Rizvi, S. J., Placenza, F. M., Nishikawa, Y., et al. (2011). Deep brain stimulation for treatment-resistant depression: follow-up after 3 to 6 years. *American Journal of Psychiatry, 168,* 502–510.

Kobayashi, K., Ikeda, Y., Sakai, A., Yamasaki, N., Haneda, E., et al. (2010). Reversal of hippocampal neuronal maturation by serotonergic antidepressants. *Proceedings of the National Academy of Sciences (USA), 107*(18), 8434–8439.

Kondziolka, D., Steinberg, G. K., Wechsler, L., Meltzer, C. C., Elder, E., et al. (2005). Neurotransplantation for patients with subcortical motor stroke: a phase 2 randomized trial. *Journal of Neurosurgery, 103,* 6–8.

Kondziolka, D., Wechsler, L., Goldstein, S., Meltzer, C., Thulborn, K. R., et al. (2000). Transplantation of cultured human neuronal cells for patients with stroke. *Neurology, 55,* 565–569.

Kong, L., Herold, C., Stieltjes, B., Essig, M., Seidl, U., et al. (2012). Reduced gray to white matter tissue intensity contrast in schizophrenia. *PLoS One, 7,* e37016.

Kuo, H.-K., Jones, R.N., Milberg, W.P., Tennstedt, S., Talbot, L., et al. (2005). Effect of blood pressure and diabetes mellitus on cognitive and physical functions in older adults: A longitudinal analysis of the advanced cognitive training for independent and vital elderly cohort. *Journal of the American Geriatrics Society, 53,* 1154–1161.

Larrabee, G. J., & Crook, T. H. (1994). Estimated prevalence of age-related memory impairment derived from standardized tests of memory function. International *Psychogeriatrics, 6,* 95–104.

Leritz, E. C., Salat, D. H., Williams, V. J., Schnyer, D. M., Rudolph, J. L., et al. (2011). Thickness of the human cerebral cortex is associated with metrics of cerebrovascular health in a normative sample of community dwelling older adults. *Neuroimage, 54,* 2659–2671.

Levin, H. S., Benton, A. L., & Grossman, R. G. (1982). *Neurobehavioral consequences of closed head injury.* New York: Oxford University Press.

Malberg, J. E., Eisch, A. J., Nestler, E. J., & Duman, R. S. (2000). Chronic antidepressant treatment increases neurogenesis in adult rat hippocampus. *Journal of Neuroscience, 20,* 9104–9110.

Martin, J. P. (1967). *The basal ganglia and posture.* London: Ritman Medical Publishing.

Mateer, C. A., & Sira, C. S. (2006). Cognitive and emotional consequences of TBI: Intervention strategies for vocational rehabilitation. *NeuroRehabilitation, 21,* 315–326

McEvoy, J. P. (2007). The costs of schizophrenia. *Journal of Clinical Psychiatry, 68,* Suppl. 14, 4–7.

McGowan, P. O., Sasaki, A., D'Alessio, A. C., Dymov, S., Labonté, B., et al. (2009). Epigenetic regulation of the glucocorticoid receptor in human brain associates with childhood abuse. *Nature Neuroscience, 12,* 342–348.

McGowan, P. O., & Szyf, M. (2010). Environmental epigenomics: Understanding the effects of parental care on the epigenome. *Essays in Biochemistry, 48,* 275–287.

Miller, B. L., Boone, K., Cummings, J. L., Read, S. L., & Mishkin, F. (2000). Functional correlates of musical and visual ability in frontotemporal dementia. *British Journal of Psychiatry, 176,* 458–463.

Murphy, M. P., & Carmine, H. (2012). Long-term health implications of individuals with TBI: A rehabilitation perspective. *NeuroRehabilitation, 31,* 1–12.

National Institute of Mental Health Web site: http://www.nimh.nih.gov/health/publications/the-numbers-count-mental-disorders-in-america/index.shtml

Niino, M. (2010). Vitamin D and its immunoregulatory role in multiple sclerosis. *Drugs Today, 46,* 279–290.

Opriş, D., Pintea, S., García-Palacios, A., Botella, C., Szamosközi, S., & David, D. (2012). Virtual reality exposure therapy in anxiety disorders: A quantitative meta-analysis. *Depression and Anxiety, 29,* 85–93.

Pannese, R., Minichino, A., Pignatelli, M., Delle Chiaie, R., Biondi, M., & Nicoletti, F. (2012). Evidences on the key role of the metabotrobic glutamatergic receptors in the pathogenesis of schizophrenia: A "breakthrough" in pharmacological treatment. *Reviews in Psychiatry, 47,* 149–169.

Politis, M., & Lindvall, O. (2012). Clinical application of stem cell therapy in Parkinson's disease. *BMC Medicine, 10,* 1.

Rajji, T. K., Liu, S. K., Frantseva, M. V., Mulsant, B. H., Thoma, J., et al. (2011). Exploring the effect of inducing long-term potentiation in the human motor cortex on motor learning. *Brain Stimulation, 4,* 137–144.

Rapoport, J. L., Giedd, J. N., & Gogtay, N. (2012). Neurodevelopmental model of schizophrenia: Update 2012. *Molecular Psychiatry.* [Epub ahead of print]

Reger, G. M., Holloway, K. M., Candy, C., Rothbaum, B. O., Difede, J., Rizzo, A. A., & Gahm, G. A. (2011). Effectiveness of virtual reality exposure therapy for active duty soldiers in a military mental health clinic. *Journal of Trauma and Stress, 24,* 93–96.

Rizzo, A., Buckwalter, J. G., John, B., Newman, B., Parsons, T., Kenny, P., & Williams, J. (2012). STRIVE: Stress Resilience In Virtual Environments: A pre-deployment VR system for training emotional coping skills and assessing chronic and acute stress responses. *Studies in Health and Technology Information, 173,* 379–385.

Rogawski, M. A., & Loscher, W. (2004). The neurobiology of antiepileptic drugs. *Nature Reviews Neuroscience, 5,* 553–564.

Sacks, O. (1998). *The man who mistook his wife for a hat: And other clinical tales.* New York: Touchstone.

Santarelli, L., Saxe, M., Gross, C., Surget, A., Battaglia, F., et al. (2003). Requirement of hippocampal neurogenesis for the behavioral effects of antidepressants. *Science, 301,* 805–809.

Sohlberg, M. M., & Mateer, C. (1989). *Introduction to cognitive rehabilitation.* New York: Guilford Press.

Steinman, L., Martin, R., Bernard, C., Conlon, P., & Oksenberg, J. R. (2002). Multiple sclerosis: Deeper understanding of its pathogenesis reveals new targets for therapy. *Annual Review of Neuroscience, 25,* 491–505.

Tsai, G., & Coyle, J. T. (1995). N-Acetylaspartate in neuropsychiatric disorders. *Progress in Neurobiology, 46,* 531–540.

Wang, J., Zhao, X., & He, M. (2012). Is BDNF a biological link between depression and type 2 diabetes mellitus? *Medical Hypotheses, 79,* 255–258.

Wassermann, E. M., & Zimmermann, T. (2012). Transcranial magnetic brain stimulation: Therapeutic promises and scientific gaps. *Pharmacology and Therapeutics, 133*(1), 9.

A

Abramowitz, J. S., 602
Ackerly, Stafford, 421
Adar, Einat, 77
Adelmann, Pamela, 425
Adolphs, Ralph, 535
Agate, Robert, 275
Alessandria, M., 385
Allen, M. J., 137
Alzheimer, Alois, 494
Ames, Elenor, 273
Anda, Robert, 259
Antle, M. C., 453, 454, 458
Argyropoulos, S., 469
Aristotle, 6, 7
Ariznavarreta, C., 451
Arnold, A. P., 349
Arnold, S. E., 598
Asanuma, Hiroshi, 391
Asarnow, J. R., 188
Aschoff, Jurgen, 447, 449
Aserinsky, Eugene, 460
Asperger, Hans, 256
Auburtin, Ernest, 213
Audero, Enrica, 246
Axel, Richard, 403

B

Bachevalier, Jocelyne, 265, 274
Bailey, Craig, 167
Bainbridge, M. N., 74
Balaban, E., 103
Balint, R., 316
Ballard, A., 162
Bano, S., 110
Bao, A.-M., 436, 437
Barbeau, A., 154
Bartholow, Roberts, 113
Bartoshuk, Linda, 404
Becker, J. B., 196, 436
Becker, Jill, 179
Beery, Alexis, 74
Beery, Joe, 74
Beery, Noah, 74
Beery, Retta, 74
Bell, Charles, 63
Berger, Hans, 213, 224
Berkman, E. T., 542
Bermudez, P., 346
Berridge, Kent, 196, 439
Bersani, F. S., 577
Berthold, A. A., 202, 205
Binkley, S., 448
Bird, Christopher, 30
Birkmayer, W., 154
Blakemore, Sarah, 392
Bliss, Timothy, 502, 504
Bodner, M., 496
Boehner, John, 2
Boria, Sonia, 537
Brankack, J., 471
Brinkman, C., 359
Broca, Paul, 213, 219, 339, 417, 544
Brodmann, Korbinian, 56, 213
Bronstone, A., 516
Brown, S. A., 454
Brunetti, M., 436
Buck, Linda, 403

Bucy, Paul, 423, 424
Bueller, Joshua, 237
Bunney, William, 278
Butler, Robert, 400, 401
Büttner, A., 200
Buxton, O. M., 444
Bygren, Lars Olov, 106
Byron, Don, 324

C

Cacucci, F., 228
Cahill, L., 501
Cain, S. W., 454, 458
Cajal, Santiago Ramón y, 76, 257, 502
Calne, Donald, 594
Calvin, William, 526
Candia, Victor, 510
Cao, Z. F., 134
Carmine, H., 584
Carr, Catherine, 537
Casile, A., 359
Caspi, Avshalom, 237
Castiello, U., 561
Caton, Richard, 113
Cha, J. H., 101
Chalfie, Martin, 75
Chalmers, David, 563
Chang, Fen-Lei, 507, 508
Changeux, Jean-Pierre, 258
Charcot, Jean-Martin, 137, 142
Charney, E., 99, 105
Chase, T. N., 374
Chen, Mary, 167
Chen, R., 256
Cho, R. Y., 598
Cho, Y., 364
Chomsky, Noam, 338
Chrétien, Jean, 63
Church, J. A., 374
Cicoria, Dr. Anthony, 347
Claes, S., 598
Clark, E. L., 478
Clarke, R. S., 271
Cole, Jonathan, 382
Comer, R. J., 187
Connolly, R. K., 577
Coristine, Marjorie, 500
Corkin, Suzanne, 391, 491, 2002
Cormack, Allan, 228
Corrigan, N. M., 603
Cowan, R. L., 200
Coyle, J. T., 583
Craighero, L., 536, 537
Crick, Francis, 561
Critchley, M., 394
Crook, T. H., 589
Crow, Timothy, 597
Cytowic, Richard, 556

D

Daly, Martin, 408
Damasio, Antonio, 422, 423
Dapretto, Mirella, 537
Daroff, R. B., 469
Darwin, Charles, 8, 9, 11, 249, 258, 349, 411
Davis, Michael, 423, 500
Day, J. J., 409, 512

de Mairan, Jean Jacques d'Ortous, 446
Deary, I. J., 27
Debanne, D., 133
DeCarolis, N. A., 187
DeCharms, R. C., 580
DeKrosky, Steven, 589
Delezie, J., 444
DeLisi, L. E., 200
Dement, William, 460, 463, 467, 468, 477
Dennett, Daniel, 12
Dere, E., 145
Descartes, René, 7, 8, 46, 110, 111, 479
Desimone, Robert, 533
Dethier, Vincent, 409
Diamond, Marion, 558
Drake-Lee, A. B., 324
Dudai, Y., 470
Duerson, Dave, 582
Duffy, Valerie, 405
Duman, Ronald, 599
Dunbar, Robin, 23
Duncan, J., 558
Durell, T. M., 190
Dylan, Bob, 101

E

Eccles, John C., 128, 129, 130
Edgerton, Robert, 193, 194
Ehrhardt, Anke, 434
Ehringer, H., 142
Eibl-Eibesfeldt, Irenäus, 5
Einstein, Albert, 29, 558
Eisch, A. J., 187
Eisenberger, Naomi, 398
Eisener-Dorman, A. F., 103
Emery, Nathan, 30
Epel, D., 36
Epstein, Herman, 265
Eriksson, P. S., 506
Evarts, Edward, 367
Everitt, B. J., 197
Everitt, Barry, 436

F

Fabbri-Destro, M., 537
Falk, Dean, 25, 26
Farah, Martha, 314
Faraone, S. V., 241
Fenno, L., 134
Fernald, R. D., 414
Field, Tiffany, 407
Fink, Bob, 320
Fins, J. J., 13
Fiorito, Graziano, 49
Flechsig, Paul, 250
Fox, Douglas, 34
Fox, Peter, 231
Fraga, Mario, 237
Fragoso, Y. D., 589
Franklin, Benjamin, 459
Fraioli, Sabina, 179
Freud, Sigmund, 58, 172, 191, 426, 433, 464, 567, 579
Friedhoff, A. J., 374
Fritsch, Gustave, 111, 365
Fromme, K., 194
Fulton, John, 344

Fung, Teresa, 429
Fuster, Joaquin, 496

G

Galaburda, Albert, 482
Galvani, Luigi, 111
Garcia, John, 409
Gardner, Alan, 9
Gardner, Beatrice, 9
Gardner, D. M., 188
Gardner, Howard, 28, 558, 559, 560
Gardner, Randy, 468
Gaulin, S. J. C., 554
Gazzaniga, Michael, 524, 550, 563
Gebicke-Haerter, P. J., 598
Geissman, Thomas, 320
Gerhig, Lou, 137
Gerrard, J. L., 470
Gershon, E. S., 178
Geschwind, Norman, 314, 482
Geuze, E., 566
Ghosh, K. K., 75
Gibb, R., 515
Giffords, Gabrielle, 2, 584
Gilbert, S. F., 36
Gilbertson, M. W., 208
Giles, Jim, 603
Gill, J. M., 103
Giussani, Carlo, 269
Gladwell, Malcolm, 29
Glickman, Steve, 399
Gogtay, N., 574
Goldsmith, Charlie, 555
Goldstein, Jill, 273, 274, 552
Golgi, Camillo, 76
Gonzalez, Claudia, 222
Goodale, Mel, 296, 312, 315, 316
Goodall, Jane, 30
Gordon, A. D., 21
Gorski, Roger, 434
Gould, Elizabeth, 506, 513
Gould, Stephen Jay, 28
Gray, J. R., 558
Gray, Stephen, 111
Graziano, M., 366
Greely, H., 172
Greenough, William, 507, 508
Griffin, J. A., 194
Guerrini, R., 277
Guilford, J. P., 559, 560
Guthrie, Woody, 101

H

Hall, Calvin, 464
Hall, Stuart, Ph.D., 462
Halliwell, C., 276
Hamilton, Roy, 285
Hamilton, T. J., 159
Hampson, Elizabeth, 205, 512
Hansen, Rick, 364
Harlow, Harry, 272, 400, 401
Harvey, T., 558
Hatcher, M. A., 45
Hebb, Donald O., 12, 164, 267, 268, 271, 275, 276, 400, 401, 507, 527, 560
Heinz, A., 603
Held, R., 287
Heninger, George, 571
Henquet, C., 201

Heron, Woodburn, 5, 400
Hertz, Heinrich Rudolph, 319
Hess, Walter, 360, 361
Hinshelwood, James, 482
Hitzig, Eduard, 113, 365
Hoang, M. T., 186
Hobson, Allan, 479
Hobson, J. Allan, 465
Hodgkin, Alan, 114, 115, 116
Horn, Gabriel, 271
Hornykiewicz, O., 142, 154
Hounsfield, Godfrey, 228
Houston, Whitney, 183
Hu, M., 196
Hubbard, Edward, 556
Hughlings-Jackson, John, 356, 357,
  358, 586
Hulshoff Pol, H. E., 232
Hunter, J. V., 583
Huttenlocher, Peter, 258
Huxley, Andrew, 114, 115, 116

I
Iacoboni, Marco, 537
Insel, T. R., 414
Isbell, Harris, 177
Iwaniuk, Andrew, 30

J
Jackson, S. R., 374
Jacob, S., 403
Jacobs, B., 510
Jacobsen, Carlyle, 424
Jacobson, Edmond, 5
James, Charmayne, 103
James, William, 533
Janes, Amy, 440
Jeannerod, Marc, 561
Jerison, Harry, 22, 302
Johansson, B. B., 347
Jones, Barbara, 474
Jouvet, Michael, 474, 478
Julien, R. M., 182
Jung, Carl, 464
Juraska, J. M., 274

K
Kaas, Jon, 390, 509
Kaati, G., 106
Kalueff, A. V., 361
Kandel, Eric, 164, 165
Kanner, Leo, 256
Karlsson, Erik, 102
Katz, Bernard, 146
Kaufer, Daniel, 589
Kelly, Mark, 2
Kempermann, Gerd, 507
Kennedy, Edward, 83
Kern, U., 385
Kigar, D. L., 558
Kim, Karl, 339
Kimura, Doreen, 205, 512, 546,
  551, 553
Kinney, Hannah, 246
Klein, Denise, 269
Kleitman, N., 447, 460, 463, 467,
  468
Kline, Don, 292
Klinger, Eric, 466
Klüver, Heinrich, 423, 424
Knight, Heather, 80f
Koch, Christof, 561
Kolb, B., 215, 219, 259, 512, 514,
  515, 518, 520, 552
Kondziolka, Douglas, 576

Korsakoff, Sergei, 498
Kosofsky, Barry, 276
Kovelman, Joyce, 278
Kreitzer, A. C., 373
Kroger, J. K., 496
Kross, Ethan, 398
Kues, W. A., 103
Kuffler, Stephen, 300, 301
Kuo, H.-K., 595
Kutcher, S., 188

L
Lamb, Mary, 457
Lamberg, L., 460
Landré, L., 208
Lane, E. L., 162
Lang, C. E., 359
Langer, N., 560
Langhorne, P., 45
Langston, J. William, 162
Larrabee, G. J., 589
Lashley, Karl, 220, 358, 490, 491
Leakey, Louis, 21
Leakey, Mary, 21
Lecours, A.-R., 262
LeDoux, Joseph, 424, 500
Legenstein, R., 133
Lehman, M. N., 452, 454
Leonard, C. M., 392
Lepage, M., 489
Leritz, E. C., 595
Lezak, Muriel, 583
Lieberman, M., 541, 542
Lindblad-Toh, K., 102
Lindvall, O., 593
Linge, Fred, 2, 12, 13
Little, William, 362
Livet, Jean, 75
Loeb, G. E., 335
Loftus, E. F., 487
Lomber, S. G., 276
Lomo, Terje, 502, 504
Lorenz, Konrad, 271
Loscher, W., 587
Loui, Psyche, 540
Lundstrom, Johan, 404

M
Maass, W., 133
MacAndrew, Craig, 193, 194
MacDonald, T. K., 193
Mackler, Scott, 354, 361
Magendie, François, 63
Magoun, Horace, 472, 473
Maguire, Eleanor, 496, 506
Mahncke, H. W., 516
Malanga, Carl, 276
Malcolm-Smith, S., 466
Malenka, R. C., 373
Maquet, Pierre, 471
Markowitsch, Hans, 489
Marler, P., 348
Marshall, P. J., 273
Martin, Alex, 488
Martin, Jean Prudin, 591
Mateer, C. A., 579
May, L., 212
McCarthy, M. M., 436
McClintock, Martha, 403
McDonald, C. L., 87
McEvoy, J. P., 597
McFarland, K. N., 101
McGaugh, James, 501
McGowan, Patrick, 208, 238

McHugh, Tommy, 603
McKinney, M. L., 26
McLure, Samuel, 542
Meaney, Michael, 238
Meesters, Y., 450
Melzack, Ronald, 386, 387
Mendel, Gregor, 9, 10, 11, 74, 97, 99,
  104, 106
Merzenich, M., 482, 516
Milberg, William, 595
Miller, B. L., 603
Miller, Courtney, 512
Milner, Brenda, 220, 490, 554
Milner, David, 296, 312, 315, 316
Milner, H. M., 490
Milner, Peter, 438
Milroy, C. M., 199
Milton, Katharine, 24
Miniaci, M. C., 167
Mishkin, Mortimer, 497, 500
Mistlberger, R. E., 454
Mizuno, Y., 594
Mlodinow, Leonard, 543
Molaison, Henry, 490
Monecke, S., 454
Money, John, 434
Moniz, Egas, 424
Moran, James, 533
Morellis, S. A., 541
Morris, Richard, 218
Moruzzi, Giuseppe, 472, 473
Mosso, Angelo, 343
Mountcastle, Vernon, 391
Mozaffarian, D., 429
Mukerjee, M., 527
Murdock, G. P., 29
Murphy, M. P., 584
Murray, E., 497
Mychasiuk, Richelle, 238, 252, 275,
  515

N
Nabokov, Vladimir, 556
Nadel, Lynn, 495
Naert, G., 85
Nelini, C., 471
Nelson, Charles, 273
Nestler, E. J., 198
Newsome, William, 528
Niemann, H., 103
Niino, M., 588
Nonneman, Arthur, 406
Nordahl, C. W., 256
Nottebohm, Fernando, 349
Nudo, Randy, 368, 508

O
O'Hare, E. D., 259
O'Keefe, John, 227, 228, 495
Oberman, L. M., 537
Ogawa, Segi, 231
Olds, James, 438
Olsson, S. B., 404
Opris, D., 580
Osorio, I., 469
Overman, William, 265, 274

P
Panagiotakos, D. B., 429
Panksepp, J., 392
Pannese, R., 598
Pannese, R., 598
Parai, J. L., 199
Parker, Elizabeth, 489
Parkinson, James, 142

Passingham, Richard, 34
Paterson, David, 246
Paulignan, Y., 561
Pavlov, Ivan, 483
Payne, B. R., 276
Penfield, Wilder, 221, 340, 341, 342,
  343, 345, 365, 366, 367, 390
Pepperberg, I. M., 527
Peretz, Isabelle, 345
Petanjek, Zdravko, 259
Piaget, Jean, 263, 264, 265
Pickering, R., 21
Pietroiusti, A., 450
Pinker, Steven, 338
Poltis, M., 593
Pons, Tim, 393
Popova, S., 184
Posner, M. I., 232, 343
Potts, Rick, 24
Prinz, J., 5
Purpura, Dominique, 278
Purves, Dale, 504

Q
Quinlan, J., 473
Quinlan, Karen Ann, 473
Quinn, P. D., 194

R
Radjenovic, J., 175
Rafferty, Mary, 113
Ragert, Patrick, 510
Raichle, M. E., 232, 343
Rajji, T. K., 577
Rakic, Pasko, 254, 258
Ralph, Martin, 452, 454
Ramachandran, V. S., 385, 509, 510,
  537
Rameson, L. T., 541
Rapoport, Judith, 573
Rasmussen, Ted, 554, 555
Rauschecker, J. P., 337
Ravel, Maurice, 345
Raven, P. H., 446
Reeve, Christopher, 363, 364
Reeve, Dana, 363
Reeve, R., 80
Reger, G. M., 566
Rego, A. C., 103
Reiter, Russel, 456
Revonsuo, Anttio, 465
Reynolds, Brent, 519
Reza, M. F., 227
Rezania, K., 130
Richter, Curt, 452, 457
Rieder, R. O., 278
Riesen, Austin, 272
Rivest, S., 85
Rizzo, A., 566
Rizzolatti, Giacomo, 536, 537
Robinson, T. E., 439
Robinson, Terry, 179, 196, 514
Robison, A. J., 198
Rogawski, M. A., 587
Roland, P. E., 360
Rothwell, John, 381
Rowland, B. A., 531
Rutter, Michael, 273

S
Sacks, Oliver, 42, 154, 315, 347,
  382, 525, 526, 590, 591
Sainsbury, Bob, 500
Saint Ambrose, 29
Salm, A. K., 272

Sanders-Bush, E., 272
Sapolsky, Robert, 206, 207, 513
Sarne, Y., 200
Savage-Rumbaugh, Sue, 9
Sawamoto, Nobukatsu, 542
Schacter, Daniel, 500
Schallert, T., 221
Scheibel, Arnold, 278, 510, 511
Schenck, C. H., 478
Schibler, U., 452
Schieber, M. H., 359, 367
Schiff, Bernard, 399
Schiff, Nicholas, 13
Scholl, M., 510
Schroeder, M., 427
Schwartz, W. J., 450
Scotto, Pietro, 49
Scoville, William, 220, 490
Seeley, Randy, 432
Seneca, 248
Sertürner, Friedrich, 188
Severus, W., 187
Shai, Iris, 429
Shakespeare, William, 338, 468
Shaw, Philip, 260
Sherry, David, 495, 496
Shimomura, Osamu, 75
Shingo, T., 252
Siegel, Jerome, 477
Silver, R., 453, 458
Sira, C. S., 579
Sirevaag, Anita, 507
Skinner, B. F., 408, 409, 485, 486
Smith, A. D., 186

Smyke, A. T., 273
Snowden, Robert, 310
Sodhi, M. S., 272
Sohlberg, M. M., 579
Sowell, E. R., 259
Spearman, Charles, 28, 557, 558
Sperry, Roger, 269, 524, 544
Starr, J. A., 45
Stedman, Hansell, 26
Steen, Eric, 204
Stein, B. E., 531
Steinman, L., 588
Stern, Karen, 403
Stewart, Jane, 512, 552
Stoleru, D., 454
Suomi, Stephen, 272
Swaab, D. F., 436, 437
Sweatt, David, 409, 512
Synofzik, M., 475
Szyf, Moshe, 238

T
Takeuchi, Hikaru, 560
Tanaka, Keiji, 308, 309
Taub, Edward, 585
Tees, Richard, 258, 270
Teie, P. U., 324
Temple, E., 482
Teng, W., 333
Thach, Tom, 376
Thomas, K., 462
Thompson, C. K., 348
Thompson, P. M., 558
Thompson, Peter, 310

Thorndike, Edward, 483, 485, 486
Tourette, Georges Gilles de la, 374
Tréatikoff, Constantin, 142
Trehub, Sandra, 347
Troscianko, Tom, 310
Tsai, G., 583
Tsien, Roger, 75
Tulving, Endel, 489
Turk, Ivan, 320

V
Valli, K., 465
Van den Heuvel, M. P., 232
Vanderwolf, Case, 473
Veldhuizen, Maria, 405
Vevelstad, M., 200
Von Békésy, George, 329
Von Helmholtz, Hermann, 113, 329
Von Holst, E., 361
Voyvodic, J. T., 504
Vul, Ed, 542

W
Wada, Jun, 555
Wager, T. D., 542
Wall, Patrick, 386, 387
Wallace, Alfred Russel, 8, 9
Wallace, P. S., 262
Wang, J., 599
Wasserman, Robert, 315
Wassermann, E. M., 577
Watson, C. B., 462
Watson, James, 97

Watson, Michael, 556
Webb, Barbara, 80
Weber, Rutger, 447, 449
Weiss, Sam, 252, 519
Weizkrantz, Lawrence, 282
Wenger, John, 178
Wenner, J. B., 477
Werker, Janet, 258
Wernicke, Karl, 340
Whishaw, Ian, 179, 215, 219, 221, 262
Whitney, D., 333
Wilson, Margot, 408
Wilson, S., 469
Wise, R. A., 438, 514
Witelson, Sandra, 555, 558
Wong-Riley, Margaret, 298
Woods, Stephen, 432
Woollett, Katherine, 506

Y
Yakovlev, P. E., 262
Yeo, Thomas, 539
Young, J. Z., 114

Z
Zaidi, F. H., 453
Zaidi, Farhan, 295
Zajonc, Robert, 425
Zamboni, P., 86
Zatorre, Robert, 343, 345, 346
Zhang, G., 22
Zihl, Josef, 315
Zlomuzica, A., 145

Note: Page numbers followed by f indicate figures; those followed by t indicate tables.

A

Abducens nerve, 60–61, 60f
Ablation studies, 220–221, 221f
Abnormal behavior. *See* Brain/behavioral disorders
Absolute pitch, 322, 540–541, 540f, 559
Absolutely refractory membrane, 124
Acetate, 152
N-Acetylaspartate, in traumatic brain injury, 583
Acetylcholine (ACh), 140, 140f, 151, 157, 158–159, 161f
    in autonomic function, 158–159
    in brain activation, 473, 473f
    identification of, 151
    in muscle contraction, 136, 136f, 158, 371, 371f
    neurotoxin effects on, 176–177, 176f
    in Renshaw loop, 151, 151f
    synthesis and breakdown of, 152, 152f
Acetylcholine psychedelics, 191
Acetylcholine receptor, curare and, 177
Acetylcholine synapse, drug action at, 176–177, 176f
Acetylcholinesterase, 161
Achromatopsia, 315
ACTH, 415t, 600
Action potential, 122–126. *See also* Nerve impulse
    all-or-nothing response of, 126
    axon hillocks and, 132–133
    blocking of, 123, 123f
    definition of, 122
    measurement of, 122, 122f
    nerve impulse and, 125–126
    phases of, 125f
    postsynaptic potentials and, 128–133
    production of, 128–133
    propagation of, 125–126, 126f, 127f, 132f
        back, 133
    refractory periods and, 124–125, 126
    single-cell recordings of, 227–228
Action test, 8
Activating systems, 159–163
Activation-synthesis hypothesis, for dreams, 465
Adderall, 172
Addiction. *See* Substance abuse
Additive color mixing, 309, 309f
Adenine, 92f, 93
Adolescence, psychiatric disorders presenting in, 275
Adrenaline (epinephrine), 141, 152, 153f. *See also* Neurotransmitters
    in stress response, 206–207, 206f
Adrenocorticotropic hormone (ACTH), 415t, 600
Affective disorders, 425–427
Affective states. *See under* Emotion; Emotional
Afferent nerves, 37, 37f
Afferent sensory signals, 356
Africa, primate evolution in, 21, 24
Age-related cognitive loss, 595. *See also* Dementia
Aggression, sex differences in, 553
Agnosia
    color, 315
    facial, 300
    visual-form, 314–315, 423–424, 531
Agonists, 175–177, 176f
Agoraphobia, 426
Akathesia, 591
Akinesia, 221, 590
Alcohol. *See also* Substance abuse
    brain damage from, 199, 497–499
    disinhibition and, 193–194
    fetal alcohol spectrum disorder and, 184
    Korsakoff's syndrome and, 497–499

    as sedative hypnotic, 182
    tolerance to, 177–179, 178f
Alcohol myopia, 194
Alleles, 98
    dominant, 98, 100f
    recessive, 98, 100f
    wild-type, 98
Alpha rhythms, 224, 460
Alzheimer's disease, 70, 161, 494, 500, 593–595
    brain abnormalities in, 85, 169, 593–594, 593f, 594f
    in Down syndrome, 102
    Parkinson's disease and, 594
Amacrine cells, retinal, 293, 293f
Amblyopia, 270
Amine neurotransmitters, 152, 152t
Amines, synthesis of, 152
Amino acid(s), 93, 94, 94f. *See also* Protein(s)
    synthesis of, 153
Amino acid neurotransmitters, 152t, 153, 153f
Amino groups, 94
Amnesia, 486. *See also* Memory deficits
    anterograde, 498
    Boswell's, 492
    definition of, 486
    episodic, 489
    postencephalitic, 492
    psychogenic, 489, 489f
    retrograde, 498
    surgically induced, 490–492, 490f
AMPA receptors, in long-term potentiation, 503–504, 503f
Amphetamine, 190
    for attention-deficit/hyperactivity disorder, 172
    brain injury from, 200
    as cognitive enhancer, 172
    dosage of, 173–174
    hallucinogenic, 200
    neuronal effects of, 513–515
    sensitization to, 179–181
Amplification cascade, 157
Amplitude, of sound waves, 322f, 323–324
Amputation, cortical reorganization after, 509
Amusia, 346
Amygdala, 58–59, 58f, 417f, 418–419. *See also* Limbic system
    in attention, 535
    in autism spectrum disorders, 256
    in eating, 418–419, 431–432
    in emotional behavior, 418–419, 423–424
    in emotional memory, 484, 500–501, 501f
    in fear conditioning, 484, 500
    neural connections of, 500–501, 501f
    in sexual behavior, 436
Amyloid plaque, in Alzheimer's disease, 494, 494f, 593
Amyotrophic lateral sclerosis (ALS), 137, 354, 361
Anabolic steroids, 205
Analgesics, 387
    abuse of, 189–190
    opioid, 154, 182t, 187–190, 387. *See also* Opioids
Anandamide psychedelics, 191–192
Anatomical orientation, of brain, 36, 39
Anatomical terms, 38–39
Androgen(s), 202, 272. *See also* Sex hormones
    behavior and, 401
    in brain development, 204–205, 250, 272–275, 433–434, 551–554
    functions of, 202, 203
    lifelong effects of, 274–275
    neuroplasticity and, 513
    sexual behavior and, 435
    in sexual differentiation, 250, 433–434

Androgen-insensitivity syndrome, 435
Androgenital syndrome, 435
Anencephaly, 277
Anesthesia, epidural, 387
Aneurysms, cerebral, 345
Angel dust, 183, 192
Angiomas, 344
Animal(s)
    auditory communication in, 348–351
    auditory processing in, 318, 329–330, 336, 336f, 337, 337f
    brain size in, 30
    chimeric, 103
    cloned, 102–103, 103f
    culture of, 29
    diurnal, 444
    emotional behavior in, 411
    evolution of, 14–19
    experimental, 102–103, 240–243. *See also* Animal research
    hearing in, 322, 323f
    language in, 8, 9, 527
    learning in, 408–409
    prey-killing behavior in, 399, 401, 406–407
    scratch reflex in, 363
    singing in, 320
    skilled movement in, 367
    sleep in, 463, 463f, 467, 467f
    species-typical movement in, 360–361, 361f
    thinking in, 527, 550
    tool use by, 29, 30
    transgenic, 103
Animal research, 102–103, 240–243
    in behavioral disorders, 571
    benefits of, 240–241
    conditioning in, 483–485
    limitations of, 571
    regulation of, 241–243, 242t
Animal Welfare Act, 242
Anions, 116. *See also* Ion(s)
Anomalous speech representation, 555
Anorexia nervosa, 275, 428
Anoxia, cerebral, 362
Antagonists, 175–177, 176f
Anterior, definition of, 38, 39
Anterior cerebral artery, 43, 43f
Anterior cingulate cortex, 419, 419f, 530–531, 542
Anterograde amnesia, 498
Antianxiety agents, 182–183, 182t, 426, 602
Anticonvulsants, 587
Antidepressants, 182t, 186–187, 188
    atypical, 186
    hippocampal neurogenesis and, 601
    mechanism of action of, 599, 601
    monoamine oxidase inhibitors, 186–187
    second-generation, 186–187
    side effects of, 578
    sleep and, 469
    tricyclic, 186–187
Antipsychotics, 182t, 184–186, 185f
    side effects of, 578
Anvil, 328, 328f
Anxiety disorders, 426, 427, 596f, 602
    adolescent-onset, 275
    classification of, 596f
    treatment of, 182–183, 182t, 426, 602
Anxiety dreams, 465–466
Anxiolytics, 182–183, 182t, 426, 602
Apes, 20. *See also* Primates, nonhuman
Aphagia, 431

Aphasia
  Broca's, 228–229, 340
  definition of, 340
  Wernicke's, 340
Aplysia californica
  habituation in, 164–165, 165f
  sensitization in, 166, 167f
Apnea, sleep, 476
Apoptosis, 199, 258
Apraxia, 394, 545–546, 545f
Arachnoid layer, 37
Arachnoid membrane, 37f
Arborization, dendritic, 255, 255f, 504–505, 505f
Arcuate fasciculus, 340, 340f
Area postrema, 174, 175f
Arousal, in basic rest-activity cycle, 467–468, 468f
Arteriovenous malformations, 344
Artificial intelligence, 80
Asperger's syndrome, 256
Association cells, 79, 79f
Association cortex, 530–538
  in attention, 533–535
  components of, 530–531, 530f
  in imitation and understanding, 536–537
  lesions in, assessment for, 539
  mirror neurons in, 536–538, 537f
  multisensory integration in, 531–532, 532f
  neural connections to, 530–531, 530f
  in object recognition, 531, 531f
  in planning, 535–536
  in spatial cognition, 532–533, 532f, 533f
Associative learning, long-term potentiation and, 502–504
Astrocytes (astroglia), 82t, 83–84, 148
  in blood–brain barrier, 174, 174f
Asymmetry, cerebral. See Cerebral asymmetry
Ataxia, optic, 316
Athletes, traumatic brain injury in, 581, 582
Atoms, 88, 89, 89f
Atonia, in REM sleep, 460, 463, 476–478
Attention, 533–535
  contralateral neglect and, 534, 535f
  deficits of, 534–535, 535f
  definition of, 533
  extinction and, 534, 535f
  selective, 533–534
  vision and, 533–534
Attention-deficit/hyperactivity disorder, 163, 172, 240–241
Attitudes, 542
Atypical antidepressants, 186. See also Antidepressants
Auditory communication, in nonhuman species, 348–351
Auditory cortex, 332–333, 530f
  association cortex and, 532f
  Broca's area in, 255f, 263, 339–344, 340f, 342f
  mapping of, 340–344
  primary, 332, 332f
  secondary, 332
  structure of, 332–334, 332f
  supplementary speech area in, 342–343, 342f
  tonotopic representation in, 334–335, 335f
  Wernicke's area in, 332, 332f, 340–344, 340f, 342
Auditory flow, 284
Auditory nerve, 331
Auditory pathways, 331–332, 331f
Auditory receptors, 328f, 329, 330–331
Auditory system. See also Hearing; Sound
  in animals, 320, 330, 336, 336f, 337, 337f
  auditory cortex in. See Auditory cortex
  auditory pathways in, 331–332, 331f
  auditory receptors in, 328f, 329, 330–331
  ear in, 328–330, 328f, 329f, 330f
  evolution of, 327

functions of, 327
insula in, 332f, 333
in language processing, 338–344
lateralization in, 332–333, 546
in movement, 332
music perception in, 325–326, 336–337
sensitivity of, 325
sound perception in, 325
speech perception in, 325–326, 336–337
structure of, 327–334
Auditory vestibular nerve, 60–61, 60f
Aura, 586
Australopithecus, 20–22, 20f–23f
Autism spectrum disorder, 256
  brain size in, 27
  mirror neurons in, 537
Autobiographical memory, 488–489, 489f
Autoimmune disease, 130, 588
Automatisms, 586
Autonomic nervous system, 36f, 37, 64
  neurochemistry of, 158–159, 159f
Autoreceptors, 146
Autosomes, 98
Aversive childhood experiences, frontal lobe development and, 260
Axoaxonic synapse, 147f, 148
Axodendritic synapse, 147f, 148
Axomuscular synapse, 147f, 148
Axon(s), 47, 47f, 76, 76f, 77–78, 78f, 90f
  dendrites and, 76f, 77, 78f, 80
  giant squid, electrical activity in, 114–115, 114f
  growth cones of, 257, 257f
  myelination of, 127–128
  nerve impulse along, 125–126, 126f, 127f
  in neural circuits, 80. See also Neural circuits
  sprouting of, 85–86
    neuroplasticity and, 504–505, 505f
Axon collaterals, 77, 78f
Axon hillocks, 77, 78f, 132–133
Axosecretory synapse, 147f, 148
Axosomatic synapse, 147f, 148
Axosynaptic synapse, 147f, 148

B

Baby connectome, 362
Back propagation, 133
Balance, vestibular system in, 388–389, 388f
Barbiturates, 182–183
Basal forebrain, in brain activation, 473–474, 473f
Basal ganglia, 55, 55f, 57–58, 58f
  anatomy of, 372–373, 372f
  functions of, 373
  in memory, 491–492, 500
  in movement, 355, 372–375, 373f
  in Parkinson's disease, 42, 57–58, 374–375
  in Tourette's syndrome, 58, 374
Bases, nucleotide, 92f, 93
Basic fibroblast growth factor, 252
Basic rest–activity cycle (BRAC), 467–468, 468f
Basilar membrane, in hearing, 328f, 329, 330, 330f
Bats, echolocation in, 350–351, 351f
Behavior, 5–6
  abnormal. See Behavioral disorders
  brain development and, 247–248
  for brain maintenance, 400–401
  causes of, 399–401
  cerebral control of. See Brain-behavior link
  chemical senses in, 401–406
  cognitive stimulation and, 400–401
  comprehension of, 536–537
  definition of, 5–6
  drinking, 418–419, 432–433
  emotional, 399, 417–419, 422–427. See also Emotional behavior

environmental influences on, 408–409
evolution of, 406–408
feeding. See Eating/feeding behavior
free will and, 399
homeostatic mechanisms and, 411
inherited, 5–6, 6f, 406–408
innate releasing mechanisms for, 406–407
learned, 5–6, 6f, 408–409. See also Learning
measurement of, 213–223
motivated. See also Motivation
  neuroanatomy of, 411–421
neural circuits and, 401
neuroanatomy and, 213–215
nonregulatory, 412
  control of, 433–437
olfaction in, 402–404, 404f
overview of, 5–6
prey-killing, 399, 401, 406–407
purposeful, 410
purposes of, 399–401
regulation of
  amygdala in, 418–419
  frontal lobe in, 419–421, 419f
  hypothalamus in, 411–417, 430–431, 431f
  limbic system in, 417–419
  prefrontal cortex in, 419–421, 419f
regulatory, 411–412
  control of, 427–437
rewarding, 399, 401, 417, 438–440
selection of, 419–421, 419f
sex differences in, 553–554
sexual, 412, 434–436
species-typical, 28–29, 360–361
Behavior modification, 579
Behavioral assessment, 539
Behavioral disorders. See also Brain/behavioral disorders; Psychiatric disorders
  research methods for, 568–572
  vs. neurological disorders, 567–568
Behavioral neuroscience
  definition of, 215
  methodology of, 215–223
    comparisons for, 239, 239t
Behavioral sensitization, 514
Behavioral stimulants, 190–191, 191f
Behavioral tests, 539
Behavioral therapy, 578–580
Beliefs, 542
Bell–Magendie law, 62, 63
Bell's palsy, 63
Benzedrine, 172, 190
Benzodiazepines, 182–183, 426
Beta rhythms, 460
Bias, 542
Biceps muscle, 371, 371f
Bilateral, definition of, 39
Bilateral symmetry, 15, 17, 46
Bilingualism, cortical areas for, 269
Binding problem, 531
Binocular vision, corpus callosum in, 302–303
Biological clocks, 444, 446. See also Biorhythms
  eating/feeding behavior and, 453, 458
  entrained, 449
  free-running rhythms and, 447–449, 449f
  neural basis of, 452–458
  resetting of, 449–451, 451
  suprachiasmatic nucleus as, 452–458, 452f
Biorhythms, 445–458
  basic rest–activity cycle and, 467–468, 468f
  biological clocks and, 446. See also Biological clocks
  circadian, 444, 446–449, 447t, 448f. See also Circadian rhythms
  circannual, 447, 447t
    pacemaking, 456–457
  definition of, 445

disturbances of, psychiatric symptoms in, 457, 457f
entrained, 449–451, 451f
free-running, 447–449
genetic factors in, 454–455
infradian, 447, 447t
innate, 454
jet lag and, 451, 451f
neural basis of, 452–458
neural transplantation and, 454, 454f
origins of, 445
periods of, 447
pineal gland in, 456
in plants, 445, 446, 446f
recording of, 446–447, 447f, 448f
retinohypothalamic tract in, 453–454, 455
seasons and, 445
suprachiasmatic, 452–458
ultradian, 447, 447t
Zeitgebers and, 449–451
Bipolar cells, retinal, 293
Bipolar disorder, 187, 599. See also Depression
   adolescent-onset, 275
   drug therapy for, 182t
   mood stabilizers for, 187
Bipolar neurons, 79, 79f
   auditory, 331
   retinal, 293f
Birdsong, 348–350
   sex differences in, 275, 350
   vs. language, 348–349
Bitter taste, 404–405
Black widow spider venom, 176
Blind spot
   retinal, 282, 289, 290–291, 290f
   in visual fields, 313, 313f
Blindness. See Vision impairment
Blindsight, 282
Blobs, 298, 298f, 311
Block span, 217
Block-tapping test, 217, 217f
Blood-brain barrier, 83–84
   drug therapy and, 174–175
Blood vessels, cerebral, 42–43, 43f
Bodily-kinesthetic intelligence, 558–559
Body segmentation, 15, 17, 61, 61f
Body size, brain size and, 22, 23f, 34
Body symmetry, 15, 17
Body temperature, regulation of, 411–412
Body weight, regulation of, 428–432. See also Eating/
      feeding behavior
Bonds, 89
   peptide, 94
Bone flute, 320
Bonobos. See Primates, nonhuman
Boswell's amnesia, 492
Botulin toxin, 176
Bradycardia, diving, 140
Braille, 549
Brain. See also under Cerebral; Cortical
   behavioral control by. See Brain-behavior link
   capillary network of, 174, 174f
   chemical composition of, 88f
   cortical organization in. See also Cortical columns
      environmental influences in, 267–279, 268f
      sex differences in, 551–554
   crossed connections in, 43, 66–67
   Einstein's, 558, 558f
   electrical activity in, 111–128. See also under
         Electrical
      measurement of, 223–228
   evolution of, 14–31, 34, 48–49, 48f. See also Brain
         development
      in animals, 14–19
      culture and, 29–31, 35–36
      in humans, 19–31

functional maintenance of, behavior and, 400–401
functions of, 35
   behavior as, 35
   localized vs. distributed, 70
   movement as, 35
   principles of, 72
   sensory processing as, 35
growth of, 250f. See also Brain development
hemispheres of, 55–56, 55f
hierarchical organization of, 50–59, 68, 68f,
      355–356
information flow in. See Information flow
integration operation of, 128–133
lateralization in, 332–333, 345–347, 538. See also
      Lateralization
levels of function and, 50–59, 67
mapping of. See Brain maps
masculinization of, 204–205, 272–274, 433–434
orientation of, 38–39
parallel processing in, 68
perceptual world created by, 66
plasticity of. See Neuroplasticity
protective structures for, 37–38
sectioning of, 39, 39f, 43–46, 43f, 46f
sex differences in, 204–205, 272–275, 274f, 551–
      554, 552f, 553f
size of
   environmental stimulation and, 507
   intelligence and, 557–558, 558f
   sex differences in, 204–205, 274f, 552, 552f
spinal cord integration with, 63
split, studies of, 524, 547–549, 550
staining of, 46–47, 47f, 75, 75f, 213, 214f
stimulation of, 35
structure of, 3–4, 4f
   internal features of, 43–47, 46f, 47f
   layers in, 47, 47f, 56–57, 57f. See also Cortical
         layers
   microscopic features of, 46–47, 46f, 47f
   surface features of, 37–43, 37f, 341f
   symmetry in, 15, 17, 46, 46f
   terminology for, 38–39
   vascular, 42–43, 43f
subjective reality and, 35
summation in, 131–132, 131f, 132f
Brain–behavior link, 3
   in brain/behavioral disorders, 568–572
   brain development and, 261–267. See also Brain
         development
   brain size and, 22, 23f, 27–29, 34
   causation vs. correlation and, 266–267
   culture and, 35–36
   dualism and, 7–8, 7f
   evolutionary aspects of, 19–26. See also Evolution
   materialism and, 8–11
   mentalism and, 7
   research methods for, 568–572
   unified theory for, 566–568
Brain/behavioral disorders. See also specific disorders
   animal models of, 571. See also Animal research
   biorhythms and, 450, 457, 457f
   brain abnormalities in, 573, 574f
   causes of, 568–569
   classification of, 572–574, 573t
   compensatory mechanisms in, 571, 603
   degenerative, 589–593, 589t
   diagnosis of, 570
   epidemiology of, 572
   epigenetics and, 568
   neurobiology of, 569–572
   neuroimaging studies in, 573–574, 574f
   neurological, 581–584
      vs. psychiatric, 567–568
   psychiatric, 597–599
      vs. neurological, 567–568

research challenges for, 570–572
research methods for, 568–572
sleep problems in, 475–476
treatment of, 574–580
   behavioral, 578–580
   electrophysiological, 576–577
   neurosurgical, 575–576
   pharmacological, 577–578. See also Drug(s)
Brain–body orientation, 38
Brain cells, 46–47, 46f, 73–107. See also Glial cells;
      Neuron(s)
Brain chemistry, measurement of, 235–237
Brain–computer–brain interface, 354
Brain–computer interface, 354
Brain connectome, 230, 538
Brain-derived neurotrophic factor, 236–237
   in depression, 599
   electroconvulsive therapy and, 576
   neuroplasticity and, 513, 513t
Brain development, 247–280
   in autism spectrum disorder, 256
   behavioral development and, 247–248, 261–267
   brain injury and, 275–276
   cell death in, 258
   cellular commitment in, 255, 255f
   cognitive development and, 263–266, 264f, 265t,
         266f
   cortical layering in, 254–255
   cortical thinning in, 259, 259f
   critical periods in, 270–271
   drug effects on, 276
   environmental influences in, 267–279
      cortical organization and, 267–269, 268f
      cultural, 29–31, 35–36
      environmental stimulation, 267–269, 268f,
            271–272, 560
      negative experiences, 260, 271–272
      neural connectivity and, 269–270
      prenatal, 269–270
   gene expression in, 252
   glial development in, 260–261
   growth spurts in, 264–265
   hormonal influences in, 250, 272–275
   imprinting in, 271, 271f
   language development and, 262–263
   masculinization in, 204–205, 272–274, 433–434
   motor development and, 261–262, 262f
   music and, 268
   myelination in, 260–261, 260f
   neoteny and, 26, 26f
   neural column formation in, 270, 270f
   neural connectivity in, 269–270
   neural Darwinism in, 258
   neural placement in, 269–270, 270f
   neurobiology of, 248–261
   neurogenesis in, 253, 253f
   neuronal development in, 253–259
   neuronal differentiation in, 253–254, 253f
   neuronal maturation in, 255–257, 257f
   neuronal migration in, 253–255, 253f, 254f
   neuroplasticity and, 279
   prenatal, stages of, 248–250, 248f–250f
   in Romanian orphans, 273
   in schizophrenia, 278
   sensory input in, 267–269, 268f, 271–272
   sex differences in, 261f
   sex hormones in, 204–205, 250, 272–275, 433–
         434, 551–554
   sexual differentiation and, 204–205, 250, 251f
   stages of, 253t
   stress and, 272
   synaptic development in, 257
   synaptic pruning in, 258–259, 258f
   tactile stimulation and, 268
   time line for, 253f

Brain imaging. *See* Imaging studies
Brain-in-a-bottle experiment, 3–4, 5
Brain injury
   astrocytes in, 84
   behavioral testing in, 539
   behavioral therapy in, 518
   compensatory mechanisms in, 571, 603
   critical periods for, 275–276
   depression in, 227
   developmental effects of, 275–276
   excitation vs. inhibition in, 70–71
   frontal lobe, 56
   functional asymmetry and, 544–546, 545f
   ischemic, 584–585
   location of, functional loss and, 70
   lost-neuron-replacement solution in, 518–519
   microglia in, 84, 84f
   minimally conscious state and, 12
   neurogenesis in, 518–520
   neuroplasticity in, 276, 368–369, 368f, 516–
      520
   neuropsychological assessment in, 539
   new circuit solution in, 518
   occipital lobe, 56
   outcome in, 2, 12
   parietal lobe, 56
   persistent vegetative state and, 12
   recovery from, 2, 12–13, 516–520
   sex differences in, 553, 553f
   in substance abuse, 199–200
   temporal lobe, 56
   teratogens in, 253, 276
   three-legged cat solution and, 517–518
   timing of, 253–254
   traumatic. *See* Traumatic brain injury
   treatment of, 13
Brain lesion studies, 220–221
Brain maps, 56, 57f. *See also* Homunculus
   cortical function, 340–344, 342f, 365–367, 366f
   cytoarchitectonic, 56, 57f
   electrical stimulation in, 221–223
   event-related potentials in, 225–227, 225f, 226f
   Flechsig's, 260
   functional movement categories in, 366–367
   for infants, 362
   place cells and, 227, 227f
   of speech/language areas, 340–344
   tonotopic, 334–335, 335f
   topographic, 286, 286f. *See also* Topographic
      organization
      of motor cortex, 365–366, 368–369, 368f
      of visual cortex, 301–302, 302f
Brain size
   in animals, 18, 18f, 30
   behavior and, 22, 23f, 27–29, 34
   body size and, 22, 23f, 34
   in chordates, 18, 18f
   climate and, 24
   diet and, 24–25
   encephalization quotient and, 22, 23f
   evolution of, 23–26
   intelligence and, 28–29, 34
   neoteny and, 26, 26f
   in neurological disorders, 27
   neuroplasticity and, 28
   in nonhuman primates, 22, 23f, 34, 34f
   radiator hypothesis and, 25–26
   skull structure and, 25–26
   variability in, 27–29
Brain-stimulation reward, 438
Brain stimulation techniques, 13, 13f, 111–113, 221–
   223, 518, 575, 592
Brain surgery. *See* Neurosurgery
Brain tumors, 83
Brainbow, 75

Brainstem, 4, 4f, 41f, 42, 51–55
   in autism spectrum disorder, 256
   in drinking behavior, 361
   in feeding behavior, 361
   function of, 51–52
   in grooming behavior, 361
   injury of
      in locked-in syndrome, 354, 361–362
      REM sleep deprivation and, 469, 479
   in sexual behavior, 361
   in species-typical movement, 360–361
   structure of, 51–55, 51f–55f
Breast cancer, 98
Bregma, 220
Broca's aphasia, 228–229, 340
Broca's area, 255f, 263, 339–344, 340f, 343f
   mapping of, 340–344, 342f
Bruit, 344

C

Caenorhabditis elegans, light-sensitive ion channels
   in, 134
Caffeine, 192
Calcitonin-gene-related protein, 158
Calcium ions. *See also under* Ion
   in learning, 165–166, 165f
   in neurotransmitter release, 145–146
Calmodulin, 145–146
cAMP (cyclic adenosine monophosphate)
   caffeine and, 192
   in learning, 166–167, 168
Canadian Council on Animal Care, 242–243, 242t
Cancer, mutations in, 98
Cannabis sativa, 191–192, 201, 237
Capillaries, in blood-brain barrier, 174, 174f
Carbamazepine, for bipolar disorder, 187
Carbon monoxide, as neurotransmitter, 155
Carbon monoxide poisoning, 314–315
Carboxyl groups, 94
Card sorting tasks, 536, 536f, 539
Cataplexy, 477, 478, 478f
Catatonic posture, 586
Cations, 116. *See also* Ion(s)
Caudal, definition of, 38, 39
Caudate nucleus, 57–58, 57f
Causation, vs. correlation, 266–267
Cell. *See also specific types*
   components of, 87–91, 90f
   electrical properties of, 111–128
   function of, protein structure and, 87
   internal structure of, 87–97
   nerve. *See* Glial cells; Neuron(s)
   nucleus of, 90f, 91, 92–93
   as protein factory, 90–91
Cell-adhesion molecules, 257
Cell assemblies, in cognition, 527, 529
Cell body, 76, 76f
Cell cultures, 75
Cell death, programmed, 199, 258
Cell membrane, 90, 91, 91f. *See also* Membrane
   absolutely refractory, 124
   depolarization of, 122–124, 125f
   electrical activity in, 118–128
   hyperpolarization of, 121f, 122–124, 125f
   ion movement across, 116–118, 117f
   permeability of, 90, 117–118, 117f
   phospholipid bilayer of, 91–92, 91f
   relatively refractory, 124
   repolarization of, 125f
   structure of, 91–92, 91f
Cell migration, 253–254
Cellular commitment, in brain development, 255,
   255f
Central nervous system, 4–5. *See also* Brain; Spinal
      cord

development of, 247–280. *See also* Brain
      development
   evolution of
      in animals, 14–19
      in humans, 19–31
   functional organization of, 36–37, 36f, 50–59, 67
   hierarchical organization of, 50–59, 68, 68f
   neurotransmission in, 159–163
   sensory and motor divisions in, 68–69
   structure of, 4–5
Central pain, 385
Central sulcus, 41f, 42, 55f, 56
Cerebellar homunculus, 375–376, 375f
Cerebellum, 4, 18
   anatomy of, 375–376, 375f
   evolution of, 18, 18f
   functions of, 52–55
   in movement, 52, 375–378, 375f–378f
   structure of, 41f, 42, 51f, 52f
   topographic organization of, 375–376, 375f
Cerebral aneurysms, 345
Cerebral anoxia, 362
Cerebral aqueduct, 44
Cerebral arteries, 43, 43f
Cerebral asymmetry, 67–68, 332–333. *See also*
      Lateralization
   anatomical, 544, 544f
   functional, 332–333
      auditory processing and, 546
      cognition and, 538, 544–550
      dichotic listening and, 546
      language and, 332–333, 538, 549–550
      music and, 345–347, 546
      in neurological patients, 544–546, 545f
      in normal brain, 546–547
      visual processing and, 546, 546f. *See also* Visual
         fields
   handedness and, 332–333, 554–556
   hypotheses for, 549–550
   inferential thinking and, 550
   movement and, 549–550
   overlapping functions and, 305, 305f, 549
   in split-brain patients, 524, 547–549, 550
Cerebral circulation, 42–43, 43f
Cerebral cortex, 40–42, 40f, 55–57, 55f. *See also*
      specific subunits (e.g., Motor cortex) *and*
      under Cortical
   connections of, 56–57
   development of, 253f. *See also* Brain development
   developmental thinning of, 259, 259f
   layers of. *See* Cortical layers
   neocortex, 55
   sensory and motor divisions in, 68–69
   structure and function of, 55–57, 55f
   top-down processing in, 57
Cerebral hemispheres, 3, 4f, 40, 40f, 54–55, 55f
   evolution of, 18, 18f
   surgical removal of, 42
   symmetry/asymmetry of, 15, 17, 46, 46f, 67–68,
      332–333, 544–550. *See also* Cerebral
      asymmetry
Cerebral ischemia, 584–585, 585f
Cerebral microbleeds, dementia and, 595
Cerebral palsy, 256, 362
Cerebral ventricles, 44, 44f
Cerebral voltammetry, 236
Cerebrospinal fluid, 44, 398
   flow of, 82–83
   functions of, 82
   in hydrocephalus, 82–83
   in meningitis, 42
   production of, 82
   shunts for, 83
Cerebrum, 3–4, 4f, 18, 18f, 41f, 42. *See also under*
      Cerebral

Cervical spine, 61, 61f
Channelrhodopsins, 134
Channels. *See* Ion channels
Chemical bonds, 89
Chemical messengers. *See* Neurotransmitters
Chemical synapses, 143–144, 144f. *See also* Synapses
Chemistry basics, 88–89
Chemoaffinity hypothesis, 269–270, 270f
Children
    brain maps for, 362
    dendritic spine density in, 259
Chimeric animals, 103
Chimpanzees, 15, 19f, 20. *See also* Primates, nonhuman
Chlamydomonas reinhardtii, light-sensitive ion channels in, 134
Chloride ions. *See also under* Ion
    movement of, 117–118, 117f
    resting potential and, 119–121, 119f
Chlorpromazine, for schizophrenia, 185f
Choline, 152
Cholinergic system, 161, 161f
Chordates, 16–18, 16f, 18f. *See also* Spinal cord
Chromosomes, 92–93, 92f, 104–105
    abnormalities of, 101–102, 102f
    histones and, 104–105
    number of, 97f
        abnormal, 101–102, 102f
    sex, 97f, 98
        sex differences and, 553
Chronic cerebrospinal venous insufficiency, multiple sclerosis and, 589
Chronic traumatic encephalopathy, 582
Chronotypes, 454, 455
Cigarette smoking, nicotine addiction in, 195–196, 440
Cilia, of hair cells, 329, 330f, 334–335
Cingulate cortex, 58, 58f
Circadian genes, mutations in, 455
Circadian rhythms, 444, 446–449, 447t. *See also* Biorhythms
    cognitive, 457–458
    emotional, 457–458
    entrained, 449–451
    free-running, 447–449, 448f
    pacemaking, 455–456
Circannual rhythms, 447, 447t
    pacemaking, 456–457
Circulation, cerebral, 42–43, 43f
Cladogram, 17f, 18
Classes, 16, 16f
Classical conditioning, 483–484, 484f. *See also* Conditioning
    in substance abuse, 178, 197
Clinical trials, 13
Cloning, 102–103, 103f
Closed head injury. *See* Traumatic brain injury
Cocaine, 172, 190–191, 191f, 200
Coccygeal spine, 61, 61f
Cochlea, 328–329, 328f–330f
    hair cells of, 328f, 329, 330, 330f, 334–335
Cochlear implants, 335, 335f
Cochlear nucleus, 331, 331f
Codominance, 98
Codons, 94
Cognition, 525–564
    age-related changes in, 595
    in animals, 527, 550
    association cortex in, 530–538
    attention in, 533–535
    cell assemblies in, 527, 529
    cerebral asymmetry in, 538, 544–550
    characteristics of, 525–526
    convergent vs. divergent thinking and, 559

definition of, 525
    emotion and, 530–531
    evolution of, 526
    imitation and, 536–538
    inferential thinking in, cerebral asymmetry and, 550
    intelligence and, 557–560
    language and, 525–526
    multisensory integration in, 531–532, 532f
    neural circuits in, 526–529, 529
    neural unit of thought in, 526–529
    object knowledge in, 531
    planning and, 535–536
    psychological constructs in, 525
    sequential thinking and, 525–526
    sex differences in, 205, 551–554, 551f–553f
    sleep deprivation and, 468–469
    social, 541–542
    spatial, 532–533, 532f
        sex differences in, 551–554
    in split-brain patients, 524, 547–549, 550
    studies of, 538–543. *See also* Cognitive neuroscience
Cognitive-behavioral therapy
    for anxiety disorders, 602
    for depression, 188, 601
Cognitive development, 263–266, 264f, 265t, 266f
    cultural aspects of, 29–31
    in nonhuman primates, 30
Cognitive enhancers, 172
Cognitive neuroscience, 538–543
    applications of, 541–543
    methods of, 538–541
Cognitive organization. *See* Brain; Cerebral asymmetry; Topographic organization
Cognitive rhythms, 457–458
Cognitive stimulation
    behavior and, 400–401
    brain development and, 272–273
    brain size and, 507
    in learning, 506–507, 507f
Cognitive therapy, 579
Cogwheel rigidity, 590–591
Colliculus, 53, 53f, 69
    inferior, 331, 331f
Color agnosia, 315
Color blindness, 310
Color constancy, 312
Color mixing
    additive, 309, 309f
    subtractive, 309, 309f
Color vision, 309–312
    afterimages in, 310–311
    blobs in, 298, 298f, 311
    impaired, 310, 315
    opponent-process theory of, 310–312, 311f
    rods and cones in, 291–293, 291f, 292f
    subtractive color mixing and, 309
    trichromatic theory of, 309–310
Columns. *See* Cortical columns
Coma, 473
    reticular activating system in, 472–473
Common ancestor, 14–15, 20, 22, 22f
Competitive inhibitors, 189
Complex cells, of visual cortex, 306, 307f
Complex partial seizures, 586
Complex tones, 324–325
Computer-brain interface, 354
Computerized tomography (CT), 228–229, 229f
COMT gene, 237
Concentration gradient, 116, 117f
Concordance rate, 236
Concrete operational stage, of cognitive development, 264, 265t
Concurrent discrimination, 266, 266f, 274

Concussion, 581, 582. *See also* Traumatic brain injury
Conditioned response, 485
Conditioned stimulus, 484
Conditioning
    classical (Pavlovian), 483–484, 484f
    eye-blink, 484, 484f
    fear, 484, 500
    in learning, 408–409
    operant (instrumental), 485
    respondent, 483–484
    in substance abuse, 178–179, 197
Cones (photoreceptors), 291–293, 291f, 292f
Congenital adrenal hyperplasia, 435
Conscious memory. *See* Memory, explicit
Consciousness, 561–563
    definition of, 561
    nonunitary nature of, 479
    sleep and, 479
Conservation, in cognitive development, 263–264, 264f, 265t
Consolidation, of memory, 469–470, 499
Contralateral, definition of, 39
Contralateral neglect, 534–535, 535f
Convergent thinking, 559
Copulatory behavior. *See* Sexual behavior
Copycat (cloned cat), 103, 130f
Cornea, 289, 289f
Coronal, definition of, 39
Coronal section, 39f, 43–44, 43f, 44f
Corpus callosum, 43f, 45, 46f
    absence of, 104, 104f
    handedness and, 555–556
    severing of, cognitive effects of, 547–549
    visual, 302–303, 303f
Correlation, vs. causation, 266–267
Corsi block-tapping test, 217, 217f
Cortex. *See under* Cerebral cortex; Cortical
Cortical columns, 270, 270f, 297
    definition of, 297
    ocular-dominance, 270, 270f, 307, 308f, 311, 311f
    orientation, 307, 311, 311f
    in temporal lobe, 308–309, 308f
    in visual cortex, 270, 270f, 307, 308f
Cortical function mapping, 340–344, 342f, 366–367
Cortical layers, 47, 47f, 56–57, 57f, 69, 357
    development of, 254–255
    in motor cortex, 56–57, 357, 357f
    in neocortex, 357, 357f
    in occipital cortex, 297–299, 298f
Cortical lobes, 55–57, 55f. *See also* Frontal lobe; Occipital lobe; Parietal lobe; Temporal lobe
Cortical mapping, 340–344, 342f, 365–366. *See also* Brain maps
    homunculus and, 365–366, 365f. *See also* Homunculus
Cortical organization
    environmental influences in, 267–269, 268f
    sex differences in, 551–554
Cortical thickness
    developmental changes in, 259, 259f
    intelligence and, 260, 260f
    language and, 259, 259f, 262–263, 262f
    musical ability and, 346
    sex differences in, 552, 552f
Corticospinal tracts, 369–370, 369f
Cortisol, in stress response, 207, 207f, 600
Crack cocaine, 191f, 200
Cranial nerves, 41f, 60–61, 60f
Creativity, hyperconnectivity and, 541
Creolization, 339
Critical period, 106, 270–271
Cross-tolerance, 182–183

Crossed neural connections, 43, 66–67
cryptochrome gene, 455
CT scans, 228–229, 229f
Cultural bias, 542
Cultural influences, on brain development, 29–31, 35–36
Cultures, cell, 75
Curare, 177
Current, electrical, 112
Cyclic adenosine monophosphate (cAMP)
  caffeine and, 192
  in learning, 166–167, 168
Cytoarchitectonic maps, 56, 57f. See also Brain maps
Cytochrome P450 enzymes, 175
Cytosine, 92f, 93

D

Daily rhythms. See Circadian rhythms
Date rape drugs, 183
Deafferentiation, 381–382
Deafness, cochlear implants for, 335, 335f
Decibel (dB), 323
Decision making, 543
Declarative memory, 486, 487t
Deep brain stimulation (DBS), 13, 13f, 222–223, 518, 575, 592
Deinstitutionalization, antipsychotics and, 184, 185f
Delayed-alternation task, 497, 497f
Delayed-matching-to-sample test, 497, 497f
Delayed-response test, 497, 497f
Delta receptors, 188–189
Delta rhythms, 460
Dementia
  Alzheimer's, 70, 161, 494, 500, 593–595
    brain abnormalities in, 85, 169, 593–594, 593f, 594f
    in Down syndrome, 102
    Parkinson's disease and, 594
  concussion and, 582
  definition of, 589
  degenerative vs. nondegenerative, 589, 589t
  marginal high blood pressure and, 595
  types of, 589t
Dementia pugilistica, 582
Dendrites, 76, 76f, 77, 78f, 80, 90f, 133
  arborization of, 255, 255f, 504–5055, 505f
  changes in, 504–505, 505f. See also Neuroplasticity
  development of, 255, 255f
Dendritic spines, 77, 78f, 90f, 169
  age-related density of, 259
  growth of, 255, 255f, 504–505, 505f
  in learning, 169
Dendrodendritic synapse, 147f, 148
Deoxyribonucleic acid (DNA), 92–93, 92f
  methylation of, 105, 105f, 252
  modification of, 105, 105f
Dependence. See Substance abuse
Depolarization, 121f, 122–124
Depolarizing potentials, 133
Depression, 163, 186–187, 188, 425–427, 599–602
  adolescent-onset, 275
  in bipolar disorder, 599
  in brain injury, 227
  major, 163, 186–187, 188
  neurobiology of, 599
  postictal, 587
  seasonal, 450
  sleep problems in, 475–476
  stress and, 600–601
  treatment of, 601
    antidepressants in, 182t, 186–187, 188, 469, 578, 599, 601. See also Antidepressants

deep brain stimulation in, 575
electroconvulsive therapy in, 188, 576–577
transcranial magnetic stimulation in, 222, 576–577, 576f
Dermatomes, 61, 61f, 356–357, 357f
Development
  brain. See Brain development
  cognitive, 263–266, 264f, 265t, 266f
  gene expression in, 252
  growth spurts in, 264
  language, 255f, 259, 262–263, 338–339
  motor, 261–262, 262f
  prenatal
    cross-species similarities in, 248–250, 248f
    sexual differentiation in, 250, 251f
Developmental disability
  brain abnormalities in, 277–279
  causes of, 277, 277t
  in fetal alcohol spectrum disorder, 184
  in phenylketonuria, 569–570
Dextroamphetamine, 172
Diabetes mellitus, 204
Diagnostic and Statistical Manual of Mental Disorders (DSM), 181, 182t, 572–573, 573t
Diaschisis, 584
Diazepam, 182
Dichotic listening, 546
Diencephalon, 48, 48f, 51, 51f, 54, 54f
Diet. See also Nutrition
  brain size and, 24–25, 25f
  depression and, 186
  epigenetic effects of, 106
  multiple sclerosis and, 86
  weight-loss, 429
Diffusion, 116–118
  in neurotransmitter deactivation, 147
Diffusion tensor imaging, 230, 230f
  in tractography, 540–541, 540f
Digestive system, eating control and, 430, 430f
Dimers, 455
Disinhibition theory, 193–194
Displacement task, 266, 266f
Distal, definition of, 39
Distributed hierarchical model, 68, 68f
Distributed reinstatement theory, 499
Diurnal animals, 444
Divergent thinking, 559
Diving bradycardia, 140
DNA, 92–93, 92f
  methylation of, 105, 105f, 252
  modification of, 105, 105f
Dr. Death, 200
Dominant alleles, 98, 100f
Domino effect, in action potential propagation, 126, 126f
Domoic acid poisoning, 199, 199f
L-Dopa, 152, 153f, 154, 577, 592
Dopa-responsive dystonia, 74
Dopamine, 152, 153f, 160f, 161–163. See also Neurotransmitters
  in attention-deficit/hyperactivity disorder, 241
  in memory, 500
  in Parkinson's disease, 142
  reward and, 438f, 439
  in schizophrenia, 162–163, 185
  stimulants and, 190
  in substance abuse, 195, 196, 197
Dopamine hypothesis, 185
Dopamine neuron transplant, 162
  for Parkinson's disease, 518–519, 592–593
Dopaminergic system, 160f, 161–163, 571f
Dorsal, definition of, 38, 39
Dorsal-root ganglion neurons, 380–382, 381f
Dorsal roots, 62, 62f, 356
Dorsal spinothalamic tract, 382, 383f

Dorsal visual stream, 69–70, 69f, 296–297, 299f, 312, 531f
  as "how" pathway, 296, 316
  injury to, 316
  location of, 531f
  secondary somatosensory cortex in, 394
  in spatial cognition, 531
Dorsolateral prefrontal cortex, 530f
  theory of mind and, 541
Dorsomedial nucleus, 54, 54f
Dorsomedial prefrontal cortex, 419, 419f
Dorsomedial thalamic nucleus, 54, 54f
Dosage, 173–174
Down syndrome, 101–102, 102f
Dreams, 463–466
  activation–synthesis hypothesis for, 465
  anxiety, 465–466
  avoidance vs. approach behavior in, 466
  content of, 464–466
  evolutionary hypothesis for, 465–466
  function of, 465–472
  hallucinations and, 479
  in memory storage, 471
  in REM sleep, 460, 463–466
Drinking behavior, 432–433
  amygdala in, 418–419
  water intoxication and, 432–433
Drosophila melanogaster, learning in, 168, 168f
Drowsy state, 460
Drug(s)
  of abuse. See Substance abuse
  adverse effects of, 578
  agonist/antagonist, 175–177, 176f
  for behavioral disorders, 577–578
  blood-brain barrier and, 174–175
  classification of, 182t
  cross-tolerance to, 182–183
  dosage of, 173–174
  effects of, variability in, 177–179
  as environmental contaminants, 175
  excretion of, 175
  groupings of, 181–192, 182t
  individual responses to, 193–196
  mechanism of action of, 175–177, 175f, 176f
  metabolism of, 175
  movement disorders due to, 577–578
  neuroplasticity and, 513–515, 514f
  prenatal exposure to, brain injury from, 276
  psychoactive, 173
    antianxiety agents, 182–183, 182t, 426, 602
    antidepressants, 182t, 186–187. See also Antidepressants
    antipsychotics, 182t, 184–186, 578
    grouping of, 181–193
    mood stabilizers, 182t, 187
    names of, 182
    opioid analgesics, 182t, 187–190
    psychedelic/hallucinogenic stimulants, 182t, 189–190
    psychotropics, 182t, 190–192
    sedative hypnotics, 182–183, 182t
  routes of administration for, 173–174, 173f
  sensitization to, 179–181, 513–515
Drug dependence, 194–196
Drug-dependency insomnia, 476
Drug-induced behavioral sensitization, 514
Drug-induced psychosis, 186, 200, 201, 237
Drug tolerance, 177–179
DSM-IV-TR, 181, 182t, 572–573, 573t
Dualism, 7–8, 7f
Dunce, 168
Dura mater, 37, 37f
Dynorphins, 188–189
Dyskinesia, neuroleptics and, 185
Dyskinesias, 373

Dyslexia, 482
Dystonia, focal hand, 510

E

E cells, 454
Ear. *See also under* Auditory
   structure of, 328–330, 328f, 329f, 330f
   in vestibular system, 388–389, 388f
Eardrum, 328, 328f
Eating disorders, 275, 428
Eating/feeding behavior
   amygdala in, 418–419, 431–432, 431f
   aversive behavior in, 409, 439, 439f
   biological clocks and, 453, 458
   cognitive factors in, 431–432
   dieting and, 429
   digestive system and, 430, 430f
   eating disorders and, 428
   environmental factors in, 428
   of fly, 410
   hedonic reactions in, 439, 439f
   hypothalamus in, 399
   prefrontal cortex in, 431–432
   regulation of, 428–432
   reward and, 439–440
Echolocation
   in bats, 350–351, 351f
   in blind people, 333
Economic decision making, 543
Ecstasy (MDMA), 192, 200
EEG. *See* Electroencephalography (EEG)
Efferent nerves, 37, 37f
Efferent sensory signals, 356
Ego, 567
Elderly
   cognitive loss in, 595. *See also* Dementia
   sleep changes in, 470, 475
Electrical activity
   in brain, measurement of, 223–228
   in cell membrane, 118–128
   early studies of, 111–115
   in giant axon of squid, 114–115
   recording of
      electroencephalographic, 110, 113–114. *See also*
         Electroencephalography (EEG)
      microelectrodes in, 115–116
      oscilloscope in, 115
      single-cell, 136f
      voltmeter in, 112, 113
   during sleep, 459f, 460–461
   stimulation of, 111–113, 112
   voltage gradient and, 116–117
Electrical charges, 116–118
Electrical potentials, 112
   action, 122–126. *See also* Action potential
   definition of, 112
   excitatory postsynaptic. *See* Excitatory
      postsynaptic potentials
   graded, 121–122, 121f
      summation of, 131–132, 131f, 132f
   inhibitory postsynaptic, 129–132, 131f, 132f
   miniature postsynaptic, 146
   resting, 119–121, 119f
   threshold, 123
Electrical recording studies, 112, 113–114
Electrical self-stimulation, 221
Electrical stimulation studies, 111–113
Electrical synapse, 144–145
Electricity, 112
   early studies of, 111–115
Electroconvulsive therapy, 188, 576
Electrocorticography (ECoG), 224
Electrodes, 112
Electroencephalography (EEG), 110, 113, 213, 224–225
   alpha rhythms in, 460

applications of, 224–225
   beta rhythms in, 460
   in brain mapping, 225–226, 225f
   definition of, 113
   delta rhythms in, 460
   desynchronized tracings in, 473
   event-related potentials and, 225–226, 225f
   in graded potential recording, 224–225, 224f
   in sleep studies, 224–225, 225f, 459–460, 459f,
      461f, 472–475
   vs. magnetoencephalography, 227
   in waking state, 459f, 460, 461f, 473–474
Electrolytic lesions, 221
Electromyography, in sleep studies, 459, 459f
Electron microscope, 143, 143f
Electrons, 88, 89, 112
Electrooculography, in sleep studies, 459, 459f
Elements, chemical, 88, 88f
Eliminative materialism, 12
Embodied language, 5
Embryonic disc, 249, 249f
Embryos
   cross-species similarities in, 248–250, 248f
   development of, 248–253, 248f–250f
   preformation and, 248
Emotion
   anterior cingulate cortex and, 530–531
   definition of, 398
   expression of, 411, 422–423, 542
   regulation of, 542
Emotional behavior, 399, 417–419, 422–427
   abnormal, 425–427
   amygdala in, 58, 58f, 59, 418–419, 423–424
   in animals, 411
   control of, 542
   expressive, 411, 422–423
   facial expressions and, 424–425
   generation of, 422–423
   James–Lange theory of, 422
   limbic system in, 58–59, 58f, 417–419, 417f, 418f
   overview of, 422–423
   prefrontal cortex in, 424–425
   somatic marker hypothesis for, 422–423
   tone of voice and, 425
Emotional disorders, 425–427
Emotional memory, 500–501, 501f
Emotional pain, 398
Emotional rhythms, 457–458
Emotional therapy, 579–580
Empathy, 541
Encephalitis, 42
   L-dopa for, 152, 153f, 154
   memory loss in, 492
Encephalization, 16
Encephalization quotient (EQ), 22, 23f, 34
End foot, 77–78, 78f
End plate, 136, 136f
Endocrine glands, 202–203. *See also* Hormones *and
   specific glands*
   regulation of, 202–203, 202f, 412–416. *See also*
      Hypothalamus
Endolymph, 389
Endomorphins, 188–189
Endoplasmic reticulum, 90f, 91, 94
Endorphins, 154, 188–189
Endothelial cells, in blood-brain barrier, 174, 174f
Enkephalins, 154, 188–189
Entorhinal cortex
   in Alzheimer's disease, 594
   in memory, 493–494, 493f
Entrainment, 449–451, 451f
Environment, epigenetics and, 11. *See also*
   Epigenetics
Environmental enrichment
   behavior and, 400–401

brain development and, 272
   brain size and, 507
   intelligence and, 560
   in learning, 506–507, 507f
Enzymes, 95
Ependymal cells, 82–83, 82t
Epidemiology, 572
Epidermal growth factor, 252
   in brain injury, 520
Epidural anesthesia, 387
Epigenetic code, 103–106
Epigenetic drift, 238
Epigenetics, 11
   applications of, 238
   autism spectrum disorders and, 256
   brain/behavioral disorders and, 568
   changes across populations and, 238
   critical periods and, 106
   DNA modification and, 105, 105f
   frontal lobe development and, 259–260, 272
   histone modification and, 105, 105f
   intelligence and, 560
   life experiences and, 236, 238
   neuroplasticity and, 515
   neurotoxins and, 198, 199
   phenotypic plasticity and, 36, 103–106
   reflective vs. reflexive systems and, 543
   RNA modification and, 105, 105f
   schizophrenia and, 598
   sex hormones and, 274–275
   sexual orientation and, 436
   stress effects and, 238, 272
   substance abuse and, 198, 199
   synaptic organization and, 560
   twin studies and, 236, 238
Epilepsy, 110, 585–589. *See also* Seizures
   brain surgery for, 342
Epinephrine (adrenaline), 140f, 141, 152, 153f. *See
   also* Neurotransmitters
   in stress response, 206–207
Episodic memory, 469, 488–489, 489f
Equilibrium, vestibular system in, 388–389, 388f
Estradiol, in brain masculinization, 434
Estrogen. *See also* Sex hormones
   activating effects of, 434–435
   in brain development, 204–205, 273–274,
      551–554
   in brain masculinization, 435
   cognitive function and, 205
   functions of, 204–205
   lifelong effects of, 274–275
   neuroplasticity and, 512, 512f
   sexual behavior and, 434–435
Event-related potentials, 225–226, 225f, 226f
   vs. magnetoencephalography, 227
Evolution
   adaptive behavior and, 30
   of behavior, 406–408
   of brain, 13–31, 34, 48–49, 48f
   brain–behavior link and, 19–26
   of cognition, 526
   common ancestor in, 14–15, 20, 22, 22f
   culture and, 29–31
   Darwin's theory of, 8–11
   dreams and, 465–466
   hierarchical organization and, 35–358
   of language, 8, 320, 337–339
   of music, 320
   natural selection and, 8–11
   neoteny and, 26
   radiator hypothesis and, 25–26
   of sex differences, 553–554
Evolutionary psychology, 406–408
Excitation, 64
   vs. inhibition, 70–71

Excitatory postsynaptic potentials, 129–132, 131f, 132f
  in habituation, 165
Excitatory synapses, 148–149, 149f
Executive functions, of frontal lobe, 419–421
Exelon (rivastigmine), 494
Expectations, 542
Experimental animals, 102–103, 240–243. *See also* Animal research
Explicit memory. *See* Memory, explicit
Extensors, 371
External ear canal, 328, 328f
Extinction, 534, 535f
Extracellular fluid, 90f, 91
Extracellular recordings, 227
Extrastriate cortex, 298, 299f
Eye. *See also under* Vision; Visual
  structure of, 287–291, 289f
Eye-blink conditioning, 484, 484f

F
Facial agnosia, 300
Facial expressions
  innate releasing mechanisms for, 407, 407f
  interpretation of, 425, 535
  loss of, 424–425
  production of, 425
Facial nerve, 60–61, 60f
Facial paralysis, 63
Families, 16, 16f
Farsightedness, 288
Fatal familial insomnia, 475
Fear, emotional memory and, 500
Fear conditioning, 484, 500
Feedback loops, hormonal, 415–416, 415f
Feeding behavior. *See* Eating/feeding behavior
Festination, 591
Fetal alcohol spectrum disorder, 184
Fight-or-flight response, 64, 158–159, 206–207, 206f
Filopodia, 169, 257, 257f
Fissures, 41f, 42
Flehmen, 404f
Flexors, 371
Flocculus, 375, 375f
Fluid homeostasis, 432–433
Fluoxetine, 187, 601
Flupentixol, sensitization to, 179
Fly, feeding behavior of, 410
fMRI, real-time, 580
Focal hand dystonia, 510
Focal seizures, 587
Follicle-stimulating hormone (FSH), 415t
Foraging, brain size and, 24–25, 25f
Forebrain, 3, 4f, 48, 48f, 54–55, 55f
  in movement initiation, 358–360, 358f, 360f
  structure of, 54–55, 55f
Foreign languages
  cortical localization for, 269
  learning of, 339
Formal operational stage, of cognitive development, 264, 265t
Fovea, retinal, 289, 289f, 290, 302, 302f
Free-running rhythms, 447–449
Frequency, of sound waves, 321–322, 322f, 324
  in music, 326, 327f
Frontal, definition of, 39
Frontal leukotomy, 424f, 425
Frontal lobe, 41f, 55f, 56
  agenesis of, 421
  in association cortex, 530, 530f
  in attention, 535–536
  in cognition, 526, 535–536
  development of, 259–260, 260f
  epigenetic effects on, 259–260, 272
  executive functions of, 419–421

injury to, 56
  loss of libido and, 437
  in memory, 496–497
  in planning, 535–536
  in schizophrenia, 278
  in sequential movement, 526
  in sexual behavior, 437
  structure and function of, 55f, 56, 419–420, 419f, 420f
Frontal lobotomy, 424f, 425
Fruit foraging, brain size and, 24–25, 25f
Functional connective magnetic resonance imaging (fcMRI), 538–540, 539f
Functional levels, 50–59, 67
Functional magnetic resonance imaging (fMRI), 231–232, 232f
  in brain mapping, 538–540, 539f
  real-time, 580
  resting-state, 232
Functional near-infrared spectroscopy (fNIRS), 212, 234, 235f
Fundamental frequency, 324, 324f, 326
Fusiform face area, 299–300

G
G factor, 557–558
G proteins, 156, 156f
GABA (gamma-aminobutyric acid), 153, 153f, 571
GABA$_A$ receptor, drug effects at, 183
Gamma-hydroxybutryic acid (GHB), 183
Ganglia, 16, 17
  basal. *See* Basal ganglia
  dorsal-root, 380–382, 381f
  parasympathetic, 64
  sympathetic, 64
Ganglion cells
  melanopsin, 455
  off-center, 304f, 305
  on-center, 304f, 305
  retinal. *See* Retinal ganglion cells
Gap junctions, 144–145, 144f, 148
Gaseous neurotransmitters, 155
Gastrointestinal system, eating control and, 430, 430f
Gate theory of pain, 386–387, 386f
Gated ion channels, 96–97, 96f, 120–121, 123f. *See also* Ion channels
Gender differences. *See* Sex differences
Gene(s), 10–11, 92–93, 97–106
  alleles and, 98, 100f
  identification of, 236–237
  transcription of, 94f
Gene expression, 10, 98, 103–104, 104f
  critical period for, 106
  definition of, 237, 252
  in development, 252
  epigenetics and, 11, 103–106, 237–238. *See also* Epigenetics
  measurement of, 237–238
  methylation and, 252, 252f
Gene knockout, 103
Gene methylation, 105, 105f, 252
General intelligence, 28, 557–558
General-purpose neurons, 252–253
General stimulants, 192
Generalized anxiety disorder, 426, 602. *See also* Anxiety disorders
Generalized seizures, 587
Genetic code, 94, 94f
Genetic diseases
  diagnosis of, 74
  inheritance of, 99–102
Genetic engineering, 102–103, 103f
Genetic mutations, 98–102
  beneficial, 98–99
  disease-causing, 99–102

learning and, 168, 168f
  in selective breeding, 102
Genetic studies, 236–238
Genetic testing, 74
Geniculate nuclei. *See* Lateral geniculate nucleus; Medial geniculate nucleus
Geniculostriate pathway/system, 294–295, 295f, 296, 296f
Genome sequencing, 74
Genotype, 10
Genus, 16, 16f
Giant axon of squid, electrical activity in, 114–115, 114f
Glabrous skin, sensory receptors in, 379
Glasgow Coma Scale, 13
Glial cells, 46, 47f, 82–87, 148
  in blood–brain barrier, 173
  development of, 260–261
  formation of, 253
  functions of, 82t
  in nerve repair, 85–87, 85f
  properties of, 82t
  radial, 254, 254f
  types of, 82–87, 82f
Glial uptake, in neurotransmitter deactivation, 147
Glioblasts, 251
Gliogenesis, 253
Gliomas, 83
Globus pallidus, 57–58, 58f
  electrode placement in, 575, 592
  in movement, 373f, 374–375
  in Parkinson's disease, 374–475, 592
Glomeruli, olfactory, 402f, 403
Glossopharyngeal nerve, 60–61, 60f
Glucocorticoids, 203. *See also* Hormones
  neurotoxicity of, 513
Glucose, in eating behavior, 430
Glutamate, 153, 153f
  in domoic acid poisoning, 199
  in long-term potentiation, 503–504, 503f
  neurotoxicity of, 199
Glutamate psychedelics, 192
Glutamate receptor, 199
Glycine, 153
Golgi body, 90f, 91, 95–96, 95f
Gollin figure test, 486, 486f
Gonadal hormones. *See* Sex hormones
Graded potentials, 121–122, 121f
  EEG recording of, 224–225, 224f
  summation of, 131–132, 131f
Grammar, universality of, 339, 526
Grand mal seizures, 587
Grasping
  development of, 262, 262f
  hapsis and, 379f, 380
  movements in, 359, 359f
  vision in, 299, 312, 315, 315f
Gray matter, 43–44, 43f, 46–47, 47f
  layers of, 56–57, 57f
  periaqueductal, 53, 53f
    in pain, 387–388
  in reticular formation, 52
  sex differences in, 552, 552f
  spinal, 62f
  thickness of, language development and, 259, 259f, 262f, 263
Great apes, 20. *See also* Primates, nonhuman
Grooming behavior, 361
Growth cones, 257, 257f
Growth factors, 84
  in brain development, 252
  neuroplasticity and, 84, 513, 513t
Growth hormone (GH), 415t
Growth spurts, 264–265
Guanine, 92f, 93

Gustation, 404–406
Gynandromorphs, 275
Gyrus, 42, 55, 55f
    Heschl's, 332, 332f, 346, 346f
Gyrus fornicatus, 38

H

Habituation, 164–168, 165f, 502
    dendritic spines in, 169
Hair cells
    in auditory system, 328f, 329, 330, 330f, 334–335
    in vestibular system, 388f, 389
Hairy skin, sensory receptors in, 379
Hallucinations
    dreams and, 479
    hypnogogic, 477
    in schizophrenia, 186, 278
    in substance abuse, 200
Hallucinogenic amphetamine, 200
Halorhodopsin, 134
Hammer, 328, 328f
Handedness, cortical organization and, 332–333,
    554–556
Hapsis, 379f, 380
Haptic-proprioceptive pathway, 382–383, 383f
Head trauma. See Traumatic brain injury
Headaches, migraine, scotoma in, 282
Hearing. See also Auditory system; Sound
    in animals, 322, 323f
    evolution of, 327
    mechanics of, 328–330, 328f–330f
    movement and, 332
    neural activity in, 334–337
    in owls, 336, 337f
Hebb synapse, 164
Hebb–Williams mazes, 267–268, 268f
Helix
    DNA, 92f, 93
    protein, 94, 94f
Hemianopia, homonymous, 313
Hemispherectomy, 42
Hemispheres. See Cerebral hemispheres
Hemorrhagic stroke, 45
Heritability. See Inheritance
Heroin, 189, 189f
    synthetic, parkinsonian symptoms from, 162
Herpes simplex encephalitis, amnesia in, 492
Hertz (Hz), 321–322
Heschl's gyrus, 332, 332f
    in music processing, 346, 346f
Heterosexuality, 436–437
Heterozygous alleles, 98
Hibernation, circannual rhythms and, 456–457
Hierarchical organization, of nervous system, 50–59,
    68, 68f
High blood pressure, marginal, dementia and, 595
Higher vocal control center, in birds, 349, 349f, 350
Hindbrain, 48, 48f, 51–53, 51f–53f
Hippocampus, 57, 58f, 417, 417f
    antidepressants and, 187
    in domoic acid poisoning, 199f
    epigenetic differences in, 238
    in learning, 215, 216
    in memory, 215, 216
        in consolidation, 499
        explicit, 499
        spatial, 495–496, 495f, 496f
    mood and, 601
    neurogenesis in, 506, 506f
        antidepressants and, 601
    place cells of, 470
    in schizophrenia, 278
    in sleep, 470
    stress and, 187, 206f, 207–208, 601
Histamine, 152

Histones, 104–105
Home environment. See also Environmental
    enrichment
    experimental effects of, 179–180
Homeostasis, definition of, 203
Homeostatic hormones, 203–204
Homeostatic mechanisms, 411, 411f
Homicide, 408
Hominids, 20–21
Homo erectus, 21, 21f, 22, 23f, 24
Homo floresiensis, 21
Homo habilis, 21, 21f, 24
Homo neanderthalensis, 21, 21f
    music-making by, 320
Homo sapiens, 16. See also Humans
    evolution of, 20–31
Homonymous hemianopia, 313
Homosexuality, 436–437
Homozygous alleles, 98
Homunculus
    cerebellar, 375–376, 375f
    motor, 365–366, 365f, 370–371
    somatosensory, 390–391, 393f
Horizontal, definition of, 39
Horizontal cells, retinal, 293, 293f
Horizontal section, 39f
Hormones, 141, 202–208
    in brain development, 250, 272–275
    classification of, 203
    cognitive function and, 205
    early studies of, 202
    eating behavior and, 430–431
    feedback loops for, 415–416, 415f
    functions of, 203
    glucocorticoid, 203
        neurotoxicity of, 513
    homeostatic, 203–204
    neuropeptide, 153–155
    neuroplasticity and, 512–513, 512f
    organizational hypothesis and, 204
    peptide, 203
    pituitary, 202, 202f, 414–416, 415t
    receptors for, 202f, 203
    regulation of, 202–203, 202f, 412–416, 413f–415f.
        See also Hypothalamus
    releasing, 414–415, 415t
    sex (gonadal). See Sex hormones
    stress, 206–208, 206f, 207f, 600–601, 600f
    target glands of, 202, 202f
    thyroid, 415, 415t
Horror autotoxicus, 588
HPA axis, stress effects on, 600–601
Humans
    apes and, 19f, 20
    as chordates, 16–18, 18f
    early ancestors of, 20–21, 21f
    evolution of, 29–31. See also Evolution
    as hominids, 20–21
    as primates, 19–20, 19f
    taxonomy of, 16–17, 16f
Huntington's disease, 100–101, 100f, 103, 373
Hydrocephalus, 83
Hydrogen bonds, 89
Hyperactivity, 163
    in attention-deficit/hyperactivity disorder, 163,
        240–241
    in brain stimulation studies, 222
Hypercomplex cells, of visual cortex, 306, 307f
Hyperconnectivity, 541
Hyperkinetic rats, 222
Hyperkinetic symptoms, 373
Hyperopia, 288
Hyperphagia, 431
Hyperpolarization, 121f, 122–124
Hyperthymestic syndrome, 489

Hypnogogic hallucinations, 477
Hypoactivity, in brain stimulation studies, 222
Hypocretin, 478
Hypoglossal nerve, 60–61, 60f
Hypoglycemia, 204
Hypokinetic rats, 222
Hypokinetic symptoms, 373
Hypothalamic-pituitary-adrenal (HPA) axis, stress
    effects on, 600–601
Hypothalamus, 54, 54f
    amygdala and, 418
    in behavior generation, 416–417, 417f
    in biorhythms, 452–454
    electrical stimulation of, 221
    in feeding behavior, 399, 430–431, 431f
    in homeostasis, 411–416, 411f
    in hormone regulation, 412–416, 413f–415f
    limbic system and, 417–419
    neurohormone secretion by, 202f, 203
    in nonregulatory behaviors, 412
    pituitary gland and, 412–416, 414f, 600–601
    in regulatory behaviors, 411–412
    in sexual behavior, 435–437, 436f
    in sexual orientation, 436–437
    in stress response, 206, 206f, 207f
    in temperature regulation, 411–412
Hypovolemic thirst, 433

I

Id, 567
Idiopathic seizures, 586
Illusions, perceptual, 286, 286f
Imagination, 559
Imaging studies, 228–234, 239, 239t
    in brain/behavioral disorders, 573–574, 574f, 583
    in brain mapping, 538–541, 539f
    dynamic, 231–234
    static, 228–230
    in traumatic brain injury, 582, 583
Imitation, 536–538
Immune system, 84
Implicit memory. See Memory, implicit
Imprinting, 271, 271f
Incentive-sensitization theory, 196–197, 197f
Induced neurogenesis, 575–576, 592–593
Infants, brain maps for, 362
Inferential thinking, cerebral asymmetry and, 550
Inferior, definition of, 38, 39
Inferior colliculus, 53, 53f, 69, 331, 331f
Information flow
    in brain, 294–295, 295f
    Descartes's theory of, 110–111
    electrical activity and, 113–114, 114f
    in somatosensory system, 356–357, 356f
    in visual system, 294–295, 295f
Infradian rhythms, 447, 447t
Inheritance
    of dominant traits, 98, 100f
    of genetic diseases, 99–102
    of intelligence, 560
    Mendelian, 99–102, 100f
    of recessive traits, 98, 100f
Inherited behavior, vs. learned behavior, 5–6, 6f
Inherited traits, 10–11, 99–102, 100f
Inhibition, vs. excitation, 64, 70–71
Inhibitory postsynaptic potential, 129–132, 131f, 132f
Inhibitory synapses, 148–149, 149f
Innate releasing mechanisms, 406–407
Insomnia, 475–476
    brainstem injury and, 479
    drug-dependency, 476
    fatal familial, 475
Instrumental (operant) conditioning, 485
Insula, 332f, 333
    in gustation, 405–406

Insulin, 204
Integration, in neurons, 128–133
Intellectual potential, 560
Intelligence, 557–560. *See also* Cognition
  animal, 527
  artificial, 80–81
  bodily-kinesthetic, 558–559
  brain size and, 28–29, 34, 557–558, 558f
  convergent, 559
  cortical thickness and, 260, 260f
  divergent, 559
  epigenetics and, 560
  evolution of, 29–31
  frontal lobe development and, 260, 260f
  general, 28, 557–558
  heritability of, 560
  intellectual potential and, 560
  interpersonal, 558–559
  intrapersonal, 558–559
  linguistic, 558–559
  logical-mathematical, 558–559
  multiple, 28, 558–559
  musical, 558–559
  network efficiency and, 560
  observed, 560
  spatial, 558–559
  synaptic organization and, 560
Intelligence A, 560
Intelligence B, 560
Intelligence tests, 559
Interblobs, 298
Intergeniculate leaflet, 452, 453, 454
International Classification of Diseases (WHO), 572
Interneurons, 79, 79f
  motor, 370
Interpersonal intelligence, 558–559
Intoxication, toxin action at synapses in, 176–177
Intracellular fluid, 90f, 91
Intracellular recordings, 227
Intracranial self-stimulation, 438
Intrapersonal intelligence, 558–559
Invertebrates, learning in, 48–49
Ion(s), 88, 89, 89f
  electrical charge of, 116–118, 117f
  movement of, 116–118
Ion channels, 96–97, 96f, 117–118, 119f
  gated (voltage-sensitive), 96–97, 96f, 120–121, 123f, 124f
  in learning, 165–166, 165f, 167f
  light-sensitive, 134
  in muscle contraction, 136, 136f
  in nerve impulse production, 128, 135f
  in sensitization, 166–167, 167f
  in sensory processing, 135–138, 135f
  stretch-sensitive, 135
  transmitter-sensitive, 136
Ion pumps, 96–97, 96f, 119–121, 119f
Ionic bonds, 89
Ionotropic receptors, 155, 155f
Ipsilateral, definition of, 39
Iris, 289, 289f
Ischemia, cerebral, 584
Ischemic stroke, 45, 584–585
Itch, 363, 385–386

J
Jacksonian focal seizures, 586
James–Lange theory, 422
Jerison's principle of proper mass, 302
Jet lag, 451, 451f

K
K-complexes, 470
Kainic acid poisoning, 199
Kappa receptors, 188–189

Ketamine, 183, 192, 199
Kingdoms, 16, 16f
Klüver–Bucy syndrome, 423–424
Knee jerk reflex, 51, 384, 384f
Knockout technology, 103
Korsakoff's syndrome, 497–499

L
L-dopa, 152, 153f, 154, 577, 592
Landmark-learning task, 218f, 219
Landmarks, visual processing of, 299, 532
Language. *See also* Speech
  in animals, 8, 10, 527
  birdsong and, 348–350
  Braille and, 549
  Broca's area for, 255f, 263, 339–340, 340f, 343f, 344
    mapping of, 340–344, 342f
  cognitive functions of, 525–526
  consciousness and, 563
  cortical development and, 259
  cortical localization of, 70, 332–333, 339–344, 538, 549–550
    mapping of, 340–344
    for second languages, 269
    sodium amobarbital test for, 554, 555
  creolization of, 339
  development of, 255f, 262–263, 338–339
  embodied, 5
  evolution of, 8, 320, 337–339
  genetic aspects of, 338–339
  handedness and, 332–333, 554–556
  lateralization for, 332–333, 538, 549–550
  movement and, 549–550
  music and, 320
  pidgin, 339
  processing of, 338–344
  properties of, 326
  second
    cortical localization for, 269
    learning of, 339
  self-regulation and, 542
  sex differences in, 551–554
  as sound, 325–326
  structural uniformity of, 339
  syntax of, 339, 525–526
  vs. birdsong, 348–349
  Wernicke's area for, 332, 332f, 340–344, 340f
    mapping of, 340–344, 342f
Language acquisition, critical period for, 339
Language deficits
  in autism spectrum disorder, 256
  in carbon monoxide poisoning, 314
Language test, 8
Lark chronotype, 454, 455
Latent content, of dreams, 464
Lateral, definition of, 38, 39
Lateral corticospinal tract, 369f, 370, 370f
Lateral fissure, 41f, 42, 55f, 56
Lateral geniculate nucleus, 54, 54f, 296, 296f, 297f, 301–302, 301f, 302f
  receptive field of, 301–302
Lateral hypothalamus, in eating, 431, 431f
Lateral sulcus, 43f
Lateral ventricles, 43f
Lateralization, 332–333. *See also* Cerebral asymmetry
  for auditory processing, 546
  for cognition, 538, 544–550
  handedness and, 332–333, 554–556
  for language, 332–333, 538, 549–550
  for music, 345–346, 546
  overlapping functions and, 305, 305f, 549
  in split-brain patients, 524, 547–549, 550
  for visual processing, 545, 546f. *See also* Visual fields

Laughter, during tickling, 392
Law of Bell and Magendie, 62, 63
Learned behavior, 5–6, 6f, 408–409
Learned taste aversion, 409
Learned tolerance, 178
Learning, 5–6, 483–485. *See also* Memory
  associative, 502
  behavior and, 408–409
  cognitive enhancers and, 172
  conditioning in, 408–409, 483–485, 484f. *See also* Conditioning
  definition of, 164, 483
  dendritic spines in, 169
  enriched experience in, 506–507, 507f
  habituation in, 164–168, 165f, 502
  hippocampus in, 215, 216
  in invertebrates, 48–49
  long-term potentiation and, 502–504
  mutations affecting, 168, 168f
  neurogenesis in, 507
  neuronal changes in, 167–168
  neuroplasticity and, 35–36, 271–272, 502–504, 510–511. *See also* Neuroplasticity
  object-reversal, 274
  preparedness and, 409
  second messengers in, 167–168, 167f
  sensitization in, 166–167, 167f
  sites of, 488
  during sleep, 468–470
  studies of, 483–485
  in substance abuse, 178–179, 197
  synapses in, 164–168
    loss or formation of, 167–168
    structural changes in, 167–168
Learning disabilities, 482
Left-handedness. *See also* Cerebral asymmetry
  cortical organization and, 332–333, 554–556
  lateralization and, 332–333
Lens, 289, 289f
  in refractive errors, 288
Lesions, brain. *See* Brain injury
Levels of function, 50–59, 67
Lewy body, 594, 594f
Liberation therapy, 589
Libido, loss of, 437
Life forms, classification of, 16–17, 16f
Light
  circadian rhythms and, 449–451
  perception of, 288
  properties of, 288, 288f
  receptors for, 284, 287–293, 289f, 291f, 292f
    retinal ganglion cells as, 295, 300–301, 453
  in seasonal affective disorder, 450
  wave form of, 288, 288f
Light pollution, 450
Light-sensitive ion channels, 134
Limbic system, 55, 57–58, 58f, 417–419
  in eating, 418–419, 431–432
  in sexual behavior, 435–436
  structure and function of, 417–419, 417f, 418f
Linguistic intelligence, 558–559
Lithium, for bipolar disorder, 187
Lobotomy, frontal, 424f, 425
Locked-in syndrome, 354, 361–362
Logical-mathematical intelligence, 558–559
Loglio vulgaris
  electrical activity in, 114–115
  giant axon of, electrical activity in, 114f
Long-term depression, 503
Long-term memory, 488
Long-term potentiation, 502–504, 503f
Longitudinal fissure, 41f, 42, 55, 55f
Lordosis, in copulation, 435
Lou Gehrig's disease, 137, 354, 361

Loudness, 322f, 323–324, 323f
  detection of, 336
  of music, 326
LSD (lysergic acid diethylamide), 192, 200
Lumbar spine, 61, 61f
Luminance contrast, 305, 305f
Luteinizing hormone (LH), 415t
Lysosomes, 90f, 91

M

M cells, 293–297, 293f, 454
Magnesium ions. See also under Ion
Magnetic resonance imaging (MRI), 229–230, 229f, 537
  in diffusion tensor imaging, 230, 230f
  functional, 231–232, 232f
    in brain mapping, 538–540, 539f
    connective, 538–540, 539f
    real-time, 580
    resting-state, 232
Magnetic resonance spectroscopy (MRS), 230, 583
Magnetoencephalography (MEG), 226–227
Magnocellular (M) cells, 293–297, 293f, 454
Magnocellular nucleus of medulla, in sleep, 474
Major depression. See Depression
Malnutrition, epigenetic effects of, 106
Mania, 163, 187, 599. See also Bipolar disorder
Manifest content, of dreams, 464
MAO inhibitors, 186–187
Maps. See Brain maps
Marginal high blood pressure, dementia and, 595
Marijuana, 191–192, 200, 201, 237
Masculinization, of brain, 204–205, 272–274, 433–434
Matching-to-place learning task, 218f, 219
Materialism, 8–11
  eliminative, 12
  religion and, 13
Mathematical ability, 558
Mathematical intelligence, 558–559
Mating behavior. See Sexual behavior
MDMA (ecstasy), 192, 200
Medial, definition of, 38, 39
Medial forebrain bundle, 413, 413f
  reward and, 438
Medial geniculate nucleus, 331, 331f
Medial lemniscus, 382, 383f
Medial pontine reticular formation (MPRF), in REM sleep, 474, 474f, 475f
Medial preoptic area, in sexual behavior, 435–436
Medial temporal region, in explicit memory, 493–499
Medial thalamus, in memory, 497–499, 500f
Median raphé, in brain activation, 473–474, 473f
Mediterranean diet, 429
Medulla, 52–53, 53f
Melanopsin, 453
Melanopsin ganglion cells, 455
Melatonin
  biorhythms and, 456–457
  in sleep, 472
Membrane
  basilar, in hearing, 328f, 329, 329f, 330, 330f
  cell. See Cell membrane
  postsynaptic, 144, 144f, 146
  presynaptic, 144, 144f, 146
  refractory, 124–125, 126
  tectorial, of inner ear, 328f, 329
Membrane potential. See Electrical potentials
Membrane proteins, 95f, 96–97, 96f
Memory. See also Learning
  accuracy of, 487–488
  autobiographical, 488–489
  basal ganglia in, 491–492, 500
  brain areas for, 70
  classification of, 485–487, 487t

consolidation of, 469–470, 499
  reconsolidation in, 470, 499
declarative, 486, 487t
definition of, 483
distribution of, 488–489, 488f
emotional, 500–501, 501f
encoding in, 487–488
episodic, 469, 488–489, 489f
explicit, 486–487, 487t
  consolidation of, 499
  deficits in, 489
  long-term potentiation and, 502–504
  Mishkin model of, 497–499, 500, 500f
  neural circuits for, 493–499, 493f, 500f
  NREM sleep and, 470
false, 488
hippocampus in, 215, 216
implicit, 469, 486–487, 487t
  deficits in, 491–492
  neural circuits for, 500, 500f
  NREM sleep and, 470
limbic system in, 57
long-term, 488
medial temporal region in, 493–499, 500f
medial thalamus in, 497–499, 500f
parahippocampal cortex in, 493–494, 493f
personal, 488–489, 489f
priming of, 487
procedural, 486, 487t
processing of, 488
short-term, 488
  tests of, 496–497, 497f
sites of, 488, 488f
spatial, hippocampus in, 495–496, 495f, 496f
storage of, 488–489
  neuroplasticity and, 502–512
  during sleep, 471
structural basis of, 502–516. See also Neuroplasticity
studies of, 485–487
terminology of, 487t
visuospatial, 493–494
Memory deficits
  in Alzheimer's disease, 494, 500. See also Alzheimer's disease
  in amnesia, 486, 488–489, 489f. See also Amnesia
  brain lesions and, 490–492
  explicit-memory, 490–491
  implicit-memory, 491–492
  in Korsakoff's syndrome, 497–499
  normal age-related, 595
  in Parkinson's disease, 491–492
  in substance abuse, 199, 200
Memory trace, 483, 490
Menarche. See Menstruation
Ménière's disease, 389
Meninges, 37, 37f
Meningiomas, 83
Meningitis, 40, 42
Menstruation
  cognitive function and, 205
  as infradian biorhythm, 447
  neuroplasticity and, 512, 512f
  synchronized cycles in, 403
Mental illness. See Psychiatric disorders
Mental retardation. See Developmental disability
Mentalism, 6–7, 12
Mescaline, 192
Mesencephalon (midbrain), 48, 48f, 51–52, 51f, 53, 53f
Mesolimbic dopaminergic system, 162–163
  reward and, 438–439, 438f
  structure and function of, 438, 438f
  in substance abuse, 197
Messenger RNA, 94, 94f

Met allele, 237
Metabolic syndrome, 444, 450–451
Metabolism, drug, 175
Metabotropic receptors, 155–157, 156f
Metaplasticity, 515–516
Metastatic brain tumors, 83
Metencephalon, 48, 48f
Methamphetamine, 190
Methylation, 105, 105f, 252
Methylphenidate, 172
Microdialysis, 236, 236f
Microelectrodes, 115–116
Microfilaments, 90f, 91
Microglia, 82t, 84–85
  in neuron repair, 85–86
Microscope, electron, 143
Microsleep, 469
Microtubules, 95–96, 95f
Microvilli, of taste buds, 405, 405f
Midbrain, 48, 48f, 51–52, 51f, 53, 53f
Middle cerebral artery, 43, 43f
Migraine, scotoma in, 282
Migration, neuronal, 253–254
Mild cognitive impairment, 595
Mind, 6–7. See also Consciousness
  Freud's theory of, 567
Mind–body problem, 8
Miniature postsynaptic potentials, 146
Minimally conscious state, 12
Minor tranquilizers, 182–183
Mirror-drawing task, 217, 217f
Mirror neurons, 359, 536–538
Mishkin model, for explicit memory, 497–499, 500, 500f
Mitochondria, 90f, 91, 145
Molecules, 89
Monkeys. See Primates, nonhuman
Monoamine oxidase (MAO) inhibitors, 186–187
Monocular blindness, 313
Monosodium glutamate (MSG), neurotoxicity of, 199
Monosynaptic reflex, 384
Mood disorders, 596f, 599–602. See also Bipolar disorder; Depression
Mood stabilizers, 182t, 187
Morphine, 154, 188–190, 189f
  abuse of, 189–190
  for pain, 387
Morris task, variations on, 218–219, 218f
Motivation
  amygdala in, 436
  definition of, 398
  limbic cortex in, 55
  neuroanatomy of, 411–421
  in nonregulatory behavior, 412
  in regulatory behavior, 411–412
  sexual, 436, 436f
Motor cortex, 365–366, 365f
  association cortex and, 531, 532f
  corticospinal tracts and, 369–370
  functional movement categories in, 366–367
  injuries of, 368–369
  layers of, 56–57, 57f, 357, 357f. See also Cortical layers
  mapping of, 365–366, 368–369
  neuroplasticity in, 507–510, 508f
  posttraumatic reorganization of, 368–369, 368f
  primary, 358f–360f, 359–360
  in skilled movement, 367
  supplementary, 365
  topographic organization of, 56–57, 365–366, 365f, 368–369
    in animals, 507–510, 508f
    changes in, 368–369, 368f, 507–510, 508f
Motor development, 261–262, 262f
Motor end plate, 136, 136f

Motor function. *See also* Movement
  separation from sensory function, 5, 52, 68–69
Motor homunculus, 365–366, 365f, 370–371
Motor neurons, 79, 79f, 370–371
  in muscle contraction, 136, 136f, 371
Motor pathways, efferent nerves in, 37, 37f
Motor sequences, 358–360, 358f, 360f
Motor skills. *See also* Movement, skilled
  learning of, explicit vs. implicit memory in,
    486–487
Motor system. *See also* Movement
  basal ganglia in, 372–375, 372f
  cerebellum in, 375–378, 375f–378f
  corticospinal tracts in, 369–370, 369f, 370f
  as efferent system, 356
  motor cortex in, 365–366, 365f
  motor neurons in, 370–371
  in muscle control, 136, 136f, 371, 371f
  organization of, 365–371
  in pianists, 510
  in skilled movements, 367
  somatosensory system and, 356–357, 390–394.
    See also Somatosensory system
Motor training, neuroplasticity and, 510
Movement, 353–396
  accuracy of, cerebellum in, 376–378, 378f
  brainstem in, 360–363
  cerebellar control of, 52–55, 52f
  cerebral asymmetry and, 549–550
  force of, basal ganglia and, 372–375, 373f
  forebrain in, 358–360, 358f, 360f
  functional categories of, 366–367
  globus pallidus in, 373f, 374–375
  hierarchical control of, 355–364
    experimental evidence for, 360
  hindbrain in, 52–53, 52f
  imitation of, 536–538
  inhibition of, 70–71
  initiation of, 358–360, 358f, 360f
  integrated control of, 63
  involuntary
    drug-induced, 578
    in Parkinson's disease, 591
  language and, 549–550
  learning and, 510
  modeling of, 366–367
  motor sequences in, 358–360, 358f, 360f
  neurotransmitters in, 157–158
  orienting, 53, 69
  within perceptual world, 66
  prefrontal cortex in, 358, 358f, 360, 360f
  premotor cortex in, 358–359, 358f–360f, 360
  primary cortex in, 358–359, 358f–360f, 360
  production of, nerve impulses in, 136, 136f
  sensory input in, 69–70, 376–378, 380–382, 390–
    394. *See also* Somatosensory system
  sequential, 526
  skilled, 367
    brain injury and, 368–369
  sleep and, 479
  sound in, 332
  species-typical, 360–361, 361f
  spinal circuits in, 363
  spinal cord in, 363–364
  timing of, cerebellum in, 376–378
  understanding meaning of, 536–538
  urge-to-action system and, 374
  ventral roots in, 62, 356
  vision in, 312, 315–316, 315f, 382
  visual perception of, 312
Movement disorders
  basal ganglia in, 373–375
  drug-induced, 578
  hyperkinetic symptoms in, 373
  hypokinetic symptoms in, 373

MPTP, parkinsonian symptoms from, 162
MRI. *See* Magnetic resonance imaging (MRI)
MRNA, 94, 94f
MSG, neurotoxicity of, 199
Mu receptors, 188–189
Müller–Lyer illusion, 286, 286f
Multimodal neurons, 418
Multiple intelligences, 28, 558–559
Multiple sclerosis, 85, 86, 587–589
Muscle contraction, nerve impulses in, 136, 136f
Muscle control, 371
Muscle end plates, 136, 136f
Muscular rigidity, in Parkinson's disease, 590–591
Music
  agnosia and, 314
  brain development and, 268
  cerebral asymmetry for, 345–347, 546
  cognitive benefits of, 326
  cortical thickness and, 346
  evolution of, 320, 337–339
  genetic aspects of, 347
  obsession with, 347
  patterns of, detection of, 347
  perfect pitch and, 322, 540–541, 540f
  processing of, 345–347, 546
  properties of, 326–327
  as sound, 325–326
  sound waves in, 326, 327f
  vs. speech, 326
Music processing, 345–347
Music therapy, 347
Musical intelligence, 558–559
Musicians, learning in, 510
Mutations. *See* Genetic mutations
Myasthenia gravis, 130, 177
Myelencephalon, 48, 48f
Myelination, 85
  axonal, 126f, 127–128
  development of, 260–261, 260f
  loss of, in multiple sclerosis, 85, 86, 587–588,
    587f
Myopia, 288–289
  alcohol, 194

N

Nalorphine, 189
Naloxone, 189
Nanotechnology, for spinal cord injuries, 364
Narcolepsy, 476
  sleep paralysis in, 476
Narcotics. *See* Opioids
Nasal retina, 294, 294f
Natural selection, 8–11. *See also* Evolution
  behavior and, 406–408
Navigational skills, 299, 532
Neanderthals, 21–22, 21f
  music-making by, 320
Near-infrared spectroscopy (NIRS), functional, 234,
    235f
Nearsightedness, 288–289
Negative pole, 112
Neglect, contralateral, 535–536
Neocortex, 54–55. *See also* Cerebral cortex
  layers in, 357, 357f
Neoteny, 26, 26f
Nerve(s), 47, 47f
  afferent, 37, 37f
  cranial, 60–61, 60f
  efferent, 37, 37f
  peripheral, 62
  regeneration of, in spinal cord injury, 364
  spinal, 61–62, 61f, 356–357, 357f
Nerve cells. *See* Neuron(s)
Nerve growth factor, neuroplasticity and, 513,
    518

Nerve impulse, 125–126. *See also* Action potential
  definition of, 125
  input from
    integration of, 128–133
    summation of, 131–132, 131f, 132f
  in muscle contraction, 136, 136f
  production of, sensory input in, 135–138, 135f
  saltatory conduction of, 127–128, 127f
Nerve injuries, repair of, 85–87, 85f, 364
Nerve net, 15, 17, 17f, 76
Nerve roots, 62, 62f
Nervous system
  autonomic, 36f, 37, 64, 158–159, 159f
  cells of, 46–47, 46f, 47f. *See* Glial cells;
    Neuron(s)
  central. *See* Central nervous system
  in chordates, 16–18, 18f
  development of, 247–280. *See also* Brain
    development
  evolution of, 48–49, 48f. *See also* Evolution
    in animals, 14–19, 15f–19f
    in humans, 19–31
  functional organization of, 36–37, 36f
  functions of, principles of, 66–71, 72
  hierarchical organization of, 50–59, 68, 68f,
    355–364
  layers of. *See also* Cortical layers
  parasympathetic, 64, 158–159, 206–207, 206f
  peripheral, 5, 36–37, 36f
  somatic, 60–64, 158
  structure of, 4–5, 4f
  sympathetic, 64
Netrins, 257
Neural circuits, 47, 47f, 76–80, 76f, 78f, 526–529
  behavior and, 401
  as cell assemblies, 527, 529
  in cognition, 526–529, 529
  crossed, 43, 66–67
  efficiency of, intelligence and, 560
  environmental influences on, 270
  excitation and inhibition in, 64, 70–71, 80
  in habituation, 165–166, 165f
  hyperconnectivity of, 541
  modification of, 504–505, 505f. *See also*
    Neuroplasticity
  in movement, 363
  novel, creation of, 505–506, 507f. *See also*
    Neurogenesis; Neuroplasticity
  in ocular-dominance columns, 307, 308f
  in sensitization, 166–167, 167f
  spinal, 363
  in visual system, 270
Neural columns. *See* Cortical columns
Neural Darwinism, 258
Neural development. *See* Brain development
Neural groove, 249, 249f
Neural organization
  afferent vs. efferent pathways in, 37, 37f
  functional, 36–37, 36f, 37f
  hierarchical, 50–59, 68, 68f, 355–364
  plastic patterns of, 35–36. *See also*
    Neuroplasticity
Neural plate, 249, 249f
Neural processing, hierarchical vs. parallel circuits
    in, 68
Neural relays, in sensory systems, 285–286
Neural shock, 584
Neural stem cells, 251–252
  transplantation of, 162, 518–519, 575–576,
    592–593
Neural streams, 69–70, 69f
Neural transplantation, biorhythms and, 454,
    454f
Neural tube, 249–250, 249f, 250f
Neural tube defects, 276–277

Neuritic plaque, in Alzheimer's disease, 494, 494f, 593
Neuroblasts, 251
Neurodegenerative disorders, 589–593. See also Alzheimer's disease; Dementia; Parkinson's disease
Neuroeconomics, 543
Neurogenesis, 253, 253f, 505–506, 507f. See also Neuroplasticity
    antidepressants and, 601
    in brain injury, 518–520
    induced, 575–576, 592–593
Neurohormones, 202f, 203
Neuroimaging studies, 228–234, 239, 239t
    in brain/behavioral disorders, 573–574, 574f, 583
    in brain mapping, 538–541, 539f
    dynamic, 231–234
    static, 228–230
    in traumatic brain injury, 582, 583
Neurological disorders, 581–584. See also Brain/behavioral disorders
    brain size and, 27
    organic, 27, 567–568
    vs. psychiatric disorders, 567–568
Neuron(s), 4, 46–47, 46f, 47f. See also Cell
    axons and, 47, 47f, 76, 76f, 77–78, 78f. See also Axon(s)
    bipolar, 79, 79f
        auditory, 331
        retinal, 293, 293f
    cell body of, 76, 76f
    components of, 87–97
    connections between, 47, 47f, 77–80, 78f, 79f. See also Neural circuits
    culture of, 75
    death of, 258
    dendrites of, 76, 76f, 77, 78f, 80
    differentiation of, 253–254, 255f
        abnormal, 276–277
    dorsal-root ganglion, 380–382, 381f
    electrical activity in, 111–128. See also under Electrical
    evolution of, 15
    excitation of, 64, 70–71
    excitatory, 80
    function of, 46, 76
        shape/size and, 79–80, 79f
    general-purpose, 252–253
    generation of. See Neurogenesis
    glial cell repair of, 85–87
    growth and development of, 253–259
    information flow through, 77–78, 79f
    inhibitory, 64, 70–71, 80
    interneurons, 79, 79f
    labeling of, 75, 213, 214f
    longevity of, 77
    maturation of, 255–257, 255f
    migration of, 253–255, 253f, 254f
    mirror, 359, 536–538
    motor, 79, 79f, 370–371
        in muscle contraction, 136, 136f, 371
    multimodal, 418
    networks of, 79–80, 79f. See also Axon(s); Neural circuits
    number of, 76
    origin of, 251–253, 251f
    plasticity of. See Neuroplasticity
    properties of, 76
    proteins in, 94
    rainbow, 75
    repair of, 85–87, 85f
    retinal, 287, 293–297, 293f
        receptive fields for, 301–302
    sensory, 79, 79f

    sex differences in, 552, 552f
    shape of, 79–80
        neuroplasticity and, 504–505
    somatosensory, 79, 79f, 380–382, 380f
    spinal-cord gray-matter, 383
    staining of, 75, 75f, 78f
    structure of, 76f, 77–78
    summation in, 131–132, 131f, 132f
    transplantation of, 162, 518–519, 575–576, 592–593
    types of, 79, 79f
    as unit of cognition, 526–529, 534
    ventrolateral thalamic, 382, 383
Neuron theory, 76
Neuronal migration, 253–255, 253f, 254f
Neuropeptides, 153–155, 153t
Neuroplasticity, 35–36, 66, 76, 164
    adolescent-onset psychiatric disorders and, 275
    in adults, 269
    axonal sprouting in, 504–505, 505f
    in brain development, 279
    in brain disorders, 571
    in brain injury, 275–276, 368f, 516–520
    brain size and, 28
    compensatory, 571, 603
    definition of, 36
    dendritic changes and, 504–505, 505f
    drug-induced, 513–515, 514f
    environmental stimulation and, 271–272, 506–507, 507f
    epigenetics and, 36, 103–106
    experience-dependent, 510–511
    frontal lobe and, 259–260
    growth factors and, 84, 513, 513t
    hormones and, 512–513, 512f
    learning and, 35–36, 271–272, 502–504, 510–511
    long-term potentiation and, 502–504, 503f
    memory storage and, 502–512
    metaplasticity and, 515–516
    modification of existing circuits in, 504–505, 505f
    of motor cortex, 368–369, 368f
    motor training and, 507–510, 508f, 510
    neurogenesis in, 505–506, 507f
    neurotrophic factors in, 513, 513t
    phenotypic plasticity and, 36
    principles of, 515–516
    restraint-induced therapy and, 369
    of somatosensory cortex, 392–393, 393f
    structural basis of, 502–516
    synaptic change in, 164, 504–505, 505f
Neuroprosthetics, 354
Neuroprotection, 37–40, 37f, 585
Neuropsychoanalysis, 568
Neuropsychological testing, 217–218, 539
Neuropsychological therapy, 579
Neuropsychology, 213
Neuroscience
    cognitive, 538–543
        applications of, 541–543
        methods of, 538–541
    social, 541–542
Neurosurgery, 575–576
    frontal lobotomy, 424f, 425
    for Parkinson's disease, 592–593
    split-brain studies and, 524, 547–549, 550
Neurotoxic lesion studies, 221
Neurotoxins
    drugs as, 199
    epigenetics and, 198, 199
    mechanism of action of, 176–177, 176f, 199t
Neurotransmission
    in central nervous system, 159–163
    drug effects on, 175f, 176–177, 176f
    in somatic nervous system, 158
    steps in, 145–147, 145f, 146f, 175–176, 175f

Neurotransmitters, 141–142. See also specific types
    actions of, 146
    activating systems for, 159–163
    amine, 152, 152t
    amino acid, 152t, 153, 153f
    autoreceptors for, 146
    behavior and, 149
    classification of, 151–155
    deactivation of, 146–147
    definition of, 141
    degradation of, 147
    diffusion of, 147
    evolution of, 149
    gaseous, 155
    glial uptake of, 147
    identification of, 150–151, 150f
    interaction of, 157–158
    in learning, 164–167, 165f, 167f
    peptide, 153–155, 153t
    properties of, 150–151
    putative, 150
    quanta of, 146
    receptors for, 155–157, 155f, 156f
        activation of, 145f, 146
    release of, 145–146, 146f
        regulation of, 148
    reuptake of, 147
    second messengers and, 156f, 157
    small-molecule, 151–153, 159–163
    synthesis and storage of, 145–146, 145f, 146f, 152–153, 152f, 153f
    types of, 150–157
Neurotrophic factors, 236–237, 252
    for brain injury, 518–520
    in depression, 599
    electroconvulsive therapy and, 576
    in induced neurogenesis, 576
    neuroplasticity and, 513, 513t
Neutrons, 89
Nicotine
    addiction to, 195–196, 440. See also Substance abuse
    in learning, 222–223
Nicotinic acetylcholine receptor, 158, 158f
Night terrors, 463
Night vision, age-related decline in, 292
Nigrostriatal dopaminergic system, 160f, 161–162
Nitric oxide, 155
NMDA receptors, in long-term potentiation, 503–504, 503f
Nociception, 379–380, 379f, 382–388, 383f
Nodes of Ranvier, 126f, 127–128
Noise, 325
Nomenclature, anatomical, 38–39
Non-REM sleep. See NREM sleep
Nonmatching-to-sample task, 266, 266f
Nonregulatory behaviors, 412
    control of, 433–437
Nontasters, 405
Noradrenergic system, 160f, 163
Norepinephrine (noradrenaline), 140f, 141, 152, 153f, 163. See also Neurotransmitters
    in depression, 600
Norepinephrine psychedelics, 192
Norrbotten (Sweden), epigenetic influences in, 106
Notes, musical, frequencies of, 326, 327f
Notochord, 16–17
NREM sleep, 460–463, 461f. See also Sleep
    disorders of, 475–476
    memory storage during, 470
    vs. REM sleep, 461–463
Nuclear membrane, 90f, 91
Nuclei, 47
    sensory, 69
Nucleotide bases, 92f, 93

Nucleus
    caudate, 57–58, 58f
    cell, 90f, 91, 93
    cochlear, 331, 331f
    dorsomedial, 54, 54f
    dorsomedial thalamic, 54, 54f
    lateral geniculate, 54, 54f, 296, 296f, 297f, 301–302, 301f, 302f
    magnocellular medullary, 474
    medial geniculate, 331, 331f
    paraventricular, in eating, 431
    raphé, 453, 454
    red, 53, 53f, 84f
    subcoerulear, 474
    suprachiasmatic, 452–458, 452f
    thalamic, 69
Nucleus accumbens, behavioral sensitization and, 514
Nucleus robustus archistriatalis, in birds, 349, 349f
Nutrition. See also Diet
    brain size and, 24–25, 25f
    depression and, 186
    epigenetic effects of, 106
    multiple sclerosis and, 86
Nystagmus, 313

O
Obesity, 428
    biologic clocks and, 450–451
    dieting and, 429
    metabolic syndrome and, 444, 450–451
Object knowledge, 531
Object location, in visual system, 300–303, 301f, 302f
Object manipulation, mental, 532–533, 532f, 533f
Object permanence, 263, 264f, 265t
Object-position test, 495, 495f
Object recognition, 69, 531
Object-reversal learning, 274
Obsessive-compulsive disorder, 602. See also Anxiety disorders
    serotonin in, 163
Occipital lobe, 41f, 55f, 56
    in association cortex, 530, 530f
    injury to, 56
    structure and function of, 55f, 56
    visual regions of, 297–299, 298f
Ocular-dominance columns, 270, 270f, 307, 308f, 311, 311f
Oculogyric crisis, 591
Oculomotor nerve, 60–61, 60f
Off-center ganglion cells, 304f, 305
Olfaction, 402–404, 402f, 404f
    in animals, 403
    in humans, 404
Olfactory bulbs, 41f, 59f, 402f, 403
Olfactory epithelium, 402–403, 402f
Olfactory nerve, 60–61, 60f
Olfactory pathways, 403
Olfactory receptors, 402–403, 402f
Olfactory system, 59, 59f, 402–404, 402f
    accessory, 403
Oligodendroglia, 82t, 85, 127, 127f
On-center ganglion cells, 304f, 305
Operant (instrumental) conditioning, 485
Opioids, 154, 182t, 187–190, 387
    abuse of, 189–190
    physical effects of, 189
    synthetic, 189
Opium, 188, 189f
Opponent-process theory, 310–312, 311f
Optic ataxia, 316
Optic chiasm, 294, 294f
Optic disc, 289, 289f, 290
    swelling of, 290
Optic flow, 284

Optic nerve, 60–61, 60f, 289, 290–291, 294, 294f
    inflammation of, 291
Optic neuritis, 291
Optical tomography, 212, 234, 235f
Optogenetics, 134, 223
Orbital prefrontal cortex, 530f
Orbitofrontal cortex, 419, 419f
    in eating, 432
    in gustation, 405–406
    in olfaction, 403, 403f
Orders, taxonomic, 16, 16f
Orexin, 478
Organ of Corti, 328, 328f
    hair cells of, 328f, 329, 330f, 334–335
Organelles, 90
Organic neurological disorders, 567–568
Organizational hypothesis, 204
Organophosphates, 177, 278
Orientation, to sound, 336, 336f
Orientation columns, 307, 311
Orientation detectors, 306–307
Orienting movements, 53, 69
Oscilloscope, 115, 115f
Osmotic thirst, 432
Ossicles, 328, 328f
Otoconia, 389
Otolith organs, 388, 388f, 389
Oval window, 328, 328f
Ovarian hormones
    activating effects of, 434–435
    in brain development, 204–205, 273–274, 551–554
    cognitive function and, 205
    functions of, 204–205
    neuroplasticity and, 512, 512f
    sexual behavior and, 434–435
Overtones, 324, 324f
Overweight, 428
Owl chronotype, 454, 455
Owls, hearing in, 336, 337f
Oxytocin, 414, 415f, 416

P
P cells, 293–297, 293f
Pacemakers. See also Biological clocks
    circadian rhythms as, 455–456
    circannual rhythms as, 456–457
Pain, 384–388, 386f–388f
    central, 385
    emotional, 398
    expectation of, 542
    gate theory of, 386–387, 386f
    perception of, 385–386, 386f
    phantom-limb, 385
    referred, 388, 388f
    response to, 386–387
    treatment of, 387–388
Pain gate, 386–387, 386f
Panic disorder, 426, 427, 602. See also Anxiety disorders
Papaver somniferum, 188, 189f
Papez circuit, 417–418, 418
Papilloedema, 290
Parahippocampal cortex, 417
    in memory, 493–494, 493f
Parallel processing, 68, 69
Paralysis, 85, 86–87, 363
    facial, 63
Paramethoxymethamphetamine (PMMA), 200
Paraplegia, 363, 364
Parasympathetic nervous system, 64
    neurotransmission in, 158–159
    in rest and digest response, 158–159, 206–207
Paraventricular nucleus, in eating, 431
Parietal lobe, 41f, 55f, 56

    in association cortex, 530, 530f
    in attention, 534
    in cognition, 530, 533, 534
    injury to, 56
    in spatial cognition, 533
    structure and function of, 55f, 56
Parkinson's disease, 42, 57–58, 71, 142, 214f, 373–375, 589–593
    Alzheimer's disease and, 594
    basal ganglia in, 373
    causes of, 591–592
    globus pallidus in, 374–375
    L-dopa for, 152, 153f, 154, 577, 592
    lesion studies of, 220–221
    Lewy bodies in, 594, 594f
    memory deficits in, 491–492, 500
    neuron transplant for, 518–519, 592–593
    neuronal/behavioral relations in, 214
    prevalence of, 589–593
    progression of, 590
    symptoms of, 221, 590–591
    toxins and, 162
    treatment of, 152, 153f, 154, 575, 577, 592–593
    tremor in, 142, 162, 590
Parrots, cognition in, 526–527, 550
Parvocellular (P) cells, 293–297, 293f
Patellar reflex, 51, 384, 384f
Pavlovian conditioning, 483–484. See also Conditioning
    in substance abuse, 197
PCP (phencyclidine), 183, 192, 199
Peptide bonds, 94
Peptide hormones, 203. See also Hormones
Peptide transmitters, 153–155, 153t
Perception, 35, 286
    subjective reality and, 35
Perceptual illusions, 286, 286f
Perceptual world, creation of, 66
Perfect pitch, 322, 540–541, 540f, 559
Periaqueductal gray matter, 53, 53f
    in pain, 387–388
Peribrachial area, in REM sleep, 474, 474f, 475f
period gene, 455
Periodic limb movement in sleep, 462
Periods, in activity cycle, 446, 449–451
Peripheral nervous system, 4f, 5, 36–37, 36f, 62
Peripheral vision, 290, 290f
Perirhinal cortex, in memory, 493–494, 493f
Perseveration, 536
Persistent vegetative state, 12
Personal memory, 488–489, 489f
PET scan, 232–234
    in auditory cortex mapping, 343–344
Petit mal seizures, 587
Peyote, 192
Phagocytosis, 84
Phantom-limb pain, 385
Phencyclidine (PCP), 183, 192, 199
Phenotype, 9, 98
Phenotypic plasticity, 36, 103–106
Phenylketonuria (PKU), 569–570, 570t
Pheromones, 403, 404f, 412
Phobias, 426, 427, 602. See also Anxiety disorders
Phospholipid bilayer, 91–92, 91f
Photoreceptors, 284, 287–293, 289f, 291f, 292f
    retinal ganglion cells as, 295, 300–301, 453
Photosensitive retinal ganglion cells, 453
Phototherapy, for seasonal affective disorder, 450
Phyla, 16, 16f
Physostigmine, 177
Pia mater, 37
Piaget's cognitive theory, 263–265, 264f, 265t
Pianists, motor skills in, 510
Picrotoxin, 183
Pidgin, 339

Pigments, in rods vs. cones, 291–292
Pincer grip, 359
  movements in, 359f
Pineal gland, 7, 7f, 46
  in biorhythms, 452, 456
  blood–brain barrier and, 174, 175f
  melatonin secretion by, 456
  in sleep, 472
Pinna, 328, 328f
Pitch, 322, 322f, 326
  perception of, 335
  perfect (absolute), 322, 540–541, 540f, 559
Pituitary gland, 202, 202f
  blood–brain barrier and, 174, 175f
  definition of, 413
  hormones of, 202, 202f, 414–415, 415t
  hypothalamus and, 412–416, 414f, 600–601
  structure and function of, 414–415, 414f
PKU (phenylketonuria), 569–570, 570t
Place cells, 227–228, 227f, 470
Place-learning task, 218f, 219
Planning, 535–536
Plants, biorhythms in, 445, 446, 446f
Planum temporale, 332, 332f
Plaque
  in Alzheimer's disease, 85, 494, 494f, 593
  in multiple sclerosis, 86
Plasphenes, 222
Plasticity
  neural. See Neuroplasticity
  phenotypic, 36, 103–106
Pleasure, reward and, 438–440
PMMA (paramethoxymethamphetamine), 200
Poisoning
  carbon monoxide, 314–315
  domoic acid, 199
  kainic acid, 199
  MPTP, 162
  toxin action at synapses in, 176–177
Polar molecule, 89
Polygraphs, 224, 224f
Polypeptide chains, 94, 94f. See also Protein(s)
Pons, 52–53, 53f
Poppy, opium, 188, 189f
Positive pole, 112
Positron emission tomography (PET), 232–234
  in auditory cortex mapping, 343–344
Posterior, definition of, 38, 39
Posterior cerebral artery, 43, 43f
Postictal depression, 587
Postsynaptic membrane, 144, 144f, 146
Postsynaptic potentials
  excitatory/inhibitory, 129–132, 131f, 132f
  miniature, 146
Posttraumatic stress disorder (PTSD), 166, 208, 427, 566, 602
Potassium ion(s). See also under Ion
  resting potential and, 119–121, 119f, 121f
Potassium ion channels, in sensitization, 167, 167f
Potentials. See Electrical potentials
Power grasp, 359f
Power grip, 359
Precursor cells, 251
Preformation, 248
Prefrontal cortex, 419–421
  association, 530–538
  in behavior selection, 420–421
  dorsolateral, 530f
    theory of mind and, 541
  dorsomedial, 419, 419f
  in eating, 432
  in emotional behavior, 424–425
  executive functions of, 358, 420–421
  in memory, 500, 500f
  in movement, 358, 358f, 360, 360f

orbital, 530f
  in planning, 358, 535
  structure of, 419–420, 419f, 420f, 530–531, 530f
  ventromedial, 419, 419f, 530f
Pregnancy, alcohol use in, 184
Premotor cortex, 419, 419f
  in memory, 500, 500f
  in movement initiation, 358–359, 358f–360f, 360
Preoperational stage, of cognitive development, 264, 265t
Preparedness, 409
Presbyopia, 288
Presynaptic membrane, 144, 144f, 146
Prey-killing behavior, 399, 401, 406–407
Primary auditory cortex, 332, 332f. See also Auditory cortex
Primary motor cortex, 358f–360f, 359–360. See also Motor cortex
Primary protein structure, 95f
Primary visual cortex. See Visual cortex, primary (striate)
Primates
  characteristics of, 19–20
  humans as, 19–20, 19f. See also Humans
  nonhuman
    brain size in, 22, 23f, 34, 34f
    classification of, 19f, 20
    cognitive development in, 30
    culture of, 30
    evolutionary link to humans and, 14–15, 20
    language in, 8
    music and, 320
  relationships among, 19f, 20
Priming, 487
Principle of proper mass, 22, 302
Procedural memory, 486, 487t
Progenitor cells, 251
Programmed cell death, 199, 258
Prolactin, 252, 415, 415t
Proper mass principle, 302
Proprioception, 379f, 380
  loss of, 382
Prosencephalon, 48, 48f
Prosody, 327, 425
Prosopagnosia, 300
Protein(s), 94–97
  amino acids in, 92, 94, 94f, 95f, 153
  cell function and, 87
  definition of, 94
  destinations of, 95–96, 95f
  enzyme, 95
  export of, 95–96, 95f
  genes coding for. See Gene(s)
  membrane, 95f, 96–97, 96f
  shape-changing, 96, 96f
  structure of, 94–95, 95f
  synthesis of, 93, 94f
  transport of, 95f, 96–97
  transporter, 145
Protein channels. See Ion channels
Protein receptors, 96, 96f
Protons, 89
Proximal, definition of, 39
Prozac, 187
Psilocybin, 192
Psyche, 6–7
Psychedelics, 191–192, 200
Psychiatric disorders, 597–599. See also Brain/behavioral disorders
  adolescent-onset, 275
  anxiety, 182–183, 182t, 426, 427, 596f, 602. See also Anxiety disorders
  mood, 596f, 599–602. See also Bipolar disorder; Depression

psychotic, 184–186, 596f, 597–599. See also Schizophrenia
  types of, 596f
  vs. neurological disorders, 567–568
Psychoactive drugs. See Drug(s); Substance abuse
Psychoanalysis, 567
Psychogenic amnesia, 489, 489f
Psychological constructs, 525
Psychology, evolutionary, 407–408
Psychomotor activation, in substance abuse, 195
Psychopathology. See Psychiatric disorders
Psychopharmacology. See also Drug(s)
  definition of, 172
  principles of, 173–181
Psychosis, 184–186, 596f, 597–599. See also Schizophrenia
  adolescent-onset, 275
  amphetamine, 185
  drug-induced, 186, 201, 237
  drug therapy for, 184–186
Psychosurgery. See Neurosurgery
Psychotherapy, 579–580
Psychotropics, 182t, 190–192
Puffer fish, 122
Pulvinar, 297
Pumps, ion, 96–97, 96f, 119–121, 119f
Pupil, 289, 289f
Pure tones, 324
Purkinje cells, 79, 79f
Pursuit-rotor task, 486, 486f
Putamen, 57–58, 58f
Putative neurotransmitter, 150
Puzzle box, 485, 485f
Pyramidal cells, 79, 79f, 80
Pyramidal tracts, 369–370, 369f, 370f
Pyriform cortex, 59

Q
Quadrantanopia, 313, 313f
Quadriplegia, 363, 364
Quanta, 146
Quaternary protein structure, 95f

R
Rabies, 417
Radial glial cells, 254, 254f
Radiator hypothesis, 25–26
Rainbow neurons, 75
Raphé nucleus, 453, 454
Rapid eye-movement sleep. See REM sleep; Sleep
Rapidly adapting receptors, 380
Rasmussen's encephalitis, 42
Rate-limiting factor, 152
Rats, behavioral analysis of, 218–219
Real-time fMRI, 580
Recency memory task, 217, 217f
Receptive fields, 284, 301–302, 301f
  of lateral geniculate nucleus, 301–302
  overlapping, 305, 305f
  of primary visual cortex, 304–306, 307f
  in shape perception, 303–309, 306f, 307f
Receptor(s), 96–97, 96f
  auditory, 328f, 329, 330–331, 330f
  autoreceptors, 146
  ionotropic, 155, 155f
  light, 287–293, 289f, 291f, 292f
  metabotropic, 155–157, 156f
  neurotransmitter, 155–157, 156f
    activation of, 145f, 146
  olfactory, 402–403, 402f
  rapidly adapting, 380
  sensory, 135, 283–285, 378–380
    density of, 284–285
    sensitivity of, 284–285
  slowly adapting, 380

Receptor(s) (*cont.*)
  somatosensory, 378–380
  taste, 405
  transmitter-activated, 146
  vestibular, 388–389, 388f
Receptor-behavior links, 571
Recessive alleles, 98, 100f
Reconsolidation, of memory, 470, 499
Recreational drugs. *See* Drug(s); Substance abuse
Red nucleus, 53, 53f, 84f
Referred pain, 388, 388f
Reflective vs. reflexive systems, 543
Reflexes
  monosynaptic, 384
  patellar tendon, 51, 384, 384f
  scratch, 363
  spinal, 363, 384, 384f
Refractive errors, 288–289
Refractory membrane, 124
Refractory periods, 124–125, 126
Regulatory behaviors, 411–412
  control of, 427–433
Reinforcers, 408
Relatively refractory membrane, 124
Releasing hormones, 414–415, 415f
Religion, science and, 13
REM sleep, 460–463. *See also* Sleep
  atonia in, 460, 463, 476–478
  basic rest-activity cycle and, 467–468
  definition of, 460
  deprivation of, 469
  disorders of, 476–478
  dreaming in, 463–466. *See also* Dreams
  implicit memory and, 471, 471f
  memory storage during, 471
  neural basis of, 474–475, 474f, 475f
  vs. NREM sleep, 461–463
  without atonia, 479
Renshaw loop, 151, 151f
Repetitive transcranial magnetic stimulation, 222, 222f, 577
Reproduction, sex hormones in, 203. *See also* Sex hormones
Research methods, 215–223, 568–572
  comparison of, 239, 239t
Respondent conditioning, 483–484
Rest–activity cycle, 467–468
Rest and digest response, 158–159, 206–207
Resting potential, 119–121, 119f
Resting-state fMRI, 232
Restless legs syndrome, 462
Restraint-induced therapy, 369
Reticular activating system, 472–473, 473f
Reticular formation, 52, 53f
Retina, 289, 289f
  blind spot in, 282, 289, 289f, 290–291, 290f
  cells of, 287–290, 293–297, 293f
  nasal, 294, 294f
  receptive fields in. *See* Receptive fields
  temporal, 294, 294f
Retinal ganglion cells, 293–297, 293f
  lateral geniculate nuclei and, 296, 296f, 297f, 301–302, 301f, 302f
  on-center/off-center, 304f, 305
  as photoreceptors, 295, 300–301, 453
  photosensitive, 453
  receptive field of, 301–302, 301f, 304–305, 305f
  in shape perception, 304–305, 305f
Retinal neurons, 287, 293–297, 293f
  receptive fields for, 301–302, 301f, 304–305, 305f
Retinohypothalamic tract, 295, 453–454, 455
Retrograde amnesia, 498
Rett syndrome, 256
Reuptake
  of neurotransmitters, 147

Reward
  behavior and, 399, 400, 417, 438–440
  in conditioning, 483–484
  wanting and liking and, 439
Rhombencephalon (hindbrain), 48, 48f, 51–53, 51f–53f
Rhythms, biological. *See* Biorhythms
Ribonucleic acid (RNA), 94, 94f
  modification of, 105, 105f
Ribosomes, 94
Right-handedness. *See also* Cerebral asymmetry
  cortical organization and, 332–333, 554–556
  lateralization and, 332–333
Rigidity, in Parkinson's disease, 590–591
Ritalin, 172
Rivastigmine (Exelon), 494
RNA, 94, 94f
  modification of, 105, 105f
Robots, 80–81, 80f
Robustus archistriatalis, in birds, 349, 349f
Rods (photoreceptors), 291, 291f
Romanian orphans, brain development in, 273
Rostral, definition of, 38, 39
Round window, 328f, 329
Routes of administration, 173–174, 173f
Rubin's vase illusion, 286, 286f
rutabaga gene, 168

S

Saccule, 388, 388f, 389
Sacral spine, 61, 61f
Sagittal, definition of, 39
Sagittal section, 39f, 45–46, 46f
Saltatory conduction, 127–128, 127f
Salts, 89. *See also under* Ion
Schizophrenia, 277, 278, 596f, 597–599
  adolescent-onset, 275
  adult-onset, 574, 574f
  biochemical changes in, 597–599, 599t
  brain abnormalities in, 573–574, 574f, 597–599, 598f
  classification of, 597
  definition of, 162–163
  diagnosis of, 597
  dopamine in, 162–163, 185
  drug therapy for, 182t, 184–186, 185f
  early-onset, 573–574, 574f
  epigenetics in, 598
  imaging studies in, 573–574, 574f
  serotonin in, 163
  symptoms of, 597
  transcranial magnetic stimulation for, 577
  type I, 597
  type II, 597
Schwann cells, 82t, 85, 85f
  in neuron repair, 85–86
Science, religion and, 13
Sclera, 289
Scotomas, 313, 313f
  migraine, 282
Scratch reflex, 363
Seasonal affective disorder, 450
Second-generation antidepressants, 186–187
Second languages
  cortical localization for, 269
  learning of, 339
Second messengers, 156f, 157
  in learning, 167–168, 167f
Secondary auditory cortex, 332. *See also* Auditory cortex
Secondary protein structure, 95f
Secondary visual cortex, 298, 298f. *See also* Visual cortex
Sections, of brain, 39f, 43–46, 43f, 46f
Sedative hypnotics, 182–183, 182t
  insomnia and, 475–476

Segmentation, 15, 17
Seizures, 110, 585–589
  brain surgery for, 342
  focal, 586
  generalized, 587
  grand mal, 587
  idiopathic, 586
  petit mal, 587
  symptomatic, 586
Selective attention, 533–534
Selective breeding, 102
Selective serotonin reuptake inhibitors (SSRIs), 187, 188. *See also* Antidepressants
  side effects of, 578
Self-concept, 542
  neural basis of, 567–568
Self-recognition, 541–542, 541f
Self-regulation, 542
Semicircular canals, 388–389, 388f
Sensation, 283–286. *See also* Sensory function *and specific senses*
Sensitization, 164–167, 167f, 168
  behavioral, 514
  dendritic spines in, 169
  drug, 179–181, 513–516
  habituation to, 164–168, 165f, 169, 502
Sensorimotor stage, of cognitive development, 264, 265t
Sensory cortex
  association cortex and, 531, 532f
  layers of, 56–57, 57f. *See also* Cortical layers
Sensory deprivation, 5, 400–401, 400f
Sensory function
  dorsal roots in, 62
  integrated control of, 63
  midbrain in, 52, 53
  motor control and, 69–70
  neural streams in, 69–70, 69f
  for object recognition, 69, 531
  separation from motor function, 5, 52, 68–69
Sensory input
  in brain development, 267–269, 268f, 271–272
  habituation to, 164–166, 165f
  integration of, 128–133
  in movement, 376–378, 380–382. *See also* Somatosensory system
  in nerve impulse production, 135, 135f
  summation of, 131–132, 131f, 132f
Sensory neurons, 79, 79f
Sensory nuclei, 69. *See also* Nucleus
Sensory pathways, afferent nerves in, 37, 37f
Sensory perception, 35
  subjective reality and, 35
Sensory processing, 283–286
  synesthesia in, 556–557
Sensory receptors, 135, 378–380. *See also specific types*
  density of, 284–285
  sensitivity of, 284–285
Sensory systems
  coding and representation in, 285–286
  distinguishing between, 285–286
  neural relays in, 285–286
  receptive fields in, 284
  receptors in, 283–285
Serotonergic system, 160f, 163
Serotonin, 153, 163. *See also* Neurotransmitters
  in brain activation, 473–474, 473f
  in depression, 600–601. *See also* Selective serotonin reuptake inhibitors (SSRIs)
  synthesis of, 153
Serotonin psychedelics, 192
Serotonin synapse, antidepressant action at, 186f, 187
Setpoint, in temperature regulation, 411

Sex chromosomes, 97f, 98
  sex differences and, 553
Sex determination, 204–205, 250, 251f, 433–434, 435
  brain development and, 204–205
  disorders of, 435
Sex differences
  in behavior, 553–554
  in birdsong, 275
  in brain development, 272–275, 274f
  in brain injury, 553, 553f
  in brain size, 552, 552f
  in cognition, 205, 551–554, 551f–553f
  in cortical thickness, 552, 552f
  evolution of, 553–554
  in language, 551–554
  in neuron structure, 552, 552f
Sex hormones, 203, 204–205. See also Hormones
  activating effects of, 434–435
  behavior and, 401
  in brain development, 204–205, 250, 272–275, 433–434, 551–554
  cognitive function and, 205, 551–554
  function of, 203
  lifelong effects of, 274–275
  neuroplasticity and, 512–513, 512f
  organizing effects of, 433–434
  sexual behavior and, 434–435
  in sexual differentiation, 204–205, 250, 251f, 433–434
  steroid, 203
  target glands of, 202, 202f
Sexual behavior, 412, 434–436
  amygdala in, 436
  cerebral cortex in, 437
  cognitive influences in, 437
  frontal lobe in, 437
  hypothalamus in, 435–437, 436f
  loss of libido and, 437
  motivational, 436, 436f
  neural circuits and, 401
  neural control of, 435–436
  sex hormones and, 436
  sexual identity and, 436–437
  sexual orientation and, 436–437
Sexual dimorphism, 250, 434. See also Sex determination
Sexual identity, 436–437
Sexual orientation, 436–437
  gender identity and, 437
Shape perception, 303–309, 304f–309f
Shift work, biorhythms and, 450–451
Shock, neural, 584
Short-term memory, 488
  tests of, 496–497, 497f
Shunts, cerebrospinal fluid, 83
Side chains, 94
Sight. See Vision
Simple cells, of visual cortex, 306, 306f
Singing, 320, 342, 345
  agnosia and, 314
  by birds, 275, 348–350
Single-cell recordings, 227–228
Skilled-reaching tasks, 219, 219f
Skin
  sensory receptors in, 378–380
  two-point sensitivity in, 379, 379f
Skinner box, 408, 408f, 409, 485
"Slave" oscillators, in suprachiasmatic nucleus, 455–456, 456f
Sleep, 458–479
  age-related changes in, 470, 475
  alpha rhythms in, 460
  in animals, 463, 463f, 467, 467f
  antidepressants and, 469

  in basic rest-activity cycle, 467–468, 468f
  as biological adaptation, 467–468
  brainstem injury and, 469, 479
  circadian rhythms and, 444, 446–449, 447t, 448f
  consciousness and, 479
  delta rhythms in, 460
  in depression, 475–476
  dreaming in, 463–466. See also Dreams
  drowsy state in, 460
  duration of, 459, 467, 467f
  electroencephalography in, 459, 472–475
  energy conservation and, 467
  function of, 466–472
  hippocampus in, 470f
  learning during, 470, 471
  medial pontine reticular formation in, 474, 474f
  melatonin in, 472
  in memory storage, 471
  microsleep and, 469
  neural basis of, 472–475
  normal variations in, 459
  NREM, 460–463, 461f
    disorders of, 475–476
    explicit memory and, 470, 470f
  peribrachial area in, 474, 474f, 475f
  pineal gland in, 472
  REM, 460–463, 467–468
    atonia in, 463, 476–478
    definition of, 460
    deprivation of, 469
    disorders of, 476–478
    memory storage during, 470, 470f
    neural basis of, 474–475, 474f, 475f
    without atonia, 479
  as restorative process, 468–469
  reticular activating system in, 472–473
  slow-wave, 460, 461f
  stages of, 459–460, 461f
  waking and, 473–474
Sleep aids, insomnia and, 476
Sleep apnea, 476
Sleep deprivation
  brainstem injury and, 479
  REM-sleep, 469
Sleep disorders, 462, 475–478
  genetic factors in, 455
  of NREM sleep, 475–476
  of REM sleep, 476–478
Sleep paralysis, 476
Sleep-producing substance, 472
Sleep spindles, 470
Sleep studies, 224–225, 225f, 459–460, 459f, 461f, 472–475
  electroencephalography in, 472–475
  electromyography in, 459, 459f
  electrooculography in, 459, 459f
Sleeping sickness, 42
Sleepwalking, 462–463
Slow-wave sleep, 460, 461f. See also NREM sleep
Slowly adapting receptors, 380
Small-molecule neurotransmitters, 151–153, 152t, 159–163
Smell. See under Olfaction; Olfactory
Smoking, nicotine addiction in, 195–196, 440
Social cognition, 541–542
  mirror neurons in, 536–538
Social neuroscience, 541–542
Social phobia, 602. See also Anxiety disorders
Sodium amobarbital test, 554, 555
Sodium ions. See also under Ion
  resting potential and, 119–121, 119f, 121f
Sodium-potassium pump, 120
Solitary tract, 405
Soma, 76, 76f
Somasomatic connections, 148

Somatic marker hypothesis, 422–423
Somatic nervous system, 36f, 37, 60–64
  connections of, 62, 62f
  cranial nerves in, 60–61, 60f
  neurotransmission in, 158
  sensory and motor divisions of, 68–69
  spinal nerves in, 61–62, 61f
Somatosensory cortex, 390–394
  anatomy of, 390, 391f
  in gustation, 405
  hierarchical organization of, 390–391
  homunculus of, 390–391, 393f
  injury of, 392–393
  plasticity of, 392–393, 393f
  primary, 390
  secondary, 390, 394
  topographic organization of, 390–391, 391f, 393f
Somatosensory neurons, 79, 79f
Somatosensory receptors, 378–380, 388f
  distribution and density of, 379–380
  haptic, 379f, 380
  nociceptive, 379f, 380
  proprioceptive, 379f, 380
  rapidly adapting, 380
  slowly adapting, 380
  types of, 388f
Somatosensory system, 378–394
  as afferent system, 356
  in balance, 388–389
  dorsal-root ganglion neurons in, 380–382, 381f
  dorsal spinothalamic tract in, 382, 383f
  functions of, 378
  in hapsis, 380
  medial lemniscus in, 382, 383f
  motor system and, 356–357, 390–394
  in nociception, 380, 384–388
  overview of, 378
  in pain perception, 387–388
  pathways to brain in, 382–383
  in perception and movement, 378–394
  in pianists, 510
  in proprioception, 380, 382
  receptor distribution in, 379–380
  segregation and synthesis in, 391
  somatosensory cortex in, 390–394. See also Somatosensory cortex
  spinal reflexes and, 384, 384f
  ventral spinothalamic tract in, 383, 383f
  ventrolateral thalamus in, 382, 383, 383f
  vestibular system in, 388–389, 388f
Sorting tasks, 536, 536f, 539
Sound. See also Auditory system; Hearing
  language as, 325–326. See also Language
  localization of
    in bats, 350–351, 350f
    in humans, 336, 336f
  loudness of, 322f, 323–324, 323f, 326
    detection of, 336
  music as, 325–326. See also Music
  patterns of, detection of, 336
  perception of, 325
  pitch of, 322, 322f, 327
    perception of, 335
  properties of, 321–325, 322f
  source of, detection of, 336, 336f
  timbre (quality) of, 322f, 327
Sound waves, 321–327
  amplitude of, 322f, 323–324, 323f. See also Loudness
  complexity of, 322f, 324–325
  curves of, 334–335, 334f
  cycles of, 321, 322f
  definition of, 321
  frequency of, 321–322, 322f, 324. See also Pitch
  production of, 321, 321f, 322f

Spatial cognition, 532–533, 532f, 533f
  sex differences in, 553–554
Spatial intelligence, 558–559
Spatial localization, visual, 300–303, 301f, 302f
Spatial memory, hippocampus in, 495–496, 495f
Spatial orientation, of brain, 38
Spatial summation, 131, 132f
Special K (ketamine), 183, 192, 199
Species, 9, 16, 16f
  between-species vs. within-species comparisons
    and, 28–29
Species-typical behavior, 28–29
  motor, 360–361
Specific phobias, 602. See also Anxiety disorders
Speech. See also Language
  acquisition of, 255f, 262–263
  Broca's area for, 255f, 263, 339–344, 340f, 343f
  cortical localization of, 339–344
    mapping of, 339–344
  lateralization for, 332–333, 538, 549–550
  motor aspects of, 549–550
  motor sequences in, 358, 358f
  musicality of, 327
  perception of, 325–326
  production of, 340, 340f, 342–343
  rate of, 326
  vs. music, 326
  Wernicke's model of, 340, 340f
Speech arrest, 342
Speech discrimination, 336–337
Speech patterns, detection of, 336–337
Speech-sound discrimination task, in brain mapping,
  343f, 344
Speed (methamphetamine), 190
Spider venom, 176
Spina bifida, 276–277
Spinal accessory nerve, 60–61, 60f
Spinal anesthesia, 387
Spinal cord, 4, 50–51
  dorsal-root ganglia in, 380–382, 381f
  evolution of, 16–17, 48–49, 48f
  functions of, 363
    integrated, 63
  interneurons in, 370
  motor neurons in, 370–371
  orientation of, 38
  segments of, 61–62, 61f, 356–357, 357f
  spinal nerves and, 61–62, 61f
  structure of, 61–62, 61f, 356–357
Spinal-cord gray-matter neurons, 383
Spinal cord injury, 85, 86–87, 363–364
  dorsal vs. ventral root damage in, 62, 62f
  emotion in, 422, 422f
  nerve regeneration in, 364
  permanence of, 85, 86–87
  treatment of, 364
  unilateral, somatosensory deficits in, 383,
    383f
Spinal nerves, 61–62, 61f, 356–357, 357f
Spinal reflexes, 363, 384, 384f
Spine
  segments of, 61–62, 61f, 356–357, 357f
  structure of, 356–357, 357f
Spinothalamic tracts, 382–383, 383f
Split brain studies, 524, 547–549, 550
SRY gene, 204
SSRIs (selective serotonin reuptake inhibitors), 187,
  188. See also Antidepressants
Staining, brain tissue, 46–47, 47f, 75, 75f, 78f, 213,
  214f
Stellate cells, 79, 79f
Stem cells, neural, 251–252
  transplantation of, 162, 518–519, 575–576,
    592–593
Stereotaxic apparatus, 220–221, 221f

Steroid hormones, 203. See also Hormones
  anabolic, 205–206
  neurotoxicity of, 513
Stimulants, 190–191, 191f
  brain damage from, 200
  general, 192
  psychedelic/hallucinogenic, 189–190, 200
Stimulus
  conditioned, 484
  unconditioned, 484
Stimulus equivalence, 308
Stirrup, 328, 328f
Storage granules, 144, 144f
Stress
  anxiety and, 602
  brain development and, 271–272
  depression and, 600–601
  epigenetic changes and, 238, 272
  frontal lobe development and, 260
  hippocampus and, 187, 206f, 207–208, 601
  hypothalamic-pituitary-adrenal axis and, 600–
    601
  neuronal death and, 513
  posttraumatic stress disorder and, 208
Stress response, 206–208, 206f, 207f
  activation of, 206–207, 206f
  hormones in, 206–208, 206f, 207f, 600–601,
    600f
  termination of, 207–208
Stretch-sensitive channels, 135
Striate cortex. See Visual cortex, primary (striate)
Striatum, 236
Stroke, 43, 45, 584–585
  mild cognitive impairment and, 595
  sleep disturbances after, 479
  stem cell transplant for, 576
Subcoerulear nucleus, in sleep, 474, 475f
Subcortical regions, 46, 46f
Subsong, 349
Substance abuse, 194–201
  addiction in, 194–196
    definition of, 195
    dopamine in, 195, 196, 197
    neural basis of, 196–198
    psychomotor activation in, 195
    sex differences in, 195–196
    treatment of, 198–199
    withdrawal in, 195
  adolescent-onset, 275
  behavior and, 193–194
  brain damage from, 199–200
  conditioning in, 178–179, 197
  criminalization of, 198–199
  definition of, 195
  disinhibition and, 193–194
  dopamine in, 195
  epigenetics and, 198
  genetic factors in, 198
  narcotics in, 189–190
  psychomotor activation in, 195
  psychotic symptoms in, 186, 201, 237
  rates of, 198, 198f
  reward and, 440, 440f
  risk factors for, 198
  sensitization in, 179–181, 513–515
  tolerance in, 177–179, 178f
  treatment of, 198–199
  twin studies of, 198
  wanting-and-liking theory of, 196–197
Substantia nigra, 42, 53, 53f
  in Alzheimer's disease, 594
  electrode placement in, 220–221, 221f
  in Parkinson's disease, 142, 591, 594
Subtractive color mixing, 309, 309f
Subventricular zone, 251

Sudden infant death syndrome (SIDS), 163, 246
Suicide, 187
Sulci, 41f, 42, 55, 55f
Summation, 131–132, 131f, 132f
  ions in, 132
  spatial, 131, 132f
  temporal, 131, 131f
Superego, 567
Superior, definition of, 38, 39
Superior colliculus, 53, 53f, 69
Superior olive, 331, 331f
Supertasters, 405
Supplementary speech area, 342–343, 342f
Suprachiasmatic nucleus
  in biorhythms, 452–458, 452f
  "slave" oscillators in, 455–456, 456f
  structure of, 453–454, 453f
  transplantation of, 454, 454f
Surgery. See Neurosurgery
Swimming pool tasks, 218–219, 218f
Sympathetic nervous system, 64
  neurotransmission in, 158–159
  in stress response, 158–159, 206–208, 206f
Symptomatic seizures, 586
Synapses, 78, 78f
  axoaxonic, 147f, 148
  axodendritic, 147f, 148
  axomuscular, 147f, 148
  axosecretory, 147f, 148
  axosomatic, 147f, 148
  axosynaptic, 147f, 148
  chemical, 143–144, 144f
  dendrodendritic, 147f, 148
  drug action at, 175f, 176–177, 186f, 187
  electrical, 144–145
  excitatory, 148–149, 149f
  formation of, 258, 504–505. See also
    Neuroplasticity
  Hebb, 164
  inhibitory, 148–149, 149f
  in learning, 164–168, 510. See also Learning
  plasticity of, 164, 504–505, 505f. See also
    Neuroplasticity
  structure of, 143–145, 144f
  types of, 148
Synaptic cleft, 143, 144f
Synaptic organization, intelligence and, 560
Synaptic pruning, 258–259, 258f
Synaptic transmission. See Neurotransmission
Synaptic vesicles, 143, 144f
Synesthesia, 286, 556–557
Syntax, 525–526
  definition of, 525
  uniformity of, 339
Systematic desensitization, 579

T
t-PA (tissue plasminogen activator), 45, 585
Tactile stimulation
  brain development and, 268
  sensory processing in, 135, 135f
Tardive dyskinesia, 578
Taste, 405–407
  reactions to, 439–440, 439f
Taste aversions
  behavior in, 439–440, 439f
  learned, 409
Taste buds, 405, 405f
Taste preferences, 409
Taxonomy, 16–17, 16f
Tay–Sachs disease, 99–100
Tectopulvinar pathway, 295, 295f, 296, 297
Tectorial membrane, of inner ear, 328f, 329
Tectum, 53, 53f
Tegmentum, 53, 53f

Telencephalon, 48, 48f
Telodendria, 77, 78f
Temperature perception, 380, 382–388
Temperature regulation, 411–412
Temporal auditory cortex, 332
    tractography of, 540, 540f
Temporal lobe, 41f, 55f, 56
    in association cortex, 530, 530f
    in cognition, 530, 531
    injury to, 56
    in memory, 497–499, 500f
    neural columns in, 308, 308f
    in shape perception, 308–309
    structure and function of, 55f, 56
    in visual system, 299–300, 299f, 308–309
Temporal retina, 294, 294f
Temporal summation, 131, 131f
Teratogens, 253, 276
Terminal button, 77–78, 78f
Terminology, anatomical, 38–39
Tertiary protein structure, 95f
Testes-determining factor, 553
Testosterone, 202. See also Sex hormones
    behavior and, 401
    in brain development, 204–205, 250, 272–275,
        433–434, 551–554
    functions of, 202, 203
    lifelong effects of, 274–275
    neuroplasticity and, 513
    sexual behavior and, 436
    in sexual differentiation, 250, 433–434, 435
Tetrahydrocannabinol (THC), 191–192, 200, 201
Tetrodotoxin, 122
Thalamic nuclei, 69
Thalamus, 54, 54f
    association cortex and, 530
    electrode placement in, 575
    in gustation, 405
    medial, in memory, 497–499, 500f
    in movement, 373f, 374
    in olfaction, 402f, 403
    ventrolateral, 382, 383, 383f
Theory of mind, 541
Thermoregulation, 411
Thiamine deficiency, in alcoholism, memory loss and,
    199, 497–499
Thinking. See Cognition; Thought
Thirst. See also Drinking behavior
    hypovolemic, 433
    osmotic, 432
Thoracic spine, 61, 61f
Thorndike's puzzle box, 485, 485f
Thought. See also Cognition
    characteristics of, 525–526
    neural unit of, 526–529, 528f, 534
    as psychological construct, 525
Three-legged cat solution, 517–518
Threshold potential, 123
Thymine, 92f, 93
Thymus gland, in myasthenia gravis, 130
Thyroid gland, 415, 415f
Thyroid hormones, 415, 415t
Thyroid-stimulating hormone (TSH), 415t
Tickling, 392
Tight junctions, in blood-brain barrier, 174, 174f
Timbre, 322f, 327
Tissue plasminogen activator (t-PA), 45, 585
Tolerance, drug/alcohol, 177–179, 178f
Tone deafness, 346
Tone of voice, 327, 425
Tongue, taste receptors on, 405, 405f
Tonotopic representation, 335, 335f
Tool use, by animals, 29, 30
Top-down processing, 57
Topographic maps, 286, 286f

Topographic organization. See also Brain maps;
        Homunculus
    of auditory cortex, 340–344
    of cerebellum, 375–376
    definition of, 366
    experience-based changes in, 368–369, 508f,
        510–511
    of motor cortex, 365–366, 365f, 368–369
        changes in, 368–369, 368f, 507–510, 508f
    of somatosensory cortex, 390–391, 391f, 393f
    of visual cortex, 301–302, 302f
Touch, sensory processing in, 135, 135f
Tourette's syndrome, 58, 71, 374
Tower of Hanoi test, 247, 247f
Tractography, 540–541, 540f
Tracts, 47, 47f, 62, 62f, 356. See also specific types
Tranquilizers
    insomnia and, 476
    minor, 182–183
Transcranial magnetic stimulation (TMS), 222, 222f,
        576–577, 577f
    repetitive, 222, 577
Transcription, 93, 94f, 105
Transcription-translation-inhibition feedback loop, 455
Transfer RNA, 94
Transgenic animals, 103
Translation, 94, 94f, 105
    in transcription-translation-inhibition feedback
        loop, 455
Transmembrane proteins, 96–97, 96f
Transmitter-activated receptors, 146
Transmitter-sensitive channels, 136
Transplantation, of neurons, 162, 518–519, 575–576,
        592–593
Transporters, 145
Transsexuality, 437
Trapezoid body, 331, 331f
Traumatic brain injury, 2, 12–13, 579–580, 581–584,
        581f. See also Brain injury
    in athletes, 581, 582
    behavioral effects of, 520–521
    chronic traumatic encephalopathy and, 582
    concussion and, 581, 582
    diagnosis of, 583
    incidence of, 581, 581f
    mechanics of, 582–583, 583f
    neuroplasticity in, 516–520
    recovery from, 516–520, 584
    symptoms of, 581–582
Tremor, in Parkinson's disease, 142, 162, 590
Triceps muscle, 371, 371f
Trichromatic theory, 309–310
Tricyclic antidepressants, 186–187. See also
        Antidepressants
Trigeminal nerve, 60–61, 60f
Trisomy 21, 101–102, 102f
TRNA, 94
Trochlear nerve, 60–61, 60f
Trophic factors, 252
    for brain injury, 518–520
    neuroplasticity and, 513, 513t
Tropic molecules, 257
Tubules, 91
Tumors, brain, 83
Tuning curves, 334–335, 334f
Tuning fork, 321, 321f, 322f
Twin studies, 236
    epigenetics and, 236, 238
    of substance abuse, 198
Two-point sensitivity, 379, 379f
Tyrosine, 152, 153f

U
Ultradian rhythms, 447, 447t
Umami receptor, 405

Unconditioned response, 484
Unconditioned stimulus, 484
Unconscious memory. See Memory, implicit
Uracil, 94
Urge-to-action system, 374
Utricle, 388, 388f, 389

V
Vagus nerve, 60–61, 60f
Val allele, 237
Valproate, for bipolar disorder, 187
Ventral, definition of, 38, 39
Ventral corticospinal tract, 369f, 370, 370f
Ventral horn, motor neurons in, 370, 370f
Ventral roots, 62, 62f, 356
Ventral spinothalamic tract, 383, 383f
Ventral thalamus, in memory, 500, 500f
Ventral visual stream, 69, 69f, 296–297, 299f, 531f
    injury to, 314–316, 315f
    location of, 531
    secondary somatosensory cortex in, 384
    as "what" pathway, 296, 314–316, 315f, 531
Ventricles, 44, 44f
Ventrolateral thalamic neurons, 383
Ventrolateral thalamus, 382, 383, 383f
Ventromedial hypothalamus
    in eating, 431, 431f
    in sexual behavior, 435
Ventromedial prefrontal cortex, 419, 419f, 530f
Verbal fluency, sex differences in, 551–554
Vertebrae, 61, 61f, 356. See also under Spinal
Vertigo, 389
Vesicles, synaptic, 143, 144f
Vestibular system, 388–389, 388f
Virtual Iraq, 566
Virtual-reality exposure therapy, 566
Vision
    attention and, 533–534
    binocular, corpus callosum in, 302–303
    color, 309–312. See also Color vision
    light in, 288
    in movement, 382
    neuropsychology of, 313–316
    night, age-related decline in, 292
    peripheral, 290, 290f
    refractive errors and, 288–289
    shape perception in, 303–309, 304f–309f
    spatial localization in, 300–303
Vision impairment, 313–316
    agnosia and, 300, 314–315
    in carbon monoxide poisoning, 314–315, 423, 531
    echolocation in, 333
    monocular blindness and, 313
    refractive errors and, 288–289
    visual-form agnosia and, 423
Visual agnosia, 314–315
Visual cortex
    association cortex and, 531, 532f
    complex cells of, 306, 307f
    hypercomplex cells of, 306, 307f
    ocular dominance columns in, 270, 271f, 307, 308f
    primary (striate), 294f, 295, 298, 298f
        processing in, 304–307
        in shape perception, 304–307
        topographic organization of, 301–302, 302f
    secondary (extrastriate), 298, 298f
    simple cells of, 306, 306f
Visual fields, 294–295, 300–303, 300f
    blindness of, 282, 313–314, 313f
    cerebral asymmetry and, 546, 546f
Visual-form agnosia, 314–315, 423, 531
Visual illuminance, 292
Visual pathways, 294–295, 295f
    geniculostriate, 294–295, 295f, 296–297, 296f
    tectopulvinar, 295, 295f, 297

Visual-recognition task, 495, 495f
Visual streams
    dorsal, 296–297, 299f, 312, 316, 316f, 384, 531f.
        See also Dorsal visual stream
    injuries of, 314–316
    ventral, 296–297, 299f, 314–316, 316f, 531, 531f.
        See also Ventral visual stream
Visual system, 281–318
    anatomy of, 287–300
    chemoaffinity in, 269–270, 270f
    coding of location in, 300–301, 301f, 302f
    corpus callosum in, 302–303
    dorsal, 312
    "how" function of, 296, 299, 312
        impairment of, 316
    lateralization in, 546, 546f. *See also* Visual fields
    in movement, 312, 315–316, 315f, 382
    movement perception and, 312
    neural connectivity in, 270
    neural streams in, 69, 69f
    neuronal activity in, 303–312
    ocular dominance columns in, 270, 271f, 307, 308f
    ocular structures in, 287–291, 289f
    parallel processing in, 68, 69, 69f
    photoreceptors in, 287–293, 289f, 292f, 295f
    primary visual cortex in, 295, 298, 301–302,
        304–307

    receptive fields in, 301–302, 301f, 304–305
    retinal neurons in, 292f, 293–297, 293f
    secondary visual cortex in, 298, 298f
    segregated visual input in, 296–297, 297f
    temporal lobe in, 299–300, 299f, 308–309
    topographic maps in, 301–302, 302f
    visual pathways in, 294–295, 295f, 296–297, 297f
    visual streams in, 296–297, 299f, 313–316, 384,
        531, 531f. *See also* Dorsal visual stream;
        Ventral visual stream
    "what" function of, 296, 299
        impairment of, 314–316
    "where" function of, 297
Visuospatial memory, 494
Vitamin B$_1$ deficiency, in alcoholism, memory loss
        and, 199, 407–499
Vitamin D deficiency
    depression and, 186
    multiple sclerosis and, 86, 588
Vocal tone, 327, 425
Volt, 112
Voltage gradient, 116–117
Voltage-sensitive ion channels, 120–121, 123f, 124,
        124f. *See also* Ion channels
Voltmeter, 113
Vomeronasal organ, 403
Vomiting, 174

W
Waking state, 460, 479. *See also* Sleep
    electroencephalogram in, 459f, 461f, 473–474
Wanting-and-liking theory, 196–197, 439
Water, chemistry of, 89
Water intoxication, 432–433
Wave effect, 113–114, 113f
Waves, sound, 321–327
Weight, regulation of, 428–432. *See also* Eating/
        feeding behavior
Weight-loss strategies, 429
Wernicke's aphasia, 340
Wernicke's area, 332, 332f, 340–344, 340f
    mapping of, 340–344, 342f
White matter, 43f, 44, 47f
    in reticular formation, 52
    spinal, 62f
Wild-type alleles, 98
Wisconsin Card Sorting Task, 536, 536f, 539
Withdrawal symptoms, 195
World Health Organization, 572

Y
Y chromosome, *SRY* gene of, 204

Z
Zeitgebers, 449–451

Witness Neuroscience Firsthand

NEUROSCIENCE TOOL KIT

Available Spring 2013 at www.worthpublishers.com/ntk

The Neuroscience Tool Kit is a powerful web-based tool for learning the core concepts of behavioral neuroscience—by witnessing them firsthand. These thirty interactive tutorials allow students to see the nervous system in action via dynamic illustrations, animations, and models that demystify the neural mechanisms behind behavior. Videos taken from actual laboratory research enhance students' understanding of how we know what we know, and carefully crafted multiple choice questions make it easy to assign and assess each activity. Based on Worth Publishers' groundbreaking *Foundations of Behavioral Neuroscience CD-ROM*, the Neuroscience Tool Kit is a valuable accompaniment to any biopsychology course.

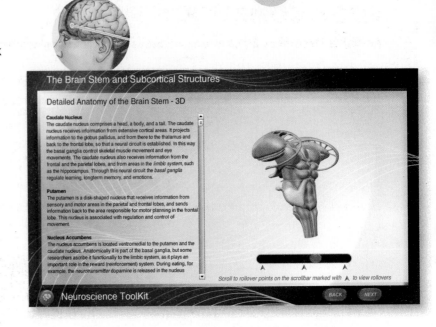

TOPICS AND ACTIVITIES

Neural Communication

Structure of a Neuron

The Membrane Potential

Conduction of the Action Potential

Synaptic Transmission

Neural Integration

Central Nervous System

Subdivisions of the Central Nervous System

Subcortical Structures

Sensory Systems–Vision

Sensory Systems–Audition

Sensory Systems–Somatosenses

Sensory Systems–Olfaction

Motor System

Limbic System

Language

The Cortex

Brain Stem

The Spinal Cord

Visual System

The Eye

Retina

Optic Chiasm

Lateral Geniculate Nucleus

Superior Colliculus

Primary Visual Cortex

Higher Order Visual Areas

Control of Movement

Organization of the Motor Systems

Muscle and Receptor Anatomy

Muscle Contraction

Spinal Reflexes

Descending Motor Tracts

Primary Motor Cortex

Higher Order Motor Cortex

Basal Ganglia

Cerebellum